The Hell Creek Formation and the Cretaceous-Tertiary boundary in the northern Great Plains: An integrated continental record of the end of the Cretaceous

Edited by

Joseph H. Hartman
Department of Geology and Geological Engineering and
Energy & Environmental Research Center
University of North Dakota, Box 8358
Grand Forks, North Dakota 58202-8358
USA

and

Kirk R. Johnson
Department of Earth Sciences
Denver Museum of Nature & Science
Denver, Colorado 80205
USA

and

Douglas J. Nichols
U.S. Geological Survey, M.S. 939, Box 25046
Denver Federal Center
Denver, Colorado 80225
USA

THE GEOLOGICAL SOCIETY OF AMERICA

Special Paper 361

3300 Penrose Place, P.O. Box 9140 ▪ Boulder, Colorado 80301-9140 USA

2002

Published by The Geological Society of America, Inc.
3300 Penrose Place, P.O. Box 9140, Boulder, Colorado 80301
www.geosociety.org

Printed in U.S.A.

GSA Books Science Editor: Abhijit Basu
Cover graphics production by Heather L. Sutphin

Library of Congress Cataloging-in-Publication Data

The Hell Creek Formation and the Cretaceous-Tertiary boundary in the northern Great Plains : an integrated continental record of the end of the Cretaceous / edited by Joseph H. Hartman, Kirk R. Johnson, and Douglas J. Nichols.

p. cm. — (Special Paper ; 361)

Includes bibliographical references and index.

ISBN 1-8137-2361-2

1. Cretaceous-Tertiary boundary—Northwestern States. 2. Geology—Northwestern States. 3. Paleontology—Northwestern States. 4. Hell Creek Formation. I. Hartman, Joseph Herbert. II. Johnson, Kirk R. III. Nichols, Douglas J. IV. Special papers (Geological Society of America) ; 361.

QE689 .H45 2002
551.7′7′0978—dc21 2002024222

Cover: Illustration is a reconstruction of Late Cretaceous (66 Ma) herbaceous vegetation from the middle part of the Hell Creek Formation. A *Neurankylus eximius* turtle encounters a *Tyrannosaurus rex* footprint that has been covered by herbs and small woody plants growing on a mid-channel bar in a large river. Hidden beneath the plants are a crocodile scute, a dinosaur claw, a conifer shoot, a bowfin vertebra, and a freshwater mussel. Artist Donna Braginetz painted this image in acrylic on Bristol board. The painting and copyright are owned by the National Science Museum, Tokyo, Japan, and it is reprinted here with permission. The concept and scientific interpretation for the painting is by Kirk Johnson with consultation by Kazuhiko Uemura and Makota Manabe. The painting is based on plant fossils from Denver Museum of Nature & Science (DMNH) localities 567, 2203, 2087, Slope County, North Dakota, and the vertebrate fauna of the Sandy Site (DMNH 1.1492), Harding County, South Dakota. Details of this small stature vegetation are provided in Johnson (Chapter 15).

10 9 8 7 6 5 4 3 2 1

Contents

Geological Society of America
Special Paper 361
2002

Preface

The Hell Creek Formation of the northern Great Plains is the most thoroughly sampled source of paleontological and geological data used to evaluate changes in nonmarine faunas and floras across the Cretaceous-Tertiary (K-T) boundary. Much of the data have come from exposures near Fort Peck Reservoir in east-central Montana. This volume provides new and well-integrated studies from Hell Creek exposures in western North Dakota and South Dakota and surrounding areas and presents new analyses of the Fort Peck record. In contrast to other studies conducted in the northern Great Plains, the patterns of biotic change throughout the Hell Creek Formation presented here are stratigraphically well documented, and numerous lithic, biotic, and paleomagnetic observations control the placement of the K-T boundary in the sections studied. The breadth of coverage is broad and includes the records of plant macrofossils, palynomorphs, mollusks, insects, and vertebrates, as well as the supporting stratigraphic framework. Much of the debate that has ensued over the K-T boundary record in the Fort Peck area has been based on an incomplete knowledge of the patterns of faunal and floral changes throughout the Hell Creek Formation. In addition, the Fort Peck sections have been plagued with difficulties associated with the reworking of Cretaceous fossil material into Paleocene rocks and the near absence of plant macrofossil and insect fossil studies. This contribution resolves these problems with new data that will place the answers to many end-Cretaceous questions on a more substantive foundation.

This volume had its conception in 1997 when Dean Pearson of the Pioneer Trails Regional Museum, Bowman, North Dakota, suggested to Kirk Johnson and Joseph Hartman that a compilation of papers based on the long-term North Dakota field studies of many of the contributors would provide a new and timely synthesis to resolve questions concerning the paleontological record of the end of the Cretaceous and the faunal and floral transition into the Paleocene in central North America. The concept of a comprehensive volume seemed particularly appropriate because many of the results were new and few of the basic data for this part of the northern Great Plains had been published. Preliminary results of this research were presented as a topical session at the 1999 annual meeting of the Geological Society of America. We hope that this volume will provide a data-rigorous foundation upon which others will ultimately build a more comprehensive picture of nonmarine and marine conditions at the end of the Cretaceous and earliest Paleocene in North America. Having put the volume together, we now realize that we have just begun to mine the information lode available on biotic patterns preserved in this part of the northern Great Plains.

STRATIGRAPHIC CONVENTIONS

The North Dakota, South Dakota, and Montana state geological surveys recognize the Hell Creek Formation as a lithostratigraphic unit, delimited below by the Fox Hills Formation and above by the Fort Union Formation or Fort Union Group. The U.S. Geological Survey (USGS) recognizes the Fort Union as a formation with member subdivisions, whereas the North Dakota Geological Survey recognizes the Fort Union Group with formally defined subdivisions as formations. In this volume, we use the nomenclature applied by the USGS to the stratigraphic units of the Williston Basin. This step was taken to provide the reader with a uniformity of usage from one chapter to the next, but does not imply that one approach is more or less valid than the other. All beds and lithofacies discussed in this volume are treated informally.

For the geologic period, the name Tertiary is considered antiquated, but will undoubtedly continue to remain in use. The use of the symbol K-T to label the Cretaceous-Tertiary boundary has extended the use of Tertiary; however, K-P for Cretaceous-Paleogene would be more meaningful and consistent with suggested chronostratigraphic standardization, although not sounding quite right. In this volume, the word contact is used in reference to lithostratigraphic units, and the word boundary is used in reference to chronostratigraphic or geochronologic unit nomenclature. Thus the K-T boundary is distinguished chronostratigraphically, whereas the Hell Creek–Fort Union formational contact is distinguished lithostratigraphically. Although this use may be well understood today,

Hartman, J.H., Johnson, K.R., and Nichols, D.J., 2002, Preface, *in* Hartman, J.H., Johnson, K.R., and Nichols, D.J., eds., The Hell Creek Formation and the Cretaceous-Tertiary boundary in the northern Great Plains: An integrated continental record of the end of the Cretaceous: Boulder, Colorado, Geological Society of America Special Paper 361, p. v–vi.

many previous arguments concerning the placement of the K-T boundary resulted from the lack of differentiation of these horizons.

ACKNOWLEDGMENTS

The volume editors thank their respective institutions for support to complete this project. Hartman gratefully acknowledges the financial support of Gerald Groenewold and Edward Steadman of the Energy & Environmental Research Center for this endeavor.

Joseph H. Hartman
Kirk R. Johnson
Douglas J. Nichols

Printed in the U.S.A.

Geological Society of America
Special Paper 361
2002

Hell Creek Formation and the early picking of the Cretaceous-Tertiary boundary in the Williston Basin

Joseph H. Hartman*
University of North Dakota Department of Geology and Geological Engineering and Energy & Environmental Research Center, Box 8358, Grand Forks, North Dakota 58202-8358, USA

ABSTRACT

The placement of the Mesozoic-Cenozoic boundary was a source of much interest long before the now well-documented theory of an extraterrestrial impact was proposed in 1980. The earlier interpretations of the stratal and biotic nonmarine record of the end of the Cretaceous in the Williston and associated basins of the Great Plains and Rocky Mountains provide us with perspective on the development of ideas and practices in recognizing and correlating events in geologic time. Arguments over the pattern and processes interpreted from this part of the fossil record were as rich and lively 90 years ago as they are today.

CRETACEOUS-TERTIARY BOUNDARY

A number of volumes (e.g., Christensen and Birkelund, 1979; Silver and Schultz, 1982; Sharpton and Ward, 1990; Ryder et al., 1996) and a multitude of papers have been written on the boundary conditions and faunal and floral patterns of the end of the Cretaceous and the Cretaceous-Tertiary (K-T) mass extinction. In their preface, Sharpton and Ward (1990) noted that there had been no loss of interest in the subject a decade after the introduction of an extraterrestrial hypothesis by Alvarez et al. (1980). More than 20 years later, enthusiasm for this topic remains unabated. The detailed geologic and paleontologic data from the K-T boundary interval have fostered additional studies that strive to provide greater resolution to the geologic record, permitting argument of process and pattern for shorter-duration phenomena. This volume is a direct outgrowth of these earlier studies, and its contributors have improved documentation to define hypotheses relating to events near and at the end of the Cretaceous.

Since the initial drumbeat begun by Alvarez et al. (1980), that the extinction of the dinosaurs and many other organisms was due to the impact of an extraterrestrial body, paleontologists have argued the pros and cons regarding this event. Some have concluded that the evidence does not support the theory (e.g., Archibald and Clemens, 1982; Archibald, 1996). Although the faunal and floral turnovers associated with the K-T boundary in the terrestrial realm have been studied by paleontologists who specialize in dinosaurs, mammals, mollusks, and plants, the impact scenario challenged the rigor of standard paleontological approaches. The iridium (Ir) anomaly represented a new standard for boundary recognition to which all other K-T criteria could be compared. Vertebrate paleontologist William Clemens and his students were the first sounding boards for Alvarez (1983) with respect to nonmarine environments and the terrestrial vertebrate record. The long-term research of Clemens in the type area of the Hell Creek Formation along the Fort Peck Reservoir, Garfield County, Montana, marked this location as a staging area for a diverse group of new studies focusing on specific K-T boundary topics (e.g., Archibald and Clemens, 1982; Fastovsky and Dott, 1986; Retallack et al., 1987). Although many of the studies in the Hell Creek badlands were conducted with the cooperation of Clemens' research group, most data sets have since been compiled independently of one another. This approach, however, resulted

*E-mail: joseph_hartman@und.nodak.edu

Hartman, J.H., 2002, Hell Creek Formation and the early picking of the Cretaceous-Tertiary boundary in the Williston Basin, *in* Hartman, J.H., Johnson, K.R., and Nichols, D.J., eds., The Hell Creek Formation and the Cretaceous-Tertiary boundary in the northern Great Plains: An integrated continental record of the end of the Cretaceous: Boulder, Colorado, Geological Society of America Special Paper 361, p. 1–7.

in different data sets being used by researchers seeking to answer related questions that required the same stratigraphic control of fossil occurrences. In contrast, this volume integrates a large stratigraphic data set across a number of fossil studies, thus obviating most of the problems associated with taxon range comparisons.

WILLISTON BASIN STUDY AREA

Stratigraphic overview

All studies presented herein were conducted in the Williston Basin, a sedimentary depocenter since the Ordovician. The stratigraphic interval covered in this volume includes the Upper Cretaceous Pierre, Fox Hills, and Hell Creek Formations overlain by the uppermost Cretaceous and Paleocene Fort Union Formation (Hartman, 1999) (Fig. 1). Most studies in this volume address the occurrence and distribution of fossils in the mostly terrestrial Hell Creek Formation of the Maastrichtian Stage and, at least in part, of the Lancian North American Land Mammal Age. The lower part of the Hell Creek Formation includes the Breien Member, which represents a short-term marine incursion of the Western Interior seaway (see Hoganson and Murphy, this volume; Hartman and Kirkland, this volume). The Fort Union Formation in the type area of the Hell Creek Formation in Montana includes, in ascending order, the Tullock, Lebo, and Tongue River Members, and in North Dakota and adjacent Montana, the Ludlow, Tongue River, and Sentinel Butte Members. The Ludlow and lower part of the Tongue River in North Dakota interfinger to the east with the marine Cannonball Formation, which represents the last transgressive-regressive cycle of the Western Interior seaway in North America. The Fort Union Formation spans nearly the entire Paleocene, from Danian to Thanetian, as correlated with Puercan (Pu1) to Clarkforkian North American Land Mammal Age biochrons.

Hell Creek Formation and bounding units

In 1902, under the loose and long-distance direction of H.F. Osborn of the American Museum of Natural History, Barnum Brown began his dinosaur study in Montana in the formation he would name the Hell Creek beds (Brown, 1907). Brown discovered dinosaurs and extensive deposits of freshwater mollusks along the southern tributaries of the Missouri River between Big Dry Creek to the east and the Mussellshell River to the west, in Garfield County, Montana, in the westernmost portion of the Williston Basin (Fig. 2). Brown successfully prospected the badlands formed by the tributaries of Gilbert, Crooked, Hell, Snow, and Seven Blackfoot Creek, which expose the Bearpaw (Pierre of Brown), Fox Hills, Hell Creek, and Fort Union Formations.

Brown (1907) differentiated the Hell Creek from the underlying marine Fox Hills Formation and overlying unnamed lignite-bearing unit that today is included in the Fort Union

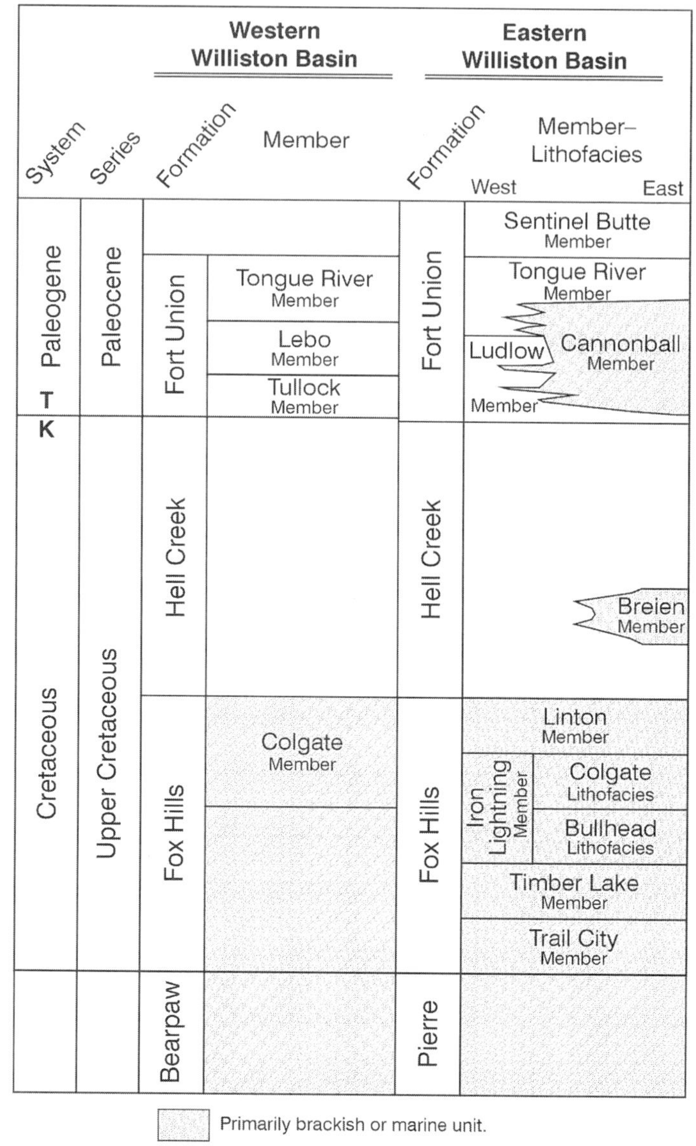

Figure 1. Stratigraphic nomenclature in Williston Basin. Diagram is intended to show various lithostratigraphic names used in this volume and does not show their complex lithostratigraphic and chronostratigraphic correlations (see individual contributions). Western Williston Basin includes Garfield, McCone, and other counties approximately west of Dawson County (contains Glendive; see Fig. 2), Montana, and eastern Williston Basin includes North and South Dakota and easternmost Montana, including Makoshika State Park at Glendive. Note that, except for Colgate Member, Fox Hills members and lithofacies are recognized primarily in southeastern portion of Williston Basin. Note that North Dakota Geological Survey (NDGS) raises rank of Fort Union members to formation and Fort Union to group. NDGS uses name Bullion Creek for Tongue River and Slope Formations for upper part of Ludlow Formation.

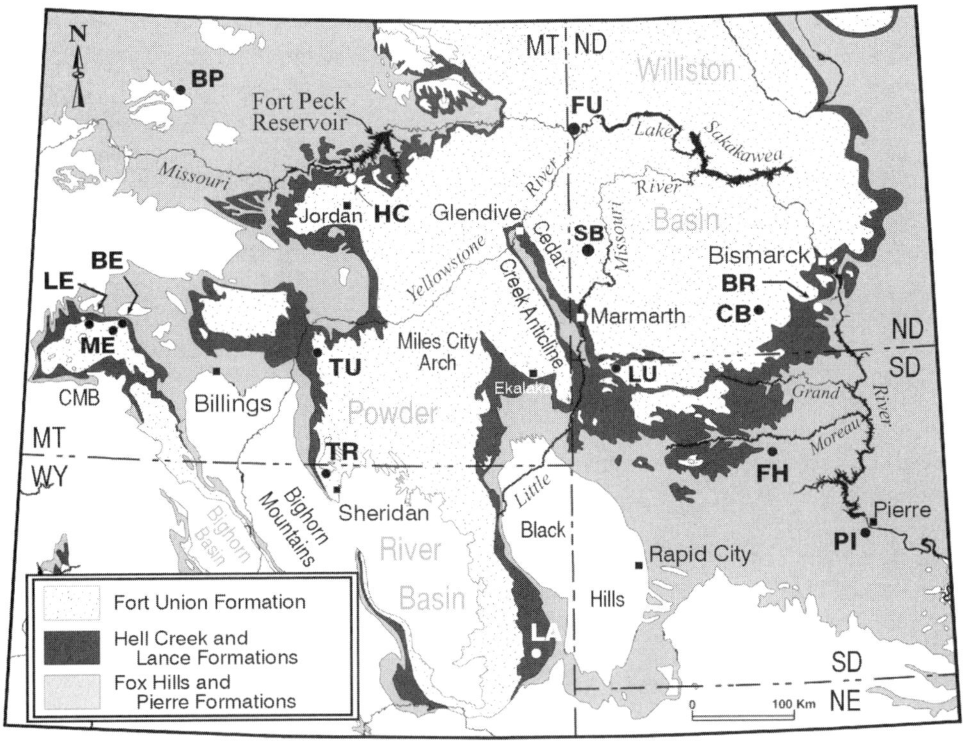

Figure 2. General geologic map of United States portion of Williston Basin and contiguous Powder River Basin. Map includes locations of type areas for following lithostratigraphic units (in ascending order): Upper Cretaceous marine Pierre (PI) Formation, marine Bearpaw (BP) Formation, predominantly marine Fox Hills (FH) Formation, predominantly terrestrial Hell Creek (HC) Formation, marine Breien (BR) Member of Hell Creek Formation; uppermost Cretaceous and predominantly Paleocene terrestrial Ludlow (LU) Member of Fort Union Formation; Paleocene Fort Union (FU) terrestrial members including Tullock (TU), Bear (BE), Lebo (LE), Melville (ME), Tongue River (TR), and Sentinel Butte (SB); and Paleocene marine Cannonball (CB) Member of Fort Union Formation. Crazy Mountains Basin (CMB) includes stratotypes of Bear, Lebo, and Melville Members.

Formation. No stratotype for the formation has ever been proposed. Uppermost Cretaceous rocks rim the Williston Basin and are exposed at the edge of the basin and along uplifts (e.g., Cedar Creek Anticline; see Fig. 2). Brown (1914) demonstrated that the previously named Lance Formation of the Powder River Basin and the Hell Creek Formation were, at least in part, equivalent in age on the basis of faunal and floral characteristics. These units merge south of the Miles City Arch, where the Lance Formation is exposed along the rim of the Powder River Basin (Fig. 2). In Canada, the Hell Creek Formation is correlated with the Frenchman Formation, where it rims the northern edge of the Williston Basin. The Fort Union Formation overlies the Hell Creek and constitutes the majority of the bedrock and surface exposures of the Williston Basin in the United States. In Canada, the Frenchman Formation is overlain by the Ravenscrag Formation, which can be lithostratigraphically and chronostratigraphically correlated with the Fort Union Formation. Similar to its placement near the Hell Creek–Fort Union formational contact in the United States, the K-T boundary is very near or at the Frenchman-Ravenscrag contact in the Canadian part of the Williston Basin (Braman et al., 1999).

K-T boundary in the Williston Basin

The following discussion brings into focus the problems faced by paleontologists who attempted to identify the Cretaceous-Tertiary or Mesozoic-Cenozoic boundary in the interior of North America prior to the proposal of the impact scenario (Alvarez et al., 1980; Alvarez, 1983). The basis of many current perceptions is rooted in the context of earlier arguments on the placement of the K-T boundary.

Meek and Hayden (1856) were the first to describe the geologic section of the Williston Basin. They used a numerical system for lithic units they interpreted to be Cretaceous and recognized the overlying strata simply as of Tertiary age. Meek and Hayden (1862) named "Formation No. 5" as the "uppermost Cretaceous Fox Hills beds" and the overlying "Eocene?" strata as the "Fort Union or Great Lignite Group." Strata overlying the Fort Union Group were of uncertain age, but would soon be assigned to the Eocene (the Paleocene in North America had yet to be delimited). Thus the Hell Creek Formation of Brown was originally included in the basal portion of the Fort Union Group of Meek and Hayden (for additional details on the Fox Hills–Hell Creek transition, see Hartman and Kirkland, this volume). Brown (1907) considered the Hell Creek beds as Cretaceous and considered the dinosaur-bearing strata of the Little Missouri River Valley to be of the same age. The Cretaceous age was based predominantly on the affinity of dinosaurs to those of the older Judith River Formation and recognition of the distinction between the freshwater mussels ("*Unio*" fauna), mammals, and plants of the Hell Creek and those of the known Tertiary ("Eocene") record.

The Fort Union Group of Meek and Hayden was also named without a stratotype being designated. The name Fort Union was taken from the fur-trading fort on the banks of the confluence of the Yellowstone and Missouri Rivers on the state line of North Dakota and Montana (Fig. 2). That the Fort Union included all of the strata overlying the Fox Hills is clear in that

Meek and Hayden reported brackish to freshwater and terrestrial mollusks in their description of the formation (Klett and Erickson, 1977; Hartman and Kirkland, this volume). The range in age represented by these strata included the uppermost Cretaceous and nearly the entire Paleocene. However, the Fort Union Formation in its type area and along most of its outcrop along the Missouri River in North Dakota is upper Paleocene (Hartman and Kihm, 1995).

With the advent of the North Dakota Geological Survey and the work of state geologist A.G. Leonard (1908), the lower part of the Fort Union Formation was referred to as the somber-colored (dark to medium gray) beds and considered to be of Tertiary age. These beds were recognized as dinosaur bearing, but they also included overlying nondinosaur-bearing strata (Ludlow Member). The somber beds were distinguished from the brighter- or lighter-colored (yellowish gray to grayish yellow) beds that would be later referred to as the Tongue River Member. The strata referred to the Hell Creek Formation were included by Lloyd and Hares (1915) in the Lance Formation, which at the time was interpreted by the U.S. Geological Survey as possibly Tertiary in age. The Fox Hills Formation was considered to be Cretaceous, and no unconformity was thought to exist between marine strata of the Fox Hills and the dinosaur-bearing rocks (but see Belt et al., 1997). Lloyd and Hares (1915) also introduced the "Ludlow lignitic member" as a unit in the upper part of the Lance Formation for exposures near Ludlow in Harding County, South Dakota (Fig. 2).

The problems associated with the placement of the K-T boundary in the section of North Dakota postdating the Fox Hills Formation were summarized by Leonard (1911). He noted that the Lance Formation was conformable with underlying and overlying strata of the Fox Hills and Fort Union, and thus deposition was continuous from the Cretaceous into the Tertiary. F.H. Knowlton's (1909, 1914) interpretation of the age of plant fossils suggested correlation of the lower member of the Lance Formation (Hell Creek) with the Tertiary, thus suggesting to both Knowlton and Leonard that this member (the Hell Creek) should be placed in the lower part of the Fort Union Formation. Leonard (1911) revised his view based on arguments presented by Stanton (1909, 1914), who suggested, like Brown, that the dinosaur and molluscan faunas should be regarded as Cretaceous.

Dinosaur bones continued to play an increasing role in efforts to define the limits of the Lance and Hell Creek Formations and the horizon of the K-T boundary. The frustrations associated with these tasks were stated by Calvert (1912), when he noted that a simple color change recognized by previous authors was not a convincing argument for the recognition of a formational contact. Calvert (1912, p. 197) went so far as to say that (author's brackets) "it is evident, therefore, that in the area mapped [in the Yellowstone drainage of eastern Montana] the upper limit of the Lance formation can not be based on lithologic grounds, but must instead be dependent upon the occurrence of fossil Ceratopsia bones. . . . The highest horizon at which these bones were found is just above [below was meant] the lowest persistent lignite bed, but there is certainly nothing in the character of the overlying strata to suggest that similar bones do not occur therein up through a stratigraphic distance of perhaps 500 feet [152 m]." Calvert's misstatements (discussed in the following) notwithstanding, his decision to tie the placement of the formational contact to the presence or absence of fossils would make the traditional lithic discrimination of map units less significant, put into question the importance of distinguishing between the Lance and/or Hell Creek–Fort Union Formations, and require the redrawing of geologic maps whenever new fossils were found. This approach of using fossils to correlate rocks was not unusual for the time, but it resulted in rock names being inappropriately used when chronostratigraphic correlation was actually being conducted (e.g., use of Wasatch Formation to represent the early Eocene).

T.W. Stanton, a U.S. Geological Survey (USGS) paleontologist who specialized in invertebrates, was both part of the solution and part of the problem in determining the placement of the K-T boundary and certain formational contacts. Although Stanton (1914) used the Lance Formation to represent the "Ceratops beds" and knew of the absence of dinosaurs in the upper part of the Lance, he was convinced that the marine Cannonball "formation" and its tongues in western North Dakota had to be of Cretaceous age because of certain faunal similarities of the mollusks of the Cannonball and those of the Fox Hills Formation. Stanton also recognized the distinctiveness of the fauna that would be assigned to the Hell Creek Formation in the Williston Basin, but like Knowlton, overinterpreted the apparent similarity of his fossils throughout the section. Stanton (1914, p. 352) interpreted the marine Cannonball mollusks as a "modified Fox Hills fauna" and thus determined them to be of Cretaceous age. Knowlton (1914) considered the plants of the Lance and Fort Union to be overwhelmingly similar and thought the Lance to be of Eocene age (Paleocene of later usage).

Hares (1928) correlated the Hell Creek Member, exposed along the Cedar Creek Anticline near Glendive, Montana, with strata in the Little Missouri River Valley of southwestern North Dakota and thus provided a name for the lower, previously unnamed member of the Lance Formation. Hares' placement of the contact between the Hell Creek and overlying "Ludlow lignitic member" resulted in the lower part of the mostly Paleocene Ludlow Member (as recognized today) being included in the Hell Creek Formation. In the type Hell Creek Formation in Montana, Collier and Knechtel (1939) also included the Hell Creek "member" in the Lance Formation, maintaining the early Eocene(?) age assignment for both the Hell Creek and overlying Tullock Member of the Lance Formation. The Tullock Member was introduced by Rogers and Lee (1923) for strata exposed along Tullock Creek in the northern Powder River Basin in southeastern Montana (Fig. 2). The Tullock was introduced as a member of the Lance Formation, with its base identified, in part, on the presence of a fairly persistent coal bed. As interpreted at the time, the fossils of the Tullock in its type area were not particularly distinctive, but C.W. Gilmore (*in* Rogers and Lee, 1923) identified a horn core of *Triceratops* 15.2 m above the base of the member (discussed in the following).

In eastern Montana, the Tullock is overlain by the Lebo Member of the Fort Union Formation. The "Lebo Andesitic or Shale Member" was introduced by Stone and Calvert (1910) for exposures on Lebo Creek on the northeast flank of the Crazy Mountains, southwest of Harlowton, Montana (Fig. 2). The Lebo Member was described as lithologically distinct from the underlying Lance and overlying Fort Union Formations, but was correlated with the Fort Union on the basis of its mammalian and floral fossil content and interpreted as earliest Tertiary in age. Stone and Calvert (1910) did not distinguish a relatively unfossiliferous unit above the Lance and below the Lebo (Fort Union) that Simpson (1937) would call the Bear Formation. In naming the Bear Formation for strata exposed on the east flank of Bear Butte (Fig. 2), Simpson essentially declined to enter the debate on the placement of the Lance and/or Hell Creek–Fort Union formational contact. He stated, however, that (1) the Crazy Mountains section included the Hell Creek, Bear, Lebo, and Melville units; (2) the Hell Creek Formation belonged to the Cretaceous based on the presence of dinosaurs; (3) the assignment of the Bear in a formation or group was subject to discussion, but that he favored assignment to the Fort Union because it lacked dinosaur bones and contained a mammalian faunule of "Tertiary aspect"; and (4) the Lebo and Melville, and thus the Fort Union as generally understood, were clearly of Paleocene age. Simpson (1929) also noted that mammals found in dinosaur-bearing strata were a distinctive assemblage (see Hunter and Archibald, this volume). Simpson (1937, p. 17) referred to undocumented observations by A.C. Silberling of "isolated, rare, and very fragmentary specimens" of dinosaurs in the lower part of the Bear Formation, as being redeposited. Given that the pre-Fort Union strata were never a focus of Silberling's field studies, these fossils more likely came from the underlying Hell Creek Formation. In any case, no dinosaur fossils have been found in subsequent extensive exploration for fossils in the Bear Formation (Hartman and Krause, 1993; see also Buckley, 1994).

With the introduction of the Paleocene as the basal series of the Tertiary and all of the major lithostratigraphic units of interest having been named in the Williston Basin, the timing was propitious for someone to review the inconsistencies apparent in the distribution of dinosaur fossils about the K-T boundary. This task was undertaken by USGS paleontologist Roland Brown in 1962 with his review of the Paleocene flora of the Rocky Mountains and Great Plains. Brown (1938, p. 422) foreshadowed his monograph with an abstract in which he reported his summary conclusions that (1) the Hell Creek Formation was uppermost Cretaceous age with a distinctive fossil record including the dinosaur *Triceratops*, ammonites, and certain species of plants (see also Dorf, 1942); (2) the Fort Union Formation was a mappable unit with its own distinctive flora and fauna that should be assigned to the Paleocene; and (3) "the thin zone of interfingering beds at the upper limit of the Hell Creek formation and the base of the immediately overlying Fort Union . . . marks the boundary between the Mesozoic and the Cenozoic." Cockerell (1915) noted the correlation between the extinction of the distinctive molluscan assemblage with that of dinosaurs at the end of the Cretaceous (see Hartman, 1998). Thus, with the preceding interpretations in place and their own extensive field studies, Cobban and Reeside (1952) constructed what would become a Western Interior standard for Cretaceous chronostratigraphic correlation as part of a series of such works published by the Geological Society of America. Cobban and Reeside (1952) recognized a *Triceratops* zonal index for the end of the Cretaceous and the Hell Creek as a distinct formation in the Williston Basin of the United States.

Brown (1962) resolved the records of dinosaurs erroneously reported from post-Hell Creek and/or Lance strata (Calvert, 1912; Rogers and Lee, 1923; Schlaikjer, 1935). With Brown's revisions, Calvert's original statement could be interpreted to locate the K-T boundary below the lowest persistent lignite bed above the highest dinosaur remains. This axiom proved remarkably useful until 1980, when impact scenarios required considerably more precision for rigorous discussion of pattern and process of fossil occurrence and faunal turnover. The stratigraphic interval between the highest dinosaur and the Ir-based K-T boundary (~ lowest lignite) would become the 3 m gap of later authors arguing for a pre–K-T boundary extinction of the dinosaurs (Archibald and Clemens, 1982; Archibald, 1996; but see Sheehan et al., 2000; Pearson et al., 2001, this volume).

Changing the definition of the end of the Cretaceous

In 1970, summaries of palynological data were presented that would change how the K-T boundary was recognized in nonmarine sections (Leffingwell, 1970; Tschudy, 1970). The GSA Special Paper in which these papers were published was the product of a decade of research contributing to the conclusion that palynomorphs could biostratigraphically zone this portion of the geologic record in North America. Leffingwell (1970) convincingly showed that the palynomorph taxa of *Gunnera, Proteacidites, Aquilapollenites*, and *Cranwellia* represented an assemblage of Maastrichtian age across the northern Great Plains (e.g., Lance and/or Hell Creek Formations). This assemblage was followed by a distinctive Danian assemblage consisting of mostly fern and moss spores and small tricolpate and tricolporate angiosperm pollen that could be used to mark the beginning of the Paleocene. This research, along with that of Tschudy (1970), would ultimately provide a means to uniquely locate the Ir-based K-T boundary using palynomorphs and correlate this horizon throughout the Western Interior of North America (see Nichols and Johnson, this volume).

SUMMARY

Contentious discussions on the pattern of extinction of life at the end of the Cretaceous that followed the publication of the now well-documented impact scenario are not unique to the placement of the K-T boundary. Paleontologists have argued, sometimes rather heatedly, over system boundaries, believing that their understanding of a fossil group provided the best means to delineate geologic time (e.g., the Cambrian-Silurian debate of Sedgwick and Murchison; Rudwick, 1976; Thackray, 1976). The placement of the Cretaceous-Tertiary system boundary is no exception, so it is not surprising that the interpretation of this boundary interval is fraught with disputes dating back to the late 1800s. In nonmarine faunas, dinosaurs were early recognized as significant in determining where the K-T boundary should be placed. Many of the pre-1960s arguments on placing the boundary were flawed, however, in part because of the inability of some researchers to distinguish between rock and time correlation and to define the actual pattern of extinction (or survival) across the Hell Creek–Fort Union formational contact. However, it is important to note that the placement of the era boundary in North American terrestrial sediments has been known within a 100 m interval since the turn of the twentieth century and that refinement of its stratigraphic horizon became possible with a more accurate knowledge of the stratigraphic distribution of plant and animal fossils. With the new model summarized by Brown (1962), the K-T event horizon was delimited to within a few meters, and by the time the search for the Ir anomaly began in terrestrial sediments of North America, palynologists had argued for a nonmarine biohorizon that would correlate with the global fallout of an extraterrestrial impact.

ACKNOWLEDGMENTS

Critiques of earlier versions of this paper provided by Ken Carpenter, Kirk Johnson, Robert Raynolds (Denver Museum of Nature & Science), William Cobban, Doug Nichols (U.S. Geological Survey), and Allen Kihm (Minot State University) were much appreciated. I also gratefully acknowledge the support of the U.S. Department of Energy through a joint research project with the Museum of the Rockies sponsored by the Energy & Environmental Research Center of the University of North Dakota.

REFERENCES CITED

Alvarez, L.W., 1983, Experimental evidence that an asteroid impact led to the extinction of many species 65 million years ago: National Academy of Sciences Proceedings, v. 80, p. 627–642.

Alvarez, L.W., Alvarez, W., Asaro, F., and Michel, H.V., 1980, Extraterrestrial cause for the Cretaceous-Tertiary extinction: Science, v. 208, p. 1095–1108.

Archibald, J.D., 1996, Dinosaur extinction and the end of an era: What the fossils say: New York, Columbia University Press, 237 p.

Archibald, J.D., and Clemens, W.A., 1982, Late Cretaceous extinctions: American Scientist, v. 70, p. 377–385.

Belt, E.S., Hicks, J.F., and Murphy, D.A., 1997, A pre-Lancian regional unconformity and its relationship to Hell Creek paleogeography in southeastern Montana: University of Wyoming Contributions to Geology, v. 31, no. 2, p. 1–26.

Braman, D.R., Sweet, A.R., and Lerbekmo, J.F., 1999, Upper Cretaceous–lower Tertiary lithostratigraphic relationships of three cores from Alberta, Saskatchewan, and Manitoba, Canada: Canadian Journal of Earth Sciences, v. 36, p. 669–683.

Brown, B., 1907, The Hell Creek beds of the Upper Cretaceous of Montana: Their relation to contiguous deposits, with faunal and floral lists, and a discussion of their correlation: American Museum of Natural History Bulletin, v. 23, p. 823–845.

Brown, B., 1914, Cretaceous-Eocene correlation in New Mexico, Wyoming, Montana, Alberta: Geological Society of America Bulletin, v. 25, p. 355–380.

Brown, R.W., 1938, The Cretaceous-Eocene boundary in Montana and North Dakota: Washington Academy of Sciences Proceedings, v. 28, p. 421–422.

Brown, R.W., 1962, Paleocene flora of the Rocky Mountains and Great Plains: U.S. Geological Survey Professional Paper 375, 119 p.

Buckley, G.A., 1994, Paleontology, geology and chronostratigraphy of Simpson Quarry (early Paleocene), Bear Formation, Crazy Mountains Basin, south-central Montana [Ph.D. thesis]: New Brunswick, New Jersey, Rutgers State University, 375 p.

Calvert, W.R., 1912, Geology of certain lignite fields in eastern Montana: U.S. Geological Survey Bulletin 471, p. 187–201.

Christensen, W.K., and Birkelund, T., eds., 1979, Symposium on Cretaceous-Tertiary boundary events, Part II, Proceedings: Copenhagen, Denmark, University of Copenhagen, 262 p.

Cobban, W.A., and Reeside, J.B., Jr., 1952, Correlation of the Cretaceous formations of the Western Interior of the United States: Geological Society of America Bulletin, v. 63, p. 1011–1044.

Cockerell, T.D.A., 1915, New species of *Unio* from the Tertiary rocks of Wyoming: American Museum of Natural History Bulletin, v. 34, p. 121–126.

Collier, A.J., and Knechtel, M.M., 1939, The coal resources of McCone County, Montana: U.S. Geological Survey Bulletin 905, p. 80.

Dorf, E., 1942, Upper Cretaceous floras of the Rocky Mountain region. 2. Flora of the Lance Formation at its type locality, Niobrara County, Wyoming: Carnegie Institution of Washington Contributions to Paleontology 508, p. 83–168.

Fastovsky, D.E., and Dott, R.H., Jr., 1986, Sedimentology, stratigraphy, and extinctions during the Cretaceous-Paleogene transition at Bug Creek, Montana: Geology, v. 14, p. 279–282.

Hares, C.J., 1928, Geology and lignite resources of the Marmarth field, southwestern North Dakota: U.S. Geological Survey Bulletin 775, 110 p.

Hartman, J.H., 1998, The biostratigraphy and paleontology of latest Cretaceous freshwater bivalves from the western Williston Basin, Montana, U.S.A., *in* Johnston, P., and Haggart, J., eds., Bivalves: An eon of evolution—Paleobiological studies honoring Norman D. Newell: Calgary, Alberta, University of Calgary Press, p. 317–345.

Hartman, J.H., 1999, The paleontologic and geologic record of North Dakota: Important sites and current interpretations, a symposium overview and geologic framework: North Dakota Academy of Science Proceedings, v. 53, p. 116–118.

Hartman, J.H., and Kihm, A.J., 1995, Age of Meek and Hayden's Fort Union Group (Paleocene), upper Missouri River, North Dakota-Montana, *in* Hunter, L.D.V., and Schalla, R.A., eds., 7th International Williston Basin Symposium, 1995 Guidebook: Billings, Montana Geological Society, p. 417–428.

Hartman, J.H., and Krause, D.W., 1993, Cretaceous and Paleocene stratigraphy and paleontology of the Shawmut Anticline and the Crazy Mountains Basin, Montana—Road log and overview of recent investigations, *in* Hunter, L.D.V., ed., Montana Geological Society Field Conference Guidebook–Energy and mineral resources of central Montana: Billings, Montana Geological Society, p. 71–84.

Klett, M.C., and Erickson, J.M., 1977, Type and reference sections for a new member of the Fox Hills Formation, Upper Cretaceous (Maestrichtian) in the Missouri Valley region, North and South Dakota: North Dakota Geological Survey Miscellaneous Series 56, 24 p.

Knowlton, F.H., 1909, The stratigraphic relations and paleontology of the "Hell Creek beds," "Ceratops beds" and equivalents, and their reference to the Fort Union Formation: Washington Academy of Sciences Proceedings, v. 11, p. 179–238.

Knowlton, F.H., 1914, The Cretaceous-Tertiary boundary in the Rocky Mountain region: Geological Society of America Bulletin, v. 25, p. 325–340.

Leffingwell, H.A., 1970, Palynology of Lance (Late Cretaceous) and Fort Union (Paleocene) Formations of the type Lance area, Wyoming, *in* Kosanke, R.M., and Cross, A.T., eds., Symposium on palynology of the Late Cretaceous and early Tertiary: Geological Society of America Special Paper 127, p. 1–64.

Leonard, A.J., 1908, Geology of southwestern North Dakota, with special reference to the coal, *in* Leonard, A.G., ed., State Geological Survey of North Dakota 5th Biennial Report: Bismarck, North Dakota, North Dakota Geological Survey, p. 29–114, pl. I–XIX.

Leonard, A.G., 1911, The Cretaceous and Tertiary formations of western North Dakota and eastern Montana: Journal of Geology, v. 19, p. 507–547.

Lloyd, E.R., and Hares, C.J., 1915, The Cannonball marine member of the Lance Formation of North and South Dakota and its bearing on the Lance-Laramie problem: Journal of Geology, v. 23, p. 523–547.

Meek, F.B., and Hayden, F.V., 1856, Descriptions of new species of Gasteropoda from the Cretaceous formations of Nebraska Territory: Philadelphia Academy of Natural Sciences Proceedings, v. 8, p. 63–69.

Meek, F.B., and Hayden, F.V., 1861 [1862], Descriptions of new Silurian (Primordial), Jurassic, Cretaceous, and Tertiary fossils, collected in Nebraska Territory, by the exploring expedition under the command of Capt. Wm. F. Raynolds, U.S. Top. Engrs., with some remarks on the rocks from which they were obtained: Philadelphia Academy of Natural Sciences Proceedings, v. 13, p. 415–448.

Pearson, D.A., Schaefer, T., Johnson, K.R., and Nichols, D.J., 2001, Palynological calibrated vertebrate record from North Dakota consistent with abrupt dinosaur extinction at the Cretaceous-Tertiary boundary: Geology, v. 29, p. 39–42.

Retallack, G.J., Leahy, G.D., and Spoon, M.D., 1987, Evidence from paleosols for ecosystem changes across the Cretaceous-Tertiary boundary in eastern Montana: Geology, v. 15, p. 1090–1093.

Rogers, G.S., and Lee, W., 1923, Geology of the Tullock Creek coal field, Rosebud and Big Horn Counties, Montana: U.S. Geological Survey Bulletin 749, 181 p.

Rudwick, M.J.S., 1976, Levels of disagreement in the Sedgwick-Murchison controversy: Journal of the Geological Society of London, v. 132, pt. 4, p. 373–375.

Ryder, G., Fastovsky, D., and Gartner, S., eds., 1996, The Cretaceous-Tertiary event and other catastrophes in Earth history: Geological Society of America Special Paper 307, 569 p.

Schlaikjer, E.M., 1935, The Torrington member of the Lance formation and a study of a new *Triceratops*: Harvard University Museum of Comparative Zoology Bulletin 76, p. 31–68.

Sharpton, V.L., and Ward, P.D., eds., 1990, Global catastrophes in Earth history: An interdisciplinary conference on impacts, volcanism, and mass mortality: Geological Society of America Special Paper 247, 631 p.

Sheehan, P.M., Fastovsky, D.E., Barreto, C., and Hoffmann, R.G., 2000, Dinosaur abundance was not declining in a "3 m gap" at the top of the Hell Creek Formation, Montana and North Dakota: Geology, v. 28, p. 523–526.

Silver, L.T., and Schultz, P.H., eds., 1982, Geological implications of impacts of large asteroids and comets on the earth: Geological Society of America Special Paper 190, 528 p.

Simpson, G.G., 1929, American Mesozoic mammals: Yale University Peabody Museum of Natural History Memoir 3, 235 p.

Simpson, G.G., 1937, The Fort Union of the Crazy Mountain Field, Montana, and its mammalian faunas: U.S. National Museum Bulletin 169, 287 p.

Stanton, T.W., 1909, The age and stratigraphic relations of the "Ceratops beds" of Wyoming and Montana: Washington Academy of Sciences Proceedings, v. 11, p. 239–293.

Stanton, T.W., 1914, Boundary between Cretaceous and Tertiary in North America as indicated by stratigraphy and invertebrate faunas: Geological Society of America Bulletin, v. 25, p. 341–354.

Stone, R.W., and Calvert, W.R., 1910, Stratigraphic relations of the Livingston formation of Montana: Economic Geology, v. 5, p. 741–764.

Thackray, J.C., 1976, The Murchison-Sedgwick controversy: Journal of the Geological Society of London, v. 132, p. 367–372.

Tschudy, R.H., 1970 [1971], Palynology of the Cretaceous-Tertiary boundary in the northern Rocky Mountain and Mississippi embayment regions, *in* Kosanke, R.M., and Cross, A.T., eds., Symposium on palynology of the Late Cretaceous and early Tertiary: Geological Society of America Special Paper 127, p. 65–111.

MANUSCRIPT ACCEPTED BY THE SOCIETY SEPTEMBER 24, 2001

Printed in the U.S.A.

Geological Society of America
Special Paper 361
2002

Lithostratigraphy of the Hell Creek Formation in North Dakota

Edward C. Murphy
John W. Hoganson
North Dakota Geological Survey, 600 East Boulevard, Bismarck, North Dakota 58505, USA
Kirk R. Johnson
Department of Earth Sciences, Denver Museum of Nature & Science, Denver, Colorado 80205, USA

ABSTRACT

In North Dakota, the Hell Creek Formation is a Laramide clastic wedge of primarily nonmarine strata intercalated with marine and brackish facies. The formation ranges in thickness from 100 to 60 m on a west to east transect across the southern portion of the state and tongues within the Hell Creek represent the youngest Cretaceous marine strata in the Western Interior. Hell Creek strata consist of poorly cemented fine-grained sandstone, siltstone, and carbonaceous-rich shale, mudstone, claystone, and rare lignite. These sedimentary rocks were deposited primarily in laterally accreting fluvial channel systems and associated flood plains.

Although localized stratigraphic patterns were recognized within the 76 sections measured for this study, no regional trends or marker horizons were observed. The marine Breien Member, formally proposed in 1942, maintains its integrity in outcrop over an area of at least 6000 km^2 in south-central North Dakota and is the only valid member of the Hell Creek Formation.

The Cantapeta Tongue, a marine brackish-water tongue characterized by *Ophiomorpha* and named herein, is present 16 m below the top of the Hell Creek Formation (40 m above the Breien Member and 20 m below the base of the Paleocene Cannonball Member [Fort Union Formation]). These occurrences of marine or brackish-water strata near both the base and top of the Hell Creek Formation in south-central North Dakota suggest that the Fox Hills Sea may not have withdrawn completely from the area before the advance of the Cannonball Sea in Paleocene time. Dinosaur remains are found throughout the Hell Creek Formation in North Dakota but are clustered within 20 m of the base of the formation in the south-central portion of the state where the fossils are concentrated a few meters above and below the Breien Member.

INTRODUCTION

The Hell Creek Formation was first described by Brown (1907) in east-central Montana, and much of the subsequent work on the formation has been done in the type area. Extensive exposures of the formation in North Dakota and South Dakota have received much less attention. Hell Creek strata were deposited during the last 1.8–2.2 m.y. of the Cretaceous, prior to 65.5 Ma (Johnson and Hickey, 1990; Hicks et al., this volume). The fossiliferous nature of the formation and its proximity to the Cretaceous-Tertiary (K-T) boundary has resulted in a significant increase in research since 1980. Collectively, the writers have more than 60 years of experience investigating Hell Creek and Fort Union strata in the Williston Basin. We

Murphy, E.C., Hoganson, J.W., and Johnson, K.R., 2002, Lithostratigraphy of the Hell Creek Formation in North Dakota, *in* Hartman, J.H., Johnson, K.R., and Nichols, D.J., eds., The Hell Creek Formation and the Cretaceous-Tertiary boundary in the northern Great Plains: An integrated continental record of the end of the Cretaceous: Boulder, Colorado, Geological Society of America Special Paper 361, p. 9–34.

have measured ~200 stratigraphic sections through the Hell Creek Formation, primarily in North Dakota, but also in northwestern South Dakota and eastern Montana. This work was done in the context of studying the K-T boundary, Late Cretaceous paleobotany, and surveys of the formation for mapping purposes. As with most geologists who have recently studied the Hell Creek Formation, we first began by referring to Charles Frye's (1967) doctoral dissertation on the formation. Frye informally proposed seven members of the Hell Creek in his thesis and formalized the nomenclature in a North Dakota Geological Survey Bulletin (Frye, 1969). We have concluded that these members are not recognizable beyond their designated type sections and should not be considered formal members of the Hell Creek Formation. The purpose of this chapter is to document the occurrence and nature of the formation in North Dakota and to demonstrate that rapid lateral facies changes within the formation do not allow its subdivision into smaller stratigraphic units. We believe that the following discussion will benefit future scientists studying Hell Creek strata by saving them valuable field time in futile attempts to follow Frye's stratigraphic usage.

Area of study

Hell Creek Formation outcrops in North Dakota are limited primarily to the Little Missouri River badlands in Slope and Bowman Counties and exposures along the Missouri River and its principal tributaries in Burleigh, Morton, Emmons, Grant, and Sioux Counties in south-central North Dakota (Fig. 1). The area of exposed Hell Creek strata in the south-central part of the state is approximately five times that in southwestern North Dakota, but the south-central outcrops are less contiguous. In addition, complete sections of Hell Creek strata are measurable at several localities in south-central North Dakota, but a complete section of Hell Creek strata is not exposed at any one locality in southwestern North Dakota (the Sunset Butte and West Marmarth localities in Bowman and Slope Counties are composite sections along 5.6 and 9 km transects, respectively). In spite of this limitation, most research on the Hell Creek Formation has taken place in southwest North Dakota. This is attributable to the expansive areas of outcrops created by badlands topography in southwest North Dakota, the greater thickness of Hell Creek strata, the abundance of fossils, and the extensive exposures of overlying Paleocene strata.

In addition to the 150 geologic sections the writers had previously measured through Hell Creek and Ludlow strata in North Dakota (Johnson, 1989; Murphy et al., 1995), an additional 76 geologic sections were measured for this project (Fig. 2). When there was sufficient outcrop control, sections were spaced relatively close together (within a few hundred meters) to facilitate tracing beds across the area. The new Hell Creek sections were measured in two areas of the state, west of Marmarth (north from Little Beaver Creek to Baldy Butte) and south of Bismarck (along the west edge of the Missouri River valley from Fort Rice to Huff). The upper Hell Creek contact, with the Ludlow Member of the Fort Union Formation, is exposed at only one locality, and the lower contact, with the Fox Hills Formation, is exposed at a few sites in the Marmarth transect, but both contacts are exposed at several localities in the Fort Rice–Huff transect. The Marmarth cross section is 10.5 km long, and the Fort Rice–Huff cross section extends for 21 km. These areas were chosen for detailed stratigraphic analysis because they include five of Frye's type member sections and encompass areas of relatively continuous Hell Creek outcrops.

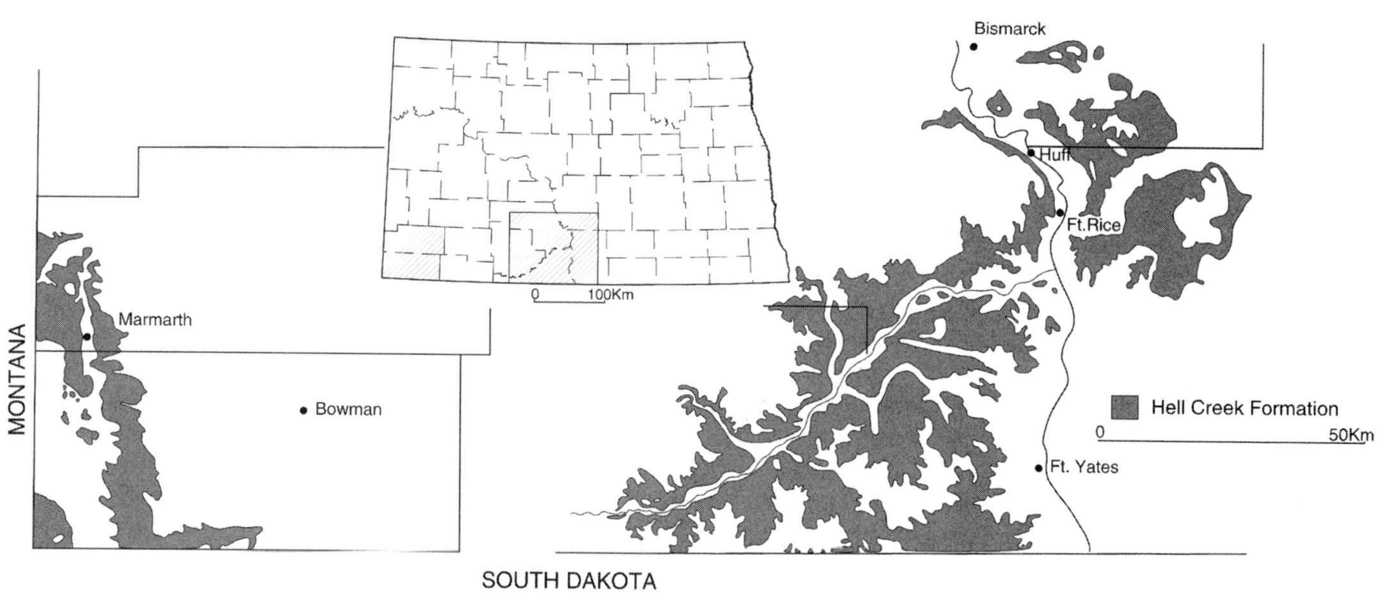

Figure 1. Areas of North Dakota where Hell Creek strata are present at surface.

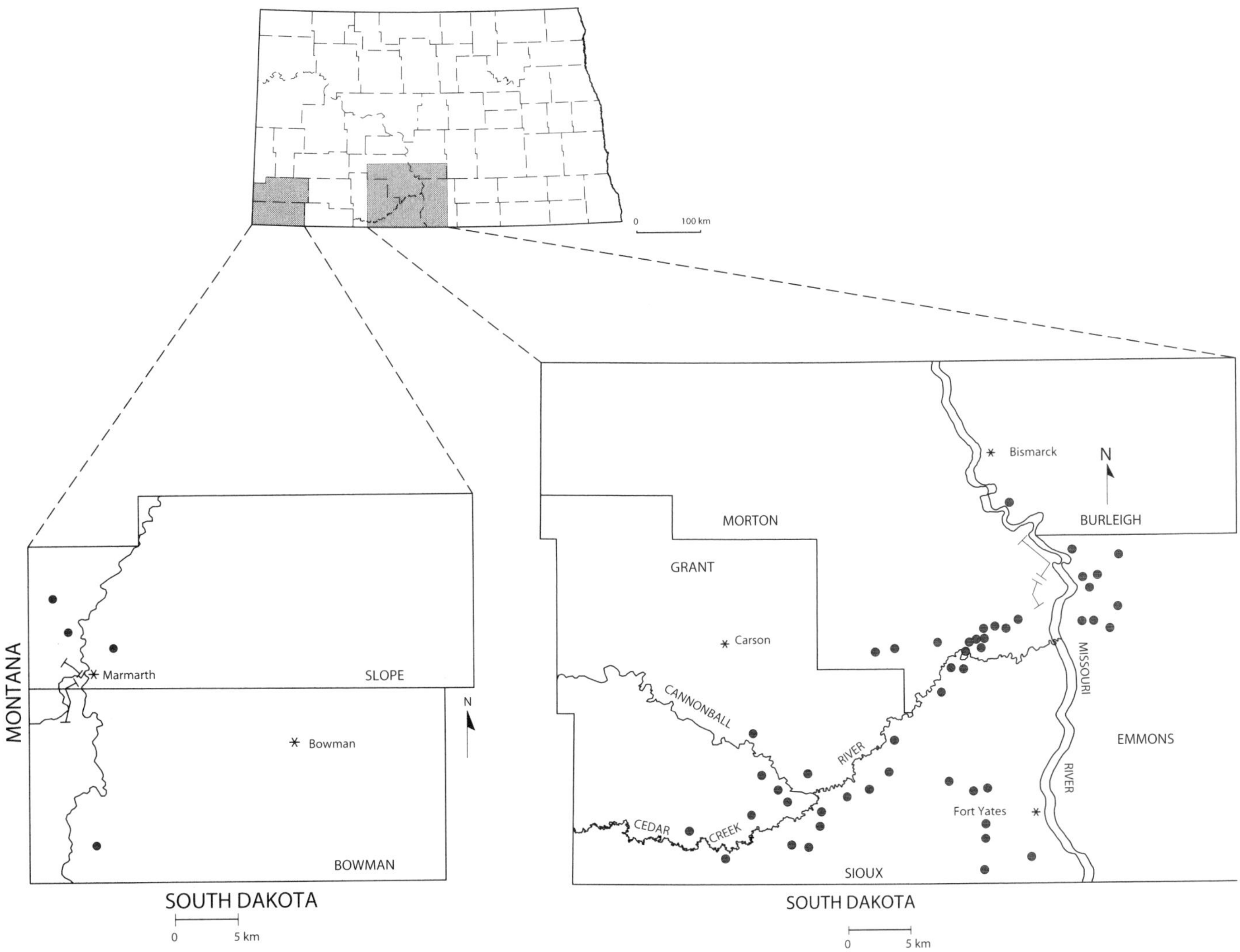

Figure 2. Location of selected measured sections (from both Murphy et al., 1995, and this study) and traces of major cross sections generated during this study in Marmarth (southwestern North Dakota) and Fort Rice–Huff (south-central North Dakota) areas.

The stratigraphic positions of vertebrate, invertebrate, and plant fossils were noted when encountered in the field. In addition to these detailed sections, three cross sections were constructed of Hell Creek strata, one in southwestern North Dakota and two in south-central North Dakota.

Hell Creek strata in North Dakota

In North Dakota, the Hell Creek Formation is a prograding clastic wedge associated with the retreat of the Western Interior Seaway that extends over an area of ~100000 km^2, roughly the western half of the state. Most Hell Creek strata in North Dakota are buried in the subsurface. The Hell Creek occurs at the surface over only ~3% of this area, 540 km^2 in the southwestern corner of the state and 2440 km^2 in south-central North Dakota (Fig. 1). Channel and crevasse-splay sandstone and siltstone, overbank mudstone and siltstone, paleosols, carbonaceous claystone, and occasional paludal lignite are the dominant Hell Creek lithologies.

Channel sandstone bodies range from a few meters to more than 30 m thick and are typically trough cross-bedded, very fine grained to fine-grained, poorly cemented, and contain thin lenses or laminae of organic debris (Fig. 3A). These channel deposits typically fine upward from sandstone with little or no clay content to interbedded sandstone and claystone (Fig. 3B). Characteristically, these sandstones contain clay derived from the diagenetic decomposition of volcanic rock fragments (Fastovsky, 1987). Hell Creek sandstone is generally light gray to light brown. In contrast to the channel sandstone bodies, sandstone deposited as crevasse splays is generally thin (less than a few meters thick), silty, and heavily rooted (Fig. 3C).

Siltstone beds are generally no more than a few tens of centimeters thick in Hell Creek strata. For this reason, in this study they were typically included with adjacent mudstones

Figure 3. (*See caption on following page*)

rather than identified as separate lithologies. Siltstone ranges from gray to brown and there are occasional shades of white. Siltstone is generally found as thin, overbank deposits, and tends to be heavily rooted. Siltstone also occasionally occurs in thicker, nonrooted deposits at the top of fining-upward channel sequences.

Mudstone is the most common and most variable lithology in the Hell Creek Formation (Fig. 3, D and E). These deposits generally contain nearly equal amounts of silt and clay and variable amounts of sand. Thin siltstone and sandstone lenses are common within the mudstone layers. Mudstone is generally very well indurated, blocky, and exhibits popcorn-weathering

Figure 3. Typical Hell Creek strata in North Dakota. A: Fining-upward sandstone in Fort Rice area (section 24). B: Interbedded sandstone and mudstone at top of channel in Fort Yates section. C: 2-m-thick, heavily rooted sandstone near base of section 25 in Marmarth area. D: Mudstone with lignitic lens in Marmarth area (section 12). E: Mottled mudstone with amber and abundant plant fragments (section 36). F: Popcorn-weathering surface on swelling claystone in middle Hell Creek strata near Marmarth (section 15). G: Carbonaceous claystone grading laterally into coal in western Morton County (T.134N, R.81W, sec. 3). H: Iron-oxide concretionary layer in Fort Rice area (section 48). I: Nodules of iron and manganese oxide near base of section 21 near Marmarth. The pick handle shown in most photos is 1 m long.

surface textures (granular surface texture created by repeated volume changes due to hydration and dehydration). The surface texture of the mudstone varies with differing concentrations of swelling clays. Mudstone beds typically range from light to dark shades of gray and brown, and occasionally purple. This lithology is typically rooted and occasionally contains thin lenses of organic material that may be lignitic. Carbonaceous mudstone is dark brown to black and often contains amber. Fastovsky and McSweeney (1987) and Retallack (1994) interpreted the majority of these rooted mudstones as young paleosols developed on wet flood plains.

Claystone and shale beds are shades of gray or brown and are moderately to well indurated (Fig. 3F). Shale is typically carbonaceous, often exhibits a paper texture, and is generally <3 m thick. Gray claystone is often relatively clean with popcorn-weathering surface textures. Brown claystone tends to be carbonaceous and rooted; it contains leaf fossil fragments, and generally does not exhibit well-developed popcorn weathering surfaces. Carbonaceous claystone was deposited in paludal environments and often contains thin lenses of organic debris. Occasionally carbonaceous claystone grades laterally and/or vertically into a thin lignite bed.

Although much of the Hell Creek mudstone and claystone is bentonitic, a subset of these layers appears to represent primary air-fall or secondarily reworked volcanic ash deposits that were deposited in lowland or lacustrine settings where the glass was later altered to bentonite. These bentonites are characterized by extreme popcorn weathering to 0.5 m thick and, in fresh surfaces, tend to exhibit a light green translucent color. Samples from several of these layers have been submitted for radiometric dating, but all have yielded mixed ages indicating either a multiple age volcanic source or depositional sorting.

Lignite beds, although sparse, may occur throughout Hell Creek strata (Fig. 3G). These lignite beds are generally thin (<1 m thick) and limited in lateral extent. A relatively persistent lignite is present near the base of the formation in south-central North Dakota. This coal does not exceed 0.5 m in thickness, but can be traced in both the surface and subsurface in southeastern Morton County and appears to be present in the subsurface in the northwestern portion of the county. This coal is also present on the east side of the Missouri River, where it was mined from underground workings at Coal Butte in Emmons County from the 1920s through the 1950s. It appears to have an areal extent of at least 1000 km^2.

Another Hell Creek coal was mined in South Dakota. The 0.2–2.1-m-thick Isabel-Firesteel coal member apparently extends over portions of four counties in north-central South Dakota (Curtiss, 1952, 1954). Lignite beds are locally abundant in the upper third of the Hell Creek Formation in the southern end of the Cedar Creek anticline just south of the North Dakota–South Dakota border.

Nodules of iron and manganese oxides and concretions cemented by iron oxide are common throughout Hell Creek strata (Fig. 3H). Iron oxide, typically limonite, most commonly forms thin (<1 m), reddish-brown, lenticular, localized beds. These beds often contain molds of conifer cones and other plant fossils, steinkerns of freshwater mollusks, and occasionally dinosaur bones, suggesting that iron in groundwater precipitated on organic-rich deposits. Purplish-black manganese nodules, often mistaken as fossils (pseudofossils), generally contain intricate surface weathering patterns that are unique to Hell Creek strata in North Dakota (Fig. 3I). Cone-in-cone concretions are locally common. Elongate ("log") concretions, cemented by calcium carbonate and oriented parallel to paleocurrent direction, are common in Hell Creek paleochannel deposits.

Exposed Precambrian, Paleozoic, and Mesozoic rocks in west-central Montana, along with ejecta from active volcanoes in that area, were probably the principle source of sediment for the Hell Creek (Cherven and Jacob, 1985). Although it has long been assumed that volcanic debris was a significant source of Hell Creek clastics, Fastovsky (1986) found little or no evidence of volcaniclastic debris in his petrographic analysis of Hell Creek strata in western North Dakota and eastern Montana.

PREVIOUS WORK

White et al. (1998) suggested that the Hell Creek is among the most fossiliferous and carefully studied Upper Cretaceous nonmarine deposits in the world. In North Dakota, most of the early work on the stratigraphy of Cretaceous and lower Tertiary rocks was by A.G. Leonard (1908). Leonard applied the term "Lance" to the dinosaur-bearing rocks in the state. Thom and Dobbin (1924) introduced the term Hell Creek to North Dakota, considering it to be the lower member of the Lance Formation. Hares (1928) reinforced the use of that terminology in southwestern North Dakota and also considered the Hell Creek to be the lower member of the Lance Formation. The Hell Creek was elevated to formation rank by Brown (1938) and has held that rank in North Dakota since.

Leonard (1912) and Hares (1928) were the first to map Hell Creek strata at relatively detailed scales. Leonard (1912) mapped Hell Creek strata (at the time termed the Lance) at a scale of 1:125 000 in a 2100 km^2 area around Bismarck. Hares (1928) mapped Hell Creek occurrences within a 4600 km^2 area in the southwestern corner of North Dakota at a scale of 1:127 000. Both commented on the lack of lignite in Hell Creek strata, and Hares (1928, p. 22) noted that the Hell Creek was "decidedly heterogeneous . . . and seemingly orderless."

Brown (1952, 1962) proposed that the lowest persistent coal bed above the highest dinosaur occurrence is a workable marker for the K-T boundary in the Great Plains. This criterion was initially used by Calvert (1912), who was working near Glendive, Montana. This same criterion for recognizing the K-T boundary had essentially been established in North Dakota by Hares (1928) when he divided the Lance into the underlying *Triceratops*-bearing beds of the "Hell Creek member" overlain by the lignite-bearing "Ludlow lignitic member."

In 1940, W.A. Cobban discovered a 10-m-thick fossiliferous marine tongue in the Hell Creek Formation between the towns of Solen and Breien on the Cannonball River (Seager et al., 1942). Cobban traced the strata in a 32-km-wide zone to the South Dakota line. Laird and Mitchell (1942) proposed the name Breien Member for this marine tongue. Although they listed four localities where the Breien Member is exposed, noting primary lithology and thickest occurrence, they did not designate a type section. In addition to naming this member, Laird and Mitchell placed the Hell Creek–Ludlow contact at the top of the highest bentonitic bed in southern Morton County. They also noted that the uppermost bentonitic beds were often overlain by a lignite or a carbonaceous horizon.

Frye (1967, 1969) followed the criterion established by Laird and Mitchell (1942) to identify the Hell Creek–Ludlow contact in the Little Missouri River and Missouri River valleys. Frye (1967) divided the formation into seven informal members based primarily upon the presence or absence and relative thicknesses of channel sandstone bodies (Fig. 4). Frye restricted three members to the Little Missouri River badlands (Little Beaver Creek, Marmarth, and Bacon Creek members), three to the Missouri River Valley (Crowghost, Breien, and Fort Rice members), and two (Huff and Pretty Butte members) were thought to extend throughout both southwestern and south-central North Dakota. Frye (1969) formally proposed these new members along with a suggested type member area for the Breien Member.

Moore (1976), as a result of his studies near the Limber Pines in northwestern Slope County, suggested that the Hell Creek and Ludlow Formations are not only separable by floral and faunal differences, but also because beds within the Ludlow Formation are generally more laterally persistent than those in the Hell Creek Formation and lignite beds are fewer, thinner, and more discontinuous in the Hell Creek Formation. Moore noted that the basal Ludlow coal is often reduced to a smut, lignitic shale, or carbonaceous clay, which, in areas of good exposures could be traced laterally into an equivalent coal. Moore found Frye's (1969) Pretty Butte member to be a valid stratigraphic unit in this area and noted that the top of this unit is often marked by a gray to blue gray, swelling clay. Although Moore found this member to be useful in the field, he questioned the utility of the other Frye members because of rapid lithofacies changes (Moore, 1998, personal commun.). Moore (1976, p. 9) noted that in Hell Creek strata, "internal continuity

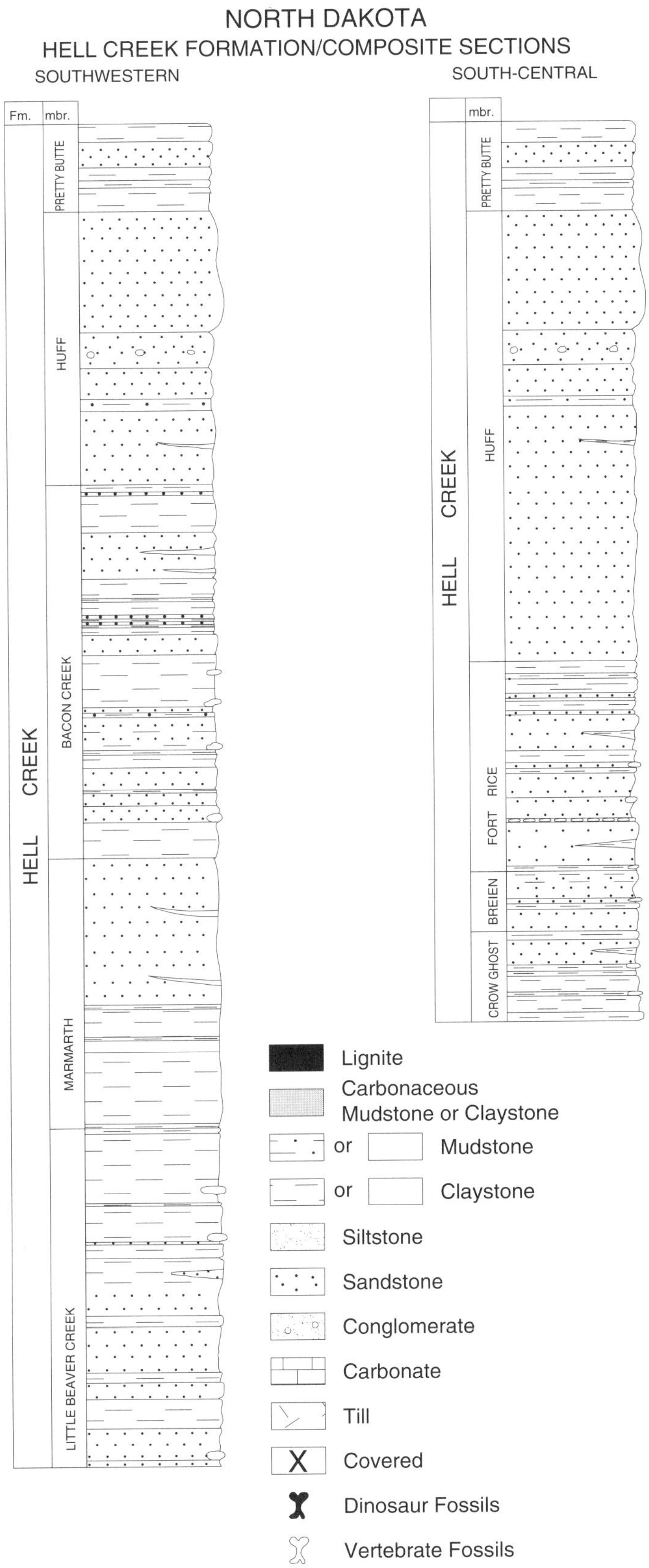

Figure 4. Generalized stratigraphic column of Hell Creek strata showing Frye's (1969) members in southwestern and south-central North Dakota. Frye's members are shown merely for reference; we do not consider them to be valid or usable.

is poor and few beds can be traced for distances in excess of a mile or more."

Butler (1980) measured 67 sections through Fox Hills, Hell Creek, and Ludlow strata in a roughly 210 km^2 area centered on Makoshika State Park near Glendive in eastern Montana. Butler applied Frye's members to Hell Creek strata in that area, the only detailed stratigraphic study to do so since these members were proposed. Member thicknesses were kept relatively constant on eight cross sections constructed for the area. The rationale for placement of member boundaries is difficult to follow on the sections because the contacts often cross lithologically distinct beds. Butler noted that Frye's members can be grouped into categories ranging from highly to slightly meandering channel sandstone facies. Although a few (Farris, 1984; Cherven and Jacob, 1985; Hunter et al., 1997) have applied Frye's members to one degree or another, Butler's thesis is the only detailed stratigraphic study to utilize them.

Cherven and Jacob (1985) compiled data, largely from published sources, and reconstructed depositional systems for Cretaceous and early Tertiary strata in eastern Montana and western North Dakota. They occasionally applied Frye's member names to generalized diagrams but apparently did not do the detailed correlation necessary to determine if these members are valid. They placed the marine-nonmarine margin just west of Bismarck in a paleogeographic reconstruction at the end of the Maastrichtian.

Fastovsky (1986) studied six Hell Creek localities, three in central Montana and three in southwestern North Dakota, as a dissertation project. Papers resulting from this work (Fastovsky, 1987; Fastovsky and McSweeney, 1987) provided the first modern sedimentological analysis of the formation, documented the ubiquity of paleosols in the formation, and provided petrographic analyses of Hell Creek sandstone and mudrock. This work also found that paleocurrent directions in the Hell Creek both in Montana and North Dakota were predominantly to the south, thus calling into question Cherven and Jacob's (1985) Sheridan Delta model.

Johnson (1989) measured 41 sections while studying floral localities throughout Hell Creek strata in southwestern North Dakota. Johnson and Hickey (1990) subdivided the Hell Creek into four megafloral zones and two subzones. Neither paper applied Frye's members because they were unrecognizable in the field. Johnson et al. (1989) documented the first K-T boundary in North Dakota to be recognized by an iridium anomaly, shocked minerals, and floral extinction. Johnson (1992) demonstrated that the K-T boundary is not always coincident with the Hell Creek–Fort Union contact and that, in many areas in Slope County, North Dakota, the K-T boundary is as much as 2.6 m above the formational contact. This work has been refined by Nichols and Johnson (this volume).

Murphy et al. (1995) measured 32 sections in south-central North Dakota across the Hell Creek–Ludlow contact. Palynomorph analysis of 12 sections revealed that the K-T boundary is nearly coincidental with the Hell Creek–Ludlow lithostratigraphic contact in this area. No iridium spike was identified and shocked quartz was not found at the K-T boundary in a section near the town of Huff that was sampled for these anomalies. For that study, Moses Attrep Jr. (Los Alamos National Laboratory) analyzed 20 samples bracketing the K-T boundary as determined by palynology. Samples were obtained at 10 cm intervals and analyzed for a series of metals, including antimony, uranium, arsenic, and iridium. Iridium ranged in concentrations from 0.01 to 0.12 ppb but did not indicate a spike at the K-T boundary. However, concentrations of the other three metals spiked at the boundary. Although several of the lithologic generalities used by Frye (1969) to subdivide the Hell Creek were noted during the Murphy et al. (1995) study, his members, other than perhaps the Pretty Butte member, could not be recognized in the field.

Belt et al. (1997) studied Hell Creek and Lance strata in eastern Montana, northern Wyoming, and the western edge of North and South Dakota. They concurred with Fastovsky (1987) that paleodrainage patterns for this stratigraphic interval indicated a predominant south-southeastward paleocurrent direction along the state line between Montana and North Dakota. They also concluded that a pre-Lancian regional unconformity exists within their four state study area, but this remains a controversial interpretation.

HELL CREEK STRATIGRAPHY

Basal contact

In the Little Missouri River and Missouri River valleys in North Dakota, the top of the Fox Hills Formation is typically a 5–15-m-thick, very fine grained to fine-grained sandstone commonly referred to as the Colgate Member (Daly, 1984; Erickson, 1992; Belt et al., 1997). The Colgate Member is distinguished in the field from overlying Hell Creek channel sandstone bodies by its light gray to white weathered surfaces and grayish-green fresh surfaces (attributable to the presence of glauconite). It is typically cleaner than overlying Hell Creek channel sandstone and exhibits fluted surface weathering patterns. The Colgate Member and Hell Creek channel sandstone bodies have similar grain sizes, contain organic and clay-rich lenses, and occasionally exhibit trough cross-bedding. The base of the Hell Creek is a 0.3–1-m-thick carbonaceous sandstone, siltstone, or mudstone. Previous workers have referred to this thin, carbonaceous bed as a lignite (e.g., Frye, 1969), but a basal lignite was not observed anywhere within the study area. The basal contact is regionally conformable but locally, such as in the Breien-Solen area of Sioux County, may show relief of a few meters, due to scouring. The base of this carbonaceous unit varies from a flat to scoured surface. There is a remarkable similarity in character of this basal carbonaceous bed (thickness and lithology) from southwestern to south-central North Dakota (Fig. 5, A and B). The basal carbonaceous bed represents the eastward progradation of paludal facies. Erickson (1992) interpreted it as a paleosol, implying that a period of quiescence, sufficient to form a soil horizon, advanced from west to east across the basin following deposition of the Fox Hills. Because it is thin, this carbonaceous bed is difficult to identify in the subsurface, and most workers have placed the Fox Hills–Hell Creek contact at the top of a log profile indicating a 5–15-m-thick, clean sandstone (the Colgate Member). In the Little Beaver Creek, Fort Rice, and Crowghost Cemetery areas, the carbonaceous bed directly overlies the Colgate Member. The depositional setting of the Colgate Member has been variously interpreted as marine, lagoonal, or back-barrier beach (Feldmann, 1972); shallow subtidal or intertidal topset deposits in environments associated with the distributary system (Waage, 1968); and nonmarine fluvial systems (Butler, 1980). As noted by Erickson (1992) and verified by the authors of this paper, thin coals are occasionally present in middle to upper portions of the Fox Hills Formation.

Figure 5. Thin (<1 m) carbonaceous mudstone at base of Hell Creek Formation in (A) southwestern North Dakota (section 26) and (B) south-central North Dakota (Crowghost Cemetery).

Erickson noted that while the previously mentioned localities are the most visited and most often cited localities in the

TABLE 1. CRITERIA FOR IDENTIFYING THE HELL CREEK-LUDLOW CONTACT

Base of the lowest persistent lignite.
Top of the highest occurring swelling "popcorn weathered" claystone.
At the color change from grayish, somber tones to light brownish or yellowish tones.
The base of yellowish strata in portions of Bowman and Slope County.
At a pronounced break or change in slope.
At an increased in vegetative cover.
At the base of laterally extensive beds.
At the base of the first occurrence of variegated siltstone and claystone beds.

literature, they are atypical of the Fox Hills–Hell Creek contact in North Dakota (J. Erickson, 1999, personal commun.). Although the Fox Hills–Hell Creek contact is lithologically remarkably similar within the study area, Erickson has argued that the contact is extremely variable in both the surface and subsurface in the surrounding region, and it may be preferable to place the contact at the first well-developed carbonaceous bed or coal. Even though it is often difficult to separate the Colgate Member from Hell Creek channel sandstone, it may still be preferable to placing the contact at underlying coals when they are present. These coals may only represent thin, nonmarine tongues and not the base of a thick, nonmarine sequence of strata. The basal Hell Creek carbonaceous bed is overlain by a variety of lithologies indicating an array of depositional environments.

The base of the Hell Creek Formation is likely older in western North Dakota and becomes progressively younger toward the east (Johnson and Hickey, 1990). This hypothesis cannot be verified because of the absence of reliable radiometric dates in the lower Hell Creek Formation. A thick, tuffaceous sandstone (Cannonball River ash) in the uppermost part of the Fox Hills Formation is exposed at several localities in the Cannonball River Valley in southeastern Morton and northeastern Sioux counties (Fig. 6). This ash also extends into the Coal Butte area in Emmons County. Frye (1967) noted the presence of the Cannonball River ash (near the town of Breien) and an additional tuff at the base of the Hell Creek in Emmons County (T.135N., R.78W., section 13). The tuff in Emmons County could not be found during this study. Frye speculated that these tuffs might represent the same volcanic eruption because anal-

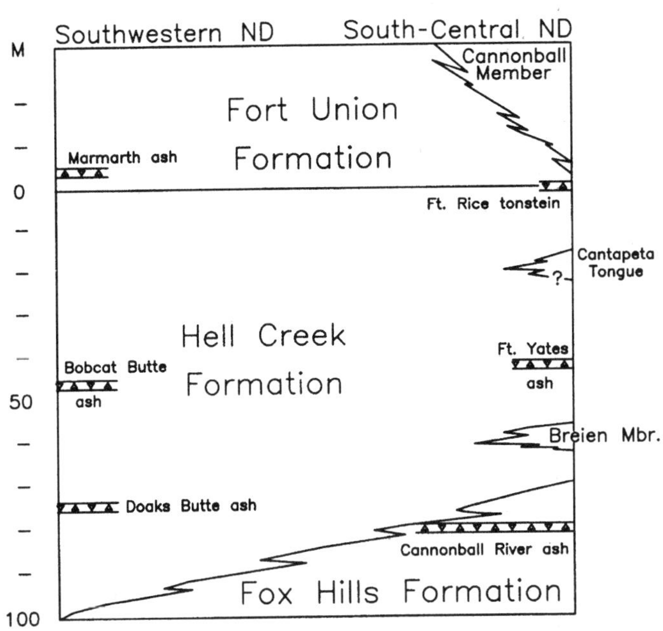

Figure 6. Relative position of known tuffaceous beds in Hell Creek and lower Fort Union Formations in southwestern and south-central North Dakota.

Figure 7. Thin coals and carbonaceous beds at base of Ludlow Member. A: Eastern Grant County (section 2 of Murphy et al., 1995). B: Fort Rice area (section 45). Note tonstein at contact in section 45.

Figure 8. Location of measured sections in Marmarth area. Marmarth is located at lat. 46°17′30″ and long. 103°55′.

ysis of glass shards from both tuffs contained identical indexes of refraction. The dating of any tuff from the base of the Hell Creek in North Dakota would provide important insight into the rate of regression of the Western Interior seaway from western North Dakota.

Upper contact

Since the early studies of Calvert (1912), workers have recognized the general absence of lignite beds in the Hell Creek Formation in Montana and North Dakota and have often placed the Hell Creek–Fort Union contact at the base of the lowest persistent lignite (Leonard, 1911; Hares, 1928; Brown, 1938, 1962; Laird and Mitchell, 1942; Frye, 1969; Moore, 1976; Butler, 1980; Fastovsky, 1987; Murphy et al., 1995). In Morton County, Laird and Mitchell (1942) placed this contact at the top of the highest bentonitic bed and noted that the bentonite was generally overlain by a lignite or a carbonaceous bed. Frye (1969) listed several criteria he found useful to identify the upper Hell Creek contact. He warned that this contact is variable and that no single criterion could be used to identify the contact throughout the Williston Basin. Fastovsky (1987) and Murphy et al. (1995) arrived at the same conclusion and suggested that often several criteria are used in combination to identify the contact (Table 1). In addition to the criteria listed in Table 1, Hell Creek strata often have pronounced erosional piping, whereas basal Fort Union strata generally do not. Carbonaceous mudstone and shale in the Hell Creek typically are reddish-brown or a mottled black and tan and contain visible pieces of amber, in contrast to carbonaceous mudstone in the Ludlow Member of the Fort Union Formation, which typically is uniform shades of dark brown and contains fragments of angiosperm leaves. Data from Murphy et al. (1995) and this report indicate that lignite is present at the Hell Creek–Ludlow contact at ~40% of the localities in south-central North Dakota; this is essentially the same percentage observed by Frye (1967). In contrast, Frye (1967) and Kirk Johnson (1999, personal commun.) placed 80%–85% of their contacts in southwestern North Dakota at coal (Fig. 7, A and B).

Breien Member

The Breien Member was proposed by Laird and Mitchell (1942). Frye (1969) designated 6 m of strata exposed at the Crowghost Cemetery locality in Sioux County as the type section, because Laird and Mitchell (1942) did not present a type section. Laird and Mitchell (1942, p. 14) characterized the Breien lithology as "two beds of grey sand separated by a grey bentonite." Frye (1969, p. 35) noted that "the lithology of the Breien is variable but is characterized by glauconitic sands containing *Ophiomorpha major*. A persistent bentonite is often found near the middle of the member . . . the Breien Member contains marine, brackish-water, fresh-water, and land fossils."

Scattered outcrops of the Breien Member suggest that it extends over an area of at least 6000 km^2 centered along the Missouri River Valley in south-central North Dakota, from the South Dakota border to the town of Huff. This marine tongue consists primarily of burrowed sandstone and mudstone and occasionally a thin bentonite and varies in thickness from 1.5 to 8.5 m. The base of the Breien Member is 1.5–9 m above the base of the Hell Creek Formation. It is often easy to identify in outcrop because of abundant iron-oxide concretions, some of which are typically *Ophiomorpha* burrows. The Breien Member is described in detail by Hoganson and Murphy (this volume).

Figure 9. Hell Creek strata in Marmarth area. A: Note mudstones at top and channel sandstone at base of section 20. B: Note carbonaceous beds in lower Hell Creek outcrop along Little Beaver Creek (sections 25–29).

HELL CREEK LITHOFACIES AND CORRELATIONS

Marmarth cross section

We measured 29 geologic sections along a 10.5 km transect west of Marmarth in southwestern North Dakota (Figs. 8–10). This area was chosen for detailed study because of excellent outcrop control and because it includes, or is in close proximity to, the type sections of Frye's (1969) Little Beaver Creek, Marmarth, and Bacon Creek members. In addition, the majority of manuscripts in this volume focus on the paleontology of the Hell Creek in the Marmarth area. To facilitate presentation, the Marmarth cross section was divided in two at a 0.8 km gap in outcrops along U.S. Highway 12. Hell Creek strata in the northern portion of the transect dip to the north-northeast at ~8 m/km. Strata along the south end of the panel dip to the north at 18 m/km due to proximity to the axis of the Cedar Creek anticline. Hell Creek strata in this area consist predominantly of

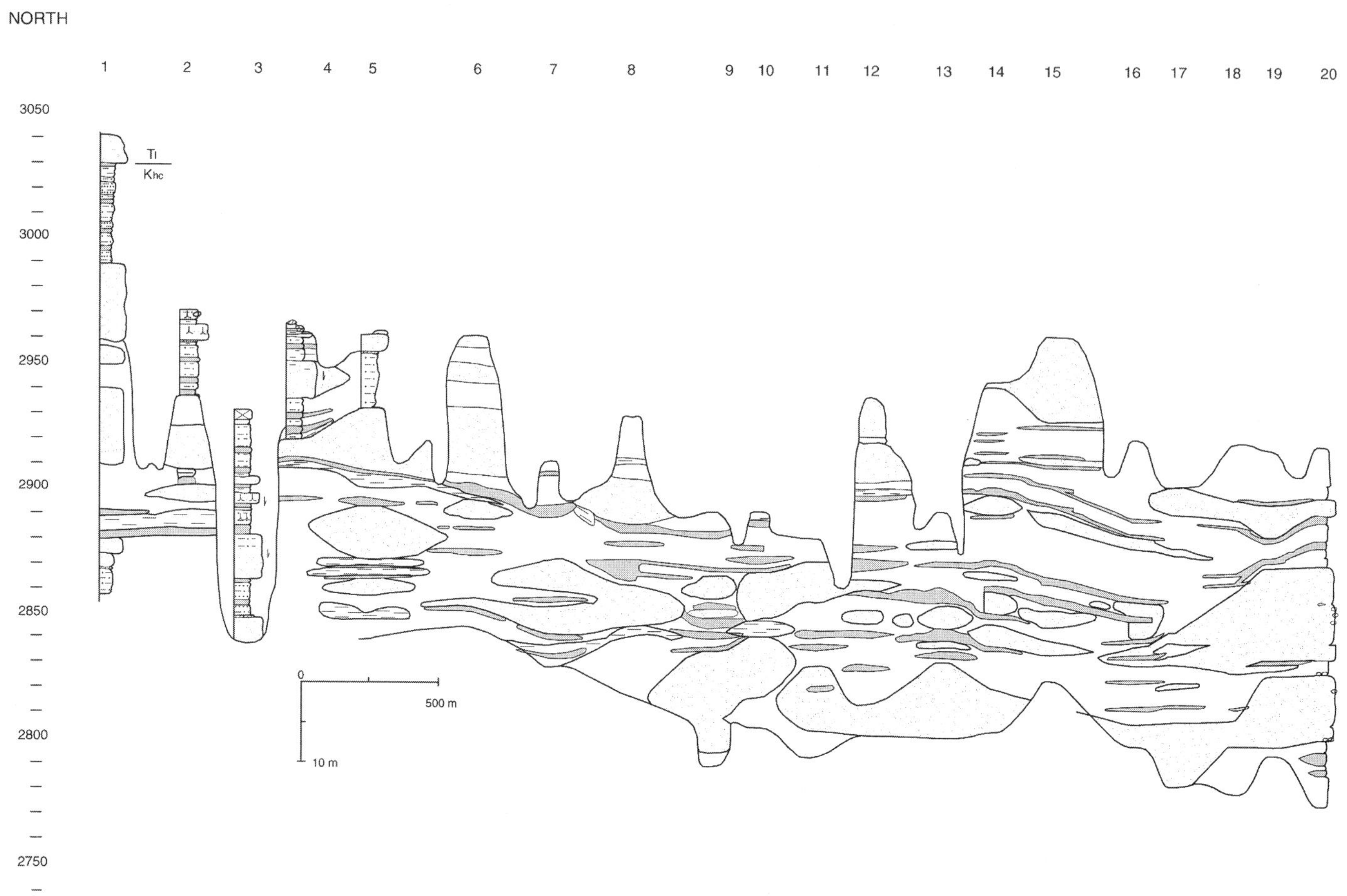

Figure 10. Cross section of Hell Creek strata in Marmarth area. See Figure 4 for legend and Figure 8 for location of measured sections. Mudstone and claystone are depicted with blank pattern to simplify diagram.

thin to thick channel sandstone beds, rooted mudstone paleosols, thin carbonaceous mudstone beds, and bentonites.

The Fox Hills–Hell Creek contact is exposed at five localities (25–29) along the southern edge of the cross section in the area adjacent to Little Beaver Creek (Fig. 10). The Colgate Member of the Fox Hills Formation is well exposed in this area and consists of at least 16 m of sandstone, which in some places is burrowed, and mudstone. The base of the Hell Creek in this area is a 1-m-thick carbonaceous siltstone or mudstone. The Hell Creek–Ludlow contact is exposed at only one locality (no. 1, Baldy Butte) in this cross section. A composite thickness of 100.6 m for the Hell Creek in this area was obtained by tracing beds across the 10.5 km transect. The lack of outcrops between sections 20 and 23 made correlation difficult and may have introduced error into this compilation. However, this thickness is surprisingly similar to the 100.3 m measured by David Fastovsky and Kirk Johnson at Sunset Butte in nearby Bowman County (Johnson, this volume).

Channel sandstone bodies are common throughout the Hell Creek Formation in the Marmarth area. Sandstone beds are very fine grained to fine-grained, poorly cemented, clean to dirty, concretionary, and contain both organic laminae and clay lenses (Fig. 10). Sandstone beds range in thickness from a few meters to 15 m. In this area, the thickest channel sandstone tends to occur in the middle to upper part of the formation (Fig. 10). However, thick sandstone beds also occur in the lower part of the Hell Creek (Fig. 10, sections 22 and 24). Most channel sandstones have a lateral extent of <500 m, but some of the larger sandstone bodies may extend for several kilometers (Fig. 10). With few exceptions, these channel sandstones are fining-upward sequences that typically grade upward from a relatively clean sandstone to an interbedded sandstone and mudstone and eventually into a mudstone. A large channel sandstone is present at the base of the outcrops in sections 12, 13, and 15–18 (Fig. 10). The outcrops in this area roughly parallel the axis of this channel as indicated by opposing meander scars.

The dominant Hell Creek lithology in the Marmarth area is mudstone. The mudstone exhibits varying degrees of popcorn-weathering surface textures. Mudstone beds range in thickness from several meters to more than 10 m; the average thickness is 3–4 m. They are slightly to heavily rooted and are carbonaceous to varying degrees. Fastovsky and McSweeney (1987)

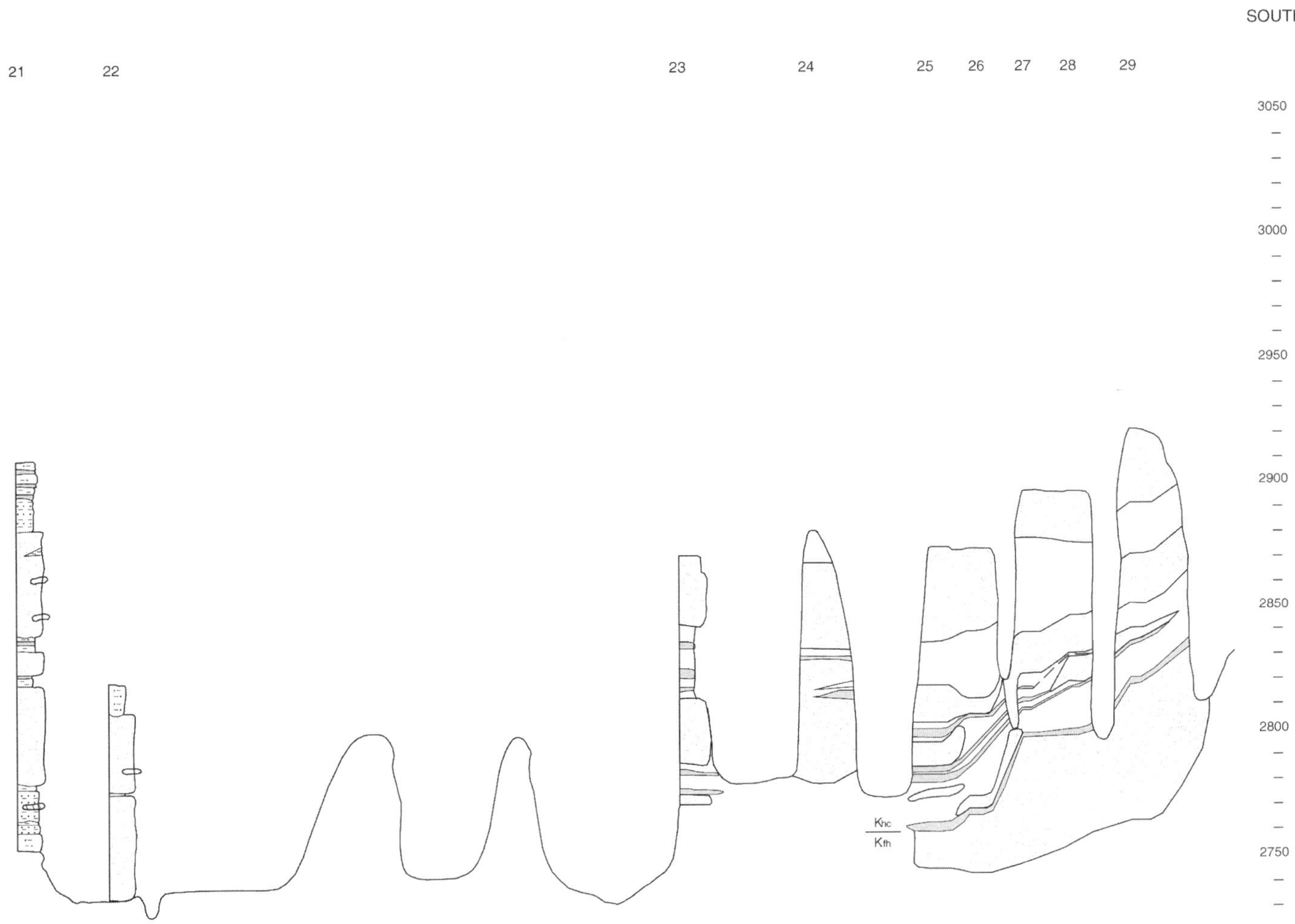

Figure 10. (*continued*)

interpreted similar lithologies to represent weakly developed paleosols.

Carbonaceous beds are common throughout Hell Creek strata in this area. These beds typically are carbonaceous mudstone or, in some places, claystones, and are rich in plant fragments and amber. They average 1 m in thickness, although they may be several meters thick. The lateral extent of these beds is extremely variable, ranging from just a few hundred meters to >1 km.

Greenish-blue bentonite beds are present in the middle to upper portion of Hell Creek strata in the northern end of this area, especially in sections 1–10 (Fig. 10). These greenish-gray to bluish-green to dark gray swelling claystone beds have extremely well developed popcorn-weathering surface textures, as compared to the bentonite beds lower in the section in the Little Beaver Creek area, and are generally less than a few meters thick. These beds are easy to trace because of their color contrast with the surrounding grayish-brown to purplish-brown mudstones. In some cases, these bentonite beds can be traced for 1 km or more, although most pinch out into adjacent strata within a few hundred meters. Another volcanic ash-related strata is a white, tuffaceous, channel sandstone that grades into a grayish-white mudstone in section 28 near the base of the Hell Creek Formation. In the Marmarth area, three volcanic ash beds (Marmarth ash, Bobcat Butte ash, Doaks Butte ash, Fig. 6) have been subjected to radiometric analysis with limited results (see Hicks et al., this volume).

Large, laterally accreting channel systems are more prevalent near the middle of the cross section in the vicinity of Highway 12, and flood basin environments (overbank, lacustrine, and paludal) are more common along the north and south ends of the transect (Fig. 10).

Although individual beds, and in some places groups of beds, can be traced across portions of this transect, the Hell Creek Formation cannot be divided into member rank lithostratigraphic units in this area. Although we were able in many places to recognize the generalized lithologic patterns upon which Frye (1967) based his subdivision of the Hell Creek, it

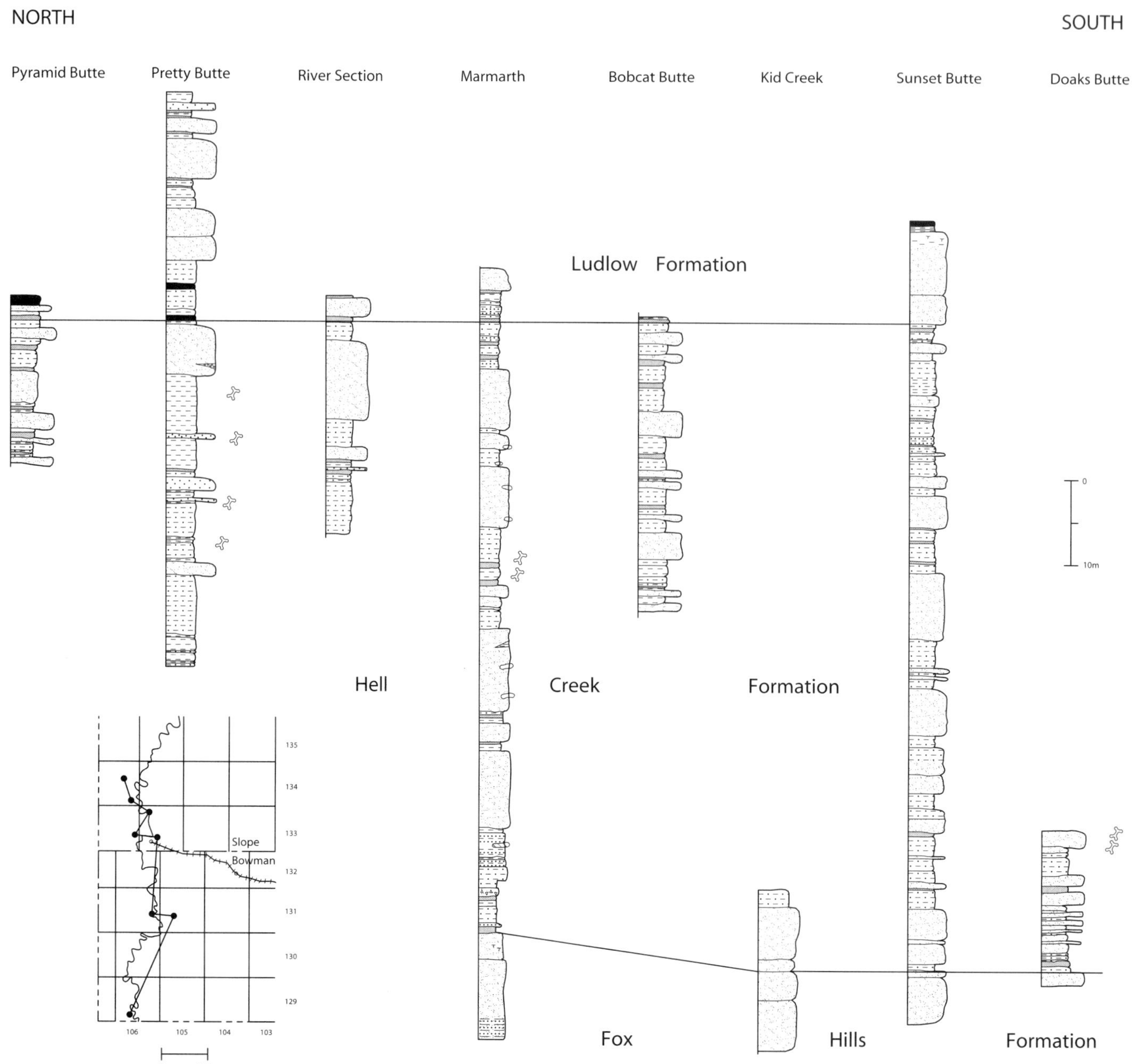

Figure 11. Selected measured sections from Bowman and Slope Counties in southwestern North Dakota. See Figure 4 for legend.

was not possible to trace these patterns for any significant distance.

Fossils

Plant fragments are common in Hell Creek mudstone and occasionally in sandstone in the Marmarth area, but leaf fossils were only observed at locality 18. Johnson (this volume) has documented more than 100 productive leaf quarries within the Hell Creek Formation in the Marmarth area. These quarries were collected between 1986 and 1999 and represent extensive prospecting efforts. Vertebrate fossils (dinosaur, turtle, and crocodilian) were found scattered throughout middle and upper Hell Creek strata in the Marmarth area, but no fossils were observed in lower Hell Creek strata (Fig. 10). Pearson et al. (this volume) presents an exhaustive vertebrate fossil survey of the Marmarth area. Fossils (primarily turtle, crocodilian, and occasionally dinosaur) were commonly found in sandstone at the base of fluvial channels. Dinosaur bones generally were found as single occurrences in mudstone. Although dinosaur fossils are scattered throughout the Hell Creek Formation in this area, there is a slight increase of fossil occurrences around the middle of the formation. A *Triceratops* skull was recently excavated by the North Dakota Geological Survey from a channel sandstone in middle Hell Creek strata, 1.6 km west of section 20 (Fig. 8).

Western North Dakota cross section

Many of the Hell Creek sections that have been previously measured in this area were measured in order to record the lithology and stratigraphic position of faunal or floral localities. As a result, many of these sections are relatively short. A 55-km-long north-south cross section from northwestern Slope County to the South Dakota border was constructed from eight relatively thick sections (Fig. 11). Only two composite sections, Marmarth (100.6 m) and Sunset Butte (100.3 m) include the entire Hell Creek Formation. Thick channel sandstone beds are present in the upper Hell Creek at several of the sections north of U.S. Highway 12 (Pretty Butte, River, and Marmarth sections). However, thick channel sandstone is absent from this interval at three other sections (Pyramid Butte, Bobcat Butte, and Sunset Butte). Johnson (this volume) presents detailed descriptions of these latter sections. Stratigraphic patterns that could be used to confidently subdivide the lower portion of the Hell Creek are also absent from the western North Dakota cross section. Channel sandstone occurrences and thicknesses are variable in this part of the section, as demonstrated by a comparison of the Doaks Butte, Sunset Butte, West Marmarth, and Kid Creek sections. No stratigraphic patterns were evident on the cross section that would enable the Hell Creek Formation to be readily subdivided into members (Fig. 11).

Fort Rice–Huff cross section

The most continuous outcrops of Hell Creek strata in south-central North Dakota occur along the west edge of the Missouri River escarpment (Fig. 12). We measured 35 sections along a 21-km-long transect from the vicinity of Fort Rice to Huff in southeastern Morton County (Fig. 13). The Fort Rice–Huff cross section is separated in two at a 2.6 km gap in an area that otherwise contains a nearly continuous transect of Hell Creek outcrops. This area was chosen for study because it contains excellent outcrops (except in the north end of the area), important vertebrate fossil localities (sections 53–56), a site that was analyzed for iridium and shocked quartz by Murphy et al. (1995) (section 67), and several of Frye's (1969) type member sections (sections 67 and 68).

The Fort Rice portion (southern) of the cross section is 4.8 km long and contains 17 measured sections, and the Huff portion (northern) consists of 18 sections measured along a 13.6 km transect along the western edge of the Missouri River escarpment. The Huff half of the panel is almost three times longer than the Fort Rice portion (Fig. 13).

Figure 12. Western edge of Missouri River escarpment near Huff. Locations of measured sections 57–64 are noted on oblique aerial photograph. View is to north-northwest.

The basal contact of the Hell Creek was identified in four measured sections (34, 36, 48, and 51) and the upper Hell Creek contact was identified in 13 others (37, 44, 45, 46, 56, 59, 60–62, 64, 65, 67, and 68). Although no individual measured section contains both the upper and lower Hell Creek contacts, composite sections from localities in relatively close proximity (within a few hundred meters) contain ~68 m of Hell Creek strata (Fig. 14). The Breien Member is exposed in more than one-third of the localities (34–38, 42, 43, 48, 51, 53, 55, 60, and 68). Thickness variations in Hell Creek strata from the top of the Breien Member to the Hell Creek–Ludlow contact range from 44.2 to 59.4 m. The top of the Breien Member is from 1.5 to 8 m above the base of the Hell Creek Formation (Hoganson and Murphy, this volume).

Throughout the Fort Rice–Huff area, the base of the Hell Creek is typically a 1–2-m-thick carbonaceous mudstone. This carbonaceous bed can be as thick as 4 m and includes a thin coal in section 51 (Fig. 14). The Fox Hills–Hell Creek contact in the Fort Rice–Huff area is very similar to the contact near the town of Breien in Morton and Sioux Counties and southwest of Marmarth in Bowman County. The lithology at the base of the Hell Creek in the Fort Rice–Huff area is a thin (<1 m), carbonaceous mudstone, siltstone, or sandstone. In the Breien area (Crowghost and State Butte localities) the basal carbonaceous bed is overlain by a thin popcorn-weathered claystone that contains numerous fossilized tree stumps in growth position. The roots of these stumps extend down into the carbonaceous bed.

Hell Creek strata below the Breien Member consist primarily of carbonaceous mudstone, thin sandstone, lignite, and occasionally swelling claystone. The lignite, up to 1.5 m thick, is present in the Huff area and extends east of the Missouri River to Coal Butte (section 71), where it was mined for domestic use in underground works at various times from the early to mid 1900s (Fig. 14). This lignite is evident on electric logs

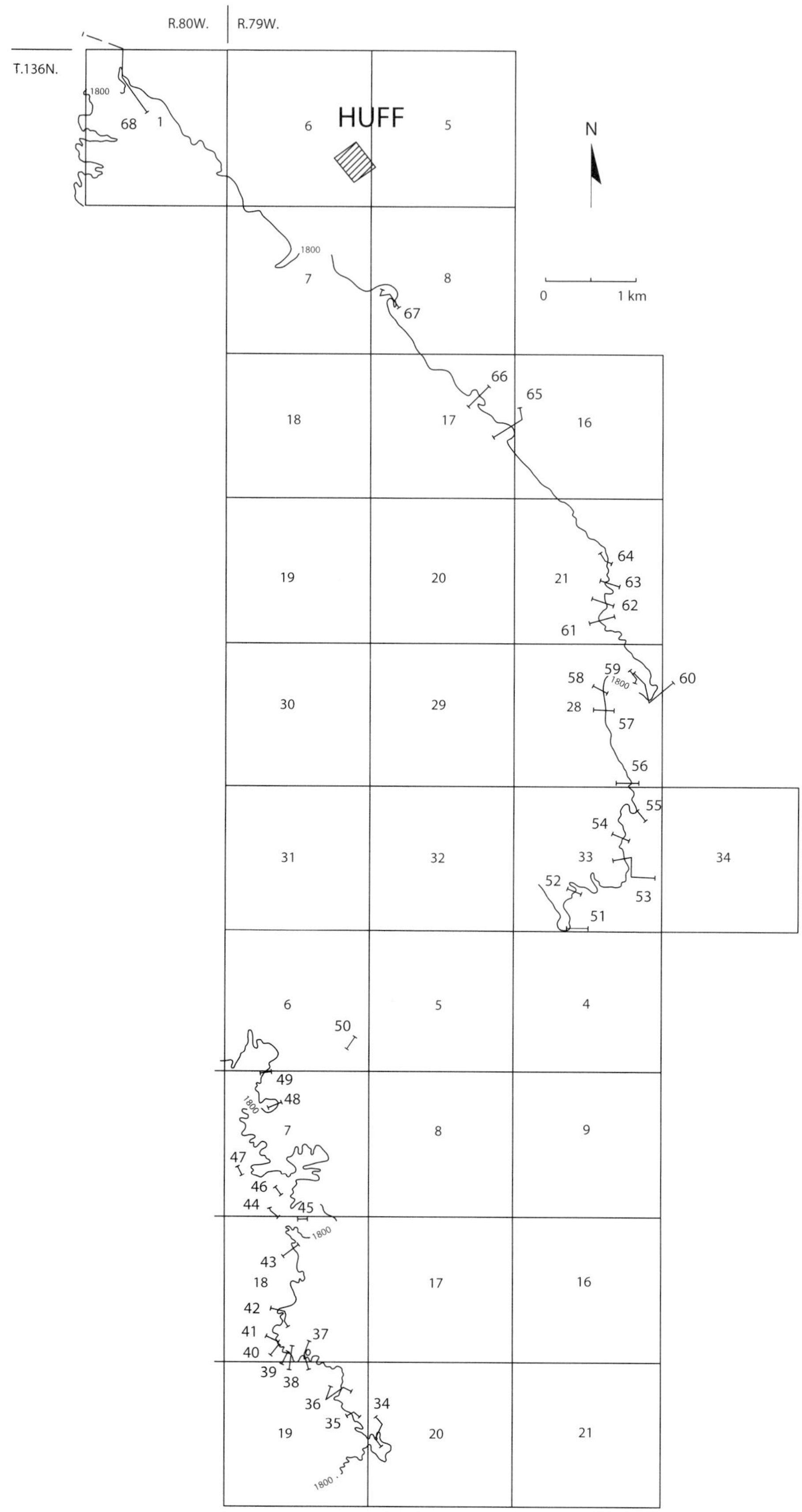

Figure 13. Location of measured sections in Fort Rice and Huff area. Huff is located at lat 46°37′30″ and long 100°39′.

and appears to have a minimum areal extent of several hundred square kilometers.

In this study, and in Hoganson and Murphy (this volume), the Breien Member has been restricted to strata that have obviously been burrowed and/or contain oyster, marine snail, and marine fish fossils. The member consists of 2.4–6.7 m of heavily iron-stained, burrowed, grayish-green sandstone and occasionally mudstone. In contrast, Frye (1967) often included an interbedded sandstone and mudstone in the upper part of this member. Although these strata are present throughout much of the Huff area (51, 53, 54, 60, and 68), they were not included in the Breien because they lacked evidence of deposition in a marine or brackish-water environment (Fig. 14). The Breien Member contains numerous iron concretions and is readily identified in surface exposures by abundant ironstone litter. This litter commonly contains rare to abundant fragments of *Ophiomorpha*. In addition to *Ophiomorpha*, Breien strata contain a mixture of marine, brackish-water, and nonmarine fossils (Hoganson and Murphy, this volume).

Hell Creek strata in the Fort Rice–Huff area are dominated by sandstone, >60% compared to 45%–48% in southwestern North Dakota. In some areas, such as section 64, the stacking of sandstone beds (multistory channels) has resulted in a section that is >80% sandstone (Fig. 14). Sandstone typically ranges from 3 to 26 m in thickness and is, on average, ~8 m thick. In general, the thicker sandstone occurs in the middle to upper portions of the Hell Creek, although, as suggested by the relatively evenly distributed sandstone exposed in complete sections (e.g., section 37), this may be a bias created by the general lack of exposures of the lower Hell Creek in this area (Fig. 14). As in the Marmarth area, sandstone is typically very fine to fine grained, clean to dirty, and poorly cemented. When bedding is apparent, sandstone beds are typically trough cross-bedded with paleocurrent directions to the east-southeast. In general, individual sandstone beds in the Fort Rice–Huff area tend to have a much greater lateral extent than those in southwestern North Dakota.

Carbonaceous mudstone beds are present throughout the Hell Creek section in the Fort Rice–Huff area and overall compose ~10% of the strata. These mudstone beds range in thickness from a few centimeters to 2.5 m. The carbonaceous beds vary considerably in their lateral extent but may extend for more than 1 km.

Swelling claystone, although common in this area, is much less pronounced than near Marmarth. These beds are generally gray to grayish-brown, unlike some of the greenish-blue bentonite in southwestern North Dakota, and tend to have relatively poorly developed popcorn-weathering surface textures. Field observations suggest that these claystone beds may, on average, contain a lower percentage of smectites than those in southwestern North Dakota. No laboratory analyses were run on these clays to test this field observation.

The top of the Hell Creek is exposed in approximately one-third of the measured sections in the Fort Rice–Huff area. The contact is easily identified in section 46, where it meets several of the criteria outlined earlier; i.e., a change in slope, an increase in vegetative cover, the first persistent coal (although at this locality the coal grades laterally into a carbonaceous claystone), a color change, and the highest occurrence of popcorn-weathering surface texture (Fig. 14). The basal Ludlow coal contains a 10-cm-thick tonstein layer at this locality. A coal is present at approximately half of the Hell Creek–Ludlow contacts (7 of 13). Where coal is absent, the contact is more difficult to pick, especially when it occurs at the base of Ludlow sandstone (i.e., sections 45 and 60).

One of Frye's (1967) measured sections (Frye no. 36) coincides with one of our sections (46) in this area (Fig. 14). He assigned all but the top 3.5 m of this section to his Huff member, which included 10 m of claystone, carbonaceous claystone, and sandstone at the base of the section. On the basis of his member criteria, these lower strata should have been placed in his Fort Rice member. This section, perhaps more than any other, demonstrates the difficulty that even Frye had in consistently recognizing his members.

Cantapeta Tongue

There is a marked color change (yellowish and/or brown tones overlying gray) at the base of a thick, laterally extensive, 12–15-m-thick sandstone in the middle to upper portion of the Hell Creek Formation near Huff. The base of this massive sandstone appears to be a scour surface that was filled with tidal-influenced deposits. This bed can be traced in outcrop over a distance of at least 21 km in the western part of the Missouri River escarpment (Fig. 14). The base of this bed occurs ~16 m below the top of the Hell Creek (between the elevations of 564 and 579 m in the Fort Rice area and between the elevations of 541 and 549 m in the Huff area (Fig. 14). In addition to a distinct color change, the base of the sandstone typically contains iron nodules and pebbles that are occasionally cemented. In section 60, the base of this bed consists of several meters of an intraformational pebble conglomerate deposited in high-angle cross-stratification. Pebbles consist primarily of locally derived mudstone and are to 8 cm in diameter. The sandstone is typically trough cross-bedded with a predominant east-south-east paleocurrent direction. At one locality (section 48), this sandstone contains several *Ophiomorpha* burrows (Fig. 15) and has a westward paleocurrent direction. This unit is also heavily burrowed in an adjacent measured section (45). This sandstone was initially interpreted to be Cannonball strata that had incised into the Hell Creek. However, upon further investigation, characteristics indicative of the Hell Creek Formation, such as popcorn-weathering claystone and dinosaur bones, were found in strata overlying this unit in sections 60 and 64 (Fig. 14). This thick sandstone was likely deposited in an interdistributary channel. This channel occasionally acted as a tidal inlet as a result of an increase in sea level or a decrease in runoff into the Western Interior Seaway. Additional marine influence in this bed is suggested by a series of distorted cross-beds in section 46 and flaser bedding in sections 60 and 61 (Figs. 14 and 16). The presence of *Ophiomorpha* and these sedimentary structures indicates a marine or brackish-water depositional environment for at least part of this sandstone, suggesting that the western edge of the Western Interior Seaway was in relatively close proximity to the area at the time of deposition. This transgression is here named the Cantapeta Tongue (for a creek in the area). Evidence for this advance is found stratigraphically in the upper one-third of the Hell Creek Formation between the Breien Member and the Cannonball Member. The Cantapeta Tongue is only 19.5 m below the Paleocene Cannonball Member of the Fort Union Formation, strongly suggesting that the Western Interior Sea never receded from the upper mid-continent at the end of the Cretaceous.

Fossils

Plant fragments are common throughout Hell Creek strata in the Fort Rice–Huff area, but leaf fossils were not discovered during the measuring of sections or tracing of beds. However, the authors of this paper discovered several leaf localities in upper Hell Creek strata ~50 km southwest of Fort Rice. Vertebrate fossils (turtle, crocodilian, and dinosaur) when present were concentrated near the base of the Hell Creek. Vertebrate fossils tend to be concentrated in carbonaceous mudstone and sandstone at specific stratigraphic levels either just above or below the Breien Member. Dinosaur taxa include ceratopsians, hadrosaurs, and tyrannosaurids (Fig. 17). Unlike the Marmarth area, vertebrate fossils were seldom observed at the base of channel sandstone beds in the Fort Rice–Huff area.

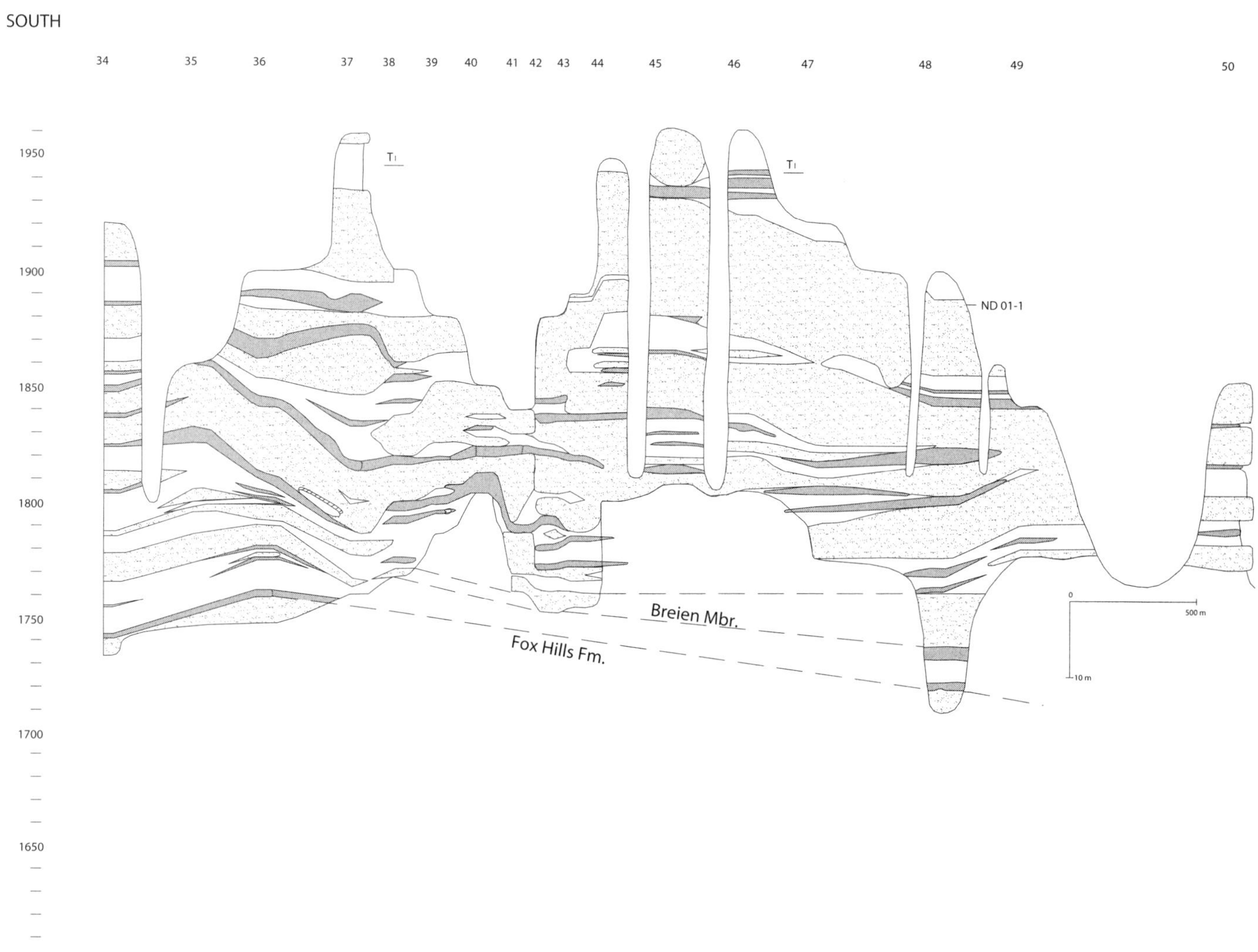

Figure 14. Profile of measured sections of Fox Hills, Hell Creek, and Fort Union strata in Fort Rice and Huff area. See Figure 4 for legend and Figure 13 for location of measured sections. Mudstone and claystone are depicted with blank pattern to simplify diagram. ND 01-1 represents stratigraphic position of *Ophiomorpha* specimens. Tl is abbreviation for Ludlow Member. (*Continued on following page.*)

A list of invertebrate and vertebrate taxa from the Hell Creek Formation of south-central North Dakota was given in Murphy et al. (1995). Hoganson and Murphy (this volume) define the vertebrate fauna of the Breien Member and discuss the significance of these fossils. Identifications of a few of the invertebrate non-Breien specimens collected and identified by Frye (1967) as possible brackish-water or marine fauna were confirmed by J.H. Hartman (1999, personal commun.), offering additional evidence of marine influence in Hell Creek strata above the Breien Member in south-central North Dakota.

South-central North Dakota cross sections

East-west and north-south cross sections were constructed from 24 sections measured by Murphy et al. (1995) in south-central North Dakota. On average, these sections are spaced 9–10 km apart and generally include only the top 25%–30% of the Hell Creek Formation. It is obvious, even from these incomplete Hell Creek sections, that the distribution of channel and nonchannel lithologies is variable across the upper portion of the formation and no consistent stratigraphic pattern is discernable. In general, there appears to be an increase in sandstone thickness in upper Hell Creek strata from west to east in south-central North Dakota (Fig. 18).

Subsurface correlation

Over the years, we have studied hundreds of electric logs that included the stratigraphic interval containing lower Paleocene and Upper Cretaceous strata. These logs were from ura-

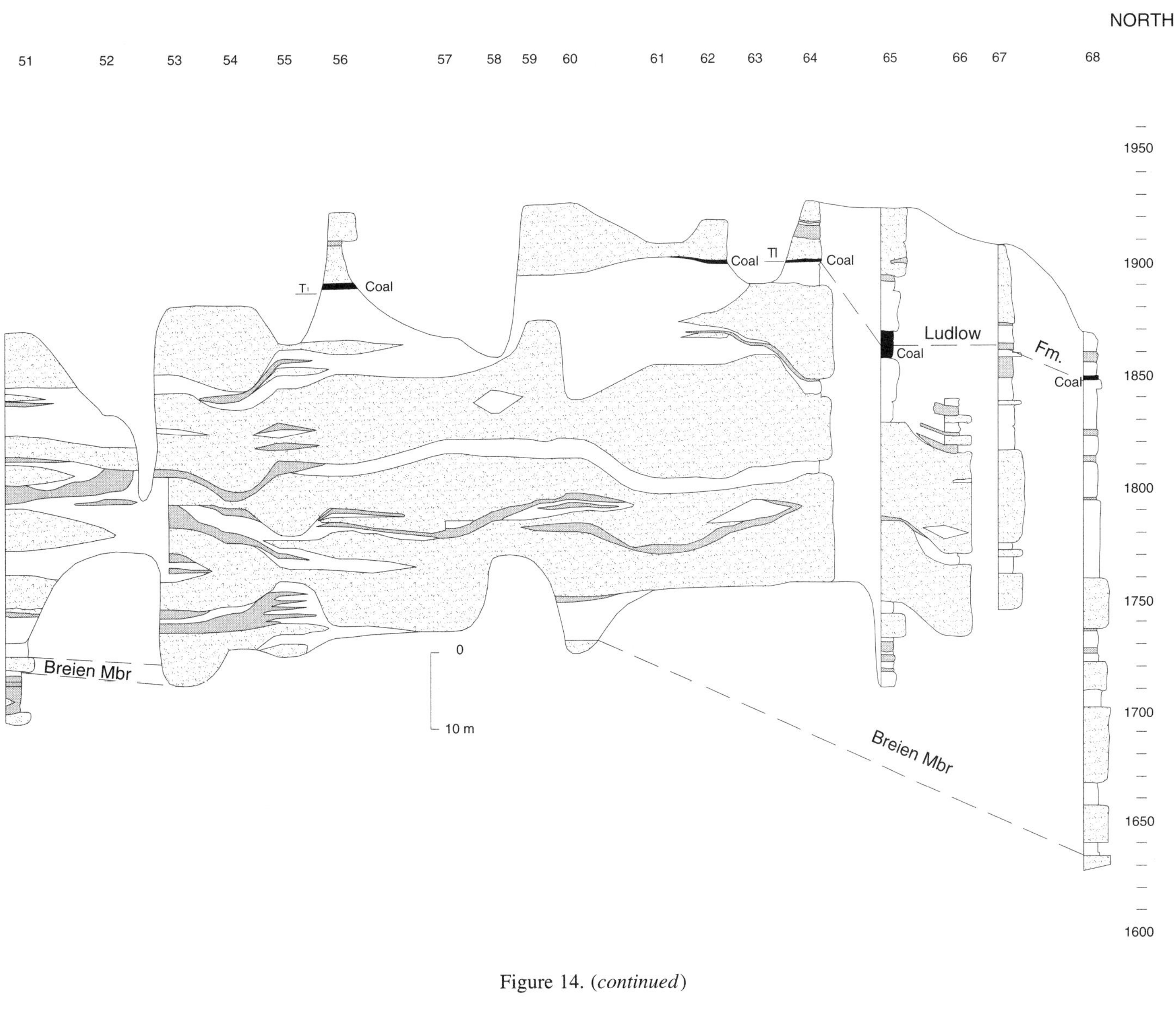

Figure 14. (*continued*)

Figure 15. A: Burrowed bed in upper Hell Creek Formation (Cantapeta Tongue) at section 48 in Fort Rice–Huff area. B: *Ophiomorpha* obtained from burrowed bed in section 48 (North Dakota State Fossil Collection catalog numbers ND01-1.1, ND01-1.2, ND01-1.3, ND01-1.4, North Dakota Heritage Center, Bismarck). Scale in cm.

Figure 16. Sandstone and mudstone exhibiting flaser and lenticular bedding in upper portion of measured section 60 south of Huff.

Figure 17. Hadrosaur femur few meters above Breien Member of Hell Creek Formation in measured section 36 south of Huff. Pick is 1 m long.

nium holes, oil wells, or North Dakota State Water Commission observation wells. This database consists mostly of gamma logs run through oil and gas surface casing. Relatively few suites of logs (gamma, spontaneous potential, resistivity, and density) are available for the Upper Cretaceous and lower Paleocene stratigraphic interval. Fox Hills, Hell Creek, and Fort Union strata are notoriously difficult to differentiate in the subsurface because these strata are lithologically similar. The stratigraphically highest contact that can be determined in the subsurface with confidence is the transition from shale to mudstone and siltstone at the Pierre–Fox Hills contact. The Fox Hills–Hell Creek contact is only obvious on gamma logs when the Colgate Member of the Fox Hills is much thicker or better sorted than the overlying channel sandstone in the Hell Creek Formation. Occasionally, a caliper log may also be useful to identify swelling claystone at or near the base of the Hell Creek Formation.

The top of the Hell Creek Formation is generally much more difficult to identify in the subsurface. Surface observations indicate that 80% of measured sections in southwestern North Dakota contained coal at the Hell Creek–Ludlow contact compared to only 40% of the sections in the south-central part of the state. When present, the basal Ludlow coal is generally <1 m thick and may be difficult to identify on oil well gamma logs at a scale of 1:1200. Selected, widely spaced logs may give a false impression that there are beds or facies that can be correlated within the Hell Creek. However, cross sections constructed from closely spaced wells in oil fields demonstrate the same lithologic variability in the Hell Creek that is obvious at the surface. For these reasons, we believe that current remote sensing technology is inadequate to accurately recognize the contacts of the Hell Creek Formation in the subsurface.

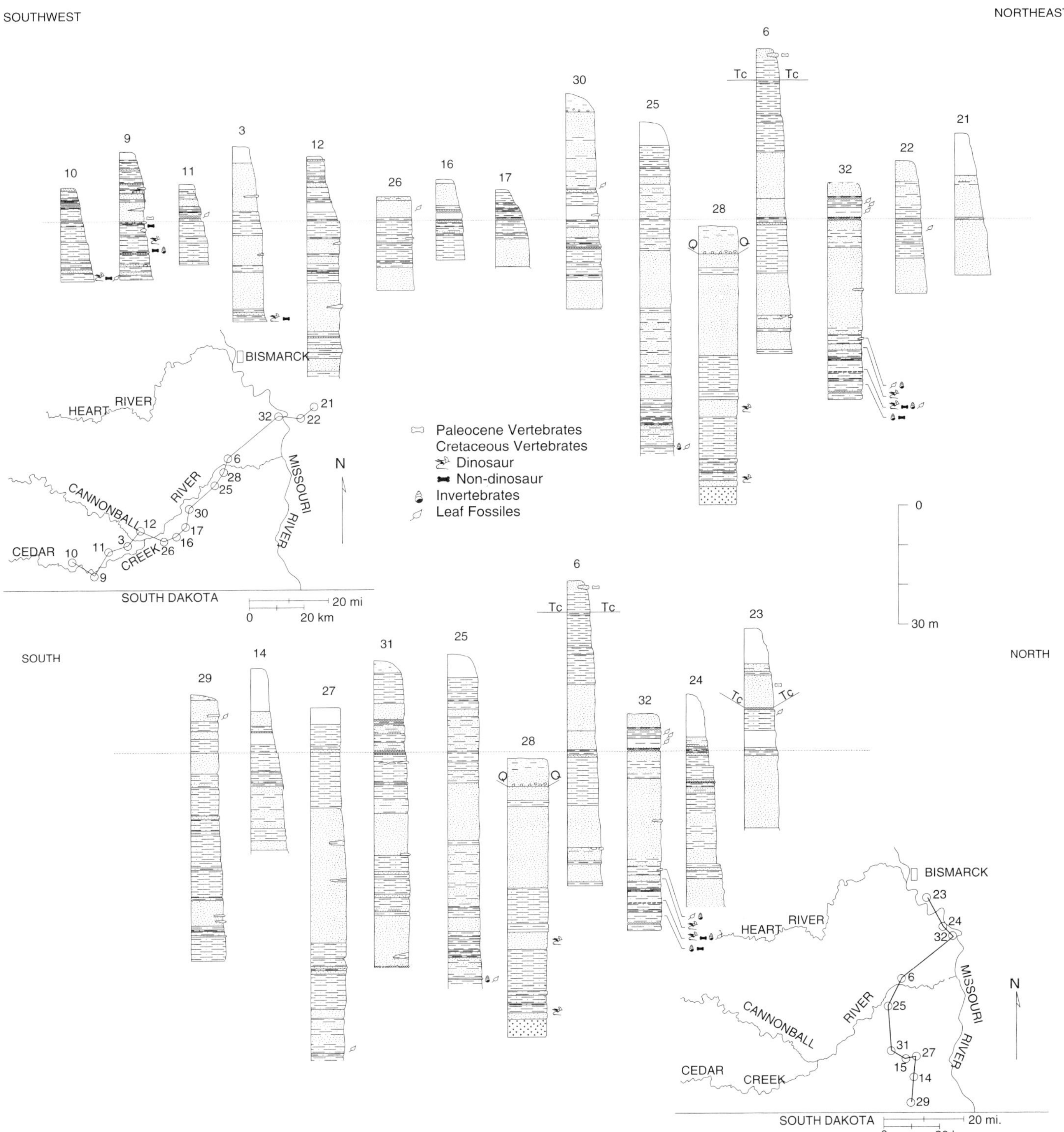

Figure 18. Cross sections of upper Hell Creek and lower Fort Union strata in south-central North Dakota. Sections, including section numbers, are from Murphy et al. (1995). See Figure 4 for legend. Datum is Hell Creek–Ludlow contact. Tc is abbreviation for Cannonball Member.

CONCLUSIONS

Subdivision of Hell Creek strata

The Hell Creek Formation thins from ~100 m (100.3 m for the Sunset Butte composite section and 100.6 m for the Marmarth composite section) in southwestern North Dakota to 47 m (Buddy Butte section) east of the Missouri River in south-central North Dakota (Fig. 19). In general, the overall character of the formation remains unchanged from western to central North Dakota, but there are some obvious differences between the two outcrop areas. Light blue and bluish-green swelling claystone beds are common in middle Hell Creek strata in the Marmarth area but are absent from Hell Creek strata in the Fort Rice–Huff area. Frye (1969) attempted to use this brightly colored swelling claystone in the Marmarth area to define a member in middle Hell Creek strata in southwestern North Dakota but, these claystone beds occur much higher in the Hell Creek section at Sunset Butte than Frye suggested.

The relative percentage of sandstone in the Hell Creek is higher in south-central North Dakota than in the Marmarth area, and sandstone geometry is quite different between these two areas (Figs. 10 and 14). Near Marmarth, sandstone is typically lenticular, in contrast to the elongated (sheet) sandstone bodies that occur above scour surfaces in upper Hell Creek strata in south-central North Dakota (Figs. 10 and 14). The difference in sandstone geometry is likely attributable to marine depositional influences in south-central North Dakota. In the Marmarth area, fluvial channels meandered over vast flood plains, whereas in the Fort Rice–Huff area, fluvial systems were focused into distributary channels along the margin of the Western Interior Seaway.

The Breien Member is a valid formal lithostratigraphic unit and can be mapped in the surface over an area of 6100 km^2 in south-central North Dakota. No other valid subdivisions of the Hell Creek Formation can be recognized. Those proposed by Frye (1969) are generally not mappable, even over relatively short distances of a few kilometers, although mappability is not a criterion for member status. Although Frye made numerous valid observations relating to lithofacies changes in the Hell Creek Formation, these changes are either too localized or too subtle to subdivide the formation into members as he did.

Workers in south-central and southwestern North Dakota have been unable to subdivide the Hell Creek for mapping purposes at scales ranging from 1:127 000 to 1:24 000 (Carlson, 1979, 1982, 1983; Murphy, 1991a, 1991b). C.J. Hares (1928, p. 28) made the following observations concerning Hell Creek stratigraphy in southwestern North Dakota: "There is the utmost variation in the rocks, both vertically and horizontally. Two detailed sections taken by the same person across the same stratigraphic interval and less than 100 feet apart would not show marked likeness, and if separated by a distance of a mile very likely they would be absolutely different." Fastovsky (1987) noted that correlation between noncontinuous exposures of Hell Creek channel sandstone is difficult not only because the lithology and petrology are indistinguishable, but also because the vertical position of a given channel varies with time. After years of working on Hell Creek stratigraphy, we agree with the observations of Hares and Fastovsky, and the stratigraphic cross sections presented herein support these conclusions.

Depositional environments

Measured sections in the Marmarth area, along with the resulting cross sections, support previous interpretations that Hell Creek strata in this area were deposited on a broad, fluvial, highly vegetated plain that was dissected by meandering rivers that often overflowed their banks (Fastovsky, 1986; Johnson, 1989). This depositional setting resulted in relatively narrow channel widths (generally <300 m) and laterally extensive (several hundred meters) but thin (<3 m) flood-plain strata (Fastovsky, 1987; Belt et al., 1997). This model is not applicable in the Fort Rice–Huff area, where channel widths in the upper part of the Hell Creek are at times an order of magnitude greater than those observed in southwestern North Dakota.

It has been suggested that the Western Interior Seaway had withdrawn significantly from North Dakota during deposition of upper Hell Creek strata (Lillegraven and Ostresh, 1990). However, the marine or brackish-water lithologies and fossil assemblages of the Breien Member near the base of the Hell Creek Formation in south-central North Dakota (Hoganson and Murphy, this volume), marine or brackish-water fossils in the upper Hell Creek Formation in south-central North Dakota (Cantapeta Tongue), and the occurrence of fossil faunas that could tolerate marine or brackish-water environments in southwestern North Dakota (Pearson et al., this volume) suggest that marine conditions existed in south-central North Dakota. A stratigraphic interval of only 19.5 m separates the *Ophiomorpha*-bearing Cantapeta Tongue from marine strata of the Cannonball Member (Fig. 20). Evidence of marine conditions during deposition of the Breien Member and the later Cantapeta Tongue in south-central North Dakota supports others that have proposed a shoreline position in the vicinity of central North Dakota near the end of the Cretaceous (Roberts and Kirschbaum, 1995; Kennedy et al., 1998; Erickson, 1992, 1999).

Dinosaur fossils

Although not the primary focus of this study, dinosaur fossils were often encountered while measuring section or mapping geology. Pearson et al. (this volume) present a detailed analysis of dinosaur occurrence in the Hell Creek Formation of southwestern North Dakota. Nonetheless, we find the observations of dinosaur remains made while analyzing the stratigraphy provide some ability to compare the occurrence of fossils in the two main study areas. Relatively few dinosaur fossils were discovered in Hell Creek strata during this study; 18 specimens from the Marmarth area and 35 specimens from the Fort

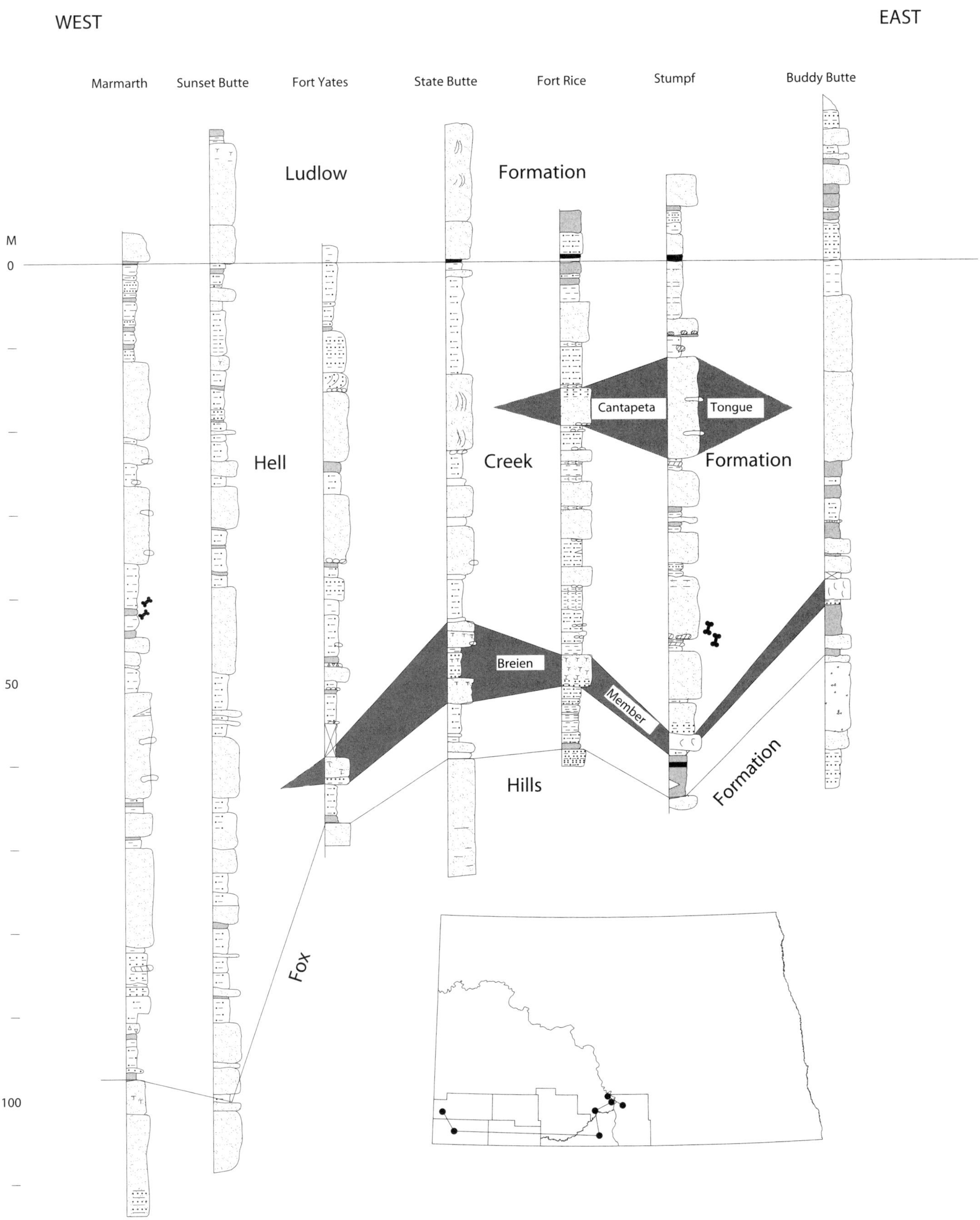

Figure 19. Cross section of select complete sections of Hell Creek strata in North Dakota. In addition to these, there are few additional sections in south-central North Dakota where entire Hell Creek Formation is exposed. See Figure 4 for legend.

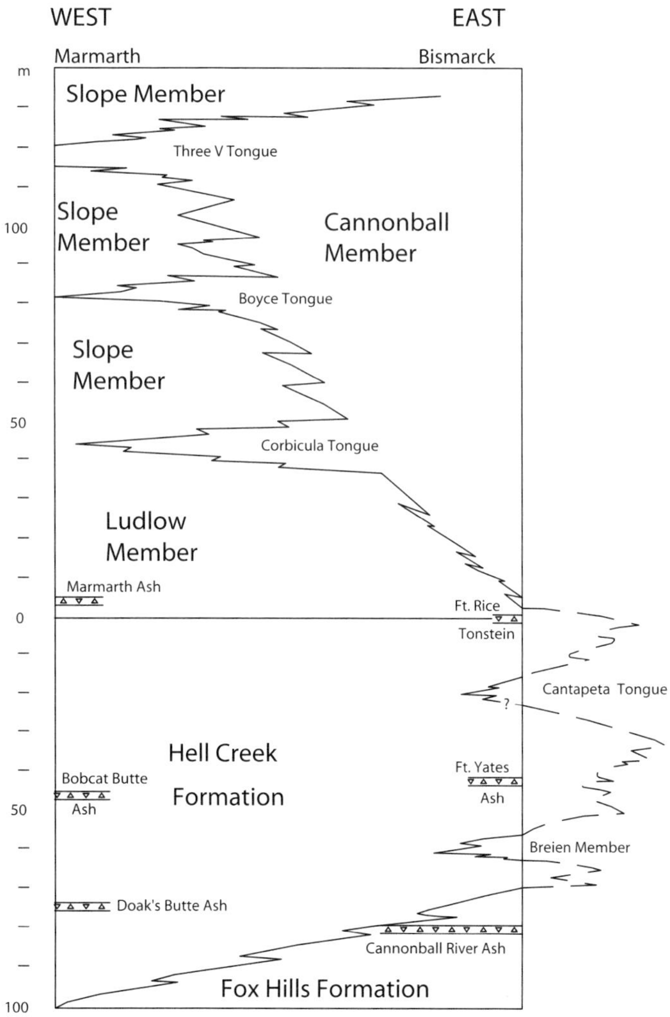

Figure 20. Profile of marine and nonmarine beds in Upper Cretaceous and Paleocene strata in North Dakota. Data on Cannonball tongues are from Hartman (1993).

Rice–Huff area. The majority of these specimens were single bone occurrences, although a few articulated bones were observed in both areas. Identical field methods were used in both areas, i.e., one individual noting the position, and when possible the taxon, of dinosaur fossils found while measuring sections and laterally tracing beds. However, even with this limited database, it is apparent that there is a significant difference in both abundance and relative position of dinosaur fossils in Hell Creek strata between the Marmarth and Fort Rice areas. Dinosaur fossils are present throughout middle to upper Hell Creek strata near Marmarth, but absent from both the lower half and upper 10 m of the formation (Fig. 21). Pearson et al. (this volume), because of their intensive sampling, recorded dinosaur remains throughout the formation in southwestern North Dakota. In contrast, there is a strong concentration of dinosaur fossils in the lower part of the Hell Creek Formation in south-central North Dakota (Fig. 21). Of the dinosaur fossils observed in south-central North Dakota, 95% were found within 10 m of the Breien Member.

Recommendations

Evidence (trace fossils, bedding changes, and green, potentially glauconitic sandstone) suggest a marine influence on deposition of strata in the upper part of the Hell Creek Formation between Fort Rice and Huff in south-central North Dakota. Petrographic analysis of sandstone in this area could verify the presence of glauconite. Field work south of the Fort Rice–Huff area, in the Fort Yates area or north-central South Dakota, may provide additional evidence to support the conclusions in this report.

It is assumed that the base of the Hell Creek Formation is progressively younger to the east and that the top of the formation is in close proximity to the K-T boundary throughout the Williston Basin. Analysis of datable minerals in tuff and

Figure 21. Position of dinosaur fossil occurrences in Hell Creek strata in southwestern and south-central North Dakota. Thicknesses (from Ludlow–Hell Creek contact to base of Hell Creek Formation) are plotted against number of individual dinosaur fossils.

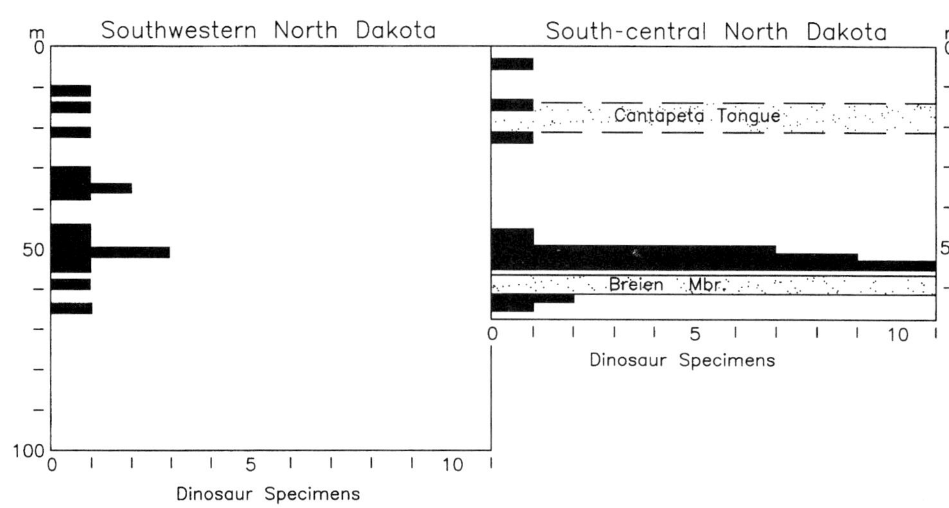

bentonite beds from both the base and top of the Hell Creek, as well as from throughout the unit, will likely support this theory and may answer other pertinent stratigraphic questions such as the position of temporally equivalent units to the Breien Member in southwestern North Dakota. Tuff beds (~6) have recently been discovered in Hell Creek strata, both in southwestern North Dakota and in the Missouri River trench area (Fig. 6). Most of these tuff deposits have not been precisely dated despite some serious effort (Hicks et al., this volume). A recent attempt failed to find datable material within the Fort Rice tonstein at the base of the Ludlow Formation in south-central North Dakota (Lisa Peters, 1999, personal commun.). Several previously reported tuff beds (the Cannonball River ash at the base of the Hell Creek Formation and the Marmarth ash at the top) are scheduled for analysis.

ACKNOWLEDGMENTS

All of the geologic sections in this report were measured by Ed Murphy except for the following: Doaks Butte—Dean Pearson, Terry Schaefer, and Jason Hicks; Pyramid and Bobcat Buttes—Kirk Johnson and Rob Myers; River Section and Sunset Butte—Dave Fastovsky and Kirk Johnson. The following landowners graciously allowed access to their land: Mike Sonsalla, Douglas Duffield, Scott and Henry Bradac, Dale Rebenitsch, Doug Lockner, Frank Armbrust, Lyle Grenz, Ruben Harriet, John Stumpf, Ken Miller, Tom Hunter, Frank Vogel, and Eugene Graner. We thank the Standing Rock Sioux Nation for allowing access to outcrop areas on this reservation. We appreciate the helpful suggestions for manuscript improvement made by David Fastovsky, James Schmitt, Joseph Hartman, and Douglas Nichols

REFERENCES CITED

Belt, E.S., Hicks, J.F., and Murphy, D.A., 1997, A pre-Lancian regional unconformity and its relationship to Hell Creek paleogeography in southeastern Montana: University of Wyoming Contributions to Geology, v. 31, no. 2, p. 1–26.

Brown, B., 1907, The Hell Creek beds of the Upper Cretaceous of Montana: American Museum of Natural History Bulletin, v. 23, p. 823–845.

Brown, R.W., 1938, The Cretaceous-Eocene boundary in Montana and North Dakota [abs.]: Washington Academy of Science Journal, v. 28, p. 421–422.

Brown, R.W., 1952, Tertiary strata in eastern Montana and western North and South Dakota: Billings Geological Society Guidebook, v. 3, p. 89–92.

Brown, R.W., 1962, Paleocene flora of the Rocky Mountains and Great Plains: U.S. Geological Survey Professional Paper 375, 119 p.

Butler, R.D., 1980, Stratigraphy, sedimentology, and depositional environments of the Hell Creek Formation (Late Cretaceous) and adjacent strata, Glendive area, Montana [Ph.D. thesis]: Grand Forks, University of North Dakota, 398 p.

Calvert, W.R., 1912, Geology of certain lignite fields in eastern Montana: U.S. Geological Survey Bulletin 471, p. 187–201.

Carlson, C.G., 1979, Geology of Adams and Bowman County, North Dakota: North Dakota Geological Survey Bulletin 65, 29 p.

Carlson, C.G., 1982, Geology of Grant and Sioux Counties, North Dakota: North Dakota Geological Survey Bulletin 67, 32 p.

Carlson, C.G., 1983, Geology of Morton County, North Dakota: North Dakota Geological Survey Bulletin 72, 37 p.

Cherven, V.B., and Jacob, A.R., 1985, Evolution of Paleogene depositional systems, Williston Basin, in response to global sea level changes, *in* Flores, M.R., and Kaplan, S.S., eds., Cenozoic paleogeography of the west-central United States: Society of Economic Paleontologists and Mineralogists, Rocky Mountain Section, p. 127–170.

Curtiss, R.E., 1952, Areal geology of the Isabel quadrangle: Rapid City, South Dakota School of Mines, scale 1:62 500.

Curtiss, R.E., 1954, Areal geology of the Firesteel Creek quadrangle: Rapid City, South Dakota School of Mines, scale 1:62 500.

Daly, D.J., 1984, Stratigraphy and depositional environment of the Fox Hills Formation (Upper Cretaceous), Bowman County, North Dakota [M.S. thesis]: Grand Forks, University of North Dakota, 200 p.

Erickson, J.M., 1992, Subsurface stratigraphy, lithofacies and paleoenvironments of the Fox Hills Formation (Maastrichtian: Late Cretaceous) adjacent to the type area, North Dakota and South Dakota: Towards a more holistic approach, *in* Erickson, J.M., and Hoganson, J.W., eds., Proceedings of the F.D. Holland, Jr., Geological Symposium: North Dakota Geological Survey Miscellaneous Series 76, p. 199–242.

Erickson, J.M., 1999, The Dakota Isthmus: Closing the late Cretaceous Western Interior seaway: North Dakota Academy of Science Proceedings, v. 53, p. 124–129.

Farris, R.A., 1984, Heavy minerals of the Cretaceous Hell Creek and Paleocene Ludlow Formations of Slope and Bowman counties, North Dakota [M.S. thesis]: Grand Forks, University of North Dakota, 109 p.

Fastovsky, D.E., 1986, Paleoenvironments of vertebrate-bearing strata at the Cretaceous-Paleogene boundary in northeastern Montana and southwestern North Dakota [Ph.D. thesis]: Madison, University of Wisconsin, 301 p.

Fastovsky, D.E., 1987, Paleoenvironments of vertebrate-bearing strata during the Cretaceous-Paleogene transition, eastern Montana and western North Dakota: Palaios, v. 2, p. 282–295.

Fastovsky, D.E., and McSweeney, K., 1987, Paleosols spanning the Cretaceous-Paleocene transition, eastern Montana and western North Dakota: Geological Society of America Bulletin, v. 99, p. 66–77.

Feldman, R.M., 1972, Stratigraphy and paleoecology of the Fox Hills Formation (Upper Cretaceous) of North Dakota: North Dakota Geological Survey Bulletin 61, 65 p.

Frye, C.I., 1967, The Hell Creek Formation in North Dakota [Ph.D. thesis]: Grand Forks, University of North Dakota, 411 p.

Frye, C.I., 1969, Stratigraphy of the Hell Creek Formation in North Dakota: North Dakota Geological Survey Bulletin 54, 65 p.

Hares, C.J., 1928, Geology and lignite resources of the Marmarth Field, southwestern North Dakota: U.S. Geological Survey Bulletin 775, 110 p.

Hartman, J.H., 1993, Road log to the type areas of the Paleocene Slope Formation and intercalated tongues of the Cannonball Formation, Slope County, North Dakota, *in* Kihm, A.J., and Hartman, J.H., eds., The Marshall Lambert Symposium: Bismarck, North Dakota Geological Society, p.78–86.

Hunter, J.P., Hartman, J.H., and Krause, D.W., 1997, Mammals, and mollusks across the Cretaceous-Tertiary boundary from Makoshika State Park and vicinity (Williston Basin), Montana: University of Wyoming Contributions to Geology, v. 32, p. 61–114.

Johnson, K.R., 1989, A high resolution megafloral biostratigraphy spanning the Cretaceous-Tertiary boundary in the Northern Great Plains [Ph.D. thesis]: New Haven, Connecticut, Yale University, 581 p.

Johnson, K.R., 1992, Leaf-fossil evidence for extensive floral extinction at the Cretaceous-Tertiary boundary, North Dakota, USA: Cretaceous Research, v. 13, p. 91–117.

Johnson, K.R., and Hickey, L.J., 1990, Megafloral change across the Cretaceous/Tertiary boundary in the northern Great Plains and Rocky Mountains, U.S.A., *in* Sharpton, V.L., and Ward, P.D., eds., Global catastrophes in Earth history: An interdisciplinary conference on impacts, volcanism, and mass mortality: Geological Society of America Special Paper 247, p. 433–444.

Johnson, K.R., Nichols, D.J., Attrep, M., Jr., and Orth, C.J., 1989, High-resolution leaf-fossil record spanning the Cretaceous/Tertiary boundary: Nature, v. 340, p. 708–711.

Kennedy, W.J., Landman, N.H., Christensen, W.K., Cobban, W.A., and Hancock, J.M., 1998, Marine connections in North America during the late Maastrichtian: Palaeogeographic and palaeobiogeographic significance of *Jeletzkytes nebrascensis* Zone cephalopod fauna from the Elk Butte Member of the Pierre Shale, southeast South Dakota and northeast Nebraska, 1998: Cretaceous Research 19, p. 745–775.

Laird, W.M., and Mitchell, R.H., 1942, The geology of the southern part of Morton County, North Dakota: North Dakota Geological Survey Bulletin 14, 42 p.

Leonard, A.G., 1908, The geology of southwestern North Dakota with special reference to the coal: North Dakota Geological Survey Fifth Annual Report, p. 1–113.

Leonard, A.G., 1911, The Cretaceous and Tertiary formations of western North Dakota and eastern Montana: Journal of Geology, v. 14, p. 507–547.

Leonard, A.G., 1912, Bismarck folio: U.S. Geological Survey Geologic Atlas of the United States, no. 181, 8 p.

Lillegraven, J.A., and Ostresh, L.M., Jr., 1990, Late Cretaceous (earliest Campanian/Maastrichtian) evolution of western shorelines of the North American Western Interior seaway in relation to known mammalian faunas, *in* Bown, T.M., and Rose, K.D., eds., Dawn of the age of mammals in the northern part of the Rocky Mountain Interior, North America: Geological Society of America Special Paper 243, p. 1–30.

Moore, W.L., 1976, The stratigraphy and environments of deposition of the Cretaceous Hell Creek Formation (reconnaissance) and the Paleocene Ludlow Formation (detailed), southwestern North Dakota: North Dakota Geological Survey Report of Investigations, v. 56, 40 p.

Murphy, E.C., 1991a, Geologic map of the Lookout Butte Quadrangle: North Dakota Geological Survey, scale 1:24000.

Murphy, E.C., 1991b, Geologic map of the Lookout Butte SE Quadrangle: North Dakota Geological Survey, scale 1:24000.

Murphy, E.C., Nichols, D.J., Hoganson, J.W., and Forsman, N.F., 1995, The Cretaceous/Tertiary boundary in south-central North Dakota: North Dakota Geological Survey Report of Investigations, v. 98, 74 p.

Retallack, G.J., 1994, A pedotype approach to latest Cretaceous and earliest Tertiary paleosols in eastern Montana: Geological Society of America Bulletin, v. 106, p. 1377–1397.

Roberts, L.N.R., and Kirschbaum, M.A., 1995, Paleogeography of the Late Cretaceous of the Western Interior of middle North America: Coal distribution and sediment accumulation: U.S. Geological Survey Professional Paper 1561, p. 1–49.

Seager, O.A., Blackstone, D.L., Jr., Cobban, W.A., Downs, G.R., Laird, W.M., and Sloss, L.L., 1942, Stratigraphy of North Dakota: Discussion: American Association of Petroleum Geologists Bulletin, v. 26, p. 1414–1423.

Thom, T.R., Jr., and Dobbin, C.E., 1924, Stratigraphy of Cretaceous-Eocene transition beds in eastern Montana and the Dakotas: Geological Society of America Bulletin, v. 35, p. 431–506.

Waage, K.M., 1968, The type Fox Hills Formation, Cretaceous (Maestrichtian), South Dakota. 1. Stratigraphy and paleoenvironments: New Haven, Connecticut, Yale University Peabody Museum of Natural History Bulletin 27, 171 p.

White, P.D., Fastovsky, D.E., and Sheehan, P.M., 1998, Taphonomy and suggested structure of the dinosaurian assemblage of the Hell Creek Formation (Maastrichtian), eastern Montana and western North Dakota: SEPM (Society for Sedimentary Geology) Research Reports, p. 41–51.

Manuscript Accepted by the Society September 24, 2001

Printed in the U.S.A.

Geological Society of America
Special Paper 361
2002

Magnetostratigraphy and geochronology of the Hell Creek and basal Fort Union Formations of southwestern North Dakota and a recalibration of the age of the Cretaceous-Tertiary boundary

Jason F. Hicks
Kirk R. Johnson
Department of Earth Sciences, Denver Museum of Nature & Science, 2001 Colorado Boulevard, Denver, Colorado 80205-5798, USA
John D. Obradovich
U.S. Geological Survey, Box 25046, Denver Federal Center, Denver, Colorado 80225-0046, USA
Lisa Tauxe
David Clark
Scripps Institution of Oceanography, Geological Research Division, 9500 Gilman Drive, La Jolla, California 92093-0220, USA

ABSTRACT

To estimate the age and duration of the Maastrichtian Hell Creek Formation, an extensive program of paleomagnetic sampling was carried out on six surface sections that span the Maastrichtian and Paleocene interval on the southern portion of the Cedar Creek anticline in Slope and Bowman Counties in North Dakota. The magnetic polarity sequence measured can be correlated with confidence to that part of the geomagnetic polarity time scale that ranges from polarity subchron C30n through C29n. A revised age estimate for the Cretaceous-Tertiary boundary of 65.51 ± 0.10 Ma has been obtained by normalizing the most recently published isotopic ages for the boundary to a standard monitor age of 28.02 Ma for the Fish Canyon Tuff and 28.32 Ma for the Taylor Creek Rhyolite. Orbital chronology gives very precise estimates for the duration of C29r that range from 570 to 673 k.y. The most recently published estimate gives a value of 603 k.y. for C29r, with a level of precision that is much higher than ages obtained by conventional dating methods. By extrapolating the measured sediment accumulation rate of the Cretaceous portion of C29r (333 k.y.) through to the base of the Hell Creek, we estimate the Hell Creek Formation to be 1.36 m.y. in duration. If the range of precessional age estimates for C29r and the differences in thickness of the reversal between the sections are taken into account, then the duration of the Hell Creek could range from 1.05 to 1.90 m.y.

Hicks, J.F., Johnson, K.R., Obradovich, J.D., Tauxe, L., and Clark, D., 2002, Magnetostratigraphy and geochronology of the Hell Creek and basal Fort Union Formations of southwestern North Dakota and a recalibration of the age of the Cretaceous-Tertiary boundary, *in* Hartman, J.H., Johnson, K.R., and Nichols, D.J., eds., The Hell Creek Formation and the Cretaceous-Tertiary boundary in the northern Great Plains: An integrated continental record of the end of the Cretaceous: Boulder, Colorado, Geological Society of America Special Paper 361, p. 35–55.

INTRODUCTION

The terrestrial strata of the Hell Creek Formation contain a rich and diverse fossil assemblage that has been a focus of numerous studies relating to the latest Cretaceous terrestrial biota and its reaction to the terminal Cretaceous event, the Cretaceous-Tertiary (K-T) boundary. The age of the terminal K-T event that occurs near the top of the formation is well known, but no studies have measured the actual duration or time span represented by the full thickness of the Hell Creek Formation. The absolute maximum age of the formation is defined by the magnetostratigraphy of the underlying Pierre Shale (Hicks et al., 1999) and its well-known and well-dated high-resolution ammonite zonation (Obradovich, 1993). However, marked lithofacies shifts and evidence for a disconformity at the top of the Fox Hills Formation (Jensen and Varnes, 1964; Belt et al., 1997) indicate that a considerable amount of time may be have been lost from the sedimentary record at the Fox Hills–Hell Creek formational contact. This indicates that the Hell Creek may be significantly shorter in duration than the bounding biostratigraphic and isotopic ages would suggest. This has direct bearing on the rate of change in biodiversity and extinction rates in the period leading to the terminal Cretaceous event.

Much of the previous work on the Hell Creek has focused on fossils collected from the exposures of the Hell Creek Formation in eastern Montana (Brown, 1907; Frye, 1969; Archibald et al., 1982; Sloan et al., 1986; Sheehan et al., 1991; Archibald, 1996; Clemens, this volume). The rocks of the upper part of the Hell Creek in Montana were correlated to the geomagnetic polarity time scale (GPTS) by Archibald et al. (1982) and Swisher et al. (1993), but the eastern exposures of the Hell Creek Formation in North Dakota and South Dakota that were measured in this study contain a number of more complete sections. The badlands of the Little Missouri River around Marmarth, southwestern North Dakota, contain superb badland exposures that crop out at the point where the northward-flowing Little Missouri River intersects the north-northwest–south-southeast-trending Cedar Creek anticline. The exposed rocks include the Pierre Shale, Fox Hills, and Hell Creek Formations of the Cretaceous to the Tertiary Ludlow Member of the Fort Union Formation (Hares, 1928; Carlson, 1979, 1983; Moore, 1976; Belt et al., 1984; Fastovsky, 1987; Johnson, 1992; Murphy et al., this volume).

In this area the Hell Creek Formation and the Ludlow Member of the Fort Union Formation have yielded abundant plant, invertebrate, and vertebrate fossil sites that allow detailed biostratigraphic correlations to other Upper Cretaceous and lower Paleocene sections and the geochronologic time scale. The flora includes abundant palynomorphs (Johnson et al., 1989; Nichols, this volume; Nichols and Johnson, this volume) and leaf fossils (Johnson et al., 1989; Johnson, 1992, this volume; Johnson and Hickey, 1990). The region is well known for its diverse vertebrate fossils (Sheehan et al., 1991; Holroyd and Hutchinson, this volume; Pearson et al., this volume) and mammal fossils (Hunter and Pearson, 1996; Hunter and Archibald, this volume). The underlying marine facies contains a high-resolution ammonite zonation (Bishop, 1965, 1973; Gill and Cobban, 1966, 1973) and numerous levels of datable bentonitic ash (Obradovich, 1993).

In Slope County, the K-T boundary has been identified biostratigraphically and geochemically, and is within the basal 3 m of the Ludlow Member (Johnson et al., 1989; Johnson, 1992; Nichols and Johnson, this volume). Farther south in Bowman County, North Dakota, the formational contact and the K-T boundary become coincident. The proximity of the K-T boundary with the upper formational contact of the Hell Creek Formation is fortuitous; the change in lithology is clearly visible in outcrop, so an approximation of the K-T boundary can be easily located. The K-T boundary is well dated and delimits the age of the end of Hell Creek deposition. Similarly, the base of the Hell Creek is readily defined in the Cedar Creek anticline at the formational contact with the distinctive white sandstone units of the Colgate Member of the Fox Hills.

The Hell Creek Formation has been the subject of several paleomagnetic studies. The first magnetostratigraphic analysis of the Hell Creek Formation was based on four sections in Garfield and McCone Counties, Montana (Archibald et al., 1982). This composite section spanned the K-T boundary, and the resulting magnetostratigraphy consisted of two zones of normal polarity surrounding a zone of reversed polarity containing the K-T boundary that was interpreted as polarity subchron C29r. Minor plotting errors in this work were corrected and the section was calibrated to the geochronologic time scale with a series of ^{40}Ar-^{39}Ar ages obtained from the Paleocene portion of the section by Swisher et al. (1993). Lerbekmo and Coulter (1984a) completed a small paleomagnetic section of the K-T boundary interval near Huff in central North Dakota. Lerbekmo and Sweet (1994) reported a thin (2 m) zone of normal polarity around the K-T boundary in the Hell Creek type area and in the three sites in Canada. In addition, Lund et al. (this volume) studied the magnetostratigraphy of four sections of the Hell Creek in eastern Montana and central North Dakota and made regional correlations of the formational boundaries of the Hell Creek Formation.

STRATIGRAPHIC SECTIONS

Six stratigraphic sections in the Marmarth area of North Dakota were sampled for magnetic orientations, and are shown in Figure 1. From north to south, these sections are Pyramid Butte, Pretty Butte, Bobcat Butte, Sunset Butte, Mud Buttes, and Doaks Butte. In each case the section was trenched to expose fresh bedrock and measured to the nearest centimeter. In five cases, the reference datum was defined as the contact of the Hell Creek and overlying Ludlow Member of the Fort Union Formation. At the Doaks Butte section the datum was the contact of the Hell Creek and underlying Fox Hills For-

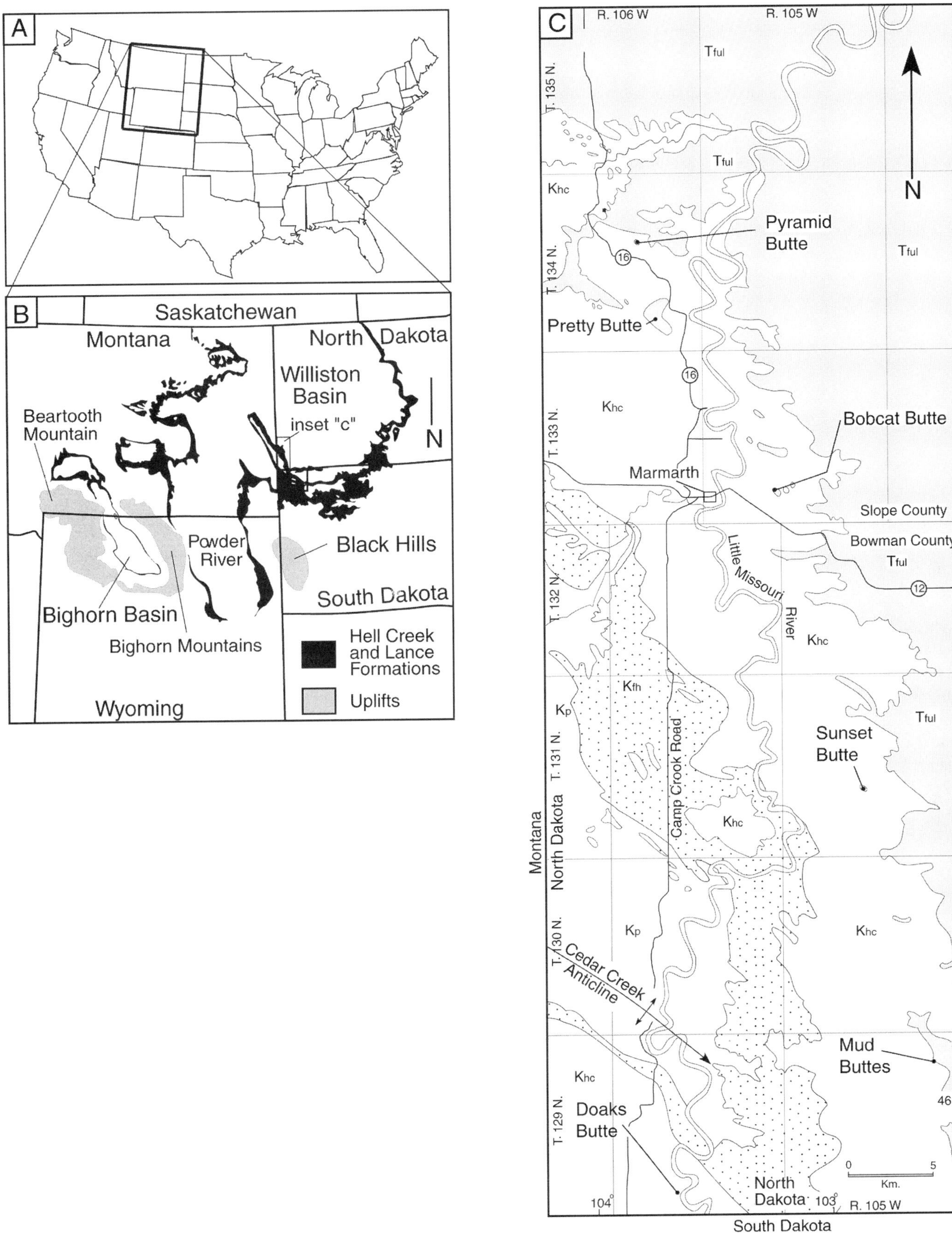

Figure 1. Map of Hell Creek Formation field area in Cedar Creek anticline, adjacent to Little Missouri River in Slope and Bowman County, North Dakota. Formational contacts between Pierre Shale (Kp), Fox Hills Formation (Kfh), Hell Creek Formation (Khc), and Ludlow Formations (Tful) are also shown.

mation. At each of these sections except Doaks Butte, palynology was used to define the K-T boundary relative to the formational contact datum (Nichols and Johnson, this volume). For more complete section descriptions see Johnson (this volume).

1. Pyramid Butte, the northernmost section, contains a K-T boundary interval that is identified by an iridium anomaly, shocked mineral grains, and palynological extinctions in lignite at the base of the Fort Union (Johnson et al., 1989). The complete section is 65 m thick; 26 m are in the Hell Creek Formation and 39 m are in the Fort Union Formation (Fig. 2).

2. At Pretty Butte the K-T boundary is defined palynologically at 112 cm above the base of the lignite that defines the base of the Fort Union Formation. A section measured up the east flank of the butte contained 50 m of Hell Creek Formation and 76 m of Fort Union Formation (Fig. 3).

3. At Bobcat Butte the palynofloral K-T boundary occurs 30 cm above the top of the 40 cm basal lignite, ~70 cm above the formational contact datum. The section is 96 m thick, comprising 58 m of Hell Creek Formation and 37 m of Fort Union strata (Fig. 4).

4. Sunset Butte is a composite section based on five smaller sections that begin on the east bank of the Little Missouri River and extend to the top of the adjacent Sunset Butte. The K-T boundary is defined palynologically (Nichols and Johnson, this volume). The complete composite section is 144 m thick, comprising 8 m of Fox Hills Formation at its base, overlain by 100 m of Hell Creek Formation and 35 m of Fort Union strata (Fig. 5).

5. Mud Buttes is located 35 km southeast of Marmarth, and east of the Little Missouri River. The K-T boundary is within 1 cm of the formational contact. The section is 61 m thick, comprising 16 m of Hell Creek Formation and 45 m of Fort Union strata (Fig. 6).

6. Doaks Butte is on the west bank of the Little Missouri River (Fig. 1). The section at Doaks Butte is the only section studied that is located on the western limb of the asymmetrical Cedar Creek anticline. The section, shown in Figure 7, encompasses the basal 25 m of the Hell Creek Formation, which overlies a thick exposure of the Colgate Member of the Fox Hills Formation (Waage, 1968).

AGE CALIBRATION

Five distinct sources have been used to calibrate and date the Hell Creek sequence of the Cedar Creek anticline: a wide range of biostratigraphic correlations, magnetostratigraphic studies, precessional chronometric age determinations for the duration of polarity subchron C29r, a reevaluation of the isotopic ages of the K-T boundary, and isotopic ages from ashes in the sedimentary sequence.

Biostratigraphy

The Hell Creek Formation in southwestern North Dakota has yielded age-diagnostic palynomorphs, megaflora, mammals, and dinosaurs. Nichols (this volume) has described a palynoflora of ~100 species, 30 of which are diagnostic of the *Wodehouseia spinata* assemblage zone of late Maastrichtian age (Nichols et al., 1982). Johnson et al. (1989) and Johnson (1992, this volume) documented a megaflora of 328 types that shares many taxa with the type Lance Formation and is considered to be late Maastrichtian. The vertebrate fauna of the Hell Creek includes mammals typical of the Lancian North American Land Mammal Age (Hunter and Pearson, 1996; Hunter this volume), and classic dinosaurs of the late Maastrichtian *Triceratops* zone (Sheehan et al., 1991; Pearson et al., this volume).

Figure 2. Line of section 86161 (Johnson, this volume) on south side of Pyramid Butte looking north across red clinker-capped Limber Pine Hills, Slope County, North Dakota. Figure shows contact between Hell Creek Formation and Ludlow Member of Fort Union. Iridium anomaly and shocked minerals were found at palynological K-T boundary in this section (Johnson et al., 1989; Nichols and Johnson, this volume). Lower part of measured section is located off lower right corner of image.

Figure 3. Pretty Butte, section 9735 (Johnson, this volume), looking west at north end of Pretty Butte in Slope County, North Dakota. Base of section is down in gully to right. Pretty Butte section measured by Murphy et al. (this volume) is located 1 km to south.

Figure 4. View of Bobcat Butte, section 86148 (Johnson, this volume), looking east in southern Slope County, North Dakota. Extensive badlands in foreground contain numerous fossil plant localities described by Johnson (this volume). Lower part of section is in gully below line of section and in second section (87148) that is located behind camera.

The overlying Ludlow Member of the Fort Union Formation has yielded a basal Paleocene (P1-P2) palynoflora (Nichols and Ott, 1978; Johnson et al., 1989; FUCAT, 1999; Nichols and Johnson, this volume). Megaflora from the Ludlow represents the widespread basal Paleocene FUI zone of Johnson and Hickey (1990). Sparse mammalian faunas reported by Hunter (1999) from the Ludlow are diagnostic of the Puercan and Torrejonian North American Land Mammal Age stages, which represent the first 4 m.y. of the Paleocene (Swisher et al., 1993). Marine molluscan and foraminferal faunas from the Cannonball Member of the Fort Union Formation (which interfingers with the Ludlow) are of Danian (early Paleocene) age (Fox and Olsson, 1969; Stanton, 1920; Cvancara and Hoganson, 1993).

In the study area the terrestrial sediments of the Hell Creek Formation (Fig. 1) directly overlie the Colgate Member of the Fox Hills Formation. Immediately beneath the Fox Hills Formation are the marine shales of the uppermost Pierre Shale containing ammonites of the *Baculites baculus* and *Baculites grandis* zones (Corbin, 1938; Bishop, 1965, 1973; Johnson and Hickey, 1990; W. Cobban, 2001, personal commun.). This level can be directly correlated with the provincial marine ammonite zonation of the Western Interior Late Cretaceous seaway, which is defined at the informal type section at Red Bird (Gill and Cobban, 1966).

Magnetostratigraphy

Hicks et al. (1999) studied the magnetostratigraphy of the Pierre Shale at Red Bird, near Lusk, Wyoming (Gill and

Figure 5. View of uppermost part of Sunset Butte composite section (DEF–071487-1, Johnson, this volume), looking to northeast in Bowman County, North Dakota. Base of composite section is located on east bank of Little Missouri River 5 km to west of Sunset Butte summit. Correlation between five legs of composite was walked out on continuously exposed marker beds. Southern face of Sunset Butte has catastrophically slumped and region behind foreground on right side of image is chaotically jumbled. Hell Creek–Fort Union contact occurs at base of 10 m sandstone bed. This sandstone is lenticular and is overlain by thick lignite (at level of dotted line). Palynological Cretaceous-Tertiary (K-T) boundary was not found on face of this butte due to lack of suitable rock for palynological sampling at base of channel. Sunset Patch section (9317, Johnson, this volume) is located 5 km to southeast. In that section sandstone channel is only 5 m thick and is underlain by fine-grained Hell Creek–Fort Union contact. Palynology of Sunset Patch section (Nichols and Johnson, this volume) places K-T boundary 170 cm above Hell Creek–Fort Union formational contact.

Figure 6. View of Mud Buttes section (88103) looking to southeast. Hell Creek–Fort Union contact and Cretaceous-Tertiary (K-T) boundary are coincident in this area and K-T boundary is marked by pollen extinction, shocked minerals, clay spherules, and iridium anomaly. Additional section that includes 15 m of Hell Creek Formation was measured behind camera view.

Cobban, 1966). This section was correlated to that part of the GPTS that ranges from the base of subchron C33n in the lower part of the Pierre Shale to the base of C31n in the Fox Hills Formation. This sequence has been calibrated independently with a series of seven isotopic ages, making the Red Bird section one of the best dated Late Cretaceous sequences in the region (see also Obradovich, 1993). The interval encompassed by the *B. baculus* and *B. grandis* range zones are especially well dated, because the C32n-C31r reversal boundary is near the base of the *B. grandis* zone, adjacent to an isotopic age of 70.15 ± 0.65 Ma. Bounded by calibrating isotopic ages, the reversal boundary C32n-C31r has been dated by interpolation from isotopic ages as 70.44 ± 0.44 Ma (Hicks et al., 1999). Direct correlation of the *B. baculus*-*B. grandis* zone boundary

Figure 7. Doaks Butte ash bed is located on flat top of this small butte that is located ~1 km south of Doaks Butte in southernmost Bowman County, North Dakota (DMNH fossil plant locality 2102). South Dakota state line is located immediately south of this butte. It is typical in Hell Creek and Fort Union Formations for volcanic ash beds to form flat-topped buttes capped with deeply weathered bentonitic ash.

underlying the Hell Creek to the Red Bird reversal sequence gives an interpolated age for the top of the Pierre Shale in the Cedar Creek anticline of ca. 70.5–70.7 Ma (Hicks et al., 1999, their Fig. 8).

Precessional chronometry

Maastrichtian to Danian deep-sea cores from the South Atlantic contain a rhythmic sedimentary cyclicity with alternations between lighter carbonate (75–85 wt% calcium carbonate), and darker marl (40–55 wt%) laminations. The alternation is caused by some combination of changes in the supply of clastic and organic material to the deep ocean, fluctuations in productivity in the surface and near-surface waters, and changes in carbonate dissolution rates (Herbert and D'Hondt, 1990; D'Hondt et al., 1992; Park et al., 1993; Herbert et al., 1995). This cyclicity has been detected over at least 20 m.y. of the Late Cretaceous and early Tertiary from the base of C33r to C28n (Deep Sea Drilling Project [DSDP] Site 516F, Herbert et al., 1995). When calibrated with isotopic age estimates for the magnetostratigraphy and biostratigraphy of this time period (in this case, to the time scale of Berggren et al., 1985), the mean periodicity of the most consistently prominent cycle was 23.5 ± 4.4 k.y. (Herbert and D'Hondt, 1990; D'Hondt et al., 1996). This agrees well with the Late Cretaceous precessional period of 20.8 k.y. predicted by Berger et al. (1989). Because the carbonate cycles are so closely spaced, this technique of orbital chronometry allows the duration of C29r and the K-T boundary event to be measured with a precision of better than 10.4 k.y., or the interval between a single precessional carbonate maximum and another. This assumes that sedimentation is constant for the time period being measured, and that all the cycles can be identified and counted.

Herbert and D'Hondt (1990) used spectral analysis to count 17 ± 2 cycles between the base of C29r and the K-T boundary in their most complete cores through this interval (cores 89, 90, Site 516F). Missing section in Sites 528, 527, and 525A caused them to give estimates of 15 ± 1, 14.5 ± 2 and 14.5 ± 2 cycles for the same time period, and complications with slumping yielded an estimate of 18 ± 2 cycles in Site 529. Their best estimate of 350 k.y. for the age of the Cretaceous part of C29r is based on a count of 17 precessional cycles with a cyclicity of 20.8 k.y. (after Berger et al., 1989). The age of the Paleocene part of C29r is a little more difficult to estimate because the pattern of carbonate cyclicity is difficult to recognize immediately above the K-T boundary, but Site 529 shows 8 cycles between the K-T boundary and the top of C29r, with an interval of 55 cm immediately above the boundary level that is estimated to represent the duration of at least 3 more cycles. The Paleocene portion of C29r yielded 11 cycles indicating an duration of ca. 229 k.y., and a total period for C29r of ~580 k.y.

In a reanalysis of these data, Herbert et al. (1995) compared the findings of the South Atlantic cores with sections from Spain (Ten Kate and Sprenger, 1993) and combined the spectral data with measurements of magnetic susceptibility. Revision of the number of cycles in C29r led them to estimate the Cretaceous part of C29r as 377 ± 20 k.y. (18.5 ± 1 cycles), and the Tertiary C29r as 296 ± 41 k.y. (14.5 ± 2 cycles). This increased their estimate of the duration of C29r from 580 k.y. to 633 ± 41 k.y. Using magnetic susceptibility data that can pick out carbonate cycles in sequences where the color contrasts are less well defined, D'Hondt et al. (1996) made another revised estimate of 603 ± 26 k.y. for the entire C29r; 333 ± 20 k.y. from the base of C29r and the K-T boundary, and 270 ± 17 k.y. for the interval from the K-T boundary to the top of the chron. Because the magnetic susceptibility method appears to

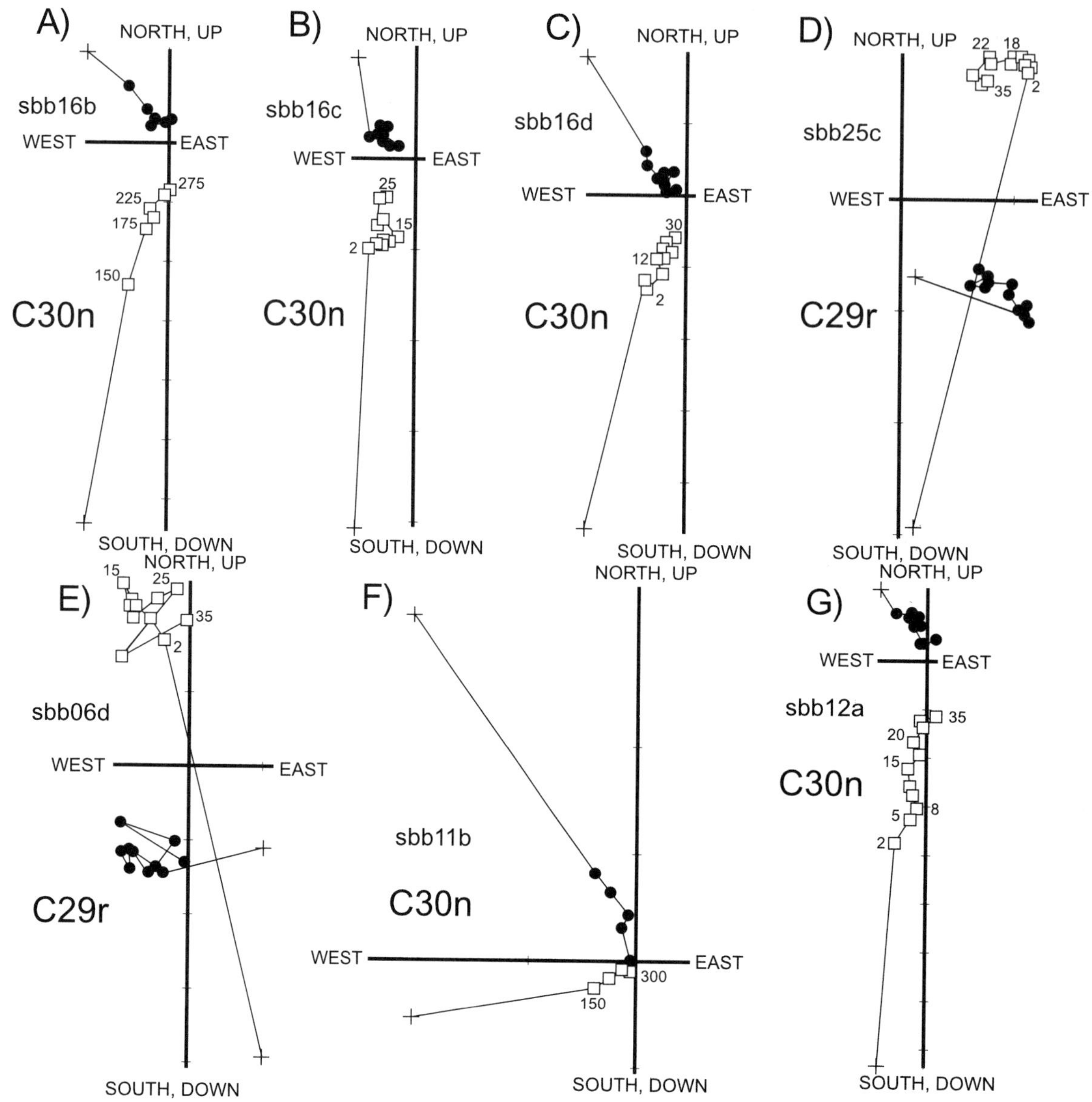

Figure 8. Representative samples from Sunset Butte section. Vector end-point diagrams (see caption Fig. 2) (A–C) display excellent sample consistency of normal polarity (N1) site sbb16. A: Thermal demagnetization. B, C: Alternating field (AF). D: AF demagnetized, multicomponent reversed polarity sample from within R1. E: Reversed AF sample from R1 showing "hovering" behavior where sample mean was calculated by Fisher average of cluster of points. F, G: Both normal polarity (N2) thermally demagnetized samples, both with quasilinear demagnetization paths.

count the precessional cycles more accurately, this latter age estimate for C29r (D'Hondt et al., 1996) is the one that we use in the course of this study.

Recalibration of the isotopic age of the K-T boundary

The time scales of Cande and Kent (1992, 1995) were based on a cubic spline solution of nine calibration points versus seafloor spreading distances of the various magnetic anomalies. The calibration points in both the 1992 and 1995 versions are a mixture of conventional K-Ar and ^{40}Ar-^{39}Ar ages and astronomical dating. The main differences between the 1992 and 1995 versions are that the age of the K-T boundary was changed from 66.0 Ma to 65.0 Ma and the astronomically dated portion of the time scale originally set at 2.60 Ma for chron C2An, was extended to 5.23 Ma for chron C3n.4n. This boundary and all younger subchrons were taken from Shackleton et al. (1990) and Hilgen (1994) with a refined astronomical age suggested for the Gauss-Matuyama boundary by Langereis et al. (1994).

A number of modifications to the original Cande and Kent

(1992) time scale have been proposed. Wuchang Wei (1995) proposed a recalibration of the Cenozoic portion using an age of 65.0 Ma for the K-T boundary. This K-T boundary age has been derived by consensus from a number of authors (Harland et al., 1989; Berggren et al., 1992; Swisher et al., 1992, 1993; Dalrymple et al., 1993). Wei also introduced a calibration age for C24n.1r of 52.8 Ma (Tauxe et al., 1994), which substitutes for the more controversial isotopic age of the −17 ash in Hole 550 near the Paleocene-Eocene boundary (Swisher and Knox, 1991).

Neither the 1992 nor 1995 Cande and Kent time scales considered the problem of interlaboratory bias in both the K-Ar and ^{40}Ar-^{39}Ar methodologies. Clarification with respect to the ^{40}Ar-^{39}Ar calibrations has only come about recently. The revised Cande and Kent (1995) time scale used the astronomical tuning of Langereis et al. (1994) for the last 5.23 Ma. Renne et al. (1994) published a series ^{40}Ar-^{39}Ar ages lined up with certain key points of the astronomically tuned time scale of Shackleton et al. (1990). An age of 28.03 Ma for the Fish Canyon Tuff sanidine monitor brings the ^{40}Ar-^{39}Ar ages into accord with those proposed by astronomical tuning (Langereis et al., 1994). The monitor age long used by the Berkeley Geochronology Center was 27.84 Ma. Moreover, the age of 65.0 Ma for the K-T boundary used by Cande and Kent (1992) was based on work by Swisher et al. (1992) that employed an age of 27.84 Ma for the Fish Canyon Tuff sanidine monitor. Another calibration point in Cande and Kent (1995) is based on a Fish Canyon Tuff monitor age of 27.84 Ma, the −17 ash in DSDP Hole 550, which has been dated as 54.5 Ma. The agreement of the last 5 m.y. of the astronomically tuned Pliocene-Pleistocene period with the age of 28.03 Ma age for the Fish Canyon Tuff sanidine suggests that the K-T boundary age should be recalibrated with this 28.03 Ma monitor age.

However, Fish Canyon Tuff sanidine is not the only monitor mineral used. The Taylor Creek Rhyolite monitor (Duffield and Dalrymple, 1990) has been used exclusively since 1990 by one of us (Obradovich) with an assigned age of 28.32 Ma normalized to an age of 520.4 Ma for MMhb-1 (Samson and Alexander, 1987). This age differs from that of 27.92 Ma assigned by Sarna-Wojcicki and Pringle (1992). The choice of 28.32 Ma was entirely pragmatic because this monitor age provided the best comparison with ages derived by Obradovich and Cobban (1975). In an intercalibration study of various monitors, Renne et al. (1998) obtained ages of 28.34 Ma for Taylor Creek Rhyolite and 28.02 Ma for Fish Canyon Tuff when calibrated against GA1550 biotite as their primary standard with an age of 98.79 Ma. This value of 28.02 Ma agrees quite well with the value of 28.03 Ma obtained through calibration based on the astronomical time scale (Renne et al., 1994). On the basis of unpublished data, one of us (Obradovich) obtained an age of 28.03 Ma for the Fish Canyon Tuff sanidine monitor of W. McIntosh (Geoscience Department, New Mexico Institute of Mining and Technology, Socorro, New Mexico), calibrated against an age of 28.32 Ma for Taylor Creek Rhyolite.

TABLE 1. $^{40}Ar/^{39}Ar$ AGES FOR THE CRETACEOUS-TERTIARY BOUNDARY AVAILABLE TO CANDE AND KENT IN 1995

Izett et al. (1991)	64.50 ± 0.10 Ma	Tektites
	64.60 ± 0.20 Ma	Sanidine Z Coal
Dalrymple et al. (1993)	64.42 ± 0.06 Ma	Tektites TF
	64.38 ± 0.18 Ma	Tektites plateau
	64.77 ± 0.07 Ma	Sanidine Z Coal
Swisher et al. (1992)	65.01 ± 0.08 Ma	Tektites Beloc
	65.07 ± 0.10 Ma	Tektites Mimbral
Swisher et al. (1993)	65.16 ± 0.04 Ma	Sanidine IrZ Coal
Obradovich (1993)	65.43 ± 0.11 Ma	Sanidine Z Coal
McWilliams (1994)	64.91 ± 0.12 Ma	Tektites
	64.71 ± 0.09 Ma	Sanidine Z Coal

TABLE 2. $^{40}Ar/^{39}Ar$ AGES FOR THE CRETACEOUS-TERTIARY BOUNDARY AFTER NORMALIZING PRIOR RESULTS (TABLE 1) TO 28.32 MA (TCR) AND 28.02 MA (FCT)

Izett et al. (1991)	65.41 ± 0.10 Ma	Tektites
	65.51 ± 0.20 Ma	Sanidine Z Coal
Dalrymple et al. (1993)	65.33 ± 0.06 Ma	Tektites TF
	65.29 ± 0.18 Ma	Tektites plateau
	65.69 ± 0.07 Ma	Sanidine Z Coal
Swisher et al. (1992)	65.42 ± 0.08 Ma	Tektites Beloc
	65.48 ± 0.10 Ma	Tektites Mimbral
Swisher et al. (1993)	65.57 ± 0.04 Ma	Sanidine IrZ Coal
Obradovich (1993)	65.43 ± 0.11 Ma	Sanidine Z Coal
McWilliams (1994)	65.83 ± 0.12 Ma	Tektites
	65.63 ± 0.09 Ma	Sanidine Z Coal

To provide the most precise time scale possible it is important that all calibration points used be somehow tied to one another. This dictates that the data used be primarily ^{40}Ar-^{39}Ar results, based on monitors with well known calibrations against one another, and that there be a certain coherency if the astronomically tuned time scale is employed for the last 5 m.y. With this in mind we present Table 1, which shows all available ^{40}Ar-^{39}Ar isotopic dates as originally published for the K-T boundary interval, along with their authors and source, and Table 2, which shows the same data after normalization based on the monitor age of 28.02 Ma for Fish Canyon Tuff and 28.32 Ma for Taylor Creek Rhyolite, which yields an average age of 65.51 ± 0.10 Ma for the K-T boundary.

Isotopic age determinations

Bentonitic ash has been sampled from several layers throughout the Hell Creek Formation. We carried out 9 single-crystal ^{40}Ar-^{39}Ar analyses on an ash from the Doaks Butte section (Fig. 7, +27.6 m above base of section, lower Hell Creek), but only 5 gave reasonable ages: 65.44, 65.46, 66.10, 66.19, 66.76 Ma. Using the time scale of Cande and Kent (1995), this age range of 1.32 m.y. spans almost the entire time period from the K-T boundary at 65.0 Ma, through to the lower part of C30n ca. 67.6 Ma. This does little to refine the age of the sequence. All the ashes so far collected from the Hell Creek contain detrital K-Feldspar and none so far have proven to be suitable as calibrating ages for the purposes of this study.

PALEOMAGNETIC ANALYSIS

Paleomagnetic sample collection

Paleomagnetic sampling was carried out in four phases. The first suite of samples was collected in 1987 from the Sunset Butte, Pyramid Butte, Mud Buttes, and Bobcat Butte sections (Fig. 1). Initial paleomagnetic analysis showed that the sediments contained a measurable paleomagnetic direction, but that there were gaps in the sampling series at critical levels. A second suite of samples was collected in 1992 from the same sections, primarily at levels where the initial results were questionable, or where a closer sampling interval was deemed necessary. To provide more information on the basal part of the Hell Creek Formation, a third suite was sampled in 1993 at Doaks Butte (Fig. 1). A fourth suite of samples was collected in 2000 from the Pretty Butte section (Fig. 1), and a number of samples were also collected from the Sunset Butte (+28 to +34 m), Mud Buttes (+19 to +21 m), and Bobcat Butte (−11 to −14 m) sections in order to place the reversal boundaries more accurately.

In all the sampling phases, the same standard sampling and data analysis procedures were used. Three to five separately oriented hand samples, evenly distributed throughout the five measured sections, were collected for paleomagnetic analysis from each stratigraphic level. A flat face was shaved on the in situ specimen with a hand rasp, and the dip and strike orientation of the face measured using a Brunton compass.

Paleomagnetic sample preparation and measurement

In the paleomagnetic laboratory at Scripps Institution of Oceanography the samples were sanded into cubes and measured using a CTF DRM-430 cryogenic magnetometer located inside a shielded room with a background magnetic field of <200 nT. The samples were thermally demagnetized in air using either a Schonstedt oven or a high-capacity custom-built oven. Step-wise alternating field (AF) demagnetization was carried out in a single axis AF demagnetizer. Where measured, the susceptibility of the samples at each thermal demagnetization step was obtained using a Bartington MS2 susceptibility meter.

All specimens were measured in either four or eight orientations per demagnetization step. Each measurement position constitutes an average of 25 individual samplings, and the standard deviations of the measurements are automatically screened. The measurement is rejected if the standard deviation exceeds 10^{-10} Am^2. Typical standard deviations are in the range of 1–2 times 10^{-11} Am^2. This procedure is repeated for all the sample orientations and is designed to compensate for the inhomogeneity of sample magnetization, shape, or (very rarely) small viscous or induced components acquired in the low magnetic field present in the measurement region of the magnetometer (~35 nT). The four or eight resulting vectors are averaged using Fisher (1953) statistics, and a circular standard deviation (CSD) is calculated. Measurements with a CSD of more than 3° were repeated and an average of all measurements was used in the following data analysis. If repeat measurements were not reproducible, the sample was eliminated from further treatment.

To determine the best method for isolating a stable characteristic remanence, two samples from each level were treated using both a step-wise AF and thermal demagnetization. Representative results are shown plotted in Figures 8 and 9 as vector end-point (Zijderveld, 1967) and equal-area diagrams. At least one sample from each polarity interval measured in these sections is shown. One sample from each level was demagnetized thermally in 25 and 50 °C steps, starting at 125 °C to a maximum temperature of 350 °C (e.g., see Figs. 8, A and F, and 9B). A second sample was subjected to AF demagnetization in a sequence of 1, 2.5, and 5 mT steps to a maximum of 45–55 mT. Comparison of three normal samples from level sbb16 at Sunset Butte (Fig. 8A, thermal demagnetization; B, C, AF demagnetization), demonstrates how the thermal and AF samples behave in a very similar and consistent fashion. In general, AF demagnetization results were similar to thermal demagnetization but appeared less scattered, so all remaining samples were subjected to step-wise AF demagnetization.

Extensive mineralogical and rock magnetic work carried out by Lund et al. (this volume) on sediments from the Hell Creek in central North Dakota has shown that the dominant ferromagnetic mineral is ilmenohematite that was probably formed in magmas during extensive Laramide volcanism to the west. Lund found that the primary sedimentary remanence in these Hell Creek sediments is between 10–40 mT and 100–250 °C. Electron microprobe and X-ray diffraction work carried out by Butler on Hell Creek rocks from the Fort Peck region (see Swisher et al., 1993) found that the dominant ferromagnetic mineral in that region was titanohematite, with characteristics similar to sediments studied by Butler and Lindsay (1985) from the San Juan Basin of New Mexico.

Thermal demagnetization indicates that the majority of samples show a maximum blocking temperature of between 200 and 350 °C (Figs. 8, A and F, and 9B), and the low-coercivity overprint seen in most samples was completely removed by AF demagnetization in fields of 10–35 mT (Figs. 8, B–E and G, and 9, A, C, and D), or rarely by fields as high as 55 mT (Fig. 9E). This behavior has been observed in sediments of Late Cretaceous to early Tertiary age from across the region (Hicks et al., 1995, 1999).

Components of magnetization for samples that display a simple quasilinear characteristic component were calculated using principal component analysis (PCA, Kirschvink, 1980). Best-fit lines were calculated when a minimum of three demagnetization points (and usually many more) had a maximum angle of deviation (MAD) of <20° and trended broadly to the origin. Examples of such specimens are shown from Sunset

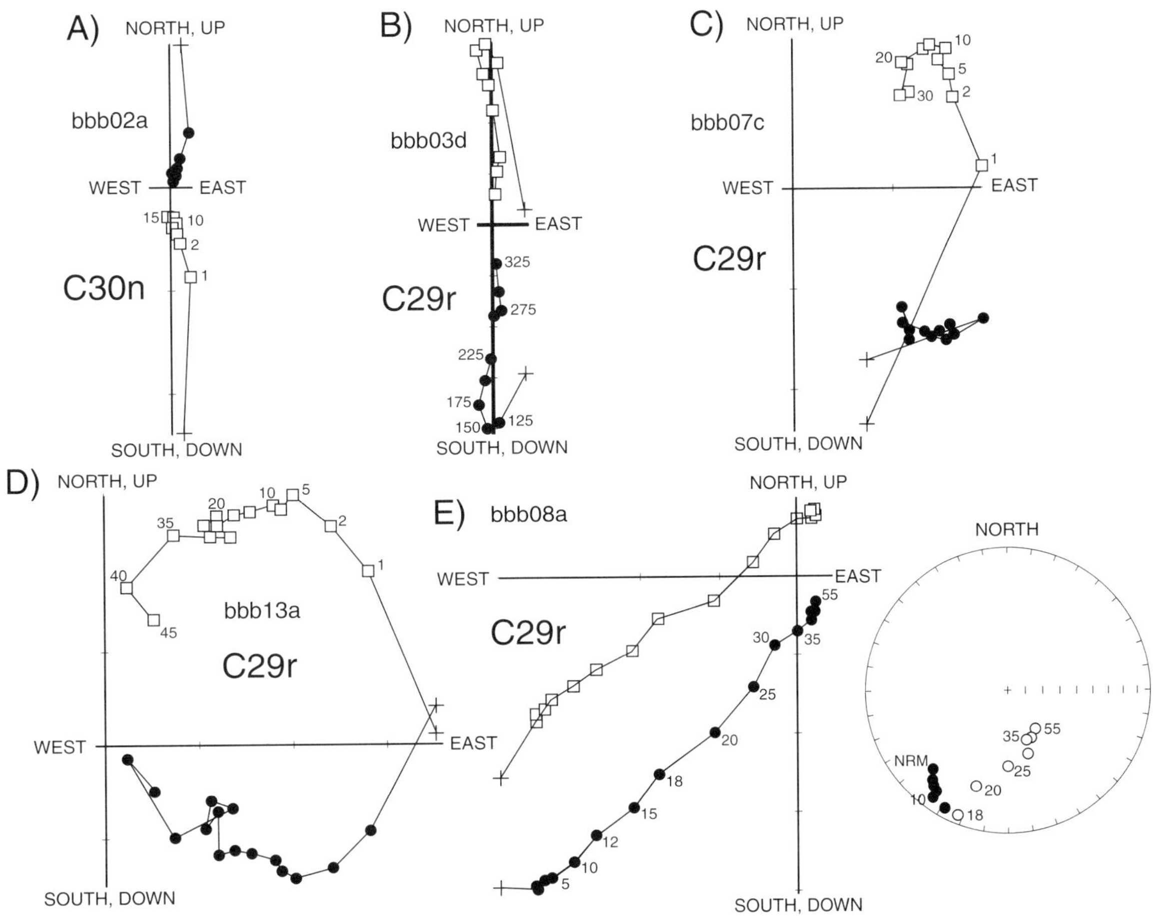

Figure 9. Representative samples from Bobcat Butte section. Vector end-point diagrams (see caption Fig. 1). A: Normal polarity sample from N1, alternating field (AF) demagnetized, quasilinear demagnetization path. B–D: Reversed polarity samples from R1, thermal (B) and AF demagnetization (C and D). E: R1, AF demagnetization, circuitous demagnetization (great circle) path that tracks to reversed polarity direction, clearly shown on right in accompanying equal-area plot.

Butte (Fig. 8, A–C and F) and from Bobcat Butte (Fig. 9, A and B). The sample data from stratigraphic levels whose specimens yielded acceptable best-fit lines were averaged using standard Fisher (1953) statistical techniques. Sites with an α_{95} >35° were discarded from the final interpretation of polarity, a cutoff based on the randomness criteria of Watson (1956).

In a number of specimens the end points tend to cluster or remain in one orientation rather than trend to the origin. This behavior is clearly shown in sample sbb06d (Fig. 8E). The overprint is completely removed by the 15 mT step, and the sample begins to trend toward the origin between 15 and 20 mT, but by 25 mT this trend stops and the sample remains in the same direction with no further decrease in intensity. In cases such as this we discard the low (natural remanent magnetization to 2 mT) and high AF (>35 mT) data and selected a set of at least five consecutive demagnetization points that remained in the same direction to define the characteristic component. Sample means calculated by Fisher statistics are considered acceptable if the α_{95} is <35°.

The demagnetization data of specimens from a number of sites did not trend to the origin or cluster in a single direction, but were within a plane. An example of this behavior is shown in an equal-area plot of the reversed sample bbb08a (Fig. 9E). If the site also has one specimen that allows calculation of either a Fisher mean or PCA, then the great circles and single directed line can be combined using the method of McFadden and McElhinny (1988) to give a mean direction.

The site means are plotted as circles in the equal-area projection shown in Figure 10 (left). Both the normal and reversed modes are Fisher distributed (Fig. 10, lower table), and the mean normal and reversed directions calculated by Fisher statistics are shown in Figure 10 (right) surrounded by their respective calculated α_{95} confidence ellipses. The direction of the present dipole field and the expected direction for the Late

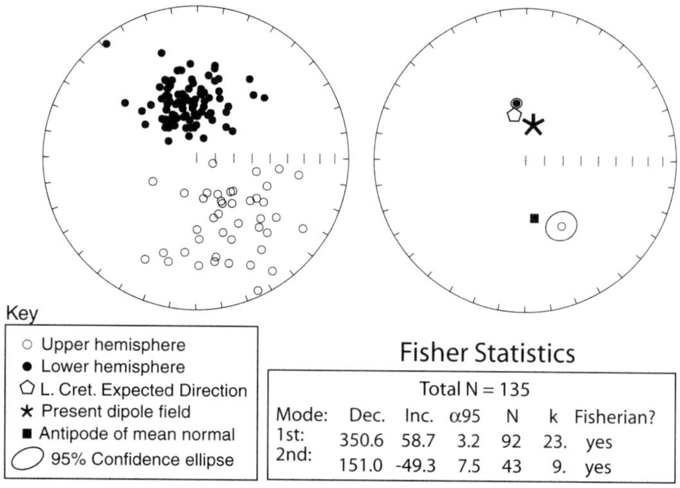

Figure 10. Left: Equal-area plot of all site mean directions measured in this study (open circles, upper hemisphere; filled circles, lower hemisphere). Right: Mean normal and reversed directions, plotted with present-day field direction and antipode of mean normal direction. Fisher statistics of site mean directions were calculated by methods of Fisher (1953).

Cretaceous calculated from the reference pole of Diehl (1991) obtained for the North American craton from the Elkhorn Mountains Volcanics of southwestern Montana are also shown. The calculated mean normal direction is very close to the Late Cretaceous expected direction, and is clearly distinct from the present dipole field. Furthermore, the antipode of the normal direction is very close to the measured reverse direction.

MAGNETOSTRATIGRAPHIC RESULTS

Reversal placement

The reversal boundaries are placed at a point stratigraphically midway between the adjacent oppositely polarized sites. The stratigraphic position of each reversal is tabulated by site in Tables 3 and 4 relative to the Hell Creek–Fort Union formational contact and the palynological K-T boundary. The difference between the formational contact and the actual K-T boundary datum level is also shown in the tables along with the distance of the reversal (chron) boundary to the K-T boundary. In the following, the stratigraphic levels of the polarity subchron reversals are reported in meters relative to the K-T boundary rather than the Hell Creek–Fort Union formational contact. Where the sediments are suitable and the section could be closely sampled, the reversals can be placed with a precision of ±0.6 m. In intervals where the lithologies are unsuitable for paleomagnetic sampling and the exposures are poor, the precision can fall to ±2.2 m.

Pretty Butte

The Pretty Butte section was extensively sampled during the 2000 field season: 27 sites were sampled through a 126 m section (Fig. 3). The mean interval between sites was 4.7 m, with a maximum of 14 m spacing between the uppermost sites in the butte where exposure was poor due to extensive weathering.

A reversed polarity interval, containing a single anomalous normal polarity site, brackets the palynological K-T boundary. The position of the K-T boundary correlates the sequence to that part of the GPTS that ranges from the upper part of C30n to C29n. The inferred stratigraphic positions of the reversal boundaries in this section are documented in Tables 3 and 4.

Bobcat Butte

We collected 34 sites from the 80 m thickness of Bobcat Butte over 3 field seasons, 1987, 1992, and 2000. The average sample interval was 2.9 m; the maximum interval was 12 m at the top of the section, and the minimum was 0.36 m. A single polarity reversal from normal to reversed is at −10.8 m ±0.7 m in the section (Table 4). The top of the reversed interval is uncontrolled; there is no overlying normal polarity interval. The position of the palynological K-T boundary within the reversed interval confirms the correlation of the upper part of the section to C29r.

Sunset Butte

Sunset Butte is a composite section that ranges through the entire thickness of the Hell Creek Formation from the underlying contact with the Colgate Member of the Fox Hills Formation. The thickness of the Hell Creek Formation has been independently confirmed by a second section logged through the same interval by Murphy et al. (this volume) from Little Beaver Creek to Baldy Butte, west of Marmarth. The two sections confirm that the thickness of the Hell Creek is 100 m with a measured difference of only ~50 cm. Further discussion of the thickness of the formation is provided in Johnson (this volume) and Murphy et al. (this volume). The total section thickness of 144 m was sampled at 54 levels in 1987 and again in 1992, with an average interval of 2.7 m, and a maximum interval of 10.3 m (Fig. 11). Due to overlapping samples collected at the same stratigraphic levels in successive field seasons, the minimum sampling interval is 0 m. The highest density of resampling was carried out at the upper reversal boundary and between −60 and −85 m to look for the very short reversed subchron C30r, which has a duration of 125 k.y. (Cande and Kent, 1995). A well-defined reversed interval was found in the Sunset Butte section. There is a single normal polarity site in the lower part of the reversal (Fig. 11), which is believed to be in an overprinted interval. The position of the palynological K-T boundary within the reversed polarity interval in the upper

TABLE 3. SOURCE DATA BRACKETING THE TOP OF POLARITY SUBCHRON C29R

	Sample number	Sample to HC-FU datum (m)	Chron boundary to datum (m)	Error (m)	K-T correction (m)	Chron boundary to K-T (m)
Sunset Butte	Upper SBB-11	31.5	30.7	0.8	1.7	29.0
	Lower SBB-26	29.9				
Mud Butte	Upper DP-6	24.5	22.9	1.6	0.0	22.9
	Lower MBB-04	21.3				
Pretty Butte	Upper PYA-19	26.6	24.4	2.2	1.1	23.3
	Lower PYA-18	22.2				
Pyramid Butte	Upper PB-02	29.5	28.9	0.6	1.0	28.0
	Lower PBB-15	28.4				

Note: K-T is Cretaceous-Tertiary.

TABLE 4. SOURCE DATA BRACKETING THE BASE OF POLARITY SUBCHRON C29R

	Sample number	Sample to HC-FU datum (m)	Chron boundary to datum (m)	Error (m)	K-T correction (m)	Chron boundary to K-T (m)
Sunset Butte	Upper SB-19	−21.4	−22.5	1.1	1.7	−24.2
	Lower SBB-04	−23.6				
Bobcat Butte	Upper BBB-03	−9.3	−10.0	0.7	0.8	−10.8
	Lower BBB-14	−10.8				
Pretty Butte	Upper PYA-09	−22.4	−23.4	1.0	1.1	−24.5
	Lower PYA-08	−24.4				

Note: K-T is Cretaceous-Tertiary.

part of the section correlates the sequence to that part of the GPTS that ranges from C30n to C29n (Fig. 11).

Mud Buttes

The Mud Buttes section was extensively sampled in 1987 and subsequently resampled in 1992 and in 2000 to improve the precision of the position of the single reversal boundary measured (Fig. 11). We sampled 37 sites in 61.6 m of section with an average spacing interval of 2.6 m, a maximum interval of 6.38 at the base of the section, and a minimum of 0.6 m. The single reversed to normal polarity change is placed with a precision of ±1.6 m at +22.9 m in the section (Table 3). The K-T boundary is in the lower part of the section in a reversed interval, so the entire sequence at Mud Buttes can be correlated to that interval of the GPTS that ranges from the base of C29r to the lower part of C29n.

Doaks Butte

Paleomagnetic samples were collected in 1993 at a very closely spaced interval through the section at Doaks Butte in an attempt to find C30r in the lower part of the Hell Creek Formation. This stratigraphic level was relatively poorly covered in the lower part of the Sunset Butte section, where there were a number of gaps in the sampling sequence (Fig. 11). The total section of 27 m was sampled at 16 levels at an average interval of 1.65 m; the maximum interval was 4.4 m at the top and the minimum was 0.4 m. Despite this sampling effort no reversal was found, and given the density of the sampling it is unlikely that the C30r reversal exists anywhere in this sequence, and the entire section is interpreted to be within the lower part of C30n.

Pyramid Butte

The 64.8 m paleomagnetic section at Pyramid Butte was sampled at 24 levels in 1987 and 1992. Samples are spaced with a maximum interval of 8.7 m, and a minimum of 1.1 m. Figure 12 shows the virtual geomagnetic pole (VGP) directions for the Pyramid Butte section. There is a clear but very short reversed polarity interval immediately below the K-T boundary, but the reversal sequence below the K-T boundary is difficult to interpret. The simplest explanation is that there has been a localized erosive episode that has almost completely cut out the Cretaceous part of C29r, leaving the K-T boundary, which here is recognized by an iridium anomaly only a few tens of centimeters above the reversal. These difficulties are compounded

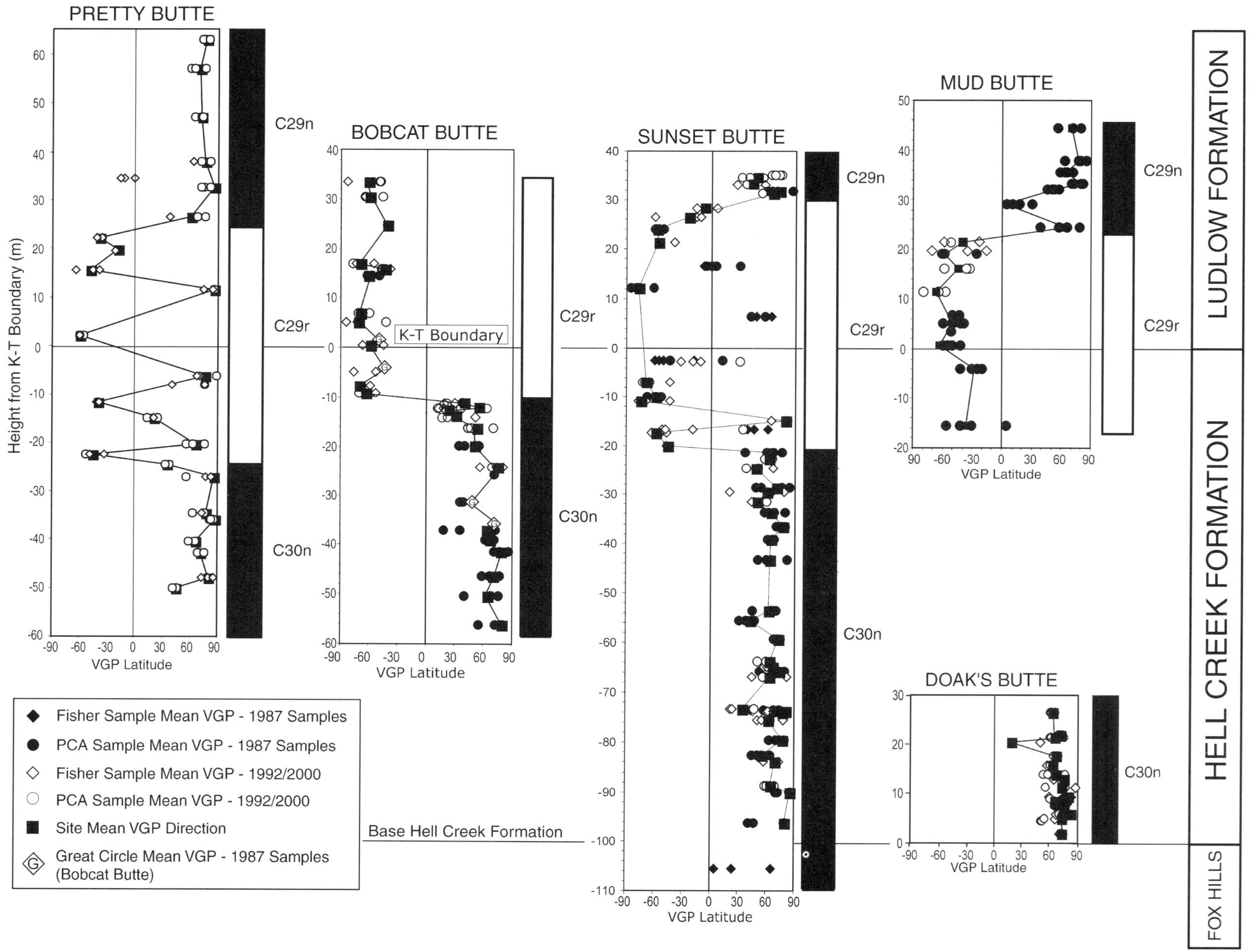

Figure 11. Virtual geomagnetic poles (VGP) versus stratigraphic height are plotted for four of five sections studied. Sections are shown in order of alignment, broadly north to south along axis of Cedar Creek anticline (see Fig. 1). Different suites of samples are clearly distinguished. Earlier data set collected in 1987 is shown with filled diamond and circle icons, data collected in 1992 and 2000 are shown with open diamond and circle icons, and site means are shown by filled black squares. No mean was calculated or plotted if mean directions of samples in site exceeded parameters outlined in text. Sections are aligned to K-T boundary datum (0 m) level. Interpreted polarity of each section is shown to right (black is normal, white is reversed). PCA is principal component analysis.

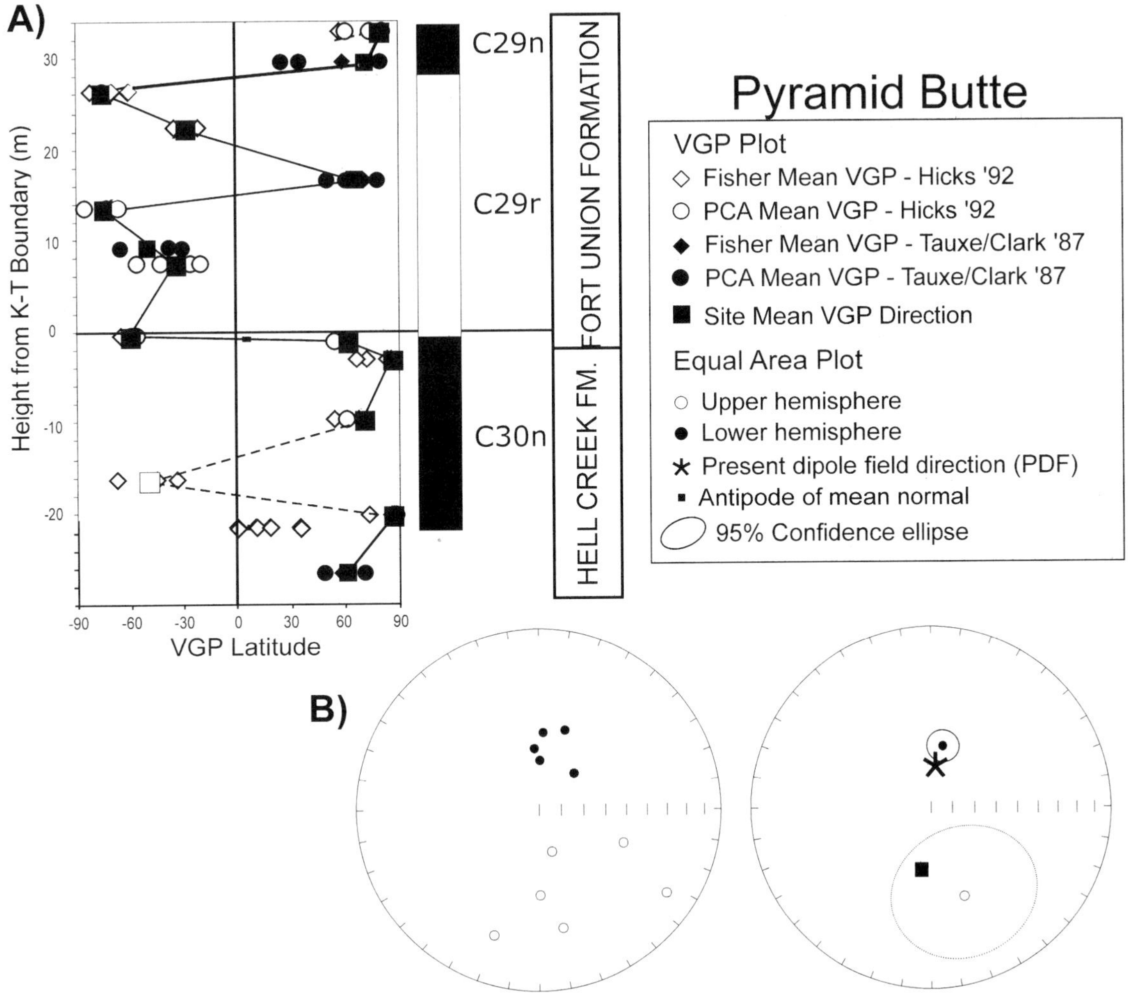

Figure 12. A: Plot of virtual geomagnetic poles (VGP) vs. stratigraphic height for Pyramid Butte section. B: Left, equal-area plot of mean sample directions obtained from Pyramid Butte section. Right, equal-area plot of mean normal and/or reversed direction with associated calculated 95% confidence ellipse. Star is present dipole field, and filled square is antipode of mean normal direction. Mean reversed direction found at − 17 m in section is believed to be anomalous and is shown as shaded filled square.

by a single anomalous reversal at − 15 m (shown as a shaded point in Fig. 12). It is difficult to reliably interpret the lower part of the sequence at Pyramid Butte, so although the VGP data are presented in Figure 12A with an equal-area plot of the sample mean directions (Fig. 12B, left), it has been left out of any further analysis. In Figure 12B (right) the calculated mean normal direction can be seen to be close to the present dipole field, but the antipode of the mean normal is within the 95% confidence ellipse of the mean reversed direction. The upper part of the section above the K-T boundary contains a single anomalous normal level (+ 16 m, Fig. 12A), but the position of the upper reversal at + 28.0 m is clear, and these data have been included in the time-estimate calculations for the Tertiary part of C29r (T-C29r, Table 3) in the next section.

Anomalous overprints

One of the difficulties encountered in this paleomagnetic study of the Hell Creek has been the occurrence of a number of anomalous single-site normal polarity intervals found in the Pretty Butte, Sunset Butte, and Pyramid Butte sections. A normal polarity subchron within C29r was reported by Lerbekmo et al. (1996) at the level of the K-T boundary. None of the single site normals that we have encountered occur near the K-T boundary interval, so this does not provide an acceptable explanation. It appears to be due to complete overprinting, probably by secondary precipitation of iron-oxide minerals such as goethite or hematite. Similar samples were found by Lerbekmo, who has carried out a large amount of work on sediments of a similar age in Canada and the northern Great Plains region of the United States. Because the age control of the Hell Creek sections is so good, with the K-T boundary clearly identified, the pattern of the C29r reversal can be clearly discerned in all cases. The actual nature and cause of the anomalous overprints is the subject of future study, but the overprints are most likely caused by the fracturing of the outcrop, penetration of water, dissolution of the magnetic carrier minerals, and the redeposition of iron-oxide minerals.

DISCUSSION

Calculated duration of the Hell Creek Formation

Given that the position of the K-T boundary is known in each section, the correlation of the reversal sequence found in these sections (Fig. 11) can be confidently correlated with the GPTS. The K-T boundary is known to be within a reversed interval, C29r, and its relative position has been fixed in the upper part of C29r at Gubbio (Alvarez et al., 1977), in deep sea cores (e.g., Herbert et al., 1995), in the published time scales (e.g., Cande and Kent, 1992), and in surface sections in the Western Interior (Lerbekmo and Coulter, 1984b; Swisher et al., 1993). The overlying normal is C29n, and the underlying normal is C30n (Fig. 11).

Combining all the data shown in Tables 3 and 4, we can calculate an average height for the top and bottom of C29r, relative to the K-T boundary. In the Bobcat Butte section the base of C29r is at −10 m, compared to the position in Sunset and Pretty Buttes, where it is at −20 and −24 m. This suggests that some strata are missing at Bobcat Butte in the Cretaceous part of C29r, probably due to a minor diastem. We used a number of different parameters to calculate a sedimentation rate for the C29r interval. The full ranges of the precessional cyclicity age estimates for the duration of C29r are shown in Table 5. All the data used in the following discussion are shown in Tables 3 through 6. These tables list the primary data required to recalculate the sedimentation rates, age, and duration of the Hell Creek Formation when revised estimates become available that supersede the current age estimates for the K-T boundary or the duration of C29r.

In calculating the duration of the Hell Creek our primary assumption is that there is no appreciable amount of sediment missing from any of the sections. This assumption does not appear to be valid in the case of the lower part of the Pyramid Butte section, nor in the case of the Bobcat Butte section, where the lower part of C29r is less than half the thickness in other adjacent sections. The duration of the Hell Creek is calculated by extrapolating downward from the calibration points that we have chosen (see Fig. 13). Our primary calibration point is the palynological K-T boundary, plotted as the zero datum in the sections, and dated at 65.51 ± 0.10 Ma. The second calibration point is derived from the precessional age of the Cretaceous part of C29r, which varies from a low of 570 k.y. (Herbert and D'Hondt, 1990) to a high of 673 k.y. (Herbert et al., 1995). From these data we can calculate the age of the Hell Creek by extrapolating down from the K-T boundary, through the C30n-C29r reversal, to the base of the Hell Creek Formation.

TABLE 5. ESTIMATES OF THE DURATION OF THE HELL CREEK FORMATION

	m.y.	m	m/m.y.	yr/m	Hell Creek duration (m.y.)
A. Calculations with all sections (including Bobcat Butte)					
D'Hondt, 1996					
T-C29R	0.27	25.8	95.6	10465	1.05
K-C29R	0.333	19.8	59.5	16818	1.68
C29R	0.603	45.6	75.6	13224	1.32
Herbert and D'Hondt, 1990					
T-C29R*	0.27	25.8	95.6	10465	1.05
K-C29R†	0.3	19.8	66.0	15152	1.52
C29R§	0.57	45.6	80.0	12500	1.25
Herbert et al., 1995					
T-C29R	0.296	25.8	87.2	11473	1.15
K-C29R	0.377	19.8	52.5	19040	1.90
C29R	0.673	45.6	67.8	14759	1.48
B. Same calculations as above but with Bobcat Butte removed					
D'Hondt, 1996					
T-C29R	0.27	25.8	95.6	10465	1.05
K-C29R	**0.333**	**24.4**	**73.3**	**13648**	**1.36**
C29R	0.603	50.2	83.3	12012	1.20
Herbert and D'Hondt, 1990					
T-C29R	0.27	25.8	95.6	10465	1.05
K-C29R	0.3	24.4	81.3	12295	1.23
C29R	0.57	50.2	88.1	11355	1.14
Herbert et al., 1995					
T-C29R	0.296	25.8	87.2	11473	1.15
K-C29R	0.377	24.4	64.7	15451	1.55
C29R	0.673	50.2	74.6	13406	1.34

Note: Line highlighted in bold is our best estimate for the duration of the Hell Creek, based on: 1. Sediment accumulation rates used are from the K-C29r interval; 2. Only use the most complete magnetostratigraphic sections (Bobcat Butte excluded); 3. Use the precessional age estimate for C29r of 333 k.y. after D'Hondt et al. (1996).
*Tertiary portion of the C29r reversal, i.e. C29r above the Cretaceous-Terriary boundary.
†K-C29R Cretaceous portion of the C29r (see note above)
§All of the chron 29 reversed polarity interval.

If we calculate the duration of the Hell Creek using all three of the precessional age estimates, both with and without the Bobcat Butte section, then a relatively close range of ages is produced (Table 5). The sedimentation rate for the Tertiary part of C29r is consistently higher than the underlying Cretaceous sequence. The Tertiary sedimentation rates labeled T-C29r (for Tertiary part of C29r) in Table 5 range from 87.2 to 95.6 m/m.y. (average 91.4 m/m.y.), whereas the rates in the equivalent Cretaceous interval (labeled K-C29r) range from 52 to 81 m/m.y. (average of 66 m/m.y.), or a difference of ~72%. The difference in sedimentation rates between the Hell Creek Formation and the Fort Union Formation is quite marked. For the purpose of calculating the age of the Hell Creek, we use the data from the Cretaceous part of C29r, and the whole of

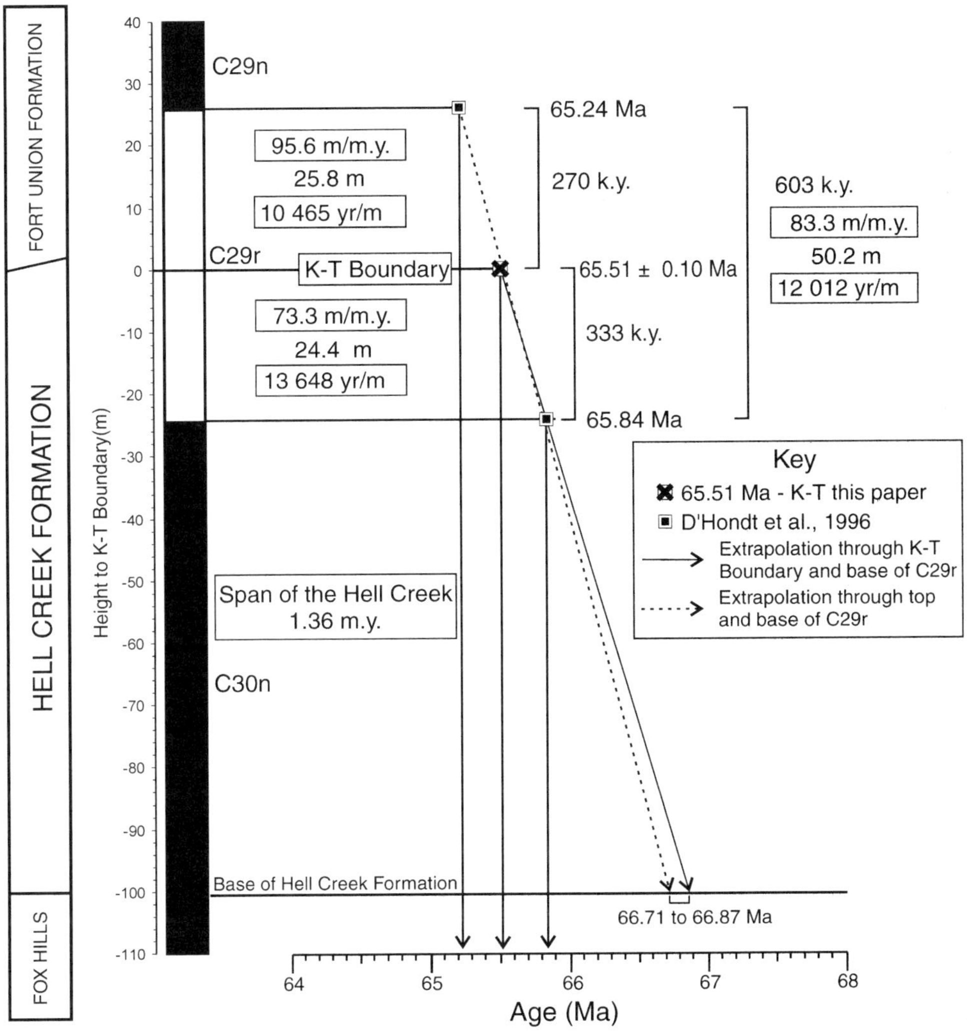

Figure 13. Composite age correlation diagram constructed using parameters outlined in text, and highlighted in Table 5. Stratigraphic position of C29r-C29n reversal boundary is average of Sunset, Mud, Pretty, and Pyramid Butte sections. Position of C30n-C29r boundary is derived from Sunset and Pretty Butte sections. Age of Cretaceous-Tertiary (K-T) boundary is plotted at 65.51 Ma. Stratigraphic thickness of Hell Creek is 100 m, and precessional ages of D'Hondt et al. (1996) are used. Extrapolation from K-T boundary through C30n-C29r reversal yields age duration for Hell Creek Formation of 1.36 m.y. Calculated sedimentation rates for each interval are shown.

the chron, and choose not to use Fort Union Formation accumulation rates (T-29r, Table 5) to estimate the duration of the Hell Creek Formation.

The longest calculated duration for the Hell Creek Formation of 1.90 m.y. is obtained from the sedimentation rate for the Cretaceous part of C29r (K-C29r) combined with the longest precessional age estimate of Herbert et al. (1995), and includes the Bobcat Butte section (which shortens the calculated distance of the K-T boundary from the C30n-C29r reversal boundary). The shortest duration of 1.05 m.y. listed in Table 5 is derived using the Tertiary sedimentation rate (T-C29r) from the shortest C29r duration of 570 k.y. (Herbert and D'Hondt, 1990) and includes the Bobcat Butte data.

Composite magnetostratigraphic section

The range of estimates for the duration of the Hell Creek Formation is relatively small, from 1.90 m.y. to 1.05 m.y., but a final determination depends on the assumptions that we make. The first assumptions when extrapolating a sedimentation rate over any stratigraphic section are that the sedimentary record is complete and that sedimentation has occurred uniformly in all the sections. This is demonstrably not true in the case of Bobcat and Pyramid Butte, so both sections are removed from the final calculation. The extrapolation is based on an average of the reversal positions for the C30n-C29r boundary from the three sections at Pretty Butte, Sunset Butte, and Mud Buttes. Because there is a marked difference in sedimentation rate above and below the K-T boundary, we disregard the higher Tertiary rates obtained from the Ludlow Member of the Fort Union and assume that the lower sedimentation rate for the Cretaceous part of C29r (K-C29r) is most applicable to Hell Creek. The precessional age we use in our calculations is the most recent one of D'Hondt et al. (1996), where the K-T boundary event is estimated to occur 333 k.y. after the C30n-C29r reversal. The age of the K-T boundary event we use is 65.51 ± 0.10 Ma, which is calculated as an average of the published ages for the impact horizon normalized to 28.32 Ma (Taylor

Creek Rhyolite) and 28.02 Ma (Fish Canyon Tuff). The thickness of the formation, which is a critical component in the extrapolation, has been directly measured as 100 m in two sections in the region (Johnson, this volume; Murphy et al., this volume).

The averages of the stratigraphic positions for the reversal boundaries C30n-C29r and C29r-C29n (Tables 3 and 4) are shown in a composite magnetostratigraphic section in Figure 13. The position of the top of C29r is plotted at +25.8 m, and the base of C29r is shown at +24.4 m. Extrapolating down from the K-T boundary, dated as 65.51 ± 0.10 Ma, through the 333 k.y. duration of the Cretaceous part of C29r, to the base of the 100-m-thick Hell Creek, gives an age estimate for the base of the formation of 66.87 Ma (Fig. 13), and a duration for the Hell Creek of ~1.36 m.y.

The parameters used for this age estimate are shown in Table 5 by the bold highlighting of the K-C29r data for the most recent precessional information from D'Hondt et al. (1996). In Figure 13 we have also shown the age estimate if all of C29r is used in the calculation, and the higher Tertiary sedimentation rates are averaged into the equation. In that case the age estimate changes very little, from 66.87 to 66.71 Ma, so the potential for error appears relatively small.

CONCLUSIONS

A number of possible estimates have been presented here that attempt to solve the problem of the age and duration of the Hell Creek Formation. Using a range of estimates for the duration of C29r we have estimated that the time span of Hell Creek sedimentation is probably no more than 1.9 m.y., may be as short as 1.05 m.y. (Table 5), and using the most likely parameters outlined here, is probably ~1.36 m.y. Despite the excellent biostratigraphic control, the precise duration of Hell Creek sedimentation cannot be estimated by any other conventional means. Ideally a single isotopic age from one of the ashes (Fig. 7) found in the sequence would solve the problem, but all the ashes sampled and analyzed so far have proven to contain a significant population of detrital K-feldspar, and have yielded a wide range of single-crystal ages. The complexity of the problem is due in part to the nature of the sedimentary system. Diastems, or localized erosional horizons, are an integral part of the terrestrial fluvial environment. Of the six magnetostratigraphic sections measured, two are foreshortened by erosion. The high density of paleomagnetic samples collected through the Hell Creek Formation as part of this study makes it unlikely that any reversals in the sequence have been missed. It is still possible that C30r has been eroded completely from the sequence, but with an estimated duration of ~125 k.y. (Cande and Kent, 1995), the reversal should be more than 9 m thick (using the 13 648 yr/m sedimentation rate highlighted in Table 5), and there is only a single gap in the composite sequence that is more than 10 m wide (Fig. 11, ~−50 m), which makes it unlikely that we have missed this chron through insufficient paleomagnetic sampling.

Given the relatively high sedimentation rates of the Hell Creek, we postulate that C30r does not occur at all in the Hell Creek, and is either somewhere in the underlying Fox Hills, or has been lost in a disconformity. The presence of such a disconformity was suggested by Brown (1907), Jensen and Varnes (1964), Reynolds (1967), and Belt et al. (1997). The earlier evidence recognized physical evidence for an erosional downcutting at the contact of the Hell Creek and the Fox Hills Formations, and lenses of extrabasinal conglomerate in the base of the Hell Creek Formation in Montana. Belt et al. (1997) discussed the age relationships of the adjacent units. A significant disconformity with relief of as much as 70 m has been reported from the base of the Frenchman Formation, a time-equivalent unit to the Hell Creek located in southwestern Saskatchewan, Canada (Lerbekmo, 1987). This would imply that the disconformity is regional in extent, and that it is certainly related to a large-scale process such as tectonism or eustasy, not just to some local downcutting.

The occurrence of *Baculites baculus* and *B. grandis* in the uppermost Pierre Shale suggests that the marine sediments immediately underlying the Fox Hills are ca. 70 Ma. At Red Bird the middle part of *B. grandis* is dated by an isotopic age of 70.15 ± 0.65 Ma (Hicks et al., 1999). In the vicinity of the Cedar Creek anticline the Fox Hills Formation is only 30–60 m thick. Bracketed by our age estimate for the base of the Hell Creek of 66.7 Ma, and by the top of the Pierre at 70 Ma, the sediments of the Fox Hills Formation, which are only 30–60 m thick, encompass ~3.3 m.y. It is at this stratigraphic level that we must look for a large regional disconformity.

This postulated missing interval of time coincides exactly with the rapid regional regression of the Western Interior Late Cretaceous seaway (Gill and Cobban, 1966, 1973; Zapp and Cobban, 1960, 1962; Haq et al., 1988; Lillegraven and Ostresh, 1990), and it is almost certainly related to the rapid fall in base levels along margins of the seaway, and the exposure and erosion of large areas of unconsolidated sediment. A similar marine regression that was followed by a period of erosion and fluvial incision into an exposed seafloor has been noted at the base of the Pierre Shale at its contact with the underlying Niobrara Formation (DeGraw, 1975). The most likely level for the postulated 3 m.y. old gap in the sedimentary record of the Fox Hills is either at the basal contact of the Hell Creek Formation, where physical evidence for a regional disconformity has been noted by previous workers, or somewhere within the Colgate Member, a white sandstone unit that may have been formed by a considerable period of weathering, reworking and erosion. If the Hell Creek is only 1.36 m.y. in duration, then the onset of Hell Creek sedimentation is closely related to the rise in base levels associated with the rising sea level at the end of the Cretaceous, as the Cannonball Sea began to transgress near the end of the Maastrichtian.

ACKNOWLEDGMENTS

This study was partly supported by Geological Society of America Research Grants to Hicks (1992); the Petroleum Research Fund, administered by the American Chemical Society (ACS-PRF, 30442-AC8), grants to Tauxe; and National Science Foundation grant EAR-9805474 to Johnson. David Fastovsky measured the original section at Sunset Butte. Dean Pearson and Terry Schaefer of the Pioneer Trails Regional Museum, Bowman, North Dakota, logged the Doaks Butte section. Russell Crane, Beth Ellis, Marshall McLean, Rob Myers, and Kevin Werth assisted in the trenching and measuring of the other sections. Dean Pearson, Terry Schaefer, Regan Dunn, Julia Sankey, Gary Staab, and Allison Sundine assisted in the collection of the paleomagnetic samples.

Note in proof: A recent paper by Busby et al. (2002) reports on a submarine landslide, purportedly triggered by the K-T boundary bolide impact. A submarine tuff associated with this event was dated by ^{40}Ar-^{39}Ar and yielded an isotopic age of 65.5 $\pm$ 0.6 Ma. This age agrees almost exactly with the isotopic age estimate for the K-T boundary interval of 65.51 Ma that is presented here.

REFERENCES CITED

Alvarez, W., Arthur, M.A., Fischer, A.G., Lowrie, W., Napoleone, G., and Silva, I.P., 1977, Upper Cretaceous-Paleocene magnetic stratigraphy at Gubbio Italy V, Type section for the Late Cretaceous-Paleocene geomagnetic reversal time scale: Geological Society of America Bulletin, v. 88, p. 383–389.

Archibald, J.D., Butler, R.F., Lindsay, E.H., Clemens, W.A., and Dingus, L., 1982, Upper Cretaceous-Paleocene biostratigraphy and magnetostratigraphy, Hell Creek and Tullock Formations, northeastern Montana: Geology, v. 10, p. 153–159.

Archibald, J.D., 1996, Dinosaur Extinction and the End of an Era: What the fossils say: Critical moments in paleobiology and earth history series: New York, Columbia University Press, 237 p.

Belt, E.S., Flores, R.M., Warwick, P.D., Conway, K.M., Johnson, K.R., and Waskowitz, R.S., 1984, Relationship of fluviodeltaic facies to coal deposition in the lower Fort Union Formation (Palaeocene), southwestern North Dakota, *in* Rahmani, R.A., and Flores, R.M., eds., Sedimentology of coal and coal-bearing sequences: International Association of Sedimentology Special Publication 7, p. 177–195.

Belt, E.S., Hick, J.F., and Murphy, D.A., 1997, A pre-Lancian regional unconformity and its relationship to Hell Creek paleogeography in southeastern Montana: University of Wyoming, Contributions to Geology, v. 31, p. 1–26.

Berger, A.L., Loutre, M.F., and Dehant, V., 1989, Influence of the changing lunar orbit on the astronomical frequencies of pre-Quaternary insolation patterns: Paleogeography, v. 4, p. 555–564.

Berggren, W.A., Kent, D.V., Flynn, J.J., and Van Couvering, J.A., 1985, Cenozoic geochronology: Geological Society of America Bulletin, v. 96, p. 1407–1418.

Berggren, W.A., Kent, D.V., Obradovich, J.D., and Swisher, C.C., III, 1992, Toward a revised Paleogene geochronology *in* Prothero, D.R., and Berggren, W.A., Eocene-Oligocene Climatic and Biotic Evolution: Princeton, New Jersey, Princeton University Press, p. 29–45.

Bishop, G.A., 1965, The paleontology and stratigraphy of the Upper Cretaceous Upper Pierre Shale on Cedar Creek anticline, Montana: Proceedings of the South Dakota Academy of Science, v. 44, p. 91–96.

Bishop, G.A., 1973, Geology, stratigraphy, and biostratigraphy of the north end of Cedar Creek anticline, Dawson County, Montana: Montana Bureau of Mines and Geology Special Publication 61, 2 p.

Brown, B., 1907, The Hell Creek Beds of the Upper Cretaceous of Montana: Bulletin of the American Museum of Natural History, v. 23, p. 823–845.

Busby, C.J., Yip, G., Blikra, L., and Renne, P., 2002, Coastal landsliding and catastrophic sedimentation triggered by Cretaceous-Tertiary bolide impact: A Pacific margin example?: Geology, v. 30, p. 687–690.

Butler, R.F., and Lindsay, E.H., 1985, Mineralogy of magnetic polarity stratigraphy of continental sediments, San Juan basin, New Mexico: Journal of Geology, v. 89, p. 299–316.

Cande, S.C., and Kent, D.V., 1992, A new geomagnetic polarity time scale for the Late Cretaceous and Cenozoic: Journal of Geophysical Research, v. 97, B10, p. 13,917–13,951.

Cande, S.C., and Kent, D.V., 1995, Revised calibration of the geomagnetic polarity time scale for the Late Cretaceous and Cenozoic: Journal of Geophysical Research, v. 100, B4, 6093–6095.

Carlson, C.G., 1979, Geology of Adams and Bowman Counties, North Dakota: North Dakota Geological Survey Bulletin, v. 65, no. 1, p. 29.

Carlson, C.G., 1983, Geology of Billings, Golden Valley, and Slope Counties: North Dakota Geological Survey Bulletin, v. 76, no. 1, p. 40.

Corbin, R.S., 1938, Stratigraphy and paleontology of the Pierre, Fox Hills, and Lance Formations in southwestern North Dakota [Bachelor's thesis]: Princeton, New Jersey, Princeton University.

Cvancara, A.M., and Hoganson, J.W., 1993, Vertebrates of the Cannonball Formation (Paleocene) in North and South Dakota: Journal of Vertebrate Paleontology, v. 13, no. 1, p. 1–23.

Dalrymple, G.B., Izett, G.A., Snee, L.W., and Obradovich, J.D., 1993, ^{40}Ar/^{39}Ar age spectra and total-fusion ages of tektites from Cretaceous-Tertiary boundary sedimentary rocks in the Beloc Formation, Haiti: U.S. Geological Survey Bulletin, 2065, 20 p.

DeGraw, H.M., 1975, The Pierre-Niobrara unconformity in western Nebraska, *in* Caldwell, W.G.E., ed., The Cretaceous system in the Western Interior of North America: Geological Association of Canada Special Publication 13, p. 589–606.

D'Hondt, S., King, L., Gibson, J.W., and Park, J., 1992, The evolution of Late Cretaceous climatic cycles at South Atlantic DSDP Site 516F: EOS (Transactions, American Geophysical Union), v. 73, p. 271.

D'Hondt, S., Herbert, T.D., King, J., and Gibson, C., 1996, Planktic foraminifera, asteroids and marine production: Death and recovery at the Cretaceous-Tertiary boundary: Boulder, Colorado, Geological Society of America Special Paper 307, p. 303–317.

Diehl, J.F., 1991. The Elkhorn Mountains revisited: New data for the Late Cretaceous paleomagnetic field of North America. Journal of Geophysical Research, v. 96, B6, p. 9887–9894.

Duffield, W.A., and Dalrymple, G.B., 1990, The Taylor Creek Rhyolite of New Mexico: a rapidly emplaced field of domes and flows: Bulletin of Volcanology, v. 52, 475–478.

Fastovsky, D.E., 1987, Paleoenvironments of vertebrate-bearing strata during the Cretaceous-Paleogene transition, eastern Montana and western North Dakota: Palaios, v. 2, p. 282–295.

Fisher, R.A., 1953, Dispersion on a sphere: Proceedings of the Royal Society, v. A217, p. 295–305.

Fox, S.K., Jr., and Olsson, R.K., 1969, Danian planktonic foraminifera from the Cannonball Formation in North Dakota: Journal of Paleontology, v. 43, p. 1397–1404.

Frye, C.I., 1969, Stratigraphy of the Hell Creek Formation in North Dakota: North Dakota Geological Survey Bulletin, v. 54, 65 p.

FUCAT (Fort Union Coal Assessment Team), 1999, 1999 Resource assessment of selected Tertiary coal beds and zones in the Northern Rocky Mountains and Great Plains region: U.S. Geological Survey Professional Paper 1625-A (CD-ROM).

Gill, J.R., and Cobban, W.A., 1966, The Red Bird Section of the Upper Cretaceous Pierre Shale in Wyoming: U.S. Geological Survey Professional Paper 393-A, 73 p.

Gill, J.R., and Cobban, W.A., 1973, Stratigraphy and geologic history of the Montana Group and equivalent rocks, Montana, Wyoming, and North and South Dakota: U.S. Geological Survey Professional Paper 776, 36 p.

Haq, B.U., Hardenbol, J., and Vail, P.R., 1988, Mesozoic and Cenozoic chronostratigraphy and cycles of sea-level change, *in* Sea-level changes—An integrated approach: Society of Economic Paleontologists and Mineralogists Special Publication 42, p. 71–108.

Hares, C.J., 1928, Geology and lignite resources of the Marmarth field, southwestern North Dakota: U.S. Geological Survey Bulletin 776, 110 p.

Harland, W.B., Armstrong, R.L., Cox, A.V., Craig, L.E., Smith, A.G., and Smith, D.G., 1989, A Geologic time scale 1989: Cambridge, Cambridge University Press, 263 p.

Herbert, T.D., and D'Hondt, S.L., 1990, Precessional climate cyclicity in Late Cretaceous–early Tertiary marine sediments: A high-resolution chronometer of Cretaceous–Tertiary boundary events: Earth and Planetary Science Letters, v. 99, p. 263–275.

Herbert, T.D., D'Hondt, S.L., Premoli-Silva, I., Erba, E., and Fischer, A. G., 1995, Orbital chronology of Cretaceous—early Paleocene marine sediments, *in* Berggren, W.A., et al., eds., Geochronology, time scales and stratigraphic correlation: Society Economic Paleontologists and Mineralogists Special Publication 54, p. 81–94.

Hicks, J.F., Obradovich, J.D., and Tauxe, L., 1995, A new calibration point for the Late Cretaceous time scale: The ^{40}Ar-^{39}Ar isotopic age of the C33r/C33n Geomagnetic reversal from the Judith River Formation (Upper Cretaceous), Elk Basin, Wyoming, U.S.A: Journal of Geology, v. 103, p. 243–256.

Hicks, J.F., Obradovich, J.D., and Tauxe, L., 1999, Magnetostratigraphy, isotopic age calibration and intercontinental correlation of the Red Bird section of the Pierre Shale, Niobrara County, Wyoming, USA: Cretaceous Research, v. 20, p. 1–27.

Hilgen, F.J., 1994, Extension of the astronomically calibrated (polarity) time scale to the Miocene/Pliocene boundary: Earth and Planetary Science Letters v. 107, p. 349–368.

Hunter, J.P., 1999, The radiation of Paleocene mammals with the demise of the dinosaurs: evidence from southwestern North Dakota: Proceedings of the North Dakota Academy of Sciences, v. 53, p.141–144.

Hunter, J.P., and Pearson, D.A., 1996, First record of Lancian (Late Cretaceous) mammals from the Hell Creek Formation of southwestern North Dakota, USA: Cretaceous Research, v. 17, p. 633–643.

Izett, G.A., Dalrymple, G.B., and Snee, L.W., 1991, ^{40}Ar-^{39}Ar age of Cretaceous-Tertiary boundary tektites from Haiti: Science, 252, p. 1539–1542.

Jensen, F.S., and Varnes, H.D., 1964, Geology of the Fort Peck area, Garfield, McCone and Valley Counties, Montana: U.S. Geological Survey Professional Paper 414F.

Johnson, K.R., 1992, Leaf-fossil evidence for extensive floral extinction at the Cretaceous-Tertiary boundary, North Dakota, USA: Cretaceous Research, v. 13, p. 91–117.

Johnson, K.R., Nichols, D.J., Attrep, M., Jr., and Orth, C.J., 1989, High-resolution leaf-fossil record spanning the Cretaceous/Tertiary boundary: Nature, 340, p. 708–711.

Johnson, K.R., and Hickey, L.J., 1990, Megafloral change across the Cretaceous/Tertiary boundary in the northern Great Plains and Rocky Mountains, U.S.A.: Boulder, Colorado, Geological Society of America Special Paper 247, p. 433–444.

Kirschvink, J.L., 1980, The least-squares line and plane and the analysis of paleomagnetic data: Royal Astronomical Society Geophysics Journal, v. 62, p. 699–718.

Langereis, G.G., van Hoof, A.A.M., and Hilgen, F.J., 1994, Steadying the rates: Nature, 369, p. 615.

Lerbekmo, J.F., 1987, Magnetostratigraphic restrictions on the age of the Frenchman Formation and the magnitude of the sub-Frenchman disconformity in southwest Saskatchewan: Bulletin of Canadian Petroleum Geology, v. 35, no. 4, p. 454–459.

Lerbekmo, J.F., and Coulter, K.C., 1984a, Magnetostratigraphic and biostratigraphic correlation of late Cretaceous to early Paleocene strata between Alberta and North Dakota, *in* Stott, D.F., and Glass, D.J., The Mesozoic of middle North America: Canadian Society of Petroleum Geologists Memoir 9, p. 313–317.

Lerbekmo, J.F., and Coulter, K.C., 1984b, Late Cretaceous to early Tertiary magnetostratigraphy of a continental sequence: Red Deer Valley, Alberta, Canada: Canadian Journal of Earth Science, v. 22, p. 567–583.

Lerbekmo, J.F., and Sweet, A.R., 1994, A short normal magnetic interval at the K/T boundary: a measure of continuity of record across the boundary and synchroneity of boundary events: Lunar and Planetary Institute Contribution no. 825, p. 66–69.

Lerbekmo, J.F., Sweet, A.R., and Duke, M.J.M., 1996, A normal polarity subchron that embraces the K-T boundary: A measure of sedimentary continuity across the boundary and synchroneity of boundary events: Boulder, Colorado, Geological Society of America, Special Paper 307, p. 465–476.

Lillegraven, J.A., and Ostresh, L.M., 1990, Late Cretaceous (earliest Campanian/Maastrichtian) evolution of western shorelines of the North American Western Interior Seaway in relation to known mammalian faunas: Boulder, Colorado, Geological Society of America Special Paper 243, p. 1–8.

McFadden, P.L., and McElhinny, M.W., 1988, The combined analysis of remagnetization circles and direct observations in paleomagnetism: Earth and Planetary Science Letters, v. 87, p. 161–172.

McWilliams, M.O., Baksi, A.K., Bohor, B.F., and Murali, A.V., 1994, High precision relative ages of K/T boundary events in North America and Deccan Trap volcanism in India: EOS (Transactions, American Geophysical Union), v. 73, 14, p. 363.

Moore, W., 1976, The stratigraphy and environments of the Cretaceous Hell Creek Formation (reconnaissance) and the Paleocene Ludlow Formation (detailed), southwestern North Dakota: North Dakota Geological Survey Report, v. 56.

Nichols, D.J., and Ott, H.L., 1978, Biostratigraphy and evolution of the *Momipites-Caryapollenites* lineage in the early Tertiary in the Wind River Basin, Wyoming: Palynology, v. 2, p. 93–112.

Nichols, D.J., Jacobson, S.R., and Tschudy, R.H., 1982, Cretaceous palynomorph biozones for the central northern Rocky Mountain Region of the United States, in Powers, R.B., ed., Geologic studies of the Cordilleran thrust belt: Rocky Mountain Association of Geologists, p. 721–734.

Obradovich, J.D., and Cobban, W.A., 1975, A time scale for the Late Cretaceous of the Western Interior of North America, *in* Caldwell, W.G.E., ed., The Cretaceous System in the Western Interior of North America: Geological Association of Canada Special Paper 13, p. 31–54.

Obradovich, J.D., 1993, A Cretaceous time scale, *in* Caldwell, W.G.E., and Kauffman, E.G., Evolution of the Western Interior Basin: Geological Association of Canada Special Paper 39, p. 379–396.

Park, J., D'Hondt, S., King, J.W., and Gibson, C., 1993, Late Cretaceous precessional cycles in double time: A warm earth Milankovitch response: Science, v. 261, p. 1431–1434.

Renne, P.R., Deino, A.L., Walter, R.C., Turrin, B.D., Swisher, C.C., Becker, T.A., Curtis, G.H., Sharp, W.D., and Jaouni, A.R., 1994, Intercalibration of astronomical and radioisotopic time: Geology, v. 22, p. 783–786.

Renne, P.R., Swisher, C.C., Deino, A.L., Karner, D.B., Owens, T.L., and DePaolo, D.J., 1998, Intercalibration of standards, absolute ages and uncertainties in ^{40}Ar-^{39}Ar dating: Chemical Geology (Isotope Geoscience Section), v. 145, p. 117–152.

Reynolds, M.W., 1967, Physical evidence for Late Cretaceous unconformity, south-central Wyoming: U.S. Geological Survey Professional Paper 575-D, D24–D28.

Samson, S.A., and Alexander, E.C., Jr., 1987, Calibration of the interlaboratory ^{40}Ar-^{39}Ar dating standard, MMhb-1: Chemical Geology (Isotope Geoscience Section), v. 66, p. 27–34.

Sarna-Wojcicki, H.M., and Pringle, M.S., Jr., 1992, Laser fusion ^{40}Ar-^{39}Ar ages of the tuff at Taylor Canyon and Bishop Tuff, E. California–W. Nevada: EOS (Transactions, American Geophysical Union), v. 73, no. 43, p. 633.

Shackleton, N.J., Berger, A., and Peltier, W.R., 1990, An alternative astronomical calibration of the lower Pleistocene time scale based on ODP Site 677: Transactions of the Royal Society of Edinburgh, Earth Sciences, v. 81, p. 251–261.

Sheehan, P.M., Fastovsky, D.E., Hoffmann, R.G., Berghaus, C.B., and Gabriel, D.L., 1991, Sudden extinction of the Dinosaurs: Latest Cretaceous, Upper Great Plains, U.S.A.: Science, v. 254, p. 835–839.

Sloan, R.E., Rigby, J.K., Jr., Van Valen, L.M., and Gabriel, D., 1986, Gradual dinosaur extinction and simultaneous Ungulate radiation in the Hell Creek Formation: Science, v. 232, p. 629–633.

Stanton, T.W., 1920, The fauna of the Cannonball marine member of the Lance formation: U.S. Geological Survey Professional Paper 128-A.

Swisher, C.C., III, and Knox, R.W.O'B., 1991, The age of the Paleocene/Eocene boundary: ^{40}Ar-^{39}Ar dating of the lower part of NP10, North Sea Basin and Denmark: Brussels, Belgium, International Geological Correlation project 308, Brussels Meeting.

Swisher, C.C., III, Grajales-Nishimura, J.M., Montanari, A., Margolis, S.V., Claeys, P., Alvarez, W., Renne, P.R., Cedillo-Pardo, E., Maurrasse, F.J.-M.R., Curtis, G.H., Smit, J., and McWilliams, M.O., 1992, Coeval ^{40}Ar-^{39}Ar ages of 65.0 million years ago from Chicxulub crater melt rock and Cretaceous-Tertiary boundary tektites. Science, v. 257, p. 954–958.

Swisher, C.C., III, Dingus, L., and Butler, R.F., 1993, ^{40}Ar-^{39}Ar dating and magnetostratigraphic correlation of the terrestrial Cretaceous-Paleogene boundary and Puercan Mammal Age, Hell Creek–Tullock formations, eastern Montana: Canadian Journal of Earth Sciences, v. 30, p. 1981–1996.

Tauxe, L., Gee, J., Gallet, Y., Pick, T., and Bown, T., 1994, Magnetostratigraphy of the Willwood Formation, Bighorn basin, Wyoming: new constraints on the location of the Paleocene/Eocene: Earth and Planetary Science Letters, v. 125, p. 159–170.

Ten Kate, W.G.H., and Sprenger, A., 1993, Orbital cyclicities above and below the Cretaceous/Paleogene boundary at Zumaya (N. Spain), Agost and Relleu (SE Spain): Sedimentary Geology, v. 87, p. 69–101.

Waage, K.M., 1968, The Type Fox Hills Formation, Cretaceous (Maestrichtian), South Dakota Part 1. Stratigraphy and paleoenvironments: Peabody Museum Natural History Bulletin, v. 27, 171 p.

Watson, G.S., 1956, A test for randomness: Royal Astronomical Society Geophysical Supplement, Monthly Notices, v. 7, p. 160–161.

Wei, W., 1995, Revised age calibration points for the geomagnetic polarity time scale: Geophysical Research Letters, v. 22, no. 8, p. 957–960.

Zapp, A.D., and Cobban, W.A., 1960, Some Late Cretaceous strand lines in northwestern Colorado and northeastern Utah: U.S. Geological Survey Professional Paper 400-B, p. B246–B249.

Zapp, A.D., and Cobban, W.A., 1962, Some Late Cretaceous strand lines in southern Wyoming: U.S. Geological Survey Professional Paper 450-D, p. D52–D55.

Zijderveld, J.D.A., 1967, A.C. Demagnetization of rocks: Analysis of results, *in* Collinson, D.W., Creer K.M., and Runcorn, S.K.: Methods in Paleomagnetism, Elsevier, Amsterdam, p. 254–286.

Manuscript Accepted by the Society September 24, 2001

Printed in the U.S.A.

Geological Society of America
Special Paper 361
2002

Magnetostratigraphy of interfingering upper Cretaceous–Paleocene marine and continental strata of the Williston Basin, North Dakota and Montana

Steve P. Lund*
Department of Earth Sciences, University of Southern California, Los Angeles, California 90089-0740, USA
Joseph H. Hartman
Department of Geology and Geological Engineering and Energy & Environmental Research Center, University of North Dakota, P.O. Box 8358, Grand Forks, North Dakota 58202-8358, USA
Subir K. Banerjee
Department of Geology and Geophysics, University of Minnesota, Minneapolis, Minnesota 55455-0219, USA

ABSTRACT

We have carried out a magnetostratigraphic study of uppermost Cretaceous and lower Paleocene strata of the Williston Basin that are exposed in central North Dakota and easternmost Montana. Five overlapping sections were sampled that extend from the marine Fox Hills Formation through the predominantly continental Hell Creek Formation and Ludlow Member of the Fort Union Formation into the overlying marine Cannonball Member of the Fort Union Formation. Rock magnetic study indicated that the detrital magnetic mineral carrying the sediment's natural remanent magnetization is an ilmeno-hematite with low Curie temperature (<225 °C) and that secondary viscous and chemical overprints are routinely present. The base of magnetic polarity Chron 29r, which contains the K-T boundary, is located at the Hell Creek–Fort Union formational contact in central North Dakota but lies within the uppermost Hell Creek Formation in eastern Montana. This is consistent with palynological placement of the K-T boundary within the Ludlow Member in central North Dakota and at the Hell Creek–Fort Union formational contact in eastern Montana. The paleomagnetic data suggest that the Hell Creek Formation in central North Dakota terminated at least 90 000 yr before the palynological K-T boundary, while the Hell Creek Formation in eastern Montana terminated synchronously with the palynological K-T boundary. This resulted from a time-transgressive change in paleoenvironmental conditions related to westward transgression of the Cannonball Sea, which was present in central North Dakota about 160 000 yr after the K-T boundary. The earlier Upper Cretaceous marine Fox Hills regression and overlying marine Breien Member (Hell Creek Formation) incursion occurred during normal magnetic polarity Chron 30n.

*E-mail: slund@usc.edu

Lund, S.P., Hartman, J.H., and Banerjee, S.K., 2002, Magnetostratigraphy of interfingering upper Cretaceous–Paleocene marine and continental strata of the Williston Basin, North Dakota and Montana, *in* Hartman, J.H., Johnson, K.R., and Nichols, D.J., eds., The Hell Creek Formation and the Cretaceous-Tertiary boundary in the northern Great Plains: An integrated continental record of the end of the Cretaceous: Boulder, Colorado, Geological Society of America Special Paper 361, p. 57–74.

INTRODUCTION

The uppermost Cretaceous and lowermost Paleocene strata of the Williston Basin of North and South Dakota (Fig. 1) record the last vestiges of a marine epeiric seaway that had inundated the Western Interior of North America during most of the Cretaceous. The Pierre and Fox Hills Formations represent the last major marine regression (Gill and Cobban, 1973) of the Western Interior Seaway during the Upper Cretaceous (Maastrichtian). There is clear evidence in the eastern portion of the Williston Basin for a number of subsequent marine transgressive-regressive cycles in the uppermost Cretaceous and into the Paleocene (Danian to Selandian) (Fig. 2). The later Cretaceous marine cycles are represented by the Maastrichtian Breien Member of the Hell Creek Formation and by other later shifts in shoreline position upsection in the Hell Creek Formation (Hoganson and Murphy, this volume; Hartman and Kirkland, this volume). Pulses of a Paleocene Western Interior Sea are represented by the Cannonball Formation and its marine tongues (Hartman and Butler, 1995; Hartman et al., 1998, 1999; Tibert et al., 2001). The post–Fox Hills marine strata in the Williston Basin contain a variety of invertebrate and vertebrate fossils, including Maastrichtian ammonoid cephalopods (Kennedy et al., 1998) and Danian planktic foraminifera and nautiloid cephalopods (Fox and Olsson, 1969; Feldmann, 1972a). The continental strata that interfinger with these marine strata contain diverse invertebrate, vertebrate, and plant assemblages (e.g., Pearson et al., this volume; Johnson, this volume) of the Lancian (Maastrichtian) North American Land Mammal Age (NALMA) and a less diverse nonmarine biota of Puercan to early Tiffanian NALMA (Danian to Selandian) age. The exact placement of the K-T boundary in these strata is currently inferred from detailed palynomorph studies in nonmarine strata (Lerbekmo and Coulter, 1984; Murphy et al., 1995; and Nichols and Johnson, this volume) and their relationship to iridium anomaly and shocked quartz horizons in western North Dakota (Johnson et al., 1987; Nichols et al., 2000).

In spite of the relatively numerous and well-documented biostratigraphic sections (e.g., Frye, 1969; Feldmann, 1972b; Cvancara, 1976a; Murphy et al., 1995) through the Fox Hills–Cannonball stratigraphic interval in two areas, there are few well-established tie points between these strata and the magnetic polarity time scale (MPTS) (e.g., Archibald et al., 1982; Lerbekmo and Coulter, 1984; Swisher et al., 1993) or the European stratotypes (Kennedy et al., 1998). This has led to significant uncertainty in the interpretation of late Maastrichtian to earliest Danian marine fluctuations in the Williston Basin, which represents the best record of marine strata for this interval of time in North America and best opportunity to correlate marine with nonmarine chronostratigraphic frameworks in the Western Interior.

This study develops a magnetic polarity stratigraphy for the Fox Hills–lowermost Cannonball interval as a means to improve our understanding of the chronology of these strata.

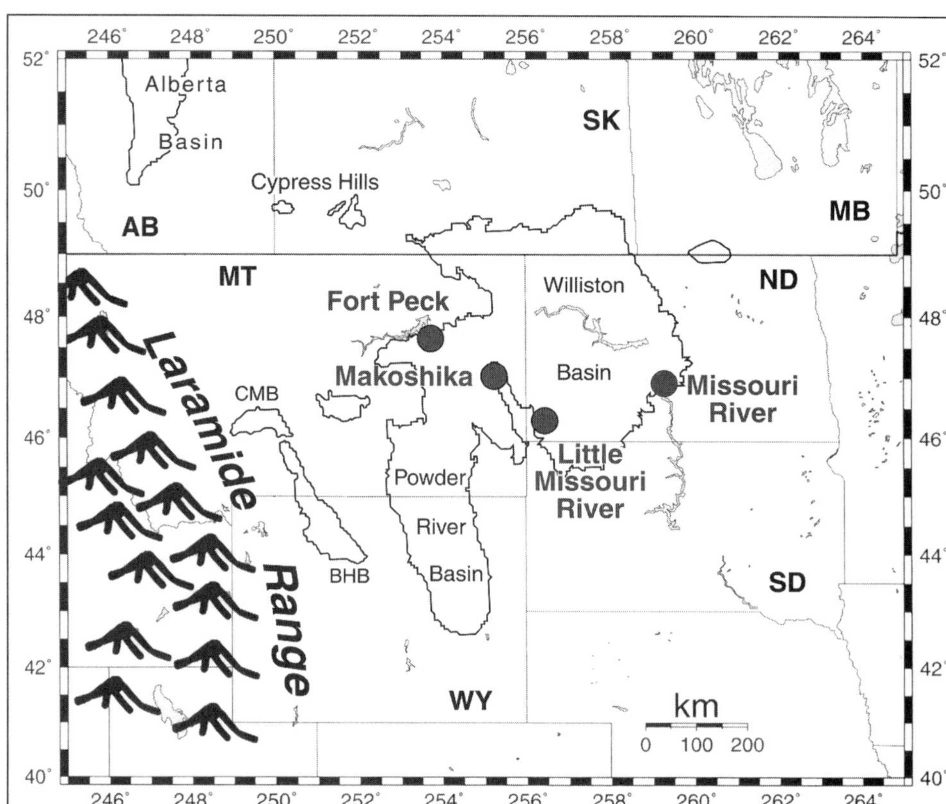

Figure 1. Location of discussed paleomagnetic sections within the Williston Basin of North Dakota and Montana. The basin outlines represent the Hell Creek–Fort Union or equivalent formational contact, which approximates the K-T boundary throughout this area (Hartman and Roth, 1998). The mountainous portion of the base map is drawn from Gill and Cobban (1973).

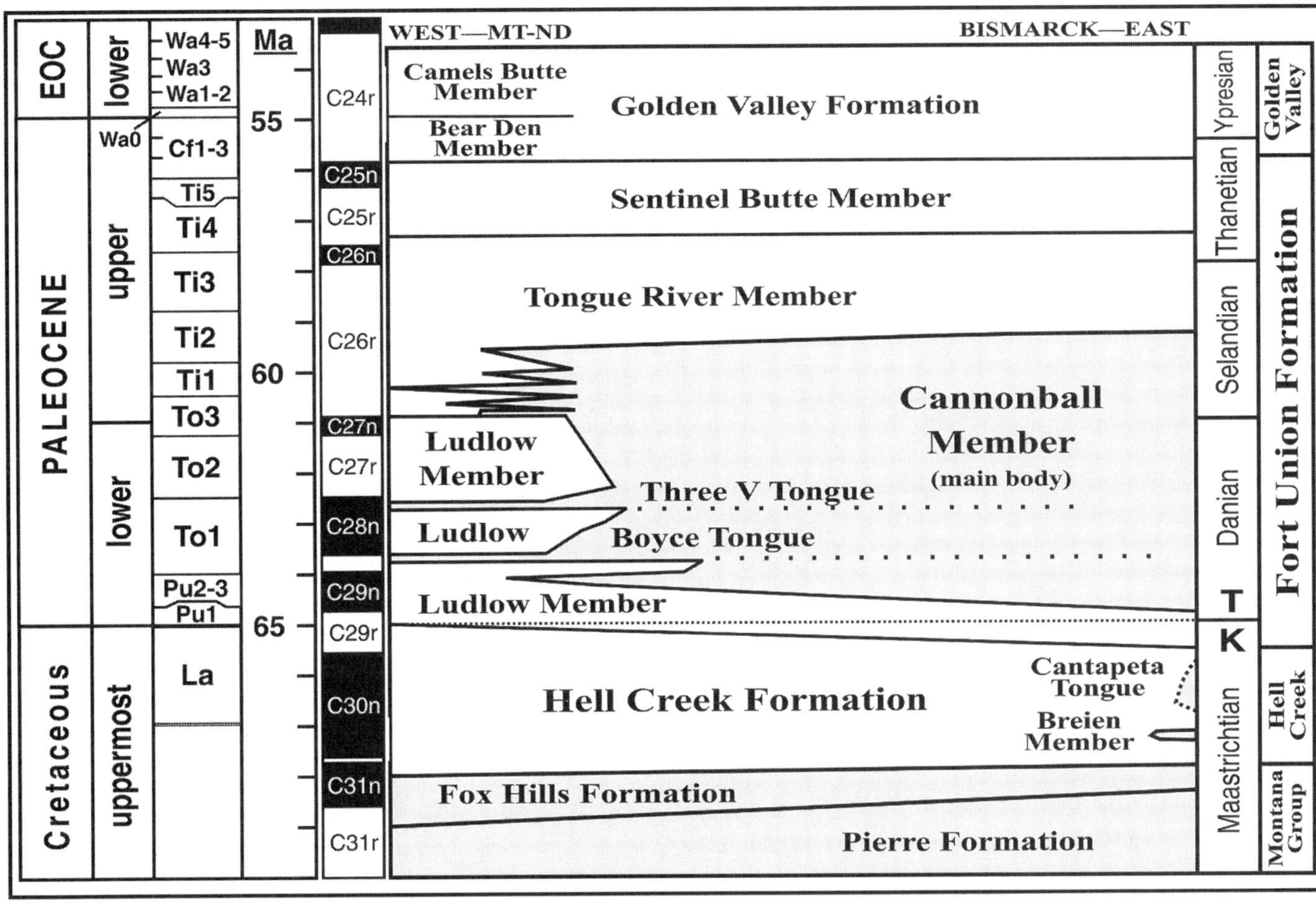

Figure 2. Chronostratigraphy of Upper Cretaceous and lower Paleogene strata in the eastern Williston Basin. This figure represents the organization of strata in the Dakotas and easternmost Montana relative to a radiometric time scale. The geomagnetic polarity time scale and associated radiometric dates are from Berggren et al. (1995), Gradstein et al. (1995), and Hardenbol et al. (1998). Except for the paleomagnetic sections addressed in this study, the North Dakota and adjacent Montana lithostratigraphic units are correlated with the polarity time scale on the basis of all available biochronologic information. The relative chronostratigraphic correlation of contacts is based on paleomagnetic sections (this study) on the bluffs of the Missouri River (east) (see Fig. 3) and Makoshika State Park (west) (see Fig. 12); mammalian local faunas (Hartman and Kihm, 1992, 1996) correlated to the NALMA chronology (after Prothero, 1995); marine faunal data (Hartman et al., 1998, 1999); palynomorph data (Nichols and Johnson, this volume; Kroeger et al., 1993); and subsurface geophysical studies (Hartman and Butler, 1995).

Our work is the outgrowth of an unpublished Master's thesis paleomagnetic study (Lund, 1976) of the Upper Cretaceous Hell Creek Formation in central North Dakota, which had the goal of identifying the K-T boundary. The results of Lund (1976) were not published, however, for no clearcut intervals of reversed polarity (assignable to Chron 29r, which contains the K-T boundary) were found in the studied section. The absence of Chron 29r implied either that the paleomagnetic results still contained undiagnosed normal-polarity overprints or that the Hell Creek–Fort Union formational contact was not the actual K-T boundary. It was not possible, at that time, to choose between these two possibilities.

STRATIGRAPHIC FRAMEWORK

Upper Cretaceous bedrock exposed in the Williston Basin (Fig. 1) consists of the marine Pierre and Fox Hills Formations and the primarily nonmarine Hell Creek Formation. These formations typically are youngest where they outcrop in central North and South Dakota (Fig. 2) (Waage, 1968; Gill and Cobban, 1973; Cvancara, 1976b). The Fox Hills Formation has been reported as notably time transgressive in west-to-east transects in eastern Montana (Tschudy, 1970; Gill and Cobban, 1973), with its final retreat from Montana dated biostratigraphically as within the *Baculites grandis* Ammonite Zone of the Western Interior (Waage, 1968; Gill and Cobban, 1973). This zone is lower Maastrichtian in age (70.22–69.69 Ma from Hardenbol et al., 1998). The youngest Fox Hills strata in South Dakota and Nebraska lie within the *Jeletzkytes nebrascensis* Ammonite Zone of the upper Maastrichtian on the basis of fossils from the Timber Lake Member. This is the youngest ammonite zone recognized in the Western Interior (Kennedy et al., 1998; <68.89 Ma from Hardenbol et al., 1998), but still younger occurrences of ammonite fossils exist higher in the section in the Linton Member of the Fox Hills (scaphitid ammonite fragments, Klett and Erickson, 1976) and in the Breien Member of the Hell Creek (*Discoscaphites* cf. *D. conradi* Morton, Kennedy et al., 1998). Thus the Fox Hills and overlying Breien Member in central North Dakota are estimated from marine biostratigraphy to range potentially from MPTS Chron 31n to Chron 29r (Fig. 2).

The overlying Hell Creek Formation in the Williston Basin represents the delta-building, continental facies, equivalent to the Fox Hills and Pierre Formations farther to the east (Butler and Hartman, 1999). The extensive channel facies and associated interbelt floodbasin mudstones of the Hell Creek contain important and distinctive assemblages of dinosaurs of the *Triceratops* Zone as used by Cobban and Reeside (1952) (see Russell and Manabe, this volume), mammals representative of the Lancian NALMA after Dorf (1938, 1942; Clemens et al., 1979; Clemens, this volume; Hunter and Archibald, this volume), plants (Johnson, this volume), palynomorphs (Nichols, this volume), and mollusks (Hartman, 1998, 1999). The Hell Creek Formation in central North Dakota also includes a marine tongue (Breien Member of Laird and Mitchell, 1942) in its lower part and subsequent oscillatory shoreline movements, as represented by trace and mollusk fossils (Hoganson and Murphy, this volume; Hartman and Kirkland, this volume) higher in the Hell Creek section that document smaller-scale transgressive-regressive cycles related to the movements of the Western Interior Seaway.

In the Williston Basin, the lignite-poor Hell Creek Formation gives way to the lignite-rich nonmarine Ludlow Member of the Fort Union Formation (Moore, 1976), with their formational contact being picked in many studies as the first persistent lignite or carbonaceous unit above the last dinosaur-bearing beds (Calvert, 1912). This placement was also traditionally used to delimit the K-T boundary (Brown, 1962; Hartman et al., this volume). The Fort Union Formation also contains the marine and marginal marine Cannonball Member, the primarily nonmarine but marine-influenced Tongue River Member, and the overlying nonmarine Sentinel Butte Member in stratigraphic succession (Fig. 2). The K-T boundary in central North Dakota is currently placed within the Ludlow Member on the basis of palynomorph changes (Lerbekmo and Coulter, 1984; Murphy et al., 1995; Nichols and Johnson, this volume). In western North Dakota and easternmost Montana, the K-T boundary is placed at the Hell Creek–Ludlow contact (Nichols and Johnson, this volume; Hunter et al., 1998, respectively).

The marine Cannonball Member interfingers with the Ludlow and Tongue River Members in the central and eastern Williston Basin. Various studies by Cvancara (1965, 1976a), his students (Hartman and Kihm, 1992, 1996; Hartman and Butler, 1995; Hartman et al., 1998, 1999), and Tibert et al. (2001) indicate that the Cannonball contains a number of relatively short-duration transgressive-regressive cycles. The Cannonball can be chronostratigraphically assigned to the Danian Stage (P1a Planktic Foraminifera Zone, Berggren et al., 1995) on the basis of its distinctive foraminifera (Fox and Olsson, 1969; Fenner, 1974a, 1974b) and to the subsequent Selandian Stage (NP4 Nannoplankton Zone, Perch-Nielsen, 1989) on the basis of preliminary documentation of *Sphenolithus primus* Perch-Nielsen (Hartman et al., 1999) from the upper part of the Cannonball. Macrofossil evidence also suggests (Cvancara, 1965, 1976a; Silfer, 1990) that part of the Cannonball is post-Danian in age. The Cannonball can be stratigraphically correlated with marginal marine or marine-influenced sediments in the lower part of the Tongue River Member (e.g., Diemer et al., 1996; Tibert et al., 2001). These strata are, in turn, overlain by nonmarine strata of the upper Tongue River and Sentinel Butte of upper Tiffanian and Clarkforkian age (upper Paleocene) (Hartman and Kihm, 1996) (Fig. 1).

Key unresolved questions that relate to these Upper Cretaceous and Paleocene strata include the following. (1) What is the age of the ammonite-bearing (post–*Jeletzkytes nebrascensis* Zone) uppermost Fox Hills Formation. No exact correlation has ever been made between these upper Cretaceous ammonites of the Western Interior and the MPTS or equivalent European stratotypes. (2) Where does MPTS Chron 29r, which contains the K-T boundary (Alvarez et al., 1980; Berggren et al., 1995), lie relative to the local biostratigraphic estimates of the K-T boundary?

PREVIOUS PALEOMAGNETIC RESULTS

Lund (1976) collected paleomagnetic samples from three sections of the Hell Creek and Fort Union Formations in central North Dakota (Fig. 3, Appendix 1), but studied only one section (Eagle Nose Butte) as part of a Master's thesis study to determine placement of the K-T boundary in central North Dakota. The sediments were relatively well-indurated mudstone and siltstone and relatively unindurated fine-grained sandstone. Block samples were recovered from 20 horizons in the Eagle Nose Butte Section, and four to six specimens ($2 \times 2 \times 2$ cm sample cubes) were cut from each block for paleomagnetic studies. Typically, the natural remanent magnetizations (NRMs) of three (or more) specimens per sample were demagnetized in alternating magnetic fields (a.f.) up to 40 millitesla (mT); the NRM of at least one specimen from each of the more indurated samples was thermally demagnetized up to 300 °C. Representative demagnetization results are displayed in Figures 4 and 5. In all cases, there was a large-intensity, low-coercivity viscous overprint, with directions similar to the local present-day field direction that was demagnetized by 10 mT and occasionally a high-coercivity overprint (with a variety of directions) that remained above 50 mT. Thermal demagnetization revealed that most NRM was lost by 250 °C, although occasionally a small-intensity, higher-temperature component was also present.

Lund (1976) carried out a detailed mineralogical and rock magnetic study to determine the magnetic mineralogy of the sediments and estimate what magnetic mineral was carrying

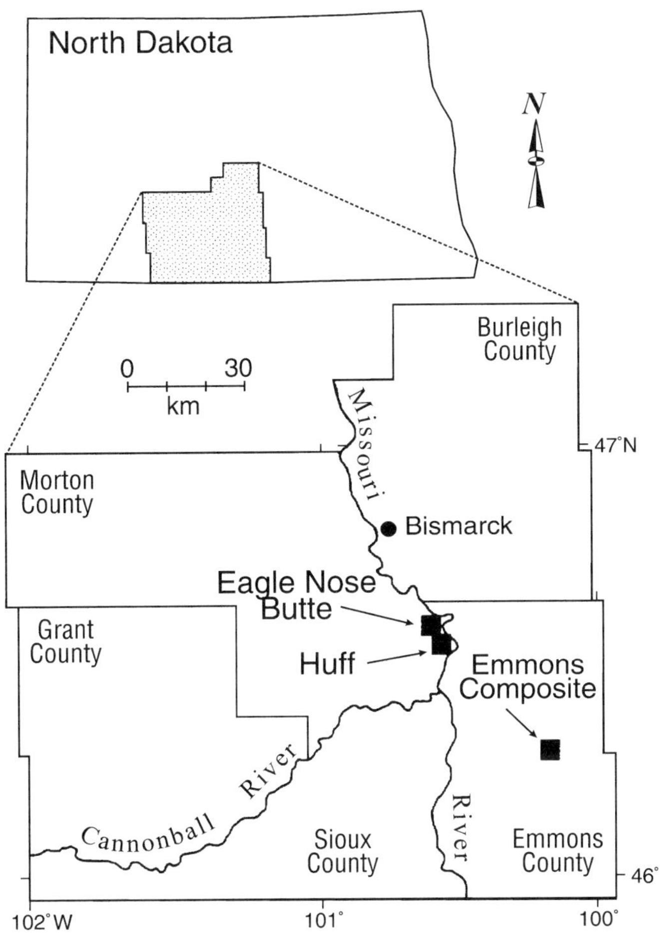

Figure 3. Site map of paleomagnetic sampling localities in central North Dakota.

most of the NRM. Magnetic minerals were separated out from six samples using a hand magnet. All magnetic separates contained abundant subhedral black detrital grains 2–30 μm in diameter. Thermomagnetic analyses (Fig. 6A) of the magnetic separates indicated that a single mineral with a Curie temperature of ~200 °C was present. Powder-camera x-ray diffraction studies determined that these grains had a rhombohedral crystal structure most like ilmenite ($FeTiO_3$), a magnetic mineral that is paramagnetic at 25 °C. Electron microprobe studies estimated that these grains were iron-titanium oxides with an average TiO_2/FeO ratio of 0.57. All of these observations suggest that the separable, detrital black magnetic grains are a single mineral made up of a solid solution of ilmenite and hematite (ilmeno-hematite) with the composition $0.70TiO_3 0.30Fe_2O_3$ ($x = 0.7$). Such phases only form during fast quenching of magmas and were probably formed during Laramide volcanism that was active to the west during the Late Cretaceous and Paleocene (Fig. 1). Yellowish iron staining and small "limonite" concretions were common in the samples. Limited thermomagnetic studies of the concretions indicated the likely presence of goethite (FeOOH) and hematite (Fe_2O_3).

Lund (1976) made artificial specimens by mixing either detrital ilmeno-hematite grains or fragments of limonite concretions with (nonmagnetic) rock salt and squeezing the mixtures in a (nonmagnetic) brass press to form $2 \times 2 \times 2$ cm cubes. Anhysteretic remanences (ARM) were applied to all artificial specimens and selected natural specimens using a 100 mT alternating field and 0.05 mT bias field. Typical alternating field demagnetizations of these specimens are displayed in Figure 6B and 6C. The coercivity spectra of the concretions and ilmeno-hematite grains were very different. The hard ARM coercivity in the concretions is likely due to fine-grained goethite or hematite, while the soft ARM coercivity is due to coarse-grained ilmeno-hematite. The fact that the NRM coercivity of all natural specimens looks most like that of the ilmeno-hematite artificial specimens suggests that the NRM is carried primarily by the ilmeno-hematite grains. This is corroborated by the similarity of the NRM thermal demagnetization results (Fig. 5) and ilmeno-hematite thermomagnetic curves (Fig. 6A). The ARMs of the natural specimens (Fig. 6B), however, indicate that an observable component of iron concretionary material (goethite, hematite) is present in most specimens and may be responsible for some of the higher-coercivity NRM.

On the basis of these results, Lund (1976) suggested that the primary sedimentary remanence resides at 10–40 mT and 100–250 °C. Viscous overprints may reside in lower coercivity or temperature ranges, and chemical overprints (associated with hematite or goethite in the concretions) may reside at higher coercivities and temperatures. Thus careful demagnetization and vector differencing (10–40 mT, 100–250 °C) were necessary to recover the original detrital remanences in these samples. On the basis of these measurements, a magnetic polarity stratigraphy was determined for the Eagle Nose Butte Section. The inclination and declination results are shown in Figure 7. Even with this careful study, however, there was no clear evidence for reversed polarity in any of the Cretaceous Hell Creek Formation samples, and the results were never published for fear that more serious overprinting had occurred than could be discerned by the study. This was a concern even though the cleaned NRM directions clearly display a Late Cretaceous pole position significantly different from the present-day field (Lund, 1976). In passing, it should be noted that the only sample studied by Lund (1976) from the Ludlow Member of the Fort Union Formation showed evidence for reversed-polarity inclinations (Fig. 7), but the sample was too weak to get a clear characteristic remanence, and the results were considered unreliable.

Lerbekmo and Coulter (1984) carried out a magnetostratigraphic study of Hell Creek, Ludlow, and Cannonball strata at the Huff Section (Fig. 3), less than 5 km south of Eagle Nose Butte Section, on the west side of the Missouri River Valley in central North Dakota. They collected three independently oriented sediment specimens from 54 sampling horizons spanning the uppermost Hell Creek Formation to middle Cannonball Member (Fig. 8). All specimens were a.f. demagnetized in fields up to 10–25 mT in order to remove a low-coercivity vis-

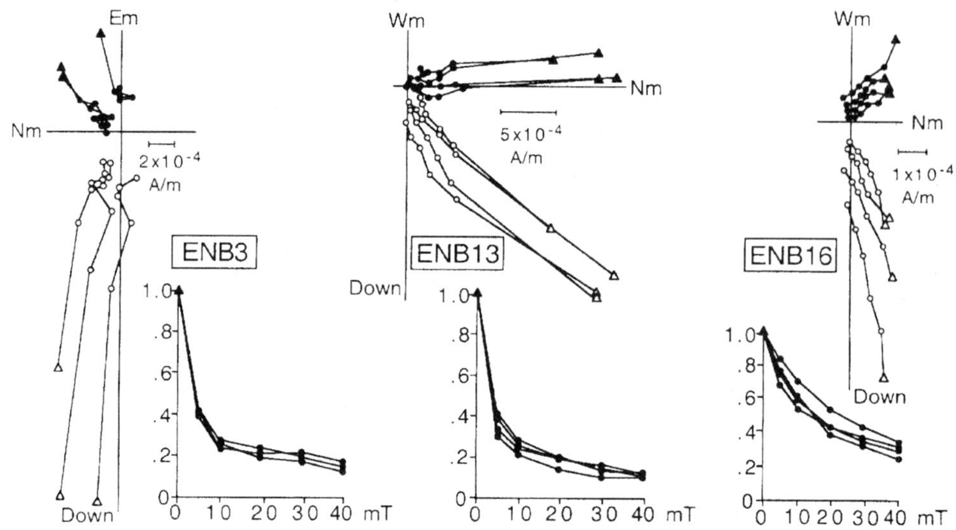

Figure 4. Zijderveld diagrams for three typical sampling horizons from the Eagle Nose Butte (ENB) Section (Appendix 1) indicating the pattern of NRM loss during a.f. demagnetization. The solid (open) triangles represent the tip of the initial NRM horizontal (vertical) component. The solid (open) circles represent the tip of the NRM horizontal (vertical) component after each step of demagnetization. For each sample, the relative loss of intensity (J/Jo) after each step of a.f. demagnetization is also plotted (from Lund, 1976).

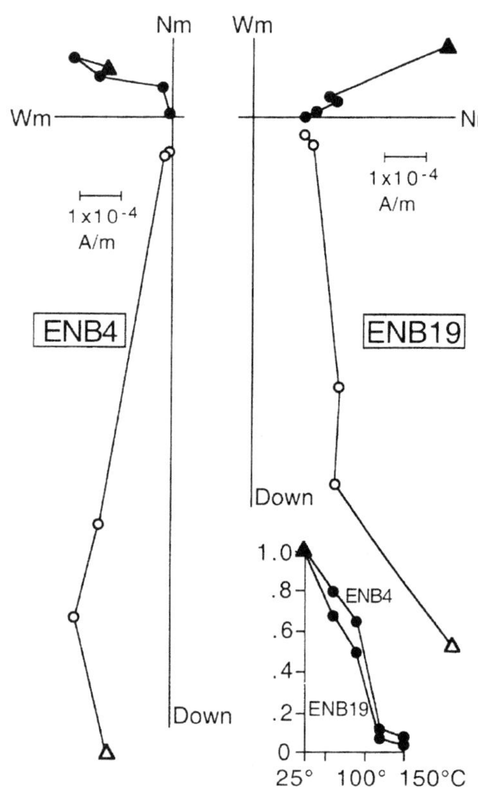

Figure 5. Zijderveld diagrams for two sampling horizons from the Eagle Nose Butte (ENB) Section (Appendix 1) indicating the pattern of NRM loss during thermal demagnetization. See Figure 4 for an explanation of the symbols (from Lund, 1976).

cous overprint. They presumed that the NRM was carried by detrital magnetite and that any NRM remaining after removal of the low-coercivity overprint should be a depositional remanence. Their final magnetostratigraphy is shown in Figure 8. Two reversed-polarity intervals were identified: one extending from the uppermost Hell Creek Formation to lowermost Cannonball Member and the other located within the middle Cannonball. The lowermost reversed-polarity interval was interpreted to be Chron 29r, which contains the K-T boundary.

Lerbekmo and Coulter (1984) also independently estimated the location of the K-T boundary on the basis of pollen occurrences. They noted the disappearance of *Aquilapollenites sp.*, a distinctive Cretaceous palynomorph, just below a 45-cm-thick coal bed in the Ludlow Member (2.3 m above the base) and the presence of distinctive Paleocene palynomorphs above the coal (Fig. 8). This placed the palynomorph K-T boundary in the middle of Chron 29r.

Archibald et al. (1982) also carried out a paleomagnetic study of the K-T boundary interval of the upper Hell Creek Formation and lower Tullock (Ludlow equivalent, in part) Member of the Fort Union Formation in the Fort Peck region of eastern Montana (Fig. 1). They assumed that the black opaque magnetic minerals in the sediments were magnetite or titanomagnetite and that a.f. demagnetization above 20 mT normally cleaned the samples and identified the primary NRM. They recovered reproducible evidence in four sections for a reversed-magnetic polarity interval bracketing the K-T boundary. They concluded that this reversed-polarity interval contained both uppermost Lancian dinosaur-bearing beds of the Hell Creek Formation and stratigraphically higher early Puercan mammal-bearing beds of the lower Tullock Member. Swisher et al. (1993) revised the magnetostratigraphic results of Archibald et al. (1982). They identified that the detrital magnetic mineral accountable for most of the sedimentary remanence is an ilmeno-hematite exactly like that noted by Lund (1976) for similar rocks of central North Dakota. Their overall magnetostratigraphy was almost identical to that of Archibald et al. (1982), with a significant interval of the Hell Creek Formation containing reversed magnetic polarity attributable to Chron 29r. This conclusion was not consistent with Lund's (1976) study of the Hell Creek Formation in central North

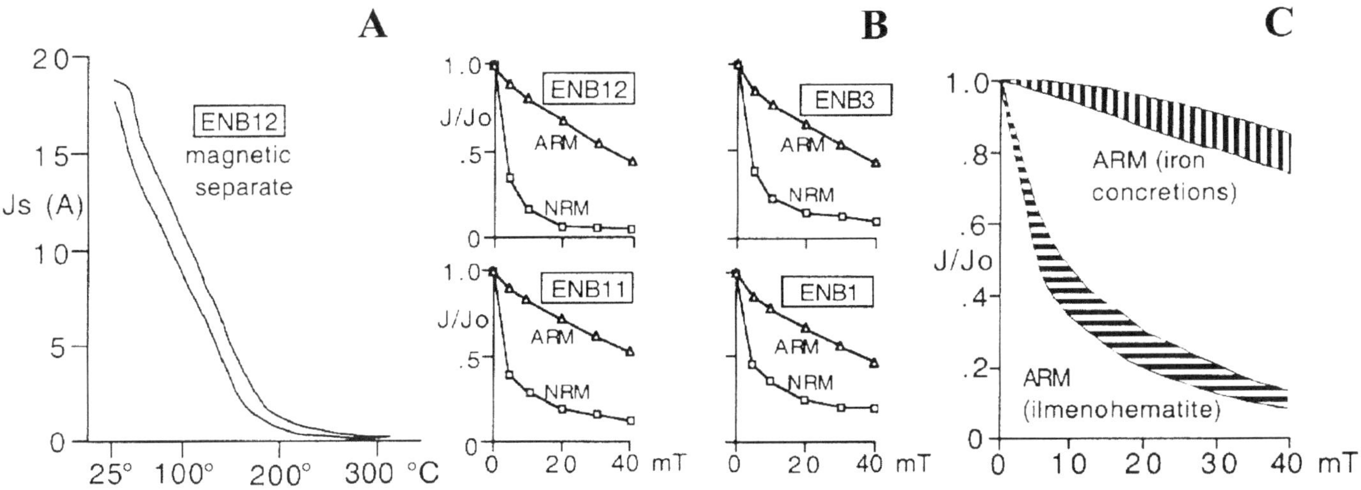

Figure 6. Rock magnetic analysis of Hell Creek Formation sediments. A: Thermomagnetic curve showing the loss of saturation magnetization (Js) as a function of temperature (in vacuum) for a typical magnetic separate. B: NRM and ARM demagnetization for four samples from the Eagle Nose Butte (ENB) Section. C: ARM demagnetization of selected artificial specimens containing either fragments of the limonite concretions or magnetic separates composed of ilmeno-hematite grains (modified from Lund, 1976).

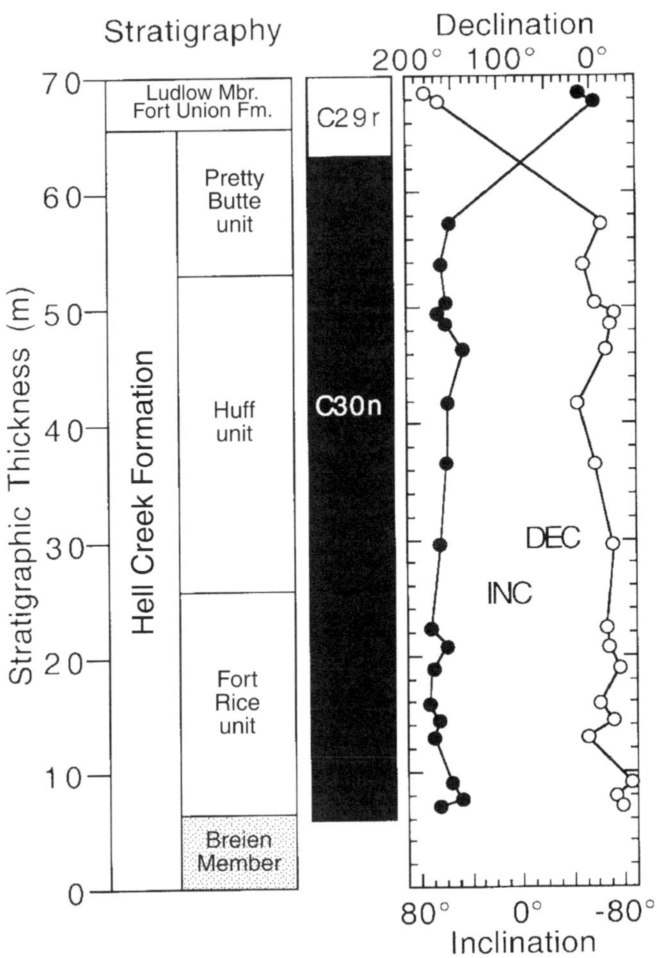

Figure 7. Average declinations and inclinations (N ≥ 3, $\alpha_{95} \leq 25°$) for 20 sampling horizons at the Eagle Nose Butte (ENB) Section and the resulting magnetic polarity stratigraphy. The Hell Creek Formation data come from Lund (1976) and the two Ludlow Member sites represent a resampling of that interval. Marine stratigraphic intervals are shown in gray. See text for further discussion.

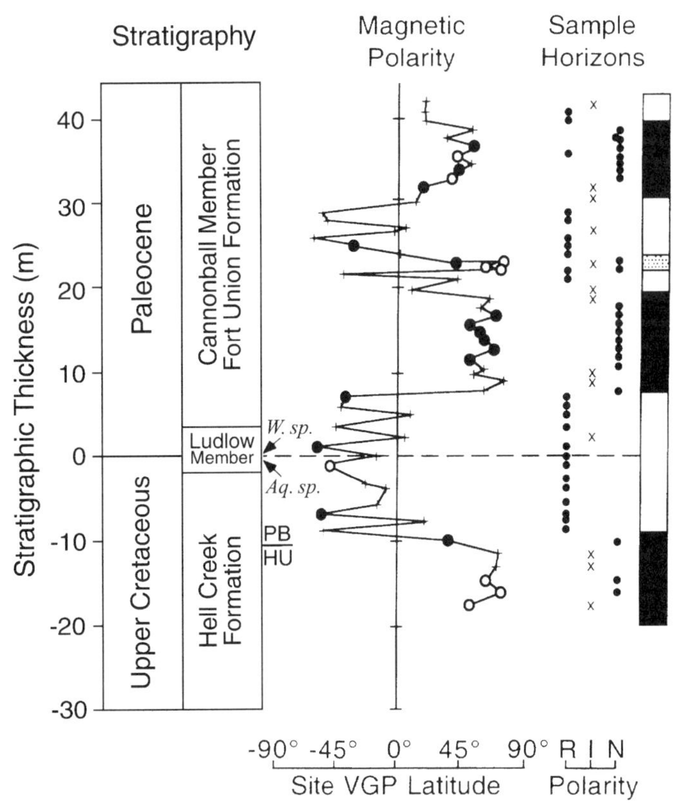

Figure 8. Magnetostratigraphy of the Huff Section as determined by Lerbekmo and Coulter (1984). The solid (open) circles indicate sites with mean VGPs which satisfy (fail) Watson's (1956) test for nonrandomness; the '+' symbols indicate sites with only one or two specimens for which no statistics were calculated. *W.* sp. indicates the lowest stratigraphic horizon with *Wodehouseia sp.*, a Paleocene palynomorph; *Aq.* indicates the highest occurrence of *Aquipollenites sp.*, a Late Cretaceous palynomorph. HU/PB indicates the contact between the lower Huff unit and the overlying Pretty Butte unit of the Hell Creek Formation (redrawn from Lerbekmo and Coulter, 1984; Frye, 1969, subsequently abandoned nomenclature, see Murphy et al., this volume).

Dakota, but it was consistent with the study of Lerbekmo and Coulter (1984).

NEW PALEOMAGNETIC RESULTS IN CENTRAL NORTH DAKOTA

After Lund's (1976) Master's thesis was completed, paleomagnetic measurements of the other two sampled sections (Emmons Composite Section and Huff Section, Appendix 1) were continued on a part-time basis in order to resolve the placement of the K-T boundary in these rocks. Special attention was given to finding a revised demagnetization procedure that would more effectively clean the specimens of unwanted overprints. Lund (1976) noted, near the end of his study, that "zero-field" (<50 nT steady magnetic field) storage of specimens for extended time intervals improved the quality of subsequent a.f. or thermal demagnetization. This observation was used to develop a modified demagnetization procedure wherein all specimens were stored in zero-field for a minimum of 6 months before any demagnetization occurred. Also, more effort was made to carry out thermal demagnetization whenever specimens were indurated enough to withstand heating. Typically, two specimens from each horizon were thermally demagnetized and two to four specimens were a.f. demagnetized.

Typical demagnetization results using the revised procedure are shown in Figure 9. In the case of both a.f. and thermal demagnetization (after extended zero-field storage, sometimes for more than 1 yr), a characteristic remanence could almost always be identified in the interval 10–50 mT or 100–300 °C. This is the interval that should demagnetize the ilmeno-hematite detrital grains. In most cases, it was clear that some small-intensity NRM, attributable to a goethite-hematite chemical remanence, remained undemagnetized above 40 mT, or 300 °C. Typically, the characteristic remanences from at least three specimens per sample were averaged to obtain a final average NRM direction. In all cases, only averaged directions with $\alpha 95$s <25° were used in the final polarity stratigraphies.

This methodology was used to first develop a magnetostratigraphy for the Emmons Composite Section (Fig. 3, Appendix 1). The paleomagnetic results are shown in Figure 10. Normal polarity is present throughout the uppermost Fox Hills and Hell Creek Formations. The Hell Creek Formation results are consistent with those of Lund (1976) and suggest that the Hell Creek Formation was deposited during Chron 30n. It is possible that the sampling process missed the short reversed-polarity interval C30r, so the lower part of the normal-polarity interval may represent some portion of Chron 31n.

When the paleomagnetic study of Lerbekmo and Coulter (1984) was published, it was apparent that the paleomagnetic study of Lund (1976) and the new unpublished study of the Emmons Composite Section (Fig. 10) differed in the apparent polarity of the uppermost Hell Creek strata. All of our paleomagnetic samples from the uppermost Hell Creek Formation

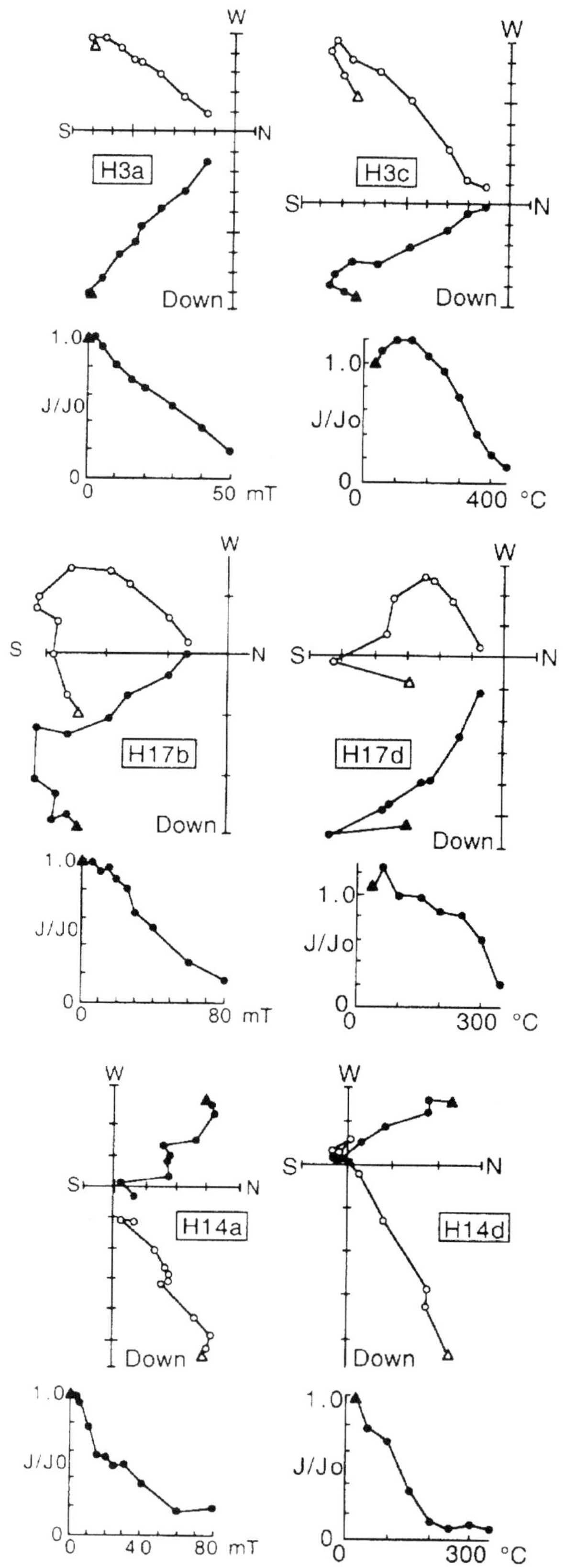

Figure 9. Zijderveld diagrams for three sampling horizons from the Huff Section (Appendix 1) indicating the pattern of NRM loss during paired a.f./thermal demagnetization. See Figure 4 for an explanation of the symbols.

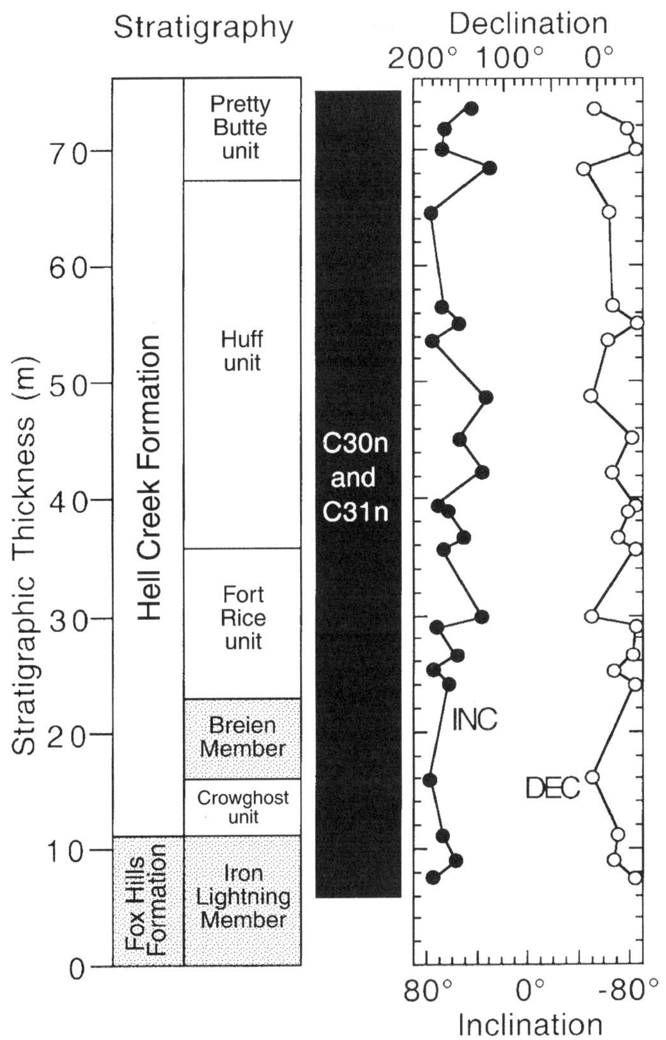

Figure 10. Average declinations and inclinations (N ≥ 3, $\alpha_{95} \leq 25°$) for 24 sampling horizons at the Emmons Composite Section and the resulting magnetic polarity stratigraphy. Marine stratigraphic intervals are shown in gray. See text for further discussion.

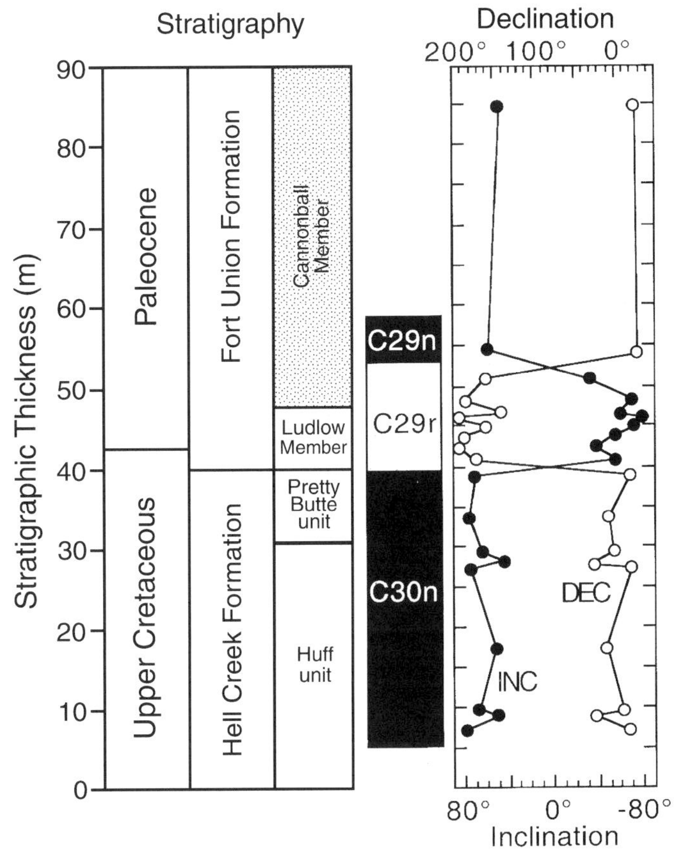

Figure 11. Average declinations and inclinations (N ≥ 3, $\alpha_{95} \leq 25°$) for 19 sampling horizons at the Huff Section and the resulting magnetic polarity stratigraphy. Marine stratigraphic intervals are shown in gray. See text for further discussion.

("Pretty Butte" unit of Frye, 1969) (Figs. 7 and 10) were normal in polarity, while those of Lerbekmo and Coulter (1984) were all reversed in polarity (Fig. 8). So, we finally measured our own samples from the Huff Section using the revised demagnetization procedure noted above to help resolve this apparent contradiction.

Our new paleomagnetic results from the Huff Section are shown in Figure 11. One clear interval of reversed polarity, about 15 m thick, is present within the Ludlow Member and lowermost Cannonball Member. This reversed interval is assignable to Chron 29r, as determined by Lerbekmo and Coulter (1984). This new magnetostratigraphy indicates that the Ludlow Member contains the K-T boundary and is, therefore, both latest Cretaceous and earliest Paleocene in age. This is consistent with Lerbekmo and Coulter's (1984) and Murphy et al.'s (1995) identification of the palynomorph-determined K-T boundary within a lignite bed near the middle of the Ludlow Member in the Huff Section (Fig. 8).

To further test our evidence that the base of Chron 29r is located at the Hell Creek–Fort Union formational contact, we resampled the Ludlow Member at two sites in the Eagle Nose Butte Section (Fig. 7). The stratigraphically lower site is the same one studied by Lund (1976), which he suggested might have reversed polarity but reached no final conclusion. Our new paleomagnetic measurements do document the presence of reversed polarity (Chron 29r) at both sites in the Ludlow Member (Fig. 7) and indicate that the base of Chron 29r is located at the Hell Creek–Fort Union formational contact in central North Dakota.

Overall, our new paleomagnetic results are generally consistent with the previous paleomagnetic studies of Lund (1976) and Lerbekmo and Coulter (1984). We disagree with Lerbekmo and Coulter (1984), however, in the placement of the base of Chron 29r. We never find reversed polarity in the uppermost strata of the Hell Creek Formation (Figs. 7, 10, and 11), while they indicate that ~8 m of the uppermost Hell Creek Formation is reversed (Fig. 8). We think that the difference is caused by chemical overprints that were not effectively identified by Lerbekmo and Coulter (1984). Lerbekmo and Coulter (1984) estimated that the NRM carrier is magnetite and that a.f. demagnetization up to 10–25 mT routinely removed a viscous overprint from the NRM-bearing magnetite. They plotted the final paleomagnetic results as virtual geomagnetic poles (VGPs) rather than inclination and declination with no zijderveld diagrams (as in Figs. 4, 5, and 9), so it is hard to be sure what NRM really remained after their limited a.f. demagnetization. However, our rock magnetic studies document clearly that magnetite is not the detrital remanence carrier and that a more careful a.f./thermal demagnetization procedure is necessary to identify the original sediment remanence direction. Lund (1976) noted weakly magnetized south- or southwest-pointing declinations (but not reversed inclinations) in selected specimens after a.f. demagnetization to 30–40 mT (Fig. 4). He viewed these as overprints in the limonite concretions. In a few specimens, our new paleomagnetic measurements also identified a weak reversed-polarity direction after a.f. or thermal demagnetization. Specimen H14d (Fig. 9) from strata at the Huff Section is one example. We think these directions are associated with a chemical remanence in limonite concretions that formed within the sediments 5–20 m below the sediment surface and a few hundred thousand years after deposition. The reversed-polarity samples of Lerbekmo and Coulter (1984) from the uppermost Hell Creek (Pretty Butte unit of Frye, 1969) are, in our opinion, just such a chemical overprint. The fact that only one of their samples of this part of the section had a statistically valid, nonrandom direction (Fig. 8) supports our suggestion that they were looking at a weakly magnetized chemical overprint. We believe that Chron 29r extends from the base of the Ludlow Member of the Fort Union Formation into the overlying Cannonball Member. Our proposed location of the upper Chron 29r boundary within the Cannonball Member is consistent with Lerbekmo and Coulter (1984). This revised placement of Chron 29r is still consistent with the location of the palynological K-T boundary (Lerbekmo and Coulter, 1984) near the middle of the Ludlow Member in the Missouri River Valley.

NEW PALEOMAGNETIC RESULTS IN EASTERN MONTANA

Our paleomagnetic results indicate that the Hell Creek Formation in central North Dakota was deposited entirely within Chron 30n. Yet, the results of Archibald et al. (1982) and Swisher et al. (1993) clearly documented that the uppermost Hell Creek Formation in eastern Montana occurred within Chron 29r. We therefore carried out a paleomagnetic study of the K-T boundary interval at Makoshika State Park, Dawson County, in easternmost Montana (Fig. 1) to further test whether or not there is a systematic time-transgressive nature to the age of the Hell Creek–Fort Union formational contact based on magnetostratigraphy. Early results of this study were reported in Hunter et al. (1997; see also Kroeger et al., 1993).

Paleomagnetic samples were collected from 40 sites in two replicate sections. The first section (M6723, 11 sites) (Fig. 12) was sampled to assess the suitability of the strata for paleomagnetic studies and estimate the location of Chron 29r. On the basis of initial paleomagnetic study of these samples, a second section (M7887b, 29 sites) was sampled in 1994 (see Hunter et al., 1997, for stratigraphic sections). An oriented hand sample was collected at each site and cut into four to six 2 × 2 × 2-cm specimen cubes in the laboratory for paleomagnetic analysis. The samples consisted of well-indurated gray mudstone and brown-to-gray siltstone and relatively unindurated brown-to-buff fine-grained sandstones.

Magnetic minerals were removed from the sediment at two sites to test whether or not the magnetic mineralogy was similar to that noted in central North Dakota. The magnetic separates both consisted of abundant black subhedral grains that were detrital in appearance. Thermomagnetic analyses of both separates revealed a loss of magnetization almost identical to that shown in Figure 6A with Curie temperatures <225 °C. No x-ray diffraction studies were made to determine the mineral crystallography. We presumed that the detrital magnetic minerals were hemo-illmenites having a Laramide volcanic provenance, just like the sediments in central North Dakota. This magnetic mineralogy is significantly different from that assumed to be present in eastern Montana by Archibald et al. (1982), but it is consistent with that documented by Swisher et al. (1993).

Typical demagnetization results for the two Makoshika State Park sections are shown in Figures 13 and 14. In the case of both a.f. and thermal demagnetization after extended zero-field storage (at least 1 month to occasionally more than 1 yr), a characteristic remanence could almost always be identified in the interval 10–50 mT or 100–300 °C. This is the interval that should demagnetize the ilmeno-hematite detrital grains. The fact that most NRM was lost by 30 mT and 250 °C indicated to us that ilmeno-hematite grains were, indeed, carrying most of the NRM. In most cases, it was clear that some small-intensity NRM, attributable to a goethite-hematite chemical remanence, remained undemagnetized above 40 mT or 300 °C. The characteristic remanences (determined by vector differencing) from at least three specimens per sample were averaged to obtain a final average NRM direction for each sample. In all cases, only averaged directions with $\alpha 95$s <25° were used in the final-polarity stratigraphic determinations.

This methodology was used to first develop a magnetostratigraphy for Section M6723. The paleomagnetic results are shown in Figure 15. Normal polarity is present in the lower Hell Creek Formation, while reversed polarity is present in the uppermost sample from the Hell Creek Formation and all of the samples from the Ludlow Member.

Our paleomagnetic results from Section M7887b are shown in Figure 16. One clear interval of reversed polarity, about 16 m thick, is present within the uppermost Hell Creek Formation and Ludlow Member of the Fort Union Formation. This reversed interval is assignable to Chron 29r. The reversed

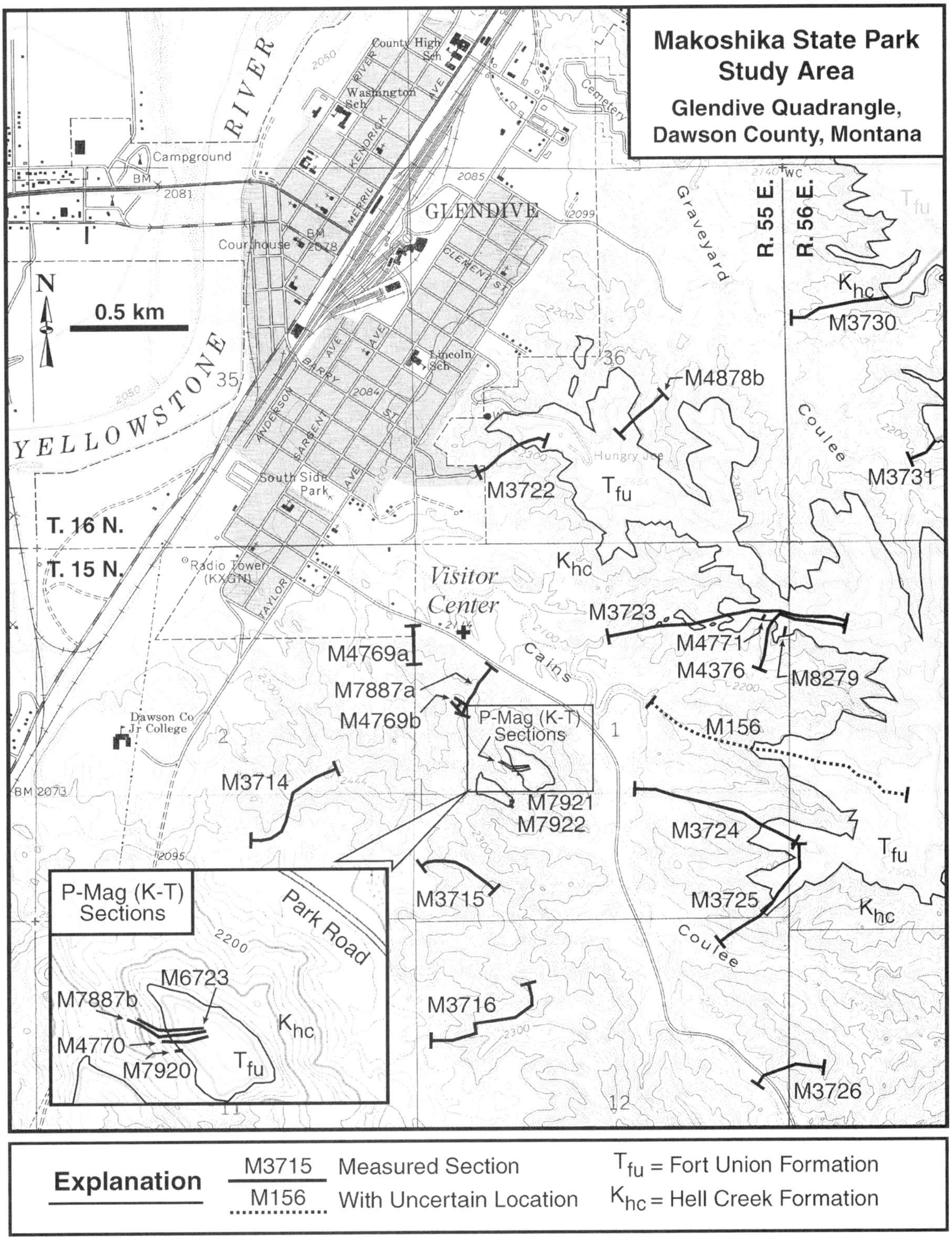

Figure 12. Makoshika State Park study area with site map of two replicate paleomagnetic sampled geologic sections (M6723 and M7887b). Base map derived from U.S. Geological Survey Glendive Quadrangle (1967), Dawson County, Montana. The contact of the Hell Creek–Ludlow Member of the Fort Union Formation is modified from Sholes and others (1989) in specific study areas (see also Hunter et al., 1997). Because of the near stratigraphic association, for the purposes of mapping, this contact represents the K-T boundary. A palynomorph-based K-T boundary and associated fern spike were identified in the paleomagnetic section (Kroeger et al., 1993; Section M4770). M-numbered lines are measured sections. Sections M3714 to M3731 were measured by Butler (1980). The transects of Sections M156 (Frye, 1967) and M4787b (Tschudy, 1970; this transect is revised from that published by Tschudy) were drawn on the basis of available information. Our interpretations of the sections by Butler, Frye, and Tschudy provided additional control on the mapped placement of the Hell Creek–Fort Union formational contact in the study area.

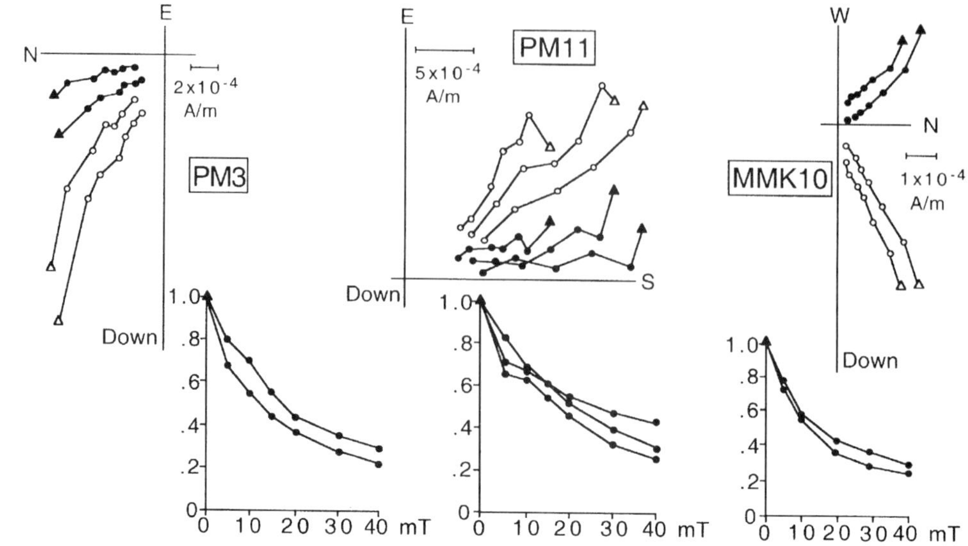

Figure 13. Zijderveld diagrams for three typical sampling horizons from Makoshika State Park Sections M6723 (PM3, PM11) and M7887b (MMK10) indicating the pattern of NRM loss during a.f. demagnetization. The solid (open) triangles represent the tip of the initial NRM horizontal (vertical) component. The solid (open) circles represent the tip of the NRM horizontal (vertical) component after each step of demagnetization. For each sample, the relative loss of intensity (J/Jo) after each step of a.f. demagnetization is also plotted.

chron placement in the two replicate sections (Figs. 15 and 16) is consistent in both stratigraphic location and interval thickness. These results are consistent with the paleomagnetic results of Archibald et al. (1982) and Swisher et al. (1993) and suggest that the Hell Creek Formation was deposited during both Chron 30n and Chron 29r.

DISCUSSION

Our new paleomagnetic data indicate that the base of Chron 29r lies at the Hell Creek–Fort Union formational contact in central North Dakota, but that it lies ~6 m below that contact in eastern Montana. These placements are consistent with palynomorph indicators of the K-T boundary and further suggest that there is a systematic time-transgressive younging of the Hell Creek–Fort Union contact as one goes from east to west.

The palynomorph changes noted by Lerbekmo and Coulter (1984) and Murphy et al. (1995) in the Huff Section in central North Dakota represent the best biostratigraphic indicator of the K-T boundary in this region. This section was first described by Laird and Mitchell (1942; Section 6) and includes the type section of the "Huff" unit of the Hell Creek Formation as defined by Frye (1969; Section 4). Lerbekmo and Coulter (1984) state that the K-T boundary palynomorph change occurs at the base of the first true lignite ~2.3 m above the base of the Ludlow Member as described by Laird and Mitchell (1942). This lignite corresponds to Unit 11 in the Appendix 1 stratigraphic column of the Huff Section (and Unit 45 in the Appendix 1 stratigraphic column of the Eagle Nose Butte Section). Stringers of lignitic shale and lignitic sand mark the base of the Ludlow Member at the Huff Section (Unit 9) and at the Eagle Nose Butte Section (Unit 43). Murphy et al. (1995) found the same palynomorph change near the base of the lignitic unit that was identified by Lerbekmo and Coulter (1984), but Murphy et al. (1995) placed the K-T boundary ~25 cm lower in the section. Thus they rejected the previous work of Laird and Mitchell (1942), Frye (1969), and Lerbekmo and Coulter (1984) and placed the basal Ludlow Member contact with the Hell Creek Formation at the base of this lignitic interval rather than ~2.3 m lower in the section as the other workers had done. We have followed the stratigraphy as defined by Laird and Mitchell (1942) and Frye (1969) in our work and place the palynological K-T boundary ~2.3 m above the Hell Creek–Fort Union formational contact in central North Dakota.

The total thickness of Chron 29r in the central North Dakota sections is ~15 m. If we assume that Chron 29r lasted ~600 k.y. (D'Hondt et al., 1996), then the average sediment accumulation rate in this interval is on the order of 1 m/40 k.y. Using this estimate and the fact that the K-T boundary lies 2.3 m above the Hell Creek–Fort Union formational contact, we suggest that Hell Creek deposition must have terminated at least 90 k.y. before the K-T boundary in central North Dakota. Similarly, the fact that at least 4 m of the basal Cannonball Member lies within Chron 29r suggests that the Cannonball Sea was present in central North Dakota within about 160 k.y. after the K-T boundary.

Archibald et al. (1982) and Swisher et al. (1993) identified Lancian faunas in the lower part of Chron 29r and Puercan faunas in the upper part of Chron 29r in eastern Montana. Hunter et al. (1997) have identified (1) a Lancian fauna just below Chron 29r, (2) the palynological K-T boundary essentially at the Hell Creek–Fort Union contact within Chron 29r (Figs. 15, 16), and (3) a Puercan-age fauna at the top of Chron 29r in Makoshika State Park.

Thus both paleomagnetic and palynological data indicate that the Hell Creek–Fort Union contact is time transgressive, being older in the east and younging as one goes west toward the axis of Laramide tectonism (Fig. 1). We interpret this to result from a time-transgressive change in paleoenvironmental conditions related to the westward transgression of the Can-

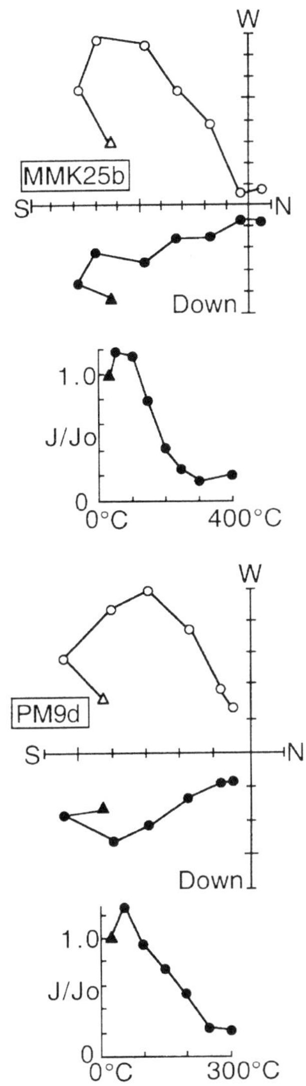

Figure 14. Zijderveld diagrams for two sampling horizons from Makoshika State Park Sections M6723 (PM9d) and M7887b (MMK25b) indicating the pattern of NRM loss during thermal demagnetization. See Figure 4 for an explanation of the symbols.

nonball Sea that occurred across the K-T boundary. In this view, the Ludlow Member lithofacies represents a nonmarine coastal-margin facies of the Cannonball marine transgression. The Cannonball marine transgression began before the palynological K-T boundary, and the Cannonball Sea was present in central North Dakota <160 k.y. after the K-T boundary. Note, however, that no Cretaceous-age strata have been found in the Cannonball Formation.

Our new paleomagnetic results from central North Dakota and eastern Montana and the paleomagnetic study of Hicks et al. (this volume) from southwestern North Dakota all indicate that the biostratigraphic K-T boundary throughout a large part of the Williston Basin occurs within the lower one-third of Chron 29r. There is a problem with this placement, however, because the marine biostratigraphic K-T boundary is located within the upper half of Chron 29r (Berggren et al., 1995). Assuming that the marine and terrestrial biotic changes are correlative (which we believe), we are left with the likelihood that sediment accumulation rates in the Williston Basin have systematically increased within Chron 29r. Hicks et al. (this volume) have argued strongly for this possibility. They think that there has been a factor-of-three increase in regional sediment accumulation rates across the K-T boundary. We agree with this assessment and suggest that such a regional change may be due to the combined influences of the westward-transgressing Cannonball Seaway and a westward shift of the Laramide foredeep depocenter, perhaps associated with Laramide tectonism.

We can also estimate the age of the Fox Hills–Hell Creek contact in central North Dakota using our paleomagnetic results. The Emmons County section (Figure 7) has at least 65 m of Hell Creek strata (the top of the section is arguably within 2 m of the Hell Creek–Ludlow contact). If we assume an average sediment accumulation rate for the pre-K-T boundary stratigraphic sections in central North Dakota comparable to that identified for Chron 29r (1 m/40 k.y.) and define the K-T boundary as 65 Ma, then the age of the Fox Hills–Hell Creek formational contact is ~67.7 Ma in the Missouri River Valley. This age estimate is near the base of Chron 30n (D'Hondt, 1996).

CONCLUSIONS

We have carried out a rock magnetic and paleomagnetic study to determine the magnetic polarity stratigraphy of uppermost Cretaceous and lower Paleocene strata of the Williston Basin that are exposed in the Missouri River Valley in central North Dakota and Makoshika State Park in eastern Montana. Five overlapping stratigraphic sections were sampled that extend from the marine Fox Hills Formation through the predominantly continental Hell Creek Formation and Ludlow Member of the Fort Union Formation and into the overlying marine Cannonball Member of the Fort Union Formation. The Hell Creek–Fort Union formational stratigraphic contact approximates the traditional, palynologically defined K-T boundary in the Williston Basin.

Rock magnetic study indicated that the primary detrital magnetic mineral carrying the sediment's natural remanent magnetization is an ilmeno-hematite with low Curie temperature (<225 °C) and that secondary viscous and chemical overprints are routinely present. This set of magnetic minerals required careful demagnetization of all samples, using both a.f. and thermal demagnetization, to recover the original magnetic polarity stratigraphy in the studied sections. Incorrect identification of the rock magnetic complexities in these sediments led Lerbekmo and Coulter (1984) to incorrectly locate the base of Chron29r in their study area.

The base of Chron 29r, which contains the K-T boundary, is located at the Hell Creek–Fort Union formational contact in

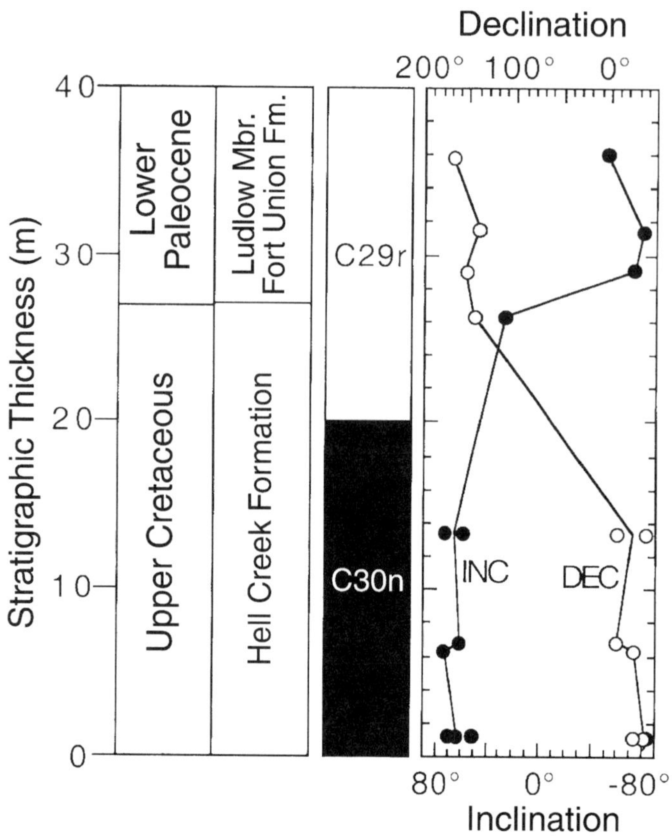

Figure 15. Average declinations and inclinations (N = 3, $\alpha_{95} < 25°$) for 11 sampling horizons at Makoshika State Park Section M6723 (see Kroeger et al., 1993; Hunter et al., 1997) and the resulting magnetic polarity stratigraphy. See text for further discussion.

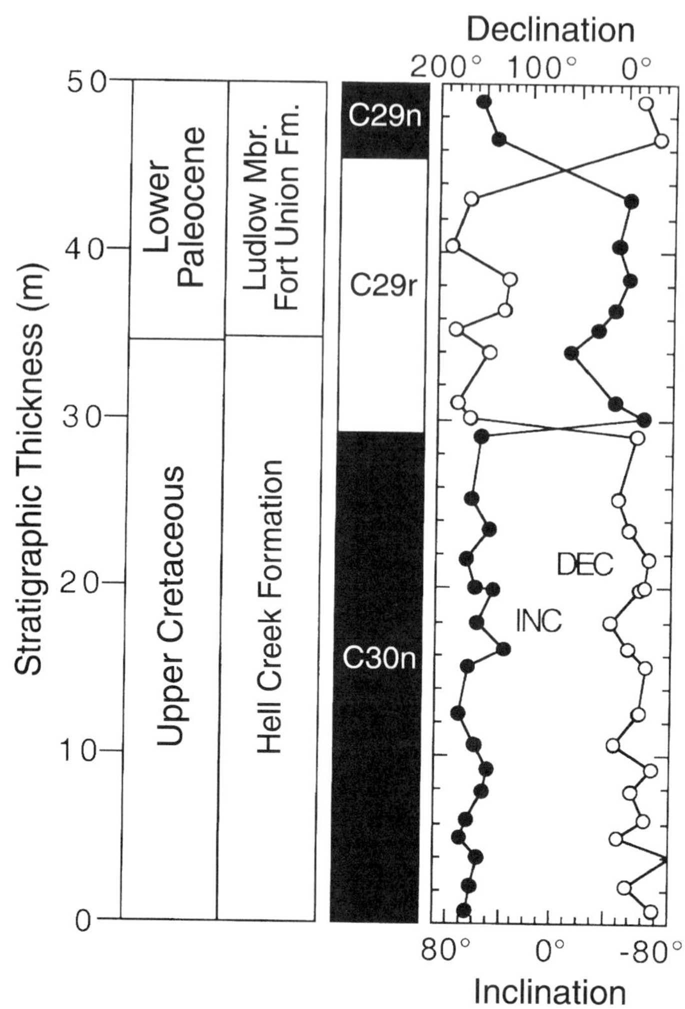

Figure 16. Average declinations and inclinations (N = 3, $\alpha_{95} < 25°$) for 29 sampling horizons at Makoshika State Park Section M7887b (see Kroeger et al., 1993; Hunter et al., 1997) and the resulting magnetic polarity stratigraphy. See text for further discussion.

central North Dakota and lies within the uppermost Hell Creek Formation in eastern Montana. This is consistent with palynological placement of the K-T boundary within the Ludlow Member in central North Dakota and at the Hell Creek–Ludlow contact in eastern Montana. Assuming no substantial sediment gaps, the paleomagnetic data suggest that the uppermost Cretaceous Hell Creek Formation in central North Dakota terminated about 90 k.y. before the palynological K-T boundary, while the Hell Creek Formation in eastern Montana terminated synchronously with the palynological K-T boundary. We interpret this to result from a time-transgressive change in paleoenvironmental conditions related to the westward transgression of the Cannonball Sea that occurred across the K-T boundary. The marine transgression represented by the Cannonball Member began before the palynological K-T boundary (although all known Cannonball strata are presently interpreted as Paleocene in age), and the Cannonball Sea was present in central North Dakota <160 k.y. after the K-T boundary.

We have located the K-T boundary in the lower third of Chron 29r in both central North Dakota and eastern Montana. Hicks et al. (this volume) have similarly identified the K-T boundary in the lower third of Chron 29r in southwestern North Dakota. By contrast, the marine biostratigraphic K-T boundary is located within the upper half of Chron 29r (Berggren et al., 1995). We assume that this is due to a significant increase in sedimentation within the Williston Basin after the K-T boundary. We suggest that such a regional change may be due to the combined influences of the westward-transgressing Cannonball Sea and a westward shift of the Laramide foredeep depocenter, perhaps associated with Laramide tectonism.

Assuming an average sediment accumulation rate for the pre–K-T boundary stratigraphic sections in central North Dakota comparable to that within Chron 29r (1 m/40 k.y.), we can also estimate the apparent age of the Fox Hills–Hell Creek formational contact to be about 67.7 Ma in the Missouri River Valley (Fig. 1). This is near the base of Chron 30n. The subsequent marine Breien (member of the Hell Creek Formation) incursion of the Western Interior Seaway also occurred during normal magnetic polarity (Chron 30n).

APPENDIX: PALEOMAGNETIC LOCALITIES IN CENTRAL NORTH DAKOTA

Eagle Nose Butte Locality

Location: N½ sec. 1, T.136 N., R. 80 W., Surgarloaf Butte (1976) Quadrangle, Morton County, North Dakota. The section extends from the top of Eagle Nose Butte on the west side of Highway 1806 to the east side of the railroad tracks on the east side of the highway. Stratigraphy: This section includes the type section of the Fort Rice Member of the Hell Creek Formation as defined by Frye (1969; Section 5). The other unit designations represent member designations by Frye in this section. Unit 46 was considered by Frye (1969) to possibly be part of the Cannonball Member of the Fort Union Formation, but we interpret it to be comparable to facies of the Ludlow Member of the Fort Union Formation of the Huff Section (see below).

EAGLE NOSE BUTTE SECTION

Unit	Lithology	Stratigraphic thickness (m)	Total thickness (m)
Fort Union Formation (Cannonball? Member)			
46	Gray to buff, brown weathering shale	1.52	70.21
Fort Union Formation (Ludlow Member)			
45	Dark gray lignitic shale	0.46	68.69
44	Buff, lignitic interbedded shale and sand	1.07	68.23
43	Dark brown lignitic interbedded shale and sand	1.52	67.16
Hell Creek Formation (Pretty Butte unit)			
42	Gray slightly bentonitic shale, siderite nodules	1.52	65.64
41	Black lignitic shale	0.30	64.12
40	Gray shale	0.46	63.82
39	Brown lignitic shale	0.46	63.36
38	Light gray sandstone	0.61	62.90
37	Brown lignitic, bentonitic sand	1.68	62.29
36	Gray-brown, bentonitic shale, siderite nodules	2.59	60.61
35	Dark purplish-brown lignitic shale	0.91	58.02
34	Gray bentonitic shale	2.29	57.11
33	Dark brown lignitic shale	0.30	54.82
32	Dark gray bentonitic shale	0.46	54.52
31	Dark brown lignitic shale	0.30	54.06
30	Dark gray bentonitic shale, siderite nodules	0.91	53.76
Hell Creek Formation (Huff unit)			
29	Medium gray shale, siderite nodules	4.88	52.85
28	Medium gray bentonitic shale	1.83	47.97
27	Brown interbedded sand and shale, siderite nodules	1.52	46.14
26	Grayish brown shale	1.52	44.62
25	Brown sand, siderite nodules	1.52	43.10
24	Brown and/or gray bentonitic shale	1.22	41.58
23	Brown and/or gray bentonitic lignitic shale	1.98	40.36
22	Brown and/or gray sand with lignitic sand partings	12.80	38.38
21	Dark brown lignitic sand	0.30	25.58
Hell Creek Formation (Fort Rice unit)			
20	Dark gray bentonitic shale	1.22	25.28
19	Dark brown lignitic shale	0.30	24.06
18	Brown bentonitic shale	1.37	23.76
17	Brown sandstone	0.30	22.39
16	Dark brown lignitic shale	0.15	22.09
15	Brown bentonitic shale	1.07	21.94
14	Brown sandstone	0.30	20.87
13	Gray sand with lignitic lenses	3.20	20.27
12	Medium gray bentonite	1.07	17.37
11	Gray sand, siderite nodules	0.46	16.30
10	Dark brown lignitic shale	0.30	15.84
9	Gray sandstone	2.44	15.54
8	Gray sand, siderite nodules	1.83	13.10
7	Siderite concretion layer	0.30	11.27
6	Brown-gray bentonitic sand and shale	4.11	10.97
5	Dark brown lignitic shale, siderite nodules	0.46	6.86
Hell Creek Formation (Breien Member)			
4	Brown and/or buff banded silt and shale	3.05	6.40
3	Dark gray bentonitic shale	1.52	3.35
2	Green glauconitic sand, siderite bed	1.37	1.83
Hell Creek Formation (Crowghost unit)			
1	Shaley lignite	0.46	0.46

EAGLE NOSE BUTTE SECTION PALEOMAGNETIC SAMPLES

Sample	Stratigraphic horizon (m)	Unit
ENB20	68.50	Unit 45, Ludlow Member
ENB18	67.82	Unit 44, Ludlow Member
ENB17	57.30	Unit 35, Pretty Butte unit
ENB16	53.68	Unit 30, Pretty Butte unit
ENB14	50.37	Unit 29, Huff unit
ENB15	49.58	Unit 29, Huff unit
ENB13	48.57	Unit 29, Huff unit
ENB12	46.44	Unit 28, Huff unit
ENB11	41.69	Unit 25, Huff unit
ENB10	36.58	Unit 22, Huff unit
ENB19	29.59	Unit 22, Huff unit
ENB09	22.38	Unit 17, Fort Rice unit
ENB05	20.75	Unit 14, Fort Rice unit
ENB04	18.77	Unit 13, Fort Rice unit
ENB03	15.82	Unit 10, Fort Rice unit
ENB02	14.32	Unit 9, Fort Rice unit
ENB01	12.87	Unit 8, Fort Rice unit
ENB06	9.09	Unit 6, Fort Rice unit
ENB07	7.97	Unit 6, Fort Rice unit
ENB08	6.97	Unit 6, Fort Rice unit

Huff Locality

Location: SW¼ sec. 8, T. 136 N., R. 79 W., Huff (1971) Quadrangle, Morton County, North Dakota. The section extends from the top of the hill at the southwest corner of sec. 8, north along the county road to the west-central edge of sec. 8, and continues eastward to the center of sec. 8. Stratigraphy: This section includes the type section of the Huff Member of the Hell Creek Formation as defined by Frye (1969; Section 4). This section was first described by Laird and Mitchell (1942; Section 6). The other unit designations represent member designations by Frye in this section.

Emmons County Locality

Location: NW¼ sec. 26, T. 136 N., R. 76 W., Dana (1975) Quadrangle (Fox Hills Formation), and SW¼ sec. 27, T. 135 N., R. 78 W., Hazelton NW (1971) Quadrangle (Hell Creek Formation), Emmons County, North Dakota. Stratigraphy: Lithostratigraphic units represented in this section include the Bullhead and Colgate lithofacies of the Iron Lightning Member of the Fox Hills Formation and subdivisions of the Hell Creek Formation correlated with the other P-MAG localities.

HUFF SECTION

Unit	Lithology	Stratigraphic thickness (m)	Total thickness (m)
Fort Union Formation (Cannonball Member)			
14	Gray to buff brown weathering shale	42.37	89.6
Fort Union Formation (Ludlow Member)			
13	Brown lignitic shale	1.83	47.23
12	Buff sandstone	2.14	45.4
11	Dark brown lignitic shale and lignite	1.52	43.26
10	Gray sandstone	1.37	41.74
9	Lignitic shale	0.61	40.37
Hell Creek Formation (Pretty Butte unit)			
8	Gray lignitic bentonitic shale, siderite nodules	9.14	39.76
Hell Creek Formation (Huff unit)			
7	Gray sands, bentonite shale layers	11.73	30.62
6	Sharpstone conglomerate, gray sandstone	3.20	18.89
5	Gray sand, siderite nodules	3.20	15.69
4	Gray bentonitic shale	0.91	12.49
3	Gray sand, calcareous lenses	7.01	11.58
Hell Creek Formation (Fort Rice unit)			
2	Gray shale, sand partings	3.05	4.57
1	Brown lignitic shale	1.52	1.52

HUFF SECTION PALEOMAGNETIC SAMPLES

Sample	Stratigraphic horizon (m)	Unit
H22	84.58	Unit 14, Cannonball Member
H04	54.41	Unit 14, Cannonball Member
H21	51.06	Unit 14, Cannonball Member
H20	48.15	Unit 14, Cannonball Member
H03	46.64	Unit 13, Ludlow Member
H02	45.86	Unit 13, Ludlow Member
H19	45.13	Unit 12, Ludlow Member
H18	43.79	Unit 12, Ludlow Member
H17	42.45	Unit 12, Ludlow Member
H07	40.86	Unit 10, Ludlow Member
H01	38.86	Unit 8, Pretty Butte unit
H16	33.56	Unit 8, Pretty Butte unit
H13	29.54	Unit 7, Huff unit
H14	27.54	Unit 7, Huff unit
H15	27.19	Unit 7, Huff unit
H12	17.30	Unit 6, Huff unit
H11	9.09	Unit 3, Huff unit
H10	8.92	Unit 3, Huff unit
H09	7.07	Unit 3, Huff unit

EMMONS COUNTY COMPOSITE SECTION

Unit	Lithology	Stratigraphic thickness (m)	Total thickness (m)
Hell Creek Formation (Pretty Butte unit)			
24	Gray bentonitic shale-silt	3.96	75.27
23	Dark brown lignitic shale	1.52	71.31
22	Dark gray bentonitic shale, siderite nodules	2.29	69.79
Hell Creek Formation (Huff unit)			
21	Gray sandstone [no thickness?]		
20	Medium gray sandy silt, siderite nodules	1.68	67.50
19	Medium gray bentonitic shale	0.30	65.82
18	Brown interbedded sand/shale, siderite nodules	1.83	65.52
17	Grayish brown shale	0.76	63.69
16	Gray lignitic shale	0.30	62.93
15	Brown-gray bentonitic shale	9.45	62.63
14	Brown-gray sand with lignitic sand partings	17.37	53.18
Hell Creek Formation (Fort Rice unit)			
13	Dark gray bentonitic shale	0.76	35.81
12	Dark brown lignitic shale	0.46	35.05
11	Brown bentonitic shale	2.74	34.59
10	Medium gray bentonite	3.66	31.85
9	Dark brown lignitic shale	0.30	28.19
8	Gray sandstone	2.44	27.89
7	Siderite concretion layer	0.30	25.45
6	Brown-gray bentonitic sandstone-shale	2.13	25.15
Hell Creek Formation (Breien Member)			
5	Covered	5.49	23.02
4	Green glauconitic sandstone, siderite bed	1.37	17.53
Hell Creek Formation (Crowghost unit)			
3	Shaley lignite	4.88	16.16
Fox Hills Formation (Colgate lithofacies)			
2	Green-gray glauconitic sandstone	3.66	11.28
Fox Hills Formation (Bullhead lithofacies)			
1	Alternating gray sandy siltstone-sandstone	7.62	7.62

EMMONS COUNTY PALEOMAGNETIC SAMPLES

Sample	Stratigraphic horizon (m)	Unit
EM54	73.28	Unit 23, Pretty Butte unit
EM53	71.91	Unit 23, Pretty Butte unit
EM52	70.01	Unit 22, Pretty Butte unit
EM51	68.46	Unit 21, Pretty Butte unit
EM50	64.38	Unit 17, Huff unit
EM49	56.57	Unit 14, Huff unit
EM48	55.15	Unit 14, Huff unit
EM47	53.85	Unit 14, Huff unit
EM46	49.28	Unit 13, Huff unit
EM45	45.22	Unit 13, Huff unit
EM44	42.17	Unit 13, Huff unit
EM43	39.35	Unit 13, Huff unit
EM42	38.99	Unit 13, Huff unit
EM41	36.60	Unit 13, Huff unit
EM40	35.86	Unit 13, Huff unit
EM39	30.09	Unit 9, Fort Rice unit
EM38	28.97	Unit 9, Fort Rice unit
EM37	26.46	Unit 7, Fort Rice unit
EM36	25.32	Unit 6, Fort Rice unit
EM35	24.05	Unit 5, Fort Rice unit
EM34	15.97	Unit 3, Crowghost unit
EM33	11.20	Unit 2, Colgate lithofacies
EM32	9.00	Unit 2, Colgate lithofacies
EM31	7.60	Unit 1, Bullhead lithofacies

REFERENCES CITED

Alvarez, L.W., Alvarez, W., Asaro, F., and Michel, H.V., 1980, Extraterrestrial cause for the Cretaceous-Tertiary extinction: Science, v. 208, no. 4448, p. 1095–1108.

Archibald, J.D., Butler, R.F., Lindsey, E.H., Clemens, W.A., and Dingus, L., 1982, Upper Cretaceous-Paleocene biostratigraphy and magnetostratigraphy, Hell Creek and Tullock Formations, northeastern Montana: Geology, v. 10, p.153–159.

Berggren, W.A., Kent, D.V., Swisher, C.C., III, and Aubry, M.-P., 1995, A revised Cenozoic geochronology and chronostratigraphy, *in* Berggren, W.A., Kent, D.V., Aubry, M.-P., and Hardenbol, J., eds., Geochronology time scales and global stratigraphic correlation: Tulsa, Oklahoma, SEPM (Society for Sedimentary Geology) Special Publication 54, p. 129–212.

Brown, R.W., 1962, Paleocene flora of the Rocky Mountains and Great Plains: U.S. Geological Survey Professional Paper 375, 119 p.

Butler, R.D., 1980, Stratigraphy, sedimentology, and depositional environments of the Hell Creek Formation (Late Cretaceous) and adjacent strata, Glendive area, Montana [Ph.D. thesis]: Grand Forks, North Dakota, University of North Dakota, 398 p.

Butler, R.D., and Hartman, J.H., 1999, Upper Cretaceous Hell Creek Formation depositional architecture in the Williston Basin: Outcrop analogs and subsurface correlations: Geological Society of America Abstracts with Programs, v. 31, no. 7, p. A71.

Calvert, W.R., 1912, Geology of certain lignite fields in eastern Montana: U.S. Geological Survey Bulletin 471, p. 187–201.

Clemens, W.A., Lillegraven, J.A., Lindsay, E.H., and Simpson, G.G., 1979, Where, when, and what: A survey of known Mesozoic mammal distribution, *in* Lillegraven, J.A., Kielan-Jaworowska, Z., and Clemens, W.A., eds., Mesozoic mammals: The first two-thirds of mammalian history: Berkeley, California, University of California Press, p. 7–58.

Cobban, W.A., and Reeside, J.B., Jr., 1952, Correlation of the Cretaceous formations of the Western Interior of the United States: Geological Society of America Bulletin, v. 63, no. 10, p. 1011–1044, 1 pl.

Cvancara, A.M., 1965, Bivalves and biostratigraphy of the Cannonball Formation (Paleocene) of North and South Dakota [Ph.D. thesis]: Ann Arbor, Michigan, University of Michigan, 470 p.

Cvancara, A.M., 1976a, Geology of the Cannonball Formation (Paleocene) in the Williston Basin, with reference to the uranium potential: North Dakota Geological Survey Report of Investigations, v. 57, 22 p.

Cvancara, A.M., 1976b, Geology of the Fox Hills Formation (Late Cretaceous) in the Williston Basin of North Dakota, with reference to uranium potential: North Dakota Geological Survey Report of Investigations, v. 55, 16 p.

Diemer, J.A., Belt, E.S., and Hartman, J.H., 1996, Base level changes as a consequence of tectonic, eustatic and autogenic processes in Late Cretaceous and Paleocene strata, western Williston Basin: Geological Society of America Abstracts with Programs, v. 28, no. 7, p. A373.

D'Hondt, S., Herbert, T.D., King, J., and Gibson, C., 1996, Planktic foraminifera, asteroids, and marine production: Death and recovery at the Cretaceous-Tertiary boundary, *in* Ryder, G., Fastovsky, D., and Gartner, S., eds., The Cretaceous-Tertiary event and other catastrophes in Earth history: Geological Society of America Special Paper 307, p. 303–317.

Dorf, E., 1942, Upper Cretaceous floras of the Rocky Mountain Region, 2, Flora of the Lance Formation at its type locality, Niobrara County, Wyoming: Washington, D.C., Carnegie Institution of Washington, Contributions to Paleontology, Publication 508, p. 83–168, 17 pls.

Feldmann, R.M., 1972a, First report of *Hercoglossa ulrichi* (White, 1882) (Cephalopoda: Nautilida) from the Cannonball Formation (Paleocene) of North Dakota, U.S.A.: Malacologia, v. 11, p. 407–413.

Feldmann, R.M., 1972b, Stratigraphy and paleoecology of the Fox Hills Formation (Upper Cretaceous) of North Dakota: North Dakota Geological Survey Bulletin 61, 65 p.

Fenner, W.E., 1974a, Foraminiferids from the Cannonball Formation (Paleocene, Danian) and their paleoenvironmental significance: Grant, Morton, and Oliver Counties, North Dakota: North Dakota Academy of Science, Proceedings, v. 28, pt. 1, p. 7.

Fenner, W.E., 1974b, The foraminiferids of the Cannonball Formation (Paleocene, Danian) and their paleoenvironmental significance: Grant, Morton, and Oliver counties, North Dakota [M.S. thesis]: Grand Forks, North Dakota, University of North Dakota, 135 p.

Fox, S.K., Jr., and Olsson, R.K., 1969, Danian planktonic Foraminifera from the Cannonball Formation of North Dakota: Journal of Paleontology, v. 43, no. 6, p. 1397–1404.

Frye, C.I., 1967, The Hell Creek Formation in North Dakota [Ph.D. thesis]: Grand Forks, North Dakota, University of North Dakota, 411 p., 8 pl.

Frye, C.I., 1969, Stratigraphy of the Hell Creek Formation in North Dakota: North Dakota Geological Survey Bulletin, no. 54, 65 p., 3 pl., geologic map, scale 1:750000, 1 sheet.

Gill, J.R., and Cobban, W.A., 1973, Stratigraphy and geologic history of the Montana Group and equivalent rocks, Montana, Wyoming, and North and South Dakota: U.S. Geological Survey Professional Paper 776, 37 p.

Gradstein, F.M., Agterberg, F.P., Ogg, J.G., Hardenbol, J., Veen, P.v., Thierry, J., and Huang, Z., 1995, A Triassic, Jurassic and Cretaceous time scale, *in* Berggren, W.A., Kent, D.V., Aubry, M.-P., and Hardenbol, J., eds., Geochronology, time scales and global stratigraphic correlation: Tulsa, Oklahoma, SEPM (Society for Sedimentary Geology) Special Publication 54, p. 95–126.

Hardenbol, J., Thierry, J., Farley, M.B., Jacquin, T., de Graciansky, P.-C., and Vail, P.R., 1998, Mesozoic and Cenozoic sequence stratigraphy of European basins, *in* Graciansky, P.-C., Hardenbol, J., Jacquin, T., Vail, P.R., eds., Mesozoic and Cenozoic sequence chronostratigraphic framework of European basins: Tulsa, Oklahoma, SEPM (Society for Sedimentary Geology) Special Publication 60, p. 3–13, with appendix, p. 763–781.

Hartman, J.H., 1998, The biostratigraphy and paleontology of latest Cretaceous freshwater bivalves from the western Williston Basin, Montana, U.S.A., *in* Johnston, P., and Haggart, J., eds., Bivalves: An eon of evolution:

Paleobiological studies honoring Norman D. Newell: Calgary, Alberta, University of Calgary Press, p. 317–345.

Hartman, J.H., 1999, End Cretaceous and early Paleocene mollusks of the Williston Basin, North Dakota and Montana: Geological Society of America Abstracts with Programs, v. 31, no. 7, p. A72.

Hartman, J.H., and Butler, R.D., 1995, Extinction and recovery of nonmarine molluscan assemblages in the Late Cretaceous and early Tertiary: Geological Society of America Abstracts with Programs, v. 27, no. 4, p. 13.

Hartman, J.H., and Kihm, A.J., 1992, Chronostratigraphy of Paleocene strata in the Williston Basin, *in* Finkelman, R.B., Tewalt, S.J., and Daly, D.J., eds., Geology and utilization of Fort Union lignites: Reston, Virginia, Environmental and Coal Associates, p. 52–75.

Hartman, J.H., and Kihm, A.J., 1996, Bio- and magnetostratigraphy of the uppermost Cretaceous and lower Tertiary strata of North Dakota, *in* Repetski, J.E., ed., Sixth North American Paleontological Convention Abstracts of Papers: Paleontological Society, Special Publications 8, p. 163.

Hartman, J.H., and Roth, B., 1998, Late Paleocene and early Eocene nonmarine molluscan faunal change in the Bighorn Basin, northwestern Wyoming and south-central Montana, *in* Aubry, M.-P., Berggren, W.A., and Lucas, S.G., eds., Late Paleocene-Early Eocene climatic and biotic events in the marine and terrestrial records: New York, Columbia University Press, p. 323–379.

Hartman, J.H., Hunter, J.P., and Krause, D.W., 1998, The last epeiric sea (Paleocene, Cannonball Formation) in North America: The age of incursions and bounding strata based on mammals, mollusks, and foraminifera: Geological Society of America Abstracts with Programs, v. 30, no. 7, p. A33.

Hartman, J.H., Collins, L.S., and Aubry, M.-P., 1999, New interpretations of the Cannonball Formation (Paleocene): North America's last interior sea: Geological Society of America Abstracts with Programs, v. 31, no. 7, p. A105.

Hunter, J.P., Hartman, J.H., and Krause, D.W., 1997 [1998], Mammals and mollusks across the Cretaceous-Tertiary boundary from Makoshika State Park and vicinity (Williston Basin) Montana, *in* Hartman, J.H., ed., A festschrift in honor of Marshall Lambert, Paleontology and geology in the northern Great Plains: Contributions to Geology (University of Wyoming), v. 32, no. 1, p. 61–114.

Johnson, K.R., Orth, C.J., and Nichols, D.J., 1987, Fossil leaf and palynomorph changes associated with an iridium anomaly at the Cretaceous-Tertiary boundary in North Dakota: Geological Society of America Abstracts with Programs, v. 19, no. 7, p. 718.

Kennedy, W.J., Landman, N.H., Christensen, W.K., Cobban, W.A., and Hancock, J.M., 1998, Marine connections in North America during the late Maastrichtian: Palaeogeographic and paleobiogeographic significance of *Jeletzkytes nebrascensis* zone cephalopod fauna from the Elk Butte Member of the Pierre Shale, SE South Dakota and NE Nebraska: Cretaceous Research, v. 19, p. 745–775.

Klett, M.C., and Erickson, J.M., 1976, Type and reference sections for a new member of the Fox Hills Formation, Upper Cretaceous (Maestrichtian) in the Missouri valley region, North and South Dakota: North Dakota Academy of Science, Proceedings, v. 28, pt. 2, p. 3–24.

Kroeger, T.J., Hartman, J.H., and Peck, W.D., 1993, Provisional palynological recognition of the fern spike at the Cretaceous-Tertiary Boundary, Makoshika State Park, Dawson County, Montana: North Dakota Academy of Science, Proceedings, v. 47, p. 48.

Laird, W.M., and Mitchell, R.H., 1942, The geology of the southern part of Morton County, North Dakota: North Dakota Geological Survey Bulletin 14, 42 p., 3 pl., geologic map, scale 1:125 000, 1 sheet.

Lerbekmo, J.F., and Coulter, K.C., 1984, Magnetostratigraphy and biostratigraphy correlations of Late Cretaceous to early Paleocene strata between Alberta and North Dakota, *in* Stott, D.F., and Glass, D.J., eds., The Mesozoic of middle North America: Calgary, Alberta, Canadian Society of Petroleum Geologists, Memoir 9, p. 313–317.

Lund, S.P., 1976, Continental sediments as paleomagnetic recorders: Evidence from the Upper Cretaceous Hell Creek Formation of the Williston Basin [M.S. thesis]: Minneapolis, Minnesota, University of Minnesota, 137 p.

Moore, W.L., 1976, The stratigraphy and environments of deposition of the Cretaceous Hell Creek Formation (reconnaissance) and the Paleocene Ludlow Formation (detailed), southwestern North Dakota: North Dakota Geological Survey Report of Investigations, v. 56, 40 p.

Murphy, E.C., Nichols, D.J., Hoganson, J.W., and Forsman, N.F., 1995, The Cretaceous-Tertiary boundary in south-central North Dakota: North Dakota Geological Survey Report of Investigations, v. 98, 74 p.

Nichols, D.J., Murphy, E.C., Johnson, K.R., and Betterton, W.J., 2000, A second K-T boundary locality in North Dakota verified by palynostratigraphy and shocked quartz: Geological Society of America Abstracts with Programs, v. 32, no. 7, p. A130.

Perch-Nielsen, K., 1989, Cenozoic calcareous nannofossils, *in* Boli, H.M., Saunders, J.B., and Perch-Nielsen, K., eds., Plankton stratigraphy: Planktic foraminifera, calcareous nannofossils, and calpionellids: Cambridge, Massachusetts, Cambridge University Press, p. 427–554.

Prothero, D.R., 1995, Geochronology and magnetostratigraphy of Paleogene North American land mammal "ages": An update, *in* Berggren, W.A., Kent, D.V., Aubry, M.-P., and Hardenbol, J., eds., Geochronology, time scales and global stratigraphic correlation: Tulsa, Oklahoma, SEPM (Society for Sedimentary Geology) Special Publication 54, p. 305–315.

Sholes, M.A., Vuke-Foster, S.M., Derkey, P.D., and Colton, R.B., 1989, Coal stratigraphy and correlation in the Glendive 30 × 60 minute Quadrangle, eastern Montana and adjacent North Dakota: Montana Bureau of Mines and Geology Geologic Map 49, 9 p., scale 1:100 000, 4 sheets.

Silfer, B.E., 1990, Neogastropods (Melongenidae, Fasciolariidae, Turridae) from the Cannonball formation (Paleocene: Thanetian?), North Dakota and South Dakota [M.S. thesis]: Grand Forks, North Dakota, University of North Dakota, 215 p.

Swisher, C.C., Dingus, L., and Butler, R.F., 1993, $^{40}Ar/^{39}Ar$ dating and magnetostratigraphic correlation of the terrestrial Cretaceous-Paleogene boundary and Puercan Land Mammal age Hell Creek–Tullock formations, eastern Montana: Canadian Journal of Earth Sciences, v. 30, p. 1981–1996.

Tibert, N.E., Curran, H.A., Hartman, J.H., Belt, E.S., and Diemer, J.A., 2001, Marine flooding intervals within the freshwater Tongue River Member (Fort Union Formation–Paleocene), western Williston Basin, U.S.A.: American Association of Petroleum Geologists, Official Program Book, v. 10, p. A200.

Tschudy, R.H., 1970 [1971], Palynology of the Cretaceous-Tertiary boundary in the northern Rocky Mountain and Mississippi embayment regions, *in* Kosanke, R.M., and Cross, A.T., eds., Symposium on palynology of the Late Cretaceous and early Tertiary: Geological Society of America Special Paper 127, p. 65–111.

Waage, K.M., 1968, The type Fox Hills Formation, Cretaceous (Maastrichtian), South Dakota: New Haven, Connecticut, Yale University Peabody Museum of Natural History Bulletin 27, 175 p.

Watson, G.S., 1956, A test for randomness of direction: Royal Astronomical Society Monthly Notices, v. 7, p. 160–161.

Manuscript Accepted by the Society September 24, 2001

Geological Society of America
Special Paper 361
2002

Chemostratigraphic correlation of four fossil-bearing sections in southwestern North Dakota

Nan Crystal Arens*
Department of Geoscience, Hobart & William Smith Colleges, Geneva, New York 14456-3397, USA
A. Hope Jahren
Morton K. Blaustein Department of Earth and Planetary Sciences, Johns Hopkins University, Baltimore, Maryland 21218, USA

ABSTRACT

A −1.5‰ to −2‰ carbon isotope shift in surface ocean dissolved inorganic carbon has been reported stratigraphically above the Cretaceous-Tertiary (K-T) boundary in the global stratotype section at El Kef, Tunisia, and in many other marine sections worldwide. Because a change in the $\delta^{13}C$ value of paleoatmospheric CO_2 would have accompanied the marine surface carbon isotope shift, the K-T carbon isotope anomaly can be used to diagnose and correlate the boundary in terrestrial sections as well. We use this shift and other secular variation in the carbon isotope signature from the Hell Creek and Fort Union Formations to correlate four sections in Slope County, North Dakota. At Pyramid Butte, the K-T boundary is marked by a −2.8‰ carbon isotope shift approximately 10 cm above the iridium-bearing impact clay. This shift is larger in magnitude than might be expected by plant vital effects or taphonomic variation. The Pyramid Butte carbon isotope shift allowed correlation to the Bobcat Butte (shift of −2.1‰), Terry's Fort Union Dinosaur (shift of −1.1‰ to −1.6‰), and the New Facet Boundary (shift of −2.3‰) sections, where an iridium enrichment was not preserved. An interval of carbon isotope fluctuation marks the transition from HCII to HCIII megafloral zones in the Bobcat Butte section, suggesting that some instability in the physical environment was correlated with the observed vegetation change. These data demonstrate that carbon isotope chemostratigraphy can provide an additional line of evidence that is largely independent from taxonomic extinction for correlating fossil-bearing terrestrial sections. Chemostratigraphic interpretations may also recognize incomplete or condensed sections, which will aid interpretation of the fossils found in them.

CHEMOSTRATIGRAPHY AS AN AID TO TERRESTRIAL CORRELATION

Testing mechanistic hypotheses of extinction and biotic change across the Cretaceous-Tertiary (K-T) boundary requires that the stratigraphic position of the boundary be placed accurately. In the global stratotype section at El Kef, Tunisia, a clay layer containing above-background concentrations of iridium records the terminal-Cretaceous bolide impact and defines the K-T boundary (Keller et al., 1995; Keller and Lindinger, 1989).

*E-mail: arens@hws.edu

Arens, N.C., and Jahren, A.H., 2002, Chemostratigraphic correlation of four fossil-bearing sections in southwestern North Dakota, *in* Hartman, J.H., Johnson, K.R., and Nichols, D.J., eds., The Hell Creek Formation and the Cretaceous-Tertiary boundary in the northern Great Plains: An integrated continental record of the end of the Cretaceous: Boulder, Colorado, Geological Society of America Special Paper 361, p. 75–93.

A similar clay layer, including impact-related shocked mineral and blast melt spherules, has been reported in terrestrial sediments and diagnoses the K-T boundary in these facies (eg., Alvarez, 1987; Bohor et al., 1984; Bohor et al., 1987a; Orth et al., 1981). However, these diagnostic components are uncommon and ephemeral in most subaerial depositional environments, making diagnosis of the K-T boundary difficult in terrestrial rocks.

To circumvent this problem, terrestrial stratigraphers have looked for other indicators associated with the K-T boundary clay, including lithostratigraphy, paleomagnetic zones, radioisotope dates, last appearances of characteristic Cretaceous pollen forms, and first appearances of typical Tertiary mammals. The first attempts to identify the K-T boundary were based on lithostratigraphy. In 1938, Roland Brown argued that both the Hell Creek–Fort Union formational contact and the chronostratigraphic K-T boundary should be placed at the base of the lowest regionally continuous lignite bed that stratigraphically overlay the last nonreworked dinosaur fossil (Brown, 1938, 1952). This approach proved unsatisfactory first because reference lignite units form diachronous complexes rather than a single isochronous datum (Archibald, 1982; Fastovsky, 1987; Lofgren, 1995), and second because a number of the basal Fort Union lignite beds have subsequently been determined to be Paleocene in age (see Clemens, this volume; Nichols and Johnson, this volume; Pearson et al., this volume; Johnson et al., 1989). Paleomagnetism documents periodic reversals of Earth's magnetic field. In the Western Interior region of North America, the K-T boundary occurs within the 29r magnetochron (Archibald et al., 1982; Butler and Lindsay, 1985; Johnson and Hickey, 1990; Johnson et al., 1989; Lerbekmo, 1985), which lasted less than one million years (Harland et al., 1989).

In terrestrial sections with high sedimentation rates, Lerbekmo et al. (1996) reported a brief (~50 k.y.) normal polarity subchron at the K-T boundary in Alberta and Saskatchewan. However, this subchron has not been reported in all sections studied, perhaps because of the discontinuous nature of terrestrial sedimentation in most locations. Radioisotope dates from volcanic ash provide isochronic markers within some sections (Swisher et al., 1993). Several ash beds were preserved within the Fort Union Formation in northeastern Montana, but no suitable volcanic ashes have been identified in southwestern North Dakota. Palynological and mammalian biostratigraphic markers diagnose the boundary by the last appearance of characteristic Maastrichtian pollen types (Hotton, 1988; Johnson et al., 1989; Nichols, 1990; Nichols and Fleming, 1990; Nichols et al., 1986; Nichols and Sweet, 1993; Orth et al., 1981; Tschudy et al., 1984) or first appearance of Paleocene mammals (Archibald et al., 1987; Lofgren, 1995; Woodburne, 1977; Woodburne and Swisher, 1995).

Biostratigraphy can leave zones of ambiguity and gaps where fossils were not preserved (Archibald, 1982) or give confusing results when fossils were reworked (Lofgren et al., 1990). Interpretation of biostratigraphic markers can also be complicated by biogeographic variation (Nichols and Sweet, 1993). First- and last-occurrence data are also insensitive to unconformities, which may be important to interpreting rates of biotic change. More significantly, biostratigraphic time markers introduce circularity to inference of extinction and biotic change within an interval, particularly when one uses the same fossils both to define zones and to discuss biotic patterns. Therefore no biostratigraphic method alone provides the biologically independent, high-resolution correlations across the wide range of terrestrial facies needed to test hypotheses of pattern and process associated with biotic change before and at the K-T boundary.

Carbon isotope chemostratigraphy may provide an additional line of evidence for correlating terrestrial sections. As outlined by Williams et al. (1988), isotope chemostratigraphy relies on temporal variability in the primary isotope composition of a chemical system. To develop a correlation, the pattern of this temporal variability is matched among stratigraphic sections. Williams et al. (1988) defined five general features of isotope records that provide useful correlation tie points: (1) cycles, which may be of uniform or variable duration, are repeating quasiperiodic components of an average signal; (2) events are unique excursions distinctly above or below the average signal; (3) shifts are abrupt (<200 k.y.) deviations from the average signal, which may be reversible; (4) long-term trends are steady directional changes in isotope composition through time; and (5) absolute magnitude may establish boundary conditions, distinguish among shifts, or identify events. Williams et al. (1988) proposed that correlations initially be developed by matching secular patterns among sections. When secular variation for an interval is well understood, new sections may be matched to a regional or global reference curve.

Isotope chemostratigraphy has been extensively applied to marine sediments. For example, Adabi (1997) compared single carbonate carbon isotope absolute magnitudes from the Renison Formation, Tasmania, to global compilation curves (Derry et al., 1992; Kaufman and Knoll, 1995) to conclude that these rocks were Neoproterozoic, not Cambrian, in age. Bartley et al. (1998) used organic carbon isotope curves to correlate the Platonovskaya Formation of Russia with other Siberian Platform sections. Body and trace fossils could bracket Platonovskaya sedimentation only between Vendian and earliest Cambrian. Comparison of both isotope shifts and absolute magnitudes through the section allowed correlation to the Precambrian-Cambrian (Yudomian-Nemakit–Daldynian) boundary and highlighted the probable presence of an unconformity at that horizon. At the Permian-Triassic boundary in Spitsbergen, Wignall et al. (1998) used organic carbon isotope data to diagnose the P-T boundary and to recognize that the sequence was indeed conformable, contrary to previous conclusions based on biostratigraphy (Nakamura et al., 1987; Stemmerik, 1988; Szaniawski and Malkowski, 1979). At the Cenomanian-Turonian boundary, Gale et al. (1993) demonstrated the synchronous nature of biostratigraphic and carbon isotope chemostratigraphic markers in both Sussex, England, and Pueblo, Colorado. Their data confirm the global character of late Cenomanian and early Turonian carbon isotope trends. Koch et al. (1992) extended the approach to terrestrial sediments by correlating the marine carbon isotope curve at the Paleocene-Eocene boundary with that developed from mammalian tooth enamel. At the K-T boundary global stratotype section in El Kef, Tunisia, a reversible carbon isotope shift has been documented in association with the iridium anomaly (Keller et al., 1995; Keller and Lindinger, 1989). This shift, along with other features of the latest Cretaceous and Paleocene (Shackleton, 1985; Shackleton and Hall, 1984) carbon isotope curve may provide similar resolution to correlations both at the K-T boundary and within the Hell Creek and Fort Union Formations.

K-T BOUNDARY CARBON ISOTOPE SHIFT IN THE ATMOSPHERE LINKS MARINE AND TERRESTRIAL RECORDS

A −1.5‰ to −2.0‰ reversible carbon isotope shift in surface ocean dissolved inorganic carbon has been documented in association with the K-T boundary impact at the global stratotype section (Keller et al., 1995; Keller and Lindinger, 1989) and at other marine sections worldwide (D'Hondt et al., 1998; Hsü et al., 1982; Perch-Nielsen et al., 1982; Robin et al., 1991; Smit, 1982; Stott and Kennett, 1989, 1990; Zachos and Arthur, 1986; Zachos et al., 1989; Zachos et al., 1992). Most workers attribute homogenization of the normally positive surface-to-deep water carbon isotope gradient to breakdown of marine primary productivity in the wake of the bolide impact (D'Hondt et al., 1998; Hsü et al., 1982; Zachos and Arthur, 1986). However, homogenization of marine carbon isotope values alone does not fully account for the negative carbon isotope shift observed in planktonic foraminifera after the boundary impact. The negative shift requires input of ^{12}C from another source. Based on modeling results, Kump (1991) concluded that fluvial input into a postimpact ocean devoid of life would cause the surface ocean to come into isotope equilibrium with weathering input over non-steady-state time scales ($<10^5$ years). This would produce a −5‰ carbon isotope shift, much larger than that observed in marine sediments (Kump, 1991). Also based on model results, Ivany and Salawitch (1993) proposed that burning ~25% of terrestrial biomass in postimpact wildfires introduced $^{12}CO_2$ into the atmosphere, lowering atmospheric carbon isotope values and raising pCO_2. However, apart from soot found in several K-T boundary sections (Wolbach et al., 1985), there is little independent evidence for a conflagration of this magnitude. Whatever the source, $^{12}CO_2$ added to the atmosphere would rapidly come into mass and isotope equilibrium with the surface ocean (Schlesinger, 1997; Siegenthaler and Sarmiento, 1993), thus lowering marine carbon isotope values, and producing the observed K-T boundary shift. Such a mechanism would produce carbon isotope shifts of similar sign and magnitude in both the surface ocean and atmosphere. Since land plants sample atmospheric CO_2 directly and discriminate in a known way during C_3 carbon fixation (Farquhar et al., 1989), the isotope composition of plant tissue reflects the composition of the atmosphere under which it was fixed (Arens et al., 2000; Medina et al., 1986; Van der Merwe and Medina, 1989), with a quantifiable error due to physiological and ecological vital effects (Arens et al., 2000). Therefore, the post-Cretaceous carbon isotope shift recorded in surface ocean dissolved inorganic carbon should also be preserved in the plant tissue and terrestrial organic carbon derived from it.

A reversible carbon isotope shift has been reported in terrestrial K-T boundary sections in northeastern Montana and southwestern North Dakota (Arens and Jahren, 2000). At the Hell Creek Road and Hauso Flat localities in Garfield County, Montana, carbon isotope shifts of −1.6‰ and −2.5‰ have been reported ~2–3 cm above the iridium-enriched K-T boundary clay (Arens and Jahren, 2000). At the Pyramid Butte locality in North Dakota, a shift of −2.8‰ was recorded 5 cm above the K-T boundary clay (Arens and Jahren, 2000). A similar isotope shift was reported at several localities in the Raton Basin of New Mexico and Colorado (Holmes and Bohor, 1994; Lomax et al., 1998; Schimmelmann and DeNiro, 1984). At York Canyon, New Mexico, a −1.8‰ carbon isotope shift occurs 10 cm above the peak iridium enrichment (there is no clear boundary clay preserved at this locality; Schimmelmann and DeNiro, 1984). At two other localities, reversible negative carbon isotope shifts were accompanied by anomalies in sulfur isotopes, interpreted as sulfuric acid rain in the aftermath of impact (Holmes and Bohor, 1994; Lomax et al., 1998). A negative shift of similar magnitude (−1.4‰ to −1.8‰) was also reported in extracted n-C_{29} alkane, a land plant biomarker, from marine rocks in Caravaca, Spain (Arinobu et al., 1999).

Therefore, an atmospheric carbon isotope shift stratigraphically above the K-T boundary offers a tool for correlating among terrestrial sections, particularly in the absence of an iridium-enriched boundary clay. This tool provides a time-stratigraphic framework in which to correlate fossil-bearing sections using data that are not autocorrelated to the biotic hypotheses that one may wish to test. Because plant-derived organic carbon is common in terrestrial rocks of this region, the technique can be applied to most fossil-bearing sections. More important, the absence of a carbon isotope shift within a suspected K-T boundary section offers powerful evidence of paraconformity—a common point of debate in many discussions of extinction and biotic change (Lofgren, 1995). Furthermore, a carbon isotope anomaly of similar magnitude and sign has been documented at the global K-T stratotype section at El Kef (Keller and Lindinger, 1989), thus making the carbon isotope signature directly diagnostic of the boundary event.

METHODS

Sediment samples for carbon isotope analysis were collected in July 1998 at localities where stratigraphic sections had been measured by K.R. Johnson. Sediment samples for bulk carbon isotope analysis were identified only by generic collection numbers in order to preclude any possibility of bias because of stratigraphy or field location. For bulk organic carbon measurements, a 1–5 g sample of the rock was acidified in 1M HCl overnight to remove carbonate. A standard weight sample of the remaining bulk organic residue was analyzed for $\delta^{13}C$ value in triplicate. Uncertainty in each carbon isotope measurement associated with mass spectrometry is $\pm 0.05‰$. All samples were prepared for stable isotope analysis by combustion in sealed tubes containing Cu, CuO, and Ag (Minagawa et al., 1984). Released CO_2 was purified cryogenically and collected for $^{13}C/^{12}C$ measurement on an Optima mass spectrometer. All isotope values are reported in the $\delta^{13}C$ notation:

$$\delta = \frac{\text{Rsample} - \text{Rstandard}}{\text{Rstandard}} \times 1000\ [‰]$$

where the reported standard is PDB limestone with Rstandard $= {}^{13}C/^{12}C = 0.011237$. Weight percent organic carbon content of sediment samples was calculated from these data.

Plant cuticle for carbon isotope analysis was picked by hand from demineralized (54% HF for three days) sediment; cuticle morphology, including the characteristic pattern of epidermal cell imprints, was verified under $100\times$ magnification. Cuticle is the waterproof coating on aerial surfaces of vascular land plants. Although it is of mixed chemical composition (De Leeuw et al., 1995), carbon isotope values of plant cuticle parallel those of whole leaf tissue with a standard error of less than $\pm 1‰$ (Upchurch et al., 1997).

SECTIONS STUDIED

To date, we have analyzed four K-T boundary crossing sections in the vicinity of Marmarth, North Dakota (Fig. 1). Stratigraphic sections figured below were based on Johnson's unpublished data and additional observations in the field. Stratigraphic positions are described with reference to the Hell Creek–Fort Union formational contact.

The Pyramid Butte section (KRJ Sections 86160 and 86161, Slope County, North Dakota; Johnson et al., 1989) has been extensively studied because it is the only section in southwestern North Dakota to yield an iridium (0.72 ppb Ir) anomaly at the K-T boundary. At Pyramid Butte, the iridium-bearing boundary clay overlies a lignite, the base of which marks the contact between the Hell Creek Formation and the Ludlow Member of the Fort Union Formation. Characteristic Cretaceous pollen taxa including *Wodehouseia spinata* and several species of *Aquilapollenites* become rare in the 20 cm above the iridium-bearing clay (see also Nichols and Johnson, this volume; Johnson et al., 1989). Plant macrofossils typical of the FUI floral zone (Johnson and Hickey, 1990), including "*Populus*" *nebrascensis, Dicotylophyllum anomalum*, and *Paranymphaea crassifolia*, are common in the mudstone immediately overlying the iridium-bearing clay (Johnson et al., 1989). We sampled for carbon isotope stratigraphy at 5 cm intervals through the 2 m that embrace the iridium-bearing boundary clay. The complete section at Pyramid Butte is ~60 m thick. Outside the recognized boundary interval, we sampled at 50 cm to 100 cm intervals in the Hell Creek Formation and, where possible (commonly 2–3 m intervals), in the overlying Ludlow Member. At Pyramid Butte, the Ludlow Member of the Fort Union Formation was relatively sandy and highly weathered; sandy sediments preserve little organic matter, and extensive weathering may alter carbon isotope values. Therefore, we focused our sampling on organic or silt-rich horizons within the Ludlow Member.

The Bobcat Butte section (KRJ Section 86148, Slope County, North Dakota) offered a long, continuous, well-exposed Hell Creek Formation section. A K-T boundary interval in the mudstone overlying the lowest lignite bed of the Fort Union Formation has been recognized by the loss of characteristic Cretaceous pollen types of the *Wodehouseia spinata* Assemblage Zone (see Nichols and Johnson, this volume). No iridium anomaly has been identified. Bobcat Butte also offered direct stratigraphic ties to the HCII-HCIII megafloral transition (see Johnson, this volume). Across the palynological K-T boundary transition, samples for carbon isotope analysis were taken at 3–10 cm intervals. Throughout the Hell Creek portion of the section, we sampled at 1 m intervals except between the highest reported HCII floral locality and the lowest HCIII horizon, where we sampled at 50 cm intervals. In the Ludlow Member, we were again hampered by poor exposure and sampled primarily at organic-rich horizons.

Terry's Fort Union Dinosaur section (KRJ Section 9653 = Dean Pearson Section 128 = DJN-96-7, Slope County, North Dakota; see also Nichols and Johnson, this volume) was of particular interest because a partially articulated ceratopsian frill, horn, and rib pieces were preserved in this section at +80 cm (KRJ Section 96530), which is stratigraphically within the Fort Union Formation. Precise positioning of the chronostratigraphic K-T boundary is particularly important in this section because many workers in the dinosaur-rich lands of northeastern Montana, eastern Wyoming, and southern Canada have noted a 2–3 m gap between the highest stratigraphic occurrence of dinosaur remains and the chronostratigraphic K-T boundary (Archibald, 1982). This observation led some to hypothesize that dinosaurs may have been extirpated in this region before the terminal Cretaceous catastrophe (Archibald, 1987). We sampled at 1 cm intervals through the exposed fossil-bearing section.

New Facet Boundary (KRJ Section 9847, Slope County, North Dakota; see also Nichols and Johnson, this volume) was newly opened during the 1998 field season. We sampled at 10 cm intervals across the Hell Creek-Ludlow contact to determine whether and where the chemostratigraphic K-T boundary was preserved.

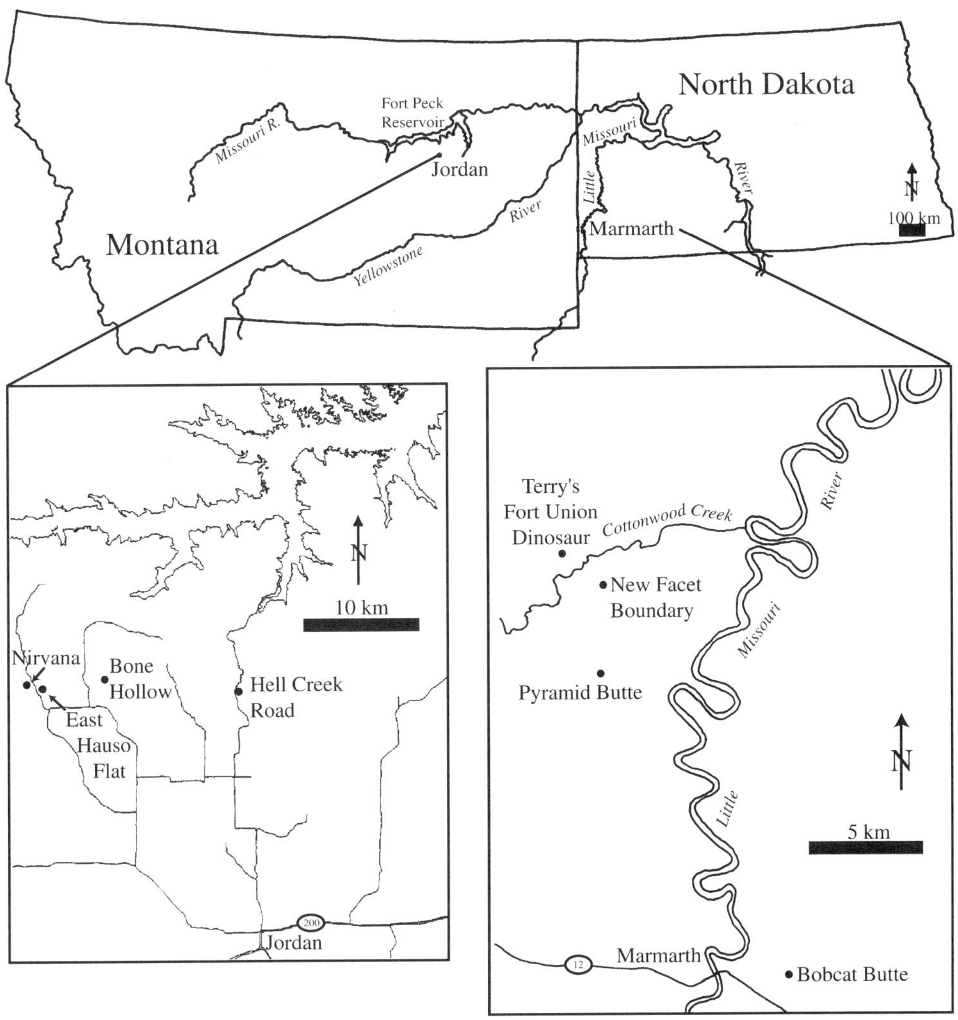

Figure 1. Location of sections discussed in this paper.

CARBON ISOTOPE VARIATION IN C3 PLANTS

Plant vital effects

A variety of plant vital effects influence the carbon isotope composition of C_3 plant tissue (see related discussion in Arens et al., 2000) and could, therefore, influence chemostratigraphic interpretation of secular variation in terrestrial organic carbon isotopes. For example, extreme light limitation can drive plant tissue $\delta^{13}C$ values as much as 5‰–6‰ lower than those of plants growing in full sun (Ehleringer et al., 1986; Zimmerman and Ehleringer, 1990). Water stress can push plant tissue to higher values, with $\delta^{13}C$ values in desert plants up to 6‰ greater than those in mesic environments (Ehleringer et al., 1992; Ismail et al., 1994; Johnson and Tieszen, 1993; Tieszen, 1991). Nutrient stress (Toft et al., 1989) and low temperature (Tieszen, 1991) can decrease leaf $\delta^{13}C$ values by as much as 4‰. Juvenile woody plants tend to have lower $\delta^{13}C$ values than do reproductive-age plants (Donovan and Ehleringer, 1991). Brooks et al. (1997a) found that growth form influenced plant $\delta^{13}C$ values, particularly when different growth forms encounter differing proportions of respired CO_2 (Broadmeadow et al., 1992; Brooks et al., 1997b; Medina et al., 1986). However, analysis of a larger data set failed to show differences in $\delta^{13}C$ value among growth forms (Arens et al., 2000). Some lineage effects emerge. The average isotope composition of conifers is about 1‰ greater than that of angiosperm trees and herbs, which have similar values (Arens et al., 2000). Among woody angiosperms, deciduous forms have lower average $\delta^{13}C$ values (<1‰) than evergreen taxa (Marshall and Zhang, 1994). However, this effect may also be related to source CO_2 because all of the carbon used to build deciduous leaves in the mid latitudes (where much of the relevant research was conducted) comes from the early spring atmosphere, which has a $\delta^{13}C$ value about 1‰ lower than the annual average (Ciais et al., 1995). The annual average atmospheric $\delta^{13}C$ value characterizes evergreen leaves, which build tissue throughout the growing season (this effect is discussed in more detail in Arens et al., 2000). C_3 grasses have average $\delta^{13}C$ values up to 2‰ greater than those of herbaceous dicots (Arens et al., 2000; Mooney et al., 1989).

However, Poaceae (grasses) does not appear in the fossil record until the late Paleocene (Crepet and Feldman, 1991).

Clearly, variation in carbon isotope composition because of plant vital effects may be large for individual plants under extreme conditions. However, organic carbon from many individual plants of different species is likely to be averaged as organic material enters the sedimentary record. Therefore, single-species data or that from plants under extreme conditions gives only partial insight into the vital effect variation that might be expected in the sedimentary record. A more realistic approach considers ranges of variation in carbon isotopic composition of many species within a single community. To quantify within-community variation in plant $\delta^{13}C$ value, we analyzed data from two modern communities in which $\delta^{13}C$ values were measured on a variety of species. Mooney et al. (1989) measured $\delta^{13}C$ values of leaf tissue from 18 species of angiosperm tree and shrub from a tropical woodland in the state of Jalisco, western Mexico. Both deciduous and evergreen species were represented in their sample and the ecosystem experienced seasonal drought. Smedley et al. (1991) studied 36 species of dicot herbs in a temperate grassland in Utah. These species showed $\delta^{13}C$ value variation due to lineage, life history (annual versus perennial) and phenology (Smedley et al., 1991). This community also experienced seasonal drought. We chose these communities because they differed in basic structure and composition and, therefore, encompass a wide range of possible vital effects.

The tropical woodland community displayed a mean $\delta^{13}C$ value of $-27.3‰$ with a standard deviation of 0.8‰ (Fig. 2, data from Mooney et al., 1989); the temperate grassland community had an average $\delta^{13}C$ value of $-26.8‰$ with a standard deviation of 0.8‰ (data from Smedley et al., 1991). An F-test for equality of variances failed to reject the null hypothesis that the variances in these samples are equal ($p = 0.46$). Residuals for both samples were normally distributed (Lilliefors $p = 0.19$ for the grassland community and $p = 0.15$ for the tropical woodland). Analysis of variance showed that the mean $\delta^{13}C$ values for these communities were not significantly different ($p = 0.06$). Given the similarity in both mean and standard deviation for these different plant communities, we propose the sample standard deviation ($\pm 0.8‰$) as a reasonable figure for expected vital effect variation within a plant community. This degree of variation might also be expected in $\delta^{13}C$ value for plant-derived organic carbon in the sedimentary record. Therefore, in chemostratigraphic interpretations, we disregard variation in the carbon isotope curve of $<\sim 1‰$, as this could be reasonably attributed to vital effects within the ancient plant community.

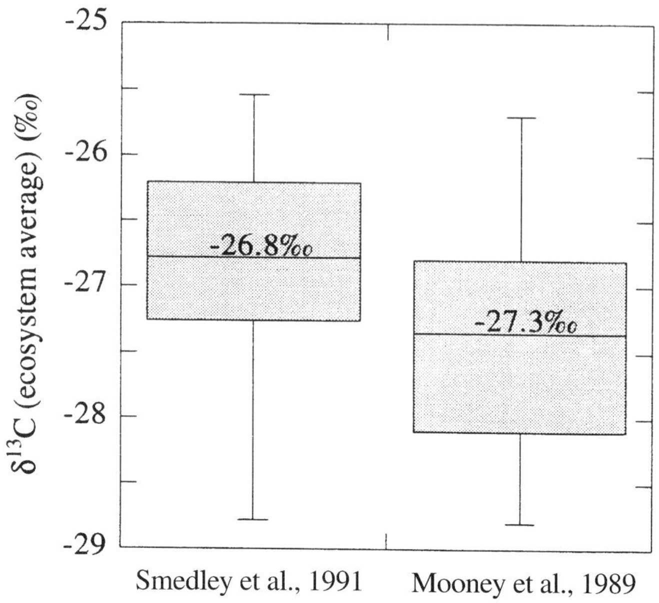

Figure 2. A comparison of whole-leaf $\delta^{13}C$ measurements from 18 species of tree and shrub from a tropical woodland in the state of Jalisco, western Mexico (Mooney et al., 1989), and 36 species of dicot angiosperms from a temperate grassland community in Utah (Smedley et al., 1991). The mean $\delta^{13}C$ value for each community is represented by the line bisecting each box, magnitude of the mean is noted above this line; one standard deviation is included within the gray boxes, this value is 0.8‰ for both communities; two standard deviations are denoted by brackets. An analysis of variance showed that these means do not differ significantly.

Choice of biogeochemical substrate

In using variation in carbon isotopic composition of C_3 plants to chart variation in the $\delta^{13}C$ value of paleoatmospheric CO_2, we must consider what organic carbon substrate to measure. Whole-fossil carbon isotopic values may be difficult to interpret because early diagenetic changes may alter the carbon isotope signal up to 4‰ (Benner et al., 1987). A variety of plant-derived macromolecules can be isolated from sedimentary rocks, and several have been proposed as useful substrates. Isolated cellulose, which ranges from 0.3‰ to 3.0‰ greater than whole tissue $\delta^{13}C$ values (Marino and McElroy, 1991; O'Leary, 1988), is preferred by many workers studying Pleistocene and Holocene materials (eg., Dupouey et al., 1993; Marino et al., 1992; Van de Water et al., 1994). Lignin ranges from 4.2‰ to 2.0‰ less than whole plant tissue values (Benner et al., 1987). Lipids, although popular for biogeochemical analysis, can be highly isotopically variable, depending on the tissue and synthetic pathway from which they were derived. In *Fagus sylvatica*, the $\delta^{13}C$ value of n-C_{27} alkane varied by more than 1‰ with leaf orientation (Lockheart et al., 1997). For sitosterol, the difference between north- and south-facing leaves was 3‰. In *Quercus robur* sun leaves, the $\delta^{13}C$ value of n-C_{27} alkane and n-C_{29} alkane was $\sim$6‰ less than whole leaf tissue values. Values from shade leaves varied between 4‰ and 6‰ less than those of bulk tissue (Lockheart et al., 1997). Individual leaves also showed seasonal variation of an additional 3‰ in n-C_{27} and n-C_{29} alkanes (Lockheart et al., 1997). Bulk tissue was less variable in all cases (Lockheart et al., 1997). Despite this eco-phenotypic variability, a −1.4‰ to −1.8‰ shift in n-C_{29} alkane has been interpreted as the K-T boundary in Spain (Arinobu et al., 1999).

We have chosen to evaluate isolated cuticle because it appears to directly parallel whole leaf carbon isotope values with low variation (<1‰, Upchurch et al., 1997). However, cuticle cannot be isolated from all sediment samples. Therefore, we must ask how well bulk organic carbon in sediment reflects the isotope composition of plants growing on the landscape. Medina et al. (1986) noted that the average $\delta^{13}C$ value of forest litter equaled that of canopy leaves, which were consistently 2‰ less than the associated soil organic matter. These data suggest that organic matter in soil has a consistent isotope relationship with the associated vegetation. To test this hypothesis, in the rock record, we regressed bulk sedimentary organic carbon isotope values onto the $\delta^{13}C$ value of a mixed sample of plant cuticle isolated from the same sediment ($n = 68$). We recovered a significant relationship between the $\delta^{13}C$ values of bulk sedimentary organic carbon and those of isolated cuticle (Fig. 3, $R^2 = 0.69$, $p < 0.001$). The residual average for this analysis was ±0.4‰. Residuals for this analysis include variation due to both plant vital effects and taphonomic variation. A slope approximately equal to one was statistically significant ($p < 0.001$), but the associated y-intercept (∼−2) was not ($p = 0.2$). However, the y-intercept was similar to the difference between the carbon isotope composition of canopy leaf tissue and that of soil organic material reported by Medina et al. (1986), suggesting that the actualistic relationship holds for ancient sediment. This shows that the bulk carbon isotope signature in sediments was related to the isotope composition of cuticle and chemically related decay-resistant materials, rather than by within-sediment carbon recycling or isotopically different compounds as previously suggested (Brooks et al., 1997b; Ke and Lin, 1997; Wedin et al., 1995). We further conclude that bulk carbon isotope measurements reflect the average signal of the ancient vegetation and, therefore, the best approximation of the paleoatmosphere. Therefore, bulk sedimentary organic carbon can be used for chemostratigraphic reconstruction in terrestrial rocks. However, we have consistently avoided sampling well-developed paleosols in the field to avoid data

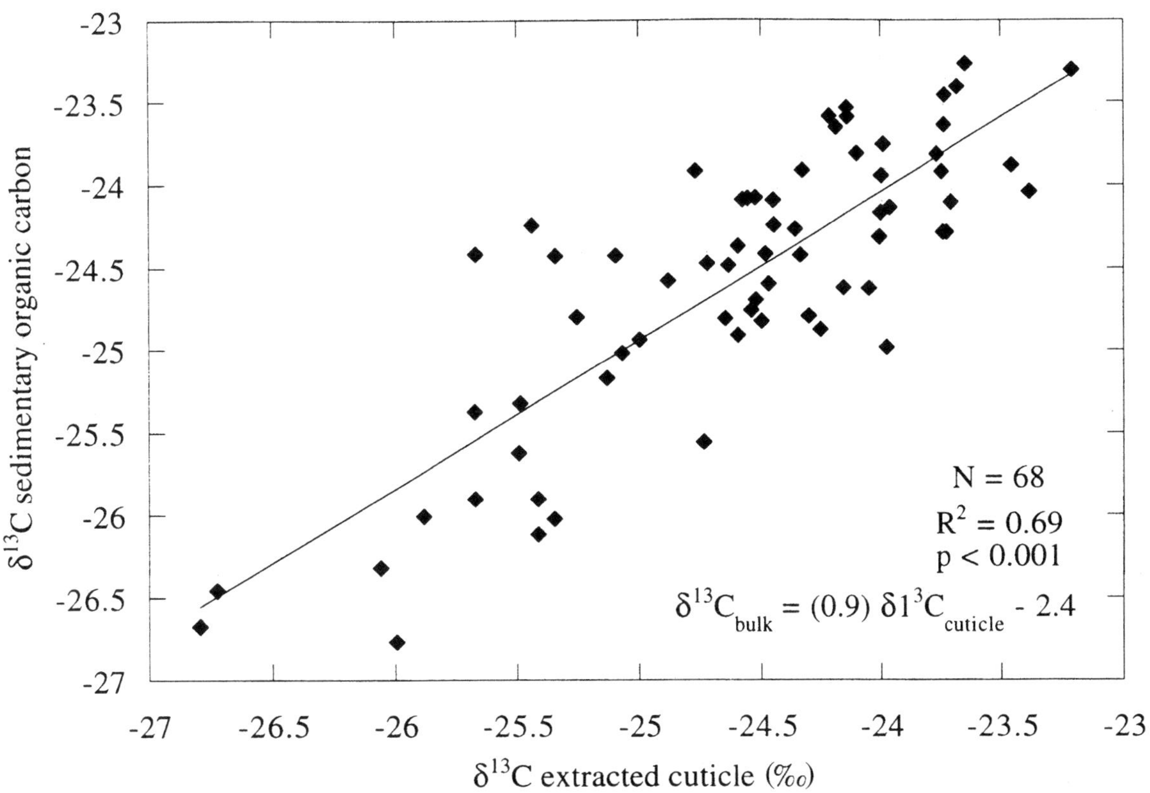

Figure 3. Regression of $\delta^{13}C$ value of sedimentary organic material onto $\delta^{13}C$ value of a mixed sample of vascular plant cuticle isolated from the same sediment. The relationship was statistically significant ($R^2 = 0.69$, $p < 0.001$). Slope ($p < 0.001$) was approximately equal to one, suggesting that bulk sedimentary organic material varied in tandem with associated plant cuticle. The regression intercept was not statistically significant ($p = 0.2$), but was approximately equal to 2‰, which was the difference between the carbon isotope composition of soil organic material and canopy leaves observed in ever-wet forests (Medina et al., 1986). This result suggests that bulk sedimentary organic material reliably parallels the isotope composition of plant tissue and thus can be used to evaluate the isotope composition of paleoatmospheric CO_2.

biased by extensive within-soil carbon recycling (Wedin et al., 1995).

TAPHONOMIC CONSIDERATIONS

Depositional environment and rock type

As discussed above, water status influences the $\delta^{13}C$ value of C_3 plants (eg., Beerling and Woodward, 1995; Farquhar and Richards, 1984; Marshall and Zhang, 1994). Consequently, one might suspect depositional environment to bias the $\delta^{13}C$ value of sedimentary organic carbon. Specifically, plants growing in better-drained substrates might experience more water stress than those growing in wetlands. Several authors have also suggested that differential diagenetic oxidation of organic material could systematically alter carbon isotopic ratios in fossils and sediment (Bartley et al., 1998; Benner et al., 1987). To test whether organic content of sediment systematically affects carbon isotope composition, we regressed the $\delta^{13}C$ values of bulk organic carbon in sediments upon weight percent organic material from the same samples (Fig. 4, $n = 411$). Weight percent organic carbon variation roughly paralleled rock-type variation observed in the field, ranging from mudstone and sandstone samples with very low organic content to lignite and carbonaceous shale samples with higher carbon content. Weight percent carbon content also serves as a crude proxy for facies variation in this context, with higher-carbon sediments corresponding to wetlands and low-carbon samples to better-drained substrates. Unlike the vital effect tests discussed above, this analysis is not designed to document a range of variation in carbon isotope composition because of facies variation. Because the isotope composition of the atmosphere is assumed to vary among samples in this analysis, the regression only tests whether there is a systematic relationship between $\delta^{13}C$ value and organic carbon content of the rock. Correlation between carbon isotope composition and percent organic carbon content ($R = 0.07$, $p = 0.1$) was not statistically significant. Bartley et al. (1998) suggested that when diagenetic alteration of the carbon isotopic signal is minimal, $\delta^{13}C$ values should be uncorrelated with total organic carbon values as in Figure 4. This result demonstrates that organic carbon content of terrestrial rocks does not systematically bias $\delta^{13}C$ values. We can, therefore, compare $\delta^{13}C$ values across this range of terrestrial depositional environments without introducing systematic bias.

Another way to approach the question of facies variation is to examine rock type. To detect a relationship between rock type and carbon isotope value, we coded each $\delta^{13}C$ value for rock type (lignite, carbonaceous shale, claystone-mudstone, siltstone, and sandstone) based on field descriptions by Arens. Since each rock type was represented by a different number of

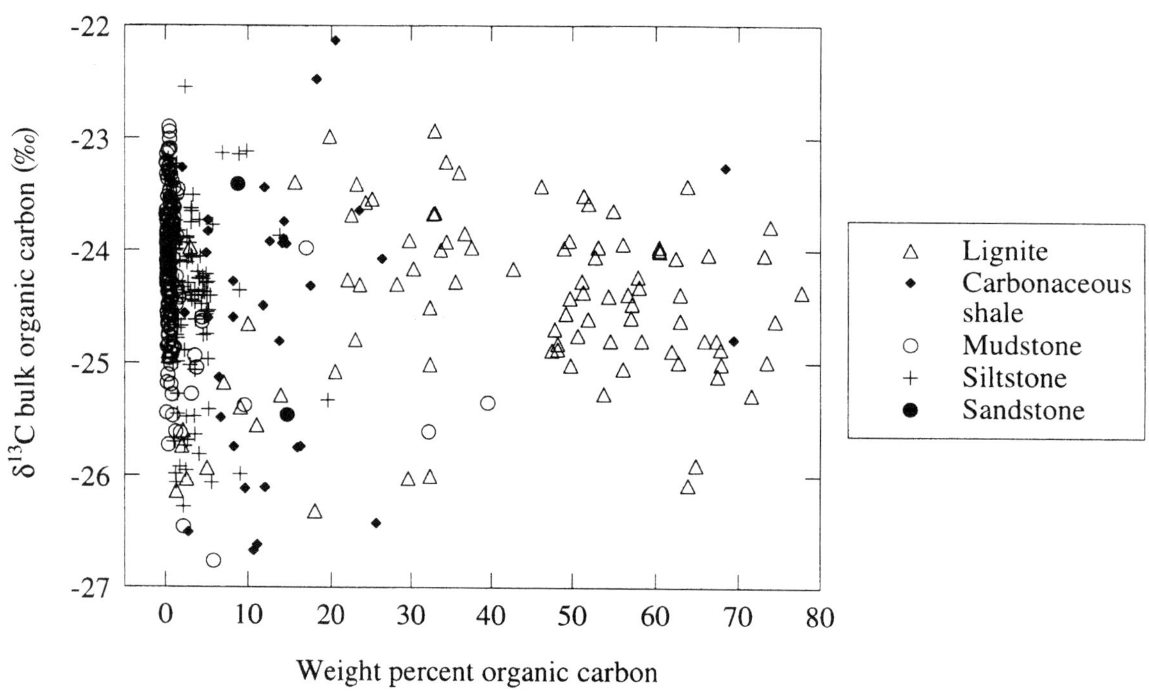

Figure 4. Regression of $\delta^{13}C$ values of bulk sedimentary organic material onto percent organic carbon content by weight. Data from several Hell Creek and Fort Union Formation localities in eastern Montana and southwestern North Dakota. There was no significant correlation ($R = 0.07$, $p = 0.1$) between these variables, suggesting that carbon isotope values of sedimentary organic material are not systematically influenced by the amount of organic carbon preserved. As weight-percent organic content was a proxy for facies variation, this result also shows that $\delta^{13}C$ values are not biased by depositional environment.

measurements, we randomly selected a balanced sample (n = 39) of lignite, carbonaceous shale, claystone-mudstone and siltstone values for analysis. We chose the nonparametric Kruskal-Wallis test for this analysis because carbonaceous shale failed the homogeneity of variance requirement for parametric ANOVA. The Kruskal-Wallis test failed to reject ($p = 0.219$) the null hypothesis that all measurements were sampled from the same population. Analysis of five randomly selected samples of lignite, carbonaceous shale, claystone-mudstone, siltstone, and sandstone yielded similar results ($p = 0.160$). We conclude that carbon isotopic value does not vary systematically with rock type (see further discussion in Arens and Jahren, 2000). Together, these analyses show that there is no significant or systematic taphonomic bias in carbon isotopic measurements of terrestrial organic material. Therefore, secular variation observed in $\delta^{13}C$ values can be attributed primarily to the carbon isotopic signature of plant-derived organic material and thus the paleoatmosphere.

Lateral variation in carbon isotope value

If the carbon isotopic composition of C_3 plant-derived sedimentary organic material faithfully parallels the carbon isotopic composition of paleoatmospheric CO_2, one would expect that samples taken from the same chronostratigraphic horizon in different locations would have similar absolute magnitudes. To test this hypothesis, we evaluated four K-T boundary-crossing sections, three in northeastern Montana and one in southwestern North Dakota (Table 1). In each of these sections, the K-T boundary is identified by an iridium-bearing boundary clay that is interpreted as the fallout of the terminal Cretaceous bolide impact (Alvarez, 1983). The assumption of temporal coincidence of the impact clay horizon at these localities hinges on the iridium-enriched clay representing a single, instantaneous event (Alvarez et al., 1980; Alvarez et al., 1995). Table 1 shows that the $\delta^{13}C$ values in the sample underlying the iridium-enriched impact clay are similar, regardless of rock type and location. Even across considerable distance, the absolute magnitude of $\delta^{13}C$ measurements from the same chronostratigraphic horizon correspond. This suggests that despite all of the possible plant vital effect and taphonomic variability, a meaningful chemostratigraphic signal emerges from these data.

TABLE 1. CARBON ISOTOPE DATA FROM SAMPLES TAKEN IMMEDIATELY STRATIGRAPHICALLY BELOW IRIDIUM-ENRICHED CRETACEOUS-TERTIARY BOUNDARY IMPACT CLAYS

Locality	Iridium (ppb)	$\delta^{13}C$ organic (‰)	Rock type
Hell Creek Road, Montana	0.57	−24.4	Mudstone
East Hauso Flat, Montana	2.22 and 11.7	−24.3	Carbonaceous shale
Nirvana, Montana	2.22 and 11.7	−24.6	Silty mudstone
Pyramid Butte, North Dakota	0.72	−24.5	Lignite

Note: The iridium anomaly at Hell Creek Road, Montana (Baadsgaard et al., 1988) is coincident with the palynological Cretaceous-Tertiary (K-T) boundary (Hotton, 1988). Iridium and palynological K-T boundaries also co-occur at Pyramid Butte, North Dakota (Johnson et al., 1989). The Nirvana and East Hauso Flats sections are ~0.5 and 1 km, respectively, east of the section discussed by Alvarez (1983, Iridium Hill, Ir = 2.2 ppb) and that of Smit and Van der Kaars (1984, Herpijunk Promontory, Ir = 11.7 ppb), where both iridium and palynological K-T boundaries have been documented. Strata are continuously exposed over this distance and the impact clay was directly traced laterally from the previous sample sites to the localities discussed here. Rock type descriptions were made in the field.

Taken together, these analyses show that bulk sedimentary organic $\delta^{13}C$ values retain the carbon isotope composition of paleoatmospheric CO_2. Therefore, carbon isotope cycles, trends, and shifts greater than that associated with vital effects (0.8‰ discussed above) can be used to correlate between terrestrial sections. Because plant-derived carbon is ubiquitous in the floodplain sediments that characterize terrestrial deposition in the Hell Creek and Fort Union Formations, this method will ostensibly work for most fossil-bearing sections. Because the method does not rely on biostratigraphic distribution of fossil forms, it offers an independent correlation, including placement of the K-T boundary, which will enhance the inference drawn from fossil distributions in space and time.

RESULTS

Pyramid Butte section

Because the Pyramid Butte section contains an iridium-bearing K-T boundary clay in addition to biostratigraphic markers (Johnson et al., 1989), it served as the reference for regional carbon isotope correlation. Figure 5 suggests cyclic variation of −24.0‰ to −25.5‰ in $\delta^{13}C$ value of bulk sedimentary organic material between −1500 cm and −200 cm (KRJ Section 86161, −365 cm to −1235 cm). We have observed similar quasiperiodic variation at the Bone Hollow locality (Los Angeles County Museum locality 7245, Garfield County, Montana, Trumbo Ranch 7.5′ Quadrangle, sec. 34, T. 21 N., R. 37 E.) in northeastern Montana (Fig. 1, Arens and Jahren, unpublished data). Cyclic variation in $\delta^{13}C$ value between −23.5‰ and −25‰ continues through the lignite at the base of the Fort Union Formation (Fig. 6, KRJ Section 86161, 0 cm to +90 cm). This lignite immediately underlies the iridium-bearing K-T boundary clay (KRJ Section 86161, +95 cm). $\delta^{13}C$ values from the 10 cm immediately below the K-T boundary clay (KRJ Section 86161, +90 cm to +80 cm) range from −23.5‰ at +80 cm to −24.5‰ at +90 cm.

At Pyramid Butte, the iridium anomaly and palynological K-T boundary co-occur at +95 cm (Table 1, Fig. 6, Nichols and Johnson, this volume; Johnson et al., 1989). Carbon isotope values in the centimeter below and centimeter above the iridium-bearing boundary clay were −24.5‰ (Table 1) and −24.4 ‰, respectively. Similar $\delta^{13}C$ values immediately above and below a K-T iridium anomaly have been observed at the

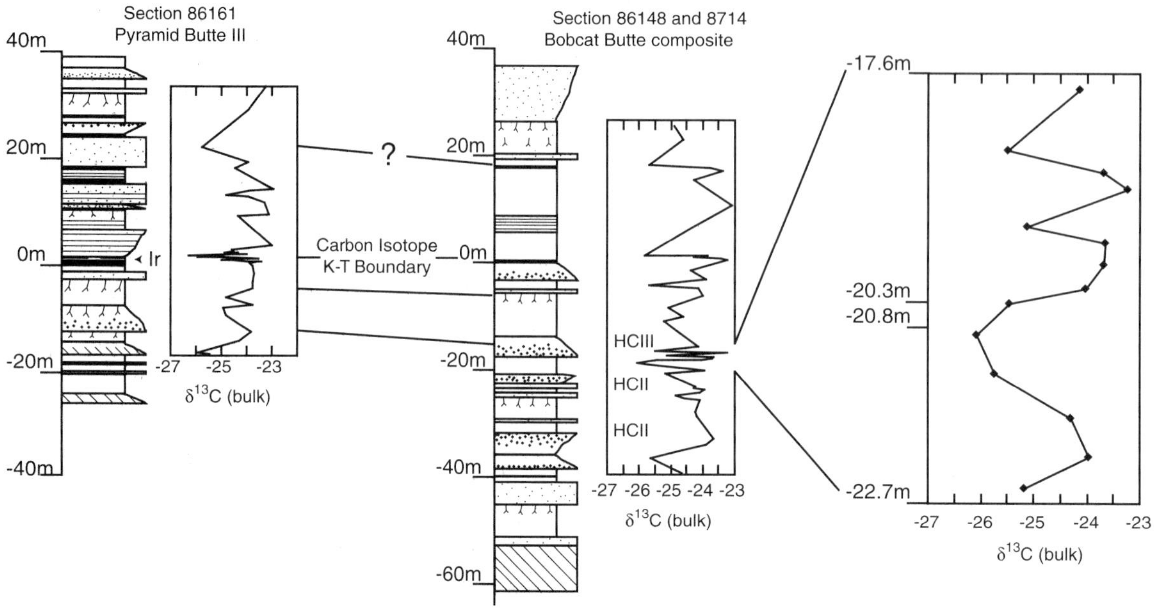

Figure 5. Carbon isotope stratigraphy for the Pyramid Butte and Bobcat Butte sections. Pyramid Butte (KRJ Section 86161), Slope County, North Dakota, Pretty Butte 7.5′ Quadrangle, Hell Creek and Fort Union Formations. Stratigraphic position of the iridium anomaly in this section is noted by "Ir". The section was measured on 23 July 1987 by K.R. Johnson and R. Myers; sediment samples for carbon isotope analysis were taken on 23 and 25 July 1998 by N.C.A., H. Pogrel and A. Thompson. Top of the section 650 m S., 1000 m E. of northwest corner sec. 15, T. 134 N., R. 106 W. (UTM 5140888 m N., 578673 m E., Zone 13); base of the section 800 m S., 1020 m E. of the northwest corner sec. 15, T. 134 N., R. 106 W. (UTM 5140744 m N., 578697 m E., Zone 13). Graphical logs correspond to KRJ Section 86161. Bobcat Butte section (KRJ Section 86148), Slope County, North Dakota, Marmarth 7.5′ Quadrangle, Hell Creek and Fort Union Formations. The section was measured on 17 July 1987 by K.R. Johnson and R. Myers; sediment samples for carbon isotope analysis were taken on 24, 25, and 27 July 1998 by N.C.A., H. Pogrel and A. Thompson. Top of the section 1610 m E. of the northwest corner sec. 28, T. 133 N., R. 105 W. (UTM 5127261 m N., 586495 m E., Zone 13). Base of the section 1520 m S., 550 m E. of northwest corner sec. 28, T. 133 N., R. 105 W. (UTM 512717 m N., 586287 m E., Zone 13). HCII and HCII megafloral zones are discussed by Johnson and Hickey (1990). Graphical logs correspond to KRJ Section 86148. The carbon isotope K-T boundary and several other chemostratigraphic tie point are noted with lines connecting the sections.

Hell Creek Road locality (Fig. 1, Garfield County, Montana, Maloney Hill 7.5′ Quadrangle, sec. 1, T. 20 N., R. 37 E.; Lerbekmo, 1985; Swisher et al., 1993) and the nearby East Hauso Flat and Nirvana localities (Fig. 1, Garfield County, Montana, Hell Hollow 7.5′ Quadrangle, sec. 36, T. 21 N., R. 35 E.; Smit and Van der Kaars, 1984). We interpret $\delta^{13}C$ values of −24.4‰ immediately above the iridium-bearing clay as the carbon isotope signature of vegetation destroyed by impact-related disturbance. Because these plants fixed their carbon under the latest Cretaceous atmosphere, their tissue retains the latest Cretaceous atmospheric signature. However, samples taken at +104 cm (10 cm above the impact clay) have $\delta^{13}C$ values decreasing to −25.9‰ and −26.3‰ at +124 cm (Fig. 6). The absolute value of the K-T carbon isotope shift at Pyramid Butte was similar to those observed at other iridium-bearing K-T boundary sections in Montana, which range from −25.9‰ at the Hell Creek Road locality to −26.8‰ at East Hauso Flat. The similarity of the carbon isotope absolute magnitudes and the stratigraphic relationship with the K-T boundary iridium anomaly suggest that we are observing the same geochemical signal in all three localities.

At Pyramid Butte, $\delta^{13}C$ values recovered to −24.4‰ at +150 cm (55 cm above the K-T impact clay). In the Hell Creek Road section in northeastern Montana, where we have fine time-stratigraphic control, $\delta^{13}C$ values recovered from a K-T shift within 22 cm above the impact clay—at least 30 cm below the 65.00 ± 0.05 million year (Swisher et al., 1993) volcanic ash preserved in this section. Therefore, carbon isotope recovery can be radiometrically bracketed between 65.16 ± 0.04 Ma (the iridium-bearing impact bed) and 65.00 ± 0.05 Ma volcanic ash (Swisher et al., 1993)—a maximum interval of 250 k.y. Thus the carbon isotope shift and recovery observed in these terrestrial sections was not only similar in magnitude, but also in duration to that observed in the "early recovery" phase of the surface ocean signature (D'Hondt et al., 1998; Hsü et al., 1982; Keller and Lindinger, 1989). If one accepts the correlation between the carbon isotope shifts observed at Pyramid Butte, North Dakota, and Hell Creek Road, Montana, a similar

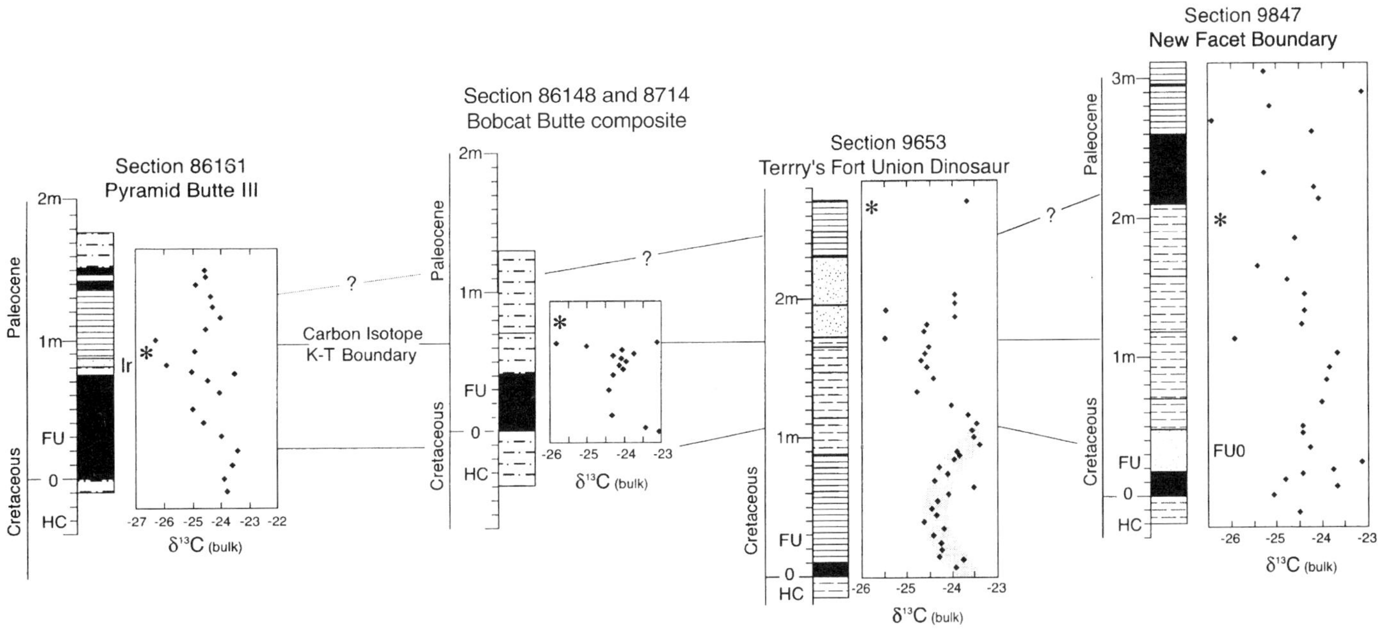

Figure 6. Detailed carbon isotope stratigraphy for the interval surrounding the Hell Creek-Fort Union formational contact and K-T boundary for the four sections analyzed in this study. The carbon isotope K-T boundary and several possible chemostratigraphic tie points are noted with lines connecting the sections. The palynological K-T boundary in denoted by asterisk (*). Stratigraphic position of the iridium anomaly in the Pyramid Butte (locality information same as Fig. 5) section is noted by "Ir". Bobcat Butte locality information same as Figure 5. Terry's Fort Union Dinosaur section (KRJ Section 9653), Slope County, North Dakota, Badland Draw 7.5′ Quadrangle, Hell Creek and Fort Union Formations. The section was measured on 16 October 1996 by K.R. Johnson, D. and K. Nichols, D. Pearson and S. Bowman; sediment samples for carbon isotope analysis were taken on 26 July by N.C.A., K.R. Johnson, T. Farnham, H. Pogrel and A. Thompson. Section located SE ¼ NW ¼ NW ¼ sec. 33, T. 135 N., R. 106 W. (UTM 5146220 m N., 756179 m E., Zone 13). Graphical log corresponds to KRJ Section 9653. Gray shading denotes the range of carbon isotope variation (0.6‰) observed in detailed sampling through this portion of the section. New Facet Butte Boundary section (KRJ Section 9847), Slope County, North Dakota, Pretty Butte 7.5′ Quadrangle, Hell Creek and Fort Union Formations. The section was measured on 25 July 1998 by K.R. Johnson and T. Farnham; sediment samples for carbon isotope analysis were taken on 25 July 1998 by N.C.A. and A. Thompson. Section located sec. 34, T. 135 N., R. 106 W. (UTM 5145400 m N., 578100 m E., Zone 13). Graphical log corresponds to KRJ Section 9847.

time scale can be inferred for terrestrial sedimentation at Pyramid Butte.

Bobcat Butte section

The long section of uppermost Cretaceous sediments at Bobcat Butte permits a direct tie between the carbon isotope stratigraphy in the Hell Creek Formation and megafloral zone transitions recognized by Johnson and Hickey (1990). Figure 5 shows bulk organic sedimentary $\delta^{13}C$ values of −25.5‰ at −4020 (KRJ Section 86148), with a trend toward higher values upsection. Between −2267 and −1756 cm (KRJ Section 86148), our data showed an interval of fluctuation in carbon isotope signature. This fluctuation (Fig. 5) is characterized by a positive shift at −2090 cm (−25.2‰ to −23.9‰), followed by a negative shift at −2050 cm (to −26.0‰), a positive shift at −2000 cm (to −24.3‰), a negative shift at −1950 cm (to −25.5‰), a positive shift at −1900 cm (to −23.7‰), a negative shift at −1800 cm (to −25.1‰), and a positive shift at −1750 cm (to −23.2‰). These shifts range in magnitude from 1.2‰ to 2.1‰, exceeding the variation that might be expected because of vital effects alone. We observed no indications of unconformities or depositional hiatuses (eg., erosive contacts, well-developed paleosols, sedimentary lags, concretion horizons, etc.) to suggest extensive missing time. We also do not recognize a significant facies change in this interval; rocks in this interval were buff to light brown siltstone or sandy siltstone. Therefore, we interpret this interval as a geologically brief period of fluctuation in the carbon isotope composition of atmospheric CO_2. We have not observed this pattern in other sections sampled to date.

The interval of carbon isotope fluctuation in the uppermost Cretaceous at Bobcat Butte (−1756 to −2267 cm, KRJ Section 86148) could be driven by a variety of factors. Climate change or regression-related releases of methane hydrate from marine sedimentary reservoirs could drive a rapid decrease in atmospheric $\delta^{13}C$ values, followed by photosynthetic compensation. This mechanism explained the 2.5‰ variation reported in $\delta^{13}C$ data from the latest Paleocene (Dickens et al., 1995). Alternatively, an input of volcanic CO_2 could drive $\delta^{13}C$ values higher (Taylor, 1986), with biomass oxidation returning the atmosphere toward lower values. Climatic variation could also drive carbon isotope fluctuation in the atmosphere by changing the size of the terrestrial biomass/soil carbon reservoir. For example, a shift of up to 1.5‰ has been reported between Pleistocene glacial and interglacial cycles (Marino et al., 1992). At this point, we lack additional lines of evidence needed to distinguish between these hypotheses. However, we note that the transition between HCII and HCIII floral zones (Johnson and Hickey, 1990) occurred at the carbon isotope instability interval (Fig. 5). Therefore, we hypothesize that factors in the physical environment may have promoted latest Cretaceous vegetation change. To test this hypothesis, we suggest a closer look at the stratigraphic interval between floral zones at other sites within the region, and a hunt for additional lines of evidence designed to falsify the various mechanisms of carbon isotope variation proposed above. What seems clear at this point is that upper Hell Creek time was not characterized by environmental stability, at least in this portion of the Western Interior region.

At −580 cm, we note a shift in $\delta^{13}C$ value from −24.1‰ to −25.7‰ to −23.9‰. This shift is similar in magnitude to the K-T boundary shift, but occurs in dinosaur-bearing beds, which also contain Maastrichtian palynomorphs and megaflora (Nichols and Johnson, this volume). We also observed this latest Cretaceous shift in terrestrial sections at the Bone Hollow locality in Montana and in some high-resolution marine sections (Keller and Lindinger, 1989). This feature is not apparent at Pyramid Butte.

No iridium anomaly has been recognized in the Bobcat Butte section. However, a −2.1‰ carbon isotope shift (latest Cretaceous $\delta^{13}C$ value of −23.7‰, overlain by a boundary shift $\delta^{13}C$ value of −25.8‰) at +62 cm (KRJ Section 86148) suggested that the carbon isotope K-T boundary can be recognized at the top of the carbonaceous shale overlying the basal lignite of the Fort Union Formation at this locality. As at Pyramid Butte, the palynological K-T boundary also overlies the basal Fort Union lignite bed (+75 cm, Table 1, see also Nichols and Johnson, this volume). Extensive weathering and poor exposure prohibited detailed sampling into the Fort Union Formation at this locality.

Terry's Fort Union Dinosaur section

This approximately 3 m section preserved a partially articulated frill, horn, and associated ribs of a ceratopsian dinosaur in a 78-cm-thick silty mudstone (Fig. 6, +80 cm, see also Pearson et al., this volume), which overlies the basal lignite bed of the Fort Union Formation. Although it would not be unusual for the palynological K-T boundary to occur in the Fort Union Formation (eg., Pyramid Butte and Bobcat Butte Sections, see also Nichols and Johnson, this volume), the position of the chronostratigraphic boundary relative to the dinosaur remains is critical to discussions of the timing of dinosaur extirpation in the region.

Detailed sampling through the section revealed apparently cyclic variation between −24.5‰ and −23.5‰ (Fig. 6). This variation was similar to that observed in the uppermost Hell Creek Formation at Pyramid Butte (KRJ Section 86161, −365 cm to −1235 cm), but at Terry's Fort Union Dinosaur, the pattern is condensed over a much shorter section (0 cm to +165 cm). Sampling density in the latter section also allows us to note a range of variation (in gray, Fig. 6) around the secular trend. This variation is ~0.6‰, which is slightly less than the range of variation associated with plant vital effects predicted both by the actualistic data set (Arens et al., 2000) and the modern ecosystem data sets discussed above. This observation suggests that although vital effects and taphonomic alteration produce some variation in bulk carbon isotopic measurements, variation >1‰ probably represents change in the isotope composition of the source CO_2 used to build ancient plant tissue. Therefore, shifts, trends, and cycles greater than this magnitude can be interpreted chemostratigraphically, as suggested by Williams et al. (1988).

We observed low $\delta^{13}C$ values, suggestive of the K-T carbon isotope shift, at +170 cm and +190 cm (KRJ 9653). These values (both −25.5‰) were higher than that observed at the Pyramid Butte K-T boundary (−26.3‰), but similar to that observed at Bobcat Butte (−25.8‰). Therefore, we place the carbon isotope K-T boundary ~1 m above the dinosaur remains. The palynological K-T boundary occurs at approximately +260 cm, ~180 cm above the dinosaur fossil.

Using the Pyramid Butte section (Fig. 5) as an interpretive template, we correlate $\delta^{13}C$ values of −24.0‰ and −24.4‰ at +70 cm at Pyramid Butte (KRJ Section 86161) with values of −24.4‰ to −24.6‰ +130 cm to +150 cm at Terry's Fort Union Dinosaur (KRJ Section 9653). We also propose to correlate higher $\delta^{13}C$ values (−23.4‰ to −23.6‰) observed at 0 cm to +10 cm at Pyramid Butte (KRJ Section 86161) with similar values (−23.5‰ to −23.9‰) at +100 cm at Terry's Fort Union Dinosaur (KRJ Section 9653). The trend toward lower values between +70 cm and +90 cm at Terry's Fort Union Dinosaur is best correlated with a similar trend between −360 cm and 0 cm at Pyramid Butte (KRJ Section 86161). Low $\delta^{13}C$ values at −425 cm (KRJ Section 86161) at Pyramid Butte (−23.8‰) and +62 cm (KRJ Section 9653) at Terry's Fort Union Dinosaur (−23.5‰) may also be correlated. Therefore, the base of the Terry's Fort Union Dinosaur section is approximately correlated with −745 cm at Pyramid Butte (KRJ Section 86161). Approximately 9 m of sediment at Pyramid Butte are condensed into ~70 cm at Terry's Fort Union Dinosaur. This interpretation highlights the differential sedimentation rates common to terrestrial deposition. This example also illustrates another power of the chemostratigraphic curve to provide isochronous links between sections within a region.

New Facet Boundary section

Reliable chemostratigraphic correlation requires that several features of the carbon isotope curves be matched among sections. The New Facet Boundary section (Fig. 6) was difficult to interpret because it lacked features characteristic of other curves where the K-T boundary could be diagnosed based on the presence of iridium and shocked mineral. At +110 cm, we recorded a carbon isotope value of −25.9‰, which overlaid values of −23.7‰ and −23.9‰ (Fig. 6). Another low value (−26.4‰) was recorded at +270 cm. This value concluded a series of values that decrease up section through the stratigraphic interval +210 cm to +270 cm. This interval of decreased carbon isotope values is underlain by values no higher than −24.1‰ at ~+210 cm to +220 cm. Palynological data placed the K-T boundary at +199 cm (see also Nichols and Johnson, this volume).

Two interpretations of these data are possible. First, the lower, −25.9‰ excursion may be interpreted as the K-T boundary carbon isotope anomaly. In this interpretation, carbon isotope values stratigraphically below the excursion are somewhat higher than those typically reported at iridium-bearing K-T boundary sites in the region (Table 1). This interpretation placed the palynological transition 90 cm above the carbon isotope anomaly.

Alternatively, the low value at +270 m might be interpreted as the K-T boundary carbon isotope excursion. We observe a similar pattern of steady decrease in carbon isotope values to a low K-T boundary value in the iridium-bearing Pyramid Butte section. However, at Pyramid Butte, the decrease occurs over a shorter stratigraphic interval (approx. 40 cm) compared to the New Facet Boundary section (approx. 70 cm). This could be explained by higher sedimentation rates at the New Facet Boundary site. Carbon isotope values underlying the +270 cm low are similar to those from iridium-bearing reference sections (Table 1). Correlating the +270 cm low with the K-T boundary carbon isotope excursion placed the palynological K-T boundary approximately 70 cm below the carbon isotope minimum attributed to the K-T boundary. This is similar to the pattern at Pyramid Butte, where the palynological K-T boundary occurs 10 cm below the minimum carbon isotope value (Table 2). Once again, higher sedimentation rate must be invoked to explain this difference. Under this interpretation, the −25.9‰ value at +110 cm is without parallel in any previously reported K-T boundary-crossing section. A carbon isotope value of −25.7‰ was recorded at −580 cm in the Bobcat Butte section. However, correlation of the −25.9‰ excursion at +110 cm at New Facet boundary with the −580 cm excursion at Bobcat Butte seems improbably. Such an interpretation would require significantly lower rates of sedimentation or a significant sedimentary haitus at New Facet Butte. There is no evidence of significant disconformity in the New Facet Boundary section and low sedimentation rates are at odds with the palynological data. Furthermore, such a correlation would require much greater time transgression of the Hell Creek–Fort Union contact than has been reported previously in this region.

TABLE 2. COMPARISON OF THE STRATIGRAPHIC POSITION OF THE PALYNOLOGICAL AND CARBON ISOTOPE MARKERS FOR THE CRETACEOUS-TERTIARY (K-T) BOUNDARY AND THE STRATIGRAPHIC DIFFERENCE BETWEEN THE TWO

Locality	Carbon isotope K-T boundary (cm)	Palynological K-T boundary (cm)	Difference (cm)
Pyramid Butte	+105	+95	−10
Bobcat Butte	+62	+75	+13
Terry's Fort Union Dinosaur	+170	+261	+91
New Facet Boundary	+110	+184	+74

Note: Stratigraphic position given relative to the Hell Creek–Fort Union formational contact. See Nichols and Johnson (this volume) for further discussion of the palynologically based K-T boundary at these localities. The iridium anomaly in the Pyramid Butte section occurs at +95 cm.

Despite our inability to develop a more definitive correlation for the New Facet Boundary section, both the carbon isotope and palynological data suggest that the FU0 flora (see also Johnson, this volume) at +30 cm (KRJ Section 9847) is of Cretaceous, rather than Paleocene, age. Difficulty in developing a carbon isotope correlation for New Facet Butte may also suggest that the section may be incomplete, as discussed below.

FEATURES OF THE CARBON ISOTOPE CURVES THAT CAN BE USED AS CHEMOSTRATIGRAPHIC TIE POINTS

Reliable chemostratigraphic correlation seldom hinges on single features. Greater confidence can be placed in correlations that match several tie points among curves. In Figures 5 and 6, we indicate several such points of correlation. In these sections, one of the most relevant tie points is the K-T boundary carbon isotope shift. As discussed above, this marker has been found in association with the K-T iridium anomaly in the global stratotype section at El Kef, Tunisia (Keller and Lindinger, 1989), and at other marine (Arinobu et al., 1999; Arthur et al., 1979; D'Hondt et al., 1998; Shackleton and Hall, 1984; Zachos et al., 1992) and terrestrial (Holmes and Bohor, 1994; Lomax et al., 1998; Schimmelmann and DeNiro, 1984) sections. Furthermore, this carbon isotope shift is backed by a widely accepted mechanism that explains the link between terrestrial and marine data for the interval (Arinobu et al., 1999; Arthur et al., 1985; Caldeira et al., 1990; D'Hondt et al., 1998; Ivany and Salawitch, 1993; Kasting et al., 1986; Kump, 1991; Stott and Kennett, 1989, 1990; Tappan, 1968; Williams et al., 1988; Wolbach et al., 1985; Zachos and Arthur, 1986; Zachos et al., 1989). Because the negative carbon isotope shift at the K-T boundary is not unique in the Hell Creek Formation, we have used other data, particularly biostratigraphy, to help distinguish the K-T negative shift from others of similar magnitude within the section (eg., −580 cm at Bobcat Butte, Fig. 5, KRJ Section 86148). Williams et al. (1988) and Gale et al. (1993) recommend this combined approach, when possible.

A second potential chemostratigraphic tie point between the Pyramid Butte and Bobcat Butte sections (Fig. 5) is the shift to lower, then higher, isotopic values recognized in the HCIII floral zone (Johnson, this volume; Johnson and Hickey, 1990).

In the Pyramid Butte section (KRJ Section 86161) this interval occurs between −400 cm and −900 cm, and between −950 cm and −1350 cm at Bobcat Butte (KRJ Section 86148). In this case, the pattern, absolute magnitude, and biostratigraphic position of these features suggest their correlation.

A third potentially useful chemostratigraphic feature is the interval of carbon isotope fluctuation identified between −1750 cm and −2270 cm in the Bobcat Butte section. This interval straddles the boundary between HCII and HCIII floral zones (Johnson, this volume; Johnson and Hickey, 1990) and is unusual in the Hell Creek Formation. We have not yet recognized this feature in any other terrestrial sections in North Dakota and Montana. However, zones of carbon isotopic instability have been reported in the uppermost Cretaceous marine record (Shackleton and Hall, 1984), suggesting that this is indeed a real feature of the carbon biogeochemical record.

A fourth chemostratigraphic feature that may provide correlative power is the trend in earliest Paleocene toward lower carbon isotopic values, which has been reported in a number of marine sections (D'Hondt et al., 1998; Shackleton and Hall, 1984). This trend is seen generally in both the Pyramid Butte (KRJ Section 86161) and Bobcat Butte (KRJ Section 86148) sections (Fig. 5). However, in both of these sections, poor exposure and extensive weathering did not permit adequate sampling to develop a detailed picture of these features in the sections sampled to date.

Together, these features of the carbon isotope curves provide several points of correlation for terrestrial sections in the region. Although much attention has been given to the K-T boundary negative carbon isotopic shift because of its importance for a variety of paleobiological questions, this feature cannot be interpreted alone. The K-T boundary shift is best interpreted in the overall context of secular variation in the carbon isotope signal in long and apparently complete sections such as that of Bobcat Butte and as part of a correlation scheme that includes many different lines of independent evidence.

COMPARISON OF K-T BOUNDARY MARKERS

A number of lines of evidence have been used to diagnose the K-T boundary. An iridium- and shocked mineral-bearing boundary clay is the preferred, but perhaps least common marker of the K-T boundary in terrestrial rocks. In the absence of an iridium anomaly, biostratigraphy is most commonly used. A carbon isotope chemostratigraphic approach can provide additional information that will aid the interpretation of terrestrial sections. We will briefly discuss the stratigraphic relationships between these different lines of evidence.

In the high-stratigraphic-resolution marine section at El Kef, Tunisia, the negative carbon isotope shift begins in the sample that also records the lowest horizon of elevated iridium abundance (Keller et al., 1995) and coincides with an increase in organic carbon deposition (Keller et al., 1995; Keller and Lindinger, 1989). At El Kef, the carbon isotope shift begins 2 cm below the level of peak iridium enrichments and reaches its lowest values ~50 cm above the maximum iridium enrichment. Unusually low $\delta^{13}C$ values are maintained for ~3 m above the iridium-defined boundary (Keller and Lindinger, 1989). Thus, although the marine carbon isotope shift begins coincident with iridium enrichment, the maximum shift does not co-occur with the iridium-enriched boundary clay. This observation is consistent with the mechanism proposed for the negative carbon isotope shift. If the marine shift begins with the cessation of marine productivity in the surface ocean because of postimpact environmental disturbance, it should co-occur with the iridium anomaly. However, we assert above that an additional input of $^{12}CO_2$ into the postimpact atmosphere is required to account for the observed magnitude of the negative carbon isotope shift. Whether because of terrestrial wildfires (Ivany and Salawitch, 1993; Wolbach et al., 1985) or other sources, a K-T related $^{12}CO_2$ input would have occurred after deposition of the impact-related fallout and thus the iridium anomaly. This lag explains why the maximum shift in $\delta^{13}C$ value is observed above the iridium anomaly at El Kef. Since our data from terrestrial sections record only atmospheric variation in carbon isotope composition, it is the postimpact maximum values, not the initial shift, that are recorded in terrestrial organic carbon. This mechanism predicts that the terrestrial K-T carbon isotope shift will always occur above an iridium-bearing boundary clay. We observe the predicted pattern at Pyramid Butte, where the minimum carbon isotope values for the shift occur 10 cm above the iridium-bearing boundary clay. The distance between the K-T boundary impact clay and minimum carbon isotope values will vary from section to section depending on depositional rate at a given site. However, it is important to note that we have correlated the lowest $\delta^{13}C$ values in Figures 5 and 6. While these lowest values are temporally coincident and thus, suitable for correlation tie points, they are not identical to the moment of impact, as recorded in the iridium-enriched boundary clay. At the Hell Creek Road locality in northeastern Montana, the terrestrial carbon isotope shift and recovery lasted significantly less than 250 k.y. Consequently, the maximum carbon isotope anomaly may still be a suitable marker for the K-T boundary for questions of biotic change when an iridium-bearing boundary clay is not preserved.

Pollen and spores have been extensively used to identify the K-T boundary in terrestrial sediments. Recognition of the palynological K-T boundary relies on the last appearance of some members of the *Wodehouseia spinata* assemblage (Nichols and Sweet, 1993). Tschudy and Tschudy (1986) stated that "The palynological Cretaceous/Tertiary boundary in the Western Interior of North America consistently occurs precisely at the horizon at which a boundary clay layer yielding anomalously high iridium and other elements is found" (p. 667). This coincidence, in fact, allowed Tschudy to successfully guide iridium sampling in the Raton Basin or New Mexico (Orth et al., 1981). However, the extinction of characteristic Cretaceous forms may occur over some stratigraphic distance. At Lance

Creek, the K-T boundary can be bracketed only within several meters using pollen and spores (Leffingwell, 1971), although a K-T iridium anomaly was subsequently identified (Bohor et al., 1987b). At Pyramid Butte, the last appearances of *Wodehouseia spinata* assemblage palynomorphs is distributed over approximately 50 cm bracketing the iridium-defined boundary (Johnson et al., 1989).

The mechanism underlying palynostratigraphic recognition of the K-T boundary is the extirpation or extinction of typical Cretaceous pollen and spore taxa due to impact-related disturbance (Nichols and Fleming, 1990; Smit and Van der Kaars, 1984; Tschudy et al., 1984). Consequently, we would expect the palynostratigraphic K-T boundary to be coincident with evidence of impact. Since the carbon isotope shift associated with the boundary should always occur stratigraphically above the impact clay because of a temporal lag between the impact and its effect on the atmosphere, the carbon isotope shift should similarly occur stratigraphically above the palynostratigraphic K-T boundary. In the four North Dakota sections studied to date, this occurs only at Pyramid Butte, where the palynological pick for the K-T boundary coincides with the iridium anomaly (Fig. 6, Table 2). In each of the other sections studied, the palynological pick for the K-T boundary (Nichols and Johnson, this volume) occurs from about 10 cm to nearly one meter above the minimum $\delta^{13}C$ value interpreted as the K-T carbon isotopic shift (Fig. 6, Table 2). Based on the few sections studied to date, we are reluctant to interpret this pattern; however, it should be noted that the last occurrence of several *Wodehouseia spinata* assemblage taxa also occurs stratigraphically above the carbon isotopic shift at Pyramid Butte (Johnson et al., 1989).

Beyond locating a stratigraphic horizon such as the K-T boundary or correlating among sections, another important application of carbon isotope stratigraphy in terrestrial sections may be the recognition of incomplete or paraconformable sections. Shackleton and Hall (1984) suggested this application for marine sediments. They proposed that for questions that required complete sections across the K-T boundary, the presence of the characteristic carbon isotope excursion offered the best check for section completeness (Shackleton and Hall, 1984). To illustrate this method, Figure 7 compares a complete hypothetical stratigraphic section with one in which the 2 m surrounding the boundary clay are missing. In both cases, a biostratigraphic transition would be recognized. However, comparison of the carbon isotope curves reveals a different pattern. In the incomplete section, the reversible carbon isotope shift associated with the biostratigraphic transition is missing. In its place, we see a non-reversible shift to larger $\delta^{13}C$ values. In a real-world example, we concluded above that the latest Cretaceous portion of the New Facet Boundary section may be incomplete, because its carbon isotope curve lacked features observed in other sections in the area. Because the relative completeness of fossil-bearing sections may be particularly important to the interpretation of the first and last occurrences that comprise our data

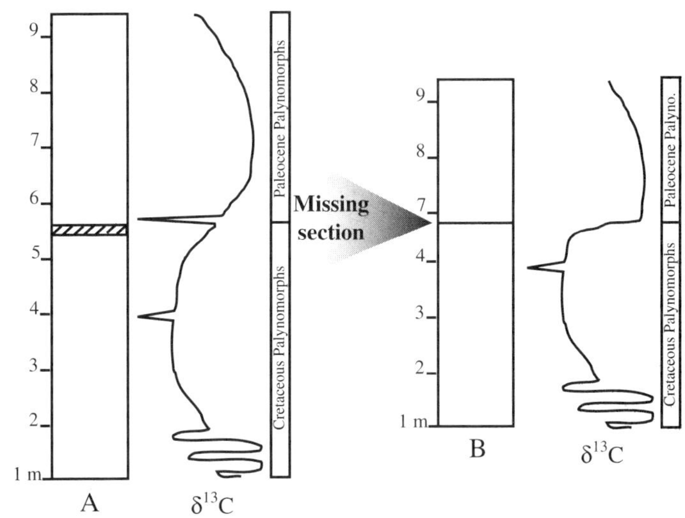

Figure 7. Carbon isotope stratigraphy can facilitate the recognition of incomplete sections. A: Hypothetical complete K-T boundary crossing section showing an iridium anomaly (hashed bar), carbon isotope curve and palynological distributions. B: The same section with the 2-m interval between 4 m and 6 m not represented by sediment. Note that vertical scales correspond in both sections. In both cases, a palynological boundary can be assigned in the correct stratigraphic position. However, differing shapes of the carbon isotope curves betray missing section.

for extinction and biotic change in the fossil record (Dingus, 1984), this application of carbon isotope stratigraphy may be the most significant.

SUMMARY AND CONCLUSIONS

When asking questions about the pattern and timing of extinction and biotic change across the K-T boundary, we are faced with several challenges: (1) diagnosing the chronostratigraphic boundary in the majority of terrestrial sections where an iridium anomaly is not preserved, (2) correlating in detail among fossil-bearing sections, and (3) recognizing incomplete and condensed sections. In the absence of impact-related iridium anomalies, lithostratigraphy, palynostratigraphy, and land mammals have been used extensively to place the K-T boundary. Although this approach works well, any biostratigraphic method will introduce circularity to questions of extinction and biotic change when the same fossils are used both to place extinction horizons and document the extinction itself. Furthermore, both palynomorphs and land mammals have been unable to provide detailed correlations within the Hell Creek and Fort Union Formations on a local scale. Biostratigraphy also commonly fails to recognize incomplete or condensed sections, a particular problem in terrestrial facies.

A $-1.5‰$ to $-2.5‰$ carbon isotope shift has been recognized above the K-T boundary in many marine sections. The marine $\delta^{13}C$ shift has been interpreted as a cessation of primary productivity in the surface ocean following the terminal-

Cretaceous impact, coupled with a change in the $\delta^{13}C$ value of atmospheric CO_2, which then equilibrated with the surface ocean. Because vascular land plants sample atmospheric CO_2 directly during photosynthesis, their tissue preserves the carbon isotope signature of the atmosphere under which they grew (Arens et al., 2000). Therefore, an atmospheric carbon isotope shift could be used to correlate in terrestrial, as well as marine, sediments.

Using this approach, we have developed a carbon isotope stratigraphy for four fossil-bearing sections from the Marmarth region of Slope County, North Dakota. We use the Pyramid Butte section, in which an iridium anomaly marks the chronostratigraphic K-T boundary, as the basis for regional correlation. At Pyramid Butte, we observed a $-2.8‰$ carbon isotope shift beginning in the 5 cm that immediately overlies the iridium-enriched horizon and peaking ~30 cm above the iridium anomaly. Carbon isotope shifts also marked the K-T boundary at Bobcat Butte and Terry's Fort Union Dinosaur localities and tentatively, at New Facet Boundary. All of these sections lack iridium-enriched boundary clays.

At Pyramid Butte, we also observed secular variation in the carbon isotope composition of sedimentary organic material in the upper Hell Creek Formation, suggesting cyclic variation in the $\delta^{13}C$ value of the latest Cretaceous atmosphere. Similar patterns, which we used to correlate between these sections, were observed at both Bobcat Butte and Terry's Fort Union Dinosaur localities. These data suggest that some environmental instability characterized the latest Cretaceous in this region. This conclusion agreed with interpretations based on leaf size and leaf margin analysis that documented changing climate through this interval (Johnson and Hickey, 1990). The long section at Bobcat Butte also allowed us to tie changes in the Hell Creek megaflora directly to an interval of fluctuating carbon isotope values. At Bobcat Butte, the transition from HCII to HCIII leaf floras was marked by ~6 m within which $\delta^{13}C$ values fluctuated over a magnitude greater than might be expected by plant vital effect or taphonomic variation alone. The carbon isotope data document rapid and substantial change in global carbon cycling. Although our data do not permit us to pinpoint a particular mechanism for the observed fluctuation, such environmental disruption may have driven the observed change in vegetation. Further investigation will show whether this pattern stands in other sections and whether similar carbon isotope instability marks the HCI-HCII floristic change as well. It seems clear at this point, however, that the latest Cretaceous was not a time of environmental stability in this portion of the Western Interior region of North America. This result should be factored into any hypothesis of extinction and biotic change.

ACKNOWLEDGMENTS

Field work and laboratory analyses were supported, in part, by the University of California Museum of Paleontology (Arens), U.C. Regent's Junior Faculty Fellowship (Arens), U.C. Berkeley Committee on Research (Arens), and the Petroleum Research Fund Type AC grant from the American Chemical Society (Arens). K.R. Johnson, T. Farnham, A. Thompson, and H. Pogrel facilitated and assisted with field work. W.M. Hagopian and A. Frumes performed invaluable service in the laboratory. This manuscript was significantly improved through conversations with, comments from, or formal reviews by D.J. Beerling, W.A. Clemens, W.E. Dean, J.H. Hartman, K.R. Johnson, J.P. Montoya, D.J. Nichols, S. Nostratinia, K. Padian, C.A.E. Strömberg, A. Thompson, and an anonymous reviewer. This is U.C.M.P. contribution 1701.

REFERENCES CITED

Adabi, M.H., 1997, Application of carbon isotope chemostratigraphy to the Renison dolomites, Tasmania: A Neoproterozoic age: Australian Journal of Earth Sciences, v. 44, p. 767–775.

Alvarez, L.W., 1983, Experimental evidence that an asteroid impact led to the extinction of many species: Proceedings of the National Academy of Sciences of the United States of America, v. 80, p. 627–642.

Alvarez, L.W., 1987, Mass extinctions caused by large bolide impacts: Physics Today, v. 40, p. 24–33.

Alvarez, L.W., Alvarez, W., Asaro, F., and Michel, H.V., 1980, Extraterrestrial cause for the Cretaceous–Tertiary extinction: Science, v. 208, p. 1095–1108.

Alvarez, W., Claeys, P., and Kieffer, S.W., 1995, Emplacement of Cretaceous–Tertiary boundary shocked quartz from Chicxulub Crater: Science, v. 269, p. 930–935.

Archibald, J.D., 1982, A study of mammalia and geology across the Cretaceous-Tertiary boundary in Garfield County, Montana: University of California Publications in Geological Sciences, v. 122, p. 1–286.

Archibald, J.D., 1987, Stepwise and non-catastrophic Late Cretaceous terrestrial extinctions in the Western Interior of North America: Testing observations in the context of an historical science: Mémoires de la Société Géologique de France, v. 150, p. 45–52.

Archibald, J.D., Butler, R.F., Lindsay, E.H., Clemens, W.A., and Dingus, L., 1982, Upper Cretaceous-Paleocene biostratigraphy and magnetostratigraphy, Hell Creek and Tullock formations, northeastern Montana: Geology, v. 10, p. 153–159.

Archibald, J.D., Gingerich, P.D., Lindsay, E.H., Clemens, W.A., Krause, D.W., and Rose, K.D., 1987, First North American land mammal ages of the Cenozoic Era, *in* Woodburne, M.O., ed., Cenozoic mammals of North America: Geochronology and biostratigraphy: Berkeley, University of California Press, p. 336.

Arens, N.C., and Jahren, A.H., 2000, Carbon isotopic excursion in atmospheric CO_2 at the Cretaceous-Tertiary boundary: Evidence from terrestrial sediments: Palaios, v. 15, p. 314–322.

Arens, N.C., Jahren, A.H., and Amundson, R., 2000, Can C3 plants faithfully record the isotopic composition of atmospheric carbon dioxide?: Paleobiology, v. 26, p. 137–164.

Arinobu, T., Ishiwatari, R., Kaiho, K., and Lamolda, M.A., 1999, Spike of pyrosynthetic polycyclic aromatic hydrocarbons associated with an abrupt decrease in $\delta^{13}C$ of a terrestrial biomarker at the Cretaceous-Tertiary boundary in Caravaca, Spain: Geology, v. 27, p. 723–726.

Arthur, M.A., Dean, W.E., and Schlanger, S.O., 1985, Variations in the global carbon cycle during the Cretaceous related to climate, volcanism, and changes in atmospheric CO_2: Geophysical Monograph, v. 32, p. 504–529.

Arthur, M.A., Scholle, P.A., and Hasson, P., 1979, Stable isotopes of oxygen and carbon in carbonates from sites 398 and 116 of the Deep Sea Drilling Project, *in* Laughter, F.H., and Fagerberg, E.M., eds., Initial reports of the

Deep Sea Drilling Project, Volume 47: Washington, D.C., U.S. Government Printing Office, pt. 2, p. 477–491.

Baadsgaard, H., Lerbekmo, J.F., and McDougall, I., 1988, A radiometric age for the Cretaceous-Tertiary boundary based on K-Ar, Rb-Sr, and U-Pb ages of bentonites from Alberta, Saskatchewan and Montana: Canadian Journal of Earth Sciences, v. 25, p. 1088–1097.

Bartley, J.K., Pope, M., Knoll, A.K., Semikhatov, M.A., and Petrov, P.Y., 1998, A Vendian-Cambrian boundary succession from the northwestern margin of the Siberian Platform: Stratigraphy, palaeontology, chemostratigraphy and correlation: Geological Magazine, v. 135, p. 473–494.

Beerling, D.J., and Woodward, F.I., 1995, Leaf stable carbon isotope composition records increased water-use efficiency of C_3 plants in response to atmospheric CO_2 enrichment: Functional Ecology, v. 9, p. 394–401.

Benner, R., Fogel, M.L., Sprague, E.K., and Hodson, R.E., 1987, Depletion of ^{13}C in lignin and its implication for stable carbon isotope studies: Nature, v. 329, p. 708–710.

Bohor, B.F., Foord, E.E., Modreski, P.J., and Triplehorn, D.M., 1984, Mineralogic evidence for an impact event at the Cretaceous-Tertiary boundary: Science, v. 224, p. 867–869.

Bohor, B.F., Modreski, P.J., and Foord, E.E., 1987a, Shocked quartz in the Cretaceous-Tertiary boundary clays: Evidence for a global distribution: Science, v. 236, p. 705–709.

Bohor, B.F., Triplehorn, D.M., Nichols, D.J., and Millard, H.T., Jr., 1987b, Dinosaurs, spherules, and the "magic layer": A new K-T boundary clay site in Wyoming: Geology, v. 15, p. 896–899.

Broadmeadow, M.S.J., Griffiths, H., Maxwell, C., and Borland, A.M., 1992, The carbon isotope ratio of plant organic material reflects temporal and spatial variations in CO_2 within tropical forest formations in Trinidad: Oecologia, v. 89, p. 435–441.

Brooks, J.R., Flanagan, L.B., Buchmann, N., and Ehleringer, J.R., 1997a, Carbon isotope composition of boreal plants: Functional grouping of life forms: Oecologia, v. 110, p. 301–311.

Brooks, J.R., Flanagan, L.B., Varney, G.T., and Ehleringer, J.R., 1997b, Vertical gradients in photosynthetic gas exchange characteristics and refixation of respired CO_2 within boreal forest canopies: Tree Physiology, v. 17, p. 1–12.

Brown, R., 1938, The Cretaceous-Eocene boundary in Montana and South Dakota: Washington Academy of Science Journal, v. 28, p. 421–422.

Brown, R., 1952, Tertiary strata in eastern Montana and western North and South Dakota: Billings Geological Society Guidebook, v. 3, p. 89–92.

Butler, R.F., and Lindsay, E.H., 1985, Mineralogy of magnetic minerals and revised magnetic polarity stratigraphy of continental sediments, San-Juan Basin, New-Mexico: Journal of Geology, v. 93, p. 535–554.

Caldeira, K.G., Rampino, M.R., Volk, T., and Zachos, J.C., 1990, Biogeochemical modeling at mass extinction boundaries: Atmospheric carbon dioxide and ocean alkalinity at the K-T boundary, *in* Kauffman, E.G., and Walliser, O.H., eds., Extinction events in Earth history: Berlin, Springer-Verlag, p. 333–345.

Ciais, P., Tans, P.P., Trolier, M., White, J.W.C., and Francey, R.J., 1995, A large Northern Hemisphere terrestrial CO_2 sink indicated by the $^{13}C/^{12}C$ ratios of atmospheric CO_2: Science, v. 269, p. 1098–1102.

Crepet, W.L., and Feldman, G.D., 1991, The earliest remains of grasses in the fossil record: American Journal of Botany, v. 78, p. 1010–1014.

D'Hondt, S., Donaghay, P., Zachos, J.C., Luttenberg, D., and Lindinger, M., 1998, Organic carbon fluxes and ecological recovery from the Cretaceous-Tertiary mass extinction: Science, v. 282, p. 276–279.

De Leeuw, J.W., Frewin, L., Van Bergen, P.F., Sinninghe Damasté, J.S., and Collinson, M.E., 1995, Organic carbon as a palaeoenvironmental indicator in the marine realm: Marine Palaeoenvironmental Analysis from Fossils, v. 83, p. 43–71.

Derry, L.A., Kaufman, A.J., and Jocobsen, S.B., 1992, Sedimentary cycling and environmental change in the Late Proterozoic: Evidence from stable and radiogenic isotopes: Geochimica et Cosmochimica Acta, v. 56, p. 1317–1329.

Dickens, G.R., O'Neil, J.R., Rea, D.K., and Owen, R.M., 1995, Dissociation of oceanic methane hydrate as a cause of the carbon isotope excursion at the end of the Paleocene: Paleoceanography, v. 10, p. 965–971.

Dingus, L.W., 1984, Effects of stratigraphic completeness on interpretations of extinction rates across the Cretaceous-Tertiary boundary: Paleobiology, v. 10, p. 420–438.

Donovan, L.A., and Ehleringer, J.R., 1991, Ecophysiological differences among juvenile and reproductive plants of several woody species: Oecologia, v. 86, p. 594–597.

Dupouey, J.-L., Leavitt, S., Choisnel, E., and Jourdain, S., 1993, Modeling carbon isotope fractionation in tree rings based on effective evapotranspiration and soil water status: Plant, Cell and Environment, v. 16, p. 939–947.

Ehleringer, J.R., Field, C.B., Lin, Z.-F., and Kuo, C.-Y., 1986, Leaf carbon isotope ratio and mineral composition in subtropical plants along an irradiance cline: Oecologia, v. 70, p. 520–526.

Ehleringer, J.R., Phillips, S.L., and Comstock, J.P., 1992, Seasonal variation in the carbon isotopic composition of desert plants: Functional Ecology, v. 6, p. 396–404.

Farquhar, G.D., Ehleringer, J.R., and Hubick, K.T., 1989, Carbon isotope discrimination and photosynthesis: Annual Review of Plant Physiology and Plant Molecular Biology, v. 40, p. 503–538.

Farquhar, G.D., and Richards, R.A., 1984, Isotopic composition of plant carbon correlates with water-use efficiency of wheat (*Triticum aestivum*) genotypes: Australian Journal of Plant Physiology, v. 11, p. 539–552.

Fastovsky, D.E., 1987, Paleoenvironments of vertebrate-bearing strata at the Cretaceous-Paleogene transition in eastern Montana and western North Dakota: Palaios, v. 2, p. 282–295.

Gale, A.S., Kenkyns, H.C., Kennedy, W.J., and Corfield, R.M., 1993, Chemostratigraphy versus biostratigraphy: Data from around the Cenomanian-Turonian boundary: Journal of the Geological Society of London, v. 150, p. 29–32.

Harland, W.B., Armstrong, R.L., Cox, A.V., Craig, L.E., Smith, A.G., and Smith, D.G., 1989, A geologic time scale: Cambridge, Cambridge University Press, 263 p.

Holmes, C.W., and Bohor, B.F., 1994, Stable isotope (C, S, N) distributions in coals spanning the Cretaceous/Tertiary boundary in the Raton Basin, Colorado and New Mexico: U.S. Geological Survey Circular, v. 1107, p. 1–141.

Hotton, C.L., 1988, Palynology of the Cretaceous-Tertiary boundary in Central Montana, U.S.A., and its implications for extraterrestrial impact [Ph.D. thesis]: Davis, University of California, 610 p.

Hsü, K.J., He, Q., McKenzie, J.A., Weissert, H., Perch-Nielsen, K., Oberhansli, H., Kelts, K., LaBrecque, J., Tauxe, L., Krahenbuhl, U., Percival, S.F., Wright, R., Karpoff, A.M., Petersen, N., Tucker, P., Poore, R.Z., Gombos, A.M., Pisciotto, K.A., Carman, M.F., and Schreiber, E., 1982, Mass mortality and its environmental and evolutionary consequences: Science, v. 216, p. 249–256.

Ismail, A.M., Hall, A.E., and Bray, E.A., 1994, Drought and pot size effects on transpiration efficiency and carbon isotope discrimination of cowpea accessions and hybrids: Australian Journal of Plant Physiology, v. 21, p. 23–35.

Ivany, L.C., and Salawitch, R.J., 1993, Carbon isotopic evidence for biomass burning at the K-T boundary: Geology, v. 21, p. 487–490.

Johnson, K.R., and Hickey, L.J., 1990, Megafloral change across the Cretaceous/Tertiary boundary in the northern Great Plains and Rocky Mountains, U.S.A., *in* Sharpton, V.L., and Ward, P.D., eds., 1990, Global catastrophes in Earth history: An interdisciplinary conference on impacts, volcanism, and mass mortality: Geological Society of America Special Paper 247, p. 433–444.

Johnson, K.R., Nichols, D.J., Attrep, M., Jr., and Orth, C.J., 1989, High-resolution leaf-fossil record spanning the Cretaceous/Tertiary boundary: Nature, v. 340, p. 708–711.

Johnson, R.C., and Tieszen, L.L., 1993, Carbon isotope discrimination, water relations, and gas exchange in temperate grass species and accessions, *in* Ehleringer, J.R., Hall, A.E., and Farquhar, G.D., eds., Stable isotopes and plant carbon-water relations: San Diego, California, Academic Press, Incorporated, p. 281–296.

Kasting, J.F., Richardson, S.M., Pollack, J.B., and Toon, O.B., 1986, A hybrid model of the CO_2 geochemical cycles and its application to large impact events: American Journal of Science, v. 286, p. 361–389.

Kaufman, A.J., and Knoll, A.H., 1995, Neoproterozoic variations in the C-isotopic composition of seawater: Stratigraphic and biogeochemical implications: Precambrian Research, v. 73, p. 27–49.

Ke, Y., and Lin, G., 1997, Isotopic changes of atmospheric carbon dioxide in two soil-vegetation systems under controlled environments [abs.]: Bulletin of the Ecological Society of America, v. 78, p. 267.

Keller, G., Li, L., and MacLeod, N., 1995, The Cretaceous/Tertiary boundary stratotype section at El Kef, Tunisia: How catastrophic was the mass extinction?: Palaeogeography, Palaeoclimatology, Palaeoecology, v. 119, p. 221–254.

Keller, G., and Lindinger, M., 1989, Stable isotope, TOC and $CaCO_3$ record across the Cretaceous/Tertiary boundary at El Kef, Tunisia: Palaeogeography, Palaeoclimatology, Palaeoecology, v. 73, p. 243–266.

Koch, P.L., Zachos, J.C., and Gingerich, P.D., 1992, Correlation between isotope records in marine and continental carbon reservoirs near the Palaeocene/Eocene boundary: Nature, v. 358, p. 319–322.

Kump, L., 1991, Interpreting carbon-isotope excursions: Strangelove oceans: Geology, v. 19, p. 299–302.

Leffingwell, H.A., 1971, Palynology of the Lance (Late Cretaceous) and Fort Union (Paleocene) Formations of the type Lance area, Wyoming, *in* Kosanke, R.M., and Cross, A.T., eds., Symposium on palynology of the Late Cretaceous and early Tertiary: Geological Society of America Special Paper 127, p. 1–64.

Lerbekmo, J.F., 1985, Magnetostratigraphic and biostratigraphic correlations of Maastrichtian to early Paleocene strata between south-central Alberta and southwestern Saskatchewan: Bulletin of Canadian Petroleum Geology, v. 33, p. 213–226.

Lerbekmo, J.F., Sweet, A.R., and Duke, M.J.M., 1996, A normal polarity subchron that embraces the K-T boundary: A measure of sedimentary continuity across the boundary and synchroneity of boundary events, *in* Ryder, G., Fastovsky, D., and Gartner, S., eds., The Cretaceous-Tertiary event and other catastrophes in Earth history: Geological Society of America Special Paper 307, p. 465–476.

Lockheart, M.J., Van Bergen, P.F., and Evershed, R.P., 1997, Variations in the stable carbon isotope compositions of individual lipids from the leaves of modern angiosperms: Implications for the study of higher land plant-derived sedimentary organic matter: Organic Geochemistry, v. 26, p. 137–153.

Lofgren, D.L., 1995, The Bug Creek problem and the Cretaceous-Tertiary transition at McGuire Creek, Montana: University of California Publications in Geological Sciences, v. 140, p. 1–185.

Lofgren, D.L., Hotton, C.L., and Runkle, A., 1990, Reworking of Cretaceous dinosaurs into Paleocene channel deposits, upper Hell Creek Formation, Montana: Geology, v. 18, p. 874–877.

Lomax, B., Beerling, D., and Upchurch, G., 1998, Acid rain and Strangelove oceans over the Cretaceous Tertiary boundary [abs.]: Palaeontological Newsletter, v. 39, p. 15.

Marino, B.D., and McElroy, M.B., 1991, Isotopic composition of atmospheric CO_2 inferred from carbon in C_4 plant cellulose: Nature, v. 349, p. 127–131.

Marino, B.D., McElroy, M.B., Salawitch, R.J., and Spaulding, W.G., 1992, Glacial-to-interglacial variations in the carbon isotopic composition of atmospheric CO_2: Nature, v. 357, p. 461–466.

Marshall, J.D., and Zhang, J., 1994, Carbon isotope discrimination and water-use efficiency in native plants of the north-central Rockies: Ecology, v. 75, p. 1887–1895.

Medina, E., Montes, G., Cuevas, E., and Rokzandic, Z., 1986, Profiles of CO_2 concentration and $\delta^{13}C$ values in tropical rainforests of the upper Rio Negro Basin, Venezuela: Journal of Tropical Ecology, v. 2, p. 207–217.

Minagawa, M., Winter, D.A., and Kaplan, I.R., 1984, Comparison of Kjelkahl and combustion tube methods for measurements of nitrogen isotope ratios in organic-matter: Analytical Chemistry, v. 56, p. 1859–1861.

Mooney, H.A., Bullock, S.H., and Ehleringer, J.R., 1989, Carbon isotope ratios of plants of a tropical dry forest in Mexico: Functional Ecology, v. 3, p. 137–142.

Nakamura, K., Kirmura, G., and Winsnes, T.S., 1987, Brachipod zonation and age of the Permian Kapp Starostin Formation (Central Spitsbergen): Polar Research, v. 5, p. 207–220.

Nichols, D.J., 1990, Geologic and biostratigraphic framework of the non-marine Cretaceous-Tertiary boundary interval in western North America: Review of Palaeobotany and Palynology, v. 65, p. 75–84.

Nichols, D.J., and Fleming, R.F., 1990, Plant microfossil record of the terminal Cretaceous event in the Western United States and Canada, *in* Sharpton, V.L., and Ward, P.D., eds., 1990, Global catastrophes in Earth history: An interdisciplinary conference on impacts, volcanism, and mass mortality: Geological Society of America Special Paper 247, p. 445–455.

Nichols, D.J., Jarzen, D.M., Orth, C.J., and Oliver, P.Q., 1986, Palynological and iridium anomalies at Cretaceous-Tertiary boundary, south-central Saskatchewan: Science, v. 231, p. 714–717.

Nichols, D.J., and Sweet, A.R., 1993, Biostratigraphy of Upper Cretaceous non-marine palynofloras in a north-south transect of the Western Interior Basin, *in* Caldwell, W.G.E., and Kauffman, E.G., eds., Evolution of the Western Interior Basin: Toronto, Geological Association of Canada Special Paper 39, p. 539–584.

O'Leary, M.H., 1988, Carbon isotopes in photosynthesis: BioScience, v. 38, p. 328–336.

Orth, C.J., Gilmore, J.S., Knight, J.D., Pillmore, C.L., Tschudy, R.H., and Fassett, J.E., 1981, An iridium anomaly at the palynological Cretaceous/Tertiary boundary in northern New Mexico: Science, v. 214, p. 1341–1343.

Perch-Nielsen, K., McKenzie, J., and He, Q., 1982, Biostratigraphy and isotope stratigraphy and the "catastrophic" extinction of calcareous nannoplankton at the Cretaceous/Tertiary boundary, *in* Silver, L.T., and Schultz, P.H., eds., 1982, Geological implications of impacts of large asteroids and comets on the Earth: Geological Society of America Special Paper 190, p. 353–371.

Robin, E., Boclet, D., Bonté, P., Froget, L., Jéhanno, C., and Rocchia, R., 1991, The stratigraphic distribution of Ni-rich spinels in Cretaceous-Tertiary boundary rocks at El Kef (Tunisia), Caravaca (Spain), and Hole 761C (Leg 122): Earth and Planetary Science Letters, v. 107, p. 715–721.

Schimmelmann, A., and DeNiro, M.J., 1984, Elemental and stable isotope variations of organic matter from a terrestrial sequence containing the Cretaceous/Tertiary boundary at York Canyon, New Mexico: Earth and Planetary Science Letters, v. 68, p. 392–398.

Schlesinger, W.H., 1997, Biogeochemistry: An analysis of global change: San Diego, California, Academic Press, Incorporated, 588 p.

Shackleton, N.J., 1985, Oceanic carbon isotope constraints on oxygen and carbon dioxide in the Cenozoic atmosphere: Geophysical Monographs, v. 32, p. 412–417.

Shackleton, N.J., and Hall, M.A., 1984, Carbon isotope data from Leg 74 sediments, *in* Blakeslee, J.H., and Lee, M., eds., Initial reports of the Deep Sea Drilling Project, Volume 74: College Station, Texas, USA, Texas A&M University Ocena Drilling Program, p. 613–619.

Siegenthaler, U., and Sarmiento, J.L., 1993, Atmospheric carbon dioxide and the ocean: Nature, v. 365, p. 119–125.

Smedley, M.P., Dawson, T.E., Comstock, J.P., Donovan, L.A., Sherrill, D.E., Cook, C.S., and Ehleringer, J.R., 1991, Seasonal carbon isotope discrimination in a grassland community: Oecologia, v. 85, p. 314–320.

Smit, J., 1982, Extinction and evolution of planktonic foraminifera after a major impact at the Cretaceous/Tertiary boundary, *in* Silver, L.T., and Schultz, P.H., eds., Geological implications of impacts of large asteroids and comets on the Earth: Geological Society of America Special Paper 190, p. 329–352

Smit, J., and Van der Kaars, S., 1984, Terminal Cretaceous extinctions in the Hell Creek area, Montana: Compatible with catastrophic extinction: Science, v. 223, p. 1177–1179.

Stemmerik, L., 1988, Brachiopod zonation and age of the Permian Kapp Starostin Formation (Central Spitsbergen): Discussion: Polar Research, v. 6, p. 179–180.

Stott, L.D., and Kennett, J.P., 1989, New constraints on early Tertiary palaeoproductivity from carbon isotopes in foraminifera: Nature, v. 342, p. 526–529.

Stott, L.D., and Kennett, J.P., 1990, Scientific results, The paleoceanographic and paleoclimatic signature of the Cretaceous/Paleogene boundary in the Antarctic: Stable isotopic results from ODP Leg 113: College Station, Texas, Ocean Drilling Program, p. 829–848.

Swisher, C.C., Dingus, L., and Butler, R.F., 1993, $^{40}Ar/^{39}Ar$ dating and magnetostratigraphic correlation of the terrestrial Cretaceous—Paleogene boundary and Puercan Mammal Age, Hell Creek-Tullock formations, eastern Montana: Canadian Journal of Earth Sciences, v. 30, p. 1981–1996.

Szaniawski, H., and Malkowski, K., 1979, Conodonts from the Kapp Starostin Formation (Permian) of Spitsbergen: Acta Paleontologica Polonica, v. 24, p. 231–264.

Tappan, H., 1968, Primary production, isotopes, extinctions and the atmosphere: Palaeogeography, Palaeoclimatology, Palaeoecology, v. 4, p. 187–210.

Taylor, B.E., 1986, Magmatic volatiles: Isotopic variation of C, H, and S: Reviews in Mineralogy, v. 16, p. 185–225.

Tieszen, L.L., 1991, Natural variations in the carbon isotope values of plants: Implications for archaeology, ecology, and paleoecology: Journal of Archaeological Science, v. 18, p. 227–248.

Toft, N.L., Anderson, J.E., and Nowak, R.S., 1989, Water use efficiency and carbon isotope composition of plants in a cold desert environment: Oecologia, v. 80, p. 11–18.

Tschudy, R.H., Pillmore, C.L., Orth, C.J., Gilmore, C.J., and Knight, J.D., 1984, Disruption of the terrestrial plant ecosystem at the Cretaceous-Tertiary boundary, Western Interior: Science, v. 225, p. 1030–1032.

Tschudy, R.H., and Tschudy, B.D., 1986, Extinction and survival of plant life following the Cretaceous-Tertiary boundary event, Western Interior, North America: Geology, v. 14, p. 667–670.

Upchurch, G.R., Marino, B.D., Mone, W.E., and McElroy, M.B., 1997, Carbon isotope ratios in extant and fossil plant cuticle [abs.]: American Journal of Botany, v. 84, no. 6 supplement, p. 143–144.

Van de Water, P.K., Leavitt, S.W., and Betancourt, J.L., 1994, Trends in stomatal density and $^{13}C/^{12}C$ ratios of *Pinus flexilis* needles during last glacial-interglacial cycle: Science, v. 264, p. 239–243.

Van der Merwe, N.J., and Medina, E., 1989, Photosynthesis and $^{13}C/^{12}C$ ratios in Amazonian rain forests: Geochimica et Cosmochimica Acta, v. 53, p. 1091–1094.

Wedin, D.A., Tieszen, L.L., Dewey, B., and Pastor, J., 1995, Carbon isotope dynamics during grass decomposition and soil organic matter formation: Ecology, v. 76, p. 1383–1392.

Wignall, P.B., Morante, R., and Newton, R., 1998, The Permo-Triassic transition in Spitsbergen: $\delta^{13}C_{org}$ chemostratigraphy, Fe and S geochemistry, facies, fauna and trace fossils: Geological Magazine, v. 135, p. 47–62.

Williams, D.F., Lerche, I., and Full, W.E., 1988, Isotope chronostratigraphy: Theory and methods: San Diego, California, Academic Press, Incorporated, 345 p.

Wolbach, W.S., Lewis, R.S., and Anders, E., 1985, Cretaceous extinctions: Evidence for wildfires and search for meteoric material: Science, v. 230, p. 167–170.

Woodburne, M.O., 1977, Definition and characterization in mammalian chronostratigraphy: Journal of Paleontology, v. 51, p. 220–234.

Woodburne, M.O., and Swisher, C.C., 1995, Land mammal high-resolution geochronology, intercontinental overland dispersals, sea level, climate, and vicariance, *in* Berggren, W.A., Kent, D.V., Aubry, M.-P., and Hardenbol, J., eds., Geochronology, time scales and global stratigraphic correlation: Tulsa, Oklahoma, SEPM (Society for Sedimentary Geology) Special Publication 54, p. 335–364.

Zachos, J.C., and Arthur, M.A., 1986, Paleoceanography of the Cretaceous/Tertiary boundary event: Inferences from stable isotopic and other data: Paleoceanography, v. 1, p. 5–26.

Zachos, J.C., Arthur, M.A., and Dean, W.E., 1989, Geochemical evidence for suppression of pelagic marine productivity at the Cretaceous/Tertiary boundary: Nature, v. 337, p. 61–64.

Zachos, J.C., Aubry, M.-P., Berggren, W.A., Ehrendorfer, T., Heider, F., and Lohmann, K.C., 1992, Scientific Results, Chemobiostratigraphy of the Cretaceous/Paleocene boundary at Site 750, southern Kerguelen Plateau: College Station, Texas, Ocean Drilling Program, v. 120, p. 961–977.

Zimmerman, J.K., and Ehleringer, J.R., 1990, Carbon isotope ratios are correlated with irradiance levels in the Panamanian orchid Catasetum viridiflavum: Oecologia, v. 83, p. 247–249.

Manuscript Accepted by the Society September 24, 2001

Printed in the U.S.A.

Geological Society of America
Special Paper 361
2002

Palynology and microstratigraphy of Cretaceous-Tertiary boundary sections in southwestern North Dakota

Douglas J. Nichols*
U.S. Geological Survey, MS 939, Box 25046, Denver Federal Center, Denver, Colorado 80225, USA
Kirk R. Johnson
Department of Earth Sciences, Denver Museum of Nature & Science, Denver, Colorado 80205, USA

ABSTRACT

Palynology is used to bracket or pinpoint the Cretaceous-Tertiary (K-T) boundary in 17 measured sections near the contact of the Hell Creek Formation and the Ludlow Member of the Fort Union Formation in southwestern North Dakota. Palynostratigraphy is the most reliable method for locating the K-T extinction horizon—which defines the K-T boundary—in nonmarine rocks. The palynological database includes 110 taxa for which relative abundance or presence or absence data were recorded in more than 350 samples based on surveys of more than 700 000 specimens. These data from laterally extensive outcrops in the badlands along the Little Missouri River provide a temporal framework for concurrent studies in the area on megafossil paleobotany, vertebrate paleontology, lithostratigraphy, magnetostratigraphy, and chemostratigraphy.

Palynology demonstrates extinction of 30% or more of the Maastrichtian palynoflora, including characteristic Maastrichtian taxa ("K taxa"), at the K-T boundary. Most of the palynomorph taxa discussed probably represent higher-level plant taxa (botanical genera or families). The K-T boundary is shown to be coincident with the Hell Creek–Fort Union formational contact at only two localities; it is as much as 2.7 m above the base of the Fort Union Formation at others. Thus, a distinctive interval of Fort Union strata of Cretaceous age is recognized that is characterized by occurrences of numerous K taxa, usually in low percentage abundance, up to the K-T boundary. This interval documents a regional paleoenvironmental change that is independent of the extinction event and that is important to understanding the K-T transition throughout western North America.

INTRODUCTION

Purpose of this study

The Cretaceous-Tertiary (K-T) boundary is extensively exposed near the contact of the Hell Creek Formation and the Ludlow Member of the Fort Union Formation in the badlands of southwestern North Dakota. Palynology has been used to bracket the K-T boundary in 17 measured sections discussed in this report. In addition, seven other localities close to the boundary are discussed that provide additional data on the nature of the K-T transition in this area. Only two of these sections have been described in publication previously. The purpose of this report is to document the position of the K-T boundary in southwestern North Dakota and to elucidate the pattern of palyno-

*E-mail: nichols@usgs.gov

Nichols, D.J., and Johnson, K.R., 2002, Palynology and microstratigraphy of Cretaceous-Tertiary boundary sections in southwestern North Dakota, *in* Hartman, J.H., Johnson, K.R., and Nichols, D.J., eds., The Hell Creek Formation and the Cretaceous-Tertiary boundary in the northern Great Plains: An integrated continental record of the end of the Cretaceous: Boulder, Colorado, Geological Society of America Special Paper 361, p. 95–143.

floral extinction across the boundary in this area. The K-T boundary is a precise timeline, a rarity in the stratigraphic record. The well-documented association of the palynologically defined K-T boundary with physical evidence of the impact of an extraterrestrial body at the end of Cretaceous time (Alvarez et al., 1980; Bohor et al., 1984; Izett, 1990) in our study area and elsewhere in the Western Interior region of North America demonstrates that palynology is the most practical means of recognizing the boundary in nonmarine strata. Thus, the utility of this report is that it provides a biostratigraphic framework based on numerous sections that permits other research in the same study area to be directly linked to the K-T boundary time line (see Arens and Jahren, this volume; Hicks et al., this volume; Johnson, this volume; Labandeira et al., this volume; and Pearson et al., this volume). This report makes reference to the description of the palynoflora of the Hell Creek Formation by Nichols (this volume) and summarizes new data on the palynoflora of the Ludlow Member of the Fort Union Formation developed during 15 years of study in southwestern North Dakota by Nichols and Johnson.

Study area

The study area extends in the north-south outcrop belt along the Little Missouri River from T. 135 N. in Slope County to T. 129 N. in Bowman County, southwestern North Dakota (Fig. 1), and includes the most extensive outcrops of the Hell Creek and Fort Union Formations in North Dakota. The badlands afford excellent, often continuous outcrops of these formations and laterally extensive outcrops of the K-T boundary interval (Fig. 2). One of us (Johnson) measured numerous reference sections from 1986 to 1999, within the Hell Creek Formation, across the Hell Creek–Fort Union formational contact, and into the Fort Union Formation. We collected hundreds of samples from these sections, from the base of the Hell Creek Formation to as high as 68 m above the base of the Ludlow Member, but concentrated on shorter intervals spanning the formation contact and the K-T boundary.

Previous studies

North Dakota. The K-T boundary in southwestern North Dakota was described previously by Johnson et al. (1989). That report documented in detail the K-T boundary in the section at Pyramid Butte, which is in the study area of the present report. The now well-known anomalous concentration of iridium (hereinafter referred to as the Ir anomaly) and shock-metamorphosed minerals (hereinafter referred to as shocked quartz) are present at the Pyramid Butte locality at the level of the palynologically defined K-T boundary. Shocked quartz of impact origin at Mud Buttes, a second palynologically defined K-T boundary section in our study area, was reported by Nichols et al. (2000). The plant megafossil record across the K-T boundary in southwestern North Dakota was discussed by Johnson et al. (1989); see

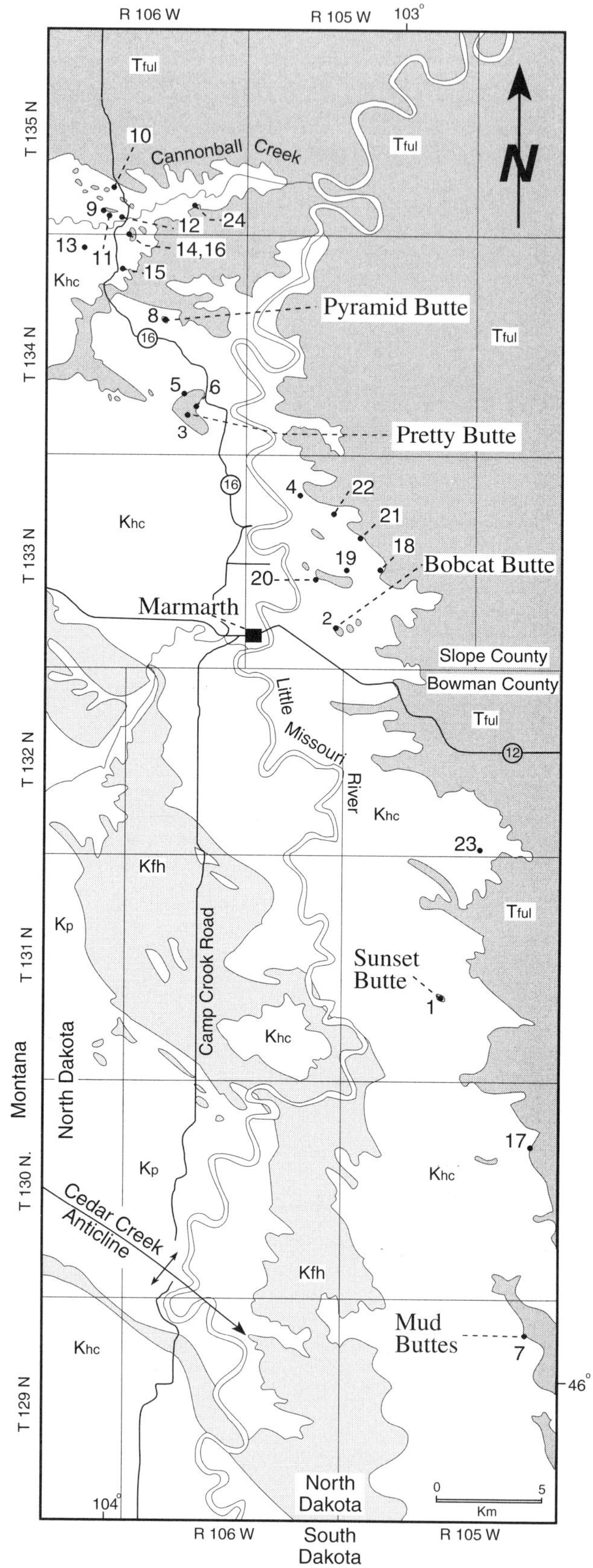

Figure 1. Geologic map of the study area in southwestern North Dakota showing major features and locations of measured sections. Geology modified from Carlson (1979, 1983). Geologic units: Kp—Pierre Shale, Kfh—Fox Hills Sandstone, Khc—Hell Creek Formation, Tful—Ludlow Member of the Fort Union Formation. Sections: 1—Sunset Butte, 2—Bobcat Butte, 3—Pretty Butte, 4—River Section, 5—Big Boundary, 6—Pretty Butte North, 7—Mud Buttes, 8—Pyramid Butte, 9—Terry's Fort Union Dinosaur, 10—Mikey's Delite, 11—Dean's High Dinosaur, 12—Cannonball Creek, 13—Scorpion Stung Kristian, 14—Vertical Doug, 15—Badland Draw, 16—New Facet Boundary, 17—Torosaurus Section, 18—Barbecue, 19—Battleship, 20—Dean's NM1, 21—NDAS Section, 22—No Leaf Minors, 23—Swanke's Nose, 24—Tonstein To Die For.

also Johnson and Hickey (1990), Johnson (1992), and Johnson (this volume).

Lerbekmo and Coulter (1984) conducted a magnetostratigraphic and palynostratigraphic study of a section near Huff, North Dakota, and compared it with a K-T boundary section in Alberta, Canada. They found that the K-T boundary in this section is within geomagnetic polarity subchron C29r. Palynological analyses reported in that study were performed by Chaitanya Singh of the Alberta Geological Survey.

Murphy et al. (1995) discussed 12 localities in the Missouri River valley in south-central North Dakota at which the boundary was bracketed palynologically. That report also presented extensive data on the lithostratigraphy of the Hell Creek and Fort Union Formations at 32 measured sections, and illustrated and discussed palynomorphs and vertebrate megafossils from that study area.

Nichols and Fleming (1990) discussed the palynology of the boundary interval at the Pyramid Butte locality and compared it with that at three other K-T boundary localities in western North America. Nichols (1990) briefly discussed the geologic and biostratigraphic framework of the K-T boundary interval in southwestern North Dakota, comparing it with that at other North American localities.

Aside from these reports that specifically concern the K-T boundary and the report on the palynoflora of the Hell Creek Formation by Nichols (this volume), there are no previously published, detailed palynological studies in North Dakota. We have published several abstracts about the palynology of the K-T boundary in the study area, however: Johnson et al. (1990), Nichols (1991), Nichols (1993), Johnson and Nichols (1994), and Nichols (1996a).

Adjacent areas. The earliest study comparing the palynology of the Hell Creek and Fort Union Formations is that of Stanley (1965) in northwestern South Dakota. Stanley's localities were geographically and stratigraphically separated, however, so his cannot be considered a study of the K-T boundary. Nonetheless, the strong floristic differences between the Hell Creek and Fort Union palynofloras are evident from his work.

Eastern Montana has been the scene of numerous studies of the palynology of the K-T boundary interval. Hall and Norton (1967), Norton and Hall (1967, 1969), and Oltz (1969, 1971) conducted a series of investigations in the Hell Creek area. Of these, the most detailed analysis is that of Norton and Hall (1969). Tschudy (1970) discussed characteristic species of the Hell Creek Formation and Tullock Member of the Fort Union Formation in a study of the K-T interval in the Hell Creek, Seven Blackfoot Creek, Brockton, and Glendive areas of eastern Montana. The extraterrestrial impact theory accounting for the K-T boundary event was not proposed until later (Alvarez et al., 1980), so none of these early studies used it to account for palynological changes observed across the boundary. Smit and Van der Kaars (1984) and Bohor et al. (1984) discussed effects of the impact on the Hell Creek area. In a major, unpublished dissertation, Hotton (1988) conducted a detailed study of the K-T boundary in six sections in the Hell Creek area (see also Hotton, this volume). Hotton's data and interpretations support the impact theory.

The palynology of the K-T boundary in southern Saskatchewan, just north of the Hell Creek area of Montana, was first described by Nichols et al. (1986). There the boundary is at the contact between the Frenchman and Ravenscrag Formations, the stratigraphic equivalents, respectively, of the Hell Creek and Fort Union Formations. Several other K-T boundary sections in this and adjacent areas in southern Canada were the subject of detailed palynological investigations by Sweet and Braman (1992) and Sweet et al. (1999).

Other areas. Several studies of the K-T boundary have been conducted in eastern Wyoming. The first was that of Leffingwell (1970) in the Lance Creek area, on the eastern flank of the Powder River Basin. The study by Leffingwell was a landmark investigation that included both taxonomy and biostratigraphy. The taxa he described and illustrated are present also in Montana and North Dakota; thus, his work provides a basis for the present study. The K-T boundary was later precisely located near Leffingwell's locality by Bohor et al. (1987), who found the Ir anomaly and shocked quartz associated with the palynological changes. Wolfe (1991) and Nichols et al. (1992) each described the K-T boundary at two separate localities on the western flank of the Powder River Basin.

Palynological studies of the K-T boundary originated in the Raton Basin of northern New Mexico and southern Colorado with the seminal research of Orth et al. (1981). Other references on the palynology of the boundary in this area are Pillmore et al. (1984), Tschudy et al. (1984), Tschudy and Tschudy (1986), Fleming (1990), and Izett (1990).

In addition to the localities in Saskatchewan previously mentioned, the palynology of the K-T boundary has been studied at several localities in Alberta and the Northwest Territories, Canada. Key references to this research are Jerzykiewicz and Sweet (1986), Lerbekmo et al. (1987), Sweet et al. (1990), Sweet and Braman (1992), Sweet (1994), Sweet et al. (1999), and Sweet and Braman (2001).

Field and laboratory procedures

Samples discussed in this report were collected from stratigraphic sections measured and described at the centimeter scale.

Figure 2. Outcrops of the Hell Creek Formation (near distance) and Ludlow Member of the Fort Union Formation (far distance) along the Little Missouri River in southwestern North Dakota.

Fresh surfaces were exposed by digging well into the outcrop. To avoid contamination of samples in the field, intervals to be sampled were brushed clean, collecting tools were carefully cleaned between samples, and samples were stored in heavy-weight plastic bags labeled with field numbers. Most intervals sampled varied in thickness from 2.5 to 5.0 cm. Stratigraphic positions of all samples were determined relative to the Hell Creek–Fort Union formational contact and are cited in this report as "–(number) cm" to indicate the number of centimeters below the formational contact or as "+(number) cm" to indicate the number of centimeters above the contact.

Field numbers applied to samples were eventually replaced by U.S. Geological Survey locality numbers. Unique four-digit locality numbers preceded by the letter "D" and followed by a hyphen and one or two additional letters distinguish individual samples. The D-numbers were then used to label all laboratory preparations, microscope slides, unmounted material, and archival splits of unprocessed samples. The repository for all microscope slides, unmounted material, and splits of samples discussed in this report is the Central Energy Resources palynology laboratory at the U.S. Geological Survey in Denver, Colorado. Standard palynological laboratory procedures were used to prepare samples for analysis (Doher, 1980).

Microscopic analyses were of a routine nature except for the procedure adopted to provide a comprehensive compilation of the diversity and abundance of palynomorphs present in some samples. Optimally, the procedure was to count 200 specimens per slide to determine relative abundance of each taxon present (as percent) and then to continue to scan the slide until a minimum of 2000 specimens had been seen. Additional scanning inevitably revealed the presence of rare species. Their presence was recorded, but they did not change the percentage numbers already determined. This procedure, which tends to emphasize presence or absence data over relative abundance data, was adopted because it became evident early on in the study that the biostratigraphically important "K taxa" may not be common in certain samples, but their presence is critical to locating the position of the K-T boundary. K taxa are here defined as those species or genera that biostratigraphically characterize and distinguish uppermost Cretaceous palynomorph assemblages and the rocks that contain them; by definition, K taxa do not occur in situ in Tertiary rocks. Relative abundance data were not obtained for all taxa in all samples, however. For long sections and those sampled in reconnaissance fashion, only the total numbers of K taxa in each sample were determined, in most cases on the basis of scans of 1000–3000 specimens. Relative abundance data were obtained for all taxa in all samples in most of the closely sampled boundary intervals.

The position of the K-T boundary is placed in each measured section above the sample having the highest occurrence of K taxa. In most sections, this horizon is clearly marked by the disappearance of several K taxa. In two sections the boundary is also marked by presence of impact ejecta (iridium and shocked quartz). Percentage extinction of palynomorph taxa across the K-T boundary is determined by counting all taxa present in the sampled interval (in long sections, only those taxa within the interval just below the boundary are used in the calculations). From that total, the number of K taxa is determined, and taxa that are present only above the boundary are subtracted. Percentage extinction is the number of K taxa divided by the remaining number of taxa, times 100.

LITHOSTRATIGRAPHY VERSUS GEOCHRONOLOGY

The Hell Creek–Fort Union contact

A simple but fundamental principle about the difference between lithostratigraphy and geochronology bears review because of a traditional tendency for the two to be confused with particular reference to the K-T boundary. In the northern Great Plains region, the K-T boundary (a nonphysical, geochronologic horizon or "time line") happens to be near—sometimes at—the physical contact between the Hell Creek and Fort Union Formations (Fig. 3). The formational contact is a lithostratigraphic boundary separating these two stratigraphic units, which differ in depositional origin and composition. Because of the near-coincidence of the temporal boundary between the Cretaceous and Tertiary Periods and the lithostratigraphic contact between the Hell Creek and Fort Union Formations in the northern Great Plains, the important point often overlooked is that they are not the same, either in concept or—in most places—in reality.

In the study area the contact between the Hell Creek and Fort Union Formations is most often placed at the base of the lowest laterally extensive lignite bed above typical Hell Creek deposits (Fastovsky, 1987; Murphy et al., this volume). There is no lignite bed that extends throughout the region, however. In places where a basal Fort Union lignite is absent, the formation contact is placed at the facies change from the underlying smectite-rich, "popcorn" weathering, gray mudstone (Hell Creek Formation) to the overlying carbonaceous shale, flat-laminated yellowish mudstone, or yellowish sandstone typical of the Ludlow Member of the Fort Union Formation. In practice, it may be somewhat difficult to pick this contact at some localities because the facies change is not as obvious as this brief description may indicate. This fact has led to some disagreement among geologists in the field, but such disagreements can be resolved when the basic lithostratigraphic concepts and specific unit definitions are applied. Real problems arise, however, when care is not taken to recognize that this formation contact is not the same as the K-T boundary. Note that in this report and elsewhere in this volume, the two words "contact" and "boundary" are used differently, and purposefully.

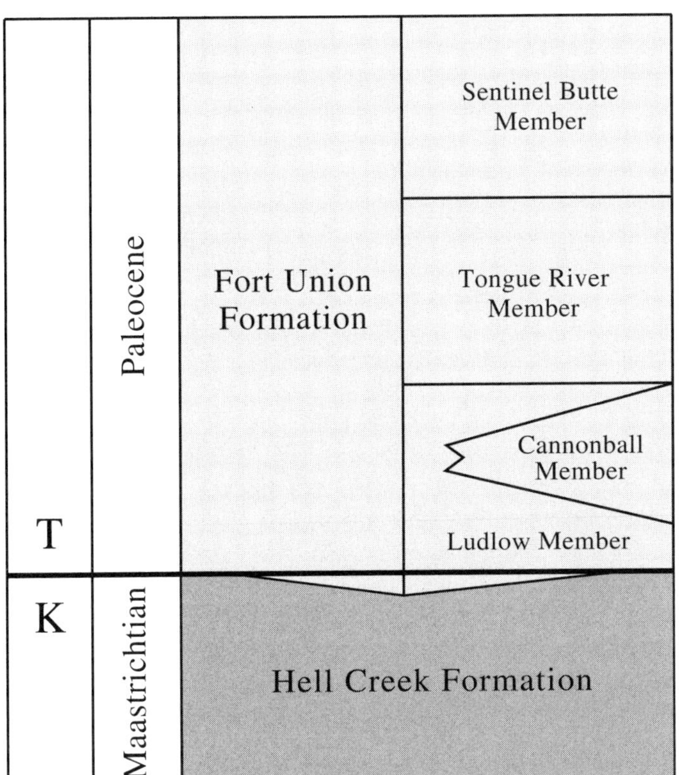

Figure 3. Chart showing stratigraphic nomenclature for the uppermost Cretaceous and lower Tertiary in the study area. The terminology is that of the U.S. Geological Survey; the North Dakota Geological Survey uses different formation names, in part.

The Cretaceous-Tertiary boundary

Brown (1952), following Calvert (1912), enunciated what has come to be known as "Brown's formula" for locating the K-T boundary in the field. This formula holds that the K-T boundary is at the level of the first laterally extensive lignite bed above the stratigraphically highest dinosaur bone. This convention, which is of some use in field mapping, nonetheless fundamentally confuses the geochronologic and lithostratigraphic concepts by equating the temporal horizon with the formation contact. As mentioned, a conveniently located basal Fort Union lignite bed is not present everywhere, and dinosaur bones, although relatively common in the Hell Creek Formation, are not ubiquitous. The boundary between the Cretaceous and Tertiary Periods, indeed between the Mesozoic and Cenozoic Eras, is not a physical or material horizon that can be identified with precision or accuracy in the field. The most reliable indications of the position of the K-T boundary in a measured section are based on biostratigraphy. Biostratigraphic placement of the boundary may be verified by presence of debris from the K-T impact event.

Biostratigraphy

It is well known that fossils provide the most practical and useful means of measuring geologic time and establishing temporal boundaries in sedimentary sequences. The most reliable fossils for K-T boundary studies in the nonmarine rocks in western North America are palynomorphs, especially angiosperm pollen. In this study and several others in this volume, the position of the K-T boundary is determined palynologically by the disappearance of the K taxa, all but one of which are angiosperms.

Our data, which are presented later in this report, show that the Hell Creek–Fort Union formational contact is diachronous in western North Dakota. In many places the K taxa that characterize the upper part of the Hell Creek Formation persist into the Fort Union Formation, despite the strong change in lithology that distinguishes these stratigraphic units. That the palynological extinction horizon marks the true K-T boundary is verified by presence of impact debris at that horizon in key sections (Johnson et al., 1989; Nichols et al., 2000), and is further supported by magnetostratigraphy (Hicks et al., this volume). It is unequivocally clear that the palynological biostratigraphy on which we rely is, for all practical purposes, independent of sedimentary facies in our study area, as well as elsewhere in western North America (see also Nichols, 1990).

PALYNOLOGY

The Maastrichtian palynoflora

Nichols (this volume) described the Maastrichtian palynoflora of the Hell Creek Formation in North Dakota. Those data are summarized here for purposes of comparison with data on the palynoflora of the Ludlow Member of the Fort Union Formation and to provide a basis for discussion of changes across the K-T boundary in the study area. The Maastrichtian palynoflora comprises 100 or more palynomorph species, of which 98 are described by Nichols (this volume). Among these 98 taxa, 68 are angiosperm pollen, eight are gymnosperm pollen, 19 are pteridophyte spores, and three are bryophyte spores. Of these, 33 are regarded as characteristically uppermost Cretaceous taxa (the K taxa), and 36 are species that are part of the *Wodehouseia spinata* assemblage sensu Nichols and Sweet (1993). The list of taxa identified in the Maastrichtian palynoflora in North Dakota is given in Table 1. The numbers shown are also used in the figures illustrating the distribution of palynomorphs at the K-T boundary localities described later.

The Paleocene palynoflora

The Paleocene palynoflora of North Dakota has not been formally described. Study of this palynoflora is ongoing, however (Nichols, 1998; Nichols, unpublished data). In the course of investigations of the palynology of the K-T boundary in North Dakota described in the present report, the palynoflora of the Ludlow Member of the Fort Union Formation was studied in considerable detail. The data are summarized here for comparison with the palynoflora of the Hell Creek Formation and to provide a basis for discussion of changes across the K-T boundary in the study area.

The Paleocene palynoflora comprises 71 taxa, of which all but ten are present also in the Maastrichtian palynoflora. Among these, 43 are angiosperm pollen, seven are gymnosperm pollen, 18 are pteridophyte spores, and seven are bryophyte spores. The palynoflora is dominated by angiosperm taxa although pteridophytes, especially ferns (13 taxa), are well represented. At least part of the Paleocene palynoflora of the Rocky Mountains and northern Great Plains region exclusive of North Dakota has been described previously by Stanley (1965) in South Dakota, Norton and Hall (1969) in eastern Montana, Leffingwell (1970) in Wyoming, and Nichols and Brown (1992) in Wyoming. Furthermore, most of the species present in the Ludlow Member of the Fort Union Formation in North Dakota are the same as those in the Hell Creek Formation, and thus are treated in the report on the Hell Creek palynoflora (Nichols, this volume). These references can be consulted for descriptions and illustrations of most of the Paleocene palynomorph species mentioned in the present report.

The list of taxa identified in the Paleocene palynoflora in North Dakota is given in Table 2. The numbers shown are also used in the figures illustrating the distribution of palynomorphs at the K-T boundary localities described later. Taxa that occur also in the Maastrichtian palynoflora are numbered as in Table 1, and taxa that occur only in the Paleocene palynoflora are listed and numbered separately. Ten taxa are listed as "(genus name) spp." Each of these categories includes two or more species that were not differentiated. Thus, the Paleocene palynoflora actually includes 85 or more species. The species that are found only in the Paleocene and not in the Maastrichtian in North Dakota are: *Brevicolporites colpella, Eucommia? leopoldae, Jarzenipollenites trinus, Momipites tenuipolus, Podocarpidites* sp., *Thomsonipollis magnificus, Triatriopollenites subtriangulus, Triporopollenites granilabratus, T. infrequens*, and *Wodehouseia fimbriata*.

Palynological definition of the K-T boundary

In southwestern North Dakota, as elsewhere in the Western Interior region of the United States and Canada, the K-T boundary is defined by the disappearance (local or total extinction) of certain palynomorph taxa. In the study area, these taxa include ~30% of the total palynoflora present in the uppermost Maastrichtian Hell Creek Formation. As a group, these palynomorphs are referred to as K taxa, because they are characteristic of uppermost Maastrichtian rocks in the area. The K taxa whose extinction defines the K-T boundary are marked with asterisks in Table 1. In this study, the K-T boundary was determined on the basis of palynology to be between the highest sample that yielded K taxa and the next sample above that lacks K taxa.

Collectively the K taxa that disappear at the K-T boundary make up as much as 35% of palynomorph assemblages in individual samples from the Hell Creek Formation. Assemblages from a single sample may include as many as 16 K taxa. In many sections in the study area, the basal part of the Ludlow Member of the Fort Union Formation is found to be of Cretaceous age, based on the presence of K taxa. In most samples

TABLE 1. MAASTRICHTIAN PALYNOFLORA

Number	Taxon
1.	**Anacolosidites rotundus* Stanley 1965
2.	**Aquilapollenites attenuatus* Funkhouser 1961
3.	*Aquilapollenites bertillonites* Funkhouser 1961
4.	*Aquilapollenites catenireticulatus* Srivastava 1968
5.	**Aquilapollenites collaris* (Tschudy and Leopold 1971) Nichols 1994
6.	**Aquilapollenites conatus* Norton 1965
7.	**Aquilapollenites delicatus* Stanley 1961
8.	**Aquilapollenites marmarthensis* sp. nov.
9.	*Aquilapollenites pyriformis* Norton 1965
10.	**Aquilapollenites quadricretaeus* Chlonova 1961
11.	**Aquilapollenites quadrilobus* Rouse 1957
12.	**Aquilapollenites reductus* Norton 1965
13.	*Aquilapollenites reticulatus* (Mtchedlishvili 1961) Tschudy and Leopold 1971
14.	**Aquilapollenites senonicus* (Mtchedlishvili 1961) Tschudy and Leopold 1971
15.	**Aquilapollenites turbidus* Tschudy and Leopold 1971
16.	**Aquilapollenites vulsus* Sweet 1986
17.	*Arecipites columellus* Leffingwell 1971
18.	*Arecipites tenuiexinous* Leffingwell 1971
19.	*Arecipites* spp. (undifferentiated)
20.	*Azolla circinata* Oltz and Hall in Hall 1968
21.	*Azolla cretacea* Stanley 1965
22.	*Boehlensipollis minimus* (Leffingwell 1971) Pocknall and Nichols 1996
23.	*Cercidiphyllites* sp.
24.	*Chenopodipollis* sp.
25.	*Cicatricosisporites* spp.
26.	*Corollina* sp.
27.	**Cranwellia rumseyensis* Srivastava 1967
28.	**Cranwellia/Striatellipollis* (undifferentiated)
29.	*Cupanieidites inaequalis* Leffingwell 1971
30.	*Cupuliferoidaepollenites* spp.
31.	*Cupuliferoipollenites* sp.
32.	*Cyathidites diaphana* (Wilson and Webster 1946) Nichols and Brown 1992
33.	*Cyathidites* sp.
34.	*Deltoidospora* sp.
35.	*Discoidites parvistriatus* (Norton in Norton and Hall 1967) Nichols and Brown 1992
36.	*Dyadonapites reticulatus* Tschudy 1973
37.	*Echitricolpites* sp.
38.	*Ephedripites jansonii* (Pocock 1964) Muller 1968
39.	**Ephedripites multipartitus* (Chlonova 1961) Yu, Guo and Mao, 1981
40.	*Ephedripites ovatus* (Pierce 1961) Norton in Norton and Hall 1969
41.	*Erdtmanipollis cretaceus* (Stanley 1965) Norton in Norton and Hall 1969
42.	*Foraminisporis undulatus* Leffingwell 1971
43.	*Fraxinoipollenites variabilis* Stanley 1965
44.	*Ghoshispora bella* (Kondinskaya 1966) Srivastava 1978
45.	*Gleicheniidites* spp.
46.	*Hamulatisporis amplus* Stanley 1965
47.	*Hamulatisporis hamulatis* Krutzsch 1959
48.	*Inaperturopollenites* sp. G
49.	*Inaperturotetradites scabratus* Tschudy 1973
50.	*Kurtzipites circularis* (Norton in Norton and Hall 1969) Srivastava 1981
51.	*Kurtzipites trispissatus* Anderson 1960
52.	*Laevigatosporites* spp.
53.	*Lanagiopollis lihoka* (Srivastava 1972) Frederiksen 1988

(*continued*)

TABLE 1. MAASTRICHTIAN PALYNOFLORA (continued)

Number	Taxon
54.	*Leptolepidites tenuis* Stanley 1965
55.	**Leptopecopites pocockii* (Srivastava 1967) Srivastava 1978
56.	**Libopollis jarzenii* Farabee et al. 1984
57.	**Liliacidites altimurus* Leffingwell 1971
58.	**Liliacidites complexus* (Stanley 1965) Leffingwell 1971
59.	*Liliacidites* spp.
60.	*Lygodiumsporites* sp.
61.	**Marsypiletes cretacea* Robertson 1973 emend. Robertson and Elsik 1978
62.	*Momipites inaequalis* Anderson 1960
63.	*Monocolpopollenites* sp.
64.	**Myrtipites scabratus* Norton in Norton and Hall 1969
65.	*Nyssapollenites* spp.
66.	**Orbiculapollis lucidus* Chlonova 1961
67.	*Osmundacidites stanleyi* sp. nov.
68.	*Palaeoisoetes subengelmannii* (Elsik 1968) gen. nov., comb. nov.
69.	*Pandaniidites typicus* (Norton in Norton and Hall 1969) Sweet 1986
70.	*Penetetrapites inconspicuus* Sweet 1986
71.	*Pityosporites* spp.
72.	**Polyatriopollenites levis* (Stanley 1965) comb. nov.
73.	*Polycingulatisporites* sp.
74.	*Porosipollis porosus* (Mtchedlishvili in Samoilovitch et al. 1961) Krutzsch 1969
75.	**Racemonocolpites formosus* Sweet 1986
76.	**Retibrevitricolporites beccus* Sweet 1986
77.	*Reticuloidosporites pseudomurii* Elsik 1968
78.	*Retitrescolpites anguloluminosus* (Anderson 1960) Frederiksen 1979
79.	*Retitriletes* sp.
80.	*Rhoipites* spp.
81.	*Simplicepollis rallus* (Stanley 1965) Nichols and Brown 1992
82.	*Sparganiaceaepollenites* sp.
83.	*Stereisporites* spp.
84.	**Striatellipollis striatellus* (Mtchedlishvili in Samoilovitch et al. 1961) Krutzsch 1969
85.	*Striatopollis tectatus* Leffingwell 1971
86.	**Styxpollenites calamitas* gen. et sp. nov.
87.	*Taxodiaceaepollenites hiatus* (Potonié 1931) Kremp 1949
88.	*Toroisporis major* (Pflug in Thomson and Pflug 1953) Stanley 1965
89.	*Triatriopollenites granulatus* (Simpson 1961) Leffingwell 1971
90.	*Tricolpites hians* Stanley 1965
91.	**Tricolpites microreticulatus* Belsky, Boltenhagen, and Potonié 1965
92.	*Tricolpites* spp.
93.	*Triporopollenites plektosus* Anderson 1960
94.	*Triporopollenites triplicatus* (Anderson 1960) comb. nov.
95.	*Triporopollenites* spp.
96.	**Tschudypollis retusus* (Anderson 1960) gen. et comb. nov.
97.	**Tschudypollis* spp.
98.	*Ulmipollenites krempii* (Anderson 1960) Frederiksen 1979
99.	*Wodehouseia spinata* Stanley 1961
100.	*Zlivisporis novomexicanum* (Anderson 1960) Leffingwell 1971

Note: The taxa *Arecipites* spp. (undifferentiated) and *Cranwellia/Striatellipollis* (undifferentiated) are not separate taxa but categories of convenience used when scanning slides. Maastrichtian taxa whose extinction marks the Cretaceous-Tertiary (K-T) boundary at some localities are marked with an asterisk. Other K taxa included in the list disappear well below the K-T boundary.

TABLE 2. PALEOCENE PALYNOFLORA

Number	Taxon
13.	*Aquilapollenites reticulatus* (Mtchedlishvili 1961) Tschudy and Leopold 1971
17.	*Arecipites columellus* Leffingwell 1971
18.	*Arecipites tenuiexinous* Leffingwell 1971
20.	*Azolla circinata* Oltz and Hall in Hall 1968
21.	*Azolla cretacea* Stanley 1965
22.	*Boehlensipollis minimus* (Leffingwell 1971) Pocknall and Nichols 1996
24.	*Chenopodipollis* sp.
25.	*Cicatricosisporites* spp.
26.	*Corollina* sp.
29.	*Cupanieidites inaequalis*
30.	*Cupuliferoidaepollenites* sp.
31.	*Cupuliferoipollenites* sp.
32.	*Cyathidites diaphana* (Wilson and Webster 1946) Nichols and Brown 1992
33.	*Cyathidites* sp.
34.	*Deltoidospora* sp.
35.	*Discoidites parvistriatus* (Norton in Norton and Hall 1967) Nichols and Brown 1992
36.	*Dyadonapites reticulatus* Tschudy 1973
38.	*Ephedripites jansonii* (Pocock 1964) Muller 1968
40.	*Ephedripites ovatus* (Pierce 1961) Norton in Norton and Hall 1969
41.	*Erdtmanipollis cretaceus* (Stanley 1961) Norton in Norton and Hall 1969
42.	*Foraminisporis undulatus* Leffingwell 1971
43.	*Fraxinoipollenites variabilis* Stanley 1965
44.	*Ghoshispora bella* (Kondinskaya 1966) Srivastava 1978
45.	*Gleicheniidites* spp.
46.	*Hamulatisporis amplus* Stanley 1965
47.	*Hamulatisporis hamulatis* Krutzsch 1959
48.	*Inaperturopollenites* sp. G
49.	*Inaperturotetradites scabratus* Tschudy 1973
50.	*Kurtzipites circularis* (Norton in Norton and Hall 1969) Srivastava 1981
51.	*Kurtzipites trispissatus* Anderson 1960
52.	*Laevigatosporites* spp.
53.	*Lanagiopollis lihoka* (Srivastava 1972) Frederiksen 1988
54.	*Leptolepidites tenuis* Stanley 1965
59.	*Liliacidites* spp.
60.	*Lygodiumsporites* sp.
62.	*Momipites inaequalis* Anderson 1960
65.	*Nyssapollenites* spp.
67.	*Osmundacidites stanleyi* sp. nov.
68.	*Palaeoisoetes subengelmannii* (Elsik 1968) Nichols 2001
69.	*Pandaniidites typicus* (Norton in Norton and Hall 1969) Sweet 1986
70.	*Penetetrapites inconspicuus* Sweet 1986
71.	*Pityosporites* spp.
77.	*Reticuloidosporites pseudomurii* Elsik 1968
78.	*Retitrescolpites anguloluminosus* (Anderson 1960) Frederiksen 1979
79.	*Retitriletes* sp.
80.	*Rhoipites* spp.
81.	*Simplicepollis rallus* (Stanley 1965) Nichols and Brown 1992
82.	*Sparganiaceaepollenites* sp.
83.	*Stereisporites* spp.
85.	*Striatopollis tectatus* Leffingwell 1971
87.	*Taxodiaceaepollenites hiatus* (Potonié 1931) Kremp 1949
88.	*Toroisporis major* (Pflug in Thomson and Pflug 1953) Stanley 1965
89.	*Triatriopollenites granulatus* (Simpson 1961) Leffingwell 1971
90.	*Tricolpites hians* Stanley 1965
92.	*Tricolpites* spp.
93.	*Triporopollenites plektosus* Anderson 1960
94.	*Triporopollenites triplicatus* (Anderson 1960) Nichols 2001
95.	*Triporopollenites* spp.
98.	*Ulmipollenites krempii* (Anderson 1960) Frederiksen 1979
99.	*Wodehouseia spinata* Stanley 1961
100.	*Zlivisporis novomexicanum* (Anderson 1960) Leffingwell 1971
101.	*Brevicolporites colpella* Anderson 1960
102.	*Eucommia leopoldae* Frederiksen 1983
103.	*Jarzenipollenites trinus* (Stanley 1965) Kedves 1980
104.	*Momipites tenuipolus* Anderson 1960
105.	*Podocarpidites* sp.
106.	*Thomsonipollis magnificus* (Pflug and Thomson in Thomson and Pflug 1953) Krutzsch 1960
107.	*Triatriopollenites subtriangulus* (Stanley 1965) Frederiksen 1979
108.	*Triporopollenites granilabratus* (Stanley 1965) Nichols and Brown 1992
109.	*Triporopollenites infrequens* (Stanley 1965) Frederiksen 1988
110.	*Wodehouseia fimbriata* Stanley 1961

Note: The first group of taxa listed (1–100) occurs also in the Hell Creek Formation; individual taxa are numbered as in Table 1. The second group (101–110) occurs only in the Fort Union Formation.

from this interval, the numbers and percentage of K taxa decline greatly from values observed in typical Hell Creek samples. As will be shown, it is evident that these reductions are a facies effect resulting from a change in plant communities associated with the transition from Hell Creek to Fort Union paleoenvironments. A critical point derived from this observation is that the position of the K-T boundary is not controlled by the change in lithology, despite the facies effect. Although there is environmental change and associated palynofloral variation, the K-T boundary extinction is clearly visible in the palynological record in most sections in North Dakota.

All but one of the palynomorph taxa that became extinct at the K-T boundary are angiosperms. Sweet et al. (1990) and Sweet and Braman (2001) hypothesized that most of the angiosperm taxa that became extinct—especially species of *Aquilapollenites*—were produced by entomophilous (insect-pollinated) plants. By that hypothesis, it is implicit that perhaps it was the extinction of insects at the K-T boundary that led to the decimation of the flowering plants. We note that this hypothesis could be a cart-before-the-horse conjecture. Labandeira et al. (this volume) present extensive data on the actual extinctions among insects at the K-T boundary. Their results make it obvious that loss of certain plant taxa would inevitably lead to loss of insect taxa dependent on them, rather than vice versa. Aside from that, we point out that an argument against the hypothesis of Sweet and coauthors is that entomophilous plants are known to produce pollen in much less quantity than anemophilous (wind-pollinated) plants, to the extent that in the pollen record, entomophilous plants tend to be underrepresented; yet pollen of *Aquilapollenites* spp. is quite abundant below the K-T boundary in North Dakota—as much as 34% of an assemblage (shale 5–10 cm below the K-T boundary at a locality discussed later). Therefore, either pollen of this genus was not in fact produced by entomophilous plants, or such plants were even more prevalent in Maastrichtian environments than they would seem to be. If the latter is true, the abrupt extinction of these communities at the K-T boundary is all the more impressive.

Although the strong change in palynofloras that marks the K-T boundary is clear in most sections, there are two phenomena that may tend to blur recognition of the palynological K-T boundary. One is that the last-occurrence data for rare K taxa are often stratigraphically below the position of the K-T boundary. This is an aspect of their general scarcity, and thus is an expression of the so-called Signor-Lipps effect (Signor and Lipps, 1982). This sampling effect causes fossil taxa that occur sporadically within a section to seem to disappear before their actual extinction. It is important to note that such last occurrences below the K-T boundary are not evidence of extinction of taxa prior to the K-T boundary event.

Conversely, rare specimens of common K taxa are sometimes found above the K-T boundary. Such specimens are interpreted as reworked.

Reworking of specimens

Rare specimens of K taxa are found above the K-T boundary in two of our sections where physical evidence of the K-T impact (the Ir anomaly and shocked quartz) supplements the palynological data. They force the question: do they represent survivors of the K-T boundary extinction event, or are they reworked? Reworking of palynomorphs from older deposits into slightly younger deposits certainly can take place easily during the deposition of clastic sediments, especially when the stratigraphic distance being considered is only centimeters. If anything, it is somewhat surprising that this does not appear to happen more often than it does. Because reworked palynomorphs, especially those that are stratigraphically out of place by centimeters, seldom if ever show any signs of reworking such as physical damage, it is virtually impossible to recognize reworking with certainty. That does not mean that isolated specimens must inevitably be interpreted as survivors, however. Because of the potential for reworking of individual palynomorphs to take place over short stratigraphic distances, rare and isolated occurrences (i.e., single specimens among many thousands) probably should be interpreted as reworked. Nonetheless, the question remains open, the answer one of interpretation rather than fact.

Palynology of Hell Creek lignite beds

Although lignite beds are more characteristic of the Fort Union Formation, a few thin beds are known within the Hell Creek Formation in the study area. These lignite beds received special attention because of the insight they can provide on the nature of floras that inhabited peat mires in Maastrichtian time and the floristic changes that occurred in such paleoenvironments across the K-T boundary.

The stratigraphically lowest Hell Creek lignite bed analyzed is near the middle of the formation, ~53 m below the top (locality D7644 of Nichols, this volume). The bed is 25 cm thick; it is present but not exposed in the cut-bank outcrop in the foreground of Figure 2. The most abundant palynomorphs in the assemblage from the lower part of this bed are "TCT" pollen (pollen of gymnosperms having affinity with the modern families Taxodiaceae, Cupressaceae, or Taxaceae; specifically in the North Dakota samples *Taxodiaceaepollenites hiatus* and *Inaperturopollenites* sp. G). Together these species account for 44% of the assemblage from the lower part of the bed. The abundance of TCT pollen drops to 18.5% in the upper part of the bed, however, and this assemblage includes more fern spores (25.5%, compared with 15.5% in the lower part). K taxa are common in this lignite bed, also; eight K taxa account for 11.5% of the assemblage in the lower part, and ten K taxa account for 21% in the upper part.

A second Hell Creek lignite bed analyzed is ~23.8–23.9 m below the top of the formation (locality D9007 of Nichols, this volume). The bed is 22 cm thick and woody in its lower part. The upper part of the bed carries a diverse palynoflora of more than 40 species, including 13 K taxa that account for 19.5% of the assemblage. There are no obvious numerical dominants among the taxa present, although the most abundant single taxon in the assemblage is the fern-spore genus *Laevigatosporites* (20.5%).

The third Hell Creek lignite bed analyzed is the one in the River Section (Appendix 1), which is described later. It is at about the same stratigraphic level as the second one discussed here, 23.5 m below the top of the formation, and is ~5 km to the southeast of it. Fern spores numerically dominate the assemblage from this bed, three species totaling 52.5%. TCT pollen is present but not abundant. Five K taxa are present that account for 15.5% of the assemblage.

The fourth Hell Creek lignite bed analyzed is the one in the Pretty Butte section (Appendix 1), which is described later. At 23.6 m below the top of the formation and 17 cm in thickness, it is possibly correlative with the beds at the second and third localities described above, although it is much farther away, and its palynomorph assemblage differs from both of the others. TCT pollen and spores of ferns (*Laevigatosporites*) and sphagnum moss (*Stereisporites*) are low in abundance. The assemblage includes eight K taxa, two of which (*Liliacidites complexus* and *Myrtipites scabratus*) account for almost 50% of the total.

Based on the palynology of such a limited number of samples, few generalizations can be made, except that the mire floras of the Hell Creek lignite beds known in the study area were varied in their floristic composition. In each deposit, TCT pollen is present, but with one exception, not notably abundant. Spores of ferns and sphagnum moss are common in some, but not all. About half of the K taxa present in the Hell Creek palynoflora are known from Hell Creek lignite. The most commonly occurring are *Liliacidites complexus* and *Myrtipites scabratus*, which thus may be considered characteristic of lignite of Maastrichtian age (although both species may occur in other facies, as well). The thin lignite beds sampled at the second, third, and fourth localities discussed may all be associated with a minor marine-transgressive event within Hell Creek time (see Murphy et al., this volume, and Hartman and Kirkland, this volume). The floras of the mires that produced these beds were not closely similar in composition, however. Given that the deposits are thin and probably laterally discontinuous, it is possible that they represent essentially contemporaneous but isolated mires.

Palynology of Fort Union lignite beds of Paleocene age

Palynological assemblages from 23 samples of lignite of Paleocene age from six measured sections in the Fort Union Formation were analyzed for comparison with those from Hell Creek lignite. The sections, and assemblages from other samples in them, are discussed in more detail later in this report.

The basal Fort Union lignite bed in the Mud Buttes section (Appendix 1) is only ~1 cm thick. Two fern-spore taxa, *Laevigatosporites* spp. and *Cyathidites diaphana*, which together account for 56.5%, numerically dominate the assemblage from this bed. This bed is Paleocene in age and no K taxa are present.

A lowermost Fort Union lignite bed 1.8 m thick is present at the Tonstein To Die For locality (Appendix 1). Ten samples in stratigraphic succession were analyzed. Fern spores of the genus *Laevigatosporites* are common in all samples, and are numerically dominant in the sample from the base of the bed. Several species of *Stereisporites* (sphagnum) are present, and these spores are notably abundant at some levels within the lignite. From this it is evident that ferns and sphagnum moss were the primary inhabitants of the peat mire that formed this thick lignite bed.

In the Badland Draw section (Appendix 1), the basal Fort Union lignite bed is Maastrichtian in age. Three thin lignite beds are present higher in the section. Assemblages from four samples collected from these beds were analyzed. The sample from the base of the lower bed is numerically dominated by TCT pollen (50.5%). TCT pollen drops in abundance to 21% in the sample from the top of this bed, and fern spores become numerically dominant (three species totaling 57.5%). In the assemblage from the sample at the base of the next higher lignite bed, the most abundant species is angiosperm pollen (*Kurtzipollenites trispissatus*, 48.5%), fern spores total 29%, and TCT pollen is present but negligible in abundance. In the assemblage from the uppermost lignite bed, pollen of another angiosperm (an unidentified tricolpate species) is numerically dominant (51%), and both fern spores and TCT pollen are present but in low percentages.

As in the Badland Draw section, the basal lignite bed of the Fort Union Formation in the Cannonball Creek section (Appendix 1) is Maastrichtian in age. The assemblage in a Paleocene lignite bed 212 cm above the base of the formation is similar to that from the base of the lowermost Paleocene lignite in the Badland Draw section. The Cannonball Creek Paleocene lignite assemblage has high abundance of TCT pollen (33.5%). Fern spores are present but not common.

The K-T boundary falls within a lignite bed in the Vertical Doug section (Appendix 1). The composition of the basal part of this bed will be discussed later. Three lignite samples of Paleocene age were analyzed. The lowermost of these has an assemblage numerically dominated by *Laevigatosporites* spores (56%). The next sample, 15 cm stratigraphically above, has only 21% *Laevigatosporites* spp., but a second fern spore taxon, *Gleicheniidites* spp., is present at 22.5%. Angiosperm pollen (*Tricolpites* spp.) is third in relative abundance in this sample at 12.5%. The assemblage from the uppermost part of this lignite bed consists largely of *Tricolpites* spp. pollen (33%) together with palm pollen (*Arecipites columellus*, 20%). These assemblages appear to reflect ecological succession in the developing mire in earliest Paleocene time.

Assemblages in samples from three thin lignite beds of Paleocene age separated by intervals of carbonaceous shale in the River Section (Appendix 1) were analyzed. The first of these lignite beds is 4.7 m above the base of the Fort Union Formation. The most abundant palynomorphs in this sample are fern spores (*Laevigatosporites* spp., 40%); second in abundance are bryophyte spores (*Stereisporites* spp., 20%). The second lignite bed is 5.6 m above the base of the formation. The most abundant palynomorphs in this bed are TCT pollen (36.5%), but at 22.5%, lycopod spores (*Palaeoisoetes subengelmannii*) are common, as well. The third bed is 6.6 m above the base of the formation. Its assemblage resembles that of the bed ~2 m below it in being dominated by *Laevigatosporites* spp. (56.5%). Bryophyte spores are rare in this bed, however, and a second fern species is second in relative abundance (*Deltoidospora* sp., 23.5%). Mire evolution cannot be perceived in this sequence of lignite beds, but differing mire plant communities are evident.

Finally, the composition of a mire assemblage from a Paleocene lignite bed more than 10 m above the base of the Fort Union Formation can be considered. This bed is the second in the succession within the formation in the Big Boundary section (Appendix 1), where the basal Fort Union lignite is Maastrichtian in age. Fern spores are the most abundant palynomorphs (*Laevigatosporites* spp., 19.5%, and *Gleicheniidites* spp., 15.5%). Next in abundance is palm pollen (*Arecipites* spp., 10%). TCT pollen is present but rare in this lignite.

To summarize, fern spores tend to dominate palynomorph assemblages in 70% or more of samples of lignite of Paleocene age analyzed. Thus, ferns appear to have been significantly abundant components of the vegetation in coal-forming peat mires in earliest Paleocene time in the study area. The data indicate that gymnosperms producing TCT pollen were present in substantial abundance in or near some of these mires and may have been involved in changing ecological successions, as were certain angiosperm species. The presence of significant amounts of TCT pollen in some basal Paleocene lignite beds in our study area contrasts with palynological data presented by Sweet et al. (1999), who observed an abrupt reduction in such pollen in assemblages just above the K-T boundary in southern Canada. McIver (1999) presented paleobotanical (leaf cuticle) evidence documenting a mass-kill of taxodiaceous forest vegetation at the K-T boundary in a coal-forming swamp. This apparent effect of the K-T impact event seems to have affected swamp vegetation to a similar but less severe extent in North Dakota.

The fern dominance of mire floras of the Ludlow Member of the Fort Union Formation in North Dakota makes them similar to mire floras of the Tullock Member of the Fort Union Formation in Wyoming and Montana (Nichols and Brown, 1992; Nichols, 1995). Most of the lowermost Paleocene lignite beds in those areas evidently developed in mires largely inhabited by ferns. This stands in striking contrast to coal beds in the Tongue River Member of the Fort Union Formation (upper Paleocene) in the region, which evidently formed in mires dominated by arboreal gymnosperms producing TCT pollen. Palynological data show those coal beds to be numerically dominated by TCT pollen, both in Wyoming and Montana (Pocknall and Flores, 1987; Nichols, 1995) and in North Dakota (Nichols, 1999).

Palynology of Fort Union lignite beds of Maastrichtian age

As mentioned in the foregoing, there are Fort Union lignite beds of Maastrichtian age in several sections in the study area. They are discussed in greater detail in the following part of this report, and their significance within the context of other Fort Union strata below the K-T boundary is discussed at length later. The palynology of five Fort Union lignite beds of Maastrichtian age (23 samples) is summarized here.

Prior to development of the data presented in this report, the best-known Fort Union lignite bed of Maastrichtian age was that in the Pyramid Butte section (Johnson et al., 1989; Appendix 1). Previously unpublished data on its palynofloristic composition are discussed here. Four samples in stratigraphic succession were analyzed. Palynomorph assemblages from all samples are diverse and no single species or group is numerically dominant. Based on palynomorphs, the only indication of ecological succession within the peat mire is a subtle shift in abundance from ferns and sphagnum moss earlier in its history to TCT-pollen-producing gymnosperms later. In all, ten different K taxa are present in low relative abundance (2.5%–7.5%).

A basal Fort Union lignite bed of Maastrichtian age 43 cm thick is present in the Battleship section (Appendix 1). Assemblages from four samples of this bed are diverse and not characterized by numerical dominance of one species or group. The most striking characteristic of assemblages from this bed is the diversity and abundance of K taxa. In all, 14 K taxa are present in abundances as high as 27%.

A lignite bed of Maastrichtian age 43 cm thick is present in the Barbecue section (Appendix 1). As in the Battleship section, this bed is the basal lignite of the Fort Union Formation. Three samples were analyzed. The assemblage from the lowermost sample has 41.5% TCT pollen and 19.5% other gymnosperm pollen. The relative abundance of gymnosperm pollen in the uppermost sample drops to about half of those values and fern spores correspondingly double in number. Eleven different K taxa are present in abundances as high as 18.5%.

A basal Fort Union lignite bed 55 cm thick is present in the Big Boundary section (Appendix 1). Eleven samples were collected in 5 cm intervals from bottom to top of this bed. Assemblages are fairly diverse throughout the lignite, with TCT pollen tending to dominate numerically (22.0%–77.5%), except in the uppermost 10 cm. Assemblages from the top of the bed consist predominantly of a species of the angiosperm genus *Triporopollenites*. Fern spores attain maximum abundances totaling ~50% near the middle of the bed. K taxa are present throughout, with as many as nine different species in a single sample and in abundances as high 27%. However, both the diversity and relative abundance of K taxa are highest in the lower part of the bed and are greatly reduced in the upper part. Thus, some ecological changes associated with mire evolution are evident.

In the Vertical Doug section, the lower part of the lignite bed is Maastrichtian in age, and the K-T boundary is within the lignite (the upper part of this bed was discussed previously). Only three K taxa are present in this sample, in very low relative abundance (0.5%). The most abundant palynomorphs in this sample are TCT pollen (40%). This assemblage together with those from the Paleocene samples from this bed suggest a rapid ecological succession within the coal-forming mire. Alternatively, the sudden increase shift from gymnosperm pollen to fern spores (56% *Laevigatosporites*) across the K-T boundary in this section may be analogous to that reported in K-T boundary sections in Canada (Sweet et al., 1999), although the relative abundance of fern spores here is less than that reported within the so-called fern-spore spike described by Tschudy et al. (1984).

In summary, Fort Union lignite beds of Maastrichtian age are somewhat more difficult to characterize palynofloristically than are Fort Union lignite beds of Paleocene age, because the Maastrichtian samples tend to be more diverse. TCT pollen is generally more common in assemblages from these lignite beds of Maastrichtian age than it is in most lignite beds of Paleocene age, however. The diverse communities of the Maastrichtian coal-forming mires included plants that produced the K taxa. The most commonly occurring K taxa are *Aquilapollenites marmarthensis, Liliacidites complexus*, and *Tschudypollis* spp. The pollen species *Myrtipites scabratus*, which is one of the most common of the K taxa in Hell Creek lignite, is also common.

Thus, comparison of palynofloral assemblages from three categories of lignite deposits—Maastrichtian Hell Creek,

Maastrichtian Fort Union, and Paleocene Fort Union—reveals little major and consistent difference among them, aside from the presence or absence of K taxa. The poorly developed mires that produced a few thin beds of Hell Creek lignite and the comparatively well-developed and common mires that produced Fort Union lignite tended to be inhabited by the same major groups of plants—ferns or taxodiaceous/cupressaceous/taxaceous gymnosperms—with certain angiosperms sometimes present in significant numbers. These observations are important because they bear on the question of possible sedimentary facies control of palynological assemblages near and across the K-T boundary, as recorded in the measured sections in the study area.

MEASURED SECTIONS

Patterns and symbols used in all diagrams that follow are shown in Figure 4. Names of palynomorph taxa are abbreviated across the top of each diagram; numbers are keyed to Tables 1 and 2, where full names of palynomorph taxa are given. Detailed descriptions of the sections are given in Appendix 1. The K-T boundary sections provide either broad or closer views of palynological changes across the boundary; they are supplemented by a few near-boundary sections that provide additional insight into the palynology of either the uppermost Cretaceous or lowermost Tertiary strata and facies in the study area.

K-T boundary sections

Sunset Butte. The K-T boundary in the Sunset Butte composite section (Fig. 5; number 1 in Fig. 1) is estimated to be 170 ± 103 cm above the Hell Creek–Fort Union formational contact. A more accurate position for the boundary in the Sunset Butte area is not possible to determine because of a lack of lithologies suitable for palynological analysis in proximity to the formational contact. See Johnson (this volume) for a discussion of the Sunset Patch section.

The analysis of the Sunset Butte composite section includes data on occurrences of 93 taxa in 34 samples collected within a total interval of ~112 m, which includes the entire thickness of the Hell Creek Formation and the lower part of the Fort Union Formation (Fig. 6). Not all individual sample levels are shown in Figure 6 because of the scale of the diagram and the close spacing of some samples. Of the 93 taxa present, 90 were recorded in samples below the K-T boundary, or both below and above it, and three were recorded only above the boundary; only the occurrences of angiosperm pollen taxa are shown in Figure 6. Including the ephedralean gymnosperm *Ephedripites multipartitus*, 28 K taxa were recorded. Seven of these K taxa are only present low in the section (more than 4 m below the top of the Hell Creek Formation); they were excluded from the calculation of percentage extinction. Only the 75 taxa that occur within 4 m of the boundary were considered in determining percentage extinction. On this basis, extinction across

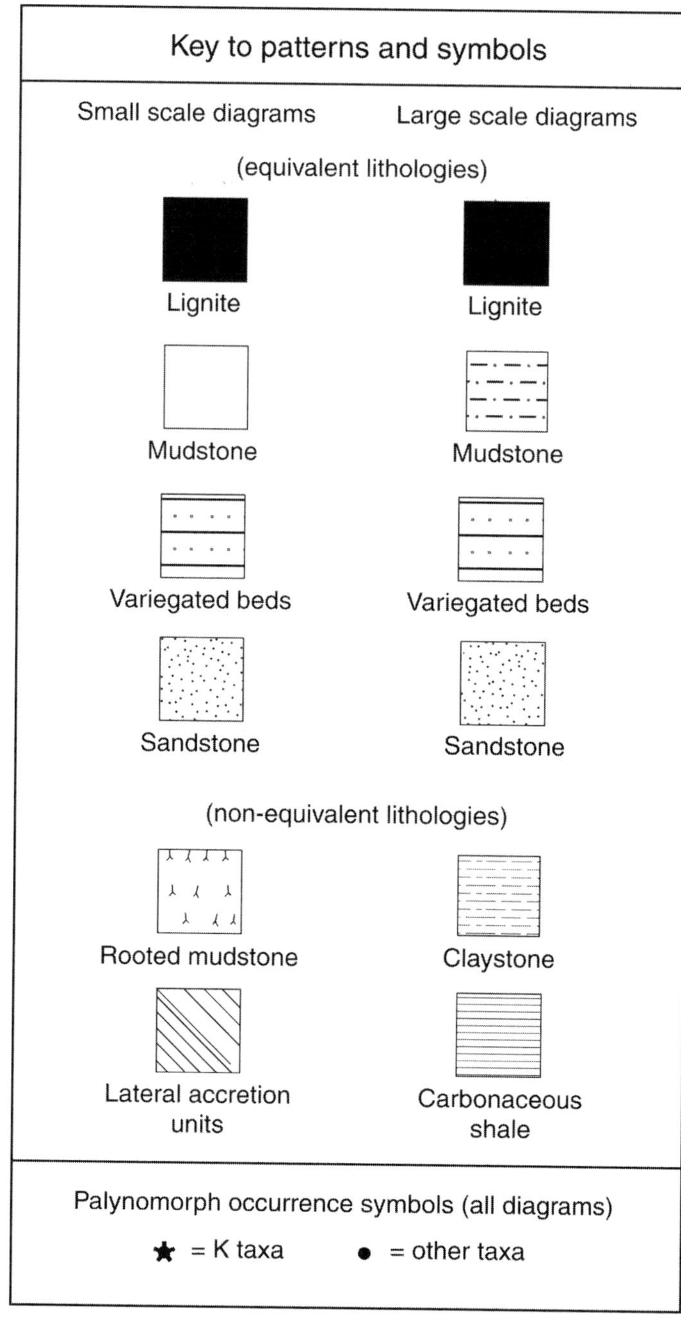

Figure 4. Key to patterns and symbols representing lithologies in Figures 4–36. K taxa = characteristic uppermost Cretaceous palynomorphs.

the K-T boundary in the Sunset Butte composite section is 28%. Among angiosperm taxa only, extinction is ~46% within this interval in this section.

Relative abundance data are available for only a few samples from the Sunset Butte composite section and are not tabulated here. The highest relative abundances of all K taxa present recorded in individual samples are 30% (Sunset Butte sample D9003-B, ~86 m below the top of the Hell Creek

Figure 5. Sunset Butte, the upper part of the Sunset Butte composite section. The position of the Hell Creek (HC)–Fort Union (FU) formational contact is indicated.

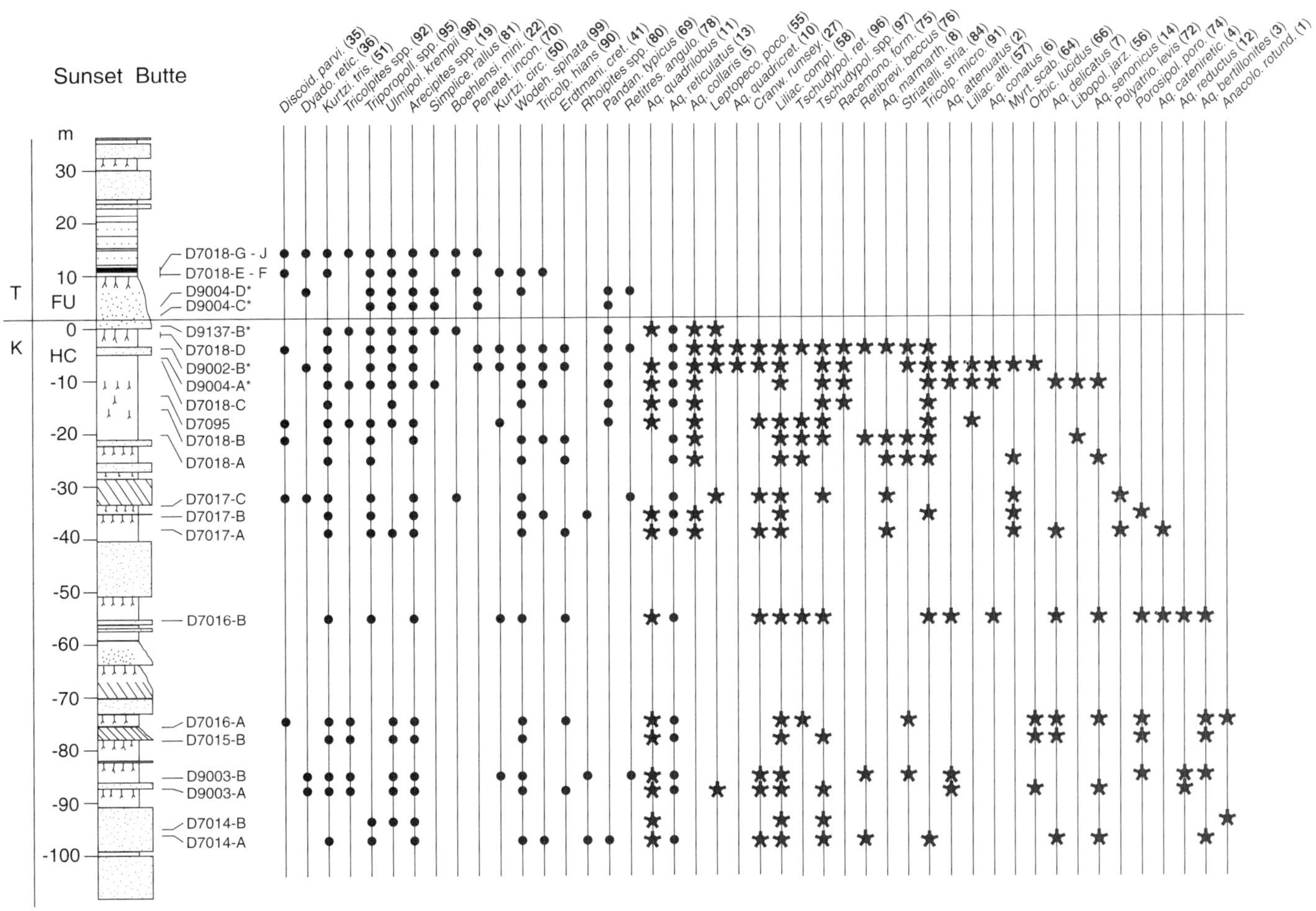

Figure 6. The K-T boundary at Sunset Butte. The entire thickness of the Hell Creek Formation is shown. HC = Hell Creek Formation, FU = Ludlow Member of the Fort Union Formation; position in the measured section is relative to the formational contact; patterns and symbols as in Figure 4. Asterisks mark samples from the Sunset Patch locality projected into this section. Occurrences of only 45 of 62 angiosperm taxa are shown, those most commonly occurring; excluded are taxa occurring in fewer than five samples, except that all K taxa are shown. The gymnosperm K taxon *Ephedripites multipartitus* occurs below the K-T boundary at this locality but is not shown. Closely spaced samples are composited. Gaps between sample levels are due to lithologies unfavorable for palynology (samples from the Sunset Patch locality fill some gaps near the boundary).

Formation, Fig. 6) and 35% (Sunset Patch sample D9002-A, ~6.5 m below the top of the formation, not shown in Figure 6 but adjacent to D7018-C). The greatest diversity in K taxa recorded in individual samples is 15 (Sunset Butte sample D7016-B, ~58 m below the top of the Hell Creek Formation, Fig. 6) and 16 taxa (Sunset Patch sample D9002-B, 3.8 m below the top of the Hell Creek, Fig. 6). Only some of the same taxa are included in these tallies from different parts of the composite measured section. Seven different K taxa are present in samples from the lowermost part of the Fort Union Formation in the Sunset Patch section, below the K-T boundary.

Bobcat Butte. The K-T boundary in the Bobcat Butte section (number 2 in Fig. 1) is 75 cm above the Hell Creek–Fort Union formational contact; the boundary is bracketed by two stratigraphically adjacent samples. No stratigraphic discontinuity or layer of different lithology was noted at this horizon, which is within a monotonous mudstone interval above a lignite bed in the Fort Union Formation.

The analysis of the Bobcat Butte section includes data on occurrences of 82 taxa in 29 samples within a total interval of ~76 m; only part of the section and 18 of the samples are shown in Figure 7. Of the total taxa, 75 were recorded in samples below the K-T boundary, or both below and above it, and seven were recorded only above the boundary. In all, 21 K taxa were recorded (occurrences of 16 are shown in Fig. 7; the others occur more than 20 m lower in the section). Single occurrences of three K taxa were recorded above the established position of the K-T boundary in a sample not shown in Figure 7. These specimens probably are reworked, but possibly they could be regarded as representing survivors of the K-T extinction event. If they are survivors, they cannot be counted among K taxa used to determine percentage extinction. If these anomalous records are treated as reworked specimens, extinction is ~23% within 0.5 m below the K-T boundary; if they are considered survivors and excluded from the calculation, extinction is 19%. Among angiosperm taxa only, extinction is ~33% within this interval just below the K-T boundary.

The late C.J. Orth of Los Alamos National Laboratory analyzed contiguous 5 cm samples collected across the formational contact for iridium in 1986, but no anomaly was detected. The position of the K-T boundary was determined palynologically at a later date and found to be well above the formational contact.

No relative abundance data are available for samples from the Bobcat Butte section. The highest diversity of K taxa in the section (12 taxa) was recorded from sample D6957-A, which is 50 m below the top of the Hell Creek Formation (not shown in the partial section illustrated in Fig. 7). Sample D6956-E from the lower part of the basal lignite bed of the Fort Union Formation in the section (Fig. 7) yielded ten K taxa. Diversity of K taxa decreases upward from this level to the K-T boundary, ~65 cm higher in the section, although seven K taxa are still present 10 cm below the boundary in sample D7103-J.

Pretty Butte. The K-T boundary in the Pretty Butte section (Fig. 8; number 3 in Fig. 1) is approximately 112 ± 26 cm above the Hell Creek–Fort Union formational contact. Samples are widely spaced stratigraphically throughout most of this long measured section, especially within the Hell Creek Formation.

The analysis of the Pretty Butte section includes data on occurrences of 77 taxa in 17 samples within a total interval of ~60 m. Only the occurrences of angiosperm pollen taxa are plotted in Figure 9. Of the 77 taxa present, 71 were recorded in samples below the K-T boundary, or both below and above it, and six were recorded only above the boundary. In all, 15 K taxa were recorded. However, a meaningful percentage extinction figure cannot be calculated for the Pretty Butte section because of a lack of samples from the upper part of the Hell Creek Formation.

River Section. The K-T boundary in the River Section (number 4 in Fig. 1, southeast of Pretty Butte) is 259.5 ± 199.5 cm above the Hell Creek–Fort Union formational contact. This is the least constrained K-T boundary section in our study area. The position of the boundary cannot be refined because unsampled intervals consist of lithologies unlikely to yield palynomorphs; all the intervals appropriate for palynology have been sampled.

The analysis of the River Section includes data on occurrences of 54 taxa in 16 samples within a total interval of ~30 m. Positions of only some of the samples and occurrences of only the angiosperm pollen taxa are plotted in Figure 10. Of the 54 taxa, 52 were recorded in samples below the K-T boundary, or both below and above it, and two were recorded only above the boundary. In all, 14 K taxa were recorded. Taxa that are present only 3 m or more below the estimated position of the K-T boundary were excluded from the calculation to determine percentage extinction. On this basis, extinction is ~21% within 3 m below the estimated position of the boundary. Sparse sampling renders this estimate somewhat unreliable.

Diversity and relative abundance data for all taxa recorded in ten samples from the River Section (those plotted in Fig. 10) are presented in Table 3, grouped by major botanical categories. Some samples that are composited in Figure 10 are listed individually in the table.

Big Boundary. The K-T boundary in this section (number 5 in Fig. 1, at Pretty Butte) is picked at 96 ± 10 cm above the Hell Creek–Fort Union formational contact, approximately 50 cm above the top of a basal Fort Union lignite bed. Most of the samples collected were in contiguous 5 cm intervals.

The analysis of the Big Boundary section includes data on occurrences of 71 taxa in 22 samples within a total interval of ~16 m (only the 3 m interval near the K-T boundary is shown in Fig. 11). Of these taxa, 65 were recorded in samples below the K-T boundary, or both below and above it, and six were recorded only above the boundary. Only the occurrences of angiosperm taxa are plotted in Figure 11. In all, 16 K taxa were recorded, including at least two species of *Tschudypollis*. Ex-

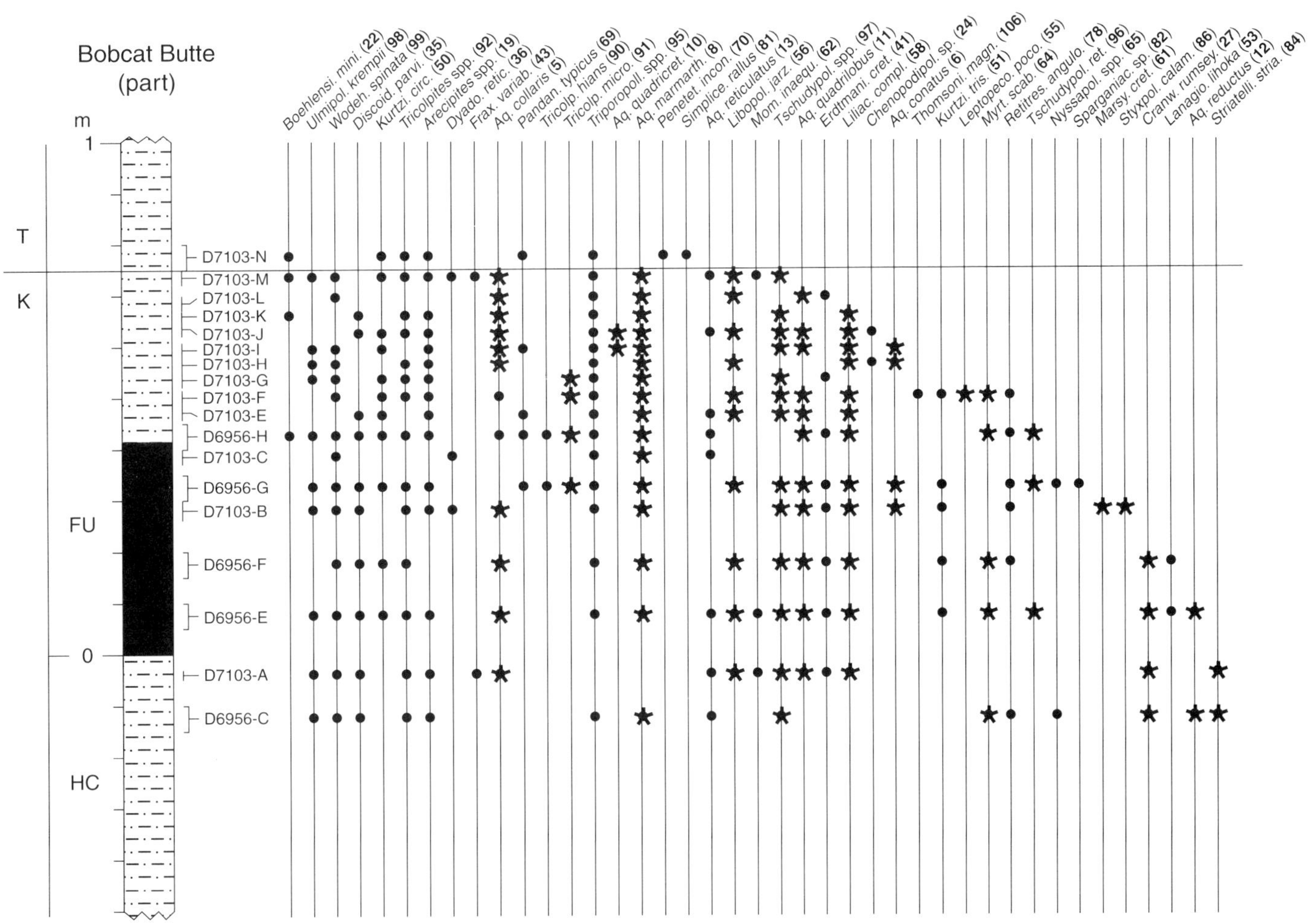

Figure 7. The K-T boundary at Bobcat Butte. HC = Hell Creek Formation, FU = Ludlow Member of the Fort Union Formation; position in the measured section is relative to the formational contact; patterns and symbols as in Figure 4. Only part of the complete measured section is shown, and only occurrences of angiosperm taxa are shown.

Figure 8. Pretty Butte, viewed from a position due east of the butte. The Hell Creek Formation is exposed in the lower part of the butte and the Fort Union Formation is exposed in the upper part. The Hell Creek (HC)–Fort Union (FU) formational contact is marked in two places, near the main section measured by E.C. Murphy (left), and near the supplementary section measured by K.R. Johnson (right). Dashed line delineates a large slump block.

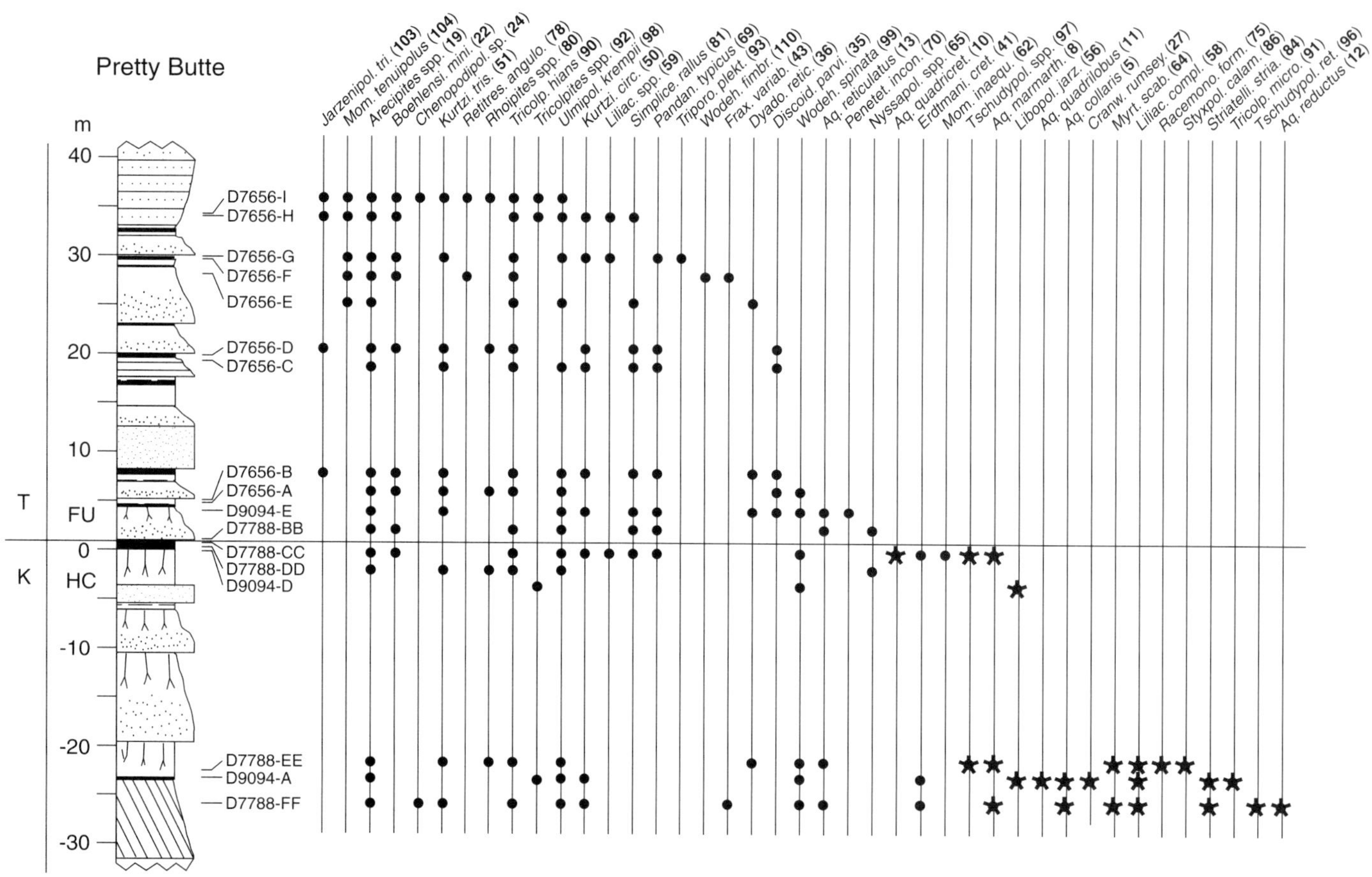

Figure 9. The K-T boundary at Pretty Butte. HC = Hell Creek Formation, FU = Ludlow Member of the Fort Union Formation; position in the measured section is relative to the formational contact; patterns and symbols as in Figure 4. Only occurrences of angiosperm taxa are shown. Gaps between sample levels are due to lithologies unfavorable for palynology. Samples marked D9094 are projected into this section from the supplementary section at Pretty Butte.

tinction is ~25% within 1 m below the K-T boundary. Among angiosperm taxa only, as shown in Figure 11, extinction is 39% within this interval.

The number of K taxa per sample decreases from a high of nine near the base of the section to two in the uppermost Cretaceous sample, through an interval of less than 1 m. With some fluctuation, relative abundance of K taxa decreases from 15% to 1% in this interval, which is within the carbonaceous and lignitic facies of the Fort Union Formation. Relative abundance data are summarized in Table 4 and shown graphically in Figure 12.

Pretty Butte North. Given the stratigraphic control provided by samples collected and analyzed, the K-T boundary in the Pretty Butte North section (number 6 in Fig. 1, at Pretty Butte) is estimated to be 248.5 ± 178.5 cm above the Hell Creek–Fort Union formational contact. No samples are available that might refine this determination. The position of the boundary determined at the nearby Pretty Butte and Big Boundary sections is considerably closer to the formational contact, ~1 m above it.

The analysis of the Pretty Butte North section includes data on occurrences of 52 taxa in 11 samples within a total interval of ~7 m within which there is a large gap (Fig. 13). Of these taxa, 47 were recorded in samples below the K-T boundary, or both below and above it, and five were recorded only above it. Occurrences of all taxa present are plotted in Figure 13. In all, 16 K taxa were recorded. Based on available data, extinction is 34%, but the wide spacing of samples bracketing the K-T boundary renders these results uncertain.

No relative abundance data are available for samples from the Pretty Butte North section. Tallies of numbers of K taxa present in individual samples show no clear pattern of difference in diversity across the formational contact within the sampled interval. Diversity ranges from four to seven K taxa in Hell Creek samples and from three to nine K taxa in Fort Union samples.

Mud Buttes. The K-T boundary in the Mud Buttes section (Fig. 14; number 7 in Fig. 1) is at the Hell Creek–Fort Union formational contact; this is one of only two sections in the study area where this is so. In addition to eight K taxa, the uppermost Cretaceous sample contains debris from the K-T impact including shocked quartz (Nichols et al., 2000), smectitic spher-

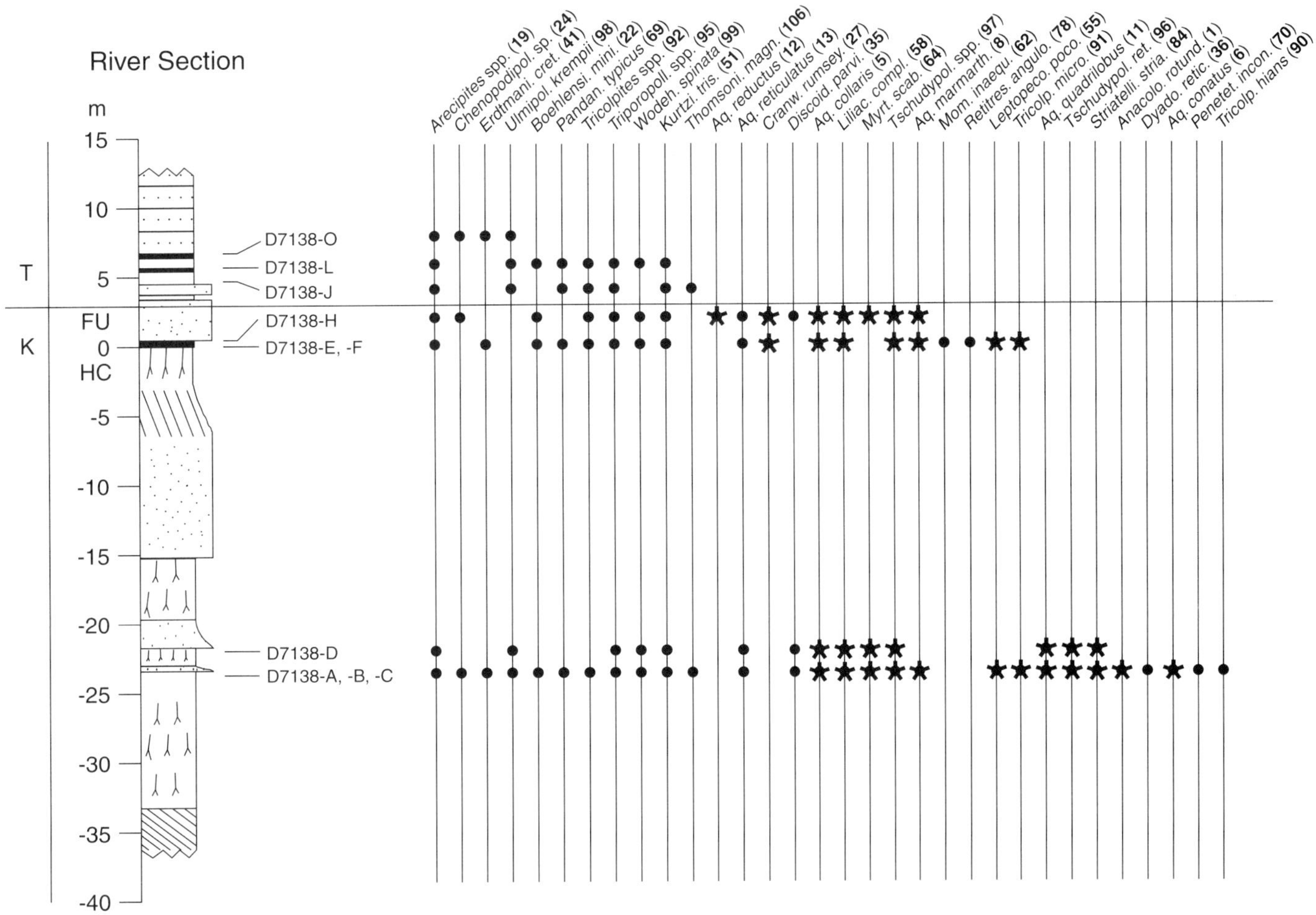

Figure 10. The K-T boundary at River Section. HC = Hell Creek Formation, FU = Ludlow Member of the Fort Union Formation; position in the measured section is relative to the formational contact; patterns and symbols as in Figure 4. Only occurrences of angiosperm taxa are shown. Closely spaced samples are composited. The large gap between sample levels is due to lithologies unfavorable for palynology.

TABLE 3. RIVER SECTION SAMPLE DATA

Sample	Position	Facies	K taxa	% K	% Angiosperms	% Gymno.	% Pterido.	% Bryo.
D7138-O	+658	FU lignite	0		8.5	4.0	84.5	3.0
D7138-L	+563	FU lignite	0		15.0	40.5	43.5	1.0
D7138-J	+469	FU lignite	0		20.5	2.5	57.5	19.5
D7138-H	+60	FU mudstone	7	6.0	26.5	9.0	38.5	26.0
D7138-F	+5	FU lignite	3	11.5	21.5	31.5	40.5	6.5
D7138-E	−5	HC mudstone	6	6.0	25.5	32.0	34.0	8.5
D7138-D	−2185	HC mudstone	7	26.0	41.0	17.0	42.0	0
D7138-C	−2341	HC mudstone	11	32.0	45.5	12.5	24.5	17.5
D7138-B	−2350	HC lignite	5	15.5	17.5	15.5	55.5	11.5
D7138-A	−2360	HC mudstone	8	17.5	28.0	5.0	65.5	1.5

Note: HC—Hell Creek Formation, FU—Ludlow Member of the Fort Union Formation; sample positions in centimeters relative to the formational contact. K taxa are also included in % Angiosperms.

ules altered from tektite glass, and an Ir anomaly of 1.38 ppb (J.R. Budahn, U.S. Geological Survey, 2000, written commun.). The K-T boundary layer at Mud Buttes is shown in Figure 15, and shocked quartz from this locality is shown in Figure 16.

The analysis of the Mud Buttes section includes data on occurrences of 69 taxa in 16 samples within a total interval ~13 m in thickness; positions of 11 samples within 6 m of the section are shown in Figure 17. Of these 69 taxa, 57 were recorded in samples below the K-T boundary, or both below and above it, and 12 were recorded only above the boundary. Only the occurrences of angiosperm taxa are plotted in Figure 17. In all, 17 K taxa were recorded. A single occurrence of the K taxon

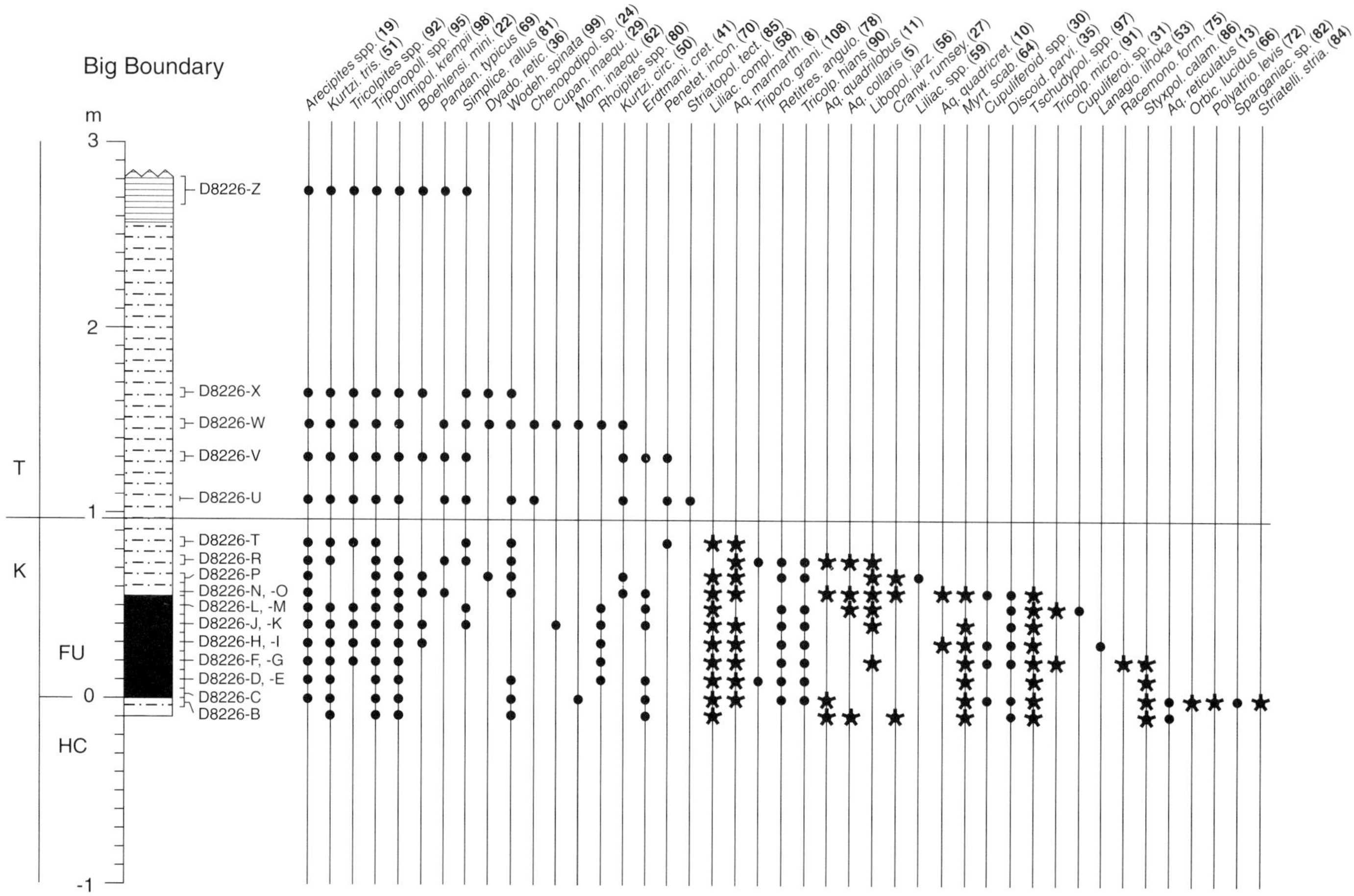

Figure 11. The K-T boundary at Big Boundary. HC = Hell Creek Formation, FU = Ludlow Member of the Fort Union Formation; position in the measured section is relative to the formational contact; patterns and symbols as in Figure 4. Only part of the entire section is shown, and only occurrences of angiosperm taxa are shown. Pairs of closely spaced samples are composited.

Liliacidites complexus was recorded ~2 cm above the K-T boundary. This specimen probably is reworked, but possibly could be regarded as representing a survivor of the K-T extinction event. If it is a survivor, it cannot be counted among K taxa used to determine percentage extinction. If this specimen is regarded as reworked, extinction is ~30% within <0.5 m of the K-T boundary; if it is regarded as a survivor and excluded from the calculation, extinction is 28%. Among angiosperm taxa only (Fig. 17), extinction is ~40% within this interval.

There is no appreciable decline in numbers or relative abundance of K taxa approaching the K-T boundary in the Mud Buttes section. The numerical data are summarized in Table 5, and data from the five samples not shown in Figure 17 are included. Relative abundance data from part of the boundary interval are shown in Figure 18.

Pyramid Butte. On the basis of the palynological samples bracketing the K-T boundary in the Pyramid Butte section (Fig. 19; number 8 in Fig. 1), the boundary would be picked at 102.5 ± 7.5 cm above the Hell Creek–Fort Union formational contact. However, the uppermost Cretaceous sample at Pyramid Butte also yielded the Ir anomaly and shocked quartz that characterize the K-T boundary around the world and that are evidence of the extraterrestrial impact at the end of Cretaceous time (Johnson et al., 1989). Thus, the actual position of the boundary is 95.0 ± 0 cm above the top of the Hell Creek Formation, just above the iridium-bearing carbonaceous claystone, which overlies a lignite bed at the base of the Fort Union Formation (Fig. 20). Shocked quartz from Pyramid Butte is shown in Figure 21.

The analysis of the section at the Pyramid Butte locality reported here is based on a reevaluation of the original samples. The new analysis includes data on occurrences of 63 taxa in 12 samples within a total interval of ~2 m (Fig. 22). Of these, 50 were recorded in samples below the K-T boundary, or both below and above it, and eight were recorded only above the boundary. The gymnosperm K taxon *Ephedripites multipartitus* occurs below the K-T boundary in this section but is not shown in Figure 22; only occurrences of angiosperm taxa are plotted in Figure 22. In all, 12 K taxa were recorded. Single occurrences of three K taxa were recorded above the established

TABLE 4. BIG BOUNDARY SECTION SAMPLE DATA

Sample	Position	Facies	K taxa	% K	% Angiosperm	% Gymno.	% Pterido.	% Bryo.
D8226-CC	+1064	FU lignite	0		28.5	12.0	50.0	9.5
D8226-Z	+267–282	FU carb shale	0		59.0	10.5	22.0	8.5
D8266-X	+162–167	FU shale	0		11.0	17.5	67.5	4.0
D8226-W	+145–150	FU mudstone	0		24.5	32.5	41.5	1.5
D8226-V	+127–132	FU mudstone	0		64.5	23.0	12.0	0.5
D8226-U	+106–109	FU claystone	0		63.0	13.5	22.5	1.0
D8226-T	+82–86	FU mudstone	2	1.0	23.0	24.0	3.5	15.5
D8226-R	+72–77	FU mudstone	4	2.5	8.0	45.0	42.0	5.0
D8226-P	+62–67	FU mudstone	3	4.5	30.0	20.0	46.0	4.0
D8226-O	+57–62	FU mudstone	6	4.0	33.0	14.0	53.0	0
D8226-N	+55–57	FU mudstone	7	4.0	27.0	19.0	54.0	0
D8226-M	+50–55	FU lignite	3	0	48.5	5.5	32.0	14.0
D8226-L	+45–50	FU lignite	3	0	42.0	10.5	31.5	16.0
D8226-K	+40–45	FU lignite	4	6.0	21.5	60.0	18.5	0
D8226-J	+35–40	FU lignite	3	2.0	13.5	25.0	48.5	13.0
D8226-I	+30–35	FU lignite	4	0.5	12.0	22.5	50.5	15.0
D8226-H	+25–30	FU lignite	2	2.0	10.5	78.5	11.0	0
D8226-G	+20–25	FU lignite	6	3.0	12.5	73.0	13.5	1.0
D8226-F	+15–20	FU lignite	6	27.0	35.0	37.0	24.0	4.0
D8226-E	+10–15	FU lignite	4	7.5	17.5	64.5	17.0	1.0
D8226-D	+5–10	FU lignite	5	8.0	25.5	37.0	34.5	3.0
D8226-C	+0–5	FU lignite	9	15.0	48.0	45.5	6.5	0
D8226-B	−0–5	HC claystone	8	5.0	24.5	0	42.0	33.5

Note: HC—Hell Creek Formation, FU—Ludlow Member of the Fort Union Formation; sample positions in centimeters relative to the formational contact. K taxa found after counts of 200 specimens per sample are recorded with 0% abundance. K taxa are also included in % Angiosperms.

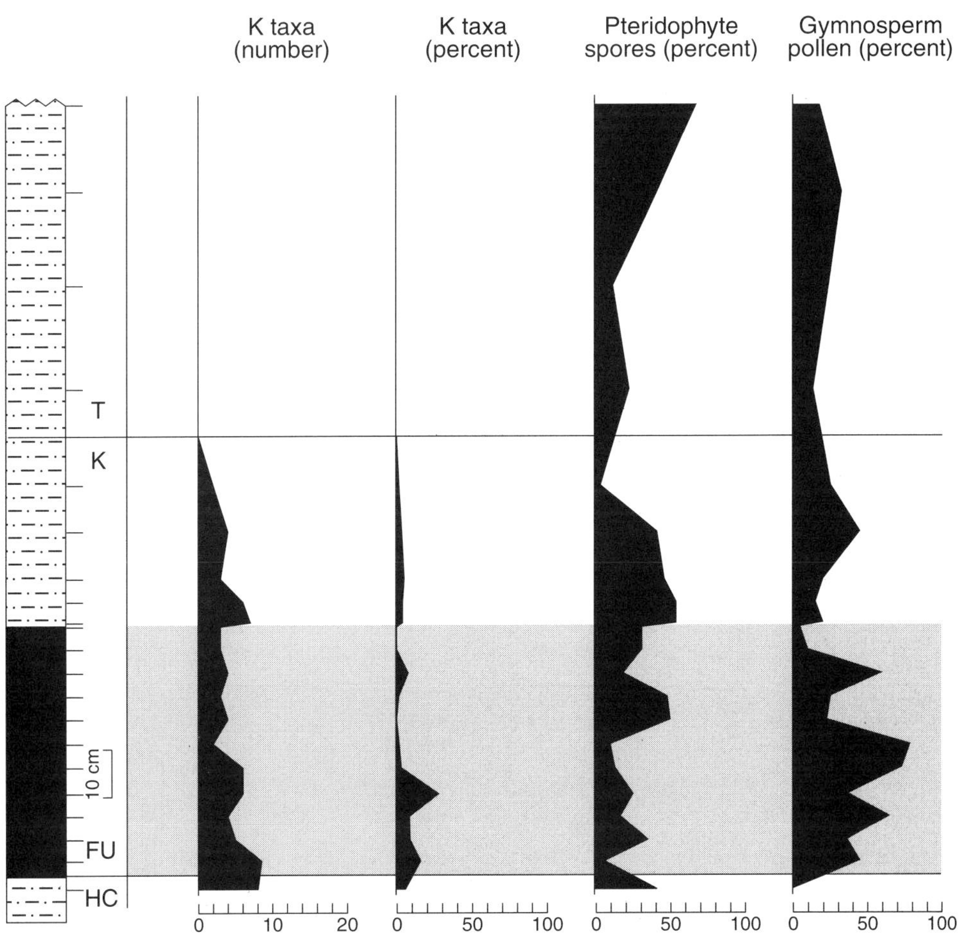

Figure 12. Diagram depicting numbers of K taxa and relative abundance of K taxa, pteridophyte spores, and gymnosperm pollen in part of the Big Boundary section.

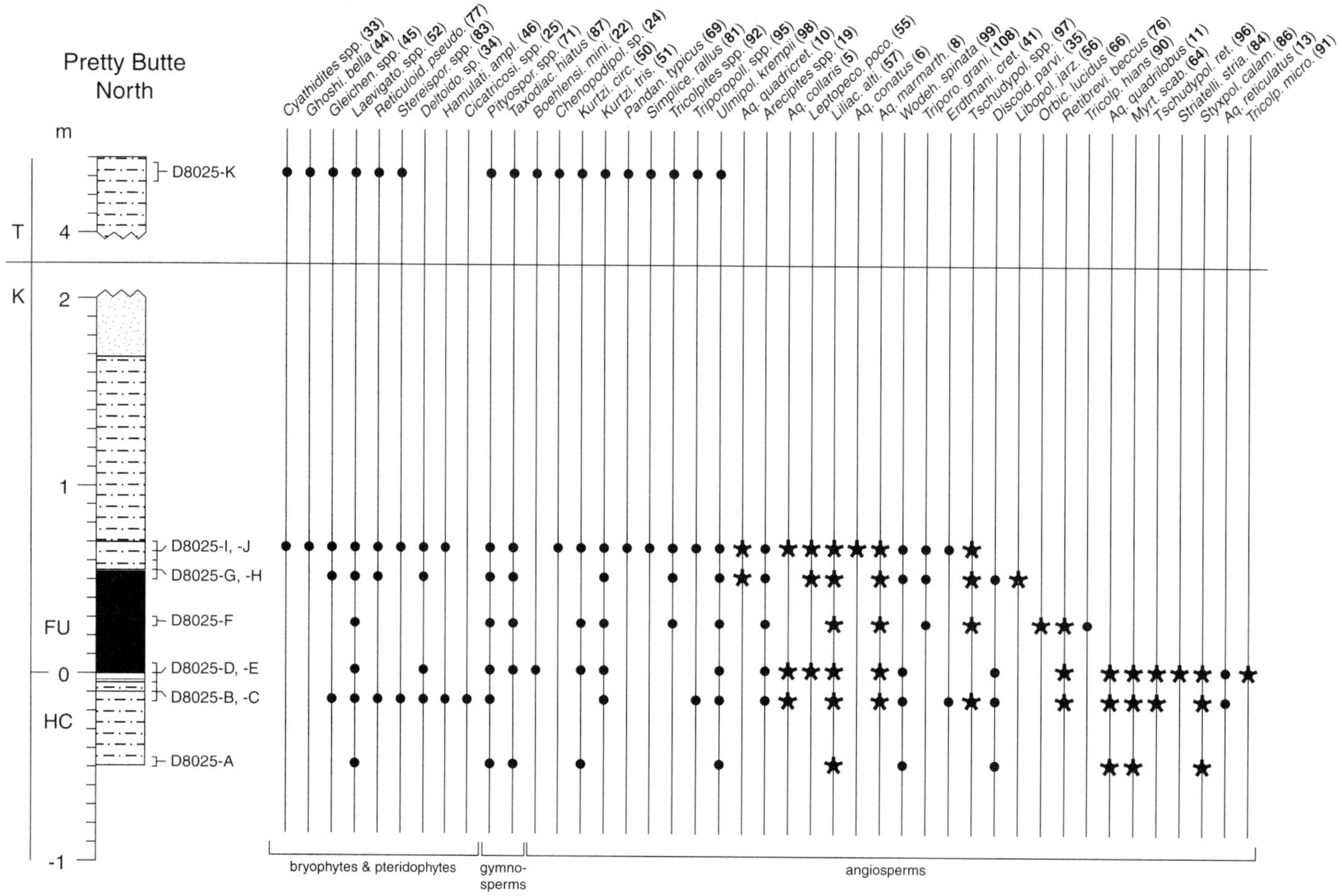

Figure 13. The K-T boundary at Pretty Butte North. HC = Hell Creek Formation, FU = Ludlow Member of the Fort Union Formation; position in the measured section is relative to the formational contact; patterns and symbols as in Figure 4. Occurrences of all taxa recorded in more than one sample are shown (except that all K taxa are shown); taxa are grouped by botanical affinity. Pairs of closely spaced samples are composited. The large gap between sample levels is due to sampling having been of a reconnaissance nature except near the formational contact. The position of the K-T boundary is uncertain because of the sample spacing.

position of the K-T boundary. These specimens probably are reworked, but possibly could be regarded as representing survivors of the K-T extinction event. If they are survivors, they cannot be counted among K taxa used to determine percentage extinction. If these specimens are regarded as reworked, extinction is 22%; if they are regarded as survivors and excluded from the calculation, extinction is 16%. Among angiosperm taxa only (Fig. 22), extinction is ~34% at Pyramid Butte. These percentage extinction figures, which are among the lowest from any K-T boundary section in the study area, are due to the lack of productive Hell Creek samples for comparison.

Relative abundance data from the Pyramid Butte section are summarized in Table 6. Percentages are based on counts of 200 specimens, and numbers of K taxa are based on scans of 2000 specimens, per sample.

Terry's Fort Union Dinosaur. The K-T boundary in the Terry's Fort Union Dinosaur section (Fig. 23; number 9 in Fig. 1, in Cannonball Creek area) is 260.5 ± 2.5 cm above the Hell Creek–Fort Union formational contact. The boundary is 176.5 ± 2.5 cm above the level of a partial *Triceratops* skeleton (Pioneer Trails Regional Museum V96016; see Pearson et al., this volume).

The analysis of the Terry's Fort Union Dinosaur section includes data on occurrences of 80 taxa in 17 samples within a total interval of less than 3 m (Fig. 24). Of these, 78 were recorded in samples below the K-T boundary, or both below and above it, and two were recorded only above the boundary. Only the occurrences of angiosperm taxa are plotted in Figure 24. In all, 26 K taxa were recorded, including the ephedralean gymnosperm *Ephedripites multipartitus* (not shown in Fig. 24) and at least two species of *Tschudypollis*. Extinction is 33% within 2.6 m of the K-T boundary. Among angiosperm taxa only, extinction is 50% across the K-T boundary in the Terry's Fort Union Dinosaur section.

Diversity of K taxa decreases upward in the section at Terry's Fort Union Dinosaur, through >2.5 m of Fort Union lignite-bearing facies. Relative abundance data are summarized

Figure 14. Outcrops of the Hell Creek and Fort Union Formations in the Mud Buttes area. The people are at an exposure of the K-T boundary, which can be traced laterally along the contact of the Hell Creek and Fort Union Formations. The boundary claystone layer is visible even at this distance as a thin, light-colored bed marked by the tip of the arrow.

Figure 15. The K-T boundary at Mud Buttes. The scale (15 cm in length) is at the level of the boundary claystone layer, which is composed mostly of kaolinitic spherules derived from impact ejecta. Iridium and shocked quartz are present in a thin layer at the top. A thin lignite bed lies just above these boundary layers.

in Table 7 and shown graphically in Figure 25. Although K taxa persist, they appear to be reduced in number and relative abundance because other palynomorphs—those produced by plants that thrived in Fort Union–style environments—are increasingly more common. These palynomorphs include various species of fern spores.

Mikey's Delite. The K-T boundary in the Mikey's Delite section (number 10 in Fig. 1, Cannonball Creek area) is 248.5 $\pm$ 1.5 cm above the Hell Creek–Fort Union formational contact. The formational contact in this section is marked by a facies change from light gray, fine-grained sandstone to dark gray, carbonaceous shale.

The analysis of the Mikey's Delite section includes data on occurrences of 63 taxa in seven samples within a total interval of $<$3 m (Fig. 26). Of these, 58 were recorded in samples below the K-T boundary, or both below and above it, and five were

Figure 16. Scanning-electron micrographs of etched grains of shocked quartz from the K-T boundary at Mud Buttes. Individual grains measure 110 to 190 micrometers. B is an enlargement of part of A. Three sets of parallel deformation features can be seen in these grains, including the one represented by the face parallel to the plane of the photograph.

recorded only above the boundary. Only the occurrences of angiosperm taxa are plotted in Fig. 26. In all, 20 K taxa were recorded. Extinction is 34.5% within 2.5 m of the K-T boundary; among angiosperm taxa only (Fig. 26), extinction is 50%.

Occurrences of angiosperm pollen shown in Figure 26 appear to show a decrease in diversity upward through the section at Mikey's Delite, but the data show no clear pattern of change in diversity among the K taxa. For example, sample D9166-I, which is <30 cm below the K-T boundary, yielded ten K taxa that account for 15.5% of the assemblage. Relative abundance data are not available for all taxa in all samples, and none are tabulated here.

Dean's High Dinosaur. The K-T boundary in the Dean's High Dinosaur section (number 11 in Fig. 1, Cannonball Creek area) is 214 ± 3 cm above the Hell Creek-Fort Union formational contact. The section includes *Triceratops* bones (Pioneer Trails Regional Museum V95014) just a meter below the base of the Fort Union Formation. Palynological sampling originally centered on the lower of two lignite beds in the Fort Union Formation in the section, but it proved to be Maastrichtian in age, and the dinosaur bones are more than 3 m below the K-T boundary.

The analysis of the Dean's High Dinosaur section includes data on occurrences of 74 taxa in 13 samples closely spaced in two groups, within a total interval of less than 4 m (Fig. 27). Of these, 70 were recorded in samples below the K-T boundary, or both below and above it, and four were recorded only above the boundary. In all, 21 K taxa are present including the gymnosperm *Ephedripites multipartitus* (not shown in Fig. 27). Extinction is 30% within 2.5 m of the K-T boundary; among angiosperm taxa only (Fig. 27), extinction is ~47%. If the analysis involves only the upper group of samples near the K-T boundary, extinction among all taxa present in that interval is ~33%.

No relative abundance data are available for samples from the Dean's High Dinosaur section. Tallies of numbers of K taxa in each sample based on scans of 2000 or more specimens show that the greatest diversity (nine K taxa) is in sample D9092-D, ~30 cm below the K-T boundary (see Fig. 27).

Cannonball Creek. The K-T boundary in the Cannonball Creek section (number 12 in Fig. 1, Cannonball Creek area) is 170.5 ± 7.5 cm above the Hell Creek–Fort Union formational contact. The Cannonball Creek measured section is near the Dean's High Dinosaur section and has a similar stratigraphic setting. The K-T boundary is in a mudstone interval between two lignite beds present in the Fort Union, close to the base of the upper lignite bed.

The analysis of the Cannonball Creek section includes data on occurrences of 62 taxa in nine samples within a total interval of a little more than 2 m (Fig. 28). Of these, 61 were recorded in samples below the K-T boundary, or both below and above it, and two were recorded only above the boundary. In all, 17

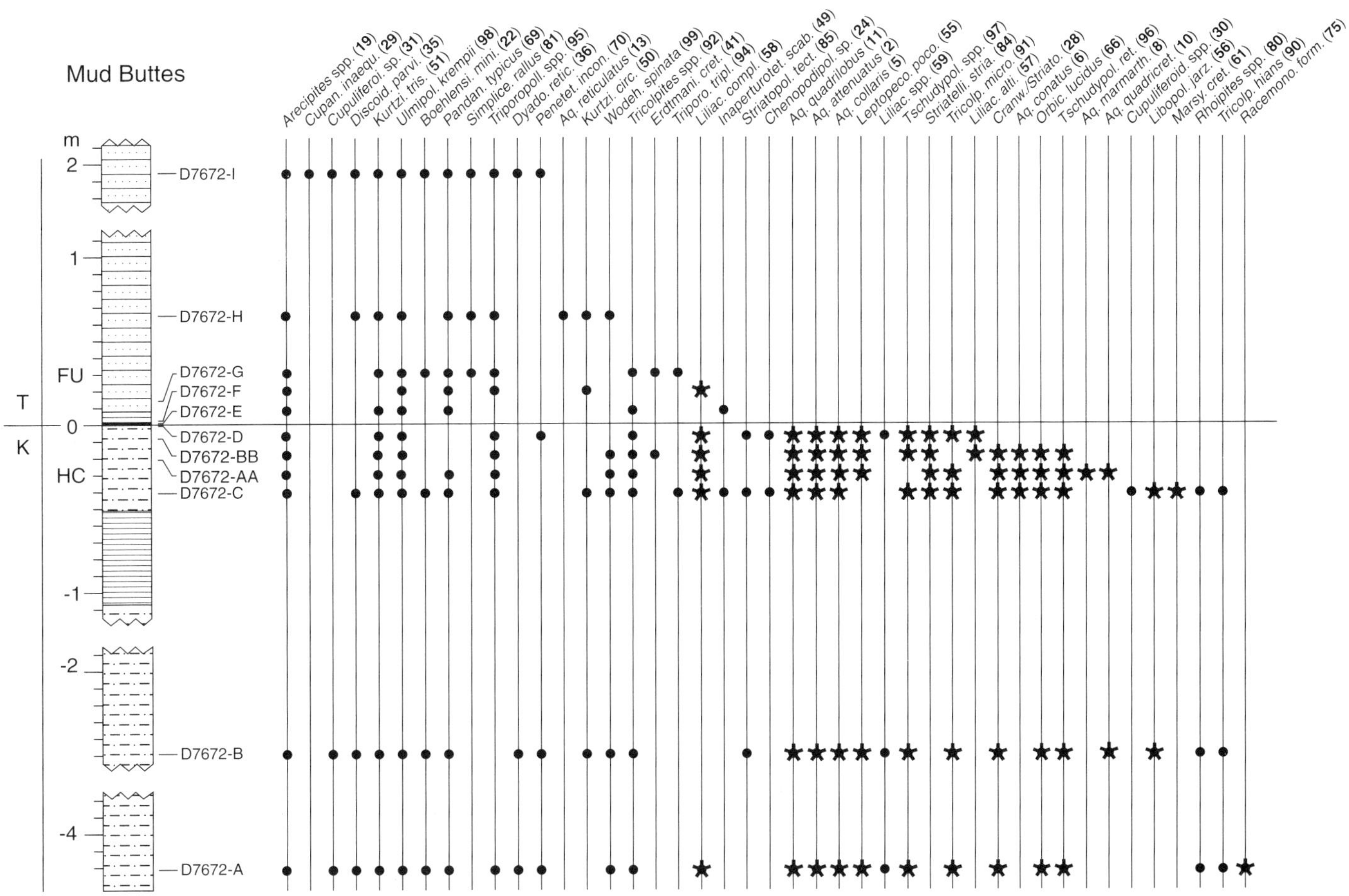

Figure 17. The K-T boundary at Mud Buttes. HC = Hell Creek Formation, FU = Ludlow Member of the Fort Union Formation; position in the measured section is relative to the formational contact; patterns and symbols as in Figure 4. Only part of the complete measured section is shown, and only occurrences of angiosperm taxa are shown. Gaps between sample levels are due to sampling having been of a reconnaissance nature except near the formational contact. This is one of only two sections in the study area where the K-T boundary is at the Hell Creek–Fort Union formational contact.

TABLE 5. MUD BUTTES SECTION SAMPLE DATA

Sample	Position	Facies	K taxa	% K	% Angio.	% Gymno.	% Pterido.	% Bryo.
D7415-D	+839	FU mudstone	0		39.5	27.5	28.0	5.0
D7415-C	+513	FU shale	0		5.5	7.5	86.5	0
D7415-B	+355	FU shale	0		23.5	42.5	25.5	8.5
D7276-I	+194	FU carb shale	0		35.5	11.0	50.0	3.5
D7276-H	+65	FU carb shale	0		11.0	7.5	79.0	2.5
D7416	+20	FU mudstone	0		25.5	13.0	43.5	18.0
D7276-G	+14	FU mudstone	0		44.0	16.0	38.0	2.0
D7415-A	+10	FU shale	0		14.5	13.5	67.0	5.0
D7672-F	+2	FU shale	1	0	12.0	0	88.0	0
D7672-E	+0–1	FU lignite	0		43.0	0.5	56.5	0
D7672-D	−1	HC claystone	7	13.5	29.5	11.0	59.5	0
D7672-BB	−8–10	HC mudstone	12	60.0	71.5	4.0	23.5	1.0
D7672-AA	−20–22	HC mudstone	13	33.0	65.0	8.5	25.5	1.0
D7672-C	−42	HC mudstone	10	34.5	83.0	5.0	11.5	0.5
D7672-B	−249	HC mudstone	11	28.0	86.5	2.5	10.5	0.5
D7672-A	−422	HC mudstone	10	17.5	79.0	11.0	10.0	0

Note: HC—Hell Creek Formation, FU—Ludlow Member of the Fort Union Formation; sample positions in centimeters relative to the formational contact. K taxon found after count of 200 specimens is recorded with 0% abundance. K taxa are also included in % Angiosperms.

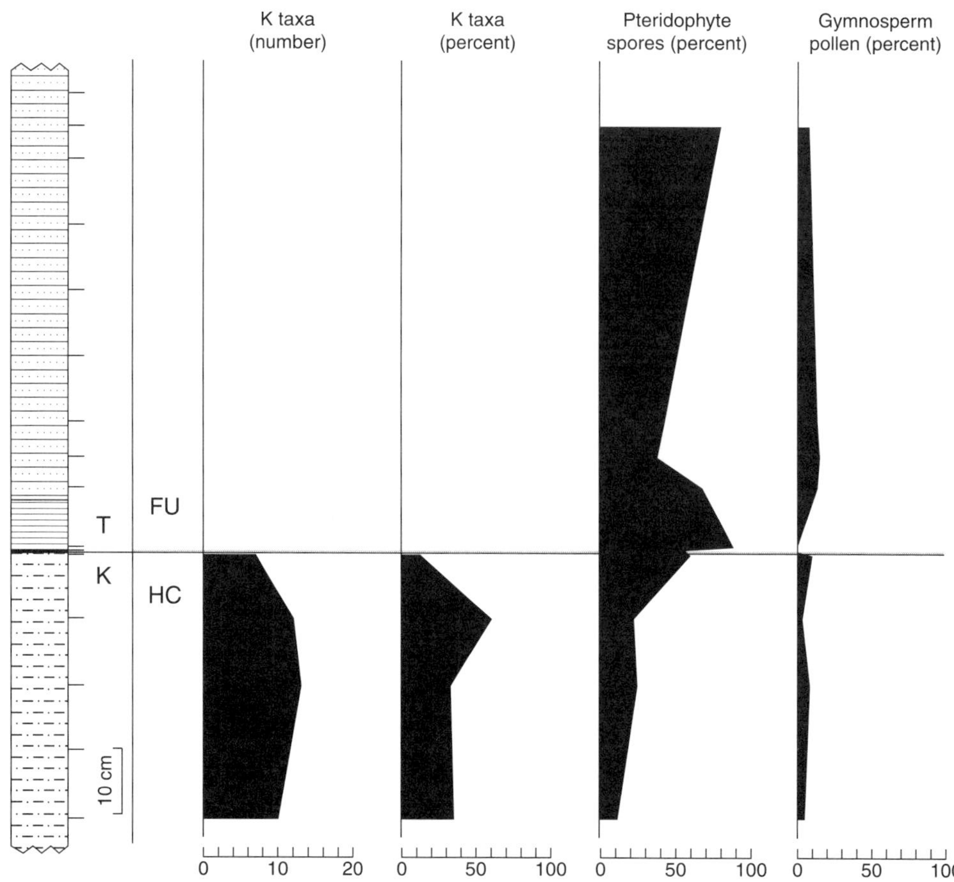

Figure 18. Diagram depicting numbers of K taxa and relative abundance of K taxa, pteridophyte spores, and gymnosperm pollen in part of the Mud Buttes section.

Figure 19. Pyramid Butte. The K-T boundary is about two-thirds the distance to the top of the butte.

K taxa were recorded. Extinction is 28% within 2.0 m of the K-T boundary; among angiosperm taxa only (Fig. 28), extinction is ~44%.

Relative abundance data from the Cannonball Creek section are summarized in Table 8. Given the close proximity and similar stratigraphic setting, these data may be taken as representative of the Dean's High Dinosaur section as well. There is a decrease in the number and relative abundance of K taxa upward through the carbonaceous and lignitic facies of the Cretaceous part of the Fort Union Formation.

Scorpion Stung Kristian. The K-T boundary in the Scorpion Stung Kristian section (number 13 in Fig. 1, Cannonball

Figure 20. The K-T boundary interval at Pyramid Butte. The fingers bracket the mudstone layer containing the Ir anomaly and shocked quartz.

Creek area) is 269 ± 4 cm above the Hell Creek–Fort Union formational contact. This is the highest position of the K-T boundary above the top of the Hell Creek Formation anywhere in the study area.

The analysis of the Scorpion Stung Kristian section includes data on occurrences of 47 taxa in eight samples within a total interval of ~3.5 m (Fig. 29). All of these taxa were recorded in samples below the K-T boundary, or both below and above it; none was recorded only above the boundary. All taxa present are plotted in Figure 29, although some individual species are combined within their genus as "spp." to facilitate drafting the figure. The taxa are sorted into major botanical groups. In all, 12 K taxa were recorded. Extinction is 25% within the interval shown in Figure 29; among the angiosperm taxa only, extinction is 40%.

Detailed relative abundance data are available for only a few samples from the Scorpion Stung Kristian section, and none are tabulated here. Even in the absence of numerical data for each sample, a decline in the diversity of K taxa in upward in the section is evident in Figure 29. The complete absence of K taxa from sample D9162-A (none was observed in a scan of 3200 specimens) is interesting, given that the sample is bracketed by two samples that yielded K taxa and thus are Maastrichtian in age. The explanation is that the palynomorph assemblage from D9162-A is numerically dominated by specimens of the isoetalean spore species *Palaeoisoetes subengelmannii* (83.5%), whose presence in such abundance suppresses the counts of all other taxa present. This species was produced by quillworts, which inhabit aquatic depositional environments. The sample represents Fort Union beds that, despite their Maastrichtian age, are characterized by few if any K taxa.

Vertical Doug. The K-T boundary in the Vertical Doug section (number 1 in Fig. 14, Cannonball Creek area) is 224.5 ± 5.5 cm above the Hell Creek–Fort Union formational contact. The sampled interval in this section includes only Fort Union strata.

The analysis of the Vertical Doug section includes data on occurrences of 70 taxa in 11 samples within a total interval of <3 m (Fig. 30). Of these, 68 were recorded in samples below the K-T boundary, or both below and above it, and two were recorded only above the boundary. In all, 22 K taxa were recorded. Extinction is 31% within 2.5 m of the K-T boundary; among angiosperm taxa only (Fig. 30), extinction is 48%.

There is a marked decrease in the diversity and relative abundance of K taxa within the section as shown in Figure 30, enumerated in Table 9, and shown graphically in Figure 31. As in the Cannonball Creek section, the presence of typical Fort Union carbonaceous and lignitic facies appears to increasingly suppress the characteristic Cretaceous components of the flora upward in the section. In the Vertical Doug section, the K-T boundary occurs within a lignite bed, whereas in the Cannonball Creek section, the boundary is in the mudstone facies below a lignite bed. These relationships indicate that although sedimentary facies can influence the relative abundance of K taxa, variable facies do not overwhelmingly control the persistent occurrence of these taxa, especially when presence or absence data are considered.

Badland Draw. The K-T boundary is 122.5 ± 7.5 cm above the Hell Creek–Fort Union formational contact. It is near or at the base of the second of three lignite beds present in the Ludlow Member of the Fort Union Formation in this measured section.

The analysis of the Badland Draw section (number 15 in Fig. 1, Cannonball Creek area) includes data on occurrences of 66 taxa in 11 samples within an interval of ~3 m (Fig. 32). Of these, 63 were recorded in samples below the K-T boundary, or both below and above it, and three were recorded only above the boundary. In all, 16 K taxa were recorded. Extinction is 25% within 2 m of the K-T boundary; among angiosperm taxa only (Fig. 32), extinction is 39%.

Relative abundance data from the Badland Draw section are summarized in Table 10 and shown graphically in Figure 33. The decrease in relative abundance of K taxa seen in other sections where the K-T boundary occurs within the Fort Union Formation is noticeable in this section, as well, although diversity of K taxa remains high in some Fort Union samples of Cretaceous age.

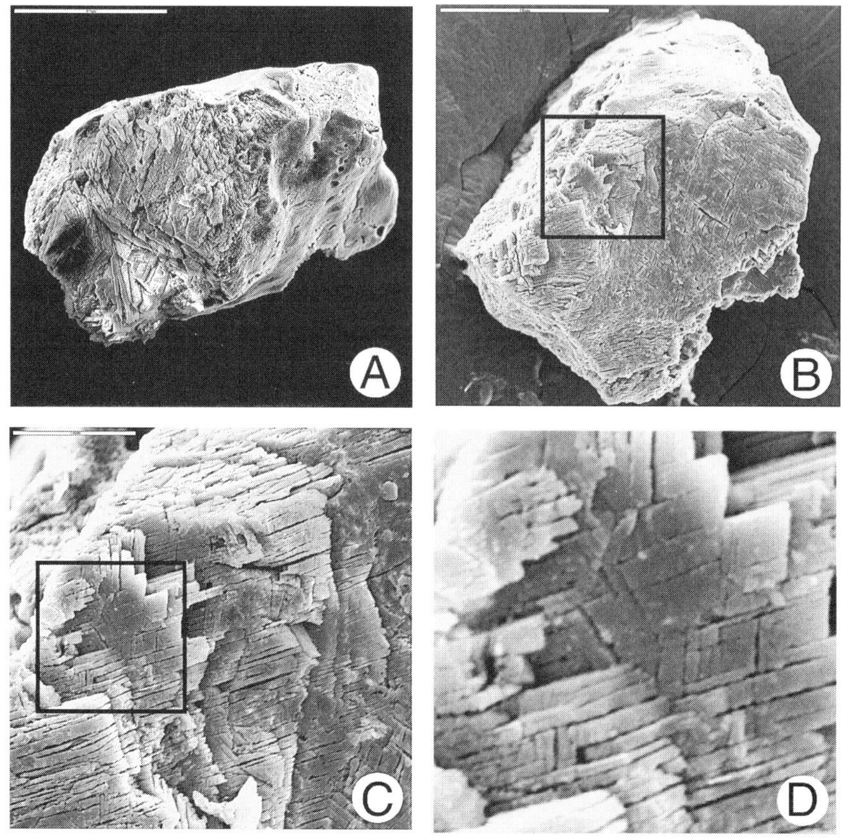

Figure 21. Scanning-electron micrographs of etched grains of shocked quartz from the K-T boundary at Pyramid Butte. Scale bars in A and B equal 50 micrometers. C is a magnified view of the area within the box in B, and D is a magnified view of the area in the box in C. Four sets of parallel deformation features can be seen in D, including the one represented by the face parallel to the plane of the photograph.

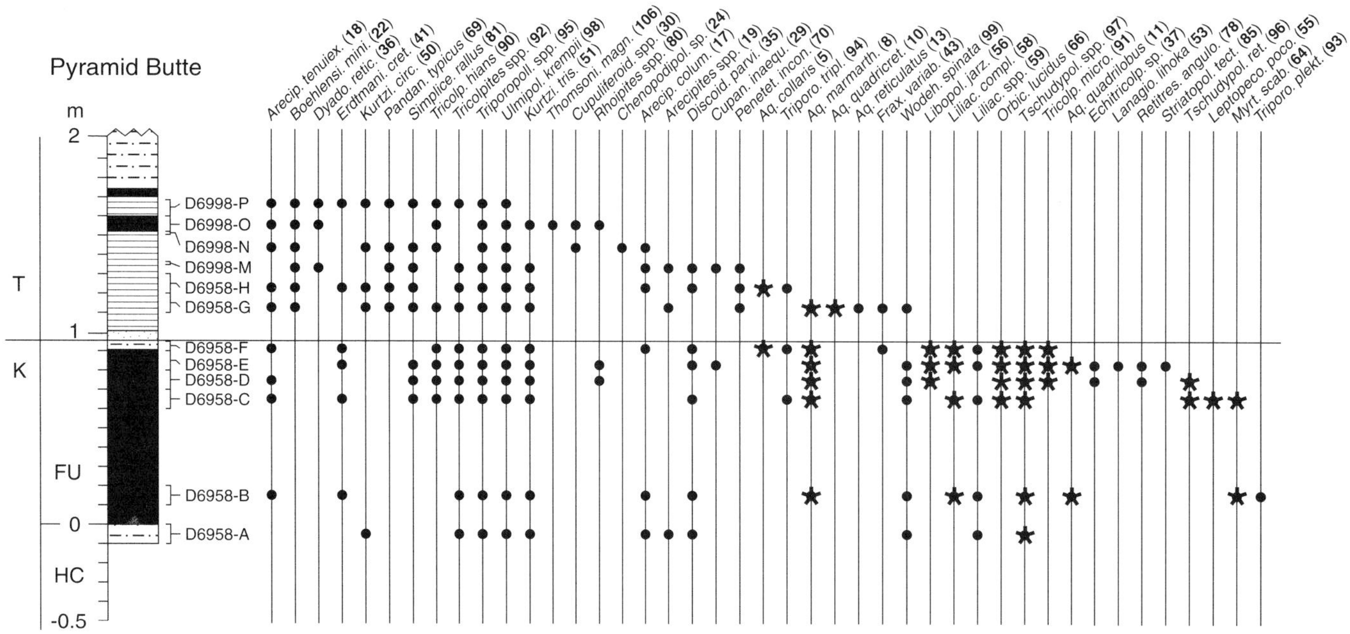

Figure 22. The K-T boundary at Pyramid Butte. HC = Hell Creek Formation, FU = Ludlow Member of the Fort Union Formation; position in the measured section is relative to the formational contact; patterns and symbols as in Figure 4. The gymnosperm K taxon *Ephedripites multipartitus* occurs below the K-T boundary at this locality but is not shown; only occurrences of angiosperm taxa are shown.

TABLE 6. PYRAMID BUTTE SECTION SAMPLE DATA

Sample	Position	Facies	K taxa	% K	% Angio.	% Gymno.	% Pterido.	% Bryo.
D6998-P	+160–170	FU carb shale	0					
D6998-O	+152–160	FU lignite	0		64.0	13.0	20.0	3.0
D6998-N	+150–152	FU carb shale	0		87.5	5.0	7.0	0.5
D6998-M	+136–137	FU carb shale	0		69.0	20.5	8.0	2.5
D6958-H	+120–130	FU carb shale	1	0	83.5	5.5	10.5	0.5
D6958-G	+110–120	FU carb shale	2	0	67.5	8.5	22.5	1.5
D6958-F	+90–95	FU mudstone	8	3.0	29.5	19.0	46.0	5.5
D6958-E	+80–90	FU lignite	9	6.5	41.5	28.0	24.5	6.0
D6958-D	+70–80	FU lignite	6	7.0	37.0	44.0	13.5	5.5
D6958-C	+60–70	FU lignite	8	2.5	49.5	14.0	16.5	20.0
D6958-B	+10–20	FU lignite	7	7.5	31.5	26.0	27.0	15.5
D6958-A	−0–10	HC claystone	3	0.5	5.0	0	75.5	19.5

Note: HC—Hell Creek Formation, FU—Ludlow Member of the Fort Union Formation; sample positions in centimeters relative to the formational contact. K taxa found after counts of 200 specimens per sample are recorded with 0% abundance. Relative abundance data unavailable for D6998-P.

Figure 23. The K-T boundary interval in the Terry's Fort Union Dinosaur section showing the position of the associated partial skeleton of a *Triceratops* (PTRM V96019) within Fort Union strata of Cretaceous age. The Hell Creek–Fort Union formational contact is at the level of the pickax head, at the base of a lignite bed; the K-T boundary is near the top of the exposure shown in the photograph. Pioneer Trails Regional Museum paleontologist Dean Pearson is kneeling by the plaster jacket containing the dinosaur bones.

New Facet Boundary. The K-T boundary in the New Facet Boundary section (number 16 in Fig. 1, Cannonball Creek area) is 199 ± 1 cm above the Hell Creek–Fort Union formational contact. It is within a claystone interval between two Fort Union lignite beds, near the base of the upper bed. A distinctive mire megaflora is present in the interval between the lignite beds.

The analysis of the New Facet Boundary section includes data on occurrences of 54 taxa in 14 samples within a total interval of ~3 m (Fig. 34). All taxa were recorded in samples below the K-T boundary or both below and above it; none was recorded only above the boundary. All taxa present are plotted in Figure 34, although some individual species are combined within their genus as "spp." to facilitate drafting the figure. In all, 16 K taxa were recorded including the ephedralean gymnosperm *Ephedripites multipartitus*, at least two species of *Tschudypollis*, and two species whose occurrences are combined in Figure 34 as *Cranwellia/Striatellipollis*. Extinction is 30% within 2 m of the K-T boundary; among the angiosperm taxa only, extinction is 50%.

Relative abundance data are available for only a few samples from the New Facet Boundary section and none are tabulated here. Relative abundance of K taxa is 7% or less in samples just below the K-T boundary, an expression of the facies-related phenomenon previously noted. This section includes DMNH loc. 2103, one of several "FU0" megafloral localities known in this vicinity (see Johnson, this volume). Its position is bracketed by samples bearing K taxa palynomorphs. The megaflora of zone FU0 (Cretaceous) is more similar in composition to the overlying "FUI" (Paleocene) megaflora than it is to the underlying "HCIII" (Cretaceous) megaflora. It represents leaf-bearing mire deposits not found in the Hell Creek Formation. Thus, the FU0 megafloras represent the mire vegetation of the Cretaceous and provide the insight that megafloral species inhabiting mires were preferential survivors of the K-T boundary event. See Johnson (this volume) for a more complete discussion.

Torosaurus Section. The K-T boundary in the *Torosaurus* Section (number 17 in Fig. 1, north of Mud Buttes) is at the

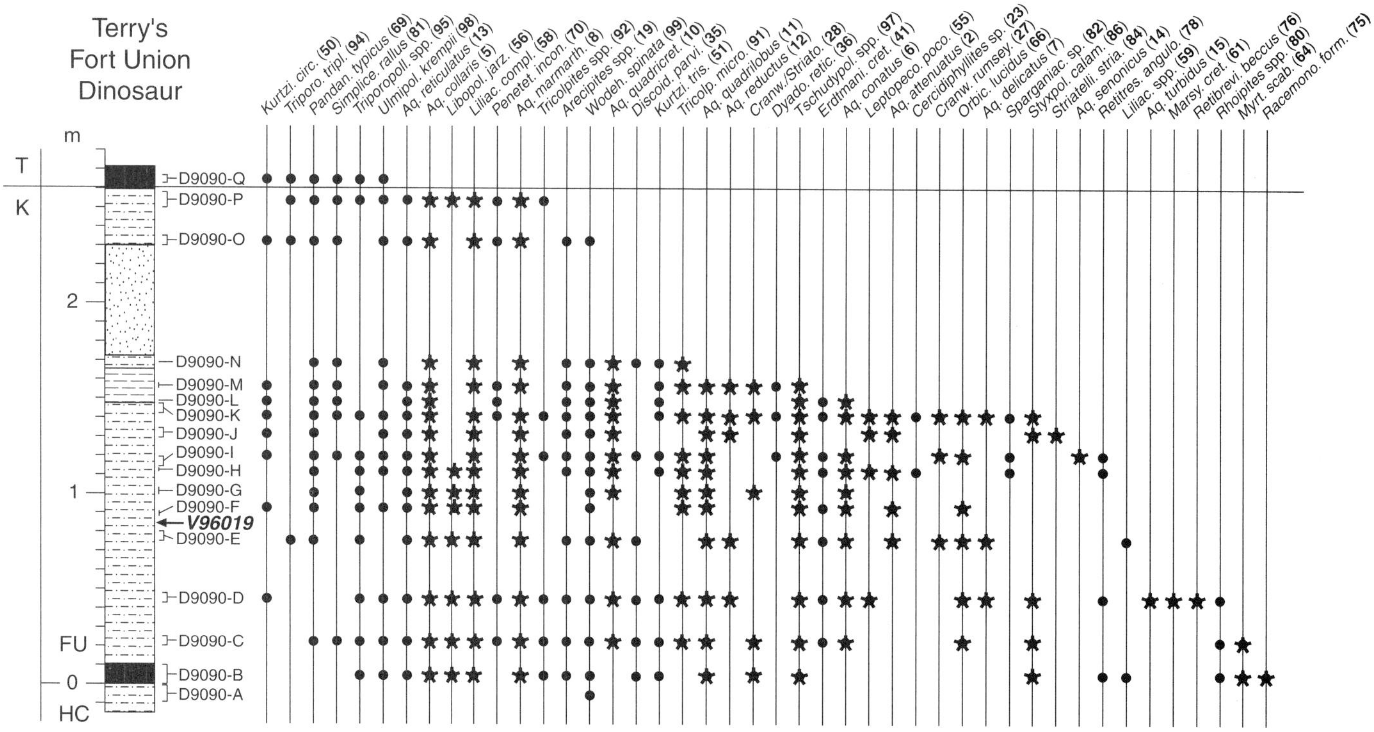

Figure 24. The K-T boundary at Terry's Fort Union Dinosaur. HC = Hell Creek Formation, FU = Ludlow Member of the Fort Union Formation; position in the measured section is relative to the formational contact; patterns and symbols as in Figure 4. Sample interval V96019 shows the position of a partial *Triceratops* skeleton. The gymnosperm K taxon *Ephedripites multipartitus* occurs below the K-T boundary at this locality but is not shown; only occurrences of angiosperm taxa are shown. The gap between sample levels just below the K-T boundary is due to the presence of sandstone, a lithology unfavorable for palynology.

TABLE 7. TERRY'S FORT UNION DINOSAUR SECTION SAMPLE DATA

Sample	Position	Facies	K taxa	% K	% Angio.	% Gymno.	% Pterido.	% Bryo.
D9090-Q	+263–267	FU lignite	0		94.0	6.0	0	0
D9090-P	+250–258	FU mudstone	4	1.0	18.0	26.5	55.5	0
D9090-O	+230–235	FU lignite	4	3.0	22.5	16.0	61.5	0
D9090-N	+168–169	FU mudstone	6	5.0	18.0	25.5	56.5	0
D9090-M	+155–158	FU claystone	12	5.0	23.0	26.0	51.0	0
D9090-L	+147–148	FU claystone	6	1.5	21.5	20.0	58.5	0
D9090-K	+142–147	FU mudstone	11	5.5	12.0	20.5	66.0	1.5
D9090-J	+129–134	FU mudstone	11	2.5	13.5	12.0	74.5	0
D9090-I	+114–119	FU mudstone	6	4.0	9.5	6.5	84.0	0
D9090-H	+112–114	FU mudstone	12	10.5	15.5	19.0	65.0	0.5
D9090-G	+99–103	FU mudstone	11	13.0	15.0	7.5	77.5	0
D9090-F	+88–91	FU mudstone	11	10.5	13.5	6.5	78.0	0
D9090-E	+75–80	FU mudstone	15	29.0	40.5	4.0	55.5	0
D9090-D	+43–48	FU mudstone	18	32.5	46.5	9.0	42.5	2.0
D9090-C	+20–25	FU mudstone	14	16.0	33.0	6.5	55.0	5.5
D9090-B	+0–10	FU lignite	14	14.5	40.0	19.5	32.5	8.0
D9090-A	−0–8	HC mudstone	0		0	0	43.0	57.0

Note: HC—Hell Creek Formation, FU—Ludlow Member of the Fort Union Formation; sample positions in centimeters relative to the formational contact. K taxa are also included in % Angiosperms.

Hell Creek–Fort Union formational contact. This is the second of only two sections in the study area where this is so; both are located at the southern end of the study area (localities 7 and 17 in Fig. 1). The section was collected for palynological analysis to determine the position of the K-T boundary because a partial skeleton of the dinosaur *Torosaurus* had been collected here by the Pioneer Trails Regional Museum, Bowman, North Dakota.

The analysis of the *Torosaurus* Section includes data on occurrences of 37 taxa in nine samples within a total interval of ~4.5 m; only ~3 m of the section and eight samples are shown in Figure 35. Of all of these taxa, 35 were recorded in

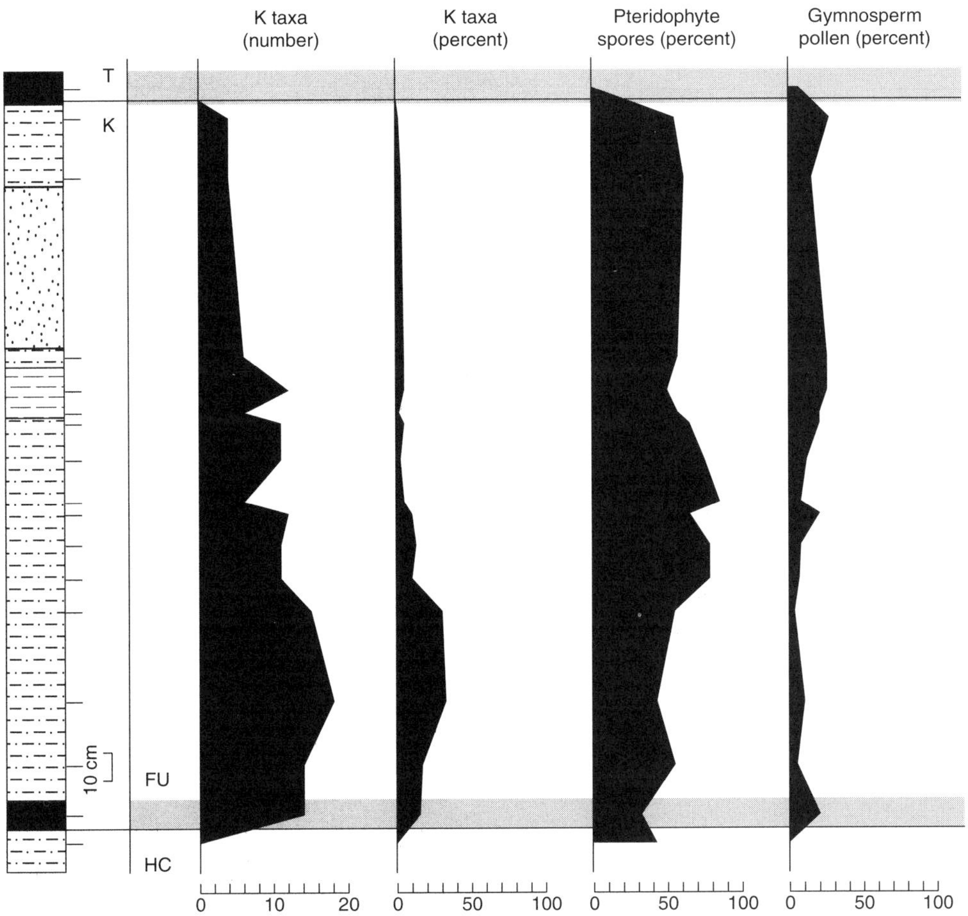

Figure 25. Diagram depicting numbers of K taxa and relative abundance of K taxa, pteridophyte spores, and gymnosperm pollen in part of the Terry's Fort Union Dinosaur section.

samples below the K-T boundary, or both below and above it, and two were recorded only above it; occurrences of all taxa recorded are shown in Figure 35. In all, 11 K taxa were recorded. Excluding the lowermost sample shown in Figure 35, extinction is 31% within 0.5 m of the K-T boundary; among angiosperm taxa only, extinction is ~42% within this interval. Relative abundance data are available for only some of the samples from the Torosaurus Section and none are tabulated here.

Near-boundary sections

Barbecue. The sampled interval in the Barbecue section (number 18 in Fig. 1, northeast of Bobcat Butte) spans the Hell Creek–Fort Union formational contact, which is marked by a basal Fort Union lignite bed. Eight samples within an interval of ~1 m were analyzed in an early attempt to locate the K-T boundary. At that time it was assumed that the boundary position would be at or near the formational contact. The entire sampled interval proved to be Cretaceous in age, however, and the position of the K-T boundary was not determined.

Occurrences of 16 different K taxa were recorded in this section, 11 in Hell Creek samples and 14 in Fort Union samples. The greatest diversity and abundance of K taxa occurs in the uppermost Hell Creek sample; diversity and abundance of K taxa are less in the Fort Union part of the section, although not especially within the basal Fort Union lignite bed. Lowest numbers for diversity and abundance were recorded from mudstone beds overlying the lignite. Relative abundance data are summarized in Table 11.

Battleship. As in the Barbecue section, the sampled interval in the Battleship section (number 19 in Fig. 1, north of Bobcat Butte) spans the Hell Creek–Fort Union formational contact, which is marked by a basal Fort Union lignite bed. Seven samples within a 75-cm-thick interval were analyzed. The entire sampled interval is Maastrichtian in age and the position of the K-T boundary has not been determined. This section is significant nonetheless because it is above the stratigraphically highest well-sampled Hell Creek megafloral locality (DMNH loc. 900), which is 2.8 m below the Hell Creek–Fort Union formational contact (Johnson, this volume).

Occurrences of 15 different K taxa were recorded in this section. Numbers of K taxa per sample range from three to nine. There is no clear pattern of occurrence of K taxa that is related either to lithofacies or to position below or above the formational contact. The highest numbers of K taxa recorded are in assemblages from the lignite bed. A sample from the base of

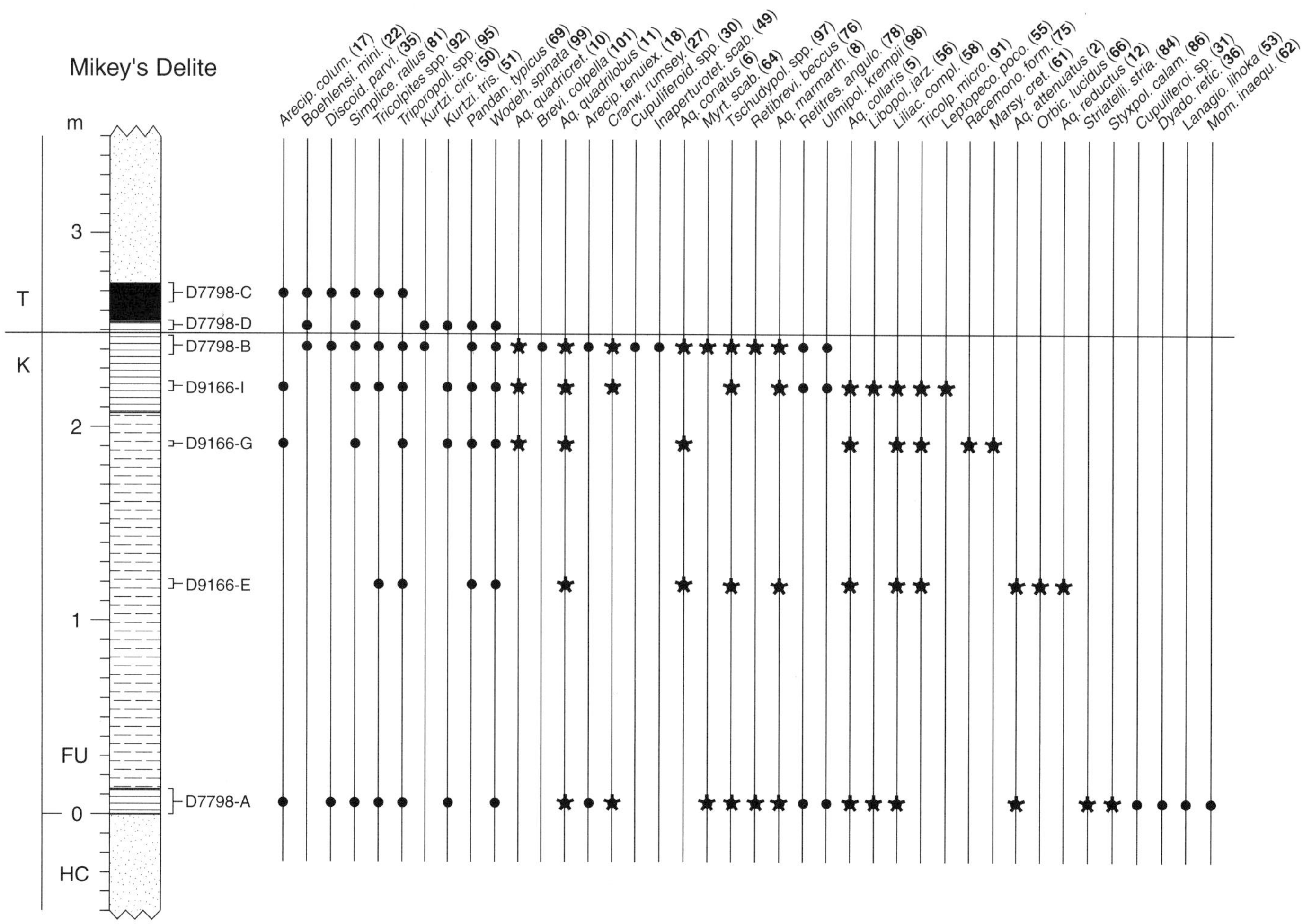

Figure 26. The K-T boundary at Mikey's Delite. HC = Hell Creek Formation, FU = Ludlow Member of the Fort Union Formation; position in the measured section is relative to the formational contact; patterns and symbols as in Figure 4. Only occurrences of angiosperm taxa are shown.

this bed yielded eight K taxa accounting for 27% of the assemblage; a sample from the top of the bed yielded nine K taxa (only five in common with the lower sample) accounting for 24.5%. Both diversity and relative abundance of K taxa are much lower in non-coal samples from both the Hell Creek and the Fort Union Formations in this section.

Dean's NM1. This measured section (number 20 in Fig. 1, north of Bobcat Butte) is near PTRM loc. V86002, one of the "vertebrate microsites" (localities from which vertebrate microfossils were collected) discovered by D. A. Pearson of the Pioneer Trails Regional Museum (see Pearson et al., this volume). Both the measured section and the sampled interval span the Hell Creek–Fort Union formational contact, which is marked by a thin lignite bed. Six samples were analyzed, two from the Hell Creek and four from the Fort Union. The entire sampled interval is Maastrichtian in age.

Occurrences of 16 different K taxa were recorded in samples from this section. Numbers of K taxa per sample range from three to nine. A sample from Fort Union carbonaceous shale 118–123 cm above the base of the formation yielded the nine K taxa, which account for 23% of the assemblage. K taxa persist in reduced numbers to the top of the sampled interval, 159 cm above the base of the Fort Union.

NDAS Section. Preparation and analysis of samples from this section (number 21 in Figure 1, north of Bobcat Butte) was supported by a North Dakota Academy of Sciences (NDAS) grant to D.A. Pearson of the Pioneer Trails Regional Museum. Although the sampled interval extends from the base of the Fort Union Formation to 237 cm above, it does not include the K-T boundary. All samples analyzed yielded K taxa and are Maastrichtian in age.

Occurrences of 18 different K taxa were recorded in samples from the NDAS Section. Sphagnum spores (*Stereisporites* spp.) numerically dominate the lowermost sample, accounting for 56.5% of the assemblage. Consequently, this assemblage has low diversity and relative abundance of K taxa: two taxa

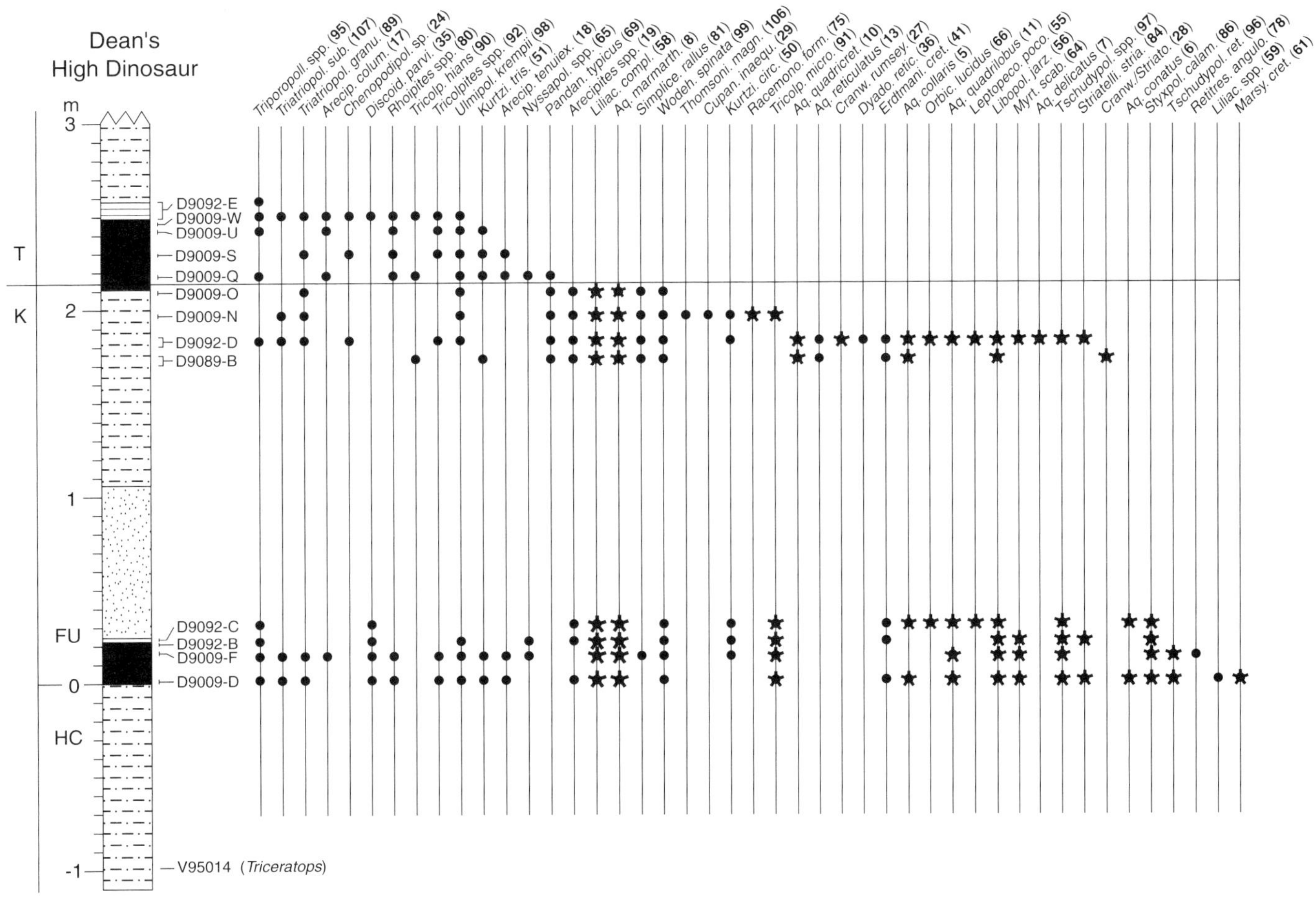

Figure 27. The K-T boundary at Dean's High Dinosaur. HC = Hell Creek Formation, FU = Ludlow Member of the Fort Union Formation; position in the measured section is relative to the formational contact; patterns and symbols as in Figure 4. Sample V95014 is a partial *Triceratops* skeleton. The gymnosperm K taxon *Ephedripites multipartitus* occurs below the K-T boundary at this locality but is not shown; only occurrences of angiosperm taxa are shown.

totaling 0.5%. Peak diversity and abundance of K taxa is in a sample of mudstone 79–84 cm above the base of the Fort Union Formation (15 taxa, 25.0%). Mudstone 184–189 cm above the formational contact datum yielded 11 K taxa totaling 12.0%. Diversity and relative abundance are considerably lower in mudstone 204–206 cm and 214–216 cm above the datum (six taxa in each sample, 4.0 and 6.0%, respectively). Assemblages from both of these samples are numerically dominated by fern spores (78.5% and 72.0%, respectively), whose presence in such abundance suppresses records of occurrence of other taxa. K taxa persist to the top of the sampled interval 237 cm above the base of the Fort Union (five taxa, 2.5%). The NDAS Section provides excellent documentation of the palynology of Fort Union strata of Cretaceous age. Relative abundance data are summarized in Table 12.

No Leaf Minors. The Hell Creek–Fort Union formational contact in the No Leaf Minors section (number 22 in Fig. 1, north of Bobcat Butte) is marked by a lignite bed 44 cm thick. Three samples collected were analyzed for palynology, one from the Hell Creek just below the lignite and two from the Fort Union just above the lignite. All three are Maastrichtian in age. The section is significant because it includes DMNH megafloral loc. 2098, which is the stratigraphically highest Cretaceous pond flora, an assemblage containing no taxa that survive the K-T boundary event (see Johnson, this volume).

Occurrences of a total of 12 K taxa were recorded in this section. Although only three samples were analyzed, their assemblages show a decline in the numbers of K taxa; this decline evidently is related to the change in lithofacies at the formational contact. The Hell Creek sample (mudstone 0–10 cm below the contact) yielded eight K taxa. The lower of the Fort Union samples (claystone 44–54 cm above the contact) yielded six K taxa, and the upper Fort Union sample (claystone 64–69 cm above the contact) yielded only two K taxa.

Swanke's Nose. This section (number 23 in Fig. 1, southeast of Marmarth) includes an interval of 80 cm spanning the Hell Creek–Fort Union formational contact at a location where there is no lignite at or near the base of the Fort Union

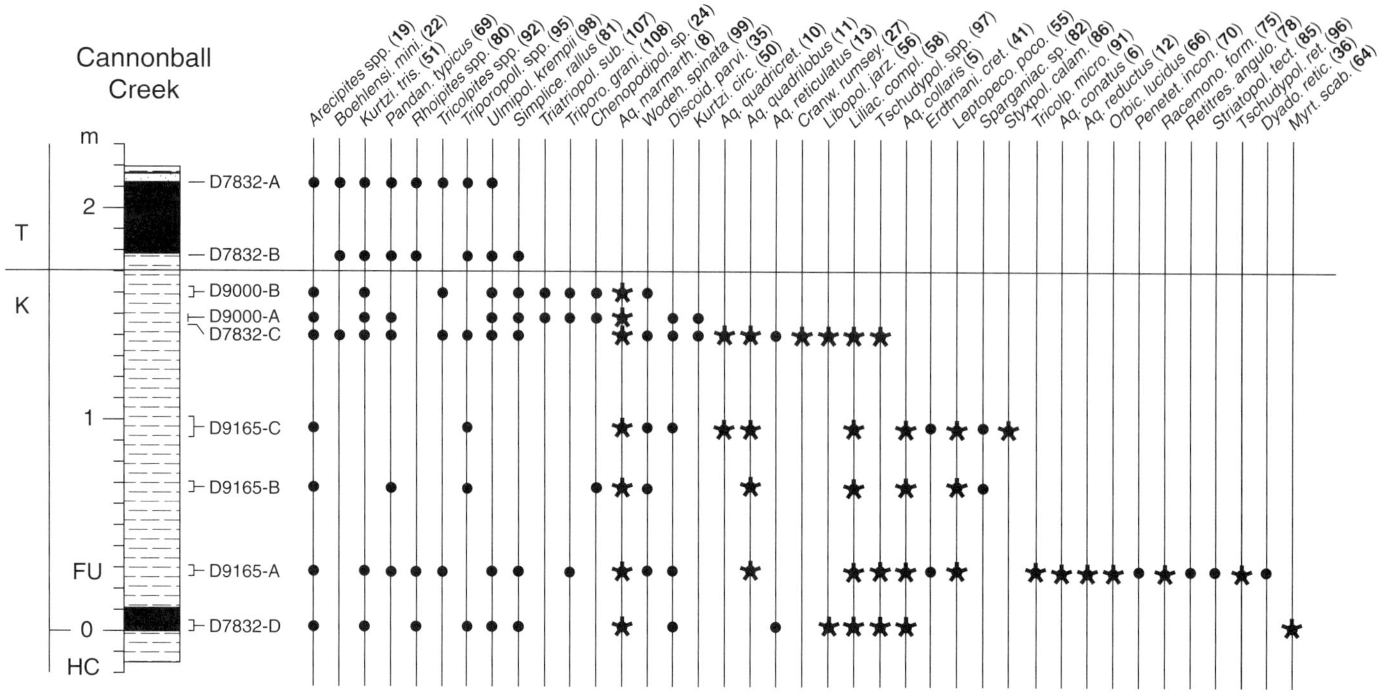

Figure 28. The K-T boundary at Cannonball Creek. HC = Hell Creek Formation, FU = Ludlow Member of the Fort Union Formation; position in the measured section is relative to the formational contact; patterns and symbols as in Figure 4. Only occurrences of angiosperm taxa are shown.

TABLE 8. CANNONBALL CREEK SECTION SAMPLE DATA

Sample	Position	Facies	K taxa	% K	% Angio.	% Gymno.	% Pterido.	% Bryo.
D7832-A	+212	FU lignite	0		43.0	35.0	15.5	6.5
D7832-B	+178	FU mudstone	0		24.5	22.0	53.5	0
D9009-B	+158–163	FU mudstone	1	0.5	15.0	24.5	60.5	0
D9009-A	+146–150	FU mudstone	1	0.5	67.0	9.0	23.5	0.5
D7832-C	+146	FU mudstone	7	4.0	39.5	26.5	34.0	0
D9165-C	+91–101	FU mudstone	8	2.0	5.5	9.5	84.0	1.0
D9165-B	+66–71	FU mudstone	5	1.0	3.0	9.5	86.5	1.0
D9165-A	+26–31	FU mudstone	13	9.5	26.5	12.5	55.0	6.0
D7832-D	+0–5	FU lignite	6	2.5	9.0	79.0	7.5	4.5

Note: HC = Hell Creek Formation, FU = Ludlow Member of the Fort Union Formation; sample positions in centimeters relative to the formational contact. K taxa are also included in % Angiosperms.

Formation. The contact is marked by a change in facies within clastic deposits, from dark gray shale with "popcorn" weathering at the surface to brown mudstone that weathers to a buff color with no "popcorn" on the weathered surface. Ten evenly spaced samples were collected through this interval, five from the Hell Creek Formation and five from the Fort Union. Only four of these were analyzed, however, including two from the Hell Creek and two from the Fort Union. All are Maastrichtian in age.

Significantly, there is no pattern of decline in occurrence of K taxa across the formational contact in this section, presumably because the basal part of the Fort Union in this local area is not lignitic, and thus does not represent mire paleoenvironments. Occurrences of 11 different K taxa were recorded. Assemblages from the Hell Creek samples (shale 5–10 cm and 35–40 cm below the formational contact) each yielded nine taxa that account for 26.0% and 30.0% of their assemblages, respectively. Those from the Fort Union samples (mudstone 5–10 and 35–40 cm above the contact) yielded seven and six taxa that account for 24.0% and 31.0%, respectively, of those assemblages. Two K taxa, *Aquilapollenites collaris* and *A. quadrilobus*, are the most common K taxa in three of the four samples, and *A. collaris* is second in abundance in the fourth. Thus, it is evident that in non-lignitic facies, typical diversities and abundances of K taxa persist into strata of Paleocene age.

Tonstein To Die For. This measured section (number 24 in Fig. 1, Cannonball Creek area) was named for the presence of an intriguing ash layer (tonstein) within a 1.8-m-thick Fort Union lignite bed. Twelve lignite samples were analyzed, all of which proved to be Paleocene in age. Thus, the tonstein is not

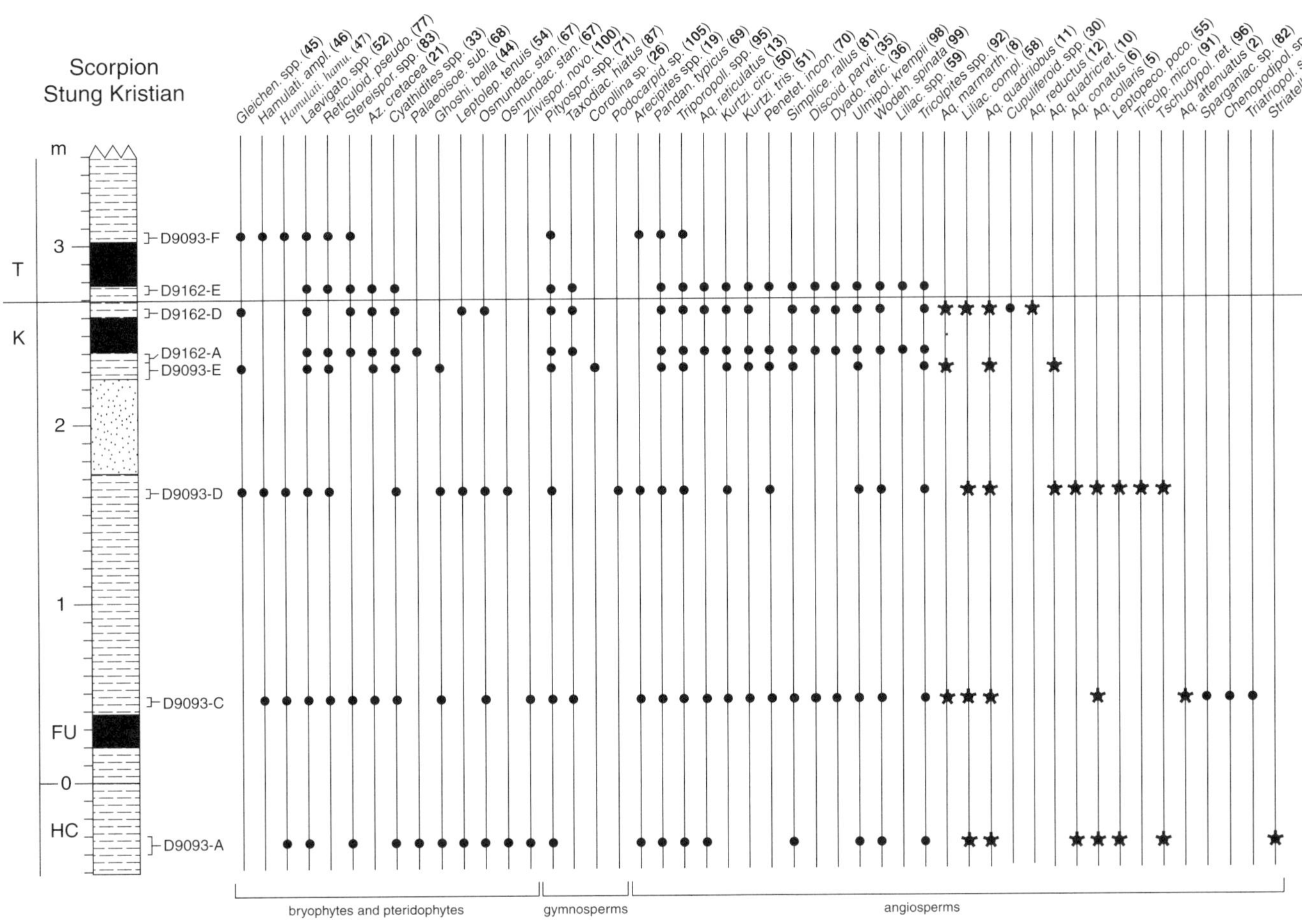

Figure 29. The K-T boundary at Scorpion Stung Kristian. HC = Hell Creek Formation, FU = Ludlow Member of the Fort Union Formation; position in the measured section is relative to the formational contact; patterns and symbols as in Figure 4. Occurrences of all taxa are shown; taxa are grouped by botanical affinity.

associated with the K-T boundary. This site is nonetheless valuable for the insight it provides about the palynological composition of a typical Fort Union lignite of Paleocene age.

No K taxa were recorded from any samples from this section. Fern spores of the genus *Laevigatosporites* are common in all samples and numerically dominant in a sample from the base of the lignite bed. Several species of *Stereisporites* (sphagnum) are present and are notably abundant at some levels within the lignite bed. From this it is evident that ferns and sphagnum moss were the primary inhabitants of the mire in which the peat that formed this lignite deposit accumulated. This is the only section in the study area that is entirely Tertiary in age.

FORT UNION STRATA OF CRETACEOUS AGE

It is clearly evident from the foregoing analyses that the Hell Creek–Fort Union formational contact is only rarely coincident with the K-T boundary in the study area. Of the 24 sections described in this report (17 K-T boundary and seven near-boundary), the contact and the boundary are coincident at only two. Twenty-one sections reveal the presence of Fort Union strata of Cretaceous age at or near the K-T boundary. The pinpointed or bracketed position of the boundary is above the formational contact in 15 of the 17 K-T boundary sections. And, in six of the seven near-boundary sections discussed, samples were collected through intervals within the lower part of the Fort Union Formation in efforts to find the boundary, but all samples proved to be of Maastrichtian age; thus the boundary is above the formational contact in unsampled intervals in these six sections, as well. (The seventh near-boundary section was entirely Tertiary in age.) K-T boundary positions for the 17 complete sections are summarized in Table 13.

It follows that results of our analyses in 21 measured sections reveal the existence of a palynologically distinctive stratigraphic unit: Fort Union strata of Cretaceous age. In Figure 3, Fort Union strata of Cretaceous age are represented by the space between the line marking the top of the Hell Creek Formation and the line marking the K-T boundary. Figure 36 shows correlation of eight measured sections in and near the Cannonball Creek valley, an area of eight square kilometers, using the K-T boundary as the datum. The relationship of the K-T boundary

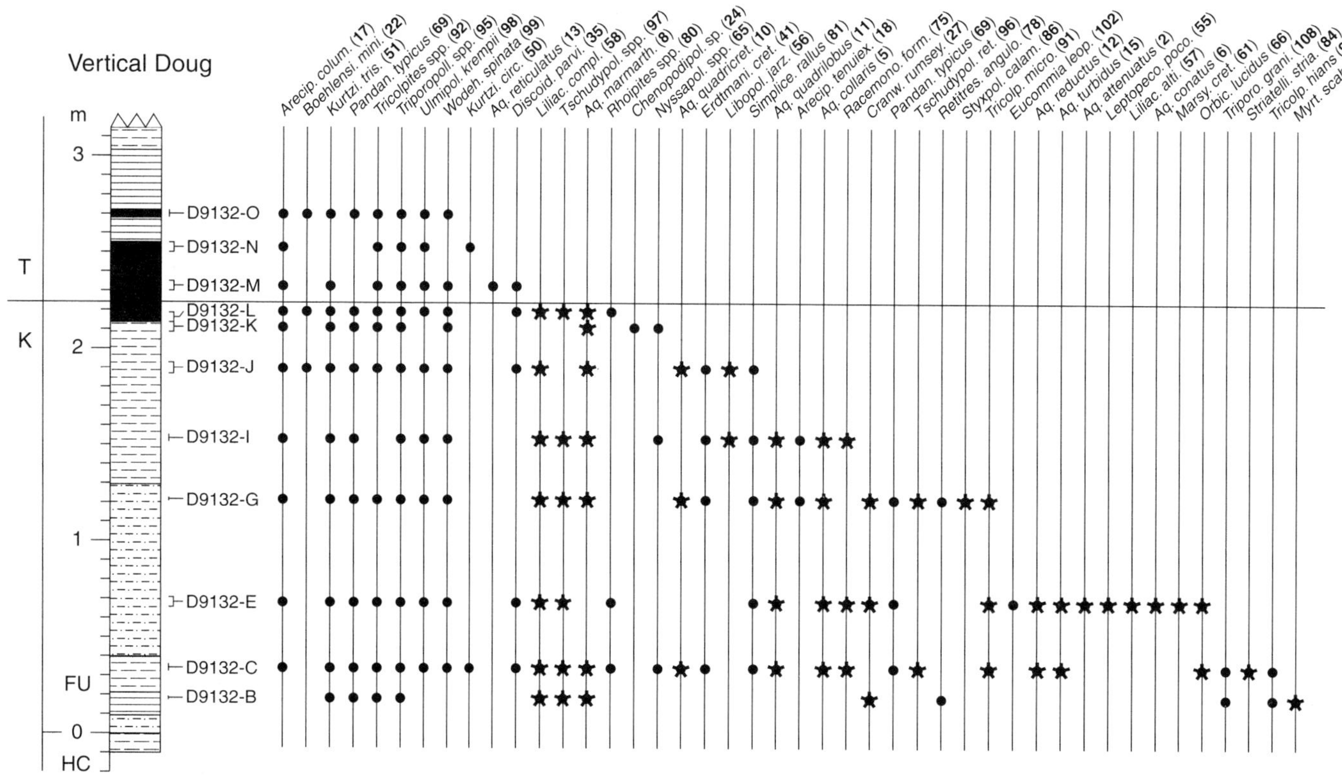

Figure 30. The K-T boundary at Vertical Doug. HC = Hell Creek Formation, FU = Ludlow Member of the Fort Union Formation; position in the measured section is relative to the formational contact; patterns and symbols as in Figure 4. Occurrences of 44 of 47 angiosperm taxa occurring in more than one sample are shown, except that all K taxa shown.

TABLE 9. VERTICAL DOUG SAMPLE DATA

Sample	Position	Facies	K taxa	% K	% Angio.	% Gymno.	% Pterido.	% Bryo.
D9132-O	+268–272	FU lignite	0	0	70.0	20.5	9.5	0
D9132-N	+250–255	FU lignite	0	0	31.5	19.0	46.5	3.0
D9132-M	+230–235	FU lignite	0	0	20.0	12.0	66.5	1.5
D9132-L	+214–219	FU lignite	3	0.5	40.0	41.5	18.0	0.5
D9132-K	+209–214	FU carb shale	1	0.5	8.5	25.0	62.5	4.0
D9132-J	+187–193	FU carb shale	4	1.0	10.0	31.5	58.0	0.5
D9132-I	+152–156	FU claystone	7	4.5	14.5	28.0	45.5	12.0
D9132-G	+121–123	FU mudstone	10	5.5	13.5	37.5	45.0	4.0
D9132-E	+66–71	FU mudstone	15	24.0	49.5	17.5	32.0	1.0
D9132-C	+33–36	FU claystone	9	11.5	25.0	20.0	52.5	2.5
D9132-B	+17–19	FU carb shale	5	2.5	6.5	81.0	12.5	0

Note: HC—Hell Creek Formation, FU—Ludlow Member of the Fort Union Formation; sample positions in centimeters relative to the formational contact. K taxa are also included in % Angiosperms.

to the Hell Creek–Fort Union formational contact is indicated; the interval between the boundary and the contact comprises Fort Union strata of Cretaceous age. This distinctive unit is characterized lithologically by deposits typical of the Ludlow Member of the Fort Union Formation, including lignite, carbonaceous shale, brown claystone and mudstone that weather to buff shades, and variegated beds; and palynologically by assemblages that include typically Maastrichtian K taxa. Of 30 K taxa recorded in the upper part of the Hell Creek Formation, 27 (90%) are present in Fort Union strata of Cretaceous age in different sections. The number of K taxa in 149 individual samples of this unit ranges from one to 13 and averages about six. These numbers are comparable to those for the Hell Creek Formation, which are a range of one to 16 and an average of about eight. The most commonly occurring K taxa in Fort Union strata of Cretaceous age are *Aquilapollenites collaris, A. marmarthensis, A. quadricretaeus, A. quadrilobus*, and *Tschudypollis* spp., all of which are present in >80% of the sections that contain these strata.

Although palynological assemblages from the Fort Union

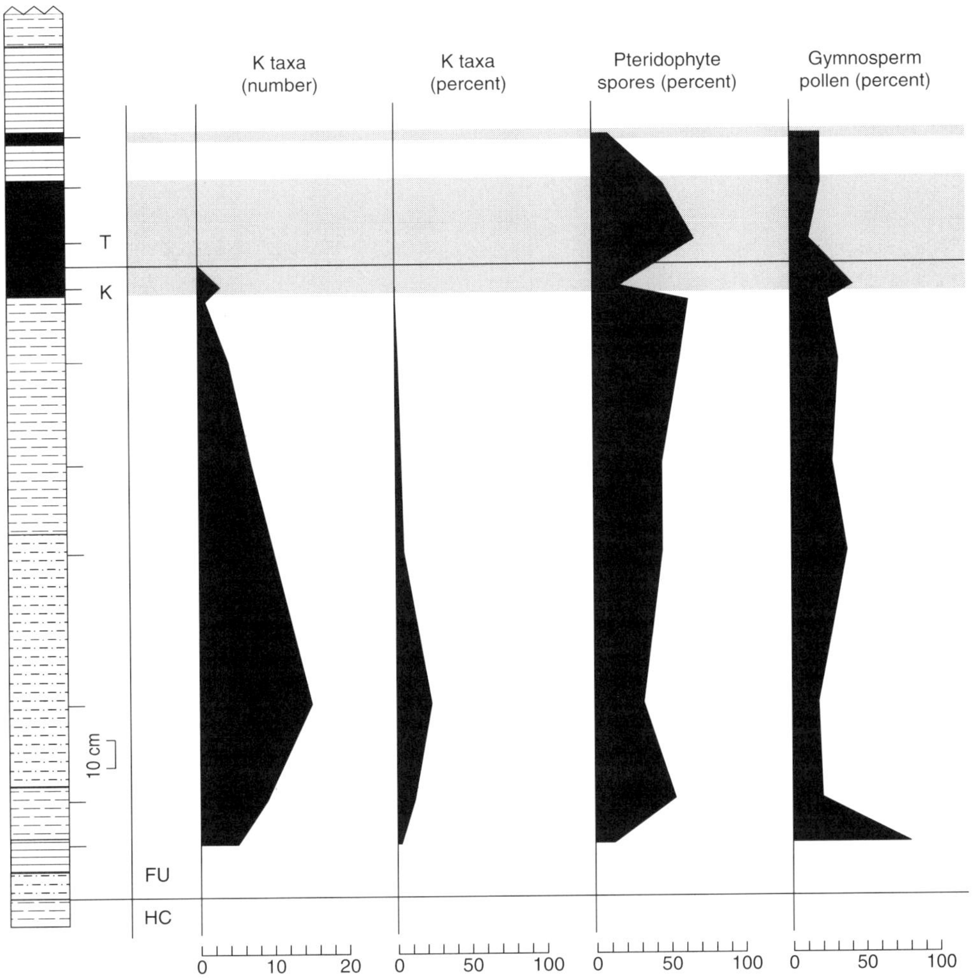

Figure 31. Diagram depicting numbers of K taxa and relative abundance of K taxa, pteridophyte spores, and gymnosperm pollen in part of the Vertical Doug section.

strata of Cretaceous age include numerous K taxa, they differ from Hell Creek assemblages in having reduced relative abundances of these typically Maastrichtian species. Relative abundance of K taxa in typical assemblages from the Hell Creek Formation averages 25.5%, and relative abundance of K taxa in typical assemblages from Fort Union strata of Cretaceous age averages ~6% (Tables 3–12). These relative abundance numbers are useful guides, but as with most geologic phenomena, there are exceptions. Hell Creek assemblages that have <1% K taxa are known, as at Pyramid Butte, and some Fort Union samples of Cretaceous age yielded assemblages having percentage abundances of K taxa that are within the range of typical Hell Creek assemblages, as at Battleship and Terry's Fort Union Dinosaur. The more important observation is that most samples of Fort Union strata of Cretaceous age contain a low percentage of K taxa.

In some sections, the Fort Union strata of Cretaceous age are further characterized palynologically by the presence of taxa that are typically Paleocene. There are four such taxa that particularly appear to presage the arrival of the Paleocene flora: *Brevicolporites colpella, Eucommia leopoldae, Podocarpidites* sp., and *Thomsonipollis magnificus*. These species evidently are strongly associated with typical Ludlow paleoenvironments and lithofacies. This pattern is even stronger for FU0 megafloral taxa, which have a marked Paleocene aspect (Johnson, this volume). Thus, Fort Union strata of Cretaceous age represent Paleocene-style depositional environments that developed locally a short time prior to the end of the Cretaceous. These environments were inhabited by some plants (represented especially by the four taxa mentioned) associated with pond and mire environments of a kind that were rare in the area in Maastrichtian time but common to dominant in Paleocene time.

Aside from K taxa and the typically Paleocene taxa, assemblages from the Fort Union strata of Cretaceous age are characterized by high percentages of pteridophyte spores, sometimes in abundances that constitute numerical dominance. The most abundant taxa are fern spores of the genera *Laevigatosporites* and *Gleicheniidites* and, in a few samples, lycopod spores of the genus *Palaeoisoetes*. The plants that produced these spores evidently inhabited certain waterlogged paleoenvironments in great numbers, including but not restricted to peat mires. The relative abundance of these spores tends to

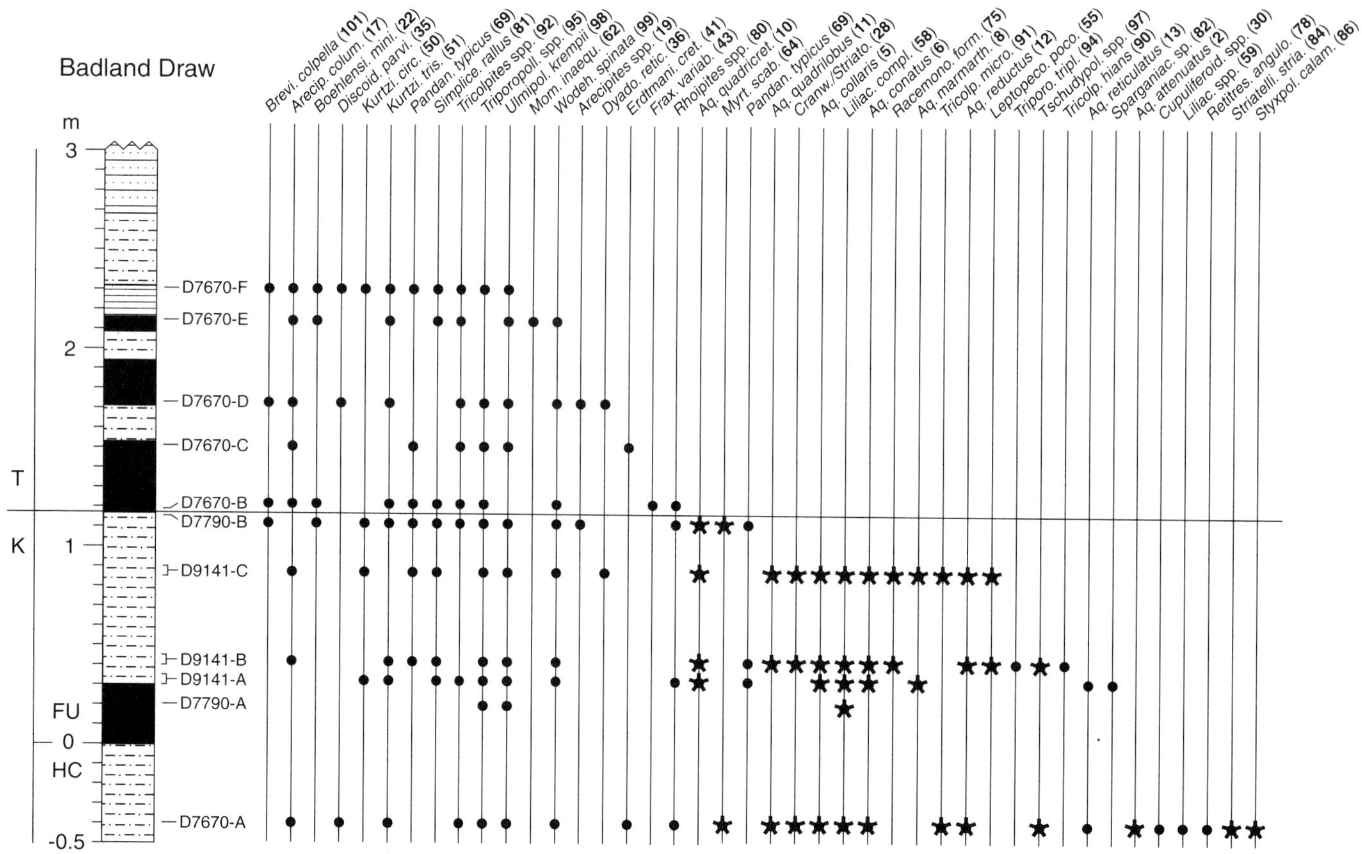

Figure 32. The K-T boundary at Badland Draw. HC = Hell Creek Formation, FU = Ludlow Member of the Fort Union Formation; position in the measured section is relative to the formational contact; patterns and symbols as in Figure 4. Only occurrences of angiosperm taxa are shown.

TABLE 10. BADLAND DRAW SECTION SAMPLE DATA

Sample	Position	Facies	K taxa	% K	% Angio.	% Gymno.	% Pterido.	% Bryo.
D7670-F	+231	FU carb shale	0		53.5	36.0	10.0	0.5
D7670-E	+220	FU lignite	0		60.0	19.0	20.0	1.0
D7670-D	+185	FU lignite	0		57.0	7.5	34.5	1.0
D7670-C	+157	FU lignite	0		12.5	7.5	59.0	5.5
D7670-B	+130	FU lignite	0		31.0	51.0	16.0	2.0
D7790-B	+115	FU siltstone	1	0	63.0	17.5	19.5	0
D9141-C	+85–90	FU shale	9	4.5	13.0	24.0	62.0	1.0
D9141-B	+40–45	FU shale	10	2.5	20.5	35.5	36.5	7.5
D9141-A	+30–35	FU shale	5	4.0	12.5	18.5	62.5	6.5
D7790-A	+20	FU lignite	2	15.5	21.0	68.5	7.0	3.5
D7670-A	−40	HC claystone	13	17.0	46.0	1.0	27.5	25.5

Note: HC—Hell Creek Formation, FU—Ludlow Member of the Fort Union Formation; sample positions in centimeters relative to the formational contact. K taxon found after count of 200 specimens per sample recorded with 0% abundance. K taxa are also included in % Angiosperms.

overwhelm the numbers of K taxa present, which contributes significantly to the reduced percentages of K taxa observed. Some samples of Fort Union strata of Cretaceous age are characterized by high relative abundances of pollen of the genera *Taxodiaceaepollenites* and *Inaperturopollenites*, instead of pteridophyte spores. This abundant pollen was produced by gymnosperms of taxodiaceous/cupressaceous/taxaceous affinity. Great abundances of gymnosperm pollen similarly suppress the percentages of K taxa recorded in such samples. These gymnosperm taxa represent plants that inhabited mires and associated paleoenvironments. Although commonly associated with bog environments, spores of the bryophyte genus *Stereisporites* (sphagnum) are not necessarily significant components of these assemblages. In several sections these spores are more abundant in the Hell Creek Formation than in the Fort Union (see Tables 4, 7, 10, 11).

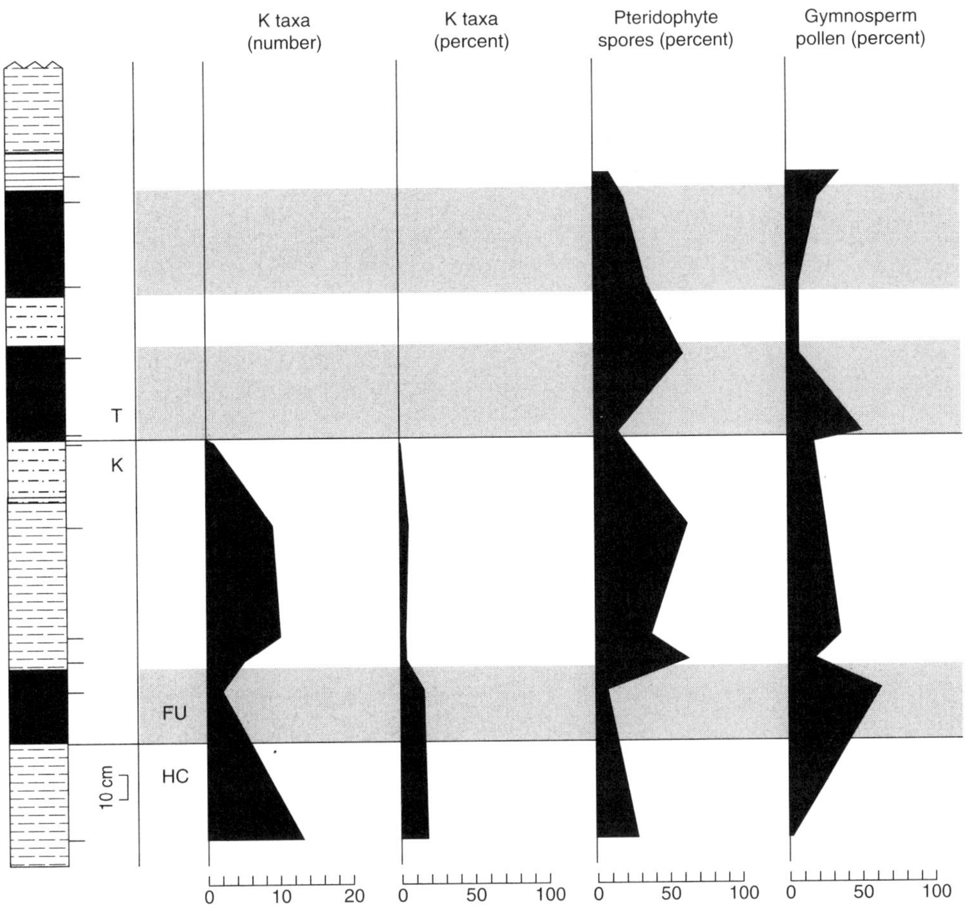

Figure 33. Diagram depicting numbers of K taxa and relative abundance of K taxa, pteridophyte spores, and gymnosperm pollen in part of the Badland Draw section.

From a palynological perspective, however, it is not the presence of coal-forming peat mires per se that constitutes the critical change in depositional environment in the transition from the Hell Creek Formation to the Fort Union Formation. This is evident from patterns of diversity and relative abundance of K taxa in lignite and non-lignite facies shown in Figures 12, 18, 25, 31, and 33. In some sections, basal Fort Union lignite beds yield assemblages characterized by high diversity and abundances of K taxa. Examples include the Battleship section (three samples with six to nine K taxa accounting for 13.5 to 27.0% of the assemblages); Barbecue (two samples with five to six K taxa each, accounting for 15.0% to 18.0%); Big Boundary (two samples with six to nine taxa accounting for 15.0% to 27.0%); and Terry's Fort Union Dinosaur (one sample with 12 K taxa accounting for 14.5%). The critical change in paleoenvironment appears to be the replacement of widespread channel and crevasse-splay deposits and their associated plant communities in the Hell Creek Formation with a prevalence of overbank fluvial and pond deposits and their associated plant communities in the Ludlow Member of the Fort Union Formation. Paleoenvironments characterized by the presence of abundant pteridophytes tend to include fewer and less abundant K taxa angiosperms (see Figs. 12, 18, 25, 31, and 33). The Fort Union strata of Cretaceous age are critical to understanding the palynological record across the K-T boundary in western North Dakota, and by implication the record of the K-T transition in all areas where the boundary lies within rocks representing peat mires and associated paludal paleoenvironments.

COMPARISONS WITH OTHER K-T BOUNDARY LOCALITIES

In the foregoing discussions, the numerous K-T boundary sections in southwestern North Dakota have been compared and contrasted with each other. The question remaining is: how do these localities compare and contrast with others in the Western Interior of North America? A brief review of the characteristics of these other localities that is based on previously published reports and two unpublished Ph.D. dissertations provides answers to this question. The similarities and differences revealed are the basis of further interpretations of the significance of the K-T boundary sections in our study area.

South-central North Dakota

In the outcrop section near the town of Huff, North Dakota, studied by Lerbekmo and Coulter (1984), the K-T boundary is marked by disappearance of *Aquilapollenites* spp., "*Proteaci-*

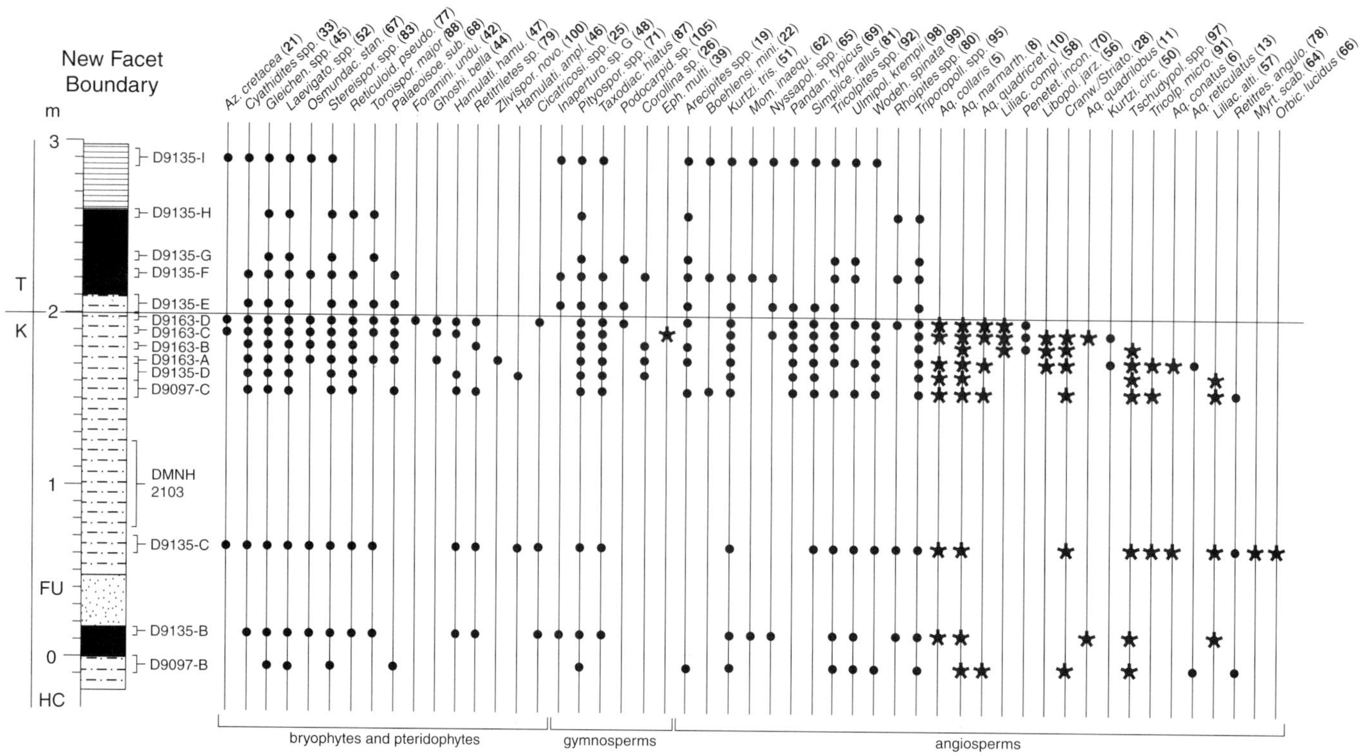

Figure 34. The K-T boundary at New Facet Boundary. HC = Hell Creek Formation, FU = Ludlow Member of the Fort Union Formation; position in the measured section is relative to the formational contact; patterns and symbols as in Figure 4. Occurrences of all taxa recorded in more than one sample are shown (except that all K taxa are shown); taxa are grouped by botanical affinity. A Maastrichtian mire megaflora is present in the interval designated DMNH 2213.

dites" (= *Tschudypollis* of this report), and *Tricolpites microreticulatus* just below the lowest lignite bed in the Ludlow Member of the Fort Union Formation. The K-T boundary interval was determined by these authors to be within magnetostratigraphic subchron C29r.

Murphy et al. (1995) bracketed the K-T boundary in 12 sections in south-central North Dakota, including the one near Huff. In eight of their 12 sections the interval in which the boundary is bracketed is 30 cm or less in thickness. The boundary occurs at or near the Hell Creek–Fort Union formational contact in four sections, above it in five sections, and below it in two (in one section the bracketing interval was too thick to permit an accurate determination of its position). Murphy et al. stated that a K-T boundary claystone of the kind known at other localities in western North America is not present in their study area. Iridium analyses were conducted at one site but a significant Ir anomaly was not detected. These results could indicate that their sections are incomplete to the extent that they do not preserve the thin deposit of impact ejecta known elsewhere. The palynomorphs named and illustrated (Murphy et al., 1995, Fig. 7) show that the Maastrichtian and Paleocene assemblages in south-central North Dakota are the same as those in our study area in the southwestern part of the state. Megafossil paleobotanical studies were not conducted in south-central North Dakota although some collections were made, and floras typical of uppermost Maastrichtian zone HCIII (see Johnson, this volume) were found in the uppermost Cretaceous deposits.

Thus, the data from south-central North Dakota are comparable with those from our study area. The palynofloras are the same, extinction of several K taxa marks the K-T boundary within subchron C29r, and the K-T boundary varies in stratigraphic position with relation to the Hell Creek–Fort Union formational contact.

Southern Saskatchewan

Nichols et al. (1986) originally described the K-T boundary at the contact of the Frenchman and Ravenscrag Formations (Hell Creek and Fort Union Formation equivalents) in what they called the Morgan Creek area, which is now known as the Rock Creek area (Sweet et al., 1999). Only 29 palynomorph taxa were listed in this brief report (Nichols et al., 1986, Fig. 2), but close palynofloristic similarity to the stratigraphically equivalent assemblages from North Dakota is evident; an extinction of ~30% of the Maastrichtian palynoflora was noted. The palynological extinction is closely associated with an Ir anomaly of 3 ppb over a background averaging 0.028 ppb, and shocked quartz is present in a distinctive claystone layer at the palynologically defined K-T boundary. Nichols et al. (1986) called special attention to the survival of species of *Arecipites*

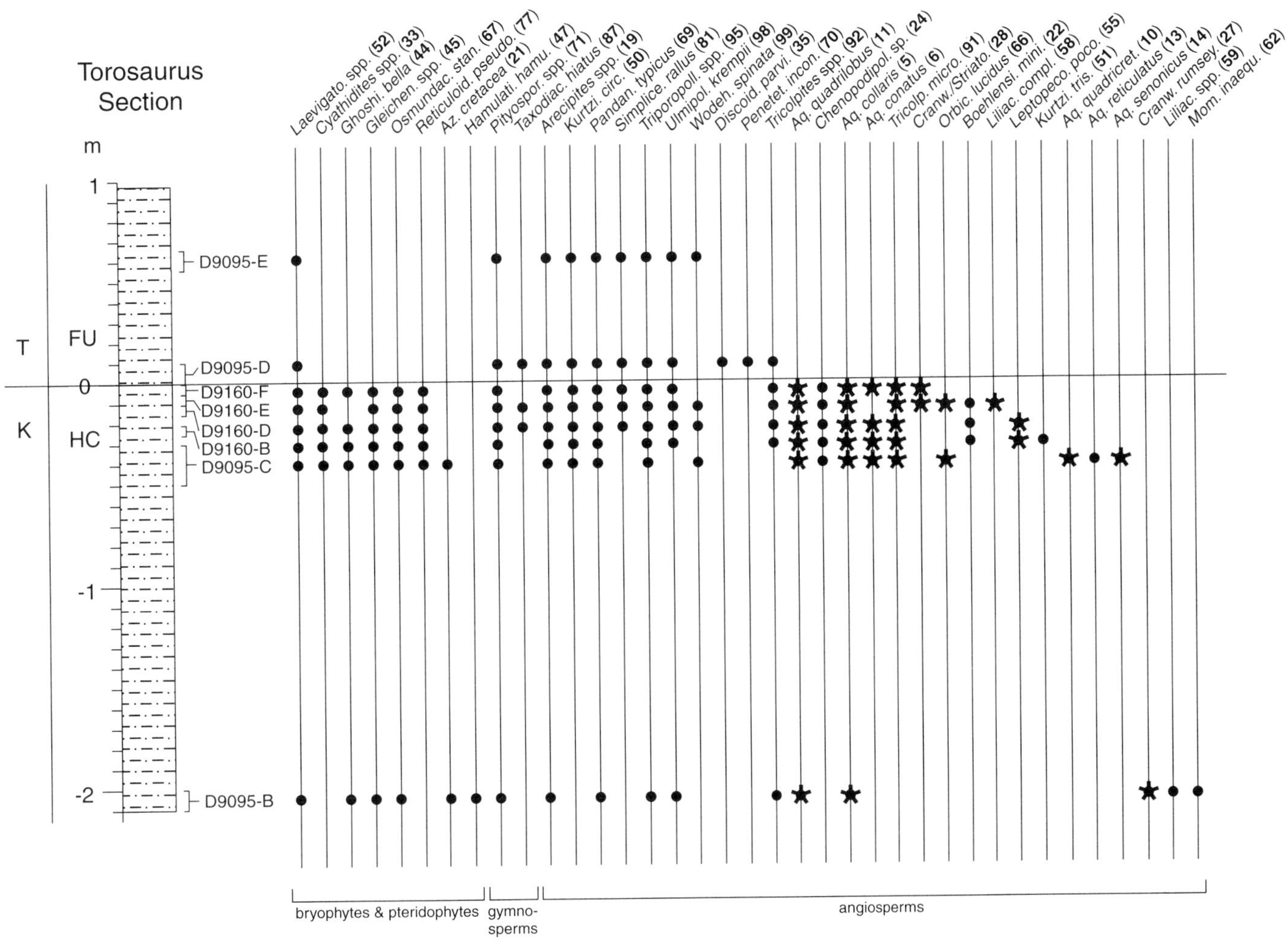

Figure 35. The K-T boundary at Torosaurus Section. HC = Hell Creek Formation, FU = Ludlow Member of the Fort Union Formation; position in the measured section is relative to the formational contact; patterns and symbols as in Figure 4. Occurrences of all taxa are shown; taxa are grouped by botanical affinity. This is one of only two sections in the study area where the K-T boundary is at the Hell Creek–Fort Union formational contact.

TABLE 11. BARBECUE SAMPLE SECTION DATA

Sample	Position	Facies	K taxa	% K	% Angio.	% Gymno.	% Pterido.	% Bryo.
D8224-H	+56–60	FU mudstone	4	4.0	29.5	26.0	43.5	1.0
D8224-G	+46–50	FU mudstone	5	5.0	24.0	33.0	41.0	2.0
D8224-F	+37–42	FU lignite	5	18.0	44.5	32.5	21.5	1.5
D8224-E	+20–30	FU lignite	6	15.0	48.0	9.0	36.0	7.0
D8224-D	+0–5	FU lignite	9	6.5	26.0	61.0	12.0	1.0
D8224-C	−0–5	HC mudstone	9	28.0	47.0	13.0	31.5	8.5
D8224-B	−12–15	HC mudstone	8	16.0	43.5	5.5	36.5	14.5
D8224-A	−40–42	HC mudstone	8	5.0	21.5	0.5	66.0	12.0

Note: HC—Hell Creek Formation, FU—Ludlow Member of the Fort Union Formation; sample positions in centimeters relative to the formational contact. K taxa are also included in % Angiosperms.

and *Pandaniidites* across the K-T boundary. They interpreted these taxa to represent thermophilic plants such as palms whose persistence is evidence against a climate change involving onset of winter frosts (hence, the advent of cold winters in Tertiary time can be ruled out as the cause of plant extinctions at the K-T boundary).

Recently, Sweet et al. (1999) published detailed analyses of the K-T boundary at two localities in the Rock Creek area

TABLE 12. NDAS SECTION SAMPLE DATA

Sample	Position	Facies	K taxa	% K	% Angio.	% Gymno.	% Pterido.	% Bryo.
D9164-O	+235–237	FU claystone	5	2.5	17.5	31.0	40.0	11.5
D9164-N	+224–226	FU claystone	6	3.0	16.5	35.0	40.0	8.5
D9162-M	+214–216	FU claystone	6	7.0	19.0	12.0	63.0	6.0
D9164-L	+204–206	FU claystone	6	4.0	7.0	5.0	86.5	1.5
D9164-K	+184–189	FU claystone	11	12.0	28.0	17.5	54.5	0
D9164-J	+143–145	FU mudstone	13	11.0	27.0	21.0	49.5	2.5
D9164-H	+111–113	FU mudstone	12	18.5	36.5	10.5	50.5	2.5
D9164-F	+79–84	FU mudstone	15	25.0	37.0	11.0	45.5	6.5
D9164-D	+56–61	FU mudstone	10	10.5	38.5	11.5	46.0	4.0
D9164-A	+0–5	FU claystone	1	0.5	2.0	0	41.5	56.4

Note: FU—Ludlow Member of the Fort Union Formation; sample positions in centimeters above the Hell Creek–Fort Union formational contact. K taxa are also included in % Angiosperms.

TABLE 13. CRETACEOUS-TERTIARY BOUNDARY POSITIONS

Locality	Position	Error	Samples
1. Pyramid Butte (KJ86161)	96.0 cm above contact	±0 cm	D6958-F/D6958-G
2. Mikey's Delite (KJ9142)	248.5 cm above contact	±1.5 cm	D7798-B/D7798-D
3. Terry's FU Dino (KJ9653)	260.5 cm above contact	±2.5 cm	D9090-P/D9090-Q
4. Dean's High Dino (KJ9612)	214 cm above contact	±3 cm	D9009-O/D9009-Q
5. Cannonball Creek (KJ9144)	170.5 cm above contact	±7.5 cm	D9000-B/D7832-B
6. Scorpion Stung . . . (KJ9734)	269 cm above contact	±4 cm	D9162-B/D9162-E
7. Vertical Doug (KJ9840)	224.5 cm above contact	±5.5 cm	D9132-L/D9132-M
8. Badland Draw (KJ9909)	117.5 cm above contact	±2.5 cm	D7790-B/D7670-B
9. New Facet (KJ9847)	199 cm above contact	±1 cm	D9163-D/D9135-E
10. Big Boundary (KJ9328)	96 cm above contact	±10 cm	D8226-T/D8226-U
11. Pretty Butte North (KJ9749)	248.5 cm above contact*	±178.5 cm	D8025-J/D8025-K
12. Pretty Butte (Murphy)	112 cm above contact	±26 cm	D7788-CC/D7788-BB
13. River Section (KJ87147)	259.5 cm above contact*	±199.5 cm	D7138-H/D7138-I
14. Bobcat Butte (KJ86148)	75 cm above contact	±0 cm	D7103-M/D7103-N
15. Sunset Butte (Fastovsky)	170 cm above contact*	±103 cm	D9161-G/D9004-C
16. Torosaurus Section (KJ9738)	At the formation contact	±0 cm	D9160-F/D9095-D
17. Mud Buttes (KJ9905)	At the formation contact	±0 cm	D7672-D/D7672-E

Note: Some locality names are abbreviated. Localities 2–9 are all in the Cannonball Creek valley; map positions of all localities are shown in Figure 1 (but with different numbers). Asterisks mark sections in which the position of the Cretaceous-Tertiary boundary is not well defined but where no refinement is possible on the basis of existing samples.

of southern Saskatchewan (among others described or reinterpreted in that report). They called attention to an abrupt drop in relative abundance of TCT pollen from 60% to 80% below the K-T boundary to near zero above. From this they interpreted that the K-T boundary event caused destruction of arboreal gymnosperms. McIver (1999) substantiated this conclusion using paleobotanical data. A parallel reduction in TCT pollen across the boundary is not evident in our sections in North Dakota, however. In seven sections for which adequate data are available to evaluate this phenomenon (Tables 3–6 and 8–10), we observe averages of 9%–37% TCT pollen below the boundary and 7%–28% above. In three of these sections a reduction of ~50% across the boundary is noted, but in the other four there is negligible difference. As mentioned previously in the

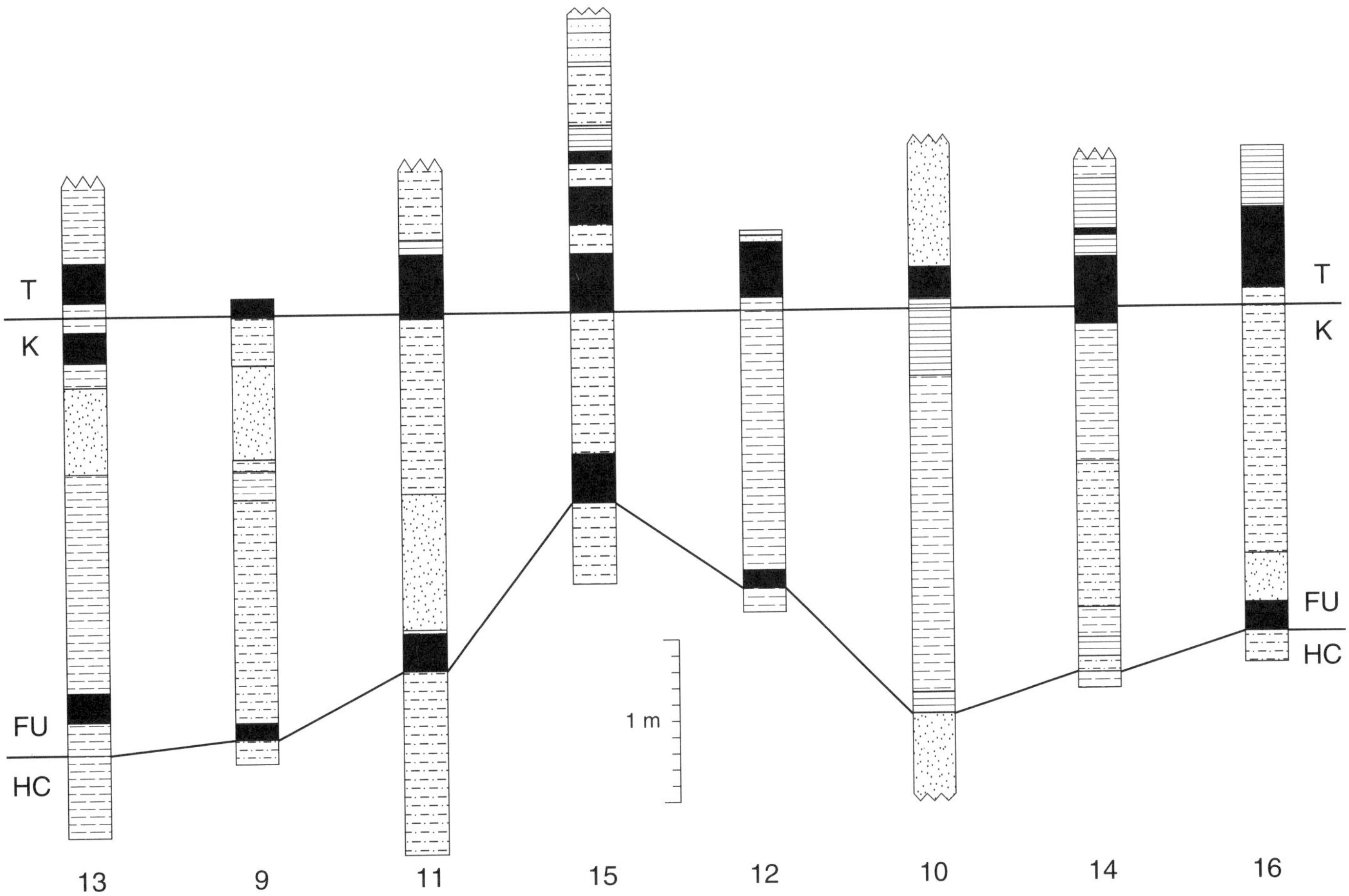

Figure 36. Diagram showing correlation of eight measured sections in and near the Cannonball Creek valley (Fig. 1) and the relationship of the K-T boundary to the Hell Creek–Fort Union formational contact. Datum is the K-T boundary; HC = Hell Creek Formation, FU = Ludlow Member of the Fort Union Formation; patterns and symbols as in Figure 4. Sections are (13) Scorpion Stung Kristian, (9) Terry's Fort Union Dinosaur, (11) Dean's High Dinosaur, (15) Badland Draw, (12) Cannonball Creek, (10) Mikey's Delite, (14) Vertical Doug, (16) New Facet Boundary. The interval between the formational contact and the K-T boundary comprises a distinctive stratigraphic unit, Fort Union strata of Cretaceous age.

section on the palynology of lignite beds of Paleocene age, assemblages from three such deposits in three different sections are composed of 33%–50% TCT pollen.

Eastern Montana

The most extensive and detailed study of the K-T boundary in eastern Montana is that of Hotton (1988; see also Hotton, this volume). Hotton's analyses involved 52 samples from the Hell Creek Formation and 103 from the Tullock Member of the Fort Union Formation collected from six localities in the Hell Creek area. She described 313 palynomorph taxa, including 291 species of pollen and spores. This palynofloral database is about three times as large as ours because Hotton conducted meticulous and detailed taxonomic analyses as a major part of her dissertation research. Her results provide strong support for the extraterrestrial-impact theory of the K-T extinctions (Hotton, 1984; 1988; this volume). Her data and interpretations are in close agreement with ours, in general. Specifically, they are closely parallel with our results in the Mud Buttes section, which, in lacking an interval of Fort Union strata of Cretaceous age below the K-T boundary, has a geologic setting like Hotton's in eastern Montana.

Eastern Wyoming

In his K-T boundary section in the Lance Creek area in the southeastern part of the Powder River Basin, Leffingwell (1970) analyzed 18 samples and recorded the relative abundances and stratigraphic ranges of 38 taxa within an interval of 38 m. His section was not sampled in the detail necessary to make valid comparisons with data from our North Dakota sections, however. Bohor et al. (1987) located the K-T boundary in the same area using 17 samples within an interval of 60 cm. An Ir anomaly, shocked quartz, and impact-derived spherules were found in the K-T boundary claystone layer. The palynological data from the Lance Creek area ("Dogie Creek" locality of Bohor et al., 1987) show that the K-T boundary is near the top of the Lance Formation (the Hell Creek Formation equivalent), several centimeters below a thin lignite bed that marks the base of the Tullock Member of the Fort Union Formation. The palynological analysis produced a list of ~50 palynomorph species. The Maastrichtian assemblage included ten K taxa, all species that are now also known in the Maastrichtian of southwestern North Dakota. Extinction was calculated at 29% (this was not mentioned in the published report), and was described as abrupt because all ten of the K taxa occurred within 6 cm below the K-T boundary and seven were present in the sample 0–2 cm below the boundary.

Nichols et al. (1992) described the K-T boundary in a measured section near the abandoned town of Sussex, Wyoming, on the western edge of the Power River Basin. The K-T boundary occurs at the formational contact between the Lance and Fort Union Formations, just below the lowest lignite bed of the Tullock Member of the Fort Union Formation. A large iridium anomaly (26 ppb above a background of 0.012–0.30) was measured, and shocked quartz was found. The palynologic analysis was based on 31 closely spaced samples within an interval of 1.2 m; 74 palynomorph taxa were identified. The Maastrichtian palynoflora includes most of the species we recorded in North Dakota, but there are some differences. For example, Nichols et al. (1992) identified seven species of *Aquilapollenites*; in contrast, we list twice that number (Table 1). In all, 16 K taxa are present, and the magnitude of extinction at the K-T boundary is 34%. The extinction is abrupt; 14 of the K taxa occur within 5.5 cm below the boundary and 11 of them within 2 cm below.

Thus, the sections in eastern Wyoming are quite similar to one another. They differ from those in North Dakota somewhat in the species composition of the Maastrichtian assemblages. These differences are an aspect of variation in Maastrichtian assemblages with paleolatitude. (See Nichols, 1984; Nichols et al., 1990; Sweet et al., 1990; Nichols and Sweet, 1993; and Nichols, 1996b for further discussions of this paleolatitudinal variation in palynofloral assemblages.) The sections in eastern Wyoming differ from our sections in which an interval of Fort Union strata of Cretaceous age is present below the K-T boundary in showing abrupt extinction of almost all of the K taxa present within a few centimeters, and are most similar to the Mud Buttes section in this respect. The explanation is that there is no facies effect from the presence of lignitic or paludal facies in the Wyoming sections, or at Mud Buttes.

Raton Basin, Colorado and New Mexico

In the Raton Basin of southern Colorado and northern New Mexico, the K-T boundary occurs within the coal-bearing Raton Formation. Investigations began here with the work of Orth et al. (1981) and have continued up to the present. More than 25 localities are now known in the basin (Pillmore et al., 1999). Key references to the palynostratigraphy of the K-T boundary in the Raton Basin are Orth et al. (1981), Tschudy et al. (1984), Tschudy and Tschudy (1986), Fleming (1990), and Pillmore et al. (1984, 1999). The unpublished dissertation by Fleming (1990) is the most extensive study.

Variations in floristic composition of palynologic assemblages due to differences in paleolatitude render the Maastrichtian and lower Paleocene palynofloras of the Raton Basin significantly different from those in North Dakota. For example, *Aquilapollenites* is exceedingly rare in the Raton Basin, and many other K taxa known in North Dakota are absent; correspondingly, some of the Raton Basin K taxa are unknown in North Dakota. Fleming (1990) listed only seven taxa as becoming extinct within 5 cm or less of the K-T boundary, and he calculated extinction averaging 17% at the three localities he studied.

Thus, K-T boundary sections in the Raton Basin differ from those in our study area in southwestern North Dakota in the composition of the palynofloras and in the percentage of K taxa that become extinct. The palynofloristic differences are due to paleolatitude (cf. Nichols and Sweet, 1993). The differences in magnitude of the extinction appear to be due to the geologic setting. The K-T boundary in the Raton Basin is within the coal-bearing Raton Formation; there is no facies change at or near the boundary at any locality in southern Colorado or northern New Mexico that is comparable to that between the Hell Creek and the Fort Union Formation in North Dakota, or between the equivalents of these formations in Saskatchewan, Montana, and Wyoming. The facies containing the boundary interval in the Raton Basin are directly comparable to the Fort Union strata of Cretaceous age we describe in our study area. The effect of the ubiquitous presence of the paludal facies in the Raton Basin appears to be analogous to what we observe at most localities in our study area: reduced diversity and abundance of K taxa present below the K-T boundary. There is less palynofloral change where there is no facies change across the K-T boundary, although the precise stratigraphic level of the palynological extinction is clear.

Alberta and the Northwest Territories, Canada

Several K-T boundary sections in western Canada were described by Jerzykiewicz and Sweet (1986), Lerbekmo et al. (1987), Sweet et al. (1990), Sweet and Braman (1992), Sweet (1994), and Sweet et al. (1999). See Sweet et al. (1990) and Sweet (1994) for summaries, and Sweet et al. (1999) and Sweet and Braman (2001) for the most recent data and interpretations. The more northern of these localities differ from those in our study area in that they exhibit some degree of floristic difference in Maastrichtian and Paleocene palynomorph assemblages due to latitudinal variation (cf. Nichols and Sweet, 1993).

The data and interpretations that have come from these sections and the studies cited above have contributed a great deal to the understanding of vegetational history in western

North America during the Cretaceous-Tertiary transition. The most important difference in the interpretation of these sections in comparison with interpretations we would emphasize from our study is that they have been seen as documenting gradual or step-wise extinction of the Maastrichtian flora through an interval that includes the K-T boundary, as opposed to an abrupt and profound extinction at the boundary (see MacLeod et al., 1997; but see also Hudson, 1998). What seems to have been overlooked by MacLeod and his coauthors is that Sweet and his coauthors have never denied that the palynological record in western Canada provides evidence of plant extinction at the K-T boundary, but only that they have argued that this event was one of several events of lesser magnitude affecting the flora of that region at the end of the Cretaceous and beginning of the Tertiary (see especially Sweet and Braman, 2001). At least some of the lesser events evidently were driven by changes in depositional environment, especially those associated with a regional increase in "wetness" (Sweet and Braman, 1992) leading to development of facies analogous to what we describe here as the Fort Union strata of Cretaceous age. We find this phenomenon to be more local than regional when considering all of western North America, but agree that it had an effect on the composition and relative abundance of elements of both Maastrichtian and Paleocene floras. We assert that the main event at the end of the Cretaceous was a simultaneous, continent-wide extinction of components of regionally varied plant communities inhabiting a variety of environments, and believe that Sweet and his coauthors would concur.

Canadian sections and the hypothesis of a regional unconformity

On the basis of the absence of a clearly identifiable K-T boundary claystone layer and associated physical evidence of impact ejecta in one of their sections, Catuneanu and Sweet (1999) proposed that the region east of a tectonic hinge line extending generally northwestward from the Raton Basin to southern Alberta has a disconformity at the K-T boundary involving the uppermost Maastrichtian and the K-T boundary itself. According to this hypothesis, debris from the K-T impact event could be preserved in the rock record to the west of the hinge line but not to the east of it. This hypothesis was reiterated and somewhat refined by Sweet and Braman (2001). The presence of impact debris at Pyramid Butte did not fit this hypothesis and constituted a strong challenge to it with respect to southwestern North Dakota, but it was disregarded by Catuneanu and Sweet (1999). The presence of impact debris including the Ir anomaly, shocked quartz, and spherules in a well-defined K-T boundary layer at Mud Buttes (which was unknown to Catuneanu and Sweet and to Braman and Sweet when their papers were written) falsifies their hypothesis.

In another test of the hypothesis of Catuneanu and Sweet (1999), which appears to rest largely on palynostratigraphic data presented by Braman and Sweet (1999), we reassessed unpublished data of Murphy et al. (1995) from south-central North Dakota. Critical to the recognition of missing section in a core hole at Turtle Mountain, Manitoba, is the absence of five palynomorph species (Braman and Sweet, 1999, p. 735). The Turtle Mountain area is north of the Murphy et al. study area and east of the Catuneanu and Sweet hinge line. The missing species are *Aquilapollenites quadricretaeus, A. collaris* ("*A. delicatus* var. *collaris*" of Braman and Sweet, 1999), *Leptopecopites pocockii, Marsypiletes cretacea*, and *Myrtipites scabratus*.

The unpublished palynostratigraphic data on which the Murphy et al. (1995) paper was based show that in eight of their closely bracketed K-T boundary sections, there is a composite assemblage of 16 K taxa within 15 cm of the K-T boundary. Among these species are *Aquilapollenites collaris* and *Leptopecopites pocockii*. At the Snyder site (locality 22 of Murphy et al., 1995), where the K-T boundary is bracketed by two samples separated by an interval of 5 cm, *A. collaris* and *L. pocockii* together account for almost 5% of the assemblage in the uppermost Maastrichtian sample. Thus, there is no palynologic evidence supporting the hypothesis of Catuneanu and Sweet (1999) as it might apply to south-central North Dakota.

In our study area in southwestern North Dakota, palynostratigraphic data support our contention that our sections are not affected by the hypothetical regional unconformity. These palynostratigraphic data are in addition to the presence of impact debris at Pyramid Butte and Mud Buttes. All five of the supposedly critical species are present in our study area. *Aquilapollenites quadricretaeus* is present in 19 of our boundary and near-boundary sections; the highest stratigraphic occurrence is at New Facet Boundary, within the interval 1–3 cm below the boundary. *Aquilapollenites collaris* is present in 20 of our sections; the highest stratigraphic occurrence is at Mud Buttes, where it is present 1 cm below the boundary (it is also present at Bobcat Butte in the interval 0–3 cm below, and at New Facet Boundary in the interval 1–3 cm below). *Leptopecopites pocockii* is present in 12 of our sections; the highest stratigraphic occurrence is at Mud Buttes, where it is present 1 cm below the boundary. *Marsypiletes cretacea* is present in six of our sections; its highest stratigraphic occurrence is at Mud Buttes, 41 cm below the K-T boundary. *Myrtipites scabratus* is present in 15 of our sections; the highest stratigraphic occurrence is at Badland Draw, where it is present 7.5 cm below the K-T boundary. Therefore, by the criterion of the absence of these species being an indication of missing section of Maastrichtian age (Catuneanu and Sweet, 1999; Braman and Sweet, 1999), there is no such evidence in our study area.

CONCLUSIONS

Palynostratigraphy, the most practical and reliable method for locating the K-T boundary in nonmarine rocks, brackets or pinpoints the boundary in 17 measured sections in southwestern North Dakota. Iridium anomalies and shocked quartz are known from two of these sections, those that were subjected to appropriate sampling and analysis. This physical evidence of extraterrestrial impact verifies that the global K-T boundary horizon is present in the study area, as it is in previously described uppermost Cretaceous and lowermost Tertiary intervals in adjacent areas and elsewhere in western North America. Palynostratigraphy provides a framework to evaluate other investigations of the K-T boundary in the study area, including those based on lithostratigraphy, magnetostratigraphy, chemostratigraphy, paleobotany, invertebrate paleontology, and vertebrate paleontology.

The K-T boundary is near—but rarely coincident with—the Hell Creek–Fort Union formational contact. A distinctive unit of Fort Union strata of Cretaceous age is recognized at many localities. There is a sedimentary facies effect on palynological assemblages occurring within the Fort Union strata of Cretaceous age that tends to reduce the number and relative abundance of K taxa (characteristic uppermost Cretaceous palynomorphs), but the extinction of K taxa at the K-T boundary was essentially independent of paleoenvironment. The actual effect of the K-T boundary event on the palynoflora can be seen as the total change in the composition of the flora less the component of change attributable to the facies effect. Where there are no Fort Union strata of Cretaceous age, as at Mud Buttes, there is no significant disappearance of palynomorph taxa approaching the boundary, and the extinction can be seen to be more abrupt.

It has long been understood from studies of the morphology of pollen of living plants that, with few exceptions, pollen produced by individual species within a genus cannot be distinguished on the basis of transmitted light microscopy. In some cases the same is true of entire families. From this it follows that "species" of fossil pollen usually represent rather broad botanical categories with respect to neobotanical classification. Thus, palynology demonstrates a major extinction of 30% or more of higher-level plant taxa (botanical genera or families) at the K-T boundary in North Dakota. All but one of the palynomorph taxa that became extinct at the K-T boundary in this area are angiosperms (the single exception is the ephedralean gymnosperm *Ephedripites multipartitus*). In the Pyramid Butte and Mud Buttes sections the palynological extinction is inextricably linked to the occurrence of geochemical and mineralogical evidence of a large-body impact on the Earth, which was the primary cause of the extinction of the plants that produced this pollen.

ACKNOWLEDGMENTS

We thank R.F. Fleming, C.L. Pillmore, and A.R. Sweet for critical reviews of the manuscript. Each contributed materially to its improvement. W.J. Betterton prepared and photographed shocked quartz from Mud Buttes and Pyramid Butte.

APPENDIX 1. K-T BOUNDARY SECTIONS

In the following descriptions of sections measured and sampled, the designation "KJ" followed by a number refers to sections cataloged by Johnson, and the designation "D" followed by a number refers to U.S. Geological Survey localities recorded in the Denver laboratory and sampled for palynology by Nichols. Different D-numbers may designate different suites of samples collected from the same section. Johnson measured all sections unless otherwise noted. Sections are listed alphabetically by name.

Badland Draw

Localities D7670, D7790, and D9141 (KJ9909), Badland Draw quadrangle, Slope County, North Dakota, SW ¼ SE ¼ SW ¼ sec. 4, T. 134 N., R. 106 W., lat. 46° 26′ 24″ N, long. 104° 0′ 09″ W., elev. 847–850 m (no. 15 in Fig. 1). Complete measured section: Hell Creek Formation (0.40 m) and Fort Union Formation (3.03 m); sampled interval: Hell Creek Formation (0.40 m) and Fort Union Formation (2.31 m).

The Badland Draw section, which is only ~3.5 m in thickness, includes the uppermost part of the Hell Creek Formation and the lower part of the Fort Union Formation. There are three lignite beds within this interval; the lower one is Maastrichtian in age and the upper two are Paleocene.

Big Boundary

Locality D8226 (KJ9328), Pretty Butte quadrangle, Slope County, North Dakota, SW ¼ NE ¼ NW ¼ sec. 26, T. 134 N., R. 106 W., lat. 46° 23′ 30″ – 46° 23′ 32″ N., long. 103° 5′ 41″ – 103° 5′ 37″ W., elev. 863–879 m (no. 5 in Fig. 1). Complete measured section: Hell Creek Formation (0.10 m) and Fort Union Formation (25.56 m); sampled interval: Hell Creek Formation (0.10 m) and Fort Union Formation (15.8 m).

The Big Boundary section is located on the north side of Pretty Butte; it was named for the numerous, closely spaced palynological samples collected. The complete measured section includes ~26 m of the uppermost Hell Creek Formation and the lower part of the Fort Union Formation. All but two of the palynology samples are from the lower 16 m of the Fort Union.

Bobcat Butte

Composite section incorporating localities D6956, D6957, and D7103 (KJ86148), Marmarth quadrangle, Slope County, North Dakota, sec. 28, T. 133 N., R. 105 W., lat. 46° 17′ 41″ – 46° 17° 43″ N., long. 103° 5′ 47″ – 103° 5′ 37″ W., elev. 841–917 m (no. 2 in Fig. 1). Complete measured section: Hell Creek Formation (52.77) and Fort Union Formation (36.72 m); sampled interval: Hell Creek Formation (50.00 m, estimated) and Fort Union Formation (26.23 m).

Bobcat Butte is ~3 km east of Marmarth, North Dakota. The complete measured section includes just under 53 m of the Hell Creek Formation and just under 37 m of the Fort Union Formation. The sampled interval includes ~50 m of the Hell Creek and ~26 m of the Fort Union (the uppermost sample is from a unit extending from 25.53 to 26.23 m above the formational contact). In all, 29 samples were used in the analysis. A lignite bed 40 cm thick that marks the Hell Creek–Fort Union formational contact is present within the measured section, but this lignite bed is Maastrichtian in age.

Cannonball Creek

Composite section incorporating localities D7832 and D9000 (KJ9144), Badland Draw quadrangle, Slope County, North Dakota, sec. 33, T. 135 N., R. 106 W., lat. 46° 27′ 38″ – 46° 20′ 48″ N., long. 104° 0′ 9″–103° 5′ 43″ W., elev. 850–852 m (no. 12 in Fig. 1). Complete measured section: Fort Union Formation (2.34 m); sampled interval: Fort Union Formation (2.12 m).

This measured section includes two lignite beds in the Ludlow Member of the Fort Union Formation; six samples were analyzed. The lower lignite bed is Maastrichtian in age and the upper one is Paleocene. The K-T boundary is ~2 m above the Hell Creek–Fort Union formational contact. The boundary is bracketed by two samples separated by ~15 cm.

Dean's High Dinosaur

Composite section incorporating localities D9009, D9089, and D9092 (KJ9612), Badland Draw quadrangle, Slope County, North Dakota, SE ¼ SW ¼ NW ¼ sec. 33, T. 135 N., R. 106 W., lat. 46° 27′ 40″ – 46° 23′ 32″ N., long. 104° 0′ 22″ – 103° 5′ 37″ W., elev. 847–850 m (no. 11 in Fig. 1). Complete measured section: Hell Creek Formation (0.98 m) and Fort Union Formation (3.86 m); sampled interval: Hell Creek Formation (0.18 m) and Fort Union Formation (2.60 m).

D.A. Pearson of the Pioneer Trails Regional Museum collected bones of two dinosaurs from this measured section. There are two lignite beds in the interval, a basal Fort Union lignite bed that is 22 cm thick and a second bed ~2 m above the first that is 38 cm thick. The lower lignite bed is Maastrichtian in age and the upper one is mostly or entirely Paleocene. In all, 35 samples were collected for palynology, although only 13 of them were analyzed for this study. The K-T boundary is either just below the upper lignite bed or in the basal part of it. The boundary is bracketed by two samples separated by 6 cm.

Mikey's Delite

Locality D7798 (KJ9142), Pretty Butte quadrangle, Slope County, North Dakota, SW ¼ SW ¼ SE ¼ sec. 28, T. 135 N., R. 106 W., lat. 46° 28′ 10″ – 46° 20′ 48″ N., long. 103° 5′ 44″ – 103° 5′ 43″ W., elev. 861–864 m (no. 10 in Fig. 1). Complete measured section: Fort Union Formation (14 + m); sampled interval: Fort Union Formation (274 cm).

This measured section is entirely within the Ludlow Member of the Fort Union Formation. The K-T boundary is ~2.5 m above the Hell Creek–Fort Union formational contact and is closely bracketed by two samples that are separated by 3 cm.

Mud Buttes

Localities D7415, D7416, and D7672 (KJ9905), Mud Buttes quadrangle, Bowman County, North Dakota, NW ¼ sec. 1, T. 130 N., R. 105 W., lat. 46° 1′6″ – 46° 1′ 8″ N., long. 103° 4′ 37″–103° 4′ 31″ W., elev. 954–967 m (no. 7 in Fig. 1). Complete measured section: Hell Creek Formation (15.35 m) and Fort Union Formation (45.40 m); sampled interval: Hell Creek Formation (4.22 m) and Fort Union Formation (8.39 m).

Mud Buttes, the southernmost section in the study area, is ~32 km southeast of Marmarth, North Dakota. The measured section includes ~15 m of the Hell Creek Formation and ~45 m of the Ludlow Member of the Fort Union Formation. The sampled interval includes ~4 m of the Hell Creek and ~8 m of the Fort Union; 16 samples were collected and analyzed. A basal Fort Union lignite bed of Paleocene age is present, although it is only ~1 cm thick. The K-T boundary is quite closely bracketed at the Hell Creek–Fort Union formational contact; it is bracketed by two stratigraphically juxtaposed samples.

New Facet Boundary

Localities D9097 and D9135 (KJ9754 and KJ9847), Pretty Butte quadrangle, Slope County, North Dakota, NW ¼ SW ¼ NE ¼ SW ¼ sec. 34, T. 135 N., R. 106 W., lat. 46° 27′ 35″ N., long. 103° 5′ 59″ W., elev. 835 m (no. 16 in Fig. 1). Measured section supplemented by samples from adjacent locality D9097 (KJ9754). Complete measured section: Hell Creek Formation (8.28 m) and Fort Union Formation (44.22 m); sampled interval: Hell Creek Formation (0.20 m) and Fort Union Formation (2.98 m).

Fourteen samples were analyzed from this locality. These samples were supplemented by two from an adjacent locality called Facet Butte, which is a few meters lateral to it in the same section, township, and range. Together these 16 samples encompass ~3 m of section. There are two thin lignite beds within the section, in the Ludlow Member of the Fort Union Formation. The lower lignite bed is Maastrichtian in age and the upper one is Paleocene. The K-T boundary is ~2 m above the Hell Creek–Fort Union formational contact. It is bracketed by two samples separated by 2 cm. A megafossil flora was collected from this locality by Johnson that includes taxa of Paleocene aspect but Maastrichtian age.

Pretty Butte

Composite section incorporating localities D7656 and D7788 and supplemented by D9094 and D9134 (KJ9735), Pretty Butte quadrangle, Slope County, North Dakota, SW ¼ SE ¼ SE ¼ sec. 35, T. 134 N., R. 106 W., lat. 46° 22′ 33–35″ N, long. 103° 56′ 52″–103° 57′ 08″ W., elev. 835–854 m (main section, measured by E. C. Murphy), and sec. 35, T. 134 N., R. 106 W., lat. 46° 22′ 41–43″ N., long. 103° 5′ 5–52″ W., elev. 836–845 m (supplementary section); (both sections at no. 3 in Fig. 1). Complete measured section: Hell Creek Formation (52.33 m) and Fort Union Formation (78.28); sampled interval: Hell Creek Formation (26.12 m) and Fort Union Formation (34.54 m).

This composite reference section was measured and sampled on Pretty Butte, a well-known geomorphologic feature ~13 km north of Marmarth, North Dakota. The measured section includes ~26 m of the Hell Creek Formation and 68 m of the Fort Union Formation. Only the lower 30.5 m part of the Fort Union is included in this analysis. The analysis is based on data from 22 samples collected from both units at both the main and supplementary sections. The K-T boundary is located in the lower part of the Ludlow Member, ~1 m above the formational contact; the boundary is bracketed by two samples in the main measured section that are separated by an interval of ~0.5 m.

Pretty Butte North

Locality D8025 (KJ9749), Pretty Butte quadrangle, Slope County, North Dakota, SW ¼ SE ¼ NE ¼ sec. 26, T. 134 N., R. 105 W., lat. 46° 23′ 5″–46° 22′ 41″, long. 103° 5′ 7″–103° 5′ 5″ W., elev. 890–897 m (no. 6 in Fig. 1). Complete measured section: Hell Creek Formation (0.5 m) and Fort Union Formation (7.0); sampled interval: Hell Creek Formation (0.5 m) and Fort Union Formation (6.5 m).

This section is located at the north end of Pretty Butte; only a few samples were analyzed. The K-T boundary is poorly constrained at this locality. It is well above the basal Fort Union lignite bed, but its exact position has not been determined; it is bracketed by two samples separated by ~3.5 m.

Pyramid Butte

Localities D6958 and D6998 (KJ86161), Pretty Butte quadrangle, Slope County, North Dakota, NW ¼ sec. 15, T. 134 N., R. 106 W., lat. 46° 25′ 4″ N., long. 103° 5′ 33″ W., elev. 857–859 m (no. 8 in Fig. 1). Complete measured section: Hell Creek Formation (23.94 m) and Fort Union Formation (39.05 m); sampled interval: Hell Creek Formation (0.10 m) and Fort Union Formation (1.78 m).

As reported by Johnson et al. (1989), 12 samples for palynological analysis were collected from part of the Pyramid Butte measured section. The sampled interval includes a basal Fort Union lignite bed 90 cm thick that is Maastrichtian in age. The K-T boundary is ~1 m above the Hell Creek–Fort Union formational contact, just above the basal Fort Union lignite bed. The boundary is bracketed by two samples separated by 15 cm. The exact position of the boundary between the bracketing samples is marked by shocked quartz and an Ir anomaly of 0.72 ppb. A thin bed of sandstone just above the boundary indicates that there was a minor episode of erosion and redeposition just after the K-T event. This erosion may account for the presence of rare specimens of K taxa above the documented position of the K-T boundary in this section, and perhaps also for the magnitude of the iridium anomaly, which is smaller than those recorded at most other K-T boundary localities in western North America.

River Section

Locality D7138 (KJ87147), Marmarth quadrangle, Slope County, North Dakota., sec. 8, T. 133 N., R. 105 W., lat. 46° 20′ 57″–46° 20′ 58″ N., long. 103° 5′ 49″–103° 5′ 42″ W., elev. 841–871 m (no. 4 in Fig. 1). Complete measured section: Hell Creek Formation (33.31 m) and Fort Union Formation (12.39 m); sampled interval: Hell Creek Formation (23.60 m) and Fort Union Formation (6.74 m).

The River Section is ~8 km northeast of Marmarth, North Dakota. The measured section includes ~33 m of the Hell Creek Formation and 12 m of the Ludlow Member of the Fort Union Formation. The sampled interval includes ~24 m of the Hell Creek and ~7 m of the Fort Union; 10 samples were collected and analyzed. The sampled interval includes one lignite bed in the Hell Creek Formation and three in the Fort Union. The K-T boundary is not well constrained in this section due to a lack of suitable lithologies for palynological analysis. It is apparently within the Ludlow Member, ~2.5 m above the Hell Creek–Fort Union formational contact; it is bracketed by two samples separated by an interval of ~4 m.

Scorpion Stung Kristian

Locality D9093 (KJ9734), Badland Draw quadrangle, Slope County, North Dakota, sec. 5, T. 134 N., R. 106 W., lat. 46° 27′ 1″–46° 19′ 6″ N., long. 104° 1′ 27″–103° 5′ 30″ W., elev. 878–881 m (no. 13 in Fig. 1). Complete measured section: Hell Creek Formation (0.40 m) and Fort Union Formation (4.48 m); sampled interval: Hell Creek Formation (0.40 m) and Fort Union Formation (3.08 m).

This measured section includes a sampled interval of ~3 m, which was sampled only in reconnaissance fashion. Eight samples were collected for palynology. The K-T boundary is 2.7 m above the Hell Creek–Fort Union formational contact; it is bracketed by two samples separated by 8 cm.

Sunset Butte

Composite section incorporating localities D7014, D7015, D7016, D7017, D7018, D7095, D7099, D9001, D9002, D9003, D9004, D9137, and D9161, Cedar Hills quadrangle, Bowman County, North Dakota, sec. 30 to sec. 21 and sec. 34, T. 131 N., R. 105 W., lat. 46° 8′ 11″–46° 8′ 29″ N., long. 103° 5′ 30″–103° 4′ 4″ W., elev. 847–959 m (main section, measured by D.E. Fastovsky); lat. 46° 6′ 12″–46° 7.′ 50″ N., long. 103° 4′ 14″–103° 4′ 8″ W., elev. 908–933 m (supplementary section); (both sections at no. 1 in Fig. 1). Complete measured section: Hell Creek Formation (100.05 m) and Fort Union Formation (36.02 m); sampled interval: Hell Creek Formation (99.65 m) and Fort Union Formation (11.90 m).

The Sunset Butte composite reference section includes the entire thickness of the Hell Creek Formation (~100 m) and ~36 m of the Ludlow Member of the Fort Union Formation; only part of the Fort Union interval was sampled. The main Sunset Butte measured section is supplemented by an additional short section called the Sunset Patch, here included as part of the composite section. The Sunset Patch section includes 6.6 m of the Hell Creek and 9.5 m of the Fort Union Formations (see Johnson, this volume, for a more complete description and discussion). In all, 22 samples from the Sunset Butte section and 12 samples from the Sunset Patch section were used in the analysis. The K-T boundary is above the formational contact; it is bracketed by samples from the Sunset Patch section that are separated by an interval of ~2 m. The precise position of the boundary in the Sunset Butte area is difficult to determine because of the presence of a thick channel-form sandstone deposit in the main measured section and the lack of lithologies suitable for palynological analysis in the supplementary section. Thus, although the Sunset Butte composite section is important because it encompasses all of the Hell Creek Formation and a substantial part of the Ludlow Member of the Fort Union Formation, it is not one of the better sections for precise location of the K-T boundary.

Terry's Fort Union Dinosaur

Locality D9090 (KJ9653; Pearson locality 128), Badland Draw quadrangle, Slope County, North Dakota, SE ¼ NW ¼ NW ¼ sec. 33, T. 135 N., R. 106 W., lat. 46° 28′ 2″–46° 27′ 37″ N., long. 104° 0′ 28″–104° 0′ 19″ W., elev. 857–860 m (no. 9 in Fig. 1). Measured section and sampled interval: Hell Creek Formation (0.15 m) and Fort Union Formation (2.71 m).

A partial associated skeleton of a *Triceratops* was discovered at this locality by Terry Schaefer of the Pioneer Trails Regional Museum, Bowman, North Dakota, 84 cm above the base of the Fort Union Formation, in facies of Maastrichtian age. The base of the Fort Union is marked by a lignite bed 10 cm thick. The section measured here includes this bed and a second lignite bed more than 2 m above it. The K-T boundary evidently lies near the base of the second lignite, ~2.5 m above the Hell Creek–Fort Union contact and ~1.7 m above the dinosaur bones. The boundary is bracketed by two samples separated by 5 cm.

Torosaurus Section

Locality D9095 (KJ9738), Mud Buttes quadrangle, Bowman County, North Dakota, NW ¼ SW ¼ SW ¼ sec. 12, T. 130 N., R. 105 W., lat. 46° 22′ 41″ N., long. 103° 5′ 5″ W., elev. 922–927 m (no. 17 in Fig. 1). Complete measured section: Hell Creek Formation (10.52 m) and Fort Union Formation (14.46 m); sampled interval: Hell Creek Formation (3.80 m) and Fort Union Formation (0.96 m).

The skeleton of a *Torosaurus* was collected at this locality by the Pioneer Trails Regional Museum (PTRM V92007). Eight samples were collected and analyzed for palynology. They verify the Maastrichtian age of the dinosaur. The boundary is closely bracketed at the Hell Creek–Fort Union formational contact.

Vertical Doug

Locality D9132 (KJ9840), Pretty Butte quadrangle, Slope County, North Dakota, SW ¼ NW ¼ SW ¼ sec. 34, T. 135 N., R. 106 W., lat. 46° 27′ 32″–46° 20′ 55″ N. long. 103° 5′ 22-24″ W., elev. 838–841 m (no. 14 in Fig. 1). Complete measured section: Fort Union Formation (3.53 m); sampled interval: Fort Union Formation (2.66 m).

This measured section includes only the Ludlow Member of the Fort Union Formation. The stratigraphically lowest lignite bed in the Fort Union Formation at this locality is more than 2 m above the Hell Creek–Fort Union formational contact. The formational contact is marked by a facies change, from olive-gray claystone with "popcorn" weathering on the surface to dark gray-brown mudstone containing root traces and charcoal. In all, 15 samples were collected, of which 11 were analyzed. The K-T boundary is bracketed by two samples separated by 11 cm.

APPENDIX 2. NEAR-BOUNDARY SECTIONS

Barbecue

Locality D8224 (KJ9324), Marmarth SE quadrangle, Slope County, North Dakota, SW ¼ NE ¼ NW ¼ sec. 22, T. 133 N., R. 105 W., lat. 46° 19′ 12″ N., long. 103° 5′ 17″ W., elev. 873 m (no. 18 in Fig. 1). Complete measured section: Hell Creek Formation (0.42 m) and Fort Union Formation (3.22 m); sampled interval: Hell Creek Formation (0.42 cm) and Fort Union Formation (0.60 cm).

Battleship

Locality D8149 (KJ9860), Marmarth SE quadrangle, Slope County, North Dakota, NE ¼ NE ¼ sec. 21, T. 133 N., R. 105 W., lat. 46° 19′ 15″–46° 20′ 51″ N., long. 103° 5′ 58″–103° 5′ 34″ W., elev. 860 m (no. 19 in Fig. 1). Complete measured section: Hell Creek Formation (2.80) and Fort Union Formation (6.10 m); sampled interval: Hell Creek Formation (0.12 m) and Fort Union Formation (0.73 m).

Dean's NM1

Locality D7629 (KJ88119), Marmarth quadrangle, Slope County, North Dakota, NW ¼ sec. 20, T. 133 N., R. 106 W., lat. 46° 19′ 6″ N., long. 103° 5′ 30″ W., elev. 884–886 m (no. 20 in Fig. 1). Complete measured section: Hell Creek Formation (53.07 m) and Fort Union Formation (4.16 m); sampled interval: Hell Creek Formation (0.47 m) and Fort Union Formation (1.59 m).

Ndas Section

Locality D9164 (KJ0042), Marmarth SE quadrangle, Slope County, North Dakota, SW ¼ NE ¼ SE ¼ NE ¼ sec. 16, T. 133 N., R. 105 W., lat. 46° 19′ 59.7″ N., long. 103° 51′ 59″ W., elev. 863 m (no. 21 in Fig. 1). Complete measured section: Hell Creek Formation (5.02 m) and Fort Union Formation (3.09 m); sampled interval: Fort Union Formation (2.4 m).

No Leaf Minors

Locality D9096 (KJ9750), Marmarth SE quadrangle, Slope County, North Dakota, NE ¼ SE ¼ SW ¼ NE ¼ sec. 9, T. 133 N., R. 105 W., lat. 46° 20′ 49-50″ N., long. 103° 5′ 13″ W., elev. 847–848 m (no. 22 in Fig. 1). Complete measured section: Hell Creek Formation (3.88 m) and Fort Union Formation (12.74 m); sampled interval: Hell Creek Formation (0.10 m) and Fort Union Formation (0.69 m).

Swanke's Nose

Locality D9006, Cedar Hills quadrangle, Bowman County, North Dakota, NE ¼ SW ¼ SE ¼ sec. 34, T. 132 N., R. 105 W., lat. 46° 12′ 47″–46° 23′ 32″ N., long. 103° 4′ 34″–103° 5′ 37″ W., elev. 884–885 m (no. 23 in Fig. 1). Measured section and sampled interval: Hell Creek Formation (40 cm) and Fort Union Formation (40 cm).

Tonstein To Die For

Locality D9133 (KJ9842), Pretty Butte quadrangle, Slope County, North Dakota, SW ¼ SE ¼ NE ¼ sec. 25, T. 135 N., R. 106 W., lat. 46° 29′ 5″–46° 20′ 55″ N., long. 103° 5′ 54″–103° 5′ 24″ W., elev. 800–803 m (no. 24 in Fig. 1). Complete measured section: Fort Union Formation (4.46 m); sampled interval: Fort Union Formation (2.33 m).

REFERENCES CITED

Alvarez, L.W., Alvarez, W., Asaro, F., and Michel, H.V., 1980, Extraterrestrial cause for the Cretaceous-Tertiary extinction: Science, v. 208, p. 1095–1108.

Bohor, B.F., Foord, E.E., Modreski, P.J., and Triplehorn, D.M., 1984, Mineralogic evidence for an extraterrestrial impact event at the Cretaceous-Tertiary boundary: Science, v. 224, p. 867–869.

Bohor, B.F., Triplehorn, D.M., Nichols, D.J., and Millard, H.T., 1987, Dinosaurs, spherules and the "magic layer"; a new K–T boundary clay site in Wyoming: Geology, v. 15, p. 896–899.

Braman, D.R., and Sweet, A.R., 1999, Terrestrial palynomorph biostratigraphy of the Cypress Hills, Wood Mountain, and Turtle Mountain areas (Upper Cretaceous-Paleocene) of western Canada: Canadian Journal of Earth Sciences, v. 36, p. 725–741.

Brown, R.W., 1952, Tertiary strata in eastern Montana and western North and South Dakota: Billings Geological Society Third Annual Field Conference, p. 89–92.

Calvert, W.R., 1912, Geology of certain lignite fields in eastern Montana: U.S. Geological Survey Bulletin 471, p. 187–201.

Carlson, C.G., 1979, Geology of Adams and Bowman Counties, North Dakota: North Dakota Geological Survey Bulletin 65, no. 1, 29 p.

Carlson, C.G., 1983, Geology of Billings, Golden Valley, and Slope Counties, North Dakota: North Dakota Geological Survey Bulletin, v. 76, no. 1, 40 p.

Catuneanu, O., and Sweet, A.R., 1999, Maastrichtian–Paleocene foreland-basin stratigraphies, western Canada: A reciprocal sequence architecture: Canadian Journal of Earth Sciences, v. 36, p. 685–703.

Doher, L.I., 1980, Palynomorph preparation procedures currently used in the paleontology and stratigraphy laboratories, U.S. Geological Survey: U.S. Geological Survey Circular 830, 29 p.

Fastovsky, D.E., 1987, Paleoenvironments of vertebrate-bearing strata during the Cretaceous–Paleogene transition, eastern Montana and western North Dakota: Palaios, v. 2, p. 282–295.

Fleming, R.F., 1990, Palynology of the Cretaceous–Tertiary boundary interval and Paleocene part of the Raton Formation, Colorado and New Mexico [Ph.D. thesis]: Boulder, University of Colorado, 283 p.

Hall, J.W., and Norton, N.J., 1967, Palynological evidence of floristic change across the Cretaceous–Tertiary boundary in eastern Montana (U.S.A.): Palaeogeography, Palaeoclimatology, Palaeoecology, v. 3, p. 121–131.

Hotton, C.L., 1984, Palynofloral changes across the Cretaceous–Tertiary boundary in east central Montana [abs.]: Palynology, v. 8, p. 242.

Hotton, C.L., 1988, Palynology of the Cretaceous–Tertiary boundary in central Montana, U.S.A., and its implications for extraterrestrial impact [Ph.D. thesis]: Davis, University of California, 732 p.

Hudson, J.D., 1998, Discussion on the Cretaceous–Tertiary biotic transition: Journal of the Geological Society of London, v. 155, p. 413–419.

Izett, G.A., 1990, The Cretaceous/Tertiary boundary interval, Raton Basin, Colorado and New Mexico, and its content of shock-metamorphosed minerals: Evidence relevant to the K-T boundary impact-extinction theory: Geological Society of America Special Paper 249, 100 p.

Jerzykiewicz, T., and Sweet, A.R., 1986, The Cretaceous–Tertiary boundary in the central Alberta Foothills: Stratigraphy: Canadian Journal of Earth Sciences, v. 23, p. 1356–1374.

Johnson, K.R., 1992, Leaf-fossil evidence for extensive floral extinction at the Cretaceous–Tertiary boundary, North Dakota, USA: Cretaceous Research, v. 13, p. 91–117.

Johnson, K.R., and Hickey, L.J., 1990, Megafloral change across the Cretaceous/Tertiary boundary in the northern Great Plains and Rocky Mountains, U.S.A., *in* Sharpton, V.L., and Ward, P.D., eds., Global catastrophes in Earth history: An interdisciplinary conference on impacts, volcanism, and mass mortality: Geological Society of America Special Paper 247, p. 433–444.

Johnson, K.R., and Nichols, D.J., 1994, Fossil plants and the Cretaceous–Tertiary boundary in the northern Great Plains: The best terrestrial argument for an abrupt catastrophe: Geological Society of America Abstracts with Programs, v. 26, no. 7, p. 395.

Johnson, K.R., Nichols, D.J., Attrep, M., Jr., and Orth, C.J., 1989, High-resolution leaf-fossil record spanning the Cretaceous/Tertiary boundary: Nature, v. 340, no. 6236, p. 708–711.

Johnson, K.R., Nichols, D.J., Tauxe, L., and Clark, D., 1990, Floral zonation and magnetostratigraphy of the Hell Creek (late Maastrichtian) and lower Fort Union (early Paleocene) Formations, North Dakota: Geological Society of America Abstracts with Programs, v. 22, no. 7, p. 323.

Leffingwell, H.A., 1970, Palynology of the Lance (Late Cretaceous) and Fort Union (Paleocene) Formations of the type Lance area, Wyoming, *in* Kosanke, R.M., and Cross, A.T., eds., Symposium on palynology of the Late Cretaceous and early Tertiary: Geological Society of America Special Paper 127, p. 1–64.

Lerbekmo, J.F., and Coulter, K.C., 1984, Magnetostratigraphic and biostratigraphic correlations of Late Cretaceous to early Paleocene strata between Alberta and North Dakota, *in* Stott, D.F., and Glass, D.J., eds., The Mesozoic of middle North America: Calgary, Alberta, Canadian Society of Petroleum Geologists, Memoir 9, p. 313–317.

Lerbekmo, J.F., Sweet, A.R., and St. Louis, R.M., 1987, The relationship between the iridium anomaly and palynological floral events at three Cretaceous–Tertiary boundary localities in western Canada: Geological Society of America Bulletin, v. 99, p. 325–330.

MacLeod, N., Rawson, P.F., Forey, P.L., Banner, F.T., BouDagher-Fadel, M.K., Bown, P.R., Burnett, J.A., Chambers, P., Culver, S., Evans, S.E., Jeffrey, C., Kaminski, M.A., Lord, A.R., Milner, A.C., Milner, A.R., Morris, N., Owen, E., Rosen, B.R., Smith, A.B., Taylor, P.D., Urquhart, E., and Young, J.R., 1997, The Cretaceous-Tertiary biotic transition: Journal of the Geological Society of London, v. 154, p. 265–292.

McIver, E.E., 1999, Paleobotanical evidence for ecosystem disruption at the Cretaceous–Tertiary boundary from Wood Mountain, Saskatchewan, Canada: Canadian Journal of Earth Sciences, v. 36, p. 775–789.

Murphy, E.C., Nichols, D.J., Hoganson, J.W., and Forsman, N.F., 1995, The Cretaceous–Tertiary boundary in south-central North Dakota: North Dakota Geological Survey Report of Investigations, v. 98, 73 p.

Nichols, D.J., 1984, Latitudinal variation in terrestrial palynomorph assemblages from Maestrichtian deposits in western North America: Sixth International Palynological Conference, Calgary, Abstracts, p. 117.

Nichols, D.J., 1990, Geologic and biostratigraphic framework of the nonmarine Cretaceous–Tertiary boundary interval in western North America: Review of Palaeobotany and Palynology, v. 65, p. 75–84.

Nichols, D.J., 1991, Palynology supports catastrophic event at Cretaceous–Tertiary boundary: An overview of ten years of research in the United States: Geological Society of America Abstracts with Programs, v. 23, no. 5, p. 357.

Nichols, D.J., 1993, Palynological evidence on the nature of the terminal Cretaceous event in the northern Great Plains of the U.S.: Geological Association of Canada—Mineralogical Association of Canada, Joint Annual Meeting, Program and Abstracts, p. A76.

Nichols, D.J., 1995, The role of palynology in paleoecological analyses of Tertiary coals: International Journal of Coal Geology, v. 28, p. 139–159.

Nichols, D.J., 1996a, Palynology of the Cretaceous–Tertiary boundary interval, Western Interior, U.S.A.: Ninth International Palynological Congress, Program and Abstracts: Houston, Texas, USA, American Association of Stratigraphic Palynologists, p. 115–116.

Nichols, D.J., 1996b, Vegetational history in Western Interior North America during the Cretaceous–Tertiary transition, Chapter 29D, *in* Jansonius, J., and McGregor, D.C., eds., Palynology: Principles and applications: Dallas, Texas, American Association of Stratigraphic Palynologists, v. 3, p. 1189–1195.

Nichols, D.J., 1998, Palynostratigraphy of the Fort Union Formation (Paleocene) in the North Dakota part of the Williston Basin and its application to coal resource assessment: Geological Society of America Abstracts with Programs, v. 30, no. 7, p. 366.

Nichols, D.J., 1999, Biostratigraphy, Williston Basin, *in* 1999 Resource Assessment of selected Tertiary coal beds and zones in the northern Rocky Mountains and Great Plains region: U.S. Geological Survey Professional Paper 1625A, p. WB1–WB3, 9 figs.

Nichols, D.J., and Brown, J.L., 1992, Palynostratigraphy of the Tullock Member (lower Paleocene) of the Fort Union Formation in the Powder River Basin, Montana and Wyoming: U.S. Geological Survey Bulletin 1917-F, 35 p.

Nichols, D.J., and Fleming, R.F., 1990, Plant microfossil record of the terminal Cretaceous event in the western United States and Canada, *in* Sharpton, V.L., and Ward, P.D., eds., Global catastrophes in Earth history: An interdisciplinary conference on impacts, volcanism, and mass mortality: Geological Society of America Special Paper 247, p. 445–455.

Nichols, D.J., and Sweet, A.R., 1993, Biostratigraphy of Upper Cretaceous nonmarine palynofloras in a north-south transect of the Western Interior Basin, *in* Caldwell, W.G.E., and Kauffman, E.G., eds., Evolution of the Western Interior Basin: Geological Society of Canada Special Paper 39, p. 539–584.

Nichols, D.J., Brown, J.L., Attrep, M., Jr., and Orth, C.J., 1992, A new Cretaceous–Tertiary boundary locality in the western Powder River Basin, Wyoming: Biological and geological implications: Cretaceous Research, v. 13, p. 3–30.

Nichols, D.J., Fleming, R.F., and Frederiksen, N.O., 1990, Palynological evidence of effects of the terminal Cretaceous event in western North America, *in* Kauffman, E.G., and Walliser, O.H., eds., Extinction events in Earth history, Lecture Notes in Earth Sciences, Volume 30: Berlin, Springer-Verlag, p. 351–364.

Nichols, D.J., Jacobson, S.R., and Tschudy, R.H., 1982, Cretaceous palynomorph biozones for the central and northern Rocky Mountain region of the United States, *in* Powers, R.B., ed., Geologic studies of the Cordilleran thrust belt: Denver, Colorado, Rocky Mountain Association of Geologists, v. 2, p. 721–733.

Nichols, D.J., Jarzen, D.M., Orth, C.J., and Oliver, P.Q., 1986, Palynological and iridium anomalies at Cretaceous–Tertiary boundary, south-central Saskatchewan: Science, v. 231, p. 714–717.

Nichols, D.J., Murphy, E.C., Johnson, K.R., and Betterton, W.J., 2000, A second K–T boundary locality in North Dakota verified by palynostratigraphy and shocked quartz: Geological Society of America Abstracts and Programs, v. 32, no. 7, p. 130.

Norton, N.J., and Hall, J.W., 1967, Guide sporomorphae in the Upper Cretaceous–Lower Tertiary of eastern Montana (U.S.A.): Review of Palaeobotany and Palynology, v. 2, p. 99–110.

Norton, N.J., and Hall, J.W., 1969, Palynology of the Upper Cretaceous and lower Tertiary in the type locality of the Hell Creek Formation, Montana, U.S.A.: Palaeontographica, Abteilung B, v. 125, p. 1–64.

Oltz, D.F., 1969, Numerical analyses of palynological data from Cretaceous and early Tertiary sediments in east central Montana: Palaeontographica. Abteilung B: Palaophytologie, v. 128, p. 90–166.

Oltz, D.F., 1971, Cluster analyses of Late Cretaceous–Early Tertiary pollen and spore data: Micropaleontology, v. 17, p. 221–232.

Orth, C.J., Gilmore, J.S., Knight, J.D., Pillmore, C.L., Tschudy, R.H., and Fassett, J.E., 1981, An iridium anomaly at the Cretaceous–Tertiary boundary in northern New Mexico: Science, v. 214, p. 1341–1343.

Pillmore, C.L., Tschudy, R.H., Orth, C.J., Gilmore, J.S., and Knight, J.D., 1984, Geologic framework of nonmarine Cretaceous–Tertiary boundary sites, Raton Basin, New Mexico and Colorado: Science, v. 223, p. 1180–1183.

Pillmore, C.L., Nichols, D.J., and Fleming, R.F., 1999, Field guide to the continental Cretaceous–Tertiary boundary in the Raton Basin, Colorado and New Mexico, *in* Lageson, D.R., Lester, A.P., and Trudgill, B.D., eds., Colorado and adjacent areas: Boulder, Colorado, Geological Society of America Field Guide 1, p. 135–155.

Pocknall, D.T., and Flores, R.M., 1987, Coal palynology and sedimentology in the Tongue River Member, Fort Union Formation, Powder River Basin, Wyoming: Palaios, v. 2, p. 133–145.

Signor, P.W., III, and Lipps, J.H., 1982, Sampling bias, gradual extinction patterns and catastrophes in the fossil record, *in* Silver, L.T., and Schultz, P.H., eds., Geological implications of impacts of large asteroids and comets on the Earth: Geological Society of America Special Paper 190, p. 291–296.

Smit, J., and van der Kaars, S., 1984, Terminal Cretaceous extinctions in the Hell Creek area, Montana: Compatible with catastrophic extinction: Science, v. 223, p. 1177–1179.

Stanley, E.A., 1965, Upper Cretaceous and Paleocene plant microfossils and Paleocene dinoflagellates and hystrichosphaerids from northwestern South Dakota: Bulletins of American Paleontology, v. 49, no. 222, p. 177–384.

Sweet, A.R., 1994, Relationships between depositional environments and changes in palynofloras across the K-T boundary interval, *in* Traverse, A., ed., Sedimentation of organic particles: Cambridge, Cambridge University Press, p. 461–488.

Sweet, A.R., and Braman, D.R., 1992, The K-T boundary and contiguous strata in western Canada: Interactions between paleoenvironments and palynological assemblages: Cretaceous Research, v. 13, p. 31–79.

Sweet, A.R., and Braman, D.R., 2001, Cretaceous-Tertiary palynofloral perturbations and extinctions within the *Aquilapollenites* phytogeographic province: Canadian Journal of Earth Sciences, v. 38, p. 249–269.

Sweet, A.R., Braman, D.R., and Lerbekmo, J.F., 1990, Palynofloral response to K-T boundary events: A transitory interruption within a dynamic system, *in* Sharpton, V.L., and Ward, P.D., eds., Global catastrophes in Earth history: An interdisciplinary conference on impacts, volcanism, and mass mortality: Geological Society of America Special Paper 247, p. 457–469.

Sweet, A.R., Braman, D.R., and Lerbekmo, J.F., 1999, Sequential palynological changes across the composite Cretaceous-Tertiary (K-T) boundary claystone and contiguous strata, western Canada and Montana, U.S.A.: Canadian Journal of Earth Sciences, v. 36, p. 743–768.

Tschudy, R.H., 1970 [1971], Palynology of the Cretaceous–Tertiary boundary in the northern Rocky Mountain and Mississippi Embayment regions, *in* Kosanke, R.M., and Cross, A.T., eds., Symposium on palynology of the Late Cretaceous and early Tertiary: Geological Society of America Special Paper 127, p. 65–111.

Tschudy, R.H., Pillmore, C.L., Orth, C.J., Gilmore, J.S., and Knight, J.D., 1984, Disruption of the terrestrial plant ecosystem at the Cretaceous–Tertiary boundary, western North America: Science, v. 225, p. 1030–1032.

Tschudy, R.H., and Tschudy, B.D., 1986, Extinction and survival of plant life following the Cretaceous/Tertiary boundary event, Western Interior, North America: Geology, v. 14, p. 667–670.

Wolfe, J.A., 1991, Palaeobotanical evidence for a June 'impact winter' at the Cretaceous–Tertiary boundary: Nature, v. 352, p. 420–423.

MANUSCRIPT ACCEPTED BY THE SOCIETY SEPTEMBER 24, 2001

Printed in the U.S.A.

Geological Society of America
Special Paper 361
2002

Vertebrate biostratigraphy of the Hell Creek Formation in southwestern North Dakota and northwestern South Dakota

Dean A. Pearson
Terry Schaefer
Pioneer Trails Regional Museum, Bowman, North Dakota 58623, USA
Kirk R. Johnson
Department of Earth Sciences, Denver Museum of Nature & Science, Denver, Colorado 80205, USA
Douglas J. Nichols
U.S. Geological Survey, Denver Federal Center, Denver, Colorado 80225, USA
John P. Hunter
Department of Anatomy, New York College of Osteopathic Medicine, New York Institute of Technology, Old Westbury, New York 11568, USA

ABSTRACT

Field surveys of the Upper Cretaceous Hell Creek Formation in southwestern North Dakota since 1986 have produced a total of 10 124 specimens from 42 vertebrate microsites and an additional 41 dinosaur skulls or partial skeletons or skulls from separate sites representing 61 taxa of vertebrates dominated by fish, dinosaurs, turtles, and crocodilians. Common elements of this diverse fauna occur to within 2.37 m of the Cretaceous-Tertiary (K-T) boundary. The stratigraphically highest fossil in the study is a partial ceratopsian skeleton 1.77 m below the K-T boundary in the basal Fort Union Formation. All dinosaurs that occur at more than two sites also occur at the highest level that yielded more than 500 specimens (8.40 m below the K-T boundary). The fine-grained uppermost 2 m of the Hell Creek is nearly devoid of all fossils, including taxa known to occur in the overlying formation. The absence of channel deposits in this part of the formation may be the reason for the absence of fossil localities. The presence of marine-tolerant taxa suggests that the study area may have been adjacent to a previously unidentified seaway of latest Cretaceous age. Rarefaction analysis indicates no evidence for a decline in vertebrate diversity through the formation or dinosaurian diversity in the 3 m below the K-T boundary. Our results are not compatible with gradual vertebrate extinction at the end of the Cretaceous.

INTRODUCTION

Despite years of work on the vertebrate faunas of the Hell Creek Formation and their central role in the Cretaceous-Tertiary (K-T) boundary extinction debate, information is still elusive on the last known occurrence of vertebrate species relative to the K-T boundary. Perhaps the most common reason that these data were not collected or published is that the K-T

Pearson, D.A., Schaefer, T., Johnson, K.R., Nichols, D.J., and Hunter, J.P., 2002, Vertebrate biostratigraphy of the Hell Creek Formation in southwestern North Dakota and northwestern South Dakota, *in* Hartman, J.H., Johnson, K.R., and Nichols, D.J., eds., The Hell Creek Formation and the Cretaceous-Tertiary boundary in the northern Great Plains: An integrated continental record of the end of the Cretaceous: Boulder, Colorado, Geological Society of America Special Paper 361, p. 145–167.

boundary cannot be recognized in the field and must be determined by laboratory analysis of palynology or geochemistry. The contact of the Hell Creek Formation with the overlying Fort Union Formation has long been recognized to occur near the K-T boundary (Brown, 1952). Because the Hell Creek–Fort Union formational contact is defined by changing depositional environments (Fastovsky, 1987), it is readily recognized in the field by experienced geologists. Given its easy field identification and its proximity to the K-T boundary, the formational contact has been used as a casual proxy for the K-T boundary even when workers clearly realized that this lithostratigraphic contact and the chronostratigraphic K-T boundary were genetically unrelated (Sheehan et al., 1991).

Furthermore, the lack of stratigraphically controlled collecting has led to a debate over the significance of the stratigraphic separation between the highest dinosaur and the K-T boundary (Archibald, 1982; Retallack and Leahy, 1986; Retallack et al., 1987; Bryant, 1989). Archibald (1996a, p. 43) discussed this "10 foot gap" at length and other authors have even suggested that the gap implied that dinosaurs were already extinct before the K-T boundary (Clemens et al., 1981; Williams, 1994). Sheehan et al. (2000) presented data from western North Dakota and eastern Montana that dinosaurs occurred within this gap in the same abundance as elsewhere in the formation, but Pearson et al. (2001) found no vertebrate fossils of any kind in the uppermost 1.7 m of Cretaceous strata. This ongoing debate clearly defines a need to stratigraphically locate vertebrate fossils in the body of the Hell Creek Formation and to determine their occurrence throughout the formation to ascertain if the strata near the K-T boundary are typical or anomalous with respect to their vertebrate fauna.

The purpose of this study is to document the presence and abundance of vertebrate taxa at specific stratigraphic levels within the Hell Creek Formation in southwestern North Dakota and extreme northwestern South Dakota, and to relate these occurrences to the palynologically and geochemically defined K-T boundary. Fossil occurrences are recorded in two types of localities: vertebrate microsites (localities from which vertebrate microfossils were collected) and isolated in situ dinosaur occurrences. We use these data to evaluate trends in vertebrate diversity through the formation.

PREVIOUS STUDIES

The Hell Creek fauna in the northern Great Plains of the United States is one of the best-sampled uppermost Cretaceous vertebrate assemblages in the world (Bryant, 1989; Fastovsky, 1990; Archibald and Bryant, 1990; Keller and Barrera, 1990; Sheehan and Fastovsky, 1992; Williams, 1994; Archibald, 1996a, 1996b; Fastovsky and Sheehan, 1997; Hunter et al., 1997). These assemblages and those in equivalent strata have been extensively studied and documented (Estes, 1964; Sloan and Van Valen, 1965; Estes et al., 1969; Estes and Berberian, 1970; Greenwald, 1971; Carpenter, 1979; Breithaupt, 1982; Whitmore and Martin, 1986; Hutchison and Archibald, 1986; Rigby et al., 1987; Sloan, 1987; Bryant, 1989; Archibald and Bryant, 1990; Sheehan et al., 1991). Some of these studies were based upon the extremely productive Bug Creek sequence in northeastern Montana (Sloan and Van Valen, 1965; Estes et al., 1969; Estes and Berberian, 1970; Sloan, 1987; Rigby et al., 1987). Although rich in specimens, the Bug Creek sites are stratigraphically ambiguous because they have been documented as containing a mixture of Late Cretaceous and early Tertiary specimens created by Paleocene streams downcutting into Cretaceous strata (Fastovsky and Dott, 1986; Lofgren, 1995). In many of the studies cited, taxonomy was the priority, and while some consideration was given to placing the collected sites in a stratigraphic framework relative to K-T boundary, the nature of the deposits made this a difficult goal to achieve.

Five of these studies (Estes et al., 1969; Estes and Berberian, 1970; Carpenter, 1979; Breithaupt, 1982; Whitmore and Martin, 1986) dealt with only a single site or the majority of the data were from a single small study area. These studies are useful for recording assemblages at a single stratigraphic interval, but are less useful for elucidating trends in vertebrate evolution and extinction through time.

Four of the studies (Estes, 1964; Greenwald, 1971; Bryant, 1989; Archibald and Bryant, 1990) considered more than one site or study area in their analyses. Estes (1964) utilized sites collected from the Lance Formation in Wyoming, although he did not list the sites by sedimentary facies type or indicate stratigraphic positions that could be used to identify trends in occurrence of vertebrates within that formation. Greenwald (1971) indicated that his data were from seven closely spaced sites, but did not give a complete faunal list for each site or its stratigraphic position. Bryant (1989) and Archibald and Bryant (1990) presented vertebrate lists from the extensive collections at the University of California Museum of Paleontology. Bryant (1989) studied two geographic areas in northeastern Montana that produced 49 localities from the Hell Creek Formation. These localities are identified by sedimentary facies type, but they are not listed by stratigraphic position within the formation. Trends dealing with vertebrate extinctions reported by Bryant (1989) are based upon a presence or absence comparison between the Hell Creek and Fort Union Formations. As Bryant (1989) indicated, no attempt was made to place these localities in a stratigraphic framework relative to the K-T boundary. Thus her data indicate whether a species survived the K-T boundary, but do not reveal trends within the formation leading up to the boundary. Archibald and Bryant (1990) attempted to track the complete Lancian vertebrate assemblage through the K-T transition. They indicated that the stratigraphic position of the fauna within the Hell Creek Formation was not considered because they were examining only the overall survivorship from the uppermost Cretaceous into the Paleocene. In their study, there is no way to determine the level in the Hell Creek Formation where vertebrate taxa first appeared or last disappeared.

Driven by the perception that stratigraphic position and

sedimentary facies were critical to understanding vertebrate distribution relative to the K-T boundary, Sheehan, Fastovsky, and colleagues organized a large field effort in the late 1980s (Fastovsky, 1987; Sheehan et al., 1991, 2000; White et al., 1998). They used a large volunteer force from the Milwaukee Public Museum's Dig-a-Dinosaur program to prospect the Hell Creek Formation in Dawson County, Montana, and Slope and Bowman Counties, North Dakota. The resulting database contained records of the stratigraphic position, identification, and sedimentological facies of 556 individual dinosaur finds representing eight dinosaur families. They used these data to argue that there was no significant change in dinosaurian diversity through the time represented by the Hell Creek Formation (Sheehan et al., 1991). Although generally well conceived, this research suffered several shortcomings during execution. By using only isolated large dinosaurian fragments and not vertebrate microsites, they limited the taxonomic potential of the study, and they were later criticized for presenting their data only at the family level (Hurlbert and Archibald, 1995). In addition, most of these specimens were left in the field, negating the potential to check identifications. The data were initially split into three stratigraphic levels, the bottom, middle, and top thirds of the Hell Creek Formation (Sheehan et al., 1991), thus decreasing their stratigraphic resolution (although this resolution was increased to 5 m intervals in Sheehan et al., 2000). These three stratigraphic units were then treated as unordered categories in an analysis of variance (Sheehan et al., 1991) rather than as items in a time series. Perhaps the most significant weakness of their study was an incomplete knowledge of the position of the K-T boundary relative to the Hell Creek–Fort Union formational contact. At the time of their study, the K-T boundary was only known at one site in North Dakota (Johnson et al., 1989). Subsequent work in southwestern North Dakota has shown the Hell Creek–Fort Union formational contact to be diachronous relative to the K-T boundary, with a separation of as much as 2.7 m between them (Johnson, 1992; Nichols and Johnson, this volume). Sheehan et al. (2000) used their data from the late 1980s to argue for the occurrence of dinosaurs in the upper 3 m of the Cretaceous strata in Montana and North Dakota. One of us (Johnson) checked Sheehan and Fastovsky's primary data from North Dakota and verified the dinosaurs they recorded from the Mud Buttes area in Bowman County where the K-T boundary and the Hell Creek–Fort Union formational contact are now known to be coincident (Nichols et al., 2000; Nichols and Johnson, this volume). The work of Fastovsky and Sheehan sets the stage for a more detailed study utilizing precise stratigraphic position, microsites, and individual dinosaurs, vouchered specimens, and a well-defined K-T boundary.

GEOLOGIC AND STRATIGRAPHIC FRAMEWORK

The study area is along the southwestern margin of the Williston Basin in southwestern North Dakota and northwestern South Dakota and is centered in the valley of the Little Missouri River to the east of the North Dakota–Montana border. This river has cut across the northwest-southeast–trending Cedar Creek Anticline, exposing the Hell Creek Formation in a series of erosional badlands (Figs. 1 and 2).

The Hell Creek Formation in this area is a 100-m-thick interval of flat-lying, root-bioturbated mudstone beds intercalated with lenticular sandstone and muddy sandstone beds (Frye, 1969; Moore, 1976; Fastovsky, 1987; Johnson, 1992; Murphy et al., this volume). The mudstone units have been interpreted as poorly drained flood-plain paleosols (Fastovsky and McSweeney, 1987), and the lenticular deposits have been interpreted as channel point bars, oxbow lakes, or channel thalwegs (Fastovsky, 1987; Johnson, 1992).

The lithostratigraphic contact between the Hell Creek and Fort Union Formations is placed at the base of the first laterally continuous lignite above the typical Hell Creek deposits. Where the basal Fort Union lignite is not present, the basal contact of the Fort Union Formation is defined by the juxtaposition of underlying smectite-rich, "popcorn-weathered," (having a surface texture bearing a resemblance to that of popcorn) gray mudstone typical of the Hell Creek and overlying carbonaceous shale, flat-laminated yellow mudstone, or yellow sandstone typical of the Ludlow Member of the Fort Union Formation (Fastovsky, 1987; Johnson, 1992; Murphy et al., this volume).

The Ludlow Member of the Fort Union Formation is composed of tabular and lenticular sand bodies, flat-laminated variegated mudstone beds, carbonaceous shale, and lignite. These deposits have been interpreted as originating in ponds, swamps, crevasse splays, and meandering and distributary channels (Moore, 1976; Belt et al., 1984; Fastovsky, 1987; Johnson, 1992). The Ludlow Member is generally more deeply weathered than the Hell Creek Formation and, as a result, usually supports vegetation, allowing the formations to be readily distinguished (Johnson, 1992).

The K-T boundary in the northern Great Plains of the United States has traditionally been placed at the first laterally continuous lignite above the highest dinosaur fossils, a procedure advocated by Roland Brown (1952) and now known as "Brown's formula" (Archibald, 1982, p. 13). This technique was first established as only a field approximation (Brown, 1952), and has been superseded by the development and use of palynological methods and the recognition of geochemical and mineralogical indicators of debris from extraterrestrial impact. The palynologically defined K-T boundary is based upon the disappearance of certain palynomorphs of the *Wodehouseia spinata* assemblage zone of Nichols et al. (1982) (see Nichols and Johnson, this volume).

In our study area, the K-T boundary is near the Hell Creek–Fort Union formational contact but may be as much as 3 m above it (Nichols and Johnson, this volume). At Pyramid Butte, 9 km north of Marmarth, North Dakota, the K-T boundary is 0.9 m above the formational contact and is defined by palynomorph extinctions coincident with an iridium anomaly and the presence of shocked mineral grains (Johnson et al., 1989).

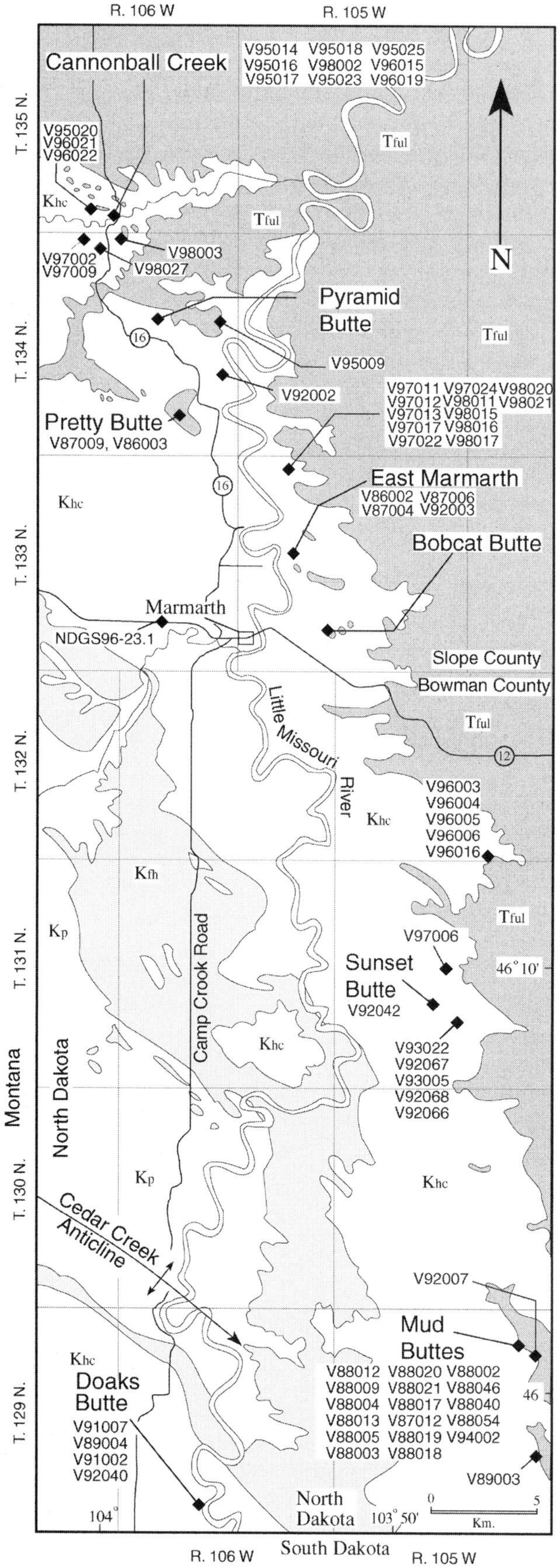

Figure 1. Locality map of study area in Williston Basin in southwestern North Dakota showing locations of Pioneer Trails Regional Museum (PTRM) vertebrate fossil sites. South Dakota sites V91010, V91014, V92025, V92027, and V92029 are not shown; they are within 2 km south of state line in Doaks Butte area. Specific site data are on file at PTRM and are available to qualified researchers. Left margin of map is Montana–North Dakota border, and bottom margin is North Dakota–South Dakota border. Kp—Pierre Shale, Kfh—Fox Hills Sandstone, Khc—Hell Creek Formation, and Tful—Ludlow Member of Fort Union Formation. Geology is modified from Carlson (1979, 1983).

At Mud Buttes, 27 km south of Rhame, North Dakota, the K-T boundary is coincident with the formational contact (Pearson et al., 2001) and is associated with shocked mineral grains (Nichols et al., 2000). Nichols and Johnson (this volume) have used palynology to define the K-T boundary at 17 sections in Slope and Bowman Counties. These sections allow us to convert the stratigraphic position of a vertebrate site relative to the formational contact and to its position relative to the K-T boundary.

METHODOLOGY

Fossil sites and stratigraphy

Vertebrate microsites and isolated in situ dinosaur occurrences represent two different methods of sampling vertebrate occurrence and diversity in the Hell Creek Formation. Both types of sites were located by systematic surface prospecting. Microsites were collected by crawling the surface and collecting all identifiable fragments. The sites that produced large numbers of specimens were repeatedly sampled over the years to achieve a representative collection approaching the original diversity of the assemblage at that stratigraphic interval (Fig. 3). Only in a few cases were microsites collected by conventional screening techniques. Isolated in situ dinosaur occurrences were either collected or, in a few cases, recorded and left on the outcrop.

Each site was placed within the stratigraphic framework by the identification of the horizon producing the specimens, its lithology, physical dimensions, and distance below the Hell Creek–Fort Union formational contact (Table 1). The Hell Creek–Fort Union formational contact was used as the field datum because it could be easily recognized throughout the study area. If the formational contact was not located close to the vertebrate site, we traced the fossiliferous horizon laterally to an adjacent section containing the contact or estimated its location using elevation. Stratigraphic sections were measured in each new collection area (Murphy et al., this volume; Johnson, this volume). These measurements were made in collaboration with Nichols and Johnson (this volume), who used palynology to define the K-T boundary in or near these sections. The depositional environment of each site was identified using the criteria supplied by Fastovsky (1987) and Johnson (1992).

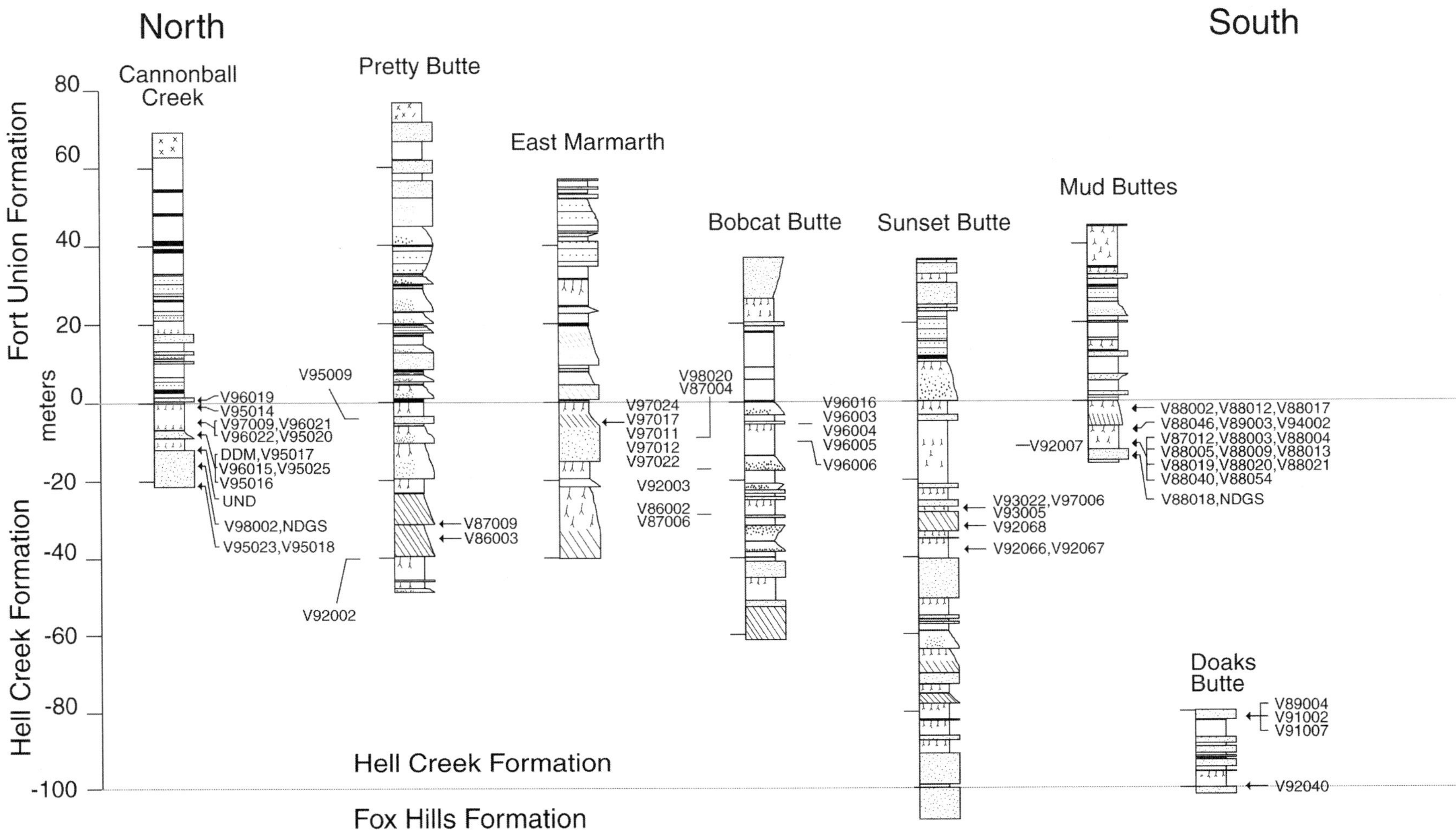

Figure 2. Stratigraphic cross section showing key sections arranged from north to south. Stratigraphic positions of Pioneer Trails Regional Museum (PTRM) vertebrate fossil sites were plotted against sections where it was possible to correlate locality directly to measured section and plotted between sections where site was correlated to section by measurement to nonillustrated stratigraphic section. Stratigraphic sections are further described in Murphy et al. (this volume), Johnson (this volume), and Hicks et al. (this volume).

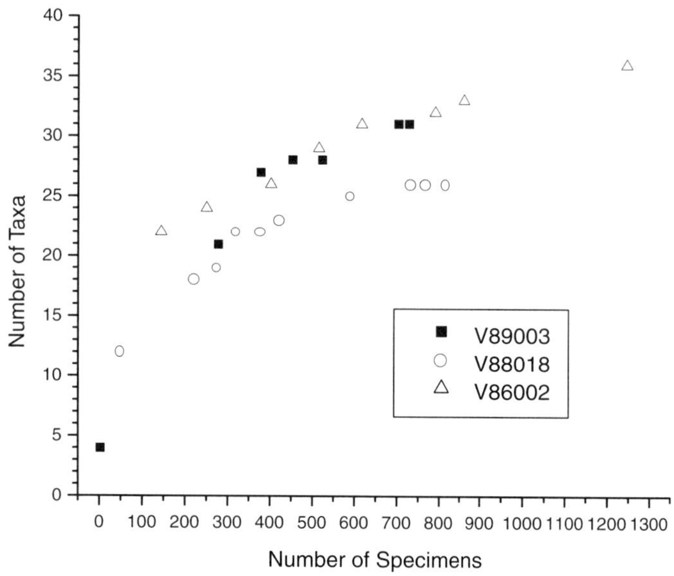

Figure 3. "Collector's curves," cumulative numbers of specimens and taxa over series of time-successive (yearly) samples, for three representative vertebrate localities in Hell Creek Formation. Although number of taxa recovered continued to increase with cumulative sampling, ~70% of vertebrate taxa recovered from each site are represented in first 300 specimens. Another locality not shown here, Pioneer Trails Regional Museum (PTRM) V92067, has yielded 3146 specimens representing only 35 vertebrate taxa, with no new taxa added after first 2693 specimens, suggesting that sampling at this locality may have been exhaustive. Although sampling more than few hundred specimens yields diminishing returns, these returns are meaningful in terms of recovering less common vertebrate taxa.

Identification of vertebrate fossils

All specimens were identified to the highest resolution possible on the basis of the materials available. We followed the protocol proposed by Badgley (1986) and utilized by Brinkman et al. (1998) for estimating the paleofaunal relative abundance in assemblages from which the probability of association among individual elements is low, and each identifiable specimen was counted as a single element. The specimens have all been identified to genus or to species where possible, and in all other cases, they have been identified to the finest level possible. The familial level was used when elements are present that are not diagnostic to a lower level, but the presence of these taxa provides a measure of ecological composition. For example, there are two genera of ceratopsians in the study area, *Torosaurus* and *Triceratops*, and ceratopsians have been reported to be the most abundant dinosaur in the faunal assemblage (White et al., 1998). However, the shed teeth of these dinosaurs found in vertebrate microsites are not taxonomically distinct (Coombs and Galton, 1988). If we had disregarded the isolated, shed ceratopsian teeth found at all of our vertebrate microsites simply because they could not be identified below the family level, we would have biased the study by eliminating the reportedly most abundant type of dinosaurs on the basis of inadequate taxonomic resolution. Thus, these shed teeth are recorded as Ceratopsidae indeterminate. A similar situation exists for the shed teeth of hadrosaurs found at vertebrate microsites. Coombs (1988) and Horner (1990) showed that isolated teeth of hadrosaurs are diagnostic only to the subfamily level (see also Fiorillo and Currie, 1994).

TABLE 1. DINOSAUR SITES AND VERTEBRATE MICROSITES

PTRM locality	Site type	Datum (m)	K-T section	Correction (m)	K-T (m)
V96019	Dinosaur	0.84	9653	2.61	−1.77
V88012	Microsite	−2.37	9906	0.00	−2.37
V88002	Dinosaur	−2.72	9906	0.00	−2.72
V95014	Dinosaur	−0.98	9612	2.14	−3.12
V88017	Microsite	−3.17	9906	0.00	−3.17
V96016	Dinosaur	−2.76	86148	0.75	−3.51
V97024	Microsite	−3.40	86148	0.75	−4.15
V95009	Dinosaur	−4.00	86161	0.96	−4.96
V96003	Microsite	−4.50	86148	0.75	−5.25
V96004	Microsite	−4.50	86148	0.75	−5.25
V97017	Dinosaur	−4.65	86148	0.75	−5.40
V97022	Microsite	−4.81	86148	0.75	−5.56
V94002	Dinosaur	−6.00	9906	0.00	−6.00
V98003	Dinosaur	−4.35	9144	1.71	−6.06
V97011	Dinosaur	−5.40	86148	0.75	−6.15
V96005	Microsite	−5.50	86148	0.75	−6.25
V97012	Dinosaur	−5.60	86148	0.75	−6.35
V98011	Microsite	−5.90	86148	0.75	−6.65
V97009	Dinosaur	−4.50	9734	2.69	−7.19
V96022	Microsite	−4.90	9653	2.61	−7.51
V98017	Dinosaur	−7.12	86148	0.75	−7.87
V88046	Dinosaur	−8.00	9906	0.00	−8.00
V95020	Microsite	−5.50	9653	2.61	−8.11
V96021	Dinosaur	−5.50	9653	2.61	−8.11
V88009	Microsite	−8.29	9906	0.00	−8.29
V89003	Microsite	−8.40	9906	0.00	−8.40
V98020	Both	−8.80	86148	0.75	−9.55
V98015	Dinosaur	−8.85	86148	0.75	−9.60
V87004	Microsite	−9.00	86148	0.75	−9.75
V96006	Dinosaur	−9.00	86148	0.75	−9.75
ND92-19.1	Dinosaur	−7.16	9653	2.61	−9.77
V87012	Microsite	−9.85	9906	0.00	−9.85
V88005	Dinosaur	−10.00	9906	0.00	−10.00
V88020	Microsite	−10.00	9906	0.00	−10.00
V88021	Microsite	−10.00	9906	0.00	−10.00
V88040	Dinosaur	−10.00	9906	0.00	−10.00
V96015	Microsite	−8.41	9144	1.71	−10.12
V95017	Dinosaur	−8.00	9612	2.14	−10.14
V88003	Dinosaur	−10.50	9906	0.00	−10.50
V88004	Microsite	−10.61	9906	0.00	−10.61
V88013	Microsite	−10.62	9906	0.00	−10.62
V95025	Microsite	−8.50	9612	2.14	−10.64
V95016	Microsite	−8.85	9612	2.14	−10.99
V98016	Dinosaur	−10.68	86148	0.75	−11.43
V92007	Dinosaur	−11.49	9738	0.00	−11.49
V88019	Microsite	−12.00	9906	0.00	−12.00
V88054	Dinosaur	−12.00	9906	0.00	−12.00
V98027	Microsite	−10.62	9909	1.45	−12.07
V98021	Dinosaur	−11.34	86148	0.75	−12.09
V88018	Microsite	−14.27	9906	0.00	−14.27
UND3000	Dinosaur	−12.00	9653	2.61	−14.61
MPM51016	Dinosaur	−15.00	9906	0.00	−15.00
V97013	Dinosaur	−14.50	86148	0.75	−15.25
V97002	Dinosaur	−13.00	9734	2.69	−15.69
V92003	Microsite	−17.00	86148	0.75	−17.75
V98002	Microsite	−16.00	9840	2.25	−18.25

(*continued*)

TABLE 1. DINOSAUR SITES AND VERTEBRATE MICROSITES (continued)

PTRM locality	Site type	Datum (m)	K-T section	Correction (m)	K-T (m)
ND95-117.1	Dinosaur	−18.00	9653	2.61	−20.61
V95023	Dinosaur	−20.00	9612	2.14	−22.14
V95018	Dinosaur	−21.30	9612	2.14	−23.44
V93022	Microsite	−25.30	9317	1.70	−27.00
V93005	Dinosaur	−25.50	9317	1.70	−27.20
V97006	Microsite	−27.00	9317	1.70	−28.70
V87006	Microsite	−28.00	86148	0.75	−28.75
V86002	Microsite	−29.18	86148	0.75	−29.93
V87009	Microsite	−31.43	9735	1.12	−32.55
V92068	Dinosaur	−31.18	9317	1.70	−32.88
V86003	Microsite	−35.05	9735	1.12	−36.17
V92066	Dinosaur	−37.00	9317	1.70	−38.70
V92067	Microsite	−37.18	9317	1.70	−38.88
V92002	Microsite	−40.00	9735	1.12	−41.12
V92042	Dinosaur	−40.00	9317	1.70	−41.70
V92029	Microsite	−60.00	9906	0.00	−60.00
V92027	Microsite	−67.00	9906	0.00	−67.00
V92025	Microsite	−73.00	9906	0.00	−73.00
V91010	Dinosaur	−75.00	9906	0.00	−75.00
V91012	Dinosaur	−75.00	9906	0.00	−75.00
V91014	Microsite	−75.00	9906	0.00	−75.00
V91007	Microsite	−80.16	9906	0.00	−80.16
V89004	Microsite	−81.68	9906	0.00	−81.68
V91002	Dinosaur	−82.68	9906	0.00	−82.68
ND96-23.1	Dinosaur	−85.00	86148	0.75	−85.75
V92040	Dinosaur	−100.00	9906	0.00	−100.00

Notes: PTRM—Pioneer Trails Regional Museum; Datum—distance to Hell Creek–Fort Union formational contact; K-T section—palynological K-T section (Johnson, this volume, Table A2) used to adjust from datum to K-T; Correction—distance from datum to K-T at nearest K-T section; K-T—distance to K-T boundary.

Among dinosaurs, only the teeth of the small theropods seem to provide reliable criteria for generic and specific identification (Currie et al., 1990; Fiorillo and Currie, 1994). Some other theropods are known only from their teeth and are commonly referred to as dental taxa. Although previous publications have rejected these taxa because their phylogenetic relationships cannot be assessed (e.g., Sullivan, 1987), others (Currie et al., 1990) acknowledge their importance and assign them to genera with familial affinities unknown. We have followed Currie et al. (1990) for our placement of the dental taxa *Paronychodon* and *Richardoestesia*, which have been found only at vertebrate microsites. Edentulous dinosaurs, such as Caenagnathidae and ornithomimids, were identified only to family because the recovered postcranial elements did not permit generic identifications.

All fossils that could not be identified at the Pioneer Trails Regional Museum (PTRM) were forwarded to specialists at other institutions for comparative analysis and identification. Specimens are stored at the PTRM and are available for study to qualified researchers.

The abundance of identifiable elements was tabulated for each of the vertebrate microsites. Because of the fragmentary nature of turtle shells, we indicated a turtle's presence at a vertebrate microsite if an identifiable element assignable to that taxon was found, but we did not count the individual shell fragments. Other common and repeated body parts, such as gar scales and crocodilian osteoscutes, were not included in specimen counts for the same reason. These isolated elements are often the most numerous elements found and, if included, likely would result in an overestimate of the actual abundance of individuals.

RESULTS

Nearly 14 000 specimens were collected for this study. Of these, we were able to identify 10 165 specimens, which form the database for the following analysis. A total of 42 vertebrate microsites yielded 10 124 specimens (Table 2). An additional 41 sites produced in situ dinosaur skulls or partial skeletons (Table 3). One of these localities (V98020) appears in both databases, so the total number of localities sampled for this study is 82 rather than 83.

The fauna of the Hell Creek Formation based on these collections includes fish (including sharks and rays), amphibians, lizards, turtles, crocodilians, champsosaurs, dinosaurs, birds, pterosaurs, and mammals. We recognize a total of 61 taxa (Table 4) including 13 dinosaurs, nine turtles, nine fish, and six mammals. The stratigraphic ranges of 56 of these taxa are plotted in Figure 4. These taxa are ranked in Table 5 by number of specimens and by the number of localities where they occur. The fish *Kindleia* is the most abundant fossil. The shed teeth of hadrosaurs are the most abundant dinosaur specimens, and Ceratopsian dinosaurs are present at more sites than any other dinosaur. The dinosaurs *Ricardoestesia* and the Caenagnathidae have not been previously reported from the Hell Creek Formation.

Stratigraphic distribution of localities

Fossil localities are concentrated near the top of the Hell Creek Formation; 59 of the 82 localities (72%) occur in the upper 25 m of the formation. The 25–50 m level yielded 12 localities, the 50–75 m level had only two localities, and the basal 25 m of the formation produced nine localities. The high concentration of sites in the upper part of the formation is largely a historical artifact caused by outcrop availability and a deliberate emphasis on sites in close proximity to the formational contact and K-T boundary.

The majority of localities found in the lower part of the Hell Creek Formation are associated with channel deposits (Tables 2 and 3). These channels are prominent features and are easily recognized because most contain calcareous sandstone lenses and weather to a high relief (Frye, 1969). Higher in the section, more localities are associated with channel margin deposits than with the channels. These deposits are small-scale, discontinuous deposits of sandstone associated with siltstone and macerated plant material. In the uppermost 15 m of the formation, vertebrate localities in flood-plain deposits become more abundant than lower in the formation. These flood-plain

TABLE 2. THE PIONEER TRAILS REGIONAL MUSEUM (PTRM) VERTEBRATE MICROSITE DATABASE

PTRM locality	Lith	Depositional environment	Datum (m)	K-T (m)	Number of taxa	Number of specimens	Number of specimens from major vertebrate groups								
							F	A	L	T	C	P	D	B	M
V88012	ZS	Flood plain	−2.37	−2.37	9	58	17	1	0	1	15	0	24	0	0
V88017	Carb	Flood plain	−3.17	−3.17	18	123	22	3	0	2	17	0	78	0	1
V97024	Carb	Flood plain	−3.40	−4.15	21	50	11	1	1	6	19	0	11	1	0
V96003	ZMS	Flood plain	−4.50	−5.25	9	10	2	0	0	5	2	0	1	0	0
V96004	ZMS	Flood plain	−4.50	−5.25	12	39	20	0	0	3	6	0	10	0	0
V97022	ZMS	Flood plain	−4.81	−5.56	6	7	2	0	0	4	1	0	0	0	0
V96005	ZMS	Channel	−5.50	−6.25	11	53	45	1	0	6	0	0	1	0	0
V98011	ZMS	Channel	−5.90	−6.65	12	19	3	0	0	4	4	0	7	0	1
V96022	SS	Channel	−4.90	−7.51	4	5	0	2	0	2	0	0	1	0	0
V95020	ZMS	Channel	−5.50	−8.11	14	32	15	1	0	4	4	0	8	0	0
V88009	SZS	Crevasse splay	−8.29	−8.29	8	10	4	0	0	3	3	0	0	0	0
V89003	SS	Crevasse splay	−8.40	−8.40	29	715	377	25	5	0	129	1	172	0	6
V98020	SS	Channel	−8.80	−9.55	3	3	0	0	0	1	1	0	1	0	0
V87004	SS	Channel	−9.00	−9.75	4	5	0	0	0	1	1	0	3	0	0
V87012	ZMS	Flood plain	−9.85	−9.85	35	566	277	19	17	6	121	0	115	4	7
V88020	SS	Channel	−10.00	−10.00	11	38	9	0	4	2	12	0	11	0	0
V88021	SZS	Crevasse splay	−10.00	−10.00	6	6	0	1	1	1	1	0	2	0	0
V96015	ZMS	Flood plain	−8.41	−10.12	14	42	30	0	0	6	2	0	4	0	0
V88004	ZMS	Flood plain	−10.61	−10.61	29	304	180	18	10	7	38	0	50	0	1
V88013	SS	Channel	−10.62	−10.62	12	31	12	2	1	3	8	0	5	0	0
V95025	SS	Crevasse splay	−8.50	−10.64	8	16	3	1	0	3	0	0	9	0	0
V95016	SS	Channel	−8.85	−10.99	20	110	30	1	0	6	20	0	51	1	1
V88019	SZS	Crevasse splay	−12.00	−12.00	15	68	27	3	2	3	21	0	10	2	0
V98027	SS	Channel	−10.62	−12.07	19	136	81	3	2	4	12	0	34	0	0
V88018	Carb	Flood plain	−14.27	−14.27	27	799	307	53	20	6	182	0	229	1	1
V97002	SS	Chanel	−13.00	−15.69	5	5	0	0	0	4	0	0	1	0	0
V92003	SS	Channel	−17.00	−17.75	5	7	0	0	0	3	0	0	4	0	0
V98002	Carb	Crevasse splay	−16.00	−18.25	9	29	14	0	0	4	1	0	10	0	0
V93022	SS	Crevasse splay	−25.30	−27.00	8	11	3	1	0	0	3	0	1	0	3
V97006	SS	Channel	−27.00	−28.70	12	32	18	0	0	3	5	0	5	1	0
V87006	SS	Channel	−28.00	−28.75	26	186	42	4	0	9	36	0	95	0	0
V86002	SS	Channel	−29.18	−29.93	41	1180	476	41	16	9	297	1	328	4	8
V87009	SS	Channel	−31.43	−32.55	17	97	48	4	0	3	17	0	22	0	3
V86003	SS	Channel	−35.05	−36.17	27	186	79	14	9	6	24	0	49	0	5
V92067	SS	Crevasse splay	−37.18	−38.88	49	3243	2122	106	40	7	307	0	550	2	109
V92002	SS	Channel	−40.00	−41.12	25	446	291	4	0	5	114	0	29	1	2
V92029	SS	Channel	−60.00	−60.00	9	12	0	0	0	4	2	0	6	0	0
V92027	SS	Channel	−67.00	−67.00	7	12	3	0	0	2	3	0	4	0	0
V92025	SS	Channel	−73.00	−73.00	26	215	81	0	2	8	47	0	76	0	1
V91014	SS	Crevasse splay	−75.00	−75.00	8	8	2	0	0	4	1	0	1	0	0
V91007	SS	Channel	−80.16	−80.16	17	230	178	2	1	0	28	0	17	1	3
V89004	SS	Channel	−81.68	−81.68	44	980	606	13	3	7	178	0	157	1	15
				Totals		10124	5437	324	134	167	1682	2	2192	19	167

Notes: n = 42 sites. Major vertebrate groups: F—fish; A—amphibians; L—lizards; T—turtles; C—crocodilians; P—pterosaurs; D—dinosaurs; B—birds; M—mammals. Lith—lithology: Z—silt; Carb—carbonaceous shale; M—mud, S—sand (e.g., ZMS—silty mudstone). K-T—Cretaceous-Tertiary.

deposits consist of laminated mudstone, silty mudstone, and depositional horizons of unrecognizable plant debris that are laterally traceable on the surface of the outcrop as dark horizons.

The vertebrate microsite database and the number of different taxa found at each locality identified by environment type are shown in Table 2. A change in the environments that preserve vertebrate microsites is evident upsection. There are, on average, 17.5 taxa per locality from vertebrate microsites in the flood-plain environments, 16 taxa in the channel margin environments, and 15.9 taxa per site from the channels. It can also be seen in Table 2 that dinosaurs occur and are the most common fossils in the highest vertebrate microsite (2.37 m below the K-T boundary).

Table 3 shows the stratigraphic occurrence of individual dinosaur sites. These data reveal the presence of dinosaurs throughout the depositional sequence of the Hell Creek Formation and up into the basal part of the Fort Union Formation. The highest dinosaur site recorded, PTRM locality V96019, is 0.84 m above the formational contact in the basal Fort Union Formation, indicating that the basal part of the Fort Union Formation is, in places, Cretaceous in age (Pearson, 1997). This is concordant with the palynology of this site, which places the K-T boundary at a position 1.77 m above this dinosaur. The site contains associated skull elements consisting of a brow horn, nose horn, frill, and right maxillary from a ceratopsian dinosaur preserved in claystone. This association indicates that

TABLE 3. PIONEER TRAILS REGIONAL MUSEUM (PTRM) INDIVIDUAL DINOSAUR DATABASE

PTRM locality	Dinosaur taxon	Skeletal elements preserved	Lithology	Depositional environment	Datum (m)	K-T (m)
V96019	Ceratopsian	Frill, horns, maxillary	CS	Flood plain	0.84	−1.77
V88002	Hadrosaur	Vertebrae and ribs	ZMS	Flood plain	−2.72	−2.72
V95014	Ceratopsian	Frill, femur	ZMS	Flood plain	−0.98	−3.12
V96016	Ceratopsian	Skull, large bone frag.	ZMS	Flood plain	−2.76	−3.51
V95009	Ceratopsian	Skull frag. and frill frags.	ZMS	Flood plain	−4.00	−4.96
V97017	Ceratopsian	Associated skeleton	RMS	Flood plain	−4.65	−5.40
V92004	Ceratopsian	Skull (commercial collector)			−6.00	−6.00
V98003	Ceratopsian	Frill (?parietal)	ZMS	Flood plain	−4.35	−6.06
V97011	Hadrosaur	Vertebrates, ribs, and femur	SZS	Crevasse splay	−5.40	−6.15
V97012	Hadrosaur	Vertebrae	ZMS	Flood plain	−5.60	−6.35
V97009	Ceratopsian	Frill	carb	Flood plain	−4.50	−7.19
V98017	Ceratopsian	Vertebrae	SS	Channel	−7.12	−7.87
V88046	Ceratopsian	Frill			−8.00	−8.00
V96021	Ceratopsian	Frill and large bones	ZS	Flood plain	−5.50	−8.11
V98020	Hadrosaur	?juv. illium	SS	Channel	−8.80	−9.55
V98015	Ceratopsian	Frill	ZMS	Flood plain	−8.85	−9.60
V96006	Ceratopsian	Frill fragments	SZS	Channel	−9.00	−9.75
ND92-19.1	*Triceratops*	Skull and skeleton	SS		−7.16	−9.77
V88005	Ceratopsian	Frill and large bone fragments	SS	Channel	−10.00	−10.00
V88040	Ceratopsian	Frill	ZMS	Flood plain	−10.00	−10.00
V95017	Ceratopsian	Frill fragments	SS	Channel	−8.00	−10.14
V88003	Ceratopsian	Frill and large bone fragments			−10.50	−10.50
V98016	Ceratopsian	Parietal	SZS	Crevasse splay	−10.68	−11.43
V92007	*Torosaurus*	Associated skull and post cranial	MS	Flood plain	−11.49	−11.49
V88054	Ceratopsian	Frill			−12.00	−12.00
V98021	Hadrosaur	Femur	SZS	Crevasse splay	−11.34	−12.09
UND3000	*Triceratops*	Skull	MS		−12.00	−14.61
MPM51016	*Triceratops*	Skull and skeleton			−15.00	−15.00
V97013	Hadrosaur	Vertebra and ribs	SS	Channel	−14.50	−15.25
ND95-117.1	*Triceratops*	Skull			−18.00	−20.61
V95023	Ceratopsian	Skull and frill fragments	SS	Channel	−20.00	−22.14
V95018	Hadrosaur	Associated skeleton	SS	Channel	−21.30	−23.44
V93005	Ceratopsian	Pubis and large bone fragments	SS	Channel	−25.50	−27.20
V92068	*Tyrannosaurus*	Associated skeleton	MS	Flood plain	−31.18	−32.88
V92066	Ceratopsian	Skull fragments w/brow horn	SS	Channel	−37.00	−38.70
V92042	Ceratopsian	Frill and rib fragments			−40.00	−41.70
V91010	Hadrosaur	R. dentary and vertebra	SS	Channel	−75.00	−75.00
V91012	Ceratopsian	Frill fragments and ribs	SS	Channel	−75.00	−75.00
V91002	Ceratopsian	L. brow horn and squasmosal	ZS	Flood plain	−82.68	−82.68
ND96-23.1	*Triceratops*	Skull and associated parts	SS	Channel	−85.00	−85.75
V92040	Ceratopsian	Vertebra	SS	Channel	−100.00	−100.00

Note: n = 41 sites. K-T, Cretaceous-Tertiary. Major vertebrate groups: F—fish; A—amphibians; L—lizards; T—turtles; C—crocodilians; P—pterosaurs; D—dinosaurs; B—birds; M—mammals. Lith—lithology: Z—silt; Carb—carbonaceous shale; M—mud, S—sand (e.g., ZMS—silty mudstone). K-T—Cretaceous-Tertiary.

this fossil did not get deposited by stream action, but rather was lying in place on the flood plain presumably at or near the place of death. This dinosaur is the geologically youngest dinosaur to be recorded to date from the study area (Pearson et al., 2001).

We tracked taxonomic diversity throughout the formation by recording the presence of all taxa at each stratigraphic level and charted the ranges of the taxa to determine the point of last occurrence relative to the Hell Creek–Fort Union formational contact and the K-T boundary. Figure 4 illustrates the observed stratigraphic ranges of the taxa throughout the formation. Of the 56 taxa whose ranges have been plotted, most occur throughout the formation. Only 12 taxonomic appearances and five disappearances occur at levels other than the base and the top of the formation, respectively.

Sample size rather than stratigraphic position may be the primary factor in determining the presence of taxa in the assemblage (see also Rarefaction analysis, below). Table 2 lists the size of the collection recorded for each site by number of specimens belonging to each taxon. It is interesting to note from this table that all dinosaurs that occur at more than two sites also occur at the highest stratigraphic level, yielding more than 500 specimens. This observation implies that moderately large sample sizes may be needed to sample the less common elements of the fauna.

Figure 5 illustrates all of the sites within 15 m of the K-T boundary plotted against the position of the palynologically defined K-T boundary. The first 15 m below the K-T boundary have produced a total of 53 sites, including both individual dinosaur sites and vertebrate microsites. Dinosaurs, which here are assumed to have become extinct at the end of the Cretaceous,

TABLE 4. SYSTEMATIC LIST OF ALL 61 VERTEBRATE TAXA RECOGNIZED IN THE HELL CREEK FORMATION IN SOUTHWESTERN NORTH DAKOTA AND NORTHWESTERN SOUTH DAKOTA

Class	Order	Family	Genus species or lowest taxon
CHONDRICTHYES			
	BATOIDEA		
		ANACORACIDAE	
			Myledaphus bipartitus
	ELASMOBRANCHII (check)		
		HYBODONTIDAE	
			Lissodus selachos
		ORECTOLOBIDAE	
			Squatirhina americana
		SCLERORHYNCHIDAE	
			Ischyrhiza avonicola
OSTEICHTHYES			
	ACIPENSERIFORMES		
		ACIPENSERIDAE	
			cf. *"Acipenser"* sp.
	AMIIFORMES		
		AMIIDAE	
			Kindleia fragosa
			Melvius thomasi
	LEPISOSTEIFORMES		
		LEPISOSTEIDAE	
			Lepisosteus occidentalis
	TELEOSTEI		
		ASPIDORHYNCHIDAE	
			Belonostomus longirostris
AMPHIBIA			
	ANURA		
		Anura Indeterminate	
	CAUDATA		
		BATRACHOSAUROIDIDAE	
			Opisthotriton kayi
		SCAPHERPETODONTIDAE	
			Scapherpeton tectum
		SIRENIDAE	
			Habrosaurus dilatus
		Caudata indeterminate	
REPTILIA			
	TESTUDINES		
		BAENIDAE	
			Plesiobaena antiqua
			Baenidae indeterminate
		CHELYDRIDAE	
			Chelydridae indeterminate
		DERMATEMYDIDAE	
			Adocus sp.
			Basilemys sinuosa
		KINOSTERNIDAE	
			Kinosternidae indeterminate
		MACROBAENIDAE	
			Macrobaenidae indeterminate
		TRIONYCHIDAE	
			Helopanoplia distincta
			Trionychidae indeterminate
		Incertae sedis	
			Compsemys victa
	LEPIDOSAURIA		
			Lepidosauria indeterminate
	CHORISTODERA		
		CHAMPSOSAURIDAE	
			Champsosaurus sp.
	CROCODILIA		
		CROCODYLIDAE	
		ALLIGATORINAE	
			Brachychampsa montana
REPTILIA (continued)			
		CROCODYLINAE	
			Leidyosuchus sternbergi
		THORACOSAURINAE	
			Thoracosaurus neocesariensis
	PTEROSAURIA		
		Pterosauria indeterminate	
	ORNITHISCHIA		
		CERATOPSIDAE	
			Torosaurus latus
			Triceratops horridus
			Ceratopsidae indeterminate
		HADROSAURINAE	
			Hadrosaurinae indeterminate
		HYPSILOPHODONTIDAE	
			Thescelosaurus neglectens
	SAURISCHIA		
		AVISAURIDAE	
			cf. *Avisaurus archibaldi*
		CAENAGNATHIDAE	
			Caenagnathidae indeterminate
		ORNITHOMIMIDAE	
			Ornithomimidae indeterminate
		TROODONTIDAE	
			Troodon sp.
		TYRANNOSAURIDAE	
			Tyrannosaurus rex
		VELOCIRAPTORINAE	
			Saurornitholestes sp.
		COELUROSAURIDAE	
			Coelurosauridae indeterminate
		Incertae sedis	
			Paronychodon lacustris
			Richardoestesia sp.
AVES			
		Aves indeterminate	
		Hesperornithiformes indeterminate	
MAMMALIA			
	MULTITUBERCULATA		
		CIMILODONTIDAE	
			Cimolodon nitidus
			Cimolodon sp.
		CIMOLOMYIDAE	
			Meniscoessus robustus
			Meniscoessus sp.
			? *Meniscoessus* sp.
		NEOPLAGIAULACIDAE	
			Mesodma thompsoni
	EUTHERIA		
		?LEPTICTIDAE	
			Gypsonictops illuminatus
		PALAEORYCTIDAE	
			Cimolestes magnus
	METATHERIA		
		STAGODONTIDAE	
			Didelphodon vorax
			Didelphodon sp.
			? *Didelphodon* sp.
		PEDIOMYIDAE	
			Pediomys florencae
			Pediomys cf. *P. florencae*
		PERADECTIDAE	
			Alphadon marshi
	Mammalia indeterminate		

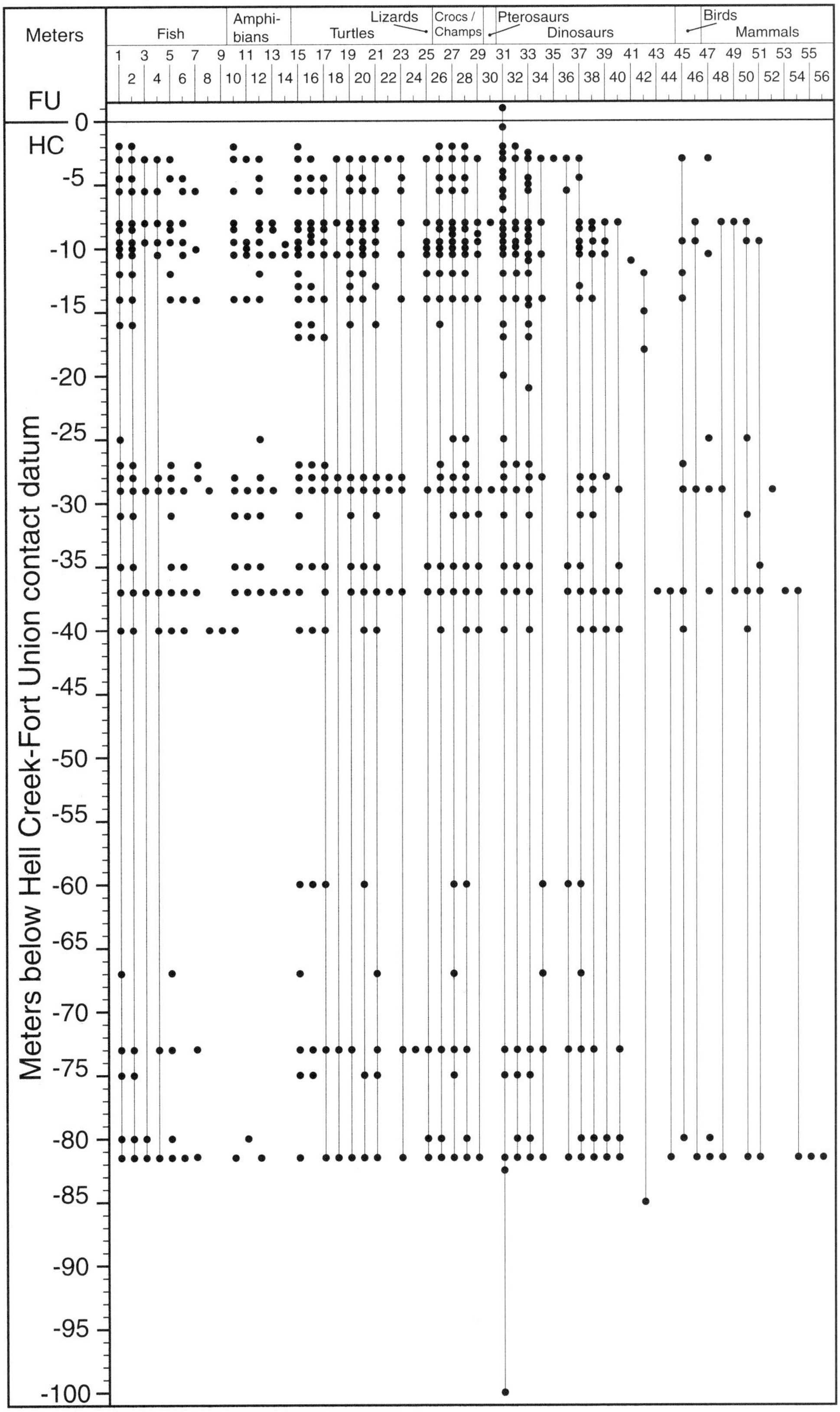

Figure 4. Stratigraphic range chart prepared with data on all localities (vertebrate microsite and dinosaur) and all taxa (some large systematic categories deleted) sorted first by higher taxonomic group and then by "tops" (last occurrence datum) relative to Hell Creek–Fort Union (HC-FU) contact datum. Plotted taxa: 1—*Kindleia fragosa*, 2—*Lepisosteus occidentalis*, 3—*Belonostomus longirostris*, 4—*Melvius thomasi*, 5—*Myledaphus bipartitus*, 6—*Lissodus selachos*, 7—cf. "*Acipenser*," 8—*Squatirhina americana*, 9—*Ischyrhiza avonicola*, 10—Caudata indet., 11—*Opisthotriton kayi*, 12—*Scapherpeton tectum*, 13—*Habrosaurus dilatus*, 14—Anura indet., 15—Trionychidae indet., 16—*Adocus* sp., 17—Baenidae indet., 18—*Basilemys sinuosa*, 19—Chelydridae indet., 20—*Compsemys victa*, 21—*Helopanoplia distincta*, 22—Kinosternidae indet., 23—*Plesiobaena antiqua*, 24—Macrobaenidae indet., 25—Lepidosauria indet., 26—*Brachychampsa montana*, 27—*Champsosaurus* sp., 28—*Leidyosuchus sternbergi*, 29—*Thoracosaurus neocesariensis*, 30—Pterosauria indet., 31—Ceratopsidae indet., 32—*Ricardoestesia isolosletes*, 33—Hadrosaurinae indet., 34—Caenagnathidae indet., 35—Coelurisauridae indet., 36—Ornithomimidae indet., 37—*Tyrannosaurus rex*, 38—*Paranychodon lacustris*, 39—*Saurornitholestes* sp., 40—*Thescelosaurus neglectens*, 41—*Torosaurus latus*, 42—*Triceratops horridus*, 43—cf. *Avisaurus archibaldi*, 44—*Troodon* sp., 45—Aves indet., 46—Hesperinthifomes indet., 47—Theria indet., 48—*Cimolodon nitidus*, 49—*Didelphodon* sp., 50—*Meniscoessus robustus*, 51—*Didelphodon vorax*, 52—*Cimolestes magnus*, 53—*Gysonictops illuminatus*, 54—*Mesodma thompsoni*, 55—*Alphadon marshi*, 56—*Pediomys florencae*.

TABLE 5. RANK ORDER OF VERTEBRATE TAXA AND THE NUMBER OF LOCALITIES WHERE THEY OCCUR

Vertebrate taxon	Rank by specimens	Percentage	Vertebrate taxon	Rank by localities	Percentage
Kindleia fragosa	2610	25.68	*Ceratopsidae* indet.	59	71.95
Myledaphus bipartitus	1677	16.50	*Trionychidae* indet.	36	43.90
Hadrosaurinae indet.	1006	9.90	*Kindleia fragosa*	34	41.46
Lepisosteus occidentalis	938	9.23	*Hadrosaurinae* indet.	32	39.02
Ceratopsidae indet.	845	8.31	*Lepisosteus occidentalis*	31	37.80
Leidyosuchus sternbergi	841	8.27	*Leidyosuchus sternbergi*	31	37.80
Champsosaurus sp.	430	4.23	*Adocus* sp.	29	35.37
Brachychampsa montana	354	3.48	*Champsosaurus* sp.	27	32.93
Caudata indet.	149	1.47	*Brachychampsa montana*	25	30.49
Scapherpeton tectum	144	1.42	*Tyrannosaurus rex*	24	29.27
Lepidosauria indet.	134	1.32	*Compsemys victa*	24	29.27
Tyrannosaurus rex	121	1.19	*Myledaphus bipartitus*	22	26.83
Ricardoestesia sp.	73	0.72	*Scapherpeton tectum*	18	21.95
Melvius thomasi	67	0.66	*Ricardoestesia* sp.	18	21.95
Paronychodon lacustris	64	0.63	*Baenidae* indet.	18	21.95
Thoracosaurus neocesariensis	57	0.56	*Helopanoplia distincta*	18	21.95
Thescelosaurus neglectens	50	0.49	*Chelydridae* indet.	17	20.73
Teleost indet.	48	0.47	*Caudata* indet.	16	19.51
Lissodus selachos	41	0.40	*Lepidosauria* indet.	16	19.51
Trionychidae indet.	36	0.35	*Melvius thomasi*	14	17.07
Mammalia indet.	35	0.34	*Paronychodon lacustris*	14	17.07
Saurornithoiestes sp.	34	0.33	*Lissodus selachos*	13	15.85
Meniscoessus robustus	30	0.30	*Thoracosaurus neocesariensis*	13	15.85
Adocus sp.	29	0.29	*Plesiobaena antiqua*	13	15.85
Belonostomus longirostris	28	0.28	*Opisthotriton kayi*	9	10.98
Compsemys victa	24	0.24	*Caenagnathidae* indet.	9	10.98
Theria indet.	24	0.24	*Aves* indet.	9	10.98
Meniscoessus sp.	23	0.23	cf. "*Acipenser*"	8	9.76
Opisthotriton kayi	22	0.22	*Belonostomus longirostris*	8	9.76
cf. "*Acipenser*"	18	0.18	*Teleost* indet.	8	9.76
Baenidae indet.	18	0.18	*Thescelosaurus neglectens*	8	9.76
Helopanoplia distincta	18	0.18	*Saurornitholestes* sp.	8	9.76
Chelydridae indet.	17	0.17	*Mammalia* indet.	8	9.76
Ornithomimidae indet.	15	0.15	*Ornithomimidae* indet.	7	8.54
Aves indet.	15	0.15	*Basilemys sinuosa*	7	8.54
Plesiobaena antiqua	13	0.13	*Meniscoessus robustus*	7	8.54
Caenagnathidae indet.	12	0.12	*Theria* indet.	7	8.54
Didelphodon vorax	11	0.11	*Meniscoessus* sp.	6	7.32
?*Mammalia* indet.	11	0.11	*Habrosaurus dilatus*	4	4.88
Multituberculata indet.	10	0.10	*Triceratops horridus*	4	4.88
Basilemys sinuosa	7	0.07	*Kinosternidae* indet.	4	4.88
Habrosaurus dilatus	6	0.06	*Didelphodon vorax*	4	4.88
Squatirhina americana	5	0.05	*Hesperorinthifomes* indet.	4	4.88
Ischyrhiza avonicola	5	0.05	*Anura* indet.	3	3.66
Troodon sp.	5	0.05	*Cimolodon nitidus*	3	3.66
Triceratops horridus	4	0.04	*Multituberculata* indet.	3	3.66
Kinosternidae indet.	4	0.04	*Squatirhina americana*	2	2.44
Hesperorinthifomes indet.	4	0.04	*Troodon* sp.	2	2.44
Anura indet.	3	0.03	*Cimolodon* sp.	2	2.44
Cimolodon nitidus	3	0.03	?*Meniscoessus* sp.	2	2.44
Marsupialia indet.	3	0.03	?*Multituberculata* indet.	2	2.44
Coelurisauridae indet.	2	0.02	*Mesodma thompsoni*	2	2.44
Cimolodon sp.	2	0.02	*Didelphodon* sp.	2	2.44
?*Meniscoessus* sp.	2	0.02	?*Didelphodon* sp.	2	2.44
?*Multituberculata* indet.	2	0.02	*Marsupialia* indet.	2	2.44
Mesodma thompsoni	2	0.02	?*Mammalia* indet.	2	2.44
Didelphodon sp.	2	0.02	*Pterosauria* indet.	2	2.44
?*Didelphodon* sp.	2	0.02	*Ischyrhiza avonicola*	1	1.22
Pterosauria indet.	2	0.02	*Torosaurus latus*	1	1.22
Torosaurus latus	1	0.01	cf.*Avisaurus archibaldi*	1	1.22
cf. *Avisaurus archibaldi*	1	0.01	*Coelurisauridae* indet.	1	1.22
Macrobaenidae indet.	1	0.01	*Macrobaenidae* indet.	1	1.22
Gysonictops illuminatus	1	0.01	*Gysonictops illuminatus*	1	1.22
Cimolestes magnus	1	0.01	*Cimolestes magnus*	1	1.22
Pediomys florencae	1	0.01	*Pediomys florencae*	1	1.22
Pediomys cf. *P. florencae*	1	0.01	*Pediomys* cf. *P. florencae*	1	1.22
Alphadon marshi.	1	0.01	*Alphadon marshi.*	1	1.22

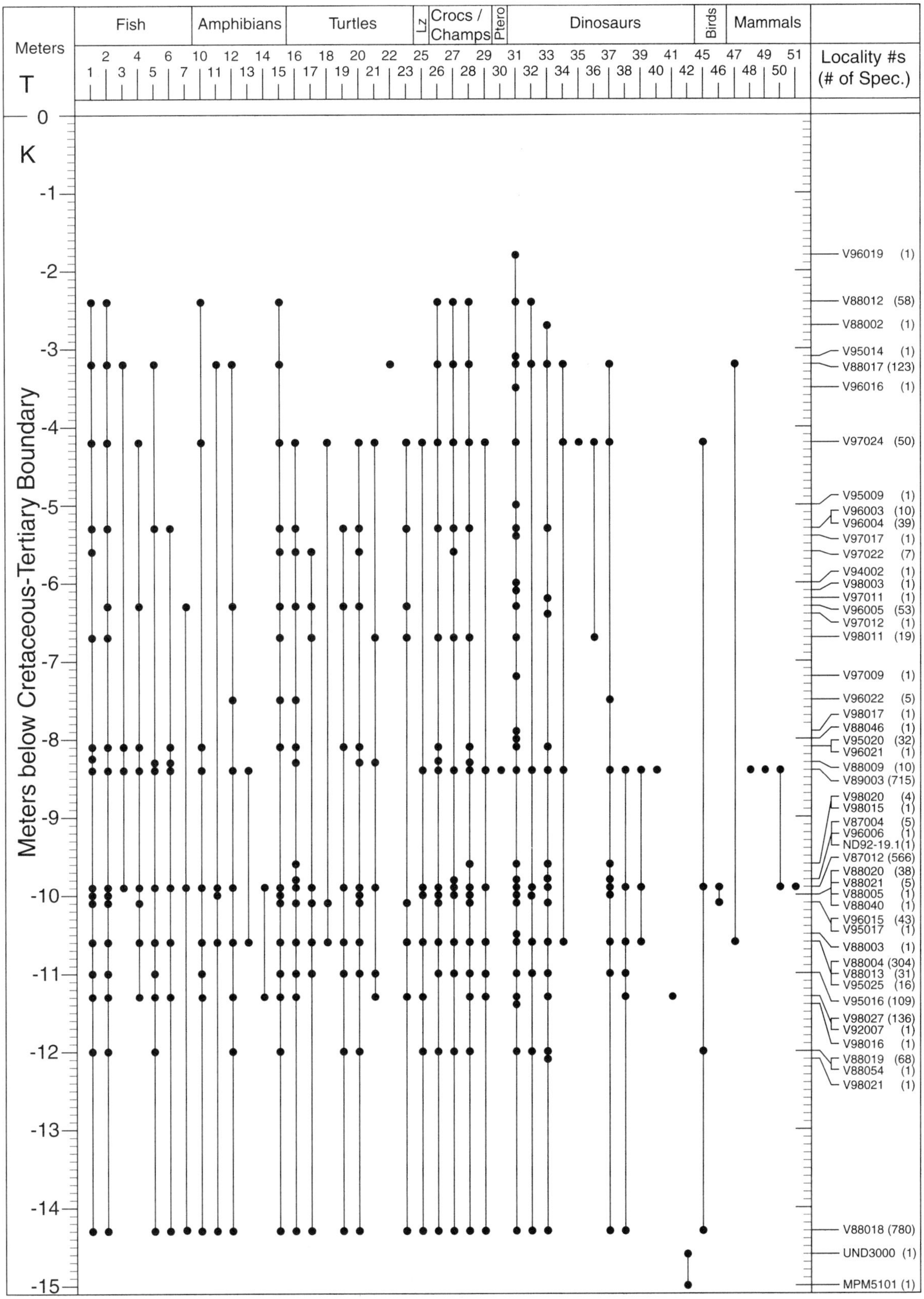

Figure 5. Stratigraphic ranges of vertebrate taxa prepared with data from all localities (vertebrate microsite and dinosaur) located within 15 m of Cretaceous-Tertiary (K-T) boundary. Each locality was measured to Hell Creek–Fort Union formational contact datum, then position of K-T boundary was determined palynologically at that section, and locality was recalibrated to K-T boundary and plotted (see Table 1 and Nichols and Johnson, this volume). Localities were plotted at same level if were within same 10 cm interval. Numbers for each taxon are same as those listed in caption for Figure 4.

occur to within 1.77 m of the K-T boundary. There are three dinosaur taxa within 3 m of the K-T boundary and seven dinosaur taxa within 4 m of the boundary. Dinosaurs are also present in the highest samples that produce vertebrates. We are in general agreement with Williams (1994) that the presence of dinosaur remains closer to the boundary than previously documented suggests that there is nothing wrong with the capacity of the sediments to preserve bones. We found a single vertebrate locality (V95014) consisting of a partial ceratopsian skeleton in the uppermost 1 m of the Hell Creek Formation, but this was at a site where the K-T boundary is located 2.14 m above the top of the formation. Other than this specimen, and despite extensive prospecting, we recovered no other vertebrate fossils in the uppermost 2 m of the Hell Creek Formation. Even species that are known to survive the K-T boundary and that are present in the overlying Fort Union Formation are absent from this interval. The reason for this gap in the record is uncertain, but our data in Figure 5 show that the highest well-sampled site (V88017, 3.17 m below the K-T boundary), at which 17 taxa are present, is as diverse as the average for all 42 vertebrate microsites within the Hell Creek Formation (16.3 taxa, data from Table 2).

Stratigraphic distribution of vertebrate taxa

The small shark *Squatirhina* and the sawfish *Ischyrhiza* are rare in the assemblage. Their presence is noted only at sites between 29 and 40 m below the formational contact, which positions them at the approximate stratigraphic position of the marine Breien Member in south-central North Dakota. Extant orectolobid sharks are found in intertidal zones, along sandy beaches, and along reefs to depths of 70 m, and the modern sawfish occurs in shallow seas and near the mouths of rivers. The absence of these taxa at other stratigraphic intervals may indicate that their preferred habitats were no longer present in the area. We acknowledge, however, that the dental remains of these taxa are extremely small, which may have contributed to their not being found elsewhere. The hybodont shark *Lissodus* occurs throughout the formation.

Bryant (1989) indicated that the large amiid *Melvius* had become extinct by the end of the Cretaceous, and that the stratigraphically highest record of this genus was at University of California Museum of Paleontology (UCMP) locality V73087 (Flat Creek Five) ~5 m below the top of the formation (Bryant, 1987a). The occurrence of this taxon at PTRM locality V97024 extends its last known occurrence to 4.15 m below the K-T boundary. The amiid *Kindleia*, the ray *Myledaphus*, and the gar *Lepisosteus* occur throughout the formation.

All of the salamanders found in the study area have been previously identified as being primarily aquatic forms (Estes, 1964; Bryant, 1989). Specimens representing the taxon *Scapherpeton* seem to vary in relative abundance more closely with lizards than with the other salamanders. At localities where the abundance of lizards is high, *Scapherpeton* abundance was similarly high. This corresponding change in taxonomic relative abundance suggests that both may have been responding to the same environmental stimuli. Estes (1964) indicated that *Scapherpeton* had large limb elements, well-developed rib-bearers, and ribs suggesting a quadrepedal animal more similar to a lizard than to a neotenic salamander. Bryant (1989) also noted that *Scapherpeton* had an elongated body suggestive of aquatic adaptation, but not necessarily swimming locomotion. It is possible that *Scapherpeton* was a more terrestrial animal than previously thought, inhabiting the stream borders more like lizards than like salamanders. The large salamander *Habrosaurus* is also present in the assemblage, although rare. Many fragments of salamander vertebrae were catalogued as Caudata indeterminate.

Our study shows that lizard remains may be found in any sedimentary environment, but that they are also a rare component in the assemblage. Because of the limited taxonomic utility of isolated vertebra, the lizard material was not identified to a more precise level. Bryant (1989) also reported that lizards are relatively rare in the Hell Creek Formation and that they appear to be concentrated in sandy lithologies.

The crocodilian *Leidyosuchus* and the alligator *Brachychampsa* occur throughout the formation. The crocodilian *Thoracosaurus* is also present, although rare. This taxon is more commonly found in deposits in the eastern and southeastern United States (Troxell, 1925; Carpenter, 1983; Gallagher, 1993). The aquatic *Champsosaurus* is present throughout the formation, but the collected remains consist primarily of isolated, dissociated, or fragmentary, postcranial elements.

Turtle fossils are ubiquitous through the formation (cf. Bryant, 1989) and are usually found as fragmentary shell material. Chelonian faunas based upon fragments of the shell are difficult if not impossible in some cases to distinguish at the species level. Archibald and Bryant (1990) indicated that all of the turtles that occur in the Hell Creek area cross the K-T boundary, with the exception of the trionychid *Helopanoplia* and the dermatemydid *Basilemys*, which are reported to become extinct at the boundary. *Helopanoplia* and *Basilemys* are present to within 4.15 m of the K-T boundary in our study area.

Ceratopsian and hadrosaur shed teeth are commonly found in all sedimentary environments, suggesting that they were present throughout the deposition of the formation. Brinkman et al. (1998) indicated that ceratopsian and hadrosaur teeth are similar in size and shape and that any differences in the abundance of their shed teeth could likely be a reflection of their actual abundance in the original communities. They also reported that ceratopsian dinosaurs were members of the coastal community during the Campanian, and that they were more abundant in these areas than in inland areas during these times.

By the end of the Cretaceous, if ceratopsians still preferred a more coastal environment (Brinkman et al., 1998) and the shorelines had receded from the area (see following discussion of the Western Interior seaway), one would expect a decrease in their abundance approaching the K-T boundary. This is not the case. Either ceratopsians had evolved since the Campanian to be more independent of coastal areas, or the coastal areas were not that far removed from the study area at the time of the deposition of the upper part of the Hell Creek Formation. The latter scenario is favored by sedimentological evidence (Murphy et al., this volume), faunal evidence (Hartman, this volume; see following), and recent studies of the coastline of the Western Interior seaway (Kennedy et al., 1998).

Specimens from the hypsilophodont dinosaur *Thescelosaurus* are found more commonly associated with channel or near-channel environments. However, *Thescelosaurus* has not been found higher than 8.4 m below the K-T boundary in our study area. The environments that produce the majority of specimens above this stratigraphic horizon are predominantly flood plains, and this may contribute to the absence of hypsilophodonts. Carpenter and Breithaupt (1986) stated that *Thescelosaurus* is known by at least one specimen from the University of California Museum of Paleontology collections from within 3 m of the Z coal, which they used to approximate the K-T boundary.

The small theropod *Troodon* is a rare component of the dinosaurian fauna. No specimens of this dinosaur have been found in the upper 36 m of the formation. The teeth of this animal are tiny (~ 5 mm), and so its absence in the upper part of the formation may reflect a sampling bias.

A single skull fragment assignable to the pachycephalosaur *Stygimoloch* was found in the study area in 1987 by the Milwaukee Public Museum, and a skull and postcranial skeleton of a pachycephalosaur were found by a commercial collector in the middle Hell Creek Formation in northwestern South Dakota in 1995 (Triebold, 1997). No other specimens of pachycephalosaurs have been found in the study area. Furthermore, no specimens assignable to the ankylosaurs or nodosaurs have been identified from the study area.

The Caenagnathidae and Ornithomimidae are represented by postcranial elements that do not provide for more precise taxonomic resolution than the familial level. The taxon cf. *Avisaurus* is represented by the distal end of the metatarsal II (P. Currie, 1997, written commun.) and is known only from a single occurrence at one locality 37.18 m below the formational contact. The small velociraptorine *Saurornitholestes* is present to within 8.4 m of the K-T boundary.

Of the theropods known only by teeth, *Paronychodon* is present to within 8.40 m of the K-T boundary and *Richardoestesia* is found within 2.37 m of the K-T boundary. Sankey (2001) named the Hell Creek specimens *Richardoestesia isolosletes* and indicated that *R. gilmorei* is absent in the Hell Creek Formation.

Tyrannosaurus remains occur throughout the formation, and, on the basis of their small size, a significant number of teeth collected were identified as juveniles (P. Curry, 1997, personal commun.). We have found no evidence from the study area suggesting that these tyrannosaurs were nesting in the immediate vicinity, but the presence of juvenile teeth implies that young were in the area soon after birth. The small size of some of these Tyrannosaur teeth (1–2 cm in length) indicates a stature probably incapable of long migratory treks. Peter Larson (2000, personal commun.) suggested that some of these small tyrannosaurid teeth may represent *Nanotyrannus*. Specimens assigned to *Tyrannosaurus* are found within 3.17 m of the K-T boundary.

There have been no reports of eggshell from the study area. It is unknown whether this lack of eggshell indicates the absence of either nesting grounds or of conditions favorable to their preservation (Carpenter, 1982; Retallack and Leahy, 1986), but the poor condition of shelly invertebrate remains in the area suggests that the latter might be so.

Bird remains representing 18 individuals occur at 12 sites ranging to within 3.4 m below the K-T boundary. The Hesperornithiformes are well represented in the collection (T. Tokaryk, 1999, written commun.). Tentative identification of these specimens indicates they may belong to a small-bodied Hesperornithiform known as *Potamornis* (Elzanowski et al., 2001). Hesperornithiformes are reported to be sensitive to fluctuation of shoreline positions (Tokaryk et al., 1997) and are believed to be generally typical of marine habitats (Tokaryk, 1997). The presence of these marine taxa in our study area suggests that the marine environment may have been closer to the study area than previously thought or that the shoreline habitats may have been extended geographically, so that the habitat that they preferred may have extended farther inland.

Two specimens identified as "cf. pterosaur" (P. Currie, 1997, personal commun.) occur at 8.4 m and 29.9 m below the K-T boundary. Subsequent review of these specimens by K. Padian and T. Stidham (2000, personal commun.) suggested that these specimens are neither pterosaur nor avian.

Mammalian taxa from 16 sites represent only 1.4% of the total assemblage. Single specimens were produced from six sites, 10 sites produced multiple specimens, and one site produced 111 specimens (Hunter and Pearson, 1996). The mammalian fauna are described and discussed by Hunter and Archibald (this volume).

Table 6 lists all of the nonmammalian vertebrate taxa by last appearance data. Included on this table are PTRM specimen numbers for a voucher specimen for each taxon and its last occurrence datum relative to the K-T boundary. It is interesting to note that 40 of the 56 taxa whose stratigraphic ranges were plotted in Figure 4 are present to within 10 m of the K-T boundary. All taxa that occur at more than two localities also occur to within 10 m of the K-T boundary.

Rarefaction analysis

To determine whether there are any trends in vertebrate diversity through the formation, we explored the use of rarefaction analysis (Sanders, 1968; Hurlbert, 1971; Simberloff, 1972; Raup, 1975) to mitigate problems due to varying sampling intensity. Rarefaction reduces the nuisance variation in diversity that is due solely to variation in sample size (Hayek and Buzas, 1997). This technique is useful in comparing samples that differ in size, such as fossils collected at different stratigraphic horizons. Sheehan et al. (1991) also used rarefaction to test for differences in dinosaur diversity at different stratigraphic levels of the Hell Creek Formation (lower, middle, and upper thirds). We follow up on their study using the dataset of this paper, treating diversity at successive stratigraphic horizons as a time series, which preserves the sequence of our stratigraphic units.

Rarefied diversity, the number of taxa expected in a sample of some size smaller than the original sample, can be calculated in several different ways, including randomization (see Simberloff, 1972; Foote, 1992; Alroy, 1998; Jernvall and Wright, 1998), and explicit formulation, assuming that abundances follow either a hypergeometric or a multinomial distribution (Hurlbert, 1971; Heck et al., 1975). Exploratory analysis with this dataset revealed that all three of these methods produce similar results, but that computer processing was much faster using the multinomial formulation, so we carried out rarefaction using the latter method (Heck et al., 1975). Confidence limits (95%) on this estimate derive from the standard deviation of expected diversity. We then compared rarefied vertebrate diversity at different stratigraphic levels of the Hell Creek Formation (Figs. 6 and 7).

If extinction were gradual over the 2 m.y. represented by the Hell Creek Formation, one would predict an overall decreasing trend through the entire series. No such negative trend is evident (Figs. 6 and 7), nor did we detect a negative trend by either runs test (up and down) or linear regression. This generalization holds for all levels of resolution that we explored by varying the size of the stratigraphic bin (20 m, 10 m, 5 m, and localities separately). Only the results using 20 m and 5 m bins are shown in Figures 6 and 7. It is important to emphasize that we have not demonstrated that a decreasing trend in vertebrate diversity, as predicted by the gradual extinction hypothesis, did not exist outside of either the range or scale of our data. For example, we cannot reject the possibility of a gradual decline in vertebrate diversity on an ecological time scale, out-

TABLE 6. LAST APPEARANCE DATA OF NONMAMMALIAN HELL CREEK VERTEBRATES RELATIVE TO THE CRETACEOUS-TERTIARY BOUNDARY

Vertebrate taxon	Voucher #	LAD (m)
Ceratopsidae indeterminate	PTRM-4919	−1.77
Richardoestesia sp.	PTRM-1059	−2.37
Kindleia fragosa	PTRM-4404	−2.37
Lepisosteus occidentalis	PTRM-4002	−2.37
Caudata indeterminate	PTRM-4156	−2.37
Trionychidae indeterminate	PTRM-4914	−2.37
Champsosaurus sp.	PTRM-4916	−2.37
Brachychampsa Montana	PTRM-4917	−2.37
Leidyosuchus sternbergi	PTRM-4243	−2.37
Hadrosaurinae indeterminate	PTRM-3920	−2.72
Tyrannosaurus rex	PTRM-2514	−3.17
Caenagnathidae indeterminate	PTRM-3694	−3.17
Myledaphus bipartitus	PTRM-3197	−3.17
Belonostomus longirostris	PTRM-3086	−3.17
Opisthotriton kayi	PTRM-3387	−3.17
Kinosternidae indeterminate	PTRM-2920	−3.17
Scapherpeton tectum	PTRM-4363	−3.17
Ornithomimidae indeterminate	PTRM-3266	−4.15
Coelurosauridae indeterminate	PTRM-3779	−4.15
Melvius thomasi	PTRM-3091	−4.15
Plesiobaena antiqua	PTRM-2948	−4.15
Helopanoplia distincta	PTRM-4913	−4.15
Compsemys victa	PTRM-4915	−4.15
Lepidosauria indeterminate	PTRM-3276	−4.15
Adocus sp.	PTRM-4912	−4.15
Thoracosaurus neocesariensis	PTRM-3773	−4.15
Basilemys sinuosa	PTRM-4911	−4.15
Aves indeterminate	PTRM-3662	−4.15
Lissodus selachos	PTRM-3098	−5.25
Chelydridae indeterminate	PTRM-2955	−5.25
cf. *"Acipenser"* sp.	PTRM-3243	−6.25
Baenidae indeterminate	PTRM-2919	−6.25
Thescelosaurus neglectens	PTRM-3717	−8.40
Saurornitholestes sp.	PTRM-2878	−8.40
Paronychodon lacustris	PTRM-1214	−8.40
Pterosauria indeterminate	PTRM-3784	−8.40
Habrosaurus dilatus	PTRM-3381	−8.40
Triceratops horridus	ND92-19.1	−9.77
Anura indeterminate	PTRM-4303	−9.85
Hesperornithiformes indeterminate	PTRM-3425	−9.85
Torosaurus latus	PTRM-4918	−11.49
Squatirhina Americana	PTRM-3055	−29.93
Troodon sp.	PTRM-2906	−38.88
cf. *Avisaurus archibaldi*	PTRM-2971	−38.88
Ischyrhiza avonicola	PTRM-3075	−41.12
Macrobaenidae indeterminate	PTRM-2969	−73.00

PTRM—Pioneer Trails Regional Museum.

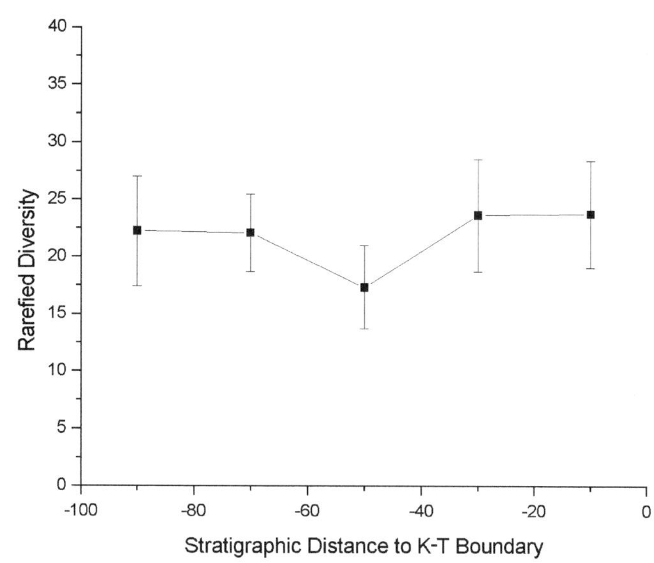

Figure 6. Each point is expected diversity, or rarefied diversity, at n = 200, and error bars are 95% confidence limits. Specimens are pooled into five 20 m stratigraphic bins. We used regression only after establishing that rarefied diversity was not significantly autocorrelated, i.e., each value was independent of previous value in series. Autocorrelation can be a problem in time series because regression assumes that data points are independent of each other (Sokal and Rohlf, 1995). Number of runs = 2, not significant (NS); autocorrelation r = −0.242, NS; regression constant = 22.928 (standard error, S.E. = 2.637), slope = 0.023 (S.E. = 0.046), NS.

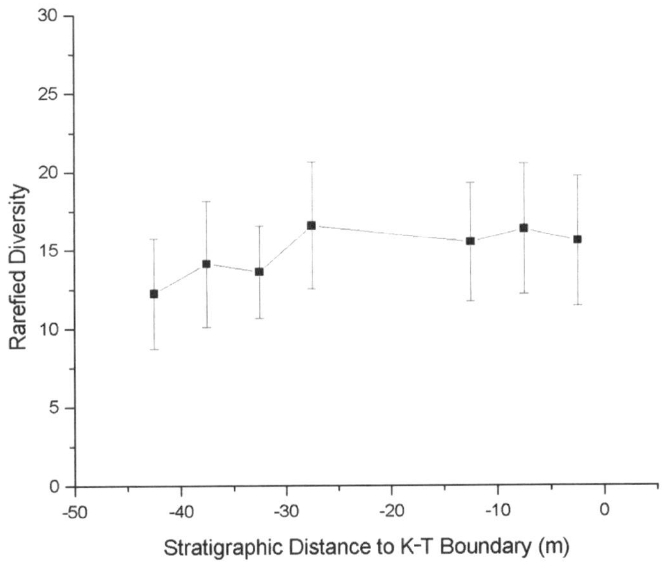

Figure 7. As for Figure 6, except that specimens are pooled into seven 5 m stratigraphic bins, and new standard sample size n = 75. Stratigraphic bins 15–20 m and 20–25 m had fewer than 50 specimens each and were discarded. Number of runs = 6, not significant (NS); autocorrelation r = 0.407, NS; regression constant = 16.583 (standard error, S.E. = 0.827), slope = 0.075 (S.E. = 0.030), p = 0.056. Note that marginally significant, positive regression slope is in opposite direction of that predicted by gradual extinction hypothesis.

side of the available resolution. Rather, we simply cannot demonstrate a decreasing trend in vertebrate diversity through the Hell Creek with these data and at the level of resolution available to us because we cannot reject the null hypothesis of no change through the formation. Vertebrate diversity appears to stay constant through the formation.

By adopting a similar approach for just the dinosaurs, it is possible to test the hypothesis of a decline in dinosaur diversity at the end of the Cretaceous corresponding to the so-called "3-meter gap" below the K-T boundary, which includes the basal part of the Ludlow Member of the Fort Union Formation. In our two databases, vertebrate microsites and single occurrences, there are 2233 dinosaurian specimens representing 14 taxa, including the single occurrence of a dinosaur in the basal Fort Union Formation. In the 3 m interval below the K-T boundary, there are 26 dinosaurian specimens representing just three taxa. Rarefaction, again using the explicit multinomial formulation, yields an expected number of taxa of 5.09 (standard deviation = 1.18), when n is 26 specimens, with upper and lower 95% confidence limits of 7.40 and 2.78, respectively. Sampling just three taxa in the upper 3 m of the Cretaceous is thus within expectation, assuming no change in diversity through the formation. There is, therefore, no evidence in our data for a decline in dinosaurian diversity below the K-T boundary. Similarly, Sheehan et al. (2000), using a different dataset, found no evidence for a decline in dinosaurian abundance in the top 3 m of the Hell Creek Formation.

DISCUSSION

Vertebrate occurrences and sedimentary environments

The distribution of taxa throughout the Hell Creek Formation may be related to the sedimentological facies in which they occur. Large channels ranging from 15 to 30 m thick are common near the base of the Hell Creek Formation (Belt et al., 1997; Murphy et al., this volume), but they decrease in size toward the top of the formation. Fastovsky (1987) indicated that 50%–60% of the upper Hell Creek Formation is composed of siltstone beds, which are interpreted as flood-plain deposits.

White et al. (1998) posed the taphonomic question: do stream channels effectively sample the flood-plain biota? They contrasted two different views regarding how to best sample the paleocommunity of ancient fluvial systems. One view (Bown and Kraus, 1981; Winkler, 1983; Retallack; 1988) holds that fossils found in flood-plain paleosols best represent the overall paleocommunity because such fossils were deposited with little time averaging and have been recovered in place, making flood-plain assemblages geographic and temporal point samples. In contrast, the other view (Behrensmeyer, 1988) argues that channel-lag deposits provide the best sampling because channels sample from the flood plains along their course and concentrate fossils, making channel assemblages geographically and temporally averaged, but more representative of larger-scale geographic and temporal units than flood-plain assemblages. White et al. (1998) used a previously published database (Sheehan et al., 1991) consisting of 649 dinosaur individuals to address the taphonomic question of whether channels concentrate fossils. They concluded that the mean density of elements in channel and flood-plain deposits do not differ, implying that the channels in the Hell Creek did not concentrate fossils above the density found on flood plains. These results contradicted earlier published findings (Fastovsky, 1987). On this basis, White et al. (1998) also stated that "contrary to the predictions of Behrensmeyer (1988), flood-plain deposits are the best source of paleontological data on large terrestrial vertebrates in the Hell Creek Formation." Also discussed by White et al. (1998), the database used by Sheehan et al. (1991) is biased toward large specimens, because their study concentrated on finding and recording in-place dinosaurs (i.e., on flood plains) and did not include vertebrate microsites (predominantly channel and crevasse-splay deposits). Therefore, the database used by White et al. (1998) was not designed to take into account geographic averaging by channels.

To determine whether vertebrate microsites indicate similar results for dinosaurs, we used the 2233 dinosaurian elements in the PTRM collection (Tables 2 and 3). These dinosaur elements were separated into three sets, specimens collected from flood-plain, channel, and crevasse-splay environments (Table 7).

Omitting localities of uncertain environmental facies, there are 759 specimens that were collected from 12 crevasse-splay localities, 929 specimens from 37 channel localities, and 535

TABLE 7. DINOSAUR TAXA RELATIVE TO STRATIGRAPHIC POSITION AND DEPOSITIONAL ENVIRONMENT

PTRM locality #	Datum (m)	Depositional environment	TOR	TRI	CER	HAD	THE	AVI	CAE	ORN	TRO	TYR	SAU	PAR	RIC	COE	Number of specimens	Taxa
V96019	−1.77	Flood plain			1												1	1
V88012	−2.37	Flood plain			21										3		24	2
V88002	−2.72	Crevasse splay				1											1	1
V95014	−3.12	Flood plain			1												1	1
V88017	−3.17	Flood plain			55	20			1			1			1		78	5
V96016	−3.51	Flood plain			1												1	1
V97024	−4.15	Flood plain			6				1	1		1				2	11	5
V95009	−4.96	Flood plain			1												1	1
V96003	−5.25	Flood plain			1	0											1	1
V96004	−5.25	Flood plain			8	2											10	2
V97017	−5.40	Flood plain			1												1	1
V97022	−5.56	Flood plain															0	0
V94002	−6.00	?			1												1	1
V98003	−6.06	Flood plain			1												1	1
V97011	−6.15	Crevasse splay				1											1	1
V96005	−6.25	Channel			1												1	1
V97012	−6.35	Flood plain				1											1	1
V98011	−6.65	Channel			6					1							7	2
V97009	−7.19	Flood plain			1												1	1
V96022	−7.51	Channel										1					1	1
V98017	−7.87	Channel			1												1	1
V88046	−8.00	?			1												1	1
V95020	−8.11	Channel			7	1				0							8	2
V96021	−8.11	Flood plain			1												1	1
V88009	−8.29	Crevasse splay															0	0
V89003	−8.40	Crevasse splay			29	130	1		1			4	1	3	3		172	8
V98020	−9.55	Channel				1						1					2	2
V98015	−9.60	Flood plain			1												1	1
V87004	−9.75	Channel				2				0		1					3	2
V96006	−9.75	Channel			1												1	1
ND92-19.1	−9.77	?			1												1	1
V87012	−9.85	Flood plain			49	47						6	2	2	9		115	6
V88005	−10.00	Channel			1												1	1
V88020	−10.00	Channel			8							1			2		11	3
V88021	−10.00	Crevasse splay			1										1		2	2
V88040	−10.00	Flood plain			1												1	1
V96015	−10.12	Flood plain			3	1											4	2
V95017	−10.14	Channel			1												1	1
V88003	−10.50	?			1												1	1
V88004	−10.61	Flood plain			28	7			2			3	3	1	6		50	7
V88013	−10.62	Channel			5												5	1
V95025	−10.64	Crevasse splay			8							1					9	2
V95016	−10.99	Channel			24	21						2		1	3		51	5
V98016	−11.43	Crevasse splay			1												1	1
V92007	−11.49	Flood plain	1														1	1
V88019	−12.00	Crevasse splay			7	1									2		10	3
V88054	−12.00	?			1												1	1
V98027	−12.07	Channel			9	24								1			34	3
V98021	−12.09	Crevasse splay				1											1	1
V88018	−14.27	Flood plain			193	17						7		3	9		229	5
UND3000	−14.61	?		1													1	1
MPM51016	−15.00	?		1													1	1
V97013	−15.25	Channel				1											1	1
V97002	−15.69	Channel										1					1	1
V92003	−17.75	Channel			2	2											4	2
V98002	−18.25	Crevasse splay			8	2											10	2
ND95-117.1	−20.61	?		1													1	1
V95023	−22.14	Channel			1												1	1
V95018	−23.44	Channel				1											1	1
V93022	−27.00	Crevasse splay			1												1	1
V93005	−27.20	Channel			1												1	1
V97006	−28.70	Channel			1	3									1	5	3	
V87006	−28.75	Channel			42	38			2			11	1	1			95	6
V86002	−29.93	Channel			81	205	2					21		8	11		328	6
V87009	−32.55	Channel			13	6						0	0	3			22	3
V92068	−32.88	Flood plain										1					1	1
V86003	−36.17	Channel			32	10	1			1		4			1		49	6

(*continued*)

TABLE 7. DINOSAUR TAXA RELATIVE TO STRATIGRAPHIC POSITION AND DEPOSITIONAL ENVIRONMENT (continued)

PTRM locality #	Datum (m)	Depositional environment	TOR	TRI	CER	HAD	THE	AVI	CAE	ORN	TRO	TYR	SAU	PAR	RIC	COE	Number of specimens	Taxa
V92066	−38.70	Channel			1												1	1
V92067	−38.88	Crevasse splay			118	341	28	1		1	3	15	14	21	8		550	10
V92002	−41.12	Channel			5	13	2					4	2	3			29	6
V92042	−41.70	Channel			1												1	1
V92029	−60.00	Channel							2	3		1					6	3
V92027	−67.00	Channel							1			3					4	2
V92025	−73.00	Channel			22	32	3		1	3		13		1	1		76	8
V91010	−75.00	Channel				1											1	1
V91012	−75.00	Channel			1												1	1
V91014	−75.00	Crevasse splay													1		1	1
V91007	−80.16	Channel				11	1					1	1	2	1		17	6
V89004	−81.68	Channel			24	62	12		1	5	2	17	10	14	10		157	10
V91002	−82.68	Flood plain			1												1	1
ND96-23.1	−85.75	Channel		1													1	1
V92040	−100.00	Flood plain			1												1	1
			1	4	845	1006	50	1	12	15	5	121	34	64	73	2	2233	

Note: TOR—*Torosaurus latus*; TRI—*Triceratops horridus*; CER—Ceratopsidae indet.; HAD—Hadrosaurinae indet.; THE—*Thescelosaurus neglectens*; AVI—cf. *Avisaurus archibaldi*; CAE—Caenagnathidae indet.; ORN—Ornithomimidae indet.; TRO—*Troodon* sp.; TYR—*Tyrannosaurus rex*; SAU—*saurornith-olestes* sp.; PAR—*Paronychodon lacustris*; RIC—*Ricardoestesia* sp.; COE—Coelurisauridae indet. PTRM—Pioneer Trails Regional Museum.

specimens from 24 flood-plain localities. Although sampling effort varied among localities, the average number of specimens per locality provides an approximate measure of specimen density. By this measure, crevasse-splay deposits (63 specimens per locality) yielded almost three times the specimen density as either channel deposits (25 specimens per locality) or flood-plain deposits (22 specimens per locality). Evidence for concentration of fossils is strongest for crevasse-splay deposits, some of which could conceivably represent multiple depositional events. Table 7 also lists the number of dinosaurian taxa collected at each site by facies type. When these taxa are averaged for each facies type, the results are very similar among facies: 2.1 taxa per site average for flood-plain facies; 2.6 taxa per site for channel margin facies; and 2.8 taxa per site for channel facies. Nevertheless, of the 13 dinosaurian taxa listed in Table 7, 11 occur in channel deposits, 11 in crevasse splay deposits, but only nine in flood-plain deposits. Most notably, some of the less common dinosaurs (*Thescelosaurus, Avisaurus*, and *Troodon*) have not yet been recovered from flood-plain deposits. Channel and crevasse-splay deposits thus contribute meaningfully to overall sampling of the Hell Creek Formation by recovering less common members of the vertebrate fauna. Possible explanations as to why channel and near-channel deposits may be more likely to sample rarer taxa than the flood-plain deposits are: (1) localities in channel deposits are more numerous than localities in flood-plain deposits; (2) localities in crevasse-splay deposits produce larger numbers of specimens than localities in flood-plain deposits; (3) localities in channel and near-channel deposits sample a broader geographic area and more environments than flood plains; and (4) channels accumulate bones and teeth by the erosion of the channel margin.

The data in Table 7 indicate that the most taxonomically complete sampling for representation of these paleocommunities is found within channel and near-channel assemblages because they incise through and collect flood-plain deposits, according to Fastovsky (1987) and Behrensmeyer (1988). Flood-plain deposits, however, are the best source for associated and/or articulated dinosaurs because dinosaur remains in place here have not been disturbed by the hydraulic and erosional action of the streams. Channels represent an assemblage more indicative of the entire paleocommunity, rather than of single individuals, but more often disarticulated elements occur there.

The change from channel-dominated to flood-plain–dominated strata upsection may be an indication of increasing distance from the shoreline. This change in depositional environment as the horizon of preservation is indicated in Table 2. As the size of a stream becomes smaller, its capacity for accumulating a representative sampling of the flood plain becomes diminished. Such streams were no longer able to accumulate the larger specimens or to sample the entire flood plain; thus, their channels contain a biased assemblage based upon the erosional and hydraulic properties of the stream. The nearer one gets to the K-T boundary, the greater the constraints placed upon the streams for accumulating specimens from the flood plain. It is in these last few meters that the final Cretaceous assemblage must reside, but the small streams were unable to concentrate the flood-plain assemblage to the same extent as seen in the larger channels lower in the formation. The physical dimensions of the last few meters limit the size of Cretaceous streams to smaller, shallower channels having the capability to accumulate only the smaller specimens, not large dinosaur bones. A channel large enough to concentrate the original flood-plain biota cannot be constrained within the sediments of the highest few meters of the Cretaceous. Such large channels would have either eroded more deeply into Cretaceous rock, resulting in a lower observed stratigraphic position, or would have eroded down into the uppermost Cretaceous rock from above, thus representing both Cretaceous and Paleocene time. Large channels found at this high stratigraphic position are almost invariably Paleocene in age with reworked Cretaceous specimens like those seen in the Bug Creek area of northeastern Montana. Large channels found at a lower stratigraphic level in the Hell Creek Formation are probably just as time averaged, but do not seem to span an interval of faunal turnover. There are no Cretaceous stream channels known to exist in the 3 m below the K-T boundary in our study area.

Proximity of the Western Interior seaway

Although the aquatic ecosystems preserved in the Hell Creek Formation were primarily of freshwater origin (Bryant, 1989; Sheehan and Fastovsky, 1992; Archibald, 1993, 1996a), taxa occur in our collections that have been reported to have a tolerance to marine environments (Myers, 1951). Bryant (1987b, 1989) and Archibald and Bryant (1990) indicate the taxa *Belonostomus, "Acipenser,"* and *Melvius* are all in that category. These taxa are present to within the uppermost 5.5 m of the Hell Creek Formation. Shark and ray specimens also occur in our collections from sites as near to the formational contact as 4.5 m and 3.17 m, respectively (see Fig. 4). The occurrence of these marine-tolerant taxa throughout the formation may indicate the nearby presence of a marine environment. Hesperornithiformes also occur in the assemblage, and their presence suggests the possibility that marine conditions may not have been far removed. Bryant (1989) cited the presence of some of these taxa as indirect evidence that the Western Interior seaway remained nearby until almost the end of the Maastrichtian, and that it was probably in the vicinity of the middle of North Dakota or South Dakota at that time. However, Roberts and Kirchbaum (1995), Archibald (1996a), and Belt et al. (1997) suggested that the southern receding coastline of the Western Interior seaway was in or near Texas by the end of the Cretaceous. Kennedy et al. (1998), through the record of a juvenile scaphitid ammonite from Niobrara County, Wyoming, suggested that the latest Maastrichtian shoreline was farther to the north of conventional estimates at that time. Murphy et al. (1995, this volume) have documented the marine Breien Member of the Hell Creek Formation at 42 m below the formational contact at the Stumpf site in south-central North Dakota. This is the marine deposit of Cretaceous age nearest to our study area. Murphy et al. (this volume) have found possible evidence of additional marine influences in some of the sandstone beds from the middle and upper parts of the Hell Creek Formation above the Breien Member along the Missouri Trench in south-central North Dakota. The close geographic proximity of the Breien Member to our study area, the possibility of marine influence in the upper Hell Creek Formation, and the presence of marine-tolerant taxa in our assemblages suggest that the study area may have been adjacent to marine environments and responding to marine influence.

CONCLUSIONS

The Hell Creek Formation in southwestern North Dakota differs from the type Hell Creek in the Fort Peck region of eastern Montana in that it lacks the large Paleocene channels that cut through the K-T boundary, causing mixing and reworking of Cretaceous and Tertiary faunas (Lofgren, 1995). In southwestern North Dakota, most vertebrate sites in the lower part of the Hell Creek Formation occur in channel deposits, whereas those in the upper part of the formation more commonly occur in flood-plain deposits. The nearer one gets to the K-T boundary, the greater the constraints that are placed upon the concentration of flood-plain faunas by channels. It is within these last few meters that the final Cretaceous assemblage must reside, but the physical dimensions of the last few meters limit the size of the streams that concentrate the assemblages. These smaller streams have a diminished capability for collecting fossils of small size and even less for the larger dinosaur material.

Sharks occur within 5.25 m of the K-T boundary and the Hesperornithiformes occur within 10.12 m. This observation and other data presented in this chapter, and others in this volume, suggest that there was possibly more marine influence in the Hell Creek Formation than previously thought.

Dinosaurs are found in the stratigraphically highest vertebrate sites, and one ceratopsian partial skeleton was found in the overlying Fort Union Formation (but within 1.77 m below the K-T boundary). Our data indicate that dinosaur taxa occurrences are averaged among the three environments (flood plain, channel, and crevasse splay), but the less common taxa were found only in the channel or near-channel environments. The most taxonomically complete representation of paleocommunities is therefore found within channel and near-channel deposits because they incise through and concentrate teeth and bones from the flood plains; however, flood-plain deposits are the best source for associated and/or articulated dinosaurs. Our data also indicate that sample size is the primary factor that determines the presence of taxa in a vertebrate site. All dinosaurs that occurred at more than two sites also occurred at the highest level that yielded more than 500 specimens (8.4 m below the K-T boundary). In this region, common vertebrate taxa range through the entirety of the Hell Creek Formation, and a major faunal extinction is coincident with megafloral and palynological extinctions. These results are in accord with previous work on dinosaurian diversity (Sheehan et al., 1991) and abundance (Sheehan et al., 2000) using different data and methods.

When variation in vertebrate diversity due to sample size is minimized using rarefaction, there is no evidence for decreasing diversity during the time represented by the Hell Creek Formation. Rarefaction of the dinosaurian specimens reveals that dinosaurian diversity within the so-called "3-meter gap" below the K-T boundary is not significantly lower than in the formation as a whole. These results are not compatible with gradual vertebrate extinction at the end of the Cretaceous.

ACKNOWLEDGMENTS

This study has been supported by the Paleontology Department of the Pioneer Trails Regional Museum and a grant from the Bowman County Economic Development Corporation to Pearson. We thank D.B. Brinkman, P.J. Currie, J.H. Hutchison, J. Horner, J. Hoganson, B. Erickson, K. Carpenter, W. Coombs, D.W. Krause, A. Sanders, and M. Williams for assisting with specimen identifications and suggestions. We also thank D.E. Fastovsky, K. Carlson, J.H. Hartman, J.F. Hicks, D. Brinkman, and E.C. Murphy for the stratigraphic placements and suggestions. We are deeply indebted to the volunteers from the Pioneer Trails Regional Museum, Denver Museum of Nature & Science, Minot State University, and the State University of New York at Stony Brook for collecting the specimens, and the landowners of the property from which they were collected. Susan Passmore, Rich Barclay, and Michele Reynolds helped to prepare the manuscript and figures. We are grateful to Marshall Lambert of Ekalaka, Montana, and the late Diane Gabriel, who gave us the desire and the drive to see this project to completion.

REFERENCES CITED

Alroy, J., 1998, Equilibrial diversity dynamics in North American mammals, *in* McKinney, M.L., ed., Biodiversity dynamics: Turnover of populations, taxa and communities: New York, Columbia University Press, p. 232–287.

Archibald, J.D., 1982, Fossil Mammalia and Testudines of the Hell Creek Formation, and the geology of the Tullock and Hell Creek formations, Garfield County, Montana: University of California Publications in Earth Sciences, v. 122, 286 p.

Archibald, J.D., 1993, Major extinctions of land-dwelling vertebrates at the Cretaceous–Tertiary boundary, eastern Montana: Comment: Geology, v. 21, p. 90–92.

Archibald, J.D., 1996a, Dinosaur extinction and the end of an era: What the fossils say: New York, Columbia University Press, 237 p.

Archibald, J.D., 1996b, Testing extinction theories at the Cretaceous–Tertiary boundary using the vertebrate fossil record, *in* MacLeod, N., and Keller, G., eds., Cretaceous-Tertiary mass extinctions: Biotic and environmental changes: New York, W.W. Norton and Company, p. 373–397.

Archibald, J.D., and Bryant, L.J., 1990, Differential Cretaceous–Tertiary extinctions of non-marine vertebrates: Evidence from northeastern Montana, *in* Sharpton, V.L., and Ward, P., eds., Global catastrophes in Earth history: An interdisciplinary conference on impacts, volcanism, and mass mortality: Boulder, Colorado, Geological Society of America Special Paper 247, p. 549–562.

Badgley, C., 1986, Counting individuals in mammalian fossil assemblages from fluvial environments: Palaios, v. 1, p. 328–338.

Behrensmeyer, A.K., 1988, Vertebrate preservation in fluvial channels: Paleogeography, Paleoclimatology, Paleoecology, v. 63, p. 183–199.

Belt, E.S., Conway, K.M., Flores, R.M., Johnson, K.R., Warwick, P.D., and Waskowitz, R.S., 1984, Relationships of fluvio-deltaic facies to coal deposition in the lower Fort Union Formation (Paleocene), southwestern North Dakota, *in* Rahmani, R., and Flores, R.M., eds., Sedimentology of coal deposits: International Association of Sedimentologists Special Publication 7, p. 317–335.

Belt, E.S., Hicks, J.F., and Murphy, D.A., 1997, A pre-Lancian regional unconformity and its relationship to Hell Creek paleogeography in southwestern Montana: University of Wyoming Contributions to Geology, v. 31, 26 p.

Bown, T.M., and Kraus, M.J., 1981, Vertebrate fossil-bearing paleosol units (Willwood Formation, northwest Wyoming, USA): Implications for taphonomy, biostratigraphy, and assemblage analysis: Paleogeography, Paleoclimatology, Paleoecology, v. 34, p. 31–56.

Breithaupt, B.H., 1982, Paleontology and paleoecology of the Lance Formation (Maastrichtian) east flank of Rock Springs uplift, Sweetwater County, Wyoming: University of Wyoming Contributions to Geology, v. 21, p. 123–151.

Brinkman, D.B, Ryan, M.J., and Eberth, D.A., 1998, The paleogeographic and stratigraphic distribution of Ceratopsids (Ornithischia) in the upper Judith River Group of western Canada: Palaios, v. 13, p. 160–169.

Brown, R.W., 1952, Tertiary strata in eastern Montana and western North and South Dakota, *in* Sonnenberg, F.P., ed., Billings Geological Society guidebook, Third Annual Field Conference Billings, Montana: Billings Geological Society, p. 89–92.

Bryant, L.J., 1987a, A new genus and species of Amiidae (Holostei: Osteichthyes) from the Late Cretaceous of North America, with comments on the phylogeny of the Amiidae: Journal of Paleontology, v. 7, p. 349–361.

Bryant, L.J., 1987b, *Belonostomus* (Teleostei: Aspidorhnychidae) from the late Paleocene of North Dakota: PaleoBios, v. 43, 3 p. 3.

Bryant, L.J., 1989, Non-Dinosaurian lower vertebrates across the Cretaceous–Tertiary boundary in northeast Montana: University of California Publications in Geological Sciences, v. 134, p. 1–107.

Carlson, C.G., 1979, Geology of Adams and Bowman Counties, North Dakota: North Dakota Geological Survey Bulletin 65, 29 p.

Carlson, C.G., 1983, Geology of Billings, Golden Valley, and Slope Counties, North Dakota: North Dakota Geological Survey Bulletin 76, 40 p.

Carpenter, K., 1979, Vertebrate fauna of the Laramie Formation (Maastrichtian), Weld County, Colorado: University of Wyoming Contributions to Geology, v. 17, p. 37–49.

Carpenter, K., 1982, Baby dinosaurs from the Late Cretaceous Lance and Hell Creek Formations and a description of a new species of theropod: University of Wyoming Contributions to Geology, v. 20, p. 123–134.

Carpenter, K., 1983, *Thoracosaurus neocesariensis* (DeKay, 1842) (Crocodylia: Crocodylidae) from the Late Cretaceous Ripley Formation of Mississippi: Mississippi Geology, v. 4, 10 p.

Carpenter, K., and Breithaupt, B.H., 1986, Latest Cretaceous occurrence of nodosaurid ankylosaurs (Dinosauria: Ornithischia), in western North America and the gradual extinction of the dinosaurs: Journal of Vertebrate Paleontology, v. 6, p. 251–257.

Clemens, W.A., Archibald, J.D., and Hickey, L.J., 1981, Out with a whimper, not a bang: Paleobiology, v. 7, p. 293–298.

Coombs, W.P., 1988, The status of the dinosaurian genus *Diclonius* and the taxonomic utility of hadrosaurian teeth: Journal of Paleontology, v. 62, p. 812–817.

Coombs, W.P., and Galton, P.M., 1988, *Dysganus*, an indeterminate ceratopsian dinosaur: Journal of Paleontology, v. 62, p. 818–821.

Currie, P.J., Rigby, J.K., Jr., and Sloan, R.E., 1990, Theropod teeth from the Judith River Formation of southern Alberta, Canada, *in* Carpenter, K., and Currie, P.J., eds., Dinosaur systematics, perspectives and approaches: New York, Cambridge University Press, p. 107–125.

Elzanowski, A., Paul, G.S., and Stidham, T.A., 2001, An avian quadrate from the Lance Formation of Wyoming: Journal of Vertebrate Paleontology, v. 20, p. 712–719.

Estes, R., 1964, Fossil vertebrates from the Cretaceous Lance Formation, eastern Wyoming: University of California Publications in Geological Sciences, v. 49, p. 1–187.

Estes, R., and Berberian, P., 1970, Paleoecology of a Late Cretaceous vertebrate community from Montana: Breviora, v. 343, p. 1–35.

Estes, R., Berberian, P., and Meszoely, C.A.M., 1969, Lower vertebrates from the Late Cretaceous Hell Creek Formation, McCone County, Montana: Brevoria, no. 337, p. 1–33.

Fastovsky, D.E., 1987, Paleoenvironments of vertebrate-bearing strata during the Cretaceous-Paleogene transition, eastern Montana and western North Dakota: Palaios, v. 2, p. 282–295.

Fastovsky, D.E., 1990, Rocks, resolution, and the record: A review of depositional constraints on fossil vertebrate assemblages at the terrestrial Cretaceous/Paleogene boundary, eastern Montana and western North Dakota, *in* Sharpton, V.L., and Ward, P., eds., Global catastrophes in Earth history: An interdisciplinary conference on impacts, volcanism, and mass mortality: Boulder, Colorado, Geological Society of America Special Paper 247, p. 541–548.

Fastovsky, D.E., and Dott, R.H., 1986, Sedimentology, stratigraphy, and extinctions during the Cretaceous-Paleogene transition at Bug Creek, Montana: Geology, v. 14, p. 279–282.

Fastovsky, D.E., and McSweeney, K., 1987, Paleosols spanning the Cretaceous-Paleogene transition, eastern Montana and western North Dakota: Geological Society of America Bulletin, v. 99, p. 66–77.

Fastovsky, D.E., and Sheehan, P.M., 1997, Demythicizing dinosaur extinctions at the Cretaceous-Tertiary boundary, *in* Wolberg, D., et al., eds., Dinofest International: Philadelphia, Pennsylvania, Philadelphia Academy of Natural Sciences, p. 527–531.

Fiorillo, A.R., and Currie, P.J., 1994, Theropod teeth from the Judith River Formation (Upper Cretaceous) of south-central Montana: Journal of Vertebrate Paleontology, v. 14, p. 74–80.

Foote, M., 1992, Rarefaction of morphological and taxonomic diversity: Paleobiology, v. 18, p. 1–16.

Frye, C.I., 1969, Stratigraphy of the Hell Creek Formation in North Dakota: North Dakota Geological Survey Bulletin 54, 65 p.

Gallagher, W.B., 1993, The Cretaceous/Tertiary mass extinction event in the North Atlantic coastal plain: The Mosasaur: Delaware Valley Paleontological Society Journal, v. 5, p. 75–154.

Greenwald, M.T., 1971, The lower vertebrates of the Hell Creek Formation, Harding County, South Dakota [M.S. thesis]: Rapid City, South Dakota, USA, South Dakota School of Mines and Technology, 76 p.

Hayek, L.-A.C., and Buzas, M.A., 1997, Surveying natural populations: New York, Columbia University Press, 563 p.

Heck, K.L., Jr., van Bell, G., and Simberloff, D., 1975, Explicit calculation of the rarefaction diversity measurement and the determination of sufficient sample size: Ecology, v. 56, p. 1459–1461.

Horner, J., 1990, Evidence of diphyletic origin of the hadrosaurian (Reptilia: Ornithischia) dinosaurs, *in* Carpenter, K., and Currie, P.J., eds., Dinosaur systematics, perspectives and approaches: New York, Cambridge University Press, p. 179–187.

Hunter, J.P., and Pearson, D.A., 1996, First record of Lancian (Late Cretaceous) mammals from the Hell Creek formation of southwestern North Dakota USA: Cretaceous Research, v. 17, p. 633–643.

Hunter, J.P., Hartman, J.H., and Krause, D.W., 1997, Mammals and mollusks across the Cretaceous-Tertiary boundary from Makoshika State Park and vicinity (Williston Basin), Montana: University of Wyoming Contributions to Geology, v. 32, p. 61–114.

Hurlbert, S.H., 1971, The nonconcept of species diversity: A critique and alternative parameters: Ecology, v. 52, p. 577–586.

Hurlbert, S.H., and Archibald, J.D., 1995, No statistical support for sudden (or gradual) extinction of dinosaurs: Geology, v. 23, p. 958–959.

Hutchison, J.H., and Archibald, J.D., 1986, Diversity of turtles across the Cretaceous/Tertiary boundary in northeastern Montana: Paleogeography, Paleoclimatology, Paleoecology, v. 55, p. 1–22.

Jernvall, J., and Wright, P.C., 1998, Diversity components of impending primate extinctions: National Academy of Sciences Proceedings, v. 95, p. 11 279–11 283.

Johnson, K.R., 1992, Leaf fossil evidence for extensive floral extinction at the Cretaceous-Tertiary boundary, North Dakota: Cretaceous Research, v. 13, p. 91–117.

Johnson, K.R., Nichols, D.J., Attrep, M., Jr., and Orth, C.J., 1989, High-resolution leaf-fossil record spanning the Cretaceous/Tertiary boundary: Nature v. 340, p. 708–711.

Keller, G., and Barrera, E., 1990, The Cretaceous-Tertiary boundary impact hypothesis and the paleontological record, *in* Sharpton, V.L., and Ward, P., eds., Global catastrophes in Earth history: An interdisciplinary conference on impacts, volcanism, and mass mortality: Boulder, Colorado, Geological Society of America Special Paper 247, p. 563–575.

Kennedy, W.J., Landman, N.H., Christensen, W.K., Cobban, W.A., and Hancock, J.M., 1998, Marine connections in North America during the late Maastrichtian: Paleogeographic and paleobiogeographic significance of *Jeletzkytes nebrascensis* zone cephalopod fauna from the Elk Butte Member of the Pierre Shale, southeast South Dakota and northeast Nebraska: Cretaceous Research, v. 19, p. 745–775.

Lofgren, D.L., 1995, The Bug Creek problem and the Cretaceous-Tertiary transition at McGuire Creek, Montana: University of California Publications in Geological Sciences, v. 140, 185 p.

Moore, W.L., 1976, The stratigraphy and environment of deposition of the Hell Creek Formation (reconnaissance) and the Paleocene Ludlow Formation (detailed), southwestern North Dakota: North Dakota Geological Survey Report of Investigations, v. 56, 40 p.

Murphy, E.C., Nichols, D.J., Hoganson, J.H., and Forsman, N.F., 1995, The Cretaceous/Tertiary boundary in south-central North Dakota: North Dakota Geological Survey Report of Investigations, v. 98, 74 p.

Myers, G.S., 1951, Freshwater fishes and East Indian zoogeography: Stanford Ichthyological Bulletin 4, p. 11–21.

Nichols, D.J., Jacobson, S.R., and Tschudy, R.H., 1982, Cretaceous palynomorph biozones for the central and northern Rocky Mountain region of the United States, *in* Powers, R.B., ed., Geologic studies of the Cordilleran thrust belt: Denver, Colorado, Rocky Mountain Association of Geologists, p. 721–733.

Nichols, D.J., Murphy, E.C., Johnson, K.R., and Betterton, W.J., 2000, A second K-T boundary locality in North Dakota verified by palynostratigraphy and shocked quartz: Geological Society of America Abstracts with Programs, v. 32, no. 7, p. 130.

Pearson, D.A., 1997, Searching for the last dinosaur, a systematic survey in the latest Cretaceous rocks of southwestern North Dakota: North Dakota Academy of Science Proceedings, v. 51, p. 213.

Pearson, D.A., Schaefer, T., Johnson, K.R., and Nichols, D.J., 2001, Palynologically calibrated vertebrate record from North Dakota consistent with abrupt dinosaur extinction at the K-T boundary: Geology, v. 29, p. 39–42.

Raup, D.M., 1975, Taxonomic diversity estimation using rarefaction: Paleobiology, v. 1, p. 333–342.

Retallack, G.J., 1988, Down-to-Earth approaches to vertebrate paleontology: Palaios, v. 3, p. 335–344.

Retallack, G.J., and Leahy, G.D., 1986, Cretaceous-Tertiary dinosaur extinctions: Science, v. 234, p. 1170–1171.

Retallack, G.J., Leahy, G.D., and Spoon, M.D., 1987, Evidence from paleosols for ecosystem changes across the Cretaceous-Tertiary boundary in eastern Montana: Geology, v. 15, p. 1090–1093.

Rigby, J.K., Jr., Newman, K.R., Smit, J., Van Der Kaars, S., Sloan, R.E., and Rigby, J.K., 1987, Dinosaurs from the Paleocene part of the Hell Creek Formation, McCone County, Montana: Palaios, v. 2, p. 296–302.

Roberts, L.R.N., and Kirschbaum, M.A., 1995, Paleogeography of the Late Cretaceous of the Western Interior of middle North America: Coal distribution and sediment accumulation: U.S. Geological Survey Professional Paper 1561, 115 p.

Sanders, H.L., 1968, Marine benthic diversity: A comparative study: American Naturalist, v. 102, p. 243–282.

Sankey, J.T., 2001, Late Campanian southern dinosaurs, Aguja Formation, Big Bend, Texas: Journal of Paleontology, v. 75, p. 208–215.

Sheehan, P.M., and Fastovsky, D.E., 1992, Major extinctions of land-dwelling vertebrates at the Cretaceous-Tertiary boundary, eastern Montana: Geology, v. 20, p. 556–560.

Sheehan, P.M., Fastovsky, D.E., Hoffman, R.G., Berghaus, C.B., and Gabriel, D.L., 1991, Sudden extinction of the dinosaurs: Latest Cretaceous, upper Great Plains, USA: Science, v. 254, p. 835–839.

Sheehan, P.M., Fastovsky, D.E., Barreto, C., and Hoffman, R.G., 2000, Dinosaur abundance was not declining in a "3 m gap" at the top of the Hell Creek Formation, Montana and North Dakota: Geology, v. 28, p. 523–526.

Simberloff, D., 1972, Properties of the rarefaction diversity measurement: American Naturalist, v. 106, p. 414–418.

Sloan, R.E., 1987, Upper Cretaceous-Paleocene sequence, Bug Creek area, northeastern Montana, *in* Beus, S.S., ed., Geological Society of America Centennial Field Guide, 38th Annual Rocky Mountain Section Meeting: Boulder, Colorado, Geological Society of America, v. 2, p. 45–48.

Sloan, R.E., and Van Valen, L., 1965, Cretaceous mammals from Montana: Science, v. 148, p. 220–227.

Sokal, R.R., and Rohlf, F.J., 1995, Biometry: Principles and practice of statistics in biological research: New York, W.H. Freeman and Company, 887 p.

Sullivan, R., 1987, A reassessment of the reptilian diversity across the Cretaceous–Tertiary boundary: Natural History Museum of Los Angeles County Contributions in Science, no. 391, 26 p.

Tokaryk, T., 1997, Preliminary review of the non-mammalian vertebrates from the Frenchman Formation (late Maastrichtian) of Saskatchewan, *in* McKenzie-McAnally, L., eds., Canadian Paleontology Conference Field Trip Guidebook No. 6., Upper Cretaceous and Tertiary stratigraphy and paleontology of southern Saskatchewan: Waterloo, Ontario, Geological Association of Canada, p. 34–44.

Tokaryk, T., Cumbaa, S.L., and Storer, J.E., 1997, Early Late Cretaceous birds from Saskatchewan: Canada: The oldest diverse avifauna known from North America: Journal of Vertebrate Paleontology, v. 17, p. 172–176.

Triebold, M., 1997, The sandy site: Small dinosaurs from the Hell Creek Formation of South Dakota, *in* Wolberg, D., et al., eds., Dinofest International: Philadelphia, Pennsylvania, Philadelphia Academy of Natural Sciences, p. 245–248.

Troxell, E.L., 1925, Thoracosaurus, a Cretaceous crocodile: American Journal of Science, ser. 5, v. 10, p. 219–233.

White, P.D., Fastovsky, D.E., and Sheehan, P.M., 1998, Taphonomy and suggested structure of the dinosaurian assemblage of the Hell Creek Formation (Maastrichtian), eastern Montana and Western North Dakota: Palaios, v. 13, p. 41–51.

Whitmore, J.L., and Martin J.E., 1986, Vertebrate fossils from the Greasewood Creek locality in the Late Cretaceous Lance Formation of Niobrara County, Wyoming: South Dakota Academy of Science Proceedings, v. 65, p. 33–50.

Williams, M.E., 1994, Catastrophic versus non-catastrophic extinction of the dinosaurs: Testing, falsifiability, and the burden of proof: Journal of Paleontology, v. 68, p. 183–190.

Winkler, D.A., 1983, Paleoecology of an early Eocene mammalian fauna from paleosols in the Clarks Fork Basin, northeastern Wyoming (U.S.A.): Palaeogeography, Palaeoclimatology, Palaeoecology, v. 43, p. 261–289.

Manuscript Accepted by the Society September 24, 2001

Printed in the U.S.A.

Geological Society of America
Special Paper 361
2002

Synopsis of the Hell Creek (uppermost Cretaceous) dinosaur assemblage

Dale A. Russell*
North Carolina State University and North Carolina State Museum of Natural Sciences, Box 8208, Raleigh, North Carolina 27695-8203, USA
Makoto Manabe
National Science Museum, 3-23-1 Hyakunin-cho, Shinjuku-ku, Tokyo 169-0073, Japan

ABSTRACT

The characteristics of the late Maastrichtian Hell Creek dinosaur assemblage can be understood as resulting in part from the influences of an unusually broad, rich coastal plain with limited niches available for large herbivorous dinosaurs. They could further stem from a regional vegetational change from gymnosperm- to angiosperm-dominated forests and the displacement toward the north of a low-diversity, southerly chasmosaurine-hadrosaurine-dominated assemblage due to a regional warming trend.

INTRODUCTION

Near the close of Cretaceous time, a great sheet of alluvium filled the former western margin of the Western Interior Sea. Termed the Hell Creek Formation on the northern plains of the United States, these strata have been known for nearly a century to contain the remains of the last dinosaurs to inhabit the region (*Triceratops* fauna). The alluvial sequence is typically 70–100 m thick and was deposited during the last 1.36 m.y. of the Cretaceous (Hicks et al., this volume).

In 1902, Barnum Brown of the American Museum of Natural History began collecting in the region of Hell Creek, Montana. He discovered many skeletal fragments, including those forming the nucleus for the famous standing mounts of *Tyrannosaurus, Triceratops*, and "*Trachodon*" (*Edmontosaurus*) displayed in the museum in New York (for a review of dinosaur collecting in the Hell Creek Formation, see Bryant, 1989). The general environmental setting of the formation was described in Lehman (1987), Russell (1989), Johnson and Hickey (1990), and Roberts and Kirschbaum (1995). Here we briefly review the attributes of the Hell Creek dinosaurian assemblage, which have long been overshadowed by the debate on the extinction of the dinosaurs.

HELL CREEK ASSEMBLAGE

The list of Hell Creek dinosaur taxa (Table 1) exhibits a strongly bimodal distribution in body size of various taxa known to be present in the assemblage. Of the 22 genera tabulated, 15 represent small forms and 7 represent large forms. The diversity of small dinosaurs is probably underestimated in view of the six different morphotypes of small ornithischian caudal vertebrae cited in Triebold (1997, p. 247) and now in the collections of the National Science Museum in Tokyo. Several large dinosaurs (and two of intermediate proportions) were among the largest members of their clades (e.g., *Tyrannosaurus, Caenagnathus, Ankylosaurus, Pachycephalosaurus, Triceratops*, and *Torosaurus*). Large size is generally indicative of abundant fodder (Coe et al., 1976). Conversely, material referred to *Dromaeosaurus* is unusually small (Baszio, 1997).

Brown (1907) noted that ceratopsids were the most abundant group represented in the Hell Creek region of eastern

*E-mail: dale_russell@ncsu.edu

Russell, D.A., and Manabe, M., 2002, Synopsis of the Hell Creek (uppermost Cretaceous) dinosaur assemblage, *in* Hartman, J.H., Johnson, K.R., and Nichols, D.J., eds., The Hell Creek Formation and the Cretaceous-Tertiary boundary in the northern Great Plains: An integrated continental record of the end of the Cretaceous: Boulder, Colorado, Geological Society of America Special Paper 361, p. 169–176.

Montana, followed by hadrosaurids. He further observed that hadrosaurid skeletons were usually articulated. Field records on file in the American Museum of Natural History indicate that most of the hadrosaurid skeletons collected by Brown were associated with stream channel environments, unlike occurrences farther southeast (cf. White et al., 1998, p. 48). Vertebrate remains do not occur in fossil soils typical of swamps or old growth forests in the Hell Creek region, possibly because the soils were too acidic. Soils that supported shrubs and small pole trees and intermediate-growth forests are associated with the greatest abundance of bones (Retallack, 1994). These forests would contain relatively accessible and abundant forage for herbivorous dinosaurs.

A preliminary attempt is made to assess the relative abundance of dinosaurian families throughout eastern Montana and the western part of the Dakotas, based on associated skeletal parts of single individuals in various collections (cf. Béland and Russell, 1978). The results (Table 2) suggest that ceratopsids are relatively less common than previously thought, but small herbivorous dinosaurs are more common.

In the vicinity of Glendive, Montana, and Marmarth, North Dakota, stream channel facies are more abundant near the bottom of the Hell Creek Formation and tend to be replaced by flood-plain facies higher in the section. Dinosaurian recovery is relatively abundant and taxonomically similar in both facies and is not known to vary significantly through the section (Sheehan et al., 1991, 1996; Sheehan and Fastovsky, 1992; Hurlbert and Archibald, 1995, 1996; White et al., 1998). It is possible that infrequent occurrences of nodosaurids may be restricted to lower levels in the formation (Larson, 1985; Carpenter and Breithaupt, 1986; Russell, 1987).

Bonebed occurrences are not uncommon in the Hell Creek Formation. A bonebed high in the section and dominated by *Triceratops* is currently under investigation in Montana (J.R. Horner, *in* Tokaryk, 1997; Horner, 2001, personal commun.). A multispecies bonebed with abundant hadrosaur remains occurs near the base of the Hell Creek in North Dakota (Hoganson et al., 1994; Murphy et al., 1995). Another, from South Dakota, has yielded over 3000 scattered elements predominantly of small raptorial dinosaurs, oviraptorosaurs, pachycephalosaurs, and thescelosaurs (Triebold, 1997; Bartlett, 1999). A bonebed 1 m above the contact of the transitional marine Fox Hills Sand-

TABLE 1. HELL CREEK DINOSAUR ASSEMBLAGE

Tyrannosauridae
- *Aublysodon* cf. *A. mirandus*
- *Tyrannosaurus rex* (including *Nanotyrannus lancensis*, Carr, 1999)

Ornithomimidae
- *Ornithomimus* sp. (Russell *in* Triebold, 1997)
- *Struthiomimus* sp. (Russell *in* Triebold, 1997)

Dromaeosauridae
- *Dromaeosaurus albertensis* (Baszio, 1997)
- cf. *Saurornitholestes langstoni* (Baszio, 1997)

Caenagnathidae
- *Caenagnathus* sp. (Currie et al., 1993)

Troodontidae
- *Troodon formosus*

Family unknown (Maniraptora)
- *Richardoestesia gilmorei*
- *Richardoestesia*, undescribed species (Baszio, 1997)

Theropoda, family unknown
- *Paronychodon lacustris* (Baszio, 1997)

Nodosauridae
- *Edmontonia* sp.

Ankylosauridae
- *Ankylosaurus magniventris*

Thescelosauridae
- *Bugenasaura infernalis* (Galton, 1995)
- *Thescelosaurus neglectus*
- *Thescelosaurus garbanii* (Morris, 1976)

Hadrosauridae
- *Anatotitan copei*
- *Edmontosaurus annectens*

Pachycephalosauridae
- Cf. Homalocephalinae, new genus and species (H. Galiano, 2001, personal commun.)
- *Pachycephalosaurus wyomingensis*
- *Stegoceras validus*
- *Stygimoloch spinifer*

Ceratopsidae
- *Torosaurus latus*
- *Triceratops horridus*
- *Triceratops prorsus* (Forster, 1996)

Note: Late Maastrichtian, after Archibald (1996) except where noted. Citations following names of taxa in this and following tables are recent references supporting the validity of the taxon and do not necessarily indicate authorship of the taxon. They may be consulted for summaries of previous taxonomic opinions. An element formerly considered to be an astragalus (Welles and Long, 1974), and suspected of therizinosaurian affinities by the first author, has been correctly identified as the right quadrate of a *Tyrannosaurus* by Carpenter (1992).

TABLE 2. ARTICULATED SPECIMENS FROM THE HELL CREEK FAUNA

	This paper %	A	A, B
Ceratopsidae	33	61	73
Hadrosauridae	27	23	10
Tyrannosauridae	16	4	12
Thescelosauridae	9	3	1
Pachycephalosauridae	8	1	1
Ornithomimidae	3	5	0
Caenagnathidae	2	0	0
Nodosauridae	1	0	0
Ankylosauridae	1	0	3
Dromaeosauridae	0	2	0
Troodontidae	0	1	0

Note: Specimen lists of articulated specimens in the Carnegie Museum of Natural History, Cleveland Museum of Natural History, Field Museum of Natural History, Chicago, Milwaukee Public Museum, Science Museum of Minnesota, St. Paul Museum, University of Michigan, Ann Arbor, and the University of California Museum of Paleontology prior to 1984 were collated by Robert Long and Dale Russell. They form the basis of the relative abundances of family groups, supplemented by specimens cited in Brown and Schlaikjer (1943); Bump (1939); Carpenter and Breithaupt (1986); Colbert and Bump (1947); Currie et al. (1993); Frye (1969); Galton (1995); Gartska and Burnham (1997); John S. McIntosh (1969, personal commun.); Larson (1997); Loomis (1929); Lull (1933, p. 117); Lull and Wright (1942); and Triebold (1997; 2000, personal commun.). The 123 articulated specimens identified are probably representative of the abundances of family groups in existing collections, both public and private. The percentages are compared to those given by (A) White et al. (1998; largely based on isolated elements), and (B) Lehman (1987).

stone is estimated to contain disarticulated skeletal elements of 2000 individuals of *Edmontosaurus*, as well as indications of the presence of tyrannosaurs, small theropods, nodosaurids, pachycephalosaurids, and ceratopsians (Larson, 1985; Christians, 1992). Bonebed and other associations of disarticulated bones belonging to several or numerous individual dinosaurs occur throughout South Dakota and are currently under review (J.W. Hoganson, R.J. Jacobson, K. Johnson, J. Nelson, S.D. Sroka, personal communs., 2001).

Available collections do not reveal clear geographic trends in the relative abundance of dinosaur taxa across the area of outcrop of the Hell Creek Formation in eastern Montana and western North Dakota and South Dakota. Remains of *Caenagnathus* (M. Triebold, 1999, personal commun.) and *Tyrannosaurus* (Larson, 1997) may be somewhat more abundant in South Dakota, although five additional *Tyrannosaurus* skeletons have recently been discovered in Montana (J.R. Horner, 2001, personal commun.).

Farther southwest, in Wyoming, the dinosaur assemblage from the Lance Formation closely resembles that of the Hell Creek (Lehman, 1987). A few taxa, including *Leptoceratops* (Ostrom, 1978), *Diceratops* (Forster, 1996), and possibly, the puzzling diplodocoid *Dyslocosaurus* (McIntosh et al., 1992; the specimen may have been derived from the Morrison Formation) occur in the Lance Formation but have not yet been recorded in the Hell Creek Formation. It has been reported (Estes and Berberian, 1970) that dinosaurian microfossils are much more abundantly represented in Lancian microsites than at a microsite (Bug Creek Anthills) in western Montana. In the Hell Creek site, dinosaurian remains appear to be 40 times less abundant, according to the minimum number of individuals represented, and 16 times less abundant by gross weight. These differences have been attributed to the greater prevalence of aquatic environments in the Hell Creek (Estes and Berberian, 1970), but are more likely the result of the reworking of Cretaceous microfossils into channel sands of basal Tertiary age (Archibald and Bryant, 1990; Lofgren, 1995). Thus a terminal Cretaceous interval of direct competition between dinosaurian and small mammalian herbivores (cf. Sloan et al., 1986) may never have occurred. Far to the northwest of the area of outcrop of the Hell Creek Formation in Montana, equivalent strata in central Alberta have yielded a small assemblage dominated by remains of ankylosaurids and protoceratopsids (Lehman, 1987).

COMPARISONS

Antedating the deposition of Hell Creek strata by ~14 m.y., a series of similar but less extensive deltas had spread into the western margin of the interior seaway (Lehman, 1997; Hicks et al., 1995). In southern Alberta and western Montana, late Campanian strata have yielded abundant skeletal remains of dinosaurs. In Dinosaur Provincial Park, Alberta, the average number of articulated specimens collected per square kilometer is 3.8, and the average number of hours needed to locate a specimen is 385 (Béland and Russell, 1978, their Table 2 and p. 1017). In two areas of badlands exposed in the Hell Creek Formation in Montana and North Dakota, these numbers are, respectively, 3.2 and 405 (White et al., 1998, p. 47). It is evident that the older and younger formations are of comparable richness in dinosaur specimens. Bonebeds, or mass-death horizons, appear to be more abundant in the Dinosaur Park (e.g., Eberth, 1996a; Getty et al., 1997; Ryan et al., 1999) and Two Medicine Formations (e.g., Rogers, 1991; Varricchio and Horner, 1993; Varricchio, 1995), although their frequency of occurrence in the Hell Creek Formation of South Dakota may prove to be comparable.

Dinosaurian diversity in the Hell Creek Formation has often been contrasted (cf. Sloan et al., 1986) to that in strata of the Dinosaur Park Formation exposed in Dinosaur Provincial Park (cf. Tables 1 and 3). The diversity of small dinosaurs is similar in the two units. However, the diversity of large dinosaurs in the Hell Creek is less than half that documented in the older assemblage, because of a lesser variety of hadrosaurids and the absence of centrosaurine ceratopsians. Regional differences in large herbivore taxa were also greater across the area that has become the northern interior plains during late Campanian time than during late Maastrichtian time (cf. Tables 1, 3, and 4). Environmental factors may have been responsible for the lower diversity of large Hell Creek herbivores. An important floral change occurred earlier in the Maastrichtian prior to Hell Creek deposition, whereby gymnosperm-dominated forests were regionally and unprecedentedly replaced by those dominated by angiosperms (Johnson, 1992; Wing et al., 1993). The relatively greater expansion of the Hell Creek flood plain probably reduced environmental and topographic gradients, rendering the region more uniform (cf. Roberts and Kirschbaum, 1995).

A prominent coastal to interior gradient in the relative abundances of large herbivorous dinosaurs is present in a previous Campanian delta (Brinkman et al., 1998). In eastern Alberta, near the coastline, the ratio of ceratopsian:hadrosaur recovery is an identical 42%:42% of the total number of collected specimens. In Dinosaur Provincial Park (toward the mountains but still within the range of tidal influences in fluvial systems, Eberth, 1996b) the ratio is 22%:41%. Ankylosaurs and pachycepalosaurids increase in relative abundance as ceratopsians decrease inland (cf. the relative abundance of ankylosaurids in inland late Maastrichtian environments, Lehman, 1987). However, there is no evidence that the ceratopsian:hadrosaur ratio departed from the Hell Creek average of the approximately equal 33%:27% anywhere in eastern Montana and North and South Dakota, thereby suggesting relatively low environmental gradients throughout the region.

Lehman (1997, p. 236) and Baszio (1997) postulated that differences between the Dinosaur Provincial Park and Hell Creek assemblages, both located near the present 49th parallel, were due to latitudinal displacements of dinosaur assemblages linked to regional shifts in climate. Abundant *Richardoestesia, Paronychodon*, pachycephalosaur, and protoceratopsid teeth

TABLE 3. DINOSAUR PARK ASSEMBLAGE

Tyrannosauridae
Aublysodon mirandus: Currie et al. (1990)
Albertosaurus libratus
Daspletosaurus, new species (P.J. Currie, 1999, personal commun.)
Ornithomimidae
Dromiceomimus samueli
Ornithomimus edmontonensis
Struthiomimus altus
Caenagnathidae: Sues (1997)
Chirostenotes pergracilis
Chirostenotes elegans
Troodontidae
Troodon formosus
Dromaeosauridae
Dromaeosaurus albertensis
Dromaeosaurus, new tooth morphotype (Sankey, 2000)
Saurornitholestes langstoni
Saurornitholestes, three new tooth morphotypes (Sankey, 2000)
Maniraptora, Family unknown: Baszio (1997)
Richardoestesia gilmorei
Richardoestesia, new species: Sankey (2000)
Theropoda, family unknown
Paronychodon, two new tooth morphotypes: Sankey (2000)
Hadrosauridae (Hadrosaurinae)
Brachylophosaurus canadensis
Gryposaurus notabilis
Gryposaurus incurvimanus (Horner, 1992)
Prosaurolophus maximus
cf. *Maiasaura* sp.: P.J. Currie (1999, personal commun.)
Hadrosaurldae (Lambeosaurinae)
Corythosaurus casuarius
Lambeosaurus lambei
Lambeosaurus magnicristatum
Parasaurolophus walkeri
Pachycephalosauridae
Gravitholus sternbergi: including *Gravitholus albertae* cf. Sullivan (2000)
Pachycephalosaurus sp.
Stegoceras validum: including *Ornatotholus browni* cf. Sullivan (2000)
"Stegoceras" lambei: Sullivan (2000)
Protoceratopsidae (Ryan and Currie, 1998)
Cf. *Leptoceratops* sp.
Ceratopsidae (Chasmosaurinae)
Chasmosaurus belli
Chasmosaurus canadensis
Chasmosaurus russelli
Chasmosaurus, new species: Holmes et al., (1999); P.J. Currie (1999, personal commun.)
Ceratopsidae (Centrosaurinae)
Centrosaurus apertus
Monoclonius crassus
Styracosaurus albertensis
Nodosauridae
Edmontonia longiceps
Edmontonia rugosidens
Panoplosaurus mirus
Ankylosauridae
Euoplocephalus tutus

Note: Late Campanian, after Weishampel (1990) except where noted.

TABLE 4. ASSEMBLAGE FROM THE UPPERMOST 80 M OF THE TWO MEDICINE ASSEMBLAGE

Tyrannosauridae
cf. *Daspletosaurus* sp.: Horner et al. (1992)
cf. *Albertosaurus* sp.
Ornithomimidae
Genus undetermined: Varricchio (1995)
Dromaeosauridae
Saurornitholestes sp. (Varricchio, 1995)
Troodontidae
Troodon formosus (Varricchio, 1995)
Family undetermined
Richardoestesia sp. (Varricchio, 1995)
Nodosauridae
Edmontonia rugosidens (Gilmore, 1930)
Ankylosauridae
Euoplocephalus sp. (Gilmore, 1930)
Pachycephalosauridae
Stegoceras sp. (Goodwin, 1992)
cf. *Pachycephalosaurus* sp.: Goodwin (1992)
Hypsilophodontidae
cf. *Orodromeus* sp.
Hadrosauridae ("Hadrosaurinae")
Prosaurolophus blackfeetensis (Horner, 1992)
Gryposaurus sp. (Varricchio, 1995)
Hadrosauridae (Lambeosaurinae)
Hypacrosaurus sp. (Varricchio, 1995)
Ceratopsidae (Centrosaurinae)
Einiosaurus procurvicornis (Sampson et al., 1997)
Achelousaurus horneri (Sampson et al., 1997)
Unnamed taxon
Brachyceratops montanensis, nomen dubium (Sampson et al., 1997)

Note: Late Campanian, as confirmed by John Horner (2000, personal commun.), with references.

and abundant skeletal remains of hadrosaurines and chasmosaurines characterized a southerly (Hell Creek) assemblage. Less abundant *Richardoestesia, Paronychodon*, pachycephalosaur, and protoceratopsid teeth and abundant skeletal remains of lambeosaurines and centrosaurines characterized the northerly (Dinosaur Park) assemblage. *Richardoestesia* is conspicuous by its absence in Alaskan tooth assemblages (Fiorillo and Gangloff, 2000). Through the last 20 m.y. of the Cretaceous, southerly faunas were displaced to the north (Milk River assemblage, Baszio, 1997), then retreated to the south in front of northerly faunas (Dinosaur Park assemblage) and finally returned to the north again during Hell Creek time. The displacements were ~1200 km, or 10° latitude, or 5°C in mean annual temperature (cf. Lehman, 1997, p. 236 and Fig. 11, and Lehman, 1987, Fig. 7). A warming trend and eustatic fall in sea level approximately coincide with the onset of Deccan Trap volcanism during latest Maastrichtian time (Wignall, 2001; Huber et al., 2002). The coincidences pose the possibility that the northward migration and expansion of the Hell Creek ecosystem was related to the formation of a large igneous province that has been suggested as the cause of the ecosystem's destruction (Wignall, 2001).

That faunal displacements occurred as a result of latitudinally controlled environmental displacements appears to be well founded, at least on a regional scale. However, smaller

Figure 1. *Triceratops horridus* (NSM PV20379) on display at the National Science Museum, Tokyo, Japan. The specimen has been named Raymond (Garstka and Burnham, 1997), and is one of the most complete articulated skeletons of *Triceratops* known.

scale environmental factors may also have played a role. For example, lambeosaurine hadrosaurids have not been recorded in the Hell Creek assemblage. Nevertheless, lambeosaurine bonebeds occur in eastern Asia associated with palynofloral indicators of Hell Creek time and regional vegetation (Godefroit et al., 2000) and at comparable paleolatitudes (C.R. Scotese, 2001, personal commun.). The lambeosaurine:hadrosaurine ratio appears to increase substantially over 20 km from east to west in Dinosaur Provincial Park (Béland and Russell, 1978; their Table 4) and vertically from the bottom to the top of a 180 m interval in the regressive lower part of the Horseshoe Canyon Formation in central Alberta (Russell and Potter Chamney, 1967). Thus the absence of records in the Hell Creek Formation could be taken as evidence of uniform coastal wetland environments usually avoided by lambeosaurines.

The relatively organic, anoxic, and slow-moving waters of coastal wetlands may provide unusual opportunities for preservation. For example, cartilaginous structures such as beaks (Morris, 1970), extensions of vertebral spines (Horner, 1984), and ribs (Fisher et al., 2000) have been reported in Hell Creek dinosaur specimens, as well as evidence of vascular structures (Fisher et al., 2000; Happ and Morrow, 2000). Spectacular skeletal specimens encased in skin impressions have long been known to occur in the contiguous Lance Formation in Wyoming (Osborn, 1912; Versluys, 1923). Evidence of soft tissues should be sought during the process of excavating well-preserved skeletal specimens of Hell Creek dinosaurs.

Low-energy, relatively anoxic environments may also have favored the preservation of infant dinosaurs. Two such specimens have been collected, a hadrosaurine from Montana (University of California Museum of Paleontology no. 128181; the length of a posterior dorsal vertebral centrum is 14 mm, that of a femur is 149 mm) and a thescelosaur from Saskatchewan (Canadian Museum of Nature no. 22039; lengths of the same ossifications are 11 mm and 92 mm). Scaling from adult specimens of *Edmontosaurus* and *Thescelosaurus* implies body weights, respectively, of 6 and 2 kg. In the case of the hadrosaurine, an infant having just left the nest is suggested (cf. Horner et al., 2000). It is thus probable that both genera nested in coastal lowlands.

CONCLUSIONS

The Hell Creek dinosaur assemblage was probably adapted to broad, homogeneous coastal plains in a warm, midlatitudinal environment and had remained generally recognizable for more than 20 m.y., through Campanian and Maastrichtian time. It was also unusual in the abundance of small, birdlike dinosaurs, which, under normal circumstances, would have been ancestral to a subsequent radiation of diverse and active vertebrates capable of increasingly complex behavior (Russell, 1996). The Hell Creek assemblage, together with that of the Lance Formation and equivalent strata in Wyoming, represents the last, best ecosystem known to have been dominated by dinosaurs. It

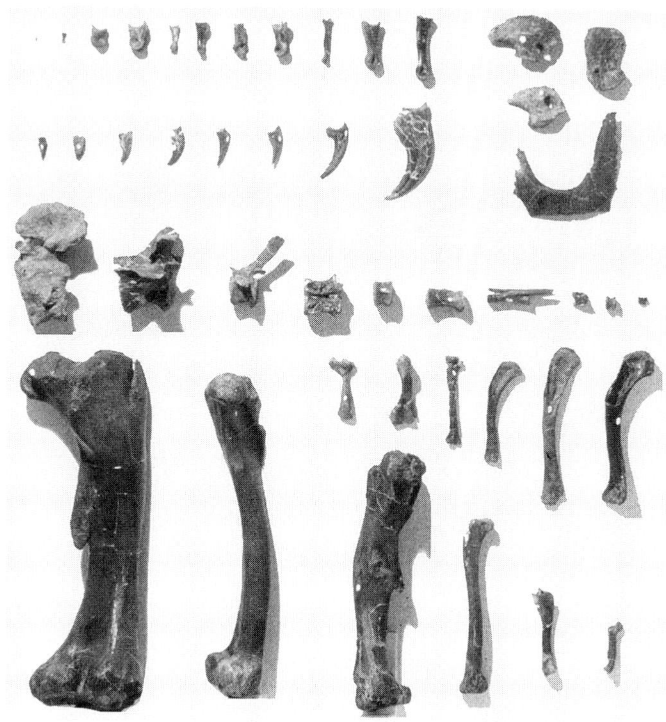

Figure 2. Display of the ecology and environment of the Hell Creek Sandy Site in the National Science Museum, Tokyo, Japan. The cabinet is designed to accommodate the results of ongoing investigations of Hell Creek biota at the Museum. Vertebrate materials are from the Sandy Site in South Dakota (Triebold, 1997), now in museum collections. The specimens will figure in future research projects.

is a pleasure to note that this ecosystem is attracting the careful and extended study it deserves.

ACKNOWLEDGMENTS

We thank Kirk Johnson for affording us the opportunity to present this abbreviated overview of the Hell Creek dinosaur assemblage and for his editorial support. The manuscript was greatly improved through the constructive criticism of Kenneth Carpenter and Thomas Lehman. We are grateful to John Hoganson, Russell Jacobson, Kirk Johnson, Peter Larson, Jim Nelson, Steven Sroka, and Michael Triebold for information on dinosaur occurrences in South Dakota and to John Horner for information on Montana occurrences and for editing the list of dinosaurian taxa from the Two Medicine Formation. Kevin Padian generously supplied archival data on the infant hadrosaurine, Christopher Scotese confirmed the paleolatitudinal position of the Hell Creek assemblage and Asian lambeosaurine bonebeds, and Henry Galiano provided information on an undescribed homalocephaline pachycephalosaur. The contents of Table 2 were largely compiled by Robert Long and Russell in the course of visiting collections of Hell Creek dinosaurs many years ago.

REFERENCES CITED

Archibald, J.D., 1996, Dinosaur extinction and the end of an era: What the fossils say: New York, Columbia University Press, 237 p.

Archibald, J.D., and Bryant, L.J., 1990, Differential Cretaceous/Tertiary extinctions of nonmarine vertebrates: Evidence from northeastern Montana, *in* Sharpton, V.L., and Ward, P., eds., Global catastrophes in Earth history: An interdisciplinary conference on impacts, volcanism, and mass mortality: Geological Society of America Special Paper 247, p. 549–562.

Bartlett, J., 1999, Taphonomy and characteristics of a diverse Maastrichtian assemblage: The Sandy Site, Hell Creek Formation, South Dakota: Journal of Vertebrate Paleontology, v. 19, supplement to no. 3, p. 31A.

Baszio, S., 1997, Investigations on Canadian dinosaurs: Courier Forschungsinstitut Senckenberg, v. 196, p. 1–77.

Béland, P., and Russell, D.A., 1978, Paleoecology of Dinosaur Provincial Park (Cretaceous), Alberta, interpreted from the distribution of articulated vertebrate remains: Canadian Journal of Earth Sciences, v. 15, p. 1012–1024.

Brinkman, D.B., Ryan, M.J., and Eberth, D.A., 1998, The paleogeographic and stratigraphic distribution of ceratopsids (Ornithischia) in the Upper Judith River Group of western Canada: Palaios, v. 13, p. 160–169.

Brown, B., 1907, The Hell Creek Beds of the Upper Cretaceous of Montana: American Museum of Natural History Bulletin, v. 23, p. 823–845.

Brown, B., and Schlaikjer, E.M., 1943, A study of troodont dinosaurs with the description of a new genus and four new species: American Museum of Natural History Bulletin, v. 82, p. 121–149.

Bryant, L.J., 1989, Non-dinosaurian lower vertebrates across the Cretaceous-Tertiary boundary in northeastern Montana: University of California Publications in Geological Sciences, v. 134, 107 p.

Bump, J.D., 1939, Dinosaurs collected by the School of Mines (South Dakota): Black Hills Engineer, v. 25, p. 228–229.

Carpenter, K., 1992, Tyrannosaurids (Dinosauria) of Asia and North America, *in* Mateer, N., and Chen, P.J., eds., Aspects of nonmarine cretaceous geology: Beijing, China Ocean Press, p. 250–268.

Carpenter, K., and Breithaupt, B., 1986, Latest Cretaceous occurrence of nodosaurid ankylosaurs (Dinosauria: Ornithischia) in western North America and the gradual extinction of the dinosaurs: Journal of Vertebrate Paleontology, v. 6, p. 251–257.

Carr, T.D., 1999, Craniofacial ontogeny in Tyrannosauridae (Dinosauria, Coelurosauria): Journal of Vertebrate Paleontology, v. 19, p. 497–520.

Christians, J.P., 1992, Taphonomy and sedimentology of the Mason Dinosaur Quarry, Hell Creek Formation (Upper Cretaceous), South Dakota [M.S. thesis]: Madison, University of Wisconsin, 91 p.

Coe, M.C., Cumming, D.H., and Phillipson, J., 1976, Biomass and production of large African herbivores in relation to rainfall and primary production: Oecologia, v. 22, p. 341–354.

Colbert, E.H., and Bump, J.D., 1947, A skull of *Torosaurus* from South Dakota and a revision of the genus: Philadelphia, Pennsylvania, Philadelphia, Academy of Natural Sciences, v. 99, p. 93–106.

Currie, P.J., Rigby, J.K., and Sloan, R.E., 1990, Theropod teeth from the Judith River Formation of southern Alberta, Canada, *in* Carpenter, K., and Currie, P.J., eds., Dinosaur systematics: Approaches and perspectives: Cambridge, Cambridge University Press, p. 107–125.

Currie, P.J., Godfrey, S.J., and Nessov, L., 1993, New caenagnathid (Dinosauria: Theropoda) specimens from the Upper Cretaceous of North America and Asia: Canadian Journal of Earth Sciences, v. 30, p. 2255–2272.

Eberth, D.A., 1996a, Ceratopsian bonebeds in the Dinosaur Park Formation (Campanian) of southern Alberta: Bigger than we thought?: Journal of Vertebrate Paleontology, v. 16, supplement to no. 3, p. 32A.

Eberth, D.A., 1996b, Origin and significance of mud-filled incised valleys (Upper Cretaceous) in southern Alberta, Canada: Sedimentology, v. 43, p. 459–477.

Estes, R.D., and Berberian, P., 1970, Paleoecology of a Late Cretaceous vertebrate community from Montana: Breviora, Museum of Comparative Zoology, v. 343, 35 p.

Fiorillo, A.R., and Gangloff, R.A., 2000, Theropod teeth from the Prince Creek Formation (Cretaceous) of northern Alaska, with speculations on Arctic dinosaur paleoecology: Journal of Vertebrate Paleontology, v. 20, p. 675–682.

Fisher, P.E., Russell, D.A., Stoskopf, M.K., Barrick, R.E., Hammer, M., and Kuzmitz, A.A., 2000, Cardiovascular evidence for an intermediate or higher metabolic rate in an ornithischian dinosaur: Science, v. 288, p. 503–505.

Forster, C., 1996, Species resolution in *Triceratops*: Cladistic and morphometric approaches: Journal of Vertebrate Paleontology, v. 16, p. 259–270.

Frye, C.I., 1969, Stratigraphy of the Hell Creek Formation in North Dakota: North Dakota Geological Survey Bulletin 54, 65 p.

Galton, P.M., 1995, The species of the basal hypsilophodontid dinosaur *Thescelosaurus* Gilmore (Ornithischia: Ornithopoda) from the Late Cretaceous of North America: Neues Jahrbuch für Geologie und Paläontologie, Abhandlungen, v. 198, p. 297–311.

Gartska, W.R., and Burnham, D.A., 1997, Posture and stance of *Triceratops*: Evidence of a digitigrade manus and cantilever vertebral column, *in* Wolberg, D.L., et al., eds., Dinofest International: Philadelphia, Pennsylvania, Philadelphia Academy of Natural Sciences, p. 385–391.

Getty, M., Eberth, D.A., Brinkman, D.B., Tanke, R.M., and Vickaryous, M., 1997, Taphonomy of two *Centrosaurus* bonebeds in the Dinosaur Park Formation, Alberta, Canada: Journal of Vertebrate Paleontology, v. 17, supplement to no. 3, p. 48A.

Gilmore, C.W., 1930, On dinosaurian reptiles from the Two Medicine Formation of Montana: U.S. National Museum Proceedings, v. 77, p. 1–21.

Godefroit, P., Zan, S.Q., and Jin, L.Y., 2000, Charonosaurus jiayinensis n. g., n. sp. Lambeosaurine dinosaur from the Late Maastrichtian of northeastern China: Comptes Rendus de l'Academie des Sciences, Serie 2, Sciences de la Terre et des Planètes, v. 330, p. 875–882.

Goodwin, M.B., 1992, Pachycephalosaurid dinosaurs from the Two Medicine Formation (Campanian) of northwestern Montana: Journal of Vertebrate Paleontology, v. 12, supplement to no. 3, p. 30A.

Happ, J.W., and Morrow, C.M., 2000, Evidence of soft tissue associated with nasal and supraorbital horn cores, rostral and exoccipital of *Triceratops*: Journal of Vertebrate Paleontology, v. 20, supplement to no. 3, p. 47A.

Hicks, J.F., Obradovich, J.D., and Tauxe, L., 1995, A new calibration point for the Late Cretaceous time scale: The ^{40}Ar/^{39}Ar isotopic age of the C33r/C33n geomagnetic reversal from the Judith River Formation (Upper Cretaceous), Elk Basin, Wyoming: Journal of Geology, v. 103, p. 243–256.

Hoganson, J.W., Campbell, J.M., and Murphy, E.C., 1994, Stratigraphy and paleontology of the Cretaceous Hell Creek Formation, Stumpf Site, Morton County, North Dakota: North Dakota Academy of Science Proceedings, v. 48, p. 95.

Holmes, R., Shepherd, K., and Forster, C., 1999, An unusual chamosaurine ceratopsid from Alberta: Journal of Vertebrate Paleontology, v. 19, supplement to no. 3, p. 52A

Horner, J.R., 1984, A "segmented" epidermal tail frill in a species of hadrosaurian dinosaur: Journal of Paleontology, v. 58, p. 270–271.

Horner, J.R., 1992, Cranial morphology of *Prosaurolopus* (Ornithischia: Hadrosauridae), with descriptions of two new hadrosaurid species and an evaluation of hadrosaurid phylogenetic relationships: Museum of the Rockies Occasional Paper 2, 119 p.

Horner, J.R., Varricchio, D.J., and Goodwin, M.B., 1992, Marine transgressions and the evolution of Cretaceous dinosaurs: Nature, v. 358, p. 59–61.

Horner, J.R., De Riqlès, A., and Padian, K., 2000, Long bone histology of the hadrosaurid dinosaur *Maiasaura peeblesorum*: growth dynamics and physiology based on an ontogenetic series of skeletal elements: Journal of Vertebrate Paleontology, v. 20, p. 115–129.

Huber, B.T., Norris, R.D., and MacLeod, K.G., 2002, Deep-sea paleotemperature record of extreme warmth during the Cretaceous: Geology, v. 30, p. 123–126.

Hurlbert, S.H., and Archibald, J.D., 1995, No statistical support for sudden (or gradual) extinction of dinosaurs: Geology, v. 23, p. 881–884.

Hurlbert, S.H., and Archibald, J.D., 1996, No statistical support for sudden (or gradual) extinction of dinosaurs: Reply: Geology, v. 24, p. 958–959.

Johnson, K.R., 1992, Leaf-fossil evidence for extensive floral extinction at the Cretaceous-Tertiary boundary: North Dakota, USA: Cretaceous Research, v. 13, p. 91–117.

Johnson, K.R., and Hickey, L.J., 1990, Megafloral change across the Cretaceous/Tertiary boundary in the northern Great Plains and Rocky Mountains, U.S.A., *in* Sharpton, V.L., and Ward, P., eds., Global catastrophes in Earth history: An interdisciplinary conference on impacts, volcanism, and mass mortality: Geological Society of America Special Paper 247, p. 433–444.

Larson, P.L., 1985, A preliminary paleontologic and geologic investigation of the Ruth Mason dinosaur quarry (Maestrichtian age), South Dakota: Annual Meeting of the Society of Vertebrate Paleontology, Abstracts, p. 5.

Larson, P.L., 1997, The king's new clothes: A fresh look at *Tyrannosaurus rex*, *in* Wolberg, D.L., et al., eds., Dinofest International: Philadelphia, Pennsylvania, Philadelphia Academy of Natural Sciences, p. 65–71.

Lehman, T.M., 1987, Late Maastrichtian paleoenvironments and dinosaur biogeography in the Western Interior of North America: Palaeogeography, Palaeoclimatology, Palaeoecology, v. 60, p. 189–217.

Lehman, T.M., 1997, Late Campanian dinosaur biogeography in the Western Interior of the United States, *in* Wolberg, D.L., et al., eds., Dinofest International: Philadelphia, Pennsylvania, Philadelphia Academy of Natural Sciences, p. 223–240.

Lofgren, D.L., 1995, The Bug Creek problem and the Cretaceous-Tertiary transition at McGuire Creek, Montana: University of California Publications in Geological Sciences, v. 140, 185 p.

Loomis, F.B., 1929, Report of the Geology Department of Amherst College to the Alumni Visiting Committee of the Alumni Council, November 16, 1928: Amherst Alumni Council News, v. 2, no. 3, p. 5–8.

Lull, R.S., 1933, A revision of the Ceratopsia, or horned dinosaurs: Peabody Museum of Natural History Bulletin, v. 3, p. 1–175.

Lull, R.S., and Wright, N.E., 1942, The hadrosaurian dinosaurs of North America, *in* The hadrosaurian dinosaurs of North America: Boulder, Colorado, Geological Society of America Special Paper 40, p. 1–242.

McIntosh, J.S., Coombs, W.P., and Russell, D.A., 1992, A new diplodocid sauropod (Dinosauria) from Wyoming, U.S.A.: Journal of Vertebrate Paleontology, v. 12, p. 158–167.

Morris, W.J., 1970, Hadrosaurian dinosaur bills: Morphology and function: Natural History Museum of Los Angeles County Contributions in Science, no. 193, p. 1–14.

Morris, W.J., 1976, Hypsilophodont dinosaurs: A new species and comments on their systematics, *in* Churcher, C.S., ed., Athlon: Essays on paleobiology in honour of Loris Shano Russell: Royal Ontario Museum Life Sciences, Miscellaneous Publications, p. 93–113.

Murphy, E.C., Nichols, D.J., Hoganson, J.W., and Forsman, N.F., 1995, The Cretaceous/Tertiary boundary in south-central North Dakota: North Dakota Geological Survey Report of Investigations, no. 98, p. 1–73.

Osborn, H.F., 1912, Integument of the iguanodont dinosaur *Trachodon*: American Museum of Natural History Memoirs, v. 1, p. 33–54.

Ostrom, J.H., 1978, *Leptoceratops gracilis* from the "Lance" Formation of Wyoming: Journal of Paleontology, v. 52, p. 697–704.

Retallack, G.J., 1994, A pedotype approach to latest Cretaceous and earliest Tertiary paleosols in eastern Montana: Geological Society of America Bulletin, v. 106, p. 1377–1397.

Roberts, L.N.R., and Kirschbaum, M.A., 1995, Paleogeography of the Late Cretaceous of the Western Interior of middle North America: Coal distribution and sediment accumulation: U.S. Geological Survey Professional Paper 1561, 115 p.

Rogers, R.R., 1991, Taphonomy of three dinosaur bone beds in the Upper Cretaceous Two Medicine Formation of northwestern Montana: Evidence for drought-related mortality: Palaios, v. 53, p. 94–413.

Russell, D.A., 1987, A note on the terminal Cretaceous nodosaurids of North America: Journal of Vertebrate Paleontology, v. 7, p. 102.

Russell, D.A., 1989, An odyssey in time: The dinosaurs of North America: Toronto, University of Toronto Press, 240 p.

Russell, D.A., 1996, The meaning of the extinction of the dinosaurs, *in* Ryder, G., et al., eds., 1996, The Cretaceous-Tertiary event and other catastrophes in Earth history: Geological Society of America Special Paper 307, p. 381–388.

Russell, D.A., and Potter Chamney, T., 1967, Notes on the biostratigraphy of dinosaurian and microfossil faunas in the Edmonton Formation (Cretaceous), Alberta: National Museum of Canada Natural History Paper 35, 22 p.

Ryan, M.J., and Currie, P.J., 1998, First report of protoceratopsians (Neoceratopsia) from the Late Cretaceous Judith River Group, Alberta, Canada: Canadian Journal of Earth Sciences, v. 35, p. 820–826.

Ryan, M.J., Russell, A.P., Getty, M., Eberth, D.A., and Brinkman, D.B., 1999, A new centrosaurine ceratopsid from the Oldman Formation of Alberta, Canada: Journal of Vertebrate Paleontology, v. 19, supplement to no. 3, p. 72A.

Sampson, S., Ryan, M.J., and Tanke, D.H., 1997, Craniofacial ontogeny in centrosaurine dinosaurs (Ornithischia: Ceratopsidae): Taxonomic and behavioral implications: Linnaean Society Zoological Journal, v. 121, p. 293–337.

Sankey, J.T., 2000, New theropod and bird teeth from the Late Cretaceous (Campanian) Judith River Group, Alberta: Journal of Vertebrate Paleontology, v. 20, supplement to no. 3, p. 67A.

Sheehan, P.M., and Fastovksy, D.E., 1992, Major extinctions of land-dwelling vertebrates at the Cretaceous-Tertiary boundary, eastern Montana: Geology, v. 20, p. 556–560.

Sheehan, P.M., Fastovsky, D.E., Hoffmann, R.G., Berghaus, C.B., and Gabriel, D.L., 1991, Sudden extinction of the dinosaurs: Latest Cretaceous, upper Great Plains, U.S.A.: Science, v. 254, p. 835–839.

Sheehan, P.M., Fastovsky, D.E., Hoffman, R.G., and Barreto, C.B., 1996, No statistical support for sudden (or gradual) extinction of dinosaurs: Comment: Geology, v. 24, p. 957–958.

Sloan, R.E., Rigby, J.K., Van Valen, L.M., and Gabriel, D.L., 1986, Gradual dinosaur extinction and simultaneous ungulate radiation in the Hell Creek Formation: Science, v. 232, p. 629–633.

Sues, H.D., 1997, On *Chirostenotes*, a Late Cretaceous oviraptorosaur (Dinosauria: Theropoda) from western North America: Journal of Vertebrate Paleontology, v. 17, p. 698–716.

Sullivan, R.M., 2000, *Stegoceras* revisited: Journal of Vertebrate Paleontology, v. 20, supplement to no. 3, p. 72A.

Tokaryk, T.T., 1997, First evidence of juvenile ceratopsians (Reptilia: Ornithischia) from the Frenchman Formation (late Maastrichtian) of Saskatchewan: Canadian Journal of Earth Sciences, v. 34, p. 1401–1404.

Triebold, M., 1997, The Sandy Site: Small dinosaurs from the Hell Creek Formation of South Dakota, *in* Wolberg, D.L., et al., eds., Dinofest International: Philadelphia, Pennsylvania, Philadelphia Academy of Natural Sciences, p. 245–248.

Varricchio, D.J., 1995, Taphonomy of Jack's Birthday Site, a diverse dinosaur bonebed from the Upper Cretaceous Two Medicine Formation of Montana: Palaeogeography, Palaeoclimatology, Palaeoecology, v. 114, p. 297–323.

Varricchio, D.J., and Horner, J.R., 1993, Hadrosaurid and lambeosaurid bone beds from the Upper Cretaceous Two Medicine Formation of Montana: Taphonomic and biologic implications: Canadian Journal of Earth Sciences, v. 30, p. 997–1006.

Versluys, J., 1923, Der Schädel des Skelettes von *Trachodon annectens* im Senckenberg-Museum: Abhandlungen Senckenbergischen Naturforschenden Gesellschaft, v. 38, p. 1–19.

Weishampel, D.B., 1990, Dinosaur distributions, *in* Weishampel, D.B., et al., eds., The dinosauria: Berkeley, University of California Press, p. 63–139.

Welles, S.P., and Long, R.A., 1974, The tarsus of theropod dinosaurs: South African Museum Annals, v. 64, p. 191–218.

White, P.D., Fastovsky, D.E., and Sheehan, P.M., 1998, Taphonomy and suggested structure of the dinosaurian assemblage of the Hell Creek Formation (Maastrichtian), eastern Montana and western North Dakota: Palaios, v. 13, p. 41–51.

Wignall, P.B., 2001, Large igneous provinces and mass extinctions: Earth-Science Reviews, v. 53, p. 1–33.

Wing, S.L., Hickey, L.J., and Swisher, C.C., 1993, Implications of an exceptional fossil flora for Late Cretaceous vegetation: Nature, v. 363, p. 342–344.

Manuscript Accepted by the Society September 24, 2001

Geological Society of America
Special Paper 361
2002

Patterns of geographic variation in latest Cretaceous vertebrates: Evidence from the turtle component

Patricia A. Holroyd
J. Howard Hutchison
Museum of Paleontology, University of California, Berkeley, California 94720, USA

ABSTRACT

Turtles are among the most widespread, abundant, and diverse fossil vertebrate groups in the latest Cretaceous. On the basis of new data from the Hell Creek Formation (North Dakota) and Lance and Ferris Formations (Wyoming) and previously published data from Montana, we examined latest Cretaceous intraregional variability in this group, comparing these three areas to a large data set previously published from the Hell Creek Formation of Montana. Using discrete locality occurrences as an index of relative abundance, significant differences were found. Trionychids are the only group that evinces no differences among the four study areas. Baenids and kinosternoids are more common than expected in North Dakota. The pleurosternid *Compsemys*, chelydrids, and *Adocus* are notably rare in the Ferris Formation, and the terrestrial turtle *Basilemys* is more abundant there than expected. The Lance Formation differs in the greater relative abundance of uncommon taxa, i.e., chelydrids, the macrobaenid "*Clemmys*" *backmani*, kinosternoids, and *Basilemys*.

These differences are statistically independent of simple lithologic differences and thus appear to reflect underlying heterogeneity in the aquatic habitats of the latest Cretaceous that may not have obtained to the same degree in terrestrial habitats. The idea of heterogeneous aquatic habitats is consistent with depositional models indicating that the northern Great Plains area was dominated by dynamic fluvial systems that created an unstable landscape. Adaptation to fluctuating environments was probably a feature of latest Cretaceous aquatic vertebrate ecology and may be a factor in their greater relative success in surviving Cretaceous-Tertiary boundary events.

INTRODUCTION

Much research on latest Cretaceous faunas has focused on events leading up to and coincident with the Cretaceous-Tertiary boundary and their effects on vertebrate diversity. Extinction scenarios were crafted from analyses of changes in latest Cretaceous ecologic (e.g., Sheehan et al., 1991) and taxonomic (e.g., Archibald and Bryant, 1990) diversity, drawing upon abundance and richness data compiled from study of the records preserved in the Hell Creek Formation of North Dakota and Montana. Loss of primary productivity (Sheehan and Fastovsky, 1992), loss of range, habitat fragmentation, and seaway regression (Archibald and Bryant, 1990) have been identified as possible mechanisms for change in these scenarios. A shortcoming of these discussions, however, is a limited understanding of possible differences in occurrences or abundances on local to regional scales in the latest Cretaceous.

Sloan (1969) and Lehman (1987) examined dinosaurian

Holroyd, P.A., and Hutchison, J.H., 2002, Patterns of geographic variation in latest Cretaceous vertebrates: Evidence from the turtle component, *in* Hartman, J.H., Johnson, K.R., and Nichols, D.J., eds., The Hell Creek Formation and the Cretaceous-Tertiary boundary in the northern Great Plains: An integrated continental record of the end of the Cretaceous: Boulder, Colorado, Geological Society of America Special Paper 361, p. 177–190.

biogeography from the Arctic to the southern United States and recognized three distinct latest Cretaceous dinosaurian faunas. They recognized the northern Great Plains as a faunal unit distinct from that of the southern Rocky Mountain region and the Arctic based on the faunal dominance of *Triceratops*. Comparable regional variation in the latest Cretaceous flora owing to latitudinal gradients has been noted based on palynomorphs (Sweet et al., 1990; Nichols and Fleming, 1990; Nichols et al., 1992). How these floral communities relate to vertebrate faunal distribution is not known, but the larger scale groupings correspond well with those recognized in the dinosaurian faunas. Variability within these regions is reported only anecdotally (e.g., Nichols et al., 1992) or figuratively (e.g., Lehman, 1987, his Figs. 6 and 9) without discussion. A detailed characterization of faunal or floral variability within the Late Cretaceous of the northern Great Plains is lacking. These types of data are essential if we are to evaluate to what degree habitat change (e.g., range truncation, seaway regression) can be implicated among the mechanisms by which vertebrate diversity was affected along with impact-related effects.

Turtles are an excellent group in which to begin an exploration of biogeographic variability. Smaller scale biogeographic differences due to ecology can be documented in the fossil record based on either abundance counts or occurrence data (Hutchison, 1980; Bryant, 1989; Holroyd and Hutchison, 2000; Holroyd et al., 2001). Turtle fossils are virtually ubiquitous in Late Cretaceous deposits of the northern Great Plains. They occur in a variety of lithologies, and thus are less subject to the recovery biases inherent in studies of dinosaurians or small-bodied taxa such as mammals, amphibians, and lizards. Most elements of the shell and skull are diagnostic to family and typically to genus, providing greater taxonomic content than is available for studies based on dinosaurians (e.g., Sheehan et al., 1991). Although only some of the Cretaceous taxa have modern representatives (trionychids, chelydrids, kinosternoids), most others are adequately known to be able to infer their habitat preferences based on morphological criteria (Hutchison, 1984; Hutchison and Archibald, 1986).

In order to explore possible geographic variation, we examine the testudine faunas of four areas in the Late Cretaceous: the Hell Creek Formation in eastern Montana, the Hell Creek Formation in western North Dakota, the Lance Formation of eastern Wyoming, and the Ferris Formation of southeastern Wyoming. Data have been published for the Montana faunas (Hutchison and Archibald, 1986; Bryant, 1989) and for the Ferris Formation (Lillegraven and Eberle, 1999), and Estes (1964) briefly reported on the turtles of the Lance Formation. Here we provide an updated discussion of latest Cretaceous turtle systematics and ecology, expanded and updated data for the Ferris Formation and Lance Formation localities, and the first reports of the fauna from selected North Dakota localities, discuss the materials and methods used to compile abundance data, examine taphonomic biases, and analyze patterns of turtle abundance.

MATERIALS

Localities

This analysis is based on collections housed at the University of California Museum of Paleontology (UCMP) and collections of the Department of Geology and Geophysics, University of Wyoming (UW), made in four study areas shown in Figure 1: the Hell Creek area in Garfield and McCone Counties, northeastern Montana (UCMP); near Marmarth, Slope County, western North Dakota (UCMP); the Lance Creek area, Niobrara County, eastern Wyoming (UCMP); and the Hanna basin, Carbon County, southeastern Wyoming (UW). During the latest Cretaceous these areas were in the flood plains west of the Western Interior seaway (e.g., Lillegraven and Ostresh, 1988) that upsection became progressively farther inland as the sea retreated. Only UCMP or UW localities were used in the study; other collections are not comparable due to differences in collecting methodologies (e.g., see Sheehan et al., 1991). All areas were surface collected; limited quarrying was done in areas of bone abundance. In addition, some localities in both the Lance Formation and Hell Creek Formation in Montana were subjected to intensive screen washing, resulting in the recovery of a greater number of specimens of small body size. This latter bias has no discernable effect on turtle diversity.

The stratigraphy and vertebrate fauna of the Hell Creek area in Montana are previously published (e.g., Archibald, 1982; Hutchison and Archibald, 1986; Bryant, 1989). In Montana, only Late Cretaceous localities in the Hell Creek Formation are included; potentially Paleocene localities in the Bug Creek channels are not (see Lofgren, 1995, for a discussion of the Cretaceous-Tertiary [K-T] boundary in that area). The Montana data are from the same localities included in a diversity study by Hutchison and Archibald (1986). Lance Formation localities are those originally described by Estes (1964) and Clemens (1963, 1966, 1973). Localities and specimens from the Ferris Formation of the Hanna Basin are those described by Lillegraven and Eberle (1999). The vertebrate localities in North Dakota are those noted by Fastovsky (1987) from the Bradac Ranch, Marmarth Road, and Pretty Butte northeast areas or from sections described by Johnson et al. (this volume). Stratigraphic data for these localities, where available, are included in Table 1. Detailed locality information for these areas is on file at the appropriate institutions.

Because the faunas of these Hell Creek Formation localities in North Dakota have not been previously reported, lists and voucher specimens for these localities in North Dakota are provided (Table 2). In overall composition, the fauna recovered from these localities is similar to that from the better known Montana sections, although some genera seem to occur more commonly than in collections from Montana. The most common faunal elements after turtles are those of fish, typically preserved only as vertebra. Lepisosteids (gar) are common in the Hell Creek Formation. Both *Atractosteus* Rafinesque 1820

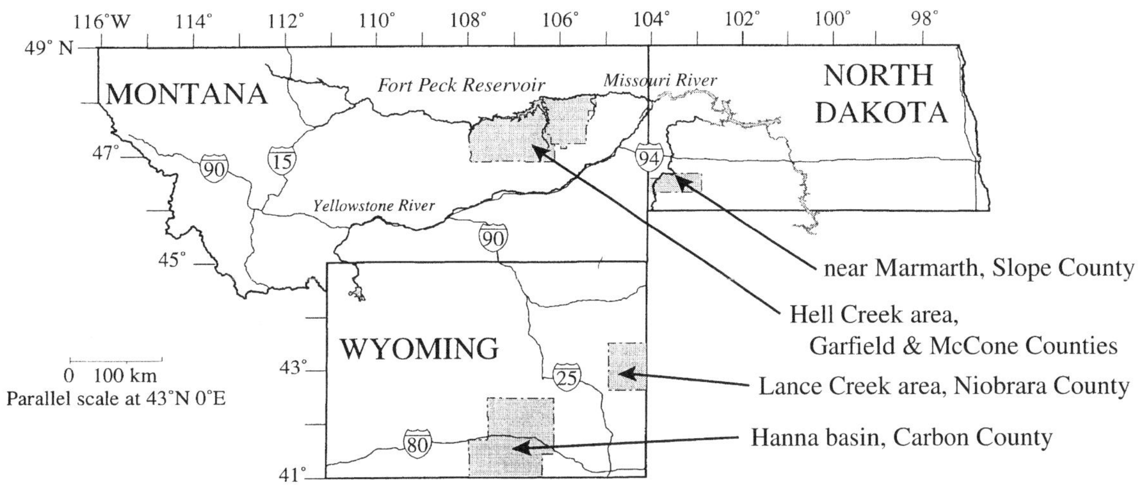

Figure 1. Map showing four study areas discussed here.

TABLE 1. STRATIGRAPHIC POSITIONS OF UCMP NORTH DAKOTA VERTEBRATE LOCALITIES

UCMP Loc. No.	Elevation (ft)	Elevation (m)	Datum elevation (ft)	Nearest KJ section	Position relative to formation contact (m)
V 86048				Dino Overlook	
V 86049	2760	841	2850	Pretty Butte	−27.4
V 86050				Conning Tower West	
V 86051				Conning Tower	
V 86063			2850	Pretty Butte	
V 86065				Conning Tower North	
V 86066				Conning Tower Northwest	
V 86071	2810	857	2900	Bobcat Butte	−24.4
V 86072	2815	858	2900	Bobcat Butte	−24.4
V 86073	2810	857	2900	Bobcat Butte	−24.4
V 86074	2800	854	2900	Bobcat Butte	−33.8
V 86075	2780	848	2900	River Section	−36.5
V 86076	2810	857	2900	Bobcat Butte	~ −24.4
V 86095	2860	872	2900	Bobcat Butte	−12
V 86096	2780	848	2850	Pretty Butte	−32.99
V 86097	2780	848	2850	Pretty Butte	−47.59
V 86098	2760	841	2850	Pretty Butte	−27.4
V 86099	2780	848	2850	Pretty Butte	−22
V 86100	2780	848	2850	Pretty Butte	−22
V 86101	2800	854		Pretty Butte	−44.59
V 91091	2840	866	2900	Bobcat Butte	−18.3
V 91092	2810	857		West of Town	−80
V 98035	2800	854	2900	Bobcat Butte	−33.8

Note: UCMP—University of California Museum of Paleontology. Meter levels from K. Johnson [KJ] (written commun.) and correlated to his stratigraphic sections (Johnson, this volume).

and *Lepisosteus* Lacepède 1803 are recorded from the formation (Bryant, 1989). Wiley (1976) recognized these taxa as distinct genera, but they are generally regarded by neoichthyologists (Eschmeyer, 1990) as separate subgenera of the genus *Lepisosteus* that are principally differentiated by body size. The generally large size of specimens from the Hell Creek of North Dakota suggests that these are probably all or nearly all *Lepisosteus* (*Atractosteous*). In addition to lepistosteids, amiids (bowfins) are relatively common fossils, and most are attributable to the large-bodied *Melvius* Bryant 1987. The smaller amiid *Cyclurus* Agassiz 1844 (= *Kindleia* Jordan 1927, see Grande and Bemis, 1998) is also known. All of the fish taxa recovered are from freshwater groups that possess a modified air bladder, in addition to gills, that enable them to use surface air and to live in polluted or stagnant water unsuitable for most other fish.

Crocodylian remains are also relatively common in the Marmarth, North Dakota, area, and are typically represented by isolated teeth or osteoderms. Much of the identifiable remains are attributable to the alligatoroid *Brachychampsa montana* Gilmore 1911 including a virtually complete skeleton (UCMP 133901) whose skull is described in Norell et al. (1994). A

TABLE 2. NONMAMMALIAN VERTEBRATE OCCURRENCES IN THE HELL CREEK FORMATION OF WESTERN NORTH DAKOTA BY LOCALITY

Locality	Occurrences
V 85051:	177965, Saurischia
V 86048:	177967–177968, *Adocus*; 177966, *Neurankylus*; 177969–177970, Trionychidae; 172452, Saurischia
V 86049:	177971–177972, Lepisosteidae; 177998–177999, *Champsosaurus*; 133901, 177996–177997, *Brachychampsa montana*; 177985–177994, *Adocus*; 177977–177984, Baenidae; 177973, 177976, *Compsemys victa*; 177974–177975, 178002, Trionychidae; 172451, Ceratopsidae; 178000, Saurischia
V 86050:	177920–177924, *Lepisosteus*; 177925, *Melvius*; 177935, *Champsosaurus*; 177933–177934, Crocodylia indet.; 177931–177932, *Adocus*; 177930, Baenidae; 177928, *Helopanoplia distincta*; 177926–177927, Trionychidae indet.; 177929, Kinosternoidae; 177937, Ceratopsidae; 177936, Hadrosauridae; 177938, Saurischia
V 86051:	177939, *Atractosteus*; 177940–177942, *Melvius thomasi*; 177961, Crocodylia indet.; 177949–177955, *Adocus*; 177956–177957, *Neurankylus*; 177958–177959, *Compsemys victa*; 177946–177948, *Helopanoplia distincta*; 177943–177945, Trionychidae indet.; 177960, Kinosternoidae; 177962, Hadrosauridae; 177963, Saurischia
V 86063:	172263, 172296–172297, 172327, 172351, *Myledaphus*; 172265, 172290, 172298, 172341, *Melvius*; 172299, 172328 Amiidae indet.; 172261–172262, 172300, 172342, Lepisosteidae; 172264, 172322, 172349, Urodela; 172259, 172266, 172289, 172301, 172332, 172336, *Champsosaurus*; 172302, 172333, Alligatoridae indet.; 172260, 172292, 172303, 172325, 172334, 172343, Crocodylia indet.; 172287, *Adocus*; 172257, *Palatobaena*; 136554, *Plesiobaena antiquus*; 172285, 172337, *Neurankylus*; 172256–172267, 172280–172283, 172313–172319, 172329, 172338, 172348, 172353, Baenidae indet.; 172258, *Protochelydra*; 172284, 172308, 172347, Chelydridae indet.; 172307, "*Clemmys*" *backmani*; 172268, 172293, 172339, *Compsemys victa*; 172286, 172294–172295, 172324, *Basilemys*; 172255, 172269, 172288, 172309–172311, 172330–172331, 172340, 172346, Trionychidae; 172312, Chelonia indet.; 172271, 172274, 172305, Ceratopsidae; 172272–3, 172304, 172321, 172335, 177912, Hadrosauridae; 172270, 172275–9, 172306, 172320, 172344–5, Saurischia indet.
V 86064:	131769, *Leidyosuchus sternbergi*
V 86065:	172119, *Cyclurus*; 172110, *Melvius thomasi*; 172115, 172120, Lepisosteidae; 172109, Urodela; 172121, *Champsosaurus*; 172122, Crocodylia; 172111, *Adocus*; 172123, Chelydridae; 172112, *Compsemys victa*; 172113–172114, 172124–172125, Trionychidae; 172126, Chelonia indet; 172127, Ceratopsidae; 172116, Hadrosauridae; 172117, Saurischia
V 86066:	172104, *Atractosteus*; 172107, *Adocus*; 172106, Baenidae; 172108, Trionychidae
V 86071:	172103, Lepisosteidae; 172100, *Champsosaurus*; 172099, *Brachychampsa montana*; 172096–172097, *Protochelydra*; 172094–172095, Chelydridae; 172091–172093, Trionychidae; 172098, Chelonia indet.; 172090, *Triceratops*; 172101, Ceratopsidae; 172102, Saurischia
V 86072:	172131, Amiidae; 172130, Caudata; 172132, *Champsosaurus*; 172133, *Brachychampsa montana*; 172134, Chelydridae; 172135–6, Trionychidae; 172137, Ceratopsidae
V 86073:	172138, *Cyclurus*; 172139, *Atractosteus*; 172141, *Champsosaurus*; 172140, Crocodylia; 177915, *Neurankylus*; 172142, Baenidae; 172143, *Protochelydra*; 172144, *Helopanoplia distincta*; 172145, 172148, Trionychidae; 172146, Ceratopsidae; 172147, Pachycephalosauridae
V 86074:	172239, Amiidae; 172240, Lepisosteidae; 172234, 172242, 177964, *Champsosaurus*; 172233, 172241, Crocodylia; 172238, 172245, Baenidae; 172237, *Protochelydra*; 172246, Chelydridae; 172235–172236, 172247–172248, Trionychidae; 172232, Ceratopsidae; 172230–172231, 172244, Hadrosauridae; 172243, Saurischia;
V 86075:	172410, *Cyclurus*; 131681, 131707–131708, 172411, 172434, *Melvius thomasi*; 172409, 172435, 172450, Lepisosteidae; 172405, 172427, *Palaeosaniwa canadensis*; 172402, 172449, *Champsosaurus*; 172399, Alligatoridae; 172401, 172428, 172448, Crocodylia; 172396, 172418, 172442, 177913, *Adocus*; 172420, *Emarginochelys cretacea*; 131705, 132057–132058, 136031, 172444, 177914, *Neurankylus*; 131703, 172397, 172421, 172445, 177917–177918, Baenidae; 172390, 172412, *Protochelydra*; 172425–172426, 172446, Chelydridae; 172395, 172417, *Basilemys*; 172398, 172422, 172441, *Compsemys victa*; 172391, 172413, 177919, *Aspiderites*; 177916, *Helopanoplia distincta*; 172392–172394, 172416, 172423–172424, 172443, Trionychidae; 172388–172389, Chelonia indet.; 172415, 172419, 172436–172440; 172403, 172433, Ceratopsidae; 172404, 172432, Hadrosauridae; 172400, Pachycephalosauridae; 172429, Theropoda; 172406–172407, 172414, 172430–172431, 172447, Saurischia
V 86076:	172251, *Champsosaurus*; 172252, Crocodylia; 172229, Baenidae; 172253–172254, Trionychidae; 172250, Saurischia
V 86095:	172185, *Melvius*; 177883, Amiidae; 172199, 177879, Lepisosteidae; 177882, *Piceoerpeton*; 172186, 177884, Urodela; 172184, 177907, Lacertilia; 172194, *Brachychampsa montana*; 172191, 177908, Crocodylia; 172189, 172197, 177905, 177906, *Adocus*; 172465, *Emarginochelys cretacea*; 177904, *Neurankylus*; 172187–172188, 177899–177903, Baenidae; 177889–177890, Chelydridae; 172198, 177893–177897, *Compsemys victa*; 172195, *Helopanoplia distincta*; 172190, 177885–177888, 177898 Trionychidae; 172196, 177880–177881, 177891–177892, Kinosternoidae, 172193; 172192, Hadrosauridae; 177911, Saurischia
V 86096:	172183, *Cyclurus*; 172182, 172201, *Melvius*; 172218, Amiidae; 172181, 172200, Lepisosteidae; 172220, Urodela; 172211, 172224, *Champsosaurus*; 172207, Crocodylia; 172177, 172204, *Adocus*; 172214, *Plesiobaena*; 172179, 172203, 172213, Baenidae; 172175, 172205, 172216, Chelydridae; 172178, *Basilemys*; 172212, "*Clemmys*" *backmani*; 172180, 172202, 172221, *Compsemys victa*; 172176, 172206, 172215, 172222–172223, Trionychidae; 172208–172209, Chelonia; 172171, 172210, Ceratopsidae; 172172, 172217, Hadrosauridae; 172173, Saurischia
V 86097:	172227, *Brachychampsa montana*; 172226, Baenidae; 172225, Saurischia; 172228, *Tyrannosaurus rex*
V 86098:	177878, Crocodylia; 177877, *Adocus*; 177876, *Aspiderites*
V 86099:	177851, *Cyclurus*; 177850, *Melvius*; 177852, Lepisosteidae; 177853, Caudata; 172374, 177870, *Champsosaurus*; 177871, *Brachychampsa montana*; 172372, Crocodylia; 177856, *Adocus*; 177864, *Neurankylus*; 172373, 177854, *Basilemys*; 177855, *Compsemys victa*; 177857–177860, 177862–177863, Chelydridae; 172371, 172479–172481, 177865–177867, Trionychidae; 177861, Chelonia; 172375, 177868–177869, 177875, Ceratopsidae; 172482, Pterosauria
V 86100:	172159, *Melvius thomasi*; 172158, *Brachychampsa montana*; 172157, Trionychidae
V 86101:	172358, *Myledaphus*; 172160, 172359, *Melvius thomasi*; 172360, 172383, Amiidae; 172161, Lepisosteidae; 172355, Urodela; 137357, Anura; 172356, *Champsosaurus*; 172162, 172378, Crocodylia; 172166, *Adocus*; 137358, *Palatobaena*; 137356, 172164–172165, 172361–172362, 172381, Baenidae; 172168, *Basilemys*; 172167, 172382, *Compsemys victa*; 172169, 172354, 172379, Trionychidae; 137359, 137360, Kinosternoidae; 172163, Chelonia; 172363, *Triceratops*; 172170, 172376, Ceratopsidae; 172369, Hadrosauridae; 172366, *Tyrannosaurus rex*; 172364–172365, 172368, Saurischia; 172367, 172377, Aves
V 91091:	172384, Baenidae; 172385, *Protochelydra*; 172386–7, Trionychidae
V 91092:	172456, Amiidae; 172454, *Neurankylus*; 172453, 177846, Baenidae; 177845, *Helopanoplia distincta*; 172455, Trionychidae;
V 98035:	172151, *Myledaphus bipartitus*; 172152, *Champsosaurus*; 177848, *Neurankylus*; 172153, Baenidae; 177849, Chelydridae; 177847, *Aspiderites*; 172154, Trionychidae; 172155, Saurischia;
V 98036:	172150, *Melvius thomasi*; 172149, *Basilemys*

Note: All voucher specimens (indicated by specimen numbers) are from the collections of the University of California Museum of Paleontology (UCMP).

general survey of the Montana and North Dakota collections in the UCMP suggests that *Brachychampsa* is relatively more common in the North Dakota section, while the slightly smaller crocodilian *Borealosuchus* Brochu 1997 is more common in Montana. The greater abundance of the large baenid *Neurankylus* Lambe 1902, large amiids, and the large crocodylian *Brachychampsa* may reflect the greater proximity of the North Dakota deposits to base level and, consequently, increased frequency of open bodies of water.

Fragmentary remains of rays, amphibians, saurians, dinosaurians, a bird, and a pterosaur were also recovered. However, few of these are diagnostic and are poorly represented in our surface collections. A more detailed accounting of these taxa from North Dakota is presented elsewhere (Pearson et al., this volume).

Late Cretaceous turtle fauna, taxonomy, and ecology

The turtle taxa encompassed by this study and their inferred ecological preferences are summarized in Tables 3 and 4, and illustrated in Figure 2. Since last detailed (Hutchison and Archibald, 1986; Bryant, 1989) the systematics of Cretaceous turtles have been considerably revised in light of a better understanding of phylogenetic relationships. These changes and their effect on ecologic interpretations are summarized here.

Pleurosternidae. The genus *Compsemys* Leidy 1856 is the sole representative of this family in the Late Cretaceous of North America. It was originally placed in the Dermatemydidae by Hay (1908), Baenidae by Gaffney (1972), and incertae sedis by Hutchison and Archibald (1986). Hutchison (1984) followed Gilmore (1919) and placed it in the Pleurosternidae as a specialized and terminal member. The shell (Fig. 2A) is moderate sized, high domed, and has a fine punctuate sculpture and full plastron. The shell shape generally resembles that of the extant emydid *Emydoidea* Gray 1870, but the skull functionally resembles the extant platysternid *Platysternon* Gray 1831, a carnivorous turtle. The Montana specimen indicates a carapace length of about 23 cm, but shell fragments from the Hell Creek Formation of North Dakota indicate a significantly larger size.

TABLE 3. TURTLE TAXA KNOWN FROM LATEST CRETACEOUS OF THE NORTHERN GREAT PLAINS WITH UPDATED TAXONOMIC PLACEMENT

- Cryptodira
  - Pleurosternidae
    - *Compsemys victa* Leidy 1856—F, L, M, ND (Whitmore and Martin, 1986)
  - Baenidae
    - Neurankylinae
      - *Hayemys latifrons* (Hay 1908) (Hell Creek Fm. of Wyoming)
      - *Neurankylus* cf. *N. eximius* Lambe 1902—F, M
    - Baeninae
      - *Plesiobaena antiqua* (Lambe 1902)—L, M
      - *Stygiochelys estesi* Gaffney 1972—M
      - *Eubaena cephalica* (Hay 1908)—L, M
      - *Palatobaena* sp. nov. (= *P. bairdi* of Gaffney, 1972, in part); Archibald and Hutchison, 1979)—M, ND (this paper)
      - *Thescelus insiliens* Hay 1908—L, M
      - "*Baena*" *hatcheri* Hay 1901—L
      - "*Baena*" *marshi* Hay 1904—L
      - "*Baena*" *hayi* Gilmore 1916—L
  - Kinosternoidae incertae sedis
      - Kinosternoidae, 2 genera—M, ND
      - *Emarginochelys cretacea* Whetstone 1978—M, ND
  - Adocidae
    - *Adocus* sp.—F, L, M, ND
  - Nanhsuingchelyidae
    - *Basilemys sinuosa* Riggs 1906—F, L, M, ND
  - Trionychidae
    - Trionychinae
      - *Apalone* sp.—M
      - "*Aspideretes*" s. l.—L, M, ND
    - Plastomeninae
      - *Helopanoplia distincta* Hay 1908—L (Whitmore and Martin, 1986), M, ND
      - Plastomenine A—M, ND
      - Plastomenine C—M, ND
    - Trionychidae indent.—L (Whitmore and Martin, 1986)
  - Macrobaenidae
    - "*Clemmys*" *backmani* Russell 1935—L, M, ND
  - Chelydridae
    - *Protochelydra* sp. (Chelydridae gen. indet., in part of Hutchison and Archibald, 1986)—M, ND

Note: Geographic data from this report, Hutchison and Archibald (1986), or as noted. Abbreviations: F = Ferris Formation, Wyoming; L = Lance Formation, Wyoming; M = Hell Creek Formation, Montana; ND—Hell Creek Formation, North Dakota.

TABLE 4. ADULT CARAPACE LENGTH AND INFERRED HABITAT AND DIET

Taxon	Carapace Length (cm)	Habitat	Diet
Compsemys victa	23	Aquatic	Carnivorous
cf. *Neurankylus*	43.5	Aquatic	Omnivorous?
Thescelus insiliens	40*	Aquatic, channel	Carnivorous?
Plesiobaena antigua	29*	Aquatic, channel	Carnivorous
Eubaena cephalica	31	Aquatic, channel	Carnivorous
Stygiochelys estesi	30*	Aquatic, channel	Carnivorous
Palatobaena sp.	?	Aquatic, channel	Molluscivorous
Emarginochelys cretacea	35	Aquatic, channel	Carnivorous
Kinosternia (*Hoplochelys-Agomphus* morph)	30*	Aquatic	Carnivorous?
Kinosternia (Kinosternid-morph)	15*	Aquatic	Carnivorous?
Adocus sp.	58*	Aquatic	Herbivorous
Basilemys sinuosa	92	Terrestrial	Herbivorous
Apalone sp.	29.5	Aquatic	Carnivorous
"*Aspideretes*" s. l.	59*	Aquatic	Carnivorous
Helopanoplia distincta	48*	Aquatic	Carnivorous
Plastomeninae type A	16.5	Aquatic, shallow	Carnivorous
Plastomeninae type C	18	Aquatic, shallow	Carnivorous
Protochelydra sp.	21*	Aquatic	Carnivorous
"*Clemmys*" *backmani*	60	Aquatic	Herbivorous

Note: *Estimated values; complete adult shells not known.

Only one species, *Compsemys victa* Leidy 1856, is currently recognized (Gaffney, 1972) and ranges from Campanian to Torrejonian. It is unclear whether the size difference observed within the Hell Creek Formation is due to ecological or taxonomic distinctions.

Baenidae. This extinct family is the most diverse represented in the latest Cretaceous of the Great Plains area, comprising six to seven genera. All of these appear to be carnivorous or molluscivorous. We choose to recognize two major groups within this family, the Neurankylinae and the Baeninae.

Neurankylinae. This group was proposed as a family by Williams (1950, p.555) to include the genus *Neurankylus* Lambe 1902, then known from a shell and some postcranial elements. Gaffney (1972), in his revision of the Baenidae, included it within the Baenidae, as did Hutchison and Archibald (1986) and Brinkman and Nicholls (1993). Gaffney and Meylan (1988) placed it in the basal dichotomy of the Baeninae (on the basis of *Hayemys* Gaffney 1972 based on *Eubaena latifrons* Hay 1908), and Brinkman and Nicholls (1993) altered this assessment only by placing *Neurankylus* as more basal, although that rests on a single dubious character state (i.e., greater emargination of the temporal, a feature not supported by the figures of *Hayemys* and *Neurankylus* skulls; cf. Brinkman and Nicholls, 1993, their Fig. 1, with Fig. 6 of Gaffney, 1972). Because the Baeninae, excluding *Hayemys* and *Neurankylus* (Baenodd of Gaffney and Meylan 1988), form a clearly contemporaneous monophyletic group, we choose to recognize the subfamily group name Neurankylinae (Neurankylidae of Williams, 1950) for these generally more primitive but vexatious turtles. Hutchison and Archibald (1986) provisionally included *Hayemys*, known from a single isolated skull, in *Neurankylus* based on *Neurankylus*-like shells with *Hayemys*-like skull fragments. Brinkman and Nicholls (1993) clarified the systematics of the type species *Neurankylus eximius* Lambe 1902 based upon subsequently discovered skull material and noted that the skull of *Hayemys* cannot be referred to *N. eximius*. However, they noted, as Hutchison and Archibald (1986) suggested, that these two taxa are closely related. The question remains as to whether some or all the material referred to *N.* cf. *N. eximius* by Hutchison and Archibald (1986) represents *Neurankylus*, *Hayemys*, or both. This question will remain unresolved pending a review of the Hell Creek neurankylines. Neurankylines are moderately large turtles (Fig. 2B), the skull and jaws of which display relatively simple triturating surfaces, suggesting a carnivorous or omnivorous diet.

Baeninae. This group is the Baenodd of Gaffney and Meylan (1988). Five genera are represented in the Hell Creek Formation by single species, and three species placed in the form-genus "*Baena*" are additionally known from the Lance Formation. *Thescelus insiliens* Hay 1908 is rare and represented only by its shell, and its systematic position within baenids is unclear. *Plesiobaena antiqua* Lambe 1902 (Fig. 2C) is known from both cranial and shell material and is considered the sister taxon of the tribe Baenini (Gaffney and Meylan, 1988). Three eubaenines (subtribe Eubaenina of Gaffney and Meylan, 1988) occur in the formation: *Eubaena cephalica* Hay 1904 and *Stygiochelys estesi* Gaffney 1972 are known from skeletons, whereas *Palatobaena* Gaffney 1972 is known only from cranial material, although some nominal shell taxa may eventually be assignable to this taxon. Gaffney (1972), Archibald and Hutchison (1979), and Hutchison and Archibald (1986) originally assigned the *Palatobaena* species to *P. bairdi* Gaffney 1972, but more complete material (UCMP 131953) suggests that the Cretaceous species is probably distinct from *P. bairdi*. "*Baena*" *hatcheri* Hay 1901 (Fig. 2E), "*Baena*" *marshi* Hay 1904, and "*Baena*" *hayi* Gilmore 1916 are known only from shells. The

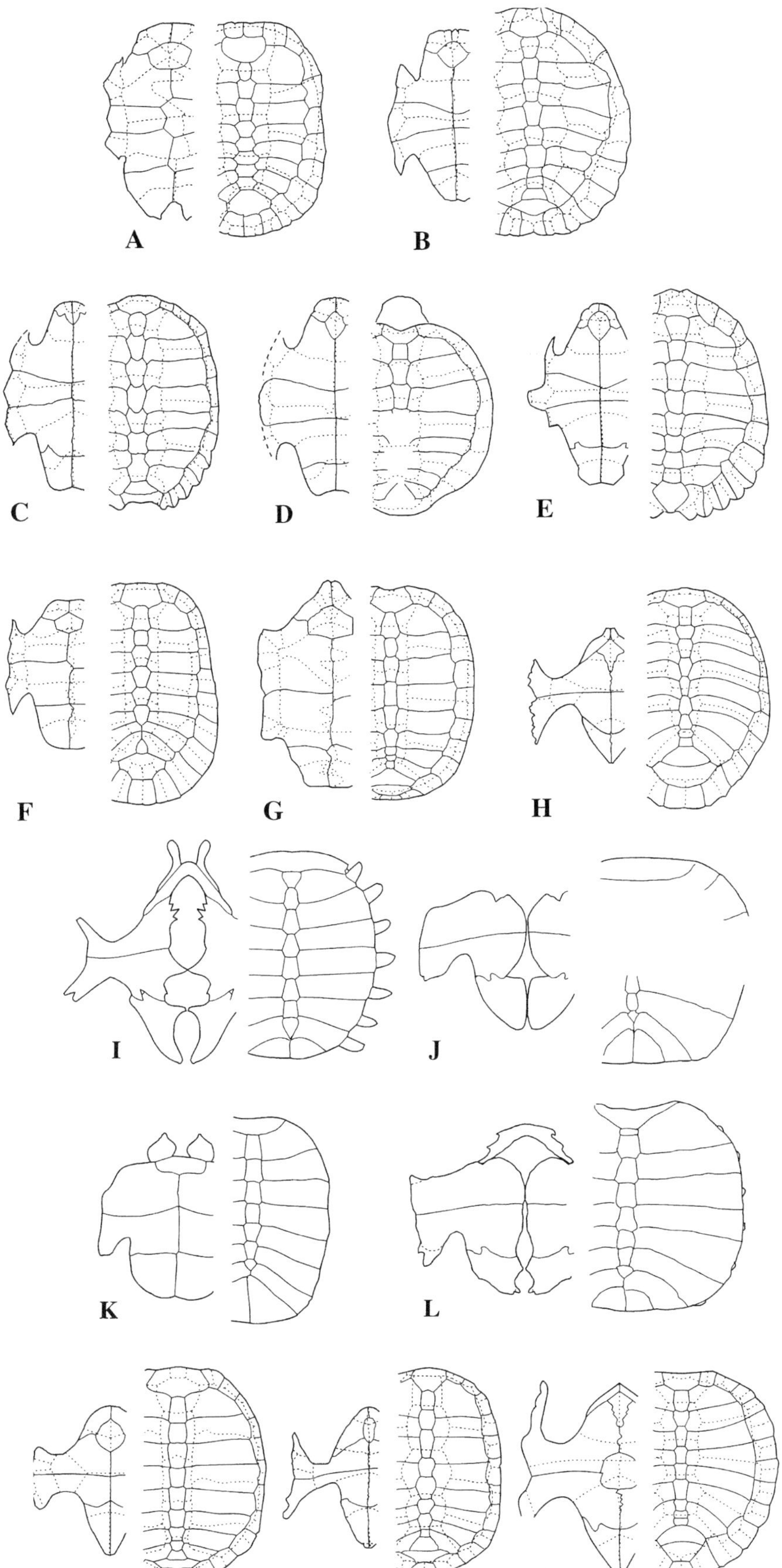

Figure 2. Ventral views of right half of plastron (left) and dorsal views of right half of carapace of selected latest Cretaceous turtles representing families recovered from Hell Creek and Lance Formations. Images sized to standardized carapace length; see Table 4 for actual lengths. A: Pleurosternidae, *Compsemys victa* after Gaffney (1972). B: Baenidae, *Neurankylus eximius* after Gaffney (1972) and University of California Museum of Paleontology (UCMP) 117103. C: Baenidae, *Plesiobaena antiqua* modified from Gaffney (1972) and after UCMP 117101. D: Baenidae, *Thescelus insiliens* after Hay (1908). E: Baenidae, eubaenine "*Baena*" *hatcheri* modified after Gaffney (1972). F: Adocidae, *Adocus punctatus* after Hay (1908). G: Nanhsuingchelyidae, *Basilemys sinuousa* Riggs, 1906, after Hay, 1908. H: Chelydridae, *Protochelydra* sp., based on *P. zangerli* (Paleocene species), Hutchison (1998), and UCMP 129693 and 136481. I: Trionychidae, *Aspideretoides foveatus* after Gardner et al. (1995). J: Trionychidae, *Helopanoplia distincta* based on UCMP 128875, 130112, 130113, 131614, 134502, and 172470. K: Trionychidae, plastomenine A, based on UCMP 130000. L: Trionychidae, plastomenine C, based on UCMP 130100 and 133900. M: Kinosternia, family incertae sedis, *Hoplochelys*, based on Paleocene species *H. bicarinata*, after Gilmore (1919). N: Kinosternia, family incertae sedis, *Emarginochelys cretacea* after Whetstone, 1978, and University of Kansas 23488. O, Macrobaenidae, "*Clemmys*" *backmani* after Russell (1935).

distribution of the Baeninae indicates a distinct preference for stream channels (Hutchison, 1984).

Kinosternoidae. This group includes two extant families (Dermatemydidae and Kinosternidae) and a number of extinct taxa that form the sister taxon of the Trionychoidae (Meylan and Gaffney, 1988). At least three members of this group are represented in the Hell Creek and Lance Formations but only one is known from a complete shell. *Emarginochelys cretacea* Whetstone 1978 is the largest member (Table 4) of this group in the Hell Creek Formation. Whetstone (1978) incorrectly reconstructed the shell because of lateral crushing; the shell (Fig. 2M) generally resembles that of chelydrids, with which it was originally classified. Meylan and Gaffney (1988) removed it to the Kinosternoidae as a basal member. The skull morphology indicates a carnivorous turtle.

An unnamed, moderate-sized turtle with a thick and high-domed carapace resembling that of *Hoplochelys* Hay 1908 and *Agomphus* Cope 1871 is only represented by shell fragments. The former genus is well known from complete shells (e.g., Fig. 2N) in Paleocene freshwater deposits of the southwest (Gilmore, 1919) and the latter appears to be a brackish or salt-water turtle (Hutchison and Weems, 1999). Skulls are unknown from any of these, so little can be deduced about feeding preference except to note that other known members of the group are carnivorous.

The smallest representative of the Kinosternoidae (Kinosternidae gen. indet. of Hutchison and Archibald, 1986) is also represented only by shell fragments and remains unnamed. These fragments lack sculpturing, and their shape indicates a shell with a cruciform plastron similar to the Eocene kinosternid *Baltemys* Hutchison, 1991. These lack at least one autapomorphy of the crown group Kinosternidae (Kinosterninae and Staurotypinae), and therefore are here regarded as representing a sister taxon of the crown group. All known kinosternids are primarily carnivorous (piscivores, insectivores, molluscivores, scavengers).

Adocidae. This extinct family is represented by *Adocus* Cope 1868, a large turtle (Table 4) with a hydrodynamic, delicately sculptured carapace and broad plastron (Fig. 2F). The skull has complex triturating ridges (Meylan and Gaffney, 1989, their Figs. 3–5) and the shell resembles the extant genus *Dermatemys* Gray 1870, with which it was previously placed (Hay, 1908). Meylan and Gaffney (1989) amply demonstrated that *Adocus* is convergent with *Dermatemys* and retained the genus within its own family. The functional resembles of *Adocus* and *Dermatemys* indicate that *Adocus* was a fully aquatic herbivore.

Nanhsuingchelyidae. The only member of this extinct family in the Hell Creek Formation is *Basilemys sinuosa* (including *B. praeclara* Hay 1910). *Basilemys* Hay 1902 is the largest turtle, as well as the only terrestrial one in the latest Cretaceous (Table 4). The shell (Fig. 2G) strongly resembles that of large testudinids in general shape (Langston, 1956); the limbs are armored (UCMP 130378), and the skull has complex triturating surfaces resembling tortoises (Brinkman, 1998). The latter indicates a herbivorous diet. Shell fragments of *Basilemys* are readily identifiable due to their distinctive surface sculpture. *Basilemys* is notable among the Late Cretaceous turtle fauna as one of the two taxa to become extinct at the K-T boundary (Hutchison and Archibald, 1986).

Trionychidae. This is the second most diverse family in the Hell Creek Formation (Hutchison and Archibald, 1986), and includes at least six taxa. These soft-shelled turtles exhibit a wide range of size and shape, although all known trionychids are carnivorous. Meylan (1987) revised the extant members of the family and partitioned taxa previously included with *Trionyx* Geoffroy Saint Hilaire 1809 into distinct genera. Gardner et al. (1995) clarified the systematics of the North American Campanian species and erected a new genus, *Aspideretoides*, for some species previously included in *Aspideretes*, and included others in the extant genus, *Apalone* Rafinesque 1832. While Meylan (1987) restricted the use of *Aspideretes* Hay 1904 to a few extant African species, the genus is used here as a form genus and enclosed in quotations. Specimens referred to *Aspideretes* in Hutchison and Archibald (1986) and *"Aspideretes"* sensu lato are here considered to belong within a group including *"Aspideretes"* and *Aspideretoides*. The systematics of the Late Cretaceous North American representatives of this group need revision. Functional morphology in this group is similar to that in extant *Apalone*, and *Aspideretoides* (Fig. 2I) is shown as an example.

Only one specimen from Montana reported by Hutchison and Archibald (1986) was referred to *Trionyx* (*Trionyx*). This remains the only specimen in the Hell Creek Formation that we recognize in this group. However, following Meylan's (1987) revision of the modern taxa, it may, on geographic bases, be better to refer it to *Apalone*, the only extant North American genus. *Apalone* is a fully aquatic, carnivorous, subaquatic burrowing turtle.

The extinct genus *Helopanoplia* Hay 1908 is represented only by its type species, *H. distincta* Hay 1908. The general conformation of the shell (Fig. 2J) resembles *"Aspideretes"* sensu lato and *Apalone*, but has a distinctive punctate sculpture on the plastron and more fully ossified plastron, thus showing some resemblance to plastomenines. This genus and species is the only aquatic turtle in the samples that is known to become extinct at the K-T boundary.

Two small unnamed plastomenines are also represented in the fauna. These are characterized by a well-ossified plastron with akinetic xiphiplastra and entoplastra and relatively thick shells with little or no free rib extensions from the costals. "Plastomenine" type A (Fig. 2K) is unusually high domed, ovate, and well ossified, with peripheral ossification in the area of the plastron-carapace contact. "Plastomenine" type C (Fig. 2L) is more circular, lower domed, and less extremely ossified.

Macrobaenidae. This extinct and primarily Asiatic family is represented in the Hell Creek and Lance Formations by *"Clemmys" backmani* Russell 1935. These are large turtles (Table 4) with a relatively thin and unsculptured carapace and cruciform plastron (Fig. 2O). The hydrodynamic shell, incompletely ossified cruciform plastron, and long digits indicate a fully aquatic turtle. The presence of sharp, parallel triturating ridges on the dentary indicate a probable herbivore.

Chelydridae. This extant family is represented by at least one genus close to or within the concept of *Protochelydra* Erickson 1973. The shell (Fig. 2H) resembles that of the macrobaenid *"Clemmys" backmani* in general shape, but is smaller (Table 4), more fully ossified, and relatively thicker. Chelydridae (snapping turtles) are generally carnivorous.

Occurrences

Data on occurrences of turtles in the Hell Creek Formation are taken from Hutchison and Archibald (1989) for Montana and from Table 2 for North Dakota. For each locality all turtle specimens were examined and identified to genus or the lowest possible taxonomic level above genus. In many cases, identification to genus is not possible due to the incompleteness or fragmentary nature of the specimens. Identification only to family level is often the case with trionychids, baenids, and kinosternians. In the case of trionychids and baenids, the high diversity and generally conservative shell morphology make identification of isolated elements to genus difficult. In the case of kinosternians, most genera are not yet described, and assignment of isolated elements would be premature. New turtle faunal lists were generated from existing collections of the Lance Formation (Table 5); some of these records were reported in Estes (1964) and are here updated. Most represent new site records. For the Ferris Formation, one of us (Hutchison) reexamined the voucher specimens published by Lillegraven and Eberle (1999) and additional material. Some material was reidentified; Table 6 is an updated and expanded list for the Ferris Formation. Although a large number of turtle taxa are known from the latest Cretaceous, the turtle fauna is fairly homogeneous in the four study areas when examined at this level. Species-level or even genus-level differences in distribution may exist, but were not detected. Thus, the principal way in which variation in turtle distribution can be examined is through differences in relative abundance.

ANALYSIS AND RESULTS

Relative abundance can be measured in a number of ways. For this study, simple counts of locality occurrences (i.e., presence and absence) were compiled for the four study areas. These occurrence data provide an index of how widely distributed and common a taxon was in the area sampled. Under other conditions, counts of minimum number of individuals (MNI) may give a better indication of assemblage structure (Gilinsky and Bennington, 1994); however, MNI counts in turtles can reflect strong local biases. Most living turtles congregate in ponds or oxbow lakes, and single-species–dominated fossil assemblages in highly localized areas are not uncommon (personal observation). Such assemblages can skew the interpretation of MNI counts. MNI counts for turtles also seem to be more strongly affected by taphonomic factors (see following)

TABLE 5. VOUCHER SPECIMENS FOR TURTLES FROM LANCE FORMATION LOCALITIES

Locality	Vouchers
V 5003:	173393, *Compsemys*; 173394, Trionychidae; 72932, cf. *Helopanoplia*; 95910, Chelydridae; 173410, *Adocus*; 173392, *Basilemys*
V 5616:	72895, *Compsemys*; 72904, cf. *Stygiochelys*; 73027, Trionychidae; 173369, *Aspiderites*; 72907, Chelydridae; 72902, *Basilemys*; 72935, Baenidae
V 5617:	173371, Chelydridae; 173370, Trionychidae
V 5620:	72897, *Compsemys*; 49759, *Plesiobaena*; 73028–29, Trionychidae; 173441, Plastomeninae; 173440, *Helopanoplia*; 125776–788, Chelydridae; 173439, *Adocus*; 72901, *Basilemys*; 125762, Kinosternoidae; 72936, Baenidae
V 5621:	173446, Trionychidae
V 5709:	173374, *Compsemys*; 173375, Trionychidae; 173409, *Helopanoplia*; 173373, Chelydridae
V 5711:	72898, *Compsemys*; 72894, *Eubaena*; 73030, Trionychidae; 173438, *Aspiderites*; 173418, Plastomeninae; 125773–775, Chelydridae; 125768–772, *Emarginochelys*; 173412, *Adocus*; 72903, *Basilemys*, 173415, cf. Macrobaenidae; 173416, Kinosternoidae
V 5714:	72913, Trionychidae; 173376, *Helopanoplia*;
V 5718:	72900, *Basilemys*
V 5815:	72896, *Compsemys*; 73026, Trionychidae; 173420, Chelydridae; 173422, Kinosternoidae; 173424, Baenidae
V 5817:	173370, Trionychidae; 173426, Plastomeninae; 170378, Chelydridae;
V 5819:	54215, "*Clemmys*" *backmani*; 124395, Baenidae
V 5820:	173437, Trionychidae; 173435, Plastomenine A; 173436, *Helopanoplia*; 173433, Chelydridae; 173434, *Basilemys*; 173432, Baenidae
V 5875:	62940, Baenidae
V 65238:	72899, *Compsemys*; 73032, Trionychidae; 73042, *Aspiderites*; 173431, Plastomenine A; 173379, *Helopanoplia*; 125789, Chelydridae; 173428, *Adocus*; 73088, *Basilemys*, 173427, "*Clemmys*" *backmani*; 73039, Kinosternoidae, 173380, Baenidae

Note: Vouchers listed by locality. All voucher specimens (indicated by specimen numbers) are from the collections of the University of California Museum of Paleontology (UCMP).

TABLE 6. IDENTIFICATIONS FOR TURTLE DATA FROM FERRIS FORMATION, HANNA BASIN

University of Wyoming locality/specimen number	Identification
V91009/27381	*Adocus*, Trionychidae
V91009/30528	*Basilemys*
V91009/27383	Trionychidae, cf. Plastomeninae
V91009/25180	cf. Plastomeninae, Trionychidae, *Adocus*
V92010/27222	cf. Plastomeninae
V92010/27362	Trionychidae
V92010/27282	Trionychidae
V92018/27384	Trionychidae
V92018/27218	Trionychidae
V92027/30604	Trionychidae
V92032/26585	Baenidae, possibly *Neurankylus*
V92036/26592	*Basilemys*, Trionychidae
V92036/26581	Trionychidae
V92036/26621	Baenidae
V92036/27279	*Compsemys*
V92036/26593	*Basilemys*, Trionychidae, Baenidae
V92040/27374	Baenidae
V92040/26477	*Neurankylus*
V92044/26586	*Basilemys*
V92044/27281	Trionychidae, Baenidae
V92046/25585	*Compsemys*, Trionychidae
V93002/27273	Baenidae
V93002/27220	*Basilemys*
V93002/27390	Trionychidae
V93003/25196	*Basilemys*
V93003/27278	*Compsemys*
V93003/27385	cf. *Helopanoplia*
V93003/27388	cf. Plastomeninae
V93003/27241	Trionychidae
V93004/30610	cf. Trionychidae, *Compsemys*, cf. Baenidae
V93007/30622	Baenidae
V93007/27387	Trionychidae, Baenidae
V93007/27389	cf. Plastomeninae
V93008/30636	Plastomeninae, Trionychidae
V93009/27380	cf. Plastomeninae
V93009/27382	Lepisosteidae
V93012/26587	*Basilemys*
V93012/27376	cf. Baenidae
V93012/30587	Baenidae, Trionychidae
V93016/30591	Trionychidae
V93018/30630	*Compsemys*
V93018/30631	Trionychidae
V93026/29027	*Basilemys*, Baenidae
V93026/30593	Trionychidae
V93030/27235	*Basilemys*
V93030/27386	Baenidae
V93030/26597	Trionychidae
V93034/30599	Trionychidae
V93036/26596	*Adocus*, Trionychidae
V93036/27221	Trionychidae
V93036/27377	Baenidae
V93036/26595	*Compsemys*
V93037/30641	Trionychidae
V94002/25556	Trionychidae

Note: revised from Lillegraven and Eberle, 1999.

than other vertebrates in the Hell Creek assemblage (e.g., dinosaurians, White et al., 1998). MNI counts that are pooled across localities can also be weighted by one or two localities with disproportionately more individuals represented or simply larger sample sizes. For example, among the localities examined from North Dakota, most have 10 or fewer individuals represented, while two localities yielded more than 40 individuals. Interpretations based on pooled MNI would be biased toward those two localities. Nonetheless, counts of occurrences or MNI constitute similar indices of relative abundance and are highly correlated with one another ($r^2 = 0.90$, $p < 0.01$) (this study).

Taphonomic factors affecting recovery of turtle fossils

Before examining intraregional variability in latest Cretaceous faunas, it is necessary to understand underlying taphonomic biases that might affect observed occurrences and abundances. Fastovsky (1987) and White et al. (1998) demonstrated that differences in sedimentary architectural elements affect recovery of certain dinosaurian taxa. Turtles present a slightly different set of possible biases, principally due to the fact that certain turtles are hypothesized to be more likely to occur in some depositional environments as a consequence of their preferred habitats in life. Bryant (1989) observed some differences in representation of turtle taxa between sandstone and mudstone sites in the Hell Hollow and Dry Arm areas of eastern Montana. Hutchison (1984) and Hutchison and Archibald (1986) also noted that articulated baenids occurred disproportionately in channel deposits, suggesting that this was their preferred habitat. Most of the differences Bryant (1989) quantified in turtle representation are minor and inconsistent between areas, although a consistently greater abundance of *Helopanoplia* in mudstone versus sandstone was observed. The absence of more marked and/or consistent differences in preservation due to lithology is consistent with White et al.'s (1998) observation that avulsion played an important role in mixing aquatic and more terrestrial elements of the fauna during deposition of the Hell Creek Formation. The problems of differential preservation common in terrestrial vertebrates due to morphological differences (e.g., bone density, body size) are largely absent from analyses of Cretaceous turtles. Most of the study taxa achieved fairly large body size, and shells are composed of dense bone.

In order to examine possible biases due to habitat preferences in life, we reanalyzed the turtle abundance (MNI) and occurrence data compiled by Bryant (1989) for the Hell Creek Formation in the Dry Arm and Hell Hollow areas of eastern Montana, in which localities were subdivided into two groups: sandstone, comprising channel fills and splays, and mudstone, representing flood-plain deposits (Table 7). We specifically tested the following hypotheses, derived from the idea that the habitat preferences of bottom-walking (baenids) or subaquatic (trionychids) turtles may cause them to be preserved in greater abundances in certain sandstones:

1. Are abundances as measured by MNI independent of architectural element?
2. Are occurrences as measured by counts of locality occurrences independent of architectural element?
3. Are baenids or trionychids more common in sandstone (a proxy for stream environments)?
4. Does *Helopanoplia* occur disproportionately in mudstone?

A chi-square analysis of abundances as calculated by MNI (Table 7) indicated that MNI distributions were not independent of lithology ($\chi^2 = 25.69$; $p < 0.01$), but locality occurrences were independent of lithology ($\chi^2 = 8.56$; NS, not significant). These results demonstrate that for turtles (and probably for other aquatic vertebrates) the MNI count is not a reliable indicator of abundance in the absence of an isotaphonomic approach to such comparisons. However, for the purposes of broader regional comparisons, we can infer that locality occurrence data are a potentially useful measure of local differences and do not strongly reflect taphonomic biases.

Therefore, using locality occurrence data (Table 8), we tested the hypothesis that the four study areas (Hanna basin, Lance Creek, Hell Creek, and Marmarth) are homogeneous. A heterogeneity chi squared test (Zar, 1999) reveals that these areas are not homogeneous ($\chi^2 = 139.6$, $p < 0.01$), i.e., the areas differ from one another in relative abundance. In order to see which taxa might be responsible for these differences and to see if the differences are merely due to the smaller number of localities sampled outside Montana, we ran bootstrap analyses (e.g., see Good, 1999; using Resampling Stats 5.0.2; ©Julian L. Simon, Resampling Stats, Inc.) to see if the observed values from North Dakota and the Wyoming formations could have occurred simply due to chance. For most values, there was no significant difference. However, several taxa stood out as unusually abundant or unusually rare when compared to the

TABLE 7. OCCURRENCES IN SANDSTONE VERSUS MUDSTONE FOR THE HELL HOLLOW AND DRY ARM AREAS OF EASTERN MONTANA

	Baenidae	*Helopanoplia*	Other Trionychidae	*Compsemys*	Chelydridae	*Adocus*	*Basilemys*	"*Clemmys*" *backmani*	Kinosternia
Locality occurrence counts									
ss	18/14.79	7/11.62	32/27.47	13/16.38	14/13.74	16/16.91	5/4.75	1	1
ms	10/13.21	15/10.38	20/24.53	18/14.62	12/12.26	16/15.09	4/4.25	0	1
Minimum number of individual (MNI) counts									
ss	39/29.06	8/14.26	97/79.25	22/29.06	20/20.60	29/29.58	7/5.81	1	1
ms	16/25.94	19/12.74	53/70.75	33/25.94	19/18.40	27/26.42	4/5.19	0	1

Note: Data are from Bryant (1989). Sandstone (ss) data are based on 28 localities; mudstone (ms) data based on 25 localities. Values in the upper row of each lithologic type are counts of locality occurrences; those in the lower row are MNI counts. All values shown as observed over expected values. Expected values are derived from an extrinsic hypothesis, testing whether the observed occurrences are equally likely in each lithology [expected value = observed value/(total localities/n localities of particular lithology)]. Data for kinosternoids and "*Clemmys*" *backmani* were excluded from the statistical analysis discussed in the text because of their very low values, and are presented only for reference. Chi-square values are for individual taxon comparisons between lithologies (see text); * indicates $p < 0.05$, ** indicates $p < 0.01$.

TABLE 8. LOCALITY OCCURRENCE COUNTS FOR TURTLES IN LATE CRETACEOUS FORMATIONS OF NORTH DAKOTA, MONTANA, AND WYOMING

	Baenids	*Helopanoplia*	Trionychids	*Compsemys*	Chelydrids	*Adocus*	*Basilemys*	*"C." backmani*	Kinosternia
Montana	142.0	63.0	291.0	174.0	122.0	127.0	38.0	11.0	4.0
North Dakota	18.0***	6.0	17.0	8.0	12.0	13.0	6.0	2.0	5.0***
Lance	8.0	6.0	12.0	7.0	10.0**	4.0	7.0***	3.0**	4.0***
Ferris	11.0	1.0	22.0	6.0**	0.0***	2.0***	8.0***	0.0	0.0

Note: Counts that differ significantly from the Montana data set are indicated by asterisks. *** indicates $p < 0.01$; ** indicates $p < 0.05$. Based on 349 sites in Montana, 24 localities each in North Dakota and the Ferris Formation, and 15 localities in the Lance Formation.

Hell Creek area of Montana. These values are indicated by asterisks in Table 8.

The values from the Hell Creek Formation of North Dakota differ least from those in Montana. However, both baenids and kinosternoids were more common in North Dakota than expected. The greater abundance of baenids may be attributable to the fact that channels tend to increase in size closer to base level, and it is the largest baenid, *Neurankylus*, which is fairly common in that area. The Lance Formation localities differ from the Hell Creek area of Montana in the greater abundance of chelydrids, *Basilemys*, the macrobaenid *"Clemmys" backmani*, and Kinosternoidae. The Ferris Formation is notable for the comparative lack of *Compsemys*, chelydrids, and *Adocus* and the abundance of *Basilemys*. These results suggest that both chelydrids and *Basilemys* were more common in the southern part of the region and hint at the type of latitudinal variation suggested by Sloan (1969) and Lehman (1987).

There is no clear explanation for the intraregional differences reported here, but the observation of such differences highlights a number of areas of inquiry. Environmental gradients could be a factor. Sloan (1969) and Lehman (1987) suggested the presence of latitudinal gradients in latest Cretaceous dinosaurian abundance. However, a clear mechanism (e.g., temperature, altitude, or rainfall) controlling such a gradient is not evident, and the absence of additional, geographically intermediate samples leaves any hypothesis of a latitudinal gradient untestable. The differential distribution of herbivorous taxa (*Adocus* and *Basilemys*) may reflect differences in the plant communities on which they depend. The existence of a southern palynofloristic province that encompasses the southern Powder River Basin and the Lance Creek area was reported by Nichols et al. (1992). Unfortunately, extrapolating from palynofloras to herbivorous turtle dietary preferences is too optimistic a task for available data. Similarly, differences in the carnivorous or molluscivorous taxa (Kinosternoidae, Chelydridae, Baenidae, Pleurosternidae) may correlate with differences in prey distribution. However, comparable data for these organisms are not yet available.

This variability in relative abundance could also reflect fundamental differences in the structure of the local landscapes being sampled. The latest Cretaceous flood plains were formed by dynamic aggrading fluvial systems with a high, fluctuating water table (Fastovsky, 1987, 1990) that created an unstable landscape altered by periodic flooding and channel migration (Fastovsky and McSweeney, 1987). A greater proportion of baenids could reflect a greater incidence of larger stream channels, and the greater numbers of *Basilemys* in the Lance Formation could reflect more terrestrial environments being sampled. The issue of sampling is further complicated by the fact that few fossils in these formations are not reworked. Rather, they represent accumulations of animals that both lived and died in the environment sampled (e.g., a river channel) as well as those deposited in the sediments through which the channel traveled and eroded (e.g., Fastovsky, 1987; Bryant, 1989). Thus, with the exception of articulated remains found in place, we cannot say with confidence whether the animal represented by the fossil occurred in the habitat represented by a particular lithologic unit. In the case of Late Cretaceous turtles, the vast majority of fossils are not articulated, but are represented only by individual skeletal elements.

Another as yet untestable hypothesis is that these areas represent slightly different temporal spans of the latest Cretaceous and that the relative differences in occurrence represent temporal trends. Typically, the four study areas are regarded as broadly contemporaneous (e.g., Lillegraven and Ostresh, 1988; Sheehan et al., 1991), representing the Lancian North American Land Mammal Age or *Triceratops* zone. However, each local section represents different thicknesses and spans of time; correlation between them is complicated by complex fluvial architecture, lack of extensive or reliable marker beds, and depositional hiatuses are clearly present (e.g., Fastovsky, 1987; Lofgren, 1995). In particular, the large samples from Montana (pooled for the purposes of this analysis) potentially comprise a longer span of time than those from North Dakota or Wyoming, given the greater stratigraphic thickness of the Hell Creek Formation in eastern Montana. Although the temporal component within the latest Cretaceous has received little attention (except as it relates to the K-T boundary), the span of time represented in these formations is more than adequate to permit dramatic changes in relative abundance. In studies of Eocene turtles, using methods similar to those employed here, comparable and even greater changes in relative abundance and among depositional basins have been documented on scales of 250–500 k.y. (Holroyd and Hutchison, 2000; Holroyd et al., 2001). Thus, the differences observed between Wyoming, Montana, and North Dakota may be attributable to changes in relative abundance through time, but a fine-scaled framework in which to test such a hypothesis does not yet exist.

DISCUSSION

The idea of a geographically widespread Hell Creek or Lancian vertebrate fauna in the northern Great Plains is widely accepted. However, the possibility and significance of small-scale intraregional differences (exclusive of taphonomic effects) in latest Cretaceous faunas has not previously been explored in detail. Based on an analysis of the very common testudine component of the fauna, there were distinct differences between the Hell Creek Formation in eastern Montana and western North Dakota and the Lance and Ferris Formations in eastern Wyoming. These differences probably reflect ecologic variability in aquatic habitats, but comparable comparative data for other organisms are lacking.

Although we can provide no ready explanations for the variation in relative abundance reported here, documenting and trying to understand such differences is critical to developing a framework in which we can better assess the effects of seaway regression, climate change, and catastrophic events on the latest Cretaceous biota. Differences in survival patterns across the K-T boundary between the aquatic and terrestrial components of the latest Cretaceous faunas are well documented (Archibald and Bryant, 1990; Sheehan and Fastovsky, 1992), but the mechanisms that controlled this differential pattern of extinction are still poorly understood. The data presented here showing biogeographic differences in aquatic vertebrate representation in the latest Cretaceous suggest that the aquatic environments of the northern Great Plains may have been more heterogeneous on local scales than the terrestrial landscape. Such heterogeneity is consistent with depositional models for latest Cretaceous formations (e.g., Fastovsky, 1990) and suggests that adaptation to fluctuating environments may have been a characteristic of Late Cretaceous aquatic vertebrates. Such adaptation may be a key factor in explaining the greater relative success of aquatic vertebrates compared to their terrestrial counterparts in surviving ecologic changes at the K-T boundary.

ACKNOWLEDGMENTS

We thank the organizers of this symposium for inviting us to participate, and especially K. Johnson for his encouragement of a more paleoecological approach to the data, and helpful discussion with W.A. Clemens, J.F. Parham, and D.J. Nichols. We thank L.J. Bryant and D.E. Fastovsky for their roles in making the collections from North Dakota, and M. Luten and K. Johnson for field assistance. Field work was funded through the Annie M. Alexander Endowment to the University of California Museum of Paleontology (UCMP) and National Science Foundation grant EAR-9505841 and prior grants to W.A. Clemens. K.T. Smith and R.E. Strasser ably assisted in the timely curation of voucher material. Eugene Gaffney and Marshal Lambert made important specimens readily available, and J.A. Lillegraven and M. Cassiliano graciously provided access to the University of Wyoming collections. This is UCMP publication 1734.

REFERENCES CITED

Archibald, J.D., 1982, A study of Mammalia and geology across the Cretaceous-Tertiary boundary in Garfield County, Montana: University of California Publications in Geological Sciences, v. 122, p. 1–286.

Archibald, J.D., and Bryant, L.J., 1990, Differential Cretaceous/Tertiary extinctions of nonmarine vertebrates: Evidence from northeastern Montana, *in* Sharpton, V.L., and Ward, P.D., eds., Global catastrophes in Earth History: An interdisciplinary conference on impacts, volcanism, and mass mortality: Geological Society of America Special Paper 247, p. 549–562.

Archibald, J.D., and Hutchison, J.H., 1979, Revision of the genus *Palatobaena* (Testudines, Baenidae), with the description of a new species: Postilla, v. 177, p. 1–19.

Brinkman, D.B., 1998, The skull and neck of the Cretaceous turtle *Basilemys* (Trionychoidea, Nanhsuingchelyidae) and the interrelationships of the genus: Paludicola, v. 1, p. 150–157.

Brinkman, D.B., and Nicholls, E.J., 1993, The skull of *Neurankylus eximius* (Testudines: Baenidae) and a reinterpretation of the relationships of this taxon: Journal of Vertebrate Paleontology, v. 13, p. 273–281.

Bryant, L.J., 1989, Non-dinosaurian vertebrates across the Cretaceous-Tertiary boundary in northeastern Montana: University of California Publications in Geological Sciences, v. 134, p. 1–107.

Clemens, W.A., 1963, Fossil mammals of the type Lance Formation, Wyoming. 1. Introduction and Multituberculata: University of California Publications in Geological Sciences, v. 48, p. 1–150.

Clemens, W.A., 1966, Fossil mammals of the type Lance Formation, Wyoming. 2. Marsupialia: University of California Publications in Geological Sciences, v. 62, p. 1–122.

Clemens, W.A., 1973, Fossil mammals of the type Lance Formation, Wyoming. 3. Eutheria and summary: University of California Publications in Geological Sciences, v. 94, p. 1–102.

Eschmeyer, W.N., 1990, Catalog of the genera of recent fishes: San Francisco, California Academy of Sciences, 697 p.

Estes, R., 1964, Fossil vertebrates from the Late Cretaceous Lance Formation, eastern Wyoming: University of California Publications in Geological Sciences, v. 49, p. 1–180.

Fastovsky, D.E., 1987, Paleoenvironments of vertebrate-bearing strata during the Cretaceous-Paleogene transition, eastern Montana and western North Dakota: Palaios, v. 2, p. 282–295.

Fastovsky, D.E., 1990, Rocks, resolution, and the record: A review of depositional constraints on fossil vertebrate assemblages at the terrestrial Cretaceous/Paleogene boundary, eastern Montana and western North Dakota, *in* Sharpton, V.L., and Ward, P.D., eds., Global catastrophes in Earth history: An interdisciplinary conference on impacts, volcanism, and mass mortality: Geological Society of America Special Paper 247, p. 541–547.

Fastovsky, D.E., and McSweeney, K., 1987, Paleosols spanning the Cretaceous-Paleogene transition, eastern Montana and western North Dakota: Geological Society of America Bulletin, v. 99, p. 66–77.

Gaffney, E.S., 1972, The systematics of the North American family Baenidae (Reptilia, Cryptodira): American Museum of Natural History Bulletin, v. 147, p. 241–320.

Gaffney, E.S., and Meylan, P.A., 1988, A phylogeny of turtles: Systematics Association Special Volume 35A, p. 157–219.

Gardner, J.D., Russell, A.P., and Brinkman, D.B., 1995, Soft-shelled turtles (Family Trionychidae) from the Judith River Group (Campanian) of North America: Canadian Journal of Earth Sciences, v. 32, p. 631–643.

Gilinsky, N.L., and Bennington, J.B., 1994, Estimating numbers of whole individuals from collections of body parts: A taphonomic limitation of the paleontological record: Paleobiology, v. 20, p. 245–258.

Gilmore, C.W., 1919, Reptilian faunas of the Torrejon, Puerco and underlying

Upper Cretaceous formations of San Juan County, New Mexico: U.S. Geological Survey Professional Paper 119, p. 1–17.

Good, P.I., 1999, Resampling methods: A practical guide to data analysis: Boston, Massachusetts, Birkhauser, 336 p.

Grande, L., and Bemis, W.E., 1998, A comprehensive phylogenetic study of amiid fishes (Amiidae) based on comparative skeletal anatomy: An empirical search for interconnected patterns of natural history: Society of Vertebrate Paleontology Memoir 4: Journal of Vertebrate Paleontology, v. 18, p. 1–690.

Hay, O.P., 1908, The fossil turtles of North America: Carnegie Institution, Publication 75, 568 p.

Holroyd, P.A., and Hutchison, J.H., 2000, Proximate causes for changes in vertebrate diversity in the early Paleogene: An example from turtles in the Western Interior of North America: GFF, v. 122, p. 75–76.

Holroyd, P.A., Hutchison, J.H., and Strait, S.G., 2001, Turtle diversity and abundance through the lower Eocene Willwood Formation of the southern Bighorn Basin: University of Michigan Papers in Paleontology, v. 33, p. 97–108.

Hutchison, J.H., 1980, Turtle stratigraphy of the Willwood Formation, Wyoming: Preliminary results: University of Michigan Papers in Paleontology, v. 24, p. 115–118.

Hutchison, J.H., 1984, Determinate growth in the Baenidae (Testudines): Taxonomic, ecologic, and stratigraphic significance: Journal of Vertebrate Paleontology, v. 3, p. 148–151.

Hutchison, J.H., 1991, Early Kinosternidae (Reptilia: Testudines) and their phylogenetic significance: Journal of Vertebrate Paleontology, v. 11, p. 145–167.

Hutchison, J.H., and Archibald, J.D., 1986, Diversity of turtles across the Cretaceous/Tertiary boundary in northeastern Montana: Palaeogeography, Palaeoclimatology, Palaeoecology. v. 55, p. 1–22.

Hutchison, J.H., and Weems, R.E., 1999, Paleocene Turtle remains from South Carolina, *in* Sanders, A.E., ed., Paleobiology of the Williamsburg Formation (Black Mingo Group; Paleocene) of South Carolina, U.S.A.: American Philosophical Society Transactions, v. 88, p.165–195.

Langston, W., Jr., 1956, The shell of *Basilemys varialosa* Cope: National Museum of Canada Bulletin, v. 147, p. 155–163.

Lehman, T.M., 1987, Late Maastrichtian paleoenvironments and dinosaur biogeography in the Western Interior of North America: Palaeogeography, Palaeoclimatology, Palaeoecology, v. 60, p. 189–217.

Lillegraven, J.A., and Eberle, J.J., 1999, Vertebrate faunal changes through Lancian and Puercan time in southern Wyoming: Journal of Paleontology, v. 73, p. 691–710.

Lillegraven, J.A., and Ostresh, L.M., Jr., 1988, Evolution of Wyoming's early Cenozoic topography and drainage patterns: National Geographic Journal of Research, v. 4, p. 303–327.

Lofgren, D.L., 1995, The Bug Creek problem and the Cretaceous-Tertiary transition at McGuire Creek, Montana: University of California Publications in Geological Sciences, v. 140, p. 1–185.

Meylan, P.A., 1987, The phylogenetic relationships of the soft-shelled turtles (family Trionychidae): American Museum of Natural History Bulletin, v. 186, p. 1–101.

Meylan, P.A., and Gaffney, E.S., 1989, The skeletal morphology of the Cretaceous cryptodiran turtle, *Adocus*, and the relationships of the Trionychoidea: American Museum Novitates, no. 2941, p. 1–60.

Nichols, D.J., and Fleming, R.F., 1990, Plant microfossil record of the terminal Cretaceous event in the Western United States and Canada, *in* Sharpton, V.L., and Ward, P.D., eds., Global catastrophes in Earth History: An interdisciplinary conference on impacts, volcanism, and mass mortality: Geological Society of America Special Paper 247, p. 445–455.

Nichols, D.J., Brown, J.L., Attrep, M., Jr., and Orth, C.J., 1992, A new Cretaceous-Tertiary boundary locality in the western Powder River basin, Wyoming: Biological and geological implications: Cretaceous Research, v. 13, p. 3–30.

Norell, M.A., Clark, J.M., and Hutchison, J.H., 1994, The Late Cretaceous alligatoroid *Brachychampsa montana* (Crocodylia): New material and putative relationships: American Museum Novitates, no. 3116, p. 1–26.

Russell, L.S., 1935, Fossil turtles from Saskatchewan and Alberta: Royal Society of Canada Transactions, v. 4, p. 101–111.

Sheehan, P.M., and Fastovsky, D.E., 1992, Major extinctions of land-dwelling vertebrates at the Cretaceous-Tertiary boundary, eastern Montana: Geology, v. 20, p. 556–560.

Sheehan, P.M., Fastovsky, D.E., Hoffman, R.G., Berghaus, C.B., and Gabriel, D.L., 1991, Sudden extinction of the dinosaurs: Latest Cretaceous, Upper Great Plains, U.S.A.: Science, v. 254, p. 835–839.

Sloan, R.E., 1969, Cretaceous and Paleocene terrestrial communities of western North America: North American Paleontological Convention Proceedings, p. 427–453.

Sweet, A.R., Braman, D.R., and Lerbekmo, J.F., 1990, Palynofloral response to K-T boundary events: A transitory interruption within a dynamic system, *in* Sharpton, V.L., and Ward, P.D., eds., Global catastrophes in Earth history: An interdisciplinary conference on impacts, volcanism, and mass mortality: Geological Society of America Special Paper 247, p. 457–469.

Whetstone, K.N., 1978, A new genus of cryptodiran turtles (Testudinoidea, Chelydridae) from the Upper Cretaceous Hell Creek Formation of Montana: University of Kansas Science Bulletin, v. 51, p. 539–563.

White, P.D., Fastovsky, D.E., and Sheehan, P.M., 1998, Taphonomy and suggested structure of the dinosaurian assemblage of the Hell Creek Formation (Maastrichtian), eastern Montana and Western North Dakota: Palaios, v. 13, p. 41–51.

Whitmore, J.L., and Martin, J.E., 1986, Vertebrate fossils from the Greasewood Creek locality in the Late Cretaceous Lance Formation of Niobrara County, Wyoming: South Dakota Academy of Science Proceedings, v. 65, p. 33–50.

Wiley, E.O., 1976, The phylogeny and biogeography of fossil and recent gars (Actinopterygii: Lepisosteidae): University of Kansas Museum of Natural History Miscellaneous Publication 64, p. 1–111.

Williams, E.E., 1950, Variation and selection in the cervical central articulations of living turtles: American Museum of Natural History Bulletin, v. 94, p. 505–562.

Zar, J.H., 1999, Biostatistical analysis (fourth edition): Englewood Cliffs, New Jersey, USA, Prentice-Hall, 620 p.

Manuscript Accepted by the Society September 24, 2001

Printed in the U.S.A.

Geological Society of America
Special Paper 361
2002

Mammals from the end of the age of dinosaurs in North Dakota and southeastern Montana, with a reappraisal of geographic differentiation among Lancian mammals

John P. Hunter*
Department of Anatomy, New York College of Osteopathic Medicine of New York Institute of Technology, Old Westbury, New York 11568, USA
J. David Archibald*
Department of Biology, San Diego State University, San Diego, California 92182, USA

ABSTRACT

An end-Cretaceous nonavian dinosaur extinction and an early Paleocene mammalian radiation is documented primarily in stratigraphic sequences in eastern Montana. To determine how representative these sequences are, we extended investigation of this Cretaceous-Tertiary (K-T) transition to new areas. Studies in southwestern North Dakota and southeastern Montana provide new records of mammals through the last 1.32–1.68 million years of the Cretaceous and extending into the Paleocene, allowing us to evaluate mammalian faunal differentiation across the geographic landscape of the North American Western Interior in the latest Cretaceous. In North Dakota, mammals occur at 15 horizons in the Hell Creek Formation from ~80 to 3 m below the Hell Creek–Fort Union formational contact. In southeastern Montana, where the Hell Creek Formation is ~150 m thick, mammals occur from 85 to 15 m below the formational contact, with well-sampled local faunas at 65 and 61 m. In faunal comparisons, the new study areas are closely similar to contemporaneous local faunas, but differ from those at higher latitudes in several ways. Although mammalian faunas of the northwestern coastal plain of the Western Interior Seaway (i.e., the Hell Creek Formation and its lateral equivalents) show little differentiation, differences that exist are strongly associated with geographic distance rather than latitude per se. The overlying Fort Union Formation in North Dakota and southeastern Montana has produced a few early Paleocene mammalian specimens, promising that future work in the area will contribute to knowledge of mammalian radiation after the K-T boundary.

INTRODUCTION

The end of the Cretaceous through the beginning of the Tertiary (Paleogene) was a critical time in the evolutionary history of mammals. During this interval, mammalian communities underwent profound changes in composition. In the Western Interior of North America, a Tertiary community with diverse eutherian mammals replaced a Cretaceous community dominated by primitive marsupials and the now extinct multituberculates (Archibald, 1996; Hunter, 1999). The highest rates

*E-mail: Hunter, jphunter@iris.nyit.edu; Archibald, darchibald@sunstroke.sdsu.edu

Hunter, J.P., and Archibald, J.D., 2002, Mammals from the end of the age of dinosaurs in North Dakota and southeastern Montana, with a reappraisal of geographic differentiation among Lancian mammals, *in* Hartman, J.H., Johnson, K.R., and Nichols, D.J., eds., The Hell Creek Formation and the Cretaceous-Tertiary boundary in the northern Great Plains: An integrated continental record of the end of the Cretaceous: Boulder, Colorado, Geological Society of America Special Paper 361, p. 191–216.

of extinction and origination, as well as the largest single increase in average body mass observed in the North American mammalian fossil record (Alroy, 1999), are associated with the Cretaceous-Tertiary (K-T) boundary Most extant orders of placental mammals first appeared in the fossil record in a dramatic Paleogene evolutionary radiation following the extinction of dinosaurs at the K-T boundary (McKenna and Bell, 1997). The temporal clustering of these ordinal appearance events (Archibald and Deutschman, 2001) and estimated preservation potential of Late Cretaceous mammals (Foote et al., 1999) demonstrate that the origin and diversification of modern placental orders deep in the Cretaceous, that is, long before the K-T boundary, is unlikely.

Unfortunately, our knowledge of the history of mammals leading up to the K-T boundary, arguably the single greatest change in mammalian communities, is restricted geographically. Although there are isolated occurrences in eastern North America (Grandstaff et al., 1992), Europe (Grigorescu and Hahn, 1987; Le Loeuff and Buffetaut, 1995), and India (Prasad et al., 1995; Krause et al., 1997), mammals from the end of the Cretaceous are best known from the North American Western Interior (Lancian North American Land Mammal Age [NALMA]) from Alberta (Lillegraven, 1969) to New Mexico (Flynn, 1986). Well-sampled stratigraphic sequences that include fossil mammals on both sides of the K-T boundary are even more restricted latitudinally to eastern Montana (Sloan and Van Valen, 1965; Archibald, 1982; Clemens, this volume) and southern Saskatchewan (Johnston, 1980; Johnston and Fox, 1984; Fox, 1989, 1997). Although similar sequences of localities occur farther south near Glendive, Montana (Hunter et al., 1997) and in the Hanna Basin of southern Wyoming (Eberle and Lillegraven, 1998a, 1998b; Lillegraven and Eberle, 1999), very earliest Paleogene mammals (Pu0 or Pu1; Archibald and Lofgren, 1990) have not been recovered near Glendive, and only a few Cretaceous mammals from the Hanna Basin.

In spite of the geographically small window of observation through which we view mammalian evolution leading up to the K-T boundary, scenarios that seek to account for mammalian turnover at the boundary have incorporated a geographic component. Differences in faunal composition among Lancian sites, e.g., an initially greater observed diversity of eutherians at higher latitudes in Lancian time (Lillegraven, 1969), have been interpreted to indicate a north-to-south sequence in the transition from Cretaceous-aspect to Paleocene-aspect vertebrate faunas, as expected with an invasion of North America by Asian immigrants (Sloan, 1969; Russell, 1975). Although some subsequent discoveries reaffirmed this interpretation (Archibald, 1982), others have rendered it less straightforward (Fox, 1989). Separating the effects of geographic, ecological, and temporal differences on the composition of Lancian mammalian local faunas has remained a problem. This difficulty has stemmed largely from the fact that no field area could boast a sequence of well-documented localities through Lancian time, by means of which at least temporal differences could be factored out (Archibald, 1982). The overall goals of this study are to move toward providing such stratigraphic sequences and to reconsider the meaning of patterns of faunal similarity among Lancian faunas.

The first specific objective of this study is to augment the known fossil record of mammals at the end of the Cretaceous, both in geographic extent and stratigraphic resolution. To this end, we first describe Late Cretaceous (Lancian) mammals from the Hell Creek Formation and Paleocene (Puercan) mammals from the Ludlow Member of the Fort Union Formation in southwestern North Dakota and adjacent South Dakota being studied by one of us (Hunter). These discoveries, primarily from North Dakota, contribute to knowledge of Lancian mammals by providing the first record of mammalian occurrences through the entire vertical extent of the Hell Creek Formation, and thus through the last ~1.32–1.68 m.y. of the Cretaceous (see Hicks et al., this volume, for estimated duration of the Hell Creek Formation of southwestern North Dakota based on a linear interpolation of sedimentation rates and a new estimate of the age of the K-T boundary). The finds derive from ongoing efforts to document Late Cretaceous and Paleocene mammalian evolution in the area (Hunter and Pearson, 1996; Hunter et al., 1997; Hunter, 1999). Second, we report on work in progress (by Archibald) on fossil mammals of the Hell Creek Formation of extreme southeastern Montana, near the town of Ekalaka in Carter County. Although these localities in Montana do not cover the stratigraphic breadth of the record in North Dakota, they nevertheless have been exhaustively sampled and likely well represent the local fauna. The third specific objective of this study is to reappraise differentiation among the well-sampled Lancian faunas, including the new ones reported here.

In this study, we follow Fox (1989, 1997) and Lillegraven and Eberle (1999) in provisionally considering the Lancian-Puercan boundary to coincide, to a first approximation, with the K-T boundary. Purportedly transitional Bug Creek assemblages in eastern Montana (Sloan and Van Valen, 1965; Lofgren, 1995) are probably best viewed as Paleocene with a reworked Late Cretaceous (i.e., Lancian) component (Lofgren, 1995). Although transitional assemblages in Saskatchewan (Fr-1 and MHBT Quarry, Long Fall horizon) might be considered Puercan on the basis of the occurrence the condylarth *Protungulatum* (Archibald and Lofgren, 1990), palynological evidence suggests that at least Fr-1 is Maastrichtian, and thus possibly in part contemporaneous with Lancian assemblages elsewhere (Fox, 1989, 1997). Although arguments based upon stratigraphy make reworking an unlikely explanation for the transitional nature of the assemblage at Long Fall (i.e., there is no obvious source for Late Cretaceous terrestrial fossils in underlying rocks nearby; Fox, 1989, 1997), interpretations other than a Late Cretaceous age remain conceivable. One possibility is that the Long Fall horizon could be Paleocene and include a component of mammalian and dinosaurian survivors from the Late Cretaceous. This scenario seems unlikely, however, because none of the dinosaurs and only one of the Lancian mammals (*Mesodma thompsoni*) present at Long Fall is argued to have survived the K-T boundary elsewhere (Archibald and Bryant, 1990). More probable is that the Long Fall assemblage is a time-averaged sample, as expected in any channel deposit (Behrensmeyer, 1988), and happens to span the K-T boundary. Even following Fox (1989, 1997) in considering Fr-1 and Long Fall to be Late Cretaceous (Lancian), these sites are arguably younger than other Lancian sites, and were treated as such in the Late Cretaceous mammalian time scale prepared by Hunter for use in Foote et al. (1999; see supplementary data at *www.sciencemag.org/feature/data/985988.shl.*). Accordingly, we omit both assemblages from the analytical faunal comparisons, and Fr-1 because of its small sample size. We instead include more typical and well-sampled Lancian assemblages from the same geographic area (Gryde and Wounded Knee localities).

GEOGRAPHIC AND GEOLOGIC CONTEXT

Other chapters in this volume (e.g., Murphy et al.) describe the geology of the Hell Creek Formation in the Great Plains in general and in North Dakota in particular. Nichols and Johnson (this volume) provide the palynological basis for recognizing the K-T boundary in North Dakota. The mammals described below derive from 17 localities distributed across southwestern North Dakota (Fig. 1) and an immediately adjacent area of South Dakota: 15 of these localities are in the Hell Creek Formation (~80–3 m below the K-T boundary), and two are in the Ludlow Member of the Fort Union Formation (~5.6–7.1 m above the K-T boundary; see Appendix). Each of these localities has also produced nonmammalian vertebrates. Pearson et al. (this volume) detail the collection and curation of the vertebrate specimens. See Hicks et al. (this volume), Johnson (this volume), and Nichols and Johnson (this volume) for the stratigraphic correlation of the localities from which these specimens derive. All 17 Pioneer Trails Regional Museum (PTRM) localities have been sampled by surface collection, but PTRM locality V92067 (Hunter and Pearson, 1996) has also been sampled by screen washing.

Belt et al. (1997) discussed the geology, particularly the stratigraphy, of the Hell Creek Formation of southeastern Montana, from which we report five mammalian localities comprising two local faunas, ~65 and ~61 m below the formational contact, respectively (see Appendix). All five UCMP localities have been sampled by screen washing.

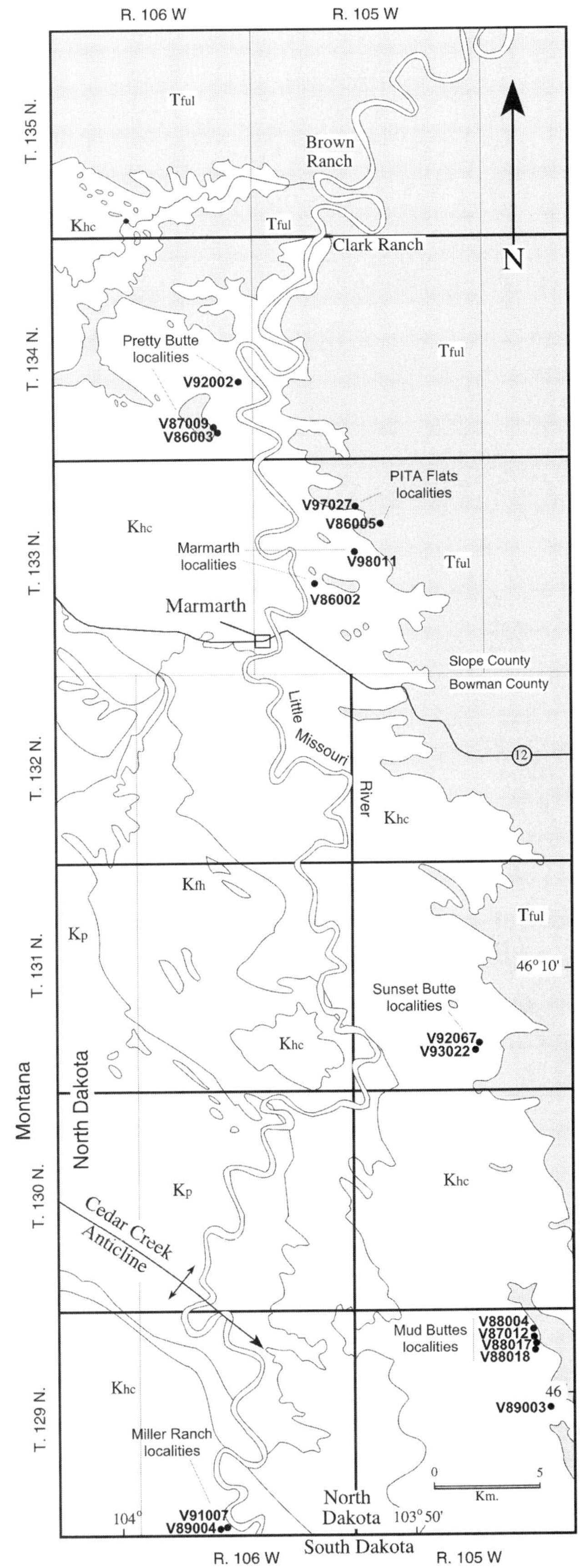

Figure 1. Map showing distribution of fossil mammal localities across study area in extreme southwestern North Dakota. Base map is adapted from Johnson (1992). All localities are identified by Pioneer Trails Regional Museum (PTRM) number. Localities in South Dakota (PTRM locality V92025) and Montana (all UCMP localities) are beyond range of map. Kfh, Fox Hills Formation; Khc, Hell Creek Formation; Kp, Pierre Formation; Tfu1, Ludlow Member of Fort Union Formation.

METHODS

Measurements (in millimeters) were taken either directly through a stereomicroscope using an optical reticle or from digital images from a Kodak MDS 120 and viewed in Adobe Photoshop. Table 1 lists the abbreviations used here. Tooth position within a tooth class (e.g., molars) is denoted by a number or by an X if position could not be ascertained.

SYSTEMATICS

Note that only mammalian specimens from the PTRM localities in southwestern North Dakota and South Dakota studied by Hunter are described in the following. Specimens from the UCMP localities in southeastern Montana studied by Archibald will be described elsewhere.

Class Mammalia
Order Multituberculata
Family Neoplagiaulacidae

Mesodma Jepsen 1940

Mesodma thompsoni Clemens 1964
Fig. 2, A–C

Nomenclatural summary. *Mesodma thompsoni* Clemens 1964 (p. 39).

Holotype. UCMP 47217, left p4.

Type locality. UCMP locality V5620, Lance Formation; Niobrara County, Wyoming.

Distribution. Lance Formation (Lancian), Wyoming; Hell Creek Formation (Lancian), Montana, North Dakota, and South Dakota; Scollard Formation (Lancian), Alberta. Ravenscrag Formation (Lancian and/or Puercan?), Saskatchewan. Bear Formation (Puercan), Montana; Nacimiento Formation (Puercan), New Mexico; Ravenscrag Formation (Puercan), Saskatchewan; Tornillo Formation (Puercan), Texas. Possibly also Kaiparowits Formation (Judithian), Utah; St. Mary River Formation (?Edmontonian), Alberta; Fox Hills Formation (Lancian), South Dakota; and Frenchman Formation (Lancian), Saskatchewan.

Referred specimens. Left dentary with p3–p4 (PTRM 1119) from PTRM locality V89004; left p4 (PTRM 1227) from PTRM locality V92067.

Discussion. The p4s of these specimens possess the low, arched profile with a sharp change in the slope of the serrate margin anteriorly that is typical of *Mesodma*. Specific assignment to *Mesodma thompsoni* is on the basis of size (p4 [PTRM 1119]: L = 4.33, W = 1.60), serration number (12), and profile shape (lower arched than in *M. formosa* [Marsh, 1889b] or *M. hensleighi* Lillegraven 1969).

Mesodma sp.

Referred specimens. Left P4 (PTRM 1274) and left p4 fragment (PTRM 1281) from PTRM locality V86005.

Discussion. Although these specimens are in the size range of *Mesodma thompsoni* and probably belong to that taxon, they have been identified only to the generic level because they are so fragmentary.

Family Cimolodontidae

Cimolodon Marsh 1889

Cimolodon nitidus Marsh 1889
Fig. 2, D–L

Nomenclatural summary. *Cimolodon nitidus* Marsh (1889a, p. 84–85, plate II, Figs. 5–8). See Clemens (1964, p. 56) for synonyms and a revised diagnosis.

Holotype. YPM 11776, left ml.

Type locality. Mammal locality 1 of Lull (1915); UCMP locality V5003, Lance Formation, Wyoming.

Known distribution. St. Mary River Formation (?Edmontonian), Alberta. Fox Hills Formation (Lancian), South Dakota; Scollard Formation (Lancian), Alberta; Frenchman Formation (Lancian), Saskatchewan; Hell Creek Formation (Lancian), Montana, North Dakota, and South Dakota; Lance and Ferris Formations (Lancian), Wyoming. Ravenscrag Formation (Lancian and/or Puercan?), Saskatchewan. Possibly also Prince Creek Formation (Lancian), Alaska.

Referred specimens. Left P4 (PTRM 614) from PTRM locality V89004; left M1 (PTRM 602) from PTRM locality V89003; left dentary with p4 (PTRM 3609) from PTRM locality V86002.

Discussion. These specimens (P4: L = 3.43, W = 1.69; M1: L = 4.32, W = 2.33; p4: L = 4.79, W = 2.00) are in the size range of *Cimolodon nitidus*, the only known Lancian multituberculate intermediate in size between the smaller neoplagiaulacids and the larger cimolomyids. The p4 (PTRM 3609) possesses the characteristic symmetrical arch of *C. nitidus*. P4 (PTRM 614) is bulky with a pronounced anterobuccal bulge, as in *C. nitidus*.

TABLE 1. ABBREVIATIONS

AMNH	=	American Museum of Natural History, New York
PTRM	=	Pioneer Trails Regional Museum, Bowman, North Dakota
UA	=	University of Alberta
UCMP	=	University of California Museum of Paleontology, Berkeley
V	=	vertebrate
YPM	=	Yale Peabody Museum, New Haven, Connecticut
L	=	length
TALW	=	talonid width
TRGW	=	trigonid width
W	=	width
C,c	=	upper or lower canine, respectively
I,i	=	upper or lower incisor, respectively
M,m	=	upper or lower molar, respectively
P,p	=	upper or lower premolar, respectively
gen.	=	genus
indet.	=	indeterminate
sp.	=	species (singular)

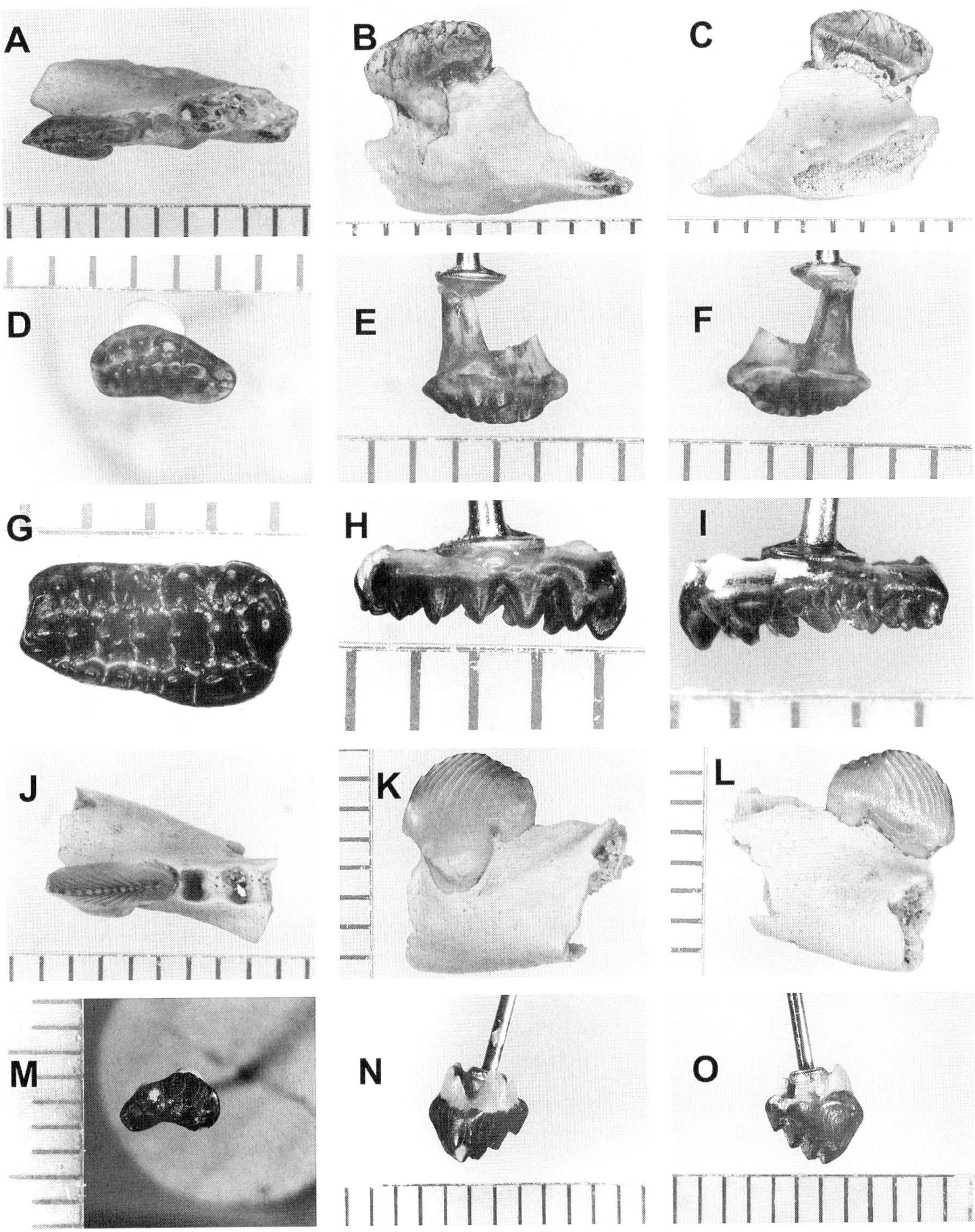

Figure 2. Photographic plate showing specimens of Late Cretaceous multituberculates from North Dakota. *Mesodma thompsoni*, left dentary with p3–p4 (PTRM 1119) from locality PTRM V89004 in occlusal (A), labial (B), and lingual (C) views. *Cimolodon nitidus*, left P4 (PTRM 614) from PTRM locality V89004 in occlusal (D), labial (E), and lingual (F) views; left M1 (PTRM 602) from PTRM locality V89003 in occlusal (G), labial (H), and lingual (I) views; left dentary with p4 (PTRM 3609) from PTRM locality V86002 in occlusal (J), labial (K), and lingual (L) views. *Meniscoessus robustus*, right P4 (PTRM 603) from PTRM locality 92002 in occlusal (M), labial (N), and lingual (O) views. Scale ruled in millimeters. PTRM is Pioneer Trails Regional Museum; for abbreviations, see Table 1.

Cimolodon sp.

Referred specimens. Left M2 (PTRM 1223) from PTRM locality V92067.

Discussion. Although this specimen is in the size range of *Cimolodon nitidus* and probably belongs to that taxon, it has been identified only to the generic level because it is so poorly preserved.

Family Cimolomyidae

Meniscoessus Cope 1882

Meniscoessus robustus (Marsh 1889)
Fig. 2, M–O; Fig. 3, A and B

Nomenclatural summary. *Dipriodon robustus* Marsh (1889a, p. 85, Plate II, Figs. 13–15). See Archibald (1982, p. 75) for synonyms and a revised diagnosis.

Holotype. YPM 11234, right m2.

Type locality. Mammal locality 2 of Lull (1915); UCMP locality V5815, Lance Formation, Wyoming.

Distribution. St. Mary River Formation, Alberta (?Edmontonian). Frenchman Formation (Lancian), Saskatchewan; Hell Creek Formation (Lancian), Montana, North Dakota, and South Dakota; Lance Formation (Lancian), Wyoming. Possibly also Frenchman and Ravenscrag Formations (Lancian and/or Puercan?), Saskatchewan.

Referred specimens. Right m2 (PTRM 1275) from PTRM locality 89004; right P4 (PTRM 603) from PTRM locality 92002; I2 (PTRM 1268), left P4 (PTRM 3041), two left M1s (PTRM 3001 and 3559), three right M2s (PTRM 619, 1288, and 2116), three left M2s (PTRM 1249, 1253, and 1256), right lower incisor (PTRM 3000), right p4 (PTRM 2111), two left p4s (PTRM 2119 and 2112), right m1 (PTRM 2115), two left m1s (PTRM 1248 and 1269), and four left m2s (PTRM 613, 1228, 1229, and 1252) from PTRM locality V92067; right p4 (PTRM 600) from PTRM locality V87009; left M1 (PTRM 1283) from PTRM locality V93022; left m2 (PTRM 3563) from PTRM locality 89003; right M1 (PTRM 3556), left M1 (PTRM 2117), and right m1 (PTRM 696) from PTRM locality V87012.

Discussion. *Meniscoessus robustus* from the Hell Creek Formation, North Dakota, is indistinguishable from samples of *Meniscoessus robustus* from other Lancian sites from North America but extends both the upper and lower ends of the size range of this taxon (see Hunter and Pearson, 1996).

Meniscoessus sp.

Referred specimens. Fragments of teeth from PTRM localities V89004 (PTRM 606 and 1291), V92067 (PTRM 621, 623, 624, 1226, 1231, 1232, 1247, 1264, 1287, 2120–2122, 2124, 3002, and 3042), V86003 (PTRM 3005, 3006, and 3046), V86002 (PTRM 3603), V89003 (PTRM 615), and V92025 (PTRM 1273).

Discussion. Many dental fragments most likely belong to *Meniscoessus robustus* on the basis of large size and cusp shape (e.g., extremely crescentic molar cusps), but they have been identified only to the generic level if they were too fragmentary or poorly preserved.

?*Meniscoessus* sp.

Referred specimens. Left petrosal (PTRM 1276) from PTRM locality V89004; PX fragment (PTRM 1115) from PTRM locality V86003.

Discussion. The isolated petrosal is not associated with any identifiable dental remains. Nevertheless, it is similar to isolated petrosals from the Lance Creek and Bug Creek local faunas that other workers (Kielan-Jaworowska et al., 1986; Luo, 1989; Rougier et al., 1992) have identified as belonging to either one of the large multituberculates *Meniscoessus* or *Catopsalis* Cope 1882. This petrosal will be described in a subsequent paper. The PX fragment is assigned to ?*Meniscoessus* on the basis of its large size.

Infraclass Metatheria
Order Marsupialia
Family Alphadontidae

Alphadon Simpson 1927

Alphadon marshi Simpson 1927
Fig. 3, C–E

Nomenclatural summary. *Alphadon marshi* Simpson 1927 (p. 125).

Holotype. YPM 13659, right m3.

Type locality. Lance Formation, Niobrara County, Wyoming.

Distribution. Scollard Formation (Lancian), Alberta; Lance Formation (Lancian), Wyoming; Hell Creek Formation (Lancian), Montana and North Dakota; Kirtland Formation (Lancian), New Mexico. Reported (material not yet described) from the Fox Hills and Hell Creek Formations (Lancian) of South Dakota.

Referred specimen. Right m1 (PTRM 608) from PTRM locality V89004.

Discussion. PTRM 608 possesses the twinned entoconid–hypoconulid and buccal postcingulid of primitive marsupials, and its cristid obliqua meets the middle of the posterior face of the trigonid as in other Lancian alphadontids (the cristid obliqua meets the posterior face of the protoconid in stagodontids and "pediomyids"). In size (L = 2.15, TRGW = 1.25, TALW = 1.4), the specimen is closest to *Alphadon marshi* from the Flat Creek local fauna (observed ranges: L = 2.09–2.25, TRGW = 1.05–1.35, TALW = 1.22–1.42; Archibald, 1982, Table 20), i.e., larger than known samples of *A. wilsoni* (see Fox, 1989, p. 23).

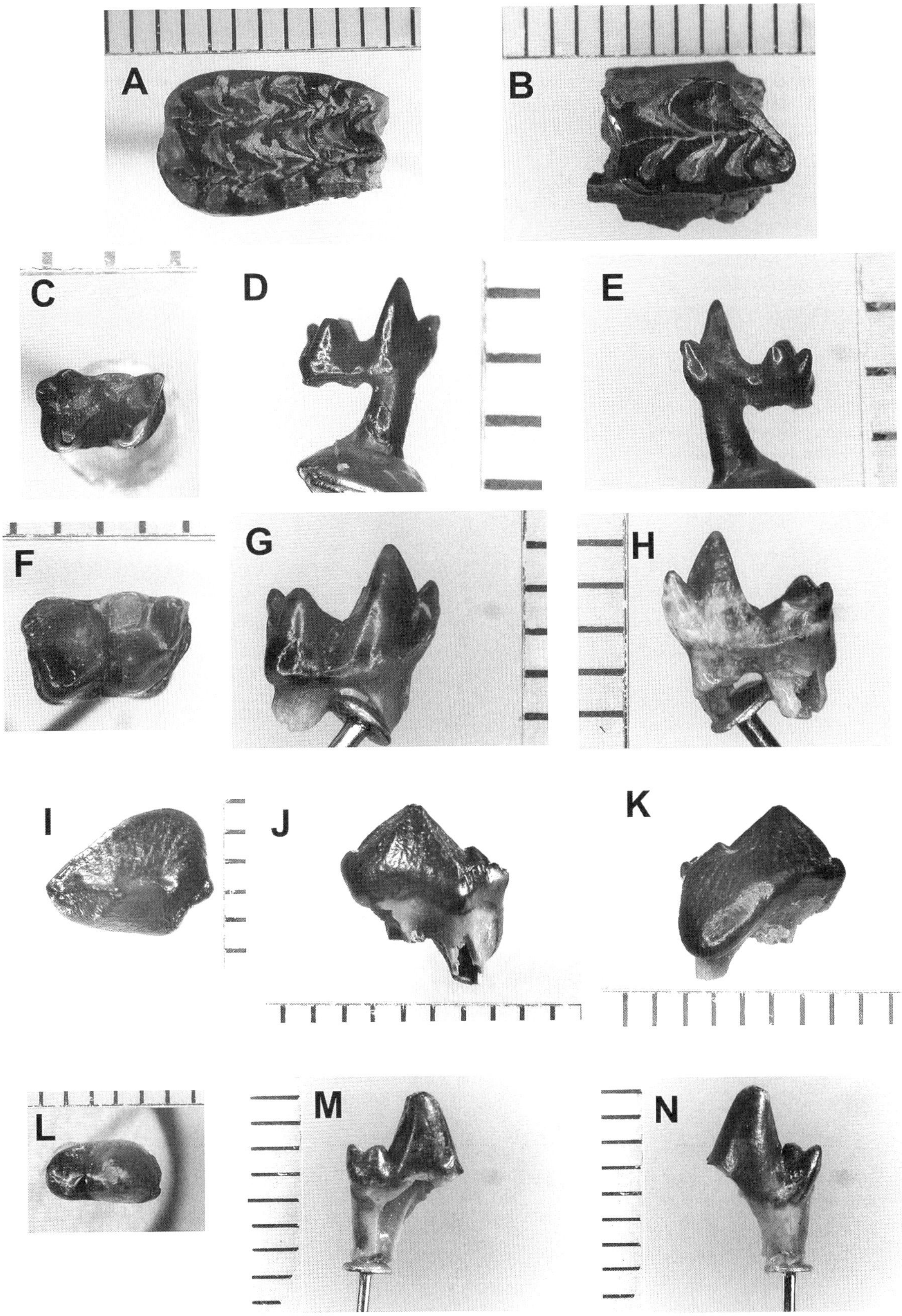

Figure 3. Photographic plate showing specimens of Late Cretaceous mammals from North Dakota. *Meniscoessus robustus*, right M1 (PTRM 3556) from PTRM locality V87012 in occlusal view (A) and left m2 (PTRM 3563) from PTRM locality 89003 in occlusal view (B). *Alphadon marshi*, right m1 (PTRM 608) from PTRM locality V89004 in occlusal (C), labial (D), and lingual (E) views. "*Pediomys*" *florencae*, right m2 (PTRM 605) from PTRM locality V89004 in occlusal (F), labial (G), and lingual (H) views. *Didelphodon vorax*, left p2 (PTRM 3012) from PTRM locality V89004 in occlusal (I), labial (J), and lingual (K) views. *Cimolestes magnus*, right p4 (PTRM 695) from PTRM locality V86002 in occlusal (L), labial (M), and lingual (N) views. Scale ruled in millimeters. PTRM is Pioneer Trails Regional Museum; for abbreviations, see Table 1.

TABLE 2. MAMMALIAN FAUNA OF THE HELL CREEK FORMATION, SOUTHWESTERN NORTH DAKOTA, AND ADJACENT SOUTH DAKOTA

Multituberculata
Mesodma thompsoni Clemens
Cimolodon nitidus Marsh
Cimolodon sp.
Meniscoessus robustus (Marsh)
Meniscoessus sp.
?*Meniscoessus* sp.
Metatheria
Alphadon marshi Simpson
Pediomys florencae Clemens
Pediomys sp. cf. *P. florencae* Clemens
Didelphodon vorax Marsh
Didelphodon sp.
?*Didelphodon* sp.
Eutheria
Gypsonictops illuminatus Lillegraven
Cimolestes magnus Clemens and Russell

Family "Pediomyidae"

"*Pediomys*" Marsh 1889

"*Pediomys*" *florencae* Clemens 1966
Fig. 3, F–H

Nomenclatural summary. *Pediomys florencae* Clemens 1966 (p. 50).

Holotype. UCMP 51440, left maxilla with M2 and M3.

Type locality. UCMP locality V5820, Lance Formation, Wyoming.

Distribution. Hell Creek Formation (Lancian), Montana, North Dakota, and South Dakota; Lance Formation (Lancian), Wyoming.

Referred specimen. Right m2 (PTRM 605) from PTRM locality V89004.

Discussion. PTRM 605 possesses the labial cristid obliqua of "pediomyids" and stagodontids, but lacks the mesiodistally compressed trigonid of stagodontids. The metaconid has been broken, but was clearly relatively larger than the metaconid of stagodontids. In size (L = 4.2, TRGW = 2.7, TALW = 2.8) it is closest to "*Pediomys*" *florencae*, the largest known Lancian "pediomyid."

"*Pediomys*" sp. cf. "*P*". *florencae*

Referred specimen. Right dentary with p3, m3 (PTRM 1255) from locality V92067.

Discussion. See Hunter and Pearson (1996).

Family Stagodontidae

Didelphodon Marsh 1889

Didelphodon vorax Marsh 1889
Fig. 3, I–K

Nomenclatural summary. *Didelphodon vorax* Marsh 1889 (1889a, p. 88, Plate IV, Figs. 1–3). See Clemens (1966, p. 60) for synonyms and a revised diagnosis.

Holotype. YPM 11827, left M2.

Type locality. Mammal locality 1 of Lull (1915); 1 UCMP locality V5003, Lance Formation, Wyoming.

Distribution. Scollard Formation (Lancian), Alberta; Frenchman Formation (Lancian), Saskatchewan; Hell Creek Formation (Lancian), Montana, and North Dakota; Lance Formation (Lancian), Wyoming. Possibly also Fox Hills and Hell Creek Formations (Lancian), South Dakota.

Referred specimens. Left p2 (PTRM 3012) from PTRM locality V89004; right M2 (PTRM 599), left M2 (PTRM 1230), left M3 (PTRM 3023), right M3 fragment (PTRM 1262), left dp3 (PTRM 2106), right p3 (PTRM 2114), right mX talonid (PTRM 1250), and left mX fragment (PTRM 2118) from PTRM locality V92067; right maxillary fragment with M4 (PTRM 1290) from PTRM locality V86003; right m4 (PTRM 700) from PTRM locality V87012.

Discussion. These specimens are indistinguishable from material of *Didelphodon vorax* from other Lancian sites in North America. Unfortunately, a P3 has not been recovered from these sites in North Dakota. P3 is the tooth by which *D. vorax* and the poorly known *D. padanicus* (Cope 1892) are known to differ in morphology. Therefore, assignment of these specimens to *D. padanicus*, unlikely as it may be, cannot be completely ruled out.

Didelphodon sp.

Referred specimens. Right mX fragment (PTRM 3264) from locality PTRM V92067; PX fragment (PTRM 3675) from PTRM locality V89003.

Discussion. Although these specimens also probably belong to *Didelphodon vorax*, they have been identified only to the generic level because they are so fragmentary.

?*Didelphodon* sp.

Referred specimens. Edentulous jaw (PTRM 1289) from PTRM locality V87009; lower canine (PTRM 697) from PTRM locality V87012.

Discussion. Because of their large size, these specimens, which clearly belong to a therian mammal, are assigned to ?*Didelphodon*, the largest known Lancian therian mammal.

Infraclass Eutheria
Order Proteutheria
Family Gypsonictopidae

Gypsonictops Simpson 1927

Gypsonictops illuminatus Lillegraven 1969

Nomenclatural summary. *Gypsonictops illuminatus* Lillegraven 1969 (p. 51).

Holotype. UA 2447, right maxilla with P3–M3.

Type locality. Locality KUA-1, Scollard Formation, Alberta.

Distribution. Scollard Formation (Lancian), Alberta; Frenchman Formation (Lancian), Saskatchewan; Hell Creek Formation (Lancian), Montana and North Dakota. Ravenscrag Formation (Lancian and/or Puercan?), Saskatchewan.

Referred specimen. Right dentary with p3–m3 (PTRM 1254) from PTRM locality V92067.

Discussion. See Hunter and Pearson (1996).

Family Palaeoryctidae

Cimolestes Marsh 1889

Cimolestes magnus Clemens and Russell 1965
Fig. 3, L–N

Nomenclatural summary. *Cimolestes magnus* Clemens and Russell 1965 (p. 33). See Lillegraven (1969, p. 73) for a revised diagnosis.

Holotype. UA 622, left dentary with p4–m3.

Type locality. Locality KUA-1, Scollard Formation, Alberta.

Distribution. Scollard Formation (Lancian), Alberta; Frenchman Formation (Lancian), Saskatchewan; Hell Creek Formation (Lancian), Montana and North Dakota; Lance Formation (Lancian), Wyoming.

Referred specimen. Right p4 (PTRM 695) from PTRM locality V86002.

Discussion. PTRM 695 is similar in size (L = 5.5, W = 2.5, both estimated because the mesial end is missing) and shape to p4 of *Cimolestes magnus*: massive protoconid and small posterior accessory cusp. The extra cuspule posterolingual to the posterior accessory cusp is better developed than in other specimens of this species. Wear extends to the tip of the extra cuspule, not just along the posterior ridge of the protoconid and crests of the posterior accessory cusp (Lillegraven, 1969).

Order Condylarthra
Family Arctocyonidae

Protungulatum Sloan and Van Valen 1965

Protungulatum sp.
Fig. 4, A–C

Referred specimen. Left mX trigonid (PTRM 1277) from locality PTRM V86005.

Discussion. Little can be gleaned from PTRM 1277 except that its paraconid is fully lingual in position and the cusps are moderately bulbous, as in *Protungulatum donnae*.

Oxyprimus Van Valen 1978

Oxyprimus galadrielae Van Valen 1978
Fig. 4, D–F

Nomenclatural summary. *Oxyprimus galadrielae* Van Valen 1978 (p. 53, Plate I, Fig. 1. See Lofgren (1995, p. 137) for a revised diagnosis.

Holotype. PU 16866, left maxilla with P3–M3.

Type locality. Mantua Lentil locality, Polecat Bench Formation, Wyoming.

Distribution. Polecat Bench and Ferris Formations (Puercan), Wyoming; Fort Union Formation (Puercan), North Dakota.

Referred specimen. Right m3 (PTRM 3564) from PTRM locality V97027.

Discussion. L = 4.4, TRGW = 2.9, TALW = 2.3. This specimen is nearly identical to the m3 of *Oxyprimus galadrielae* (by comparison with cast of PU 16713), from the Mantua Lentil locality. PTRM 3564 differs from the material from the type locality primarily in that its talonid notch (formed on the lingual side of the tooth between the entoconid and the base of the metaconid) is slightly more closed. PTRM 3564 differs from *Protungulatum donnae* Sloan and Van Valen 1965 in its closed talonid notch, more lingual paraconid, and more vertical hypoconulid. PTRM 3564 differs from *O. erikseni* Van Valen 1978 in its wider talonid and its more complete separation of the paraconid from the metaconid.

Arctocyonidae, gen. and sp. indet.
Fig. 4, G and H

Referred specimen. Left mX trigonid (PTRM 2102) from PTRM locality V86005.

Discussion. A second species of arctocyonid occurs at locality V86005 and different from and larger than *Protungulatum* sp. (above). Its cusps are more inflated than in *Protungulatum*, comparable to the degree of inflation seen in small species of *Baioconodon* Gazin 1941 and *Loxolophus* Cope 1885.

Family Periptychidae

Oxyacodon Osborn and Earle 1895

Oxyacodon priscilla Matthew 1937
Fig. 4, I–K

Nomenclatural summary. *Oxyacodon priscilla* Matthew 1937 (p. 207–208). See Archibald et al. (1983, p. 62–63) for a revised diagnosis.

Holotype. AMNH 3547a, right dentary with p2–m3.

Type locality. Nacimiento Formation, San Juan County, New Mexico.

Distribution. Nacimiento Formation (Puercan), New Mexico; Ferris Formation (Puercan), Wyoming; Fort Union Formation (Puercan), North Dakota.

Referred specimen. Left m2 (PTRM 1396) from PTRM locality V86005.

Discussion. PTRM 1396 shares with other *Oxyacodon* m2s lack of a lingual cingulid, a poorly developed labial cingulid, a median paraconid that is reduced in size, and moderately convergent cusp tips (more than in *Mimatuta* Van Valen 1978, but less than in *Conacodon* Matthew 1897). Based on size (L = 3.3, TRGW = 2.6, TALW = 2.7), this specimen is most likely *O. priscilla* or less likely *O. ferronensis* Archibald et al. (1983).

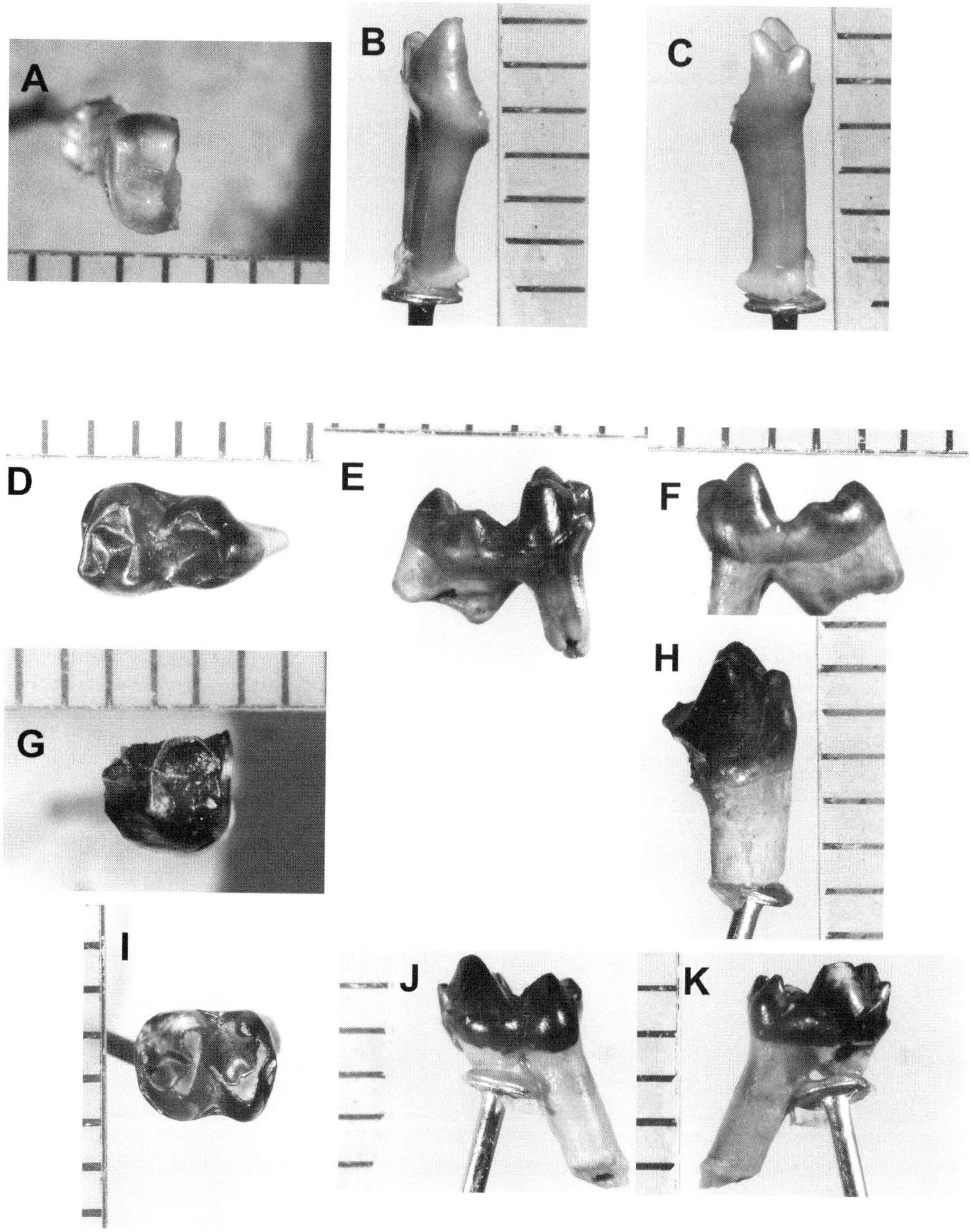

Figure 4. Photographic plate showing specimens of early Paleocene mammals from North Dakota. *Protungulatum* sp., left mX trigonid (PTRM 1277) from PTRM locality V86005 in occlusal (A), labial (B), and lingual (C) views. *Oxyprimus galadrielae*, right m3 (PTRM 3564) from PTRM locality V97027 in occlusal (D), labial (E), and lingual (F) views. Arctocyonidae, gen. and sp. indet., left mX trigonid (PTRM 2102) from PTRM locality V86005 in occlusal (G) and lingual (H) views. *Oxyacodon priscilla*, left m2 (PTRM 1396) from PTRM locality V86005 in occlusal (I), labial (J), and lingual (K) views. Scale ruled in millimeters. PTRM is Pioneer Trails Regional Museum; for abbreviations, see Table 1.

MAMMALS FROM THE HELL CREEK AND FORT UNION FORMATIONS OF NORTH DAKOTA AND ADJACENT SOUTH DAKOTA

Table 2 lists the mammalian taxa recovered from the Hell Creek Formation in southwestern North Dakota and adjacent South Dakota. Table 3 lists the mammalian taxa recovered from the Ludlow Member of the Fort Union Formation of the same area. Collecting efforts have yielded 166 mammalian specimens recovered from 15 localities in the Hell Creek and 9 specimens from 2 localities in the Ludlow Member. (In this context, a mammalian specimen is any fossil that can be identified as mammalian. For teeth, further identification is usually possible to the specific or generic level. For some cranial and postcranial fragments, identification might be possible only to the ordinal level or higher.) The locality register (see Appendix) breaks down these faunal lists by locality. Figure 5 depicts the observed stratigraphic occurrences and ranges of the identified mammalian species. In southwestern North Dakota, mammalian localities occur throughout most of the vertical extent of the Hell Creek Formation, making it difficult to define isolated local faunas, and even localities that are geographically close may be separated by considerable vertical (i.e., stratigraphic) distance. We therefore name no Cretaceous local faunas here. Nevertheless, some geographic-stratigraphic clusters of localities are discernible (see Fig. 1 for placement):

1. Pretty Butte localities (40–31 m below top): PTRM localities V92002, V87009, and V86003.
2. Marmarth localities (29–6 m below top): PTRM localities V98011 and V86002.
3. Sunset Butte localities (37–25 m below top): PTRM localities V92067 and V93022.
4. Mud Buttes localities (14–3 m below top): PTRM localities V88004, V87012, V88017, and V88018.
5. Miller Ranch localities (82–73 m below top): PTRM localities V91007, V89004, and V92025.

One of the Miller Ranch sites, PTRM locality V92025, is in South Dakota outside the range of the map used in Figure 1. PTRM locality V89003 (~8 m below the K-T boundary) is treated as an isolated occurrence southeast of the nearby Mud Buttes area.

Three species of mammals (*Alphadon marshi, Gypsonictops illuminatus*, and *Cimolestes magnus*) of the eight that have been found in the Hell Creek Formation in southwestern North Dakota occur at a single stratigraphic horizon each. Because only one specimen of each of these species has been recovered from the study area, however, little can be said except that these species occur. The recovery of these species may be largely a result of their occurrence in well-sampled vertebrate faunules: *Alphadon marshi* at PTRM locality V89004 (980 vertebrate specimens), *Gypsonictops illuminatus* at PTRM locality V92067 (3243 vertebrate specimens), and *Cimolestes magnus* at PTRM locality 86002 (1180 vertebrate specimens). (In this context, a vertebrate specimen is any vertebrate fossil that can be identified to the specific, generic, or familial level, exclusive of repeated elements such as gar scales, turtle shell fragments, and crocodilian osteoscutes; see Pearson et al., this volume.) Although the metatherian species *"Pediomys" florencae* is also known locally from only one specimen recovered ~82 m below the K-T boundary (PTRM locality V89004), material referred to "*Pediomys*" sp. cf. *P. florencae* ~37 m below the K-T boundary (PTRM locality V92067) may also belong to the same species (Hunter and Pearson, 1996).

Slightly more commonly recovered are two species of multituberculates (*Cimolodon nitidus* and *Mesodma thompsoni*) that have extensive stratigraphic ranges; however, their distributions within those ranges have been strongly influenced by the vagaries of sampling. Although *Cimolodon nitidus* is known locally with certainty from only three specimens recovered from better sampled faunules (one each from PTRM localities V89004, V86002, and V89003), its stratigraphic range spans most of the (locally 100 m thick) Hell Creek Formation from ~82 m (PTRM locality V89004) to ~8 m below the K-T boundary (PTRM locality V89003). Similarly, *Mesodma thompsoni*, a species known to have survived the K-T boundary elsewhere in the Western Interior (Archibald and Bryant, 1990), occurs as low as ~82 m below the K-T boundary and may extend as high as ~5.6–7.1 m above the boundary in the study area.

The two species of mammals most commonly recovered locally, *Didelphodon vorax* and *Meniscoessus robustus*, but not known to have survived the K-T boundary (Archibald and Bryant, 1990), occur throughout the Hell Creek Formation. To some extent, these two species may be the most commonly recovered because they are the two largest mammals in the sample and because surface collections are almost certainly biased toward recovering large specimens. Nevertheless, these two species are relatively rare at many other Lancian local faunas, particularly *Meniscoessus robustus* at higher latitudes (Lillegraven, 1969; Fox, 1989; Storer, 1991), and so their numerical dominance locally in the Hell Creek Formation of southeastern North Dakota, where *Meniscoessus robustus* composes ~58% and *Didelphodon vorax* composes ~22% of the mammalian fauna, is an unusual and significant observation.

Early Paleocene (Puercan NALMA) mammals recovered from the Ludlow Member of the Fort Union Formation occur at two localities within a single channel system and stratigraphic horizon. This assemblage of fossils has been named the

TABLE 3. MAMMALIAN FAUNA OF THE LUDLOW MEMBER OF THE FORT UNION FORMATION, SOUTHWESTERN NORTH DAKOTA

Multituberculata
Mesodma sp.
Eutheria
Protungulatum sp.
Oxyprimus galadrielae Van Valen
Arctocyonidae, gen & sp. indet.
Oxyacodon priscilla Matthew

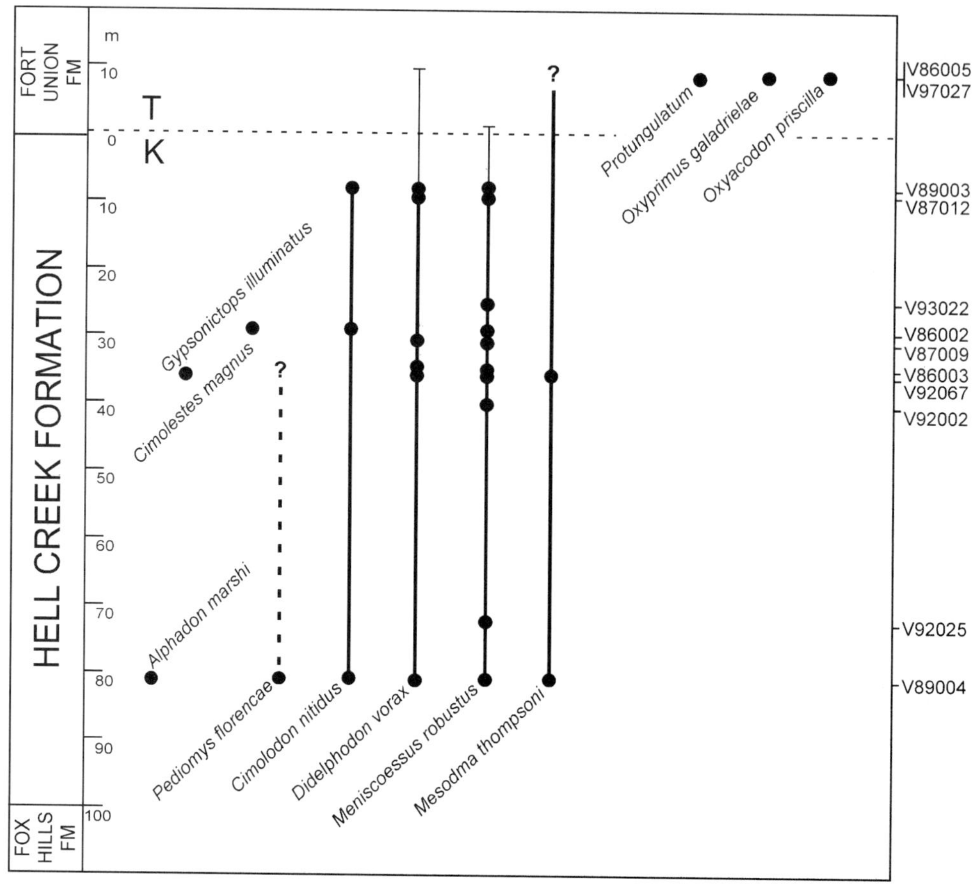

Figure 5. Chart of stratigraphic ranges of mammalian taxa in Hell Creek and Fort Union Formations (FM), southwestern North Dakota. T, Tertiary, K, Cretaceous. Local observed range of "*Pediomys*" *florencae* extends further if "*Pediomys*" sp. cf. "*P.*" *florencae* belongs to this species. *Mesodma thompsoni* may be present in Fort Union locally, but this occurrence is not unexpected given that species is known to occur in Paleocene strata outside of study area (Archibald and Bryant, 1990). One-tailed classical confidence limits (Strauss and Sadler, 1989; Marshall, 1990) have been added (thin lines) to observed ranges of *Didelphodon vorax* and *Meniscoessus robustus* to indicate degree of confidence that can be placed on highest observed local occurrences of these taxa. Only 12 localities that have produced mammalian fossils that are diagnostic to generic or specific level have been included.

PITA Flats local fauna on the basis of the faunule at PTRM locality V86005 (Hunter, 1999). This locality occurs ~8 m above the contact between the Hell Creek and Fort Union Formations. Although the K-T boundary has not been precisely identified near PTRM locality V86005, Cretaceous pollen occurs in the lowest 0.86 m of the Ludlow in the PITA Flats measured section, whereas ~0.8 km away Cretaceous pollen occurs in the lowest 2.37 m of the Ludlow in the North Dakota Academy of Science measured section (Johnson, this volume, Table A2). Given that the K-T boundary must occur at least this high, an estimate of ~5.6–7.1 m above the K-T boundary for PITA Flats seems reasonable.

Originally, one of us (Hunter, 1999, p. 142) advanced a Pu2–Pu3 age for this local fauna based on the occurrence of the condylarth *Oxyacodon priscilla*, which is known only from the Pu2 and Pu3 interval zones of the Puercan NALMA (Archibald et al., 1987; Williamson, 1996; Eberle and Lillegraven, 1998b). Two lines of evidence suggest that the PITA Flats local fauna may be older than other local faunas assigned to the Pu2–Pu3 interval zones. First, a second site (PTRM locality V97027), within the same channel system and almost assuredly the same stratigraphic horizon as PTRM locality V86005, has produced a specimen of the condylarth *Oxyprimus galadrielae*, which is know elsewhere only from the Pu1 interval zone (Van Valen, 1978; Eberle and Lillegraven, 1998b). Second, the stratigraphic position of both PITA Flats localities ~5.6–7.1 m above the K-T boundary places the PITA Flats local fauna within magnetochron 29R based on stratigraphic thickness (see Hicks et al., this volume). Known local faunas of Pu1 age elsewhere occur within C29R, whereas those of Pu2 and Pu3 age occur within the overlying C29N (Archibald et al., 1987; Swisher et al., 1993). On current evidence, the PITA Flats local fauna is clearly close to the transition between the Pu1 and Pu2 interval zones.

MAMMALS FROM THE HELL CREEK FORMATION, SOUTHEASTERN MONTANA

Table 4 lists the mammalian species recovered from the Hell Creek Formation of extreme southeastern Montana, Carter County, in the vicinity of the town of Ekalaka. Collecting efforts have yielded 874 mammalian specimens recovered from five localities. The locality register (see Appendix) breaks down this faunal list by locality. Unlike the localities in North Dakota, those in southeastern Montana are clustered stratigraphically and geographically into a Claw Butte Anthills local fauna (UCMP localities V86081–4) ~65 m and a Spigot-Bottle local fauna (UCMP locality V88007) ~61 m below the Hell Creek–

TABLE 4. MAMMALIAN FAUNA OF THE HELL CREEK FORMATION, CARTER COUNTY, SOUTHEASTERN MONTANA

Multituberculata
- *Cimolodon nitidus* Marsh
- *Cimolomys gracilis* Marsh
- *Essonodon browni* Simpson
- *Meniscoessus robustus* (Marsh)
- *Mesodma formosa* (Marsh)
- *Mesodma hensleighi* Lillegraven
- *Mesodma thompsoni* Clemens

Metatheria
- *Alphadon marshi* Simpson
- *Alphadon wilsoni* Lillegraven
- *Didelphodon vorax* Marsh
- *Glasbius twitchelli* Archibald
- *Pediomys cooki* Clemens
- *Pediomys elegans* Marsh
- *Pediomys florencae* Clemens
- *Pediomys hatcheri* Osborn
- *Pediomys krejcii* Clemens
- *Protalphadon lulli* (Clemens)
- *Turgidodon rhaister* (Clemens)

Eutheria
- *Batodon tenuis* Marsh
- *Cimolestes incisus* Marsh
- *Cimolestes propalaeoryctes* Lillegraven
- *Cimolestes stirtoni* Clemens
- *Gypsonictops hypoconus* Simpson

Fort Union formational contact. Note, however, that the Hell Creek Formation is considerably thicker near Ekalaka (~150 m) than it is in southwestern North Dakota (~100 m) (Belt et al., 1997), and so vertical distance below the formational contact cannot be compared directly between the two regions. The actual fossil material from these localities will be described elsewhere.

The Claw Butte Anthills local fauna, named for the anthills at the top of Claw Butte, was recovered from localities at the top of the Claw Butte channel belt described by Belt et al. (1997, their Fig. 12). Most material was screen washed from an in situ channel fill composed of silty sand with a very small proportion also obtained by screening the anthills. Although sediments do not directly overlie Claw Butte, Belt et al. (1997, their Fig. 7) were able to trace the beds at Claw Butte, on the basis of sedimentology, to the Blacktail Creek area ~10 km to the northeast, where the overlying formational and K-T boundaries are found. Therefore, the Claw Butte Anthill local fauna is estimated to be 65 m below the Hell Creek–Fort Union contact. The Spigot-Bottle local fauna, named for a nearby spigot and broken bottles littering the surface, was recovered from localities near the top of the Toe Bone meander belt described by Belt et al. (1997, their Fig. 12). Although there are several Spigot localities, they are far less rich than Spigot-Bottle and thus are not described further. As with Claw Butte, the sediment at Spigot-Bottle is silty sand. Although the preliminary faunal analysis presented here indicates that these two local faunas are not clearly separable on the basis of fauna, they were first named at a time when their faunal content and stratigraphic content were not well known. Accordingly, we retain them as separate local faunas but consider them as a single unit in the much broader analysis presented in this chapter.

In addition to the taxa listed in Table 4, two other records of Late Cretaceous or early Paleocene mammals from the Ekalaka area, but not from either the Claw Butte Anthills or Spigot-Bottle local faunas, are notable. Clemens (1973) reported the occurrence of *Cimolestes magnus* in the Hell Creek Formation near Ekalaka, and we have added this species to the overall faunal list for the Hell Creek Formation, Ekalaka area, used in the following faunal comparisons. Simmons (1987) described a new species of multituberculate, *Taeniolabis lamberti*, from the Fort Union Formation of the Ekalaka area, documenting the occurrence of early Paleocene mammals there.

COMPARISONS AMONG LANCIAN MAMMALIAN FAUNAS

The discovery of Lancian mammals in southwestern North Dakota and southeastern Montana augments the density of the geographic record of Lancian North American mammals and provides an opportunity to reevaluate patterns of similarity among approximately contemporaneous mammalian faunas. Previous workers have commented on putative differences in faunal composition among Lancian mammalian faunas, arguably associated with latitude. Lillegraven (1969) pointed out a higher diversity of eutherians, particularly palaeoryctoids, in the Trochu local fauna in central Alberta (1 in Fig. 6) than were known in the Lance local fauna in eastern Wyoming (7 in Fig. 6). Sloan (1969) and Russell (1975) suggested that this difference might indicate an invasion of North America by eutherian immigrants from Asia, which had not yet been completed by Lancian time. This view assumes that the Trochu and Lance local faunas were contemporaneous. Alternatively, the slightly more progressive aspect of the Trochu fauna, given that eutherians underwent a marked evolutionary radiation after the Cretaceous, could also be a result of temporal differences, the Trochu local fauna being somewhat younger than the Lance fauna. A third possibility is that local ecological differences could account for the differences in composition. Archibald (1982) described mammals from the Hell Creek Formation of eastern Montana, noting that at this intermediate latitude (i.e., between the Trochu local fauna to the north and the Lance local fauna to the south) the fauna was intermediate in taxonomic composition as well. Others have since described more Lancian faunas at intermediate latitudes (Fig. 6), and Fox (1989) and Hunter et al. (1997) discussed the general problem of separating the effects of time, latitude, and local ecology on the composition of Lancian local faunas.

Although we are not yet able to completely resolve these issues, we discuss here the more modest goal of establishing whether there are geographic patterns in the composition and abundance of Lancian mammals. To simplify, we assume that all the described Lancian local faunas are approximately contemporaneous, and therefore time has not been important in

Figure 6. Map showing distribution of well-sampled Lancian mammalian faunas compared here. Base map is adapted from Archibald et al. (1987). 1, Trochu local fauna, Scollard Formation, Alberta; 2, Gryde local fauna, Frenchman Formation, Saskatchewan; 3, Wounded Knee local fauna, Frenchman Formation, Saskatchewan; 4, Hell Creek fauna, Garfield County, Montana; 5, Hell Creek fauna, Carter County, Montana; 6, Hell Creek fauna, southwestern North Dakota; 7, Lance local fauna, Wyoming; 8, Muddy Tork local fauna, Hell Creek Formation, Dawson County, Montana; 9, Hell Creek fauna, South Dakota; 10, Red Owl local fauna, Fox Hills Formation, South Dakota.

establishing faunal differences. This is a reasonable assumption for two reasons. First, all of these local faunas occur in approximately laterally equivalent lithostratigraphic units (Gill and Cobban, 1973; Cherven and Jacob, 1985; Belt et al., 1997), the upper bounds of which in some cases approximate the K-T boundary (to within a few meters; e.g., Swisher et al., 1993; Hunter et al., 1997; Nichols and Johnson, this volume). Although Lancian faunas can be correlated approximately to upper Maastrichtian marine ammonite zones (Lillegraven and Ostresh, 1990) that represent approximately the last 4 m.y. of the Cretaceous (Gradstein et al., 1995), the finds reported here from southwestern North Dakota seem to be restricted to the last 1.32–1.68 m.y. of the Cretaceous (Hicks et al., this volume), and a similar restriction may be valid for other faunas in our comparisons (but see below regarding the Fox Hills Formation of South Dakota). Second, in the course of compiling faunal data from the literature, localities at all stratigraphic levels were pooled within the same formation and area of study to arrive at a single faunal list and set of abundances for each formation study area (Tables 5 and 6). Thus we use analytically time averaged and approximately contemporaneous (or at least temporally overlapping) samples, which should minimize any temporal pattern (i.e., difference in faunal composition correlated with difference in time) among the samples. This procedure is further justified by the fact that we observed no mammalian biostratigraphic changes through the Hell Creek Formation in our respective study areas. Nevertheless, the mammalian occurrences in the Fox Hills Formation of South Dakota (the Red Owl locality) are probably stratigraphically lower than the other occurrences and therefore probably older (Foote et al., 1999, supplementary data, *www.sciencemag.org/feature/data/985988.shl*).

In compiling data on the taxonomic composition of Lancian faunas, as presented in Tables 5 and 6, we made several decisions. First, in order to include only those faunas that have been sampled moderately well to thoroughly, so that we could assume that the known samples reflect local composition, we omitted faunas that include fewer than 8 species or 40 specimens. Unfortunately, this decision had the undesired effect of restricting the geographic scope of this study. We were forced to disregard poorly sampled Lancian, or possibly Lancian, local faunas from Alaska (Clemens and Nelms, 1993), Wyoming (Breithaupt, 1982; Eberle and Lillegraven, 1998a, 1998b; Lillegraven and Eberle, 1999), Colorado (Carpenter, 1979), Utah (Cifelli et al., 1999), and New Mexico (Flynn, 1986). We also omitted assemblages Fr-1 and the Long Fall horizon of the MHBT Quarry in Saskatchewan (Johnston, 1980; Johnston and Fox, 1984; Fox, 1989, 1997) because of their age and transitional nature, and Fr-1 further because of its small sample size, but we were able to include the more typical and well-sampled Gryde and Wounded Knee faunas from the same geographic area.

Second, we were able to include local faunas from South Dakota (Wilson, 1983, 1987) in the comparisons using presence-absence data (Table 6), but not in the comparisons using abundance data (Table 5) because only taxonomic faunal lists for the localities in the Hell Creek Formation of South Dakota have been published (Wilson, 1983) and because only multituberculates, but not therians, have been described from the Fox Hills Formation (Wilson, 1987). Moreover, precise geographic coordinates have not been published for any of these localities in South Dakota. Because of the limited geographic information available about these sites, faunal lists were pooled into two master lists, one for the Hell Creek and another for the Fox Hills, and their geographic placements (Table 7) were approximated as the towns of Buffalo and Red Owl, South Dakota, respectively. We encountered a similar problem inferring the geographic position of the Wounded Knee locality (Fox, 1989) because precise coordinates have not been published. Therefore, we assigned geographic coordinates to Wounded Knee corresponding to 3.4 km north of the Gryde locality, as described by Storer (1991).

Third, taxonomic problems in discriminating among small-bodied species of the marsupial *Alphadon* (i.e., *A. marshi*, *A. wilsoni*, and *A. jasoni*) somewhat complicate determining the abundance and even the occurrence of these species at Lancian

TABLE 5. NUMBER OF IDENTIFIABLE SPECIMENS (ABUNDANCE) OF MAMMALIAN SPECIES AT LANCIAN-AGE LOCAL FAUNAS IN THE WESTERN INTERIOR

	Scollard*	Gryde†	WK§	HC Garfield#	HC Ekalaka**	HC ND††	Lance§§	HC MT##
Alostera saskatachewanensis	1	1	3	0	0	0	0	0
Alphadon jasoni	27	309	44	13	0	0	3	0
Alphadon marshi (incl. *wilsoni*)	26	0	3	3	33	1	17	0
Batodon tenuis	9	7	0	4	3	0	1	0
Cimexomys minor	0	0	0	0	0	0	7	0
Cimolestes cerberoides	88	0	0	1	0	0	0	0
Cimolestes incisus	0	28	4	3	10	0	35	0
Cimolestes magnus	61	11	0	0	1	1	4	0
Cimolestes propalaeoryctes	19	0	2	1	4	0	0	0
Cimolestes stirtoni	0	0	2	3	5	0	13	0
Cimolodon nitidus	66	4	3	27	36	3	450	0
Cimolomys gracilis	7	18	0	7	6	0	89	0
Cimolomys trochuus	3	0	0	0	0	0	0	0
Clemensodon megaloba	0	0	0	0	0	0	3	0
Didelphodon vorax	11	3	1	16	9	11	196	6
Essonodon browni	0	0	1	10	6	0	1	0
Glasbius intricatus	0	0	0	0	0	0	16	0
Glasbius twitchelli	0	0	1	35	3	0	0	0
Gypsonictops hypoconus	8	0	0	0	78	0	314	0
Gypsonictops illuminatus	102	102	15	18	0	1	0	0
Meniscoessus robustus	0	2	2	41	55	29	186	24
Mesodma formosa	105	341	10	18	148	0	170	0
Mesodma hensleighi	34	92	0	38	34	0	26	0
Mesodma thompsoni	29	40	4	19	207	2	117	3
Neoplagiaulax burgessi	0	0	0	5	0	0	0	0
Paracimexomys priscus	9	1	2	15	0	0	0	0
Parectypodus foxi	0	4	0	0	0	0	0	0
Pediomys cooki	0	0	0	1	26	0	130	0
Pediomys elegans	9	30	3	4	106	0	152	2
Pediomys florencae	0	0	0	10	27	2	79	3
Pediomys hatcheri	2	6	0	3	33	0	73	2
Pediomys krejcii	26	28	0	9	1	0	26	1
Protalphadon lulli	0	0	0	3	3	0	19	0
Turgidodon petiminis	0	17	0	0	0	0	0	0
Turgidodon rhaister	8	0	0	4	12	0	14	1

*Trochu local fauna, Scollard Formation, Alberta: Lillegraven (1969), Lillegraven and McKenna (1986) Fox (1989, 1997).
†Gryde local fauna, Frenchman Formation, Saskatchewan: Storer (1991).
§Wounded Knee local fauna, Frenchman Formation, Saskatchewan: Fox (1989, 1997).
#Hell Creek Formation, Garfield County, near Jordan, Montana (incl. Flat Creek local fauna): Archibald (1982), Lillegraven and McKenna (1986), Johanson (1996).
**Hell Creek Formation, Carter County, near Ekalaka, Montana: this paper.
††Hell Creek Formation, southwestern North Dakota: Hunter and Pearson (1996) and this paper.
§§Lance Formation, Wyoming: Clemens (1964, 1966, 1973), Lillegraven and McKenna (1986), Krause (1992).
##Muddy Tork local fauna, Hell Creek Formation, Dawson County, Makoshika State Park, Glendive, Montana: Hunter et al. (1997).

sites. Clemens (1966) originally recognized only *A. marshi* in the Lance local fauna. After Lillegraven (1969) had named *A. wilsoni* based on material in the Trochu local fauna, Clemens (1973) reported that *A. wilsoni* also occurs in the Lance local fauna but did not report the exact number of *A. wilsoni* specimens recovered. Storer (1991) named *A. jasoni*, documenting its presence at the Gryde locality and also suggesting that it also occurred at the Wounded Knee and Bug Creek Anthills localities. Thus, as originally conceived, *A. jasoni* would have a clearly northern distribution, possibly replacing *A. marshi* and *A. wilsoni* locally. However, Johanson (1996), in revising *Alphadon*, argued that *A. wilsoni* is a junior synonym of *A. marshi* and that *A. jasoni* includes some specimens formerly identified as *A. marshi* and *A. wilsoni*. If correct, it would mean that *A. jasoni* occurs in several additional local faunas (including the Trochu, Flat Creek, and Lance) in the presence of at least *A. marshi*. Regardless of the actual status of *A. wilsoni*, we had to pool this species with *A. marshi* in order to include the Lance local fauna because published accounts do not distinguish between the two species in terms of the number of actual specimens recovered. In addition, we provisionally follow Johanson (1996) regarding specimen assignments to *A. jasoni* because they result in greater overlap between the geographic distributions of *A. jasoni* and the pooled sample of *A. marshi* + *A. wilsoni*, which is conservative relative to the hypothesis of faunal differentiation. We treated less certainly identified species (e.g., *Meniscoessus* sp. cf. *M. robustus*) as actual occurrences of these species for the sake of comparison.

Because the originally hypothesized difference among Lancian local faunas was that of a greater diversity of eutherians

TABLE 6. PRESENCE (1)/ABSENCE (0) OF MAMMALIAN SPECIES AT LANCIAN LOCAL FAUNAS IN THE WESTERN INTERIOR

	Scollard*	Gryde†	WK§	HC Garfield#	HC Ekalaka**	HC ND††	Lance§§	HC MT##	HC SD***	Fox Hills†††
Alostera saskatachewanensis	1	1	1	0	0	0	0	0	0	0
Alphadon jasoni	1	1	1	1	0	0	1	0	0	0
Alphadon marshi(incl. *wilsoni*)	1	0	1	1	1	1	1	0	1	1
Batodon tenuis	1	1	0	1	1	0	1	0	0	0
Cimexomys minor	0	0	0	0	0	0	1	0	0	0
Cimolestes cerberoides	1	0	0	1	0	0	0	0	0	0
Cimolestes incisus	0	1	1	1	1	0	1	0	0	0
Cimolestes magnus	1	1	0	0	1	1	1	0	0	0
Cimolestes propalaeoryctes	1	0	1	1	1	0	0	0	0	0
Cimolestes stirtoni	0	0	1	1	1	0	1	0	0	0
Cimolodon nitidus	1	1	1	1	1	1	1	0	1	1
Cimolomys gracilis	1	1	0	1	1	0	1	0	1	0
Cimolomys trochuus	1	0	0	0	0	0	0	0	0	0
Clemensodon megaloba	0	0	0	0	0	0	1	0	0	0
Didelphodon vorax	1	1	1	1	1	1	1	1	1	1
Essonodon browni	0	0	1	1	1	0	1	0	0	0
Glasbius intricatus	0	0	0	0	0	0	1	0	0	0
Glasbius twitchelli	0	0	1	1	1	0	0	0	0	0
Gypsonictops hypoconus	1	0	0	0	1	0	1	0	1	1
Gypsonictops illuminatus	1	1	1	1	0	1	0	0	0	0
Meniscoessus robustus	0	1	1	1	1	1	1	1	1	1
Mesodma formosa	1	1	1	1	1	0	1	0	0	1
Mesodma hensleighi	1	1	0	1	1	0	1	0	0	1
Mesodma thompsoni	1	1	1	1	1	1	1	1	0	1
Neoplagiaulax burgessi	0	0	0	1	0	0	0	0	0	0
Paracimexomys priscus	1	1	1	1	0	0	0	0	0	0
Parectypodus foxi	0	1	0	0	0	0	0	0	0	0
Pediomys cooki	0	0	0	1	1	0	1	0	0	1
Pediomys elegans	1	1	1	1	1	0	1	1	1	0
Pediomys florencae	0	0	0	1	1	1	1	1	1	0
Pediomys hatcheri	1	1	0	1	1	0	1	1	1	1
Pediomys krejcii	1	1	0	1	1	0	1	1	1	1
Protalphadon lulli	0	0	0	1	1	0	1	0	0	1
Turgidodon petiminis	0	1	0	0	0	0	0	0	0	0
Turgidodon rhaister	1	0	0	1	1	0	1	1	0	0

*Trochu local fauna, Scollard Formation, Alberta: Lillegraven (1969), Lillegraven and McKenna (1986) Fox (1989, 1997).
†Gryde local fauna, Frenchman Formation, Saskatchewan: Storer (1991).
§Wounded Knee local fauna, Frenchman Formation, Saskatchewan: Fox (1989, 1997).
#Hell Creek Formation, Garfield County, near Jordan, Montana (incl. Flat Creek local fauna): Archibald (1982), Lillegraven and McKenna (1986), Johanson (1996).
**Hell Creek Formation, Carter County, near Ekalaka, Montana: this paper.
††Hell Creek Formation, southwestern North Dakota: Hunter and Pearson (1996) and this paper.
§§Lance Formation, Wyoming: Clemens (1964, 1966, 1973), Lillegraven and McKenna (1986), Krause (1992).
##Muddy Tork local fauna, Hell Creek Formation, Dawson County, Makoshika State Park, Glendive, Montana: Hunter et al. (1997).
***Hell Creek Formation, Eureka Quarry and vicinity, and Joe Painter Quarry, South Dakota: Wilson (1983).
†††Fox Hills Formation, Red Owl local fauna, South Dakota: Wilson (1983, 1987).

TABLE 7. APPROXIMATE GEOGRAPHIC COORDINATES OF LANCIAN-AGE LOCAL FAUNAS

Local Fauna	Latitude	Longitude
Scollard	51.92	112.95
Gryde	49.31	108.42
Wounded Knee	49.34	108.42
Hell Creek Garfield	47.52	106.41
Hell Creek Ekalaka	45.86	104.87
Hell Creek North Dakota	46.14	103.80
Lance	43.20	104.55
Muddy Tork	47.09	104.72
Hell Creek South Dakota	45.35	103.32
Fox Hills, South Dakota	44.43	102.31

at higher latitudes, we first examine the number of species of multituberculates, metatherians, and eutherians among the faunas (Table 8). A simple G-test (Sokal and Rohlf, 1995, p. 724–743) establishes that there is no significant difference in the proportion of species within each higher taxon across all of the samples ($G = 12.6359$, $p = 0.8127$, degrees of freedom = 18). Moreover, results presented here on the distribution of individual species suggest that, with few exceptions, eutherians do not contribute to significant geographic patterning among Lancian faunas.

In order to summarize the variation in composition among the faunas, we examined the abundance (Table 5) and presence and/or absence data (Table 6) with correspondence analysis (CA) using NTSYS-pc (Rohlf, 1999). CA is better suited to

TABLE 8. NUMBERS OF SPECIES IN THE THREE MAJOR HIGHER TAXA OF MAMMALS AT LANCIAN-AGE LOCAL FAUNAS IN THE WESTERN INTERIOR

	Scollard	Gryde	Wounded Knee	Hell Creek Garfield	Hell Creek Ekalaka	Hell Creek North Dakota	Lance	Muddy Tork	Hell Creek South Dakota	Fox Hills
Multituberculata	7	8	6	9	7	3	9	2	3	5
Metatheria	7	6	5	11	10	3	11	6	6	6
Eutheria	7	5	5	6	6	2	5	0	1	1

species data than other ordination techniques (e.g., principal component analysis) because it characterizes relationships among rows and columns (here, species and faunal lists) using the chi-square (χ^2) distance, the metric preserved in contingency tables (Legendre and Legendre, 1998, p. 451–452). The χ^2 distance excludes double-zeros, i.e., the absence of a species from both of a pair of sites, thereby avoiding the species abundance paradox that plagues many distance metrics, including euclidean distance (Legendre and Legendre, 1998, p. 276–286). CA is equally suitable for abundance and presence and/or absence data on species. In the abundance data, rows and columns (species and faunal lists) are strongly nonindependent across the overall data table ($\chi^2 = 6328.187$, degrees of freedom = 238, $p < 0.0001$). The first CA axis explains 43.35% of the variance (eigenvalue = 0.5291), the second axis explains 21.17% (eigenvalue = 0.2584), the third axis explains 15.38% (eigenvalue = 0.1878), and the fourth axis explains 10.81% (eigenvalue = 0.1319) for a cumulative 90.17% of the variance explained by the first four axes. Figure 7 shows the projections for eight Lancian sites, for which abundance data were available, along the first two CA axes. We do not show the projections for species because of the large number (N = 35) involved. CA axis 1 recovers geographically correlated variation in pattern of species abundance, specifically correlated with latitude; localities at higher latitudes score higher along CA axis 1 than those at lower latitudes. This observation is significant by Spearman's rank order correlation (r_S, = 0.7857, $0.05 > p$; see Siegel, 1956, Table P, for critical values of r_S for $n \leq 30$) and by an exact test, systematically enumerating all 8! permutations of latitude, each time recalculating the correlation coefficient ($r = 0.7769$, $p = 0.0119$).

Unlike the abundance data, in the presence and/or absence data, independence of the rows and columns (species and sites) could not be rejected across the data table ($\chi^2 = 154.464$, not significant, degrees of freedom = 306). Accordingly, CA was also less effective in capturing variation overall in the presence and/or absence data than in the abundance data. The first CA axis explains 23.37% of the variance (eigenvalue = 0.2148), the second axis explains 17.42% (eigenvalue = 0.1602), the third axis explains 14.73% (eigenvalue = 0.1354), and the fourth axis explains 12.05% (eigenvalue = 0.1108) for a cumulative 67.56% of the variance explained by the first four axes. Figure 8 shows the projections for 10 Lancian sites along the first two CA axes, again omitting the species projections. Although CA has been less satisfactory as an ordination technique for the presence and/or absence data, CA axis 1 appears to have captured some geographically correlated variation in faunal composition, as localities at higher latitudes score higher along CA axis 1 than those at lower latitudes. This result is significant by Spearman's rank order correlation ($r_S = 0.9030$, $p < 0.01$; see Siegel, 1956, Table P, for critical values of r_S when $n \leq 30$) and by a randomization test ($r = 0.7769$, $p = 0.0123$ in 10,000 trials, 9999 random permutations of latitude plus observed).

In order to test for spatial patterning among sites in faunal abundance and composition, as suggested by the CA ordination among sites, we performed a series of Mantel tests between matrices of geographic distance and faunal distance between sites. We treated geographic distance initially both as surface distance by great circle route (Table 9) and latitudinal difference between sites. Although surface distance and latitudinal difference are obviously correlated (they are both derived from the same geographic coordinates), they are not completely redundant; there is some east-west spread among the sites. In addition, because the original hypothesis of faunal differentiation was phrased in terms of latitude, we present results for the overall faunal comparisons employing both geographical distances. Subsequently, we study the partial correlation structure of each of the measures of geographic distance with faunal distance and argue that faunal composition varies in ways more directly related to simple geographic separation than to latitude (see following). Rather than presenting actual Mantel statistics, z, which vary from problem to problem, we list the more familiar coefficient of matrix correlation (Smouse et al., 1986) and evaluate its significance by randomization test. In comparisons using the abundance data, which were available for eight faunas, testing is by systematic enumeration of all 8! permutations of one of the faunal distance matrix, recalculating z each time, with p being the proportion of z statistics obtained by randomization greater than or equal to observed z, using Matlab (Mathworks, 1999). Because exact tests would have been impractical in the presence and/or absence comparisons for all 10! permutations of the faunal distance matrix among 10 faunas, p was determined by 10 000 trials, 9999 random permutations and one observed. All tests are one-tailed and in the positive direction because negative correlation in this context, corresponding to a "repulsed" distribution across the landscape, lacks an obvious biological interpretation. We set the arbitrary threshold for sig-

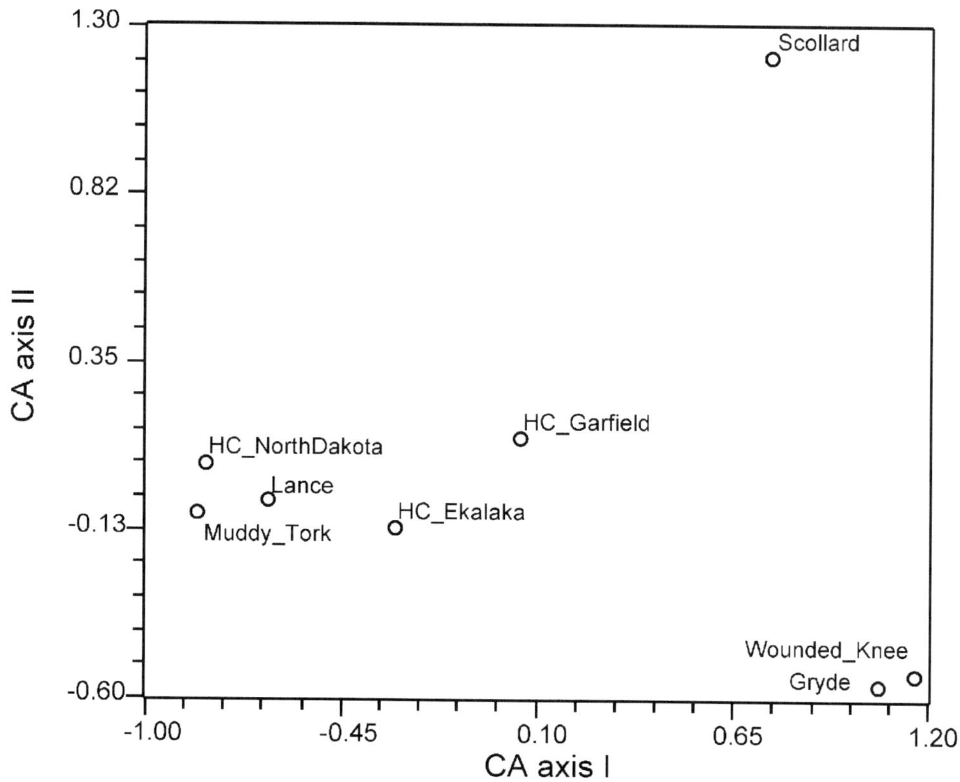

Figure 7. Ordination of Lancian mammalian faunas along first two correspondence analysis (CA) axes, comparing patterns of species abundance. See text for explanation and Table 5 for abbreviations.

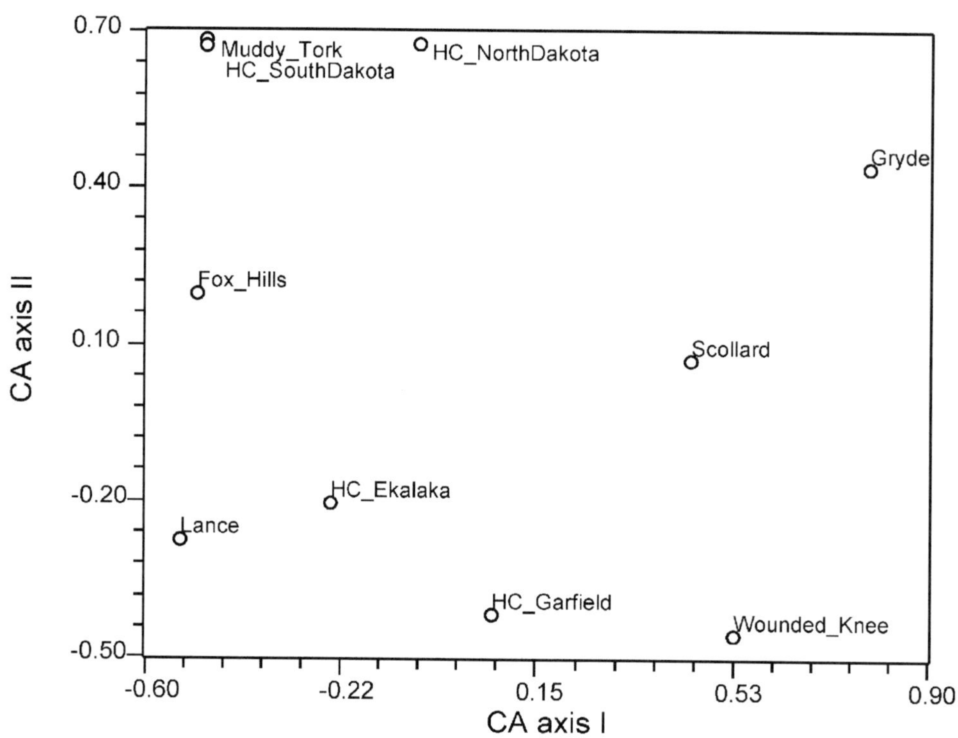

Figure 8. Ordination of Lancian mammalian faunas along first two correspondence analysis (CA) axes, comparing patterns of species presence and absence. See text for explanation and Table 6 for abbreviations.

TABLE 9. GEOGRAPHIC DISTANCES BETWEEN PAIRS OF LANCIAN-AGE LOCAL FAUNAS

	Scollard	Gryde	Wounded Knee	HC Garfield	HC Ekalaka	HC North Dakota	Lance	Muddy Tork	HC South Dakota
Gryde	431.3								
Wounded Knee	428.9	3.4							
HC Garfield	677.7	248.2	251.0						
HC Ekalaka	895.0	467.2	469.9	219.1					
HC North Dakota	925.5	494.2	496.6	251.7	88.8				
Lance	1154.9	742.0	745.1	502.3	296.9	332.2			
Muddy Tork	800.6	369.6	371.8	136.7	137.0	126.8	432.5		
HC South Dakota	1015.2	584.3	586.8	338.1	133.1	94.8	258.5	220.7	
Fox Hills, SD	1144.5	713.7	716.2	467.2	255.9	222.2	226.2	349.0	129.4

Note: Distances in kilometers, HC is Hell Creek. SD is South Dakota.

nificance at the conventional $p < 0.05$ level, but given the exploratory nature of this study and the small sample sizes ($N = 8$ or 10 faunas), we consider any $p < 0.1$ as marginally significant.

The upper half of Table 10 shows the results for χ^2 distance as the faunal distance metric between sites, taking into account variation in all 35 species. None of the tests of association between χ^2 distance based on all species and geographic distance was significant. We interpret this result to indicate great similarity among Lancian faunas in composition, with little differentiation or geographic patterning overall. Nevertheless, the existence of some latitudinal patterning is suggested by the CA ordination, at least along CA axis 1. Therefore, we also investigated faunal distance within the reduced dimensionality of the CA space. The lower half of Table 10 shows the results for a distance simply called faunal dissimilarity that we define as the euclidean distance among sites within the ordination space delimited by the first four CA axes. Such distances in the space of the principal axes should be euclidean after rotation (Legendre and Legendre, 1998, p. 452), and proper scaling of the axes results in weighting projections along the first axis higher than the second, the second higher than the third, and so forth. All of the tests of association between geographic distance and faunal dissimilarity, thus defined, are significant (Table 10, lower half). In spite of the great overall similarity in composition among these faunas, the major axes of variation among them, as summarized by CA, are related to geography.

We decided to study the strength of the association of each of these two measures of geographic distance (surface distance and latitudinal difference) with faunal dissimilarity by studying their partial correlation structure (Smouse et al., 1986). Faunal dissimilarity (*f*, calculated on the abundance data) is correlated both with surface distance, *s* ($r_{fs} = 0.5298$, $p = 0.0055$), and with latitudinal difference, *l* ($r_{fl} = 0.4248$, $p = 0.0169$), between sites (Table 10). Furthermore, the partial correlation of faunal dissimilarity and surface distance, with latitudinal difference held constant, is significantly positive (partial correlation, $r_{fs\,l} = 0.5358$, $p = 0.0076$) by an exact test, permutating faunal dissimilarity while keeping the geographic distance matrices fixed, by the method described by Smouse et al. (1986).

TABLE 10. MANTEL TESTS OF ASSOCIATION BETWEEN GEOGRAPHIC OR LATITUDINAL (DEGREES) DISTANCE, AND CHI SQUARED DISTANCE OR FAUNAL DISSIMILARITY* BETWEEN LOCALITIES

	r_M†	p§
Chi Squared Distance between Localities		
Abundance Data		
X Geographic Distance	0.1667	0.1869
X Latitudinal Distance	0.0401	0.4244
Presence and/or Absence Data		
X Geographic Distance	0.10318	0.2298
X Latitudinal Distance	0.05282	0.3190
Faunal Dissimilarity between Localities		
Abundance Data		
X Geographic Distance	0.5298	0.0055
X Latitudinal Distance	0.4248	0.0169
Presence and/or Absence Data		
X Geographic Distance	0.43494	0.0104
X Latitudinal Distance	0.42181	0.0106

Note: Distance in kilometers.
*Faunal dissimilarity: Euclidean distance between projections of objects (localities) along the first four correspondence analysis axes (see text for explanation)
†r_M, coefficient of matrix correlation
§p, one-tailed probability of obtaining random Z ≥ observed Z, from systematic enumeration of all 8! permutations of the faunal distances (abundance data) or from 10 000 enumerations (9999 random permutations + observed; presence/absence data).

In contrast, the partial correlation of faunal dissimilarity and latitudinal difference, with surface distance held constant, is actually negative (partial correlation, $r_{fl\,s} = -0.4332$). Negative correlation here lacks an obvious biological explanation, and we simply consider it to be nonsignificant. We interpret these results to indicate that considerable residual association remains between overall surface distance and faunal distance after correcting for the effect of latitude, presumably caused by some east-west differentiation. No association remains, however, between faunal distance and latitudinal difference after correcting for overall surface distance, meaning that the observed latitudinal patterning in the faunal distances (Table 10) is accounted for by simple geographic separation. For this reason, remaining comparisons are made using only surface distance.

In order to determine which species distributions are most affected by geographic separation, we performed a series of Mantel tests between geographic and faunal distance between sites, but with each species considered separately (Table 11). Because we could not calculate χ^2 distance without complete column totals (i.e., without other species from the same fauna), we instead calculated faunal distance as the Bray-Curtis distance, for the log (Y + 1) transformed abundance data, and as 1—the simple matching coefficient (SM), for the presence and/or absence data. Although SM as a similarity measure is not recommended for species data because double-zeros contribute to it (Sokal and Sneath, 1963; Legendre and Legendre, 1998), such double-zeros contribute 0 distance to the dissimilarity measure 1—SM. In the context of examining individual species distributions, registering a distance of 0 in such cases seems more reasonable than treating such cases as missing data at a consequential loss in power. We omitted species that occur at only a single locality among the faunas considered here, and we could not analyze the presence and/or absence distribution of the metatherian *Didelphodon* across the landscape because this mammal occurs at every locality.

Most species distributions do not vary significantly with geographic distance within this data set (Table 11). This observation is true of most species of eutherians including all "palaeoryctoid" species (*Cimolestes* and *Batodon*), with the possible exception of *Cimolestes propalaeoryctes*. Even this species, however, shows only marginally significant spatial patterning in the presence and/or absence comparison, probably because of its absence from the Lance local fauna, and no significant spatial patterning in the abundance comparison. Therefore, there is little evidence for geographic patterning in the evolutionary radiation of palaeoryctoid eutherians in Lancian time. Three species (*Alostera saskatachewanensis, Paracimexomys priscus*, and *Pediomys florencae*) have distributions that varied significantly with geography in both comparisons (Table 11), *Alostera saskatachewanensis* and *Paracimexomys priscus* being characteristic of northern latitudes and *Pediomys florencae* being characteristic of southern latitudes during Lancian time (Tables 5 and 6). *Meniscoessus robustus* showed significant, geographically related variation in abundance, but not in occurrence (Table 11), being present at all latitudes (Table 6) but consistently more abundant at lower latitudes (Table 5). *Gypsonictops illuminatus* is more abundant (Table 5) and occurs more consistently (Table 6) at higher latitudes, and accordingly its distribution shows significant geographic patterning in the presence and/or absence data, although the correlation

TABLE 11. MATRIX CORRELATIONS BETWEEN INDIVIDUAL SPECIES DISTRIBUTIONS* AND GEOGRAPHIC DISTANCE (KM) BETWEEN FAUNAS

	Abundance data		Presence and/or absence data	
	r_M	p	r_M	p
Alostera saskatachewanensis	**0.6098**	**0.0179**	**0.70928**	**0.0082**
Alphadon jasoni	**0.2297**	**0.0661**	**0.30076**	**0.0344**
Alphadon marshi (incl. *wilsoni*)	−0.2915	0.9585	−0.13538	0.6243
Batodon tenuis	−0.0663	0.6229	−0.01266	0.3414
Cimolestes cerberoides	0.2793	0.2143	0.34310	0.1123
Cimolestes incisus	−0.0573	0.6033	−0.01655	0.3809
Cimolestes magnus	0.0614	0.2497	−0.08233	0.8485
Cimolestes propalaeoryctes	0.0176	0.3869	**0.16703**	**0.0815**
Cimolestes stirtoni	−0.0610	0.6006	−0.12367	0.8187
Cimolodon nitidus	−0.1813	0.7253	−0.20295	0.8996
Cimolomys gracilis	−0.1692	0.8622	−0.14550	0.8882
Didelphodon vorax	0.0848	0.3754	—	—
Essonodon browni	−0.1300	0.8143	−0.12367	0.8129
Glasbius twitchelli	−0.3252	0.9881	−0.16596	0.7925
Gypsonictops hypoconus	**0.3144**	**0.0625**	0.09502	0.1904
Gypsonictops illuminatus	**0.2914**	**0.0554**	**0.34681**	**0.0266**
Meniscoessus robustus	**0.7986**	**0.0005**	0.67589	0.1002
Mesodma formosa	−0.1873	0.7067	−0.13341	0.7157
Mesodma hensleighi	−0.2120	0.8985	−0.16000	0.9189
Mesodma thompsoni	−0.1586	0.7921	−0.12218	0.5012
Paracimexomys priscus	**0.4852**	**0.0119**	**0.61717**	**0.0048**
Pediomys cooki	0.0551	0.3988	0.06117	0.2759
Pediomys elegans	−0.1032	0.4850	−0.04566	0.4486
Pediomys florencae	**0.5915**	**0.0046**	**0.35848**	**0.0300**
Pediomys hatcheri	−0.1437	0.6680	−0.15972	0.7102
Pediomys krejcii	−0.2525	0.8899	−0.15972	0.7153
Protalphadon lulli	0.0813	0.3452	0.06117	0.2753
Turgidodon rhaister	−0.1463	0.8315	−0.06661	0.6924

*Species Distribution measured as either the Bray-Curtis distance (abundance data) or 1-simple matching coefficient (presence/absence data) between localities for each species individually (see text for explanation).

was only marginally significant in the abundance comparison (Table 11). Its congener, *G. hypoconus*, also shows marginally significant geographic patterning in the abundance comparison (Table 11), and these two species may have disjunct distributions (see following). *Alphadon jasoni* is also probably more abundant (Table 5) and occurs more frequently (Table 6) at northern latitudes, but its distribution shows significant geographic patterning only for the presence and/or absence data, the abundance comparison being only marginally significant (Table 11).

Of those species distributions that vary significantly with geography, three kinds of patterns are apparent, taking into account other information about these species. First, there are species without close relatives surviving into Lancian time, with restricted geographic distributions during the Lancian, but whose ancestors may have been more widespread at earlier times. In this first category are the eutherian *Alostera saskatachewanensis* and the multituberculate *Paracimexomys priscus*. A recent interpretation of *Alostera*, an unusual and problematic eutherian (Fox, 1989), places it among the "zhelestids," a primarily Asian group of mammals from which modern ungulates may have arisen (Nessov et al., 1998). Other possible North American zhelestids, according to Nessov et al. (1998), are *Avitotherium* and *Gallolestes*, which occurred in pre-Lancian time at more southern latitudes in Utah (Cifelli, 1990), Texas (Rowe et al., 1992; Cifelli, 1994), and Baja California, Mexico (Lillegraven, 1976; Clemens, 1980). If *Alostera* is part of a North American radiation of zhelestids, then its restricted northern distribution in Lancian time may be viewed as relict. Similarly, the restricted northern distribution of the multituberculate *P. priscus* in Lancian time may also be relict, because this species occurred at least in Wyoming in pre-Lancian time (Lillegraven and McKenna, 1986). Moreover, *P. priscus* appears to be a late surviving member of a group of multituberulates that radiated much earlier in the Cretaceous (Eaton and Nelson, 1991; Eaton, 1995).

A second category of species with restricted distributions includes those with close, contemporaneous relatives of similar size and morphology (and presumably similar ecology) present during Lancian time, but that occur, if not completely allopatrically, then at least in a disjunct way. In this second category are the metatherian *Alphadon jasoni* and the eutherian *Gypsonictops illuminatus*. *A. jasoni* is undoubtedly closely related to *A. marshi* and *A. wilsoni*, either of which can be easily confused with *A. jasoni*, particularly when only lower dentitions are compared. Moreover, the distribution of *A. jasoni* seems to be negatively associated with that of the pooled sample of *A. marshi* and *A. wilsoni*, i.e., *A. jasoni* seems to be abundant when *A. marshi* + *A. wilsoni* is rare and vice versa (Table 5). This negative association, however, is only marginally significant by an exact test ($r = -0.4148$, $p = 0.0746$, 7! permutations, excluding Muddy Tork, where neither taxon has been sampled [a double-zero]). Interpretation of the distribution of *Alphadon* species is further hampered by problems in their alpha taxonomy (Johanson, 1996). More systematic work is needed in order to gain confidence in estimates of the abundance of these species. Particular attention needs to be paid to identifying lower dentitions, which cannot at present be placed with confidence.

Likewise, the eutherian *Gypsonictops illuminatus* is also extremely similar to its congener *G. hypoconus*. These two species also appear to have a negative association in abundance (Table 5), but this difference is not significant by an exact test ($r = -0.3879$, $p = 0.1048$ from 7! permutations, excluding Muddy Tork [a double-zero]). Because *G. hypoconus* is well sampled only at the southern extreme of the area considered in these comparisons and because this species shows only marginally significant spatial patterning (Table 11), it is possible that failure to find a significant negative correlation is from geographically patchy sampling of *G. hypoconus*.

The third pattern is demonstrated by *Meniscoessus robustus*, by far the largest multituberculate and one of the two largest mammals (the other being the metatherian *Didelphodon*) known from the Lancian. Because of its large size and unique dental morphology, with highly crescentic molar cusps, it is unlikely that *Meniscoessus* was closely similar ecologically to other contemporaneous and sympatric mammals. The highly significant geographic patterning of *Meniscoessus* abundance (Table 11), abundant in the south and rare farther north (Table 5), in the absence of obvious competitors, suggests that the distribution of this taxon may have simply reflected its own intrinsic ecological tolerances.

To summarize this comparison among Lancian mammalian faunas, it must be emphasized that all of these faunas are extremely similar in composition and do not vary greatly with distance, as evidenced by the nonsignificant correlations between χ^2 distance among the faunas and geographic distance (Table 10, upper half). Nevertheless, the variation among these faunas is associated with the geographic distance, as evidenced by the significant correlation between the distances between faunas in the reduced ordination space defined by correspondence analysis and geographic distance defined as either surface distance or latitudinal separation (Table 10, lower half). The partial correlation structure of each of these measures of geographic separation with faunal dissimilarity suggests that faunas differ in ways more closely related to simple geographic separation, than to latitude. Only a few species seem to show significant geographic patterning in their distributions, only two of these are eutherians, and neither of them is a palaeoryctoid (Table 11). Therefore, the hypothesis of geographic patterning to the evolutionary radiation of eutherians in the Lancian is not supported. Instead, certain species show geographic patterning that is understandable in terms of their replacement across the landscape by closely related and, presumably, ecologically similar species. Others show a pattern suggestive of contraction of a formerly more widespread distribution. Rather than historical effects, such as incomplete invasion of eutherian lineages into North America from Asia, simple spatial gradients can account for variation among mammalian faunas across this geographically small window into the latest Cretaceous.

CONCLUSIONS

The goals of this chapter have been to describe mammalian fossils recovered from the Hell Creek and Fort Union Formations of southwestern North Dakota and adjacent South Dakota, to report work in progress on the mammalian fauna of the Hell Creek Formation of southeastern Montana, and to reconsider geographic differentiation among Lancian mammalian faunas. Mammalian fossils occur throughout the vertical extent of the Hell Creek Formation in North Dakota, but at no one stratigraphic horizon are they yet exhaustively sampled. Although two well-sampled local faunas occur in the Hell Creek Formation of southeastern Montana, the distribution of fossil mammals there is geographically and stratigraphically irregular. The mammalian fauna recovered from these relatively new field areas is closely similar to other Lancian faunas, particularly those at more southern latitudes, such as the Lance fauna. We did not find evidence for geographic patterning of a latest Cretaceous evolutionary radiation of eutherians. Rather, simple spatial differences, and possibly replacement of ecologically similar species across the landscape, account for observed differences in composition among Lancian faunas. Separating the effects of geographic separation from those of time, however, may become possible in the future with more intense vertical sampling through the Hell Creek Formation and its lateral equivalents, to estimate a time factor, or by including faunas before and after the Lancian. Separating the effects of geographic separation from ecology may become possible with more explicit paleoenvironmental reconstruction to establish distances along paleoecological gradients. As these data become available, it should be possible to partition their effects by extending the partial matrix correspondence approach employed here.

ACKNOWLEDGMENTS

We thank Joseph Hartman, Kirk Johnson, and Douglas Nichols for inviting us to contribute to this volume, for providing invaluable assistance both in the field in North Dakota and in the laboratory, and for making their paleontologic, stratigraphic, and geologic data accessible. We thank the editors, Rich Cifelli and Bill Clemens, and an anonymous reader for reviews that pointed out ways to improve this manuscript. Kirk Johnson also provided the base map used to prepare Figure 1. We thank Dean Pearson, who found many of the localities reported here, collected many of the specimens described here, and has continued to supply logistical support to Hunter's field studies in southwestern North Dakota over the past several years. We also thank Robert Sokal and Michael Rosenberg for discussion of spatial statistics. Michael Rosenberg made available his spatial analysis program Passage (Rosenberg, 2000), which we used initially to calculate partial Mantel correlations, although we later replaced the significance tests with exact permutation tests using Matlab. A Geological Society of America research grant and a Patterson Award from the Society of Vertebrate Paleontology (to Hunter) supported field studies in southwestern North Dakota and South Dakota. National Science Foundation (NSF) grant EAR-84075507 (to Archibald) supported the original collection of the Ekalaka specimens and NSF grant EAR-9804771 (to Archibald) helped support their analysis.

APPENDIX: LOCALITY REGISTER

The person whose name follows each taxonomic assignment identified the fossil material. PTRM is Pioneer Trails Regional Museum.

PTRM locality V89004

Bowman County, North Dakota, Doaks Butte Quadrangle, 0.97 km south of Doaks Butte, sec. 33, T.129N, R.106W, 45°56′51″N, 103°56′46″W, elevation ~899 m. Hell Creek Formation, 81.68 m below top.

Mesodma thompsoni Clemens; Hunter
Cimolodon nitidus Marsh; Hunter
Meniscoessus robustus (Marsh); Hunter
Meniscoessus sp.; Hunter
?*Meniscoessus* sp.; Hunter
Multituberculata indet.; Hunter
Alphadon marshi Simpson; Hunter
Pediomys florencae Clemens; Hunter
Didelphodon vorax Marsh; Hunter
Marsupialia indet.; Hunter
Theria indet.; Hunter
Mammalia indet.; Hunter

PTRM locality V91007

Bowman County, North Dakota, Doaks Butte Quadrangle, 0.97 km south of Doaks Butte and 0.32 km east of V89004, sec. 33, T.129N, R.106W, 45°56′53″N, 103°56′29″W, elevation ~893 m. Hell Creek Formation, 80.16 m below top.

Mammalia indet.; Hunter
Theria indet.; Hunter

PTRM locality V92025

Harding County, South Dakota, Doaks Butte Quadrangle, 1.9 km southeast of Doaks Butte, sec. 30, T.23N, R.2 E., 45°55′47″N, 103°55′11″W, elevation ~905 m. Hell Creek Formation (lower one-third), 73 m below top.

Meniscoessus sp.; Hunter

PTRM locality V92002

Slope County, North Dakota, Pretty Butte Quadrangle, 2.7 km northeast of Pretty Butte, sec. 24, T.134N, R.106W, 46°23′47″N, 103°56′08″W, elevation ~823 m. Hell Creek Formation, 42.60 m below top.

Meniscoessus robustus (Marsh); Hunter
Multituberculata indet.; Hunter

PTRM locality V92067

Bowman County, North Dakota, Cedar Hills Quadrangle, 1.6 km southeast of Sunset Butte, sec. 27, T.131N, R.105W, 46°08′11″N,

103°47′49″W, elevation ~887 m (corrected from Hunter and Pearson, 1996). Hell Creek Formation, 37.18 m below top (corrected from Hunter and Pearson, 1996).
Mesodma thompsoni Clemens; Hunter
Cimolodon sp.; Hunter
Meniscoessus robustus (Marsh); Hunter
Meniscoessus sp.; Hunter
Multituberculata indet.; Hunter
?Multituberculata indet.; Hunter
Pediomys sp. cf. *P. florencae* Clemens; Hunter
Didelphodon vorax Marsh; Hunter
Didelphodon sp.; Hunter
Marsupialia indet.; Hunter
Gypsonictops illuminatus Lillegraven; Hunter
Theria indet.; Hunter
Mammalia indet.; Hunter
?Mammalia indet.; Hunter

PTRM locality V86003
Slope County, North Dakota, Pretty Butte Quadrangle, at base of Pretty Butte, sec. 35, T.134N, R.106W, 46°22′47″N, 103°56′53″W, elevation ~841 m. Hell Creek Formation, 35.05 m below top.
Meniscoessus sp.; Hunter
?*Meniscoessus* sp.; Hunter
Didelphodon sp.; Hunter

PTRM locality V87009
Slope County, North Dakota, Pretty Butte Quadrangle, sec. 35, T.134N, R.106W, 46°22′48″N, 103°56′53″W, elevation ~841 m. Hell Creek Formation, 31.43 m below top.
Meniscoessus robustus (Marsh); Hunter
?*Didelphodon* sp.; Hunter
Mammalia indet.; Hunter

PTRM locality V86002
Slope County, North Dakota, Marmarth Quadrangle, 2.9 km northeast of Marmarth, sec. 20, T.133N, R.105 W, 46°18′34″N, 103°53′33″W, elevation ~853 m. Hell Creek Formation, 29.18 m below top.
Cimolodon nitidus (Marsh); Hunter
Meniscoessus sp.; Hunter
Cimolestes magnus Clemens and Russell; Hunter
Theria indet.; Hunter

PTRM locality V93022
Bowman County, North Dakota, Cedar Hills Quadrangle, 1.9 km southeast of Sunset Butte, sec. 27, T.131N, R.105W, 46°08′04″N, 103°47′52″W, elevation ~893 m. Hell Creek Formation, 25.3 m below top.
Meniscoessus robustus (Marsh); Hunter
Theria indet.; Hunter
Mammalia indet.; Hunter

PTRM locality V88018
Bowman County, North Dakota, Mud Buttes Quadrangle, sec. 11, T.129N, R.105W, 46°01′00″N, 103°45′50″W, elevation ~945 m. Hell Creek Formation, 14.27 m below top.
Mammalia indet.; Hunter

PTRM locality V88004
Bowman County, North Dakota, Mud Buttes Quadrangle, sec. 2, T.129N, R.105W, 46°01′32″N, 103°45′54″W, elevation ~951 m. Hell Creek Formation, 10.61 m below top.
Theria indet.; Hunter

PTRM locality V87012
Bowman County, North Dakota, Mud Buttes Quadrangle, sec. 2, T.129N, R.105W, 46°01′08″N, 103°45′52″W, elevation ~945 m. Hell Creek Formation, 9.85 m below top. *Meniscoessus robustus* (Marsh); Hunter
Didelphodon sp.; Hunter
?*Didelphodon* sp.; Hunter
?Mammalia indet.; Hunter

PTRM locality V89003
Bowman County, North Dakota, Dogie Butte Quadrangle, sec. 13, T.129N, R.105W, 45°59′46″N, 103°45′29″W, elevation ~945 m. Hell Creek Formation, 8.4 m below top.
Cimolodon nitidus Marsh; Hunter
Meniscoessus robustus (Marsh); Hunter
Meniscoessus sp.; Hunter
Didelphodon sp.; Hunter
Mammalia indet.; Hunter

PTRM locality V98011
Slope County, North Dakota, Marmarth SE Quadrangle, sec. 16, T.133N, R.105W, 46°20′2″N, 103°52′37″W, elevation ~875 m. Hell Creek Formation, 5.90 m below top.
Mammalia indet.; Hunter

PTRM locality V88017
Bowman County, North Dakota, Mud Buttes Quadrangle, sec. 2, T.129N, R.105W, 46°01′08″N, 103°45′50″W, elevation ~951 m. Hell Creek Formation, 3.17 m below top.
Theria indet.; Hunter

UCMP locality V86081
Claw Butte Anthill No. 1. Carter County, MT, Blacktail Creek SW Quad, sec. 16, T.1 S., R.55 E., 45°45′43″N, 104°55′26″W, elevation ~1001 m. Hell Creek Formation, ~65 m below top.
Mesodma formosa (Marsh); Archibald
?*Mesodma formosa* (Marsh); Archibald
Mesodma thompsoni Clemens; Archibald
?*Mesodma thompsoni* Clemens; Archibald
Mesodma sp.; Archibald
?*Mesodma* sp.; Archibald
Cimolodon nitidus Marsh; Archibald
?*Essonodon browni* Simpson; Archibald
Meniscoessus robustus (Marsh); Archibald
Alphadon marshi Simpson; Archibald
Pediomys elegans Marsh; Archibald
Pediomys florencae Clemens; Archibald
Pediomys hatcheri Osborn; Archibald
Gypsonictops sp.; Archibald
Mammalia indet.; Archibald

UCMP locality V86082
Claw Butte Anthill No. 2. Carter County, MT, Blacktail Creek SW Quad, sec. 16, T.1 S., R.55 E., 45°45′43.5″N, 104°55′25″W, elevation ~1001 m. Hell Creek Formation, ~65 m below top.
Mesodma hensleighi Lillegraven; Archibald
Mesodma formosa (Marsh); Archibald

Mesodma thompsoni Clemens; Archibald
Mesodma sp.; Archibald
?*Mesodma* sp.; Archibald
Cimolodon nitidus Marsh; Archibald
Essonodon browni Simpson; Archibald
Meniscoessus robustus (Marsh); Archibald
Alphadon marshi Simpson; Archibald
Pediomys cooki Clemens; Archibald
Pediomys elegans Marsh; Archibald
Pediomys elegans Marsh or *P. cooki* Clemens; Archibald
Pediomys florencae Clemens; Archibald
Pediomys hatcheri Osborn; Archibald
Pediomys sp. cf. *P. hatcheri* Osborn; Archibald
Gypsonictops sp.; Archibald
Mammalia indet.; Archibald

UCMP locality V86083

Claw Butte Anthill No. 3. Carter County, MT, Blacktail Creek SW Quad, sec. 16, T.1 S., R.55 E., 45°45′43.5″N, 104°55′25″W, elevation ~1001 m. Hell Creek Formation, ~65 m below top.
Mesodma formosa (Marsh); Archibald
Mesodma hensleighi Lillegraven; Archibald
Mesodma thompsoni Clemens; Archibald
Cimolodon nitidus Marsh; Archibald
Meniscoessus robustus (Marsh); Archibald
Protalphadon lulli (Clemens); Archibald
Alphadon marshi Simpson; Archibald
Glasbius twitchelli Archibald; Archibald
Glasbius sp. cf. *G. twitchelli* Archibald; Archibald
Pediomys cooki Clemens; Archibald
Pediomys florencae Clemens; Archibald
Pediomys hatcheri Osborn; Archibald
Pediomys krejcii Clemens; Archibald
Gypsonictops hypoconus Simpson; Archibald
Gypsonictops sp.; Archibald
Mammalia indet.; Archibald

UCMP locality V86084

Claw Butte Anthill No. 4. Carter County, MT, Blacktail Creek SW Quad, sec. 16, T.1 S., R.55 E., 45°45′44″N, 104°55′24″W, elevation ~1001 m. Hell Creek Formation, ~65 m below top.
Mesodma formosa (Marsh); Archibald
Mesodma hensleighi Lillegraven; Archibald
Mesodma thompsoni Clemens; Archibald
Cimolodon nitidus Marsh; Archibald
Essonodon browni Simpson; Archibald
Meniscoessus robustus (Marsh); Archibald
?*Alphadon marshi* Simpson; Archibald
?*Alphadon wilsoni* Lillegraven; Archibald
Protalphadon lulli (Clemens); Archibald
Alphadon marshi Simpson; Archibald
Glasbius twitchelli Archibald; Archibald
Pediomys sp. cf. *P. hatcheri* Osborn; Archibald
Pediomys cooki Clemens; Archibald
Pediomys elegans Marsh; Archibald
Pediomys florencae Clemens; Archibald
Pediomys hatcheri Osborn; Archibald
Turgidodon rhaister (Clemens); Archibald
Batodon tenuis Marsh; Archibald
Gypsonictops hypoconus Simpson; Archibald
Mammalia indet.; Archibald

UCMP locality V88007

Spigot-Bottle. Carter County, MT, Blacktail Creek SE Quad, sec. 3, T.1N, R.55 E., 45°51′29″N, 104°52′28″W, elevation ~939 m. Hell Creek Formation, ~61 m below top.
Mesodma formosa (Marsh); Archibald
Mesodma hensleighi Lillegraven; Archibald
Mesodma thompsoni Clemens; Archibald
Cimolodon nitidus Marsh; Archibald
Cimolomys gracilis Marsh; Archibald
Cimolodon nitidus Marsh; Archibald
Meniscoessus robustus (Marsh); Archibald
Protalphadon lulli (Clemens); Archibald
Alphadon marshi Simpson; Archibald
Alphadon wilsoni Lillegraven; Archibald
Pediomys cooki Clemens; Archibald
Pediomys elegans Marsh; Archibald
Pediomys florencae Clemens; Archibald
Pediomys hatcheri Osborn; Archibald
Pediomys sp. cf. *P. hatcheri* Osborn; Archibald
Pediomys krejcii Clemens; Archibald
Pediomys sp. cf. *P. krejcii* Clemens; Archibald
Pediomys sp.; Archibald
Turgidodon rhaister (Clemens); Archibald
Didelphodon vorax Marsh; Archibald
Metatheria indet.; Archibald
Batodon tenuis Marsh; Archibald
Cimolestes incisus Marsh; Archibald
Cimolestes propalaeoryctes Lillegraven; Archibald
Cimolestes stirtoni Clemens; Archibald
Gypsonictops sp.; Archibald
cf. *Paranyctoides* sp.; Archibald
"Insectivora" indet.; Archibald

PTRM locality V86005

PITA Flats locality, PTRM locality P86001, NM-4. Slope County North Dakota, Marmarth SE Quadrangle, sec. 10, T.133N, R.105W, 46°20′22″N, 103°51′36″W, elevation ~881 m. Fort Union Formation, Ludlow Member, ~8 m above base.
Mesodma sp.; Hunter
Multituberculata indet.; Hunter
Oxyacodon priscilla Matthew; Hunter
?*Protungulatum* sp.; Hunter
Condylarthra indet.; Hunter
Mammalia indet.; Hunter

PTRM locality V97027

Slope County, North Dakota, Marmarth SE Quadrangle, sec. 9, T.133N, R.105 W, 46°20′55″N, 103°52′36″W, elevation ~860 m. Fort Union Formation, Ludlow Member, ~8 m above base.
Oxyprimus galadrielae Van Valen; Hunter

REFERENCES CITED

Alroy, J., 1999, The fossil record of North American mammals: Evidence for a Paleocene evolutionary radiation: Systematic Biology, v. 48, p. 107–118.

Archibald, J.D., 1982, A study of Mammalia and geology across the Cretaceous-Tertiary boundary in Garfield County, Montana: University of California Publications in Geological Sciences, v. 122, p. 1–286.

Archibald, J.D., 1996, Dinosaur extinction and the end of an era: What the fossils say: New York, Columbia University Press, 237 p.

Archibald, J.D., and Bryant, L.J., 1990, Differential Cretaceous/Tertiary extinctions of nonmarine vertebrates: Evidence from northeastern Montana, *in* Sharpton, V.L., and Ward, P.D., eds., Global catastrophes in Earth History: An interdisciplinary conference on impacts, volcanism, and mass mortality: Geological Society of America Special Paper 247, p. 549–562.

Archibald, J.D., and Deutschman, D.H., 2001, Quantitative analysis of the timing of the origination and diversification of extant placental orders: Journal of Mammalian Evolution, v. 8, p. 107–124.

Archibald, J.D., and Lofgren, D.L., 1990, Mammalian zonation near the Cretaceous-Tertiary boundary, *in* Bown, T.M., and Rose, K.D., eds., Dawn of the Age of Mammals in the northern part of the Rocky Mountain Interior, North America: Boulder, Colorado, Geological Society of America Special Paper 243, p. 31–50.

Archibald, J.D., Rigby, J.K., Jr., and Robison, S.F., 1983, Systematic revision of *Oxyacodon* (Condylarthra, Periptychidae) and a description of *O. ferronensis* n. sp: Journal of Paleontology, v. 57, p. 53–72.

Archibald, J.D., Gingerich, P.D., Lindsay, E.H., Clemens, W.A., Krause, D.W., and Rose, K.D., 1987, First North American land mammal ages of the Cenozoic era, *in* Woodbourne, M.O., ed., Cenozoic mammals of North America: Berkeley, University of California Press, p. 24–76.

Behrensmeyer, A.K., 1988, Vertebrate preservation in fluvial channels: Palaeogeography, Palaeoclimatology, Palaeoecology, v. 63, p. 183–199.

Belt, E.S., Hicks, J.F., and Murphy, D.A., 1997, A pre-Lancian regional unconformity and its relationship to Hell Creek paleogeography in southeastern Montana: University of Wyoming Contributions to Geology, v. 31, p. 1–26.

Breithaupt, B.H., 1982, Paleontology and paleoecology of the Lance Formation (Maastrichtian) east flank of Rock Springs uplift, Sweetwater County, Wyoming: University of Wyoming Contributions to Geology, v. 21, p. 123–151.

Carpenter, K., 1979, Vertebrate fauna of the Laramie Formation (Maestrichtian), Weld County, Colorado: University of Wyoming Contributions to Geology, v. 17, p. 37–49.

Cherven, V.B., and Jacob, A.F., 1985, Evolution of Paleogene depositional systems, Williston Basin, in response to global sea level changes, *in* Flores, R.M., and Kaplan, S.S., eds., Cenozoic paleogeography of west-central United States: Society of Economic Paleontologists and Mineralogists, Rocky Mountain Section, p. 127–170.

Cifelli, R.L., 1990, Cretaceous mammals of southern Utah. 4. Eutherian mammals from the Wahweap (Aquilan) and Kaiparowits (Judithian) Formations: Journal of Vertebrate Paleontology, v. 10, p. 346–360.

Cifelli, R.L., 1994, Therian mammals of the Terlingua local fauna (Judithian), Aguja Formation, Big Bend of the Rio Grande, Texas: University of Wyoming Contributions to Geology, v. 30, p. 117–136.

Cifelli, R.L., Nydam, R.L., Eaton, J.G., Gardner, J.D., and Kirkland, J.I., 1999, Vertebrate faunas of the North Horn Formation (Upper Cretaceous–lower Paleocene), Emery and Sampete counties, *in* Gillette, D.D., ed., Vertebrate paleontology in Utah: Utah Geological Survey Special Publication 99-1, p. 377–388.

Clemens, W.A., 1964, Fossil mammals of the type Lance Formation, Wyoming. 1. Introduction and Multituberculata: University of California Publications in Geological Sciences, v. 48, p. 1–105.

Clemens, W.A., 1966, Fossil mammals of the type Lance Formation, Wyoming. 2. Marsupialia: University of California Publications in Geological Sciences, v. 62, p. 1–122.

Clemens, W.A., 1973, Fossil mammals of the type Lance Formation, Wyoming. 3. Eutheria and summary: University of California Publications in Geological Sciences, v. 94, p. 1–102.

Clemens, W.A., 1980, *Gallolestes pachymandibularis* (Theria, incertae sedis; Mammalia) from Late Cretaceous deposits in Baja California del Norte, Mexico: PaleoBios, v. 33, p. 1–10.

Clemens, W.A., and Nelms, L.G., 1993, Paleoecological implications of Alaskan terrestrial vertebrate fauna in latest Cretaceous time at high paleolatitudes: Geology, v. 21, p. 503–506.

Clemens, W.A., Jr., and Russell, L.S., 1965, Mammalian fossils from the upper Edmonton Formation: University of Alberta Geology Bulletin, v. 2, p. 32–40.

Cope, E.D., 1882, A second genus of Eocene Plagiaulacidae: American Naturalist, v. 16, p. 416–417.

Cope, E.D., 1885, The oldest Tertiary Mammalia: American Naturalist, v. 19, p. 385–387.

Eaton, J.G., 1995, Cenomanian and Turonian (early Late Cretaceous) multituberculate mammals from southwestern Utah: Journal of Vertebrate Paleontology, v. 15, p. 761–784.

Eaton, J.G., and Nelson, M.E., 1991, Multituberculate mammals from the Lower Cretaceous Cedar Mountain Formation, San Rafael Swell, Utah: University of Wyoming Contributions to Geology, v. 29, p. 1–12.

Eberle, J.J., and Lillegraven, J.A., 1998a, A new important record of earliest Cenozoic mammalian history: Geologic setting, Multituberculata, and Peradectia: Rocky Mountain Geology, v. 33, p. 1–47.

Eberle, J.J., and Lillegraven, J.A., 1998b, A new important record of earliest Cenozoic mammalian history: Eutheria and paleogeographic/biostratigraphic summaries: Rocky Mountain Geology, v. 33, p. 49–117.

Flynn, L.J., 1986, Late Cretaceous mammal horizons from the San Juan Basin, New Mexico: American Museum Novitates, no. 2845, p. 1–30.

Foote, M., Hunter, J.P., Janis, C.M., and Sepkoski, J.J., Jr., 1999, Evolutionary and preservational constraints on origins of biologic groups: Divergence times of eutherian mammals: Science, v. 283, p. 1310–1314.

Fox, R.C., 1989, The Wounded Knee local fauna and mammalian evolution near the Cretaceous-Tertiary boundary, Saskatchewan, Canada: Palaeontographica, v. 208, p. 11–59.

Fox, R.C., 1997, Late Cretaceous and Paleocene mammals, Cypress Hills region, Saskatchewan, and mammalian evolution across the Cretaceous-Tertiary boundary, *in* McKenzie-McAnally, L., ed., Upper Cretaceous and Tertiary stratigraphy and paleontology of southern Saskatchewan, Paper 6 of Canadian Paleontology Conference Field Trip Guidebook: St. John's, Newfoundland, Geological Association of Canada, p. 70–85.

Gill, J.R., and Cobban, W.A., 1973, Stratigraphy and geologic history of the Montana Group and equivalent rocks, Montana, Wyoming, and North and South Dakota: U.S. Geological Survey Professional Paper 776, p. 1–37.

Gradstein, F.M., Agterberg, F.P., Ogg, J.G., Hardenbol, J., van Veen, P., Thierry, J., and Huang, Z., 1995, A Triassic, Jurassic, and Cretaceous time scale, *in* Berggren, W.A., et al., eds., Geochronology, time scales and global stratigraphic correlation: Tulsa, Oklahoma, SEPM (Society for Sedimentary Geology), p. 95–126.

Grandstaff, B.S., Parris, D.C., Denton, R.K., Jr., and Gallagher, W.B., 1992, *Alphadon* (Marsupialia) and Multituberculata (Allotheria) in the Cretaceous of eastern North America: Journal of Vertebrate Paleontology, v. 12, p. 217–222.

Grigorescu, D., and Hahn, G., 1987, The first multituberculate teeth from the Upper Cretaceous of Europe (Romania): Geologica and Palaeontologica, v. 21, p. 237–243.

Hunter, J.P., 1999, The radiation of Paleocene mammals with the demise of the dinosaurs: Evidence from southwestern North Dakota: North Dakota Academy of Science Proceedings, v. 53, p. 141–144.

Hunter, J.P., and Pearson, D.A., 1996, First record of Lancian (Late Cretaceous) mammals from the Hell Creek Formation of southwestern North Dakota, USA: Cretaceous Research, v. 17, p. 633–643.

Hunter, J.P., Hartman, J.H., and Krause, D.W., 1997, Mammals and mollusks across the Cretaceous-Tertiary boundary from Makoshika State Park and vicinity (Williston Basin), Montana: University of Wyoming Contributions to Geology, v. 32, p. 61–114.

Johanson, Z., 1996, Revision of the Late Cretaceous North American genus *Alphadon*: Palaeontographica, v. 242, p. 127–184.

Johnson, K.R., 1992, Leaf-fossil evidence for extensive floral extinction at the Cretaceous-Tertiary boundary, North Dakota, USA: Cretaceous Research v. 13, p. 91–117.

Johnston, P.A., 1980, First record of Mesozoic mammals from Saskatchewan: Canadian Journal of Earth Sciences, v. 17, p. 512–519.

Johnston, P.A., and Fox, R.C., 1984, Paleocene and Late Cretaceous mammals from Saskatchewan, Canada: Palaeontographica, v. 186, p. 163–222.

Kielan-Jaworowska, Z., Presley, R., and Poplin, C., 1986, The cranial vasculature system in taeniolabidoid multituberculate mammals: Royal Society of London Philosophical Transactions, ser. B: Biological Sciences, v. 313, p. 525–602.

Krause, D.W., 1992, *Clemensodon megaloba*, a new genus and species of Multituberculata (Mammalia) from the Upper Cretaceous type Lance Formation, Powder River Basin, Wyoming: PaleoBios, v. 14, p. 1–8.

Krause, D.W., Prasad, G.V.R., Koenigswald, W., Sahni, A., and Grine, F.E., 1997, Cosmopolitanism among Gondwanan Late Cretaceous mammals: Nature, v. 390, p. 504–507.

Legendre, P., and Legendre, L., 1998, Numerical ecology (second edition): Amsterdam, Elsevier, 853 p.

Le Loeuff, J., and Buffetaut, E., 1995, The evolution of Late Cretaceous nonmarine vertebrate fauna in Europe, *in* Sun, A., and Wang, Y., eds., Short papers, Sixth Symposium on Mesozoic Terrestrial Ecosystems and Biota: Beijing, China Ocean Press, p. 181–184.

Lillegraven, J.A., 1969, Latest Cretaceous mammals of upper part of Edmonton Formation of Alberta, Canada, and review of marsupial-placental dichotomy in mammalian evolution: University of Kansas Paleontological Contributions, v. 50, p. 1–122.

Lillegraven, J.A., 1976, A new genus of therian mammal from the Late Cretaceous "El Gallo Formation," Baja California, Mexico: Journal of Paleontology, v. 50, p. 437–443.

Lillegraven, J.A., and Eberle, J.J., 1999, Vertebrate faunal changes through Lancian and Puercan time in southern Wyoming: Journal of Paleontology, v. 73, p. 691–710.

Lillegraven, J.A., and McKenna, M.C., 1986, Fossil mammals from the "Mesaverde" Formation (Late Cretaceous, Judithian) of the Bighorn and Wind River Basins, Wyoming, with definitions of Late Cretaceous North American land-mammal "ages": American Museum Novitates, no. 2840, p. 1–68.

Lillegraven, J.A., and Ostresh, L.M., Jr., 1990, Late Cretaceous (earliest Campanian/Maastrichtian) evolution of western shorelines of the North American Western Interior Seaway in relation to known mammalian faunas, *in* Bown, T.M., and Rose, K.D., eds., Dawn of the Age of Mammals in the northern part of the Rocky Mountain Interior, North America: Boulder, Colorado, Geological Society of America Special Paper 243, p. 1–30.

Lofgren, D.L., 1995, The Bug Creek problem and the Cretaceous-Tertiary transition at McGuire Creek, Montana: University of California Publications in Geological Sciences, v. 140, p. 1–185.

Lull, R.S., 1915, The mammals and horned dinosaurs of the Lance Formation, Niobrara County, Wyoming: American Journal of Science, v. 40, p. 319–348.

Luo, Z., 1989, Structure of the Petrosals of Multituberculata (Mammalia) and morphology of the molars of Early Arctocyonids (Condylarthra, Mammalia) [Ph.D. thesis]: Berkeley, University of California, 426 p.

Marsh, O.C., 1889a, Discovery of Cretaceous Mammalia: American Journal of Science, v. 3, p. 81–92.

Marsh, O.C., 1889b, Discovery of Cretaceous Mammalia, Part 2: American Journal of Science, v. 3, p. 177–180.

Marshall, C.R., 1990, Confidence intervals on stratigraphic ranges: Paleobiology, v. 16, p. 1–10.

Mathworks, Inc., 1999, Matlab: The language for technical computing, Version 5.3.

Matthew, W.D., 1937, Paleocene faunas of the San Juan Basin: American Philosophical Society Transactions, v. 30, p. 1–510.

McKenna, M.C., and Bell, S.K., 1997, Classification of mammals above the species level: New York, Columbia University Press, 640 p.

Nessov, L.A., Archibald, J.D., and Kielan-Jaworowska, Z., 1998, Ungulate-like mammals from the Late Cretaceous of Uzbekistan and a phylogenetic analysis of Ungulatomorpha: Carnegie Museum of Natural History Bulletin, v. 34, p. 40–88.

Prasad, G.V.R., Khajuria, C.K., and Manhas, B.K., 1995, Palaeobiogeographic significance of the Deccan infra- and intertrappean biota from peninsular India: Historical Biology, v. 9, p. 319–334.

Rohlf, F.J., 1999, NTSYS-pc: Numerical taxonomy and multivariate analysis system, Version 2.02k: Setauket, New York, Exeter Software.

Rosenberg, M.S., 2000, Passage: Pattern analysis, spatial statistics, and geographic exegesis: Beta Version 0.1.4.15 (available on request from M.S. Rosenberg, Arizona State University, msr@asu.edu).

Rougier, G.W., Wible, J.R., and Hopson, J.A., 1992, Reconstruction of the cranial vessels in the Early Cretaceous mammal *Vincelestes neuquenianus*: Implications for the evolution of the mammalian cranial vascular system: Journal of Vertebrate Paleontology, v. 12, no. 2, p. 188–216.

Rowe, T., Cifelli, R.L., Lehman, T.M., and Weil, A., 1992, The Campanian Terlingua local fauna, with a summary of other vertebrates from the Aguja Formation, Trans-Peco, Texas: Journal of Vertebrate Paleontology, v. 12, p. 472–493.

Russell, L., 1975, Mammalian faunal succession in the Cretaceous System of western North America, *in* Caldwell, W.G.E., ed., The Cretaceous system in the Western Interior of North America: Geological Association of Canada Special Paper 13, p. 137–161.

Siegel, S., 1956, Nonparametric statistics for the behavioral sciences: New York, McGraw-Hill, 312 p.

Simmons, N.B., 1987, A revision of *Taeniolabis* (Mammalia: Multituberculata) with a new species from the Puercan of eastern Montana: Journal of Paleontology, v. 61, p. 794–808.

Simpson, G.G., 1927, Mesozoic Mammalia. 8. Genera of Lance mammals other than multituberculates: American Journal of Science, v. 14, p. 121–130.

Sloan, R.E., 1969, Cretaceous and Paleocene terrestrial communities of western North America: North American Paleontological Convention Proceedings, part E, p. 427–453.

Sloan, R.E., and Van Valen, L.M., 1965, Cretaceous mammals from Montana: Science, v. 148, p. 220–227.

Smouse, P.E., Long, J.C., and Sokal, R.R., 1986, Multiple regression and correlation extensions of the Mantel test of matrix correspondence: Systematic Zoology, v. 35, p. 627–632.

Sokal, R.R., and Rohlf, F.J., 1995, Biometry (third edition): New York, W.H. Freeman and Company, 887 p.

Sokal, R.R., and Sneath, P.H.A., 1963, Principles of numerical taxonomy: San Francisco, W.H. Freeman and Company, 359 p.

Storer, J.E., 1991, The mammals of the Gryde Local Fauna, Frenchman Formation (Maastichtian: Lancian), Saskatchewan: Journal of Vertebrate Paleontology, v. 11, p. 350–369.

Strauss, D., and Sadler, P.M., 1989, Classical confidence intervals and Bayesian probability estimates for ends of local taxon ranges: Mathematical Geology, v. 21, p. 411–427.

Swisher, C.C., III, Dingus, L., and Butler, R.F., 1993, $^{40}Ar/^{39}Ar$ dating and magnetostratigraphic correlation of the terrestrial Cretaceous-Paleogene boundary and Puercan Mammal Age, Hell Creek–Tullock formations, eastern Montana: Canadian Journal of Earth Sciences, v. 30, p. 1981–1996.

Van Valen, L.M., 1978, The beginning of the age of mammals: Evolutionary Theory, v. 4, p. 45–80.

Williamson, T.E., 1996, The beginning of the age of mammals in the San Juan Basin, New Mexico: Biostratigraphy and evolution of Paleocene mammals of the Nacimiento Formation: New Mexico Museum of Natural History and Science Bulletin, v. 8, p. 1–141.

Wilson, R.W., 1983, Late Cretaceous mammals of western South Dakota, 1974: National Geographic Society Research Reports, v. 15, p. 749–752.

Wilson, R.W., 1987, Late Cretaceous (Fox Hills) multituberculates from the Red Owl local fauna of western South Dakota, *in* Martin, J.E., and Ostrander, G.E., eds., Papers in vertebrate paleontology in honor of Morton Green: Rapid City, South Dakota School of Mines and Technology Museum of Geology, p. 118–122.

Manuscript Accepted by the Society September 24, 2001

Printed in the U.S.A.

Geological Society of America
Special Paper 361
2002

Evolution of the mammalian fauna across the Cretaceous-Tertiary boundary in northeastern Montana and other areas of the Western Interior

William A. Clemens
University of California Museum of Paleontology, Berkeley, California 94720, USA

ABSTRACT

Paleontological and geological studies of the Hell Creek Formation and Tullock Member of the Fort Union Formation in the northern Western Interior of North America provide a globally unique, detailed record of the evolution of terrestrial vertebrates during the transition from the latest Cretaceous into the earliest Paleocene. In the area south of the Fort Peck Reservoir, northeastern Montana, high-resolution lithostratigraphic and biostratigraphic studies, abetted by $^{40}Ar/^{39}Ar$ radiometric age determinations and paleomagnetic correlations, have clarified the local record of faunal evolution. This research demonstrates that the lithostratigraphic boundary between the Hell Creek and Tullock in northeastern Montana usually is coincident with, but can be younger than, the Cretaceous-Tertiary boundary.

A nonrandom pattern of survival and extinction of mammals and other terrestrial vertebrates characterizes faunal evolution across the Cretaceous-Tertiary boundary in northeastern Montana. Paleontological and geological data support hypotheses that a variety of long- and short-term factors influenced the course of vertebrate evolution. Recovery of the terrestrial fauna of northeastern Montana during the first million years of the Paleocene was not simply the product of evolutionary radiations stemming from some local survivors. In addition, waves of immigration of mammals from outside the northern Western Interior are recorded in the early (Pu1) and undifferentiated middle and late (Pu2-Pu3) Puercan local faunas of the area. They document both the important role of dispersal in recovery of the mammalian fauna of northeastern Montana during the Puercan as well as the biogeographic differentiation of Lancian and Puercan local faunas in more distant areas.

INTRODUCTION

For more than 90 yr, northeastern Montana has been a major source of geological and paleontological data documenting the evolution of the terrestrial biota prior to, during, and after the Cretaceous-Tertiary (K-T) boundary. Significant discoveries also have been made in other parts of the northern Western Interior of North America, which includes eastern areas of Colorado, Wyoming, and Montana; adjacent parts of North and South Dakota; and southern areas of the Canadian provinces of Alberta and Saskatchewan (Fig. 1). The goal of this chapter is to provide a resume and analysis of aspects of published research and work in progress in northeastern Montana that can serve as a basis for comparisons in research projects in North Dakota, the focus of this symposium, as well as other areas. In the long term, synthesis of these studies and parallel projects in

Clemens, W.A., 2002, Evolution of the mammalian fauna across the Cretaceous-Tertiary boundary in northeastern Montana and other areas of the Western Interior, *in* Hartman, J.H., Johnson, K.R., and Nichols, D.J., eds., The Hell Creek Formation and the Cretaceous-Tertiary boundary in the northern Great Plains: An integrated continental record of the end of the Cretaceous: Boulder, Colorado, Geological Society of America Special Paper 361, p. 217–245.

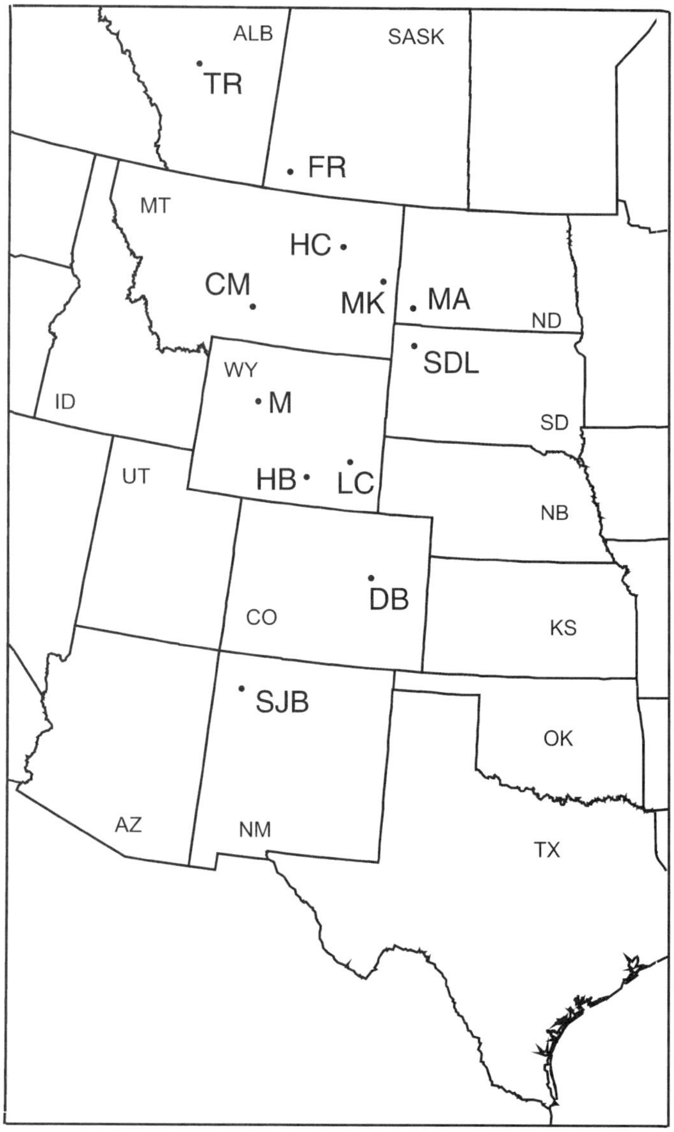

Figure 1. Index map of Western Interior showing locations of areas yielding selected Lancian and Puercan North American Land Mammal Age local faunas cited in text. Abbreviations: CM—Crazy Mountains basin, Montana; DB—Denver Basin, Colorado; FR—Frenchman and Ravenscrag Formations, Saskatchewan; HB—Hanna basin, Wyoming; HC—sites in Garfield and McCone Counties, Montana; LC—Lance Creek, Wyoming; M—Mantua Lentil, Bighorn Basin, Wyoming; MA—Sites in Marmarth area, North Dakota; MK—Makoshika State Park, Montana; SDL—sites in northwestern South Dakota; SJB—San Juan Basin, New Mexico; TR—Trochu, valley of Red Deer River, Alberta.

more distant areas will provide a much clearer understanding of the evolutionary patterns and processes involved in terrestrial biotic change across the K-T boundary.

Paleontological and geological research in northeastern Montana began in 1902 when Barnum Brown of the American Museum of Natural History, New York, started collecting vertebrate fossils in the valley of Hell Creek, north of the town of Jordan in what is now Garfield County (Fig. 2). The first vertebrate fossils he discovered are now parts of the type specimen of *Tyrannosaurus rex*. In following years, Brown worked eastward, primarily along the southern side of the valley of the Missouri River, now flooded by the Fort Peck Reservoir, into the valley of Big Dry Creek. During the summers of 1907–1908, he collected another, relatively complete skeleton of *T. rex* near the mouth of the valley of Bug Creek, an eastern tributary of Big Dry Creek (see Horner and Lessum, 1993). Through the course of this field work, Brown and his crews collected skulls and partial skeletons of ceratopsians, hadrosaurs, and other species of dinosaurs, as well as other large vertebrates including crocodilians, champsosaurs, and turtles. Combing through concentrations of fossils on the nests of harvester ants and the residues of underwater screening of sediments through burlap bags provided some mammalian and other microvertebrate remains. In addition, they made significant collections of freshwater invertebrates (Hartman, 1999). For almost 50 yr, Brown's collections remained the major and essentially only sample of the area's latest Cretaceous and early Paleocene faunas available for research (e.g., Simpson, 1927, 1929).

A few collections of vertebrates from the Hell Creek Formation were made during the 1920s and 1930s, particularly in association with the construction of the dam at Fort Peck, but their scientific significance would not be realized until several decades later. The current pulse of vertebrate paleontological research in northeastern Montana began in the 1960s. Following the now abandoned road from Jordan to the Hell Creek Recreation Area, J.A. Dorr Jr. (University of Michigan), B.R. Erickson (Science Museum of Minnesota), and R.E. Sloan (University of Minnesota) collected fossil vertebrates in the vicinity of Brownie Butte, a prominent landmark in the upper reaches of Hell Creek.

In addition to their work around Brownie Butte, Sloan and Van Valen (1965), then at the American Museum of Natural History, were drawn to McCone County by a small collection of vertebrate fossils on display in the museum in the Fort Peck Reservoir power plant. The collection included a mixture of Cretaceous dinosaurian remains as well as teeth and jaws of groups of mammals that were known only from the Paleocene. These fossils had been collected in the 1930s by Darwin Harbicht and other workers involved in construction of the Fort Peck Dam. Newell F. Joyner (National Park Service), Donald C. Beckman, and Eugene Kuszmaul (U.S. Army Corps of Engineers), and their families, who were involved in development of the museum at Fort Peck, relocated the area where Harbicht had found localities producing this intriguingly mixed assemblage. They led Sloan and Van Valen (1965) to a remarkable concentration of vertebrate fossils, the Bug Creek Anthills locality. For many years, starting in 1962, Sloan and his field crews collected at this and other localities in the valley of Bug Creek and adjacent areas, particularly to the north and east. Other paleontologists made collections at Bug Creek Anthills

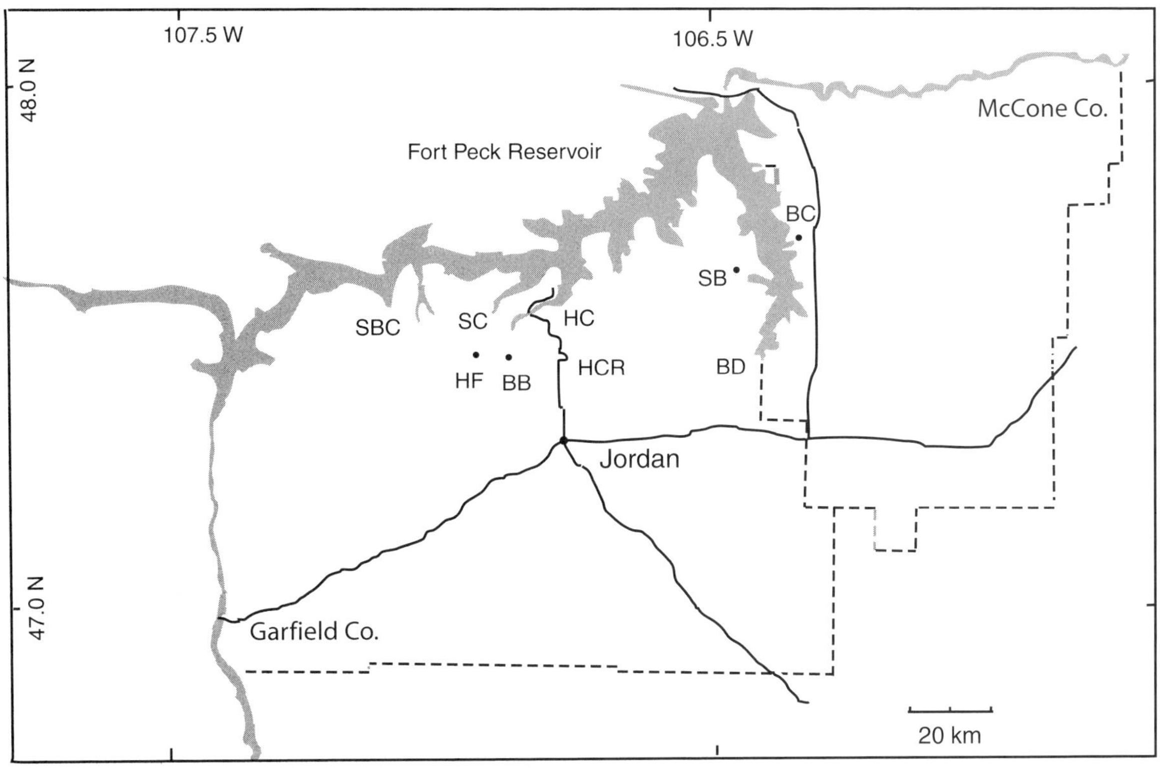

Figure 2. Index map of Garfield and McCone Counties, eastern Montana. Abbreviations: BB—Brownie Butte, BC—Bug Creek, BD—Big Dry Creek arm of Fort Peck Reservoir, HC—Hell Creek, HF—Hauso Flats, HCR—road to Hell Creek Recreation Area, SB—Sandstone Basin, SBC—Seven Blackfoot Creek, SC—Snow Creek.

that are now dispersed in many museums in different parts of the world. Subsequently, J.K. Rigby Jr. initiated an extensive research project in the area (see Rigby and Rigby, 1990, and references cited therein).

In 1965, H.J. Garbani, then an employee of the Los Angeles County Museum, began what was to be a most successful series of field seasons in the valley of Hell Creek. The following year, he discovered both another skeleton of *Tyrannosaurus rex* and a concentration of early Paleocene vertebrate fossils in a channel filling now known as the Garbani Channel. At the invitation of the staff of the Los Angeles County Museum, in 1971, the author and other paleontologists from the University of California Museum of Paleontology (UCMP) started work in the area. Since then field parties from Museum of Paleontology and, more recently, the Raymond M. Alf Museum, Webb School, Claremont, California, have worked in eastern Montana almost every summer, amassing large collections of fossils as well as undertaking a variety of other paleontological and geological studies.

Among the fossil vertebrates from northeastern Montana, its dinosaurs have attracted considerable scientific and popular attention. Weishampel (1990) published a compilation of this dinosaurian fauna. Archibald (1982), Bryant (1989), and Lofgren (1995) described and analyzed much of the available nondinosaurian, latest Cretaceous and Paleocene vertebrate paleontological record, but field and laboratory research continues to add significant new records and interpretations (e.g., Weil, 1998; Clemens, 1999; Hutchison, 2000). In general, the latest Cretaceous and earliest early Paleocene (Pu1, see discussion below) terrestrial faunas have been sampled with modern techniques and the resulting collections thoroughly studied. Analyses of large samples of middle and late early Paleocene and early middle Paleocene terrestrial faunas (Pu2-Pu3 and To1) have not progressed as far.

Following Brown's (1907) notes on occurrences of fossil leaves in the Hell Creek and Tullock, paleobotanists at the University of Minnesota undertook pioneering studies of the floras preserved in these formations. On the basis of several small but significant samples, Shoemaker (1966) analyzed the megafloral record. Subsequent collecting, reported by Johnson and Hickey (1990), yielded only scattered megafloral remains from the Hell Creek but produced samples of numerous FU I megafloras from the Tullock. Recent discoveries and initial studies of numerous concentrations of fossil leaves give promise of significantly expanding our knowledge of the megafloral records of the Hell Creek (N.C. Arens, 2000, personal commun.). Concurrently, the first studies of the palynological record across the formational contact in Garfield and McCone Counties were completed (e.g., Hall and Norton, 1967; Norton and Hall, 1969; Oltz, 1969). Hotton (1989, this volume) undertook a more extensive analysis of the palynological record. Freshwater invertebrate faunas of the area continue to be studied by Hartman (1999, this volume).

The history of geological research in Garfield and McCone Counties, like the history of paleontological research, is characterized by a series of pulses of activity. After several years of field work, which began in 1902, Barnum Brown (1907) formally established the Hell Creek Formation and published his geological observations. Brown's contributions toward documenting and analyzing the latest Cretaceous and earliest Paleocene stratigraphy and the vertebrate faunas of the Western Interior were prodigious. Beginning in 1900 with field work in the Lance Formation, Wyoming, he went on to work in the San Juan Basin, New Mexico, in the valley of the Red Deer River, Alberta, as well as in northeastern Montana. The scope of his contributions was outlined in a brief article that Brown (1914) contributed to a symposium sponsored by the Paleontological Society.

The next major contribution to knowledge of the Hell Creek and Tullock (or Fort Union Formation) of northeastern Montana was the analysis of the geology of McCone County by Collier and Knechtel (1939). As part of their studies in the area, they established the foundation for the system of lithostratigraphic subdivison of the Tullock employed today.

The current period of geological research started with the work of Bell (1965) in northwestern McCone County. Then, after a gap of almost two decades, a wealth of different geological studies began to appear. Detailed studies of the stratigraphy of the Hell Creek and Tullock were completed in the valley of Hell Creek, Garfield County (Archibald, 1982; Dingus, 1983), and the valley of McGuire Creek, McCone County (Lofgren, 1995). Smit et al. (1987) as well as Rigby and Rigby (1990) published their findings on the geology of the Bug Creek and Sand Arroyo areas in McCone County. Fastovsky (1987; Fastovsky and Dott, 1986) reported on detailed stratigraphic and sedimentological studies of sections including the contact of the Hell Creek and Tullock. Retallack et al. (1987; Retallack, 1994) analyzed paleosols in one of these, the Bug Creek–Russell Basin section.

New research techniques continue to provide greater resolution in the history of biotic change documented in the area. In the 1980s, Bohor et al. (1984) discovered a layer of sediment containing an anomalously high concentration of iridium and shocked minerals ~0.9 km east of Brownie Butte. They interpreted the layer as the signature of the impact of an exceptionally large extraterrestrial bolide. The potential for use of this layer as a means of precise correlation continues to be evaluated. In their first report on a continuing project in this area, Swisher et al. (1993) provided $^{40}Ar/^{39}Ar$ age determinations and magnetostratigraphic correlations of the terrestrial K-T boundary and the Puercan North American Land Mammal Age (NALMA) of northeastern Montana. On the basis of geochemical data drawn from eastern Montana, Landis et al. (1996) as well as Arens and Jahren (1997, 1998, this volume) investigated aspects of atmospheric change during the transition from the Cretaceous into the Tertiary and its influence on evolutionary processes. In addition, Arens and Jahren (this volume) are exploring the potential of application of carbon isotope chemostratigraphy in correlation and in evaluating the completeness of geological sections.

Field and subsequent laboratory research have contributed to extensive documentation of terrestrial biotic change in northeastern Montana prior to and during the K-T mass extinction as well as recovery of the biota in the early Paleocene. This documentation and a variety of derivative analyses are not without some significant gaps and biases that must be considered in comparative studies. Research on the fossil and geological record of Garfield and McCone Counties is far from completed. New discoveries and applications of new research techniques continue to add to the knowledge of the pattern and processes involved in biotic evolution in this area during the transition from the Cretaceous to the Tertiary.

GEOLOGY

At the beginning of the Maastrichtian, North America was bisected by a seaway that extended from the Arctic Ocean to the Gulf of Mexico. In the late Maastrichtian, Laramide orogenic pulses coupled with rapid eustatic regression were primary causes of the relatively rapid expansion of the western coastal lowlands of this seaway (Roberts and Kirschbaum, 1995). This was not a geographically simple eastward expansion of the coastal lowlands. The Sheridan delta (Gill and Cobban, 1973) brought sediments from areas of Laramide uplift eastward across southern Montana and then swung northward, north of a highland in the region of the modern Black Hills. Volcanic areas in southwestern Montana and possibly others to the west were active during the latest Cretaceous and early Paleocene, providing both water and airborne clastics to northeastern Montana. Farther to the east, the seas shallowed and a lowland, the Dakota isthmus, emerged, possibly dividing the Western Interior Sea into northern and southern embayments (Erickson, 1999). In Garfield and McCone Counties, the Bearpaw Shale and Fox Hills Sandstone, which varies in thickness from ~10 to 45 m (Jensen and Varnes, 1964), are, respectively, the last marine and the strand line and/or beach deposits formed along the western side of the regressing Cretaceous seaway.

Hell Creek Formation

The Hell Creek Formation, named and typified from sections exposed along "Hell Creek and nearby tributaries of the Missouri River" by Brown (1907, p. 829), overlies the Fox Hills Sandstone. Hartman (1999) found substantial topography on the contact of the two formations, reflecting the erosional surface of the Fox Hills. Others (e.g., Rigby and Rigby, 1990) noted areas in which the contact between the Fox Hills and Hell Creek is transitional. This passage from deposition in shallow-marine and strand-line to continental environments was associated with the advance of the Sheridan delta. Deposition of the volcanic-rich, gray to gray-green, very fine grained to medium-grained sandstone and siltstone of the Hell Creek Formation occurred on a broad fluvial plain that was dissected by meandering rivers. Moderately gleyed soils were products of a high, fluctuating water table (Fastovsky, 1987). Vegetation was abundant, but, possibly because of some combination of landscape instability, rapid input of inorganic materials, or insufficient bacterial decomposition, humification was inhibited, and in only a few areas thin beds of lignite were formed (Fastovsky, 1987).

Many investigators (e.g., Brown, 1907; Hartman, 1999) reported that the Hell Creek Formation is ~95 m thick in Garfield and McCone Counties, but Thom and Dobbin (1924) reported a thickness of ~170 m at an unspecified locality in Garfield County. Citing the work of Bell (1965), Rigby and Rigby (1990) reported that the Hell Creek Formation reached a thickness of ~114 m at the mouth of Hell Creek and tended to thicken to the west and south.

In northeastern Montana, vertebrate fossils have been found throughout the Hell Creek Formation, but sampling of the vertebrate fauna in Garfield County is stratigraphically biased. As noted by Brown (1907), commonly in the primary collecting areas, the lower part of the formation usually contains a higher proportion of sandstone relative to siltstone than the upper part, and it weathers to form cliffs and a generally rugged landscape. This greater topographic relief lowers the chances of discovery and increases the difficulty of collecting vertebrate fossils. Because of its greater area of exposure and less precipitous outcrops, the upper part (~50 m) of the formation has been the source of the vast majority of the collections of fossil vertebrates.

In the upper part of the Hell Creek Formation, many vertebrate fossils were preserved in fine-grained overbank deposits, but this was not the only mode of accumulation. The type locality of *Tyrannosaurus rex*, for example, which yielded a partial skeleton, is in a multistoried, sandy channel filling ~30 m in thickness exposed in the headwaters of Hell Creek (see Fastovsky, 1987, locality 1). Other large, sandy channel fillings within the Hell Creek Formation (e.g., Sandstone Basin, UCMP locality V80092, eastern Garfield County, Fig. 2) preserve concentrations of usually disarticulated skeletal elements of small through large species of latest Cretaceous vertebrates. Although large channel fillings have provided some scientifically significant material, the majority of the Hell Creek vertebrates was collected from overbank and pond deposits or small channel fillings. Archibald (1982, his Table 1) and Lofgren (1995) studied collections from more than 60 vertebrate fossil localities. Approximately a third of these localities were sampled with screen-washing techniques in order to obtain large collections of microvertebrate fossils.

Attempts to develop a regional lithostratigraphic subdivision of the Hell Creek Formation in Garfield and McCone Counties have yet to be successful. In most areas, the upper part of the formation lacks distinctive strata that can be traced over long distances. A possible exception is a thin lignite containing a pink tonstein, called the Tonstein lignite by Lofgren (1995), that occurs ~30 m below the Z lignite (MCZ of Lofgren, 1995) in the McGuire Creek area of McCone County. Possible correlatives of this tonstein-bearing lignite have been found in eastern Garfield County. It might be a correlative of the Null coal, which lacks an included tonstein, mapped by Rigby and Rigby (1990) in the Bug Creek region and northward. The Tonstein and Null lignites could prove to be useful lithostratigraphic markers within the Hell Creek Formation in eastern Garfield and western McCone Counties, but this hypothesis has yet to be fully tested.

Current procedure in attempts to determine the relative stratigraphic positions of fossil localities within the Hell Creek Formation is to measure their position below the local Hell Creek–Tullock formational contact. Many Hell Creek fossil localities are 1 km or more from the nearest exposure of this contact, and their positions can only be crudely estimated from differences in current elevation of the site and the closest outcrop of the contact. Where the stratigraphic position of a major fossil locality relative to the contact could be measured directly, the localities ranged from 5.2 to 45 m below the local top of the formation. As discussed in the following, detailed studies show that the Hell Creek–Tullock contact is not a synchronous datum. The varying age of the contact, coupled with the episodic and heterogeneous pattern of accumulation of the Hell Creek Formation, limits the use of relative stratigraphic position below the local formational boundary in detailed regional correlations (see Dingus, 1984; Fastovsky, 1987).

Because of the absence of deposits of volcanic ashes suitable for radiometric age determination at or near the base of the formation, the interval of deposition of the Hell Creek Formation in Garfield and McCone Counties cannot be directly determined. Making use of a variety of biostratigraphic, radiometric, geochemical, and magnetostratigraphic techniques, Hicks et al. (1999, this volume) estimated that the duration of deposition of the Hell Creek Formation in southwestern North Dakota was ~1.75 m.y. In the Cedar Creek anticline, near their study area, and in Garfield and McCone Counties, the youngest ammonite zone is the *Baculites grandis* zone (Cobban, *in* Jensen and Varnes, 1964; Sloan et al., 1986; Rigby and Rigby, 1990). This suggests that the estimate of ~1.75 m.y. for the period of deposition of the Hell Creek Formation in North Dakota is probably applicable to the Hell Creek Formation in northeastern Montana.

Tullock Formation

In Garfield and McCone Counties deposition of the Hell Creek Formation came to an end with a shift to accumulation of much more regularly stratified sandstone, siltstone, and mudstone interbedded with lignite or coal. In the many articles in

TABLE 1. GRAPHICAL COMPARISON OF COMPOSITE LANCIAN AND PUERCAN MAMMALIAN FAUNAL LISTS, GARFIELD AND McCONE COUNTIES, MONTANA

Lancian NALMA fauna	Pu1 NALMA fauna	Pu2/3 NALMA fauna
"Residents"	***"Residents"***	**"Residents"**
Meniscoessus robustus	*Cimexomys minor*	*Mesodma thompsoni*
Essonodon browni	*Cimexomys gratus*	*Mesodma* sp.
Cimolodon nitidus	*Mesodma garfieldensis*	*Microcosmodon arcuatus*
Cimolomys gracilis	*Peradectes* cf. *P. pusillus*	*Microcosmodon harleyi*
Paracimexomys priscus	*Procerberus formicarium*	*Stygimys* cf. *camptorhiza*
Mesodma hensleighi		*Eucosmodon* cf. *americanus*
Mesodma formosa	***"Aliens"***	*Cimexomys* cf. *C. minor*
Mesodma thompsoni	*Acheronodon garbani*	*Cimexomys* cf. *C. gratus*
Cimexomys minor	*Catopsalis* cf. *foliatus*	*Catopsalis* sp.
Alphadon marshi	*Stygimys kuszmauli*	*Peradectes* sp.
Alphadon wilsoni	?*Prodiacodon* sp.	*Prodiacodon crustulum*
Protalphadon lulli	*Protungulatum donnae*	*Cimolestes* cf. *C. simpsoni*
Turgidodon rhaister	*Oxyprimus erikseni*	*Procerberus* n. sp.
Pediomys elegans	*Baioconodon ?harbichti*	*Loxolophus schizophrenus*
Pediomys cooki	*Baioconodon engdahli*	*Baioconodon* sp.
Pediomys krejcii	*Baioconodon nordicum*	*Anisonchus* sp.
Pediomys hatcheri	*Mimatuta morgath*	*Oxyacodon ferronensis*
Pediomys florencae	*Mimatuta minuial*	
Didelphodon vorax	?Periptychidae, gen et sp. nov.	**"Aliens"**
Gypsonictops hypoconus		*Ptilodus* cf. *P. tsosiensis?*
Gypsonictops illuminatus		*Ectypodus* sp.
		Parectypodus armstrongi
"Aliens"		?*Parectypodus* sp.
Glasbius twitchelli		*Taeniolabis lamberti*
Batodon tenuis		*Purgatorius unio*
		Purgatorius janisae
"Unknowns"		*Pandemonium dis*
Cimolestes incisus		*Carcinodon aquilonius*
Cimolestes cerberoides		*Oxyclaenus corax*
Cimolestes propaleoryctes		*Litomylus orthronepius*
Cimolestes stirtoni		*Bubogonia saskia*
		?*Litomylus* sp.
		Eoconodon nidhoggi
		Eoconodon gaudrianus
		"Unknowns"
		Neoplagiaulax kremnus
		Neoplagiaulax spp.
		?*Xyronomys* sp.
		Multituberculata, gen et sp. nov

Note: NALMA—North American Land Mammal Ages.

this volume dealing with the stratigraphy of North Dakota, the lithostratigraphically correlative unit is identified as the Tullock Member of the Fort Union Formation. The majority of scientists who have recently published the results of their research in Garfield and McCone Counties have followed Wilmarth (1938) and accorded the Tullock full formational status (e.g., Rigby and Rigby, 1990; Retallack, 1994; Lofgren, 1995). The difference in rank accorded the Tullock reflects a difference in local usage, not disagreement on lithostratigraphic correlation.

As demonstrated by Fastovsky (1987) and amplified by other workers (e.g., Retallack et al., 1987), the change in depositional environment from the Hell Creek Formation to the Tullock Formation reflects a rise of the water table and a dramatic increase in the amount of standing water. In consequence, peat swamps and large ponds appeared in topographically low areas. The resulting strata of the Tullock consist of drab yellow, interbedded, very fine grained to medium-grained siltstone, sandstone, and mudstone of fluvial and ponded origin. The primary, if not sole, causes of change in pattern of deposition are thought to have been an increase in rainfall and decrease in annual temperature, although contributions from reduction in rate of subsidence and/or initiation of the transgression of the Cannonball Sea cannot be excluded (Fastovsky, 1987; Eberth and O'Connell, 1995).

Erosion of the Laramide uplands and continued volcanic activity to the west, as well as extensive reworking of strata of the Hell Creek Formation (Lofgren, 1995), contributed the clastics. At irregular intervals, deposition of these clastics was complicated by formation of extensive peat swamps that now are represented by deposits of lignite or coal. In the region of Hell Creek, Archibald (1982) measured thicknesses of ~88–90 m in the Tullock Formation, and noted that the formation appears to thin eastward. Collier and Knechtel (1939) reported a thickness of ~50 m in McCone County. Radiometric age determinations from the volcanic ashes in the basal coals of the Tullock (the IrZ and MCZ, see following or Fig. 3) and the basal coal

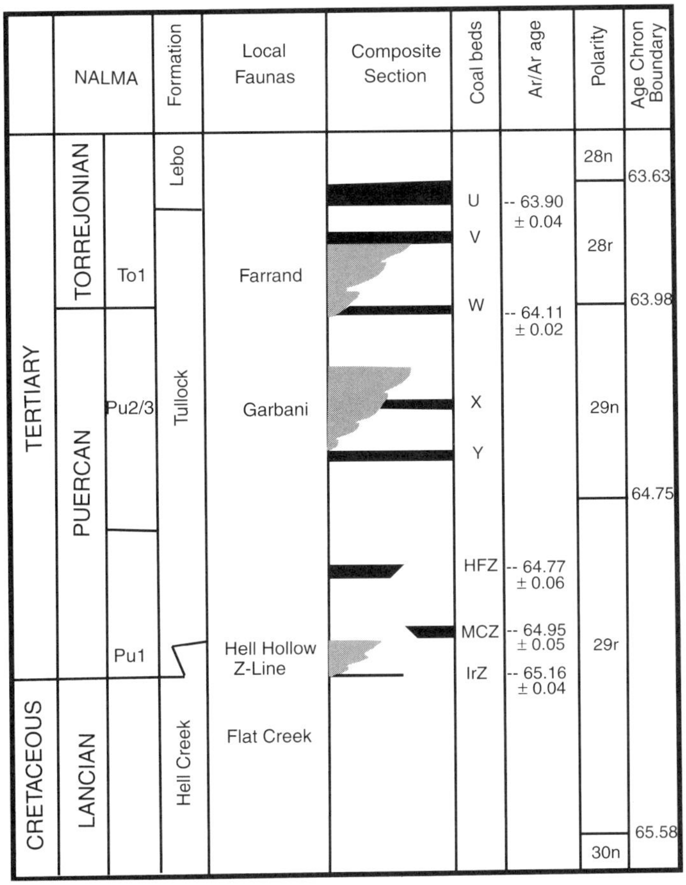

Figure 3. Composite geological section of Hell Creek Formation and Tullock Member of Fort Union Formation with occurrences of major vertebrate local faunas; biostratigraphic, chronostratigraphic, and magnetostratigraphic correlations (Berggren et al., 1995); and radiometric age determinations (Swisher et al., 1993). NALMA is North American Land Mammal Age.

of the overlying Lebo Formation (the U coal) delimit an ~1 m.y. period during which the Tullock Formation was deposited.

In their early study of the Tullock in McCone County, Collier and Knechtel (1939) documented the reoccurring pattern of deposition of beds of lignite and coal. As they stressed, these are not single, continuous strata. Rather, in McCone County and adjacent areas, these carbonaceous units vary in composition, thickness, and patterns of development of clastic interbeds. Collier and Knechtel (1939) established a system of alphabetical designations for stratigraphic intervals where lignite or coal deposition predominated. These intervals were termed coals, regardless of the fact that many were composed primarily or entirely of lignite. The stratigraphically lowest interval was dubbed the Z coal. Overlying intervals were designated in sequence up to the U coal. Collier and Knechtel (1939, p. 18) placed the contact between the Hell Creek and Tullock at the top of the Z coal, clearly stating that it "is probably not a single continuous bed but rather a succession of lenses of coal in about the same stratigraphic position." Later workers shifted the Hell Creek–Tullock contact to the base of the Z coal, and some incorrectly treated the base of this series of lignites and coals as a regionally continuous, synchronous unit (see Lofgren, 1995).

Paleobotanical studies of Tullock floras in Garfield and McCone Counties have had a checkered history. Brown (1907) noted occurrences of fossil leaves and wood in the Tullock strata. A limited study by Shoemaker (1966), an initial report on additional collections by Johnson and Hickey (1990), and the analysis of a sequence of five "petrified forests" represented by silicified stumps of *Metasequoia* (Sloan, 1976), have begun to expand on Brown's discoveries. The palynological record, particularly of the lower part of the Tullock Formation, was analyzed in pioneering studies by Hall and Norton (1967), Norton and Hall (1969), and Oltz (1969). Hotton (1989, this volume) analyzed the Tullock palynofloras in much greater detail.

Although Brown (1907) reported occurrences of fossil turtles, crocodilians, and champsosaurs in the Tullock, initially few attempts were made to sample the formation's vertebrate paleontological record. For the most part, the lignite and interbedded sediment proved to be sparsely fossiliferous at best, and vertebrate paleontologists were easily attracted to the riches of the Hell Creek Formation. In the 1960s, relatively rare channel fillings within the Tullock Formation, e.g., the Purgatory Hill channel in McCone County (Van Valen and Sloan, 1965) and the Garbani Channel in Garfield County, were discovered to be richly fossiliferous. This opened the possibility of studying the evolution of the early Paleocene faunas of the area. The question of why some channel fillings are richly fossiliferous while adjacent (contemporaneous) swamp, pond, and overbank deposits are not, has yet to be addressed. An obvious step toward answering this question would be undertaking a thorough taphonomic study of these relatively rare fossiliferous deposits. A hypothesis to be tested is that they are the products of hydrodynamic concentration in streams with a sufficient rate of flow to flush away humic and other acids, which destroyed the vertebrate remains that accumulated in overbank environments.

The relatively rare concentrations of vertebrate fossils in stream-channel fillings in the Tullock Formation frequently are bracketed between the alphabetically designated coals. Initially, preparation of composite sections describing the relative stratigraphic positions of these early Paleocene fossil localities was based on the assumption that the named coals are synchronous units over the hundreds of square kilometers of exposures in Garfield and McCone Counties. Subsequently, it was discovered that these coals preserve volcanic ashes suitable for obtaining $^{40}Ar/^{39}Ar$ age determinations. A program of analyzing these ashes is underway. First results of these and correlated magnetostratigraphic studies (Swisher et al., 1993) show that, on a time scale calibrated in 100 k.y., the Y through U coals can be treated as broadly synchronous units. They can be used to subdivide the ~1 m.y. history of deposition of the Tullock Formation. The age determinations also provide an independent temporal framework within which the record of regional waxing and waning of the coal swamps is being analyzed (see Swisher et al., 1993).

RECOGNITION OF THE K-T BOUNDARY

The Alvarez et al. (1980) hypothesis that the impact of an asteroid was the causal factor of terminal Cretaceous extinctions within marine and terrestrial biotas stimulated a wide variety of paleontological and geological research (see Glenn, 1994). The extensive record of evolutionary change of the terrestrial biota across the K-T boundary in northeastern Montana, which documented the local extinctions of nonavian dinosaurs and some other groups traditionally used to mark the end of the Cretaceous, quickly became a focus of attention.

Initially, our primary research goal in northeastern Montana was to develop increasingly higher resolution in knowledge of the lithostratigraphy and biostratigraphy of the upper part of the Hell Creek Formation and lower part of the Tullock Formation. Globally, the northern Western Interior was, and essentially remains, the only area in which the pattern of evolution of the terrestrial biota can be traced across the K-T boundary. Late Cretaceous, nonmarine vertebrate faunas have been found in other areas of North America and some other continents (Clemens et al., 1979). Worldwide, terrestrial faunas of Paleocene age—particularly early Paleocene age—are rare (Savage and Russell, 1983). Current work in China (e.g., Wang et al., 1998), Argentina (e.g., Bonaparte, 1990; Pascual, 1998; Pascual et al., 1999), and Europe (e.g., LeLoeuff, 1991; López-Martinez et al., 1999) gives promise of providing a similarly detailed record of faunal change across the K-T boundary. However, we currently have one window through which this episode in the evolution of terrestrial vertebrates can be studied. In contrast, many studies, too numerous to cite here, have focused on elucidating contemporaneous patterns of change in marine biotas. (For an introduction to this literature, see MacLeod et al., 1997; MacLeod, 1998.)

A key to synthesizing these studies of terrestrial and marine biotic change is recognition of a common standard for the K-T (Cretaceous-Paleogene) boundary. In 1982, the C-P (K-T) Boundary Working Group was established and began its deliberations; after a series of site visits and votes, it made its recommendations to the International Commission on Stratigraphy (Cowie et al., 1989). The section at El Kef, Tunisia, was recommended by a majority of the members of the working group as the boundary stratotype, the Global Stratotype Point being placed at the base of the 50-cm-thick boundary clay. The boundary clay contains an anomalous concentration of iridium and shocked mineral grains. This geochemical and mineralogical signature, interpreted as the product of an impact of an extraterrestrial body, offers the potential for high-resolution, global correlation of the Cretaceous-Tertiary boundary in both marine and nonmarine deposits. These recommendations were approved at the International Geological Congress in 1989 (see Keller et al., 1995). This action established a datum for the K-T boundary that has been recognized since 1989, but analysis of the history of concepts and criteria for recognition of a K-T boundary in terrestrial deposits of the Western Interior is required for interpretation of the results of earlier research.

Coal, dinosaurs, and Roland Brown's formula

The early geological investigation of continental deposits in the Western Interior is punctuated with accounts of lively debates concerning the recognition of the position of the K-T boundary. Foundations of the procedures for identification of this boundary in the Western Interior employed in the 1960s through the 1980s stem from the work of the paleobotanist Roland Brown. He argued that at any locality in the northern Western Interior, the boundary between the Hell Creek Formation and the Fort Union Formation should be placed at the first coal zone, no matter how thin, above the stratigraphically highest occurrence of dinosaurian remains and that the formational boundary approximated the K-T boundary (Brown, 1938, 1952). This formula for recognition of both the top of the Hell Creek Formation and the K-T boundary was widely adopted by geologists and paleontologists working in the area.

As attempts were made to maximize resolution in documentation of the lithostratigraphic and biostratigraphic records in Garfield and McCone Counties, the limitations of Brown's formula became increasingly apparent. In studies of the record, the lignite beds composing the Z coal could be treated as a unit, even though Collier and Knechtel (1939, p. 18) clearly noted that the Z coal was not a single bed, but a "succession of lenses of coal in about the same stratigraphic position."

Archibald (1982) reported results of the first detailed study of the lithostratigraphy of the contact of the Hell Creek and Tullock Formations in the valleys of Hell Creek and Snow Creek in Garfield County. There, as in McCone County, the Z coal is not a simple, uniform bed, but a complex of lignite beds and interbedded silt and sandstone that vary laterally in composition. Dingus (1983) and Fastovsky (1987) provided a more detailed analysis of this variation in Garfield County.

Lofgren (1995) analyzed variation of the Z coal complex (his MCZ coal) in the McGuire and Bug Creek regions of western McCone County and amplified Collier and Knechtel's (1939) observations. There, around the Big Dry Creek arm of Fort Peck Reservoir (Fig. 2), the uppermost beds of the Hell Creek Formation lithologically resemble those exposed in the valley of Hell Creek. In some areas, latest Cretaceous, Lancian NALMA local faunas have been discovered in the upper part of the formation. For example, a series of closely approximated sites in small channel and pond deposits in eastern Garfield County, ~5 m below the local Hell Creek–Tullock contact, has yielded the typically Lancian, Flat Creek local fauna (UCMP localities V73083 through V73087, Archibald, 1982).

In addition, other localities in the uppermost Hell Creek Formation in eastern Garfield County (Constenius local fauna, UCMP locality V96268) and western McCone County (Z-line channel local fauna, UCMP localities V84193 and V84194; Lofgren, 1995) yield earliest early Puercan (Pu1) local faunas. The biostratigraphic evidence that the uppermost strata of the Hell Creek Formation around the Big Dry Creek arm of the Fort Peck Reservoir include strata of earliest Paleocene age is supported by $^{40}Ar/^{39}Ar$ determinations of the age of the basal Tullock Z lignite (MCZ, Fig. 3) that are ~200 k.y. younger than those from the basal Tullock Z lignite (IrZ, Fig. 3) in the valley of Hell Creek (Swisher et al., 1993).

In light of increasingly resolved biostratigraphic and radiometric data (discussed in the following), it is clear that inception of deposition of the lignites composing the Z coal complex in various areas of Garfield and McCone Counties took place during an interval of at least 100 k.y. if not 200 k.y. Data provided by Johnson et al. (1989, see other chapters in this volume) demonstrate that inception of lignite deposition in the northern Western Interior was not an instantaneous event on a refined time scale.

In summary, the formula for recognition of the Hell Creek–Fort Union contact and the K-T boundary proposed by Roland Brown made a positive contribution to early investigations in the northern Western Interior. Subsequent work, however, has highlighted its limitations for detailed temporal correlations. Although it still might provide a guide for placing a local formational contact, the formula does not provide a high-resolution basis for recognition of the K-T boundary in the northern Western Interior: other data must be employed.

Extinction of nonavian dinosaurs

The extinction of nonavian dinosaurs has been adopted as a diagnostic hallmark of the end of the Cretaceous in the popular press and in many scientific papers. A prediction based on the Alvarez hypothesis is that, somewhere, remains of nonavian dinosaurs would be found covered by a layer of clay containing an anomalous concentration of iridium and shocked mineral grains, hallmarks of an asteroid impact (see Williams, 1994). Probably the record would not be a massive bone-bed accumulation (Hunter, 1994; Cutler and Behrensmeyer, 1996), but the remains of a nonavian dinosaurian so entombed by extraterrestrial debris have not yet been found.

Brown (1938, 1952) recognized that strata lacking vertebrate fossils regularly separated the stratigraphically highest occurrence of dinosaurian remains from the first regionally extensive lignite, which he interpreted as the marker of the K-T boundary. In Garfield and McCone Counties, fossil localities yielding a diverse sample of the vertebrates comprising the typical latest Cretaceous (Lancian NALMA) fauna have been discovered as much as 5.2 m below the local Hell Creek–Tullock contact (Archibald, 1987). Occurrences of isolated dinosaur bones or other vertebrates have been found stratigraphically higher. These extend the stratigraphic range to ~2 m below the local contact of the Hell Creek and Tullock Formations (Archibald, 1987). Similar gaps between the stratigraphically highest record of a group and the basal lignite of the Z coal complex have been found in some sections studied palynologically and in the record of freshwater invertebrates (Hartman, 1999).

The causes of this gap in the documented fossil record in the Western Interior have been the subject of abundant speculation and hypothesizing. It has been argued that the gap in the vertebrate record is a product of inadequate collecting (e.g., Sheehan et al., 1999) or, more charitably, the Signor-Lipps effect (Signor and Lipps, 1982). Now, after almost 30 yr of intensive collecting in Garfield and McCone Counties, much of it focused on strata of the uppermost Hell Creek Formation and lower part of the Tullock Formation, the gap in the fossil record remains. On the basis of a study of the section in the valley of Bug Creek, Retallack (1994) suggested that the gap in the record was the product of leaching because of intense acid rain produced by an impact. This hypothesis has yet to be supported by observations in other sections, and the possibility of acid rain of such great predicted intensity has been seriously challenged (Weil, 1994). Continuing studies of variation in carbon isotopic signatures of atmospheric change during deposition of the uppermost part of the Hell Creek Formation document a period of environmental instability in the Western Interior prior to the K-T boundary (Arens and Jahren, 1997; this volume). The possibility that the gaps in the vertebrate, palynological, and invertebrate records just below the boundary were products of this environmental instability is being investigated.

In a related discussion, there is debate over the possibility that the taxonomic and/or ecological diversity of nonavian dinosaurs decreased prior to their extinction (see Hurlbert and Archibald, 1995, 1996; Sheehan et al., 1996; and references cited therein). Comparison of faunal lists of nonavian dinosaurs for the northern Western Interior (Weishampel, 1990) clearly indicates that during the last 5–7 m.y. of the Cretaceous, taxonomic diversity of nonavian dinosaurs, particularly ornithischians, decreased from Judithian NALMA levels to those characteristic of the Lancian NALMA. On the basis of a series of detailed studies of the fossil record of the Hell Creek Formation in southwestern North Dakota, Sheehan et al. (1991, 1996, and references therein) argued that the ecological diversity of nonavian dinosaurs did not decrease over a significantly shorter period of time, probably <2 m.y., prior to their extinction. Their conclusions were challenged by Hurlbert and Archibald (1995, 1996).

Palynological criteria

Lofgren (1995), like many earlier workers, relied on a palynological datum to place the K-T boundary in sections lacking evidence of an iridium anomaly or boundary clay. This datum is the top of the *Wodehouseia spinata* assemblage zone, the youngest of the Late Cretaceous palynological zones of the Western Interior established by Nichols et al. (1982; see Nichols, 1994, and references cited therein). Its type section is in exposures of the Hell Creek Formation in Seven Blackfoot

Coulee, Garfield County (Fig. 2). The top of the zone is marked by the last record of *Proteacidites* spp., which is one of many, apparently contemporaneous, palynofloral extinctions used to mark the K-T boundary. The boundary between the *Wodehouseia spinata* assemblage zone and older *Aquilapollenites quadrilobatus* interval zone is placed at the first occurrence of *Wodehouseia spinata* that, in the type area, is coincident with the local contact between the nonmarine Hell Creek Formation and the underlying, partly marine Fox Hills Sandstone. Where correlation has been possible, Lancian NALMA faunas have been found to be within the *W. spinata* assemblage zone and in sediments deposited during the magnetic polarity chron 29r. Although much more widely applicable than the last occurrence of nonavian dinosaurs, the last appearance datum (LAD) provided by the top of the *Wodehouseia spinata* assemblage is similarly problematic. In some sections in the Hell Creek region, the palynologically determined LAD is separated by several meters from the bed containing an anomalous concentration of iridium and shocked mineral grains (Hotton, 1989).

Iridium anomalies, shocked mineral grains, and radiometric age determinations

In theory, the use of minute but anomalously high concentrations of iridium in combination with the presence of shocked mineral grains (interpreted as evidence of an impact of an extraterrestrial body) should provide a practical, globally applicable method of recognition of the K-T boundary in marine and nonmarine sections. Similarly, radiometric age determination of the time of deposition of these or adjacent beds through application of $^{40}Ar/^{39}Ar$ methods offers the potential of accurate, very precise correlations. In practice, in Garfield and McCone Counties, these methods have proven to be of limited utility.

Detailed studies of sections along the drainage of Hell Creek in northern Garfield County (e.g., Archibald, 1982; Dingus, 1983; Fastovsky, 1987) have documented the complex composition of the Z coal beds. In the area along the road connecting the town of Jordan with the Hell Creek Recreation Area (Fig. 2) and westward into the headwaters of Hell Creek and Snow Creek, a distance of ~35 km, the base of the Z coal complex is formed by a thin lignite, the iridium lignite (IrZ; Fig. 3). At several sites, anomalously high concentrations of iridium were discovered in the IrZ, often associated with shocked mineral grains. These anomalies vary from 0.57 ppb (Baadsgaard et al., 1988) at the Lerbekmo locality adjacent to the Hell Creek Recreation Area road (Fig. 2) and 1.00 ppb (Bohor et al., 1984) at Brownie Butte (BB, Fig. 2) to 2.22 ppb (Swisher et al., 1993) and 11.7 ppb (Smit and van der Kaars, 1984) in Hauso Flats (Fig. 2). Single crystals of sanidine from a bentonite preserved in the IrZ at several of these localities yielded weighted mean ages of 65.16 ± 0.04 Ma (Swisher et al., 1993).

In contrast, in spite of intensive search, neither anomalously high concentrations of iridium nor evidence of shocked mineral grains have yet been found in sections in eastern Garfield or western McCone Counties. There the basal lignite of the Tullock is the McGuire Creek Z coal (Fig. 3) in the terminology employed by Lofgren (1995) or Z coal in the terminology used by Swisher et al. (1993). Radiometric age determinations by the $^{40}Ar/^{39}Ar$ method on sanidines yielded a mean age of 64.95 ± 0.05 Ma, ~200 k.y. younger than the IrZ (65.16 ± 0.04 Ma). This is the area in which vertebrate local faunas of Pu1 age have been discovered in the uppermost beds of the Hell Creek Formation. These data indicate that local peat swamps had to have been present at the K-T boundary in order to preserve anomalous concentrations of iridium, shocked mineral grains, or volcanic ashes suitable for analysis by the $^{40}Ar/^{39}Ar$ method. Note that unless otherwise indicated, I have herein used the age determinations obtained at the Berkeley Geochronology Center and reported by Swisher et al. (1993). These and other relevant age determinations were recalculated by Obradovich and Hicks (1999), who reported a slightly older age of 65.5 ± 0.1 for the K-T boundary.

Current techniques

Establishment of a Global Stratigraphic Section and Point in marine sediments at El Kef, Tunisia, for the K-T boundary provided a standard reference point for correlation. Evaluation of the long-used formula for recognition of the approximate position of the K-T boundary in the northern Western Interior established by Roland Brown, i.e., the base of the first lignite above the stratigraphically highest remains of dinosaurs, showed it to be flawed. Our research in northeastern Montana and studies in other areas (Pearson et al., 1999, this volume) show that the inception of deposition of lignite characteristic of the Tullock Member of the Fort Union Formation began at different times in different areas.

Attempts to make use of biostratigraphic criteria for recognition of the K-T boundary and establishing precise correlations throughout the northern Western Interior have added resolution but have significant limitations. Adoption of the last record of nonavian dinosaurs or the extinction of the palynologically defined species characterizing the *Wodehouseia spinata* assemblage zone has the common limitation of a LAD. This is particularly true in many sections in Garfield and McCone Counties where the last occurrences of nonavian dinosaurs are overlain by strata lacking records of vertebrates or pollen. Is the last record of these organisms in the northern Western Interior a record of local extirpation, global extinction, a hiatus in sedimentation, or a taphonomic factor?

Where preserved, records of the K-T boundary impact event—an anomalous concentration of iridium or shocked mineral grains—or bentonites suitable for analysis by the $^{40}Ar/^{39}Ar$ method provide a basis for high-resolution correlation with the global standard for the K-T boundary. In the Western Interior, preservation of these records appears to have been dependent on the presence of peat swamps, and, in many areas, they were not developed at this time.

In the valley of Hell Creek and adjacent areas in Garfield County, the IrZ coal can be correlated with the K-T boundary on the basis of geochemical and mineralogical data. Resembling the records of nonavian dinosaurs, in this area the last records of Lancian NALMA local faunas are 2 m or more below the IrZ. The stratigraphically lowest record of a Puercan NALMA local fauna (UCMP locality V77129, Herpijunk Promontory) is in a channel filling capped by strata ~3.5 m above the IrZ.

Other methods that could provide high-resolution correlations have been proposed and can be evaluated. Lerbekmo et al. (1996) reported the discovery of a brief normal polarity subchron including the K-T boundary within the magnetostratigraphic chron 29r. They reported this subchron, which is estimated to have lasted at least 50 k.y., at localities in Alberta and Saskatchewan as well as at one locality ~0.8 km west of the road to the Hell Creek Recreation Area (Fig. 2) in Garfield County. In addition, variations in carbon isotopic signatures, a $-1.5‰$ to $-2.0‰$ excursion in $\delta^{13}C$ just after the K-T boundary (Arens and Jahren, 1997, 1998; this volume), may provide a widely applicable tool for precise correlation at a degree of resolution suitable for testing paleobiological hypotheses. In addition, these studies of carbon isotopic signatures could serve as a basis for determining the completeness of geological sections.

BUG CREEK PROBLEM

Understanding patterns of survival and extinction as well as recovery of the terrestrial vertebrate fauna after the terminal Cretaceous extinction has been clouded by differences of opinion concerning the origins of the so-called Bug Creek facies faunas. The pioneering work of Sloan and Van Valen (1965) and their associates resulted in the discovery and extensive sampling of a series of highly fossiliferous, sandy channel fillings in the valley of Bug Creek and adjacent areas (Fig. 2). Many of these fillings contain concentrations of fossils including remains of species previously thought to be limited to the latest Cretaceous and others thought to be limited to the early Paleocene.

The Bug Creek facies faunas—better termed assemblages—served as the basis of hypotheses advocating a pattern of gradual or stepwise faunal change in the terrestrial vertebrate fauna from the latest Cretaceous into the Paleocene (see Archibald, 1996a, and references cited therein). Evidence supporting these hypotheses initially was provided by samples from a sequence of three localities—Bug Creek Anthills, Bug Creek West, and Harbicht Hill—that were first thought to be of Cretaceous age (Sloan and Van Valen, 1965). The relative stratigraphic positions of these channel fillings could not be established by tracing the channel fillings laterally into sections where they interfingered with underlying and overlying strata. Instead, they were temporally ordered on the basis of faunal composition and stage of evolution of members of some mammalian lineages.

The taphonomy and biostratigraphic significance of the remarkably rich fossil localities in the Bug Creek region have been extensively and, at times, acrimoniously debated (Lofgren, 1995). Detailed field and laboratory studies have resolved many of these issues. Large channel fillings in the Hell Creek and Tullock record the presence of large rivers crossing the area during the latest Cretaceous and early Paleocene. Vertebrate fossils preserved in these fillings were derived from two sources: some represent animals that died just before or at the time of formation of the channel filling, and others are older fossils eroded from their entombing strata by the rivers and reworked into younger deposits.

The role of reworking and the resulting time averaging of fossil assemblages (see Kowalenski, 1996) was underestimated in early studies, as exemplified by an analysis of a sample of the multituberculate *Mesodma* from Bug Creek Anthills (Novacek and Clemens, 1977). *Mesodma* is represented by three species (*M. hensleighi, M. formosa*, and *M. thompsoni*) in Lancian NALMA local faunas in Wyoming, Alberta, and Montana. Novacek and Clemens (1977) found that the Bug Creek Anthills assemblage of posterior premolars of this genus shows a pattern of variation suggestive of one species with a range of variation that encompasses those of the three Lancian species as well as the Pu1 species, *Mesodma garfieldensis*. In view of the current recognition of the role of reworking in formation of the Bug Creek fossil assemblages, the Bug Creek Anthills species of *Mesodma* probably was based on a hydrodynamically sorted sample of all four species. A vivid example of reworking the remains of dinosaurs preserved in lumps of clay palynologically determined to be of Late Cretaceous age into an early Paleocene channel filling was described by Lofgren et al. (1990).

Lofgren (1995) addressed the general problem of the age of the Bug Creek facies assemblages. He demonstrated that, where they could be traced laterally, channel fillings containing these assemblages, which include dinosaurian remains, interfinger with flood-plain deposits that lack records of nonavian dinosaurs. Many workers (e.g., Lofgren, 1995; Nessov et al., 1998, and references cited therein) argue that these and other localities yielding Bug Creek-type assemblages were deposited during the early Paleocene and contain a mixture of fossils representing animals that lived during the latest Cretaceous and early Paleocene. Fiorillo (1998) conservatively estimated that 10%–15% of the specimens in samples of some of these Bug Creek assemblages were reworked from strata of Lancian age. Despite their taphonomic interest, for the purposes of biostratigraphic analyses and studies of the pattern of vertebrate evolution, these time-averaged assemblages must be set aside. This done, a series of fossil localities in small channel and overbank deposits remains and provides a well-documented history of faunal evolution.

What can be done with the large, well-preserved samples

of fossils in the Bug Creek assemblages? As a number of workers (e.g., Argast et al., 1987) have demonstrated, the degree of physical abrasion, particularly of small fossils, cannot be used to distinguish reworked fossils in the Bug Creek assemblages from those representing animals extant at the time of deposition. Although the fossils record the presence in the area of taxa at some time within or throughout a latest Lancian-Pu1 interval, they cannot be used to precisely establish first or last occurrences. The possibility of using the large samples of taxa that, from other data, can be demonstrated to have strictly Lancian or Pu1 ranges appears to have been successful in some instances (e.g., Luo, 1991).

SURVIVAL AND EXTINCTION OF VERTEBRATES AT THE K-T BOUNDARY; EVIDENCE FROM NORTHEASTERN MONTANA

Hypothesizing, speculating, and otherwise writing about the nature and causal factors of extinctions at the K-T boundary continues, and contributions to the scientific literature continue to multiply. Some studies focus on the history of change in ancient oceans (see MacLeod, 1998, and references cited therein), others consider evolution of vertebrate lineages, both marine and terrestrial (see Milner, 1998, and references cited therein), and fewer attempt to analyze change in the entire biota (see MacLeod et al., 1997). Individual articles and symposia continue to report debates on interpretations of the causal factors of the extinctions at the K-T boundary (e.g., Ryder et al., 1996, this volume). Similarly the scientific and popular press reports accounts of new discoveries and interpretations of events that occurred ca. 65 Ma. Although currently the rate of appearance of such articles has been decreasing, the *Science Times* of the *New York Times* continues to publish commentary on nonavian dinosaurs, their life, and their extinction. Historians of science and sociologists have found the debate about what happened at the K-T boundary an area worthy of scholarly investigation (e.g., Clemens, 1986, 1994; Galvin, 1998; Glen, 1994). An attempt to summarize and analyze the data and conclusions presented in this rapidly growing body of literature is well beyond the bounds of this article. The following discussion is limited to an evaluation of the strengths and limitations of the vertebrate fossil record of Garfield and McCone Counties and adjacent areas of the Western Interior and its contributions to current discussions of evolution around the K-T boundary.

The collections from the northern Western Interior (eastern areas of Colorado, Wyoming, and Montana; adjacent parts of North and South Dakota; and southern areas of the Canadian provinces of Alberta and Saskatchewan) constitute a globally unique sample of the course of terrestrial vertebrate evolution from the latest Cretaceous into the Paleocene. In no other area is there yet a similarly detailed record of the evolution of terrestrial vertebrates across the K-T boundary. The record from the northern Western Interior documents changes in the terrestrial vertebrate fauna of the eastern side of a peninsula, an apparently topographically diverse, straplike appendage of the Asian continent. It was bounded by the Western Interior seaway to the east, the Pacific Ocean to the west, and, south of modern Guatemala, an oceanic gulf separated the southern end of the peninsula from South America (Scotese, 1997; Hay et al., 1999). Patterns of biological change recorded in the northern Western Interior and summarized here cannot be confidently extended to characterize what happened in other areas ca. 65 Ma. There is growing evidence that the extinction and recovery of terrestrial faunas across the K-T boundary was significantly influenced by biogeographic diversity.

The patterns of survival and extinction of terrestrial vertebrates across the K-T boundary in what are now Garfield and McCone Counties were studied in detail by Archibald and Bryant (1990; see Archibald, 1996a). They undertook analyses of these patterns on the basis of the large collections amassed since 1972 by University of California Museum of Paleontology field parties (>150 000 specimens representing 111 species). Because of the rarity and fragmentary nature of avian fossils, the records of birds were not considered in their analyses. Archibald and Bryant (1990) concluded that, at the end of the Cretaceous, the terrestrial vertebrate fauna of eastern Montana was modified by extinctions of lineages concentrated in a few major groups, local survival of some lineages, and local extirpation of others through shifts in biogeographic range. The chances that this pattern of extinction was the result of chance alone appear to be nil; for some reasons, the nonavian dinosaurs were particularly vulnerable (Raup, 1991).

On a long-term (millions of years) time scale, modification of the terrestrial biota (e.g., cladogenic origins of new species and immigration events) appears to reflect changes in global environments possibly triggered by marine regression or other factors prior to the end of the Cretaceous. Short-term, ecological changes (from increased volcanic activity and impact of extraterrestrial bodies at the end of the Cretaceous) probably contributed to the alteration of the physical environment that exceeded the range of tolerance of some species and thus caused their extinction.

The survival of many lineages of amphibians, birds, and nondinosaurian reptiles indicated that, if environmental changes occurred, they were not short-term, catastrophic events. The "Dante's Inferno" class of hypotheses (see Alvarez, 1986, 1987), attributing the extinction of dinosaurs (other than birds) to extended periods (months or years) of global darkness and cold or intense deluges of acid rain of very low PH, is falsified by the survival of these lineages (e.g., see Archibald and Bryant, 1990; Clemens and Nelms, 1993; Weil, 1994). The weight of the vertebrate paleontological evidence suggests that the causal factors of terminal Cretaceous extinctions were complex, involving both long-term and short-term environmental changes.

Subsequent field and laboratory research on the vertebrate fossil record of Garfield and McCone Counties has added new data supporting the hypotheses advanced by Archibald and Bryant (1990). In part, this research derives from studies of patterns of recovery of the vertebrate fauna in the earliest Paleocene (discussed in the following). In addition, several workers have undertaken studies of the fragmentary records of latest Cretaceous and early Paleocene birds, a group not considered by Archibald and Bryant (1990).

The depositional environments of the Hell Creek and Tullock Formations, combined with the fragility of the skeletons of birds, resulted in rare and fragmentary fossil records, but these pieces can preserve characters supporting systematically detailed identifications. Study of this material is part of a rapidly intensifying interest in the origin and early evolution of birds (see Padian and Chiappe, 1998). In general, it appears that the Late Cretaceous avifauna of the Western Interior consisted of representatives of five major clades. Four are known to have been represented in Lancian avifaunas. The most primitive clade includes the curious, flightless alvarezsaurids. Hutchinson and Chiappe (1998) described a new alvarezsaurid from the Hell Creek Formation of eastern Garfield County, which added to the already documented South American and Mongolian occurrences of the group. *Avisaurus*, typified on a tarsometatarsus collected at another Hell Creek locality in Garfield County (Brett-Surman and Paul, 1985), is a representative of the Enantiornithes, which currently constitutes the most diverse and widely distributed group of Late Cretaceous birds. The Hesperornithiformes and Ichthyornithiformes are best known from pre-Maastrichtian marine deposits of the Western Interior. Ichthyorniformes are unknown in Lancian avifaunas, but the existence of a Lancian, riverine, hesperornithiform group has been announced on the basis of material from both the Hell Creek and Lance Formations (Hutchinson et al., 1997; Stidham, 1998a). These four groups of Lancian birds appear to have become extinct at or near the end of the Cretaceous.

The fifth major group of Late Cretaceous birds is the Neornithes, the clade including modern birds and their immediate ancestors. On the basis of the current fossil record, the Neornithes was the numerically dominant group of birds in the nearshore region of the Western Interior (Hope, 1998; Stidham, 1998a). Specimens from the Hell Creek Formation, northeastern Montana, and the Lance Formation, eastern Wyoming, already record the presence of multiple species of Anseriformes and Charadriiformes, which were dominantly shore and aquatic birds (see Brodkorb, 1963; Stidham, 1998b). Stidham (1998c) described a specimen of a parrot from the Lance Formation, Wyoming. This and a record of a galliform add two taxa of nonaquatic birds to the documented avifauna. Stidham, Hope, and others are studying additional material from the Hell Creek and Lance Formations that gives promise of significantly increasing documented Lancian neornithine diversity.

These new discoveries lend support to molecular studies (e.g., Cooper and Penny, 1997) suggesting that most if not all modern bird orders had differentiated before the end of the Cretaceous. Cooper and Penny (1997, p. 1109) call neornithine evolution across the K-T boundary an example of "mass survival." However, extinction of the more primitive alvarezsaurids, enantiornithines, hesperornithiforms, and, earlier, ichthyornithiforms adds to evidence of the selectivity in avian and other vertebrate survival and extinction across the K-T boundary.

DISTRIBUTION AND CORRELATION OF LATE CRETACEOUS–EARLY PALEOCENE LOCAL FAUNAS WITHIN THE NORTHERN WESTERN INTERIOR

Three topics must be addressed to set the stage for an analysis of research on patterns of recovery of the terrestrial vertebrate fauna of the northern Western Interior during approximately the first million years after the K-T extinction. Current bases for the fine-scaled correlations of terrestrial vertebrate local faunas are analyzed, recent studies of the nature of biogeographic provincialism of Lancian and Puercan vertebrate faunas are summarized, and the nature of the database—number of localities, geographic distribution, stratigraphic sequence, and related topics—is briefly evaluated.

Historically, biostratigraphic methods have played the predominant role in correlations of Late Cretaceous and early Cenozoic terrestrial vertebrate local faunas of the Western Interior. In recent decades, a time scale for the Cenozoic based on first, last, and unique occurrences of mammals calibrated in NALMA was developed and continues to be refined (see discussions in Woodburne, 1987). In this scale, the early Paleocene Puercan NALMA is divided into three units, Pu1 (the oldest) through Pu3. A fourth unit, Pu0, based on the Bug Creek assemblages that, most probably, are time-averaged assemblages of the remains of Lancian and Pu1 vertebrates, is not employed here (see preceding discussion; Lofgren, 1995; Eberle and Lillegraven, 1998a, 1998b). The middle Paleocene Torrejonian NALMA is divided into three units, To1 to To3 (Archibald et al., 1987; Archibald and Lofgren, 1990; Lofgren, 1995), or more (Williamson, 1996). A comparable system of units was proposed for the Late Cretaceous by Russell (1964) and refined by Lillegraven and McKenna (1986), who recast some of the units proposed by Russell into the format of NALMA. The youngest of these is the Lancian NALMA (Fig. 3). In some areas the biostratigraphic correlations have been strengthened through magnetostratigraphic correlations and radiometric age determinations, but for many areas these data are not available (Archibald et al., 1987; Swisher et al., 1993; Williamson, 1996).

Recent research (see Lillegraven and McKenna, 1986; Lehman, 1997; Weil, 1997, 1999a, 1999b) suggests that Judithian mammalian faunas from parts of the northern Western Interior (eastern areas of Colorado, Wyoming, and Montana; adjacent parts of North and South Dakota; and southern areas of the Canadian provinces of Alberta and Saskatchewan) are samples of local faunas from a single biogeographic province. In the Lancian NALMA, mammalian local faunas from the northern Western Interior show a high level of similarity, indicating that they probably sample the fauna of a single biogeographic province. Although Lancian mammals from southern sites are not well known, patterns of floral and nonmammalian vertebrate distributions suggest that the provinciality of the Judithian was maintained into the Lancian (see Lehman, 1987; Weil, 1999a, 1999b). Samples of mammalian local faunas are known in the Puercan from southern Canada to New Mexico. Biogeographic analyses indicate that differentiation of mammalian faunas into northern and southern provinces, first seen in the Judithian, continued on into the Puercan (see Anthony and Maas, 1990; Eberle and Lillegraven, 1998b; Weil, 1999a, 1999b).

A complete catalog of all currently known Lancian and Puercan localities yielding one or more mammalian specimens is beyond the scope of this chapter. In this and the following paragraphs citations are limited to recognition of areas that have produced large samples and a few less productive sites that help define the limits of the geographic area being sampled. The northernmost local faunas certainly of Lancian age are in central Alberta (TR, Fig. 1) and southwestern Saskatchewan (FR, Fig. 1) and have been studied by a number of paleontologists. Overviews of their research were provided by Clemens et al. (1979) and, with a focus on the record from Saskatchewan, by Fox (1997). The implications of the presence of a greater diversity of eutherians in these faunas than in apparently contemporaneous Lancian local faunas to the south were discussed by Fox (1997) and Eberle and Lillegraven (1998b). In Montana, in addition to the local faunas from Garfield and McCone Counties, an important Lancian local fauna has been reported from the vicinity of Makoshika State Park (Fig. 1; Hunter et al., 1997). Lancian local faunas have been discovered in other areas of southeastern Montana (Hunter and Archibald, 1999, this volume). Samples of Lancian local faunas have been obtained from southwestern North Dakota (M, Fig. 1; Hunter and Pearson, 1996; Hunter and Archibald, 1999; this volume) and western South Dakota (Fig. 1; Wilson, 1983, 1987). The primary Lancian local faunas of Wyoming come from the Lance Creek area (Fig. 1; Clemens 1973a, and references cited therein) and the Hanna basin of Wyoming (Fig. 1; Eberle and Lillegraven, 1998a, 1998b). A few Lancian vertebrates have been found farther south, e.g.,, in Weld County, northern Denver Basin, Colorado (Fig. 1; Carpenter, 1979) and the San Juan Basin of New Mexico (Fig. 1; Lehman, 1984; Flynn, 1986), but these samples are not large enough to adequately document local faunal composition.

The Mantua Lentil local fauna from northwestern Wyoming (Fig. 1) was the first of earliest Puercan (Pu1) age to be discovered (see Archibald et al., 1987, and references cited therein). Current studies of local faunas from southwestern Saskatchewan have not resulted in discovery of a Pu1 local fauna, but problems of correlation of localities around the K-T boundary have yet to be resolved (see Fox, 1997). Pu1 local faunas also are known from sites in Garfield and McCone Counties (Archibald, 1982; Lofgren, 1995). A Pu1 local fauna has been described from the Hanna basin, south-central Wyoming (Eberle and Lillegraven, 1998a, 1998b). Remarkably well preserved mammalian remains from the Denver Basin, Colorado, document the Littleton local fauna, which appears to be intermediate in age between the Pu1 Mantua Lentil local fauna and the Pu2 fauna of the San Juan Basin (Middleton, 1982, 1983; Archibald et al., 1987). Where magnetostratigraphic studies have been carried out, all Pu1 local faunas occur in chron 29r. Radiometric age determinations in Garfield County show that the well-studied Hell's Hollow local fauna (Archibald, 1982) and other Pu1 local faunas are preserved in rocks deposited between 64.77 and 65.16 Ma (Fig. 3).

The pattern of sampling of middle and late Puercan (Pu2 and Pu3) local faunas differs from that of the earliest Puercan. Pu1 local faunas are unknown in the San Juan Basin of New Mexico (Fig. 1). The classic Puercan local faunas of the basin, discovered in the nineteenth century and most recently analyzed by Williamson (1996; see Williamson and Lucas, 1993), can be segregated into the stratigraphically lower *Hemithlaeus kowalevskianus-Taeniolabis taoensis* assemblage zone and superjacent *Taeniolabis taoensis-Periptychus carinidens* assemblage zone. All occurrences of these zones are in strata deposited in chron 29n. The faunas of these assemblage zones provide the characterizing taxa for Pu2 and Pu3, respectively (see Archibald et al., 1987). Outside the San Juan Basin Pu2 and Pu3 local faunas have been recognized in the Hanna basin, Wyoming (Eberle and Lillegraven, 1998a, 1998b) and, possibly, in the Denver Basin (Middleton, 1983; Archibald et al., 1987).

To the north, the Hiatt local fauna from Makoshika State Park, southeastern Montana (Fig. 1; Hunter et al., 1997), has been interpreted as being of Pu2 age but might have lived during chron 29r. Two major localities apparently of Pu2 age have been reported from Saskatchewan (FR, Fig. 1; Fox, 1997). Because of biogeographic provinciality (Anthony and Maas, 1990) and, possibly, differences in taphonomy, Pu2 and Pu3 cannot yet be distinguished in the local faunas of Garfield and McCone Counties where post-Pu1 Puercan faunas are referred to as Pu2-Pu3 undifferentiated. A local fauna from the Crazy Mountain Field, south-central Montana (Fig. 1; Buckley, 1997, and references cited therein) has been correlated with the Pu2-Pu3 local faunas of Garfield and McCone Counties. Where magnetostratigraphic analyses have been successfully undertaken, with the possible exception of the Hiatt local fauna, all Pu2, Pu3, or Pu2-Pu3 undifferentiated local faunas are preserved in rocks deposited during chron 29n. In Garfield and McCone Counties, the Pu2-Pu3 localities are between 64.77 and 64.11 Ma.

In Garfield and McCone Counties the most recent local faunas preserved in the Tullock are correlated with the early Torrejonian (To1) NALMA. There, the To1 local fauna, represented by the largest sample from the Farrand Channel (UCMP localities V76170 and V76171, Fig. 3), is between 64.11 and 63.90 Ma.

RECOVERY OF THE TERRESTRIAL VERTEBRATE FAUNA

In the past decade, a number of workers have undertaken research on the evolution of vertebrates across the K-T boundary. Some of these studies are primarily concerned with documenting the history and elucidating the evolutionary processes involved in recovery of the mammalian fauna after the K-T mass extinction (e.g., Archibald, 1993; Alroy, 1999). Others emphasize what the fossil record can tell us about preextinction diversity of mammals (e.g., Weil and Clemens, 1998; Foote et al., 1999a). A few deal with the evolutionary histories of other groups of terrestrial vertebrates (e.g., Hutchison and Archibald, 1986; Holroyd and Hutchison, this volume). The following sections of this paper present analyses, some preliminary, of the record of mammalian evolution in Garfield and McCone Counties and what it tells us about the patterns and processes of evolution after the mass extinction marking the K-T boundary. In turn, data from analyses of the Puercan faunas serve as a basis for assessing the completeness of sampling of Late Cretaceous mammalian faunas.

Mammalian faunal database

In order to test current hypotheses concerning patterns and processes of mammalian evolution during the Late Cretaceous and early Paleocene, A. Weil and I are developing a mammalian faunal database for the Lancian, Pu1, and Pu2-Pu3 faunas of Garfield and McCone Counties. The database on which the preliminary analyses presented in Weil and Clemens (1998) were founded, revised to reflect results of ongoing research, is given in Table 1. The Lancian faunal list is a compilation of the taxa found in the Flat Creek localities in Garfield County (UCMP localities V73083–V73085, Archibald, 1982) and the K-Mark locality in McCone County (UCMP locality V85092, Lofgren, 1995). The Pu1 faunal list is a compilation of the taxa found in the Hell's Hollow Channel (UCMP V74110), Worm Coulee 1 (UCMP 74111), and McKeever Ranch (UCMP V72210 and V77087) localities in Garfield County (Archibald, 1982) and the Z-Line (UCMP V84193 and V84194) localities in McCone County (Lofgren, 1995). In preparing the compilations (Table 1) for the Lancian and Pu1 faunas, some changes in original identifications (Archibald, 1982; Lofgren, 1995) have been made to reflect subsequent changes in taxonomy.

The list for the Pu2-Pu3 fauna of Garfield and McCone Counties (Table 1) must be taken as a preliminary assessment of taxonomic diversity. Study of the very large collections from the Garbani Channel deposits in Garfield County is under way. In compilation of the Pu2-Pu3 list for Garfield and McCone Counties, we took advantage of the work of P.A. Johnston, who studied the collections made in the first years of our work in Garfield County in his analysis of the Rav W-1 local fauna of Saskatchewan (Johnston and Fox, 1984) and found many taxa in common. Subsequently, larger samples of some taxa in the Garbani Channel local faunas have been studied in greater detail (e.g., Simmons, 1987; Weil, 1998, 1999a, 1999b; Clemens, 1974, 1999), and the conclusions of these analyses have been used to amend the initial faunal list. In order to avoid duplication of records, we have not added several new taxa from the Pu2-Pu3 Purgatory Hill local fauna in McCone County proposed by Sloan and Van Valen (1965; Van Valen, 1978). Some already may be recognized in our preliminary list (Table 1); e.g., *Procerberus* n. sp. might well be referable to *Procerberus plutonis* (Van Valen, 1978), but the necessary comparisons have yet to be made. Where comparative studies have been made, taxa first recognized in the Purgatory Hill local fauna have been identified in the Garbani Channel local faunas; e.g., both *Purgatorius unio* (Van Valen and Sloan, 1965), typified on material from Purgatory Hill, and *P. janisae* (Van Valen, 1994), typified on material from the Garbani Channel deposits, are recognized as members of the Pu2-Pu3 fauna. On the basis of the trend for recognition of new taxa established in the detailed studies completed to date, the provisional faunal list for Pu2-Pu3 mammals in Garfield and McCone Counties presented in Table 1 may underrepresent the taxonomic diversity of this fauna.

Role of immigration in the evolution of the mammalian fauna of Garfield and McCone Counties

Archibald (1993) focused on the importance of phylogenetic analysis in assessments of species turnover, and provided analyses of the evolution of three groups of mammals—taeniodont eutherians, taeniolabidoid multituberculates, and periptychid eutherian ungulates—during the Paleocene. His study highlights the importance of clearly differentiating between pseudoextinctions of metaspecies, groups of organisms lacking characterizing autapomorphies, and true lineage extinctions. Among the taeniodonts, taeniolabidoids, and periptychids, almost 25% of the disappearances of species from the fossil record are products of pseudoextinctions.

In a parallel fashion, recognizing what was known of their biogeographic ranges and phylogenetic relationships, Weil and Clemens (1998) separated the mammalian taxa in Lancian, Pu1, and Pu2-Pu3 local faunas in northeastern Montana into three categories: residents, aliens, and unknowns. In this analysis, species were classified as residents if they were present, or their lineage was represented by a closely related sister species, in one or more local faunas of the preceding NALMA in the study area.

The study areas used in these comparisons vary somewhat. In the analysis of the Lancian fauna of Garfield and McCone Counties, records from older Edmontonian and Judithian NALMA faunas of the northern Western Interior (specifically, Edmontonian and Judithian local faunas in Wyoming, Montana, and Alberta) were considered. For Pu1 species, comparisons were made with Lancian local faunas in Colorado, Wyoming,

Montana, and the Dakotas, as well as some local faunas in Alberta and Saskatchewan. Only the local faunas from Canada that are clearly of Lancian age were considered in analysis of the Pu1 fauna of Garfield and McCone Counties; the transitional assemblages identified by Fox (1997) were excluded at this stage. In analysis of Pu2-Pu3 species, comparisons were made with Pu1 local faunas in the northern Western Interior; the Canadian transitional assemblages (Fox, 1997) were excluded in the first stage of analysis.

In all comparisons, species whose ancestors or closely related sister taxa did not precede them in the geographic areas specified were classified as aliens. The numbers of aliens and residents during a temporal interval record the total documented taxonomic diversity of each category. It is not asserted that all alien species of a fauna individually entered the area; some could have been species that evolved in Garfield and McCone Counties from immigrant stocks. Similarly, total diversity of residents could reflect local speciation during the temporal interval. Because of uncertainties concerning their phylogenetic relationships or records of occurrence, several taxa were classified as unknowns.

Lancian fauna of Garfield and McCone Counties. Weil and Clemens' (1998) analysis showed that the mammalian Lancian fauna of Garfield and McCone Counties was composed of a great majority of resident taxa. Subsequent research, reflected in Table 1, continues to support their interpretation of the dominance of residents, but two Lancian eutherian genera, *Cimolestes* and *Batodon*, require special comment. Records of *Cimolestes* (usually classified as a member of the Palaeoryctidae but placed in the Cimolestidae by McKenna and Bell, 1997), or other palaeoryctids, which could not be identified at the generic level, have been reported from Edmontonian and Judithian local faunas of both the northern and southern Western Interior. In the north, Russell (1975) mentioned the occurrence of *Cimolestes* in the Edmontonian Scabby Butte local fauna of Alberta, but the specimens on which this record was based have yet to be described. Fox (1979, p. 123–124) described an isolated premolar from the Judithian Oldman Formation and identified it as "Palaeoryctidae, genus and species undetermined." In an abstract, Carrano et al. (1997) reported the occurrence of both *Cimolestes* and *Batodon* in local faunas of the type area of the Judith River Formation, Montana. Through the courtesy of M.T. Carrano and R.W. Blob, I had the opportunity to study the material on which these identifications were based. In my opinion, some of these specimens are referable to *Paranyctoides*; another appears to document a new eutherian species. None are referable to *Cimolestes* or *Batodon*.

Occurrences of as yet undescribed species of *Cimolestes* also have been noted in the southern Western Interior. In a faunal list, Eaton et al. (1999) reported the occurrence of *Cimolestes* sp. in the fauna of the Judithian Kaiparowits Formation, Utah. Clemens (1973b) and Flynn (1986) noted the presence of palaeoryctid-like teeth in the Judithian, lower Hunters Wash fauna of the San Juan Basin, New Mexico. Rigby and Wolberg (1987) described *Cimolestes lucasi* from the geographically nearby and approximately contemporaneous Fossil Forest local fauna of the San Juan Basin. I agree with Cifelli (1990) that this species is based on a lower molar of *Alphadon*, a marsupial.

Currently, the only fully described, pre-Lancian record of a palaeoryctid is an isolated premolar from the Oldman Formation (Fox, 1979). The several passing references to occurrences of palaeoryctids in Alberta and in sites in Utah and New Mexico suggest that *Cimolestes* or *Cimolestes*-like palaeoryctids were present in these areas prior to the Lancian. However, the lack of a record of *Cimolestes* or another palaeoryctid among the ~800 identifiable mammalian teeth so far recovered from the Judith River Formation in Montana (see Sahni, 1972; Montellano, 1992; Carrano et al., 1997) suggests that palaeoryctids were very rare in or absent from these Judithian faunas. It appears best to recognize the current lack of documentation by listing the Lancian species of *Cimolestes* as unknowns (Table 1). Future studies of the currently incompletely documented occurrences will probably establish the presence of *Cimolestes* or a closely related palaeoryctid in Edmontonian or Judithian faunas of the Western Interior.

In summary, *Batodon* and *Glasbius* are the only members of the Lancian mammalian fauna of Garfield and McCone Counties that lack records in older faunas of the northern Western Interior and are classified as aliens. As a result of the change in classification of the species of *Cimolestes*, the mammalian fauna of Garfield and McCone Counties is now interpreted as having consisted of 21 resident species, 2 alien species, and 4 unknowns (Table 1). Thus the Lancian fauna of northeastern Montana appears to have been the product of evolutionary differentiation of lineages present during the Edmontonian or Judithian of the northern Western Interior. Immigration played a minor role in the evolution of this fauna.

Pu1 fauna of Garfield and McCone Counties. Extinctions at the K-T boundary reduced the total diversity of mammalian fauna of Garfield and McCone Counties to ~60% of its Lancian level (from 27 to 17 species). As noted by Archibald and Bryant (1990), most of the lineages of Lancian marsupials in northeastern Montana became extinct. Only one species, *Peradectes* cf. *P. pusillus*, is known in the Pu1 fauna of eastern Montana. Consideration of the phylogenetic relationships of all members of the Pu1 mammalian fauna shows that local extinction at the K-T boundary cut even more deeply than would be concluded from comparisons of tabulations of total diversity. Lofgren (1995) suggested survival of only 10% of the Lancian lineages across the K-T boundary; our figures suggest survival of ~18% (5 of 27). Of the mammalian species in the Pu1 fauna of Garfield and McCone Counties ~70% (12 of 17) were aliens, species representing lineages unknown in the Lancian faunas of the northern Western Interior (excluding the transitional assemblages of Fox, 1997).

The large proportion of alien species in the Pu1 fauna of Garfield and McCone Counties highlights two points. Analysis of their phylogenetic relationships shows that evolutionary radiations of some groups of multituberculates and eutherians had begun in the Cretaceous. It also indicates that Lancian mammalian faunas of areas outside the northern Western Interior, but not necessarily outside western North America, had distinctly different compositions.

Where were the homelands of these immigrants? The limited Late Cretaceous mammalian fossil record from outside the northern Western Interior offers some clues. One of the abundantly represented immigrant multituberculates in the Pu1 faunas of northeastern Montana is the eucosmodontid *Stygimys kuszmauli*. The genus and family are not certainly represented in earlier faunas of the northern Western Interior. The earliest known North American eucosmodontid, ?*Stygimys* sp., appears in the Campanian of Baja California del Norte (Lillegraven, 1972, 1976). This occurrence suggests that the *Stygimys* lineage had differentiated from other multituberculate lineages by at least the Judithian in an area outside the northern Western Interior and dispersed into northeastern Montana at the beginning of the Paleocene (Weil and Clemens, 1998; Weil, 1999a).

Another example of the significance of immigration in origins of the Pu1 mammalian fauna of Garfield and McCone Counties is the locally abrupt appearance of species of *Protungulatum, Oxyprimus, Baioconodon*, and *Mimatuta*, genera of primitive ungulates. All four genera are represented in the Pu1 Hell's Hollow local fauna in Garfield County (Archibald, 1982). They are also common members of Pu1 local faunas in the McGuire Creek area, McCone County (Lofgren, 1995). These local faunas are of particular interest because many of them are preserved in deposits stratigraphically overlain by the MCZ coal (e.g., the Z-line channel local fauna, UCMP localities V84193 and V84194; Lofgren, 1995). A volcanic ash in the MCZ coal has yielded an age determination of 64.95 ± 0.05 Ma (Fig. 3; Swisher et al., 1993), indicating that these local faunas inhabited northeastern Montana at times during the first 100 to 200 k.y. of the Paleocene. All four genera of primitive ungulates found in Pu1 local faunas of Garfield County are represented in one or more of these local faunas in McCone County.

Luo (1991) analyzed the dental morphology and phylogenetic relationships of the time-averaged assemblage of primitive ungulates in the Bug Creek Anthills channel filling. Given the absence of primitive ungulates in Lancian local faunas of Garfield and McCone Counties, probably all the material in his sample is of Pu1 age. Luo (1991) recognized three species in the Bug Creek Anthills assemblage, *Oxyprimus erikseni, Mimatuta morgoth*, and *Protungulatum donnae*, and concluded that *O. erikseni* was the most primitive species of the group. In a recent classification, McKenna and Bell (1997) treated *Protungulatum* as an ungulate, incertae sedis; *Oxyprimus* was placed in the Oxyclaenidae; and *Mimatuta* was referred to the Periptychidae. A fourth genus of primitive ungulates found in Pu1 local faunas of Garfield and McCone Counties, *Baioconodon*, was referred to the Arctocyonidae by McKenna and Bell (1997). The allocation of these genera to different families supports the hypothesis that an evolutionary radiation of primitive ungulates had begun in the Cretaceous in an area outside Garfield and McCone Counties.

The ancestry of these primitive ungulates is unclear. Recent field work in Turonian-Coniacian deposits in Asia by Archibald (1996b) and Nessov et al. (1998) revealed the presence of a diverse group of primitive ungulate-like forms, the zhelestids. They were probably distant sister taxa of the primitive ungulates that appeared in northeastern Montana during Pu1 time. In addition to these Asian forms, several North American and European species were allied in the zhelestid group. They include the Campanian genera *Gallolestes* from Baja California del Norte (Lillegraven, 1976) and Texas (Cifelli, 1994) and *Avitotherium* from the Kaiparowits Formation of south-central Utah (Cifelli, 1990). In addition, Nessov et al. (1998) allied *Alostera*, known from isolated teeth found in two Lancian local faunas in Saskatchewan (Gryde local fauna, Storer, 1991; Wounded Knee local fauna, Fox, 1997, and references cited therein), and the assemblage from the Bug Creek Anthills channel filling in McCone County (Fox, 1989). Because of reworking of fossils during deposition of this channel filling, the specimen of *Alostera* from the Bug Creek Anthills channel could represent either a Lancian or a Puercan mammal.

Other than the occurrence of *Alostera* in McCone County, which might be of Lancian age, all Lancian local faunas of Montana, western North Dakota and South Dakota, Wyoming, and Colorado lack representatives of the zhelestids and primitive ungulates. To the north, in Canada, the situation was different. As recognized by Lillegraven (1969), Russell (1975), and Fox (1997), in comparison to Lancian local faunas in the United States, the relative abundance and taxonomic diversity of eutherians was greater in Lancian local faunas of Alberta and Saskatchewan. In a recent summary and analysis, Fox (1997) dubbed two Late Cretaceous local faunas, the Fr-1 and MHBT Quarry (Long Fall), from southwestern Saskatchewan transitional assemblages. In addition to mammals that are typical of Lancian NALMA local faunas, the Fr-1 local fauna includes a record of the multituberculate *Catopsalis* and representatives of two lineages of primitive ungulates that do not appear in more southern local faunas until the Puercan (Fox, 1997, and references cited therein). The MHBT Quarry (Long Fall) local fauna includes five taxa of primitive ungulates (Fox, 1997, and references cited therein). The greater abundance and diversity of eutherians in Canadian Late Cretaceous local faunas and composition of these transitional assemblages suggest that diversification of some groups of eutherians was well under way in Canada in the Lancian. Their abrupt appearance in typical Pu1 local faunas in the United States might have been the product of southward dispersal. However, the Lancian age determination of the transitional assemblages has been challenged by some (e.g., Lerbekmo, 1985), who argue that they represent time-averaged Lancian and Puercan assemblages.

In addition, in a review of research on mammalian evolution across the K-T boundary, Fox (1997, p. 76) reported the discovery of a eutherian with " . . . progressive, taeniodont-like features of the dentition." in the Lancian KUA-1 fauna of Alberta. Eberle (1999) analyzed a cimolestid, *Alveugena*, which was closely related to the taeniodonts, from a Pu2 local fauna in the Hanna basin, Wyoming. Currently, the oldest record of taeniodonts is from Pu2 faunas of the San Juan Basin, New Mexico (see Williamson, 1996). These data suggest that differentiation of the lineage leading to taeniodonts had begun in the Lancian. Origin and evolutionary diversification of the group might have occurred then, or during Pu1 times, outside the areas from which local faunas are known.

The examples provided by the known histories of evolution of the multituberculate *Stygimys*, primitive ungulates, and taeniodont-like eutherians are stark reminders of the incompleteness of our sampling of the diversity and evolutionary histories of the Late Cretaceous mammals from areas outside of the northern Western Interior of North America. In addition, the high relative abundance of eutherians in the Lancian local faunas of Alberta and Saskatchewan and the composition of the transitional assemblages (Fox, 1997) suggest that some of the Pu1 eutherian aliens of the northern Western Interior originated in areas to the north and dispersed southward after the K-T mass extinction.

Pu2-Pu3 fauna of Garfield and McCone Counties. Immigration continued to play a significant role in recovery of the mammalian fauna of Garfield and McCone Counties during Pu2-Pu3 time. Stressing that the listing of the Pu2-Pu3 fauna (Table 1) is a preliminary compilation, Weil and Clemens (1998) noted that mammalian taxonomic diversity more than doubled (17 to 36 species) during this interval. A significant proportion of the Pu2-Pu3 fauna, ~40%, is made up of alien multituberculates and eutherians.

A recent study of one of the alien Pu2-Pu3 taxa, the triisodont *Eoconodon*, illustrates both the biogeographic diversity of Pu1 faunas and the role of immigration (Clemens, 1999). In addition, the phylogenetic relationships of *Eoconodon* are of interest in research on the origin of whales. O'Leary (1998), in her analysis of the phylogenetic relationships of the Cete (a clade including the terrestrial mesonychids and the amphibious and aquatic whales), recognized the triisodont *Eoconodon* as one of the early branches within the clade, a branch postdating the evolutionary divergence of Cete and Artiodactyla. On the basis of this phylogenetic interpretation, which was derived from analysis of dental characters, the occurrence of *Eoconodon* in Puercan faunas would indicate that differentiation of the Cete and Artiodactyla took place sometime in the Cretaceous. A more recent analysis of the phylogenetic relationships of the Cete by O'Leary and Geisler (1999) considered a broader spectrum of mammals and was based on dental, cranial, and postcranial characters of the extinct taxa. In their preferred analysis, *Eoconodon* is well outside a clade including the Cete, Artiodactyla, and Perissodactyla. As recognized by its authors, the distant phylogenetic relationships of *Arctocyon, Eoconodon*, and *Andrewsarchus* in the preferred analysis (O'Leary and Geisler, 1999, their Fig. 2) are at variance with other studies (e.g., McKenna and Bell, 1997). Whether *Eoconodon* provides the earliest record of the Cete in the Puercan or represents a primitive ungulate lineage characterized by convergence in dental morphology on the mesonychids is unknown.

Whatever its phylogenetic relationships, the fossil record of the species of *Eoconodon* documents its diversification and dispersal across the Western Interior during the Puercan. The oldest records of the genus are from the Pu1 Mantua Lentil local fauna of northwestern Wyoming and contemporaneous deposits in the Hanna basin, south-central Wyoming (Eberle and Lillegraven, 1998b). The genus is not represented in the well-sampled Pu1 faunas of Garfield and McCone Counties. Later in the Puercan, two species of *Eoconodon* appeared in northeastern Montana; one, *E. gaudrianus*, is one of the largest mammals in the Pu2-Pu3 fauna of Garfield and McCone Counties. To the south, in the San Juan Basin, New Mexico, *Eoconodon* is represented by three species in the Pu2 fauna; one of these, *E. coryphaeus*, survived in Pu3 fauna (Williamson, 1996). Unfortunately, in the San Juan Basin, Lancian local faunas currently are very poorly documented and Pu1 local faunas are unknown; the time of appearance of *Eocondon* in the area cannot be determined.

In summary, the Pu2-Pu3 fauna of Garfield and McCone Counties was more than twice as diverse as the Pu1 fauna of the area. Resembling the Pu1 fauna, a significant proportion of the Pu2-Pu3 mammals are immigrants, alien species. The profound shift from the Lancian, when mammalian diversity appears to have been dominantly the product of evolutionary diversification of resident taxa, to the Pu1, when immigration of new lineages played a major role in contributing to faunal diversity, was maintained into the Pu2-Pu3 in Garfield and McCone Counties.

Discussion. In a series of papers, Alroy (1999, and references cited therein) analyzed several patterns of North American mammalian evolution during the Late Cretaceous and Cenozoic. These highlighted the exceptional change in taxonomic diversity and estimated body mass of mammals during the first million years of the Paleocene, essentially the Puercan NALMA. The results of these studies were compared with those obtained in analyses of evolution of the Lancian and Puercan mammalian faunas in Garfield and McCone Counties. In evaluating the results of these comparisons, differences in databases must be emphasized.

Alroy's (1999) analyses are based on a comprehensive North American database; the ongoing studies by Weil and Clemens (1998) are based on samples from the northern Western Interior. The difference is significant; research by Weil (1999a, 1999b), which incorporates the work of others (e.g., Eberle and Lillegraven, 1998b), indicates distinct biogeographical provincialism in the Western Interior during the Puercan and, possibly, the Lancian. During this interval in the southern Western Interior, the Lancian mammalian fauna is poorly represented and a Pu1 fauna has yet to be discovered. In contrast,

Pu2 and Pu3 faunas are well known from the San Juan Basin, New Mexico (e.g., Archibald et al., 1987; Williamson, 1996). In the northern Western Interior, Lancian and Pu1 faunas are documented in Garfield and McCone Counties and adjacent areas. Except in the Hanna basin of Wyoming (Eberle and Lillegraven, 1998a, 1998b), Pu2 and Pu3 faunas cannot yet be differentiated and are treated as Pu2-Pu3 undifferentiated. By limiting research to the northern Western Interior, with particular emphasis on the faunas of Garfield and McCone Counties, the ongoing research (Weil and Clemens, 1998) focuses on the pattern of change in taxonomic diversity within one biogeographic province. The database used by Alroy (1999) does not recognize this biogeographic provincialism.

Similarly, studies of evolution of the mammalian fauna in the northern Western Interior are calibrated on a different time scale. First, Alroy's (1999) study was based on 2.5 Ma bins for Cretaceous faunas and 1 Ma bins for Cenozoic faunas. In Garfield and McCone Counties, the mammalian occurrences in Alroy's (1999) first 1 Ma bin can be divided between the Pu1, with a duration of ~400 k.y., and the Pu2-Pu3, with a duration of ~600 k.y. Second, in compilation of the database, we (Weil and Clemens, 1998) did not include records of the time-averaged Bug Creek assemblages and did not recognize the Pu0 subdivision (see preceding discussion).

Studies of change in the mammalian fauna of Garfield and McCone Counties, in particular, and the northern Western Interior, in general, support and amplify two of Alroy's (1999) major conclusions. His analysis of the North American record indicated that there was a mass extinction of mammalian lineages at the K-T boundary and that as many as two-thirds went extinct (Alroy, 1999, p. 114). The record from Garfield and McCone Counties is not only in line with Alroy's estimate but suggests that, at least in the northern Western Interior, extinction cut more deeply, to ~80% or 90%. However, total taxonomic diversity of the Pu1 fauna was not so deeply depressed below Lancian levels because of the appearance of 12 alien species.

Similarly, the record of Puercan faunal evolution in Garfield and McCone Counties supports Alroy's (1999) contention that diversification of the North American mammalian fauna was not a slow, steady process, but "... the most rapid phase of the radiation was confined to the very first 1.0 MY of the Paleocene (Alroy, 1999, p. 113)." Subdivision of the Puercan record of Garfield and McCone Counties into Pu1 and Pu2-Pu3 faunas reveals that the major burst in local taxonomic diversification did not occur immediately after the K-T mass extinction but during Pu2-Pu3 time. Consideration of the phylogenetic relationships of the members of Pu1 and Pu2-Pu3 faunas highlights the significant contribution of immigration to the increase in diversity of the mammalian faunas of Garfield and McCone Counties after the K-T mass extinction.

Evolution of mammalian body size

Data from northeastern Montana provide evidence concerning the pattern and tempo of evolution of mammalian body size during the Puercan. In temporally and biogeographically coarser grained studies, Alroy (1998, 1999) compared estimates of body sizes of Late Cretaceous and Puercan North American mammals and documented a marked increase in average body size during the first million years of the Tertiary. The magnitude of this change is greater than any recorded during other equivalent periods of North American mammalian evolution (Alroy, 1998, 1999).

Table 2 presents some preliminary results of the ongoing study of evolution of mammalian body size across the K-T boundary in Garfield and McCone Counties. This geographic limitation has the disadvantage of producing small sample sizes of the taxa but avoids introduction of possible latitudinal or other biogeographic variation in body size. However, limiting the database to Garfield and McCone Counties permits direct application of radiometric age determinations to subdivide the Puercan (Fig. 3), and makes it possible to determine the contributions of alien and resident species to the local pattern of change in mammalian body size.

As a method of making approximate comparisons among multituberculates, metatherians, or eutherians, length of M/1 is used as an index of their body sizes. A number of studies of the relationships of dental dimensions to body mass of eutherians have produced formulae that allow the estimation of body mass from the occlusal area of M/1 (see Bloch et al., 1998). Where possible, these values have been calculated using the formula for insectivores (Bloch et al., 1998) and are presented in the text. Similar studies have not been carried out for metatherians or multituberculates. At the moment, differences in length of M/1 among either metatherians or multituberculates should not be taken as anything more than rough indices of differences in body size. Comparisons of molar lengths among eutherians, metatherians, and multituberculates as estimates of differences in body size should be regarded as impracticable.

Lancian mammals of Garfield and McCone Counties. In the Lancian fauna of Garfield and McCone Counties, the largest known eutherian is *Cimolestes stirtoni*. An M/1 of this species has yet to be discovered in northeastern Montana; therefore, the length of M/2 is listed. This substitution is warranted by the observation that in other species of *Cimolestes*, e.g., *C. incisus* (Clemens, 1973a), in which both M/1 and M/2 are known, these molars are essentially of the same length. A contemporary Lancian species somewhat larger than *C. stirtoni*, *C. magnus*, is known from several specimens found in Alberta (Lillegraven, 1969), some rare, fragmentary material from Carter County, southeastern Montana, and the type area of the Lance Formation, Niobrara County, eastern Wyoming (Clemens, 1973a).

Estimates of the body mass of the largest species of *Cimolestes* were obtained through application of the formula proposed by Bloch et al. (1998) for insectivores to samples of teeth from sites outside Garfield and McCone Counties. The estimated body mass of *Cimolestes stirtoni*, based on the sample from the type Lance Formation, Wyoming (Clemens, 1973a), is ~375 g. The estimated body mass of *Cimolestes magnus*,

TABLE 2. DENTAL DIMENSIONS, LARGEST LANCIAN AND PUERCAN MULTITUBERCULATES AND THERIAN MAMMALS FROM GARFIELD AND MCCONE COUNTIES, MONTANA

Taxon and sources of data*	Lancian NALMA	Pu1 NALMA	Pu2-Pu3 NALMA
Multituberculata			
Mensicoessus robustus (1)	7.89 (N = 3)		
Catopsalis joyneri (2)		6.44 (N = 11)	
Taeniolabis sp. (3)			21.4 (N = 1)
Metatheria			
Didelphodon vorax (1)	6.74 (N = 1)		
Pediomys florencae (2)	4.74 (N = 2)		
Peradectes cf. *P. pusillus* (1.2)		2.39 (N = 10)	
Eutheria			
Cimolestes stirtoni (1)	4.24 (N = 1)		
Baioconodon nordicum (2)		4.47 (N = 7)	
Eocondon gaudrianus (4)			9.00 (N = 2)

Note: NALMA is North American Land Mammal Ages, N is number.
*Sources of data: (1) Archibald (1982), (2) Lofgren (1995), (3) Simmons (1987), (4) Clemens, personal observation.

based on the sample from the Lancian fauna of the Scollard Formation, Alberta (Lillegraven, 1969), is ~565 g.

The largest metatherians in the Lancian fauna of northeastern Montana are referable to *Didelphodon vorax*, which might not be referable to the Marsupialia (see Marshall and Kielan-Jaworowska, 1992; Fox and Naylor, 1995). The largest marsupial is *Pediomys florencae*. The length of M/1 of *Didelphodon vorax* is reported (Table 2). In northeastern Montana the lower molar dentition of *Pediomys florencae* is known only from isolated teeth, and M/1 has yet to be discovered; the average length of teeth tentatively identified as M/2 or M/3 is recorded (Table 2).

On the basis of length of M/1, *Meniscoessus robustus* was the largest multituberculate in the Lancian fauna of Garfield and McCone Counties.

Comparison of the Lancian fauna of Garfield and McCone Counties with Lancian mammalian faunas of the Scollard Formation, Alberta (Lillegraven, 1969), and the Lance Formation, Wyoming (Clemens, 1963, 1966, 1973a), reveals a generally similar pattern of mammalian size distribution. In all three, the largest eutherian was a species of *Cimolestes*; the largest metatherian was *Didelphodon vorax*. In both northeastern Montana and Wyoming, the largest species of multituberculate was *Meniscoessus robustus*. *Meniscoessus robustus* has not been reported from the Scollard Formation, but occurs in several Lancian faunas of Saskatchewan (Fox, 1997).

Puercan mammals of Garfield and McCone Counties. The largest mammals in the Pu1 fauna of Garfield and McCone Counties are aliens (Table 2). Among the eutherians, the largest species is the primitive ungulate *Baioconodon nordicum* (estimated body mass ~686 g). A second primitive ungulate, *Protungulatum gorgun*, is only slightly smaller. Most of the metatherians in Garfield and McCone Counties died out during the K-T mass extinction. The only lineage that survived is represented by *Peradectes* cf. *P. pusillus*, a relatively small form in comparison to other Pu1 mammals. Currently, the largest multituberculate recorded in Garfield and McCone Counties by M/1s is *Catopsalis joyneri*. Lofgren (1995) noted that an even larger species, *C. alexanderi*, was present, but an M/1 of this species has yet to be discovered.

The same pattern of the largest mammals being alien species is repeated in the Pu2-Pu3 fauna of Garfield and McCone Counties (Table 2). The largest multituberculate is *Taeniolabis* sp. Among the eutherians, *Eoconodon gaudrianus* is the largest species yet recorded (estimated body mass 3665 g). Marsupials continue to be represented by a relatively small species of *Peradectes*.

Discussion. As noted in the previous discussion of the role of immigration in evolution of taxonomic diversity of the mammalian fauna of Garfield and McCone Counties, the databases of Alroy's (1998, 1999) analyses and our ongoing study (Weil and Clemens, 1998) differ in significant respects. The database of our study is limited to a subset, the faunas of Garfield and McCone Counties, of Alroy's North American database. Also, the first million years of his database are subdivided to distinguish between Pu1 and Pu2-Pu3 faunas of northeastern Montana.

As with the analysis of taxonomic diversification, data from Garfield and McCone Counties suggest some amplification of Alroy's (1998, 1999) conclusions concerning the pattern of evolution of mammalian body size during the first million years of the Tertiary. In northeastern Montana, the overall increase in eutherian and multituberculate maximum body size during the Puercan was extreme, in line with that recorded in the North American database. Among Lancian eutherians, the largest estimated body masses ranged between ~375 g (*Cimolestes stirtoni*, based on a sample from the type Lance Formation) and ~565 g (*Cimolestes magnus*, based on a sample from the Scollard Formation). By the end of the Puercan, the largest eutherian in Garfield and McConc Countics was *Eoconodon gaudrianus* (estimated body mass 3665 g). A cruder index of body size, length of M/1, shows a similar pattern in a comparison of the Lancian *Meniscoessus robustus* with the largest known multituberculate in the Puercan of Garfield and McCone Counties, *Taeniolabis* sp. (Table 2). Increase in body size is not recorded immediately after the K-T mass extinction. The largest Pu1 eutherians and multituberculates in northeastern Montana are about the same size as those of the Lancian. The major increase in maximum body size in both groups appears in the Pu2-Pu3 fauna.

Alroy (1999, p. 116) posed a question concerning the significant change in body size in the Puercan, "Was the K-T shift a true evolutionary event, or was it merely the side-effect of immigration (specifically from Eurasia) in the wake of a major mass extinction?" Although recognizing that immigration played a role, he tended to minimize the contribution of immigrant species. The evidence from Garfield and McCone Counties suggests that, at least in this area, immigration played a major role. In both the Pu1 and Pu2-Pu3 faunas of Garfield and McCone Counties, the largest eutherians and multituberculates are alien species. Although some of the immigrant species might have Eurasian origins, the sparse records of North American Judithian and Lancian faunas from outside the northern Western Interior indicate distinct biogeographic diversity and possible sources for the alien species that appear in the Pu1 fauna of Garfield and McCone Counties.

It is of interest to note that the largest Pu1 species, all alien species, in Garfield and McCone Counties are not larger than the largest mammalian species of the Lancian fauna of the area. This suggests that they differentiated in another area during the Cretaceous when there appears to have been an evolutionary cap on maximum mammalian body size (Alroy, 1999).

Sampling the pattern of mammalian evolution in the Late Cretaceous and early Paleocene

Currently, two aspects of mammalian evolution in the Late Cretaceous and Paleocene are foci of investigation and discussion: completeness of the available fossil record of mammalian, particularly therian, evolution during this interval and timing of differentiation of the major ordinal clades of eutherian mammals. Studies of the evolution of the mammalian faunas of Garfield and McCone Counties add limited but pertinent data to these discussions.

From a global perspective, Jurassic and, particularly, Cretaceous mammalian faunas of the Northern Hemisphere and Southern Hemisphere were the products of long histories of evolution in isolation. The mammalian faunas of Australia, Antarctica, and South America appear to have consisted entirely of nontherian mammals; no unequivocal records of therians have been recovered (Bonaparte, 1990; Pascual, 1998; Pascual et al., 1999). Within this area, only in South America is the pattern of faunal change across the K-T boundary beginning to be documented. There, the transition from the Cretaceous into the Tertiary was characterized by the extinction of lineages of nontherian mammals and dispersal of marsupials and eutherians from North America or northern South America (Pascual et al., 1999). The pattern of mammalian evolution during this interval on other southern continents, Australia and Africa (except Madagascar), for example, is unknown.

The available fossil record suggests that the initial diversification of marsupials and eutherians occurred during the Cretaceous on continents that are now part of the Northern Hemisphere. How well is this phase of mammalian evolution sampled? The Eurasian record of Cretaceous mammalian evolution is dominated by a series of remarkably productive sites in Mongolia. Several of these have yielded specimens, now including numerous partial and complete skeletons, of a variety of multituberculates and therian mammals (Novacek et al., 1997, and references cited therein). The therians include members of at least two eutherian stem groups (Asioryctitheria and Zalambdalestidae), representatives of an early eutherian radiation outside the crown group Eutheria (Novacek et al., 1997). Multituberculates from these Mongolian sites also appear to document an endemic radiation. Many, but possibly not all, of these lineages do not appear to be closely related to the multituberculates that dominated Late Cretaceous (Judithian through Lancian) faunas of North America (Kielan-Jaworowska and Hurum, 1997; Rougier et al., 1997). Similarly, most of the metatherians from Mongolia currently are thought to represent a sister group (Deltatheroida) to the crown group Marsupialia. It has been argued, however, that the Mongolian mammal *Asiatherium* represents an endemic family of marsupials (Szalay and Trofimov, 1996, and references cited therein). The preponderance of endemic lineages comprising the Mongolian faunas suggests that these faunas were dominated by species particularly adapted to the desert environments of the area.

Determination of the ages of these remarkably productive Mongolian fossil localities, which cannot be directly correlated with marine deposits and occur in sections lacking rocks suitable for radiometric age determinations, has been discussed (Jerzykiewicz and Russell, 1991; Kielan-Jaworowska and Hurum, 1997, and references cited therein). Currently, none of the Mongolian faunas is recognized as a correlative of Lancian local faunas of the Western Interior. The youngest Late Cretaceous Mongolian local faunas are derived from the Nemegt Formation. They might be of Maastrichtian age, intermediate between the oldest Judithian and the Lancian local faunas of the Western Interior, i.e., contemporaneous with the poorly sampled Edmontonian interval (Jerzykiewicz and Russell, 1991; Kielan-Jaworowska and Hurum, 1997). Although there are problems and uncertainties in correlation, the available Mongolian fossil record does not appear to include latest Cretaceous or early Paleocene vertebrate faunas.

The wealth of mammalian material from Mongolia has tended to overshadow occurrences of Late Cretaceous mammals in other parts of Eurasia. Although represented by rarer and not as complete fossils, Late Cretaceous faunas from Uzbekistan and adjacent areas (Nessov et al., 1994, and references cited therein) are beginning to illuminate the biogeographic and phylogenetic histories of middle Asian mammals. Many of these faunas can be correlated with marine strata and, on this basis, are thought to range from Turonian to Campanian in age (Nessov et al., 1994). In general, the multituberculates present in these faunas appear to have been allied with lineages also known from Mongolia. In addition to deltatheridians, which currently are regarded as metatherians but not marsupials, *Marsasia* (Coniacian, Uzbekistan) and *Asiatherium* (probably Campanian, Mongolia) might be the earliest records of marsupials in Asia (Averianov and Kielan-Jaworowska, 1999, and references cited therein). The most abundant eutherians in these faunas are the zhelistids, eutherians thought to be a closely related sister group to the ungulates (Nessov et al., 1998). In addition, some dentally primitive eutherians, which might be related to palaeoryctids or lipotyphlans, have been reported (see Nessov et al., 1994). The youngest mammalian fauna from middle Asia is from Kazakhstan, and is Campanian (Averianov and Nessov, 1995), and, as is the case in Mongolia, the evolution of mammals across the K-T boundary is not documented in middle Asia.

Glimpses of the biogeographic diversification of mammals during the Campanian and Maastrichtian in Europe, which was an archipelago at this time, are provided by a growing number of sites in Spain and France (Gheerbrant and Astiba, 1994, and references cited therein) and Romania (Râdulescu and Samson, 1997, and references cited therein). The small samples currently available indicate that the mammalian faunas of these islands included groups of endemic, primitive cimolodontan multituberculates (Râdulescu and Samson, 1997). The reported presence of marsupials has been questioned (Gheerbrant and Astiba, 1994). Eutherians reported to date appear to represent two groups; one is referable to the zhelestids (Nessov et al., 1998). Isolated teeth appear to represent primitive eutherians of either lipotyphlan or palaeorcytoid affinities (Gheerbrant and Astiba, 1994).

Fragmentary fossils from a Late Cretaceous site in India represent *Deccanolestes*, which originally was referred to the Palaeoryctidae, but shows dental similarities to the kennalestids, which are best known from Mongolia. Godinot and Prasad (1994), however, have suggested that *Deccanolestes* is allied to the Archonta, a group including primates, bats, and their allies. Whether India was biogeographically part of a southern Gondwanan or Asian continent in the latest Cretaceous remains unknown (see Krause et al., 1997, and references cited therein).

The limited available fossil record from the Late Cretaceous of Eurasia, extending from a major landmass in the east to the archipelago that formed what is now Europe in the west, already is beginning to document the presence of biogeographically diverse mammalian faunas. An early review of the Eurasian Late Cretaceous record of mammalian evolution (Clemens et al., 1979) highlighted the significantly greater sampling of contemporaneous North American faunas. Although many significant discoveries have been made subsequently in Eurasia, the cursory review of the Eurasian record given here indicates that our knowledge of mammalian evolution in this area remains far from complete. Phylogenetic analyses of these Eurasian mammals suggest that many were members of endemic groups. As discussed in the following, others appear to have been related to mammalian lineages that dispersed to North America in the Late Cretaceous or early Tertiary.

On a global basis, the Western Interior of North America currently provides by far the most detailed record of mammalian evolution during the close of the Cretaceous (Judithian through Lancian NALMAs) and the beginning of the Tertiary (Puercan NALMA). The results of analyses by Weil (1999a) and others (e.g., Lehman, 1997) of the documented diversity of faunas in this area reveal the presence of biogeographic provinces from the Judithian to the Puercan. The dominance of immigrant taxa comprising Pu1 and Pu2-Pu3 faunas of the northern Western Interior points to evolutionary diversification of mammals during the Late Cretaceous in other areas. The composition of some North American Late Cretaceous local faunas from areas outside the northern Western Interior supports this interpretation.

The Judithian local faunas of Baja California del Norte and southern Texas contain significant records of mammals unknown in contemporaneous faunas of the northern Western Interior. The eucosmodontid, *Stygimys* (see Weil, 1999a), which is known from Baja California del Norte, is one of the common alien taxa in the Pu1 fauna of Garfield and McCone Counties. Local faunas from both Baja California del Norte and Texas include species of *Gallolestes*, a zhelestid (Cifelli, 1994). These occurrences raise questions about the origins of the primitive ungulates, *Oxyprimus, Protungulatum, Baioconodon*, and *Mimatuta*, which were abundantly represented immigrants in Pu1 faunas of the northern Western Interior. The possibility that the differentiation of these lineages took place in western North America rather than Asia must be entertained.

Recent discoveries in Utah, particularly on the Kaiparowits Plateau (Eaton et al., 1999, and references cited therein) and adjacent plateaus, are beginning to fill in a major gap in the pre-Campanian Late Cretaceous record of mammalian evolution. Although sharing some taxa with contemporaneous faunas in the northern Western Interior, ongoing studies of Judithian faunas of Utah are revealing endemic species. These include *Symmetrodontoides*, the last record of the ancient holotherian symmetrodonts, as well as derived therians, including the marsupials *Aenigmadelphys archeri* and *Turgidodon madseni* and the eutherian *Avitotherium utahensis*, which has been included in the zhelestids (Nessov et al., 1998). These discoveries enhance the picture of Judithian biogeographic diversity in the Western Interior.

As discussed in preceding sections, evidence provided by Lancian faunas of Alberta and Saskatchewan, in which eutherians were more abundant and diverse than in faunas from the south, suggests that an evolutionary radiation of eutherians was under way in the Late Cretaceous. If the Late Cretaceous age of the transitional assemblages in Saskatchewan is confirmed, they indicate that many of the alien taxa of the Pu1 fauna of Garfield and McCone Counties may have been immigrants from what is now southern Canada.

In summary, analyses of the record of Late Cretaceous and early Tertiary (Judithian, Lancian, and Puercan) mammalian diversity in the northern Western Interior suggest that it is relatively well sampled (e.g., Weil, 1999a). The results of these studies are supported by analyses focusing on biogeographic patterns and phylogenetic relationships reported here. However, these findings should not be misinterpreted as indicating that the faunas of the northern Western Interior completely document contemporaneous mammalian diversity in other parts of western North America. Many of the immigrant taxa that formed major parts of the Pu1 and Pu2-Pu3 faunas of Garfield and McCone Counties might have originated in and dispersed from other areas in western North America. Similarly, the limited but growing records of Late Cretaceous mammalian evolution in Eurasia are beginning to document significant biogeographic diversity. These studies demonstrate that endemic Mongolian mammals adapted to the desert environments of the area dominated its Cretaceous faunas. Growing records from other areas of Asia and the archipelago of islands in what is now Europe reveal greater mammalian diversity.

A related topic of current discussion is determination of the times of origin and diversification of extant eutherian orders. Several studies have suggested that the differentiation of extant orders of eutherian mammals did not occur until after the K-T mass extinction (Alroy, 1999; Foote et al., 1999a, 1999b). Others, on the basis of morphological and molecular data (e.g., Novacek, 1992, 1999, and references cited therein), suggested that a radiation of eutherians was under way during the Late Cretaceous, but recognized that most extant orders did not appear in the fossil record until after the K-T boundary. Students of molecular evolution (e.g., Kumar and Hedges, 1998) have argued for much more ancient (possibly Early Cretaceous) differentiation of lineages of modern orders of eutherian mammals.

As pointed out by Archibald (1999), any attempt to evaluate these competing hypotheses must distinguish between the origin of stem lineages of supraordinal groups and the origin of extant orders. The foregoing analysis of the Lancian and Puercan faunas of Garfield and McCone Counties provides evidence pertinent to times of origin of some, but not all, extant orders and supraordinal groups of eutherians. In particular, the fossil record from this area sheds light on the origins of the orders Insectivora and Carnivora as well as the supraordinal groups including primates and ungulates.

Only two known Late Cretaceous eutherians might be representatives of the extant order Lipotyphla (Insectivora). These are *Paranyctoides*, a member of many Judithian faunas of the northern and southern Western Interior, and *Batodon*, represented in Lancian faunas of the northern Western Interior. McKenna and Bell (1997) referred both to the extant Order Soricomorpha, thus placing them firmly within a crown group Lipotyphla. Several other workers have expressed less certainty about the relationships of these taxa, which are known primarily from isolated teeth (e.g., Foote et al., 1999a). At the moment, it would appear best to take the records of *Paranyctoides* and *Batodon* as clues that the stem lineage of the Lipotyphla had differentiated before the end of the Cretaceous.

A second extant eutherian order that might have differentiated in the Late Cretaceous is the Carnivora. Many workers have suggested close phylogenetic relationships between the Carnivora and several extinct ordinal groups, e.g., Creodonta, Tillodonta, and Taeniodonta (Lillegraven, 1969; Clemens, 1973a; McKenna and Bell, 1997). Some of these studies and others (e.g., Hunt and Tedford, 1993) have allied primitive members of the Carnivora with Lancian species of *Cimolestes*. Fox and Youzwyshyn (1994), however, argued that the known cimolestids were too derived to include the ancestry of carnivorans, which must be sought in dentally more primitive eutherians. Currently, there is no unequivocal evidence of a Late Cretaceous member of the Carnivora. However, the hypothesis that a supraordinal group including the cimolestids and stem group carnivorans had differentiated in the Late Cretaceous is warranted.

The Archonta is the third supraordinal group that may have originated in the Late Cretaceous. Equivocal evidence comes from the work of Godinot and Prasad (1994), who suggested that *Deccanolestes*, from the Late Cretaceous of India, had archontan affinities. Beard (1998), in a biogeographic analysis of the origination of the Primates (excluding the plesiadapiforms), argued that this order differentiated in Asia but did not voice an opinion on the area of origin of its sister group, the plesiadapiforms. The appearance of the immigrant plesiadapiforms *Purgatorius* and *Pandemonium* in the Pu2-Pu3 fauna of Garfield and McCone Counties suggests that the differentiation of Primates and plesiadapiforms had occurred outside the northern Western Interior in Pu1 time or, more likely, earlier.

The fourth supraordinal group that contributed to the evolution of the Puercan faunas of Garfield and McCone Counties was the Ungulatomorpha (Nessov et al., 1998). A wide variety of evidence documents differentiation of the zhelestids, a basal stock of the supraordinal ungulatatomorpha group (Nessov et al., 1998), in the Late Cretaceous and its dispersal through the Eurasian and North American continents prior to the K-T boundary. Most zhelestid lineages appear to have become extinct prior to the end of the Cretaceous, with the possible exception of *Alostera*. Most of the ungulates of Pu1 faunas of the northern Western Interior are referable to primitive ungulate groups of condylarthran grade. With the intriguing exception of the apparently Late Cretaceous transitional assemblages of Saskatchewan (Fox, 1997), these taxa are unknown in Lancian faunas of the Western Interior. Current evidence suggests the diversification of these primitive ungulate groups began in the Cretaceous in areas yet to be sampled. The occurrence of *Eoconodon* in some but not all Pu1 faunas of the northern Western Interior requires notice. If, as one recent analysis (O'Leary,

1998) indicated, *Eoconodon* is a sister group to the mesonychids and whales, the Pu1 occurrence of this genus would indicate that the lineages leading to whales and artiodactyls had differentiated in the Cretaceous. A subsequent analysis (O'Leary and Geisler, 1999) removed Eoconodon from the mesonychid + whale clade, but did not clarify its phylogenetic relationships.

The Late Cretaceous fossil record of mammalian evolution suggests that the early diversification of therian mammals and particularly the origin of the metatherian and eutherian clades occurred in the Northern Hemisphere. Available samples of the fossil record of this phase of mammalian evolution are biased in favor of the North American Western Interior. A smaller but growing number of faunas sample the Cretaceous faunas of Eurasia. The known fossil record documents evolution of significant biogeographic diversity during the Cretaceous. There is no unequivocal evidence of the differentiation of extant eutherian orders in the Cretaceous, but differentiation of major supraordinal eutherian groups certainly was under way prior to the K-T boundary.

CONCLUSIONS

In almost a century of research, the fossiliferous deposits of the Hell Creek Formation and Tullock Member of the Fort Union Formation have continued to yield information concerning the course of evolution of the terrestrial biota across the K-T boundary. The collections from McCone and Garfield Counties constitute a diverse and abundant record of terrestrial vertebrate evolution during this significant period of Earth history. Considered in context with the results of past and ongoing research in adjacent areas, on a global basis, the northern Western Interior provides the most detailed record of the evolution of the terrestrial biota across the K-T boundary.

Results of studies of the record from the North American Western Interior have come to be regarded by some workers as exemplifying the global pattern of change in terrestrial vertebrates and particularly the mammalian fauna across the K-T boundary. This assumption has been falsified. The composition of the Late Cretaceous mammalian faunas of the Northern and Southern Hemispheres was significantly different. Origin and diversification of metatherian and eutherian mammals appears to have taken place in the Cretaceous in the Northern Hemisphere. Although not as extensive, the known Cretaceous fossil record from outside the North American Western Interior already provides clear records of biogeographic heterogeneity. However, the data derived from the northern Western Interior can be used in tests of hypotheses invoking global environmental change at the K-T boundary.

Increasingly detailed stratigraphic studies, abetted by application of recently developed techniques of radiometric age determination as well as magnetostratigraphy and chemostratigraphy, continue to refine the resolution of knowledge of the pattern of evolution across the K-T boundary in the northern Western Interior. This increasingly detailed record of terrestrial vertebrate evolution across the K-T boundary indicates to me, and to many others, that the causal factors of this extinction involved a complex of long- and short-term causal factors.

Recovery of the terrestrial vertebrate fauna of Garfield and McCone Counties after the K-T boundary was not simply the product of an explosive evolutionary radiation of a few surviving stocks. Immigration from other areas played a major role in reconstitution of mammalian diversity.

ACKNOWLEDGMENTS

The contributions to the studies of the faunas of the Hell Creek and Tullock Formations by my colleagues, including many former and current students, with whom I enjoyed working in the field and in the laboratory, are gratefully acknowledged. In particular I thank L.B. Albright, J.D. Archibald, N.C. Arens, R.W. Blob, L. Bryant, M.T. Caranno, D. Clemens-Knott, L. Dingus, D.E. Fastovsky, J.H. Hartman, S. Hope, C.L. Hotton, A.H. Jahren, J.A. Lillegraven, D.L. Lofgren, Z. Luo, M. Novacek, N.B. Simmons, T.A. Stidham, C.C. Swisher III, A. Weil, and G.P. Wilson for their participation in the field or laboratory work, contributions to discussions of Cretaceous-Tertiary boundary phenomena, and/or the opportunity to refer to the results of their studies, which are cited throughout this chapter. Harley J. Garbani deserves special recognition for his remarkable discoveries and the many other ways in which he contributed to the successes of our field work. I also thank the Engdahl, Hauso, McKeever, Trumbo, and Twitchell families and many other residents of Garfield and McCone Counties who gave us access to their lands and/or provided significant help in other ways. Financial support from the National Science Foundation came through a series of grants (most recently EAR-9505841, part of a cooperative research project involving W.A. Clemens, D.L. Lofgren, C.C. Swisher III, and L. Dingus). In addition, I gratefully acknowledge the University of California Museum of Paleontology for underwriting my research and my students' research in northeastern Montana. Sincere thanks are due to Don and Marge Beckman as well as various members of the Bureau of Land Management and the Charles M. Russell National Wildlife Refuge, who assisted us in many ways over the past several decades and facilitated this research. I also gratefully acknowledge the constructive reviews of earlier drafts of this manuscript by N.C. Arens, J.J. Eberle, J.H. Hartman, J.P. Hunter, J.A. Lillegraven, D.J. Nichols, and A. Weil. This is University of California Museum of Paleontology Contribution 1715.

REFERENCES CITED

Alroy, J., 1998, Cope's rule and the dynamics of body mass evolution in North American fossil mammals: Science, v. 280, p. 731–734.

Alroy, J., 1999, The fossil record of North American mammals: Evidence for a Paleocene evolutionary radiation: Systematic Biology, v. 48, p. 107–118.

Alvarez, L.W., Alvarez, W., Asaro, F., and Michel, H.V., 1980, Extraterrestrial cause for the Cretaceous–Tertiary extinction: Science, v. 208, p. 1095–1108.

Alvarez, W., 1986, Toward a theory of impact crises: EOS (Transactions, American Geophysical Union), v. 67, p. 649–658.

Alvarez, W., 1987, Mass extinctions caused by large bolide impacts: Physics Today, v. 40, p. 24–33.

Anthony, M.R.L., and Maas, M.C., 1990, Biogeographic provinciality in North American Paleocene mammalian faunas: Journal of Vertebrate Paleontology, v. 9 (supplement to no. 3), p. 12A.

Archibald, J.D., 1982, Fossil Mammalia and Testudines of the Hell Creek Formation, and the geology of the Tullock and Hell Creek Formations, Garfield County, Montana: University of California Publications in Earth Sciences, v. 122, 286 p.

Archibald, J.D., 1987, Stepwise and non-catastrophic Late Cretaceous terrestrial extinctions in the Western Interior of North America: Testing observations in the context of an historical science: Mémoires de la Société Géologique de France, Nouvelle Série, v. 150, p. 45–52.

Archibald, J.D., 1993, The importance of phylogenetic analysis for the assessment of species turnover: A case history of Paleocene mammals in North America: Paleobiology, v. 19, p. 1–27.

Archibald, J.D., 1996a, Dinosaur extinction and the end of an era: What the fossils say: New York, Columbia University Press, Critical moments in Paleobiology and Earth History Series, 237 p.

Archibald, J.D., 1996b, Fossil evidence for a Late Cretaceous origin of "hoofed" mammals: Science, v. 272, p. 1150–1153.

Archibald, J.D., 1999, Divergence times of eutherian mammals: Technical Comments, Science, v. 285, p. 2031a, *www.sciencemag.org*

Archibald, J.D., and Bryant, L.J., 1990, Differential Cretaceous-Tertiary extinctions of non-marine vertebrates: Evidence from northeastern Montana, *in* Sharpton, V.L., and Ward, P.D., eds., Global catastrophes in Earth History: An interdisciplinary conference on impacts, volcanism, and mass mortality: Geological Society of America Special Paper 247, p. 549–562.

Archibald, J.D., Clemens, W.A., Gingerich, P.D., Krause, D.W., Lindsay, E.H., and Rose, K.D., 1987. First North American land mammal ages of the Cenozoic, *in* Woodburne, M.O., ed., Cenozoic mammals of North America: Berkeley, California, University of California Press, p. 24–76.

Archibald, J.D., and Lofgren, D.L., 1990, Mammalian zonation near the Cretaceous-Tertiary boundary, *in* Bown, T.M., and Rose, K.D., eds., Dawn of the age of mammals in the northern part of the Rocky Mountain Interior, North America: Geological Society of America Special Paper 243, p. 31–50.

Arens, N.C., and Jahren, A.H., 1997, An empirical model for reconstructing the carbon isotopic composition of atmospheric CO_2 in geologic time: Geological Society of America Abstracts with Programs, v. 29, no. 6, p. A395.

Arens, N.C., and Jahren, A.H., 1998, Vascular plants provide clues to global carbon cycling, example from Cretaceous-Tertiary boundary: Geological Society of America Abstracts with Programs, v. 30, no. 7, p. A116.

Argast, S., Farlow, J.O., Gabet, R.M., and Brinkman, D.L., 1987, Transport-induced abrasion of fossil reptilian teeth: Implications for the existence of Tertiary dinosaurs in the Hell Creek Formation, Montana: Geology, v. 15, p. 927–930.

Averianov, A., and Kielan-Jaworowska, Z., 1999, Marsupials from the Late Cretaceous of Uzbeckistan: Acta Palaeontologica Polonica, v. 44, p. 71–81.

Averianov, A., and Nessov, L.A., 1995, A new Cretaceous mammal from the Campanian of Kazakhstan: Neues Jahrbuch für Geologie und Paläontologie, v. 1995, p. 65–74.

Baadsgaard, H., Lerbekmo, J.F., and McDonald, I., 1988, A radiometric age for the Cretaceous-Tertiary boundary based on K-Ar, Rb-Sr, and U-Pb ages of bentonites from Alberta, Saskatchewan and Montana: Canadian Journal of Earth Sciences, v. 25, p. 1088–1097.

Beard, K.C., 1998, East of Eden: Asia as an important center of taxonomic origination in mammalian evolution, *in* Beard, K.C., and Dawson, M.R., eds., Dawn of the Age of Mammals in Asia: Bulletin of the Carnegie Museum of Natural History, v. 34, p. 5–38.

Bell, R.C., 1965, Geology and stratigraphy of the Fort Peck fossil field, northwest McCone County, Montana [M.S. thesis]: Minneapolis, Minnesota, University of Minnesota, 166 p.

Berggren, W.A., Kent, D.V., Swisher, C.C., III. and Aubry, M.-P., 1995, A revised Cenozoic geochronology and chronostratigraphy, *in* Berggren, W.A., Kent, D.V., Hardenbol, J., and Scholle, P.A., eds., Geochronology, time scales, and global stratigraphic correlation: Society for Sedimentary Geology, Special Publication 54, p. 129–212.

Bloch, J.I., Rose, K.D., and Gingerich, P.D., 1998, New Species of *Batodonoides* (Lipotyphla, Geolabididae) from the early Eocene of Wyoming: Smallest known mammal: Journal of Mammalogy, v. 79, p. 804–827.

Bohor, B.F., Foord, E.E., Modreski, P.J., and Triplehorn, D.M., 1984, Mineralogic evidence for an impact event at the Cretaceous–Tertiary boundary: Science, v. 224, p. 867–869.

Bonaparte, J., 1990, New Late Cretaceous mammals from the Los Alamitos Formation, northern Patagonia: National Geographic Research, v. 6, p. 63–93.

Brett-Surman, M.K., and Paul, G.S., 1985, A new family of bird-like dinosaurs linking Laurasia and Gondwanaland: Journal of Vertebrate Paleontology, v. 5, p. 133–138.

Brodkorb, P., 1963, Birds from the Upper Cretaceous of Wyoming: Proceedings, 13th International Ornithological Congress, 1962, v. 1, p. 55–70.

Brown, B., 1907, The Hell Creek beds of the Upper Cretaceous of Montana: Bulletin of the American Museum of Natural History, v. 23, p. 823–845.

Brown, B., 1914, Cretaceous–Eocene correlations in New Mexico, Wyoming, Montana, Alberta: Geological Society of America Bulletin, v. 25, p. 355–380.

Brown, R., 1938, The Cretaceous–Eocene boundary in Montana and South Dakota: Washington Academy of Science Journal, v. 28, p. 421–422.

Brown, R., 1952, Tertiary strata in eastern Montana and western North and South Dakota: Billings Geological Society Guidebook, v. 3, p. 89–92.

Bryant, L., 1989, Non-dinosaurian lower vertebrates across the Cretaceous–Tertiary boundary in northeastern Montana: University of California Publications in Geological Sciences, v. 134, 107 p.

Buckley, G.A., 1997, A new species of *Purgatorius* (Mammalia; Primatomorpha) from the lower Paleocene Bear Formation, Crazy Mountains Basin, south-central Montana: Journal of Paleontology, v. 71, p. 149–155.

Carpenter, K., 1979, Vertebrate fauna of the Laramie Formation (Maestrichtian), Weld County, Colorado: Contributions to Geology (University of Wyoming), v. 17, p. 37–49.

Carrano, M.T., Blob, R.W., Flynn, J.J., Rogers, R.R., and Forster, C.A., 1997, The mammalian fauna of the Judith River Formation type area (Campanian, Central Montana) revisited: Journal of Vertebrate Paleontology, v. 17 (supplement to no. 3), p. 36A.

Cifelli, R.L., 1990, Cretaceous mammals of southern Utah. 4. Eutherian mammals from the Wahweap (Aquilan) and Kaiparowits (Judithian) Formations: Journal of Vertebrate Paleontology, v. 10, p. 346–360.

Cifelli, R.L., 1994, Therian mammals of the Terlingua local fauna (Judithian), Aguja Formation, Big Bend of the Rio Grande, Texas: Contributions to Geology (University of Wyoming), v. 30, p. 117–136.

Clemens, E.S., 1986, Of asteroids and dinosaurs: The role of the press in the shaping of scientific debate: Social Studies of Science, v. 16, p. 421–456.

Clemens, E.S., 1994, The impact hypothesis and popular science: Conditions and consequences of interdisciplinary debate, *in* Glen, W., ed., The mass extinction debates: How science works in a crisis: Stanford, California, Stanford University Press, p. 92–120.

Clemens, W.A., 1963, Fossil mammals of the type Lance Formation, Wyoming.

1. Introduction and Multituberculata: University of California Publications in Geological Sciences, v. 48, 105 p.

Clemens, W.A., 1966, Fossil mammals of the type Lance Formation, Wyoming. 2. Marsupialia: University of California Publications in Geological Sciences, v. 62, 122 p.

Clemens, W.A., 1973a, Fossil mammals of the type Lance Formation, Wyoming. 3. Eutheria and summary: University of California Publications in Geological Sciences, v. 94, 102 p.

Clemens, W.A., 1973b, The roles of fossil vertebrates in interpretation of Late Cretaceous stratigraphy of the San Juan Basin, New Mexico, *in* Fassett, J.E., ed., Cretaceous and Tertiary rocks of the southern Colorado Plateau: Four Corners Geological Society Memoir, p. 154–167.

Clemens, W.A., 1974, *Purgatorius*, an early paromomyid primate (Mammalia): Science, v. 184, p. 903–905.

Clemens, W.A., 1999, Biogeography and phylogenetic relationships of the Puercan mammal *Eoconodon* (Triisodontidae, Cete): Journal of Vertebrate Paleontology, v. 19 (supplement to no. 3), p. 39A.

Clemens, W.A., Lillegraven, J.A., Lindsay, E.H., and Simpson, G.G., 1979, Where, when, and what: A survey of known Mesozoic mammal distribution, *in*, Lillegraven, J.A., Kielan-Jaworowska, Z., and Clemens, W.A., eds., Mesozoic mammals: The first two-thirds of mammalian history: Berkeley, University of California Press, p. 7–58.

Clemens, W.A., and Nelms, L.G., 1993, Paleoecological implications of Alaskan terrestrial vertebrate fauna in latest Cretaceous time at high paleolatitudes: Geology, v. 21, p. 503–506.

Collier, A.J., and Knechtel, M., 1939, The coal resources of McCone County, Montana: U.S. Geological Survey Bulletin 905, 80 p.

Cooper, A., and Penny, D., 1997, Mass survival of birds across the Cretaceous–Tertiary boundary: Molecular evidence: Science, v. 275, p. 1109–1113.

Cowie, J.W., Ziegler, W., and Remane, J., 1989, Stratigraphic Commission accelerates progress, 1984 to 1989: Episodes, v. 12, p. 79–83.

Cutler, A.H., and Behrensmeyer, A.K., 1996, Models of vertebrate mass mortality events at the K/T boundary, *in* Ryder, G., Fastovsky, D., and Gartner, S., eds., The Cretaceous–Tertiary event and other catastrophes in Earth history: Geological Society of America Special Paper 307, p. 465–476.

Dingus, L., 1983, A stratigraphic review and analysis of selected sections spanning the Cretaceous–Tertiary boundary [Ph.D. thesis]: Berkeley, University of California, 156 p.

Dingus, L., 1984, Effects of stratigraphic completeness on interpretations of extinction rates across the Cretaceous–Tertiary boundary: Paleobiology, v. 10, p. 420–438.

Eaton, J.G., Cifelli, R.L., Hutchison, J.H., Kirtland, J.I., and Parrish, J.M., 1999, Cretaceous vertebrate faunas from the Kaiparowits Plateau, south-central Utah, *in* Gillette, D.D., ed., Vertebrate paleontology in Utah: Utah Geological Survey, Miscellaneous Publication 99-1, p. 345–353.

Eberle, J.J., 1999, Bridging the transition between didelphodonts and taeniodonts: Journal of Paleontology, v. 73, p. 936–944.

Eberle, J.J., and Lillegraven, J.A., 1998a, A new important record of earliest Cenozoic mammalian history: Geological setting, Multituberculata, and Peradectia: Rocky Mountain Geology, v. 33, p. 3–47.

Eberle, J.J., and Lillegraven, J.A., 1998b, A new important record of earliest Cenozoic mammalian history: Eutheria and paleogeographic/biostratigraphic summaries: Rocky Mountain Geology, v. 33, p. 49–117.

Eberth, D.A., and O'Connell, S.C., 1995, Notes on changing paleoenvironments across the Cretaceous–Tertiary boundary (Scollard Formation) in the Red Deer River valley of southern Alberta: Bulletin of Canadian Petroleum Geology, v. 43, p. 44–53.

Erickson, J.M., 1999, The Dakota Isthmus: Closing the Late Cretaceous Western Interior Seaway: North Dakota Academy of Science, Proceedings, v. 53, p. 124–129.

Fastovsky, D.E., 1987, Paleoenvironments of vertebrate-bearing strata during the Cretaceous–Paleogene transition, eastern Montana and western North Dakota: Palaios, v. 2, p. 282–295.

Fastovsky, D.E., and Dott, R.H., Jr., 1986, Sedimentology, stratigraphy, and extinctions during the Cretaceous/Paleogene transition at Bug Creek, Montana: Geology, v. 14, p. 279–282.

Fiorillo, A.R., 1998, Measuring fossil reworking within a fluvial system: An example from the Hell Creek Formation (Upper Cretaceous) of eastern Montana, *in* Tomida, Y., Flynn, L.J., and Jacobs, L.L., eds., Advances in vertebrate paleontology and geochronology: National Science Museum Monographs 14, p. 243–251.

Flynn, L.J., 1986, Late Cretaceous mammal horizons from the San Juan Basin, New Mexico: American Museum Novitates, no. 2845, 30 p.

Foote, M., Hunter, J.P., Janis, C.M, and Sepkoski, J.J., Jr., 1999a, Evolutionary and preservational constraints on origins of biologic groups: Divergence times of eutherian mammals: Science, v. 283, p. 1310–1314.

Foote, M., Hunter, J.P., Janis, C.M, and Sepkoski, J.J., Jr., 1999b, Divergence times of eutherian mammals: Technical Comments, Science, v. 285, p. 2031A (*www.sciencemag.org*).

Fox, R.C., 1979, Mammals from the Upper Cretaceous Oldman Formation, Alberta. 3. Eutheria: Canadian Journal of Earth Sciences, v. 16, p. 114–125.

Fox, R.C., 1989, The Wounded Knee local fauna and mammalian evolution near the Cretaceous–Tertiary boundary, Saskatchewan, Canada: Palaeontographica. Abteilung A: Paläozoologie-Stratigraphie, v. 208, p. 11–59.

Fox, R.C., 1997, Late Cretaceous and Paleocene mammals, Cypress Hills Region, Saskatchewan, and mammalian evolution across the Cretaceous–Tertiary boundary: Geological Association of Canada, Canadian Paleontology Conference, Field Trip Guidebook No. 6, p. 70–85.

Fox, R.C., and Naylor, B.G., 1995, The relationships of the Stagodontidae, primitive North American Late Cretaceous mammals, *in* Sun, A., and Wang, Y., eds., Short papers, Sixth Symposium on Mesozoic Terrestrial Ecosystems and Biota: Beijing, China Ocean Press, p. 247–250.

Fox, R.C., and Youzwyshyn, G.P., 1994, New primitive carnivorans (Mammalia) from the Paleocene of Western Canada, and their bearing on relationships of the order: Journal of Vertebrate Paleontology, v. 14, p. 382–404.

Galvin, C., 1998, The great dinosaur extinction controversy and the K-T Research Program in the late 20th Century: Earth Sciences History, v. 17, p. 41–55.

Gheerbrant, E., and Astibia, H., 1994, Un nouveau mammifére du Maastrichtien de Laño (Pays Basque espagnol): Comptes Rendus de l'Académie des Sciences, Série 2, Sciences de la Terre et des Planetes, Earth and Planetary Sciences, v. 318, p. 1125–1131.

Gill, J.R., and Cobban, W.A., 1973, Stratigraphy and geologic history of the Montana Group and equivalent rocks, Montana, Wyoming, and North and South Dakota: U.S. Geological Survey Professional Paper 776, 37 p.

Glen, W., 1994, The mass extinction debates: How science works in a crisis: Stanford, California, Stanford University Press, 370 p.

Godinot, M., and Prasad, G.V.R., 1994, Discovery of Cretaceous arboreal eutherians: Naturwissenschaften, v. 81, p. 79–81.

Hall, J.W., and Norton, N.J., 1967, Palynological evidence of floristic change across the Cretaceous–Tertiary boundary in eastern Montana (U.S.A.): Palaeogeography, Palaeoclimatology, and Palaeoecology, v. 3, p. 121–131.

Hartman, J.H., 1999, The biostratigraphy and paleontology of latest Cretaceous and Paleocene freshwater bivalves from the western Williston Basin, Montana, U.S.A., *in* Johnston, P.A., and Haggart, J.W., eds., Bivalves: An eon of evolution: Calgary, Alberta, University of Calgary Press, p. 317–345.

Hay, W.W., DeComo, R.M., Wold, C.N., Wilson, K.M., Voigt, S., Schulz, M., Wold-Rossby, A., Dullo, W.C., Ronov, A.B., Balukhovsky, A.N., and Söding, E., 1999, Alternative global Cretaceous paleogeography, *in* Barrera, E., and Johnson, C.C., eds., Evolution of the Cretaceous ocean climate system: Geological Society of America Special Paper 332, p. 1–47.

Hicks, J.F., Johnson, K.R., Tauxe, L., Clark, D., and Obradovich, J.D., 1999, Geochronology of the Hell Creek Formation of southwestern North Dakota: A multidisciplinary approach using biostratigraphy, isotopic dating, geochemistry, and magnetostratigraphy: Geological Society of America Abstracts with Programs, v. 31, no. 7, p. A71.

Hope, S., 1998, The Mesozoic record of the Neornithes: Journal of Vertebrate Paleontology, v. 18 (supplement to no. 3), p. 51A.

Horner, J.R., and Lessum, D., 1993, The Complete T. rex: New York, Simon and Schuster, 239 p.

Hotton, C.L., 1989, Palynology of the Cretaceous–Tertiary boundary in central Montana, U.S.A., and its implications for extraterrestrial impact [Ph.D. thesis]: Davis, University of California, 731 p.

Hunt, R.M., Jr., and Tedford, R.H., 1993, Phylogenetic relationships within the Aeluroid Carnivora and implications of their temporal and geographic distribution, *in* Szalay, F.S., Novacek, M.J., and McKenna, M.C., eds., Mammal phylogeny: Placentals: New York, Springer-Verlag, p. 53–73.

Hunter, J.P., 1994, Lack of a high body count at the K–T boundary: Journal of Paleontology, v. 68, p. 1158.

Hunter, J.P., and Archibald, J.D., 1999, Mammals from the end of the Age of Dinosaurs in North Dakota and Southeastern Montana: Geological Society of America Abstracts with Programs, v. 31, no. 7, p. A73.

Hunter, J.P., Hartman, J.H., and Krause, D.W., 1997, Mammals and mollusks across the Cretaceous–Tertiary boundary from Makoshika State Park and vicinity (Williston Basin), Montana: Contributions to Geology (University of Wyoming), v. 32, p. 61–114

Hunter, J.P., and Pearson, D.A., 1996, First record of Lancian (Latest Cretaceous) mammals from the Hell Creek Formation of southwestern North Dakota, U.S.A.: Cretaceous Research, v. 17, p. 633–643.

Hurlbert, S.H., and Archibald, J.D., 1995, No statistical support for sudden (or gradual) extinction of dinosaurs: Geology, v. 23, p. 881–884.

Hurlbert, S.H., and Archibald, J.D., 1996, No statistical support for sudden (or gradual) extinction of dinosaurs: Reply: Geology, v. 24, p. 958–959.

Hutchinson, J.R., and Chiappe, L.M., 1998, The first known alvarezsaurid (Theropoda: Aves) from North America: Journal of Vertebrate Paleontology, v. 18, p. 447–450.

Hutchinson, J.R., Stidham, T.A., and Snively, E., 1997, New birds and other Late Cretaceous theropods from Montana in the University of California Museum of Paleontology: Journal of Vertebrate Paleonotology, v. 18 (supplement to no. 3), p. 55A.

Hutchison, J.H., 2000, Diversity of Cretaceous turtle faunas of eastern Asia and their contribution to the turtle faunas of North America, *in* Lee, Y.-N. ed., Proceedings, 2000 International Dinosaur Symposium for Kosong County in Korea: Kosong, Korea, The Paleontological Society of Korea, Special Publication 4, p. 27–38.

Hutchison, J.H., and Archibald, J.D., 1986, Diversity of turtles across the Cretaceous/Tertiary boundary in northeastern Montana: Palaeogeography, Palaeoclimatology, and Palaeoecology, v. 55, p. 1–22.

Jensen, F.S., and Varnes, H.D., 1964, Geology of the Fort Peck area, Garfield, McCone, and Valley Counties, Montana: U.S. Geological Survey Professional Paper 414F, 49 p.

Jerzykiewicz, T., and Russell, D.A., 1991, Late Mesozoic stratigraphy and vertebrates from the Gobi Basin: Cretaceous Research, v. 12, p. 345–377.

Johnson, K.R., and Hickey, L.J., 1990, Megafloral change across the Cretaceous/Tertiary boundary in the northern Great Plains and Rocky Mountains, U.S.A., *in* Sharpton, V.L., and Ward, P.D., eds., Global catastrophes in Earth History: An interdisciplinary conference on impacts, volcanism, and mass mortality: Boulder, Colorado, Geological Society of America Special Paper 247, p. 433–444.

Johnson, K.R., Nichols, D.J., Attrep, M., Jr., and Orth, C.J., 1989, High-resolution leaf fossil record spanning the Cretaceous/Tertiary boundary: Nature, v. 340, p. 708–711.

Johnston, P.A., and Fox, R.C., 1984, Paleocene and Late Cretaceous mammals from Saskatchewan, Canada: Palaeontographica. Abteilung A: Pälaozoologie-Stratigraphie, v. 186, p. 163–222.

Keller, G., Li, L., and MacLeod, N., 1995, The Cretaceous/Tertiary boundary stratotype section at El Kef, Tunisia: How catastrophic was the mass extinction?: Palaeogeography, Palaeoclimatology, Palaeoecology, v. 119, p. 221–254.

Kielan-Jaworowska, Z., and Hurum, J.H., 1997, Djadochtatheria: A new suborder of multituberculate mammals: Acta Palaeontologica Polonica, v. 42, p. 201–242.

Kowalewski, M., 1996, Time-averaging, overcompleteness, and the geological record: The Journal of Geology, v. 104, p. 317–326.

Krause, D.W., Prasad, G.V.R., Koenigswald, W.v., Sahni, A., and Grine, F.E., 1997, Cosmopolitanism among Gondwanan Late Cretaceous mammals: Nature, v. 390, p. 504–507.

Kumar, S., and Hedges, S.B., 1998, A molecular timescale for vertebrate evolution: Nature, v. 392, p. 917–920.

Landis, G.P., Rigby, J.K., Jr., Sloan, R.E., Hengst, R., and Snee, L.W., 1996, Pele hypothesis: Ancient atmospheres and geologic-geochemical controls on evolution, survival and extinction, *in* MacLeod, N., and Keller, G., eds., Cretaceous-Teritary mass extinctions: Biotic and environmental changes: New York, W.W. Norton and Company, p. 519–556.

Lehman, T.M., 1984, The multituberculate *Essonodon browni* from the Upper Cretaceous Naashoibito Member of the Kirtland Shale, San Juan Basin, New Mexico: Journal of Vertebrate Paleontology, v. 4, p. 602–603.

Lehman, T.M., 1987, Late Maastrichtian paleoenvironments and dinosaur biogeography in the Western Interior of North America: Palaeogeography, Palaeoclimatology, Palaeoecology, v. 60, p. 189–217.

Lehman, T.M., 1997, Late Campanian dinosaur biogeography in the Western Interior of North America, *in* Wolberg, D., Stump, E., and Rosenberg, G.D., eds., Dinofest International: Philadelphia, Pennsylvania, The Academy of Natural Sciences of Philadelphia, p. 223–240.

LeLoeuff, J., 1991, The Campano-Maastrichtian vertebrate faunas from southern Europe and their relationships with other faunas in the world: Palaeobiogeographical considerations: Cretaceous Research, v. 12, p. 93–114.

Lerbekmo, J.F., 1985, Magnetostratigraphic and biostratigraphic correlations of Maestrichtian to early Paleocene strata between south-central Alberta and southwestern Saskatchewan: Bulletin of Canadian Petroleum Geology, v. 33, p. 213–226.

Lerbekmo, J.F., Sweet, A.R., and Duke, M.J.M., 1996, A normal polarity subchron that embraces the K/T boundary: A measure of sedimentary continuity across the boundary and synchroneity of boundary events, *in* Ryder, G., Fastovsky, D., and Gartner, S., eds., The Cretaceous–Tertiary event and other catastrophes in Earth history: Geological Society of America Special Paper 307, p. 465–476.

Lillegraven, J.A., 1969, Latest Cretaceous mammals of upper part of Edmonton Formation of Alberta, Canada, and review of marsupial-placental dichotomy in mammalian evolution: The University of Kansas Paleontological Contributions, Article 50 (Vertebrata 12), 122 p.

Lillegraven, J.A., 1972, Preliminary report on Late Cretaceous mammals from the El Gallo Formation, Baja California del Norte, Mexico: Contributions to Science, Natural History Museum of Los Angeles County, no. 232, 11 p.

Lillegraven, J.A., 1976, A new genus of therian mammal from the Late Cretaceous, "El Gallo Formation," Baja California, Mexico: Journal of Paleontology, v. 50, p. 437–443.

Lillegraven, J.A., and McKenna, M.C., 1986, Fossil mammals of the "Mesaverde" Formation (Late Cretaceous, Judithian) of the Bighorn and Wind River Basins, Wyoming with definitions of Late Cretaceous Land Mammal "Ages": American Museum Novitates, no. 2840, 68 p.

Lofgren, D.L., 1995, The Bug Creek Problem and the Cretaceous–Tertiary transition at McGuire Creek, Montana: University of California Publications in Geological Sciences, v. 140, 185 p.

Lofgren, D.L., Hotton, C., and Runkel, A.C., 1990. Reworking of Cretaceous dinosaurs into Paleocene channel deposits, upper Hell Creek Formation, Montana: Geology, v. 18, p. 874–877.

López-Martinez, N., Fernández-Marrón, M.T., and Valle, M.F., 1999, The succession of vertebrates and plants across the Cretaceous–Tertiary boundary in the Tremp Formation, Ager Valley (south-central Pyrenees, Spain): Geobios, v. 32, p. 617–627.

Luo, Z., 1991, Variability of dental morphology and the relationships of the earliest arctocyonid species: Journal of Vertebrate Paleontology, v. 11, p. 452–471.

MacLeod, N., 1998, Impacts and marine invertebrate extinction, *in* Grady, M.M., Hutchison, R., McCall, G.J.H., and Rothery, D.A., eds., Meteorites: Flux with time and impact effects: Geological Society [London] Special Publication 140, p. 217–246.

MacLeod, N., Rawson, P.F., Forey, P.L., Banner, F.T., Boudagher-Fadel, M.K., Brown, P.R., Burnett, J.A., Chambers, P., Culver, S., Evans, S.E., Jeffery, C., Kaminski, M.A., Lord, A.R., Smith, A.B., Taylor, P.D., Urquhart, E., and Young, J.R., 1997, The Cretaceous–Tertiary biotic transition: Journal of the Geological Society of London, v. 154, p. 265–292.

Marshall, L.G., and Kielan-Jaworowska, Z., 1992, Relationships of the dog-like marsupials, deltatheroidans and early tribosphenic mammals: Lethaia, v. 25, p. 361–374.

McKenna, M.C., and Bell, S., 1997, Classification of mammals above the species level: New York, Columbia University Press, 631 p.

Middleton, M.D., 1982, A new species and additional material of *Catopsalis* (Mammalia, Multituberculata) from the Western Interior of North America: Journal of Paleontology, v. 56, p. 1197–1206.

Middleton, M.D., 1983, Early Paleocene vertebrates of the Denver Basin, Colorado [Ph.D. thesis]: Boulder, University of Colorado, 403 p.

Milner, A.C., 1998, Timing and causes of vertebrate extinction across the Cretaceous–Tertiary boundary, *in* Grady, M.M., Hutchison, R., McCall, G.J.H., and Rothery, D.A., eds., Meteorites: Flux with time and impact effects: Geological Society [London] Special Publication 140, p. 247–257.

Montellano, M., 1992, Mammalian Fauna of the Judith River Formation (Late Cretaceous, Judithian), northcentral Montana: University of California Publications in Geological Sciences, v. 136, 115 p.

Nessov, L.A., Archibald, J.D., and Kielan-Jaworowska, Z., 1998, Ungulate-like mammals from the Late Cretaceous of Uzbeckistan and a phylogenetic analysis of Ungulatomorpha, *in* Beard, K.C., and Dawson, M.R., eds., Dawn of the Age of Mammals in Asia: Bulletin of the Carnegie Museum of Natural History, v. 34, p. 40–88.

Nessov, L.A., Sigogneau-Russell, D., and Russell, D.E., 1994, A survey of Cretaceous tribosphenic mammals from middle Asia (Uzbekistan, Kazakhstan and Tajikistan), of their geological setting, age and faunal environment: Palaeovertebrata, v. 23, p. 51–92.

Nichols, D.J., 1994, A revised palynostratigraphic zonation of the nonmarine Upper Cretaceous, Rocky Mountain Region, United States, *in* Caputo, M.V., Peterson, J.A., and Franczyk, K.J., eds., Mesozoic systems of the Rocky Mountain Region, USA: SEPM (Society for Sedimentary Geology), Rocky Mountain Section, p. 503–521.

Nichols, D.J., Jacobson, S.R., and Tschudy, R.H., 1982, Cretaceous palynomorph biozones for the central and northern Rocky Mountain region of the United States, *in* Powers, R.B., ed., Geologic studies of the Cordilleran Thrust Belt: Denver, Colorado, Rocky Mountain Association of Geologists, p. 721–733.

Norton, N.J., and Hall, J.W., 1969, Palynology of the Upper Cretaceous and lower Tertiary in the type locality of the Hell Creek Formation, Montana: Palaeontographica. Abteilung B: Paläophytologie v. 125, p. 1–64.

Novacek, M.J., 1992, Mammalian phylogeny: Shaking the tree: Nature, v. 356, p. 121–125.

Novacek, M.J., 1999, 100 million years of land vertebrate evolution: The Cretaceous–early Tertiary transition: Annals of the Missouri Botanical Garden, v. 86, p. 230–258.

Novacek, M.J., and Clemens, W.A., 1977, Aspects of intrageneric variation and evolution of *Mesodma* (Multituberculata, Mammalia): Journal of Paleontology, v. 51, p. 701–717.

Novacek, M.J., Rougier, G.W., Wible, J.R., McKenna, M.C., Dashzeveg, D., and Horowitz, I., 1997, Epipubic bones in eutherian mammals from the Late Cretaceous of Mongolia: Nature, v. 389, p. 483–486.

Obradovich, J.D., and Hicks, J.F., 1999, A review of the isotopic calibration points for the geomagnetic polarity time scale in the interval 83 to 33 Ma (C34N to C13N): Geological Society of America Abstracts with Programs, v. 31, no. 7, p. A71.

O'Leary, M.A., 1998, Phylogenetic and morphometric reassessment of the dental evidence for a Mesonychian and Cetacean Clade, *in* Thewissen, J.G.M., ed., The emergence of whales, evolutionary patterns in the origin of Cetacea: New York, Plenum Press, p. 133–161.

O'Leary, M.A., and Geisler, J.H., 1999, The position of Cetacea within Mammalia: Phylogenetic analysis of morphological data from extinct and extant taxa: Systematic Biology, v. 48, p. 455–490.

Oltz, D.F., Jr., 1969, Numerical analysis of palynological data from Cretaceous and early Tertiary sediments in east central Montana: Palaeontographica. Abteilung B: Paläophytologie v. 128, p. 90–166.

Padian, K., and Chiappe, L.M., 1998, The origin and early evolution of birds: Biological Reviews, v. 73, p. 1–42.

Pascual, R., 1998, The history of South American land mammals: The seminal Cretaceous–Paleocene transition: Asociación Paleontológica Argentina, Publicación Especial, no. 5, p. 9–18.

Pascual, R., Ortiz-Jaureguizar, E., and Prado, J.L., 1999, Land mammals: Paradigm for Cenozoic South American geobiotic evolution: Münchener Geowissenschaftliche Abhandlungen, v. 30, p. 265–319.

Pearson, D.A., Schaeffer, T., Johnson, K.R., and Nichols, D.J., 1999, Vertebrate biostratigraphy of the Hell Creek Formation in southwestern North Dakota: Geological Society of America Abstracts with Programs, v. 31, no. 7, p. A72–A73.

Râdulescu, C., and Samson, P.-M., 1997, Late Cretaceous Multituberculata from the Ha–eg Basin (Romania): Sargetia, v. 17, p. 247–255.

Raup, D.M., 1991, Extinction: Bad genes or bad luck?: New York, W.W. Norton & Company, Inc., 210 p.

Retallack, G.J., 1994, A pedotype approach to latest Cretaceous and earliest Tertiary paleosols in eastern Montana: Geological Society of America Bulletin, v. 106, p. 1377–1397.

Retallack, G.J., Leahy, G.D., and Spoon, M.D., 1987, Evidence from paleosols for ecosystem changes across the Cretaceous/Tertiary boundary in eastern Montana: Geology, v. 15, p. 1090–1093.

Rigby, J.K., and Rigby, J.K., Jr., 1990, Geology of the Sand Arroyo and Bug Creek Quadrangles, McCone County, Montana: Brigham Young University, Geology Studies, v. 36, p. 69–134.

Rigby, J.K., Jr., and Wolberg, D.L., 1987, The therian mammalian fauna (Campanian) of Quarry 1, Fossil Forest study area, San Juan Basin, New Mexico, *in* Fassett, J.E., and Rigby, J.K., Jr., eds., The Cretaceous–Tertiary boundary in the San Juan and Raton Basins, New Mexico and Colorado: Geological Society of America Special Paper 209, p. 51–80.

Roberts, L.N.R., and Kirschbaum, M.A., 1995, Paleogeography of the Late Cretaceous of the Western Interior of middle North America: Coal distribution and sediment accumulation: U.S. Geological Survey Professional Paper 1561, 115 p.

Rougier, G.W., Novacek, M.J., and Dashzeveg, D., 1997, A new multituberculate from the Late Cretaceous locality Ukhaa Tolgod, Mongolia: Considerations on multituberculate interrelationships: American Museum Novitates, no. 3191, 26 p.

Russell, L.S., 1964, Cretaceous non-marine faunas of northwestern North America: Royal Ontario Museum, Life Sciences, Contribution 61, 24 p.

Russell, L.S., 1975, Mammalian faunal succession in the Cretaceous System of western North America, *in* Caldwell, W.G.E., ed., The Cretaceous System in the Western Interior of North America: Geological Association of Canada Special Paper 13, p. 137–161.

Ryder, G., Fastovsky, D., and Gartner, S., editors, 1996, The Cretaceous–Tertiary event and other catastrophes in Earth history: Geological Society of America Special Paper 307, 469 p.

Sahni, A., 1972, The vertebrate fauna of the Judith River Formation: Bulletin of the American Museum of Natural History, v. 147, p. 321–412.

Savage, D.E., and Russell, D.E., 1983, Mammalian paleofaunas of the world: London, Addison-Wesley Publishing Company, 432 p.

Scotese, C.R., 1997, Paleogeographic Atlas, PALEOMAP Progress Report 90-0407: Arlington, University of Texas, Department of Geology, 45 p.

Sheehan, P.M., Fastovsky, D.E., Barretto, C.B., and Hoffmann, R.G., 1999, Does the three-meter gap at the top of the Hell Creek Formation indicate dinosaurs were declining prior to the bolide impact?: Geological Society of America Abstracts with Programs, v. 31, no. 7, p. A473.

Sheehan, P.M., Fastovsky, D.E., Hoffmann, R.G., and Barretto, C.B., 1996, No statistical support for sudden (or gradual) extinction of dinosaurs: Comment: Geology, v. 24, p. 957–958.

Sheehan, P.M., Fastovsky, D.E., Hoffmann, R.G., Berghaus, C.G., and Gabriel, D.L., 1991, Sudden extinction of the dinosaurs: Latest Cretaceous, upper Great Plains, U.S.A.: Science, v. 254, p. 835–839.

Shoemaker, R.E., 1966, Fossil leaves of the Hell Creek and Tullock formations of eastern Montana: Palaeontographica. Abteilung B: Paläophytologie, v. 110, p. 54–75.

Signor, P.W., III, and Lipps, J.H., 1982, Sampling bias, gradual extinction patterns, and catastrophes in the fossil record, *in* Silver, L.T., and Schultz, P.H., Geological implications of impacts of large asteroids and comets on the Earth: Boulder, Colorado, Geological Society of America Special Paper 190, p. 291–296.

Simmons, N.B., 1987, A revision of *Taeniolabis* (Mammalia, Multituberculata), with a new species from the Puercan of eastern Montana: Journal of Paleontology, v. 61, p. 794–808.

Simpson, G.G., 1927, Mammalian fauna of the Hell Creek Formation of Montana: American Museum Novitates, no. 267, 7 p.

Simpson, G.G., 1929, American Mesozoic Mammalia: New Haven, Connecticut, Yale University Peabody Museum Memoir, v. 3, pt. 1, 171 p.

Sloan, R.E., 1976, The ecology of dinosaur extinction, *in* Churcher, C.S., ed., Athlon: Essays in honour of Loris Shano Russell: Royal Ontario Museum, Miscellaneous Publications, p. 134–1154.

Sloan, R.E., Rigby, J.K., Jr., Van Valen, L.M., and Gabriel, D., 1986, Gradual dinosaur extinction and simultaneous ungulate radiation in the Hell Creek Formation: Science, v. 232, p. 629–633.

Sloan, R.E., and Van Valen, L., 1965, Cretaceous mammals from Montana: Science, v. 148, p. 220–227.

Smit, J., and van der Kaars, S., 1984, Terminal Cretaceous extinctions in the Hell Creek area, Montana: Compatible with catastrophic extinction: Science, v. 223, p. 1177–1179.

Smit, J., van der Kaars, W.A., and Rigby, J.K., Jr., 1987, Stratigraphic aspects of the Cretaceous–Tertiary boundary in the Bug Creek area of eastern Montana, U.S.A.: Paris, Mémoires de la Société Géologique de France, Nouvelle Série, no. 150, p. 53–73.

Stidham, T.A., 1998a, The most diverse Cretaceous avifauna in the world: Ostrich, v. 69, p. 413.

Stidham, T.A., 1998b, Phylogenetic and ecological diversification of waterfowl (Anseriformes) in the Late Cretaceous and Paleogene: Journal of Vertebrate Paleontology, v. 18 (supplement to no. 3), p. 80A.

Stidham, T.A., 1998c, A lower jaw from a Cretaceous parrot: Nature, v. 396, p. 29–30.

Storer, J.E., 1991, The mammals of the Gryde local fauna, Frenchman Formation (Maastrichtian, Lancian), Saskatchewan: Journal of Vertebrate Paleontology, v. 11, p. 350–369.

Swisher, C.C., III, Dingus, L., and Butler, R.F., 1993, $^{40}Ar/^{39}Ar$ dating and magnetostratigraphic correlation of the terrestrial Cretaceous–Paleogene boundary and Puercan Mammal Age: Canadian Journal of Earth Sciences, v. 30, p. 1981–1996.

Szalay, F.S., and Trofimov, B.A., 1996, The Mongolian Late Cretaceous *Asiatherium*, and the early phylogeny and paleobiogeography of Metatheria: Journal of Vertebrate Paleontology, v. 16, p. 474–509.

Thom, W.T., and Dobbin, C.E., 1924, Stratigraphy of Cretaceous–Eocene transition beds in eastern Montana and the Dakotas: Geological Society of America Bulletin, v. 35, p. 481–505.

Van Valen, L.M., 1978, The beginning of the Age of Mammals: Evolutionary Theory, v. 4, p. 103–121.

Van Valen, L.M., 1994, The origin of the plesiadapid primates and the nature of *Purgatorius*: Evolutionary Monographs, v. 15, 79 p.

Van Valen, L.M., and Sloan, R.E., 1965, The earliest primates: Science, v. 150, p. 743–745.

Wang, Y., Hu, Y., Chow, M., and Li, C., 1998, Chinese Paleogene mammal faunas and their correlation, *in* Beard, K.C., and Dawson, M.R., eds., Dawn of the Age of Mammals in Asia: Bulletin of the Carnegie Museum of Natural History, v. 34, p. 89–123.

Weil, A., 1994, K/T survivorship as a test of acid rain hypotheses: Geological Society of America Abstracts with Programs, v. 26, no. 7, p. A335.

Weil, A., 1997, Phylogeny, biogeography, and origin of the Eucosmodontidae: Implications for the origins of North American Early Paleocene fauna: Journal of Vertebrate Paleontology, v. 17 (supplement to no. 3), p. 84A.

Weil, A., 1998, A new species of *Microcosmodon* (Mammalia: Multituberculata) from the Paleocene Tullock Formation of Montana, and an argument for Microcosmodontinae: PaleoBios, v. 18, p. 1–15.

Weil, A., 1999a, Multituberculate phylogeny and mammalian biogeography in the Late Cretaceous and earliest Paleocene Western Interior of North America [Ph. D. thesis]: Berkeley, University of California, 242 p.

Weil, A., 1999b, Biogeographic provinciality of mammalian faunas in the Late Cretaceous and Early Paleocene Western Interior of North America: Geological Society of America Abstracts with Programs, v. 31, no. 7, p. A473.

Weil, A., and Clemens, W.A., 1998, Aliens in Montana: Phylogenetically and biogeographically diverse lineages contributed to an earliest Cenozoic community: Geological Society of America Abstracts with Programs, v. 30, no. 5, p. 69–70.

Weishampel, D.B., 1990, Dinosaurian distribution, *in* Weishampel, D.B., Dodson, P., and Osmóska, H., eds., The Dinosauria: Berkeley, California, University of California Press, p. 63–139.

Williams, M.E., 1994, Catastrophic versus noncatastrophic extinction of the dinosaurs: Testing, falsifiability, and the burden of proof: Journal of Paleontology, v. 68, p. 183–190.

Williamson, T.E., 1996, The beginning of the Age of Mammals in the San Juan Basin, New Mexico: Biostratigraphy and evolution of Paleocene mammals of the Nacimiento Formation, New Mexico: New Mexico Museum of Natural History and Science Bulletin, v. 8, p. 1–141.

Williamson, T.E., and Lucas, S.G., 1993, Paleocene vertebrate paleontology of the San Juan Basin, New Mexico: New Mexico Museum of Natural History and Science Bulletin, v. 2, p. 105–135.

Wilmarth, M.G., 1938, Lexicon of geologic names of the United States (including Alaska): U.S. Geological Survey Bulletin 895, 2396 p.

Wilson, R.W., 1983, Late Cretaceous mammals from western South Dakota, 1974: National Geographic Research Reports, v. 15, p. 749–752.

Wilson, R.W., 1987, Late Cretaceous (Fox Hills) multituberculates from the Red Owl local fauna of western South Dakota: Dakoterra, v. 3, p. 118–122.

Woodburne, M.O., 1987, Cenozoic mammals of North America: Berkeley, California, University of California Press, 336 p.

MANUSCRIPT ACCEPTED BY THE SOCIETY SEPTEMBER 24, 2001

Printed in the U.S.A.

Geological Society of America
Special Paper 361
2002

Marine Breien Member (Maastrichtian) of the Hell Creek Formation in North Dakota: Stratigraphy, vertebrate fossil record, and age

John W. Hoganson*
Edward C. Murphy
North Dakota Geological Survey, 600 East Boulevard Avenue, Bismarck, North Dakota 58505, USA

ABSTRACT

The Breien Member of the Hell Creek Formation represents the last well-documented advance of the Fox Hills Sea into the upper midcontinent during the Late Cretaceous (Maastrichtian). This member is lithologically variable and consists of burrowed, greenish-gray, glauconitic sandstone and mudstone or burrowed, interbedded sandstone and mudstone. Iron-oxide staining is usually present. Ichnofossils, most of which are *Ophiomorpha*, occur throughout the member and are often associated with *Crassostrea subtrigonalis*. The Breien Member is exposed in isolated ravine, road cut, and butte outcrops in south-central North Dakota. It ranges in thickness from 1.5 m in Emmons County to 8.5 m in Morton County. The Breien interfingers with nonmarine, dinosaur fossil-bearing Hell Creek deposits. Both upper and lower contacts of the Breien Member with nonmarine Hell Creek strata are lithologically variable and conformable. The base of the member is 1.5–9.0 m above the Fox Hills Formation and the top of the member is 46–61 m below the Hell Creek–Fort Union formational contact. Because of lithologic variability at the Breien Member's type locality and elsewhere in its outcrop area, three reference sections are designated to characterize the Breien.

Fossils are scarce in the Breien Member but indicate deposition within the Upper Cretaceous *Jeletzkytes nebrascensis* Zone and a Lancian age. The Breien vertebrate fossil assemblage consists of terrestrial and marine taxa, including fish, crocodilians, dinosaurs, and mammals. Marine chondrichthyan teeth are the most abundant vertebrate fossils. Most of these taxa corroborate a Late Cretaceous age for the member and indicate that the Breien was deposited in relatively shallow water, nearshore, marine environments. This marine chondrichthyan fauna is similar to, but less diverse than, the Maastrichtian Fox Hills Formation fauna and exhibits no similarity to the early Paleocene Cannonball Member fauna. This finding lends support for a major end-Cretaceous extinction of marine chondrichthyan species.

*E-mail: jhoganso@state.nd.us

Hoganson, J.W., and Murphy, E.C., 2002, Marine Breien Member (Maastrichtian) of the Hell Creek Formation in North Dakota: Stratigraphy, vertebrate fossil record, and age, *in* Hartman, J.H., Johnson, K.R., and Nichols, D.J., eds., The Hell Creek Formation and the Cretaceous-Tertiary boundary in the northern Great Plains: An integrated continental record of the end of the Cretaceous: Boulder, Colorado, Geological Society of America Special Paper 361, p. 247–269.

INTRODUCTION

Surface exposures within three counties in south-central North Dakota, consisting primarily of poorly consolidated glauconitic sandstone containing the trace fossil *Ophiomorpha* and *Crassostrea subtrigonalis* represent the last well-documented Late Cretaceous (Maastrichtian) marine advance into the upper midcontinent. This unit was first recognized by W.A. Cobban and was named the Breien Member of the Hell Creek Formation by Laird and Mitchell (1942). Earlier, Seager et al. (1942) had mentioned the presence of a marine unit within the Hell Creek Formation in south-central North Dakota, but did not name it. Frye (1964, 1967, 1969), Kume and Hansen (1965), Carlson (1982), Moore (1976), Bluemle (1984), Hoganson et al. (1994), Hoganson (2000), and Murphy et al. (1995) endorsed the concept of a widely identifiable marine member of the Hell Creek and perpetuated the use of Breien terminology. Murphy et al. (this volume) suggest that the Breien is the only member of the several members of the Hell Creek Formation proposed by Frye (1969) that is laterally extensive and useful enough to be retained as a viable member. Because of the scarcity and lack of study of fossils in the member, the Breien fauna is poorly known and has been mentioned in few reports (Seager et al., 1942; Frye, 1964, 1967, 1969; Laird and Mitchell, 1942; Hoganson et al., 1994; Hoganson, 2000; Murphy et al., 1995).

Recent interest in the stratigraphic framework of the Hell Creek Formation in North Dakota (e.g., Murphy et al., 1995; Murphy et al., this volume), the dynamics of marine shorelines during the latest Cretaceous and early Paleocene in the Western Interior Seaway (e.g., Lillegraven and Ostresh, 1990; Erickson, 1992, 1999; Cvancara and Hoganson, 1993; Roberts and Kirshbaum, 1995; Archibald, 1996; Belt et al., 1997; Kennedy et al., 1998), and documentation of extinction of marine vertebrate faunas in the Western Interior at the end of the Cretaceous (Hoganson and Cvancara, 1989, 1991; Cvancara and Hoganson, 1993; Hoganson et al., 1997; Hoganson, 2000) prompted this study.

METHODS

The stratigraphy and geographic extent of the Breien Member were defined through field mapping and measuring stratigraphic sections using hand level and Jacob's staff. Fossil specimens recovered during field surveys and those examined in collections were used to define the vertebrate fauna of the Breien. Screen washing for microscopic vertebrate fossil specimens was employed at the Stumpf site. Fossil specimens collected during this study are reposited at the North Dakota Heritage Center, Bismarck, and are cataloged into the State Fossil Collection. State Fossil Collection figured specimens have a prefix of ND (e.g., ND99-25.1) before the catalog number, as do fossil locality numbers (e.g., ND99-25). Figured specimens in this report collected by Charles Frye are in the paleontology collection at the University of North Dakota and have a prefix of UND-PC (e.g., UND-PC 15836) before the catalog number. University of North Dakota fossil localities referred to in this study also have a prefix of UND-PC (e.g., UND-PC A493).

STRATIGRAPHY AND GEOGRAPHIC EXTENT

Laird and Mitchell (1942) named marine beds near the base of the Hell Creek Formation exposed in the Cannonball River Valley (Morton and Sioux Counties, North Dakota) the Breien Member of the Hell Creek Formation (after the nearby town of Breien). They presented four measured sections that included the Breien Member and described the member as consisting primarily of two gray, buff-weathering, fine-grained sandstone beds separated by a gray bentonite. They suggested that the greenish color of the upper sandstone indicated a marine origin; this was confirmed by the discovery of marine fossils in the sandstone. They noted the occurrence of *Halymenites* (*Ophiomorpha*) and limonite concretions in the upper sandstone and marcasite concretions and gypsum crystals in the gray bentonite layer. They also observed that the Breien strata interfinger with undifferentiated nonmarine Hell Creek Formation deposits and that it occurs, on average, ~6 m above the base of the Hell Creek.

Frye (1964, 1969) identified several additional Breien Member localities and extended the geographic range of the member to include Morton, Sioux, and Emmons Counties in North Dakota. Laird and Mitchell (1942) did not designate a type section for the Breien Member in their initial description of the unit but expressed to Frye (1969, p. 35) in a personal communication that the exposure at the Crowghost Cemetery site (now called Red Cedar Cemetery) in Sioux County most typically represented the member (Fig. 1). Frye (1964, 1969) designated the Crowghost Cemetery site (sec. 33, T.134N., R.81W.) as the type section for the Breien Member, but erroneously gave the location as the center of sec. 33, T.134N., R.82W. Frye (1969, p. 57, 58) gave the following description of the Breien Member at its type locality.

Hell Creek Formation	Thickness (ft/m)
Lignitic shale and shaly lignite with siderite nodules weathering to limonite near the top....	2.5 (0.76)
Hell Creek Formation (Breien Member)	
Gray, clay-rich, banded, medium-grained sands grading downward into soft, greenish-brown, brown weathered, glauconitic sands with *Ophiomorpha major*	9.0 (2.75)
Siderite bed weathered to limonite underlain by moderate-grayish-brown sand	1.5 (0.45)
Moderate-gray bentonite..............................	1.5 (0.45)
Grayish-green, glauconitic sand weathering rusty brown, containing *Ophiomorpha major* and gastropod molds near the top.................	7.0 (2.15)
Hell Creek Formation	
Brown, lignitic shale..................................	2.5 (0.76)

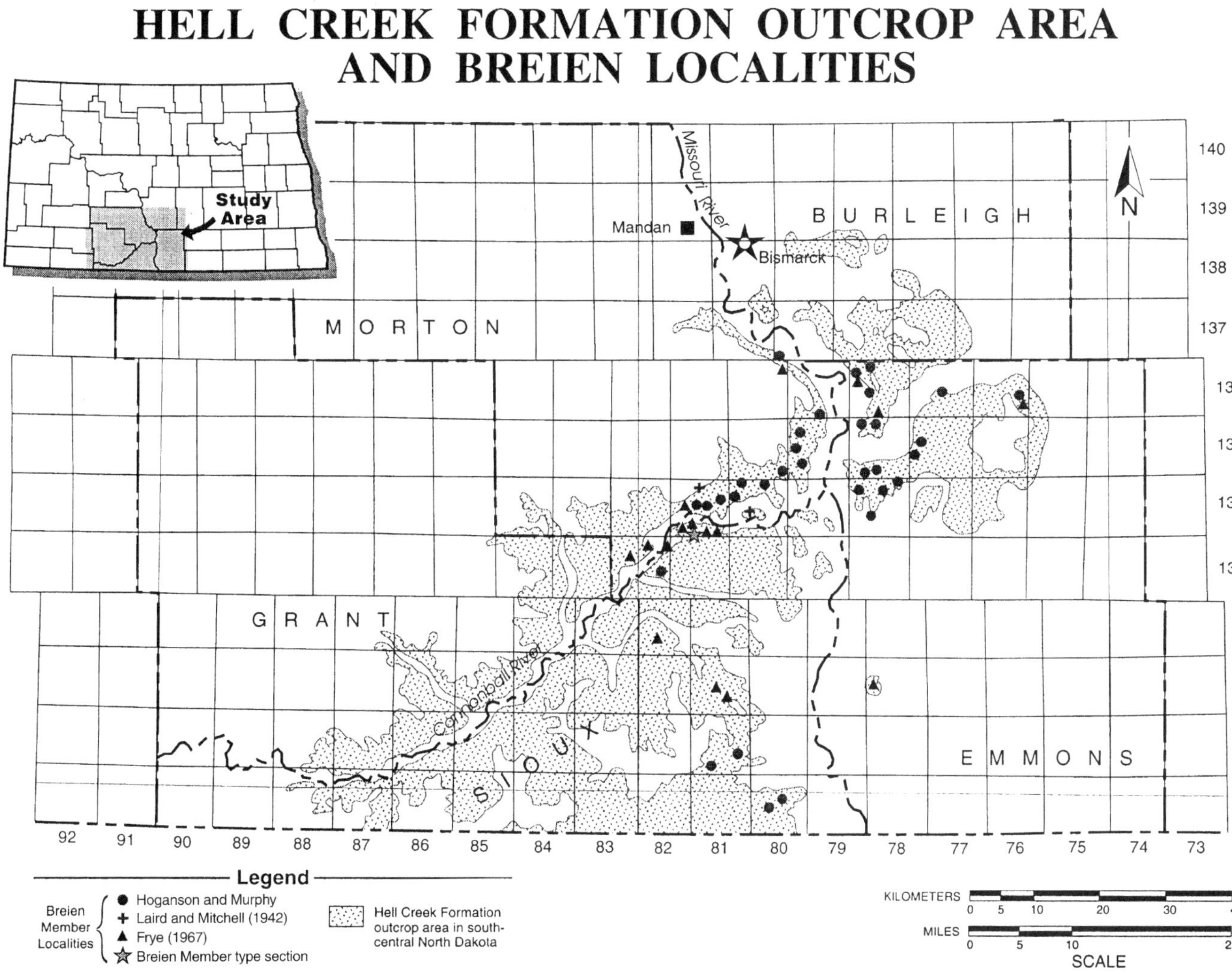

Figure 1. Hell Creek Formation outcrop area in south-central North Dakota and Breien Member localities.

Exposures of the Breien are restricted to isolated butte, stream, and road-cut exposures in Morton, Sioux, and Emmons Counties in a 6000 km^2 area of south-central North Dakota (Fig. 1). Breien outcrops are more numerous and more extensive than were known by earlier workers. More than 50 outcrops were examined and 24 stratigraphic sections were measured for this study (Figs. 1, 2, 3, 4; Table 1). Badlands terrain at the Crowghost Cemetery site, the type section of the Breien Member, provides the opportunity to view vertical and lateral lithologic variations within the member. At this site, the lower and upper contacts of the member are discernable and the contact between the Hell Creek and Fox Hills Formations is well defined (Fig. 3, section 11; also see Figs. 2 and 5). The lithostratigraphy of the member at this site is more variable and complex than indicated in Laird and Mitchell's (1942) and Frye's (1964, 1969) measured sections. Rapid facies changes and channel dissection at the site complicate interpretation of the stratigraphy of the Breien. This is evident when Frye's description of the type section is compared to the Crowghost Cemetery reference section presented in the following and in Figures 2, 3, and 5. The most obvious difference between Frye's type section and the Crowghost Cemetery reference section is that the burrowed upper sandstone described by Frye is absent at the reference section. Lithologic variation of the Breien Member throughout the outcrop area is also indicated by the several measured sections provided herein (Figs. 2, 3, 4, 5, 6, and 7).

Because of the lateral lithologic variability of the Breien Member at its type locality and throughout the outcrop area, three reference sections are presented here to further characterize the member.

CROWGHOST CEMETERY REFERENCE SECTION

NE1/4NW1/4NW1/4 sec. 33, T.134N., R.81W., Breien Quadrangle, Sioux County, North Dakota (Fig. 2, *section CG*; Fig. 3, *section 11*; Fig. 5; Table 1)

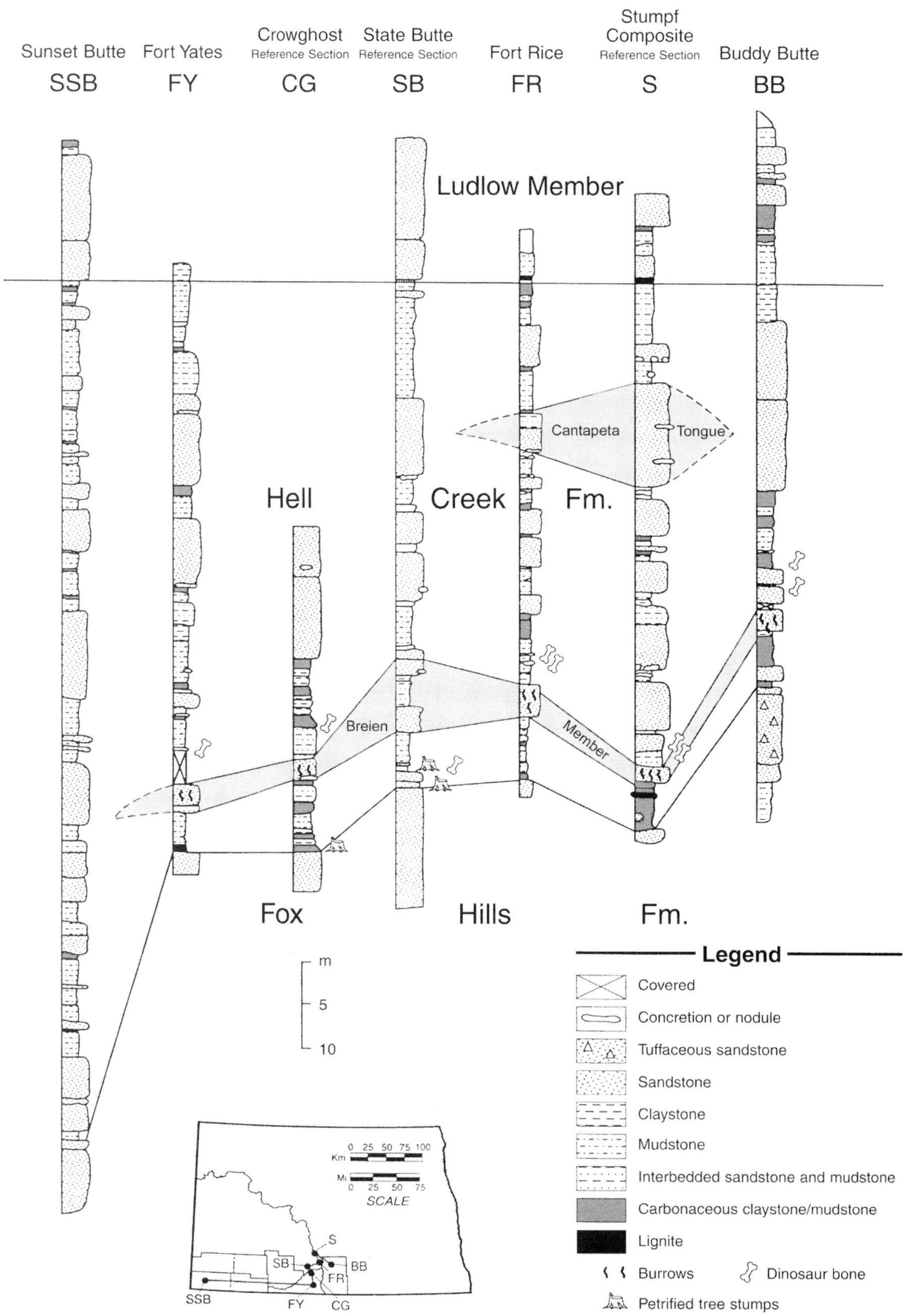

Figure 2. Measured sections showing position of Breien Member in Hell Creek Formation and its position relative to Fox Hills–Hell Creek and Hell Creek–Fort Union (Ludlow Member) formational contacts. Modified from Murphy et al. (this volume). See Table 1 for location of measured sections.

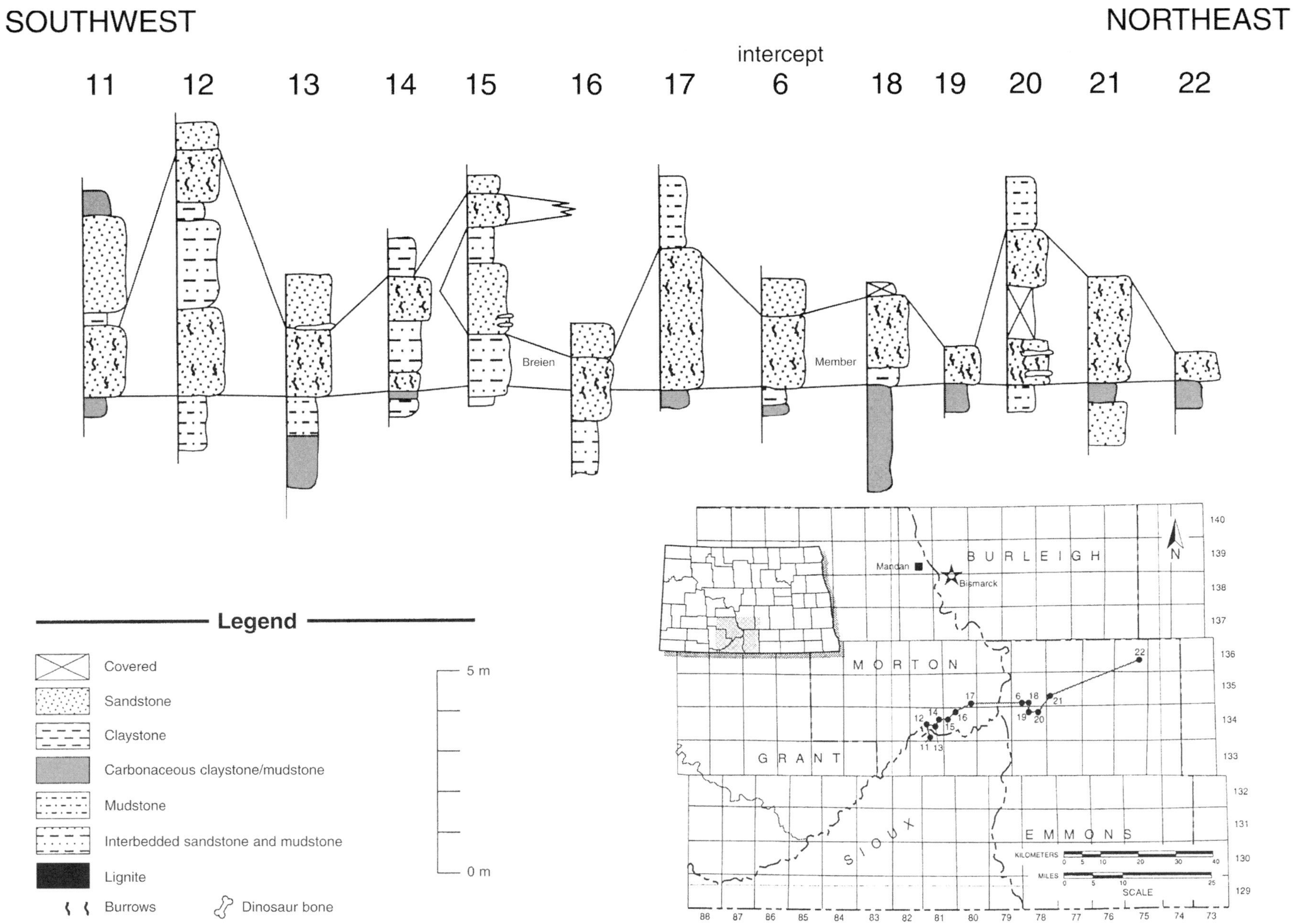

Figure 3. Southwest-northeast cross section of Breien Member measured sections. See Table 1 for measured section names and locations.

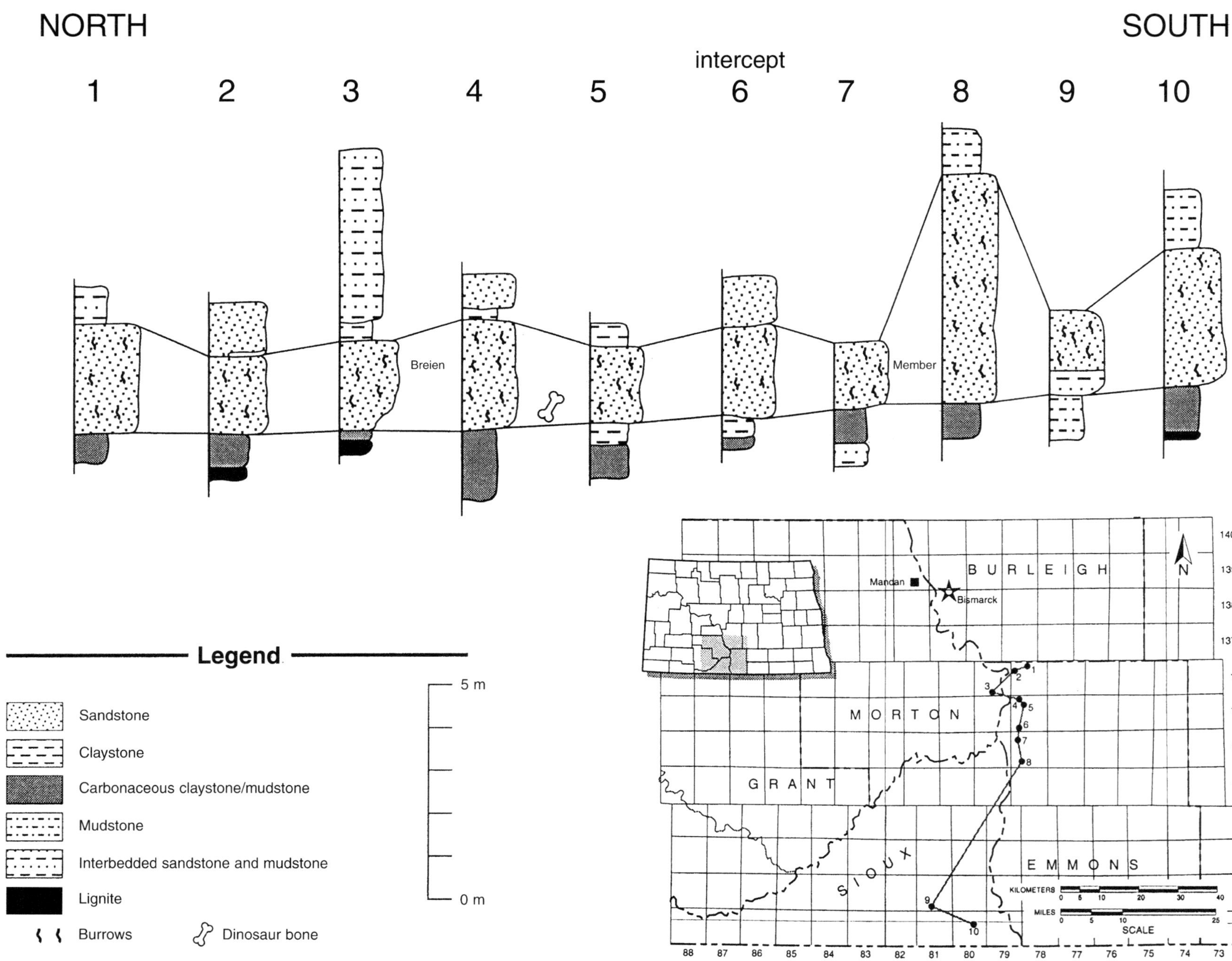

Figure 4. North-south cross section of Breien Member measured sections. See Table 1 for measured section names and locations.

TABLE 1. BREIEN MEMBER MEASURED SECTION LOCALITIES, NORTH DAKOTA

Measured section number	Locality name	NDGS Fossil site accession number	Location
1	Glencoe Creek Ravine		SE1/4 SE1/4 SW1/4, sec. 5, T.136N., R.78W. Emmons County Fort Rice Quadrangle
2	Kimball Road cut	ND99-53	SE1/4 SE1/4 SE1/4, sec. 7, T.136N., R.78W. Emmons County Fort Rice Quadrangle
3 (S)	Stumpf Composite Reference Section	ND96-4 ND96-5 ND99-15 ND99-155	SE1/4 SE1/4 sec. 28, NE1/4 SE1/4 sec. 33, SE1/4 SW1/4 sec. 34, SW1/4 SE1/4 sec. 34, T.136N., R.79 W. Morton County Fort Rice Quadrangle
4	Devils Gulch North		NW1/4 NW1/4 NW1/4 sec. 4, NE1/4 NE1/4 NE1/4 sec. 5, T. 135N., R. 78W. Emmons County Fort Rice Quadrangle
5	Devils Gulch South		SE1/4 SW1/4 NW1/4 sec. 4, T. 135N., R. 78W. Emmons County Moffit SW Quadrangle
6	Coal Butte	ND99-111	NW1/4 NE1/4 SW1/4 sec. 32, T. 135N., R. 78W. Emmons County Cannon Ball Quadrangle
7	Horstmeyer Road cut		NW1/4 NE1/4 NW1/4 sec. 8, T. 134N., R. 78W. Emmons County Cannon Ball Quadrangle
8	Sunburst Lake Road cut		SW1/4 NW1/4 NW1/4 sec. 28, T. 134N., R. 78W. Emmons County Cannon Ball Quadrangle
9	Lund		SE1/4 SE1/4 sec. 23, SW1/4 NW1/4 sec. 25, T. 130N., R. 81W. Sioux County Selfridge SE Quadrangle
10	Fourmile Creek Ravine		SE1/4 NW1/4 SW1/4 sec. 14, T. 129N., R. 80W. Sioux County Mahto NW Quadrangle
11 (CG)	Crowghost Cemetery Reference Section	ND99-25	NE1/4 NW1/4 NW1/4 sec. 33, T. 134N., R. 81W. Sioux County Breien Quadrangle
12 (SB)	State Butte Reference Section	ND00-39 ND00-40	SW1/4 sec. 16, NW1/4 sec. 21, T. 134N., R. 81W. Morton County Breien Quadrangle
13	Hunter Composite		SE1/4 sec. 22, SW1/4 NW1/4 NW1/4 sec. 23, T.134 N., R. 81W. Morton County Solen Quadrangle
14	Hunter	ND99-99 ND99-114	SE1/4 SE1/4 NE1/4 sec. 13, T. 134N., R. 81W. Morton County Solen Quadrangle
15	Solen North Road cut Composite	ND99-102 ND99-103	SW1/4 SW1/4 SW1/4 sec. 7, NW1/4 NW1/4 NW1/4 sec. 18, T. 134N., R. 80W. Morton County Solen Quadrangle
16	Cantapeta Creek Road cut	ND99-104	NE1/4 NW1/4 NW1/4 sec. 4, T. 134N., R. 80W. Morton County Solen Quadrangle
17	Armbrust Road cut	ND99-110	NE1/4 NE1/4 NW1/4 sec. 36, T. 135N., R. 80W. Morton County Cannonball NW Quadrangle
18 (BB)	Buddy Butte		NE1/4 NW1/4 NE1/4 sec. 33, T. 135N., R. 78W. Emmons County Hazelton NW Quadrangle

(*continued*)

TABLE 1. BREIEN MEMBER MEASURED SECTION LOCALITIES, NORTH DAKOTA (continued)

Measured section number	Locality name	NDGS Fossil site accession number	Location
19	Gimble Cemetery Road cut		SE1/4 SE1/4 SE1/4 sec. 3, T. 134N., R. 78W. Emmons County Hazelton NW Quadrangle
20	Horsehead Creek Road cut	ND99-55	NE1/4 SE1/4 NE1/4 sec. 1, T. 134N., R. 78W. Emmons County Hazelton NW Quadrangle
21	Hazelton Road cut	ND99-22 ND99-57	NW1/4 NE1/4 NE1/4 sec. 9, T. 135N., R. 77W. Emmons County Moffit SW Quadrangle
22	Dana Road cut		SW1/4 NW1/4 SW1/4 sec. 23, T. 136N., R. 76W. Emmons County Dana Quadrangle
SSB	Sunset Butte Composite		NW1/4 sec. 25, NW1/4 sec. 30, T. 131N., R. 105W. Bowman County Cedar Hills Quadrangle
FY	Fort Yates Composite		SW1/4 SE1/4 sec. 14, SE1/4 SE1/4 sec. 23, SW 1/4 NW 1/4 sec. 25, T. 130N., R. 81W. Sioux County Selfridge SE Quadrangle
FR	Fort Rice		NW1/4 SE1/4 SW1/4 sec. 7, SW1/4 SW1/4 NE1/4 sec. 7, T. 135N., R. 79W. Morton County Huff Quadrangle

Note: North Dakota (ND) locality numbers refer to localities of fossils in the North Dakota State Fossil Collection, North Dakota Heritage Center, Bismarck. Measured section numbers refer to sections illustrated in Figures 2, 3, and 4.

Hell Creek Formation — Thickness (ft/m)
- Interbedded sandstone, mudstone, and claystone; variegated; thin to medium beds (part) .... (part)
- Bentonitic claystone 1.5 (0.46)

Breien Member of the Hell Creek Formation
- Sandstone: gray-brown to orange (fresh), light grayish-green (weathered); very fine grained; not cemented; contains burrows (including *Ophiomorpha*); glauconitic; iron-oxide stained; contains iron-oxide-cemented concretions (<10 cm diameter) at base; contains oyster and other invertebrate fossils, and vertebrate fossils including shark and chimaeroid remains 7.9 (2.40)

Hell Creek Formation
- Claystone: slightly carbonaceous at base; highly carbonaceous at top; contains jarosite and petrified wood fragments 2.0 (0.60)
- Interbedded sandstone, mudstone, and claystone; variegated; thin to medium beds; contains root traces and petrified wood fragments 5.9 (1.80)
- Carbonaceous mudstone and claystone 4.4 (1.35)
- Mudstone containing root traces and other plant fossil fragments; iron-oxide staining at base; popcorn surface weathering 5.9 (1.80)
- Carbonaceous mudstone with paper shale 1.5 (0.46)
- Lignite 2.0 (0.60)
- Bentonite with in situ petrified tree stumps 2.0 (0.60)
- Carbonaceous mudstone 3.0 (0.90)

Fox Hills Formation (Colgate Member)
- Sandstone: gray; very fine grained to fine grained; moderately cemented; castellate surface weathering; contains in situ petrified tree stumps .... (part)

STATE BUTTE REFERENCE SECTION

SW1/4 sec. 16 and NW1/4 sec. 21, T.134N., R.81W., Breien Quadrangle, Morton County, North Dakota (Fig. 2, *section SB*; Fig. 3, *section 12*; Fig. 6; Table 1)

Hell Creek Formation — Thickness (ft/m)
- Sandstone: grayish-brown (fresh), gray (weathered); clean; very fine grained; not burrowed; gradational from below .... (part)

Breien Member of the Hell Creek Formation
- Sandstone: grayish-green (fresh), grayish-brown (weathered); very fine grained; poorly cemented; glauconitic; iron-oxide concretions (<9 cm in diameter) at base; iron-oxide chips throughout; contains burrows (including *Ophiomorpha*) throughout but especially at top, and vertebrate fossils including shark and chimaeroid remains 6.0 (1.80)
- Bentonitic claystone: popcorn surface weathering 2.0 (0.60)

Crowghost Cemetery
Reference Section
NE1/4, NW1/4, NW1/4,
Section 33, T. 134 N., R. 81 W.
Sioux County, North Dakota
Breien Quadrangle
Elevation Base of Breien = 1720 ft.

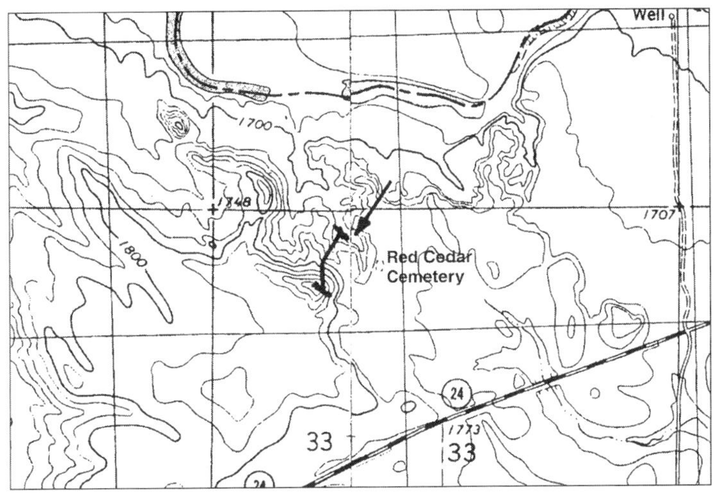

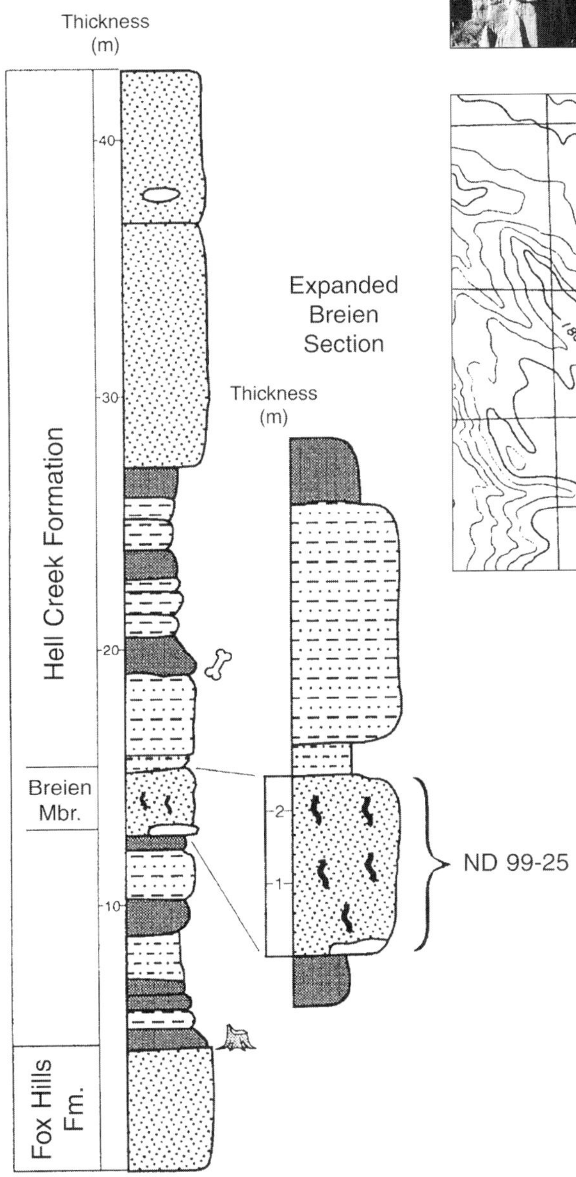

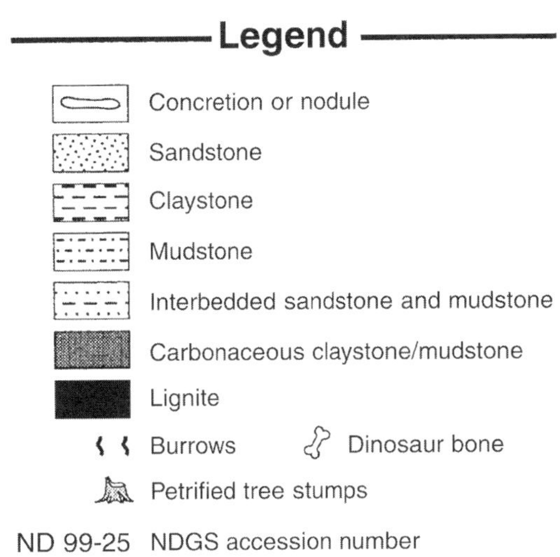

Figure 5. Detailed measured section at Crowghost Cemetery Breien Member reference section site. Photograph is of Breien Member at Crowghost Cemetery site (view is to southwest). Lines on photograph indicate upper and lower contacts of Breien Member. Map (USGS 1:24000 topographic map) shows location of measured section and arrow points in direction of photograph.

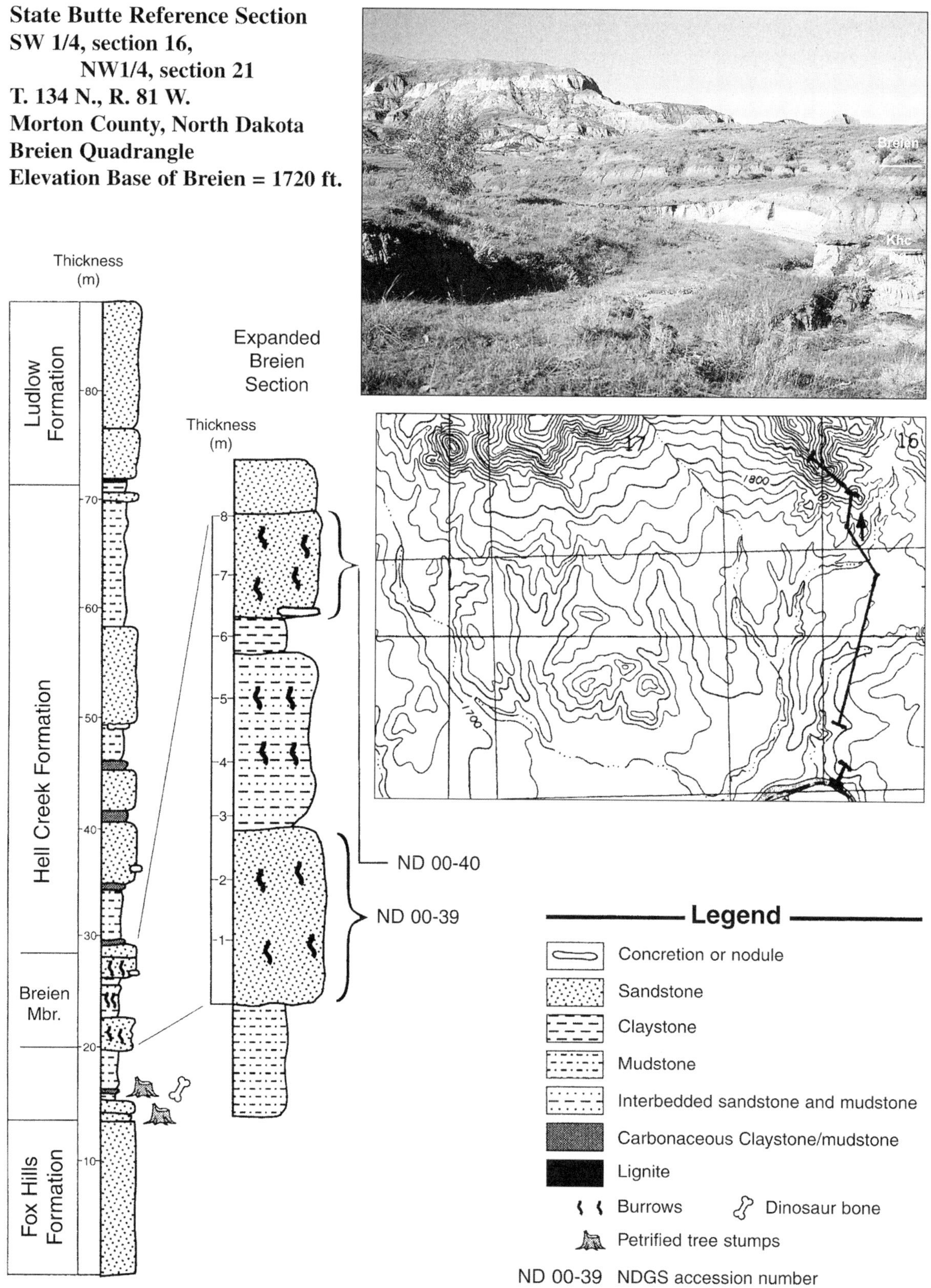

Figure 6. Detailed measured section at State Butte Breien Member reference section site. Photograph is of Breien Member at State Butte site (view is to north). Lines on photograph indicate upper and lower contacts of Breien Member. Map (USGS 1:24 000 topographic map) shows location of measured section and arrow points in direction of photograph.

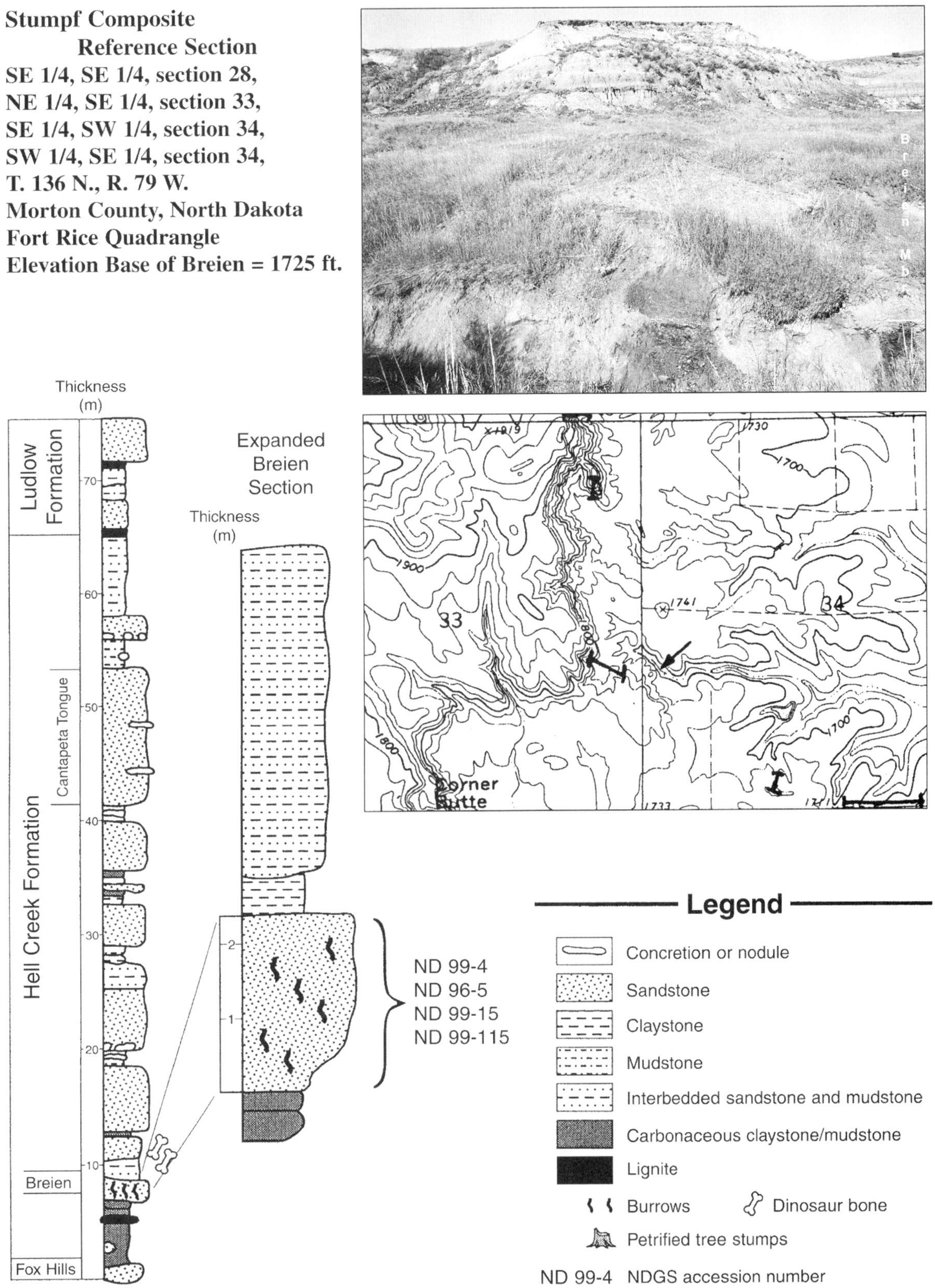

Figure 7. Detailed measured section at Stumpf composite Breien Member reference section site. Photograph is of Breien Member at Stumpf site (view is to southwest). Lines on photograph indicate upper and lower contacts of Breien Member. Map (USGS 1:24 000 topographic map) shows location of measured section and arrow points in direction of photograph.

Interbedded sandstone, mudstone, and claystone: sandstone is yellowish gray to gray (both fresh and weathered) and very fine grained; mudstone is dark gray; both sandstone and mudstone are burrowed (including *Ophiomorpha*) and top 10 cm of sandstone is heavily burrowed 9.8 (3.00)
Sandstone: grayish-green to gray (fresh), gray (weathered); very fine to fine grained; poorly cemented; glauconitic, contains iron-oxide concretions (<3 cm in diameter); contains organic-rich planar clay lenses with slight popcorn weathering; burrowed (including *Ophiomorpha*); clay drapes on sandstone 9.8 (3.00)

Hell Creek Formation
Mudstone: contains silty layers; slight popcorn weathering 12.0 (3.60)
Paper shale: contains amber, lignite lenses, petrified tree stumps, and crocodile and turtle bones 0.9 (0.30)
Bentonite: contains dinosaur bones................ 2.0 (0.60)
Sandstone: contains root traces; clay draped..... 2.8 (0.86)
Carbonaceous siltstone: contains in situ petrified tree stumps 2.8 (0.86)

Fox Hills Formation (Colgate Member)
Sandstone: grayish blue (fresh), grayish brown (weathered); very fine to fine grained; poorly to moderately cemented; organic rich; burrowed; planar bedded; clay draped .... (part)

STUMPF SITE REFERENCE SECTION

SE1/4SE1/4 sec. 28; NE1/4SE1/4 sec. 33; SE1/4SW1/4 sec. 34; SW1/4,SE1/4 sec. 34, T.136N., R.79W., Fort Rice Quadrangle, Morton County, North Dakota (Fig. 2, *section S*; Fig. 4, *section 3*; Fig. 7; Table 1)

Hell Creek Formation — Thickness (ft/m)
Mudstone: variegated (part) 12.0 (3.60)

Breien Member of the Hell Creek Formation
Sandstone: grayish-green (fresh and weathered); very fine grained; glauconitic; contains iron-oxide concretions (<6 cm in diameter); burrowed throughout (including *Ophiomorpha*); contains oysters, other invertebrate fossils, and vertebrate fossils including shark, ray, and chimaeroid remains................ 6.0 (1.80)
Mudstone: gray to dark gray (fresh and weathered); iron-oxide staining; burrowed (including *Ophiomorpha*).. 2.8 (0.86)

Hell Creek Formation
Paper shale 2.8 (0.86)
Lignite 1.0 (0.30)
Paper shale: carbonaceous; 0.86-m-thick sandstone in middle.................................... 12.0 (3.60)

Fox Hills Formation (Colgate Member)
Sandstone: greenish-gray (fresh), gray (weathered); very fine grained; poorly cemented; contains clay partings and disseminated plant fragments.................. (part)

As defined in this chapter, the Breien Member consists of a 1.5–8.5-m-thick interval of burrowed, greenish-gray glauconitic, fine-grained sandstone, mudstone, or interbedded sandstone and mudstone. The mudstone occurs only in isolated areas. Iron-oxide staining is usually present and is most pronounced along bedding surfaces. The burrows, most of which are *Ophiomorpha*, occur throughout the member, are preserved by limonite, and are often associated with the oyster *Crassostrea subtrigonalis* (Fig. 8). Depositional strike indicates that the Breien was deposited along the southeastern edge of the Williston Basin (Fig. 9). The Breien attains a maximum thickness of 8.5 m at the State Butte section in Morton County (Fig. 10). It is lost in the subsurface to the west and north in Sioux, Morton, and Burleigh Counties and pinches out to the east in eastern Emmons County (Fig. 10). Although variable, the Breien's lower contact is generally placed where the member is underlain by a brown, carbonaceous mudstone. The upper contact is placed above distinctly burrowed sandstone or mudstone. The Breien is overlain by a variety of lithologies, usually mudstone or siltstone. Both upper and lower contacts are conformable. The base of the Breien Member occurs 1.5–9.0 m above the Fox Hills–Hell Creek formational contact and 46–61 m below the Hell Creek–Fort Union formational contact (Fig. 2). The Breien Member is 60–70 m below the Paleocene marine Cannonball Member of the Fort Union Formation, and interfingers with continental deposits in the Hell Creek Formation that often contain the remains of dinosaurs, mammals, other vertebrates, freshwater bivalves and gastropods, and plants determined to be from the HC III megafloral zone by Johnson (1998, personal commun.).

PALEONTOLOGY

Except for burrows, many of which are identified as *Ophiomorpha*, and the oyster *Crassostrea subtrigonalis*, fossils are uncommon in the Breien Member (Fig. 8). Other than these oysters, invertebrate fossils are particularly scarce and consist mostly of casts and molds in limonite (Fig. 8). As a result, reports of Breien fossils are few and mostly only mentioned in conjunction with stratigraphic studies of the Hell Creek Formation. W.A. Cobban (*in* Seager et al., 1942) first identified fossils from the Breien Member and noted the presence of marine indicators such as *Halymenites* (*Ophiomorpha*), *Ostrea* (*Crassostrea*) *glabra*, and other invertebrates: Laird and Mitchell (1942) republished the list of taxa. Frye (1964, 1967, 1969) collected and documented several taxa of invertebrate fossils from the Breien for his study of the Hell Creek Formation. He did not illustrate those fossils, and many specimens have sub-

Figure 8. Invertebrate and trace fossils from Breien Member. A: ?*Pachymelania*. ND99-25.1, height 83.9 mm. B, C: *Jeletzkytes* cf. *J. nebrascensis*. partial specimen. UND-PC 15836. B is posterior view, height 33.9 mm; C is lateral view. D: *Ophiomorpha*. ND00-40.1, height 67.0 mm. E: *Crassostrea subtrigonalis*, ND95-15.3, height 133.4 mm.

sequently been lost. Hartman and Kirkland (this volume) identify and illustrate invertebrate fossils collected by Frye from the Breien. Hoganson et al. (1994) and Murphy et al. (1995) provided lists of marine invertebrate fossils recovered from the Breien at the Stumpf site, Morton County. Kennedy et al. (1998) reported the occurrence of *Discoscaphites* cf. *D. conradi* in the Breien Member.

Remains of vertebrate animals are also not common in the Breien Member. The first mention of vertebrate fossils from the Breien was by Frye (1964), who recorded the presence of skate and shark teeth. In his dissertation, he greatly expanded the list of vertebrate taxa in the Breien to include the remains of sharks, rays, bony fish, turtles, crocodiles, alligators, champsosaurs, dinosaurs, and mammals (Frye, 1967). Most of these specimens could not be located for reexamination. Lists of vertebrate taxa, including chondrichthyans and bony fish, turtles, and crocodilia from the Breien recovered at the Stumpf site, Morton County, were presented by Hoganson et al. (1994) and Murphy et al. (1995). Murphy et al. (1995) illustrated the right palatine of the chimaeroid *Ischyodus*, and a lateral tooth of the lamniform shark, *Carcharias*, from the Stumpf site.

Of the 50 Breien fossil sites identified during this study, only 16 contain the remains of vertebrate animals (Fig. 11; Table 2). Vertebrate fossils from the Breien Member consist of disarticulated skeletal elements, mostly teeth and vertebrae. No articulated skeletons were found.

Remains of fish are the most common vertebrate fossils in the Breien. We identified 13 species of chondrichthyans, including sharks, rays, and chimaeroids, and 3 species of actinopterygians (Fig. 12; Table 2). The pristid *Peyeria*, originally described from the Upper Cretaceous of Egypt, has not previously been reported from North America, although Gorden Bell (1998, personal commun.) recovered a tooth of *Peyeria* in 1985 from the Maastrichtian Ripley Formation in Wilcox County, Alabama (Fig. 12). Reptilia is represented by the remains of champsosaurs, crocodilians, and tyrannosaurid and hadrosaurid dinosaurs. One multituberculate mammalian taxon was recorded (Fig. 12). Coprolites, presumably from fish, were also found.

AGE OF THE BREIEN

The age of the Breien Member can be determined by its stratigraphic position and by paleontological information. At the Stumpf site in south-central North Dakota, nonmarine Hell Creek strata immediately adjacent to the Breien have been determined to be Lancian in age based on vertebrate fossils (Hoganson et al., 1994; Murphy et al., 1995). On average, the base of the Breien Member occurs ~7 m above the Fox Hills Formation that was deposited within the Upper Cretaceous *Jeletzkytes nebrascensis* Zone (Landman and Waage, 1993). A poorly preserved fragment of an ammonite was collected by Frye (1967) from the Breien Member in Emmons County. One of us (Hoganson) provided a cast of that specimen to W.A. Cobban, who identified it as *Discoscaphites* cf. *D. conradi* (Kennedy et al., 1998). At our request, N.H. Landman examined the actual specimen and provisionally identified it as *Jeletzkytes* cf. *J. nebrascensis* (Fig. 8). The occurrence of this specimen in the Breien suggests deposition at the time of the *Jeletzkytes nebrascensis* Zone. We were able to relocate the site where Frye collected the ammonite. Breien sandstone was observed, even though the site was badly overgrown, but no additional

CONTOUR MAP ON BASE OF BREIEN MEMBER

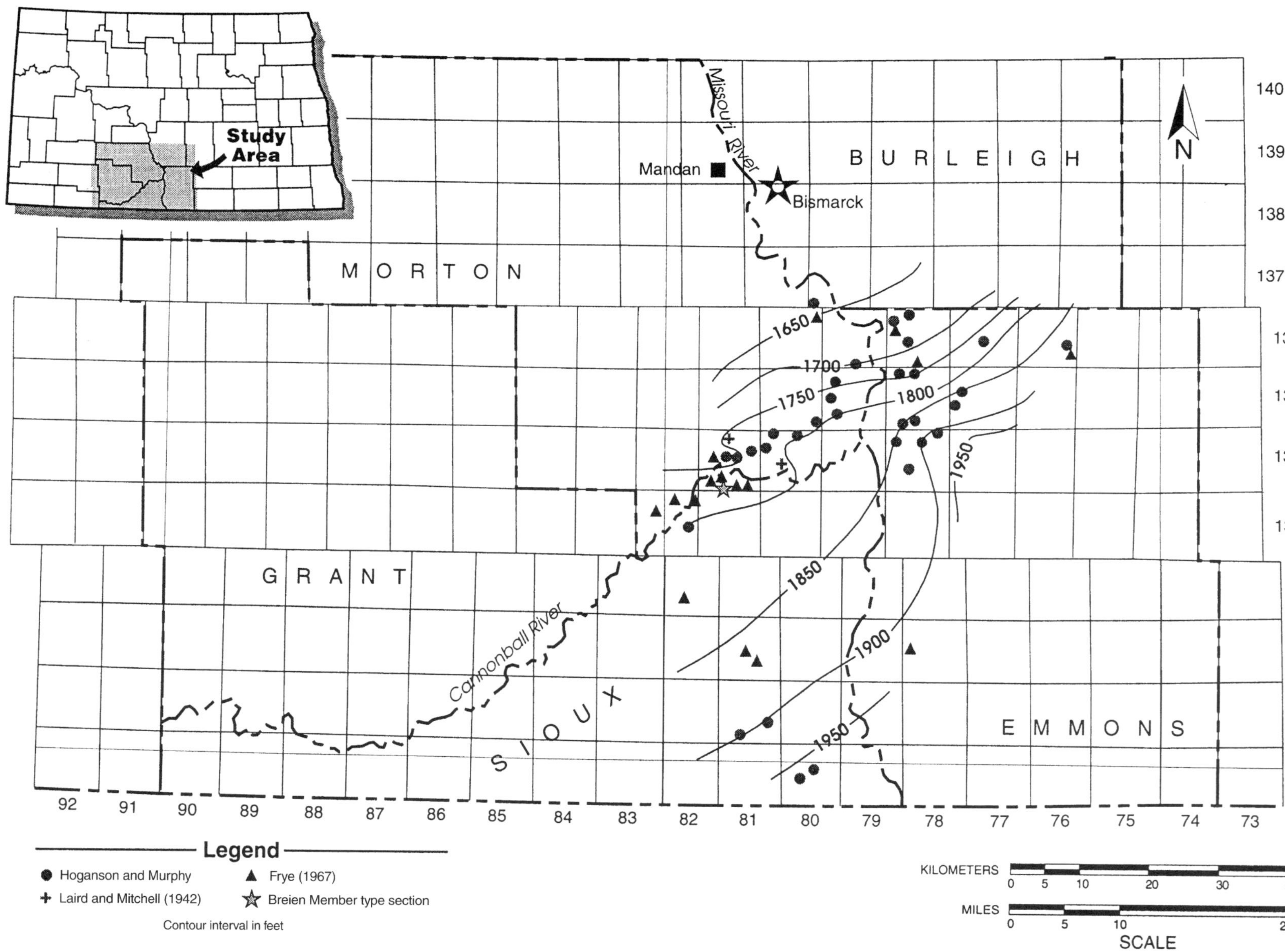

Figure 9. Contour map on base of Breien Member in south-central North Dakota showing depositional strike.

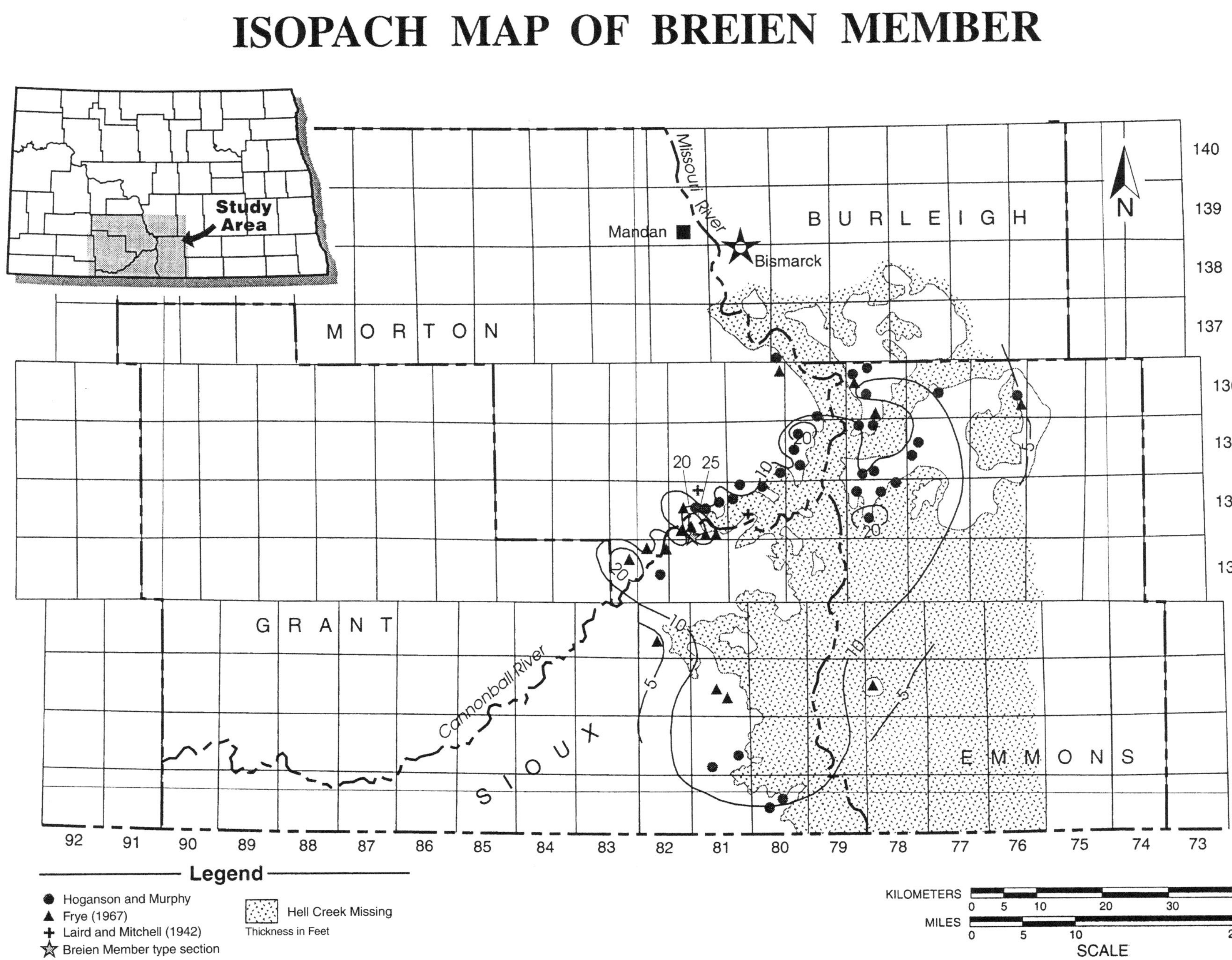

Figure 10. Isopach map of Breien Member in south-central North Dakota.

BREIEN MEMBER FOSSIL SITES

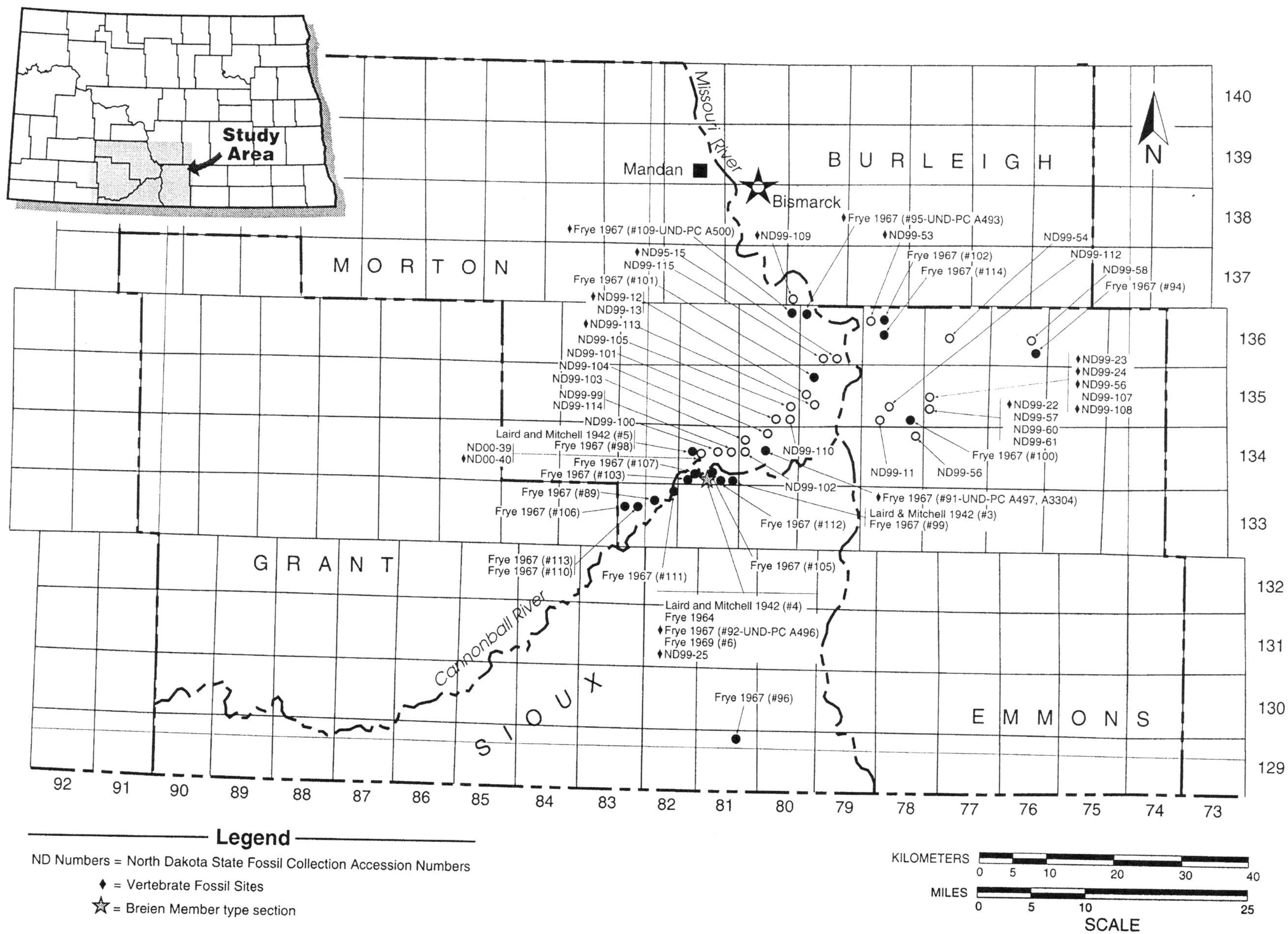

Figure 11. Map showing location of Breien fossil sites in south-central North Dakota.

TABLE 2. LIST OF VERTEBRATE TAXA FROM THE BREIEN MEMBER OF THE HELL CREEK FORMATION AND THE LOCALITIES WHERE THEY WERE FOUND

Taxon	Locality																	
	UND A493	UND A496	UND A497	UND A500	UND A3304	ND95-15	ND96-4	ND99-12	ND99-22	ND99-23	ND99-25	ND99-56	ND99-108	ND99-109	ND99-113	ND00-37	ND00-40	ND00-59
Class Chondrichthyes																		
Order Lamniformnes																		
Family Carchariidae																		
Carcharias sp.	–	–	+	–	–	+	+	+	+	+	+	+	+	–	+	–	+	–
Family Odontaspidae																		
Odontaspis sp.	–	–	–	–	–	+	–	–	–	–	–	–	–	–	–	–	–	–
Family Cretoxyrhinidae																		
Cretolamna appendiculata (Agassiz 1843)	–	–	–	–	–	+	–	–	–	–	–	–	–	–	–	–	–	–
Family Serratolamnidae																		
Serratolamna serrata (Agassiz 1843)	–	–	+	–	–	–	+	–	–	–	–	–	–	–	–	–	–	–
Family Alopiidae																		
Paranomotodon sp.	–	–	–	–	–	+	–	–	–	–	–	–	–	–	–	–	–	–
Family Triakidae																		
Galeorhinus sp.	–	–	–	–	–	+	–	–	–	–	–	–	–	–	–	–	–	–
Order Rajiformes																		
Rhinobatoidea incertae sedis																		
?Pseudohypolophus sp.	–	–	–	–	–	+	–	–	–	–	–	–	–	–	–	–	–	–
Myledaphus bipartitus (Cope 1876)	–	+	+	+	–	+	–	–	–	–	–	–	–	–	–	–	+	–
Family Sclerorhynchidae																		
Ischyrhiza sp.	–	–	–	–	–	+	–	–	–	–	–	–	–	–	–	–	–	–
Ptychotrygon cf. *P. texana* (Leriche 1940)	–	–	–	–	–	+	–	–	–	–	–	–	–	–	–	–	–	–
Family Pristidae																		
Peyeria sp.	–	–	–	–	–	+	–	–	–	–	–	–	–	–	–	–	–	–
Order Chimaeriformes																		
Family Edaphodontidae																		
Ischyodus n. sp.	–	–	–	–	–	+	–	+	–	?	–	–	–	+	–	+	–	–
Ischyodus sp.	–	–	–	–	–	+	–	–	–	–	–	–	–	–	–	–	–	–
Chimaeriformes indet.	–	–	–	–	–	+	+	–	–	–	+	–	–	+	–	–	+	+
Class Actinopterygii																		
Order Amiiformes																		
Family Amiidae																		
Melvius thomasi (Bryant 1987)	–	–	–	–	–	+	–	–	–	–	–	–	–	–	–	–	–	–
Order Aulopiformes																		
Family Enchodontidae																		
Enchodus sp.	–	–	–	–	–	+	–	–	–	–	–	–	–	–	–	–	–	–
Order Elopiformes																		
Family Phyllodontidae																		
?Paralbula sp.	–	–	–	–	–	+	–	–	–	–	–	–	–	–	–	–	–	–
Actinopterygii indet.	–	–	+	–	–	+	–	–	–	+	–	–	–	–	–	–	–	–
Class Reptilia																		
Order Christodera																		
Family Champsosauridae																		
Champsosaurus sp.	+	–	–	–	–	–	–	–	–	–	–	–	–	–	–	–	–	–
Order Crocodilia																		
Family Crocodilidae																		
?*Brachychampsa* sp.	–	–	–	+		–	–	–	–	–	–	–	–	–	–	–	–	–
Crocodilia indet.	–	–	+	–	–	+	–	–	–	–	+	–	–	–	–	–	–	–
Order Saurischia																		
Family Tyrannosauridae																		
Tyrannosaurus sp.	–	–	–	+	–	–	–	–	–	–	–	–	–	–	–	–	–	–
Family Hadrosauridae																		
Hadrosauridae indet.	–	–	–	–	?	?	–	–	–	–	–	–	–	–	–	–	–	–
Class Mammalia																		
Order Multituberculata																		
Family Cimolomyidae																		
Meniscoessus robustus (Marsh 1889)	–	–	?	–	–	–	–	–	–	–	–	–	–	–	–	–	–	–
Trace Fossils																		
Coprolite	–	–	–	–	–	+	–	–	+	–	–	–	–	–	–	–	–	–

Note: See Figure 11 for location of fossil sites.

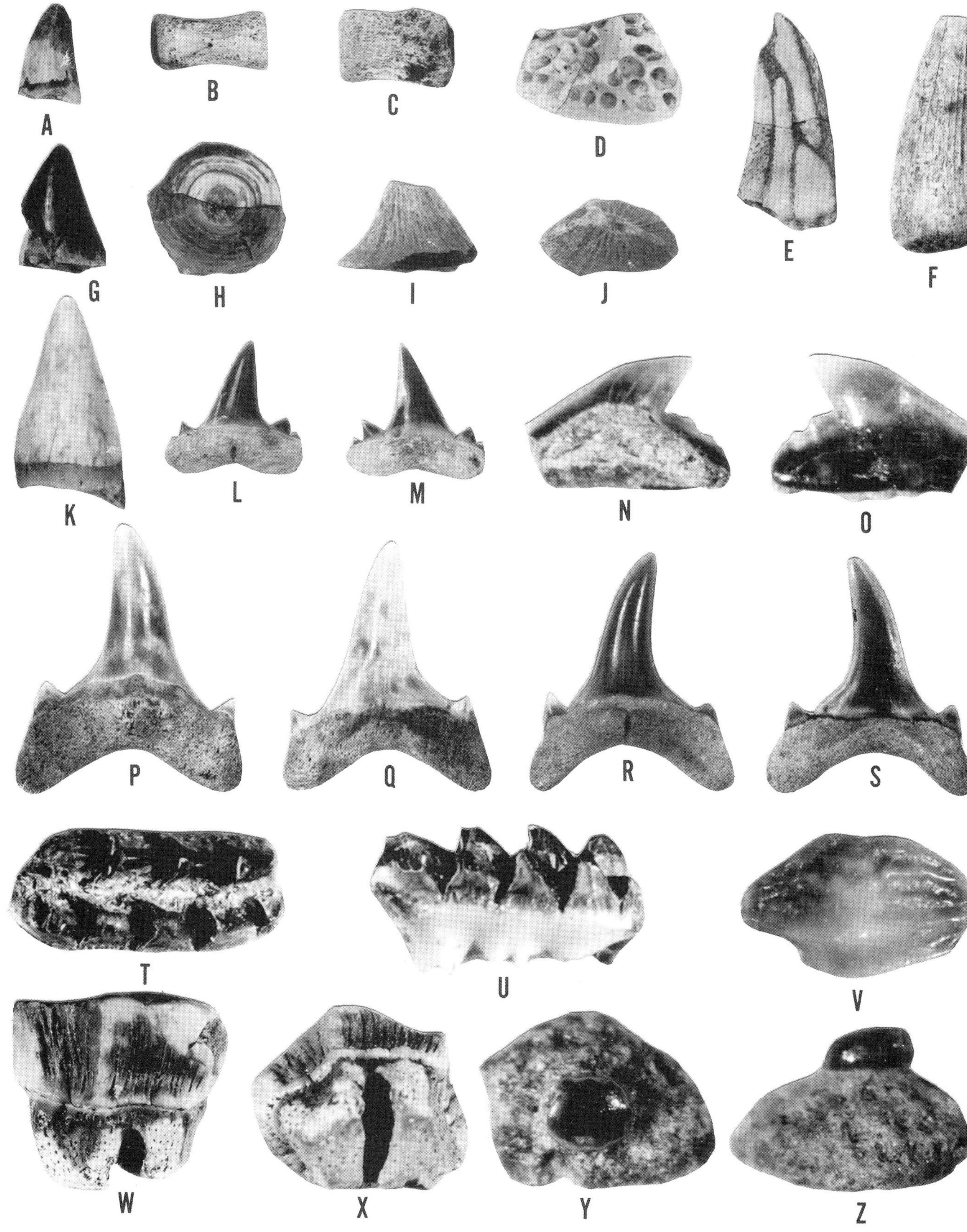
A
B
C
D
E
F
G
H
I
J
K
L
M
N
O
P
Q
R
S
T
U
V
W
X
Y
Z

Figure 12. Representative vertebrate fossils from Breien Member. A: ?*Brachychampsa*. lateral view of tooth, UND-PC 15838, height 12.8 mm. B, C: *Champsosaurus*. dorsal and lateral views of caudal vertebra, UND-PC 15837, width 13.9 mm. D: Crocodilia, dorsal view of scute, ND99-25.2, width 21.1 mm. E: *Ischyodus* n. sp. occlusal view of right palatine tooth plate, ND95-15.1, height 15.7 mm. F: *Enchodus*. lateral view of tooth, ND95-15.4, height 12.6 mm. G: *Tyrannosaurus*. lateral view of tooth, UND-PC 15840, height 15.7 mm. H: Fish vertebra. ND95-15.5, height 16.2 mm. I, J: *Peyeria*. lateral and top views of rostral tooth, ND95-15.6, width 16.2 mm. K: *Melvius thomasi*. medial view of tooth, ND95-15.7, height 7.1 mm. L, M: *Serratolamna serrata*, lingual and labial views of tooth, ND96-4.1, height 15.8 mm. N, O: *Galeorhinus*, lingual and labial views of tooth, ND95-15.8, height 2.4 mm. P, Q: *Cretolamna appendiculata*, lingual and labial views of tooth, ND95-15.12, height 13.6 mm. R, S: *Carcharias*, lingual and labial views of tooth, ND95-15.2, height 12.1 mm. T, U: *Meniscoessus robustus*, occlusal and labial views of Rm1, UND-PC 15839, width 8.4 mm. V: *Ptychotrygon* cf. *P. texana*, occlusal view of tooth, ND95-15.9, width 1.8 mm. W, X: *Myledaphus bipartitus*, lingual and basal views of tooth, ND95-15.10, width 6.6 mm. Y, Z: Elasmobranchii, dorsal and lateral views of dermal denticle, ND95-15.11, width 1.8 mm.

ammonites were found. The Breien is exposed in a road cut (Fig. 3, section 22) very near Frye's hillside ammonite site. At that road cut, the Breien is only 1.5 m above the Colgate Member of the Fox Hills Formation. As suggested by Erickson (1999), reworking of Frye's Breien ammonite from the underlying Fox Hills Formation cannot be ruled out, although it is still likely that the Breien Member was deposited at the time of the *Jeletzkytes nebrascensis* Zone.

The Breien Member interfingers with nonmarine Hell Creek strata containing vertebrate fossils that provide information about the age of the member. At the Stumpf site and other sites, nonmarine, fossiliferous Hell Creek beds are found directly above and directly below the marine fossil-bearing Breien Member. Vertebrate fossils found in those beds, including the dinosaurians *Tyrannosaurus rex* and *Triceratops* and the mammalian taxon *Meniscoessus robustus*, corroborate a late Maastrichtian age for the Breien and indicate that the Breien was deposited at the time of the Western Interior *Triceratops* Zone. The possible occurrence of *Jeletzkytes nebrascensis* in the Breien suggests that the *Jeletzkytes nebrascensis* Zone overlaps, at least in part, the *Triceratops* Zone. Frye (1967) discovered a mammalian tooth in the Breien Member that was identified by W.A. Clemens (letter to F.D. Holland, Jr., April 20, 1965) as *Meniscoessus robustus* (Fig. 12). That identification was confirmed through reexamination of the specimen (Hoganson). Frye's mammal tooth locality is still well exposed, and examination of the site indicates that the specimen could easily have weathered out of nonmarine Hell Creek strata and ended up as float on the Breien surface.

The Breien fish fossil assemblage provides some age information (Fig. 12; Table 2). *Paranomotodon, Ischyrhiza, Ptychotrygon*, and *Peyeria* are only known from the Late Cretaceous (Cappetta, 1987). As previously noted, in addition to being initially found in Upper Cretaceous deposits in Egypt, *Peyeria* has only been reported from the Maastrichtian Ripley Formation in Alabama (Gorden Bell, 1998, personal commun.). Welton and Farish (1993) gave the geologic ranges of *Cretolamna appendiculata, Serratolamna serrata, Paranomotodon, Ischyrhiza*, and *Ptychotrygon* in Texas as Late Cretaceous. The unidentified *Carcharias* species from the Breien Member is the same taxon as *Carcharias* sp. B described by Welton and Farish (1993) from the Maastrichtian of Texas, the youngest Cretaceous *Carcharias* in Texas. *Myledaphus bipartitus* is generally considered a Maastrichtian taxon. It has been reported from younger strata but is probably reworked in those instances (Bryant, 1989). The amiid *Melvius thomasi* is widespread in Maastrichtian deposits from the western margin of the Western Interior Seaway, but also occurs in earlier Cretaceous strata (Bryant, 1987).

PALEOECOLOGY AND PALEOBIOGEOGRAPHY

The dynamic delta and shallow-marine ecological setting during Breien time provided habitat for diverse communities of terrestrial, brackish water, and marine animals. This is reflected by the Breien vertebrate fossil assemblage, which consists of a mix of marine, brackish water, and terrestrial taxa (Fig. 12; Table 2). The terrestrial taxa presence in this assemblage, including champsosaurs, crocodiles, dinosaurs, and mammals, have also been reported from Hell Creek nonmarine deposits in south-central North Dakota (Hoganson et al., 1994; Murphy et al., 1995) and are indicative of the marginal marine depositional setting of the Breien Member. These taxa, typical of Upper Cretaceous terrestrial faunas in the U.S. Western Interior, also inhabited areas farther away from the Fox Hills Sea in western North Dakota (Pearson et al., this volume), South Dakota, Montana, and elsewhere. Their remains in Breien strata indicate that the geographic range of these animals extended to Fox Hills Sea shorelines.

Paleoecological information can be gained from the Breien fish fauna by comparison to the occurrence and behavior of related extant taxa and by the reported occurrence of Breien taxa elsewhere (Fig. 12; Table 2). The marginal marine, nearshore depositional setting implied by the lithology of the Breien Member and the occurrence of *Ophiomorpha, Crassostrea subtrigonalis*, and other invertebrates (Hartman and Kirkland, this volume) is further defined by the presence of shallow-water, euryhaline fish taxa in the Breien. For example, sawfish, represented by *Ischyrhiza* and *Peyeria* in the Breien fossil assemblage, today are nearshore dwellers and prefer brackish to estuarine conditions (Estes, 1964; Estes and Berberian, 1970). They often enter river systems and even breed in completely freshwater habitats. Although the rajiform *Myledaphus* was probably mostly a freshwater taxon (Estes, 1964), its occurrence in the Breien suggests that it could tolerate brackish-water conditions. *Myledaphus* and *Ischyrhiza* have also been reported from fluvial Hell Creek deposits close to Breien shoreline areas in south-central North Dakota (Murphy et al., 1995). They also occur in nonmarine Hell Creek deposits in southwestern North Dakota (Pearson et al., this volume) and eastern Montana (Bryant, 1989; Archibald, 1996) and in fluvial Lance Formation deposits in Wyoming (Estes, 1964), far removed from the Western Interior Seaway.

A nearshore, shallow-water marine depositional setting is also indicated by the presence of another euryhaline taxon, the amiid *Melvius thomasi* (Fig. 12). *Melvius thomasi* is mostly found in nonmarine Hell Creek deposits (Bryant, 1987; Pearson et al., this volume). Bryant (1987, 1989) suggested that *Melvius thomasi* probably entered marine water occasionally, but preferred living in rivers along the margin of the Western Interior Seaway. Grande and Bemis (1998) point out that *Melvius thomasi* belongs in the mostly marine group, Vidalamiinae, and inhabited both marine and freshwater habitats. Chimaeroids also live today in, and inhabited in the past, mostly shallow-water marine areas (Stahl, 1999). Estuaries and other nearshore marine habitats were likely preferred by phyllodontids (Estes, 1969; Estes and Hiatt, 1978). Triakid sharks mostly inhabit nearshore marine areas in water depths of <2 m (Cvancara and Hoganson, 1993).

Other than the triakids, most of the lamniform shark taxa recovered from the Breien Member are considered to be taxa that either inhabited shallow, coastal waters or would often venture into nearshore areas from the open ocean (Welton and Farish, 1993). The living *Carchaias taurus*, for example, is found mostly in coastal depths of 2–5 fathoms (Cvancara and Hoganson, 1993). The enchodontid *Enchodus* was apparently an open-water pelagic fish that also occasionally frequented shallow waters (Goody, 1976). Most of the living relatives of the chondrichthyan taxa present in the Breien inhabit tropical to warm temperate waters today suggesting similar water temperatures during deposition of the Breien Member.

The Breien brackish-water and marine chondrichthyan and bony fish fauna inhabited estuaries and other nearshore habitats at the western edge of the Western Interior Seaway. These fish shared habitats with the shrimp *Callianassa*, indicated by the trace fossil *Ophiomorpha*, gastropods (e.g., *Pachymelania*), and brackish-water bivalves (e.g., *Crassostrea subtrigonalis, Corbicula*) (Fig. 8). This association of invertebrates, termed the *Crassostrea–Pachymelania* estuarine faunule by Erickson (1974), is also present in the Fox Hills Formation. A more complete picture of this community is gained with inclusion of the vertebrate taxa. It is assumed that fish occupied the top of the food chain in this nearshore marine habitat, although it is interesting to speculate that nonmarine taxa, including champsosaurs, crocodilians, and theropod dinosaurs, occasionally preyed on fish in these areas. The rhinobatoids (rays), chimaeroids (ratfish), and phyllodontids possessed teeth or tooth plates adapted for crushing hard-shelled invertebrates. These animals would cruise the bottom feeding on a variety of invertebrates, including gastropods, bivalves, crustaceans, and soft-bodied animals (Welton and Farish, 1993; Stahl, 1999; Estes, 1969). It is believed that the sawfish *Ischyrhiza* was a bottom dweller that primarily preyed on other fish, but it would also use its snout to probe into the bottom to dig up shellfish (Welton and Farish, 1993). Triakids, soupfin sharks, also generally live in water depths of <2 m and feed mainly on bony fish and invertebrates (Cvancara and Hoganson, 1993). The primarily fish-eating predators that inhabited the Breien coastal waters included the large lamniform sharks, *Carcharias, Odontaspis, Cretolamna*, and *Serratolamna* (Welton and Farish, 1993; Cvancara and Hoganson, 1993), and the enchodontid *Enchodus* (Goody, 1976).

Many of the chondrichthyans present in the Breien Member were cosmopolitan during the Late Cretaceous and inhabited southern areas of the Western Interior Seaway (Welton and Farish, 1993), the Atlantic coastal plain (Cappetta and Case, 1975; Gallagher, 1993; Kent, 1994), Europe (Cappetta, 1987), and Morocco (Noubhani and Cappetta, 1997), suggesting connection of Breien shallow-water habitats with shallow-marine waters elsewhere.

PALEOGEOGRAPHIC IMPLICATIONS

The Breien Member of the Hell Creek Formation represents the last well-documented advance of the Fox Hills Sea into the upper Western Interior, probably at the time of the late Maastrichtian *Jeletzkytes nebrascensis* Zone. The Breien was thought to represent the last Cretaceous marine advance into the upper mid-continent, until Murphy et al. (this volume) discovered marine strata above the Breien in the upper part of the Hell Creek Formation in south-central North Dakota (Fig. 13). They called this younger Cretaceous incursion the Cantapeta Tongue. The dynamics of shallow-water movements at the interface between the Fox Hills Sea and Sheridan (Hell Creek) Delta (Cherven and Jacob, 1985) at the end of the Cretaceous is illustrated by the complex facies relationships observed in the Breien and Cantapeta deposits. Deposition of the marginal marine sediments of the Breien Member was influenced by subtle eustatic sea-level changes and by lateral interdistributary embayment and channel migrations along the highly dissected margin of the prograding Sheridan Delta (Erickson, 1999). This depositional setting produced a laterally restricted and discontinuous Breien outcrop pattern. The depositional pattern reflected by the Breien Member continued with deposition of Cantapeta strata and was terminated in the Paleocene by advance of the Cannonball Sea.

The documentation of marine conditions as far north as North Dakota by Breien and Cantapeta strata indicates that shallow seas existed in the upper midcontinent prior to the major Paleocene advance of the Cannonball Sea. These findings are contrary to the models proposed by Lillegraven and Ostrech (1990), Roberts and Kirschbaum (1995), Archibald (1996), and Belt et al. (1997), who suggested that the seas completely receded from the upper midcontinent and shorelines retracted far south into Texas by the end of the Maastrichtian. The marine

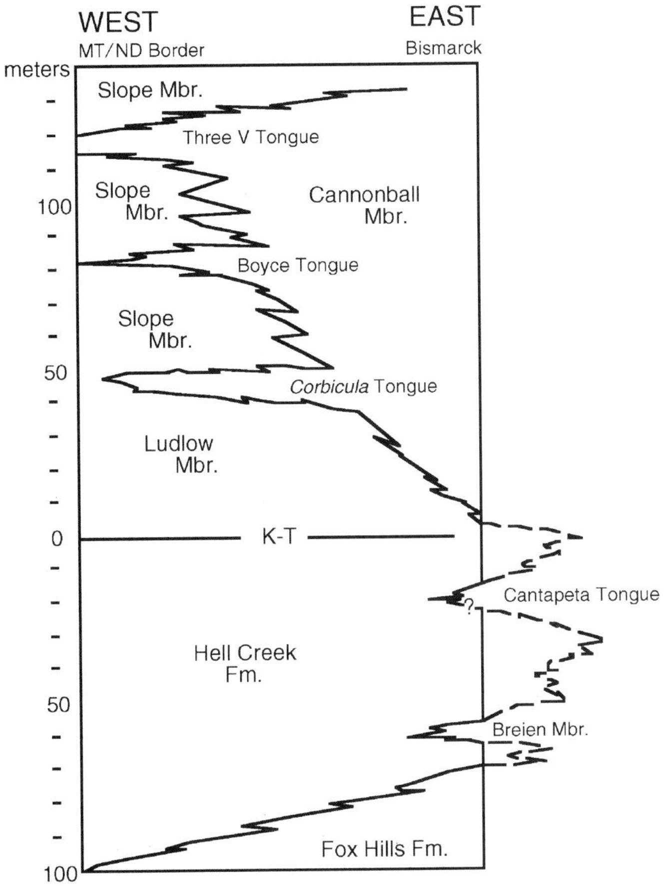

Figure 13. Diagram showing position of Breien Member in Hell Creek Formation and its stratigraphic relationship with Fox Hills Formation and Cantapeta and Cannonball tongues. Modified from Murphy et al. (this volume). MT—Montana; ND—North Dakota.

Hell Creek tongues, recognized in south-central North Dakota, support the conclusions reached by Erickson (1992, 1999) and Kennedy et al. (1998), that marine conditions lasted into the latest Cretaceous in the upper midcontinent. In this regard, the Fox Hills, Breien, Cantapeta, and Cannonball transgressive-regressive pulses should be considered a continuum of oscillating marine conditions during the Late Cretaceous and early Paleocene in the upper midcontinent prior to final withdrawal of the Cannonball Sea in the Paleocene.

EXTINCTION OF MARINE CHONDRICHTHYANS AT THE END OF THE CRETACEOUS

Although less diverse, the Breien Member marine chondrichthyan fauna exhibits close affinities to the Maastrichtian Fox Hills Formation fauna from North Dakota (Hoganson et al., 1997; Hoganson, 2000). All taxa found in the Breien, except for the sawfish *Peyeria*, have also been recovered from the Fox Hills Formation. This includes the unidentified species of sand-tiger shark, *Carcharias*, and the new species of the chimaeroid, *Ischyodus*. This species of *Ischyodus* is only known from the Breien and Fox Hills. None of the chondrichthyan species found in the Breien were reported by Hoganson and Cvancara (1989, 1991) and Cvancara and Hoganson (1993) from the Paleocene Cannonball Member of the Fort Union Formation in North and South Dakota.

Cvancara and Hoganson (1993) pointed out that all 15 chondrichthyan species present in the Cannonball Member, with the exception perhaps of *Otodus obliquus*, are restricted to Paleocene and younger strata. They interpreted their findings to lend support for significant marine chondrichthyan extinctions at the end of the Cretaceous. Hoganson et al. (1997) noted that none of the 22 chondrichthyan species identified from the Fox Hills Formation in North Dakota are found in the Cannonball Member and that none of the identified species of chondrichthyans occur in both rock units. They concluded that the end-Cretaceous extinction event had a major effect, at least at the species taxonomic level, on Western Interior Seaway shallow-water chondrichthyan faunas. The chondrichthyans that inhabited the Breien shallow-marine environment were the same species as those that existed earlier, when the Fox Hills Sea covered much of the area (Hoganson, 2000). These findings provide additional evidence that chondrichthyan species that inhabited the Western Interior Seaway during deposition of Fox Hills and Breien sediments became extinct at the end of the Cretaceous and that the Cannonball Sea was populated by new species that originated early in the Paleocene. Extinction of chondrichthyans at the end of the Cretaceous has also been documented in Hell Creek brackish-water or freshwater faunas from Montana (Bryant, 1989; Archibald and Bryant, 1990; Archibald, 1996) and southwestern North Dakota (Pearson et al., 2001; Pearson et al., this volume), in marine fossil assemblages from the United States northern Atlantic coastal plain (Gallagher, 1993) and the western United States (Eaton et al., 1989), and in European and North African marine faunas by Cappetta (1987).

CONCLUSIONS

The Breien Member of the Hell Creek Formation was deposited in interdistributary tidal channels, bays, and estuaries at the eastern edge of the prograding Sheridan Delta in south-central North Dakota. The fine-grained, glauconitic sandstone, siltstone, and mudstone of the Breien contain burrows of *Ophiomorpha*, which is often associated with the oyster *Crassostrea subtrigonalis*. Deposition of the Breien Member occurred during the Late Cretaceous, probably during the time of the *Jeletzkytes nebrascensis* Zone. Breien outcrops are scattered over a 6000 km^2 area of south-central North Dakota, a much larger area than had previously been determined. This member interfingers with nonmarine, dinosaur fossil-bearing Hell Creek strata. Fossils, other than oysters and burrows, are not common in the Breien, but include a mix of marine, brackish-water, and terrestrial taxa. The fauna includes the disarticulated remains (mostly teeth) of chondrichthyan and bony fish, crocodilians, dinosaurs, and mammals. Remains of marine chondrichthyans are the most abundant vertebrate fossils and represent several shallow-water Late Cretaceous taxa. The Breien chondrichthyan fauna contains taxa that are also found in the Upper Cretaceous Fox Hills Formation. None of the Breien chondrichthyan taxa occur in the Paleocene Cannonball Member of the Fort Union Formation. Extinction of shallow-water marine chondrichthyan species in the Western Interior Seaway at the end of the Cretaceous is supported by this finding.

ACKNOWLEDGMENTS

We thank John Stumpf, Ken Miller, Tom Hunter, Doug Lochner, Dale Rebenitsch, Lyle Grenz, Frank Armbrust, and Ruben Grenz for giving us permission to examine Breien outcrops on their property. We are grateful to the Standing Rock Sioux Nation for allowing us access to the Breien Member type locality. Johnathan Campbell and Michele Gutenkunst helped collect, prepare, and catalog fossils. We also thank Bill Cobban, Neil Landman, J. Mark Erickson, David Ward, John Hunter, Phil Currie, and Gorden Bell for examining fossils. Patricia Kelley, previously of the Department of Geology and Geological Engineering, University of North Dakota provided access to the Frye collection. We thank Brian Austin for production of the computer-generated figures. We appreciate the useful suggestions for improvement of the manuscript by J. Mark Erickson, Doug Nichols, Joseph Hartman, Kirk Johnson, and Jim Kirkland.

REFERENCES CITED

Archibald, J.D., 1996, Dinosaur extinction and the end of an era: What the fossils say: New York, Columbia University Press, 237 p.

Archibald, J.D., and Bryant, L.J., 1990, Differential Cretaceous/Tertiary extinctions of nonmarine vertebrates: Evidence from northeastern Montana, *in* Sharpton, V.L., and Ward, P.D., eds., Global catastrophes in Earth history: An interdisciplinary conference on impacts, volcanism, and mass mortality: Geological Society of America Special Paper 247, p. 549–562.

Belt, E.S., Hicks, J.F., and Murphy, D.A., 1997, A pre-Lancian regional unconformity and its relationship to Hell Creek paleogeography in southwestern Montana: Contributions to Geology (University of Wyoming), v. 31, no. 2, p. 1–26 p.

Bluemle, J.P., 1984, Geology of Emmons County: North Dakota Geological Survey Bulletin 66, 69 p.

Bryant, L.J., 1987, A new genus and species of Amiidae (Holostei; Osteichthyes) from the Late Cretaceous of North America, with comments on the phylogeny of the Amiidae: Journal of Vertebrate Paleontology, v. 7, no. 4, p. 349–361.

Bryant, L.J., 1989, Non-dinosaurian lower vertebrates across the Cretaceous–Tertiary boundary in northeastern Montana: University of California Publications in Geological Sciences, v. 134, 107 p.

Cappetta, H., 1987, Extinctions et renouvellements fauniques chez les Selaciens post-jurassiques: Mémoires de la Société Géologique de France, Nouvelle Série, no. 150, p. 113–131.

Cappetta, H., and Case, G., 1975, Contributions à l'étude des sélaciens du groupe Monmouth (Campanian-Maastrichtian) du New Jersey: Palaeontographica. Abteilung A: Palaozoologie-Stratigraphie 15, p. 1–46.

Carlson, C.G., 1982, Geology of Grant and Sioux Counties, North Dakota: North Dakota Geological Survey Bulletin, no. 67, pt. 1, 32 p.

Cherven, V.B., and Jacob, A.R., 1985, Evolution of Paleogene depositional systems, Williston Basin in response to global sea level changes, *in* Flores, R.M., and Kaplan, S.S., eds., Cenozoic paleogeography of the west-central United States: Society of Economic Paleontologists and Mineralogists, Rocky Mountain Section, p. 127–170.

Cvancara, A.M., and Hoganson, J.W., 1993, Vertebrates of the Cannonball Formation (Paleocene) in North and South Dakota: Journal of Vertebrate Paleontology, v. 13, no. 1, p. 1–23.

Eaton, J.G., Kirkland, J.I., and Doi, K., 1989, Evidence of reworked Cretaceous fossils and their bearing on the existence of Tertiary dinosaurs: Palaois, v. 4, p. 281–286.

Erickson, J.M., 1974, Revision of the Gastropoda of the Fox Hills Formation, upper Cretaceous (Maestrichtian) of North Dakota: Bulletin of American Paleontology, v. 66, no. 284, p. 131–253.

Erickson, J.M., 1992, Subsurface stratigraphy, lithofacies, and paleoenvironments of the Fox Hills Formation (Maastrichtian: Late Cretaceous) adjacent to the type area, North Dakota and South Dakota: Toward a more holistic view, *in* Erickson, J.M., and Hoganson, J.W., eds., Proceedings of the F.D. Holland, Jr., Geological Symposium: North Dakota Geological Survey, Miscellaneous Series 76, p. 199–241.

Erickson, J.M., 1999, The Dakota Isthmus: Closing the Late Cretaceous Western Interior Seaway: North Dakota Academy of Science, Proceedings, v. 53, p. 124–129.

Estes, R., 1964, Fossil vertebrates form the Late Cretaceous Lance Formation eastern Wyoming: University of California Publications in Geological Sciences, v. 49, 183 p.

Estes, R. 1969, Studies of fossil phyllodont fishes: Interrelationships and evolution in the Phyllodontidae (Albuloidei): Copeia, no. 2, p. 317–331.

Estes, R., and Berberian, P., 1970, Paleoecology of a Late Cretaceous vertebrate community from Montana: Breviora-Museum of Comparative Anatomy, no. 343, 35 p.

Estes, R., and Hiatt, R., 1978, Studies of fossil phyllodont fishes: A new species of *Phyllodus* (Elopiformes, Albuloidea) from the Late Cretaceous of Montana: PaleoBios, no. 28, p. 1–10.

Frye, C.J., 1964, Marine tongues in the Hell Creek Formation, North Dakota: The Compass, v. 41, no. 2, p. 167–171.

Frye, C.J., 1967, The Hell Creek Formation in North Dakota [Ph.D. thesis]: Grand Forks, University of North Dakota, 411 p.

Frye, C.J., 1969, Stratigraphy of the Hell Creek Formation in North Dakota: North Dakota Geological Survey Bulletin, no. 54, 65 p.

Gallagher, W.B., 1993, The Cretaceous/Tertiary mass extinction event in the northern Atlantic coastal plain: The Mosasaur, v. 5, p. 75–154.

Goody, P.C., 1976, Enchodus (Teleostei: Enchodontidae) from the Upper Cretaceous Pierre Shale of Wyoming and South Dakota with an evaluation of the North American enchodontid species: Palaeontographica. Abteilung A: Palaozoologie-Stratigraphie, v. 152, p. 91–112.

Grande, L., and Bemis, W.E., 1998, A comprehensive phylogenetic study of amiid fishes (Amiidae) based on comparative skeletal anatomy: An empirical search for interconnected patterns of natural history: Society of Vertebrate Paleontology Memoir 4, (Supplement to Journal of Vertebrate Paleontology, v. 18, no. 1), 690 p.

Hoganson, J.W., 2000, Additional evidence for the extinction of cartilaginous fish at the end of the Cretaceous from the marine Breien Member (Maastrichtian) of the Hell Creek Formation, North Dakota, USA [abs.]: Journal of Vertebrate Paleontology, v. 20, (supplement to no. 3), p. 49A.

Hoganson, J.W., and Cvancara, A.M., 1989, Vertebrate fauna of the Cannonball Sea: The last (Paleocene) transgression into central North America: Geological Society of America Abstracts with Programs, v. 21, no. 6, p. 4310.

Hoganson, J.W., and Cvancara, A.M., 1991, Biochronologic and paleobiogeographic significance of fishes from the marine Cannonball Formation (Paleocene), North and South Dakota [abs.]: Journal of Vertebrate Paleontology, v. 11, (supplement to no. 3), p. 35A.

Hoganson, J.W., Campbell, J.M., and Murphy, E.C., 1994, Stratigraphy and paleontology of the Cretaceous Hell Creek Formation, Stumpf site, Morton County, North Dakota: North Dakota Academy of Science, Proceedings, v. 48, p. 95.

Hoganson, J.W., Erickson, J.M., Cvancara, A.M., and Holland, F.D., Jr., 1997, Terminal Cretaceous extinction event documented by marine cartilaginous fishes from the Fox Hills (Maastrichtian) and Cannonball (Danian/Thanetian) Formations, North Dakota [abs.]: Journal of Vertebrate Paleontology, v. 17, (supplement to no. 3), p. 53A.

Kennedy, W.J., Landman, N.H., Christensen, W.K., Cobban, W.A., and Hancock, J.M., 1998, Marine connections in North America during the late Maastrichtian: Palaeogeographic and palaeobiogeographic significance of *Jeletzkytes nebrascensis* Zone cephalopod fauna from the Elk Butte Member of the Pierre Shale, SE South Dakota and NE Nebraska: Cretaceous Research, v. 19, p. 745-775.

Kent, B.W., 1994, Fossil sharks of the Chesapeake Bay region: Columbia, Maryland, Egan, Rees, and Boyer, Inc., 146 p.

Kume, J., and Hansen, D.E., 1965, Geology and ground water resources of Burleigh County, North Dakota. 1. Geology: North Dakota Geological Survey Bulletin, no. 42, 111 p.

Laird, W.M., and Mitchell, R.H., 1942, The geology of the southern part of Morton County, North Dakota: North Dakota Geological Survey Bulletin, no. 14, 42 p.

Landman, N.H., and Waage, K.M., 1993, Scaphitid ammonites of the upper Cretaceous (Maastrichtian) Fox Hills Formation in South Dakota and Wyoming: Bulletin of the American Museum of Natural History, v. 215, 257 p.

Lillegraven, J.A., and Ostresh, L.M., Jr., 1990, Late Cretaceous (earliest Campanian/Maastrichtian) evolution of western shorelines of the North American Western Interior Seaway in relation to known mammalian faunas, *in* Bown, T.M., and Rose, K.D., eds., Dawn of the Age of Mammals in the northern part of the Rocky Mountain Interior, North America: Geological Society of America Special Paper 243, p. 1–30.

Moore, W.L., 1976, The stratigraphy and environments of deposition of the Cretaceous Hell Creek Formation (reconnaissance) and the Paleocene Ludlow Formation (detailed), southwestern North Dakota: North Dakota Geological Survey Report of Investigations, v. 56, 40 p.

Murphy, E.C., Nichols, D.J., Hoganson, J.W., and Forsman, N.F., 1995, The Cretaceous/Tertiary boundary in south-central North Dakota: North Dakota Geological Survey Report of Investigations, v. 98, 73 p.

Noubhani, A., and Cappetta, H., 1997, Les Orectolobiformes, Carcharhiniformes et Myliobatiformes (Elasmobranchii, Neoselachii) des bassins à phosphate du Maroc (Maastrichtien-Lutétien basal): Palaeo Ichthyologica, v. 8, p. 1–327.

Pearson, D.A., Schaefer, T., Johnson, K.R., and Nichols, D. J., 2001, Palynologically calibrated vertebrate record from North Dakota consistent with abrupt dinosaur extinction at the Cretaceous–Tertiary boundary: Geology, v. 29, no. 1, p. 39–42.

Roberts, L.N.R., and Kirshbaum, M.A., 1995, Paleogeography of the Late Cretaceous of the Western Interior of middle North America: Coal distribution and sediment accumulation: U.S. Geological Survey Professional Paper 1561, 115 p.

Seager, O.A., Blackstone, D.L., Jr., Cobban, W.A., Downs, G.R., Laird, W.M., and Sloss, L.L., 1942, Stratigraphy of North Dakota: Discussion: American Association of Petroleum Geologists Bulletin, v. 26, no. 8, p. 1414–1423.

Stahl, B.J., 1999, Handbook of Paleoichthyology: Chondrichthyes 3: Holocephali: Munchen, Verlag Dr. Friedrich Pfeil, 164 p.

Welton, B.J., and Farish, R.F., 1993, The collector's guide to fossil sharks and rays from the Cretaceous of Texas: Dallas, Texas, Horton Printing Company, 204 p.

Manuscript Accepted by the Society September 24, 2001

Printed in the U.S.A.

Geological Society of America
Special Paper 361
2002

Brackish and marine mollusks of the Hell Creek Formation of North Dakota: Evidence for a persisting Cretaceous seaway

Joseph H. Hartman*
University of North Dakota, Department of Geology and Geological Engineering and Energy & Environmental Research Center, P.O. Box 8358, Grand Forks, North Dakota 58202-8358, USA
James I. Kirkland
Utah Geological Survey, P.O. Box 146100, Salt Lake City, Utah 84114-6100, USA

ABSTRACT

Molluscan species associations reflecting varying salinities in the final North American interior seaway occur in Cretaceous to Paleocene strata. Deposits of the last few million years of the Cretaceous sea record an open marine environment characterized by species-rich faunal associations documenting lowering levels of the Western Interior Sea. Relatively diverse molluscan associations at the beginning of the deposition of the Fox Hills Formation are dramatically reduced in both bivalves and gastropods in the younger brackish-water lithofacies of the formation, indicating continued loss of habitat. Marginal marine to marine conditions returned to the area of central North Dakota and South Dakota, as recorded in the transition deposits of the Fox Hills into the lower Hell Creek Formation. More fully marine (mesohaline to brachyhaline) conditions are documented by a relatively diverse molluscan fauna in south-central North Dakota, as evidenced in the Breien Member of the Hell Creek Formation. The short-lived species richness recorded in the Breien strata is overlain by evidence of isolated rare molluscan and arthropod fossils in the upper, but not uppermost, part of the Hell Creek. These molluscan faunules contain individual ammonite, scaphopod, and bivalve taxa that indicate a resurgence of westward shoreline advance or fluctuation in shoreline configuration and continued influence of sublittoral marine conditions. Thus, as reflected in the deltaic depositional history of the Hell Creek Formation through much of its thickness, the molluscan record of the last 1–2 m.y. of the Cretaceous indicates environmental settings characteristic of the presence of an adjacent interior sea.

LAST MARINE DEPOSITS OF THE CRETACEOUS

Overview

One of the basic questions of the geologic history of the Cretaceous is the movement of the shoreline of the Western Interior Seaway. There has been a long-held perception that the last Cretaceous sea is represented by the Fox Hills Formation and its constituent fossils in the northern Great Plains. The regressive Fox Hills marine and marginal marine deposits were derived from penecontemporaneous river-dominated sediments

*E-mail: joseph_hartman@und.nodak.edu

Hartman, J.H., and Kirkland, J.I., 2002, Brackish and marine mollusks of the Hell Creek Formation of North Dakota: Evidence for a persisting Cretaceous seaway, *in* Hartman, J.H., Johnson, K.R., and Nichols, D.J., eds., The Hell Creek Formation of the northern Great Plains: An integrated continental record of the end of the Cretaceous: Boulder, Colorado, Geological Society of America Special Paper 361, p. 271–296.

of the Hell Creek Formation to the west of the retreating sea. The Hell Creek Formation, along with the overlying mostly Paleocene Fort Union Formation (Fig. 1), represents a wedge of progradational, largely unconsolidated sediments derived from the uplifting Rocky Mountains near the beginning of the Laramide orogeny. Strata of the Hell Creek and Fort Union Formations thin from west to east and intercalate with marine strata from the base of the Hell Creek during the Maastrichtian up into the early late Paleocene (Selandian) (Hartman et al., 1998, 1999) and are represented by an erosional edge in the central Dakotas along the eastern edge of the Williston Basin. Although previously reported, knowledge of the continuation of marine conditions above the Fox Hills Formation is not well or widely known. The last interior marine deposits and fossils of the Cretaceous are present in the Hell Creek Formation (1) at or near its base in South Dakota and North Dakota, (2) as a relatively fossiliferous marine tongue (Breien Member) in the lower part of the Hell Creek in North Dakota, and (3) as tidally influenced distributary deposits in the upper part of the Hell Creek Formation in North Dakota.

The presence of marine waters very near or at the end of the Cretaceous seems a likely possibility. Subsurface geophysical logs interpreted by R. Butler (1999, personal commun.; see also Butler and Hartman, 1999) indicate deltaic deposits in the uppermost part of the Hell Creek near the southeastern edge of the Williston Basin in North Dakota. In addition, sediments and fossils show the return of marine waters in central North Dakota during the early Paleocene (plankton foraminifera zone P1a; Hartman et al., 1999; Lund and Hartman, 1999; Lund et al., this volume). These marine deposits are of the Cannonball Member of the Fort Union Formation and represent the last interior marine incursion of the Cenozoic. On the basis of this study and the work of Hoganson and Murphy (this volume and references cited therein), we argue that an interior sea persisted throughout much of the end of the Cretaceous, with, at most, a brief hiatus of marine deposition in the interval immediately bounding the Cretaceous-Tertiary (K-T) boundary. Thus a generally perceived dramatic regression of the Western Interior Sea and a completely nonmarine K-T stratal interval of 1–2 m.y. did not occur in the central interior of North America.

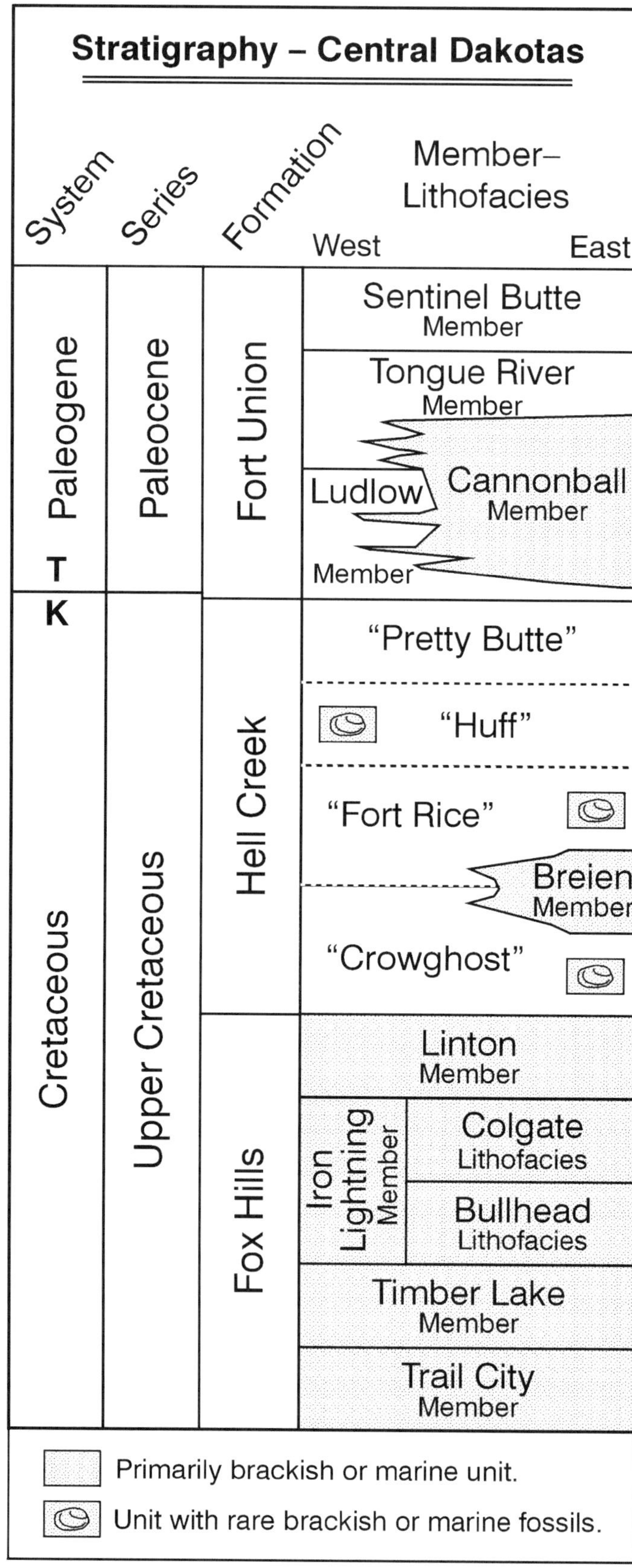

Figure 1. Stratigraphy of Upper Cretaceous and Paleocene strata in south-central North Dakota and north-central South Dakota on eastern margin of Williston Basin (not to scale). This diagram provides nomenclature and horizon of marine conditions (stippled pattern) for Fox Hills, Hell Creek, and Fort Union Formations. Stippled blocks in pre-Breien (Crowghost) and post-Breien (Fort Rice and Huff) subdivisions represent isolated evidence of brackish water or marine waters in what are otherwise interpreted as nonmarine strata. Except for Breien Member, names applied to subdivisions of Hell Creek are informal. Depiction of Paleocene strata is after Hartman (1999) and Hartman et al. (1998, 1999). K-T is Cretaceous-Tertiary.

Purpose

Brackish mollusks of the Hell Creek Formation, although known since 1942 (Seager et al., 1942), have only been reported informally (Frye, 1967) and until now have never been illustrated (see Plates 1–3). Along with another paper in this volume (Hoganson and Murphy) primarily concerning the vertebrates of the Breien Member of the Hell Creek Formation, we provide the current state of knowledge regarding the latest known marginal marine macrofossil assemblages of the Cretaceous in the Western Interior of North America. Our molluscan results are derived significantly from the collections C.I. Frye made in the early 1960s while conducting his dissertation field studies of the Hell Creek Formation in North Dakota (Appendix 1). Frye's subsequently controversial lithostratigraphic interpretations were published in 1969, but excluded his major effort to identify the fauna and flora represented in his extensive collections from the Hell Creek Formation (Fig. 2, A and B). Thus a major purpose of this article, ~30 yr after the completion of his dissertation, is to document and illustrate the brackish-water and marine mollusks collected by Frye. With new identifications, we are able to note specific trends in molluscan faunal associations in marine-influenced depositional environments during the deposition of the Hell Creek Formation to very near the end of the Cretaceous.

GENERAL STRATIGRAPHY AND LOCATION

North Dakota

The introduction of the name Hell Creek for strata between the underlying Fox Hills and the overlying Fort Union Formations by Brown (1907), and its use in North Dakota, is reviewed in the introductory chapter to this volume (Hartman et al.). Subsequently, a marine tongue intercalated with the predominantly nonmarine deposits of the Hell Creek Formation was named the Breien Member by Laird and Mitchell (1942) for exposures along the Cannonball River near the town of Breien in Morton County, North Dakota. The introduction of the Breien Member is locally useful as a means to order pre-Breien and post-Breien stratigraphic intervals of the Hell Creek (Fig. 1). Frye (1969) formally introduced a complete subdivision of the Hell Creek Formation in North Dakota and suggested lithostratigraphic correlation of certain members into east-central Montana. In south-central North Dakota, Frye (1969) delineated stratotypes and other sections for the Crowghost, Breien, Fort Rice, Huff, and Pretty Butte units, while the stratotype of the Pretty Butte unit was designated in Slope County in far southwestern North Dakota.

Butler (1980) made extensive use of Frye's subdivisions in the well-exposed badlands of the Hell Creek Formation in Makoshika State Park, Dawson County, Montana. As one of the few geologists to embrace Frye's nomenclature, Butler's dissertation concerned the depositional history of the Hell Creek Formation throughout the park and described and mapped in detail the sedimentology of each member. Subsequently, however, workers in North Dakota and eastern Montana have found the members difficult or impossible to recognize as mappable units. As a typical expression of this condition, Carlson (1982, p. 12) stated "the Hell Creek is characterized by a general lack of continuity of beds and a heterogeneity so that subdivision into members is not very useful for either surface or subsurface interpretations" and added "it is difficult to trace these units away from their type areas." Murphy et al. (this volume) strongly and effectively argue that Frye's members should be abandoned. In preliminary work, Butler and Hartman (1999) supported an effort to recognize a sedimentological framework for the Hell Creek that reflected changes in depositional architecture in response to changes in base level as recorded in lithic packages through the Hell Creek. This work was based on the interpretation of subsurface oil and gas logs, but is analogous to Waage's (1968) recognition of environmental shifts through the Fox Hills Formation in South Dakota. The depositional architecture suggested by Butler and Hartman (1999) most likely reflects broader changes in depositional patterns than were envisaged by Frye (1969).

For the purposes of this chapter, the use of Frye's lithic units represents the only method to communicate the relative stratal position of the mollusks collected by him in the Hell Creek Formation. The use of his names here is not intended as validation, but simply a means to subdivide the Hell Creek using Frye's collections that cannot as yet be referenced to a formational contact or other marker bed. Frye's names are hereafter referred to as informal lithic units (e.g., Huff unit).

In south-central North Dakota, strata beneath the well-established Breien Member of Laird and Mitchell (1942) represent the basal strata of the Hell Creek Formation. Frye (1969) assigned the name Crowghost to these strata, after Crowghost Cemetery on the Standing Rock Indian Reservation in Sioux County, North Dakota. The Crowghost unit was observed by Frye to lie either on the Colgate or Bullhead lithofacies of the Iron Lightning Member of the Fox Hills Formation (the Linton Member was not yet introduced). Frye also noted that the Crowghost and Bullhead units are intercalated at least at one location in Morton County, North Dakota. Regardless of nomenclature (Bullhead versus Linton), Frye's observations were consistent with Waage's (1968) interpretation of the locally complex Fox Hills–Hell Creek formational contact. Frye recognized the Fort Rice and Huff units overlying the Breien Member. The Fort Rice unit, named after an old fort in Morton County, is 12–18 m thick in the study area and gradational with the overlying Huff unit. The Huff unit, which is named after the town of Huff along the Missouri River in Morton County, is ~26 m thick in the study area and is gradational with the overlying Pretty Butte unit. The Pretty Butte unit, thought by Frye to be the most extensive of his units, represents the remaining 3–8 m of the Hell Creek Formation.

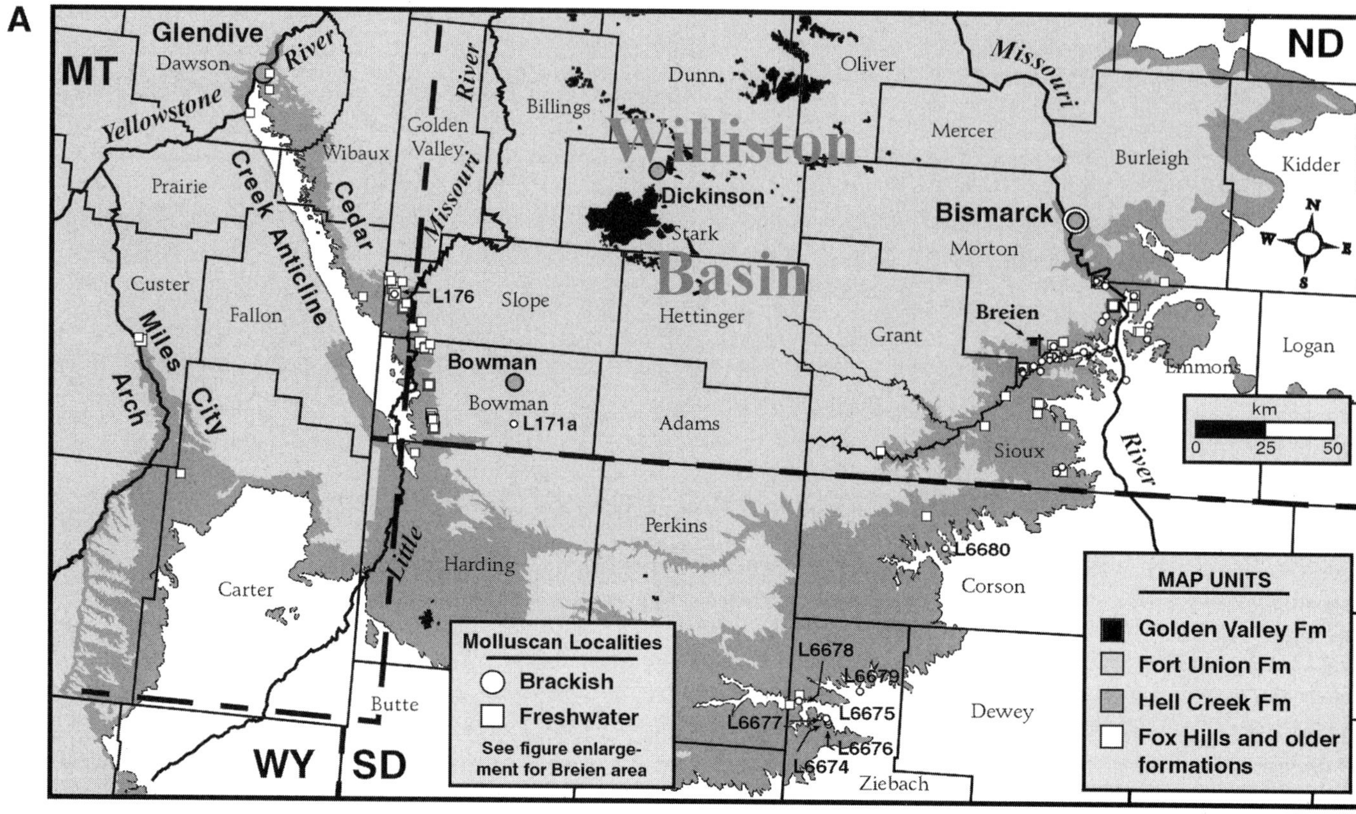

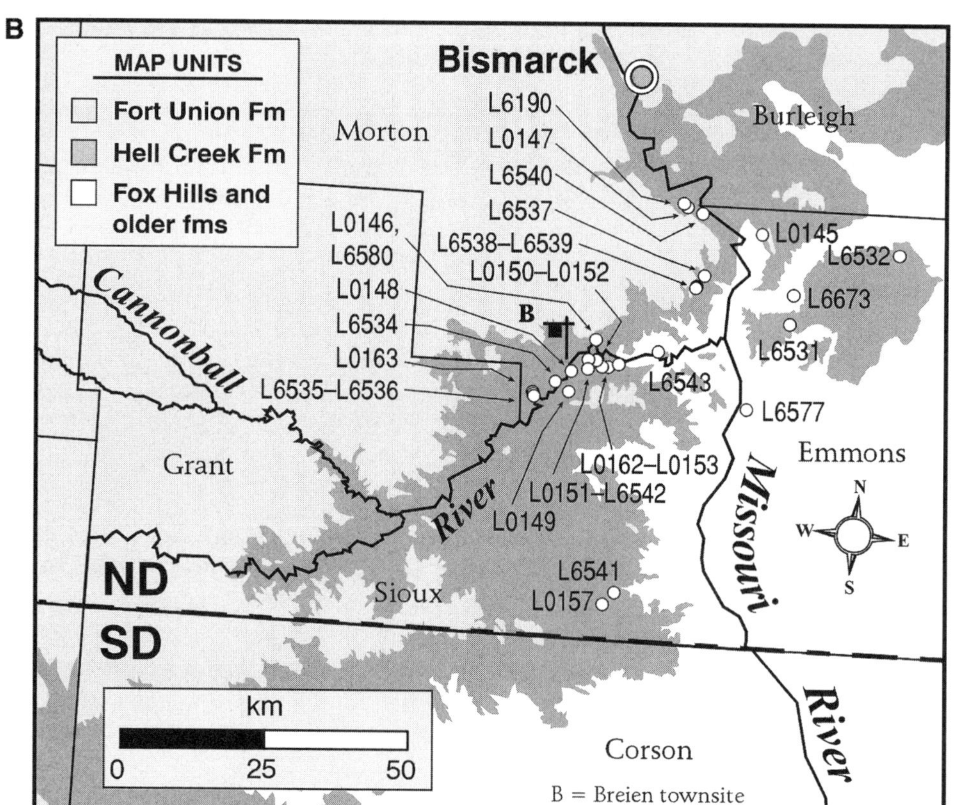

Figure 2. A: Distribution of molluscan localities in uppermost Cretaceous Hell Creek Formation of southeastern portion of Williston Basin, North Dakota, South Dakota, and Montana. Geologic base map for this figure is after Hartman and Kihm (1995; including information from S.M. Vuke (1993, 1994, written commun.) of Montana Bureau of Mines and Geology and Forster Sawyer (1994, written commun.) of South Dakota Geological Survey). This map displays distribution of brackish-water and marine molluscan localities of Hell Creek Formation in North Dakota and South Dakota in context of freshwater molluscan localities known from formation. Locality number (L number) is part of locality database system used by Hartman for North American molluscan and other collections. Additional locality numbers (e.g., L0145) are shown in B, which represents enlargement of Breien (B; location indicated by signpost) area of south-central North Dakota. Location data for brackish-water and marine localities of North and South Dakota are given in Appendices 1 and 3, respectively. B: Distribution of uppermost Cretaceous brackish-water and molluscan localities in south-central North Dakota (detail of A). Locality numbers (e.g., L0145) are documented in Appendix 1 (see A for symbols and additional explanation).

South Dakota

Although the focus of this chapter is on the record of brackish-water faunas from the Hell Creek Formation of North Dakota, discussion of the depositional history of the Fox Hills Formation is necessary to highlight the transition from shallow open marine shelf conditions to fully terrestrial environments. The lower Fox Hills Formation is relatively species rich when compared to a dramatic reduction in species diversity in the upper part of the formation. An influx of sediments into high-energy, nearshore, and brackish environments in the upper Fox Hills was replaced by complex fluvially dominated environments, apparently for the remainder of the Cretaceous.

The Fox Hills Formation represents marine, marginal marine, brackish, and less common nonmarine deposits associated with a regressing sea during the Maastrichtian. The name Fox Hills was introduced by Meek and Hayden (1861) for exposures between the Moreau and Cheyenne Rivers in Ziebach and Dewey Counties, South Dakota (Waage, 1967, his Fig. 1). Meek and Hayden did not delimit the Hell Creek Formation, but they clearly recognized the transition from marine to nonmarine environments in the northern plains in their use of the Fox Hills and Fort Union Formations.

Subdivision of the Fox Hills Formation was first undertaken using the distinctive lithofacies exposed in the breach of the Cedar Creek Anticline, Dawson County, Montana. Calvert et al. (1914) introduced the Colgate Member for a distinctively fluted, white sandstone near Colgate Station. The use of this name in North Dakota and South Dakota was redefined by Waage (1968) to represent a lithofacies within the Iron Lightning Member (discussed in the following).

Morgan and Petsch (1945) introduced the basic stratigraphic framework of the Fox Hills Formation in its type area; they named the Trail City and Timber Lake Members after towns in Corson and Dewey Counties, respectively, and recognized what would become the Timber Lake and part of the Iron Lightning Members. The Trail City Member crops out in the breaks of the Grand and Moreau Rivers, and the Timber Lake Member covers broad upland areas. Fisher (1952) first correlated these members with the North Dakota section. The Bullhead Member was introduced by Stevenson (1956) for strata interpreted as gradational between the Timber Lake and Colgate Members in exposures in the Grand River watershed near the Bullhead Indian village, Corson County, South Dakota. Waage (1968) recognized the Bullhead Member as a lithofacies of the Iron Lightning Member.

In the 1960s and early 1970s, a number of field studies reinterpreted the paleoenvironmental history of the Fox Hills Formation. Waage (1968) recognized the validity of the Trail City Member in South Dakota, but distinguished depositional environments within the member by identifying the Irish Creek and Little Eagle lithofacies. The Timber Lake Member was recognized as the sandstone-bearing unit of the lower part of the Fox Hills Formation, which could be subdivided into biofacies based on fossil content. The Bullhead and Colgate lithofacies of earlier usage were recognized by Waage (1968) as, at times, both lateral and superposed lithofacies of his Iron Lightning Member. Waage named the Iron Lightning Member for exposures along the Moreau River in Ziebach County, South Dakota. The Bullhead lithofacies of the Iron Lightning Member has a sharp, diachronous contact with the sandstones of the underlying Timber Lake Member (Waage, 1968), reflecting a shift in environmental conditions through time. A lignite, lignitic, or carbonaceous unit, or distinctive color change, representing the transition from marine to generally nonmarine environments, are typical criteria used to mark the contact between the Iron Lightning Member and the overlying Hell Creek Formation.

The Colgate lithofacies of the Iron Lightning Member represents a special stratigraphic problem (see Klett and Erickson, 1976). The Colgate in the type area of the Fox Hills Formation is a volcanic-rich, poorly sorted, white to grayish-white sandstone with distinctive fluted exposures. Thicker outcrops of the cross-bedded sandstones are near the top of the formation and are overlain by brown clays of the Hell Creek Formation. Klett and Erickson (1976) argued that the interpretation of the Colgate lithofacies by Waage (1968) in South Dakota and by Feldmann (1972) in North Dakota was too inclusive, containing lithofacies reflecting different environments and times of deposition. Klett and Erickson (1976) introduced the Linton Member for volcanic-rich, poorly sorted, light green to gray sandstones capping buttes throughout its distribution in Emmons and Sioux Counties, North Dakota, and Corson County, South Dakota. It was named for exposures in a row of buttes north of the town of Linton in Emmons County. According to Klett and Erickson (1976), the Linton Member represents the uppermost beds of the Fox Hills Formation, but the member's relation to underlying units is not well known because of poor exposures. In addition to lithic differences, the faunas of the units appear to be distinct. Neither unit is particularly fossiliferous, but the Colgate lithofacies contains extensive plant debris, a limited distribution of *Crassostrea* and *Corbicula* lenses (Waage, 1968), and local occurrences of freshwater caenogastropod and pulmonate mollusks (Hartman's own collections). The Linton Member is largely without hard-part fossils, but large specimens of *Ophiomorpha* are relatively common (Klett and Erickson, 1976).

HELL CREEK MARINE STUDIES IN NORTH DAKOTA

In the first report of marine fossils between the Fox Hills Formation and Cannonball Member of the Fort Union Formation, Seager et al. (1942, p. 1418) reported William Cobban's discovery of a "30-foot [9.1 m] marine bed . . . near the base of the Hell Creek" (locality L0152, Appendix 1). This bed contains the crustacean trace fossil *Ophiomorpha,* the oyster *Crassostrea subtrigonalis,* the heterodont bivalve *Dosiniopsis?,* and the inarticulate brachiopod *Lingula.* This faunule documents marine and brackish-water fossils overlying the Fox Hills Formation and within the Hell Creek Formation. A more extensive documentation of the marine deposition in the Hell Creek was given by Laird and Mitchell (1942). They chose the name Breien for the member to represent the stratigraphic interval exposed along the Cannonball River in the vicinity of the village of Breien in Morton County and adjacent strata in Sioux County, North Dakota.

On the basis of Cobban's section and the field studies of Laird and Mitchell and their field partners (see Laird and Mitchell, 1942), the Breien was recognized primarily as two beds of limonitic concretion-bearing, gray, fine-grained sandstone interbedded with a gray bentonite. The sandstones are *Ophiomorpha* rich, and distinctive horizons and may reflect a cyclic pattern of deposition. They recognized the relatively thin Breien as of unusual interest because it represented a brief return of marine conditions. Laird and Mitchell (1942, p. 15; brackets of present authors) stated that "as the member has a small east-west extension but is found to the south of this area [in Sioux County], it would appear that a seaway from the south or east apparently invaded this territory for a short distance [to the west] . . . This return of the sea interrupted the typical continental deposition of the Hell Creek in this area. After a comparatively short time the sea again retreated and conditions of continental deposition resumed. This marine member suggests that marine conditions were not far distant from this area from Fox Hills time until the end of the [Paleocene] Cannonball deposition." These concluding comments are supported by successive researchers of the fossil record of the Breien Member (see references cited in Erickson, 1992).

More than 20 years passed without new data or interpretations concerning the marine deposits of the Hell Creek Formation. Frye's (1967, 1969) studies at the University of North Dakota focused on the lithostratigraphic interpretation of the Hell Creek Formation along the southern rim of the Williston Basin in North Dakota. His sections (>70) represented the most comprehensive study of the Hell Creek Formation until that time. Sections were measured primarily in south-central and southwestern North Dakota; a few were measured in the type area of the Hell Creek in Garfield County, Montana. During his field studies, Frye also made an extensive collection of mollusks, vertebrates, and plants. Although the collection and documentation of his freshwater and terrestrial fossils have now largely been superseded by the work presented in this volume, Frye's collection of fossils reflects an outstanding effort, because few to no such fossils had been previously reported in North Dakota; the collections were reported in his dissertation in the form of locality faunules and lists. Frye's identifications of brackish-water molluscan taxa, although in need of revision, were sufficient to deduce marine-influenced deposition through most of the Hell Creek Formation.

While undertaking his studies, Frye (1964, p. 171) briefly reported on marine tongues in the Hell Creek Formation and made similar conclusions to those of today; he stated that the Hell Creek Formation was "deposited very close to sea level" and that the "wide extent of rather thin marine units would indicate that there was very little relief in North Dakota" during the latest Cretaceous. Frye (1964, 1967, 1969) reported that the Breien Member consists of ~5.8 m of dominantly glauconitic sandstones containing *Ophiomorpha* and interbeds of claystone and limonitic concretions, representing a sedimentary package reflecting changes in sea base level. Frye (1969) recognized that the lower contact with the basal Hell Creek Formation (Crowghost unit) is sharp, while the overlying contact with the Fort Rice unit (or main body) of the Hell Creek Formation is less distinct, but also sharp, with lignitic shales representing more fluvially dominated environments. Frye (1969, p. 36) stated that the two most common fossils in the Breien Member are *Ophiomorpha* (Frye's *O. major*) and *Crassostrea subtrigonalis* (Frye's *Ostrea glabra*), and that "these fossils were found at nearly every location and would lead one to believe that probably all of the terrestrial and most of the fresh-water fossils were washed into a marine or brackish-water environment." Although the admixture of freshwater mollusks is relatively uncommon in the Breien Member, Frye stated that the "relatively good preservation of the fresh-water [sic] and terrestrial fossils would indicate that these fossils had not been transported very far and would suggest that the Breien Member originated in an estuarine or lagoonal environment" (Frye, 1969, p. 36). On the basis of an examination of the collections, we are uncertain what faunal elements Frye considered to be of a terrestrial nature.

Frye (1964, 1969) clearly recognized a brackish-water and/or marine fauna from strata overlying the Breien Member. Frye (1964, p. 170) observed "brackish water beds near the top of the Hell Creek Formation in Slope and Bowman Counties." These beds were described as yellow-gray, medium-grained, cross-bedded channel sandstones interbedded with gray bentonites. These records are interpreted here as the Huff unit of Frye. On the basis of additional studies, Frye (1969) reported fossils from strata overlying the Breien Member (Fort Rice unit) in south-central North Dakota.

Frye (1969) reported taxa from both the post-Breien strata (Fort Rice and Huff units) characteristic of marine waters. His Fort Rice unit includes a relatively diverse nonmarine fauna, but also contains a scaphopod *Dentalium* and the ammonite *Hoploscaphites?* (Frye's *Discoscaphites*). We believe that Frye's (1967, 1969) other locality (L0157), from which he identified the brackish-water taxa, *Ostrea* and *Corbicula,* is more likely from higher in the section (Huff unit; see Appendix 1), and the specimens identified here are a possible freshwater unionid and the marine heterodont *Clisocolus,* most likely reworked from coeval deposits.

The taxa from brackish environments from strata in the valley of the Little Missouri River in southwestern North Dakota are of particular interest because of their relatively high stratigraphic position and their geographic location, inland compared to the record of the Breien marine tongue. Frye

(1964) first recognized this higher marine biofacies based on the presence of *Ophiomorpha* (Frye's *Halymenites*) and skate teeth. Frye (1964, p. 171) stated that the faunule represented "a death assemblage deposited near the mouth of a river in which salt water had invaded." He considered this occurrence distinct from the Breien Member and noted that although similar lithologies exist in south-central North Dakota south of Bismarck, similar fossils have yet to be found (discussed in the following).

With the completion of his dissertation studies, Frye (1967) recognized brackish conditions elsewhere in North Dakota on the basis of the crustacean trace fossil *Ophiomorpha* (localities L171a and L6191) in Bowman County; a possible brackish-water bivalve *Corbula* (locality L176) in Slope County; "skate" teeth of *Myledaphus* (localities L175 and L179) in Slope County; and a shark (cf. *Lamna*) (locality L178) in Slope County (see Fig. 3). The *Ophiomorpha* appears to be *O. irregulaire* (Frey et al., 1978; D.J. Daly, 2001, written commun.), and the *Corbula* is probably *Spaniorinus,* a heterodont bivalve. The skate teeth represent a freshwater ray present elsewhere in the Hell Creek in North Dakota and Montana. The records of *Lamna,* like *Myledaphus,* most likely do not indicate local marine or brackish conditions, because the associated local fauna is distinctly freshwater, but may indicate the propinquity to an as yet unmapped shoreline of the interior sea. Although limited in numbers, the molluscan occurrence and other faunal elements, along with similar inferences from fossils reported by Hoganson and Murphy (this volume) and Pearson et al. (this volume), indicate the continued presence of marine conditions lateral to the fluvial-deltaic environments of the Hell Creek Formation during the latest Cretaceous.

MARINE TRANSITION FROM THE FOX HILLS TO THE HELL CREEK

Molluscan record of the Fox Hills

The most closely related molluscan associations to those of the Hell Creek Formation are found in the subjacent Fox Hills Formation of South Dakota and North Dakota. Waage (1968) and Speden (1970) provided a detailed interpretation of the stratigraphic context for the formation and molluscan bivalve systematics in South Dakota. Feldmann (1967) and Erickson (1971) presented studies of the bivalves and gastropods, respectively, of the Fox Hills Formation in North Dakota. These and later works by these authors form a basis to interpret Cretaceous shoreline movements of the Western Interior Sea during the deposition of the Fox Hills Formation and its continuation as incursions into nonmarine Hell Creek deposition as represented by the Breien Member and other strata upsection in the upper part of the formation.

To facilitate a comparison between the Fox Hills and the basal Hell Creek faunal associations, the record of the molluscan taxa and their distribution in the Fox Hills Formation was compiled using the members and other lithic units described here (Fig. 4; see Appendix 3). Co-occurring Fox Hills and Hell Creek taxa are shown by the letter Y. Note that the vertical ordering of the members of the Fox Hills in Figure 4 oversimplifies the lithofacies relationships mentioned earlier for the upper part of the formation. The Colgate is recognized as a lithofacies within the Iron Lightning, but is more common in the upper part of the member (see Fig. 3 of Speden, 1970). The greater species richness of the Fox Hills marine record is shown by tabulating the number of Fox Hills taxa not present in the Hell Creek Formation and entering this number under "other Fox Hills bivalves, gastropods, and cephalopods" on the right side of Figure 4.

The Trail City and Timber Lake Members of the lower part of the Fox Hills Formation represent marine platform deposits that contain numerous molluscan species derived largely from limestone concretions (Waage, 1964). Each member has a faunal identity with distinctive faunal associations (Waage, 1968; Speden, 1970). Each association (except for the "Lower and Upper *nicolleti*" assemblage zones) is represented by a different dominant taxon. Thus the record of associations through the Fox Hills provides a unique upsection history of changing marine environments. From species-rich assemblages in sublittoral marine seas, a dramatic decrease in diversity of both bivalves and gastropods is evident in the upper part of the Timber Lake Member into the Iron Lightning Member (Fig. 4). The upper Timber Lake records shallow-water to intertidal conditions with the presence of the *Tancredia-Ophiomorpha* association with limonitic sediments. Although more restricted, chiefly marine conditions prevailed with the deposition of the Bullhead lithofacies of the Iron Lightning Member (Fig. 4; Waage, 1968). The loss of bivalve diversity noted here reflects changing depositional conditions. The apparent rapid deltaic sedimentation of the Bullhead lithofacies resulted in unsuitable habitats for bivalves and gastropods. Although the gastropods of this unit may not have been completely studied (J.M. Erickson, 2001, personal commun.), with no species yet recorded, there was a 90% reduction in the number of bivalve taxa in the Timber Lake Member (the presence of some of these taxa in the Breien Member suggests that marine conditions continued to exist nearby). The nearshore or intertidal marine and brackish environments in the upper part of the Timber Lake were more restricted.

Colgate lithofacies are in different levels within the Iron Lightning Member. To the west, Colgate deposits are interbedded as lenses within the typically banded Bullhead lithofacies. The Colgate also generally overlies the Bullhead in more blanket-form deposits. Waage (1967) also noted that Colgate-type sediments occur in the basal portion of the Hell Creek Formation and contain a similar brackish-water fauna. The Colgate, in its various depositional forms, is dominated by taxa considered brackish in origin. Intertidal estuaries, tidal flats, tidal channels, and lagoonal beaches represent suitable habitats of varying salinity where a small number of mollusks can flourish (discussed in the following). These types of depositional

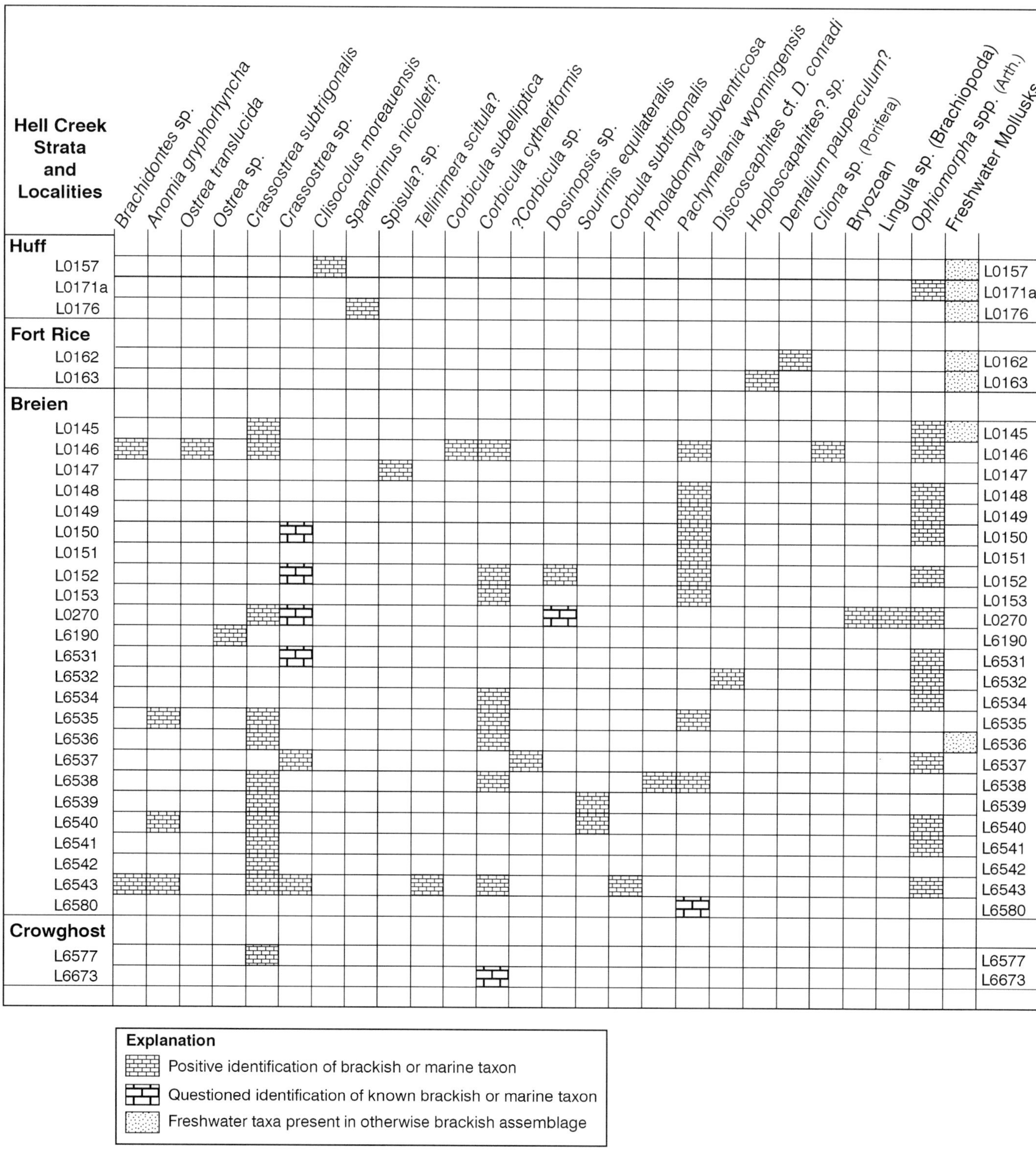

Figure 3. Stratigraphy of brackish-water and marine mollusks and related faunule taxa in Hell Creek Formation. Locality numbers (e.g., L0145) are documented in Appendix 1 and shown in Figure 2. Note that single specimen record of *Jeletzkytes* cf. *J. nebrascensis* is a reidentification by N.L. Landman (see Hoganson and Murphy, this volume) of a published identification of *Discoscaphites* cf. *D. conradi* (Kennedy et al., 1998).

Hell Creek and Fox Hills Strata	Brachidontes sp.	Anomia gryphorhyncha	Ostrea translucida	Ostrea sp.	Crassostrea subtrigonalis	Crassostrea sp.	Clisocolus moreauensis	Spaniorinus nicolleti?	Spisula? sp.	Tellinimera scitula?	Corbicula subelliptica	Corbicula cytheriformis	?Corbicula sp.	Dosinopsis sp.	Sourimis equilateralis	Corbula subtrigonalis	Pholadomya subventricosa	Pachymelania wyomingensis	Jeletskytes cf. J. nebrascensis	Hoploscapahites? sp.	Dentalium paupercullum?	Cliona sp. (Porifera)	Bryozoan	Lingula sp. (Brachiopoda)	Ophiomorpha spp. (Arth.)	Freshwater Mollusks	Other Fox Hills Bivalves	Other Fox Hills Gastropods	Other Fox Hills Cephalopods	
Huff							1	1																1	1	3				Huff
Fort Rice																				1	1					2				Fort Rice
Breien	2	3	1	1	11	4/2	0	0	1	1	1	8	1	1/1	2	1	1	9/1	1	0	0	1	1	1	14	2				Breien
Crowghost					1							0/1																		Crowghost
Hell Creek-SD		Y			Y						~	~							~							Y	3	1	4	Hell Creek-SD
Linton																									Y		2	0	1	Linton*
Colgate	Y	Y			Y				Y			Y		?				~								Y	5	3	1	Colgate
Iron Lightning			Y						Y												~						5	0	1	Iron Lightning
Timber Lake			Y		Y			Y		Y		Y		~	Y	~	~	Y				Y	Y		Y		56	21	3	Timber Lake
Trail City		Y	Y				Y	Y		Y					Y		~			~							38	9	3	Trail City

Explanation

- 8/1 Positive/questioned identification of taxon
- Y Taxon present in Fox Hills or Hell Creek of South Dakota
- * Note also other taxa that cannot be assigned to a unit at this time
- ~ Taxon of similar morphology present in Hell Creek of North Dakota

Figure 4. Abundance of brackish-water and marine mollusks and related faunule taxa in Hell Creek Formation of North Dakota and comparison with molluscan fauna of Fox Hills Formation of South and North Dakota. This figure compares diversity and presence or absence of Fox Hills and Hell Creek molluscan faunas.

environments through the transition from Fox Hills to Hell Creek locally make distinguishing between the formations difficult (see Erickson, 1992). The overlying Linton Member, likely the youngest member of the Fox Hills Formation (Klett and Erickson, 1976), contains rare bivalves and rarer ammonites in deposits to the south in Corson County, South Dakota, suggesting the continuation of marine shelf habitats. In North Dakota, however, locally abundant but widely dispersed *Ophiomorpha* trace fossils suggest brackish conditions at the mouth of tidally influenced river systems (e.g., distributary channels) (Klett and Erickson, 1976). Note that in Figure 4, the Colgate Member (particularly for localities in Corson County, South Dakota) may include taxa that would otherwise be assigned to the Linton Member of the Fox Hills because of the earlier lack of distinction between these units (see Klett and Erickson, 1976). Because the Linton is relatively unfossiliferous, the inappropriate stratal assignment of a few species does not have any immediate consequence to the paleontological interpretation of its depositional environment.

Brackish-water mollusks of the Hell Creek Formation in South Dakota

On the basis of information provided by the Yale Peabody Museum (T. White, 2001, written commun.; and Yale Peabody Museum, 2001; Appendix 2) from the studies of Waage (1967), the transition from the Fox Hills to the Hell Creek Formation in South Dakota is marked locally by a return to brackish and potentially sublittoral marine conditions. The occurrence of chiefly brackish-water taxa, including *Anomia, Crassostrea,* and *Corbicula,* in the basal Hell Creek are comparable to channel-bottom accumulations in the Colgate lithofacies. Waage (1967) noted this similarity and considered the depositional conditions to be the same. The delta front and tidally influenced distributary channel environments of the Bullhead and Colgate lithofacies of the Iron Lightning Member are also in the Hell Creek, but represent marine inundation of distributary channel and overbank environments and/or changes in shoreline configuration upsection. Nearby marine shelf conditions are evidenced by other chiefly marine bivalve species, such as *Granocardium whitei, Hiatella?* sp. A, *Leptosolen* sp., and *Lunatia* sp., and ammonites, such as *Discoscaphites rossi, Jeletzkytes nebrascensis,* and *Sphenodiscus lenticularis* (see Appendices 2 and 3).

Brackish-water mollusks of the Hell Creek Formation in North Dakota

The brackish-water mollusks of the Hell Creek Formation are known primarily from the Breien Member, but as noted here, they are also known from strata below (Crowghost unit) and above (Fort Rice and Huff units). Our reidentification of Frye's specimens, given in Appendix 1 and summarized in Figure 3, is ordered vertically by lithic unit. The classification of taxa identified in this report is given in Figure 5.

The typical limonitic (or ferruginous) preservation of North Dakota Hell Creek mollusks makes them difficult to find because of the ubiquity of ironstone concretions resembling mollusks. The localities in Appendix 1 were discovered and collected largely by Charles Frye. This paper is the first detailed record of Frye's brackish-water and marine mollusks, the documentation of which is not straightforward. Frye conducted his field studies in 1961–1963. U.S. Geological Survey topographic quadrangles were not available to Frye, so he located his collections on North Dakota county highway maps (scale 1:126720). Many of the collections are thus only known to a general location, and relocating them has been difficult. However, in 1990, Hartman employed Frye to relocate as many of his localities as his time permitted. As a result, we learned that many of Frye's localities were not accurately reported (e.g., NW¼ instead of NE¼), and that collection information was incomplete. Frye also provided field notes to better document localities and geologic sections mentioned in his dissertation (C.I. Frye, 1990, written commun.; field note volumes 1961, 1962, 1963a, 1963b; see Appendix 1).

The brackish-water and marine faunule associations of the Hell Creek Formation in North Dakota are dominated by three taxa: the bivalves *Crassostrea* and *Corbicula* and the gastropod *Pachymelania* (see specimen figures). One or more of these taxa is in almost 90% of the faunules with associations typically considered brackish. The remaining faunules consist of one bivalve species of *Clisocolus, Spisula,* or *Spaniorinus.* The oyster *Crassostrea* (Plate 1, figs. 9–12) clearly is dominant in single- and multiple-species accumulations. It occurs singularly in ~35% of the localities and in association with others in ~30%. *Pachymelania* (e.g., Plate 3, fig. 3) is the only other taxon present at more than one locality that constitutes the entire faunule. As noted by Frye, in virtually every occurrence, the crustacean trace fossil *Ophiomorpha* (e.g., Plate 3, fig. 12) also occurs with this molluscan association. Pre-Breien and post-Breien brackish-water deposits suggest varying degrees of marine influence, from estuary river distributary mouth to changes in shoreline configuration related to changing patterns of deltaic distributary sedimentation.

Comparison of molluscan assemblages

The Breien Member is the dominant brackish-water mollusk-bearing unit in the Hell Creek Formation of North Dakota. Except for the marked reduction in diversity, the fossils of the Fox Hills and Breien strata share many common species. The Breien and the Colgate units appear to have similar brackish-water *Crassostrea-Corbicula* species association. Although sparse, bivalves associated with open marine conditions also are present in the Breien Member and indicate that more normal saline conditions were nearby. The generally low quality of preservation of the brackish-water mollusks of the Hell

Creek, however, limits the likelihood of recognizing distinctive Late Cretaceous immigrants or new species. The few chiefly marine forms present in the Breien Member have records of related taxa in the shallow-marine deposits of the Trail City and Timber Lake Members. Comparable species associations are in the *Crassostrea-Pachymelania* association of the Timber Lake Member, but the Timber Lake Member also contains a much higher species richness, indicating an open shelf environment with normal marine salinity.

PALEOENVIRONMENTS AND PALEOSALINITY

Breien Member

Although the preservation of the brackish-water mollusks of the Hell Creek Formation is far from ideal, the specimens clearly indicate a return to marine conditions following the well-known regression of the Western Interior Sea and the deposition of the Fox Hills Formation. Of the various stratigraphic levels from which mollusks are known, only the relatively thin Breien Member contains a sufficiently diverse assemblage to document the return to brackish-water and marine platform conditions (Figs. 3 and 4). Isolated marine deposits below and above the Breien Member, however, are compelling evidence for the proximity of latest Cretaceous seas in the interior of North America.

Because sedimentological criteria are of little use in distinguishing oligohaline (0.5‰–5‰), mesohaline (5‰–18‰), and brachyhaline (18‰–30‰) marine waters, Fürsich (1994) compiled examples and suggested guidelines for interpreting the paleosalinity of autochthonous faunal associations. With its more or less in situ fauna, the Breien Member represents a small remnant of a larger marine environment. The paleosalinity of the Breien Member can be environmentally characterized by comparing its dominant faunal elements with known salinity-controlled benthic macroinvertebrate associations.

Oligohaline regimes are characterized by faunas of very low diversity, commonly containing some freshwater elements. Taxa are typically shallow, burrowing suspension feeders. Faunal associations in the middle and Upper Cretaceous include the bivalve *Corbicula,* the inarticulate brachiopod *Lingula,* and gastropods such as *Pyrgulifera* and *Melanopsis.* By comparison, mesohaline faunal associations are considerably more diverse. This increase in faunal diversity reflects the greater number of organisms tolerant of saline conditions within this range, along with adaptations for variable temperatures, turbidity conditions, and oxygen levels. By the Late Cretaceous, mesohaline environments were dominated largely by shallow epifaunal bivalves (specifically oysters), infaunal bivalves, epifaunal gastropods, and *Lingula.*

Oysters successfully inhabit both mesohaline and brachyhaline regimes, where they form banks or patch reefs in fine-grained lagoonal muds (Fürsich, 1994), but are represented only by a few species of variable ecotypic-dependent morphology. From the mid-Cretaceous, the oyster *Crassostrea* dominated mesohaline environments. With increasing salinity, oysters are associated with greater numbers of other taxa. In the late Early and Late Cretaceous, the corbiculid and corbulid bivalves became mesohaline faunal associates. In brachyhaline environments, overall molluscan diversity significantly increases with the addition of a greater number of epifaunal bivalves. Diagnostically, corbiculid bivalves are far less important, and taxa with broad salinity tolerance, such as corbulids, become more prominent in brachyhaline environments.

In addition to these faunal associations, distinctive ecotypic morphotypes are recognizable in salinity-controlled environments (Fürsich, 1994). The bivalves *Crassostrea, Brachidontes* (Plate 1, fig. 1), *Anomia* (Plate 1, fig. 3), *Corbicula* (Plate 2, figs. 1 and 7), and *Corbula* (Plate 2, fig. 10) are Cretaceous genera with morphological representatives back to the Jurassic. The inarticulate brachiopod *Lingula* represents an even greater geologic history of salinity-controlled distribution. All of these taxa are represented in the Breien Member of the Hell Creek Formation, *Crassostrea, Anomia,* and *Corbicula* being dominant faunal elements.

The taxa of the Breien Member of the Hell Creek Formation strongly indicate a salinity-controlled environment, individual species ranging from oligohaline to brachyhaline salinity tolerances. Typical faunal associations, including epibionts, however, suggest a mesohaline regime augmented by a few brachyhaline to euhaline taxa. These rarer faunal constituents include chiefly marine species of bivalves and ammonites that may represent species washed in from nearby normal euhaline conditions associated with a stable marine platform or taxa brought in by transgressive seas. The Breien molluscan fauna has few freshwater admixtures, suggesting a strong marine influence. Only higher in the Hell Creek section are marine faunal elements admixed with a dominant freshwater fauna (discussed in the following).

Of the dominant Breien taxa, *Anomia and Crassostrea* are cemented epifaunal suspension feeders common in estuarine environments with limited sediment influx and potentially significant tidal influence (Fürsich and Kauffman, 1984; Fürsich and Kirkland, 1986). *Crassostrea* is frequently the dominant element in mesohaline environments. High mesohaline or brachyhaline conditions are indicated by (1) the presence of sponge (*Cliona*)-bored *Crassostrea* (L0146; Plate 3, fig. 10), (2) the occurrence of bryozoans (locality L0270, Plate 3, fig. 11), (3) the larger size of some *Crassostrea,* (4) the patch reef occurrence, and (5) the association with chiefly marine bivalves. The common *Corbicula* of the Breien fauna are shallow, burrowing infaunal suspension feeders and are typically dominant elements in oligohaline to mesohaline environments (Fürsich, 1994). Although rare, *Brachidontes* and *Corbula* are adapted to similar conditions, having colonized the mesohaline regime, but also brachyhaline to more euhaline environments. However, the variability of environmental tolerances of *Corbula* in the North Dakota Laramide section is complex. For

Class Bivalvia
Subclass Pteriomorphia
Order Mytiloida
Superfamily Mytiloidea
Family Mytilidae
Brachidontes Swainson 1840
Brachidontes sp.
Localities L0146, L6543.
Order Pterioida
Suborder Pteriina
Superfamily Anomioidea
Family Anomiidae
Anomia Linnaeus 1758
Anomia gryphorhyncha Meek 1872
Localities L6535, L6540, L6543.
Suborder Ostreina
Superfamily Ostreoidea
Family Ostreidae
Ostrea Linnaeus 1758
Ostrea translucida Meek and Hayden 1857
Locality L0146.
Ostrea sp.
Locality L6190 (not found).
Crassostrea Sacco 1897
Crassostrea subtrigonalis (Evans and Shumard) 1857
Localities L0145, L0146 (with poriferan species of *Cliona*), L0270 (not found), L6535, L6536, L6538, L6539, L6540, L6541, L6542, L6543, L6577.
Crassostrea sp.
Localities L6537, L6543.
Crassostrea? sp.
Localities L0150, L0152, L0270 (with species of bryozoan), L6531.
Subclass Heterodonta
Order Veneroida
Superfamily Lucinoidea
Family Mactromyidae
Clisocolus Gabb 1868
Clisocolus moreauensis (Meek and Hayden) 1857
Locality L0157 (probably Huff unit).
Superfamily Galeommatoidea
Family Galeommatidae
Spaniorinus Dall 1899
Spaniorinus nicolleti? Speden 1970
Locality L0176.
Superfamily Mactroidea
Family Mactridae
Subfamily Mactrinae
Spisula Gray 1837
Spisula? sp.
Localities L0147, L6543 (possible juvenile of *Spisula* or *Sourimis*, but more likely *Tellinimera scitula*)
Superfamily Tellinoidea
Family Tellinidae
Subfamily Tellininae
Tellinimera Conrad 1860
Tellinimera scitula? (Meek and Hayden) 1856
Locality L6543 (or possibly juvenile *Sourimis* or *Spisula*).

Figure 5. Classification of brackish water and marine mollusks of Hell Creek Formation. Daggers (†) indicate: 1970, Ohta published a systematic revision of selected North American Cretaceous corbiculids. Ohta (1970) proposed two taxonomic changes that would impact on taxa mentioned here (see also Table 1). Ohta introduced the genus name *Hendersona,* with *Corbicula subelliptica* (*Tellina subelliptica* Meek and Hayden) as its type species distinguishing it from *Corbicula* on the basis of differences in pallial sinus. Ohta (1970) also synonymized *Corbicula cytheriformis* (*Cyrena* [*Corbicula*?]; *cytheriformis* Meek and Hayden 1860) with *corbicula occidentalis* (*Cyrena occidentalis* Meek and Hayden 1856) and placed it in the genus *Veloritina* Meek on the basis of pallial sinus, umbo, lunule, and morphometric characteristics. Ohta (1970, p. 308) stated: "I prefer at present to regard *C. cytheriformis* as synonymous with *V. occidentalis,* because the variable outline of *C. cytheriformis* is included in the range of variation of outline of *V. occidentalis,* and both species coexist with each other in many cases." Ohta (1970) emended the definition of *Veloritina* to include the presence of a lunule. Stanton and Hatcher (1905) had earlier noted the presence of "intermediate" forms between these two species in some areas, but retained the nomenclatural distinction. Additional study may prove the value of Ohta's (1970) contribution but it apparently has not been reviewed by other corbiculid workers, and evaluating his work is beyond the scope of this paper. Double daggers (‡) (Class Gastropoda) indicate that higher classification of cerithioids is in state of flux, because many new observations provide more rigorous basis for establishing monophyletic clades within Gastropoda.

Subclass Heterodonta (continued)
Superfamily Corbiculoidea
Family Corbiculidae
Corbicula Megerle 1811
Corbicula subelliptica (Meek and Hayden) 1856†
Locality L0146.
Corbicula cytheriformis (Meek and Hayden) 1860†
Localities L0146, L0152, L0153, L6534, L6535, L6536, L6538, L6543, L6673 (not found).
?*Corbicula* sp.
Locality L6537.
Superfamily Veneroidea
Family Veneridae
Subfamily Pitarinae
Dosiniopsis Conrad 1864
Dosiniopsis sp.
Locality L0152 (Laird and Mitchell, 1942; not found).
Dosiniopsis? sp.
Locality L0270 (Seager et al., 1942; Laird and Mitchell, 1942; not found).
Sourimis Speden 1970
Sourimis equilateralis (Meek and Hayden) 1856
Localities L6539, L6540.
Sourimis? sp.
Locality L6543 (possible juvenile of *Sourimis* or *Spisula*, but more likely *Tellinimera scitula*).
Order Myoida
Superfamily Myoidea
Family Corbulidae
Corbula Bruguière 1797
Corbula subtrigonalis Meek and Hayden 1856
Locality L6543.
Subclass Anomalodesmata
Order Pholadomyoida
Superfamily Pholadomyoidea
Family Pholadomyidae
Pholadomya G.B. Sowerby 1823
Pholadomya subventricosa (Meek and Hayden) 1857
Locality L6538.
Class Gastropoda
Subclass Apogastropoda (see Ponder and Lindberg, 1997‡)
Superorder Caenogastropoda (see also Ponder and Warén, 1988‡)
Order Sorbeconcha (Neotaeinogastropoda) (see Ponder and Lindberg, 1997‡)
Superfamily Cerithioidea
Family Thairidae
Pachymelania E. Smith 1893
Pachymelania wyomingensis Meek 1873
Locality L0146, L0148, L0149, L0150, L0151, L0152, L0153, L6535, L6538, L6580(?).
Class Cephalopoda
Order Ammonoidea
Suborder Lytoceratina
Superfamily Scaphitaceae
Family Scaphitidae
Discoscaphites Meek 1870
Discoscaphites cf. *D. conradi* (Morton) 1834
Locality L6532.
Hoploscaphites Nowak 1911
Hoploscaphites? sp.
Locality L0163.
Class Scaphopoda
Order Dentaliida
Family Dentaliidae
Dentalium Linnaeaus 17589
Dentalium pauperculum? Meek and Hayden 1860
Locality L0162.

Figure 5. (*continued*)

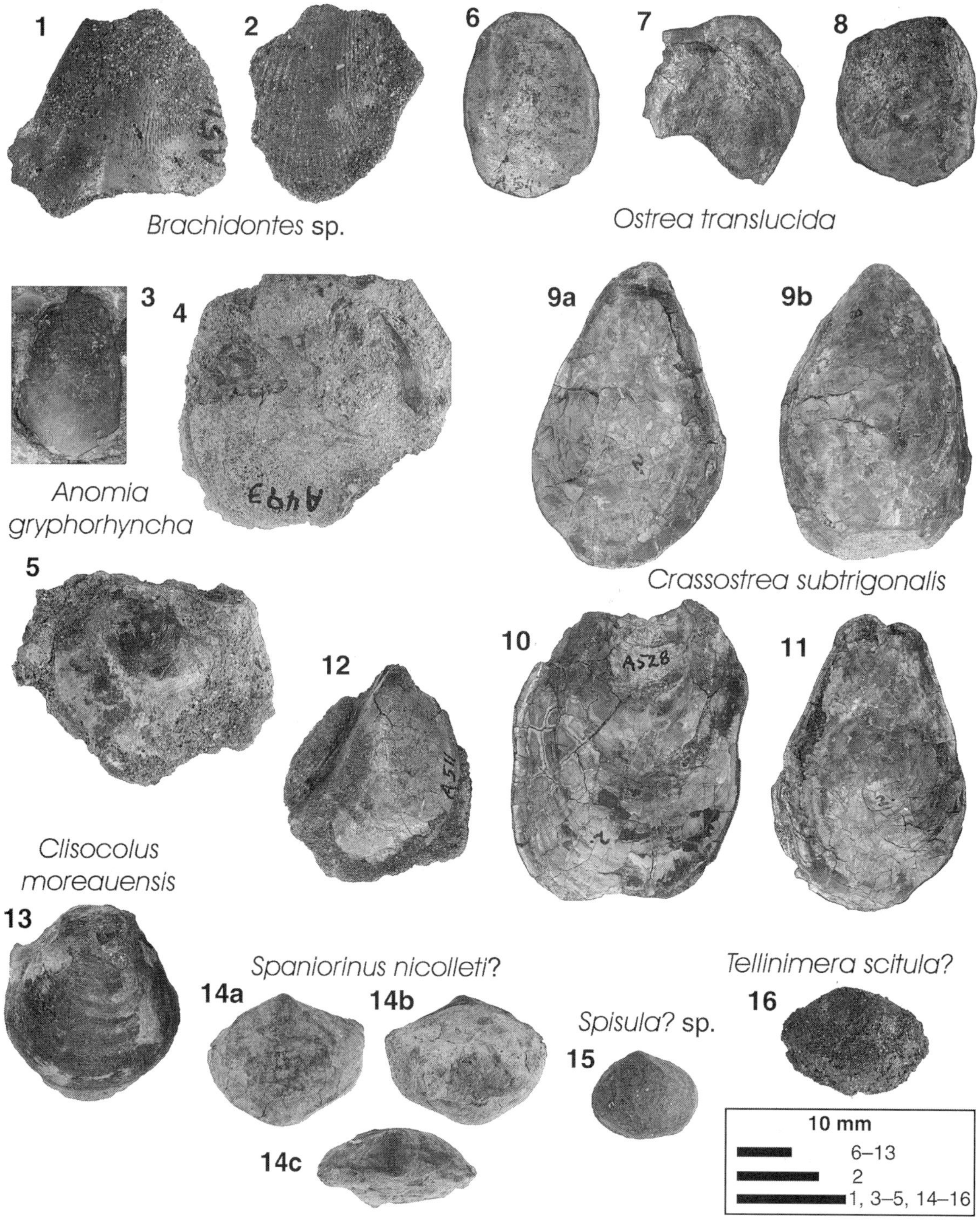

Plate 1. General quality of specimens is poor. They are preserved as limonitic steinkerns, external molds, or impressions. L-numbers (e.g., L0146) refer to localities mentioned in this report. 1, 2—*Brachidontes* sp.: 1, UND-PC 15842, L1046, internal mold; 2, UND-PC 15843, L6543, impression of external surface of valve showing curvilinear ridge sculpture. 3–5—*Anomia gryphorhyncha:* 3, UND-PC 15844, L6535, external view; 4, UND-PC 15845, L6540, external view; 5, UND-PC 15846, L6540, external view. 6–8—*Ostrea translucida:* 6, UND-PC 15847, L0146, external view of right valve; 7, UND-PC 15848, L0146, external view; 8, UND-PC 15849, L0146, external view. 9–12—*Crassostrea subtrigonalis:* 9, UND-PC 15850, L6536; a, external view of right valve, b, external view of left valve; 10, UND-PC 15851, L6536, external view of right valve; 11, UND-PC 15852, L6536, external view of right valve; 12, UND-PC 15853, L0146, internal mold with *Cliona* borings. 13—*Clisocolus moreauensis:* UND-PC 15854, L0157, internal view of right valve. 14 (a-c)—*Spaniorinus nicolleti?:* UND-PC 15855, L0176; a, external view of left valve, b, internal view (same) of hinge area, c, dorsal view (same). 15—*Spisula?* sp., UND-PC 15856, L0147, external view of left valve. 16—*Tellinimera scitula?* or possibly juvenile of species of *Sourimis* or *Spisula,* UND-PC 15857, L6543, external view of left valve with borings.

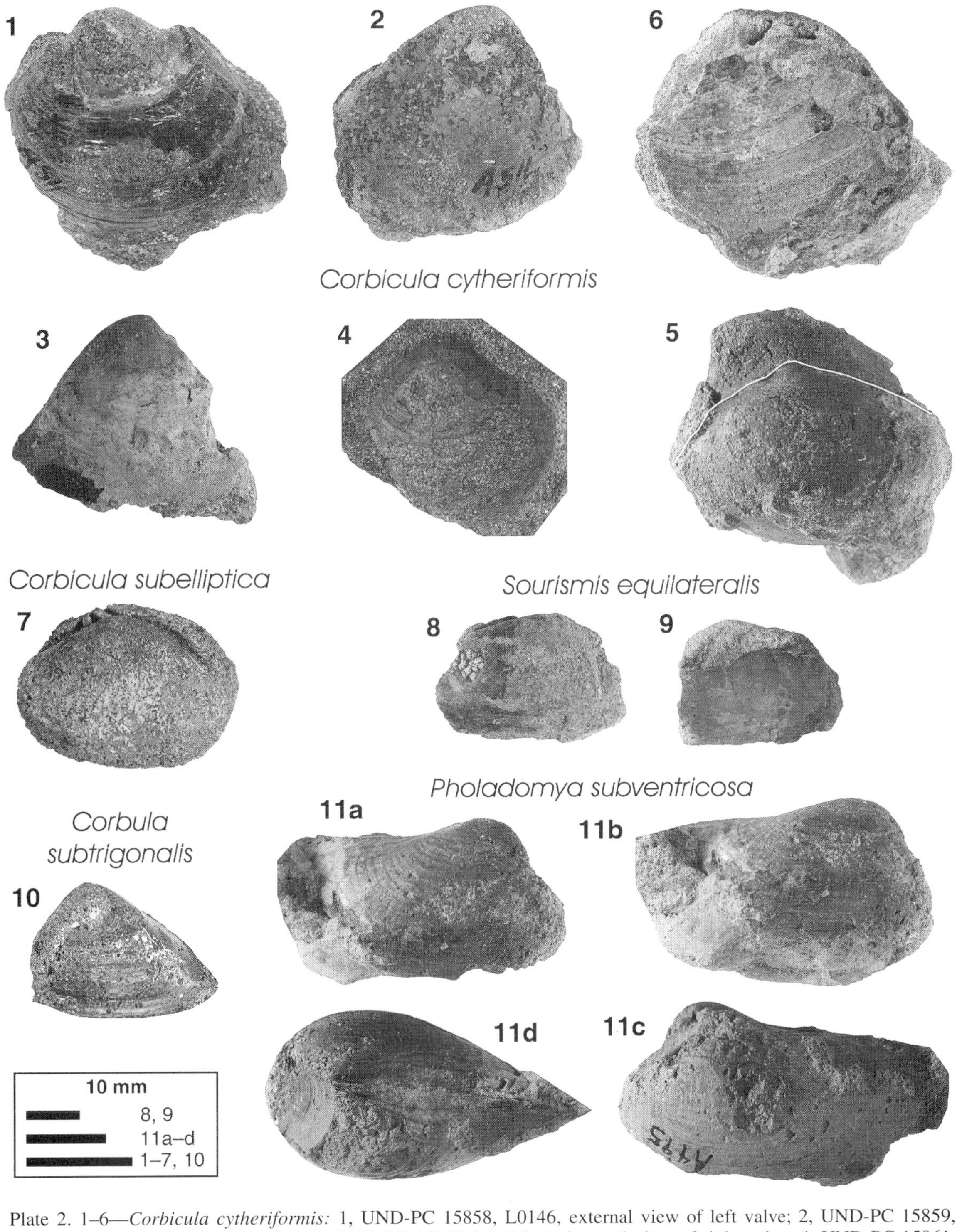

Plate 2. 1–6—*Corbicula cytheriformis:* 1, UND-PC 15858, L0146, external view of left valve; 2, UND-PC 15859, L0146, internal mold of right valve; 3, UND-PC 15860, L0146-11, internal view of right valve; 4, UND-PC 15861, L0146, external mold of left valve; 5, UND-PC 15862, L6538, steinkern of left valve, right valve behind; 6, UND-PC 15863, L0152, external mold. 7—*Corbicula subelliptica:* UND-PC 15764, L0147, internal mold of left valve with hinge teeth above. 8 and 9—*Sourimis equilateralis:* 8, UND-PC 15865, L6540, steinkern of left valve; 9, UND-PC 15866, L6539, steinkern of left valve with change in preservation color toward dorsal margin and posterodorsal matrix obscuring margin of specimen. 10—*Corbula subtrigonalis:* UND-PC 1567, L6543, external mold of left valve. 11 (a–d)—*Pholadomya subventricosa,* UND-PC 15868, L6538: a, external view of right valve, tilted for better view of posterodorsal margin, b, external view of right valve, c, external view of left valve, d, dorsal view of valves.

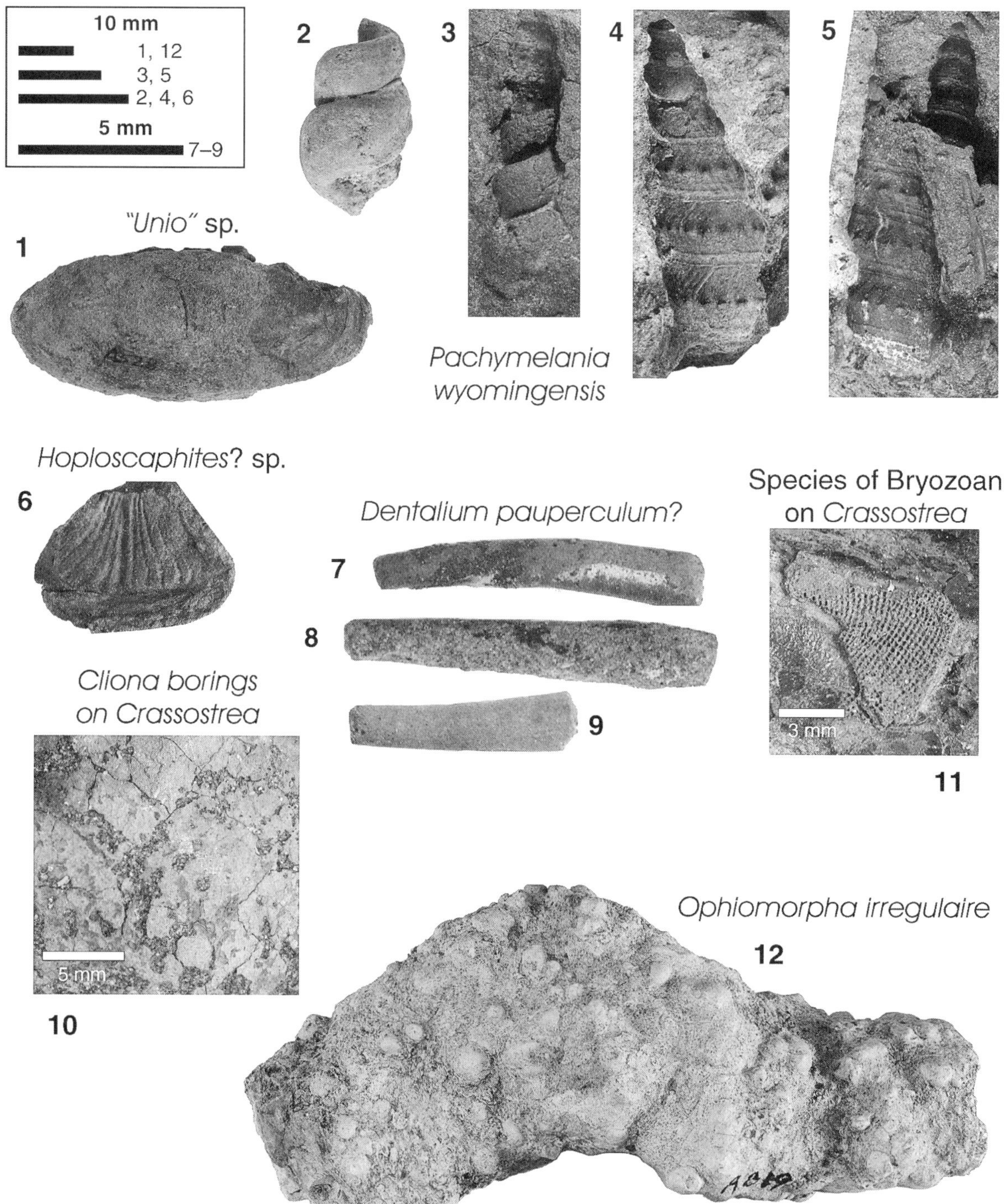

Plate 3. 1—*"Unio"* sp., UND-PC 15869, L6536. 2–5—*Pachymelania wyomingensis:* 2, UND-PC 15870, L0150, steinkern; 3, UND-PC 15871, L0146, steinkern with impression of external surface; 4, UND-PC 15872, L0153, impression of external surface; 5, UND-PC 15873, L0148, impression of external surface. 6—*Hoploscaphites?* sp. UND-PC 15874, L0163, external impression. 7–9—*Dentalium pauperculum?:* 7, UND-PC 15875, L0162, partial mold; 8, UND-PC 15876, L0162, partial mold; 9, UND-PC 15877, L0162, partial mold. 10—*Cliona* borings into surface of *Crassostrea subtrigonalis:* UND-PC 15878, L0146, enlargement of surface of oyster with borings. 11—Bryozoan on surface of *Crassostrea* sp. UND-PC 15879, L0270. 12—*Ophiomorpha irregulaire:* UND-PC 15880, L0171a, external surface.

example, *Corbula* (*Bicorbula*) *mactriformis* (Meek and Hayden) is known from a number of localities in the overlying Paleocene Cannonball Member of the Fort Union Formation (Cvancara, 1966; Hartman, 1999). This sublittoral marine platform deposit contains a diverse molluscan fauna. However, this taxon is also well known as an admixture to highly diverse freshwater molluscan assemblages in fluvial deposits and as isolated macrofossils in association with brackish-water trace fossils (Tibert et al., 2001). The former deposit is exposed near the eastward edge of predominantly nonmarine environments adjacent to the Western Interior Seaway. The latter occurrence of *Corbula* is ~200 km to the west of the deposits that record the presence of the former seaway (location of the present-day Missouri River valley), upgradient of the early late Paleocene Laramide paleotopography.

With a few exceptions, gastropods are typically not a major component (in abundance to diversity) of mesohaline faunas, although they can be locally abundant in low-diversity assemblages. Such exceptions include the cerithioids *Pyrgulifera* and *Pachymelania,* which can be significant components of oligohaline and mesohaline associations, respectively. The sole gastropod from the Breien Member is *Pachymelania wyomingensis* (Plate 3, fig. 3), which is a common component of low-diversity associations.

The fossil bivalve assemblage of the Breien Member overlaps significantly with the faunal assemblage of the Colgate lithofacies of the Iron Lightning Member (Waage, 1968; Speden, 1970; Feldmann and Palubniak, 1975) (Fig. 4). They share in common *Anomia gryphorhyncha* (Plate 1, figs. 3–5) *Crassostrea subtrigonalis* (Plate 1, figs. 9–12), and *Corbicula cytheriformis* (Plate 2, figs. 1–6E). The only relatively common taxon of the Colgate that is absent is *Modiolus galpinianus.* The other common fossil in the Breien Member is *Pachymelania,* which is also well known throughout the Western Interior in marine-influenced depositional environments (e.g., uppermost Cretaceous Laramie Formation of the Denver Basin, the type locality of *P. wyomingensis*). The Colgate lithofacies was interpreted by Waage (1967) as deltaic and as including a number of highly variable subenvironments that resulted in a limited number of complex species associations. Feldmann and Palubniak (1975) interpreted *Crassostrea glabra* (here *C. subtrigonalis*) associations in the Fox Hills as having been deposited on the landward side of barrier bars and on lagoonal beaches associated with the deposition of the Timber Lake and Colgate units, respectively. Similar environmental conditions may be represented in the Breien Member, but with the presence of the other faunal elements discussed above, the Breien represents mesohaline conditions adjacent to an accessible marine platform.

Pre-Breien and post-Breien marine fossils

Strata underlying and overlying the Breien Member include taxa that provide tantalizing clues supporting persistent marine conditions within and lateral to the Hell Creek Formation. The occurrence of *Corbicula* and *Crassostrea* in the pre-Breien (Crowghost unit) strata may reflect oligohaline conditions associated with the regressing Western Interior Sea. Although a few freshwater mollusks are known from this part of the section of the Hell Creek in South Dakota, none is known from the pre-Breien strata in North Dakota, indicating limited suitable habitats for otherwise common unionids, viviparids, and pleurocerids to colonize. The basal Hell Creek Formation in south-central North Dakota is dominated by lignitic lithologies (Frye, 1967), possibly indicating changing base-level conditions associated with the maximum retreat of the Western Interior Sea at this time.

The strata overlying the Breien Member include rare species of a more chiefly marine nature than typically found in the Breien. In south-central North Dakota, an ammonite (*Hoploscaphites?;* Plate 3, fig. 6) and scaphopod (*Dentalium;* Plate 3, figs. 7–9) are known from the Fort Rice unit. Although the ammonite is likely to be allochthonous, a *Dentalium-Lunatia* association is well documented at specific horizons at the top of the Colgate lithofacies of the Iron Lightning Member and may reflect in situ euhaline conditions. Likewise, the presence of the marine genera *Clisocolus* (L157; Plate 1, fig. 13) and *Spaniorinus* (L176; Plate 1, fig. 14) (Figure 5), as well as the presence of the crustacean trace fossil *Ophiomorpha* (L171a; Plate 3, fig. 12) ~200 km west of the Breien type locality, indicates major changes in the shoreline configuration and, possibly, a rise in sea level at that time. Although marine bivalves may have been washed into nonmarine environments or are part of a condensed section representing both freshwater and marine environments, the trace fossil is unlikely to have been reworked.

DISCUSSION OF FAUNAL DEPOSITIONAL ENVIRONMENTS

In the Breien Member, many bivalves are articulated, indicating an in situ death. Along with the faunal diversity present in the Breien Member, including the admixture of chiefly marine forms and the presence of epibionts, such as *Cliona* and colonial bryozoans, the salinity of the *Anomia-Crassostrea-Corbicula* association indicates mesohaline to brachyhaline (10‰–20‰) conditions. A shift between localities having a dominance of mobile infaunal *Corbicula* and *Pachymelania* to sessile epifaunal *Anomia* and *Crassostrea* is probably due mainly to the local or persistent habitat stability and limited sedimentation (Fürsich and Kirkland, 1986). Transgressive changes in sea level or variations in freshwater influx, noted in the cyclic nature of Breien deposition, likely allowed the incorporation of more rare, chiefly open marine taxa, such as ammonites and certain bivalves like *Pholadomya* and *Sourimis* (Plate 2, figs. 11 and 8, respectively), into a moderately diverse lagoonal (bay) fauna or into its deposits.

The common occurrence of the crustacean trace fossil *Ophiomorpha* in Breien Member molluscan assemblages does not isolate a specific environmental niche. The typical interpretation of nearshore, sandy sublittoral environments was expanded upon by Frey et al. (1978, p. 217), who stated that "burrows occur also on shoals, tidal flats, tidal stream point bars, and lagoon, bay, sound, and estuary floors . . . wherever salinities and current energy are moderately high and the substrate consists mainly of sand." Water depths of 3–4 m may be expected. The dominance of the *Anomia, Crassostrea, Corbicula,* and *Pachymelania* molluscan association, the co-occurrence of *Ophiomorpha,* and the limited admixture of chiefly marine bivalves suggest a partially protected lagoonal or bay environment.

Beyond the specific environmental interpretation of the Breien Member, deposits containing open shelf, euhaline mollusks indicate persistent marine conditions very late in the Cretaceous. The precise chronostratigraphic correlation of the strata representing the existence of this sea is uncertain. In a study of the marine connections and ammonite biostratigraphy in South Dakota, North Dakota, and Nebraska, Kennedy et al. (1998, p. 749) stated that the "*Jeletzkytes nebrascensis* zone fauna is then the highest assemblage that can be recognised in the US Western Interior, but it cannot be shown if the very highest records . . . represent this Zone, or some younger, inadequately characterised assemblage." Although not reviewed by Kennedy et al. (1998), the record of *Jeletzkytes nebrascensis* (Appendix 2) at the base of the Hell Creek Formation in South Dakota may represent the highest evidence of this zone in the Western Interior. Other ammonites at this level in South Dakota, from the Breien Member in North Dakota (*Jeletzkytes* cf. *J. nebrascensis;* see Hoganson and Murphy, this volume), from the upper part of the Hell Creek in North Dakota, and from the Lance Formation in Wyoming (Jeletzky and Clemens, 1965), suggest that additional specimens will be found that will ultimately resolve global chronostratigraphic correlation of the marine faunal elements of the Hell Creek. Such a temporal correlation will help hone the calibration of nonmarine Lancian (North American Land Mammal Age) faunas of the Hell Creek with its Maastrichtian marine faunal counterparts. The documentation of post-Breien marine mollusks and trace fossils further supports continued refinement of paleogeographic models associated with the retreat of the Cretaceous sea (Erickson, 1999, and references cited therein). The extent and development of Erickson's Dakota isthmus dividing northern and southern latest Cretaceous and Paleocene seas is also refined by providing temporal limits of 100–300 k.y. for the withdrawal of the Cretaceous Western Interior Seaway and its return as the Paleocene Cannonball Sea (Cannonball Formation) very near the beginning of the Paleocene (Hartman et al., 1999; Lund et al., this volume). Otherwise, the Western Interior Sea at the end of the Cretaceous never significantly retreated from the plains of North Dakota and South Dakota.

CONCLUSIONS

The marine and brackish-water mollusks found through much of the Hell Creek Formation of south-central North Dakota document the presence of an open marine shelf environment and lateral to marginal marine and brackish environments through nearly the latest part of the Cretaceous. Molluscan associations and isolated records indicate continual changes in shoreline configuration likely related to changes in base level. The well-documented open marine environment with species-rich associations of the lower part of the Fox Hills Formation was replaced by restricted brackish-water conditions as recorded in the upper part of the formation, associated with a regressing sea. The presence of marine and brackish-water molluscan species in the Hell Creek Formation of South Dakota records shelf and probably lagoonal conditions at a time when the Western Interior Sea is generally thought to have regressed from the central interior of North America. The basal portion of the Hell Creek Formation in North Dakota contains isolated monospecific faunules, likely of oligohaline to mesohaline waters in lagoonal environments, indicative of an interior seaway. The overlying Breien molluscan fauna represents a relatively dramatic return to mesohaline to brachyhaline conditions in a lagoonal or bay setting that was inundated by more euhaline waters that resulted in the admixture of chiefly marine mollusks. As reported by Hoganson and Murphy (this volume) and Pearson et al. (this volume), the vertebrate and other trace fossil records of Breien and overlying strata in North Dakota marine-influenced deposits indicate persistent marine conditions compared to the short-lived species richness of the Breien Member. In contrast to the older molluscan record on the eastern edge of the Williston Basin, the isolated ammonite, scaphopod, and bivalve mollusks of the upper part of the Hell Creek suggest fluctuating marine levels and/or major changes in shoreline configuration and indicate syndepositional open marine conditions in southwesternmost North Dakota. The subsequent incursion of the Cannonball Sea in the very early, but not earliest, Paleocene documents a return to open marine conditions and species-rich molluscan associations for the next 5 m.y. The retreat of the Western Interior Seaway from the area of present-day exposures of the Hell Creek Formation in North Dakota likely occurred within a few hundred thousand years of the end of the Cretaceous. The subsequent change from nonmarine conditions in the interval of the K-T boundary to marine conditions within as short a time as 100 k.y. suggests a limited regression of the sea from the interior of the continent. A working hypothesis might be that a complete withdrawal of marine conditions from the Western Interior prior to the end of the Cretaceous lacks supporting documentation in the rock record.

ACKNOWLEDGMENTS

This paper would not have been possible without the collections and observations made by Charles I. Frye (formerly of Northwest Missouri State University). We acknowledge his effort and that of his 1990 field assistant, Johnathan M. Campbell. We thank William Cobban (U.S. Geological Survey), Tim White and Copeland MacClintock (Yale Peabody Museum), and Daniel J. Daly (Energy & Environmental Research Center) for their timely contribution of specific observations or data used in this report. We also appreciated critical comments by J. Mark Erickson (St. Lawrence University), John W. Hoganson (North Dakota Geological Survey), Doug J. Nichols (U.S. Geological Survey), and Mike Lowe, Mike Hylland, and Martha Hayden (Utah Geological Survey) on preliminary drafts and Emmett Evanoff (University of Colorado, Boulder) and Steve Good (West Chester University) for their helpful review of the manuscript. This research was supported by funding from the U.S. Department of Energy through the Energy & Environmental Research Center of the University of North Dakota.

APPENDIX 1. LOCALITY REGISTER OF HELL CREEK FORMATION BRACKISH-WATER AND MARINE MOLLUSKS

Explanation

The localities in this appendix were discovered and collected almost exclusively by C.I. Frye. Much time has been spent trying to provide the most complete knowledge of Frye's important collections from the brackish-water and marine facies of the Hell Creek Formation. The documentation of Frye's collection of specimens from most of his localities is far from straightforward, and because this is the first time Frye's localities have been updated and published, we have attempted to integrate all records relating to a specific locality. In doing so, we hope to help future workers on the brackish-water and marine mollusks of the Hell Creek to better interpret new collections in light of Frye's initial effort.

Special documentation includes the use of Frye's field books for the years 1961, 1962, and 1963, which were copied and are referred to with Frye's permission. The following locality register follows the guidelines given in the introductory chapter of this volume (Hartman et al.). The register is organized by the lithic units recognized by Frye (1967). The value of these units for stratigraphic correlation is very limited (Murphy et al., this volume), thus the use of Frye's lithostratigraphy here simply references the horizon information available to us directly from Frye's studies. In ascending order, fossil localities are arranged by pre-Breien (Crowghost unit), Breien, and post-Breien (Fort Rice and Huff units). No brackish-water mollusks were reported from Frye's Pretty Butte unit from the top of the Hell Creek Formation. All taxa, except as noted, are mollusks. Locality (L) numbers (e.g., L6577) and section (M) numbers represent a Hartman system of tracking data for computer and museum curation. The numbers following the L number refer to other numbers assigned to the fossil location or collection.

Pre-Breien Member (Crowghost) mollusks

Locality L6577. Frye location 115; UND-PC A783. Documents: Frye (1967, p. 357; 1963b field notebook, p. 55). Location: Emmons County, North Dakota, Cannon Ball SE (1968) Quadrangle, NW¼ sec. 19, T.133N., R.78W. Collections: UND-PC, 1962 (7/12), C.I. Frye.

Locality L6673. Frye location 123; UND-PC A775. Documents: Frye (1967, p. 357; 1963a field notebook, p. 4; 1963b field notebook, p. 57). Location: Emmons County, North Dakota, Moffit SW (1971) Quadrangle, new highway cut, probably south edge SW¼ sec. 13, T.135N., R.78W. (this report). Elevation: 1745 ± 20 ft (525.8 ± 6.1 m). Stratigraphy: In his 1963a field notes, Frye reported: "Ash bed, 6-0 feet [1.82 m] thick. Above ash bed 1′? [0.30 m—] Gingkos plus other plants. [I]mprints of clams?" Collections: UND-PC, 1963 (6/3), C.I. Frye; no specimen(s) found [brackets of present authors].

Breien Member mollusks

The following catalog of molluscan localities from the Breien Member represents the majority of those mollusks associated with brackish environments in the Hell Creek Formation of North Dakota. Note that there is a close association of mollusks with the crustacean trace fossil *Ophiomorpha,* which is considered to normally indicate brackish-water or marine conditions. Only at location 102 of Frye (1967; L6533) was *Ophiomorpha* reported without mollusks.

Locality L0145. Frye location 114; UND-PC A524. Documents: Frye (1967, p. 359; 1962 field notebook, p. 6; 1963b field notebook, p. 54). Location: Emmons County, North Dakota, Fort Rice (1971) Quadrangle, about center sec. 17, T.136N., R.78W. (this report). Elevation: 1750 ± 30 ft (533.4 ± 9.1 m). Discussion: In his 1962 field notes, Frye provided a simple section with the brackish-water mollusks occurring in "clayey buff sand" and reported the locality from the center of sec. 17. This faunule includes *Ophiomorpha.* Collections: UND-PC, 1962 (6/21), C.I. Frye. The freshwater pleurocerid *Lioplacodes tenuicarinata* (Meek and Hayden) was reported by Frye (1967) from this locality, but this species was not found in the collection.

Locality L0146. (Frye location 98; UND-PC A511). Documents: Frye (1967, p. 360; 1961 field notebook, p. 64–66, section 7; 1962 field notebook, p. 13; 1963b field notebook, p. 49). Location: Morton County, North Dakota, Breien (1980) Quadrangle, E½ SE¼ sec. 17, T.134N., R.81W. (this report). This locality occurs in Laird and Mitchell's (1942, p. 30) section 5 (M6442), where they reported *Ophiomorpha* from "Buff, limonitic sand, with 1 foot bentonite layer near middle. *Halymenites* [*Ophiomorpha*] moderately abundant. Limonitic concretions common. The limonitic zone at the top of the Breien seems to be fairly continuous." On the basis of section 5 (Laird and Mitchell, 1942), the Breien Member occurs from ~1709 to 1726.5 ft (519.68–526.23 m). In his 1962 field notes, Frye reported "[c]oncretion zone in and above lignite above Breien present with numerous wood fossils and gastropods—again fossils are spotty but where present numerous." Collections: UND-PC, 1961 (8/?), 1962 (6/2), and 1962 (6/27), C.I. Frye. The first and last dates are from Frye (written communs., 1990; 1962 field notes), and the other is a museum catalog entry.

Locality L0147. Frye location 108 (see following); UND-PC A500. Documents: Frye (1967, p. 361; 1961 field notebook, p. 58–60, section 3; 1963b field notebook, p. 52). Location: Morton County, North Dakota, Sugarloaf Butte (1976 PR, 1962) Quadrangle, just northwest of Huff along Highway 1806 in the NE¼ NW¼ NW¼ sec. 1, T.136N., R.80W. Elevation: 1663 ± 10 ft (506.9 ± 3.05 m) (this report). Stratigraphy: "Limonitic concretions 3′ [0.91 m] above highway, both sides of road; west side primary site" (Frye, 1990, written commun.; see also L6186). Discussion: Frye (1967) inadvertently included and excluded nearby but different collections under his location 109. In his 1963b field notes, Frye included a collection from location 108 (L147) under location 109, which was not otherwise mentioned by number in his dissertation. In section 3 of Frye's 1961 field notebook, he also referenced a brackish-water fossil locality (L6190 [L147c]), but excluded it from his dissertation (Frye, 1967). Section 5 (unit 2) (M8919). Collections: UND-PC, 1961 (8/?) and 1962 (7/4), C.I. Frye. The former date was reported by Frye (1990, written commun.) and represents the date of discovery and section measuring, and the latter is a museum catalog entry for collected specimens. The collection includes a poorly preserved, undetermined gastropod.

Locality L0148. Frye location 111; UND-PC A527. Documents: Frye (1967, p. 361; 1962 field notebook, p. 23; 1963b field notebook, p. 53). Location: Sioux County, North Dakota, Solen SW (1971) Quadrangle, NE¼ sec. 1, T.133N., R.82W. Elevation 1750 ± 50 ft (533.4

± 15.2 m). Discussion: Frye (1967) omitted the header "Section 56" from his section on p. 318. This section (M8955) also omits references to mollusks. In his 1962 field notes, Frye correlates gastropods from the "South end" of the exposure to units 10 and 12 of section 56, representing "buff sands with *Ophiomorpha* near the top" and "buff sands interbedded with thin shales near top," respectively. Collections: UND-PC, 1962 (7/6) and 1962 (7/25), C.I. Frye. The former date was reported by Frye in his 1962 field notes, and the latter is a museum catalog entry.

Locality L0149. Frye location 106; UND-PC A503. Documents: Frye (1967, p. 361; 1962 field notebook, p. 16; 1963b field notebook, p. 52). Location: Sioux County, North Dakota, Solen SW (1971) Quadrangle, road cut in central portion of E½ E½ sec. 13, T.133N., R.82W. Elevation 1750 ± 15 ft (533.4 ± 4.6 m). This locality was reported by Frye (1967) as from sec. 18, but was revised to sec. 13 by Frye (1990, written commun., confirming his 1962 field note entry). Stratigraphy: In his 1962 field notes, Frye describes the locality as "In what appears to be a slump block—at least it is tilted[;] we have 2′ [0.61 m] of lignite at base + 5′ [1.52 m] clay above + 1′ [0.30 m] of sandy clay with *Halymenites* [*Ophiomorpha*] + gastropods[,] then 2–5′ [0.61–1.52 m] of gray clay (bentonitic)[,] then 5′ [1.52 m] of sand buff when weathered but greenish when fresh." This section can be correlated with Frye's section 32 (M8931). Collections: UND-PC, 1962 (7/7) and 1962 (7/6), C.I. Frye. The former date was reported by Frye (1962 field notes; 1990, written commun.), and the latter is a museum catalog entry.

Locality L0150. Frye location 105; UND-PC A504. Documents: Frye (1967, p. 362; 1962 field notebook, p. 1, with inadequate reference to collection; 1963b field notebook, p. 51). Location: Sioux County, North Dakota, Solen (1980) and Breien (1980) Quadrangles, S½ S½ SW¼ and SW¼ SE¼ sec. 28, T.134N., R.81W. (this report). Stratigraphy: In his 1962 field notes, Frye reported that "Silicified sand + lignite appears just above ash beds at several horizons. Ash deposited in lake? [N]o cross bedding[.] [A]sh interfingers with bentonite, sandy shales, clay sands, lignite, iron concretion zones, and silicified sands"; "Contains fossil silicified tree stumps 4′–20′ [1.22–6.10 m] apart and from 1′ to 2′ [0.30–0.61 m] in diameter." Discussion: "L150 and L152 [are] parts of same site" and that the location as "south edge in west 3/4 of sec. 28" (Frye, 1990, written commun.). Collections: UND-PC, 1962 (6/15) and 1962 (6/27), C.I. Frye. The former date was reported by Frye (1962 field notebook; 1990, written commun.), and the latter is a museum catalog entry. This faunule includes *Ophiomorpha.*

Locality L0151. Frye location 107; UND-PC A502. Documents: Frye (1967, p. 362; 1963b field notebook, p. 52). Location: Sioux County, North Dakota, Breien (1980) Quadrangle, W½ E½ SW¼ sec. 29 to center of sec. 32, T.134N., R.81W., along valley exposures of north-flowing tributary of the Cannonball River. Elevation: 1740 ± 40 ft [530.4 ± 12 m] (this report). Discussion: No likely reference was found to match this locality in Frye's field notes. Reported by Frye (1967) as in "sec. 29, T.134N., R.81W." Frye (1990, written commun.) added "from C[enter] of section 29 to C[enter] section 32 along north flowing gully." Collections: UND-PC, 1962 (6/18) and 1962 (6/13), C.I. Frye. The former date was reported by Frye (1990, written commun.), and the latter is a museum catalog entry.

Locality L0152. Frye location 92; UND-PC A496. Documents: Laird and Mitchell (1942, p. 28); Frye (1964, p. 170; 1967, p. 362; 1961 field notebook, p. 67, 68; 1962 field notebook, p. 3, 28; 1963b field notebook, p. 46, 47). Location: Sioux County, North Dakota, Solen (1980) Quadrangle, N½ NW¼ sec. 33, T.134N., R.81W., Crowghost (Red Cedar) Cemetery area (this report). Laird and Mitchell (1942, section 4 [M6441] reported the locality interval from the Breien Member as from 75.67 to 82.08 ft (23.06–25.02 m) above the river bed of Cannonball River. This provides an elevation estimate of 532–534 m. An elevation estimate based on Frye's (1967) section 6 (M8920) places the locality only about 12 ft [3.7 m] higher, suggesting a likely correlation between their sections. Frye (1964 [Section M8920a]) also reported fossils from this location. Frye (1990, written commun.) revised location information of locality L0152 from "C[enter] sec. 33" to "NW¼ NW¼ sec. 33" [brackets of present authors]. Stratigraphy: Type section of the Breien Member. Laird and Mitchell (1942, p. 28) reported fossils from unit 11 of section 4 (M6441): "Grey, fine-grained sand. Weathers buff grey. *Halymenites* [*Ophiomorpha*], *Dosinopsis,* and gastroped [sic] molds found." Frye (1964, p. 170, section 6 [M8920a]) reported the lithic unit as "gray-green, glauconitic sand weathering, rusty brown with *Halymenites,* and gastropod molds near the top. Occasional *Halymenites* near the bottom." Frye (1967, p. 238, section 6 [M8920]) reported the same lithic unit, but he revised the identification in his section description to "containing *Ophiomorpha major,* and gastropod molds near the top." Discussion: Laird and Mitchell (1942) and Frye (1967) reported fossils from the Crowghost Cemetery. In 1990, Frye documented this locality by photograph. Collections: 1941, collector and location of collections unknown; UND-PC, 1962 (6/18) and 1962 (7/13), C.I. Frye. The former date was reported by Frye (1962 field notes; 1990, written commun.), and the latter is a museum catalog entry.

Locality L0153. Frye location 112; UND-PC A526. Documents: Frye (1967, p. 362; 1963b field notebook, p. 53, 54). Location: Sioux County, North Dakota, Solen (1980) or Half Timber Butte (1971) Quadrangle, SW¼ sec. 34, T.134N., R.81W. (this report). Discussion: Locality L0153 was not found referenced in Frye's field notes. In 1990, Frye photographed and located this locality on a map. Frye (1967) reported the locality as "1 mile east of Crowghost Cemetery." Collections: UND-PC, 1962 (6/18) and 1962 (7/6), C.I. Frye. The former date was reported by Frye (1990, written commun.), and the latter is a museum catalog entry.

Locality L0270. Frye location 91; UND-PC A497. Documents: Frye (1967, p. 360, section 34 [M8933]; 1962 field notebook, p. 12, unnamed east-west profile). Location: Morton County, North Dakota, Cannon Ball NW (1968) Quadrangle, near center E½ SE¼ sec. 21, T.134N., R.80W. (this report). Elevation ~557.48–566.62 m (estimated using Laird and Mitchell, 1942). Stratigraphy: Hell Creek Formation, Breien Member. Frye (1967, p. 286) reported fossils from unit 9 of section 34, but omitted referring to his collection of mollusks: "Buff to gray sands with lignitic lenses, shale partings, shark teeth, skate teeth, and *Ophiomorpha major* . . . 19 ft [5.79 m]." Discussion: Laird and Mitchell (1942) also reported a Breien section (section 1 [M6439]) from this quarter section measured by W.A. Cobban. They did not mention fossils identified by Cobban in Seager et al. (1942), which are presumed to have come from unit 21: "Sandstone: greenish-grey, soft, massive; contains shaley partings and small rusty-weathering ferruginous nodules" (Laird and Mitchell, 1942, p. 26). This faunule includes *Ophiomorpha,* a bryozoan, and an inarticulate brachiopod. Collections: UND-PC, 1962 (6/26), C.I. Frye. The inarticulate brachiopod *Lingula* and bivalve *Dosinopsis?* were also reported from this location (Seager et al., 1942; Laird and Mitchell, 1942).

Locality L6190. (Frye, 1990, written commun.; field number L147c, in association with Frye location 109; 1963b field notebook, p. 52, but without reference to mollusks). Location: Morton County, North Dakota, Sugarloaf Butte (1976 PR, 1962) Quadrangle, "river side [of] small bare hill" (Frye, 1990) in NW¼ SW¼ SE¼ NW¼ SE¼ sec. 31, T.137N., R.80W. The location plotted by Frye (1990, written commun.) results in an elevation of 498.3 ± 3.05 m, which is lower than other Breien fossils reported in this area (see L0147). The locality may be on a slump block. Discussion: Frye (1990, written commun.) noted that he inadvertently omitted reporting an *Ostrea* occurrence in his dissertation (Frye, 1967). Collections: 1961 (8/?), C.I. Frye. The

Ostrea specimen(s) reported from this locality were not found in UND-PC collections.

Locality L6531. Frye location 100; UND-PC A509. Documents: Frye (1967, p. 359; 1963b field notebook, p. 50). Location: Emmons County, North Dakota, Hazelton NW (1971) Quadrangle, SE¼ sec. 35, T.135N., R.78W. Discussion: Locality L6531 was not found referenced in Frye's field notes. Collections: UND-PC, 1962 (6/24), C.I. Frye. This faunule includes *Ophiomorpha.*

Locality L6532. Frye location 94; UND-PC A494. Documents: Frye (1967, p. 359; 1963a field notebook, p. 8; 1963b field notebook, p. 47, 48); Kennedy et al. (1998); Hoganson and Murphy (this volume). Location: Emmons County, North Dakota, Dana (1975) Quadrangle, NW¼ sec. 26, T.136N., R.76W., Breien Member at top of hill. Stratigraphy: In his 1962 (p. 8, section 23) field notes, Frye reported ascending glauconite-rich units: "Green[-]gray clay, some sand . . . much limonitic stain. Cephalopod. Green sands[,] much limonite stain. *Halymenites* [*Ophiomorpha*]." Discussion: This record is the youngest, best-preserved ammonite occurrence in the northern Great Plains of North America. The published identification of *Discoscaphites* cf. *D. conradi* (Kennedy et al., 1998) was revised by N.L. Landman (*in* Hoganson and Murphy, this volume) as *Jeletskytes* cf. *J. nebrascensis.* A second, more poorly preserved ammonite was reported by Frye (1967) from locality L0163. Collections: UND-PC, 1962 (6/23), C.I. Frye.

Locality L6534. Frye location 89; UND-PC A499. Documents: Frye (1967, p. 359; 1962 field notebook, p. 42; 1963b field notebook, p. 45). Location: Morton County, North Dakota, Solen SW (1971) Quadrangle, about center east edge sec. 10, T.133N., R.82W. (this report). Elevation: 1759 ± 15 ft (536.14 ± 4.57 m). Stratigraphy: Frye reported the occurrence of Breien clams and worm tubes in his 1962 field notes, but omitted reporting this fossil from unit 2 of his dissertation section 32 (Frye, 1967, p. 283; [M8931]): "Buff sands with worm tubes . . . 1.5 feet [0.46 m]." Discussion: Reported (Frye, 1967) as from the NE¼ SE¼ sec. 10." This faunule includes *Ophiomorpha.* Collections: UND-PC, 1962 (8/8), C.I. Frye.

Locality L6535. Frye location 113; UND-PC A525. Documents: Frye (1967, p. 359; 1962 field notebook, p. 39–41; 1963b field notebook, p. 54). Location: Morton County, North Dakota, Timmer (1971) Quadrangle, S½ SE¼ sec. 17, T.133N., R.82W. (this report). Elevation: 1742 ± 20 ft (531.0 ± 6.1 m). Stratigraphy: Fossils were reported (Frye, 1967, p. 285, 296; [M8932]) from two horizons within the Breien in section 33: unit 3: "Gray sands with plant fragments, and a siderite bed weathering to limonite with oysters in the middle . . . 2.5 feet [0.76 m]" and unit 5: "Buff to green, glauconitic sands with siderite bed weathering to limonite containing oysters at the base . . . 9.0 feet [2.74 m]." Localities L6535 and L6536 correspond to these horizons and, on the basis of matrix associated with the samples, L6535 most likely corresponds to unit 3. Discussion: Frye's 1962 measured section also notes fossils collected from what can be interpreted as the uppermost part of Frye's Fort Rice unit (see L163) and base of overlying Huff unit. Collections: UND-PC, 1962 (8/7), C.I. Frye.

Locality L6536. Frye location 110; UND-PC A528. Documents: Frye (1967, p. 359, 360; 1962 field notebook, p. 39–41; 1963b field notebook, p. 53). Location: Morton County, North Dakota, Timmer (1971) Quadrangle, N½ NE¼ sec. 20, T.133N., R.82W. (this report). Elevation: 1724 ± 20 ft (525.5 ± 6.1 m). Stratigraphy: Fossils were reported (Frye, 1967, p. 285, 296) from two horizons within the Breien in section 33 (M8932). unit 3: "Gray sands with plant fragments, and a siderite bed weathering to limonite with oysters in the middle . . . 2.5 feet [0.76 m]" and unit 5: "Buff to green, glauconitic sands with siderite bed weathering to limonite containing oysters at the base . . . 9.0 feet [2.74 m]." Localities L6535 and L6536 correspond to these horizons and, on the basis of matrix associated with the samples, L6536 most likely corresponds to unit 5. Discussion: Location of locality revised based on section data. Collections: UND-PC, 1962 (8/6), C.I. Frye. This locality includes an unidentifiable freshwater unionid.

Locality L6537. Frye location 101; UND-PC A508. Documents: Frye (1967, p. 360; 1962 field notebook, p. 3; 1963b field notebook, p. 50). Location: Morton County, North Dakota, Huff (1971) Quadrangle, NW¼ sec. 8, T.135N., R.79W., near top of hill. Discussion: In his 1962 field notes, Frye reported the locality as containing *"Halymenites* [*Ophiomorpha*] + oysters near top of hill." Collections: UND-PC, 1962 (6/16), C.I. Frye.

Locality L6538. Frye location 93; UND-PC A495. Documents: Frye (1967, p. 360; 1963b field notebook, p. 47). Location: Morton County, North Dakota, Huff (1971) Quadrangle, NE¼ sec. 18, T.135N., R.79W. Discussion: Note that "sec. 8" was reported in Frye (1967), but no reference in his field notes exists to confirm this location. Thus the UND-PC catalog entry is used here. Collections: UND-PC, 1962 (6/17), C.I. Frye.

Locality L6539. Frye location 97; UND-PC A491. Documents: Frye (1967, p. 361; 1963b field notebook, p. 48). Location: Morton County, North Dakota, Huff (1971) Quadrangle, center sec. 18, T.135N., R.79W. Discussion: No field note reference can be found for this locality. Collections: UND-PC, 1962 (6/16), C.I. Frye. This faunule includes *Ophiomorpha.*

Locality L6540. Frye location 95; UND-PC A493. Documents: Frye (1967, p. 361; 1962 field notebook, p. 1; 1963b field notebook, p. 48). Location: Morton County, North Dakota, Sugarloaf Butte (1976 PR, 1962) or Huff (1971) Quadrangle, SE¼ sec. 6, T.136N., R.79W. Stratigraphy: Reported as from middle of Breien Member. Discussion: This location apparently represents an intimately mixed freshwater (L165) and brackish-water (L6540) deposit. In his 1962 field notebook, Frye "[d]iscovered in middle Breien an oyster bed [L6540]. There seems also to be other types of clams . . . Zone varies from 1″ to 12″ [2.54–30.5 cm] in thickness—may be all that is 'normal marine.'" Collections: UND-PC, 1962 (6/16), C.I. Frye. This faunule includes *Ophiomorpha.*

Locality L6541. Frye location 96; UND-PC A492. Documents: Frye (1967, p. 361; 1962 field notebook, p. 13). Location: Sioux County, North Dakota, Selfridge SE (1971) Quadrangle, NE¼ sec. 3, T.129N., R.81W., at top [of] hill. Elevation: 1940 ± 20 ft (591.3 ± 6.1 m). Discussion: Note that Frye did not mention any fossils in his report of the Breien Member at this locality in his field notebook. This faunule includes *Ophiomorpha.* Collections: UND-PC, 1962 (6/28), C.I. Frye.

Locality L6542. Frye location 103; UND-PC A506. Documents: Frye (1967, p. 362; 1963b field notebook, p. 103). Location: Sioux County, North Dakota, Solen SW (1971) or possible Breien (1980) Quadrangle, SW¼ sec. 32, T.134N., R.81W. Discussion: No record of this locality was found in Frye's field notes. The specimens from this locality represent one of his early collections (see also L0152). Collections: UND-PC, 1962 (6/18), C.I. Frye.

Locality L6543. Frye location 99; UND-PC A510. Documents: Laird and Mitchell (1942, p. 27, section 3); Frye (1967, p. 362, 363; 1961 field notebook, p. 68, 69, section 9; 1963b field notebook, p. 49, 50). Location: Sioux County, North Dakota, Solen (1980) Quadrangle, center sec. 35, T.134N., R.81W., "west of Solen railroad cut 1.3 miles [sic] [2.1 km]" (Laird and Mitchell, 1942, p. 27). Their report places the fossils between 56.58 and 65.58 ft (17.25–20.00 m) above the base of section. Elevation: 1740 ± 20 ft (530.3 ± 6.1 m). Stratigraphy: Laird and Mitchell (1942) reported fossils from unit 5 of section 3 (M6440): "Buff-weathering grey sand with some interbedded shale layers. *Halymenites* [*Ophiomorpha*] fairly abundant toward the top of the exposure. At top of the exposure is bentonitic shale which has a whitish alkali coating. Weathered marcasite concretions abundant. Gypsum crystals were also noted. Two poorly preserved gastropod molds, one *Halymenites* mold in a concretion, and one questionable pelecypod mold were found . . . 9 ft [2.74 m].” Although not reported in Frye’s dissertation, in his 1961 field notebook (p. 68), Frye measured a very similar section representing the source of his specimens: “Buff to brown limonitic sands with limonite concretions & *Halymenites* [*Ophiomorpha*].” Collections: 1941 (collector and status of collection unknown); UND-PC, 1962 (6/17), C.I. Frye.

Locality L6580. UND-PC A3305 (~A511a). Documents: Frye (1961 field notebook, p. 64, 65, section 7). Location: Morton County, North Dakota, Breien (1980) Quadrangle, SE¼ sec. 17, T.134N., R.81W. Stratigraphy: “Below bentonitic [level] in Breien Memb.-Khc” (UND-PC collection label). Discussion: This description appears to correlate with Laird and Mitchell’s section 5 (M6442). This section was also measured by Frye in 1961, but was not mentioned in his dissertation. The specimens reported here were apparently collected by Frye (label without collector’s name). In the same measured section, Frye (1967) reported a freshwater assemblage (location 88; L164) overlying the Breien in the Fort Rice unit. Collections: UND-PC, 1961 (8/18), C.I. Frye? This collection was previously unidentified and not mentioned by Frye (1967).

Post-Breien (Fort Rice) mollusks

The following catalog of molluscan localities from the Fort Rice unit includes only those that have been associated with brackish environments because of their constituent faunule elements. Note that Frye (1967) did not mention some occurrences of the trace fossils he recognized in the field (e.g., *Ophiomorpha* from locations 73 [L0154] and 85 [L0161]), which otherwise contain freshwater assemblages.

Locality L0162. Frye location 72; UND-PC A489. Documents: Frye (1967, p. 367; 1961 field notebook, p. 67, 68; 1962 field notebook, p. 28, 29; 1963b field notebook, p. 37). Location: Sioux County, North Dakota, Half Timmer Butte (1971) and Solen (1980) Quadrangles, E½ SE¼ sec. 33, T.134N., R.81W. (this report). Elevation: 1777 ± 15 ft (541.6 ± 4.6 m). Stratigraphy: Frye (1967, p. 237) did not report mollusks from section 6, but using his 1962 field notes, the scaphopods and possibly the freshwater mollusks listed below can be referred to unit 14: “Lignitic shale and shaly lignite with siderite nodules weathering to limonite near the top . . . 2.5 feet [0.76 m].” For comparison, in his 1962 field notes, Frye provided the following description for the unit overlying the Breien Member: 1′ [0.30 m] lignite + lignitic shale[,] bone fragments[,] may be up to 3′ [0.91 m] thick [overlain by] 18″ [45.72 cm] thick semicontinuous limonitic concretions with molluscan fauna[.]” Discussion: L162 was reported by Frye (1967) as from the center sec. 33, T.134N, R.81W, Crowghost Cemetery, which he (Frye, 1990, written commun.) later revised: “E½ SE¼ and adjacent portions of sec. 34 [T.134N, R.81W.] + [secs.] 4 and 3 of T.133N., R.81W . . . Extension of L0153.” Collections: UND-PC, 1962 (7/27), C.I. Frye. This collection also includes freshwater molluscan taxa.

Locality L0163. Frye location 66; UND-PC A485. Documents: Frye (1967, p. 368; 1962 field notebook, p. 39-41; 1963b field notebook, p. 34). Location: Morton County, North Dakota, Timmer (1971) Quadrangle, W½ SE¼ sec. 17, T.133N., R.82W. (this report). Elevation: 1823 ± 20 ft (555.7 ± 6.1 m). Discussion: Frye (1967) did not report fossils from his section 33 (M8932), but he did so in his 1962 field notes: “Very fossiliferous in some places” in what is likely the basal unit of his Huff unit and “1 snail bag” in the uppermost part of his underlying Fort Rice unit. The horizon from which ?*Discoscaphites* was collected cannot be determined with available information. This occurrence is likely the youngest ammonite record in the northern Great Plains of North America, because it overlies the ammonite from the Breien Member (locality L6532). Collections: UND-PC, 1962 (8/6) and 1962 (8/4), C.I. Frye. The former date was reported by Frye in his 1962 field notes, and the latter is a museum catalog entry. This collection also includes freshwater molluscan taxa. W.A. Cobban (2001, personal commun.) reidentified the ammonite as *Hoploscaphites?*

Post-Breien (Huff) mollusks

Locality L0157. Frye location 68 (part); UND-PC A483. Documents: Frye (1967, p. 365; 1962 field notebook, p. 52–54, section 33; 1963b field notebook, p. 35, 36). Location: Sioux County, North Dakota, Selfridge SE (1971) Quadrangle, center SW¼ sec. 9, T.129N., R.81W. Elevation: 2022 ± 20 ft (616.3 ± 6.1 m). Stratigraphy: Frye (1967, p. 365) reported this locality as from the Fort Rice unit. However, in his section 40 (M8939) through this interval, he reported neither fossils nor the Fort Rice unit. The molluscan horizons are referenced in section 33 of his 1962 field notes. With field information, mollusks can be referred to units 4 and 11 of section 40 (Frye, 1967). From available lithic descriptions, units 3 and 4 may be interpreted as from the Fort Rice unit, thus making the collection from unit 4 of the Fort Rice unit and the collection from unit 11 of the Huff unit. Unit 4 is “[m]oderate-brown sands and shales with small siderite nodules weathering to limonite throughout but concentrated near the base . . . 7 feet [2.13 m]” (Frye, 1967, p. 296). In his 1962 field notes, Frye noted that the plant and mollusk fossils are from the limonitic concretions. Discussion: Frye (1990, written commun.) revised the above location from “SW¼ sec. 9, sec. 10,” but he added “L0157 and L0158 [are] really parts of same site [horizon] by [which] also [occurs] in sections 16 and 15.” Location “furthest from road at foot of bluff” [brackets by present authors]. Collections: UND-PC, 1962 (8/16) and 1962 (8/18), C.I. Frye. The former date is from Frye’s 1962 field notes (also 1990, written commun.) and the latter is a museum catalog entry. This collection also includes an extensive freshwater molluscan assemblage.

Locality L0171a. Frye location 48; UND-PC A819. Documents: Frye (1967, p. 371; 1963b field notebook, p. 25, 26). Location: Bowman County, North Dakota, Mud Buttes (1973) Quadrangle, NE¼ SE¼ SE¼ NE¼ SW¼ sec. 35, T.130N., R.105W. Elevation: 3125 ± 10 ft (952.5 ± 3.0 m). Frye (1967) reported the combined locations of UND-PC A819 and A823 as from sec. 35, T.130N., R.105W. Without specifically distinguishing the locality sites, Frye (1990, written commun.) provided additional information that suggests that the above location more precisely locates the collection from L0171a and that L0171b can be found 365 m to the southwest. Stratigraphy: Hell Creek Formation, Huff unit. Discussion: Frye (1967) included two accession numbers (A819 and A823) under his location 48. For the purposes of this report, locality L0171a represents the former collection and includes the faunule given below. The remaining invertebrate taxa belong to UND-PC A823. Collections: UND-PC, 1963 (8/16), C.I. Frye. This collection includes *Ophiomorpha irregulaire* Frey and others identified by D.J. Daly (2001, written commun.) and an unidentifiable caenogastropod (possibly the freshwater taxon *Campeloma*).

Locality L0176. Frye location 62; UND-PC A787. Documents: Frye (1967, p. 375; 1963a field notebook, a nearby section is reported on p. 26–28; 1963b field notebook field notebook, p. 30). Location: Slope County, North Dakota, Pretty Butte (1980) Quadrangle, near center NE¼ sec. 16, T.134N., R.106W. Frye (1990, written commun.) placed a flag at L0176 at 1300 ft (396 m) from the west section line and 875 ft (267 m) from the south section line. This locality was reported by Frye (1967) as from the NW¼ of sec. 16, but was revised by Frye (1990, written commun.) to near the center of the NE¼. Frye (1990, written commun.) also added that the “site mostly on west edge of road.” Discussion: Frye (1990, written commun.) noted that this locality consisted of only one or two concretions producing fossils. Collections: UND-PC, 1963 (07/5), C.I. Frye. The specimen identified as *Spaniorinus nicolleti* Speden was destroyed in shipment. A unionid, caenogastropod, and pulmonate are also known from this locality.

APPENDIX 2. BASAL HELL CREEK FORMATION LOCALITIES OF SOUTH DAKOTA

Explanation

The presence of Colgate-like lithofacies in the basal strata of the Hell Creek Formation in north-central South Dakota was documented by Waage (1968), and selected molluscan taxa were reported by Speden (1970). The importance of these few taxa is in the further acknowledgment of marine or marginal marine conditions following the deposition of the Fox Hills higher in the Cretaceous section outside of North Dakota. Neither Waage or Speden provided location specifics or complete faunules of the fossil-bearing localities, so we have provided available data from Yale Peabody Museum (YPM) Invertebrate Paleontology resources through the courtesy of Tim White and Cope MacClintock. The list also includes freshwater molluscan locality records that document the occurrence of freshwater conditions on the basis of fossils and those localities in need of additional study.

Locality L6674. YPM Localities IP.A0663, IP.C0496, IP.D5248. Location: Southeast corner of the SE¼ SE¼ NE¼ sec. 32, T.14N., R.19E., and southwest corner of the SW¼ SW¼ NW¼ sec. 33, T.14N., R.19E., Redelm NE (1951) Quadrangle, Ziebach County, South Dakota. Stratigraphy: Hell Creek Formation; 4.0–5.5 m (13–18 ft) above base. The freshwater faunule includes the freshwater bivalve "*Unio*" sp. (IP.C0496, IP.D5248) and unprocessed material (IP.A0663).

Locality L6675. YPM Localities IP.C0483, IP.C0492, IP.C0493, IP.C0074; includes Waage locality 74. Location: Center east edge, sec. 32, T.14N., R.19E., Redelm NE (1951) Quadrangle, Ziebach County, South Dakota. Stratigraphy: Hell Creek Formation, ~3.4 m above base. The freshwater faunule includes "*Unio*" sp. (IP.C0483, IP.C0492) and unprocessed material (IP.C0074, IP.C0493).

Locality L6676. YPM locality IP.A0666. Location: SW¼ sec. 33, T.14N., R.19E., Redelm NE (1951) Quadrangle, Ziebach County, South Dakota. Stratigraphy: Hell Creek Formation, basal, accumulations at top of basal lignite on butte top. Faunule represents unprocessed material.

Locality L6677. YPM locality IP.A4716; includes Waage locality 304. Location: Southeast corner of the SE¼ NE¼ SW¼ sec. 29, T.14N., R.19E., Redelm NE (1951) Quadrangle, Ziebach County, South Dakota. Stratigraphy: Hell Creek Formation, basal, in and adjacent to *Unio* bed.

Locality L6678. YPM Localities IP.A0668, IP.C0075; includes Waage locality 75. Location: Southeast corner sec. 6, T.14N., R.18E., and northeast corner, sec. 7, T.14N., R.18E., Redelm NW (1951) Quadrangle, Ziebach County, South Dakota. Stratigraphy: Hell Creek Formation, lowermost. The faunule includes the following marine taxa: *Sphenodiscus lenticularis* (Owen), *Jeletzkytes nebrascensis* (Owen), *Granocardium whitei* (Dall), *Hiatella?* sp. A, *Leptosolen* sp. indet., and *Lunatia* sp.; brackish-water taxa: *Anomia gryphorhyncha* Meek and *Crassostrea subtrigonalis* (Evans and Shumard); and other specimens: undetermined caenogastropod and unprocessed material (IP.A0666).

Locality L6679. YPM locality IP.C0497; includes Waage locality 93. Location: Center SE¼ sec. 21, T.15N., R.20E., Thunder Butte Quadrangle (1951), Ziebach County, South Dakota. Stratigraphy: Hell Creek Formation, lower. Represents float from lower Hell Creek Formation on Bullhead lithofacies of Iron Lightning Member of Fox Hills Formation. The faunule includes the identification of an undetermined Mutelidae.

Locality L6680. YPM locality IP.C0478; includes Waage locality 48. Location: N½ W¼ SE¼ sec. 31, T.21N., R.23E., Black Horse NE Quadrangle (1956, PR 1978), Dewey County, South Dakota. Stratigraphy: Hell Creek Formation; 3.2 m above base. The faunule represents unprocessed material.

APPENDIX 3. MOLLUSKS OF THE FOX HILLS FORMATION OF THE EASTERN WILLISTON BASIN

Explanation

Appendix 3 provides documentation for the stratigraphic distribution of molluscan bivalves, gastropods, and cephalopods and minor other nonmolluscan groups corresponding to the summary numbers given in Figure 5 and discussed in the text of this chapter. Fox Hills Formation fossil localities include north-central South Dakota and south-central North Dakota and lower Hell Creek localities from South Dakota. The sources of biostratigraphic data and their abbreviations in Appendix 3 are Erickson (1974 [Eri74], 1978 [Eri78], 1999 [Eri99]), Feldmann (1967 [Fel67], 1972 [Fel72]), Feldmann and Pulubniak (1975 [F&P75]), Feldmann and Kammer (1976 [F&K76]), Speden (1970 [Spe70]), and the Yale Peabody Museum, Division of Invertebrate Paleontology (YPM). State postal codes include North Dakota (ND) and South Dakota (SD).

Stratigraphy and Abbreviations

Hell Creek Formation of South Dakota
Lower Hell Creek (HC)
Fox Hills Formation (FH) of South and North Dakota
Linton Member (LT); Iron Lightning Member (IL), Colgate lithofacies (CG), Bullhead lithofacies (BH); Timber Lake Member (TL); Trail City Member (TC)

Bivalves

Nucula cancellata Meek and Hayden = *Nucula (Pectinucula) cancellata* fide Fel67; TL Fel67- Fel72-ND; TL Spe70-SD; TC Spe70-SD
Nucula n.sp. e: Feldmann 1967; TL Fel67-Fel72-ND
Nucula planomarginata Meek and Hayden = *Nucula subplana* Meek and Hayden fide Speden, as used by Feldmann; IL Spe70-SD; TL Fel67-Fel72-ND; TL Spe70-SD; TC Spe70-SD
Nucula obsoletastriata Meek and Hayden; TL Fel67-Fel72-ND
Nucula percrassa Conrad; TL Spe70-SD; TC Spe70-SD
Malletia evansi (Meek and Hayden) = *Nuculana evansi* (Meek and Hayden) fide Speden, as used by Feldmann; TL Fel67-Fel72-ND; TL Spe70-SD; TC Spe70-SD
Nuculana bisulcata (Meek and Hayden); TL Fel67-Fel72-ND
Nuculana (Nuculana) grandensis Speden; TL Spe70-SD; TC Spe70-SD
Nuculana (Jupiteria) scitula (Meek and Hayden); IL Spe70 chart only, not in YPM-SD (excluded); TL Fel67-Fel72)-ND; TL Spe70-SD; TC Spe70-SD
Nuculana tarensis (Gardner); TL Fel67-Fel72-ND
Yoldia lacrima Speden: Spe70-SD; TL Spe70-SD; TC Spe70-SD
Yoldia rectanularis Speden; TL Spe70-SD; TC Spe70-SD
Solemya subplicata (Meek and Hayden); L Spe70-SD; TC Spe70-SD
Parallelodon (Cosmetodon) sulcatinus (Evans and Shumard) = *Neomodon sulcatinus* (Evans and Shumard) fide Speden, as used by Feldmann; TL Fel67-Fel72-ND; TL Spe70-SD; TC Spe70-SD
Cucullaea nebrascensis Owen = *Cucullaca (Idonearca) shumardi* (Meek and Hayden) fide Speden, as used by Feldmann; TL Fel67-Fel72-ND; TL Spe70-SD; TC Spe70-SD
Limopsis striatopunctatus Evans and Shumard; TL Fel67-Fel72-ND; TL Spe70-SD; TC Spe70-SD
Brachidontes? sp. indet.; IL (CG); Spe70-SD
Crenella elegantula Meek and Hayden; TL Fel67-Fel72-ND; TL Spe70-SD; TC Spe70-SD
?*Crenella elegantula* Meek and Hayden; TL Fel67-Fel72-ND (poor occurrence)
Modiolus galpinianus (Evans and Shumard); CG YPM fide Spe70-SD; TL Fel67-Fel72-ND; TL Spe70-SD; TC Spe70-SD

Modiolus meeki (Evans and Shumard); TL Spe70-SD; TC Spe70-SD
Modiolus n.sp. s: Feldmann 1972; TL Fel67-Fel72-ND
Modiolus aff. M. *attenuatus* (Meek and Hayden): Speden (1970); TL Spe70-SD
Tenuipteria fibrosa (Meek and Hayden); TL Spe70-SD; TC Spe70-SD
Pseudoptera subtortuosa (Meek and Hayden) = *Gervillia recta* Meek and Hayden fide Speden, as used by Feldmann; TL (basal) Fel67-Fel72-ND for both *G. subtortuosa* and *G. recta;* TL Spe70-SD; TC Spe70-SD; basal FH Eri78
Phelopteria linguaeformis (Evans and Shumard) = *Pteria lingulaeformis* fide Speden, as used by Feldmann; IL (CG) (no locality specifically mentioned) Spe70-SD; TL Fel67-Fel72-ND; TL Spe70-SD; TC Spe70-SD
Oxytoma (Hypoxytoma) nebrascana (Evans and Shumard) = *Pteria nebrascana* fide Speden, as used by Feldmann; TL Fel67-Fel72-ND; TL Spe70-SD; TC Spe70-SD; basal FH Eri78
Syncyclonema halli (Gabb) = *Pecten (Syncyclonema) halli* Gabb fide Speden, as used by Feldmann; TL Fel67-Fel72-ND; TL Spe70-SD; TC Spe70-SD
Anomia micronema Meek; CG Fel67-Fel72, F&P75-ND; FH Fel67-Fel72-ND; TL F&P75-ND
Anomia gryphorhyncha Meek; HC YPM; IL (CG) Spe70-SD; TC Spe70-SD
Ostrea translucida Meek and Hayden = *Crassostrea pellucida* (Meek and Hayden) fide Speden, as used by Feldmann; IL (not CG; chart only) Spe70-SD; TL Spe70-SD; FH Fel67-Fel72-ND; TC Spe70-SD
Crassostrea subtrigonalis (Evans and Shumard) = *Crassostrea glabra* (Meek and Hayden) fide Speden, as used by Feldmann; HC Spe70-YPM-SD; CG Fel67-Fel72-ND; CG F&P75-ND (as *C. glabra*); IL (CG) Spe70-SD; TL (top) Fel67-Fel72-ND (*C. subtrigonalis*); TL Fel67-ND; TL F&P75-ND (as *C. glabra*)
Unio sp.; HC (chart only) Spe70-YPM-SD
undet. Mutelidae; HC YPM (not in Spe70)
Epilucina cedrensis (Stanton); TL Spe70-SD
Clisocolus moreauensis (Meek and Hayden); TC Spe70-SD; TC Eri78-ND-SD; basal FH Eri78-ND
Erycina? sp. A: Speden (1970); TL Spe70-SD
Spaniorinus nicolleti Speden; TL Spe70-SD; TC Spe70-SD
Hindsiella corsonensis Speden; TL Spe70-SD; TC Spe70-SD
Crassatellina hollandi Feldmann and Kammer = *?Astarte* n.sp. h: Feldmann 1967; TL Fel67-ND; TL F&K76-ND-SD
Granocardium (Ethmocardium) aff. *G.* (*E.*) *whitei* (Dall) = *Cardium (Ethocardium) whitei* Dall, as used by Feldmann; HC YPM fide Spe70-SD; IL (BH) Spe70-SD; TL Fel67-Fel72-ND; TL Spe70-SD
Protocardia subquadrata (Evans and Shumard); IL (CG) Spe70-SD; TL Fel67-Fel72-ND; TL Spe70-SD; TC Spe70-SD; basal FH Eri78
Protocardia sp. A; TC Spe70-SD
Cymbophora warrenana (Meek and Hayden) = *Mactra warrenana* Meek and Hayden fide Speden, as used by Feldmann; IL (CG) Spe70-SD; IL Spe70-SD; TL Fel67-Fel72-ND; TL Spe70-SD; TC Spe70-SD
Mactra formosa Meek and Hayden = probably *Cymbophora* following Speden (1970); TL Fel67-Fel72-ND
Cymbophora? nitidula (Meek and Hayden); TL Spe70-SD
Spisula? sp. indet.; IL (CG) Spe70-SD; IL Spe70-SD
Tellinimera scitula (Meek and Hayden); TL Fel67-Fel72-ND; TL Spe70-SD; TC Spe70-SD; basal FH Eri78
Tellina (Eurytellina?) cheyennensis Meek and Hayden; TL Fel67-Fel72-ND
Protodonax? sp. indet.; IL (CG); Spe70-SD
Tancredia americana (Meek and Hayden); HC (but no locality given) Spe70-SD and not in YPM (excluded); IL (CG) Spe70-SD; IL (BH) Spe70-SD; TL Fel67-ND; TL F&P75-ND; TL Spe70-SD; TC Spe70-SD
Arctica cf. *A ovata* (Meek and Hayden): Erickson (1978); TL Eri78-ND
Corbicula cytheriformis (Meek and Hayden) = ? *Corbicula subelliptica* (Meek and Hayden), *Corbicula subelliptica moreauensis* (Meek and Hayden) fide Speden; CG Fel67-Fel72, F&P75-ND; TL Fel67-Fel72, F&P75-ND
Corbicula moreauensis (Meek and Hayden); TL Fel67-ND
Corbicula sp. A: Speden (1970); HC Spe70-YPM-SD; IL (CG) Spe70-SD
Dosiniopsis deweyi (Meek and Hayden); IL (CG) Spe70-SD; TL Fel67-Fel72-ND; TL Spe70-SD; TC Spe70-SD
Leptosolen sp. indet.: Speden (1970); HC Spe70-YPM-SD
Mya? sp. A; IL (CG) Spe70-SD
Varicorbula crassimarginata (Meek); IL (not CG) Spe70-SD; TL Spe70-SD; TC Spe70-SD
Corbula monmouthensis Gardner; TL Fel67-Fel72-ND
Corbulamella gregaria (Meek and Hayden) = *Corbulamella inornata* (Meek and Hayden) fide Speden, as used by Feldmann; TL Fel67-Fel72-ND; TC Spe70-SD
Cyrtodaria minuta Speden; IL Spe70-SD; TC Spe70-SD
Hiatella? sp. A: Speden (1970); HC Spe70-YPM-SD; TC Spe70-SD
Panopea occidentalis Meek and Hayden; TL Fel67-Fel72-ND; TL Spe70-SD
Parapholas sp. A: Speden (1970); Irish Creek lithofacies (TC/TL) Spe70-SD
Opertochasma cuneatum (Meek and Hayden); TL Spe70-SD; TC Spe70-SD
Ichthyosarcolites? sp. A; TC Spe70-SD
Goniomya americana Meek and Hayden; TL Fel67-Fel72-ND; TL Spe70-SD; TC Spe70-SD
Pholadomya deweyensis Speden; TL Spe70-SD; TC Spe70-SD
Periploma subgracile (Whitfield); TL Spe70-SD; TC Spe70-SD
Periploma (Aelga?) subgracile (Whitfield); TL Eri78-ND-SD
Cuspidaria moreauensis (Meek and Hayden); TL Spe70-SD; TC Spe70-SD
Cuspidaria ventricosa (Meek and Hayden); TL Spe70-SD; TL Eri78-ND-SD; TC Spe70-SD
Nymphalucina occidentalis (Morton); TL Spe70-SD; TC Spe70-SD
Vetericardia crenalirata (Conrad) [= ? *Pseudocardia;* complicated nomenclatural history, see Vokes (1980)]; TL Fel67-Fel72-ND
Sourismis equilateralis (Meek and Hayden); TL Eri78; North Dakota (and SD); TL Fel67-Fel72-ND [*Tellina equilateralis?* Meek and Hayden of Fel67)]; TL F&P75-ND; TC Spe70-SD; TC Eri78-SD; basal FH Eri78-ND

Cephalopods

Discoscaphites rossi Landman and Waage; HC YPM-SD
Jeletzkytes nebrascensis (Owen); HC YPM-SD
Sphenodiscus lenticularis (Owen); HC YPM-SD; TL Eri78-ND
Hoploscaphites nicolleti (Morton); TC Eri78-ND; basal FH Eri78
undet. Scaphitidae: Speden (1970); HC YPM-SD

Gastropods

Margaritinae gen. and sp. uncertain: Erickson (1974); TL ERI74-ND
Neritina loganensis Erickson; *Crassostrea-Pachymelania* association Eri74-ND
Oligoptycha concinna (Hall and Meek); basal FH Eri78; TC Eri74; CG F&P75-ND
Pachymelania wyomingensis (Meek); CG Eri74-ND; TL F&P75)-ND

Pachymelania inscuplta (Meek); Eri74; See Eri74-ND; LT–IL Eri99-ND (excluded); CG F&P75-ND; TL F&P75-ND
Spironema tenuilineata (Meek and Hayden); TC? Eri74-ND
Syncera? sp.: Erickson (1974); Top FH Eri74-ND
Drepanochilus evansi Cossmann; TL Eri74-ND; TC Eri74-ND; basal FH Eri78-ND
Drepanochilus nebrascensis? (Evans and Shumard); TL Eri74-ND
Vanikoropsis nebrascensis (Meek and Hayden); TL Eri74-ND; TC Eri74-ND
Lunatia sp.: Speden (not in Spe70); HC YPM-SD
Euspira rectilabrum (Conrad); CG F&P75-ND (spelled *E. reticulabrum*); TL Eri74-ND
Euspira obliquata (Conrad); TC Eri74
Euspira subcrassa (Meek and Hayden); unnamed (LT?) or CG Eri74-ND (excluded); TL F&P75-ND
Euspira? dakotensis (Henderson); CG F&P75-ND (given as *E. dakotensis*); TC Eri74-ND
Semitriton buccinoides (Meek and Hayden); FH Eri74-ND
Rhombopsis subturritus (Meek and Hayden); TC Eri74-ND
Graphidula sp.: Erickson (1978); Basal FH Eri78
Graphidula culbertsoni (Meek and Hayden); TL Eri74-ND; TC Eri74-ND
Graphidula? cretacea (Meek and Hayden); TL Eri74-ND
Piestochilus scarboroughi (Meek and Hayden); TL Eri74-ND
Piestochilus sp.: Feldmann and Pulubniak (1975); TL F&P75-ND
Piestochilus feldmanni Erickson; unnamed (LT?) Eri74-ND (excluded)
Piestochilus galpiniana? (Meek and Hayden); FH Eri74-ND
Hercorhyncus (Haplovoluta) hollandi Erickson; TL Eri74-ND
Serrifusus dakotaensis; TL Eri74-ND; basal FH Eri78
Remera? cvancarai Erickson; TC Eri74
Cryptorhytis flexicostata? (Meek and Hayden); unnamed (LT?) Eri74-ND (excluded)
Pyropsis bairdi (Meek and Hayden); TL Eri74
Perissitys? sp. A: Erickson (1974); TL? Eri74
Perissitys? sp. B: Erickson (1974); TL Eri74
Cancellaria siouxensis Erickson; TL Eri74
Cancellaria sp.: Feldmann and Pulubniak (1975); CG F&P75-ND
Amuletum (Amuletum) minor (Evans and Shumard); Basal FH Eri74-ND
Amuletum (Lutema) sp.: Erickson (1974); TL Eri74-ND
Oliogoptycha concinna (Hall and Meek); TC Eri74-ND
Ellipsoscapha cf. *E. minor* (Meek and Hayden): Erickson (1974); TC Eri74-ND
Cylichna scitula Meek and Hayden; unnamed (LT?) Eri74-ND (excluded), TL Eri74-ND
Cylichna volvaria? Meek and Hayden; TL Eri74-ND
Goniocylichna bisculpturata Wade; unnamed (LT?) Eri74-ND; TL Eri74-ND
Goniocylichna? sp.: Erickson (1974); TL Eri74-ND

Arthropoda
Ophiomorpha nodosa; TL Spe70-YPM-SD
Ophiomorpha; TL Spe70-YPM-SD

Porifera
Cliona sp.; TL F&P75-ND; CG F&P75-ND

Bryozoa
Bryozoans encrusting; TL F&P75-ND
Bryozoans boring; TL F&P75-ND; CG F&P75-ND

REFERENCES CITED

Brown, B., 1907, The Hell Creek beds of the Upper Cretaceous of Montana; their relation to contiguous deposits, with faunal and floral lists, and a discussion of their correlation: American Museum of Natural History Bulletin, v. 23, p. 823–845.

Butler, R.D., 1980, Stratigraphy, sedimentology, and depositional environments of the Hell Creek Formation (Late Cretaceous) and adjacent strata, Glendive area, Montana [Ph.D. dissertation]: Grand Forks, University of North Dakota, 398 p.

Butler, R.D., and Hartman, J.H., 1999, Upper Cretaceous Hell Creek Formation depositional architecture in the Williston Basin—Outcrop analogs and subsurface correlations: Geological Society of America Abstracts with Programs, v. 31, no. 7, p. A71.

Carlson, C.G., 1982, Geology of Grant and Sioux Counties, North Dakota: North Dakota Geological Survey Bulletin 67, 32 p.

Calvert, W.R., Beekly, A.L., Barnett, V.H., and Pishel, M.A., 1914, Geology of the Standing Rock and Cheyenne River Indian Reservation, North and South Dakota: U.S. Geological Survey Bulletin 575, 49 p.

Cvancara, A.M., 1966, Revision of the fauna of the Cannonball Formation (Paleocene) of North and South Dakota: University of Michigan Museum of Paleontology Contributions, v. 20, no. 10, 97 p.

Erickson, J.M., 1971, Gastropoda of the Fox Hills Formation (Maestrichtian) of North Dakota [Ph.D. dissertation]: Grand Forks, University of North Dakota, 247 p. E

Erickson, J.M., 1974, Revision of the Gastropoda of the Fox Hills Formation, Upper Cretaceous (Maestrichtian) of North Dakota: American Paleontology Bulletin, v. 66, p. 131–253.

Erickson, J.M., 1978, Bivalve mollusk range extensions in the Fox Hills Formation (Maestrichtian) of North and South Dakota and their implications for the Late Cretaceous geologic history of the Williston Basin: North Dakota Academy of Science Proceedings, v. 32, p. 79–89.

Erickson, J.M., 1992, Subsurface stratigraphy, lithofacies and paleoenvironments of the Fox Hills Formation (Maastrichtian: Late Cretaceous) adjacent to the type area, North Dakota and South Dakota—Toward a more holistic view, *in* Erickson, J.M., and Hoganson, J.W., eds., Proceedings of the F.D. Holland, Jr., geological symposium: North Dakota Geological Survey Miscellaneous Series 76, p. 199–241.

Erickson, J.M., 1999, The Dakota Isthmus—Closing the Late Cretaceous Western Interior Seaway: North Dakota Academy of Science Proceedings, v. 53, p. 124–129.

Feldmann, R.M., 1967, Bivalvia and paleoecology of the Fox Hills Formation (Upper Cretaceous) of North Dakota [Ph.D. dissertation]: Grand Forks, University of North Dakota, 366 p.

Feldmann, R.M., 1972, Stratigraphy and paleoecology of the Fox Hills Formation (Upper Cretaceous) of North Dakota: North Dakota Geological Survey Bulletin 61, 65 p.

Feldmann, R.M., and Kammer, T.W., 1976, *Crassatellina hollandi,* n. sp. (Bivalvia: Astartidae) from the Fox Hills Formation (Maestrichtian, Cretaceous) of North and South Dakota: Journal of Paleontology, v. 50, p. 481–487.

Feldmann, R.M., and Palubniak, D.S., 1975, Palaeoecology of Maestrichtian oyster assemblages in the Fox Hills Formation, *in* Caldwell, W.G.E., ed., The Cretaceous System in the Western Interior of North America: Geological Association of Canada Special Paper 13, p. 211–233.

Fisher, S.P., 1952, The geology of Emmons County, North Dakota: North Dakota Geological Survey Bulletin 26, p. 47 p. F.

Frey, R.W., Howard, J.D., and Pryor, W.A., 1978, *Ophiomorpha:* Its morphologic, taxonomic, and environmental significance: Palaeogeography, Palaeoclimatology, Palaeoecology, v. 23, p. 199–229.

Frye, C.I., 1964, Marine tongues in the Hell Creek Formation, North Dakota: Compass of Sigma Gamma Epsilon, v. 41, p. 167–171.

Frye, C.I., 1967, The Hell Creek Formation in North Dakota [Ph.D. dissertation]: Grand Forks, University of North Dakota, 411 p.

Frye, C.I., 1969, Stratigraphy of the Hell Creek Formation in North Dakota: North Dakota Geological Survey Bulletin 54, 65 p.

Fürsich, F.T., 1994, Paleoecology and evolution of Mesozoic salinity-controlled benthic macroinvertebrate associations: Lethaia, v. 26, p. 327–346.

Fürsich, F.T., and Kauffman, E.G., 1984, Palaeoecology of marginal marine sedimentary cycles in the Albian Bear River Formation of southwestern Wyoming: Palaeontology, v. 27, p. 501–536.

Fürsich, F.T., and Kirkland, J.I., 1986, Biostratinomy and paleoecology of a Cretaceous brackish water lagoon: Palaios, v. 1, p. 543–560.

Hartman, J.H., 1999, The paleontologic and geologic record of North Dakota—Important sites and current interpretations, a symposium overview and geologic framework: North Dakota Academy of Science Proceedings, v. 53, p. 116–118.

Hartman, J.H., and Kihm, A.J., 1995, Age of Meek and Hayden's Fort Union Group (Paleocene), upper Missouri River, North Dakota–Montana, *in* Hunter, L.D.V., and Schalla, R.A., ed., Seventh International Williston Basin Symposium, 1995 Guidebook: Billings, Montana Geological Society, p. 417–428.

Hartman, J.H., Hunter, J.P., and Krause, D.W., 1998, The last epeiric sea (Paleocene, Cannonball Formation) in North America—The age of incursions and bounding strata based on mammals, mollusks, and foraminifera: Geological Society of America Abstracts with Programs, v. 30, no. 7, p. A-33.

Hartman, J.H., Collins, L.S., and Aubry, M.-P., 1999, New interpretations of the Cannonball Formation (Paleocene)—North America's last interior sea: Geological Society of America Abstracts with Programs, v. 31, no. 7, p. A-105.

Jeletzky, J.A., and Clemens, W.A., 1965, Comments on Eutheria, Lance *Scaphites* and *Inoceramus?* ex. gr. *tegulatus:* Journal of Paleontology, v. 39, p. 952–959.

Kennedy, W.J., Landman, N.H., Christensen, W.K., Cobban, W.A., and Hancock, J.M., 1998, Marine connections in North America during the late Maastrichtian: Palaeogeographic and paleobiogeographic significance of *Jeletzkytes nebrascensis* Zone cephalopod fauna from the Elk Butte Member of the Pierre Shale, SE South Dakota and NE Nebraska: Cretaceous Research, v. 19, p. 745–775.

Klett, M.C., and Erickson, J.M., 1976, Type and reference sections for a new member of the Fox Hills Formation, Upper Cretaceous (Maestrichtian) in the Missouri Valley region, North and South Dakota: North Dakota Academy of Science Proceedings, v. 28, p. 3–24.

Laird, W.M., and Mitchell, R.H., 1942, The geology of the southern part of Morton County, North Dakota: North Dakota Geological Survey Bulletin 14, 42 p.

Lund, S.P., and Hartman, J.H., 1999, Magnetostratigraphy of Upper Cretaceous/Lower Paleocene strata in the Williston Basin—A reference chronostratigraphic framework for paleoenvironmental and paleogeographic patterns: Geological Society of America Abstracts with Programs, v. 31, no. 7, p. A-71.

Meek, F.B., and Hayden, F.V., 1862, Descriptions of new Lower Silurian (Primordial), Jurassic, Cretaceous, and Tertiary fossils, collected in Nebraska, by the exploring expeditions under the command of Capt. Wm. F. Raynolds, U.S. Top. Engrs., with some remarks on the rocks from which they were obtained: Philadelphia Academy of Natural Science of Proceedings, v. 13, p. 415–447.

Morgan, R.E., and Petsch, B.C., 1945, A geological survey in Dewey and Corson Counties, South Dakota: State Geological Survey of South Dakota Report of Investigations, v. 49, 52 p.

Ohta, Y., 1970, A review of some Cretaceous corbiculids in North America: Paleontological Society of Japan, Transactions and Proceedings, n.s., no. 79, p. 291–315, pl. 32–34.

Ponder, W.F., and Lindberg, D.R., 1997, Towards a phylogeny of gastropod molluscs—An analysis using morphological characters: Linnean Society Zoological Journal, v. 119, p. 83–265.

Ponder, W.F., and Warén, A., 1988, Classification of the Caenogastropoda and Heterostropha—A list of the family-group names and higher taxa, *in* Ponder, W.F., et al., eds., Prosobranch Phylogeny: Malacological Review, supplement 4, p. 288–317.

Seager, O.A., Blackstone, D.L., Cobban, W.A., Downs, G.R., Laird, W.M., and Sloss, L.L., 1942, Stratigraphy of North Dakota: American Association of Petroleum Geologists Bulletin, v. 26, p. 1414–1423.

Speden, I.G., 1970, The Type Fox Hills Formation, Cretaceous (Maestrichtian), South Dakota. Part 2. Systematics of the Bivalvia: Peabody Museum of Natural History Bulletin 33, 222 p.

Stanton, T.W., and Hatcher, J.B., 1905, Geology and paleontology of the Judith River beds: U.S. Geological Survey Bulletin 257, p. 1–128, pl. I–XIII.

Stevenson, R.E., 1956, Geology of the Bullhead Quadrangle: South Dakota Geological Survey, Geologic Quadrangle Series, scale, 1:62 500, 1 sheet.

Tibert, N.E., Curran, H.A., Hartman, J.H., Belt, E.S., and Diemer, J.A., 2001, Marine flooding intervals within the freshwater Tongue River Member (Fort Union Formation–Paleocene), western Williston Basin, U.S.A.: American Association of Petroleum Geologists, Official Program Book, v. 10, p. A200.

Vokes, H.E., 1980, Genera of the Bivalvia—A systematic and bibliographic catalogue (revised and updated): Ithaca, New York, Paleontological Research Institution, 307 p.

Waage, K.M., 1964, Origin of repeated fossiliferous concretion layers in the Fox Hills Formation, *in* Merriam, D.F., ed., Symposium on cyclic sedimentation: Kansas Geological Survey Bulletin 169, p. 541–563.

Waage, K.M., 1967, Cretaceous transitional environments and faunas in central South Dakota, *in* A symposium on paleoenvironments of the Cretaceous seaway in the Western Interior: Geological Society of America, Rocky Mountain Section (preprints): Golden, Colorado, Colorado School of Mines, Department of Publications, p. 237–266.

Waage, K.M., 1968, The Type Fox Hills Formation, Cretaceous (Maestrichtian), South Dakota. Part 1. Stratigraphy and paleoenvironments: Peabody Museum of Natural History Bulletin 27, 171 p.

Yale Peabody Museum, 2001, Information concerning basic data association with the studies and identifications of K.M. Waage and I.G. Speden, January 2001, Yale Peabody Museum Web site, *www.peabody.yale.edu.*

MANUSCRIPT ACCEPTED BY THE SOCIETY SEPTEMBER 24, 2001

Printed in the U.S.A.

Geological Society of America
Special Paper 361
2002

Preliminary assessment of insect herbivory across the Cretaceous-Tertiary boundary: Major extinction and minimum rebound

Conrad C. Labandeira
Department of Paleobiology, National Museum of Natural History, Smithsonian Institution, Washington, D.C. 20560-0121, USA and Department of Entomology, University of Maryland, College Park, Maryland 20742, USA
Kirk R. Johnson
Department of Earth Sciences, Denver Museum of Nature & Science, Denver, Colorado 80205, USA
Peter Lang
Web Publishing Department, Field Museum of Natural History, Chicago, Illinois 60605-2496, USA

ABSTRACT

Forty-nine insect damage types morphologically characterize plant-insect associations spanning a 136 m composite boundary interval from the uppermost Cretaceous Hell Creek Formation to the lowest Paleocene Fort Union Formation. The time duration of this interval is ~2.2 m.y., 1.4 m.y. of which is latest Cretaceous. These data originate from 80 localities from the Williston Basin of southwestern North Dakota and are allocated to four assemblage zones consisting of 385 megafloral morphotypes of bryophytes, ferns, conifers, cycads, *Ginkgo*, monocots, and 333 dicots.

All 49 foliar damage types were produced by external foliage feeding, mining, galling, and piercing and sucking insects, each of which was assigned one of the following host specificity levels: generalized, intermediate, or specialized. A distinctive pattern emerged when these damage types were plotted stratigraphically: of the 22 damage types that survive the Cretaceous-Tertiary boundary (K-T), 55% are generalized; of the 15 damage types that are extinguished immediately below the boundary, all are specialized or intermediate, as are the 12 damage types that disappear in stepped fashion somewhat before the boundary. Within this interval only two damage types are unique to the Fort Union Formation, occurring during the lowest Paleocene. These data indicate that highly and moderately specialized associations were preferentially culled at the K-T boundary, disproportionately enriching early Paleocene floras in generalized herbivores as dominant colonizers. These data are local, and the Signor-Lipps effect may partially obscure the true pattern, thus restricting extrapolation to the Western Interior of North America. Nevertheless, (1) the selective extinction of specialist herbivores at the boundary, (2) the absence of significant evidence for immigrant colonization, (3) the prolonged occurrence of species-depauperate floras during the Paleocene, and (4) the high percentage of last occurrences at and not before the boundary, despite the Signor-Lipps effect, suggest a role for end-Cretaceous abiotic perturbation and consequent biotic response in the evolution of modern plant and insect associations.

Labandeira, C.C., Johnson, K.R., and Lang, P., 2002, Preliminary assessment of insect herbivory across the Cretaceous-Tertiary boundary: Major extinction and minimum rebound, *in* Hartman, J.H., Johnson, K.R., and Nichols, D.J., eds., The Hell Creek Formation of the northern Great Plains: Boulder, Colorado, Geological Society of America Special Paper 361, p. 297–327.

INTRODUCTION

The major biotic extinction at the end of the Mesozoic was a profound event that shaped the trajectory of life during the past 65.5 m.y. This event, the Cretaceous-Tertiary (K-T) boundary, primarily has been documented in the marine realm, although finely resolved patterns of extinction and survival have been extended to the terrestrial record. Marine organisms that are identified with the end-Cretaceous event include planktonic foraminifera (Norris, 1991; D'Hondt and Zachos, 1998), ammonites (Ward, 1995), rudistid bivalves and other marine molluscs (Kauffman, 1984; Hansen et al., 1993), cartilaginous fish (Hoganson and Murphy, 1999), and large marine reptiles. Prominent continental lineages that perished at the boundary include some freshwater mollusks (Hartman, 1999), all dinosaurs (Sereno, 1999), pterosaurs and most champosaurs (Pearson et al., 1999), many lineages of birds (Chatterjee, 1997), and a broad spectrum of mammals (Hunter and Archibald, 1999). However the fates of the two most diverse clades of terrestrial macroscopic organisms—vascular plants and insects—have only been tentatively investigated, with opposite results. Although it was once thought that plant lineages were relatively immune from the K-T extinction (Hickey, 1981; Tschudy, 1984; Briggs, 1991; Hughes, 1994), current evidence from much greater sampling indicates a major extirpation of lineages, at least regionally in North America (Wolfe and Upchurch, 1987; Boulter et al., 1988; Johnson, 1989; Johnson and Hickey, 1990; Lupia, 1999; Sweet et al., 1999). Recent estimates of vascular plant extinction for the Western Interior of North America are ~70% for all megafloral morphotypes, and 20%–31% for taxonomically coarser palynomorph form genera (Hotton, 1984; Johnson, 1992, 1999; Nichols and Johnson, 1999).

Insect diversity, by contrast, has been analyzed only at the coarser family level by several authors (Whalley, 1987; Dmitriev and Zherikhin, 1988; Rasnitsyn, 1988; Jarzembowski, 1989; Labandeira, 1992; Labandeira and Sepkoski, 1993; Jarzembowski and Ross, 1993), who concluded that there was no significant decline of diversity above background levels. Nevertheless other anecdotal observations indicate that Late Cretaceous extinctions are detectable at the genus level for groups as varied as aphidoid Homoptera (Heie and Pike, 1992), scarabaeoid beetles (Scholtz and Chown, 1995; see also Chin and Gill, 1996), and perhaps weevils (Kuschel et al., 1994). By contrast, these observations differ from evidence for long-term persistence of well-defined lineages since the mid-Cretaceous, including flies (Rayner and Waters, 1990; Borkent, 1995), leaf-mining moths (Kozlov, 1988; Labandeira et al., 1994; Powell et al., 1999), pollinivorous bees (Michener and Grimaldi, 1988), and phytophagous beetles (Farrell, 1998; Windsor et al., 1999). Case studies have documented the geochronologically shorter-term persistence of intricate associations such as bees and Malpighiaceae since the late Paleocene (Taylor and Crepet, 1987), aphids and barberry since the middle Eocene (Moran, 1989), domatia-inhabiting mites and their plant associates from the late Eocene (O'Dowd et al., 1991), and leaf-mining moths and oak associations that are traceable to the late Miocene (Opler, 1973, 1974). This latter pattern of insect resiliency and immunity from extinction also is consistent with various examinations of the response of beetles to Pleistocene climatic change, which responded by biogeographical shifts rather than in situ extinction (Matthews, 1979; Coope, 1995). Studies of the terminal Pleistocene indicate that beetles—the principal taxonomic component of insect faunas—were highly immune from major environmental perturbation, unlike the very disparate pattern for the heightened extinction of vertebrates (Graham, 1986; Martin, 1986). Obvious questions issuing from these data and hypotheses include an assessment of the types of fossil data from which these varying conclusions regarding extinction are derived (e.g., Poinar et al., 1999), and more important, formulating analyses of more refined taxonomic data to test the universality of these conclusions.

To date, the predominant approach for analysis of the fossil plant and insect records across the stratigraphically narrow confines of the K-T boundary has been temporal extrapolation by examination of larger scale diversity patterns based on body-fossil occurrences. These occurrences have been documented from well-preserved and often diverse floras (Niklas et al., 1980; Niklas, 1988) and insect faunas (Ponomarenko, 1988; Rasnitsyn, 1988; Wootton, 1990; Labandeira and Sepkoski, 1993; Jarzembowski and Ross, 1993). There has been a shift, particularly in North America, toward assessment of species-level floral diversity typically from basins that are stratigraphically and geochronologically well defined (Johnson et al., 1989; Johnson, 1992). However, such a targeted and taxonomically detailed approach is unavailable for insects because of the worldwide absence of informative and continuous body-fossil sequences across the K-T boundary (Labandeira et al., 1999). Only several Late Cretaceous insect-bearing Lagerstätten are known (exceptionally preserved deposits), notably amber-bearing strata from the probable Cenomanian of Alaska (Langenheim et al., 1960), Turonian to Campanian of New Jersey (Grogan and Szadziewski, 1988; Grimaldi, 1996), the Santonian to Campanian of the Taimyr Peninsula in northern Russia (Zherikhin and Sukacheva, 1973; Botosaneanu and Wichard, 1983), and the Campanian of Alberta (McAlpine and Martin, 1969; Brown and Pike, 1990; Heie and Pike, 1992). No relevant latest Cretaceous (Maastrichtian) or earliest Paleocene (Danian) insect deposits are documented anywhere with significant diversity (Poinar, 1992; Pike, 1993; Ponomarenko, 1995). In addition, relevant insect compression deposits are limited to the mid-Cretaceous, such as those from the Cenomanian at Orapa, Botswana (Rayner and Waters, 1990), and the Czech Republic (Fritsch, 1882, 1901), or those from the Turonian of Russia (Rasnitsyn, 1988), matched in the Paleocene by the earliest significant occurrence from the Fur Formation of the Paleocene to Eocene boundary in Denmark (Andersen, 1998; Rust, 1998), ~9.5 m.y. after the K-T event.

Because of this absence of body fossils for detecting local

insect diversity anywhere across the boundary, we have used proxy data based on plant-insect associations. Our data originate from relatively continuous sequences of well-preserved foliar material across the latest Cretaceous to earliest Paleocene boundary in the Williston Basin of southwestern North Dakota. These data record patterns of plant-host use by diverse insect herbivore species, providing a detailed pattern of insect ecological diversity across an important extinction event. Notably, such data would still be valuable even if there were a continuous insect body-fossil record at this boundary, because damage-type spectra would reveal patterns of herbivore colonization of plant hosts that are difficult to unravel from the body-fossil record alone. This fine-scale approach and type of analysis for changing herbivory patterns within geochronologically short intervals has been used for the late Paleocene to early Eocene warming interval in the greater Green River Basin (Wilf and Labandeira, 1999; Coley, 1999), which demonstrated robust correlations among paleoclimate, floral turnover, and associated insect herbivory patterns. An analysis resulting from an expanded data set stemming from the current authors' research on the K-T boundary of the Williston Basin has been reported (Labandeira et al., 2002).

WILLISTON BASIN IN SOUTHWESTERN NORTH DAKOTA

Megafloras from a composite 136 m of stratigraphic section in Upper Cretaceous and lower Paleocene exposures near Marmarth, southwestern North Dakota, have been collected since 1981 and have been studied for floristic change through the K-T boundary (Belt et al., 1984; Johnson et al., 1989; Johnson and Hickey, 1990; Johnson, 1992, 1996). This sequence is paleomagnetically well calibrated (Cande and Kent, 1995; Johnson et al., 1990; Lund and Hartman, 1999; Obradovich and Hicks, 1999; Hicks et al., this volume) and increasingly is being linked to isotope records (Stott and Kennett, 1990; Arens and Jahren, 2000; Hicks et al., 1999) and foliar physiognomy (Johnson and Hickey, 1990; Johnson and Wilf, 1996) for paleoclimatic inferences. Currently 171 localities are included for this interval, representing ~25 500 specimens located principally at the Denver Museum of Nature & Science in Colorado and the Yale Peabody Museum in New Haven, Connecticut. A minor collection is present at the National Museum of Natural History in Washington, D.C.

General geological and paleoenvironmental context

The southwestern Williston Basin of the northern Great Plains consists of badlands exposures along the north-flowing Little Missouri River. These outcrops are prominent where the Little Missouri River has cut down into the northeastern flank of the northwest- to southeast-trending Cedar Creek anticline, which structurally separates the Powder River Basin of easternmost Montana from the Williston Basin of southwestern North Dakota. Progressively younger deposits are exposed toward the northeast, corresponding to the onset of transgression for the Cannonball Seaway during the early Paleocene. Deposition during the latest Maastrichtian to early Paleocene was apparently continuous, initially consisting of fluvial, lignite-poor deposits of the uppermost Colgate Member of the Fox Hills Formation and throughout the Hell Creek Formation, followed by fluviodeltaic and lignite-rich strata of the superjacent Lower Ludlow Member of the Fort Union Formation (Johnson, 1992).

The investigated late Maastrichtian section within the Williston Basin consists of a 91 m composite sequence of horizontal mudstone beds interspersed among lenticular and often argillaceous sandstone. This interval corresponds to ~1.4 m.y. (Hicks et al., this volume) and represents a flood-plain environment, as evidenced by pedogenic features such as root traces, slickenslides, and leached horizons typical of paleosols. These strata are modified by channel deposits that include lateral accretion units, cross-bedded sandstones, and planar, laminated mudstones that are considered to represent varied fluvial environments. The floras examined from this interval originate sporadically from outcrops ranging from the uppermost Fox Hills Formation below to the top of the Hell Creek Formation above, including in places as much as 3 m of the bottommost Fort Union Formation, where the K-T boundary is defined palynologically as the disappearance of the *Wodehouseia spinata* assemblage zone (Nichols et al., 1982; but see Sweet et al., 1999). Consequently, this places the K-T boundary from the top contact of the Hell Creek Formation to within the lower 3 m of the Fort Union Formation, depending on location. This boundary regionally coincides with an iridium anomaly (Alvarez et al., 1980) and is associated with shocked quartz (Bohor et al., 1987; Izett, 1990), indicating an extraterrestrial impact (Alvarez, 1997). Iridium anomalies have been found at two sites within the Williston Basin study area (Johnson et al., 1989; Nichols et al., 2000).

The lower Ludlow Member of the Fort Union Formation represents an increase in deltaic sedimentation, indicated by a greater proportion of mudstone units, lignite beds, and planar to lenticular sandstone bodies. These examined strata, corresponding an ~0.8 m.y. interval (Hicks et al., this volume), have been interpreted as ponds, backswamp deposits, crevasse splays, and interspersed meandering channels. Floras from ~44 m of this sequence above the K-T boundary and below the first marine bed that represents a transgressive event during Torrejonian time have been scrutinized. The weathering profile of the Fort Union Formation is generally better developed, supporting distinctive modern vegetation, when compared to strata of the more stark landforms of the subjacent Hell Creek Formation.

The ~2.2 m.y. represented by the Hell Creek to Fort Union succession are characterized by dramatic paleoenvironmental changes involving prolonged, ambient trends and short-term consequences of the K-T event. The geochemical and plant physiognomic data that are the source data for these trends are correlated to a known paleomagnetic chronology (Johnson et al., 1990; Hicks et al., this volume) and inserted into a vertebrate (Pearson, 1992; Pearson et al., this volume) and marine invertebrate (Gill and Cobban, 1973) record. During the earlier Maastrichtian time of this interval, mean annual temperature was 10–15 °C, followed by rapid thermal fluctuation, and then an increase to 15–19 °C during the latest Maastrichtian (Johnson and Hickey, 1990; Johnson and Wilf, 1996), a pattern corroborated by data from Antarctica (Stott and Kennett, 1990). At or immediately below the K-T boundary, there was a sharp decrease in mean annual temperature to 7–12 °C, continuing into earliest Danian time and residing at that level for the balance of the sampled interval. By comparison to the late Maastrichtian, early Paleocene precipitation increased, as probably did the mean annual range of temperature, causing biologically significant freezing winters. The Paleocene trend of relatively cool temperatures and increased precipitation persisted, with short but occasional deviations (Ellis and Johnson 1999), until the thermal maximum of the latest Paleocene to early Eocene transition (Wilf, 2000; Wing et al., 2000).

Plant hosts

Megafloras of latest Cretaceous and earliest Paleocene age have been collected for more than 100 yr from the northern Great Plains (Newberry, 1863, 1898; Lesquereux, 1874, 1878; Ward, 1887; Dorf, 1940, 1942; Brown, 1962). However, only recently has major progress been made to resolve a historically chaotic taxonomy based on the practice of assignment of Linnean names from the overall resemblance of fossil leaves to modern taxa (Upchurch and Dilcher, 1990). By establishing additional megafloral localities and relocation of older sites that are sources of known repository collections, an expanded database of 60 000 fossil leaf specimens are now known from 8 major study areas of the northern Great Plains (Johnson and Hickey, 1990). Of these study sites, the succession near Marmarth in the Williston Basin of South Dakota is the richest taxonomically (385 morphotypes) and the most heavily sampled (25 400 specimens), originating from 171 current localities (Johnson, 1999). These floras are overwhelmingly dominated by dicotyledonous angiosperms, but also include bryophytes, pteridophytes, cycads, gingkophytes, conifers, and monocotyledonous angiosperms. Currently, 385 distinctive megafloral morphotypes have been identified, 333 of which are dicotyledonous angiosperms, with links to Linnean taxa ranging in resolution from the ordinal to species rank. On the basis of stratal distributions of megafloral occurrences, this Hell Creek to Fort Union sequence was subdivided into five assemblage zones: HCI, HCII, HCIII, FU0, and FUI, from oldest to youngest; in addition, HCI and HCII were each further subdivided into two subzones (Johnson, 1989; Johnson et al., 1989; see Johnson et al., this volume). The section in the Marmarth area is the most intensely collected megafloral interval representing the K-T boundary.

The stratigraphically lowest zone, subzone HCIa, is dominated by *"Dryophyllum" subfalcatum, Leepierceia preartocarpoides, "Vitis" stantonii,* and *"Celastrus" taurenensis.* Above this is subzone Ib, characterized by small-leafed taxa that are dominant (Johnson and Hickey, 1990). Subzone IIa is dominated by *"Dryophyllum" subfalcatum* and subordinately by *Erlingdorfia montana,* whereas subzone IIb contains *"Vitis" stantonii* and subordinately *"Dryophyllum" subfalcatum.* In contrast, zone III is dominated by *Marmarthia trivialis* and *Platanites marginata,* but also harbors an abundant palm and varied lauralean taxa. Above zone III is the lowermost 1.4 m of the Fort Union Formation, Cretaceous in age, which has not yielded significant megafloral material but is represented by a diverse palynoflora up to the iridium anomaly. This interval has been assigned to zone FU0 (Johnson, this volume). The earliest Paleocene megafloral zone FUI is a distinctive assemblage of species that are rare or absent in the underlying Cretaceous. Dominants, in decreasing order of abundance, are *"Populus" nebrascensis, "Cocculus" flabella, Dicotylophyllum anomalum, Quereuxia angulata, Platanus raynoldsii, Paranymphaea crassifolia,* and *Limnobiophyllum scutatum.* The taxodiaceous conifers *Glyptostrobus europaeus* and *Metasequoia occidentalis* are minor components. The extinction of taxa between the HCIII and FUI assemblage zones ranges from ~20% to 31% for pollen and spore form-taxa, and 70% for megafloral morphotypes. This abrupt turnover corresponds to a distinctive facies change, discussed here, and is linked to poorly drained soils and meandering channels in a flood-plain environment replete with swamps, ponds, and straight to meandering deltaic river channels. Recent analyses of lowermost Fort Union floras indicate that taxa from wetland mire deposits preferentially survived the terminal Cretaceous event and formed many of the dominant elements in early Paleocene plant communities from the Williston Basin and elsewhere in the Western Interior (Gemmill and Johnson, 1997; Johnson, 1999).

Insect herbivores

Only five undetermined insect body fossils have been found throughout the examined Hell Creek and Fort Union succession, compared to more than 7000 analyzed leaf fossils. This renders body-fossil evidence of negligible consequence for understanding the dynamics of insects at the end-Cretaceous event. Because an excellent megafloral record is present throughout this interval in the Williston Basin, a unique opportunity exists for the examination of insect-mediated damage to assess the response, or lack thereof, of insect herbivores to the K-T event. Thus, this rich leaf record permits the examination of only the herbivore fraction of the total insect biota during this interval. Herbivorous taxa nevertheless constitute approximately half to somewhat more than half of all insect species (Strong et al., 1984; Wilson, 1992), a percentage that is likely to increase because of the disproportionately greater number of described insect herbivore species over nonherbivorous ones (Erwin and Scott, 1980). Such a percentage was probably true for the Late Cretaceous and Paleogene, based on the presence of a similar proportion of herbivore versus nonherbivore insect families during this interval (from Labandeira, 1994), and investigations that many major herbivore lineages have relatively deep Cretaceous or even earlier roots (Jarzembowski, 1990; Labandeira et al., 1994; Farrell, 1998). The selection of the herbivore component of the insect fauna has an additional consequence in that consumers of primary production are typically several times more abundant as individuals than the superjacent trophic level of predators (Price, 1997). Consequently, the herbivore component of the regional insect fauna is a more ubiquitous and ecologically sensitive indicator of their trophically dependent plant hosts and associated environmental perturbation than more generalist and less abundant insect predators or the more degraded fossil record of parasitoids or parasites (Labandeira, 2002a). This approach spotlights the value of trace-fossil data in assessing ecological diversity trends across intervals of major environmental perturbation.

Insect herbivores are so taxonomically diverse and ecologically pervasive that they can be characterized by the mode in which they consume plants. The functional feeding group approach is a classification of the basic ways that insects access plant tissue, and include surface fluid feeders, seed predators, borers, and spore or pollen consumers, all which have an informative fossil history that is linked to the body-fossil records of both plants and insects (Labandeira, 1998a, 2002b). However, four important functional feeding groups—external foliage feeding, leaf mining, galling, and piercing and sucking—are abundantly documented in foliar material for many deposits from the Late Pennsylvanian to the Pleistocene, where preservation is good (Chaloner et al., 1991; Scott et al., 1992) and are represented by a rich fossil record (Labandeira, 2002b). In fine-grained clastic deposits minimally affected by postdepositional diagenetic alteration, foliar material can reveal significant detail of insect-mediated damage, including bite marks from external feeders (Stephenson, 1992; Beck and Labandeira, 1998), frass trail detail of leaf miners (Kozlov, 1988; Labandeira et al., 1994), and occasionally gross structure of three-dimensional galls (Stephenson, 1992; Larew, 1992; Waggoner and Poteet, 1996). Each of these four functional feeding groups can be subdivided into categories; e.g., external foliage feeding comprises margin feeding, hole feeding, skeletonization, and special types such as bud feeding and free feeding (Becker, 1938; Kuroko, 1960; Wong and Melvin, 1976; Coulson and Witter, 1984; Kazakova, 1985; Selman, 1988; Heron and Borowiec, 1997; Heron, 1999; Togashi, 1999). These subcategories, in turn, can be objectively categorized into damage types that are the basic recognizable and irreducible units of insect-mediated damage. Thus all insect consumption or alteration of foliar tissue can be classified into discrete morphotypes, each of which represents a distinctive and diagnosable category of plant damage based on size, shape, position, or modified tissue-associated attributes. These damage morphotypes are temporally recurring, identifiable features that are present on individual leaf specimens, often specific to plant taxa or lineages, that are present within a flora or a vertical succession of floras.

METHODS

General procedure

The procedure for determining the temporal patterns of plant-insect associations across the K-T boundary was divided into four phases. The first was taxonomic morphotype identification of each megafloral specimen (Johnson, this volume); the second was a subsequent characterization of each fossil leaf according to an insect damage type, if any (Appendix 1). Third, after assignment of each leaf specimen to a plant morphotype—infrequently listed as unidentifiable if key diagnostic features were missing—specimen numbers were assigned as an identification for each specimen. Last, these data were plotted graphically (Figs. 1 and 2). For each examined flora of the K-T succession, four types of information were recorded for each specimen: locality number, specimen number, plant host morphotype, and leaf damage type, if any. Figure 1 provides a trivariate grid expressing (1) the presence and/or absence data of damage types, (2) locality with reference to the K-T boundary and the Hell Creek–Fort Union contact, and (3) the degree of host specificity discussed below. These raw data were converted to stratigraphic ranges in Figure 2, retaining the same threefold categorization of host specificity. These ranges, like their plant host taxa (Johnson et al., 1989), necessarily decay away from the K-T boundary and terminate at the beginning and end of the studied section because occurrences were not recorded beyond the confines of the Hell Creek Formation and the Ludlow Member of the Fort Union Formation. Data for each of the 80 localities, including 5178 total Cretaceous leaves (72.9%) and 1917 Paleocene leaves (27.1%), were examined and tabulated for insect damage and assigned to one of the four megafloral assemblage zones.

Plant-host morphotypes. Examination and identification of plant hosts from the K-T succession of the Williston Basin is based on the leaf morphotype system described by Johnson (1989), which borrows the terminology of Hickey (1973, 1979). This system of identification was updated and elaborated by the Leaf Architecture Working Group (1999), using leaf architectural, venational, margin-associated and other characters for elucidating Late Cretaceous and Paleogene vascular plant taxa. The motivation for establishing a seemingly parallel system to that of traditional Linnean nomenclature was the difficulty of unambiguously recognizing fossil leaves that were assigned on superficial similarities to modern leaves, rather than circumscription by objective, commonly understood, and taxonomically significant characters (Upchurch and Dilcher, 1990). Although this system is methodologically independent of previous

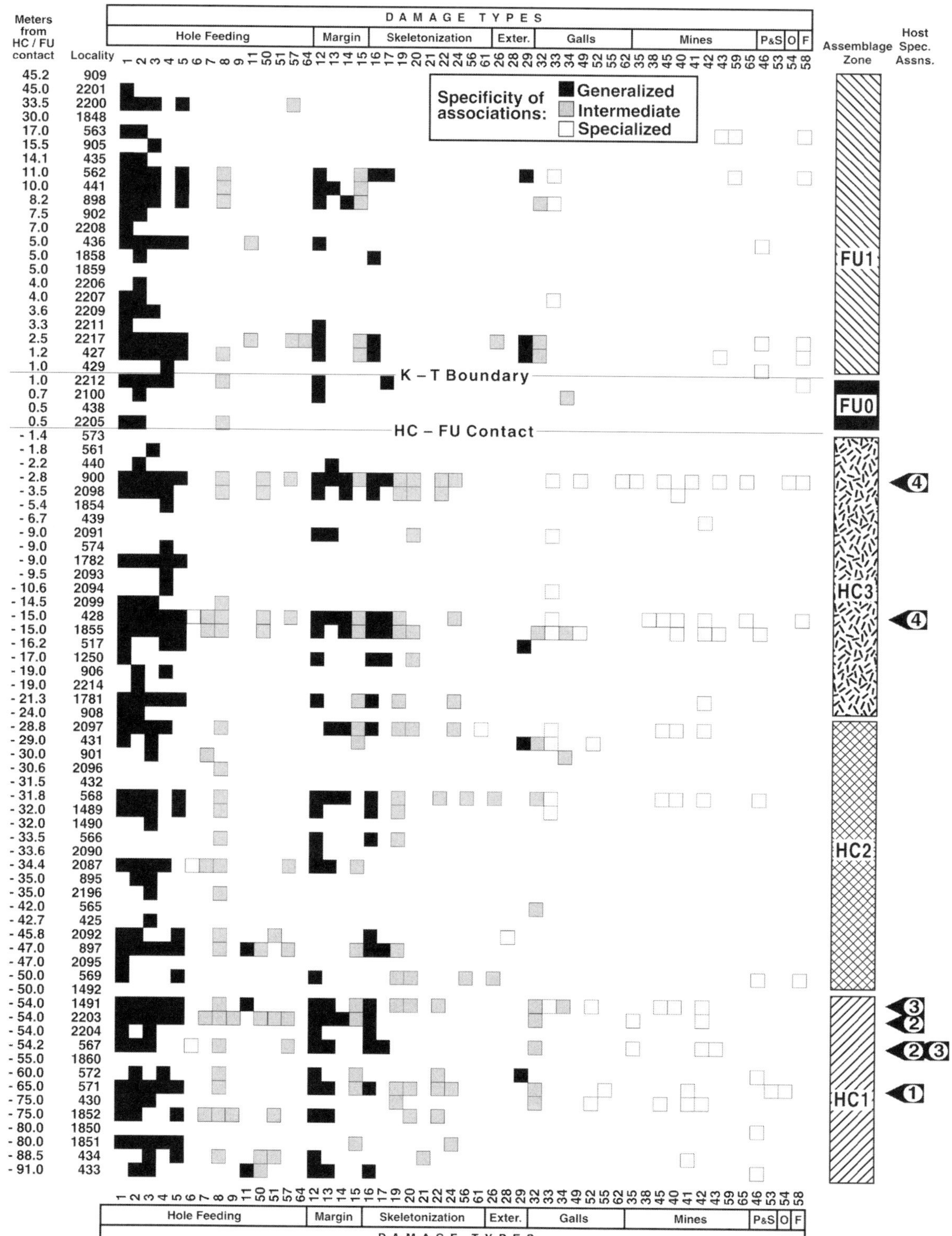

Figure 1. Matrix of damage types and inferred host specificity throughout all examined floras of upper Hell Creek (HC) to lower Fort Union (FU) formational succession in Williston Basin. Vertical stratigraphic elevations are with respect to Hell Creek–Fort Union formational contact, often not coincident with Cretaceous-Tertiary (K-T) boundary. Numbers in arrows at right refer to four host-specific associations detailed in text and in Table 1. Abbreviations: exter. = special external feeding; p&s = piercing and sucking; o = oviposition, and f = fungal damage.

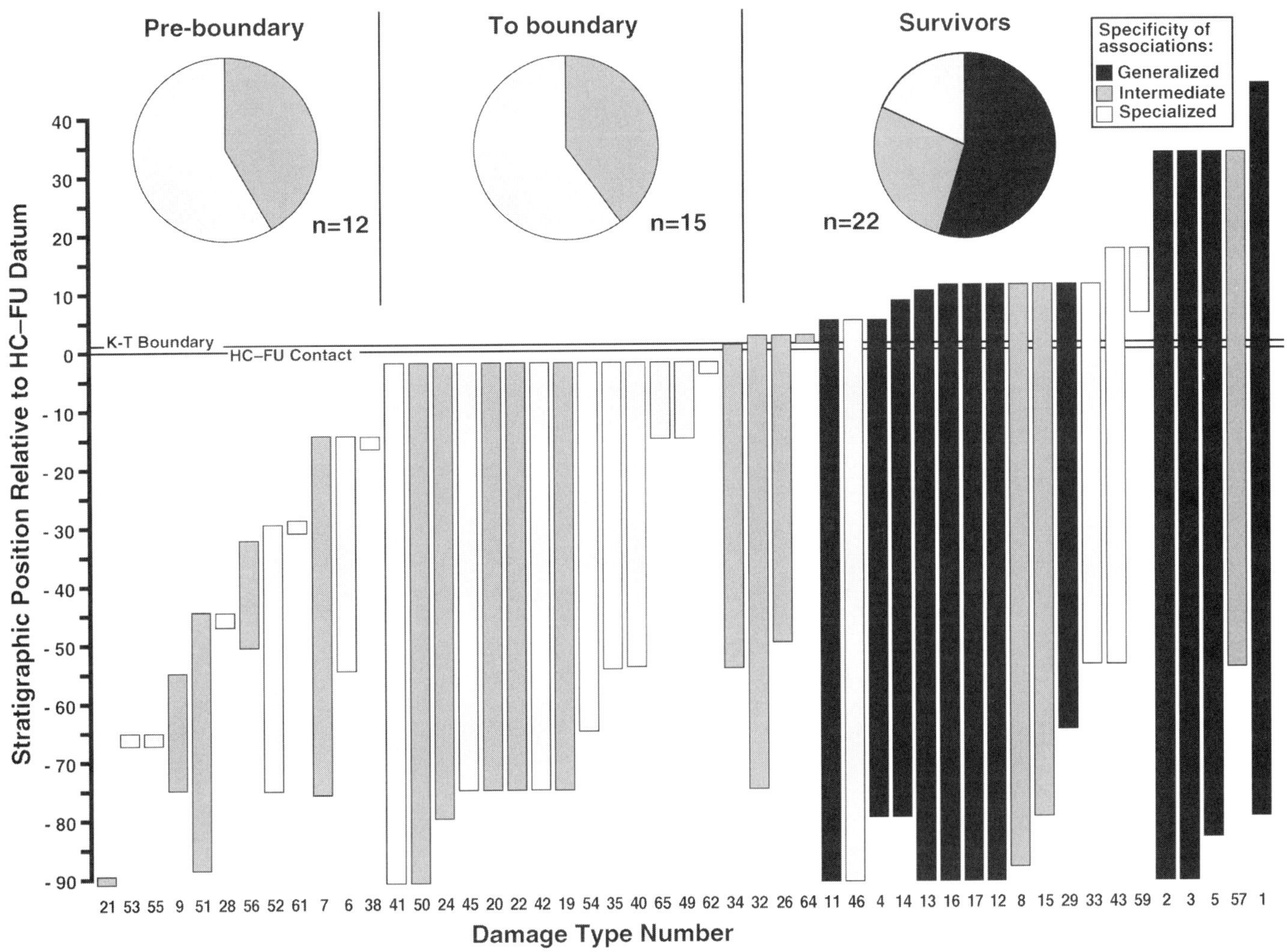

Figure 2. Summary of Figure 1, expressed as temporal ranges calibrated to standardized stratigraphic scale relative to Hell Creek–Fort Union (HC–FU) formational contact. Damage types are partitioned into three cohorts—namely those that disappear before boundary (preboundary), those that are extinguished at boundary (to boundary), and those that cross boundary (survivors). Proportions of generalized (black), intermediate (dark gray), and specialized (light gray) associations are provided for each cohort, indicating differential extinction of specialized associations before boundary and selectivity for plant host generalists among surviving taxa. K-T, Cretaceous-Tertiary.

Linnean assignments, it allows fossil leaves to be allocated to a distinctive morphotype that eventually can be upgraded to a formally described Linnean binomial once sufficient, diagnostic, and commonly understood characters are assembled from a representative suite of well-preserved material. Unlike the Linnean system, which requires a permanently designated holotype, holomorphotypes can be updated periodically as better material becomes available, providing for a reference specimen that displays more exacting criteria for eventual taxonomic placement. We acknowledge the possibility that some of these morphotypes may be taxonomically split and others amalgamated, but such reassignments would be done at low taxonomic rank and would not significantly alter our results. Although only 18% of the K-T flora has been formally and convincingly described at the species level (Johnson, 1996, 1999), it is expected that the entire flora will be converted to Linnean taxa as the quality of the morphotype reference collection is improved.

Insect damage types

A system of foliar damage types was developed by Wilf and Labandeira (1999), later elaborated (Wilf et al., 2001), explicitly to categorize recognizable insect modification of leaves in the fossil record. The initial leaf damage spectrum of 41 types, used in a study of late Paleocene to early Eocene insect herbivory of the greater Green River Basin of Wilf and Labandeira (1999), has been updated and consists of 49 damage types documented in this study. Damage type hereafter will be abbreviated DT when referring to a discussed example. An additional 16 K-T damage morphotypes not found in the previous study have been added, and 7 from the Green River list are absent in this study, resulting in a total of 49 recognizable categories for the Hell Creek to lower Fort Union succession that serves as the basis for the current study (Appendix 1). Each insect damage type from the Late Cretaceous to the early Eocene has been assigned a linked reference specimen for comparison purposes; the number of damage morphotypes is expected to increase and be replaced as better preserved and additional foliar material is examined.

These herbivore damage types can additionally be partitioned from specialized to generalized, based on knowledge of plant host specificities of modern descendants or ecologically analogous taxa, following Wilf and Labandeira (1999). Generalized associations, such as most hole feeding and margin feeding, are designated 1 (see Bernays and Chapman, 1994), whereas highly specialized associations such as leaf mines (Spencer, 1990; Hespenheide, 1991; Connor and Taverner, 1997), piercing and sucking damage (Patch, 1938; Deckle, 1965; Dixon, 1998), and galls (Gagné, 1989; Yukawa and Uechi, 1999) are assigned a 3. An intermediate assignment of 2 is given to types of damage whose modern analogs typically are not deemed either monophagously specialized or polyphagously generalized. It should be stressed that multiple, unrelated insect taxa can produce the same generalized insect damage patterns, or alternatively, the same insect species can produce different damage patterns, the latter mostly as a consequence of instar-related feeding differentiation in holometabolous insects (Buzzi, 1975). Specialized damage types, such as leaf mines, scale insect scars, and highly stereotyped hole feeding are typically synonymous with single plant-host and insect-herbivore associations, whereas intermediate associations have much less host specificity. Based on a prolific, mostly agriculturally based literature of modern plant-insect associations (summarized in Packard, 1889; Houard, 1909a, 1909b; Hering, 1951, 1967; Zerova et al., 1988, 1991; Kolomoets et al., 1989; Johnson and Lyon, 1991; Alford, 1991; and numerous other sources), generalized, intermediate, and specialized damage patterns typically can be recognized. We have taken a conservative approach, opting to characterize a damage type as specialized only if supported by widespread extant ecological observations, supplemented with patterns of host use in the investigated K-T interval. We recognize the probable inclusion of some specialist feeders possessing nondiagnostic consumption patterns within the generalized damage types of external foliage feeding. However, this inclusion would bias our results against a conclusion that recognizes the extinction of host specialists at the end of the Cretaceous.

We have established four explicit criteria by which we have based our determinations of foliar herbivory (Labandeira, 1998b), as opposed to abiotically mediated damage (Katerman, 1990; Vincent, 1990). The first criterion is presence of plant-reaction tissue adjacent to or otherwise surrounding the damaged area, including thickened callus, intervening necrotic tissue, or an encircling reaction zone some distance away (Beck and Labandeira, 1998). Second, we have used microscopic details of the damaged tissue margin, such as veinal stringers (Kazikova, 1985) or minute, cuspate tissue flaps (Gangwere, 1966) indicative of chewing activity. Third, we have noted stereotypy of the plant host damage in terms of overall pattern, position on the leaf, and size and shape of the individual damage elements (e.g., Kile, 1974; Heron and Borowiec, 1997; Heron, 1999). Last, we have noted the presence of host specificity patterns for all damage types, which would be highly atypical if the causative agent were abiotic. Single or various combinations of these four criteria in almost all instances were sufficient to separate insect-mediated from other damage, although the recognition of fungal damage (Dilcher, 1965; Sinclair et al., 1987) presented special problems in some instances. An additional complication was detection of normally delicate damage types that occur in coarser grained deposits, where the threshold of preservational detail is higher. Thus coarser sandy deposits frequently lack detail of delicate insect damage such as small leaf mines or piercing and sucking punctures that would be evident in finer grained depositional facies: in such instances, where damage was preservationally marginal but could not be documented convincingly, they were not included as data.

RESULTS

Four major results characterize this preliminary analysis of the K-T succession from the Williston Basin: (1) the greatest diversity of insect damage types so far examined explicitly from any fossil flora; (2) the occurrence of ~15 unique, host-specialized associations within the Cretaceous portion of the succession, four of which are detailed in herein (Table 1); (3) the extirpation of many associations by the K-T event, particularly specialist ones, and the subsequent preferential enrichment of generalist herbivores; and (4) the lack of a recovery phase of plant and/or insect associations during the sampled early Paleocene.

Plant host and damage-type diversity and abundance

Currently 49 diagnosable plant-insect associations have been recovered from the Hell Creek to Fort Union boundary succession of the Williston Basin. Some are abundant, particularly generalized types, but there are also some intermediate and specialized damage types; others are rare, mostly specialized damage types (Appendix 1). Based on functional feeding groups, these damage types are represented as 30 for external foliage feeding, 7 for galling, 9 for mining, 2 for piercing and sucking, and 2 extraneous types (oviposition and undifferentiated fungal damage), the latter of which was noted but not included in our analyses. We were unable to find pollen- or nectar-related activities on preserved but relatively rare flowers, although their presence in the Late Cretaceous is provided by other types of well-preserved deposits and inferences from body fossils (Crepet and Friis, 1987; Schlüter 1987; Willemstein,

1987). Nevertheless, the recorded damage types represent a broad spectrum of herbivore associations that are similar to those from extant warm-temperate to subtropical floras. Several of the plant host and insect herbivore associations are spatially and temporally unique for the fossil record, and some have been reported previously (Lang, 1996). Four selected examples are discussed here, representing the external foliage feeding, galling, piercing and sucking, and leaf-mining functional feeding groups.

A total of 7095 fossil leaves were examined for this study, 5178 of which occur in the Hell Creek Formation and represent 54 localities, and 1917 of which occur in the Fort Union Formation and represent 26 localities. Of these specimens, ~62% originate from 12 deposits with greatest leaf abundances and are distributed rather evenly throughout the K-T interval (Table 2). These 12 floras contain 8–25 damage types, although one with the least number of specimens of these is anomalously low, with 5 damage types. Of the 80 floras, 9 lacked any indication of insect-mediated damage (Fig. 1), and 18 contained only one damage type. However, there is a monotonic increase in specimens per locality from several to 135 specimens, culminating in the tally provided in Table 2. We inventoried the entire relevant collection from the K-T boundary of the Williston Basin housed at the Denver Museum of Nature & Science (as of December 2000). A subsequent analytical study has examined 6500 additional specimens inventoried at the Peabody Museum of Yale University (Labandeira et al., 2002).

With the exception of bryophytes and sphenopsids, all major taxa of plants occurring within the studied interval display evidence of interactions with insect herbivores. Several fern species (Fig. 3, A and B) and the conifer *Metasequoia* exhibit cuspate margin feeding and ovoidal hole feeding (DT 12 and DT 2, respectively), but the incidence of attack is rare. Ginkgophytes, represented by *Ginkgo adiantoides,* bear generalized skeletonization (DT 16) similar to that described by Hase (1955) (Fig. 3E), a less incised version of cuspate margin feeding (DT 12), and characteristic slot-hole feeding (DT 8) comparable to those of monocots (Selman, 1988). Cycads, present as *Nilssonia yukonensis,* are characterized by generalized window feeding (DT 29) (Fig. 3, C and D), reminiscent of modern feeding damage by the basal chrysomelid beetle *Aulacoscelis* on *Zamia fairchildiana* in Panamá and Costa Rica (Windsor et al., 1999; Cox and Windsor, 1999). No specialized associations (level 3) were found in nonangiospermous plants. Monocotyledonous angiosperms, principally the ginger *Zingiberopsis attenuata,* exhibited slot feeding (DT 8) and strip feeding inclined or perpendicular to venation (DT 28) (Fig. 3, F and G), characteristic of some extant hispine beetles (Strong, 1977; Wilf et al., 2000). A few nymphaealean and indeterminate monocots

TABLE 1. FOUR SELECTED HOST-SPECIFIC ASSOCIATIONS FROM THE HELL CREEK FORMATION, WILLISTON BASIN OF NORTH DAKOTA

DMNH localities and Hell Creek–Fort Union datum horizon (m)	Plant host and morphotype	Damage type	Functional feeding group	Text figure
900 (−2.8 m)	*Marmarthia pearsonii* (Laurales) (HC161)	DT 45	Leaf miner	12
428 (−15.0 m) 567 (−54.0 m) 1491 (−54.2 m)	Undescribed dicot (HC135)	DT 33	Galler	11
567 (−54.0 m) 2203 (−54.2 m)	Undescribed dicot (Urticales) (HC81)	DT 57	External foliage feeder	10
571 (−65.0 m)	*Erlingdorfia montana* (Platanaceae) (HC57 + HC64)	DT 53	Piercer-and-sucker	9

Note: DMNH—Denver Museum of Nature & Science.

TABLE 2. LOCALITIES WITH THE GREATEST NUMBER OF EXAMINED LEAF SPECIMENS

DMNH locality	Stratigraphic elevation to Hell Creek–Fort Union contact (m)	Number of specimens examined	Number of damage types
428	−15.0	836	20
900	−2.8	560	25
2217	+2.5	436	14
2203	−54.0	413	14
567	−54.2	396	11
2087	−34.4	378	8
571	−65.0	296	18
2097	−28.8	268	13
2200	+33.5	258	5
1491	−54.0	203	17
898	+8.2	187	10
427	+1.2	177	12
		4408	

Note: DMNH—Denver Museum of Nature & Science.

displayed rare hole feeding and two mines, although palms evidently do not show any evidence of herbivory. Dicotyledonous angiosperms, by contrast, collectively displayed virtually all of the 49 damage types described in Appendix 1. Included in this broad spectrum of insect-mediated damage is margin feeding (Fig. 4), hole feeding associated with leaf laminae (Fig. 3, B, F, G, and L), and primary veins (Fig. 5), of which characteristic slot-feeding by chrysomeloid beetles (Fig. 6) and skeletonization along primary veins indicative of leaf rollers or tiers (Fig. 7, A–N) or other external feeders (Fig. 3, M-R; Fig. 7, O and P), are very distinctive. In addition, various types of primary-vein petiolar and leaf-blade galls (Fig. 8, A–F), serpentine and

Figure 3. External feeding damage to nondicotyledonous plants; bud feeding, stereotyped hole feeding, and serpentine skeletonization from floras of Hell Creek Formation. A: Cuspate margin feeding (arrow) on unidentified filicalean fern (locality [loc.] 2087, DMNH-19962; HC241, DT 12). B: Medium hole feeding (arrows) on unidentified filicalean fern (loc. 2087, DMNH-19927b; HC241, DT 2). C: Skeletonization or surface feeding on *Nilssonia yukonensis* (Cycadales) (loc. 571, DMNH-7128; HC164, DT 16, 24). D: Detail of damage in C, exhibiting continuity of secondary venation and surrounding darker reaction rim. E: Generalized skeletonization on *Ginkgo adiantoides* (Ginkgoaceae) (loc. 568, DMNH-19979; HC114, DT 16). F: Slot hole feeding (*Cephaloleichnites*) on *Zingiberopsis attenuata* (Zingiberaceae) (loc. 2092, DMNH-19959; HC196, DT 8) (Wilf et al., 2000). G: Hole feeding and skeletonization on *Zingiberopsis attenuata* (Zingiberaceae), showing reaction rims (arrows) and continuity of secondary venation through damaged area (loc. 2092, DMNH-19957; HC36, DT 26). H: Medium polylobate hole feeding (bud feeding?) occurring approximately symmetrical on each side of midvein of unidentified dicot (loc. 900, DMNH-7240; HC-UNID, DT 3, DT 17, ?DT 6). I. Bud feeding on *Liriodendrites bradacii* (Magnoliales) (loc. 428, DMNH-20066; HC166, DT 6). J: Probable bud feeding on immature leaf of "*Cinnamomum*" *lineafolia* (Trochodendrales) (loc. 567, DMNH-19950; HC92, DT 6). K: Probable bud feeding, indicated by two arrows, on *Humulus* sp. (Cannabinaceae) (loc. 2087, DMNH-19972; HC243, DT 6). L: Polylobate hole feeding on unidentified species of Ulmaceae (loc. 2098, DMNH-20017; HC224, DT 3). M: Skeletonization of very small, immature leaf of "*Cinnamomum*" *lineafolia* (Trochodendrales) (loc. 567, DMNH-19358; HC92, DT 1). N: Serpentine skeletonization, possibly terminal phase of leaf mine (?), on *Liriodendrites bradacii* (Magnoliales) (loc. 900, DMNH-7317; HC166, DT 20). O: Serpentine skeletonization of unidentified dicot (loc. 1855; DMNH-20003a; HC254, DT 20). P: Curvilinear skeletonization on *Erlingdorfia montana* (Platanaceae) (loc. 1250, DMNH-19997; HC57, DT 20). Q: Curvilinear skeletonization associated with midvein of *Marmarthia pearsonii* (Laurales) (loc. 900, DMNH-7713; HC162, DT 20). R: Detail in Q showing feeding path and affected venation. Solid scale bars = 1.0 cm, except in G and J, where striped bars = 0.1 cm.

blotch miners (Figs. 8, G-M), piercing and sucking (Fig. 9), rare bud feeding (Fig. 3, H–K), and oviposition (Fig. 8, N and O) are also present. All major subclasses of dicots (sensu Cronquist, 1981) were herbivorized, except for Caryophyllidae and Asteridae, which were very rare in the flora; the most intensively herbivorized subclasses were Magnoliidae, especially Laurales, and Hamamelidae, particularly Platanaceae, Cercidiphyllaceae, Trochodendraceae, and Urticales.

Host specific associations

In assemblage zone HCI, at Denver Museum of Nature & Science (DMNH) locality 571, located −65.0 m from the HC-FU datum, *Erlingdorfia montana* (Brown) Johnson, a trifoliately compound member of the Platanaceae (leaf morphotype HC57), exhibits large, circular to subtly ovoidal structures 5–9.5 mm in diameter that always are situated on or adjacent to primary veins (Fig. 9). This host-specific association (DT 53) is known only from one locality that has yielded 32 instances of damage, representing a highly variable number of circular structures per leaf, including 7 on a single specimen (Fig. 9, A and B), 4 on another specimen (Fig. 9, C and D), 3 on 2 specimens, 2 on 3 specimens (Fig. 9, E, F, I, M, P), and 1 on 10 specimens (Fig. 9, G, H, N, O). In several specimens, prominent areas of discolored foliar tissue surround each or a group of these circular structures (Fig. 9, A, G, J, O, L), interpreted as a reaction response similar to chlorotic zones surrounding extensive scale infestations in modern Platanaceae (Wheeler, 1980). These circular structures lack fusain or other indication of thickened tissue, such as hyperrelief or hyporelief depression, into the matrix that would suggest a gall. We attribute these structures to scale insects (Hemiptera: Coccoidea); they are similar in size, shape, and plant host position to extant members of the Coccidae (common scale insects) that commonly occur on broadleaved dicots (Comstock, 1916; Deckle, 1965; Williams and Kosztarab, 1972). Modern Platanaceae are heavily attacked by a broad spectrum of insects (Smith, 1944; Brown and Eads, 1965; Filer et al., 1977; Thompson and Solomon, 1986; Klausnitzer, 1988; Halperin, 1990), and prominently include piercing and sucking forms such as coccids (Wunn, 1925; McConnell and Davidson, 1959; Murakami, 1970), leafhoppers (Brown and Eads, 1965; McClure, 1974), and lacebugs (Maceljski and Balarin, 1972; Rogers et al., 1982). Although the scale damage on *Erlingdorfia* is large by modern standards, it is well within the size range of some extant Margarodidae, some tropical species of which reach 40 mm in diameter (Essig, 1926; Carver et al., 1991), and Coccidae such as *Eulecanium caryae*, which occurs on hickory and occasionally on *Platanus occidentalis* (Williams and Kosztarab, 1972; Johnson and Lyons, 1991), reaching a maximum diameter of 15 mm. Notably these fossil scales cluster exclusively on primary veins—a habit that is present in many piercing and sucking folivorous insects that insert their stylets into more nutritionally rewarding vascular tissue rather than intervening epidermis and parenchyma (Sanders, 1905; Dodge, 1943; Wakgari and Giliomee, 1998). Koteja (1985) and Foldi (1997) discussed the colonization of angiosperms by scale insects during the Late Cretaceous and their resemblance to many modern forms.

A second example, this time an intermediate-specialized association, originates from zone HCI, occurring at DMNH localities 567 and 2203, at −54.2 and −54.0 m, respectively, from the HC-FU datum. Two externally feeding damage types are evident in this host: very common DT 57 consisting of triangular to ovoidal to reniform hole feeding in angles subtended by primary and secondary veins (Fig. 10), and uncommon DT 50 consisting of circular feeding along primary veins (Fig. 10, I). These two types of damage, while delicate and distinct, were probably caused by different insects. Occasional slot feeding of DT 8 also has been noted, although this feeding mode undoubtedly was caused by a species of a chrysomelid beetle or even a weevil. Hundreds of individual damage traces (DT 57) have been observed from these two floras, occurring almost exclusively on the five-lobed, serrately margined herbaceous leaf probably representing an undescribed species of

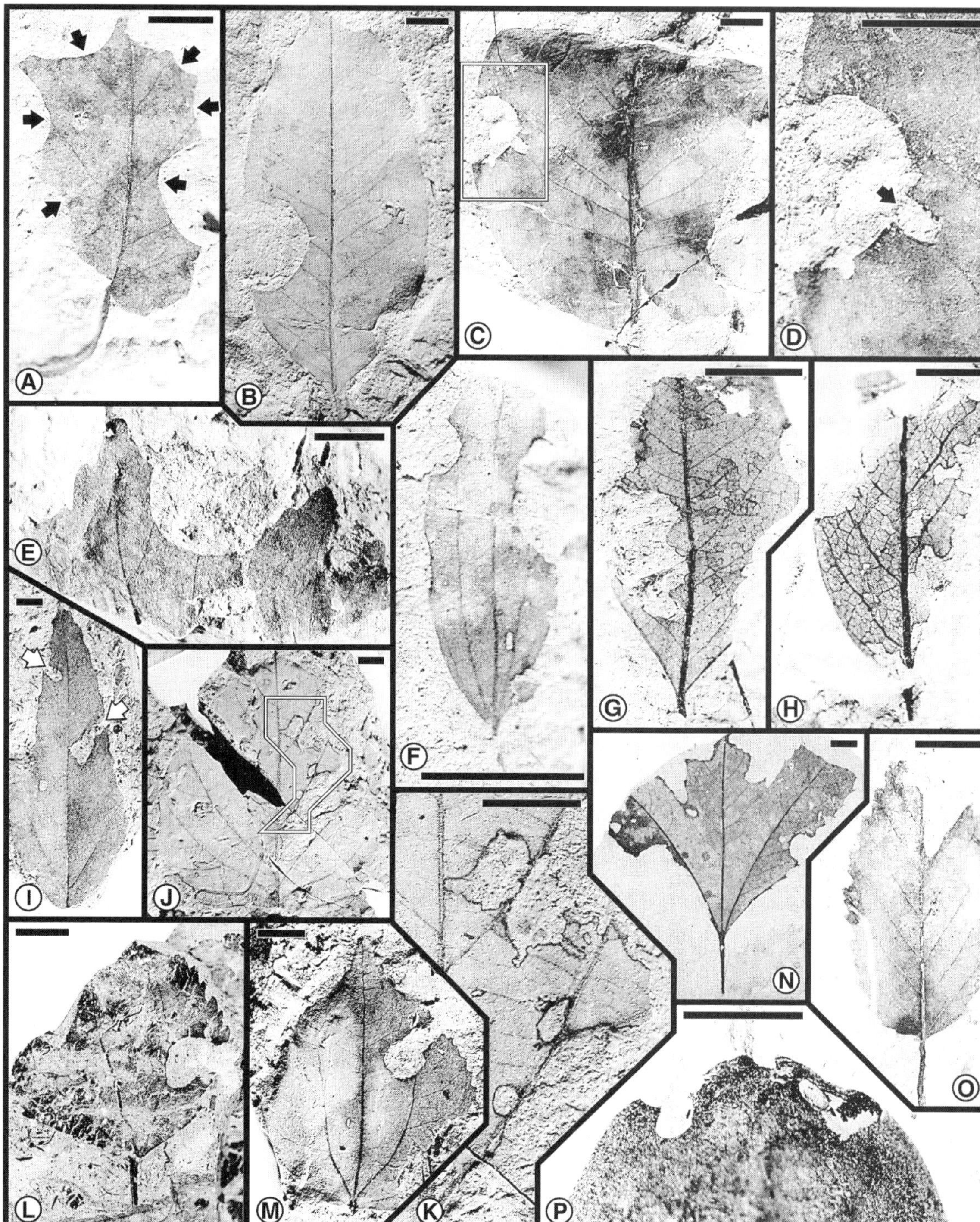

Figure 4. Varied types of insect margin feeding from Hell Creek (A-D, F, G, K-P) and Fort Union (E, J) Formations. A: Cuspate margin feeding on unidentified dicot (locality [loc.] 2091, DMNH-20058; HC-UNID, DT 12), resembling that of extant megachilid bees. Arrows indicate separate events of leaf-margin excision. B: Cuspate margin feeding on unidentified dicot (loc. 2091, DMNH-19722; HC285, DT 12). C: Cuspate and secondarily trenched margin feeding on undetermined magnoliid dicot (loc. 2091, DMNH-18874; HC66, DT 12, DT 15). D: Detail of cuspate excision in C; arrow indicating a secondary feeding event. E: Cuspate margin feeding on leaf fragment of *Dicotylophyllum anomalum,* unidentified dicot (loc. 898, DMNH-20048; FU29, DT 4, DT 14). F: Cuspate margin feeding on immature leaf of *Erlingdorfia montana* (Platanaceae) (loc. 568, DMNH-20074; HC64, DT 12). G: Margin feeding to midrib in "*Dryophyllum*" *subfalcatum,* unidentified dicot (loc. 2097, DMNH-19986; HC49, DT 14). H: Margin feeding to midrib and hole feeding in unidentified dicot "*Dryophyllum*" *subfalcatum* (loc. 2097, DMNH-19988; HC49, DT 3, DT 14). I: Trenched margin feeding on undescribed platanaceous leaf (loc. 900, DMNH-20025; HC329, DT 15). J: Varied margin feeding and hole feeding on *Pterospermites cordata* (Dilleniidae) (loc. 898, DMNH-20052; FU61, DT 2, DT 3, DT 14, DT 15). K: Detail in J, showing extent of damage. L: Trenched margin feeding on *Cercidiphyllum ellipticum* (Cercidiphyllaceae) (loc. 2087, DMNH-19967; HC212, DT 15). M: Another example of trenched margin feeding on *Cercidiphyllum ellipticum* (Cercidiphyllaceae) (loc. 1491, DMNH-19073; HC212, DT 15). N: Extensive cuspate, apical, and primary-vein associated margin feeding and hole feeding on *Erlingdorfia montana* (Platanaceae) (loc. 567, YPM-6148; HC57, DT 3, DT 12, DT 13, DT 14). O: Margin feeding associated with primary vein of unidentified ulmaceous dicot (loc. 2098, DMNH-20018; HC36, DT 14). P: Apex of unknown dicot exhibiting trenched margin feeding and medium hole feeding (loc. 1491, DMNH-18887; HC135, DT 2, DT 15). Solid scale bar = 1.0 cm.

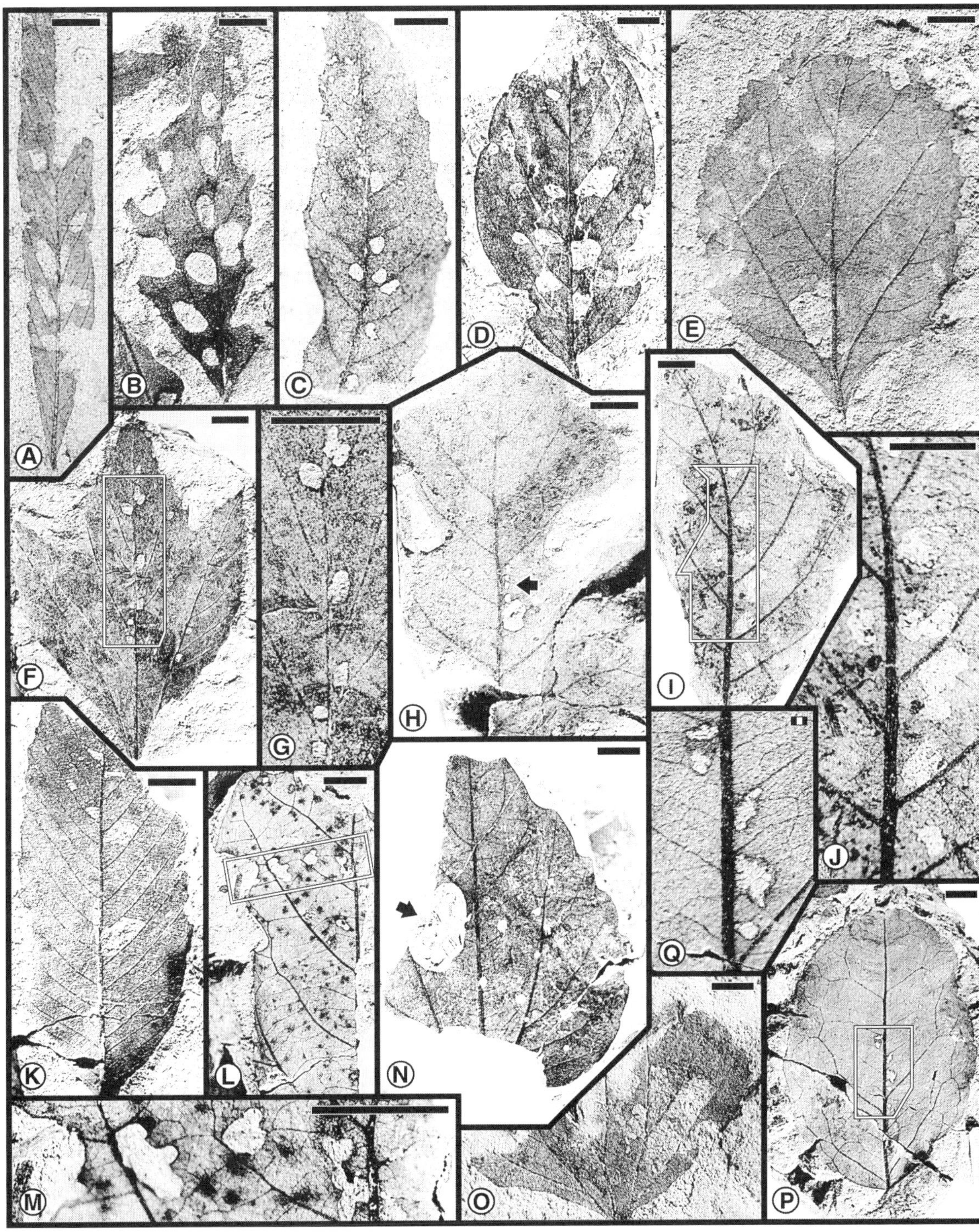

Figure 5. Examples from floras of Hell Creek Formation of mostly hole feeding and skeletonization associated with primary veins. A: Midvein-associated hole feeding on "*Dryophyllum*" *tennesseensis* (Magnoliidae) (locality [loc.] 428, DMNH-20063; HC44, DT 50). B: Extensive hole feeding and margin feeding between secondary veins on unidentified dicot (loc. 900, DMNH-7595; HC?330, DT 2, DT 14). C: Midvein-associated hole feeding on "*Dryophyllum*" *subfalcatum,* undetermined dicot (loc. 433, DMNH-13196; HC49, DT 50). D: Medium and large polylobate hole feeding on unidentified dicot (loc. 1489, DMNH-19977; HC254, DT 3, DT 5). E: Hole feeding at basal vein angles of *Grewiopsis saportana* (Platanaceae) (loc. 428, DMNH-6722; HC2, DT 57). F: Hole feeding adjacent to central primary vein on *Erlingdorfia montana* (Platanaceae) (loc. 1855, DMNH-20010; HC57, DT 50). G: Detail of hole feeding illustrated in F. H: Small and medium sized hole feeding adjacent midvein of lauralean dicot, indicated by arrow (loc. 428, DMNH-20059; HC199, DT 24, DT 39). I: Midvein associated hole feeding on unidentified dicot (loc. 428, DMNH 6789; HC-UNID, DT 50). J: Detail of damage adjacent to midvein in I, showing ovoidal and polylobate hole feeding. K: Removal of intersecondary areas by hole feeder on "*Rhamnus*" *cleburni* (Rhamnaceae) (loc. 2088, DMNH-19958; HC62, DT 63). L: Medium sized, polylobate hole feeding and margin feeding on *Bisonia niemi* (Lauraceae) (loc. 1781, DMNH-19991; HC200, DT 3). M: Detail of damage in L, perhaps showing secondary fungal infection in blackened necrotic areas. N: Large and small hole feeding on *Marmarthia trivialis* (Laurales) (loc. 1781, DMNH-19992; HC105, DT 5). O: Skeletonization of intersecondary vein areas on *Platanites marginata* (Platanaceae) (loc. 428, DMNH-7431; HC106, DT 16, DT 19). P: Midrib-associated ovoidal skeletonization of unidentified magnoliid dicot (loc. 900, DMNH-7184; HC43, DT 24). Q: Detail of skeletonization in P, showing thick reaction rims. Scale bars: solid = 1.0 cm, except in Q, where striped bar = 0.1 cm.

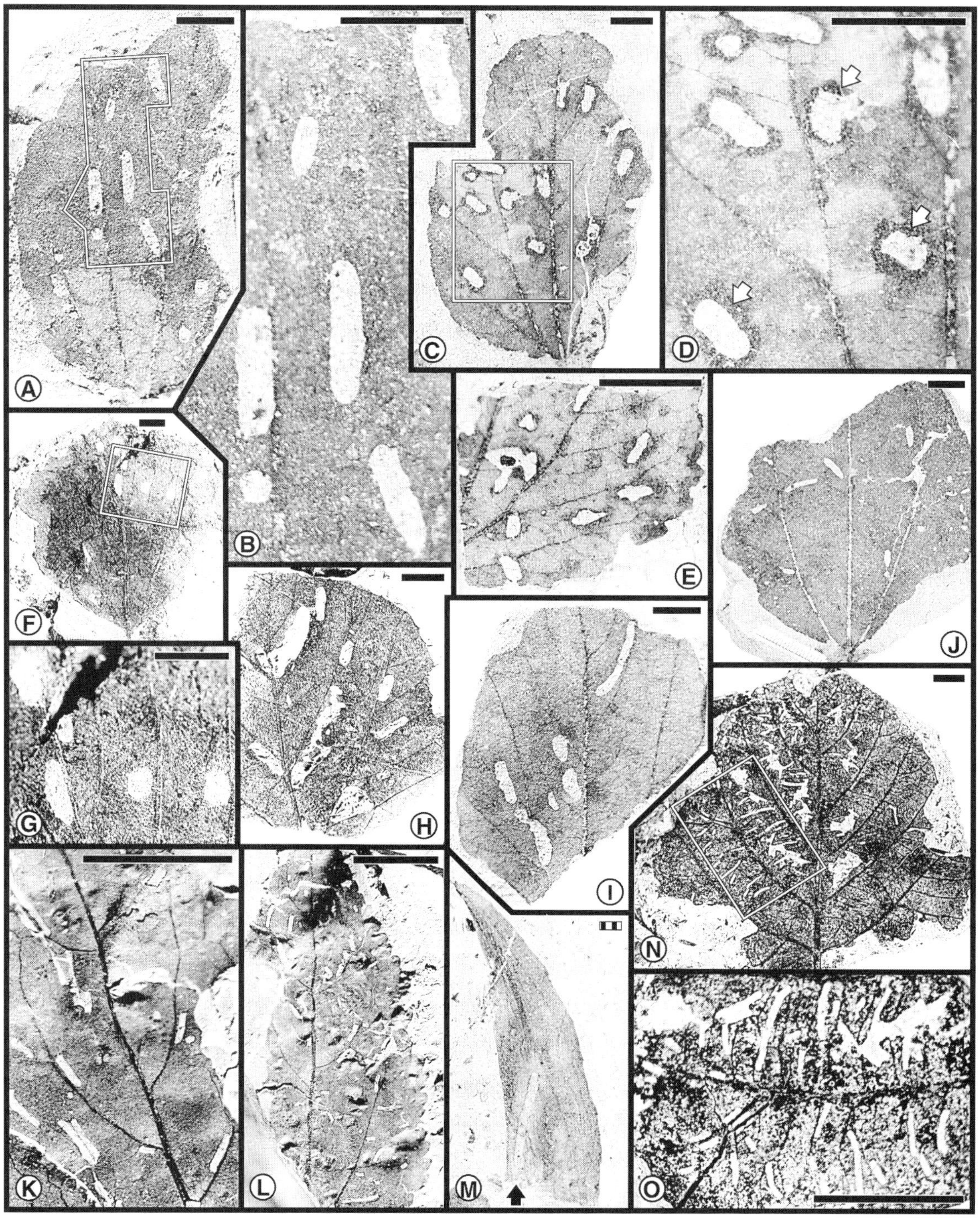

Figure 6. Examples of hole-feeding damage from floras of Hell Creek Formation, deployed principally as slots parallel to major venation. A: Slot hole feeding (top) and rectangular skeletonization (bottom) on *Marmarthia trivialis* (Laurales) (locality [loc.] 428, DMNH-20065; HC105, DT 8, DT 19). B: Detail of slot hole feeding in B. C. Slot hole feeding on *Marmarthia trivialis* (Laurales) with exceptionally thick reaction rims (loc. 428, DMNH-7495; HC105, DT 8). D: Detail of slot hole feeding in C; arrows indicate well-developed reaction rims. E: Slot and medium hole feeding on platanaceous leaf fragment, similar to C (loc. 434, DMNH-18651; HC209, DT 3). F: Medium, ovoidal hole feeding on *Platanites marginata* (Platanaceae) (loc. 906, DMNH-19995; HC106, DT 2, DT 8). G: Detail of hole feeding in F, showing avoidance of primary veins. H: Slot and large polylobate hole feeding on undetermined platanaceous leaf (loc. 897, DMNH-19955; HC98, DT 4, DT 8, DT 57). I: Slot and possible polylobate hole feeding on *Marmarthia trivialis* (Laurales) (loc. 428, DMNH-13617; HC105, DT 8). J: Another example of elongate slot hole feeding on *Marmarthia trivialis* (Laurales) (loc. 428, DMNH-6850; HC105, DT 8). K: Slot hole feeding on cercidiphyllaceous leaf fragment (loc. 2203, DMNH-19611; HC229, DT 8). L: Example of slot hole feeding on cercidiphyllaceous leaf (loc. 2203, DMNH-19550; HC229, DT 8). M: Elongate slot on unidentified dicot (loc. 567, DMNH-19386; HC131, DT 8), indicated by arrow. N: Linear and polylobate slot hole feeding parallel to tertiary venation of platanaceous leaf (loc. 434, DMNH-18659; HC209, DT 51, DT 60). O: Detail of distinctive feeding damage in N showing linear and anastomosing slots. Solid scale bars = 1.0 cm, except in M, where striped bar = 0.1 cm.

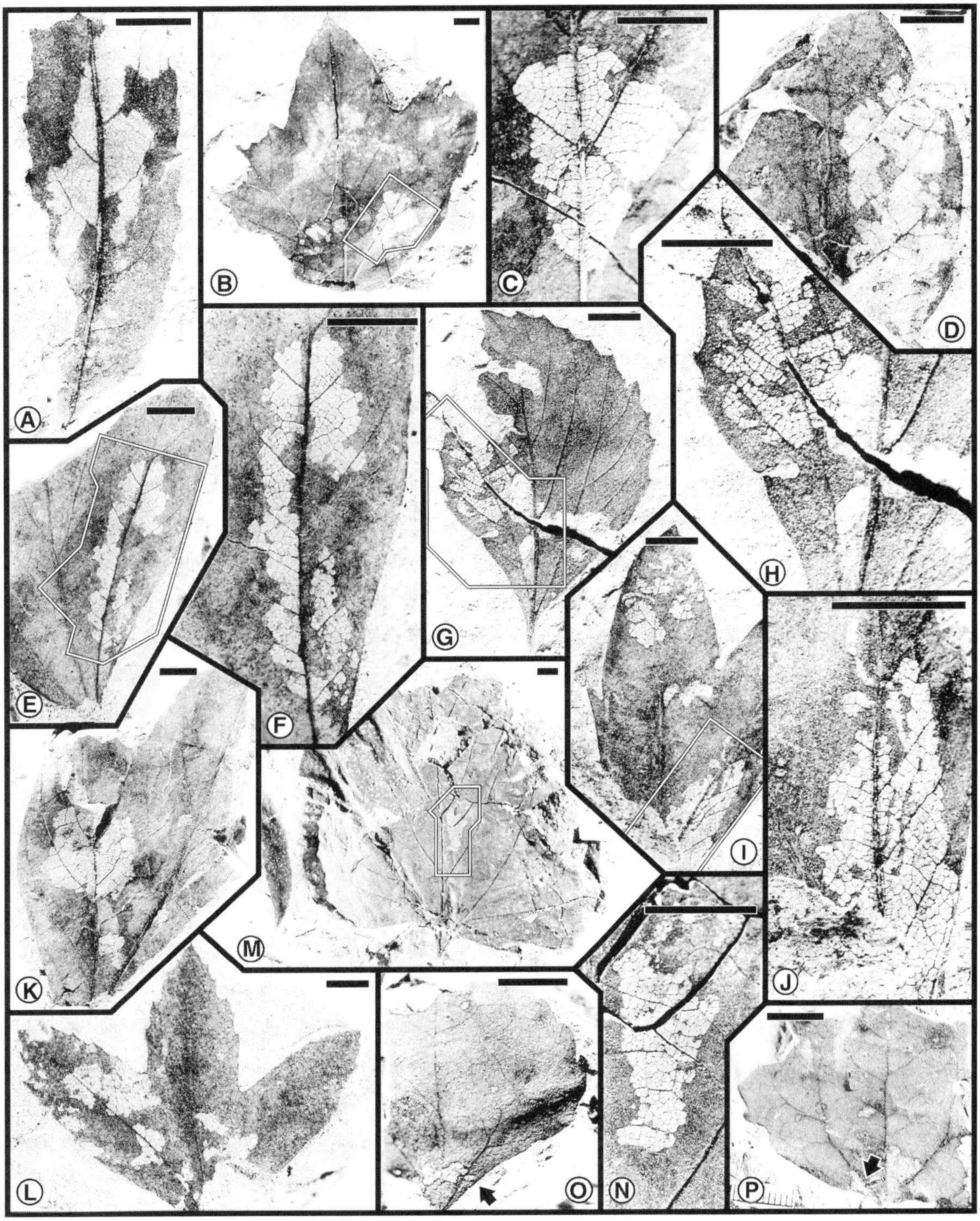

Figure 7. Leaves from floras of Hell Creek Formation, exhibiting elongate skeletonization along primary veins, suggesting folivory by leaf rollers or tiers. A: Rectangular skeletonization oriented obliquely on midrib of unidentified dicot (locality [loc.] 571, DMNH-7646; HC49, DT 19). B: Primary-vein associated skeletonization on *Erlingdorfia montana* (Platanaceae) (loc. 571, DMNH-18835; HC57, DT 16). C: Detail of skeletonized region in B, showing venational meshwork. D: Rectangular skeletonization parallel to midrib and leaf margin of unidentified dicot (loc. 1781, DMNH-19993a; HC-UNID, DT 19). E: Rectangular skeletonization associated with primary vein of *Erlingdorfia montana* (Platanaceae) (loc. 568, DMNH-20074; HC64, DT 19). F: Detail of skeletonization in E, displaying venational network. G: Skeletonization and medium hole feeding on unidentified dicot (loc. 571, DMNH-18881; HC-UNID, DT 2, DT 16). H: Detail in G, exhibiting tertiary venation at upper left and hole feeding at lower right. I: Skeletonization on lobe apex of *Erlingdorfia montana* (Platanaceae) leaf (loc. 571, DMNH-11014; HC57, DT 19). J: Detail of skeletonization associated with primary vein in I. K: Symmetrical skeletonization along primary vein of *Erlingdorfia montana* (Platanaceae) leaf (loc. 571, DMNH-11013; HC64, DT 19). L: Obliquely oriented skeletonization along primary vein of *Erlingdorfia montana* (Platanaceae) (loc. 568, DMNH-6256; HC57, DT 19). M: Elongate skeletonization parallel to midrib of undetermined platanaceous leaf (loc. 571, DMNH-8564; HC99, DT 16). N: Detail in M, illustrating tertiary and quaternary veinal network. O: Skeletonization of basal region of *Cercidiphyllum ellipticum* (Cercidiphyllaceae) leaf (loc. 1491, DMNH-19070; HC212, DT 20). P: Skeletonization of basal leaf region of *Trochodendroides nebrascensis* (Trochodendrales) leaf (loc. 569, DMNH-8490; HC103, DT 20). Solid scale bars = 1.0 mm.

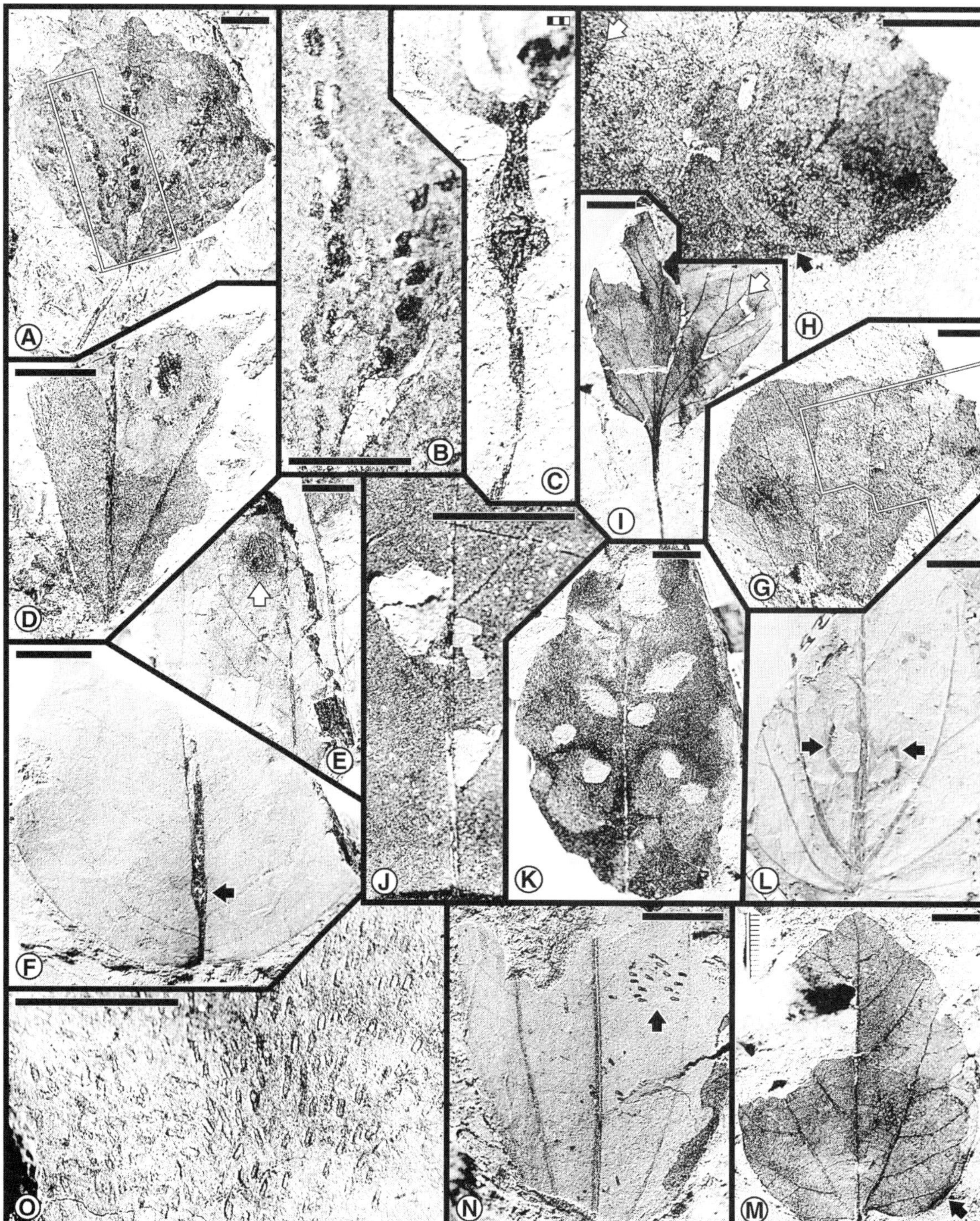

Figure 8. Examples of galls, blotch and linear leaf mines, and oviposition from floras of Hell Creek (A–K, M–O) and Fort Union (L) Formations. A: Galls on primary veins of *Trochodendroides nebrascensis* (Trochodendrales) (locality [loc.] 1489, DMNH-19976; HC103, DT 33). B: Detail of two gall-bearing veins in A; 34 galls have been recorded from this leaf. C: Petiolar gall on unidentified dicot (loc. 571, DMNH-18876; HC-UNID, DT 55). D: Gall on lamina of *Marmarthia pearsonii* (Laurales) showing reaction rim at some distance from gall core (loc. 900, DMNH-7210; HC162, DT 49). E: Large, fusain-containing gall on unidentified dicot (loc. 900, DMNH-7606; HC-UNID, DT 62). F: Incipient midrib gall (arrow) on unidentified magnoliid dicot (loc. 2091, DMNH-20011; HC43, DT 33). G: Blotch leaf mine on *Grewiopsis saportana* (Platanaceae) leaf (loc. 428, DMNH-20061; HC2, DT 35). H: Detail in G, showing tissue disruption and frass trail delineated by arrows. I: Serpentine skeletonized region on magnoliid (?Laurales) dicot, indicated by white arrow, probably representing leaf miner due to increase in damage width (loc. 2097, DMNH-19984; HC32, DT 61). J: Possible blotch mines in primary vein angles of *Erlingdorfia montana* (Platanaceae) (loc. 897, DMNH-19952; HC57, ?DT 57). K: Probable blotch mines on unidentified dicot (loc. 900; DMNH-7238; HC-UNID, DT 19); holes may represent pupation chambers with exfoliated epidermal tissue. L: Two distinctive, linear leaf mines with expansive terminal chambers (arrows) on each side of midrib of *Paranymphaea crassifolia* (?Nympheales) leaf (loc. 563, DMNH-20055; FU1, DT 59). M: Linear leaf mine along margin of "*Ficus*" *planicostata* (Laurales) (loc. 900, DMNH-7506; HC4, DT 41). N: Linearly oriented oviposition scars on *Marmarthia pearsonii* (Laurales) (loc. 900, DMNH-7265; HC162, DT 54). O: Radially oriented oviposition scars on *Erlingdorfia montana* (Platanaceae) (loc. 571, DMNH-11007; HC57, DT 54). Solid scale bars = 1.0 cm, except in C, where striped scale bar = 0.1 cm.

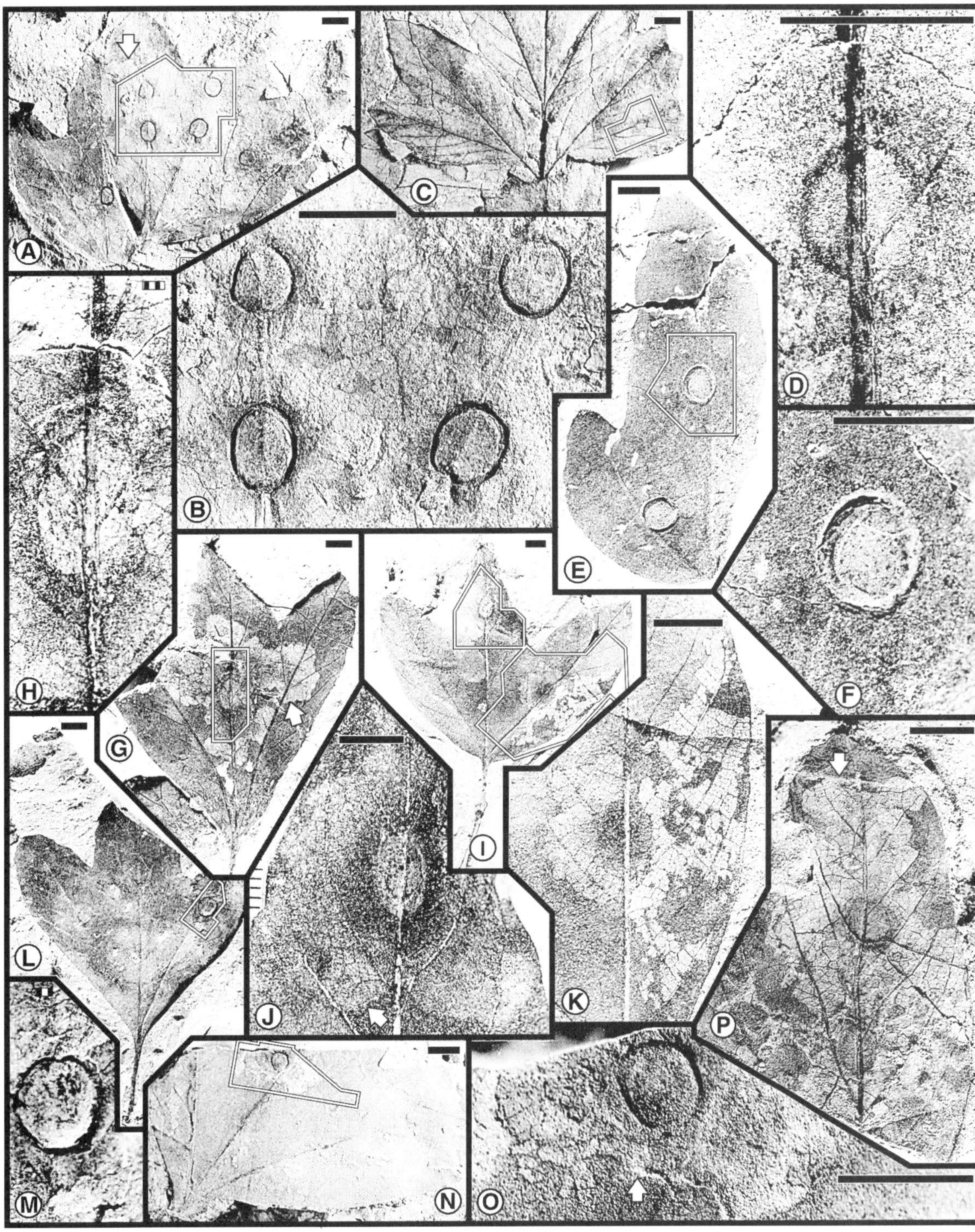

Figure 9. Host-specific association between a platanaceous host, *Erlingdorfia montana,* and a coccoid scale (Hemiptera), from Hell Creek Formation. All specimens originate from locality 571, represent damage type 53, and are found only on primary veins of compound leaf representing morphotypes HC57 (median leaves) and HC64 (lateral leaves) of *Erlingdorfia montana* (Platanaceae). A: Seven scales on simple leaf (DMNH-18829b), showing tissue imprints typical of pit scales. Note lighter hued region surrounding scales, indicating plant host-tissue response. B: Magnified detail of A, showing occurrence of four scales on two primary veins. C: Four scales on compound leaf (DMNH-18830). D: Enlargement of portion of right leaflet in C, showing scale and apparently unaffected primary vein. E: Two scales on leaf fragment, showing craterlike structure (DMNH-8587) representing opposite leaf surface to that of B. F: Detail of upper scale in E, showing well-developed rim. G: Portion of compound leaf showing central scale and associated zone of probable necrotic tissue; white arrow delineates boundary with unaffected tissue (DMNH-18836a). H: Magnified area in G, showing scale and unaffected primary vein. I: Simple leaf showing two scales (DMNH-11002). J: Detail of upper scale in I, revealing concentric ring within scale and enveloping necrotic zone (white arrow). K: Detail of lower right scale in I, showing associated necrotic zone and perhaps resulting in limited skeletonization. L: Leaf with two scales (DMNH-7641). M: Magnification of scale in L, showing structure. N: Scale on simple leaf (DMNH-6282). O: Detail of scale in N, revealing necrotic zone (white arrow) that apparently includes digitate skeletonization. P: Two scales on leaf lobe, displaying extensive necrotic zone defined by conspicuous margin (white arrow). Solid scale bar = 1.0 cm.

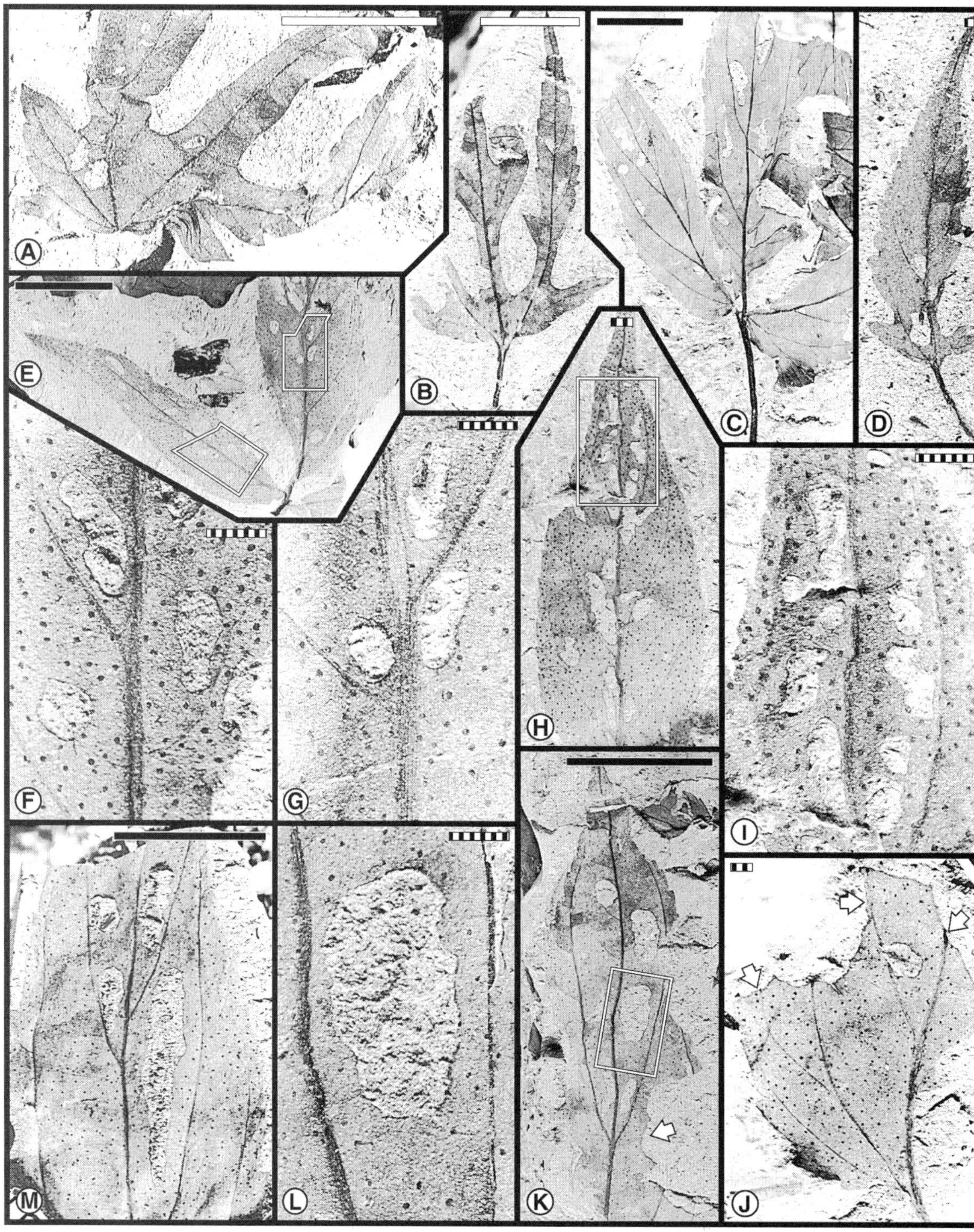

Figure 10. Host-specific association between an undescribed species of Urticales and an external foliage feeder, probably a weevil (Curculionidae) or leaf beetle (Chrysomelidae). This feeding pattern on leaf morphotype HC81 is characterized by damage type 57 and to lesser extent DT 50, consisting of triangular to ovoidal hole feeding in primary vein angles; one example of DT 12 is provided in J. This association occurs only at localities 567 and 2203. A: Almost complete, quinquepartite leaf with hole feeding at major vein angles (locality [loc.] 567, YPM-6703) B: Another relatively complete, quinquepartite, leaf with vein-angle hole feeding (loc. 567, YPM-6661). C: Tripartite leaf with damage (loc. 2203, DMNH-19206). D: Undivided leaf with margin serrations and vein-angle hole feeding (loc. 567, YPM-6661). E: Tripartite leaf with extensive hole feeding at basal vein angles (loc. 2203, DMNH-19539) F: Magnified region in E, showing vein-angle hole feeding amid mineralized epidermal structures on foliar surface, expressed as round dots. These features probably represent the silicified bases of stinging hairs. G: Closeup of E, showing juxtaposition of hole feeding and silicified structures. H: Lobe of leaf showing extensive vein-angle hole feeding (loc. 2203, DMNH-19549). I: Enlargement of region in H, illustrating consumption of nonvascular foliar tissue, including epidermis rich in presumptive stinging hair bases. J: Cuspate margin feeding (DT12), showing indiscriminate consumption of silica-containing foliar tissue (loc. 2203, DMNH-19585). Arrows point to individual feeding events revealed by crescentic margins. K: Extensive interveinal hole feeding and margin feeding (arrow) (loc. 2203, DMNH-19597). L: Closeup of region in K, showing avoidance of vascular tissue but consumption of silica-bearing intervening tissue. M: Vein-angle hole feeding on leaf lobe with extensive siliceous structures (loc. 2203, DMNH-19592). Solid scale bars = 1.0 cm; striped scale bars = 0.1 cm.

Urticales, and corresponding to leaf morphotype HC81. Although this damage type is structurally indistinguishable from that of several other leaf taxa in younger floras (Fig. 1), it is highly specific to HC81 at these two localities based on a survey of 809 leaf specimens. The sole exception is a HC228 leaf from locality 2203, which also is placed in the Urticales, the same order as HC81 (Cronquist, 1981). Lastly, an extremely small, linear leaf mine has been recorded in one specimin (DMNH 19619, not shown) that is consistent with a sap-feeding agromyzid miner occurring on such extant urticalean species as common nettle *Urtica dioica* L. and hemp *Cannabis sativa* L. from North America and Eurasia (Hering, 1932; Spencer, 1976, 1990).

The insect consumption of herbaceous HC81 provides an important occurrence in the early history of angiosperm antiherbivore defense. Under high magnification, the foliar surfaces of this taxon contains abundant, evenly spaced structures that may be calcium carbonate cystoliths in the hypodermis or more likely the silicified epidermal bases of stinging hairs (Thurston, 1974; Cronquist, 1981), features of which are amber in color and texture (Fig. 10E–10M). This leaf species only occurs in two localities and is similar to at least a few modern genera of Urticaceae (L. Hickey, 2001, personal commun.), although floral features are not present for confident assignment to any extant or extinct genus within this family. Likely herbivore coandidates, especially for DT 57, are leaf beetles of the Chrysomelidae (Essig, 1926; Jolivet and Hawkeswood, 1995; Konstantinov and Vandenberg, 1996), and less likely lepidopteran caterpillars such as members of the Pyralidae and Noctuidae (Forbes, 1923, 1960). Several modern insect taxa produce similar hole damage on native nettle, *Urtica dioica* L., including leaf beetles, weevils, and lepidopteran caterpillars (Bredemann, 1959; Carter and Hargreaves, 1986; Alford, 1991; Amelung, 1995). Implicit in this fossil association is herbivore avoidance from foliage laden with mineralic compounds, suggesting an association at 66.0 Ma similar to that of some pests on Urticaceae today, and perhaps to the deterrent effect that grass phytoliths have on extant graminivorous insects (Dravé and Laugé, 1978; Bernays and Barbehenn, 1987; Köhler et al., 2000).

The third host-specific association originates from DMNH localities 1491 and 567, situated −54.0 and −54.2 m, respectively, below the HC-FU contact, within assemblage zone HCI. This association consists of 588 galls that are thick, circular, and characteristically range from 0.5 to 3.0 mm in diameter, although the overwhelming majority are 0.75–1.25 mm in diameter (Fig. 11). They occur exclusively on the primary veins of an unassigned dicot (HC135) at both localities. These galls, assigned to DT 33, occasionally occur in clusters and are linked by necrotic foliar tissue within denser gall patches (Fig. 11, K, L, N). There is indication of a central chamber or possibly multiple chambers, revealed by the frequent presence of a central, craterlike depression. This depression is of intermediate hue between unaltered leaf and the surrounding rock matrix, indicating that there has been incomplete modification of tissue and not complete removal of the leaf that would be expected from a hole feeder (Fig. 11, B, E, J). Thus the surrounding ring of dark, thickened material is interpreted as the walls of a gall; some specimens also have an encircling zone of apparently necrotic tissue (Fig. 11, F and G). In a few instances, a light colored ovoidal structure within the gall may represent an emergence hole (Fig. 11, E and J), although in other cases differently shaped and positioned holes are interpreted as parasitoid damage or herbivory by an external feeder subsequent to gall formation (Fig. 11F). Although this association is host specific at this locality, galls of this general type occur on other plant hosts in other localities. However, this host is the only intensely galled plant within this local flora, and patterns of gall abundance, location on veinal tissue, and uniformity of size and shape among galls strongly indicate the singular targeting of this plant. This gall resembles simple dipteran galls of the family Cecidomyiidae (gall midges) that currently occur in taxonomic profusion on diverse dicotyledonous angiosperms (Gagné, 1989, 1994). The presence of gallers on this plant host is not unexpected; there is an extensive Late Cretaceous gall record on dicotyledonous angiosperms (Stephenson, 1992; Larew, 1992; Scott et al., 1994; Lang, 1996; Labandeira, 1998c).

The fourth specialized association is a lepidopteran leaf miner assigned to the family Gracillariidae, occurring on leaves of the dicot *Marmarthia pearsoni* Johnson (1996), a member of the Laurales and assigned to leaf morphotype HC162 (Fig. 12). These mines were illustrated by Lang (1996), are referable to DT 45, and occur in assemblage zone HCIII from DMNH localities 428 and 900, −15.0 and −2.8 m, respectively, from the HC-FU formational contact. Although this plant host occurred in other localities in zone HCIII, most of these deposits consist of a coarser grained, sandy matrix that probably precludes recognizable preservation. Five well-preserved and complete examples are known of this leaf mine (Fig. 12, A–E, H–L), characterized by a definable, circular oviposition site, followed by a serpentine phase representing an initial threadlike, coiled mine, followed by subsequent instar width increases that assumes a more linear trajectory, and terminating in a more expansive, elongate to ovoidal chamber. These mines were made by a minute, flattened larva that produced a median, initially linear but subsequently tightly undulatory frass trail of solid fecal material (Fig. 12, E and L). Judging from the particulate nature of the frass trail, this miner was a whole-cell feeder consuming either palisade or spongy mesophyll tissue. The mines were highly constrained by primary and secondary venation and occur between the median and lateral primary veins and rarely along the leaf margin. The prepupal instar stage occasionally crosses primary veins. A total of 31 mines, many aborted or otherwise incomplete, have been documented on 21 individuals of the plant host *M. pearsoni*. Three of these mines originate from DMNH locality 428 (Fig. 12, M–O).

Almost all modern genera of Lauraceae, particularly *Actinodaphne, Cinnamomum, Laurus, Lindera, Litsea, Parabenzoin,*

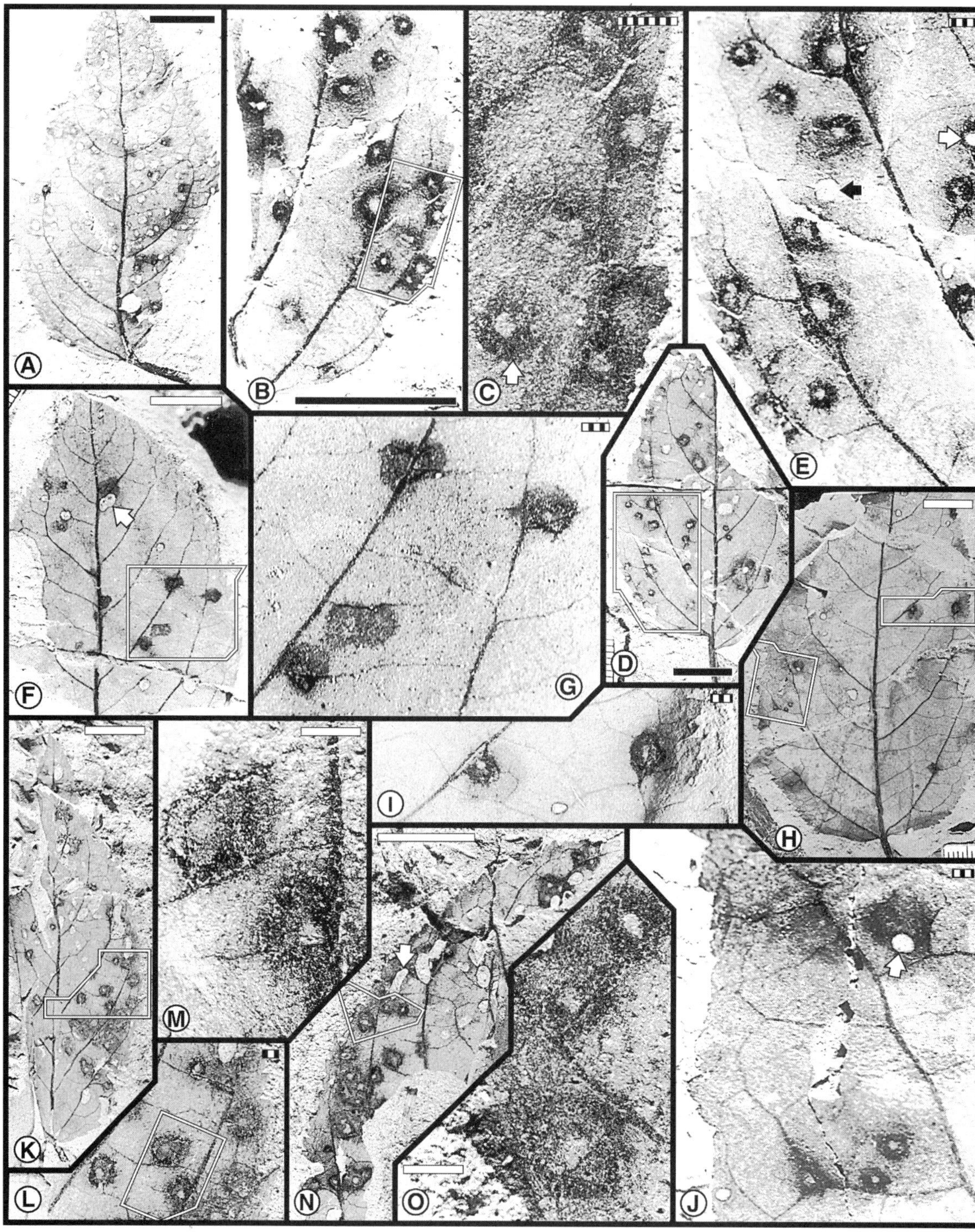

Figure 11. Small, cecidomyiid-type, cratered galls on primary veins of morphotype HC135, an unassigned dicotyledonous angiosperm from Hell Creek Formation. This host-specific association (DT 33) occurs only in localities 567 and especially 1491, latter of which supplied all illustrated examples. A: Typical example with several galls interspersed among more numerous examples of small hole feeding (DMNH-18895b). B: Extensively galled leaf fragment showing veinal association and modest variation in gall size (DMNH-19059) C: Enlargement of region in B showing five galls, each possessing pronounced fusanized rim (arrow) surrounding central depression. D: Leaf with galls and hole feeding (DMNH-18919). E: Magnified region in D illustrating random, vein-associated distribution of galls. F: Leaf with 10 galls, one of which has been partly consumed secondarily by hole feeder (arrow) after gall formation (DMNH-18894). G: Magnified region in F demonstrating absence of central cratered structures. H: Leaf with ~10 galls (DMNH-19728). I: Detail of two galls in H, with surrounding altered tissue. J: Detail of four galls, one apparently bearing emergence hole (arrow). K: Leaf with hole feeding and cluster of galls (DMNH-18914). L: Enlargement of gall cluster in K, showing central depressions and intervening altered tissue. M: Enlargement of two galls in L, illustrating outer gall wall and central cavity. N: Leaf fragment bearing cluster of galls and various hole feeding (DMNH-19040). O: Detail of three galls in N, showing outer wall, central depression, and relationship to major venation. Solid scale bars = 1.0 cm; striped scale bars = 0.1 cm.

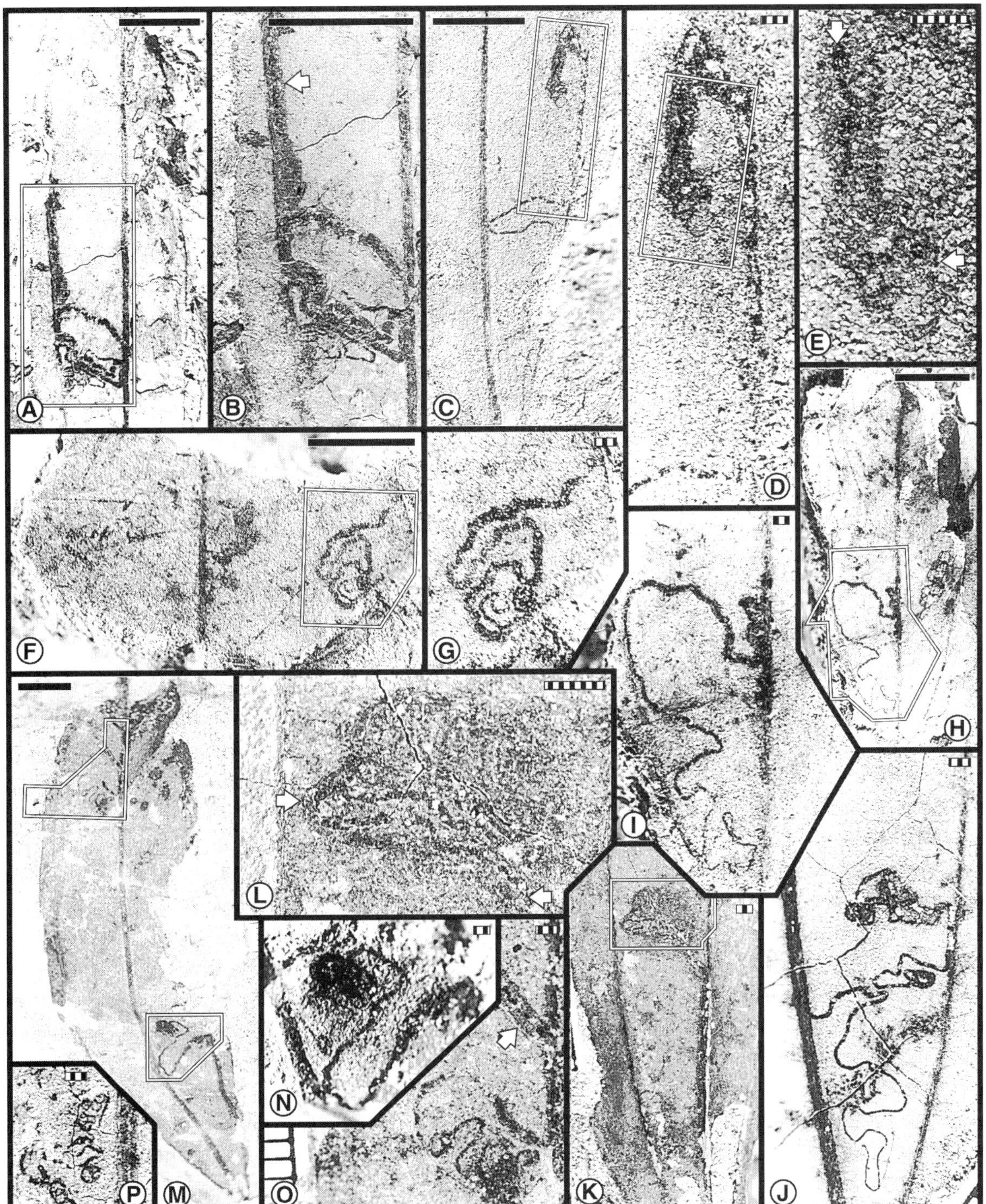

Figure 12. Host-specific association between *Marmarthia pearsonii* (Laurales) and a gracillariid leaf miner (Lepidoptera), from the uppermost Hell Creek Formation. This highly stereotyped damage type, DT 45, is a serpentine mine typically with initial coiled phase, subsequent curvilinear trajectory, and modestly expanded terminal chamber. It occurs exclusively on plant host HC162 at localities 900 and 428. A: Near-complete leaf mine following primary venation of *M. pearsonii* leaf (locality [loc.] 900, DMNH-7313). B: Enlargement of A, with dark colored, medial frass trail detectable at upper left corner, along primary vein. C: Complete leaf mine with terminal chamber at upper right (loc. 900, DMNH-7263). D: Enlargement of latter serpentine phase and terminal chamber of leaf mine in C. E: Closeup of terminal chamber displaying dark frass trail (top arrow) and chamber edge (bottom arrow). F: Fragment of leaf portion of leaf mine; note coiled early phase (loc. 900, DMNH-7199). G: Closeup of mine in F. H: Complete leaf mine bounded by median and lateral primary veins of plant host, typical for this species (loc. 900, DMNH-20023). I: Detail of mine in H, showing median frass trail at upper left. J: Three aborted leaf mines, two of which have coiled initial phase (loc. 900, DMNH-7325). K: Complete leaf mine with extensive terminal chamber illustrating frass trail (loc. 900, DMNH-7264). L: Enlargement of terminal mine chamber in K, showing path of undulatory frass trail, indicated by white arrows. M: Two aborted leaf mines with coiled initial phases (loc. 428, DMNH-7498). N: Enlargement of leaf mine at upper left in M, showing coiled (darkened) initial phase. O: Enlargement of leaf mine at lower right in M, revealing coiled (darkened) initial phase, and subsequent curvilinear phase along median primary and branching secondary veins (arrow). P: Two adjacent leaf mines aborted early in development (loc. 900, DMNH-7313). Scale bars = 1.0 cm; striped scale bars = 0.1 cm.

Persea, Sassafras, and *Umbellaria,* collectively are hosts to several genera of gracillariid leaf miners, namely species of *Acrocercops, Caloptila, Gracillaria, Lithocolletis,* and *Phyllocnistis* (Fletcher, 1920, 1933; Needham et al., 1928; Hering, 1967; Kumata, 1982; Kumata et al., 1988; Yuan and Robinson, 1993), contrary to the undocumented statement of Grimaldi (1999). These modern and latest Maastrichtian associations variously represent an old relationship between lauralean and gracillariid taxa that extends back to the early Maastrichtian of northern Wyoming (Labandeira et al., 1995; Lang, 1996), Turonian of eastern Kazakhstan (Kozlov, 1988), the earliest Cenomanian of Kansas and Nebraska (Labandeira et al., 1994; Futuyma and Mitter, 1996; Kristensen, 1997; Powell et al., 1999), and undoubtedly into the late Early Cretaceous. Several other lineages of Lepidoptera and other leaf-mining insect clades have been documented in middle and Late Cretaceous floras (Skalski, 1979; Stephenson, 1992; Labandeira et al., 1994; Lang, 1996), indicating a radiation of the leaf-mining habit by multiple lineages onto dicotyledonous hosts. The apparent extinction of *M. pearsoni* immediately before or at the K–T boundary indicates that this herbivore specialist also succumbed, although other lauraceous host lineages and their gracillariid herbivores persisted to the present.

Other host-specific associations occur, such as insect damage on the ginger *Zingiberopsis* (Zingiberaceae) (Hickey and Peterson, 1978), that are identical to modern descendant lineages of certain hispine beetles (Strong, 1977, 1984) described by Wilf et al. (2000). Additional associations include stereotyped skeletonization, such as a serpentine form occurring on the unidentified dicot HC254 (Fig. 3, Q and R), and a rectangular-elongate type found on the platanacean taxon *Erlingdorfia montana,* probably representing a leaf roller (Fig. 7, A–F, I–L). All insect damage from this K-T sequence will be qualitatively and quantitatively detailed in a forthcoming monograph.

Extinction at the boundary

These and most of the other host-specific associations occurring in Hell Creek strata do not survive the end of the Cretaceous (Fig. 2), resulting in an overall major decrease of damage types for the early Paleocene (Lang, 1996). Of the 22 damage types that remain in strata of the earliest Paleocene Fort Union Formation, 55% are generalized associations (level 1), such as nonstereotyped hole feeding, margin feeding, and some types of skeletonization. Six less generalized, moderately specific associations (level 2) survive the boundary and are more stereotyped forms of external foliage feeding. Only four highly specific associations, assigned to level 3, survive the boundary. They are two leaf mines, a gall, and a type of piercing and sucking puncture. Only two highly specific associations have been documented from earliest Paleocene deposits that do not occur previously in Hell Creek strata.

Of the 14 Cretaceous associations that are extinguished at the boundary, 6 are intermediate, 8 are highly specialized, and none are generalized. Of the 13 associations that disappear prior to but not at the boundary, 8 are plant host specialists, 5 are intermediate, and none are generalized, possibly indicating that highly host-specialized associations were declining before the terminal Cretaceous effect. Nevertheless, this geochronologically stepped pattern could be a sampling artifact, suggesting that preboundary associations may extend closer to the boundary if there were greater sampling effort in floras proximal to the HC-FU datum. However, many floras that occur throughout the 136 m interval across the boundary are highly sampled and occur throughout the entire interval (Table 2), and include DMNH locality 900, which is −2.8 m below the HC-FU contact and contains the second-most abundant number of specimens, thus making it unlikely that the pattern is only an artifact of collection strategy, which would be predicted by the Signor-Lipps effect (Signor and Lipps, 1982). However, a possible taphonomic explanation for the pattern remains, in which depositional conditions were inhospitable for preservation of floras proximal to the boundary that would extend upward damage-type ranges. Although one or any these alternatives explains the data, an objective reading of the pattern indicates that a pronounced extinction of intermediate and highly specialized associations occurred at or before the boundary, preferentially enriching the earliest Paleocene survivors in generalized herbivore taxa. Note also that the generalist category probably subsumes additional, unrecognized, host-specific plant damage that would bias against our conclusion for an extinction of host specialists at the boundary.

Lack of a recovery phase

The post-Cretaceous insect herbivore fauna, as judged from plant-related damage spectra, was not accompanied by significant immigration of new extrabasinal associations. Currently only two early Paleocene associations can be attributed to an immigrant or previously unsampled taxon, one of which is a specialized leaf miner. This suggests that the earlier extinction event at the K-T boundary was not accompanied by a short-term rebound phase to previous levels of plant-insect associations, at least on the order of tens to hundreds of thousands of years. Thus the earliest 0.8 m.y. of the Paleocene indicates depauperate levels of plant-insect associations. Additional research is needed to determine whether the 1.4 m.y. of latest Cretaceous levels of plant-insect associations reached a comparable level during the early Eocene, ~11.0 m.y. later (Wilf and Labandeira, 1999). Undoubtedly the extirpation of associations at the K-T boundary is tied to the extinction of their plant hosts as a consequence of broader ecosystem collapse at the terminal Cretaceous. However, this plant extinction did not result in an "across the board" reduction in insect associations: it preferentially affected specialized over generalized associations and also considerably curtailed the number of associations on early Paleocene survivors when compared to their latest Cretaceous conspecifics.

DISCUSSION

Three larger scale implications result from our study of regional patterns of insect herbivore diversity across the K-T boundary. First is the broader temporal context in comparison to what is known of herbivore damage trends during the Late Cretaceous and Paleocene. Second is whether our K-T boundary patterns can be explained by the biogeographical restriction of the study site and the consequent immigration of insect herbivores outside of the Williston Basin. Third is the increasing importance of insect trace-fossil data for resolving temporal trends of insect history that are not addressable by largely unavailable body-fossil data.

Larger temporal context

Judging from descriptions of insect-mediated damage (Kozlov, 1988; Lang, 1996; Stephenson, 1992; Labandeira et al., 1994, 1995), Late Cretaceous floras from the Northern Hemisphere have a comparatively elevated level of associations, characterized by a diverse spectrum of external foliage feeding, mining, galling, piercing and sucking, and oviposition damage types. This is in contrast to our preliminary survey of early Paleocene floras from the Western Interior, which lack generally the higher diversity and host specificity of associations. However, examination of insect associations for most Paleocene floras are incomplete when compared to those of the Late Cretaceous, latest Paleocene, or Eocene. Additional study of other North American Paleocene floras will be required for confirmation of this apparent trend of depauperate associational diversity. The recent documentation of an early Paleocene flora with atypical taxa indicating a warming event (Ellis and Johnson, 1999) indicates rare periodic incursions of thermophilic vegetation, possibly accompanied by distinctive herbivores. Such events would presage the warming trend that occurred during the latest Paleocene to early Eocene that was trophically connected to an increase in herbivore intensity and diversity (Wilf and Labandeira, 1999; Wilf et al., 2001). This level of herbivory continued into the later Eocene as well (Lang et al., 1995; Palmer et al., 1998; Wilf et al., 2001).

The persistence of diverse types of Late Cretaceous insect damage at elevated levels in North America currently is based on a few examined floras, principally documented from the early Cenomanian Dakota flora (98 Ma), the early Maastrichtian Meeteetsee flora (72 Ma), the middle Maastrichtian Ripley flora (ca. 68 Ma), and the late Maastrichtian Hell Creek flora (67.0 Ma) (Stephenson, 1992; Labandeira et al., 1994, 1995; Labandeira, 1998c; Lang, 1996). Herbivory in these floras variously affected all major vascular plant groups; i.e., ferns, conifers, cycadophytes, gingkophytes, monocots, and dicots. Of particular importance is the presence of extensive herbivory on herbaceous angiosperms, particularly in DMNH localities 2203 and 567 of zone HCIb. The meadow-like plants of these assemblages, consisting of low-statured and nonarborescent plants with highly lobed and dissected leaves, is also found in the early Maastrichtian Meeteetsee flora of north-central Wyoming (Wing et al., 1993; Labandeira et al., 1995), although there are few to no cooccurring lower level taxa. Undoubtedly this angiospermous life habit had earlier antecedents. Notably this Hell Creek assemblage includes the earliest megafloral representatives of the Rosaceae (HC80) and Urticales (HC81, HC229), both which harbored leaf miners (Labandeira, personal observation) and various external feeders. This colonization of herbaceous plants is complemented by observations of a variety of charcoalified floral remains from the Turonian to Santonian coastal plain sequence of eastern North America; relatively advanced plant-insect pollination mutualisms were already deployed on a diverse, lowland flora, but of woody dicotyledonous angiosperms (Crepet and Friis, 1987; Crepet and Nixon, 1996, 1998; but see Friis, 1984).

Biogeographical and ecological considerations

The demise of many of these Late Cretaceous associations essentially reset the plant-insect associational clock for the Paleogene. Thus many modern associations in the Western Interior either reevolved or possibly accumulated as immigrant herbivores colonized floras that became progressively diverse and more thermophilic during the Paleocene to Eocene transition. These two processes are demonstrated by the autochthonous radiation of Paleocene mammals (Alroy, 1999) as well as immigration of taxa from Asia during the later Paleocene (Beard, 1998). The Paleogene history of seed plants is also well documented (Manchester, 1999) and indicates that interchange of most dicotyledonous angiosperms across the Beringian or northern European land bridges did not occur until late in the Paleocene or early Eocene, although taxa such as *Nordenskioldia, Platanites,* and *Quereuxia* had earlier dispersal trajectories (Manchester, 1999). It is most likely that dispersal from Asia or Europe to North America had little effect on the character of earlier Paleocene insect herbivores, particularly because the dominant source of earliest Paleocene floras in the Williston Basin was local or regional mire-associated taxa of the latest Cretaceous (Johnson, 1999) that lacked post-Cretaceous taxonomic affinities to coeval Eurasian taxa.

Studies of modern specialist herbivorous insects and their plant hosts strongly suggest that differentiation of an ancestral species into populations that exhibit novel plant host preferences can be rapid (Jaenike, 1990). A classic example is the morphological partitioning during the past 60 yr of the soapberry bug, *Jadera haematoloma,* into three populations with distinctive beak lengths for exploiting different species of soapberry seeds (Carroll and Boyd, 1992; Bernays and Chapman, 1994). This indicates that such segregation can be a rapid process that leads to eventual species-level differentiation. A different approach has documented the extensive, species-level radiation of the weevil genus *Miocalles* onto varied plant hosts on Rapa Island during the past 10^5 yr (Paulay, 1985). These and other examples indicate that there is a significant potential for rapid evolution of herbivore colonization of novel, albeit often phylogenetically related, plants at ecological time scales (Schoonhoven et al., 1998; but see Piegler, 1986).

Our observations that largely document an absence of new associations during the 0.8 m.y. interval after the K-T event seem paradoxical. This is particularly so, given studies establishing the rapidity of host colonization by some modern herbivorous insects and the virtual absence of decreases in damage types during the late Paleocene to early Eocene warming event (Wilf and Labandeira, 1999). However, if the K-T reduction of the local insect herbivore fauna was sufficiently severe and persisted millions of years into the early Paleocene, there would have been minimal diversity of species constituting the pool for establishing plant-insect associations. Accordingly, depressed levels of earliest Paleocene plant-insect associations would have been attributable not only to the presence of a depauperate regional flora, but more important, to decimation of those insect lineages that would have been available as potential plant consumers. Thus the repertoire of present-day plant-insect associations likely is a mixture of older Mesozoic colonizations of plants (Zwölfer, 1978; Rohr et al., 1984; Thien et al., 1985; Michener and Grimaldi, 1988; Genise, 1995; Labandeira, 1998a), perhaps including such examples as the specialized leaf miners on Cenomanian Platanaceae (Labandeira et al., 1994; also see Becerra, 1997), and external foliage feeding on Maastrichtian Zingiberaceae (Wilf et al., 2000) and more recent and abundant, Paleogene originations. If correct, this temporally bimodal distribution of associations ultimately resulted from major environmental disruption and ecosystem replacement (Coley, 1998; McIver, 1999) of herbivore assemblages at the end of the Cretaceous.

The effects of extensive plant and insect herbivore extinction at the K-T boundary must have had dramatic consequences that trophically cascaded upward, affecting consumers of herbivorous insects. The superjacent trophic levels of parasites, parasitoids, and predators likely were affected dramatically by a decreased food resource base (Labandeira, 2002a). This inference would affect host-dependent parasitoids and parasites the most, because insect predators that dispatch their victims immediately are typically dietary generalists (Price, 1997). This prediction would be difficult to test, however, as sufficiently diverse body-fossil deposits proximal to the K-T boundary are unknown, and therefore appropriate before and after comparisons are unavailable. The expectation is that parasite and parasitoid diversity also suffered a major decline during the earliest Paleocene and, subsequently, reestablished newfound associations on insect hosts, approximately in tandem with the plants and herbivorous insects constituting the subjacent levels of the trophic pyramid.

Value of body-fossil versus trace-fossil data

The spatial and temporal context of this study represents a few orders of magnitude greater resolution than earlier comprehensive but qualitatively coarser grained studies of insect diversity that were based on global compendia for a 410 m.y. fossil record (Dmitriev and Zherikhin, 1988; Labandeira and Sepkoski, 1993; Jarzembowski and Ross, 1996). The relevant outcrop area within the Williston Basin represents ~0.00025 that of available terrestrial outcrop worldwide, and the 2.2 m.y. interval is ~0.005 that of the total duration of the fossil insect record. Equally important is the source of our data: we have chosen the fossil record of insect-plant associations, characterized as damage types, rather than insect body fossils evaluated at the family level. Although the insect body-fossil record is taxonomically better than previously thought (Labandeira and Sepkoski, 1993; Labandeira, 1994), it represents a series of mostly exceptionally preserved deposits that are sporadically deployed in time and space but are typically absent at important, stratigraphically confined intervals of major theoretical interest. Because of a dominantly Lagerstätten-driven record of insect occurrences, it is necessary to employ proxy data—in this case the parallel fossil record of insect damage on plants—for evaluating insect diversity across stratigraphically focused successions where insect body fossils are effectively absent but plant deposits are comparatively plentiful.

Insect herbivory patterns on fossil plant hosts are valuable for chronicling phenomena such as mass extinctions, major environmental perturbation, and the evolutionary radiations between insect and plant taxa. This is because insect-modified foliar material from the fossil record provides crucial ecological data that otherwise are unavailable from insect bodies or from pristine leaves. This approach has been underappreciated in part because the historical concern of paleoentomology has been description of body fossils from selected Lagerstätten of high preservational quality. Our study demonstrates that detailed examination of the plant-insect associational record can supply robust data for testing hypotheses (Labandeira, 2002b)—such as the response of insects to the K-T event—that otherwise would be unavailable if the only recourse was study of body fossils. In addition, for many important phytophagous taxa, the fossil record of insect-mediated plant damage reveals considerably more detail about their ecologic and even evolutionary history than the parallel body-fossil record. One example of this is the record of Cretaceous lepidopteran leaf mines (Powell et al., 1999), which provides considerably more natural historical and taxonomic insight than their corresponding depauperate record of body fossils, including those found in amber.

ACKNOWLEDGMENTS

We acknowledge the following volunteers of the Denver Museum of Nature & Science (DMNS) and the Pioneer Trails Regional Museum, who provided valuable assistance in field collection, specimen preparation, curation, and accessioning of the fossil leaf material in this study: Bill Bateman, Kristian Benson, Stephanie Bowman, Peter Buckman, Laura Carsten, Tom Collins, Beth Ellis, John Gurche, Madeline Harrell, Biz Hermsen,

APPENDIX. INSECT DAMAGE TYPES (DT)

Damage Type	Category	Physical Description	Host Specificity
1	HOLE FEEDING	Small, (<1 mm)	1
2	HOLE FEEDING	Medium (1–5 mm)	1
3	HOLE FEEDING	Medium (1–5 mm), polylobate	1
4	HOLE FEEDING	Large, (>5 mm)	1
5	HOLE FEEDING	Large, (>5 mm), polylobate	1
6	HOLE FEEDING	Bud feeding	3
7	HOLE FEEDING	Curvilinear	2
8	HOLE FEEDING	Elongated slot (length > 2.5 width)	2
9	HOLE FEEDING	Polymorphic, generally ellipses (>2 mm)	2
11	HOLE FEEDING	Thick rim of necrotic tissue	1
50	HOLE FEEDING	Adjoining midvein	2
51	HOLE FEEDING	Slots overlapping into enlarged central area and those associated by pattern	2
57	HOLE FEEDING	Pattern associated with divergence point of secondary from primary veins	2
64	HOLE FEEDING	Series of medium (<5 mm) circular holes in a margin-associated pattern	2
12	MARGIN FEEDING	Generalized, cuspate	1
13	MARGIN FEEDING	Leaf apex	1
14	MARGIN FEEDING	To primary vein	1
15	MARGIN FEEDING	Trenched or otherwise deeply incised	2
16	SKELETONIZATION	General, reaction rim weakly developed	1
17	SKELETONIZATION	General, reaction rim well developed	1
19	SKELETONIZATION	Broad, with rectangular pattern	2
20	SKELETONIZATION	Curvilinear	2
21	SKELETONIZATION	Highest order of venation removed	2
22	SKELETONIZATION	Linear pattern	2
24	SKELETONIZATION	Ovoidal, adjacent to midvein	2
56	SKELETONIZATION	At leaf base between primary veins	2
61	SKELETONIZATION	Adjoins and exactly follows primary and secondary veins, increasing in size	3
26	EXTERNAL-SPECIAL	Free feeding	2
28	EXTERNAL-SPECIAL	Strip feeding between secondary veins; *Cephaloleichnites strongi*	3
29	EXTERNAL-SPECIAL	Window feeding, generalized	1
32	GALLING	On nonveinal part of leaf	2
33	GALLING	On primary vein	3
34	GALLING	On secondary vein	2
49	GALLING	Large (core > 3 mm), with distinctive reaction rim at some distance	3
52	GALLING	Relatively large (>3 mm), with woody and radiating partitions	3
55	GALLING	On petiole	3
62	GALLING	Large, (~5 to 10 mm diameter), flat, “pock-marked,” alternating light/dark areas	3
35	MINING	Blotch, with a central chamber	3
38	MINING	Circular, with case	3
40	MINING	Semilinear, reaction rim present; frass is solid, thick and 3-D; terminal chamber	3
41	MINING	Serpentine A: long, undulatory, frass particulate	3
42	MINING	Serpentine B: medium, width rapidly increasing, irregular margin	3
43	MINING	Serpentine C: short, solid frass trail	3
45	MINING	Serpentine E: medium length, margin irregular, well-developed terminal chamber	3
59	MINING	Initially adjacent to secondary vein, then primary; thin and with minimal width increase; terminating into a 5–10 mm ovoidal/elliposoidal terminal chamber	
65	MINING	Large (>30 mm length), linear, blotchlike, with rapidly expanding width	3
46	PIERCING & SUCKING	Small scale or puncture (<2 mm), circular depression	3
53	PIERCING & SUCKING	Large scale (>4 mm), on primary veins (e.g., *Erlingdorfia*)	3
54	OVIPOSITION	Scars in arcuate rows, en echelon	3
58	FUNGAL DAMAGE	Ovoid to polylobate fungal blotches	n/a

Dennis Hopkins, Dena Hunter, Logan Ivy, Laurie Oakland, Marianne O'Connell, Michael O'Connell, Susan Passmore, Dean Pearson, Michele Reynolds, Nancy Schaefer, Terry Schaefer, Pam Schmidt, Michon Scott, Gary Staab, Patty Stoiko, Allison Sundine, Kevin Werth, and the students of the DMNS 1998 field school. Property access was granted by the Little Missouri National Grassland, the Bureau of Land Management, the East Marmarth Pasture Association, the Horse Creek Grazing Association, Scott and Henry Bradac, John and Bud Brown, and Mike Sonsella. The figures were expertly rendered by Finnegan Marsh. We thank Michele Reynolds (DMNS), who assisted Labandeira in data entry, and Logan Ivy, who made the collections available. Douglass Miller, of the U.S.D.A. Agricultural Research Service, assisted in examination of scale insect damage on HC57. Support for this research originated from the Walcott Fund of the National Museum of Natural History (to Labandeira), the Denver Museum of Nature & Science (to Johnson), and the Natural Environment Research Council (to Lang). We acknowledge the assistance of Margaret Collinson and Andrew Scott, of the University of London, who provided facilities and encouragement to Lang during his work on a dissertation, some results of which have been incorporated herein. Bill Crepet (Cornell University), Charles Mitter (University of Maryland), Peter Wilf (University of Michigan), and an anonymous reviewer provided constructive feedback of a previous version of this article. We also thank Leo Hickey (Yale University) for assistance in ascertaining the taxonomic affinities of morphotype HC81. This is contribution 71 of the Evolution of Terrestrial Ecosystems Consortium at the National Museum of Natural History.

REFERENCES CITED

Alford, D.V., 1991, A colour atlas of pests of ornamental trees, shrubs and flowers: London, Wolfe Publishing, 448 p.

Alroy, J., 1999, The fossil record of North American mammals: Evidence for a Paleocene evolutionary radiation: Systematic Biology, v. 48, p. 107–118.

Alvarez, L.W., Alvarez, W., Asaro, F., and Michel, H.V., 1980, Extraterrestrial cause for the Cretaceous-Tertiary extinction: Science, v. 208, p. 1095–1098.

Alvarez, W., 1997, *T. rex* and the crater of doom: Princeton, New Jersey, Princeton University Press, 285 p.

Amelung, D., 1995, Harmful organisms of stinging nettle and hemp: Mitteilungen aus der Biologischen Bundesanstalt für Land- und Forstwirtschaft Berlin-Dahlem, v. 310, p. 125–130.

Andersen, M.M., 1998, Water striders from the Paleogene of Denmark, with a review of the fossil record and evolution of semiaquatic bugs (Hemiptera, Gerromorpha): Biologiske Skrifter, v. 50, p. 1–157.

Arens, N.C., and Jahren, A., 2000, Carbon isotope excursion in atmospheric CO_2 at the Cretaceous-Tertiary boundary: Evidence from terrestrial sediments: Palaios, v. 15, p. 314–322.

Beard, K.C., 1998, East of Eden: Asia as an important center of taxonomic origination in mammal evolution, *in* Beard, K.C., and Dawson, M.R., eds., Dawn of the age of mammals in Asia: Carnegie Museum of Natural History Bulletin, v. 34, p. 5–39.

Becerra, J.X., 1997, Insects on plants: Macroevolutionary chemical trends in host use: Science, v. 276, p. 253–256.

Beck, A., and Labandeira, C.C., 1998, Early Permian insect folivory on a gigantopterid-dominated riparian flora from north-central Texas: Palaeogeography, Palaeoclimatology, Palaeoecology, v. 142, p. 139–173.

Becker, W.B., 1938, Leaf-feeding insects of shade trees: Massachusetts Agricultural Experiment Station Bulletin 353, 83 p.

Belt, E.S., Flores, R.M., Warwick, P.D., Conway, K.M., Johnson, K.R., and Waskowitz, R.S., 1984, Relationship of fluviodeltaic facies to coal deposition in the lower Fort Union Formation (Palaeocene), south-western North Dakota, *in* Rahmani, R.A., and Flores, R.M., eds., Sedimentology of coal and coal-bearing sequences: International Association of Sedimentologists Special Publication 7, p. 177–195.

Bernays, E.A., and Barbehenn, R. 1987, Nutritional ecology of grass foliage-chewing insects, *in* Slansky, F., Jr., and Rodriguez, J.G., eds, Nutritional ecology of insects, mites, spiders, and related invertebrates: New York, John Wiley, p. 147–175.

Bernays, E.A., and Chapman, R.F., 1994, Host-plant selection by phytophagous insects: New York, Chapman and Hall, 312 p.

Bohor, B.F., Triplehorn, D.M., Nichols, D.J., and Millard, H.T., Jr., 1987, Dinosaurs, spherules, and the "magic layer": A new K-T boundary clay site in Wyoming: Geology, v. 15, p. 896–899.

Borkent, A., 1995, Biting midges in the Cretaceous amber of North America (Diptera: Ceratopogonidae): Leiden, Backhuys, 237 p.

Botosaneanu, L., and Wichard, W., 1983, Upper Cretaceous Siberian and Canadian amber caddisflies (Insecta: Trichoptera): Bijdragen Tot de Dierkunde, v. 53, p. 187–217.

Boulter, M.C., Spicer, R.A., and Thomas, B.A., 1988, Patterns of plant extinction from some palaeobotanical evidence, *in* Larwood, G.P., ed., Extinction and survival in the fossil record: Systematics Association Special Volume 34, p. 1–36.

Bredemann, G., 1959, Die Große Brennessel *Urtica dioica* L.: Berlin, Akademie-Verlag.

Briggs, J.C., 1991, A Cretaceous-Tertiary mass extinction?: Bioscience, v. 41, p. 619–624.

Brown, L.R., and Eads, C.O., 1965, A technical study of insects affecting the sycamore tree in southern California: University of California Agricultural Experiment Station Bulletin 818, 38 p.

Brown, R.W., 1962, Paleocene flora of the Rocky Mountains and great Plains: U.S. Geological Survey Professional Paper 375, 119 p.

Brown, V.V., and Pike, E.M., 1990, Three new fossil phorid flies (Diptera: Phoridae) from Canadian Late Cretaceous amber: Canadian Journal of Earth Sciences, v. 27, p. 187–217.

Buzzi, Z.J., 1975, Redescrição e bionomia de *Anacassis cribrum* (Klug, 1829) (Coleoptera, Chrysomelidae): Acta Biólogica Paranaense, v. 4, p. 59–89.

Cande, S.C., and Kent, D.V., 1995, Revised calibration of the geomagnetic polarity time scale for the Late Cretaceous and Cenozoic: Journal of Geophysical Research, v. 100, p. 6093–6095.

Carroll, S.P., and Boyd, C., 1992, Host race radiation in the soapberry bug: Natural history with the history: Evolution, v. 46, p. 1053–1069.

Carter, D.J., and Hargreaves, B., 1986, A field guide to caterpillars of butterflies and moths in Britain and Europe: London, Collins, 296 p.

Carver, M., Gross, F.G., and Woodward, T.E., 1991, Hemiptera (bugs, leafhoppers, cicadas, aphids, scale insects, etc.), *in* Naumann, I.D., et al., eds., Insects of Australia: Ithaca, New York, Cornell University Press, p. 429–509.

Chaloner, W.G., Scott, A.C., and Stephenson, J., 1991, Fossil evidence for plant-arthropod interactions in the Palaeozoic and Mesozoic: Royal Society of London Philosophical Transactions, ser. B, v. 333, p. 177–186.

Chatterjee, S., 1997, The rise of birds: Baltimore, Maryland, Johns Hopkins University Press, 312 p.

Chin, K., and Gill, B.D., 1996, Dinosaurs, dung beetles, and conifers: Participants in a Cretaceous food web: Palaios, v. 11, p. 280–285.

Coley, P.D., 1998, Possible effects of climate change on plant/herbivore interactions in moist tropical forests: Climatic Change, v. 39, p. 455–472.

Coley, P.D., 1999, Hungry herbivores seek a warmer world: Science, v. 284, p. 2098–2099.

Comstock, J.H., 1916, Reports on scale insects: Cornell University Agricultural Experiment Station Bulletin, no. 372, p. 421–500.

Connor, E.F., and Taverner, M.P., 1997, The evolution and adaptive significance of the leaf-mining habit: Oikos, v. 79, p. 6–25.

Coope, G.R., 1995, The effects of Quaternary climatic changes on insect populations: Lessons from the past, *in* Harrington, R., and Stork, N.E., eds., Insects in a changing environment: London, Academic Press, p. 29–48.

Coulson, R.N., and Witter, J.A., 1984, Forest entomology, ecology and management: New York, Wiley Interscience, 669 p.

Cox, M.L., and Windsor, D.M., 1999, The first instar larva of *Aulacoscelis appendiculata* n. sp. (Coleoptera: Chrysomelidae: Aulacoscelinae) and its value in the placement of the Aulacoscelinae: Journal of Natural History, v. 33, p. 1049–1087.

Crepet, W.L., and Friis, E.M., 1987, The evolution of insect pollination in angiosperms, *in* Friis, E.M., et al., eds., The angiosperms and their biological consequences: Cambridge, Cambridge University Press, p. 181–201.

Crepet, W.L., and Nixon, K.C., 1996, The fossil history of stamens, *in* D'Arcy, W.G., and Keating, R.C., eds., The anther: Form, function and phylogeny: Cambridge, Cambridge University Press, p. 25–57.

Crepet, W.L., and Nixon, K.C., 1998, Fossil Clusiaceae from the Late Cretaceous (Turonian) of New Jersey and implications regarding the history of bee pollination: American Journal of Botany, v. 85, p. 1122–1133.

Cronquist, A., 1981, An integrated system of classification of flowering plants: New York, Columbia University Press, 1262 p.

Deckle, G.W., 1965, Florida armored scale insects: Arthropods of Florida, v. 3, p. 1–265.

D'Hondt, S., and Zachos, J.C., 1998, Cretaceous foraminifera and the evolutionary history of planktic photosymbiosis: Paleobiology, v. 24, p. 512–523.

Dilcher, D.L., 1965, Epiphyllous fungi from Eocene deposits in western Tennessee, U.S.A.: Palaeontographica, Abteilung B: Palaophytologie, v. 116, p. 1–54.

Dixon, A.F.G., 1998, Aphid ecology (second edition): London, Chapman and Hall, 300 p.

Dmitriev, V.Y., and Zherikhin, V.V., 1988, Changes in the familial diversity of insects and demonstration of a method of data analysis, *in* Ponomarenko, A.G., ed., The Mesozoic-Cenozoic crisis in the evolution of insects: Moscow, Academy of Sciences, p. 208–215 (in Russian).

Dodge, B.O., 1943, The sycamore plant bug: New York Botanical Garden Journal, v. 44, p. 214–215.

Dorf, E., 1940, Relationship between floras of the type Lance and Fort Union Formations: Geological Society of America Bulletin, v. 51, p. 213–236.

Dorf, E., 1942, Upper Cretaceous floras of the Rocky Mountain region. 1. Stratigraphy and paleontology of the Fox Hills and Lower Medicine Bow Formations of southern Wyoming and northwestern Colorado. 2. Flora of the Lance Formation at its type locality, Niobrara County, Wyoming: Carnegie Institution of Washington Contributions to Paleontology, no. 508, 168 p.

Dravé, E.-H., and Laugé, G., 1978, Étude de l'action de la silice sur l'usure des manibules de la pyrale duriz: *Chilo suppressalis* (F. Walker): Bulletin de la Société Entomologique de France, v. 83, p. 159–162.

Ellis, B., and Johnson, K.R., 1999, An unusually diverse megaflora from Castle Rock, Colorado: Evidence for a previously unrecognized early Paleocene warming event: Geological Society of America Abstracts with Programs, v. 31, no. 7, p. A468.

Erwin, T.L., and Scott, J.C., 1980, Seasonal and size patterns, trophic structure, and richness of Coleoptera in the tropical arboreal ecosystem: The fauna of the tree *Luehea seemannii* Triana and Planch in the Canal Zone of Panama: Coleopterists Bulletin, v. 34, p. 305–322.

Essig, E.O., 1926, Insects of western North America: Macmillan, New York, 1035 p.

Farrell, B.D., 1998, "Inordinate fondness" explained: Why are there so many beetles?: Science, v. 281, p. 555–559.

Filer, T.H., Solomon, J.D., McCracken, F.I., Oliveira, F.L., Lewis, R., Jr., Weiss, M.J., and Rogers, T.J., 1977, Sycamore pests: A guide to major insects, diseases, and air pollution: Stoneville, Mississippi, U.S. Department of Agriculture Forest Service, Southern Forest Experiment Station, 36 p.

Fletcher, T.B., 1920, Life-histories of Indian insects. Microlepidoptera. 6. Gracillariidae: Memoirs of the Department of Agriculture in India, Entomological Series, v. 6, p. 136–167.

Fletcher, T.B., 1933, Life-histories of Indian Microlepidoptera (second series)—Cosmopterygidae to Neopseustidae: Imperial Council of Agricultural Research Scientific Monograph 4, p. 1–85.

Foldi, I., 1997, Defense strategies in scale insects: Phylogenetic inference and evolutionary scenarios (Hemiptera, Coccoidea), *in* Grandcolas, P., ed., The origin of biodiversity in insects: Phylogenetic tests of evolutionary scenarios: Mémoires du Muséum National d'Histoire Naturelle, v. 173, p. 203–230.

Forbes, W.T.M., 1923, The Lepidoptera of New York and neighboring states, Part I: Primitive forms, microlepidoptera, pyraloids, Bombyces: Ithaca, New York, Cornell University Agricultural Experiment Station Memoir 68, 729 p.

Forbes, W.T.M., 1960, Lepidoptera of New York and neighboring states, Part IV: Agaristidae through Nymphalidae, including butterflies: Ithaca, New York, Cornell University Agricultural Experiment Station Memoir 371, 188 p.

Friis, E.M., 1984, Preliminary report of Upper Cretaceous angiosperm reproductive organs from Sweden and their level of organization: Annals of the Missouri Botanical Garden, v. 71, p. 403–418.

Fritsch, A., 1882, Fossile Arthropoden aus der Steinkohlen- und kreideformation Böhmens: Beiträge zur Paläontologie von Österreich-Ungarns, v. 2, p. 1–7.

Fritsch, A., 1901, Die thierischen Reste der Perucer Schichten: Archiv für die Naturwissenschaftliche Landesdurchforschung von Böhmen, v. 11, p. 163–180.

Futuyma, D., and Mitter, C., 1996, Insect-plant interactions: The evolution of component communities: Royal Society of London Philosophical Transactions, ser. B, v. 351, p. 1361–1366.

Gagné, R.J., 1989, The plant-feeding gall midges of North America: Ithaca, New York, Cornell University Press, 356 p.

Gagné, R.J., 1994, The gall midges of the Neotropical region: Ithaca, New York, Cornell University Press, 352 p.

Gangwere S.K., 1966, Relationships between the mandibles, feeding behavior, and damage inflicted on plants by the feeding of certain acridids (Orthoptera): Michigan Entomologist, v. 1, p. 13–16.

Gemmill, C.E.C., and Johnson, K.R., 1997, Paleoecology of a late Paleocene (Tiffanian) megaflora from the northern Great Divide Basin, Wyoming: Palaios, v. 12, p. 439–448.

Genise, J.F., 1995, Upper Cretaceous trace fossils in permineralized plant remains from Patagonia, Argentina: Ichnos, v. 3, p. 287–299.

Gill, J.R., and Cobban, W.A., 1973, Stratigraphy and geologic history of the Montana Group and equivalent rocks, Montana, Wyoming, and North and South Dakota: U.S. Geological Survey Professional Paper 776, p. 1–37.

Graham, R., 1986, Plant-animal interactions and Pleistocene extinctions, *in* Elliott, D.K., ed., Dynamics of extinctions: New York, John Wiley, p. 131–154.

Grimaldi, D.A., 1996, Amber: Window to the past: New York, Harry Abrams, 216 p.

Grimaldi, D.A., 1999, The co-radiations of pollinating insects and angiosperms in the Cretaceous: Missouri Botanical Garden Annals, v. 86, p. 373–406.

Grogan, W.L., Jr., and Szadziewski, R., 1988, A new biting midge from Upper Cretaceous (Cenomanian) amber of New Jersey (Diptera: Ceratopogonidae): Journal of Paleontology, v. 62, p. 808–812.

Halperin, J., 1990, Arthropod fauna and main insect pests of plane trees in Israel: Phytoparasitica, v. 18, p. 309–319.

Hansen, T.A., Farrell, B.R., and Upshaw, B., III, 1993, The first 2 million years after the Cretaceous-Tertiary boundary in east Texas: Rate and paleoecology of the molluscan recovery: Paleobiology, v. 19, p. 251–265.

Hartman, J.H., 1999, End Cretaceous and early Paleocene mollusks of the Williston Basin, North Dakota and Montana: Geological Society of America Abstracts with Programs, v. 31, no. 7, p. A72.

Hase, A., 1955, Über Insektenschäden an *Gingko biloba* L. (Kl. Ginkgoinae): Mededelingen van de Landbouwhogeschool en de Opzoekingsstations van de Staat te Gent, v. 20, p. 331–336.

Heie, O.E., and Pike, E.M., 1992, New aphids in Cretaceous amber from Alberta (Insecta, Homoptera): Canadian Entomologist, v. 124, p. 1027–1053.

Hering, E.M., 1932, Minen-Herbarium: (privately published), v. 7, 21 p.

Hering, E.M., 1951, Biology of leaf miners: S'Gravenhage, Netherlands, Junk, 1605 p.

Hering, E.M., 1967, Bestimmungstabellen der blattminen von Europa (Band 1 to 3): S'Gravenhage, Netherlands, W. Junk, 1406 p.

Heron, H., 1999, The biology of *Conchyloctenia punctata* (Fabricius): A cycloalexic cassid (Chrysomelidae: Cassidinae), *in* Cox, M.L., ed., Advances in Chrysomelidae biology 1: Leiden, Backhuys Publishers, p. 565–580.

Heron, H., and Borowiec, L., 1997, Host plants and feeding patterns of some

South African tortoise beetles (Coleoptera: Chrysomelidae: cassidoid Hispinae): Genus, v. 8, p. 625–658.

Hespenheide, H.A., 1991, Bionomics of leaf-mining insects: Annual Review of Entomology, v. 36, p. 535–600.

Hickey, L.J., 1973, Classification of the architecture of dicotyledonous leaves: American Journal of Botany, v. 60, p. 17–33.

Hickey, L.J., 1979, A revised classification of the architecture of dicotyledonous leaves, *in* Metcalf, C.R., and Chalk, L., eds., Anatomy of the dicotyledons (second edition): Oxford, Clarendon Press, p. 25–39.

Hickey, L.J., 1981, Land plant evidence compatible with gradual not catastrophic change at the end of the Cretaceous: Nature, v. 292, p. 529-531.

Hickey, L.J., and Peterson, R.K., 1978, *Zingiberopsis,* a fossil genus of the ginger family from Late Cretaceous to early Eocene sediments of Western Interior North America: Canadian Journal of Botany, v. 56, p. 1136–1152.

Hicks, J.F., Johnson, K.R., Tauxe, L., and Clark, D., 1999, Geochronology of the Hell Creek Formation of southwestern North Dakota: A multidisciplinary approach using biostratigraphy, isotopic dating, geochemistry, and magnetostratigraphy: Geological Society of America Abstracts with Programs, v. 31, no. 7, p. A71.

Hoganson, J.W., and Murphy, E.C., 1999, The marine Breien Member of the Hell Creek Formation: Stratigraphy, vertebrate paleontology, and biochronology: Geological Society of America Abstracts with Programs, v. 31, no. 7, p. A71.

Hotton, C., 1984, Palynofloral changes across the Cretaceous-Tertiary boundary in east-central Montana, U.S.A. [abs.]: International Palynological Conference, Sixth Meeting, Calgary, Alberta, p. 66.

Houard, C., 1909a, Les zoocécidies des plantes d'Europe et du Bassin de la Méditerranée [tome primaire]: Paris, A. Hermann et Fils, p. 1–572 p.

Houard, C., 1909b, Les zoocécidies des plantes d'Europe et du Bassin de la Méditerranée [tome second]: Paris, A. Hermann et Fils, p. 573–1248.

Hughes, N.F., 1994, The enigma of angiosperm origins: Cambridge, Cambridge University Press, 303 p.

Hunter, J.P., and Archibald, J., 1999, Mammals from the end of the age of dinosaurs in North Dakota and southeastern Montana: Geological Society of America Abstracts with Programs, v. 31, no. 7, p. A73.

Izett, G.A., 1990, The Cretaceous-Tertiary boundary interval, Raton Basin, Colorado and New Mexico, and its content of shocked metamorphosed minerals: Evidence relevant to the K-T boundary impact-extinction theory: Geological Society of America Special Paper 249, 100 p.

Jaenike, J., 1990, Host specialization in phytophagous insects: Annual Review of Ecology and Systematics, v. 21, p. 243–274.

Jarzembowski, E.A., 1989, Cretaceous insect extinction: Mesozoic Research, v. 2, p. 25–28.

Jarzembowski, E.A., 1990, A boring beetle from the Wealden of the Weald, *in* Boucot, A.J., ed., Evolutionary paleobiology of behavior and coevolution: Amsterdam, Elsevier, p. 373–376.

Jarzembowski, E.A., and Ross, A., 1993, Time flies: The geological record of insects: Geology Today, v. 9, p. 218–223.

Jarzembowski, E.A., and Ross, A., 1996, Insect origination and extinction in the Phanerozoic, *in* Hart, M.B., ed., Biotic recovery from mass extinction events: Geological Society [London] Special Publication 102, p. 65–78.

Johnson, K.R., 1989, A high-resolution megafloral biostratigraphy spanning the Cretaceous-Tertiary boundary in the northern Great Plains [Ph.D. thesis]: New Haven, Connecticut, Yale University, 556 p.

Johnson, K.R., 1992, Leaf-fossil evidence for extensive floral extinction at the Cretaceous-Tertiary boundary, North Dakota, USA: Cretaceous Research, v. 13, p. 91–117.

Johnson, K.R., 1996, Description of seven common fossil leaf species from the Hell Creek Formation (Upper Cretaceous: Upper Maastrichtian), North Dakota, South Dakota, and Montana: Denver Museum of Natural History Proceedings, ser. 3, no. 12, 47 p.

Johnson, K.R., 1999, The megaflora of the Hell Creek Formation, southwestern North Dakota: Biostratigraphy and paleoecology of the end-Cretaceous terrestrial vegetation: Geological Society of America Abstracts with Programs, v. 31, no. 7, p. A72.

Johnson, K.R., and Hickey, L.J., 1990, Megafloral change across the Cretaceous/Tertiary boundary in the northern Great Plains and Rocky Mountains, U.S.A., *in* Sharpton, V.L., and Ward, P.D., eds., Global catastrophes in Earth History: An interdisciplinary conference on impacts, volcanism, and mass mortality: Geological Society of America Special Paper 247, p. 433–444.

Johnson, K.R., and Wilf, P., 1996, Paleobotanical temperature curve for the Cretaceous-Tertiary boundary interval and a scenario for the revegetation of North America after the terminal Cretaceous event: Geological Society of America Abstracts with Programs, v. 28, no. 7, p. A225.

Johnson, K.R., Nichols, D.J., Attrep, M., Jr., and Orth, C.J., 1989, High-resolution leaf-fossil record spanning the Cretaceous/Tertiary boundary: Nature, v. 340, p. 708–711.

Johnson, K.R., Nichols, D.J., Tauxe, L., and Clark, D., 1990, Floral zonation and magnetostratigraphy of the Hell Creek (late Maastrichtian) and lower Fort Union (early Paleocene) Formations, North Dakota: Geological Society of America Abstracts with Programs, v. 22, no. 6, p. A323.

Johnson, W.T., and Lyons, H.H., 1991, Insects that feed on trees and shrubs (second edition): Ithaca, New York, Cornell University Press, 560 p.

Jolivet, P., and Hawkeswood, T.H., 1995, Host-plants of Chrysomelidae of the world: Leiden, Backhuys, 281 p.

Katerman, F., ed., 1990, Environmental injury to plants: San Diego, Academic Press, 290 p.

Kauffman, E.G., 1984, The fabric of Cretaceous marine extinctions, *in* Berggren, W.A., and Van Couvering, J.A., eds., Catastrophes and Earth history: The new uniformitarianism: Princeton, New Jersey, Princeton University Press, p. 151–246.

Kazikova, I.G., 1985, The character of damage to plants by Orthoptera (Insecta) linked to the structure of their mouthparts on the example of Novosibirsk Akademogorodok fauna, *in* Zolotarenko, G.S., ed., Anthropogenic influences on insect communities: Novosibirsk, Russia, Nauka, p. 122-127 (in Russian).

Kile, G.A., 1974, Insect defoliation in the eucalypt regrowth forests of southern Tasmania: Australian Forest Research, v. 6, p. 9–18.

Klausnitzer, B., 1988, Zur Kenntnis der winterlichen Insektenvergesellschaftung unter Platanenborke (Heteroptera, Coleoptera): Entomologische Nachrichten und Berichte, v. 32, p. 107–112.

Köhler, G., Jentzsch, A., and Reinhardt, K., 2000, Age-related mandible abrasion in three species of short-horned grasshoppers (Caelifera: Acrididae): Journal of Orthoptera Research, v. 9, p. 81–87.

Kolomoets, T.P., Mamaev, V.M., Zerova, M.D., Narchuk, E.P., Ermolenko, V.M., and Diakonchuk, P.A., 1989, Insect gall makers: Diptera: Kiev, Naukova Dumka, 168 p. (in Russian).

Konstantinov, A.S., and Vandenberg, N.J., 1996, Handbook of Paleoarctic flea beetles (Coleoptera: Chrysomelidae: Alticinae): Contributions on Entomology, International, v. 1, p. 233–439.

Koteja, J., 1985, Essay on the prehistory of the scale insects (Homoptera: Coccoidae): Annales Zoologici, v. 38, p. 461–503.

Kozlov, M.V., 1988, Paleontology of lepidopterans and problems in the phylogeny of the order Papilionida, *in* Ponomarenko, A.G., ed., The Cretaceous biocenotic crisis in the evolution of insects: Moscow, Academy of Sciences, p. 16-69 (in Russian).

Kristensen, N.P., 1997, Early evolution of the Lepidoptera + Trichoptera lineage: Phylogeny and the ecological scenario, *in* Grandcolas, P., ed., The origin of biodiversity in insects: Phylogenetic tests of evolutionary scenarios: Mémoires du Muséum National d'Histoire Naturelle, v. 173, p. 253–271.

Kumata, T., 1982, A taxonomic revision of the *Gracillaria* Group occurring in Japan (Lepidoptera: Gracillariidae): Insecta Matsumurana, new series, v. 26, p. 1–186.

Kumata, T., Kuroko, H., and Ermolaev, V.P., 1988, Japanese species of the

Acrocercops Group (Lepidoptera: Gracillariidae) Part 2: Insecta Matsumurana, new series, v. 40, p. 1–133.

Kuroko, H., 1960, Two species of lepidopterous larvae which make larval nests through the leaf: Kita-Kyushu no Konchu, v. 7, p. 1-3 (in Japanese).

Kuschel, G., Oberprieler, R.G., and Rayner, R.J., 1994, Cretaceous weevils from southern Africa, with description of a new genus and species and phylogenetic and zoogeographical comments (Coleoptera: Curculionoidea): Entomologica Scandinavica, v. 25, p. 137–149.

Labandeira, C.C., 1992, Diversity, diets, and disparity: Determining the effect of the terminal Cretaceous extinction on insect evolution [abs.]: Lawrence, Kansas, Paleontological Society Special Publication 6, p. 174.

Labandeira, C.C., 1994, A compendium of fossil insect families: Milwaukee Contributions in Biology and Geology, v. 88, p. 1–71.

Labandeira, C.C., 1998a, How old is the flower and the fly?: Science, v. 280, p. 57–59.

Labandeira, C.C., 1998b, Early history of arthropod and vascular plant associations: Annual Review of Earth and Planetary Sciences, v. 26, p. 329–377.

Labandeira, C.C., 1998c, The role of insects in Late Jurassic to Middle Cretaceous ecosystems, *in* Lucas, S.G., et al., eds., Lower and Middle Cretaceous terrestrial ecosystems: Filling the gap: New Mexico Museum of Natural History Bulletin, v. 14, p. 105–124.

Labandeira, C.C., 2002a, Paleobiology of predators, parasitoids, and parasites: Accomodation and death in the fossil record of terrestrial invertebrates, *in* Kowalewski, M., et al., eds., The fossil record of predation: Lawrence, Kansas, Paleontological Society Papers, v. 8 (in press).

Labandeira, C.C., 2002b, The history of associations between plants and animals, *in* Herrera, C., and Pellmyr, O., eds., Plant-animal interactions: An evolutionary approach: London, Blackwell Science, p. 26–74, 248–261.

Labandeira, C.C., and Sepkoski, J.J., Jr., 1993, Insect diversity in the fossil record: Science, v. 261, p. 310–315.

Labandeira, C.C., Dilcher, D.L., Davis, D.R., and Wagner, D.L., 1994, 97 million years of angiosperm-insect association: Paleobiological insights into the meaning of coevolution: National Academy of Sciences Proceedings, v. 91, p. 12278–12282.

Labandeira, C.C., Johnson, K.R., and Wilf, P., 2002, Impacts of the terminal Cretaceous event on plant-insect associations: National Academy of Sciences Proceedings, v. 99, p. 2061–2066.

Labandeira, C.C., Nufio, C., Wing, S., and Davis, D., 1995, Insect feeding strategies from the Late Cretaceous Big Cedar Ridge flora: Comparing the diversity and intensity of Mesozoic herbivory with the present: Geological Society of America Abstracts with Programs, v. 27, no. 6, p. A447.

Labandeira, C.C., Johnson, K.R., and Lang, P., 1999, Insect extinction at the Cretaceous/Tertiary boundary: Geological Society of America Abstracts with Programs, v. 31, no. 7, p. A72.

Lang, P.J., 1996, Fossil evidence for patterns of leaf-feeding from the Late Cretaceous and early Tertiary [Ph.D. thesis]: London, University of London, 333 p

Lang, P.J., Scott, A.C., and Stephenson, J., 1995, Evidence of plant-arthropod interactions from the Eocene Branksome Sand Formation, Bournemouth, England: Introduction and description of leaf mines: Tertiary Research, v. 15, p. 145–174.

Langenheim, R.L., Smiley, C.J., and Gray, J., 1960, Cretaceous amber from the Arctic Coastal Plain of Alaska: Geological Society of America Bulletin, v. 71, p. 1345–1356.

Larew, H.G., 1992, Fossil galls, *in* Shorthouse, J.D., and Rohfritsch, O., eds., Biology of insect-induced galls: New York, Oxford University Press, p. 50–59.

Leaf Architecture Working Group, 1999, Manual of leaf architecture: Morphological description and categorization of dicotyledonous and net-veined monocotyledonous angiosperms: Washington, D.C., Smithsonian Institution, 65 p.

Lesquereux, L., 1874, Contributions to the fossil flora of the Western Territories. 1. The Cretaceous flora: U.S. Geological and Geographical Surveys of the Territories Report, v. 6, 136 p.

Lesquereux, L., 1878, Contributions to the fossil flora of the Western Territories. 2. The Tertiary flora: U.S. Geological and Geographical Surveys of the Territories Report, v. 7, 336 p.

Lund, S., and Hartman, J.H., 1999, Magnetostratigraphy of Upper Cretaceous/Lower Paleocene strata in the Williston Basin: A reference chronostratigraphic framework for paleoenvironmental and paleogeographic patterns: Geological Society of America Abstracts with Programs, v. 31, no. 7, p. A71.

Lupia, R., 1999, Discordant morphological disparity and taxonomic diversity during the Cretaceous angiosperm radiation: North American pollen record: Paleobiology, v. 25, p. 1–28.

Maceljski, M., and Balarin, I., 1972, Novi clan stetne entomofaune u Jugoslaviji—*Corythuca ciliata* (Say), Tingidae, Heteroptera: Zastita Bilja, v. 23, p. 193–200.

Manchester, S.R., 1999, Biogeographical relationships of North American Tertiary floras: Missouri Botanical Garden Annals, v. 86, p. 472–522.

Martin, P.M., 1986, Refuting late Pleistocene extinction models, *in* Elliott, D.K., ed., Dynamics of extinction: New York, John Wiley, p. 107–130.

Matthews, J.V., Jr., 1979, Fossil beetles and the Late Cenozoic history of the tundra environment, *in* Gray, J., and Boucot, A.J., eds., Historical biogeography, plate tectonics, and the changing environment: Corvallis, Oregon State University Press, p. 371–378.

McAlpine, J.F., and Martin, J.E.H., 1969, Canadian amber: A paleontological treasure chest: Canadian Entomologist, v. 101, p. 818–838.

McClure, M.S., 1974, Biology of *Erythroneura lawsoni* (Homoptera: Cicadellidae) and coexistence in the sycamore leaf-feeding guild: Environmental Entomology, v. 3, p. 59–68.

McConnell, H.S., and Davidson, J.A., 1959, Observations on the life history and morphology of *Kermes pubescens* Bogue (Homoptera: Coccoidea: Dactylopiidae): Entomological Society of America Annals, v. 52, p. 463–468.

McIver, E.E., 1999, Paleobotanical evidence for ecosystem disruption at the Cretaceous—Tertiary boundary from Wood Mountain, Saskatchewan, Canada: Canadian Journal of Earth Sciences, v. 36, p. 775–789.

Michener, C.D., and Grimaldi, D.A., 1988, The oldest fossil bee: Apoid history, evolutionary stasis, and antiquity of social behavior: National Academy of Sciences Proceedings, v. 85, p. 6424–6426.

Moran, N.A., 1989, A 48 million-year-old aphid-host plant association and complex life cycle: Biogeographic evidence: Science, v. 245, p. 173–175.

Murakami, Y., 1970, A review of biology and ecology of diaspine scales in Japan: Mushi, v. 43, p. 65–114.

Needham, J.G., Frost, S.W., and Tothill, B.H., 1928, Leaf-mining insects: Baltimore, Maryland, Williams and Wilkins, 351 p.

Newberry, J.S., 1863, Descriptions of fossil plants collected by Mr. George Gibbs: Boston Natural History Museum Journal, v. 7, p. 506–524.

Newberry, J.S., 1898, The later extinct floras of North America: U.S. Geological Survey Monograph 35, 295 p.

Nichols, D.J., and Johnson, K.R., 1999, Palynology and microstratigraphy of the K-T boundary in southwestern North Dakota: Geological Society of America Abstracts with Programs, v. 31, no. 7, p. A72.

Nichols, D.J., Jacobsen, S.R., and Tschudy, R.H., 1982, Cretaceous palynomorph biozones for the central and northern Rocky Mountain region of the United States, *in* Powers, R.B., ed., Geologic studies of the Cordilleran thrust belt: Denver, Colorado, Rocky Mountain Association of Geologists, p. 721–733.

Nichols, D.J., Murphy, E.C., Johnson, K.R., and Betterton, W.J., 2000, A second K-T boundary locality in North Dakota verified by palynostratigraphy and shocked quartz: Geological Society of America Abstracts with Programs, v. 32, no. 7, p. A130.

Niklas, K.J., 1988, Patterns of vascular plant diversification in the fossil record: Proof and conjecture: Missouri Botanical Garden Annals, v. 75, p. 35–54.

Niklas, K.J., Tiffney, B.H., and Knoll, A.H., 1980, Apparent changes in the diversity of fossil plants: A preliminary assessment: Evolutionary Biology, v. 12, p. 1–89.

Norris, R.D., 1991, Biased extinction and evolutionary trends: Paleobiology, v. 17, p. 388–399.

Obradovich, J.D., and Hicks, J.F., 1999, A review of the isotopic calibration points for the geomagnetic polarity time scale, in the interval 83 to 33 Ma (C34N to C13N): Geological Society of America Abstracts with Programs, v. 31, no. 7, p. A71.

O'Dowd, D.J., Brew, C.R., Christophel, D.C., and Norton, R.A., 1991, Mite-plant associations from the Eocene of southern Australia: Science, v. 252, p. 99–101.

Opler, P.A., 1973, Fossil lepidopterous leaf mines demonstrate the age of some insect-plant relationships: Science, v. 179, p. 1321–1323.

Opler, P.A., 1974, Oaks as evolutionary islands for leaf-mining insects: American Scientist, v. 62, p. 67–73.

Packard, A.S., 1889, Insects injurious to forest and shade trees: U.S. Department of Agriculture Entomological Commission Report, v. 5, 955 p.

Palmer, M., Labandeira, C.C., Johnson, K., and Wehr, W., 1998, Diversity and intensity of insect herbivory on the middle Eocene Republic Flora: Comparing the fossil record with the recent: Geological Society of America Abstracts with Programs, v. 30, no. 7, p. A37.

Patch, E.M., 1938, Food-plant catalogue of the aphids of the World, including the Phylloxeridae: University of Maine Agricultural Experiment Station Bulletin, no. 393, p. 35–431.

Paulay, G., 1985, Adaptive radiation on an isolated oceanic island: The Cryptorhynchinae (Curculionidae) of Rapa revisited: Linnean Society Biological Journal, v. 26, p. 95–187.

Pearson, D.A., 1992, Microfaunal comparisons between the lower and upper portions of the Hell Creek Formation (late Maastrichtian) in southwestern North Dakota [abs.]: North Dakota Academy of Science Proceedings, v. 46, p. 84.

Pearson, D.A., Schaefer, T., Johnson, K.R., and Nichols, D.J., 1999, Vertebrate biostratigraphy of the Hell Creek Formation in southwestern North Dakota: Geological Society of America Abstracts with Programs, v. 31, no. 7, p. A72–A73.

Peigler, R.S., 1986, Worldwide predilection of resiniferous hostplants by three unrelated groups of moths in the genera *Actias, Citheronia* (Saturniidae) and the subfamily Euteliinae (Noctuidae): Tyô to Ga, v. 37, p. 45–50.

Pike, E.M., 1993, Amber taphonomy and collecting biases: Palaios, v. 8, p. 411–419.

Poinar, G.O., Jr., 1992, Life in amber: Stanford, California, Stanford University Press, 350 p.

Poinar, G.O., Jr., Archibald, B., and Brown, A., 1999, New amber deposit provides evidence of early Paleogene extinctions, paleoclimates, and past distributions: Canadian Entomologist, v. 131, p. 171–177.

Ponomarenko, A.G., ed., 1988, The Cretaceous biocenotic crisis in the evolution of insects: Moscow, Academy of Sciences, 230 p. (in Russian).

Ponomarenko, A.G., 1995, The geological history of beetles, *in* Pakaluk, J., and Slipinski, A., eds., Biology, phylogeny, and classification of Coleoptera: Papers celebrating the 80th birthday of Roy A. Crowson: Warsaw, Polish Academy of Sciences, Museum and Institute of Zoology, p. 155–171.

Powell, J.A., Mitter, C., and Farrell, B., 1999, Evolution of larval food preferences in Lepidoptera, *in* Kristensen, N.P., ed., Lepidoptera, moths and butterflies, Volume 1, Evolution, systematics, and biogeography: Handbuch der Zoologie, v. 4, no. 35, p. 403–422.

Price, P.W., 1997, Insect ecology (third edition): New York, John Wiley, 874 p.

Rasnitsyn, A.P., 1988, Problems in the global crisis in insect communities during the middle Mesozoic, *in* Ponomarenko, A.G., ed., The Mesozoic-Cenozoic crisis in the evolution of insects: Moscow, Academy of Sciences, p. 191–207 (in Russian).

Rayner, R.J., and Waters, S.B., 1990, A Cretaceous crane-fly (Diptera: Tipulidae): 93 million years of stasis: Linnean Society Zoological Journal, v. 99, p. 309–318.

Rogers, J., Locci, R., and Vescovo, P., 1982, Contribution to tree pathology. 3. On the association between *Corythucha ciliata* and saprophytic fungi in plane trees: Rivista di Patologia Vegetale, v. 4, p. 149–155.

Rohr, D.M., Boucot, A.J., Abbott, M., and Miller, J.M., 1984, Oldest termite nest from the Upper Cretaceous of west Texas: Geology, v. 14, p. 87–88.

Rust, J., 1998, Biotratinomie von Insekten aus der Fur-Formation von Dänemark (Møler, oberes Paleozän/unterers Eozän): Paläontologisches Zeitschrift, v. 71, p. 41–58.

Sanders, J.G., 1905, The cottony maple scale (*Pulvinaria innumerabilis* Rathvon.): U.S. Department of Agriculture, Bureau of Entomology Circular, no. 64, 5 p.

Schlüter, T., 1987, Die Entwicklung der coevolutiven beziehungen zwischen Angiospermen: Unter besonderer Berücksichtigung ihrer Pollen und Insekten im späten Mesozoikum: Berliner Geowissenschaften Abhandlungen, ser. A, v. 86, p. 127–143.

Scholtz, C.H., and Chown, S.L., 1995, The evolution of habitat use and diet in the Scarabaeoidea: A phylogenetic approach, *in* Pakaluk, J., and Slipinski, S.A., eds., Biology, phylogeny, and classification of Coleoptera: Papers celebrating the 80th birthday of Roy A. Crowson: Warsaw, Polish Academy of Sciences, Museum and Institute of Zoology, p. 355–374.

Schoonhoven, L.M., Jermy, T., and van Loon, J.J.A., 1998, Insect-plant biology: From physiology to evolution: London, Chapman and Hall, 409 p.

Scott, A.C., Stephenson, J., and Chaloner, W.G., 1992, Interaction and coevolution of plants and arthropods during the Palaeozoic and Mesozoic: Royal Society of London Philosophical Transactions, ser. B, v. 335, p. 129–165.

Scott, A.C., Stephenson, J., and Chaloner, W.G., 1994, The fossil record of leaves with galls, *in* Williams, M.A.J., ed., Plant galls: Organisms, interactions, populations: Oxford, Clarendon Press, p. 447–470.

Selman, B.J., 1988, Viruses and Chrysomelidae, *in* Jolivet, P., Petitpierre, E., and Hsiao, T.H., eds., Biology of Chrysomelidae: Dordrecht, Kluwer, p. 379–387.

Sereno, P.C., 1999, The evolution of dinosaurs: Science, v. 284, p. 2137–2147.

Signor, P.W., III, and Lipps, J.H., 1982, Sampling bias, gradual extinction patterns, and catastrophes in the fossil record, *in* Silver, L.T., and Schultz, P.H., eds., Geological implications of impacts of large asteroids and comets on the Earth: Geological Society of America Special Paper 190, p. 291–296.

Sinclair, W.A., Lyon, H.H., and Johnson, W.T., 1987, Diseases of trees and shrubs (second edition): Ithaca, Cornell University Press, 575 p.

Skalski, A.W., 1979, Records of oldest Lepidoptera: Nota Lepidopterologica, v. 2, p. 61–66.

Smith, R.H., 1944, Insects and mites injurious to sycamore trees (*Platanus* spp.) in western North America: Arborist's News, v. 9, p. 9–15.

Spencer, K.A., 1976, The Agromyzidae (Diptera) of Fennoscandia and Denmark: Fauna Entomologica Scandinavica, v. 5, 606. p.

Spencer, K.A., 1990, Host specialization in the world Agromyzidae (Diptera): Dordrecht, Kluwer, 440 p.

Stephenson, J., 1992, Evidence of plant/insect interactions in the Late Cretaceous and early Tertiary [Ph.D. thesis]: London, University of London, 378 p.

Stott, L.D., and Kennett, J.P., 1990, The paleoceanographic and paleoclimatic signature of the Cretaceous/Paleogene boundary in the Antarctic: Stable isotopic results from ODP Leg 113, *in* Proceedings of the Ocean Drilling Program, Scientific results, Volume 113: Washington, D.C., U.S. Government Printing Office, p. 829–848.

Strong, D.R., 1977, Rolled-leaf hispine beetles (Chrysomelidae) and their Zingiberales host plants in Middle America: Biotropica, v. 9, p. 156–169.

Strong, D.R., 1984, Banana's best friend: Natural History, v. 93, p. 50–57.

Strong, D.R., Lawton, J.H., and Southwood, T.R.E, 1984, Insects on plants: Community patterns and mechanisms: Cambridge, Massachusetts, Harvard University Press, 313 p.

Sweet, A.R., Braman, D.R., and Lerbekmo, J.F., 1999, Sequential palynological changes across the composite Cretaceous-Tertiary (K-T) boundary claystone and contiguous strata, western Canada and Montana, U.S.A.: Canadian Journal of Earth Sciences, v. 36, p. 743–768.

Taylor, D.W., and Crepet, W.L., 1987, Fossil floral evidence of Malpighiaceae and an early plant-pollinator relationship: American Journal of Botany, v. 74, p. 274–286.

Thien, L.B., Bernhardt, P., Gibbs, G.W., Pellmyr, O., Bergstrom, G., Groth, I., and McPherson, G., 1985, The pollination of *Zygogynum* (Winteraceae) by a moth, *Sabatinca* (Micropterygidae): An ancient association?: Science, v. 227, p. 540–543.

Thompson, L.C., and Solomon, J.D., 1986, Insect defoliators of young sycamore plantations: University of Arkansas Agricultural Experiment Station Bulletin. 897, 24 p.

Thurston, E.L., 1974, Morphology, fine structure, and ontogeny of the stinging emergence of *Urtica dioica:* American Journal of Botany, v. 61, p. 809–817.

Togashi, I., 1999, A new species of the genus *Caliroa* (Hymenoptera, Tenthredinidae) from Mt. Hausan, Honshu, Japan: Japanese Journal of Systematic Entomology, v. 8, p. 177–180.

Tschudy, R.H., 1984, Palynological evidence for change in continental floras at the Cretaceous-Tertiary boundary, *in* Berggren, W.A., and Van Couvering, J.A., eds., Catastrophes in Earth history: The new uniformitarianism: Princeton, New Jersey, Princeton University Press, p. 315–337.

Upchurch, G.R., Jr., and Dilcher, D.L., 1990, Cenomanian angiosperm leaf megafossils, Dakota Formation, Rose Creek locality, Jefferson County, southeastern Nebraska: U.S. Geological Survey Bulletin 1915, 57 p.

Vincent, J.F.W., 1990, Fracture properties of plants: Advances in Botanical Research, v. 17, p. 235–287.

Waggoner, B.M., and Poteet, M.F., 1996, Unusual oak leaf galls from the middle Miocene of northwestern Nevada: Journal of Paleontology, v. 70, p. 1080–1084.

Wakgari, W.M., and Giliomee, J.H., 1998, Description of the stages of the white wax-scale, *Ceroplastes destructor* Newstead (Homoptera: Coccidae): African Entomology, v. 6, p. 303–316.

Ward, L.F., 1887, Types of the Laramie flora: U.S. Geological Survey Bulletin 37, 354 p.

Ward, P.D., 1995, After the fall: Lessons and directions from the K-T debate: Palaios, v. 10, p. 530–538.

Whalley, P., 1987, Insects and Cretaceous mass extinction: Nature, v. 327, p. 562.

Wheeler, A.G., Jr., 1980, Life history of *Plagiognathus albatus* (Hemiptera: Miridae), with a description of the fifth instar: Entomological Society of America Annals, v. 73, p. 354–356.

Wilf, P., 2000, Late Paleocene—early Eocene climate changes in southwestern Wyoming: Paleobotanical analysis: Geological Society of America Bulletin, v. 112, p. 292–307.

Wilf, P., and Labandeira, C.C., 1999, Response of plant-insect associations to Paleocene-Eocene warming: Science, v. 284, p. 2153–2156.

Wilf, P., Labandeira, C.C., Johnson, K.R., Coley, P.D., and Cutter, A.D., 2001, Insect herbivory, plant defense, and early Cenozoic climate change: National Academy of Sciences Proceedings, v. 98, p. 6221–6226.

Wilf, P., Labandeira, C.C., Kress, W.J., Staines, C.L., Windsor, D.M., Allen, A.L., and Johnson, K.R., 2000, Timing the radiations of leaf beetles: Hispines on gingers from latest Cretaceous to recent: Science, v. 289, p. 291–294.

Willemstein, S.C., 1987, An evolutionary basis for pollination biology: Leiden Botanical Series, v. 10, p. 1–425.

Williams, M.L., and Kosztarab, M., 1972, Morphology and systematics of the Coccidae of Virginia, with notes on their biology (Homoptera: Coccoidea): Virginia Polytechnic Institute and State University Research Division Bulletin 74, 215 p.

Wilson, E.O., 1992, The diversity of life: Cambridge, Massachusetts, Harvard University Press, 424 p.

Windsor, D., Ness, J., Gómez, L.D., and Jolivet, P.H., 1999, Species of *Aulacoscelis* Duponchel and Chevrolat (Chrysomelidae) and *Nomotus* Gorham (Languriidae) feed on fronds of Central American cycads: Coleopterists Bulletin, v. 53, p. 217–231.

Wing, S.L., Hickey, L.J., and Swisher, C.C., 1993, Implications of an exceptional fossil flora for Late Cretaceous vegetation: Nature, v. 363, p. 342–344.

Wing, S.L., Bao, H., and Koch, P.L., 2000, An early Eocene cool period?: Evidence for continental cooling during the warmest part of the Cenozoic, *in* Huber, B.T., MacLeod, K., and Wing, S.L., eds., Warm climates in Earth history: Cambridge, Cambridge University Press, p. 197–237.

Wolfe, J.A., and Upchurch, G.R., Jr., 1987, Leaf assemblages across the Cretaceous-Tertiary boundary in the Raton Basin, New Mexico and Colorado: National Academy of Sciences Proceedings, v. 84, p. 5096–5100.

Wong, H.R., and Melvin, J.C.E., 1976, Biological observations and larval descriptions of *Enargia decolor* (Lepidoptera: Noctuidae) on trembling aspen in northern Alberta: Canadian Entomologist, v. 108, p. 1213–1220.

Wootton, R.J., 1990, Major insect radiations: Systematics Association Special Volume 42, p. 187–208.

Wunn, H., 1925, Die Coccidenfauna Badens. 6. Mitteilung über Cocciden: Zeitschrift für Angewandte Entomologie, v. 1925, p. 427–451.

Yuan, D.-C., and Robinson, G.S., 1993, *Caloptilia* leaf-miner moths (Gracillariidae) of South-East Asia: Natural History Museum of London Bulletin, v. 62, p. 1–37.

Yukawa, J., and Uechi, N., 1999, Can gallers expand the host range to alien plants within a short period of time?: Esakia, v. 39, p. 1–7.

Zerova, M.D., Diakonchuk, L.A., and Ermolenko, V.M., 1988, Insect gall makers: Hymenoptera: Kiev, Naukova Dumka, 160 p. (in Russian).

Zerova, M.D., Mamontova, V.A., Ermolenko, V.M., Diakonchuk, L.A., Sinev, S.J., and Kozlov, M.V., 1991, Insect gall makers: Homoptera, Heteroptera, Coleoptera, Lepidoptera: Kiev, Naukova Dumka, 344 p. (in Russian).

Zherikhin, V.V., and Sukacheva, I.D., 1973, On Cretaceous insect-bearing amber (retinite) from northern Siberia, *in* Bei-Benko, G.Y., Reports of the 24th Annual Lecture in Memory of N.A. Kholodkovskogo: Moscow, Academy of Sciences, p. 3–48 (in Russian).

Zwölfer, H., 1978, Mecanismen und Ergebnisse der Co-Evolution von phytophagen und entomophagen Insekten und höheren Pflanzen: Sonderbande des Naturwissenschaften Vereins in Hamburg, v. 2, p. 7–50.

Manuscript Accepted by the Society September 24, 2001

Printed in the U.S.A.

Geological Society of America
Special Paper 361
2002

Megaflora of the Hell Creek and lower Fort Union Formations in the western Dakotas: Vegetational response to climate change, the Cretaceous-Tertiary boundary event, and rapid marine transgression

Kirk R. Johnson*
Department of Earth Sciences, Denver Museum of Nature & Science, 2001 Colorado Boulevard, Denver, Colorado 80205, USA

ABSTRACT

The Hell Creek and Fort Union Formations in southwestern North Dakota and northwestern South Dakota have yielded a diverse megaflora of 380 species from 158 quarry sites. These sites are situated in a stratigraphic framework, delimited by palynology, magnetostratigraphy, and vertebrate paleontology, that contains both the Cretaceous-Tertiary (K-T) boundary event horizon and stratigraphic evidence for rapid base-level increase. The late Maastrichtian Hell Creek flora is dominated by angiosperms; ferns, fern allies, cycads, ginkgo, and conifers represent <10% of total taxa and specimens. Megafloral extinction at the K-T boundary is extensive, effectively eliminating all dominant plant taxa of the upper Hell Creek Formation. K-T survivorship appears to be greatest in plants that occupied Cretaceous mire facies. The Hell Creek megaflora is heterogeneous with respect to stratigraphic position and sedimentary facies, allowing the recognition of three superposed megafloral zones and two facies-controlled megafloras. Hell Creek vegetation represents an angiosperm-dominated woodland composed of small- to medium-sized trees, often with lobed leaves. One exception is a diverse angiosperm-dominated herbaceous vegetation associated with large paleochannels in the middle of the formation. The uppermost Hell Creek megaflora zone (HCIII) first occurs near the base of magnetic polarity subchron C29R. The HCIII megaflora is significantly more diverse than earlier Hell Creek floras, and foliar physiognomy suggests a significant climate warming during the final 300–500 k.y. of the Cretaceous. The Paleocene megaflora is depauperate, less heterogeneous than the Cretaceous megaflora, and dominated, in all facies, by taxa that were most common in Cretaceous mire facies.

INTRODUCTION

In the northern Rocky Mountain and Great Plains region of North America, the Cretaceous-Tertiary (K-T) boundary is at or near the contact of the Hell Creek Formation with the overlying lignitic Fort Union Formation. The Hell Creek Formation is widespread in the Williston and Powder River Basins of eastern Montana and western North Dakota and South Dakota. Contiguous strata in the Powder River Basin of Wyoming are known as the Lance Formation. The Hell Creek Formation

*E-mail: kjohnson@dmns.org

Johnson, K.R., 2002, Megaflora of the Hell Creek and lower Fort Union Formations in the western Dakotas: Vegetational response to climate change, the Cretaceous-Tertiary boundary event, and rapid marine transgression, *in* Hartman, J.H., Johnson, K.R., and Nichols, D.J., eds., The Hell Creek Formation and the Cretaceous-Tertiary boundary in the northern Great Plains: An integrated continental record of the end of the Cretaceous: Boulder, Colorado, Geological Society of America Special Paper 361, p. 329–391.

was named by Barnum Brown (1907), who made major discoveries of dinosaurs such as *Tyrannosaurus rex* and other vertebrates in the Fort Peck region of central Montana between 1902 and 1910. The significance of this area increased with the discovery of abundant mammalian fossils at the Bug Creek Anthills in McCone County, Montana (Sloan and Van Valen, 1965). Although there were episodic field efforts in the Hell Creek Formation in eastern Montana near the towns of Ekalaka and Glendive, most of the early paleontological work in the Hell Creek Formation occurred in the valley of the Missouri River (later dammed to create the Fort Peck Reservoir) in Garfield and McCone Counties, Montana. The Hell Creek Formation in North Dakota and South Dakota has received less study but is paleogeographically significant because it interfingers with marine strata at its eastern margin (Laird and Mitchell, 1942).

The publication of the impact hypothesis (Alvarez et al., 1980) raised questions that were not readily answered in the published literature of terrestrial paleontology, and launched or redirected several field programs to assess the effects of the K-T event on terrestrial biota (Archibald, 1996; and chapters in this volume). These efforts have resulted in extensive collections from the Hell Creek and Fort Union Formations and have begun to build a new understanding of the nature of the latest Cretaceous biota of the Rocky Mountains and Great Plains.

Although paleobotanical exploration of the western United States was a priority as early as the 1860s, two things hampered the circumscription of the Maastrichtian (latest Cretaceous) flora. The first was a controversy known as the Laramie problem, which pitted paleobotanists against vertebrate paleontologists over the placement of the K-T boundary. The second was the very real rarity of Maastrichtian fossil plants relative to Paleocene fossil plants. The Laramie problem was largely resolved before the World War II by Erling Dorf and Roland Brown, neither of whom was published prolifically, but their rivalry postponed the acceptance of their generally concordant conclusions. Dorf focused on the Cretaceous and Brown applied most of his attention to the Paleocene. Dorf's (1940) overall assessment of the Laramie problem and his monograph on the Maastrichtian megaflora of the Lance Formation (Dorf, 1942) and Brown's early papers (1938, 1939, 1952) and his (posthumous) monograph of the Paleocene megaflora of the northern Rocky Mountains and Great Plains (Brown, 1962) documented that Upper Cretaceous floras were substantially different from those of the Paleocene. Whereas Brown's work was based on hundreds of localities and tens of thousands of specimens, Dorf's work was based on a few localities and ~1000 specimens from localities around the type section of the Lance Formation in the southern portion of the Powder River Basin near Lance Creek, Wyoming. Fossil leaf sites are much more common in the Fort Union Formation than they are in the Hell Creek Formation and, as a result, studies of Hell Creek leaves have been few (Berry, 1934; Brown, 1939; Shoemaker, 1966).

The rise of palynology in the 1960s (Stanley, 1965; Leffingwell, 1970; Tschudy, 1970) provided a new stratigraphic tool that grew to supplant leaves and dinosaurs as the primary means of recognizing the K-T boundary in terrestrial rocks. In addition, the recognition that many early fossil angiosperm papers were flawed by poor taxonomy led to the waning of leaf-based studies.

Hickey and Wolfe (1975) led a revival of angiosperm leaf studies based on the two premises that angiosperm leaves contain enough taxonomic information to be systematically useful and that their physiognomy is a valid method to interpret paleoclimate (Wolfe, 1979). In addition, it was becoming clear that leaves held paleoenvironmental and paleoecological data that might not be retrievable from pollen.

By the time of the Alvarez discovery, palynology had been refined as a well-honed tool for picking the K-T boundary. When the palynological K-T boundary, defined by the decimation of the *Wodehouseia spinata* Assemblage Zone flora, was shown to coincide with shocked mineral and iridium anomalies, a link was drawn between the pollen extinctions and the K-T impact (Orth et al., 1981; Tschudy et al., 1984). Continued work by Hickey (1980) added greatly to the knowledge of the basal Paleocene floras, but Maastrichtian floras were still largely unknown by the late 1980s. The overabundance of Paleocene floras relative to uppermost Cretaceous ones caused an underestimation of the effects of the K-T boundary (Hickey, 1981, 1984) and created a need for a more thoroughly sampled uppermost Cretaceous floral record.

It became clear that an approach using both palynology and megafossil paleobotany might be a strong addition to the understanding of the K-T event, and I began to search for Upper Cretaceous floras in 1983. Preliminary prospecting in the Fort Peck region by Hickey between 1978 and 1982 suggested that Cretaceous leaves were quite rare relative to Paleocene ones. This and the stratigraphic complexities of the Bug Creek problem led me to search other areas for a K-T transition flora. In 1981, mapping of the stratigraphy of lignite-bearing strata of southwestern North Dakota by the U.S. Geological Survey (USGS) brought to light the presence of fossil leaf deposits in the Ludlow Member of the Fort Union Formation (Johnson, 1982; Belt et al., 1984). This, combined with papers by Corbin (1938) and Brown (1939), suggested that the area around Marmarth in the extreme southwestern corner of North Dakota might be a fertile spot to assess the terminal Cretaceous flora and its transition to the Tertiary. Fastovsky (1986) discovered several well-preserved megafloral sites in the course of his sedimentological field work in the Hell Creek Formation near Marmarth. My field work in 1986–1988 resulted in the discovery and collection of fossil plants from 57 Cretaceous and 30 Paleocene sites (Johnson, 1989, 1992, 1996; Johnson and Hickey, 1990). Field work under the auspices of the Denver Museum of Natural History (now the Denver Museum of Nature & Science) began in southwestern North Dakota in 1991 and has continued to the present. This chapter summarizes the results of these last 10 years of field work and integrates them with the data collected in the late 1980s.

This chapter is an attempt to provide an overview of the rich and complex megaflora from the Hell Creek Formation and the much poorer (although more common) and more uniform flora of the Paleocene portions of the overlying Ludlow Member of the Fort Union Formation. The time span represented by these rocks included three major and unrelated events: climate change, the impact of the Chicxulub bolide, and the rapid transgression of the Cannonball seaway. Each of these events left their mark on the regional landscape and vegetation. The goal of this chapter is to describe the megaflora and its stratigraphic setting in such a way that it is possible to understand the effects of each of these events.

GEOLOGIC SETTING

The badlands of the Little Missouri River in Slope and Bowman Counties, southwestern North Dakota, and Harding County, northwestern South Dakota, provide superb exposures of the Pierre Shale; the Timber Lake and Colgate Members of the Fox Hills Formation; the Hell Creek Formation; and the Ludlow, Cannonball, and Tongue River Members of the Fort Union Formation (Figs. 1 and 2) (Thom and Dobbin, 1924; Dobbin and Reeside, 1929). These outcrops have formed in the past few million years by the intersection of the north-flowing Little Missouri River with the north-northwest–south-southeast–trending Cedar Creek anticline. This asymmetrical structure plunges into the subsurface just south of the North Dakota–South Dakota state line. In eastern Harding County, the drainage of the Grand River has created a complementary complex of badland exposures.

The Hell Creek Formation and the Ludlow Member of the Fort Union Formation are the subjects of this study. Exposures of these units in North Dakota and South Dakota were first described by Leonard (1908, 1911) and Stanton (1910) and mapped by the lignite surveys of the western United States that were undertaken in the years before World War I (Bowen, 1912; Winchester et al., 1916; Bauer, 1924; Hares, 1928). Occasional geological studies in westernmost North Dakota and South Dakota were later undertaken as part of mapping projects by the South Dakota Geological Survey (Petsch, 1956; Schulte, 1957; Erickson, 1956a, 1956b) and thc North Dakota Geological Survey (Carlson, 1979, 1983; Moore, 1976), uranium prospecting in the 1950s and 1960s (Pipiringos et al., 1965), more recent surveys and assessment of lignite-bearing strata (Belt et al., 1984; FUCAT [Fort Union Coal Assessing Team], 1999), and several dissertations (Frye, 1967; Fastovsky, 1986; Johnson, 1989).

The Hell Creek Formation was first recognized in this area by Leonard (1908, 1911), who considered it a member of the Lance Formation. Laird and Mitchell (1942) recognized a marine tongue in the Hell Creek in the Missouri River Valley and named it the Breien Member. The first detailed work on paleontology of the Hell Creek in southern North Dakota was by Frye (1967, 1969), who subdivided the formation into several members: modern workers (see discussion in Murphy et al., this volume) do not recognize any of these, except the previously described Breien Member. The Hell Creek is composed primarily of rooted light gray to green to purple mudstone and fine-grained lenticular sandstone bodies (Fastovsky, 1987). The formation was deposited on a coastal plain to the west of the Cretaceous seaway. In central North Dakota, the formation shows evidence of brackish-water conditions (Hartman, this volume; Hoganson and Murphy, this volume; Murphy et al., this volume), suggesting that the marine transgression, traditionally considered to be a Paleocene event resulting in the deposition of the Paleocene Cannonball Member of the Fort Union Formation (Fox and Ross, 1942), had begun in the Late Cretaceous. The basal contact of the Hell Creek Formation and the underlying Colgate Member of the Fox Hills Sandstone is locally disconformable and is also marked by the presence of dark gray to brown carbonaceous beds in the basal few meters of the Hell Creek Formation. The Fox Hills Sandstone in the westernmost Dakotas has not yielded age-diagnostic fossils (Feldmann, 1972), but the upper portions of the underlying Pierre Shale have produced invertebrates indicative of both the *Baculites baculus* and the *Baculites grandis* zones (Bishop, 1965, 1973; Gill and Cobban, 1973). The latter has been dated as 70.15 ± 0.15 Ma (Obradovich, 1993; Hicks, 1993; Hicks et al., 1999). The Hell Creek Formation is 100 m thick in southwestern North Dakota (Murphy et al., this volume). Numerous early reports (Hares, 1928; Pipiringos et al., 1965; Erickson, 1956a) cited the thickness as ranging to 575 ft (175 m), but unlike this chapter and Murphy et al. (this volume), none of these reports based their estimates on continuous measured sections, and there has been extensive confusion in the past over the placement of the Hell Creek–Ludlow contact.

The upper contact of the Hell Creek Formation is marked by a change in depositional environments, and the overlying Ludlow Member of the Fort Union Formation is readily distinguished by a number of distinctive lithologies, including laterally extensive lignite and carbonaceous shale beds, variegated mudstone beds of lacustrine origin, and tabular sand bodies interpreted as crevasse-splay deposits (Brown, 1952; Belt et al., 1984; Fastovsky, 1987; Johnson, 1992). The facies change that marks the formational contact is consistent with a scenario of rising base level caused by the transgression of the Cannonball seaway. The Ludlow Member was first described by Lloyd and Hares (1915) from exposures on the flanks of the North Cave Hills west of the small town of Ludlow, South Dakota. The Ludlow is rich in lignite beds and these beds were extensively mined by early settlers of the region. Later, the discovery of uranium roll-front deposits in the Ludlow led to local mining efforts (Pipiringos et al., 1965). The thickness of the Ludlow ranges from 94.5 m at Lahti Butte in western Harding County to 145 m at its type area (Pipiringos et al., 1965). In 2001, we trenched the Lahti Butte section and measured the complete

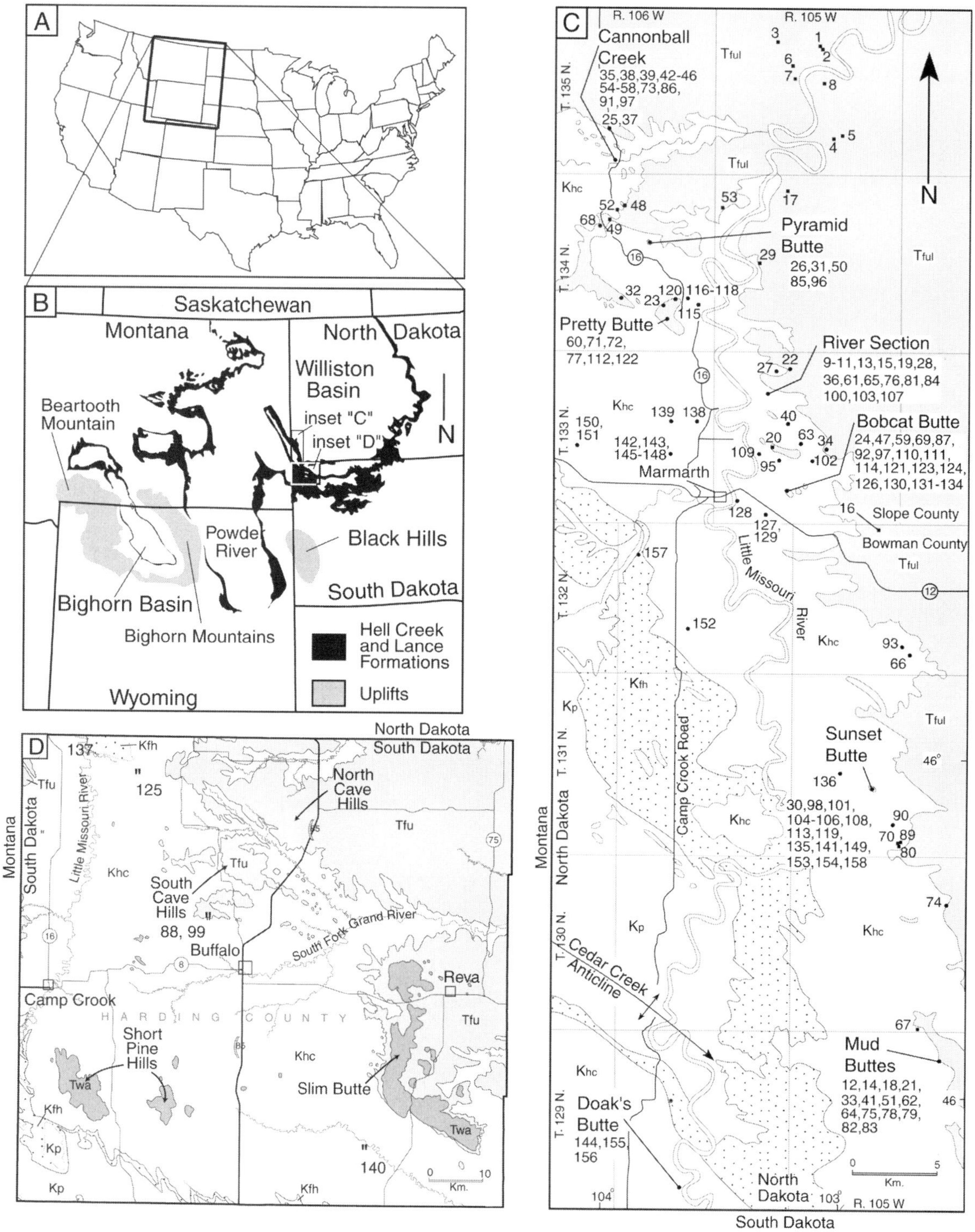

Figure 1. Locality map of study area in Williston Basin of northern Great Plains of North America showing locations of Yale Peabody Museum and Denver Museum of Nature & Science plant megafossil sites. Specific sites are listed in Table A1. Inset C is the western halves of Slope and Bowman Counties, North Dakota, and inset D is the western half of Harding County, South Dakota. Kp is Pierre Shale, Kfh is Fox Hills Sandstone, Khc is Hell Creek Formation, Tful is Ludlow Member of Fort Union Formation, Tfu is undivided Fort Union Formation, and Twa is undivided White River Group and Arikaree Formation. Geology for North Dakota modified from Carlson (1979, 1983) and for South Dakota from Ellis and Colton (1994).

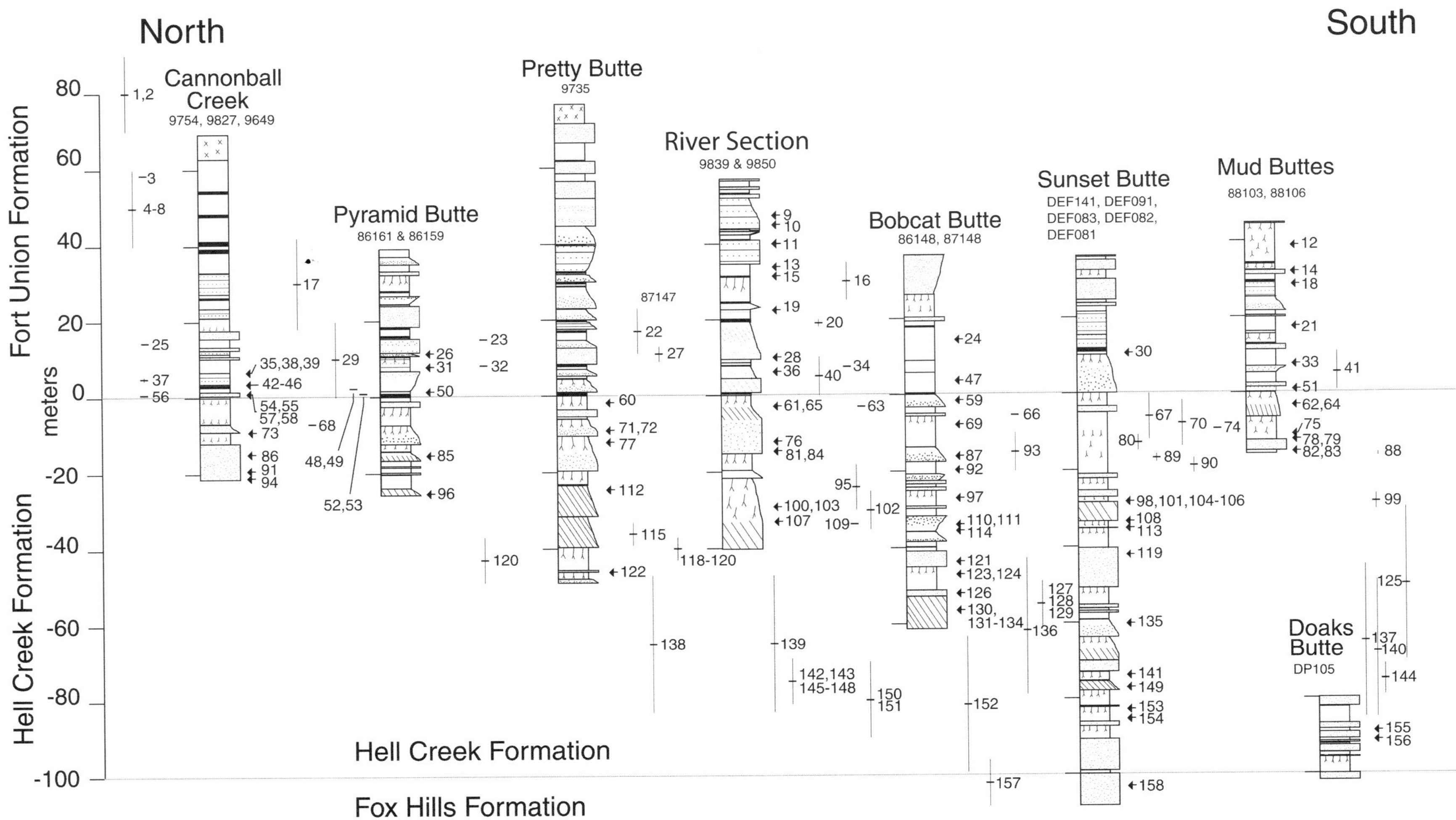

Figure 2. Stratigraphic cross section showing key sections in Slope and Bowman Counties, North Dakota, arranged from north to south. Stratigraphic positions of Yale Peabody Museum and Denver Museum of Nature & Science fossil plant sites are plotted against sections when it was possible to directly correlate locality to measured section and plotted between sections when site was correlated to section by elevation. Stratigraphic sections are further described in Murphy et al. (this volume).

thickness of the Ludlow Member as only 77 m. Complete thicknesses are more difficult to measure in northern Slope County, but Belt et al. (1984) measured 180 m of Ludlow in a series of correlated sections. Hartman (1995) reported a 183 m thickness of the Ludlow in Slope County, but cautioned that the presence of lenticular channel deposits and intercalated tongues of the Cannonball Member of the Fort Union Formation makes the thickness locally quite variable. The Ludlow is laterally equivalent to the eastward-thickening Cannonball Member, which has an early Paleocene invertebrate, vertebrate, and microfaunal marine fauna (Stanton, 1921; Fox and Ross, 1942; Fox and Olsson, 1969; Cvancara and Hoganson, 1993). The base of the Ludlow Member has yielded a sparse Puercan (Pu1-2) mammalian fauna and the middle portion of the unit has produced a sparse Torrejonian (To2) mammalian fauna, suggesting that the complete thickness of the Ludlow Member represents the first ~2–4 m.y. of the Paleocene (Hunter, 1999; this volume; Hartman, 1999; Swisher et al., 1993).

It has long been recognized that the K-T boundary is stratigraphically close to the Hell Creek–Fort Union contact (Brown, 1952). Nichols and Johnson (this volume) have documented the palynological K-T boundary in 17 sections, and physical evidence for the K-T impact has been found at two sites in southwestern North Dakota: Pyramid Butte north of Marmarth (Johnson et al., 1989), and Mud Buttes south of Marmarth (Nichols et al., 2000; Nichols and Johnson, this volume). Both sites have yielded iridium anomalies and shocked minerals coincident with palynofloral extinction. In Slope County, North Dakota, the K-T boundary is within the basal 3 m of the Ludlow Member (Johnson, 1992; Nichols and Johnson, this volume). This shows that the basal few meters of the Ludlow are Cretaceous. Farther south in Bowman County, North Dakota, the formational contact and the K-T boundary become coincident, and the Ludlow is wholly Paleocene. Preliminary data from Harding County, South Dakota, suggest that the K-T boundary may be in the uppermost portion of the Hell Creek Formation.

The K-T boundary time line and the Hell Creek–Fort Union formation contact occur within 3 m of each other in central Montana (Hotton, this volume), western North Dakota and South Dakota (Nichols and Johnson, this volume), and central North Dakota (Murphy et al., 1995), a transect of more than 500 km. Accepting that the facies change that defines the formational contact is related to Cannonball transgression, it is possible to deduce that this transgression was relatively rapid, began before the K-T boundary, and was regionally significant. For these reasons, separating the effects of the K-T boundary effect from those of the marine transgression is challenging. In this chapter, this challenge is met by precisely correlating the megafloral record with the microstratigraphic record of the K-T boundary (Nichols and Johnson, this volume).

Because the Hell Creek Formation is in the interval between *Baculites grandis* at 70.15 Ma and the K-T boundary at 65.51 Ma, its maximum possible duration is 4.6 m.y. This unlikely scenario allows no time for the deposition of the intervening 30–60-m-thick Fox Hills Formation (Feldmann, 1972; Cvancara, 1976). In addition, Belt et al. (1997) argued for the existence of a significant regional unconformity between the Fox Hills Sandstone and the Hell Creek Formation. Such an unconformity makes sense in the context of a lowered sea level caused by the regression of the Cretaceous seaway. Paleomagnetic analysis by Hicks et al. (this volume) suggests that the deposition of the Hell Creek Formation occurred in 1.36 m.y. between 66.87 and 65.51 Ma. Their analysis of the time represented in the 80 m of Ludlow strata is based on the extrapolation of sediment accumulation rates derived from the Paleocene portion of magnetic polarity subchron C29r. Their duration of 0.84 m.y. is at odds with the presence of To2 mammals between the 50 and 80 m level of the Ludlow Member at the Brown Ranch site (Hunter, 1999), arguing for a somewhat longer duration. Thus, the strata discussed in this chapter represent the last ~1.36 m.y. of the Cretaceous, and at least the first 0.84 m.y. (and perhaps more) of the Paleocene. Because both of these estimates are generated from sediment accumulation rates derived from magnetostratigraphy, there is a great need for additional radiometric control.

METHODOLOGY

Fossil leaf localities

Fossil leaves are not abundant in the Hell Creek Formation, and for many years they were considered to be extremely rare. Prospecting for these localities involves long hikes with pick axes and the excavation of hundreds of dry holes. Most fossil-leaf-bearing horizons represent rapid burial of leaves, and the result is that sites usually produce numerous leaves. Modern taphonomic studies (Johnson, 1993) strongly suggest that leaf beds represent very little time averaging, probably less than one year. When a prospective site has been found, an excavation is undertaken to remove overburden and expose a bench of the leaf-containing layer. Large blocks of matrix are split to expose the leaves, which are then roughly trimmed and wrapped. Because the matrix preferentially splits along the plane of weakness defined by the leaf, the narrow portions of the leaf (attenuated apices, bases with petioles attached, and nonentire leaf margins) often remain beneath matrix and must be prepared in the laboratory with airscribes to expose the entire preserved lamina.

Optimally, leaf quarries are excavated until 300–500 leaves have been recovered. Census counts of leaf litter and transported leaf mats from modern temperate, deciduous forests indicate that sample sizes greater than 200 are needed to reflect forest composition with some accuracy (Johnson, 1989; Burnham et al., 1992). Work in subtropical and tropical evergreen forests is in agreement with that figure (Burnham, 1989). Many sites have produced fewer specimens because of the nature of the exposure and matrix. Typical quarries cover a few meters of surface area, and the leaf-bearing horizons are often relatively thin (1–10 cm) and rarely thicker (to 1 m). The censuses in conjunction with facies analysis help to constrain taphonomic and ecological factors that affect spatial distribution of taxa. Leaf physiognomy and regional distribution are then analyzed for use in paleoclimatic interpretations.

During excavation, sedimentological features are noted, and the site is assigned to 1 of 12 different sedimentary facies categories based on the nature of the rock and presence of primary and secondary sedimentary features. These facies are described herein (and see Fig. 2 of Johnson, 1992).

Localities are numbered in the field using a two-digit annual number and a sequential site number; thus, 9735 would be the 35th site discovered in 1997. For specimens in the Yale Peabody Museum collections, these remain the only locality numbers assigned to the specimens. Denver Museum of Nature & Science (DMNH) specimens are also assigned a DMNH locality number. Thus references to DMNH localities will cite both the field and museum locality number. For example, 9735 is also DMNH locality 1222 and is noted in this chapter as 9735 (1222). Localities returned to in subsequent years receive a different field number and the same museum number, but are referred to in this chapter by the field locality number assigned when the site was first discovered. While somewhat cumbersome, referring to the site with both the primary field number and the museum number allows for comparison of the two suites of voucher fossils that are stored in separate museums. For example, specimens at DMNH from locality 88102 (428) are from the same site as those labeled 88102 in the Yale collections.

Once discovered and quarried, the sites are integrated into the stratigraphic framework described in the following section. The method requires measurement of the locality relative to the Hell Creek–Fort Union formational contact, which is designated the datum for this chapter because it is recognizable in the field, unlike the K-T boundary, which must be determined by laboratory analysis. In optimal situations, the locality is placed in a measured section and can be directly measured relative to the Hell Creek–Fort Union formational contact to an accuracy of a few centimeters. Less optimal situations occur where outcrops are not continuous and where the distance to the formational contact is >30 m. In these cases, the localities are tied to the stratigraphic framework by walking along key marker beds. In worst-case scenarios, the locality is assigned a stratigraphic position by calculating the difference between its elevation and the elevation of the nearest formational contact. Error in this method can be as large as 20 m, or 20% of the thickness of the Hell Creek Formation. In general, localities near the top or base of the formation are much better defined than those in the middle.

Stratigraphy

The well-exposed and poorly lithified bedrock badlands of the Little Missouri River and adjacent Grand River valleys provide an ideal area for stratigraphic paleontology. The structural dip on the east flank of the Cedar Creek anticline is <1° and the elevation of badland buttes provides the opportunity to measure as much as 120 m of section. The nature of the bedrock is such that most surfaces are moderately weathered and fresh rock is 20–100 cm below the surface. North-facing slopes tend to be much more deeply weathered and are often vegetated. For this reason, the best sections tend to be exposed on south-facing slopes. Even on these slopes, weathering causes contacts to migrate downslope, and small local slumps and rotated blocks are relatively common. To achieve resolution and reproducibility, most of the measured sections in this study were trenched to a depth of at least 50 cm so that measurements could be made on fresh rock with a Jacob's staff to the nearest centimeter.

Use of the Hell Creek–Fort Union formational contact as a regional datum worked well for fossil sites in the uppermost 40 m of the formation and less well for sites below that. For the lower sites, distance above the basal contact was measured and the distance to the upper contact was calculated by subtraction from the 100 m thickness of the complete Hell Creek Formation. Correlation of sections is based either on extrapolation of the formational contact datum or by physically walking along continuous marker beds.

In order to calibrate the Hell Creek–Fort Union formational contact with the palynologically defined K-T boundary, several short high-resolution sections were trenched to vertical fresh walls and measured with centimeter resolution. Palynology samples from these sections were analyzed by Nichols and Johnson (this volume). This resulted in a series of 17 K-T boundary sections that delimit the diachroneity of the formational contact relative to the K-T boundary time line. These sections are used to recalibrate the megafloral and vertebrate fossil (Pearson et al., this volume) localities and magnetostratigraphic levels (Hicks et al., this volume) from their measured stratigraphic position to their position relative to the K-T boundary.

The sections measured in this study form the stratigraphic basis for other chapters in this volume: palynology (Nichols, Nichols and Johnson), vertebrates (Pearson et al.), magnetostratigraphy (Hicks et al.), plant-insect interactions (Labandeira et al.), and chemostratigraphy (Arens and Jahren). Collectively, these other techniques strengthen the understanding of the stratigraphic placement of the leaf fossil localities and provide independent estimates of the age of the formation.

Paleobotany

Megafloral specimens tend to be relatively large and delicate and thus difficult to study and compare. Placing this material in the Linnaean taxonomic system is a time-consuming process, requiring careful comparison to both extinct and living taxa, which may require years to complete. However, there is an immediate need to abstract reproducible patterns of floral diversity, dominance, and stratigraphic range, which then can be used to elucidate trends in floral evolution, ecology, paleoclimate, and extinction in advance of a settled taxonomy. For this reason, I have developed a method that makes possible the rapid, detailed study and characterization of very large collections of plant megafossils. First used in Johnson (1989), this methodology was refined and published by the Leaf Architecture Working Group (1999). The Leaf Architecture Working Group manual builds on the leaf architectural work of Hickey (1973, 1979) and provides a methodology for sorting large samples of modern or fossil leaves.

In the morphotype analysis, fossil plant organs are placed into discrete morphological categories, or morphotypes, on the basis of their leaf architectural characters (other characters are used for nonleaf plant organs and for nonangiosperms). Each morphotype is assigned an alphanumeric identifier (e.g., HC105) and is typified by a single specimen, the holomorphotype, and all other specimens are referred to it. A morphotype quality index is then assigned to each morphotype based on its preservation and number of specimens. The holomorphotypes are sequestered for ease of comparison and digital images of the holomorphotypes are created. In this study, full-page images of each of the holomorphotypes were mounted on a large wall, allowing ready comparison of newly collected unknowns with the growing collection of types. While this method allows reproducible analysis of very large sample sizes of previously undescribed fossil plants, it is merely the first step toward full diagnosis and description of new taxa, and the method is seen as an ordered progression toward the eventual Linnaean classification of the fossil plant taxa (Johnson, 1996).

The complete suite of holomorphotype specimens for the Hell Creek and Fort Union morphotypes used in this study, as well as a representative sample of the flora from each locality, are deposited in the Yale Peabody Museum of Natural History (YPM) and the Denver Museum of Nature & Science. The specimens at the Yale Peabody Museum are sorted stratigraphically by locality using the field numbers of Johnson (1989). Those at the Denver Museum of Nature & Science are sorted in the same manner but also have DMNH locality numbers.

FOSSIL LEAF LOCALITIES

A total of 158 fossil localities in Harding, Slope, and Bowman Counties discovered and excavated for this study have yielded voucher specimens. The Ludlow Member of the Fort Union Formation produced 58 localities, and the Hell Creek Formation produced 98 localities (Figs. 1 and 2). An additional 2 localities were collected from the uppermost 3 m of the Colgate Member of the Fox Hills Formation. These two sites are combined with the Hell Creek localities in this chapter. Six of the Ludlow localities occurred in the basal meter of the Ludlow, in rocks that yielded a Cretaceous palynoflora. Nichols and Johnson (this volume) call this Cretaceous basal Ludlow "Fort Union strata of Cretaceous age." Thus, the assemblage of localities represents 106 that are latest Cretaceous (late Maastrichtian) and 52 that are earliest Paleocene (Puercan and Torrejonian) (Appendix, Table A1). The sample size at each locality was not uniform (Table 1), and the localities are more abundant in the stratigraphic interval that contains the K-T boundary (Table 2). This reflects the interest in the K-T extinction event that initiated this project. As a result, levels more than 60 m above or below the K-T are not as well sampled as those closer to the boundary.

Excavation of these leaf localities resulted in two types of collections. Voucher collections are permanently stored at the Yale Peabody Museum and Denver Museum of Nature & Science. Census collections are made on site where the majority of specimens were tallied and discarded. The former type provides a reference collection, and the latter provides ecologically sensitive census data that are not available in collector-biased museum collections. The primary thrust of this chapter is to describe the voucher collections, which now number 6801 specimens at DMNH, 6669 specimens at the Yale Peabody Museum, and 101 specimens at the Black Hills Institute of Geological

TABLE 1. SAMPLE SIZE DISTRIBUTION OF FOSSIL LEAF LOCALITIES IN THE HELL CREEK AND FORT UNION FORMATIONS

Sample size	Fort Union	Hell Creek
>500	3	2
100–499	6	27
50–99	5	8
10–49	18	40
2–9	17	22
1	3	7
Total	52	106

TABLE 2. NUMBER OF FOSSIL LEAF LOCALITIES AND NUMBER OF SPECIMENS IN 10 m INTERVALS RELATIVE TO THE CRETACEOUS-TERTIARY BOUNDARY

Level	Localities	Specimens
80–90	2	31
70–80	0	0
60–70	0	0
50–60	6	792
40–50	2	250
30–40	5	300
20–30	3	71
10–20	9	164
0–10	25	2336
−0–10	19	1251
−10–20	18	1904
−20–30	9	638
−30–40	17	1609
−40–50	9	728
−50–60	10	2015
−60–70	6	729
−70–80	9	504
−80–90	6	178
−90–100	1	26
−100–110	2	45
Total	158	13571

Note: FU0 located −0–4 below the Cretaceous-Tertiary has 6 localities and 142 specimens

Research (BHIGR) in Hill City, South Dakota. These 13571 specimens represent the largest collection of fossil plants spanning the K-T boundary.

Of the 158 localities, 157 were assigned to 1 of 12 different sedimentary facies based on the grain size, texture, primary and secondary structures, color, and enclosed fossils of the matrix at the leaf locality (Table 3). The 12 different leaf-bearing facies fall into two major categories, channel and flood plain. Although it is likely that all of the fossils in this study were transported before final deposition, there is a gradient of localities relative to transport, and sites in the channel facies are inferred to be the most transported and the flood-plain facies the least. I have previously suggested that in modern temperate forests (Johnson, 1988, 1989) and tropical forests (Christophel and Johnson, 1990), species richness of leaf litter increases with increasing amount of transportation because of increased sampling of the levee forest. Thus, forest-floor leaf litter has a significantly lower richness than litter that has only been transported tens to hundreds of meters. This becomes significant when the relative abundances of fossil leaf localities in the formations under study are compared (Table 4). The Hell Creek Formation has a significantly higher percentage of localities in

TABLE 3. FOSSIL LEAF–BEARING FACIES IN THE HELL CREEK FORMATION AND LUDLOW MEMBER OF THE FORT UNION FORMATION

Facies Categories	Description
Channel	
Channel x	Massive to cross-bedded sandstone in lenticular beds. Interpreted as lower point bar to thalweg deposits of meandering river systems. Leaves preserved at all angles to bedding and often rolled.
Channel la	Lateral accretion deposits of alternating mud and sand, often with primary depositional dip. Interpreted as upper point bar deposits of meandering river systems. Leaves preserved on bedding planes. Leaves near the top of these sequences are frequently penetrated by root traces, suggesting subaerial exposure and plant growth immediately after deposition. Very common in the Hell Creek and present in the Ludlow.
Channel	Sandy deposits that lack features of the above two categories. Interpreted as generalized river deposits.
Abandoned channel	Laminated mudstone associated with Channel la deposits, often with little or decreasing primary dip. Interpreted as the infilling of an oxbow lake within a meandering river system. Leaf cuticle sometimes preserved. Relatively rare in both formations.
Flood basin	
Splay and/or Levee	Horizontal and interbedded thin tabular sandstone and mudstone or claystone layers. Sandstone layers thinner than 20 cm. Leaves preserved as root-penetrated impressions in laminated parent rock, suggesting that leaves were deposited in a distal levee, which was later subjected to early stage pedogenic processes.
Splay	Tabular sandstone layers from 20 cm to 2 m thick with rooted zones at top. Interpreted as crevasse splay deposits on a flood plain. Leaves preserved as root-penetrated impressions.
Pond	Laminated shale or mudstone, horizontally bedded. Interpreted as ponded water on a flood plain. Associated with aquatic plants. Relatively abundant in both formations.
Pond vb	Variegated beds of mudstone and siltstone with prominent horizontal iron-oxide concretionary levels and iron-oxide staining. Interpreted as a distinctive type of laterally extensive ponded water environment. Often associated with aquatic plants and vertebrates. Common in the basal 35 m of the Ludlow and essentially absent in the Hell Creek (one example known).
Pond ls	Horizontally bedded mudstone with layers or fenses of micritic limestone. Interpreted as a alkaline ponded water environment on the flood plain. Leaves preserved in limestone layers.
Carb Shale	Organic-rich mudstone with abundant root traces, slickensides, particulate organics, and amber. Laterally or vertically transitional to lignite. Interpreted as mire deposits on a saturated flood plain. Common in the Ludlow, rare in the Hell Creek.
Flood plain soil	Gray, green, and purple mudstone layers with abundant root traces, slickensides, iron-oxide concretions. Often characterized by popcorn weathering texture due to modern weathering of smectitic clays. Interpreted as gleyed paleosols on a wet flood plain. Common in the Hell Creek, rare in the Ludlow. Leaves preserved in laminated mudstone above the heavily rooted zone.
Ash bed	Layers of altered volcanic ash that are lime green to very light gray when fresh and weather to a deep popcorn texture. Primary bedding consists of horizontal to ripple lamination. Interpreted as hydrologically reworked air-fall volcanic ash. Leaves preserved as impressions at all angles to bedding, suggesting that fluvially reworked ash was moved as a slurry.

TABLE 4. NUMBER AND RELATIVE ABUNDANCE OF FOSSIL LEAF LOCALITIES FROM DIFFERENT SEDIMENTARY FACIES, HELL CREEK FORMATION AND LUDLOW MEMBER OF THE FORT UNION FORMATION

Facies categories	Total		Hell Creek		K-Ludlow		Ludlow	
	(n)	(%)	(n)	(%)	(n)	(%)	(n)	(%)
Channel								
Channel la	63	39%	56	57%			7	13%
Channel x	18	11%	17	17%			1	4%
Channel	2	1%	1	1%			1	2%
Abandoned channel	5	3%	3	3%			2	4%
Flood basin								
Splay	13	8%	8	8%	2	33%	3	6%
Splay and/or levee	1	1%					1	2%
Pond	26	16%	7	7%	1	17%	18	35%
Pond vb	11	7%					11	21%
Pond ls	2	1%					2	4%
Carb shale	8	5%			3	50%	5	10%
Flood-plain soil	6	4%	6	6%				
Ash bed	2	1%	1	1%			1	2%
Totals	157		99		6		52	
Summary								
Channel	88	56%	77	78%	0	0%	11	21%
Flood basin	69	44%	22	22%	6	100%	41	79%

channel facies (78%) relative to the Ludlow Member (21%). This suggests that one reason the flora of the Hell Creek is more diverse than that of the Ludlow is because of increased sampling of allochthonous facies. Table 5 shows the vertical distribution of fossil-bearing localities by the two major facies categories.

Sample size is strongly correlated with species richness at a locality. Although it would have been optimal to standardize the sample size from each quarry, the reality is that some sites were well sampled and others barely collected. Census data were obtained for ~50% of the collected floras and will be discussed in subsequent papers. Despite the lack of census data in this chapter, the number of voucher specimens from each quarry is a good measure of the intensity of sampling and roughly reflects the relative abundance of taxa at a site.

STRATIGRAPHIC FRAMEWORK

The lithostratigraphy of the Hell Creek Formation in this study area is discussed by Murphy et al. (this volume) so there is no need for a detailed recapitulation here. However, it is important to present the stratigraphic framework of the fossil localities in this study. Vagaries of outcrop quality and land access defined certain areas where the field work was concentrated. An attempt was made in each of these areas to measure sections that were as long and complete as possible. In many cases, it was necessary to link adjacent sections to reconstruct the complete thickness of the strata exposed in a given area.

From north to south, the key areas and/or sections are Cannonball Creek, Pyramid Butte, Pretty Butte, River Section, Bobcat Butte, Sunset Butte, Mud Buttes, and Doaks Butte (Figs. 1 and 2; Appendix, Table A2). Photographs of most of these sections appear in Hicks et al. (this volume). In all cases except Doaks Butte, the datum was defined as the contact of the Hell Creek and overlying Ludlow Member of the Fort Union Formation. At the Doaks Butte section, the datum was the contact of the Hell Creek and underlying Colgate Member of the Fox Hills Sandstone. At each of these sections except Doaks Butte, palynology was used to define the K-T boundary relative to the formational contact datum (Tables A1 and A2; Nichols and Johnson, this volume). These sections were also sampled for magnetostratigraphy (Hicks et al., this volume), palynology (Nichols, this volume; Nichols and Johnson, this volume), vertebrate paleontology (Holroyd and Hutchison, this volume; Pearson et al., this volume), geochemistry (Arens et al., this volume), and plant-insect interactions (Labandeira, this volume).

Cannonball Creek

The valley of the Cannonball Creek, 20 km north of Marmarth and 6 km upstream from the creek's confluence with the Little Missouri River, is the site of several short K-T boundary sections described in Nichols and Johnson (this volume). A longer composite of section 9649, measured at the site of a Pioneer Trails Regional Museum (PTRM) hadrosaur quarry, and sections 9754 and 9827, measured at nearby Facet Butte, span 21 m of Hell Creek and 69 m of Fort Union Formation. The K-T boundary has been located at 8 sites in this area and ranges from 122.5 to 269 cm above the formation-contact datum. This area is significant because the K-T boundary is above the formation contact and the basal few meters of the Fort Union Formation are Cretaceous. These Cretaceous Fort Union rocks have yielded a *Wodehouseia spinata* Assemblage Zone

TABLE 5. FOSSIL LEAF LOCALITIES FROM DIFFERENT SEDIMENTARY FACIES RELATIVE TO STRATIGRAPHIC POSITION

Meters to K/T	(n)	Channel	%	Flood plain	(%)
Paleocene					
80–90	2	0	0	2	100
70–80	0	0	0	0	0
60–70	0	0	0	0	0
50–60	6	2	33	4	67
40–50	2	0	0	2	100
30–40	5	0	0	5	100
20–30	3	0	0	3	100
10–20	9	3	33	6	67
0–10	25	6	24	19	76
Cretaceous					
(FU0) −0–4	6	0	0	6	100
(HCIII) −4–10	13	5	38	8	62
−10–20	18	16	89	2	11
−20–30	9	8	89	1	11
−30–40	17	14	82	3	18
−40–50	9	7	78	2	22
−50–60	9	8	89	1	11
−60–70	6	6	100	0	0
−70–80	9	6	67	3	33
−80–90	6	4	67	2	33
−90–100	1	1	100	0	0
−100–110	2	2	100	0	0

palynoflora (Nichols and Johnson, this volume) and an unusual megaflora from six localities. This megaflora, known as FU0, are discussed in detail in the following section. This area has also yielded a partial ceratopsian dinosaur (Pearson et al., 2001, this volume) from a position 86 cm above the base of the Fort Union Formation.

Pyramid Butte

At Pyramid Butte, 15 km north of Marmarth, the K-T boundary is at the top of the 90 cm basal Fort Union lignite and is marked by the coincidence of an iridium anomaly, shocked mineral grains, and palynological extinctions (Johnson et al., 1989). Magnetostratigraphy of this section is described by Hicks et al. (this volume). The section (86159, 86160, 86161) is 65 m thick, 39 m of which are Fort Union strata and 26 m of which are Hell Creek strata. The Pyramid Butte section was the first place in North Dakota where the K-T boundary was discovered. The lowest well-sampled Paleocene leaf locality in the study area, 86107 (427), is located at Pyramid Butte and is only 20 cm above the K-T boundary.

Pretty Butte

On the west bank of the Little Missouri River, the clinker-capped Pretty Butte is the most distinctive topographic feature in the Marmarth area. It was illustrated by Brown (1962) and exposes more than 150 m of section. The butte is wooded on its northern slopes and the best exposures occur on the eastern margin. Here extensive slumping has complicated and obscured the stratigraphy, but there are two relatively undisturbed sections. The southern section is described by Murphy et al. (this volume). The northern section (9735) exposes 76 m of Fort Union Formation and at least 50 m of the Hell Creek Formation. Magnetostratigraphy of this section is described by Hicks et al. (this volume). Nichols and Johnson (this volume) have palynologically located the K-T boundary at four locations around the base of Pretty Butte: Big Boundary (9328), Pretty Butte North (9248), Pretty Butte (9735), and the section measured by Murphy et al. (this volume). The K-T boundary ranges from 96 cm above the base of the Ludlow to a level that may be as high as 248 cm, but needs refinement with more sampling.

River Section

River Section is a series of measured sections (87144, 87147, 9838, 9839, 9850) located on the east bank of the Little Missouri River in section 8, T.133N., R.106W. and extending up the north side of an unnamed drainage to the east for 4 km. This composite section spans 40 m of the Hell Creek Formation and 58 m of the Fort Union Formation. The area is very fossiliferous and includes extensive outcrops of the Marmarth ash, a reworked volcanic ash of variable thickness that occurs ~10 m above the base of the Fort Union Formation. The K-T boundary has been difficult to determine in this area because it is in the base of the Fort Union, and the highest palynology samples that produce Cretaceous palynomorphs are separated from the lowest Tertiary samples by a thick sequence of nonpalyniferous variegated mudstone. The best resolution of the K-T boundary occurs at section 87147, where the highest Cretaceous sample is 60 cm above the formation contact and the lowest Tertiary sample is 459 cm above the contact, providing an estimate of

259 ± 199.5 cm for the position of the K-T boundary. Numerous attempts to resolve this problem have been unsuccessful. However, at the nearby section 0042, Cretaceous samples extend to 237 cm above the formational contact (no Tertiary samples were obtained), suggesting a further limit on the boundary. The PITA Flats fossil mammal locality of Hunter (1999, this volume) occurs in this area at a level 10 m above the Hell Creek–Fort Union formational contact. Although represented by only a few species, the site is demonstrably Puercan in age.

Bobcat Butte

Bobcat Butte is a prominent hill located on the southern edge of section 8, T.133N., R.105W., on the north side of Highway 12 ~4 km east of Marmarth. The badlands that surround this butte are very fossiliferous and the amount of exposure is extensive. The section is 96 m thick, 37 m of which are Fort Union strata and 59 m of which are Hell Creek strata. The palynofloral K-T boundary is 75 cm above the formational contact datum, but detailed sampling at this site did not yield an Ir anomaly (C. Orth, 1987, written commun.). Magnetostratigraphy of this section is described by Hicks et al. (this volume). The megafloral zonation of Johnson and Hickey (1990) was first recognized in this area. Megafloral zones HCIb, HCIIa, HCIIb, HCIII, and FUI all occur here in superposition. In addition, this is the primary area where the small-leaved herbaceous floras that characterize zone HCIb were first discovered.

Sunset Butte

Sunset Butte, 18 km south of Marmarth, is the only place identified in southwestern North Dakota where it is possible to measure a continuous section of the entire Hell Creek Formation. The composite section is based on five smaller sections, which can be correlated by walking along key marker beds up the drainage of Bog Creek. The base of the section is located on the east bank of the Little Missouri River in section 30, T.131N., R.105W., and the top of the section is the top of Sunset Butte. The section was originally discovered and measured by David Fastovsky in 1987 and field checked by me in 1988. The section is 143 m thick, 35 m of which are Fort Union strata, 100 m of which are Hell Creek strata, and 8 m of which are Fox Hills strata. Magnetostratigraphy of this section is described by Hicks et al. (this volume). The formational contact on the south face of Sunset Butte is difficult to recognize because of the presence of a 10-m-thick sandstone layer at the base of the Fort Union Formation that has apparently cut out the carbonaceous zone at the formational contact. The basal Ludlow sand body is light gray and superficially appears to be part of the Hell Creek Formation. This situation initially caused the placement of the formation contact 10 m higher than currently recognized (Johnson, 1992). In order to obtain a suitable section for palynological analysis of the K-T, we traced the strata 3 km to the southeast of Sunset Butte where the Ludlow sandstone is only 5 m thick and is underlain by a typical fine-grained Hell Creek–Ludlow contact. The resulting section, called the Sunset Patch (9317), is where Nichols and Johnson (this volume) have defined the K-T boundary at 170 ± 103 cm above the formational contact.

Mud Buttes

The Mud Buttes section is located 35 km southeast of Marmarth and 10 km east of the Little Missouri River in sections 1, 2, 11, and 12, T.129N., R.105W., Bowman County. Here the K-T boundary is at the formational contact datum and is associated with a 2 cm claystone that contains shocked quartz, an iridium anomaly, and clay spherules (Nichols et al., 2000; Nichols and Johnson, this volume). The section is 61 m thick, 45 m of which are Fort Union strata and 16 m of which are Hell Creek strata. This area has been the site of two major surveys for vertebrate fossils (Sheehan et al., 2000; Pearson et al., 2001). Magnetostratigraphy of this section is described by Hicks et al. (this volume). Megafloral locality 88102 (428), located 15 m below the K-T boundary at the intersection of the valley floor and the surrounding buttes, is not only the richest megafloral assemblage known from the Hell Creek Formation but, with 86 taxa, is the richest known Cretaceous leaf quarry in the world.

Doaks Butte

Doaks Butte is 45 km south of Marmarth on the west bank of the Little Missouri River, just north of the North Dakota–South Dakota state line. The section is composed of 25 m of Hell Creek strata conformably overlying an exposure of the Colgate Member of the Fox Hills Formation. The magnetostratigraphy of this section is described by Hicks et al. (this volume). The Doaks Butte ash [locality 9733 (2102)], located in this section at an estimated position of 75 m below the Hell Creek–Fort Union contact datum, has been dated by Hicks et al. (this volume). This area is the southwestern margin of the Cedar Creek anticline. Mapping by Dobbin and Larsen (1934) indicates that the western margin of the anticline exhibits structural dip of as much as 30°W. Geologic maps by early USGS surveys (Winchester et al., 1916) and the South Dakota Geological Survey (Petsch, 1956; Schulte, 1957; Erickson, 1956a, 1956b) all show the Ludlow Member of the Fort Union Formation to be discontinuous between the South Cave Hills and the northwestern corner of South Dakota. This is extremely improbable and likely a geological mapping error, because several of these maps show extensive lignite beds in rocks mapped as the Hell Creek. Recent observations and palynological analysis (D.J. Nichols, 2000, written commun.) have demonstrated the presence of many previously unmapped exposures of Paleocene strata in this area. Megafloral locality 9257 (569), 2 km southwest of Doaks Butte, is the only megafloral locality in this study that occurs on the western side of the Cedar Creek anticline. Horizontal projections from the top of the Doaks Butte measured section to this locality initially suggested a position of 65 m below the formation-contact datum. The floral assemblage from this site is markedly similar to HCIII floras, rather than lower Hell Creek floras, and I now surmise that structural dip on the west side of the anticline caused me to underestimate the stratigraphic level of the locality by ~40 m.

Harding County, South Dakota

Stratigraphic work for this project in Harding County has been of a reconnaissance nature. The Hell Creek–Fort Union formational contact has the same characteristics that are seen in Slope and Bowman Counties, and the K-T boundary has been located using palynology at Lola Butte and just south of Lahti Butte at the site where a nearly complete *Tyrannosaurus rex* skeleton known as "Stan" was collected by the Black Hills Institute of Geological Research (D.J. Nichols, 2000, written commun.). The skeleton was located 16 m below the formational contact, and all palynology samples collected at or above the contact produced Tertiary pollen, suggesting that the K-T boundary is at or below the formation contact. More work needs to be done in Harding County to refine these patterns.

SUMMARY OF THE MEGAFLORA

Morphotypes and floristics

The 158 localities in this study yielded 13 571 voucher specimens, 12 150 of which were identified to one of 380 morphotypes (Appendix, Tables A3 and A4). Of this total, 221 morphotypes occur at only one quarry site each and 144 morphotypes are represented by only one specimen. The Hell Creek Formation and Cretaceous base of the Ludlow Member of the Fort Union Formation yielded a total of 328 morphotypes (Table 6) and the Paleocene portion of the Ludlow produced 85 morphotypes (Table 7). Only 33 morphotypes occurred both in the Cretaceous and Paleocene parts of the section, while 295 morphotypes were restricted to Cretaceous strata, and 52 morphotypes were found only in Tertiary strata. Thus, of the 328 Cretaceous morphotypes, only 10% occur in the overlying Paleocene.

Fossil leaves were by far the most common type of identifiable megafossil and 347 of the morphotypes are leaf taxa. The remaining 33 morphotypes include identifiable seeds, fruits, flowers, cones, roots, and fertile fronds (Tables 6 and 7). Angiosperms dominate both the Cretaceous and Paleocene floras (89% and 84% of the morphotypes, respectively). While this assessment is based on number of morphotypes, censuses of individual quarries show the same pattern for relative abundance of specimens (Johnson, 1992). The most abundant morphotypes by number of voucher specimens and by number of localities recorded are shown (Tables 8 and 9) to illustrate the important taxa in the collective flora. Note that only one species that is restricted to the Paleocene (FU1, *Paranymphaea crassifolia*) is on either list. Most of the morphotypes listed in Tables 8 and 9 are illustrated in Plates 1–17.

Systematic floral lists of the Hell Creek and Ludlow Member of the Fort Union Formation are presented in Tables 10 and 11, and a diagnostic key to the most common (and illustrated in Plates 1–17) morphotypes is presented in Table 12. The Hell Creek flora has a few rare ferns, one cycad, one ginkgo, and conifers representing the Taxodiaceae, Cupressaceae, and Araucariaceae. The identified angiosperms are dominated by the Lauraceae, Platanaceae, Magnoliaceae, Cercidiphyllaceae, Araceae, and Arecaceae (Palmae). A number of other extant families that are present in the flora include the Berberidaceae, Ranunculaceae, Cannabaceae, Urticaceae, Ulmaceae, Malvaceae, Rosaceae, Rhamnaceae, and Cornaceae. A large proportion of the flora is composed of dicotyledonous angiosperms that have not been identified to subclass. Thus the overall affinities of the flora need to be systematically studied. A comparison of this megaflora with the Hell Creek palynoflora is presented by Nichols (this volume).

The Fort Union Formation flora is much poorer and less well sampled than that of the Hell Creek Formation. It yielded only 9 localities that produced more than 100 specimens each, as opposed to the Hell Creek, which yielded 27 such localities. Whereas the 9 Ludlow sites averaged more specimens (mean

TABLE 6. MAJOR PLANT GROUPS AND ORGAN TYPES, HELL CREEK FORMATION AND CRETACEOUS INTERVAL OF THE LUDLOW MEMBER OF THE FORT UNION FORMATION

	Leaf	Root	Seed	Fruit	Cone	Sori	Sum	All (%)	Leaves only (%)
Dicots	269		3	12			284	86.6	89.1
Monocots	9						9	2.7	3.0
Conifers	10		4		4		18	5.5	3.3
Cycads	1						1	0.3	0.3
Ginkgo	1						1	0.3	0.3
Ferns	11					1	12	3.7	3.6
Bryophytes	1						1	0.3	0.3
Sphenopsids		1					1	0.3	0.0
Indeterminate			1				1	0.3	0.0
Totals	302	1	8	12	4	1	328		

TABLE 7. MAJOR PLANT GROUPS AND ORGAN TYPES, PALEOCENE INTERVAL OF THE LUDLOW MEMBER OF THE FORT UNION FORMATION

	Leaf	Root	Seed	Fruit	Cone	Sori	Sum	All (%)	Leaves only (%)
Dicots	59			8			67	78.8	78.7
Monocots	3	1					4	4.7	4.0
Conifers	4						4	4.7	5.3
Ferns	7						7	8.2	9.3
Bryophytes	2						2	2.4	2.7
Sphenopsids		1					1	1.2	0.0
Totals	75	2	0	8	0	0	85		

TABLE 8. TOP 50 MEGAFLORAL MORPHOTYPES RANKED BY NUMBER OF VOUCHERED SPECIMENS

Morphotype number	Lowest taxonomic category	Specimens	Localities	Stratigraphic occurrence
HC49	*"Dryophyllum" subfalcatum*	1003	51	K
FU2	*Quereuxia angulata*	543	15	K-T
FU4	*Glyptostrobus europaeus*	534	12	K-T
FU7	*"Populus" nebrascensis*	506	27	K-T
FU43	*Ziziphoides flabella*	439	15	K-T
FU29	*Dicotylophyllum anomalum*	373	21	K-T
HC81	Cannabaceae	365	4	K
HC57	*Erlingdorfia montana*	320	30	K
HC105	*Marmarthia trivialis*	282	26	K
HC92	*"Cinnamomum" lineafolia*	259	13	K
HC162	*Marmarthia pearsonii*	249	16	K
HC14	*"Vitis" stantonii*	234	15	K
HC80	Rosaceae/Rosales	217	3	K
FU16	*Platanus raynoldsii*	209	34	K-T
HC86	*Leepierceia preartocarpoides*	204	23	K
HC106	*Platanites marginata*	197	21	K
FU3	*Metasequoia occidentalis*	190	10	K-T
FU31	*Cornophyllum newberryi*	185	28	K-T
FU26	*Cupressinocladus interruptus*	178	15	K-T
HC62	*"Rhamnus" cleburni*	177	18	K
HC212	*Cercidiphyllum ellipticum*	162	17	K-T
HC130	Dicotyledones	157	1	K
HC166	*Liriodendrites bradacii*	156	10	K
HC94	*"Celastrus" taurenensis*	151	17	K
HC2	*Grewiopsis saportana*	134	17	K
HC64	*Erlingdorfia montana* lat. leaf	131	15	K
HC103	*Trochodendroides nebrascensis*	113	26	K
FU1	*Paranymphaea crassifolia*	108	21	T
HC135	Dicotyledones	104	3	K
FU38	*Limnobiophyllum scutatum*	99	5	K-T
HC44	cf. *"Dryophyllum" tenneseensis*	99	14	K
HC188	Berberidaceae	94	1	K
HC199	Laurales	93	2	K
HC243	*Humulus* sp.	93	1	K
HC71	*Taxodium olrikii*	92	14	K-T
HC111	*Paranymphaea hastata*	92	17	K
FU5	*Cercidiphyllum genetrix*	88	12	K-T
HC165	*Elatides longifolia*	88	9	K
FU37	*Amesoneuron* or *Sabalites*	87	24	K-T
HC114	*Ginkgo adiantoides*	84	7	K
HC7	*Palaeoaster inquirenda*	78	4	K
HC241	Filicopsida	70	4	K
HC73	Trochodendrales	69	10	K
HC84	Laurales	68	3	K
HC229	*Cercidiphyllaceae*	67	1	K
HC4	*"Ficus" planicostata*	66	13	K
HC36	Ulmaceae	65	15	K
HC43	Magnoliidae	65	14	K
HC200	*Bisonia niemii*	64	5	K
HC179	*"Artocarpus" lessigiana*	62	8	K

K-T—Cretaceous-Tertiary.

TABLE 9. TOP 50 MEGAFLORAL MORPHOTYPES RANKED BY FREQUENCY OF OCCURRENCE AT DISCRETE LOCALITIES

Morphotype number	Lowest taxonomic category	Localities	Specimens	Stratigraphic occurrence
HC49	*"Dryophyllum" subfalcatum*	51	1003	K
FU16	*Platanus raynoldsii*	34	209	K-T
HC57	*Erlingdorfia montana*	30	320	K
FU31	*Cornophyllum newberryi*	28	185	K-T
FU7	*"Populus" nebrascensis*	27	506	K-T
HC103	*Trochodendroides nebrascensis*	26	113	K
HC105	*Marmarthia trivialis*	26	282	K
FU37	*Amesoneuron* or *Sabalites*	24	87	K-T
HC86	*Leepierceia preartocarpoides*	23	204	K
FU1	*Paranymphaea crassifolia*	21	108	T
FU29	*Dicotylophyllum anomalum*	21	373	K-T
HC106	*Platanites marginata*	21	197	K
FU59	*Nyssidium arcticum*	19	49	K-T
HC62	*"Rhamnus" cleburni*	18	177	K
HC2	*Grewiopsis saportana*	17	134	K
HC94	*"Celastrus" taurenensis*	17	151	K
HC111	*Paranymphaea hastata*	17	92	K
HC212	*Cercidiphyllum ellipticum*	17	162	K-T
HC162	*Marmarthia pearsonii*	16	249	K
FU2	*Quereuxia angulata*	15	543	K-T
FU26	*Cupressinocladus interruptus*	15	178	K-T
FU43	*Ziziphoides flabella*	15	439	K-T
HC14	*"Vitis" stantonii*	15	234	K
HC36	Ulmaceae	15	65	K
HC64	*Erlingdorfia montana* lat. leaflet	15	131	K
HC1	*Platanites marginata* lat. leaflet	14	49	K
HC43	Magnoliidae	14	65	K
HC44	cf. *"Dryophyllum" tenneseensis*	14	99	K
HC71	*Taxodium olrikii*	14	92	K-T
HC4	*"Ficus" planicostata*	13	66	K
HC66	Magnoliidae	13	53	K
HC92	*"Cinnamomum" lineafolia*	13	259	K
FU4	*Glyptostrobus europaeus*	12	534	K-T
FU5	*Cercidiphyllum genetrix*	12	88	K-T
HC70	*Sequoia* sp.	12	52	K
HC137	*Fokieniopsis catenulata*	11	53	K-T
FU3	*Metasequoia occidentalis*	10	190	K-T
HC73	Trochodendrales	10	69	K
HC166	*Liriodendrites bradacii*	10	156	K
FU62	*Nelumbium montanum*	9	27	K-T
HC77	*"Pistia" corrugata*	9	61	K
HC165	*Elatides longifolia*	9	88	K
HC27	Dicotyledones	8	23	K
HC35	*Metasequoia* sp. #2	8	58	K
HC99	Platanaceae	8	34	K
HC108	aff. *"Vitis" stantonii*	8	20	K
HC123	*Marchantia pealii*	8	33	K-T
HC151	Platanaceae	8	23	K
HC174	*"Ziziphus" fibrillosus*	8	55	K
HC179	*"Artocarpus" lessigiana*	8	62	K

K-T—Cretaceous-Tertiary.

= 336) than the 27 Hell Creek sites (mean = 271), the Hell Creek had significantly more morphotypes per locality (23 versus 13). The Ludlow flora contains few ferns, no cycads or ginkgo, and a depauperate conifer assemblage composed of Taxodiaceae and Cupressaceae. The dominant angiosperm families are the Platanaceae, Cercidiphyllaceae, and Cornaceae. Less common families include the Betulaceae and Flacourtiaceae. Palms are present in the lower Fort Union, but are not present more than 20 m above the base of the formation. Aquatic plants are numerous, relatively diverse and represent primarily extinct genera including *Quereuxia* (FU2), *Limnobiophyllum* (FU38), *Ceratophyllum* (FU80), *Paleonelumbo* (FU78), *Nelumbium* (FU62), cf. *Harmsia* (FU9), *Paranymphaea* (FU1), and *Porosia* (HC12).

In the Cannonball Creek area, 6 localities in the basal 2 m of the Ludlow Member of the Fort Union Formation occur in strata that have produced Cretaceous pollen and a ceratopsian dinosaur (Pearson et al., 2001). This Cretaceous megaflora in

Plate 1. Larger plant fossils from Hell Creek Formation. 1: Compressed palm trunk from locality 9610 (1781). Trunk is 5.3 m long and 34 cm wide and contained coalified wood in center. Prominent ribs are scars left by dehisced leaf bases. Diameter of reconstructed trunk is 20 cm. Portions of trunk in Pioneer Trails Regional Museum (PTRM) collections have specimen number PTRM-4778. 2: Impression fossil of large *Sabalites* leaf, PTRM-4779, from locality 9610 (1781). 3: Detail of palm trunk in 1. 4: Cross section of buried tree trunk at locality 8102 (428). Trunk is buried in rooted fine-grained sandstone and is replaced by dark gray mudstone. 5: Typical plant fossil locality, 9322 (900). Productive horizon is dipping quarry floor, which reflects primary sedimentary dip on upper point-bar surface. Site has yielded 485 vouchered specimens.

Plate 2. Hell Creek ferns and bryophytes. 1: (HC241), DMNH-19227, locality 9727 (2087). 2: HC275, DMNH-19712, locality 88102 (428). 3: *Hydropteris pinnata* (HC129), YPM-6259, locality 86104. 4: *Marchantia pealii* (HC123), YPM-6252, locality 86102. 5: *Salvinia* sp. (HC312), DMNH-19861, locality 9610 (1781). Scale bars = 1 cm.

Plate 3. Hell Creek gymnosperms. 1: *Gingko adiantoides* (HC114), DMNH-18927, locality 86105 (571). 2: *Dammarites* sp. (HC257), DMNH-19258, locality 9610 (1781). 3: *Metasequoia* sp. #2 (HC35), YPM-6158, locality 86108. 4: *Elatides longifolia* (HC165), DMNH-19255, locality 9610 (1781). 5: *Taxodium olrikii* (HC71), YPM-6192, locality 86102. 6: *Fokieniopsis catenulata* (HC137), DMNH-19428, locality 86100 (567). 7: *Nilssonia yukonensis* (HC164), FMNH-PP45626; locality is site of *Tyrannosaurus rex* SUE near Faith, South Dakota. Scale bars = 1 cm.

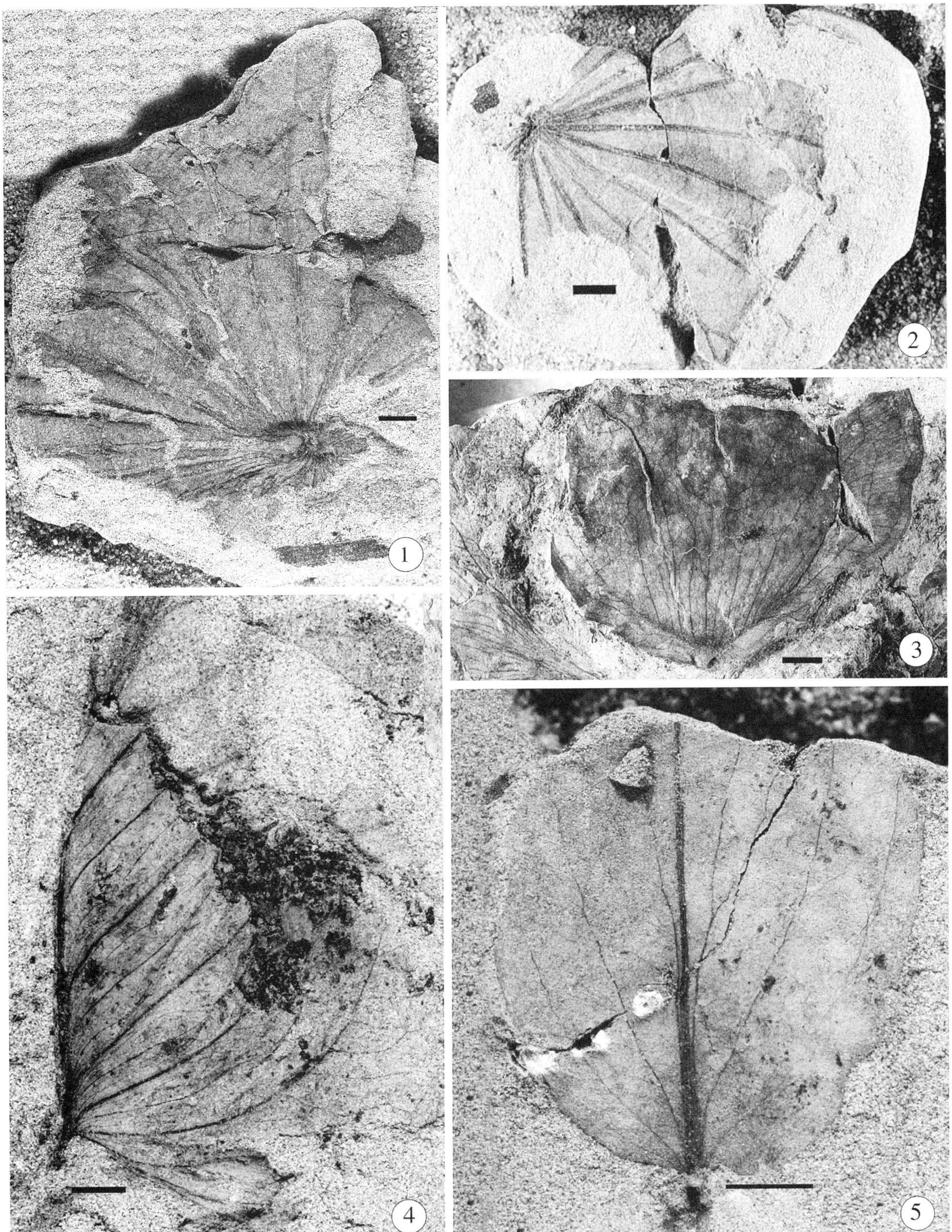

Plate 4. Cretaceous Arales and aquatics. 1: *Nelumbo* sp. (HC202), DMNH-6869, locality 88102 (428). 2: *Nelumbo* sp. (HC202), DMNH-19731, locality 88102 (428). 3: "*Pistia*" *corrugata* (HC77), Araceae, DMNH-8656b, locality 86105 (571). 4: Aff. *Philodendron* (HC201), Araceae, DMNH-6711, locality 88102 (428). 5: HC183, DMNH-6315, locality 88102 (428). Scale bars = 1 cm.

Plate 5. Hell Creek dicotyledons characteristic of megafloral zone HCIa. 1: *Leepierceia preartocarpoides* (HC86), DMNH-6359, locality 86105 (571). 2: *Cercidiphyllum ellipticum* (HC212), DMNH-19079, locality (1491). 3: "*Cinnamomum*" *lineafolia* (HC92), YPM-6220, locality 86108. 4: "*Celastrus*" *taurenensis* (HC94), YPM-6224, Dynamite Coulee locality (Corbin, 1938). Scale bars = 1 cm.

Plate 6. Herbaceous and shrubby Hell Creek flora (zone HCIb). 1: aff. *Humulus* sp. (HC243), DMNH-19217, locality 9727 (2087). 2: HC135, DMNH-18883, locality 9535 (1491). 3: HC224, DMNH-19207, locality 9824 (2203). 4: Rosaceae (HC80), YPM-6201, locality 86100 (567). Scale bars = 1 cm.

Plate 7. Herbaceous and shrubby Hell Creek flora (zone HCIb). 1: Ranunculaceae (HC226), DMNH-19201, locality 9824 (2203). 2: Urticaceae (HC93), DMNH-19229, locality 9727 (2087). 3: HC84, YPM-6207, locality 86100 (567). 4: HC211, DMNH-19037, locality 9535 (1491). 5: HC228, DMNH-19202, locality 9824 (2203). Scale bars = 1 cm.

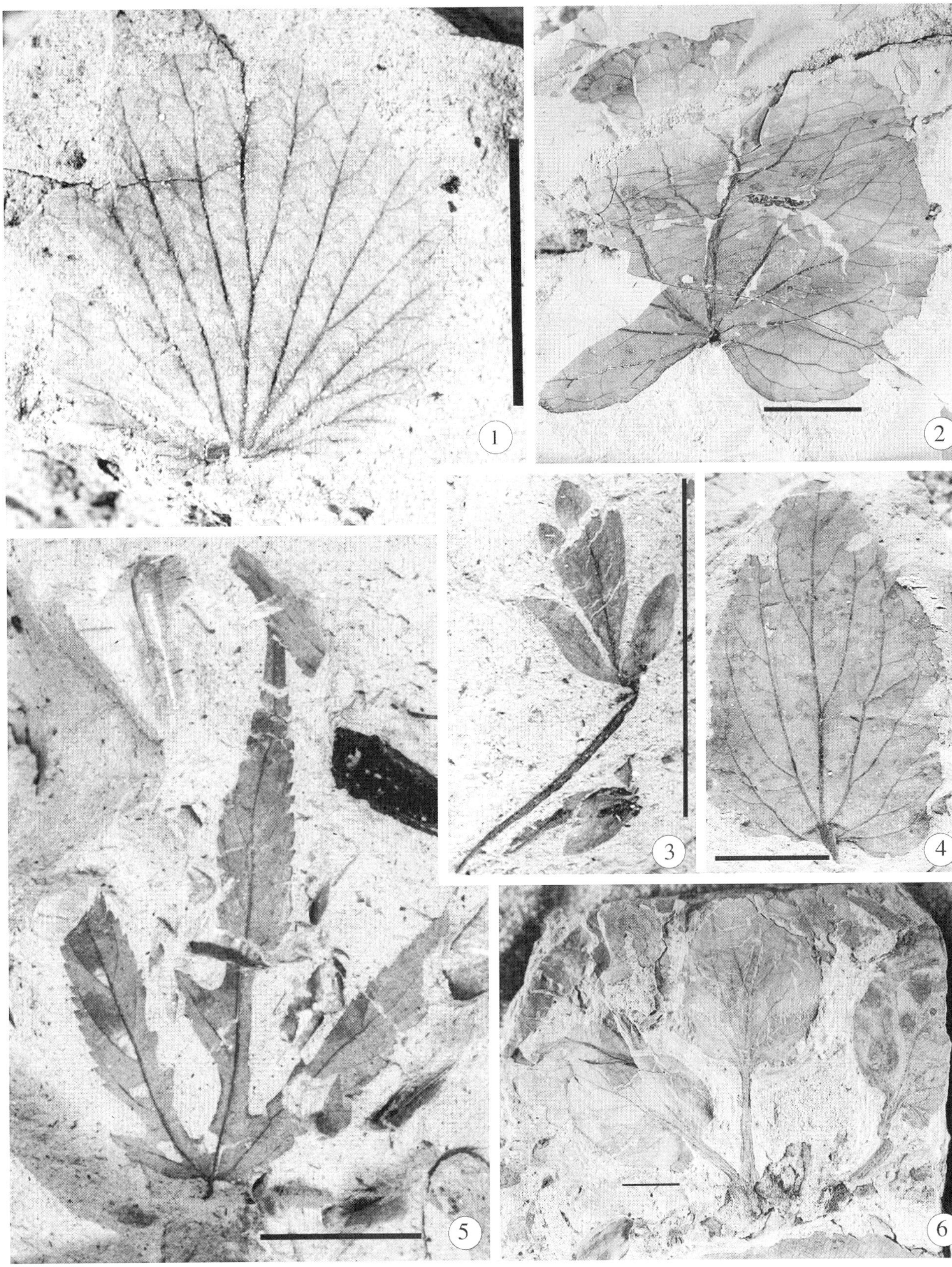

Plate 8. Herbaceous and shrubby Hell Creek flora (zone HCIb). 1: HC90, DMNH-19050, locality 9535 (1491). 2: HC230, DMNH-19205, locality 9824 (2203). 3: HC227, DMNH-19174, locality 86100 (567). 4: HC223, DMNH-19141, locality 9822 (2202). 5: Cannabaceae HC81, YPM-6205, 86100 (567). 6: Mostly complete herbaceous plant HC251, DMNH-19228, locality 9727 (2087). Scale bars = 1 cm.

Plate 9. Hell Creek dicotyledons characteristic of megafloral zone HCII. 1: *Erlingdorfia montana* (HC57-terminal leaflet; HC64-lateral leaflet), DMNH-6253, locality 86153 (568). 2: "*Vitis*" *stantoni* (HC14), YPM-6138, locality 87105. 3: "*Dryophyllum*" *subfalcatum* (HC49), YPM-6170, locality 86115. 4: HC66, YPM-6188, locality 86153. 5: "*Rhamnus*" *cleburni* (HC62), DMNH-19262, locality 87105 (2088). Scale bars = 1 cm.

Plate 10. Hell Creek dicotyledons characteristic of megafloral zone HCIII. 1: "*Ficus*" *planicostata*, (HC4) DMNH-7333, locality 9322 (900). 2: HC43, DMNH-7234, locality 9322 (900). 3: *Platanites marginata* (HC106), YPM-6234, locality 86126 (Ekalaka, Montana). 4: "*Zizyphus*" *fibrillosus* (HC174), DMNH-7463, locality 88102 (428). Scale bars = 1 cm.

Plate 11. Hell Creek dicotyledons characteristic of megafloral zone HCIII. 1: *Marmarthia trivialis* (HC105), YPM-6233, locality 86126 (Ekalaka, Montana). 2: *Liriodendrites bradacii* (HC166), YPM-6297, locality 87110. 3: *Marmarthia pearsonii* (HC162), DMNH-7713, locality 9322 (900). 4: *Paranymphaea hastata* (HC111), YPM-6438, locality 87152. 5: cf. "*Dryophyllum*"; *tennessensis* (HC44), YPM-6167, locality 87110. Scale bars = 1 cm.

Plate 12. Hell Creek lobed dicotyledons. 1: *Cissites puilosokensis* (HC180), DMNH-6240, locality 88102 (428). 2: *Grewiopsis saportana* (HC2), DMNH-21377, locality 9312 (897). 3: *Araliaephyllum polevoi* (HC60), DMNH-19233, locality 86153 (568). Scale bars = 1 cm.

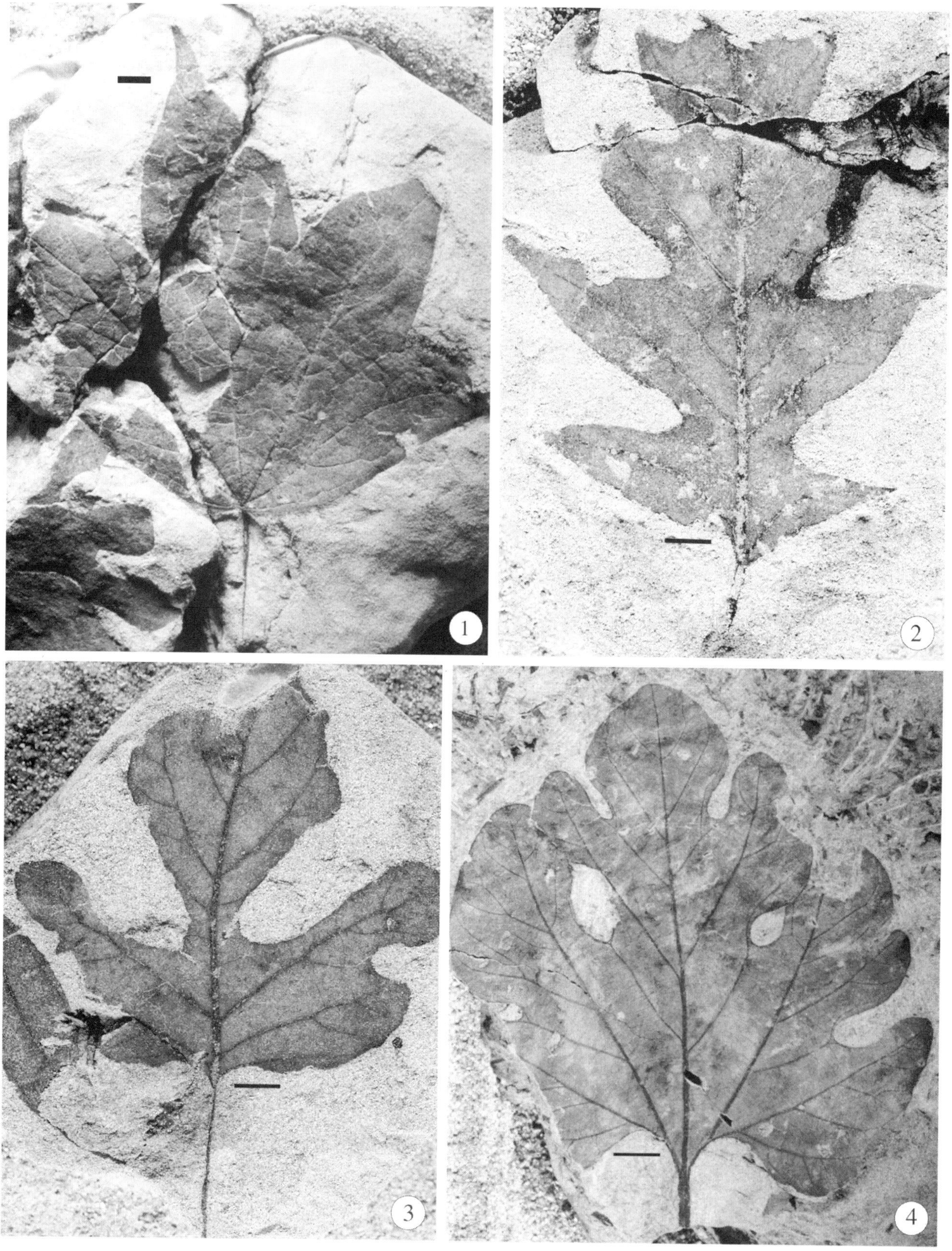

Plate 13. Hell Creek lobed dicotyledons. 1: cf. *Cissites acerifolia* (HC63), DMNH-8458, locality 86153 (568). 2: "*Artocarpus*" *lessigiana* (HC179), DMNH-13650, locality 88102 (428). 3: HC199, DMNH-13634, locality 88102 (428). 4: *Bisonia niemii* (HC200), DMNH-16694, locality 9610 (1781). Scale bars = 1 cm.

Plate 14. Hell Creek lobed dicotyledons. 1: Four-lobed *Liriodendron* (HC208), DMNH-18863, locality 9322 (900). 2: HC277, DMNH-19245, locality 9322 (900). 3: HC330, DMNH-7271, locality 9322 (900). Scale bars = 1 cm.

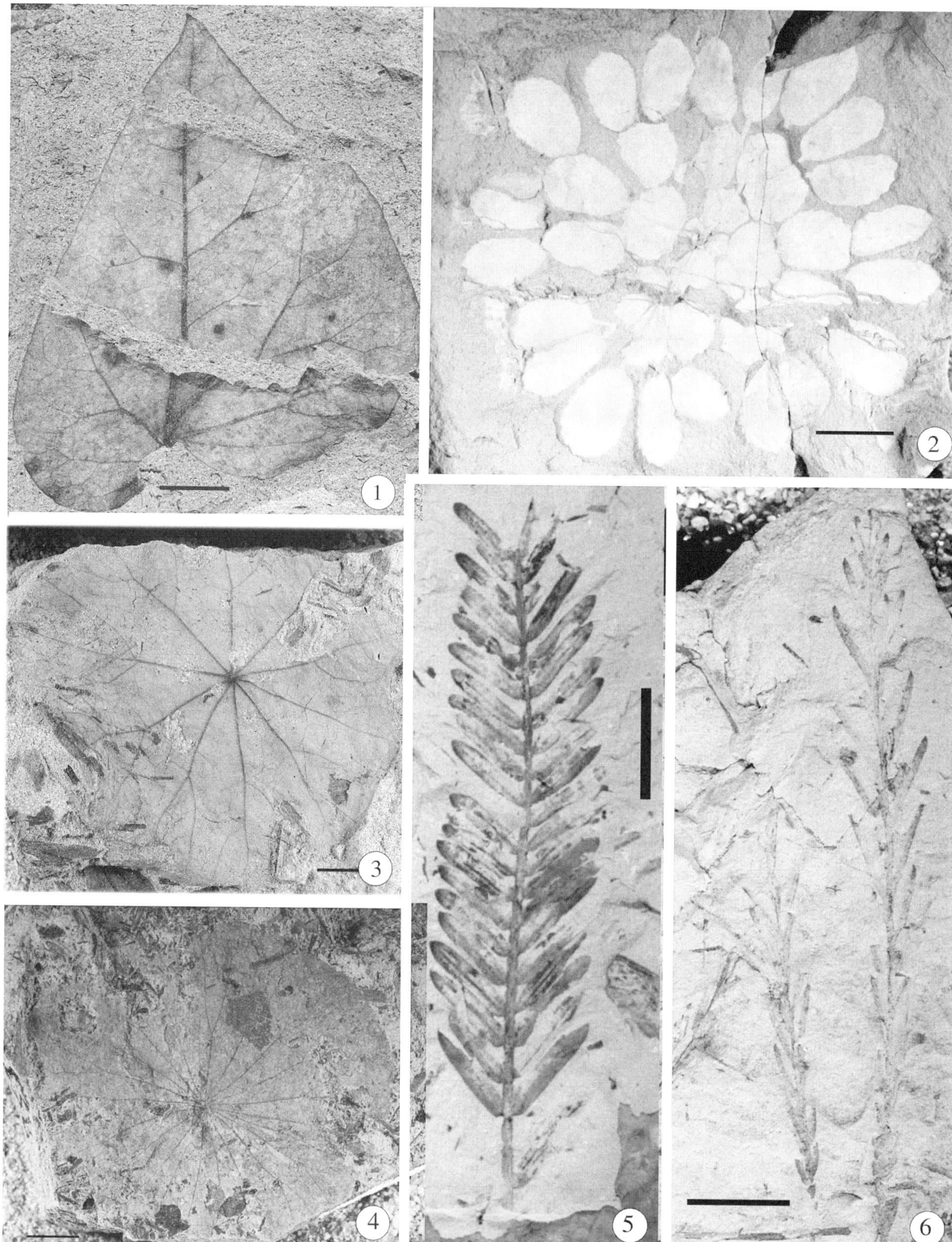

Plate 15. Paleocene and Cretaceous aquatics and gymnosperms, all but 1 occur in both Cretaceous and Paleocene. 1: *Paranymphaea crassifolia* (FU1), DMNH-19733, locality 9251(563). 2: *Quereuxia angulata* (FU2), YPM-6052, locality 87142. 3: cf. *Harmsia* (FU9), DMNH-7594, locality 9322 (900). 4: *Nelumbium montanum* (FU62), DMNH-19263, locality 9529 (1489). 5: *Metasequoia occidentalis* (FU3), YPM-6053, locality 86110. 6: *Glyptostrobus europaeus* (FU4), YPM-6054, locality 86110. Scale bars = 1 cm.

Plate 16. Paleocene and Cretaceous dicots and gymnosperms. 1: *Cupressinocladus interruptus* (FU26), DMNH-19734, locality 9858 (2217). 2: *Cercidiphyllum genetrix* (FU5), YPM-6056, locality 86110. 3: *Ziziphoides flabella* (FU43), YPM-6088, locality 86110. 4: *Penosphyllum cordatum* (FU61), YPM-6106, locality 86107 (427). Scale bars = 1 cm.

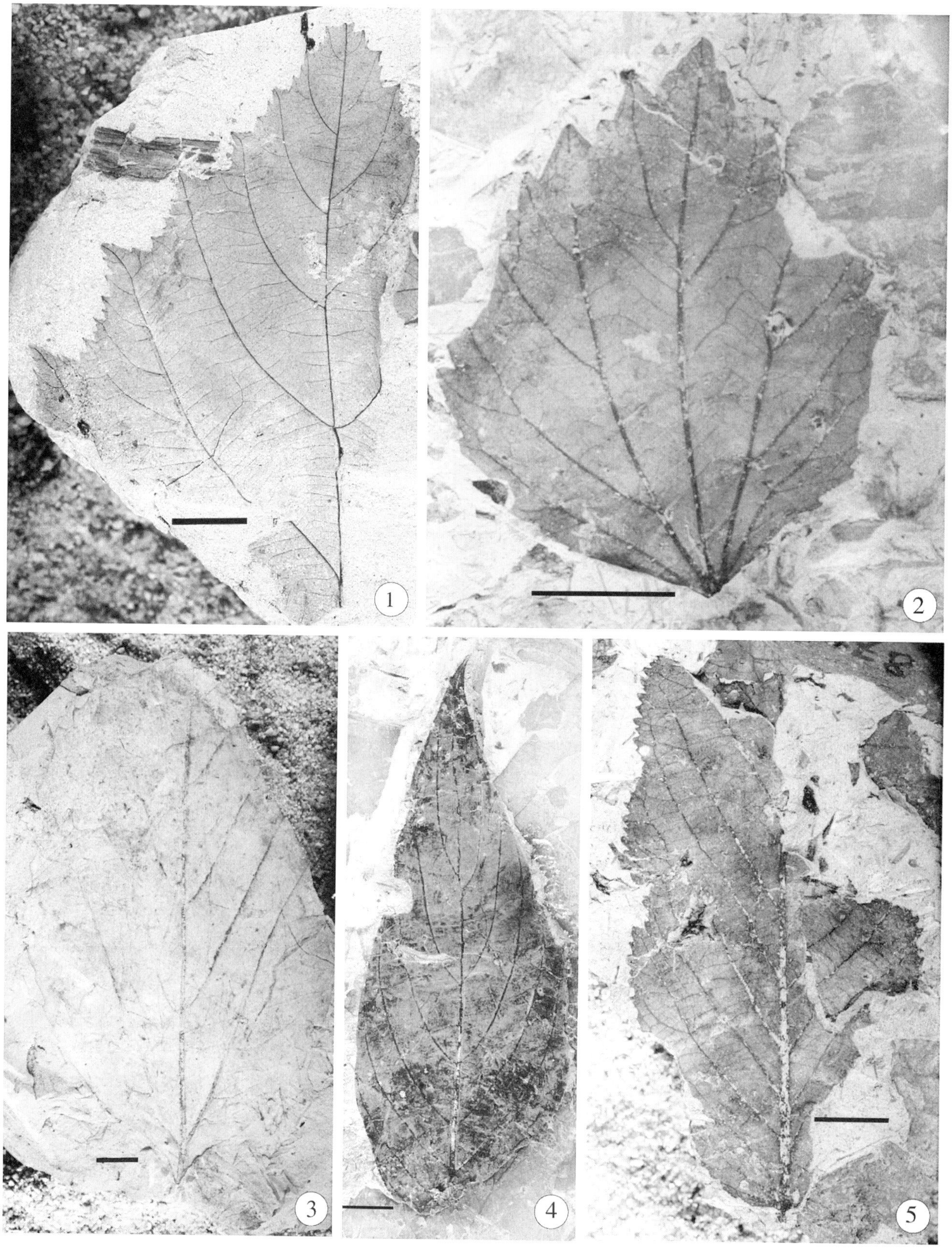

Plate 17. Paleocene and Cretaceous dicotyledons 1: *Corylus* sp. (FU79), DMNH-19732, locality 9250 (562). 2: "*Populus*" *nebrascensis* (FU7), YPM-6061, locality 86107 (427). 3: *Platanus raynoldsii* (FU16), YPM-6394, locality 86110. 4: *Cornophyllum newberryi* (FU31), YPM-6078, locality 86107 (427). 5: *Dicotylophyllum anomalum* (FU29), YPM-6076, locality 86107 (427). Scale bars = 1 cm.

TABLE 10. FLORAL LIST FOR THE HELL CREEK FORMATION

Class	Subclass	Order	Family	Genus species (morphotype number)
Bryophyta				
				Marchantia pealii (HC123)
Sphenopsida				
		Equisetales		
				Equisetum sp. (FU36)
Filicopsida				
		Hydropteridales		
			Salviniaceae	
				Salvinia sp. (HC312)
			Hydropteridaceae	
				Hydropteris pinnata (HC129)
		Filicales		
			Polypodiaceae?	
				HC241
		Incertae sedis		
				FU46, HC8, HC78, HC115, HC190, HC215, HC248, HC275, HC319
Gymnospermopsida				
		Coniferales		
			Taxodiaceae	
				Glyptostrobus europaeus (FU4) *Glyptostrobus* sp.#2 (HC9) *Metasequoia occidentalis* (FU3) *Metasequoia* sp.#2 (HC35) *Taxodium olrikii* (HC71) *Sequoia* sp. (HC70) HC339
			Cupressaceae	
				Cupressinocladus interruptus (FU26) *Fokieniopsis catenulata* (HC137)
			Araucariaceae	
				Elatides longifolia (HC165) *Dammarites* sp. (HC257) HC83, HC299
			Incertae sedis	
				HC219, HC220, HC240, HC304, HC311, HC313
		Cycadales		
				Nilssonia yukonensis (HC164)
		Ginkgoales		
				Ginkgo adiantoides (HC114)
Liliopsida				
	Arecidae			
		Arales		
			Araceae	
				"Pistia" corrugata (HC77) aff. *Philodendron* (HC201) *Limnobiophyllum scutatum* (FU38)
		Arecales		
			Arecaceae	
				Amesoneuron or *Sabalites* (FU37)
	Liliidae			
		Zingiberales		
			Zingiberaceae	
				Zingiberopsis attenuata (HC196)
	Incertae sedis			
				Porosia verrucosa (HC12) HC258, HC341
Magnoliopsida				
	Magnoliidae			
		Laurales		
			Chloranthaceae	
				HC88
			Lauraceae	
				Bisonia niemii (HC200) *Marmarthia pearsonii* (HC162) *Marmarthia trivialis* (HC105) *"Artocarpus" lessigiana* (HC179) *"Ficus" planicostata* (HC4)

(*continued*)

TABLE 10. FLORAL LIST FOR THE HELL CREEK FORMATION (continued)

Class	Subclass	Order	Family	Genus species (morphotype number)
Magnoliopsida (*continued*)				
	Magnoliidae			
		Laurales		
			Lauraceae (*continued*)	
				Araliaephyllum polevoi (HC60)
				Cissites insignis (HC186)
				Cissites lobata (HC175)
				Cissites puilosokensis (HC180)
				HC32, HC84, HC87, HC122, HC163, HC168, HC199, HC265, HC283, HC292
		Magnoliales		
			Magnoliaceae	
				Four-lobed *Liriodendron* (HC208)
				Liriodendrites bradacii (HC166)
				HC203
			Incertae sedis	
				FU25
		Nelumbonales		
			Nelumbonaceae	
				Nelumbo (HC202)
				Nelumbium montanum (FU62)
		Piperales		
			Piperaceae	
				"Ziziphus" fibrillosus (HC174)
		Incertae sedis		
				HC43, HC66
	Ranunculidae			
		Ranunculales		
			Berberidaceae	
				HC188
			Ranunculaceae	
				HC226, HC227
	Hamamelididae			
		Trochodendrales		
			Cercidiphyllaceae	
				Cercidiphyllum genetrix (FU5)
				Cercidiphyllum ellipticum (HC212)
				aff. *Cercidiphyllum* (HC124)
				Nyssidium arcticum (FU59)
				Ziziphoides flabella (FU43)
				"Populus" nebrascensis (FU7)
				Nordenskioldia borealis (FU44)
				Trochodendroides nebrascensis (HC103)
				HC73, HC229, HC355
		Fagales		
			Fagaceae	
				HC100
		Hamamelidales		
			Platanaceae	
				Erlingdorfia montana (HC57, HC64)
				Platanites marginata (HC1, HC106)
				Platanus raynoldsii (FU16)
				Leepierceia preartocarpoides (HC86)
				Grewiopsis saportana (HC2)
				"Celastrus" taurenensis (HC94)
				"Vitis" stantonii (HC14)
				aff. *"Vitis" stantonii* (HC108)
				aff. *"Vitis" stantonii* (HC189)
				HC52, FU63, HC98, HC99, HC140, HC151, HC152, HC197, HC198, HC209, HC213, HC284, HC329, HC330
		Urticales		
			Cannabaceae	
				aff. *Humulus* (HC243)
				HC81, HC225
			Ulmaceae	
				Carpites ulmiformis (HC54)
				HC36, HC218

(*continued*)

TABLE 10. FLORAL LIST FOR THE HELL CREEK FORMATION (continued)

Class	Subclass	Order	Family	Genus species (morphotype number)
Magnoliopsida (*continued*)				
			Urticaceae	
				HC18, HC93, HC228
		Incertae sedis		
				HC21, HC22, HC139
	Dilleniidae			
		Malvales		
			Malvaceae	
				Palaeoaster inquirenda (HC7)
		Incertae sedis		
				Penosphyllum cordatum (FU61) HC72, HC127, HC354
	Rosidae			
		Rosales		
			Rosaceae	
				HC80, HC288
		Rhamnales		
			Rhamnaceae	
				"Rhamnus" cleburni (HC62) HC261
		Cornales		
			Cornaceae	
				Cornophyllum newberryi (FU31)
		Incertae sedis		
				HC20, HC47, HC90, HC132, HC138, HC142, HC194, HC211, HC224, HC232, HC233, HC234, HC246
Magnoliopsida				
	Incertae sedis			
				"Dryophyllum" subfalcatum (HC49) cf. *"Dryophyllum" tenneseensis* (HC44) *"Myrica" torreyi* (HC344) cf. *Cissites acerifolia* (HC63) *Dicotylophyllum anomalum* (FU29) *Ficus ceratops* (HC256) *Harmsia hydrocotyloidea* (FU9) *Laurophyllum wardiana* (HC58) *Paranymphaea hastata* (HC111) *Quereuxia angulata* (FU2) *"Cinnamomum" lineafolia* (HC92) FU35, FU42, FU47, FU60, FU81, FU95, FU96 HC3, HC5, HC11, HC15, HC17, HC23, HC24, HC25, HC27, HC28, HC29, HC31, HC34, HC40, HC55, HC56, HC65, HC67, HC68, HC74, HC75, HC76, HC85, HC91, HC96, HC126, HC128, HC130, HC131, HC135, HC136, HC141, HC147, HC148, HC150, HC154, HC155, HC158, HC159, HC161, HC169, HC171, HC172, HC176, HC182, HC183, HC185, HC187, HC193, HC204, HC207, HC210, HC214, HC216, HC217, HC222, HC223, HC230, HC231, HC236, HC237, HC238, HC239, HC242, HC244, HC245, HC247, HC249, HC250, HC251, HC252, HC253, HC254, HC255, HC260, HC262, HC263, HC264, HC266, HC267, HC268, HC269, HC270, HC271, HC272, HC273, HC274, HC276, HC277, HC278, HC279, HC280, HC281, HC282, HC285, HC286, HC287, HC289, HC290, HC291, HC293, HC294, HC295, HC296, HC297, HC298, HC300, HC301, HC302, HC303, HC305, HC306, HC307, HC308, HC309, HC310, HC314, HC315, HC316, HC317, HC318, HC320, HC321, HC323, HC324, HC325, HC326, HC327, HC328, HC331, HC332, HC333, HC334, HC335, HC336, HC337, HC338, HC340, HC342, HC343, HC345, HC346, HC347, HC348, HC349, HC350, HC351, HC352, HC353, HC356, HC357, HC358, HC359, HC360, HC362, HC363, HC364, HC365, HC366, HC367, HC368, HC369, HC37

Note: Taxonomic arrangement based on Stewart and Rothwell (1993) for nonangiosperms and Tahktajan (1980) with modifications from Hickey and Wolfe (1975) for angiosperms.

TABLE 11. FLORAL LIST FOR THE LUDLOW MEMBER OF THE FORT UNION FORMATION

Class	Subclass	Order	Family	Genus species (morphotype number)
Bryophyta				
				Marchantia pealii (HC123) FU45
Filicopsida				
		Filicales		
			Hymenophyllaceae	
				FU21
			Polypodiaceae	
				Woodwardia gravida (FU51)
			Incertae sedis	
				FU24, FU83, FU84, FU90
Gymnospermopsida				
		Coniferales		
			Taxodiaceae	
				Metasequoia occidentalis (FU3) *Glyptostrobus europaeus* (FU4) *Taxodium olrikii* (HC71)
			Cupressaceae	
				Cupressinocladus interruptus (FU26) *Fokieniopsis catenulata* (HC137)
Sphenopsida				
		Equisetales		
				Equisetum sp. (FU36)
Liliopsida				
	Arecidae			
		Arecales		
			Arecaceae	
				Amesoneuron and *Sabalites* (FU37)
		Arales		
			Araceae	
				Limnobiophyllum scutatum (FU38)
		Nelumbonales		
			Nelumbonaceae	
				Nelumbo rhizome (FU85)
	Incertae sedis			
				Porosia verrucosa (HC12)
Magnoliopsida				
	Magnoliidae			
		Magnoliales		
				FU25
		Nelumbonales		
			Nelumbonaceae	
				Nelumbium montanum (FU62)
		Nymphaeales?		
			Ceratophyllaceae	
				Ceratophyllum sp. (FU80)
		Incertae sedis		
				Paranymphaea crassifolia (FU1)
		Incertae sedis		
				FU78

(continued)

the basal Fort Union Formation is designated megafloral zone FU0 (Figs. 3 and 4; Tables 5, A1, and A4). This zone overlies zone HCIII of Johnson and Hickey (1990) and underlies their FUI. The FU0 zone occurs between the last appearance data (LAD) of 28 HCIII megaflora taxa and the LAD of FU81 and HC49 (Fig. 3; Table A4).

The two formations are substantially different in terms of the relative abundance of plant-preserving facies (Tables 4 and 5), the Hell Creek Formation being very rich in channel and near-channel localities and the Fort Union Formation being very rich in flood-plain localities. While this clearly reflects the increased ponding of the landscape probably due to the transgression of the Cannonball seaway (Belt et al., 1984), it decreases the possibility of direct comparison of the flora of the two formations. Johnson (1989) recognized this and was somewhat successful in sampling the underrepresented facies in each formation. Nonetheless, certain facies occur only in one formation and thus could not be sampled in the other. An example of this is the variegated mudstone units in the base of the Fort Union that represent widespread shallow ponds. Whereas the abandoned channel deposits of the Hell Creek also represent ponded water, the geomorphic context is different, and it is reasonable to suspect that different vegetation types would be sampled in the different settings.

TABLE 11. FLORAL LIST FOR THE LUDLOW MEMBER OF THE FORT UNION FORMATION (continued)

Class	Subclass	Order	Family	Genus species (morphotype number)
Magnoliopsida (*continued*)				
	Hamamelididae			
		Hamamelidales		
			Platanaceae	
				Platanus raynoldsii (FU16)
				cf. *Wardiaphyllum daturaefolium* (FU48)
				FU63, FU88, FU91, HC140
			Betulaceae	
				Corylus sp. FU79
		Trochodendrales		
				Cercidiphyllum genetrix (FU5)
				"Populus" nebrascensis (FU7)
				Ziziphoides flabella (FU43)
				Nordenskioldia borealis (FU44)
				Nyssidium arcticum (FU59)
				Cercidiphyllum ellipticum (HC212)
				FU54
		Incertae sedis		
				"Ampelopsis" acerifolia (FU76)
	Dilleniidae			
		Violales		
			Flacourtiaceae	
				FU49, FU74
		Incertae sedis		
				Penosphyllum cordatum (FU61)
				FU19, FU53, FU56, FU57
	Rosidae			
		Sapindales		
				FU55
		Cornales		
			Cornaceae	
				Amersinia (FU82)
				Cornophyllum newberryi (FU31)
		Incertae sedis		
				FU23
Magnoliopsida				
	Incertae sedis			
				Quereuxia angulata (FU2)
				cf. *Harmsia hydrocotyloidea* (FU9)
				Dicotylophyllum anomalum (FU29)
				aff. *Dicotylophyllum anomalum* (FU10)
				FU12, FU14, FU15, FU17, FU18, FU20, FU35, FU39, FU40, FU42, FU47, FU52, FU60, FU64, FU65, FU69, FU70, FU72, FU73, FU75, FU77, FU81, FU86, FU87, FU89, FU92, FU93, FU94, FU97, FU98, FU46, HC231

Note: Taxonomic arrangement based on Stewart and Rothwell (1993) for nonangiosperms and Tahktajan (1980) with modifications from Hickey and Wolfe (1975) for angiosperms.

FLORAL CHANGE WITHIN THE HELL CREEK FORMATION

Johnson and Hickey (1990) described five zones (HCIa, HCIb, HCIIa, HCIIb, HCIII) in the Hell Creek Formation and one zone (FUI) in the Fort Union Formation as a reflection of floral change through time. The addition of 71 new localities and 6902 specimens to this database provides an opportunity to evaluate these zones. Figure 3 shows the ranges of stratigraphically distributed megafloral taxa. This diagram was prepared by plotting all of the morphotypes that occurred at three or more stratigraphic levels. The levels are arbitrarily defined 2 m intervals (bins) above and below the palynologically defined K-T boundary. These levels contain between one and eight localities each. Thus, the 98 morphotypes that occur at 3 or more of the 54 fossil-bearing bins are likely to be common and identifiable and, as such, should give a good overview of floral change through time. The occurrence of all 380 morphotypes is listed by locality and zone in Table A4. Because the localities are more common and more thoroughly sampled in the vicinity of the K-T boundary (Table 2), and there are no local floras known from below the Hell Creek or above the Fort Union, samples low and high in the section are less well known than those in the middle of the section. Figure 5 shows standing richness of the complete flora and number of voucher specimens relative to the K-T boundary. This shows the dramatic decrease of morphotypes at the K-T boundary as well as a low-diversity zone just below floral zone HCIII.

TABLE 12. KEY TO ILLUSTRATED HELL CREEK AND FORT UNION ANGIOSPERM LEAVES

Diagnostic characters	Morphotype	Plate	Figure
Simple unlobed leaves with pinnate primary venation			
Entire margin			
Thread-like secondary veins	HC183	4	5
Spiderweb tertiary veins	FU61	16	4
Emarginate apex	HC43	10	20
Lobate base			
Secondary veins obscure	HC111	11	4
Secondary veins prominent	HC66	9	4
Tertiary veins parallel to secondaries	HC201	4	4
Length:width ratio >3:1	HC44	11	5
Nonentire margin			
Teeth only on apical portion of leaf			
Basal secondaries eucamptodromous	FU31	17	4
Basal secondaries brochidodromous	HC86	5	1
Length:width ratio >3:1	HC49	9	3
One tooth per secondary			
Strongly percurrent tertiaries	HC62	9	5
Urticoid teeth	HC228	7	5
Secondary forks at sinus	HC211	7	4
Many teeth			
Teeth compound			
Larger teeth at secondaries	FU79	17	1
Double teeth	HC224	6	3
Teeth small			
Leaf shape oblong	FU29	17	5
Leaf shape ovate	HC135	6	2
Small teeth with rounded sinuses			
Strong basal agrophic veins	FU16	17	3
Weak basal agrophic veins	HC94	5	4
Secondaries fork near primary	HC2	12	2
Simple unlobed leaves with more than one primary vein			
Entire margin			
Basal lateral primary veins form lower leaf margin			
Prominent percurrent tertiary veins	HC4	10	1
Narrow elliptic leaf shape	HC162	11	3
Suprabasal lateral primary veins	HC84	7	3
Normal basal lateral primary veins			
Three basal primaries	HC92	5	3
More than three basal primaries			
Secondaries eucamptodromous	HC174	10	4
Secondaries brochidodromous	HC223	8	4
Secondaries festooned brochidodromous			
Tertiaries percurrent	HC230	8	2
Tertiaries irregular	FU1	15	1
Nonentire margin			
More than three basal primaries	HC90	8	1
Three basal primaries			
Craspedodromous secondaries			
Cuneate base shape	HC106	10	3
Lobate base shape	HC14	9	2
Obtuse to rounded base	FU7	17	2
Semicraspedodromous secondaries			
Urticoid teeth	HC93	7	2
Teeth apical, large, and rounded	HC105	11	1
Large chloranthoid teeth	FU43	16	3
Small chloranthoid teeth			
Leaf shape wide ovate	FU5	16	2
Leaf shape ovate	HC212	5	2

(*continued*)

Herbaceous Cretaceous: The small leaves of zone HCIb

The group of 8 last appearance data (LAD) at 36 m below the K-T boundary in Figure 3 represents the upper limit of Johnson and Hickey's (1990) floral zone HCIb, an interval zone characterized by small leaves. In 1990, HCIb was considered to be present in only 5 m of rock (50–55 m below the formational contact datum), and it was not clear if the zone was driven climatically or if it was the result of some unusual edaphic or taphonomic situation. The subsequent discovery of a number

TABLE 12. KEY TO ILLUSTRATED HELL CREEK AND FORT UNION ANGIOSPERM LEAVES (continued)

Diagnostic characters	Morphotype	Plate	Figure
Simple lobed leaves			
Palmately lobed			
Odd number of lobes			
Three lobes			
Shallow lobe sinus			
Simple teeth	HC57	9	1
Compound teeth	HC80	6	4
Deep lobe sinus			
Regular prominent secondaries	HC243	6	1
Irregular secondaries	HC81	8	5
Five lobes			
Actinodromous primaries	HC63	13	1
Palinactinodromous primaries			
Deep rounded sinuses	HC60	12	3
Shallow sinuses	HC277	14	2
Even number of lobes			
Lobes asymmetrical	HC64	9	1
Pinnately lobed			
Entire margin			
Two lobes (wide-obtuse apex)	HC166	11	2
Four lobes	HC208	14	1
More than four lobes	HC179	13	2
Nonentire margin	HC226	7	1
Complex lobes (lobes on lobes)			
Lobed apex			
Acute apical shape	HC180	12	1
Round apical shape	HC200	13	4
Unlobed apex	HC199	13	3
Leaves with peltate primary venation			
Eight to ten primary veins	FU9	15	3
More than 12 primary veins			
Leaves $<$10 cm	FU62	15	4
Leaves $>$15 cm	HC202	4	1,2
Compound leaves			
Leaves in rosettes	FU2	15	2
Leaves trifoliate			
Terminal leaflet unlobed	HC330	14	3
Terminal leaflet lobed			
Lateral leaflet bilobed	HC57/HC64	9	1
Lateral leaflet unlobed	HC227	8	3
Leaves with multistranded midveins			
Few strands	HC251	8	6
Many strands	HC77	4	3

of similar floras at the same stratigraphic level and ~15 m higher (localities scored as HCIb in Tables A1 and A4) and their sedimentological context suggest that these floras represent unusual preservation of herbaceous and shrubby vegetation on the exposed point bars of large river systems. Sedimentological evidence for this hypothesis includes the presence of these floras in large channel deposits consisting of heterolithic inclined strata. The floras occur on point-bar deposits formed by alternating mudstone and sandstone layers. Typically, the leaves occur in mudstone strata that are as much as 15 cm thick. The mudstone layers are not rooted, indicating that the floras are parautochthonous rather than autochthonous. In these localities, tree leaves are present, but rare and much less common than the small leaves.

The small leaves are thought to be herbaceous or shrubby for several reasons. Herbaceous dicots tend to have multistranded midveins, disorganized venation, and can have small leaves. Another argument for the herbaceous or shrubby nature of these local floras is their taxonomic composition. These localities contain a number of families whose living representatives are herbaceous or shrubby and whose phylogenetic placement suggests that they are ancient lineages. Examples include Urticaceae (HC93), Ranunculaceae (HC226), Rosaceae (HC80), and Cannabaceae (HC81, HC225) (Plates 6–8). Despite the phylogenetic argument that these groups likely have a Cretaceous heritage, their vegetational habit is one that is not likely to be preserved by common taphonomic processes. Herbaceous plants tend to hold their leaves and are thus relatively difficult to introduce into the fossil record because they die and dry rather than being transported and fossilized. Shrubs produce far fewer leaves than trees, and, because of their low stature, are less likely to disperse leaves for any great distance.

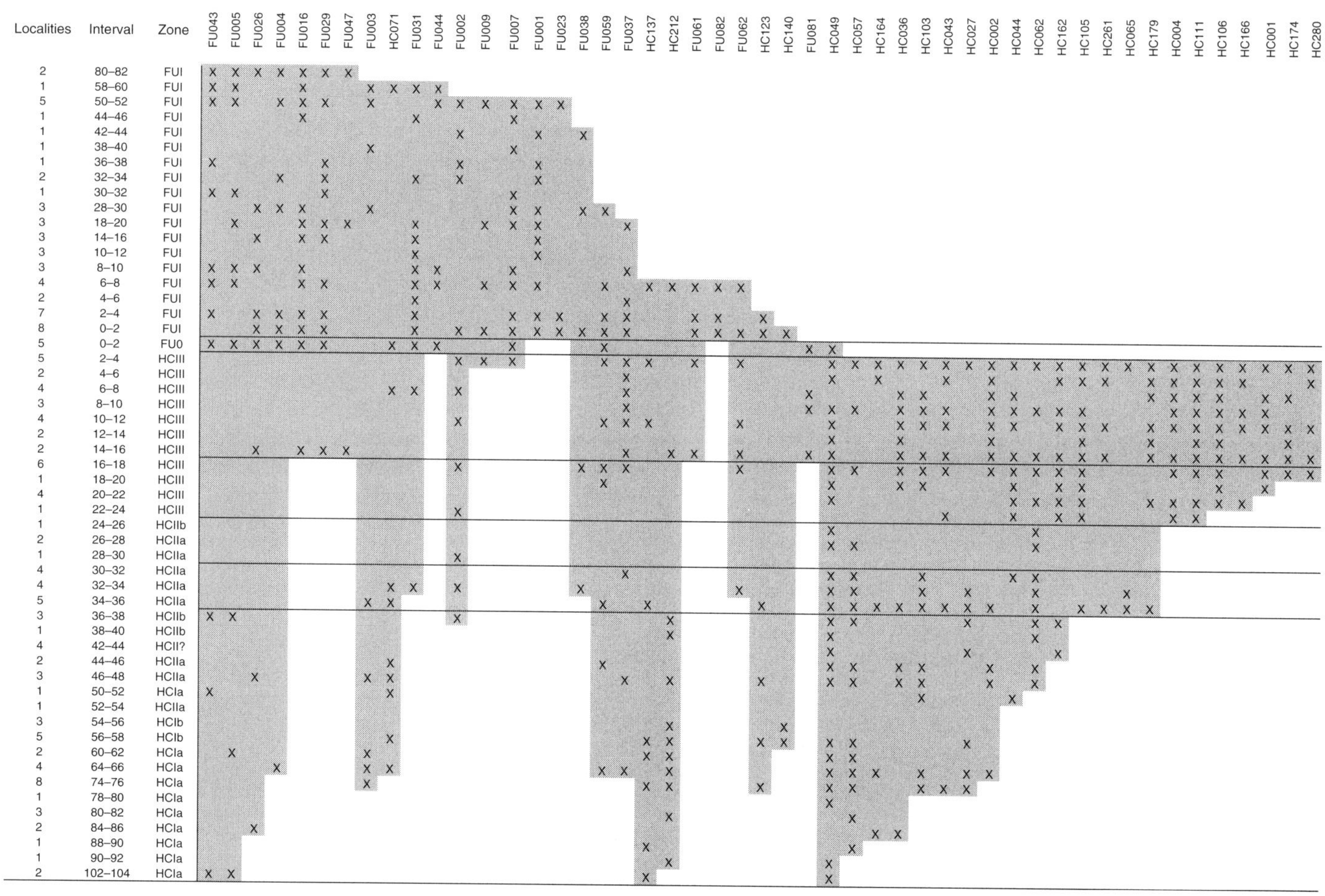

Figure 3 (on this and facing page). Stratigraphic ranges of megafloral morphotypes relative to Cretaceous-Tertiary (K-T) boundary. Individual localities were recalculated from HC-FU (Hell Creek, Fort Union) datum to K-T boundary using palynology (Table A2; Nichols and Johnson, this volume). It is not practical to plot all 380 morphotypes at all 158 localities, so data set was compressed by removing all morphotypes that occur at only one locality (n = 222). Localities were then collapsed into 2 m bins (levels) and morphotypes that occurred in only one bin were removed (n = 9). Morphotypes that occurred at only two levels were removed (n = 51). Remaining 98 stratigraphically robust morphotypes are plotted in 54 levels. Number of localities per level is specified in first column.

In contrast, the Big Cedar Ridge flora of Wing et al. (1993) is an example of an autochthonous herbaceous flora. In that case, the landscape vegetation was buried by secondarily reworked volcanic ash that overwhelmed local drainages. In the Hell Creek case, individual plants and segments of vines were apparently uprooted and deposited on point bars or mid-channel bars not far from their position of growth. I suggest that these floras grew on the point bars of large rivers during periods of low flow. They were then excavated by channel scour as flow increased and were redeposited and buried when flow slowed again. The existence of variable current flow is supported by the presence of inclined heterolithic strata in Hell Creek paleochannels (Fastovsky, 1987). These localities are characterized by excellent leaf preservation in very thin layers. At locality 9824 (2203) the productive layer is only 2 cm thick, suggesting that the deposition occurred in a single event. It is not clear if these layers represent annual variation in flow or sporadic flood events. Although several authors have suggested seasonality in the Hell Creek deposits (Derstler, 1994; Carpenter, 1987; Rigby, 1987), most of these suggestions are based on the occurrence of dinosaur mass death assemblages or dinosaur mummies, and seem to be circumstantial. There is little sedimentological evidence in the Hell Creek strata for marked seasonality or aridity (Fastovsky, 1987; Fastovsky and McSweeney, 1987).

This would suggest that zone HCIb of Johnson and Hickey (1990) represents an unusual type of deposition that preferentially preserved herbaceous and small-stature vegetation rather than representing vegetation of a specific time period. An interesting feature of HCIb floras is the rarity of tree leaves admixed with the small leaves of shrubs and herbs. This suggests that the river systems were large enough that the gallery forests were not making a major contribution to the leaf litter of the point bars in certain situations. The few tree leaves found in HCIb floras are taxa such as *Leepierceia preartocarpoides* (HC86) and "*Dryophyllum*" *subfalcatum* (HC49), which are characteristic of the underlying HCIa and overlying HCIIa zones.

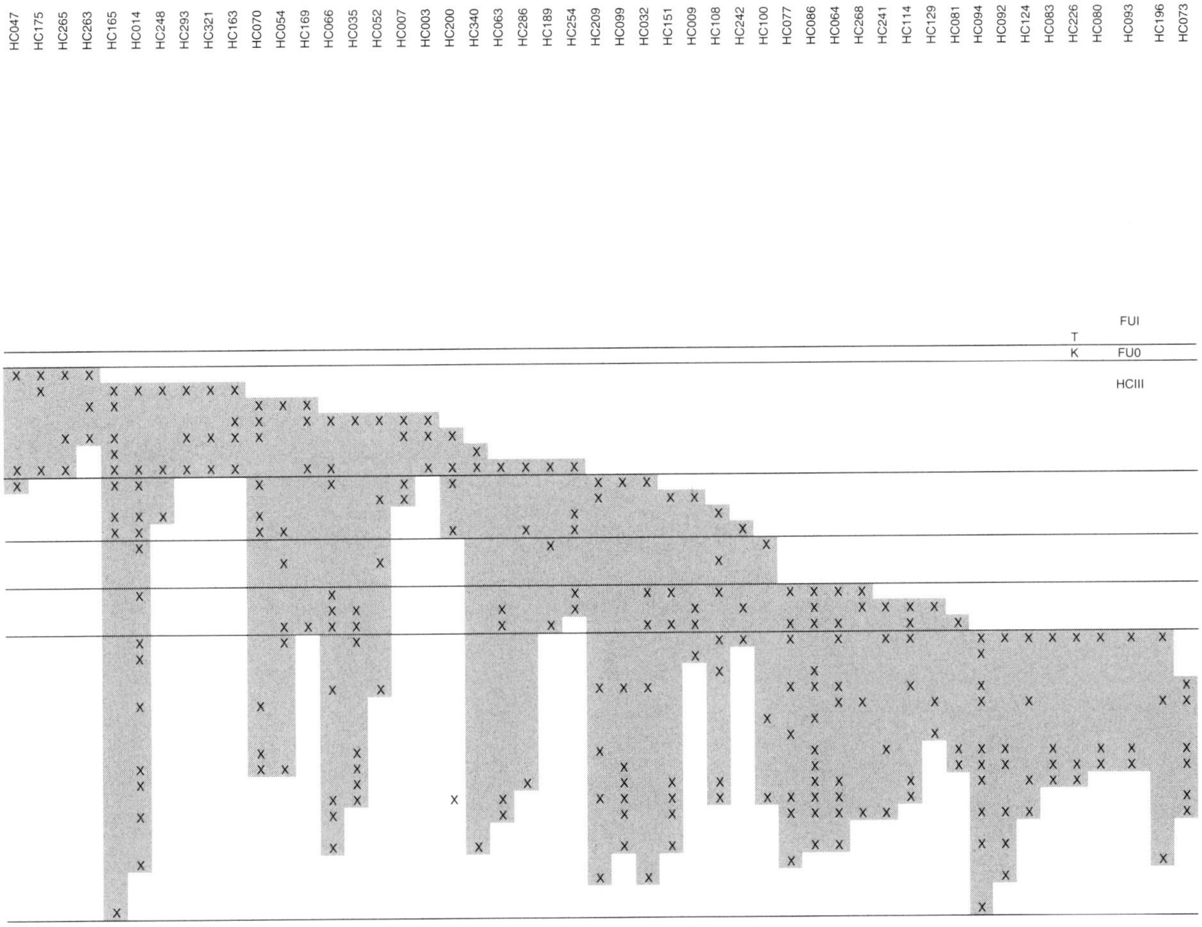

Figure 3. (*continued*)

The dominant taxa of zone HCIa all have ranges that extend into higher zones. Extinctions of common or important HCI taxa include *Leepierceia preartocarpoides* (HC86) 31.4 m below the K-T boundary, "*Celastrus*" *taurenensis* (HC94) 36.9 m below the boundary, and "*Vitis*" *stantonii* (HC14) and "*Dryophyllum*" *subfalcatum* (HC49) just below the K-T boundary (Table A4; Figs. 3 and 4, Plates 5 and 9). Although the concept of a distinct lower Hell Creek megaflora is valid, the specific upper boundary of zone HCI at 50 m (Johnson and Hickey, 1990) no longer seems particularly tenable.

Transition floras in the middle of the Hell Creek Formation

Johnson and Hickey (1990) described megafloral zone HCII as ranging from 50 m to 15–17 m below the Hell Creek–Fort Union formational contact (HCIIa 50-25 m; HCIIb 25 to 15–17 m). The addition of 21 new localities in this interval has complicated this pattern somewhat (Tables A1 and A4). The new data have shown that HCIII floras, such as locality 9610 (1781), extend as far down as 21 m, and it is clear that there is no precise stratigraphic level that marks the HCII-HCIII boundary. Rather, the HCII-HCIII contact rises and falls as much as 8–10 m. This suggests that either the HCIIb-HCIII floras are transitional, occupying different parts of the landscape at the same time, or that the time represented by a given stratigraphic thickness varies from section to section. There is reason to believe that the latter interpretation is more likely correct. Hicks et al. (this volume) show that the base of magnetic polarity subchron C29R varies from 24.5 to 10.5 m below the K-T boundary in three sections (Pretty Butte, Bobcat Butte, Sunset Butte). Because there is no example where an HCII flora overlies an HCIII flora in the same stratigraphic section, it is reasonable to posit the HCII-HCIII transition as a temporal one. Figure 6 plots all leaf localities and megafloral zones relative to the magnetostratigraphic chron boundaries. The HCIIb zone that is characterized by low species diversity and hyperabundance of "*Dryophyllum*" *subfalcatum* (HC49) and "*Vitis*" *stantoni* (HC14) also appears to represent a discrete stratigraphic interval.

The HCIIa flora appears to represent a transitional period where taxa from the lower Hell Creek are replaced by the HCIII megaflora. Figure 3 shows stepped last appearances beginning

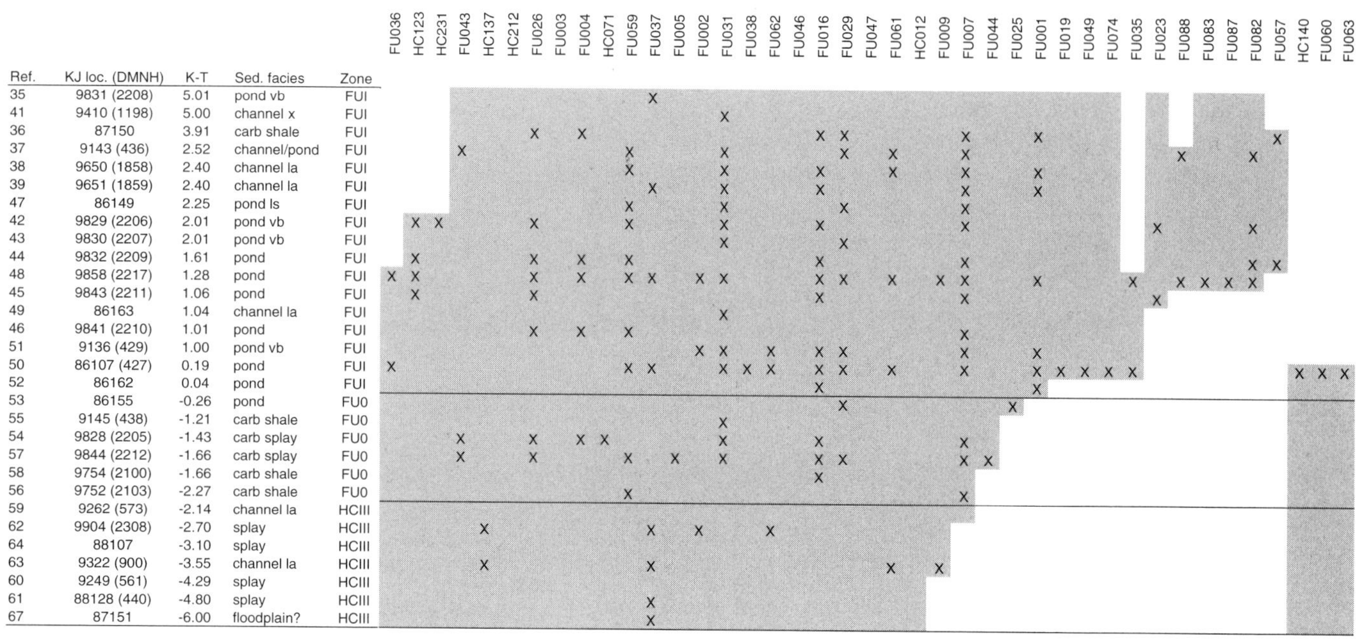

Figure 4 (on this and facing page). Stratigraphic ranges of megafloral morphotypes from 30 localities within 6 m of Cretaceous-Tertiary boundary. The six localities from Cretaceous basal portion of Fort Union Formation are designated as FU0. *Paranymphaea crassifolia* (FU1) does not occur in FU0 but appears in lowest Fort Union localities that produce Paleocene pollen.

36 m below the K-T and continuing to the boundary. This trend is balanced by a series of first appearances beginning 36 m below the K-T and culminating with a burst of 11 first appearance data (FAD) in the interval 14–16 m below the K-T boundary.

Warming in the latest Maastrichtian: The HCIII megaflora

Johnson and Hickey (1990) noted that the mean annual temperature (MAT) estimated from leaf margin analysis warmed considerably from HCII to HCIII. This climate analysis was later enhanced by Johnson and Wilf (1996), who showed that the temperature increase in the latest Cretaceous was significant, and could be roughly correlated to stable isotope results from upper Maastrichtian foraminifera in the South Atlantic that argued for climate warming between the base of polarity subchron C29R and the K-T boundary. Since 1990, 19 new HCIII sites have been added to the database and the resulting flora is extremely rich and widespread. The assemblage is characterized by a large number of very distinct and common taxa from a variety of facies.

Two localities deserve special mention, Battleship 9322 (900) and Dean Street 88102 (428). The former is located in upper point-bar deposits 2.8 m below the formation contact ~5 km northeast of Marmarth and the latter occurs in lower point-bar deposits 15 m below the formation contact at K-T at Mud Buttes. We have had some difficulty locating a complete palynological K-T boundary near Battleship and have extrapolated the K-T boundary correction from Bobcat Butte, which is 2 km to the south. This gives Battleship a position of 3.6 m below the K-T boundary. Closer sections in the vicinity of River Section (Table A2) have not yielded a well-resolved K-T boundary. Both of these sites have been extensively excavated and have yielded very rich floras that serve to characterize HCIII and thus the penultimate Cretaceous flora. Dean Street has yielded 86 morphotypes and 975 voucher specimens. Battleship has produced 42 morphotypes and 485 vouchers. These are extremely diverse floras and represent some of richest known Cretaceous megafloral quarries in the world. Some sense of the magnitude of the K-T boundary extinctions on the megaflora can be achieved by looking at what happens to the flora of these two rich sites. Of the 86 taxa at Dean Street, only 8 (9.3%) are known anywhere in Paleocene strata. In addition, those 8 taxa at Dean Street compose only 3.9% of the voucher specimens from that site, indicating that the taxa that survive were not floral dominants during the Cretaceous. A similar pattern is seen at Battleship where only 4 of 42 (9.5%) survive to the Paleocene, and those 4 only compose 2.5% of the Battleship specimens.

FLORAL CHANGE NEAR THE CRETACEOUS-TERTIARY BOUNDARY

Early studies near Marmarth in 1986 indicated that megafloral change at the K-T boundary was very dramatic. The characteristics of the transition can be summarized by the following: (1) a diverse angiosperm-rich HCIII flora is replaced by a depauperate aquatic angiosperm-rich FUI flora; (2) floral domi-

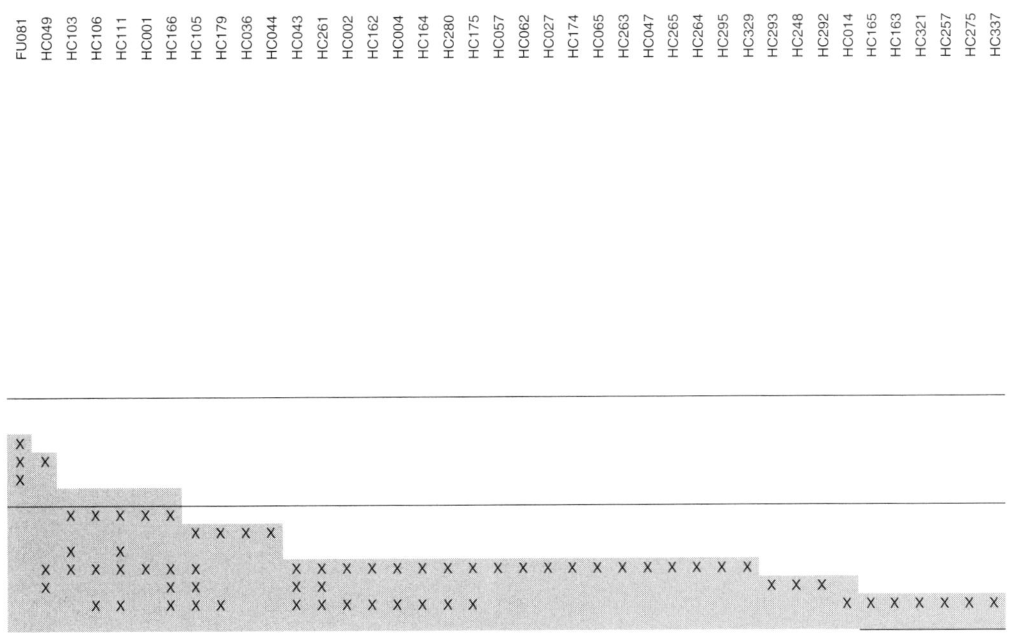

Figure 4. (*continued*)

nants in the uppermost Cretaceous rocks are absent in the basal Paleocene; (3) angiosperms that exhibit a Cretaceous "look," characterized by unusual lobation and strongly impressed tertiary veins (Plates 12–14), disappear at the K-T boundary; (4) archaic gymnosperms such as the cycad *Nilssonia* and the araucarian conifers *Elatides* and *Dammarites* disappear, and *Ginkgo* is locally extirpated; and (5) certain relatively abundant Cretaceous plants, such as palms and taxodiaceous conifers, remain relatively common into the early Paleocene. These changes are coeval with the decimation of the *Wodehouseia spinata* Assemblage Zone, within the limits of stratigraphic resolution. A key point is that palynomorphs and megafossils tend to be preserved in different rock types and facies. The dark gray mudstone, claystone, and lignite sampled for pollen rarely preserve megafossils, whereas the light gray silty mudstone and sandstone that preserve leaves rarely preserve pollen. In addition, stratigraphic resolution of palynology can be much finer than that of the megaflora, assuming that palyniferous rocks are available. Thus, despite the recognition that pollen and leaves represent the same flora filtered through different taphonomic pathways, the precise centimeter-level stratigraphy of the K-T boundary remains largely the realm of the palynologist. In addition, the ability to locate the K-T boundary is limited to sections with physical evidence for impact or high-resolution palynostratigraphy.

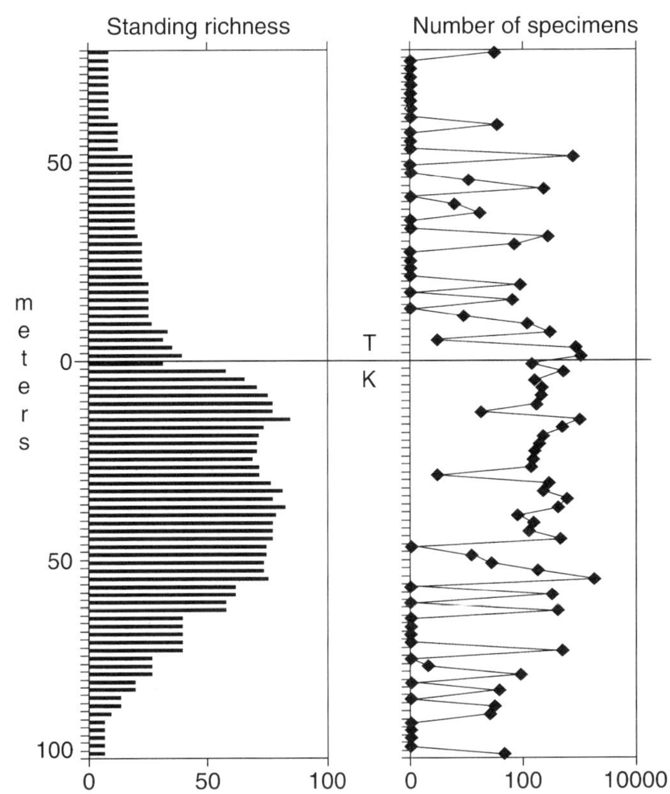

Figure 5. Standing richness of Hell Creek and Ludlow megaflora based on 2 m bins from Figure 3. Standing richness was counted as all morphotypes at each bin level plus those that occurred both below and above bin level. Number of voucher specimens per bin level is plotted (log scale) to demonstrate that dramatic decrease in standing richness is not a function of decreased sample size.

The documentation of the K-T boundary at 17 sections in the Marmarth area using palynology (Nichols and Johnson, this volume) and the discovery of an impact layer at Pyramid Butte (Johnson et al., 1989) and Mud Buttes (Nichols et al., 2000) provide a much higher resolution for interpreting the megafloral

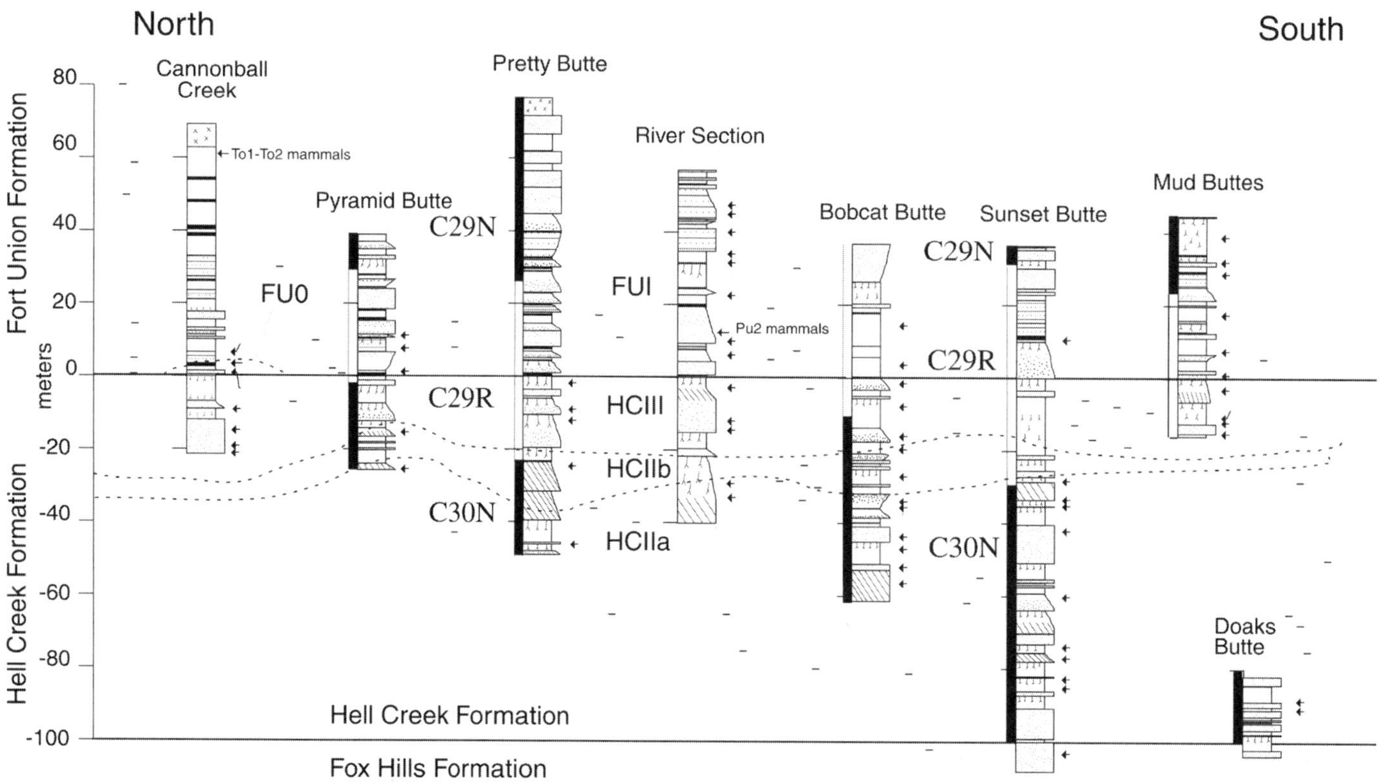

Figure 6. Key stratigraphic sections with magnetostratigraphic interpretation of Hicks et al. (this volume) and approximate boundaries of megaflora zones. Levels of megafloral localities from Figure 2 are plotted without locality identifiers simply to show how tightly they delimit megafloral zone boundaries. Note how upper limit of zone HCIIb arches upward at Pyramid Butte where base of C29R is also anomalously high. Pu marks presence of PITA Flats Puercan mammalian fauna and To marks Brown Ranch Torrejonian fauna (Hunter, 1999).

localities in the stratigraphic vicinity of the Hell Creek–Fort Union formational contact and the K-T boundary.

K-T survival of mire vegetation

Prospecting in 1998 revealed several megafloral localities in the lowermost Fort Union strata along Cannonball Creek in Slope County (designated FU0 in Tables A1 and A4). These sites occur in the lignitic facies of the Ludlow Member of the Fort Union Formation but below the palynologically defined K-T boundary (Table A2). These sites are significant because they are composed largely of species that occur in the overlying Paleocene portion of the Fort Union, rather than the species that occur in the underlying Hell Creek Formation. Johnson (1992) showed that the K-T boundary in southwestern North Dakota was characterized not only by the last occurrences of certain Cretaceous species, but also by a number of first appearance data for Paleocene species. At the time, this pattern did not agree with the pattern of palynomorph occurrence, which showed only last appearance data at the K-T boundary, a pattern more in line with the catastrophic extinction model. This difference was attributed to the different taxonomic resolution of palynomorphs and megafossils, but still suggested a rapid immigration of plants that were survivors of the K-T event in remote refugia. Discovery of Paleocene-aspect plants in the Cretaceous coaly portion of the Fort Union suggests another alternative. Coal beds are present but rare in the Hell Creek Formation. To date, none of the coaly units in the Hell Creek Formation have produced identifiable megaflora. The pattern as it exists suggests that the Paleocene megaflora was composed of Cretaceous mire vegetation that survived the K-T boundary event and that the refugia were not distant, but merely local, rare environments. Plants from the Cretaceous mires became widespread in the Paleocene and occur in all facies of the Paleocene. Not all Cretaceous mire plants survived the K-T boundary, as demonstrated by the presence of morphotypes FU81 and HC49 in the Cretaceous portion of the Fort Union and their absence higher in the section. This pattern is further reinforced by the palynological record. Certain "K taxa" used by Nichols and Johnson (this volume) to mark the end of the Cretaceous have staggered disappearances in the areas where the K-T boundary is in the Fort Union Formation. This suggests that certain Cretaceous palynomorphs were derived from plants that were not tolerant of the mire environment, although others were. This brings the interpretations of Nichols and Johnson (this volume) into concordance with earlier studies by A.R. Sweet of the Geological Survey of Canada, who had previously suggested environmental change in latest Cretaceous time rep-

Figure 7. Detail of Hell Creek diorama at Denver Museum of Nature & Science painted by Kent Pendleton. Diorama is based on HCIII megaflora from the upper 20 m of the formation and illustrates several key points about the nature of Hell Creek vegetation. The simple-leafed, thin-stemmed cycad is reconstructed based on a related cycad that was preserved as a nearly complete plant (Spicer and Herman, 1996). Aralean aff. *Philodendron* (HC201) is reconstructed using extant *Philodendron* as a model. Morphotype HC174, "*Zizyphus*" *fibrillosus* is reconstructed as vine, but could equally be weak-stemmed subshrub as reflection of modern habit of Piperaceae. Lobed leaves and slender trunks characterize canopy trees.

resented by the uppermost meter or less of Cretaceous rock (Sweet at al., 1990; Sweet and Braman, 1992).

The concept of Cretaceous mire plants as the primary survivors of the K-T event finds some resonance with the record of selective survivorship in terrestrial vertebrates. Sheehan and Fastovsky (1993) reinterpreted the work of Archibald and Bryant (1990) and argued for preferential survivorship of aquatic animals across the K-T boundary. They contended that the animals most susceptible to the K-T event were either large bodied or terrestrial. Our interpretation of the FU0 megaflora suggests that plants living in swamps were more likely to survive than were plants that lived on better drained soils. A test of this hypothesis will be the discovery of mire floras in the Hell Creek well below the K-T boundary.

Floral change in the Paleocene portion of the section

Data collected for this study, although primarily targeted at the K-T boundary, began as an assessment of the flora of the Ludlow Member of the Fort Union Formation (Johnson, 1982), and it is clear that there are trends of floral change within the Paleocene portion of the section. Palms, which are absent in the lower Hell Creek, become common in the upper 20 m of the formation and remain common through the lower 20 m of the Ludlow before disappearing again. Marine tongues of the Cannonball Member of the Fort Union Formation occur at 50 m (Boyce Tongue) and 80 m (Oyster Tongue), and it is likely that changes in the Ludlow bear some relationship to the transgression of the Cannonball seaway. The megafloral sample size is very small in the upper portions of the Ludlow and represents a future research agenda. A recently collected megaflora from the base of the Tongue River Member of the Fort Union (Williams, 1988) contains several species known from the Ludlow and suggests a certain amount of floral stasis at least through the middle Paleocene.

VEGETATION AND LANDSCAPE

Hell Creek landscapes

The Hell Creek is one of the most frequently reconstructed and painted of ancient landscapes. This is a direct reflection of the broad interest in latest Cretaceous dinosaurs such as *Triceratops, Edmontosaurus*, and *Tyrannosaurus rex* (Colagrande and Felder, 2000). In addition, interest in the K-T boundary event has generated many renderings of the terminal Cretaceous landscape, often focusing on the Hell Creek because of its charismatic megafauna and their catastrophic demise (Henderson, 2000). Despite its popularity, the Hell Creek is rarely rendered with any detailed attention to the primary data concerning its flora and vegetation. Notable exceptions to this trend are recent paintings by John Gurche of the HCIb flora (Relf, 2000) and by Brian Regal of a generalized Hell Creek forest (Fastovsky and Weishampel, 1996, Plate VIIb). As part of the exhibit "Prehistoric Journey" that opened in 1995, the Denver Museum of Nature & Science constructed a walk-through Hell Creek diorama, based on megafloral zone HCIII, that incorporates much of the primary data presented in this chapter (Figs. 7 and 8 of Johnson and Stucky, 1995; Leggitt and Johnson, 1999; Johnson, 1999).

It is clear that the Hell Creek was a forested environment. Petrified wood is rare in the Marmarth area, but rooted gley paleosols (Fastovsky and McSweeney, 1987; Retallack, 1994), limb imprints, occasional buried coalified trunks (Plate 1, Fig.

TABLE A1. MEGAFLORAL LOCALITIES FROM SLOPE AND BOWMAN COUNTIES, NORTH DAKOTA, AND HARDING COUNTY, SOUTH DAKOTA

	Locality number	Datum (m)	Error (m)	Boundary section	Cretaceous-Tertiary (m)	Morphotypes	Specimens (n=)	Sedimentary facies	Floral zone	USGS Quad	County	Latitude	Longitude
1	8408	80.00	10.0	na	80.0	4	9	Carb shale	FUI	3 V Crossing	Slope	46-30-57N	103-51-39W
2	8409	80.00	10.0	na	80.0	7	22	Carb shale	FUI	3 V Crossing	Slope	46-30-50N	103-51-34W
3	9903 (2307)	58.00	5.0	na	58.0	8	35	Channel la	FUI	Williams Lake	Slope	46-30-56N	103-52-57W
4	8403	50.00	10.0	na	50.0	12	87	Pond	FUI	Boyce Cr. West	Slope	46-28-23N	103-51-09W
5	8404	50.00	10.0	na	50.0	1	5	Channel	FUI	Boyce Cr. West	Slope	46-28-28N	103-50-35W
6	86110	50.00	10.0	na	50.0	20	652	Splay	FUI	Williams Lake	Slope	46-30-30N	103-52-34W
7	JND1	50.00	10.0	na	50.0	1	9	Pond	FUI	3 V Crossing	Slope	46-30-12N	103-52-26W
8	JND7	50.00	10.0	na	50.0	1	4	Pond	FUI	Boyce Cr. West	Slope	46-29-56N	103-51-17W
9	87143	47.50	0.0	87147	44.9	3	11	Pond	FUI	Marmarth SE	Slope	46-21-10N	103-52-15W
10	87142 (909)	45.20	0.0	87147	42.6	4	239	Pond	FUI	Marmarth SE	Slope	46-21-10N	103-52-15W
11	9819 (2201)	40.00	0.0	87147	37.4	4	17	Pond	FUI	Marmarth SE	Slope	46-21-07N	103-52-20W
12	88103l	39.00	0.0	9906	39.0	2	6	Splay/Levee	FUI	Mud Buttes	Bowman	46-01-07N	103-45-31W
13	9817 (2200)	33.50	0.5	87147	30.9	4	249	Pond	FUI	Marmarth SE	Slope	46-20-55N	103-52-28W
14	88103j	32.00	0.0	9906	32.0	1	1	Splay	FUI	Mud Buttes	Bowman	46-01-07N	103-45-31W
15	9816 (2199)	31.50	0.5	87147	28.9	1	10	Pond	FUI	Marmarth SE	Slope	46-20-55N	103-52-28W
16	9627 (1848)	30.00	5.0	86148	29.3	6	9	Pond	FUI	Marmarth SE	Slope	46-16-48N	103-48-48W
17	86113	30.00	12.0	na	30.0	5	27	Pond ls	FUI	Pretty Butte	Slope	46-26-50N	103-52-55W
18	88103i	28.80	0.0	9906	28.8	4	52	Pond vb	FUI	Mud Buttes	Bowman	46-01-07N	103-45-31W
19	87154	22.50	0.9	87147	19.9	6	30	Channel la	FUI	Marmarth SE	Slope	46-21-05N	103-52-20W
20	88120	19.10	1.0	86148	18.4	7	24	Channel la	FUI	Marmarth	Slope	46-19-00N	103-52-53W
21	88103g	18.00	0.5	9906	18.0	7	36	Splay	FUI	Mud Buttes	Bowman	46-01-07N	103-45-31W
22	9251 (563)	17.00	6.0	87147	14.4	4	34	Aban chan	FUI	Marmarth SE	Slope	46-21-27N	103-52-19W
23	9329 (905)	15.50	0.0	9328	14.5	4	10	Pond	FUI	Pretty Butte	Slope	46-23-32N	103-57-37W
24	86148q	15.50	0.2	86148	14.8	2	21	Carb shale	FUI	Marmarth	Slope	46-17-43N	103-52-42W
25	9142d (435)	14.10	0.1	9142	11.6	1	3	Pond	FUI	Pretty Butte	Slope	46-24-40N	103-59-47W
26	87130	11.50	0.0	86161	10.5	2	3	Pond vb	FUI	Pretty Butte	Slope	46-26-18N	103-54-13W
27	9250 (562)	11.00	2.0	87147	8.4	8	75	Ash bed	FUI	Marmarth SE	Slope	46-21-29N	103-52-23W
28	88122 (441)	10.00	2.0	87147	7.4	12	122	Channel la	FUI	Marmarth SE	Slope	46-20-19N	103-51-29W
29	86114	10.00	10.0	na	10.0	1	3	Pond	FUI	Pretty Butte	Slope	46-24-39N	103-54-13W
30	87140	10.00	0.0	9317	8.3	3	38	Carb shale	FUI	Cedar Hills	Bowman	46-08-59N	103-48-36W
31	86161g	10.00	0.0	86161	9.0	1	3	Pond vb	FUI	Pretty Butte	Slope	46-25-05N	103-58-35W
32	9316 (898)	8.20	0.1	9328	7.2	9	162	Pond vb	FUI	Pretty Butte	Slope	46-23-33N	103-59-43W
33	88100	7.60	0.0	9906	7.6	2	7	Pond vb	FUI	Mud Buttes	Bowman	46-01-09N	103-45-33W
34	9325 (902)	7.50	0.1	86148	6.8	5	8	Pond vb	FUI	Marmarth SE	Slope	46-19-15N	103-51-20W
35	9831 (2208)	7.00	0.0	9847	5.0	1	2	Pond vb	FUI	Pretty Butte	Slope	46-27-32N	104-59-03W
36	87150	6.50	0.0	87147	3.9	13	606	Carb shale	FUI	Marmarth	Slope	46-20-58N	103-53-45W
37	9143 (436)	5.00	0.5	9142	2.5	10	95	Aban chan	FUI	Pretty Butte	Slope	46-20-20N	103-59-39W
38	9650 (1858)	5.00	2.0	9653	2.4	6	26	Channel la	FUI	Badland Draw	Slope	46-27-52N	104-00-17W
39	9651 (1859)	5.00	2.0	9653	2.4	5	10	Channel la	FUI	Badland Draw	Slope	46-27-47N	104-00-27W
40	9855 (2216)	5.00	5.0	87147	2.4	1	1	Pond vb	FUI	Marmarth	Slope	46-19-48N	105-52-36W
41	9410 (1198)	5.00	5.0	9906	5.0	1	1	Channel x	FUI	Mud Buttes	Bowman	46-00-24N	103-45-30W
42	9829 (2206)	4.00	0.0	9847	2.0	9	53	Pond vb	FUI	Pretty Butte	Slope	46-27-34N	103-58-59W
43	9830 (2207)	4.00	0.0	9847	2.0	2	5	Pond vb	FUI	Pretty Butte	Slope	46-27-34N	103-58-59W
44	9832 (2209)	3.60	0.0	9847	1.6	8	47	Pond	FUI	Pretty Butte	Slope	46-27-34N	103-58-59W
45	9843 (2211)	3.30	0.3	9840	1.1	5	24	Pond	FUI	Pretty Butte	Slope	46-27-32N	103-58-29W
46	9841 (2210)	3.25	0.3	9840	1.0	4	6	Pond	FUI	Pretty Butte	Slope	46-27-32N	103-58-29W
47	86149	3.00	1.2	86148	2.3	4	124	Pond ls	FUI	Marmarth	Slope	46-17-43N	103-52-42W
48	9858 (2217)	2.50	0.2	9909	1.3	22	232	Pond	FUI	Badland Draw	Slope	46-26-19N	104-00-17W
49	86163	2.00	0.0	86161	1.0	1	2	Channel la	FUI	Badland Draw	Slope	46-25-56N	104-00-20W
50	86107 (427)	1.15	0.3	86161	0.2	28	640	Pond	FUI	Pretty Butte	Slope	46-25-07N	103-58-34W
51	9136 (429)	1.00	0.1	9906	1.0	7	41	Pond vb	FUI	Mud Buttes	Bowman	46-01-07N	103-45-48W
52	86162	1.00	0.0	86161	0.0	2	6	Pond	FUI	Badland Draw	Slope	46-26-08N	104-00-10W

(continued)

TABLE A1. MEGAFLORAL LOCALITIES FROM SLOPE AND BOWMAN COUNTIES, NORTH DAKOTA, AND HARDING COUNTY, SOUTH DAKOTA (continued)

	Locality number	Datum (m)	Error (m)	Boundary section	Cretaceous-Tertiary (m)	Morphotypes	Specimens (n =)	Sedimentary facies	Floral zone	USGS Quad	County	Latitude	Longitude
53	86155	0.70	0.0	86161	−0.3	3	4	Pond	FU0	Pretty Butte	Slope	46-26-20N	103-54-13W
54	9828 (2205)	0.56	0.4	9847	−1.4	8	59	Carb splay	FU0	Pretty Butte	Slope	46-27-34N	103-58-59W
55	9145 (438)	0.50	0.1	9144	−1.2	1	1	Carb shale	FU0	Badland Draw	Slope	46-27-39N	104-00-08W
56	9752 (2103)	0.34	0.1	9653	−2.3	2	2	Carb shale	FU0	Badland Draw	Slope	46-28-08N	104-00-45W
57	9844 (2212)	0.33	0.5	9847	−1.7	13	70	Carb splay	FU0	Pretty Butte	Slope	46-27-34N	103-58-59W
58	9754 (2100)	0.33	0.0	9847	−1.7	2	10	Carb shale	FU0	Pretty Butte	Slope	46-27-27N	103-59-40W
59	9262 (573)	−1.39	0.1	86148	−2.1	5	10	Channel la	HCIII	Marmarth	Slope	46-17-43N	103-52-43W
60	9249 (561)	−1.80	0.1	9248	−4.3	8	22	Splay	HCIII	Pretty Butte	Slope	46-23-23N	103-57-05W
61	88128 (440)	−2.20	0.3	87147	−4.8	21	139	Splay	HCIII	Marmarth SE	Slope	46-20-49N	103-52-28W
62	9904 (2308)	−2.70	0.1	9906	−2.7	8	15	Splay	HCIII	Mud Buttes	Bowman	46-01-08N	103-45-48W
63	9322 (900)	−2.80	0.0	86148	−3.6	42	485	Channel la	HCIII	Marmarth SE	Slope	46-19-19N	103-51-52W
64	88107	−3.10	0.0	9906	−3.1	2	7	Splay	HCIII	Mud Buttes	Bowman	46-01-03N	103-45-48W
65	9747 (2098)	−3.45	0.2	87147	−6.0	18	141	Pond	HCIII	Marmarth SE	Slope	46-20-49N	103-52-09W
66	9643 (1854)	−5.40	0.5	86148	−6.2	11	59	Splay	HCIII	Cedar Hills	Bowman	46-12-33N	103-47-47W
67	87151	−6.00	6.0	9906	−6.0	1	1	Floodplain soil	HCIII	Mud Buttes	Bowman	46-01-55N	103-46-40W
68	9146 (439)	−6.70	1.0	9909	−7.9	6	14	Channel la	HCIII	Badland Draw	Slope	46-25-57N	104-00-36W
69	87110	−7.70	0.0	86148	−8.5	25	188	Splay	HCIII	Marmarth SE	Slope	46-17-57N	103-51-45W
70	87152	−7.80	6.0	9317	−9.5	2	4	Channel x	HCIII	Cedar Hills	Bowman	46-07-30N	103-47-25W
71	9736 (2091)	−8.95	0.0	9735	−10.1	17	46	Floodplain soil	HCIII	Pretty Butte	Slope	46-22-44N	103-56-56W
72	88123 (574)	−9.00	0.0	9248	−11.5	17	49	Channel la	HCIII	Pretty Butte	Slope	46-23-23N	103-57-05W
73	9611 (1782)	−9.00	0.5	9612	−11.1	16	42	Channel x	HCIII	Badland Draw	Slope	46-27-38N	104-00-21W
74	9739 (2093)	−9.50	0.0	9738	−9.5	10	20	Channel x	HCIII	Mud Buttes	Bowman	46-05-32N	103-45-40W
75	9740 (2094)	−10.61	0.0	9906	−10.6	13	34	Channel la	HCIII	Mud Buttes	Bowman	46-01-31N	103-45-54W
76	88124	−11.80	0.0	87147	−14.4	2	6	Floodplain soil	HCIII	Marmarth SE	Slope	46-20-45N	103-52-29W
77	9729 (2089)	−11.85	0.0	9248	−14.3	1	1	Channel la	HCIII	Pretty Butte	Slope	46-23-32N	103-56-33W
78	88105	−12.40	0.0	9906	−12.4	3	4	Channel la	HCIII	Mud Buttes	Bowman	46-01-13N	103-45-52W
79	9907 (2306)	−12.40	1.0	9906	−12.4	10	14	Channel la	HCIII	Mud Buttes	Bowman	46-01-10N	103-45-43W
80	88112	−13.00	2.0	9317	−14.7	2	2	Channel	HCIII	Cedar Hills	Bowman	46-07-32N	103-47-28W
81	88127 (2099)	−14.48	0.0	87147	−17.1	13	94	Channel la	HCIII	Marmarth	Slope	46-20-46N	103-52-54W
82	88102 (428)	−15.00	1.0	9906	−15.0	86	975	Channel x	HCIII	Mud Buttes	Bowman	46-01-07N	103-45-52W
83	88104	−15.00	1.0	9906	−15.0	1	1	Channel x	HCIII	Mud Buttes	Bowman	46-01-08N	103-45-51W
84	87147a	−15.30	0.0	87147	−17.9	6	35	Channel x	HCIII	Marmarth	Slope	46-20-57N	103-53-40W
85	88117 (910)	−15.40	0.8	86161	−16.4	1	83	Channel la	HCIIb	Pretty Butte	Slope	46-25-05N	103-58-35W
86	9649a (1857)	−15.67	0.1	9612	−17.8	1	2	Channel x	HCIII	Badland Draw	Slope	46-27-35N	104-00-20W
87	86158	−16.00	0.0	86148	−16.8	14	43	Channel x	HCIII	Marmarth	Slope	46-17-42N	103-58-35W
88	9239 (517)	−16.20	0.5	9906	−16.2	20	242	Channel la	HCIII	McKenzie Butte	Harding	45-39-05N	103-37-28W
89	87153 (1250)	−17.00	1.0	9317	−18.7	19	231	Channel la	HCIII	Cedar Hills	Bowman	46-07-44N	103-47-30W
90	9331 (906)	−19.00	2.0	9317	−20.7	7	15	Channel x	HCIII	Cedar Hills	Bowman	46-07-53N	103-47-18W
91	9849 (2214)	−19.00	0.0	9612	−21.1	4	9	Channel la	HCIII	Badland Draw	Slope	46-27-07N	103-58-35W
92	87103	−19.50	0.0	86148	−20.3	2	46	Channel la	HCIIb	Marmarth SE	Slope	46-17-45N	103-52-16W
93	9645 (1855)	−20.00	5.0	86148	−20.8	12	121	Channel la	HCIII	Cedar Hills	Bowman	46-12-47N	103-47-28W
94	9610 (1781)	−21.30	1.0	9612	−23.4	27	159	Channel la	HCIII	Badland Draw	Slope	46-27-34N	104-00-19W
95	88121 (908)	−24.00	6.0	86148	−24.8	8	148	Aban chan	HCIIb	Marmarth	Slope	46-18-48N	103-52-55W
96	86115 (2215)	−25.50	0.0	86161	−26.5	8	136	Channel la	HCIIa	Pretty Butte	Slope	46-24-51N	103-58-20W
97	87102	−26.90	0.0	86148	−27.7	1	1	Channel x	HCIIb	Marmarth SE	Slope	46-17-45N	103-52-16W
98	88110	−28.50	0.0	9317	−30.2	4	26	Channel la	HCIIa	Cedar Hills	Bowman	46-08-58N	103-48-39W
99	9309 (896)	−28.70	2.5	9906	−28.7	1	3	Pond	HCIIa	Lahti Butte	Harding	45-39-05N	103-37-33W
100	9745 (2097)	−28.80	0.0	87147	−31.4	22	217	Channel la	HCIIa	Marmarth	Slope	46-20-47N	103-53-10W
101	9138 (431)	−29.00	1.0	9317	−30.7	4	35	Channel la	HCIIa	Cedar Hills	Bowman	46-08-56N	103-48-37W
102	9323 (901)	−30.00	5.0	86148	−30.8	6	13	Channel la	HCIIb	Marmarth SE	Slope	46-18-37N	103-51-42W
103	87131 (2096)	−30.60	0.0	87147	−33.2	9	110	Floodplain soil	HCIIa	Marmarth	Slope	46-20-56N	103-53-46W
104	9139 (432)	−31.50	1.0	9317	−33.2	5	7	Channel la	HCIIa	Cedar Hills	Bowman	46-08-56N	103-48-37W
105	9529 (1489)	−32.00	5.0	9317	−33.7	30	106	Channel la	HCIIa	Cedar Hills	Slope	46-09-02N	103-48-48W
106	9530 (1490)	−32.00	5.0	9317	−33.7	1	3	Channel la	HCIIa	Cedar Hills	Slope	46-09-03N	103-48-49W

TABLE A1. MEGAFLORAL LOCALITIES FROM SLOPE AND BOWMAN COUNTIES, NORTH DAKOTA, AND HARDING COUNTY, SOUTH DAKOTA (continued)

	Locality number	Datum (m)	Error (m)	Boundary section	Cretaceous-Tertiary (m)	Morphotypes	Specimens (n=)	Sedimentary facies	Floral zone	USGS Quad	County	Latitude	Longitude
107	87146	−33.30	0.0	87147	−35.9	5	42	Channel la	HCIIa	Marmarth	Slope	46-20-50N	103-53-09W
108	88111 (566)	−33.50	0.5	9317	−35.2	23	208	Floodplain soil	HCIIa	Cedar Hills	Bowman	46-08-59N	103-48-58W
109	9731 (2090)	−33.60	0.0	86148	−34.4	4	10	Channel la	HCIIa	Marmarth	Slope	46-19-16N	103-56-06W
110	86153 (568)	−33.80	0.0	86148	−34.6	35	332	Channel x	HCIIa	Marmarth	Slope	46-17-34N	103-52-33W
111	86103	−33.80	0.0	86148	−34.6	5	18	Channel x	HCIIa	Marmarth	Slope	46-17-45N	103-53-05W
112	9727 (2087)	−34.40	0.0	9248	−36.9	36	364	Channel la	HCIb	Pretty Butte	Slope	46-23-21N	103-56-30W
113	9307 (895)	−35.00	5.0	9317	−36.7	6	22	Pond	HCIIa	Cedar Hills	Bowman	46-08-08N	103-47-58W
114	87109	−35.60	6.0	86148	−36.4	5	16	Channel la	HCIIa	Marmarth SE	Slope	46-18-03N	103-51-55W
115	87105 (2088)	−36.00	3.0	9248	−38.5	8	80	Channel la	HCIIb	Pretty Butte	Slope	45-23-23N	103-56-29W
116	9811 (2195)	−40.00	3.0	9248	−42.5	1	6	Channel x	HCIIb	Pretty Butte	Slope	46-23-39N	103-56-18W
117	9812 (2196)	−40.00	3.0	9248	−42.5	1	3	Channel la	HCIIb	Pretty Butte	Slope	46-23-39N	103-56-18W
118	9813 (2197)	−40.00	3.0	9248	−42.5	4	18	Channel la	HCII?	Pretty Butte	Slope	46-23-44N	103-56-22W
119	87136 (565)	−42.00	0.0	9317	−43.7	3	118	Splay	HCIIa	Cedar Hills	Bowman	46-09-00N	103-49-00W
120	87129 (425)	−42.70	6.0	9248	−45.2	19	119	Channel x	HCIIa	Pretty Butte	Slope	46-23-53N	103-57-02W
121	88114	−43.40	0.5	86148	−44.2	5	9	Channel la	HCIIa	Marmarth	Slope	46-17-40N	103-52-48W
122	9737 (2092)	−45.77	0.0	9735	−46.9	10	75	Pond	HCIIa	Pretty Butte	Slope	46-22-45N	103-56-52W
123	9312 (897)	−47.00	2.0	86148	−47.8	16	103	Channel la	HCIIa	Marmarth	Slope	46-17-46N	103-53-16W
124	86102 (2095)	−47.00	0.0	86148	−47.8	21	277	Channel la	HCIIa	Marmarth	Slope	46-17-47N	103-53-04W
125	9536 (1492)	−50.00	20.0	9906	−50.0	7	12	Channel x	HCIa	Chimney Butte	Harding	45-52-24N	103-47-14W
126	86104	−52.00	0.0	86148	−52.8	3	28	Pond	HCIIa	Marmarth	Slope	46-17-49N	103-53-10W
127	9634 (1860)	−55.00	6.0	86148	−55.8	1	1	Channel la	HCIb	Marmarth	Bowman	46-17-00N	103-54-02W
128	86108	−55.00	6.0	86148	−55.8	12	113	Channel la	HCIb	Marmarth	Slope	46-17-46N	103-54-47W
129	Dynamite Coulee	−55.00	6.0	86148	−55.8	8	64	?	HCIb	Marmarth	Slope	46-17-00N	103-54-02W
130	86100 (567)	−56.00	0.0	86148	−56.8	39	1203	Channel la	HCIb	Marmarth	Slope	46-18-00N	103-53-30W
131	9535 (1491)	−56.00	0.0	86148	−56.8	21	260	Channel la	HCIb	Marmarth	Slope	46-18-00N	103-53-35W
132	9822 (2202)	−56.00	0.0	86148	−56.8	5	12	Channel la	HCIb	Marmarth	Slope	46-18-00N	103-53-30W
133	9824 (2203)	−56.00	0.0	86148	−56.8	19	304	Channel la	HCIb	Marmarth	Slope	46-17-59N	103-53-24W
134	9825 (2204)	−56.00	0.0	86148	−56.8	3	18	Channel la	HCIb	Marmarth	Slope	46-17-59N	103-53-26W
135	87134 (572)	−60.00	0.0	9317	−61.7	20	284	Channel la	HCIa	Cedar Hills	Bowman	46-08-27N	103-49-08W
136	9911 (2309)	−60.00	20.0	9317	−61.7	4	35	Channel la	HCIa?	Cedar Hills	Bowman	46-09-47N	103-49-54W
137	9257 (569)	−65.00	20.0	9906	−65.0	11	29	Channel la	HCIII	Doaks Butte	Harding	45-56-15N	103-56-36W
138	86105 (571)	−65.00	18.0	na	−65.0	29	360	Channel la	HCIa	Marmarth	Slope	46-19-38N	103-56-34W
139	86119	−65.00	18.0	na	−65.0	3	7	Channel la	HCIa	Marmarth	Slope	46-20-02N	103-57-21W
140	9910 (2310)	−65.00	20.0	9906	−65.0	8	14	Channel la	HCIa?	Bam Butte SE	Harding	45-17-28.5N	103-16-12.5W
141	87133	−74.00	4.0	9317	−75.7	2	8	Floodplain soil	HCIa	Cedar Hills	Bowman	46-08-12N	103-49-46W
142	87111 (430)	−75.00	18.0	na	−75.0	19	301	Channel la	HCIa	Marmarth	Slope	46-19-12N	103-57-53W
143	86142 (1852)	−75.00	18.0	na	−75.0	17	127	Channel la	HCIa	Marmarth	Slope	46-18-56N	104-57-35W
144	9733 (2102)	−75.00	5.0	na	−75.0	4	13	Volcanic ash	HCIa	Doaks Butte	Bowman	45-56-43N	103-56-27W
145	86143	−75.00	6.0	na	−75.0	1	4	Pond	HCIa	Marmarth	Slope	46-18-56N	103-57-26W
146	86144	−75.00	6.0	na	−75.0	1	6	Aban chan	HCIa	Marmarth	Slope	46-19-00N	103-57-32W
147	86145	−75.00	6.0	na	−75.0	4	5	Aban chan	HCIa	Marmarth	Slope	46-18-40N	103-57-10W
148	87112	−75.00	18.0	na	−75.0	8	38	Channel la	HCIa	Marmarth	Slope	49-19-12N	103-57-53W
149	87132	−76.80	4.0	9317	−78.5	1	2	Channel la	HCIa	Cedar Hills	Bowman	46-08-10N	103-51-24W
150	9630 (1850)	−80.00	10.0	na	−80.0	3	7	Channel la	HCIa	Waterhole Creek	Slope	46-18-52N	104-01-00W
151	9631 (1851)	−80.00	10.0	na	−80.0	9	48	Channel la	HCIa	Waterhole Creek	Slope	46-18-57N	104-01-06W
152	88139	−80.00	18.0	na	−80.0	1	35	Channel la	HCIa	Kid Creek	Bowman	46-13-42N	103-55-46W
153	88109	−83.00	4.0	9317	−84.7	4	38	Pond	HCIa	Cedar Hills	Bowman	46-08-10N	103-51-25W
154	88108	−85.10	4.0	9317	−86.8	2	19	Splay	HCIa	Cedar Hills	Bowman	46-08-11N	103-51-25W
155	9141 (434)	−88.50	5.0	na	−88.5	5	31	Channel la	HCIa	Doaks Butte	Bowman	45-56-50N	103-56-27W
156	9140 (433)	−91.00	5.0	na	−91.0	5	26	Channel la	HCIa	Doaks Butte	Bowman	45-56-53N	103-56-27W
157	87113	−102.00	6.0	na	−102.0	1	1	Channel x	HCIa	Marmarth	Bowman	46-15-51N	103-58-37W
158	88131	−103.00	6.0	na	−103.0	8	44	Channel x	HCIa	Cedar Hills	Bowman	46-08-19N	103-51-06W

Notes: Datum is the position relative to the Hell Creek–Fort Union formational contact. Error is the uncertainty of the locality with respect to the datum. Boundary section is the Cretaceous-Tertiary (K-T) boundary section used to correct the level of the locality to the level of the K-T boundary. K-T is the corrected position.

TABLE A2. STRATIGRAPHIC SECTIONS IN SLOPE AND BOWMAN COUNTIES, NORTH DAKOTA

Section Number	Section Name	K/T (cm)	± (cm)	FU (m)	HC (m)	Latitude (N)	Longitude (W)	USGS Quadrangle	County
Cannonball Creek									
9142	Mikey's Delite	248.5	1.5	14		46-28-10	103-59-43	Pretty Butte	Slope
9144	Cannonball Creek	170.5	7.5	2	0	46-27-37	104-00-08	Badland Draw	Slope
9612	Dean's High Dino	214	3	4	2	46-27-39	104-00-21	Badland Draw	Slope
9653	Terry's Ft Union Dino	260.5	2.5	3		46-28-01	104-00-28	Badland Draw	Slope
9734	Scorpion Stung Kristian	269	4	4	1	46-27-00	104-01-27	Badland Draw	Slope
9754	Facet Butte			69	0	46-27-32	103-59-00	Pretty Butte	Slope
9827	Facet Detail			35	9	46-27-32	103-59-02	Pretty Butte	Slope
9840	Vertical Doug	224.5	5.5	4	0	46-27-31	103-59-22	Pretty Butte	Slope
9847	New Facet bdy	199	1	4	0	46-27-34	103-58-58	Pretty Butte	Slope
9909	Badland Draw	122.5	7.5			46-26-25	104-00-10	Badland Draw	Slope
Pyramid Butte									
86161	Pyramid Butte III	96	0	39	22	46-25-03	103-58-33	Pretty Butte	Slope
Pretty Butte									
9328	Big Boundary	96	10	25	0	46-23-30	103-57-40	Pretty Butte	Slope
9735	Pretty Butte	112	26	76	49	46-22-43	103-56-51	Pretty Butte	Slope
9248	Pretty Butte North	248.5	179	7	1	46-23-04	103-57-06	Pretty Butte	Slope
River Section									
87144	Quereuxia Hill			38		46-21-09	103-52-20	Marmarth SE	Slope
87147	River Section	259.5	200	12	33	46-20-56	103-53-48	Marmarth	Slope
88129	Marshall's Turtle	45+		1	4	46-20-47	103-51-42	Marmarth SE	Slope
9750	No Leaf Minors	69+		9	4	46-20-49	103-52-12	Marmarth SE	Slope
9838	Lower Mt. Goat Lake			12		46-20-50	103-52-16	Marmarth SE	Slope
9839	Upper Mt. Goat Lake			38		46-20-54	103-52-27	Marmarth SE	Slope
9850	Behind Marshall's Turtle			19	1	46-20-49	103-52-34	Marmarth	Slope
9860	Battleship	73+		1	5	46-19-15	103-51-58	Marmarth SE	Slope
9913	PITA Flats	86+		21		46-20-19	103-51-36	Marmarth SE	Slope
0042	NDAS section	237+		3	5	46-20-00	103-51-59	Marmarth SE	Slope
Bobcat Butte									
86148	Bobcat Butte I	75	0	37	42	46-17-40	103-52-46	Marmarth	Slope
87148	Bobcat Butte II				26	46-17-56	103-53-29	Marmarth	Slope
Sunset Butte									
DEF081	Sunset Butte I				33	46-08-11	103-51-29	Cedar Hills	Bowman
DEF082	Sunset Butte II				33	46-08-10	103-49-41	Cedar Hills	Bowman
DEF083	Sunset Butte III				20	46-08-26	103-49-09	Cedar Hills	Bowman
DEF091	Sunset Butte IV				18	46-08-54	103-49-02	Cedar Hills	Bowman
DEF141	Sunset Butte V			36	32	46-08-59	103-48-37	Cedar Hills	Bowman
9317	Sunset Patch	170	103	13	12	46-06-11	103-47-14	Cedar Hills	Bowman
Mud Buttes									
88103	Mud Buttes I			45		46-01-06	103-45-36	Mud Buttes	Bowman
88106	Mud Buttes II			9	16	46-01-33	104-52-30	Mud Buttes	Bowman
9738	Torosaurus	0	0	3	11	46-05-33	103-45-42	Mud Buttes	Bowman
9906	Murphy Mud Buttes	0	0	2	2	46-01-11	103-45-44	Mud Buttes	Bowman
Doaks Butte									
DP 105	Doaks Butte				22	45-56-50	103-56-33	Doaks Butte	Bowman

K/T—Cretaceous-Tertiary.

4), compressed logs (Plate 1, Figs. 1 and 2), and ubiquitous tree leaves are ample evidence for forests. The nature of these forests is uncertain because so many of the species that populated them were subsequent victims of the K-T boundary extinction. The Hell Creek megaflora contains an unusually high percentage of lobed leaves. The Dean Street locality, 88102 (428), has 85 morphotypes, nearly 35% of which are lobed. This is a much higher number than is found in modern forests, where lobation occurs primarily on deciduous temperate trees or on rapidly growing secondary trees in tropical settings. These lobed leaves, combined with an extremely high angiosperm diversity, the presence of the extinct simple-leafed cycad *Nilssonia* (HC164), *Ginkgo* (HC114), aff. *Philodendron* (HC201), and several types of conifers, paint a picture of a forest that is relatively unlike any extant forest community (Fig. 7). Very few of the Hell Creek taxa represent extant genera. The abundance of large dinosaurian herbivores argues that they should have had some role in structuring the shape and density of the vegetation (Wing and Tiffney, 1987a, 1987b). The result is a forest that is similar in general appearance to a living mixed deciduous and evergreen broad-leafed forest, but differs from that general look by having palm trees and odd conifers such as *Elatides longifolia* (HC165)

TABLE A3. MEGAFLORAL MORPHOTYPES FROM THE HELL CREEK AND FORT UNION FORMATIONS

Morphotype number	Lowest Taxonomic Category	AFF	Org	M	Holomorphotype Museum #	Spec. n=	Locs. n=	Range
FU1	*Paranymphaea crassifolia*	DIC	L	E	DMNH-19733	108	21	T
FU2	*Quereuxia angulata*	DIC	L	NE	YPM-6052	543	15	K-T
FU3	*Metasequoia occidentalis*	CON	L	na	YPM-6053	190	10	K-T
FU4	*Glyptostrobus europaeus*	CON	L	na	YPM-6054	534	12	K-T
FU5	*Cercidiphyllum genetrix*	DIC	L	NE	YPM-6056	88	12	K-T
FU7	*"Populus" nebrascensis*	DIC	L	NE	YPM-6061	506	27	K-T
FU9	*Harmsia hydrocotyloidea*	DIC	L	NE	DMNH-7594	30	5	K-T
FU10	aff. *Dicotylophyllum anomalum*	DIC	L	NE	YPM-6064	6	1	T
FU12	Dicotyledones	DIC	L	NE	YPM-6391	10	2	T
FU14	Dicotyledones	DIC	L	E	YPM-6392	7	1	T
FU15	Dicotyledones	DIC	L	NE	YPM-6393	2	1	T
FU16	*Platanus raynoldsii*	DIC	L	NE	YPM-6394	209	34	K-T
FU17	Dicotyledones	DIC	L	E	YPM-6066	2	1	T
FU18	Dicotyledones	DIC	L	NE	YPM-6067	2	1	T
FU19	Pinnate Dilleniidae	DIC	L	E	YPM-6068	7	2	T
FU20	Dicotyledones	DIC	L	E	YPM-6069	4	1	T
FU21	Hymenophyllaceae	PTE	L	na	YPM-6070	4	1	T
FU23	Rosidae	DIC	L	NE	YPM-6071	13	4	T
FU24	Filicopsida	PTE	L	na	YPM-6073	1	1	T
FU25	Magnoliales	DIC	L	E	YPM-6074	3	3	K-T
FU26	*Cupressinocladus interruptus*	CON	L	na	DMNH-19734	178	15	K-T
FU29	*Dicotylophyllum anomalum*	DIC	L	NE	YPM-6076	373	21	K-T
FU31	*Cornophyllum newberryi*	DIC	L	NE	YPM-6078	185	28	K-T
FU35	Dicotyledones	DIC	L	NE	YPM-6080	19	2	K-T
FU36	*Equisetum*	SPE	R	na	YPM-6081	4	3	K-T
FU37	*Amesoneuron or Sabalites*	MON	L	na	YPM-6082	87	24	K-T
FU38	*Limnobiophyllum scutatum*	MON	L	na	YPM-6083	99	5	K-T
FU39	Dicotyledones	DIC	L	E	YPM-6084	2	1	T
FU40	Dicotyledones	DIC	L	NE	YPM-6085	3	1	T
FU42	Dicotyledones	DIC	L	E	YPM-6087	1	1	K-T
FU43	*Ziziphoides flabella*	DIC	L	NE	YPM-6088	439	15	K-T
FU44	*Nordenskioldia borealis*	DIC	F	na	YPM-6089	15	6	K-T
FU45	Bryophyta	BRY	L	na	YPM-6090	1	1	T
FU46	Filicopsida	PTE	L	na	YPM-6091	2	2	K-T
FU47	Dicotyledones	DIC	L	NE	YPM-6092	11	3	K-T
FU48	cf. *Wardiaphyllum daturaefolium*	DIC	L	NE	YPM-6093	3	1	T
FU49	Flacourticeae	DIC	L	NE	YPM-6094	5	2	T
FU51	*Woodwardia gravida*	PTE	L	na	YPM-6096	8	1	T
FU52	Dicotyledones	DIC	L	NE	YPM-6097	1	1	T
FU53	Palmate Dilleniidae	DIC	L	NE	YPM-6098	1	1	T
FU54	Trochodendrales	DIC	L	E	YPM-6099	1	1	T
FU55	Sapindales	DIC	L	NE	YPM-6100	2	1	T
FU56	Palmate Dilleniidae	DIC	L	NE	YPM-6101	3	1	T
FU57	Pinnate Dilleniidae	DIC	L	NE	YPM-6102	4	2	T
FU59	*Nyssidium arcticum*	DIC	F	na	YPM-6104	49	19	K-T
FU60	Dicotyledones	DIC	L	E	YPM-6105	5	2	K-T
FU61	*Penosphyllum cordatum*	DIC	L	E	YPM-6106	34	7	K-T
FU62	*Nelumbium montanum*	DIC	L	na	DMNH-19263	27	9	K-T
FU63	Platanaceae	DIC	L	NE	YPM-6108	3	2	K-T
FU64	Dicotyledones	DIC	L	E	YPM-6109	1	1	T
FU65	Dicotyledones	DIC	L	NE	YPM-6110	15	1	T
FU69	Dicotyledones	DIC	L	NE	YPM-6114	1	1	T
FU70	Dicotyledones	DIC	L	E	YPM-6115	1	1	T
FU72	Dicotyledones	DIC	L	NE	YPM-6117	2	1	T
FU73	Dicotyledones	DIC	L	E	YPM-6119	2	1	T
FU74	Flacourtiaceae	DIC	L	NE	YPM-6120	3	2	T
FU75	Dicotyledones	DIC	L	NE	YPM-6121	12	1	T
FU76	*"Ampelopsis" acerifolia*	DIC	L	NE	YPM-6122	6	2	T
FU77	Dicotyledones	DIC	L	E	YPM-6123	1	1	T
FU78	Magnoliidae	DIC	L	E	YPM-6124	2	2	T
FU79	*Corylus* sp.	DIC	L	NE	DMNH-19732	1	1	T
FU80	*Ceratophyllum*	DIC	L	na	DMNH-19736	4	1	T
FU81	Dicotyledones	DIC	L	E	DMNH-19735	32	6	K-T
FU82	*Amersinia* sp.	DIC	F	na	DMNH-19912	21	5	T
FU83	Filicales	PTE	L	na	DMNH-19896	27	2	T
FU84	Filicales	PTE	L	na	DMNH-19897	1	1	T
FU85	*Nelumbo* rhizome	MON	R	na	DMNH-19898	12	1	T

(*continued*)

TABLE A3. MEGAFLORAL MORPHOTYPES FROM THE HELL CREEK AND FORT UNION FORMATIONS (continued)

Morphotype number	Lowest Taxonomic Category	AFF	Org	M	Holomorphotype Museum #	Spec. n=	Locs. n=	Range
FU86	Dicotyledones	DIC	L	NE	DMNH-19899	1	1	T
FU87	Dicotyledones	DIC	L	E	DMNH-19900	2	2	T
FU88	Platanaceae	DIC	F	na	DMNH-19901	2	2	T
FU89	Dicotyledones	DIC	L	NE	DMNH-19902	2	1	T
FU90	Filicales	PTE	L	na	DMNH-19903	1	1	T
FU91	Platanaceae	DIC	F	na	DMNH-19904	1	1	T
FU92	Dicotyledones	DIC	F	na	DMNH-19905	5	1	T
FU93	Dicotyledones	DIC	F	na	DMNH-19906	1	1	T
FU94	Dicotyledones	DIC	L	E	DMNH-19907	2	1	T
FU95	Dicotyledones	DIC	L	E	DMNH-19908	2	1	K
FU96	Dicotyledones	DIC	L	NE	DMNH-19909	1	1	K
FU97	Dicotyledones	DIC	L	NE	DMNH-19910	1	1	T
FU98	Dicotyledones	DIC	F	na	DMNH-19911	1	1	T
HC1	*Platanites marginata* lat. leaflet	DIC	L	NE	YPM-6125	49	14	K
HC2	*Grewiopsis saportana*	DIC	L	NE	YPM-6126	134	17	K
HC3	Dicotyledones	DIC	L	E	YPM-6127	10	3	K
HC4	*"Ficus" planicostata*	DIC	L	E	DMNH-7333	66	13	K
HC5	Dicotyledones	DIC	L	E	YPM-6129	1	1	K
HC7	*Palaeoaster inquirenda*	DIC	F	na	YPM-6131	78	4	K
HC8	*Filicopsida*	PTE	L	na	YPM-6132	1	1	K
HC9	*Glyptostrobus* sp.#2	CON	L	na	YPM-6133	22	4	K
HC11	Dicotyledones	DIC	L	NE	YPM-6135	1	1	K
HC12	*Porosia verrucosa*	MON	L	na	YPM-6136	3	2	K-T
HC14	*"Vitis" stantonii*	DIC	L	NE	YPM-6138	234	15	K
HC15	Dicotyledones	DIC	L	NE	YPM-6139	1	1	K
HC17	Dicotyledones	DIC	L	NE	YPM-6141	1	1	K
HC18	Urticaceae	DIC	L	NE	YPM-6142	16	1	K
HC20	Rosidae	DIC	L	NE	YPM-6144	54	1	K
HC21	Dicotyledones	DIC	L	E	YPM-6145	12	1	K
HC22	Hamamelididae	DIC	L	NE	YPM-6146	1	1	K
HC23	Dicotyledones	DIC	L	E	YPM-6147	2	1	K
HC24	Dicotyledones	DIC	L	NE	YPM-6148	1	1	K
HC25	Dicotyledones	DIC	L	NE	YPM-6149	1	1	K
HC27	Dicotyledones	DIC	L	E	YPM-6340	23	8	K
HC28	Dicotyledones	DIC	L	E	YPM-6152	2	2	K
HC29	Dicotyledones	DIC	L	NE	YPM-6153	1	1	K
HC31	Dicotyledones	DIC	L	E	YPM-6155	2	1	K
HC32	Laurales?	DIC	L	E	DMNH-8428	33	6	K
HC34	Dicotyledones	DIC	L	E	YPM-6157	4	1	K
HC35	*Metasequoia* sp.#2	CON	L	na	YPM-6158	58	8	K
HC36	*Ulmaceae*	DIC	L	NE	DMNH-19252	65	15	K
HC40	Dicotyledones	DIC	L	NE	YPM-6163	2	1	K
HC43	Magnoliidae	DIC	L	E	DMNH-7234	65	14	K
HC44	cf. *"Dryophyllum" tenneseensis*	DIC	L	NE	YPM-6167	99	14	K
HC47	Rosidae	DIC	L	NE	YPM-6169	6	3	K
HC49	*"Dryophyllum" subfalcatum*	DIC	L	NE	YPM-6170	1003	51	K
HC52	Platanoids	DIC	L	NE	YPM-6172	10	4	K
HC54	*Carpites ulmiformis*	DIC	S	na	YPM-6173	9	6	K
HC55	Dicotyledones	DIC	L	NE	YPM-6174	3	2	K
HC56	Dicotyledones	DIC	L	NE	YPM-6176	2	2	K
HC57	*Erlingdorfia montana*	DIC	L	NE	YPM-6177	320	30	K
HC58	*Laurophyllum wardiana*	DIC	L	E	YPM-6178	2	2	K
HC60	*Araliaephyllum polevoi*	DIC	L	E	DMNH-19233	13	2	K
HC62	*"Rhamnus" cleburni*	DIC	L	NE	DMNH-19262	177	18	K
HC63	cf. *Cissites acerifolia*	DIC	L	NE?	DMNH-8458	29	5	K
HC64	*Erlingdorfia montana* lat. leaflet	DIC	L	NE?	DMNH-6253	131	15	K
HC65	Dicotyledones	DIC	L	E	YPM-6187	7	4	K
HC66	Magnoliidae	DIC	L	E	YPM-6188	53	13	K
HC67	Dicotyledones	DIC	L	E	YPM-6189	1	1	K
HC68	Dicotyledones	DIC	L	E	YPM-6190	2	1	K
HC70	*Sequoia* sp.	CON	L	na	YPM-6191	52	12	K
HC71	*Taxodium olrikii*	CON	L	na	YPM-6192	92	14	K-T
HC72	Palmate Dilleniidae	DIC	L	E	YPM-6193	2	1	K
HC73	Trochodendrales	DIC	L	NE	YPM-6194	69	10	K
HC74	Dicotyledones	DIC	L	E	YPM-6195	1	1	K
HC75	Dicotyledones	DIC	L	E	YPM-6196	2	1	K
HC76	Dicotyledones	DIC	L	NE	YPM-6197	4	2	K

(*continued*)

TABLE A3. MEGAFLORAL MORPHOTYPES FROM THE HELL CREEK AND FORT UNION FORMATIONS (continued)

Morphotype number	Lowest Taxonomic Category	AFF	Org	M	Holomorphotype Museum #	Spec. n=	Locs. n=	Range
HC77	*"Pistia" corrugata*	MON	L	na	DMNH-8656B	61	9	K
HC78	Filicopsida	PTE	L	na	YPM-6199	2	1	K
HC80	Rosaceae	DIC	L	NE	YPM-6201	217	3	K
HC81	Cannabaceae	DIC	L	NE	YPM-6205	365	4	K
HC83	Araucariaceae?	CON	S	na	YPM-6206	33	5	K
HC84	Laurales	DIC	L	E	YPM-6207	68	3	K
HC85	Dicotyledones	DIC	L	NE	YPM-6211	1	1	K
HC86	*Leepierceia preartocarpoides*	DIC	L	NE	DMNH-6359	204	23	K
HC87	Laurales	DIC	L	E	YPM-6213	3	2	K
HC88	Chloranthaceae	DIC	L	E	YPM-6214	10	1	K
HC90	Rosidae?	DIC	L	NE	DMNH-19050	52	5	K
HC91	Dicotyledones	DIC	L	NE	YPM-6219	1	1	K
HC92	*"Cinnamomum" lineafolia*	DIC	L	E	YPM-6220	259	13	K
HC93	Urticaceae	DIC	L	NE	DMNH-19229	34	3	K
HC94	*"Celastrus" taurenensis*	DIC	L	NE	YPM-6224	151	17	K
HC96	Dicotyledones	DIC	L	NE	YPM-6225	2	1	K
HC98	Platanaceae	DIC	L	NE	DMNH-13004	19	4	K
HC99	Platanaceae	DIC	L	E	DMNH-18925	34	8	K
HC100	Fagaceae	DIC	L	NE	YPM-6229	4	3	K
HC103	*Trochodendroides nebrascensis*	DIC	L	NE	YPM-6232	113	26	K
HC105	*Marmarthia trivialis*	DIC	L	NE	YPM-6233	282	26	K
HC106	*Platanites marginata*	DIC	L	NE	YPM-6234	197	21	K
HC108	aff. *"Vitis" stantonii*	DIC	L	NE	YPM-6236	20	8	K
HC111	*Paranymphaea hastata*	DIC	L	NE	YPM-6239	92	17	K
HC114	*Ginkgo adiantoides*	GIN	L	na	DMNH-18927	84	7	K
HC115	Filicopsida	PTE	L	na	YPM-6243	2	1	K
HC122	Laurales	DIC	L	NE	YPM-6251	3	1	K
HC123	*Marchantia pealii*	BRY	L	na	YPM-6253	33	8	K-T
HC124	aff. *Cercidiphyllum*	DIC	L	NE	YPM-6254	41	6	K
HC126	Dicotyledones	DIC	L	E	YPM-6256	1	1	K
HC127	Pinnate Dilleniidae	DIC	L	E	YPM-6257	2	1	K
HC128	Dicotyledones	DIC	L	E?	YPM-6258	1	1	K
HC129	*Hydropteris pinnata*	PTE	L	na	YPM-6259	5	4	K
HC130	Dicotyledones	DIC	S	na	YPM-6260	157	1	K
HC131	Dicotyledones	DIC	L	NE	YPM-6261	41	2	K
HC132	Rosidae	DIC	L	NE	YPM-6335	22	1	K
HC135	Dicotyledones	DIC	L	NE	DMNH-18883	104	3	K
HC136	Dicotyledones	DIC	L	NE	YPM-6266	2	2	K
HC137	*Fokieniopsis catenulata*	CON	L	na	DMNH-19428	53	11	K-T
HC138	Rosidae	DIC	L	NE	YPM-6268	1	1	K
HC139	Hamamelididae	DIC	F	na	YPM-6269	38	2	K
HC140	Platanaceae	DIC	L	NE	YPM-6270	28	3	K-T
HC141	Dicotyledones	DIC	L	E	YPM-6271	1	1	K
HC142	Rosidae	DIC	L	NE	YPM-6272	1	1	K
HC147	Dicotyledones	DIC	L	NE	YPM-6276	1	1	K
HC148	Dicotyledones	DIC	L	E	YPM-6277	1	1	K
HC150	Dicotyledones	DIC	L		DMNH-7137	3	1	K
HC151	Platanaceae	DIC	F	na	YPM-6280	23	8	K
HC152	Platanaceae	DIC	L	NE	YPM-6281	3	1	K
HC154	Dicotyledones	DIC	L	E	YPM-6284	3	1	K
HC155	Dicotyledones	DIC	L	E	YPM-6285	1	1	K
HC158	Dicotyledones	DIC	L	NE	YPM-6336	1	1	K
HC159	Dicotyledones	DIC	L	NE	YPM-6290	1	1	K
HC161	Dicotyledones	DIC	L	NE	YPM-6292	1	1	K
HC162	*Marmarthia pearsonii*	DIC	L	E	DMNH-7713	249	16	K
HC163	Laurales	DIC	L	E	YPM-6294	9	4	K
HC164	*Nilssonia yukonensis*	CYC	L	na	YPM-6295	16	5	K
HC165	*Elatides longifolia*	CON	L	na	DMNH-19255	88	9	K
HC166	*Liriodendrites bradacii*	DIC	L	E	YPM-6297	156	10	K
HC168	*Laurales*	DIC	L	E	YPM-6299	1	1	K
HC169	Dicotyledones	DIC	L	NE	DMNH-19893	25	4	K
HC171	Dicotyledones	DIC	L	E	YPM-6302	1	1	K
HC172	Dicotyledones	DIC	L	NE	YPM-6303	1	1	K
HC174	*"Ziziphus" fibrillosus*	DIC	L	E	DMNH-7463	55	8	K
HC175	*Cissites lobata*	DIC	L	NE	DMNH-6853	9	5	K
HC176	Dicotyledones	DIC	L	E	YPM-6307	1	1	K
HC179	*"Artocarpus" lessigiana*	DIC	L	E	DMNH-13650	62	8	K

(*continued*)

TABLE A3. MEGAFLORAL MORPHOTYPES FROM THE HELL CREEK AND FORT UNION FORMATIONS (continued)

Morphotype number	Lowest Taxonomic Category	AFF	Org	M	Holomorphotype Museum #	Spec. n=	Locs. n=	Range
HC180	*Cissites puilosokensis*	DIC	L	NE	DMNH-6240	16	1	K
HC182	Dicotyledones	DIC	L	E	YPM-6314	5	1	K
HC183	cf. Saururaceae	DIC	L	E	YPM-6315	1	1	K
HC185	Dicotyledones	DIC	L	NE	YPM-6317	1	1	K
HC186	*Cissites insignis*	DIC	L	E	YPM-6318	1	1	K
HC187	Dicotyledones	DIC	L	NE	YPM-6319	3	2	K
HC188	Berberidaceae	DIC	L	NE	YPM-6320	94	1	K
HC189	aff. *"Vitis" stantonii*	DIC	L	E	YPM-6321	14	3	K
HC190	Filicopsida	PTE	L	na	YPM-6322	10	2	K
HC193	Dicotyledones	DIC	L	E	YPM-6325	1	1	K
HC194	Rosidae	DIC	L	NE	YPM-6326	1	1	K
HC196	*Zingiberopsis attenuata*	MON	L	na	YPM-6328	11	3	K
HC197	Platanaceae	DIC	L	E	YPM-6329	2	1	K
HC198	Platanaceae	DIC	L	E	YPM-6330	1	1	K
HC199	Laurales	DIC	L	E	DMNH-13634	93	2	K
HC200	*Bisonia niemii*	DIC	L	E	DMNH-16694	64	5	K
HC201	aff. *Philodendron*	MON	L	E	DMNH-6711	1	1	K
HC202	*Nelumbo*	MON	L	E	DMNH-6869	4	1	K
HC203	Magnoliaceae	DIC	L	E	DMNH-6770	8	1	K
HC204	Dicotyledones	DIC	L	E	DMNH-6862	2	2	K
HC207	Dicotyledones	DIC	L	E	DMNH-6863	1	1	K
HC208	aff. *Liriodendron*	DIC	L	E	DMNH-18863	8	1	K
HC209	Platanaceae	DIC	L	NE	DMNH-13003	20	7	K
HC210	Dicotyledones	DIC	L	?	DMNH-7133	1	1	K
HC211	Rosidae	DIC	L	NE	DMNH-19037	33	1	K
HC212	*Cercidiphyllum ellipticum*	DIC	L	NE	DMNH-19079	162	17	K-T
HC213	Platanaceae	DIC	L	NE	DMNH-19057	2	1	K
HC214	Dicotyledones	DIC	L	NE	DMNH-19084	1	1	K
HC215	Filicopsida (fertile frond)	PTE	So	na	DMNH-19088	1	1	K
HC216	Dicotyledones	DIC	L	NE	DMNH-19085	1	1	K
HC217	Dicotyledones	DIC	L	NE	DMNH-19089	2	2	K
HC218	Urticales	DIC	S	na	DMNH-19090	1	1	K
HC219	Coniferales	CON	S	na	DMNH-19086	1	1	K
HC220	Coniferales	CON	S	na	DMNH-19089	1	1	K
HC222	Dicotyledones	DIC	F	na	DMNH-19103	2	1	K
HC223	Dicotyledones	DIC	L	E	DMNH-19141	1	1	K
HC224	Rosidae	DIC	L	NE	DMNH-19207	27	2	K
HC225	Cannabaceae	DIC	L	NE	DMNH-13200	10	2	K
HC226	Ranunculaceae	DIC	L	NE	DMNH-19201	26	4	K
HC227	Ranunculidae	DIC	L	?	DMNH-19174	2	2	K
HC228	Urticaceae	DIC	L	NE	DMNH-19202	13	1	K
HC229	Cercidiphyllales	DIC	L	NE	DMNH-19204	67	1	K
HC230	Dicotyledones	DIC	L	NE	DMNH-19205	1	1	K
HC231	Dicotyledones	DIC	L	NE	DMNH-19206	2	2	K-T
HC232	Rosidae	DIC	L	NE	DMNH-19209	1	1	K
HC233	Rosidae	DIC	L	NE	DMNH-19210	1	1	K
HC234	Rosidae	DIC	L	NE	DMNH-19211	1	1	K
HC236	Dicotyledones	DIC	F	na	DMNH-11040	1	1	K
HC237	Dicotyledones	DIC	L	E?	DMNH-19213	1	1	K
HC238	Dicotyledones	DIC	L	E	DMNH-19215	1	1	K
HC239	Dicotyledones	DIC	L	E	DMNH-19216	1	1	K
HC240	Coniferales	CON	S	na	DMNH-19175	2	2	K
HC241	Filicopsida	PTE	L	na	DMNH-19227	70	4	K
HC242	Dicotyledones	DIC	L	E	DMNH-19220	7	3	K
HC243	Cannabaceae	DIC	L	NE	DMNH-19217	93	1	K
HC244	Dicotyledones	DIC	L	E	DMNH-19219	4	1	K
HC245	Dicotyledones	DIC	L	NE	DMNH-19223	1	1	K
HC246	Rosidae	DIC	L	NE	DMNH-19221	13	1	K
HC247	Dicotyledones	DIC	L	E	DMNH-19246	3	2	K
HC248	Filicopsida	PTE	L	na	DMNH-19234	16	3	K
HC249	Dicotyledones	DIC	L	E	DMNH-19236	4	1	K
HC250	Dicotyledones	DIC	L	E	DMNH-19237	2	1	K
HC251	Dicotyledones	DIC	L	E	DMNH-19228	1	1	K
HC252	Dicotyledones	DIC	F	na	DMNH-19222	30	1	K
HC253	Dicotyledones	DIC	L	E	DMNH-19247	2	2	K
HC254	Dicotyledones	DIC	L	E	DMNH-19238	60	5	K
HC255	Dicotyledones	DIC	L	E	DMNH-19235	1	1	K

(*continued*)

TABLE A3. MEGAFLORAL MORPHOTYPES FROM THE HELL CREEK AND FORT UNION FORMATIONS (continued)

Morphotype number	Lowest Taxonomic Category	AFF	Org	M	Holomorphotype Museum #	Spec. n=	Locs. n=	Range
HC256	*Ficus ceratops*	DIC	F	na	DMNH-19259	7	1	K
HC257	*Dammarites* sp.	CON	C	na	DMNH-19258	19	2	K
HC258	Monocotyledones	MON	L	na	DMNH-19256	19	1	K
HC260	Dicotyledones	DIC	L	E	DMNH-19714	1	1	K
HC261	Rhamnaceae	DIC	L	E	DMNH-19248	27	6	K
HC262	Dicotyledones	DIC	L	E	DMNH-19261	4	3	K
HC263	Dicotyledones	DIC	L	E	DMNH-19240	11	4	K
HC264	Dicotyledones	DIC	L	E	DMNH-19244	25	2	K
HC265	Laurales	DIC	L	NE	DMNH-7629	19	3	K
HC266	Dicotyledones	DIC	L	E	DMNH-19250	41	1	K
HC267	Dicotyledones	DIC	L	E	DMNH-19717	9	1	K
HC268	Dicotyledones	DIC	L	NE	DMNH-19725	28	4	K
HC269	Dicotyledones	DIC	L	NE	DMNH-19724	1	1	K
HC270	Dicotyledones	DIC	L	E	DMNH-19718	1	1	K
HC271	Dicotyledones	DIC	L	E	DMNH-19226	28	1	K
HC272	Dicotyledones	DIC	L	E	DMNH-19715	3	1	K
HC273	Dicotyledones	DIC	L	E	DMNH-19723	1	1	K
HC274	Dicotyledones	DIC	L	NE	DMNH-19709	1	1	K
HC275	Filicales	PTE	L	na	DMNH-19712	10	2	K
HC276	Dicotyledones	DIC	F	na	DMNH-19730	2	2	K
HC277	Dicotyledones	DIC	L	NE	DMNH-19245	1	1	K
HC278	Dicotyledones	DIC	L	E	DMNH-19242	1	1	K
HC279	Dicotyledones	DIC	L	E	DMNH-19251	7	1	K
HC280	Dicotyledones	DIC	L	E	DMNH-19243	43	5	K
HC281	Dicotyledones	DIC	L	E	DMNH-19708	1	1	K
HC282	Dicotyledones	DIC	L	NE	DMNH-19232	1	1	K
HC283	Laurales	DIC	L	E	DMNH-19253	6	4	K
HC284	Platanaceae	DIC	L	NE	DMNH-7331	1	1	K
HC285	Dicotyledones	DIC	L	E	DMNH-19722	2	2	K
HC286	Dicotyledones	DIC	L	E	DMNH-19720	5	3	K
HC287	Dicotyledones	DIC	L	NE	DMNH-19719	2	2	K
HC288	Rosidae	DIC	L	NE	DMNH-19711	1	1	K
HC289	Dicotyledones	DIC	L	NE	DMNH-19716	1	1	K
HC290	Dicotyledones	DIC	L	E	DMNH-19249	29	2	K
HC291	Dicotyledones	DIC	L	E	DMNH-19727	6	2	K
HC292	Laurales	DIC	L	E	DMNH-7406	5	2	K
HC293	Dicotyledones	DIC	L	E	DMNH-13542	5	4	K
HC294	Dicotyledones	DIC	L	E	DMNH-7700	1	1	K
HC295	Dicotyledones	DIC	L	E	DMNH-13541	2	2	K
HC296	Dicotyledones	DIC	L	NE	DMNH-19737	1	1	K
HC297	Dicotyledones	DIC	L	NE	DMNH-19738	2	1	K
HC298	Dicotyledones	DIC	L	NE	DMNH-19739	1	1	K
HC299	Araucariaceae	CON	L	na	DMNH-19740	1	1	K
HC300	Dicotyledones	DIC	L	NE	DMNH-19741	7	2	K
HC301	Dicotyledones	DIC	L	E	DMNH-19742	1	1	K
HC302	Dicotyledones	DIC	L	NE	DMNH-19743	1	1	K
HC303	Dicotyledones	DIC	L	NE	DMNH-19768	1	1	K
HC304	Coniferales	CON	C	na	DMNH-19769	1	1	K
HC305	Dicotyledones	DIC	F	na	DMNH-19770	2	1	K
HC306	Dicotyledones	DIC	L	E	DMNH-19772	1	1	K
HC307	Dicotyledones	DIC	L	NE	DMNH-19771	1	1	K
HC308	Dicotyledones	DIC	L	E	DMNH-19773	1	1	K
HC309	Dicotyledones	DIC	L	E	DMNH-19774	1	1	K
HC310	Dicotyledones	DIC	L	E	DMNH-19775	1	1	K
HC311	Indeterminate	IND	F	na	DMNH-19860	1	1	K
HC312	*Salvinia* sp.	PTE	L	na	DMNH-19861	1	1	K
HC313	Coniferales	CON	C	na	DMNH-19862	1	1	K
HC314	Dicotyledones	DIC	L	E	DMNH-19863	1	1	K
HC315	Dicotyledones	DIC	L	E	DMNH-19864	6	2	K
HC316	Dicotyledones	DIC	L	E	DMNH-19865	1	1	K
HC317	Dicotyledones	DIC	L	E	DMNH-19866	2	1	K
HC318	Dicotyledones	DIC	L	NE	DMNH-19867	2	2	K
HC319	Filicales	PTE	L	na	DMNH-19868	1	1	K
HC320	Dicotyledones	DIC	L	NE	DMNH-19869	1	1	K
HC321	Dicotyledones	DIC	L	E	DMNH-19870	23	3	K
HC323	Dicotyledones	DIC	L	NE	DMNH-19872	1	1	K
HC324	Dicotyledones	DIC	L	E	DMNH-19873	2	1	K

(*continued*)

TABLE A3. MEGAFLORAL MORPHOTYPES FROM THE HELL CREEK AND FORT UNION FORMATIONS (continued)

Morphotype number	Lowest Taxonomic Category	AFF	Org	M	Holomorphotype Museum #	Spec. n=	Locs. n=	Range
HC325	Dicotyledones	DIC	L	E	DMNH-19874	1	1	K
HC326	Dicotyledones	DIC	L	E	DMNH-19875	1	1	K
HC327	Dicotyledones	DIC	L	E	DMNH-19876	1	1	K
HC328	Dicotyledones	DIC	F	na	DMNH-19877	1	1	K
HC329	Platanaceae	DIC	L	NE	DMNH-7304	22	2	K
HC330	Platanaceae	DIC	L	NE	DMNH-7271	6	1	K
HC331	Dicotyledones	DIC	L	E	DMNH-19878	1	1	K
HC332	Dicotyledones	DIC	L	NE	DMNH-7273	1	1	K
HC333	Dicotyledones	DIC	L	E	DMNH-7715	1	1	K
HC334	Dicotyledones	DIC	L	E	DMNH-19879	1	1	K
HC335	Dicotyledones	DIC	L	E	DMNH-7185	1	1	K
HC336	Dicotyledones	DIC	L	E	DMNH-8701	1	1	K
HC337	Dicotyledones	DIC	L	NE	DMNH-18709	2	2	K
HC338	Dicotyledones	DIC	L	E	DMNH-19880	1	1	K
HC339	Taxodiaceae	CON	C	na	DMNH-11044	1	1	K
HC340	Dicotyledones	DIC	L	E	DMNH-18712	3	3	K
HC341	Monocotyledones	MON	L	na	DMNH-19881	1	1	K
HC342	Dicotyledones	DIC	L	E	DMNH-19882	1	1	K
HC343	Dicotyledones	DIC	L	E	DMNH-19883	6	1	K
HC344	*"Myrica" torreyi*	DIC	L	NE	DMNH-7552	2	1	K
HC345	Dicotyledones	DIC	L	E	DMNH-19884	1	1	K
HC346	Dicotyledones	DIC	L	E	DMNH-19885	1	1	K
HC347	Dicotyledones	DIC	L	E	DMNH-7556	1	1	K
HC348	Dicotyledones	DIC	L	E	DMNH-7414	1	1	K
HC349	Dicotyledones	DIC	L	E	DMNH-7405	1	1	K
HC350	Dicotyledones	DIC	L	E	DMNH-7460	1	1	K
HC351	Dicotyledones	DIC	L	E	DMNH-19886	1	1	K
HC352	Dicotyledones	DIC	L	E	DMNH-7579	1	1	K
HC353	Dicotyledones	DIC	L	E	DMNH-7592	1	1	K
HC354	Dilleniidae	DIC	L	E	DMNH-19887	4	1	K
HC355	Cercidiphyllaceae	DIC	L	NE	DMNH-19888	1	1	K
HC356	Dicotyledones	DIC	L	NE	DMNH-7476	1	1	K
HC357	Dicotyledones	DIC	L	NE	DMNH-7421	1	1	K
HC358	Dicotyledones	DIC	L	E	DMNH-7475	1	1	K
HC359	Dicotyledones	DIC	L	E	DMNH-19889	1	1	K
HC360	Dicotyledones	DIC	L	NE	DMNH-19890	1	1	K
HC362	Dicotyledones	DIC	L	E	DMNH-7531	1	1	K
HC363	Dicotyledones	DIC	L	NE?	DMNH-6897	2	1	K
HC364	Dicotyledones	DIC	L	NE	DMNH-19891	1	1	K
HC365	Dicotyledones	DIC	L	E	DMNH-18762	1	1	K
HC366	Dicotyledones	DIC	L	E	DMNH-18760	1	1	K
HC367	Dicotyledones	DIC	L	NE	DMNH-13183	1	1	K
HC368	Dicotyledones	DIC	L	E	DMNH-19892	1	1	K
HC369	Dicotyledones	DIC	L	E	DMNH-19894	2	1	K
HC370	Dicotyledones	DIC	L	E	DMNH-19217	1	1	K

Notes: AFF—affinity in larger taxonomic category: DIC—dicotyledones; MON—monocotyledones; CON—coniferales; PTE—pteridophytes; SPE—sphenophytes; BRY—bryophytes; GIN—ginkgophyte; CYC—cycadophyte; IND—indeterminate. Org—organ type: L—leaf; R—root or rhizome; F—flower or fruit; S—seed; So—sori; C—cone. M—dicotyledon angiosperm leaf margin type: E—entire; NE—nonentire. Spec.—number of voucher specimens at DMNH and YPM. Locs.—numbers of localities where the morphotype is found. Range—stratigraphic distribution of the morphotype: T—Tertiary only: K-T—both Cretaceous and Tertiary; K—Cretaceous only.

and *Fokieniopsis catenulata* (HC137). There is certainly no justification for illustrating *Araucaria araucana*, the monkey puzzle tree, a common element in almost all dinosaur landscape reconstructions, in the Hell Creek landscape.

Beyond these generalizations, it is clear that the Hell Creek Formation has several different floras, depending on stratigraphic position and sedimentary environment. The occurrence of small-leaved herbaceous floras of zone HCIb is certainly controlled by sedimentary environment, and it remains to be seen if they are also stratigraphically controlled. The variety of vegetation and landscapes offers a wealth of new information about the world of the last dinosaurs.

The sedimentology and physiographic setting of the Hell Creek Formation in southwestern North Dakota provide further information toward a reconstruction of the landscape. Data provided by Hoganson and Murphy (this volume) and Murphy et al. (this volume) place the shoreline of the Cretaceous sea somewhere east of the position of the present Missouri River. Their conclusion that a seaway was present in North Dakota and South Dakota through the end of the Cretaceous counters cur-

TABLE A4. MEGAFLORAL MORPHOTYPE OCCURRENCE BY LOCALITY AND MEGAFLORAL ZONE

FUI MEGAFLORAL ZONE

1. 8408 (80 m; carb shale): FU004, FU016, FU025, FU043
2. 8409 (80 m; carb shale): FU004, FU005, FU016, FU025, FU026, FU029, FU047
3. 9903 (2307), Kenny's Crib, (58 m; channel la): FU003, FU005, FU016, FU031, FU031?, FU043, FU044, FU097, HC071
4. 8403 (50 m; pond): FU003, FU004, FU005, FU012, FU016, FU021, FU023, FU024, FU029, FU043, FU044, FU046
5. 8404 (50 m; channel): FU016
6. 86110 (50 m; splay): FU002, FU003, FU004, FU005, FU007, FU009, FU010, FU012, FU014, FU015, FU016, FU017, FU018, FU019, FU020, FU023, FU029, FU043, FU044, FU045
7. JND1 (50 m; pond): FU001
8. JND7 (50 m; pond): FU002
9. 87143 (47.5 m; pond): FU007, FU016, FU031
10. 87142 (909), Quereuxia Hill, (45.2 m, pond): FU001, FU002, FU038, FU048
11. 9819 (2201), Jack, (40 m; pond): FU001, FU002, FU029, FU043
12. 88103l (39 m; splay/levee): FU003, FU007
13. 9817 (2200), Mt. Goat Lake, (33.5 m; pond): FU001, FU002, FU004, FU029
14. 88103j (32 m; splay): FU031
15. 9816 (2199), Sore Heel Aquatics, (31.5 m; pond): FU038
16. 9627 (1848), Fort Dilts Sunset, (30 m; pond): FU004, FU007, FU016, FU059, FU083, FU090
17. 86113 (30 m; pond ls): FU005, FU007, FU029, FU043, FU049
18. 88103l (28.8 m; pond vb): FU001, FU003, FU026, FU075
19. 87154 (22.5 m; channel la): FU001, FU005, FU007, FU016, FU029, FU047, FU049?, HC174?
20. 88120 (19.1 m; channel la): FU001, FU007, FU016, FU029, FU031, FU059, FU076
21. 88103g (18 m; splay): FU001, FU009, FU025?, FU029, FU031, FU037, FU037, FU077, FU078
22. 9251 (563), Paleocene Leaf, (17 m; aban chan): FU001, FU016, FU029?, FU031, FU082?, FU089
23. 9329 (905), Paranymphaea Heaven, (15.5 m; pond): FU001, FU016, FU029, FU031
24. 86148q (15.5 m; carb shale): FU016, FU026
25. 9142d (435), Mikey's Delite Pollen D, (14.1 m; pond): FU001
26. 87130 (11.5 m; pond vb): FU001, FU031
27. 9250 (562), Marmarth Ash, (11 m; ash bed): FU005, FU007, FU016, FU026, FU031, FU043, FU044, FU091
28. 88122 (441), Dean's NM04 Microsite/PITA Flats, (10 m; channel la): FU005, FU007, FU016, FU031, FU043, FU044, FU074, FU076, FU077?, FU078, FU078?, FU087, HC137, HC212
29. 86114 (10 m; pond): FU001
30. 87140 (10 m; carb shale): FU007?, FU016, FU037, HC012
31. 86161g (10 m; pond vb): FU016
32. 9316 (898), Pretty Ridge Paleocene, (8.2 m; pond vb): FU001, FU007, FU007?, FU009, FU016, FU029, FU031, FU059, FU061, FU082
33. 88100 (7.6 m; pond vb): FU007, FU016, FU035?
34. 9325 (902), New Paleocene above BBQ, (7.5 m; pond vb): FU007, FU016, FU031, FU062, FU080
35. 9831 (2208), Burnt Palms, (7 m; pond vb): FU037
36. 87150 (6.5 m; carb shale): FU001, FU004, FU005?, FU007, FU016, FU026, FU029, FU051, FU052, FU053, FU054, FU055, FU056, FU057
37. 9143 (436), Mikey's Delite, (5 m; channel/pond): FU007, FU016?, FU029, FU031, FU043, FU059, FU061, FU079, FU082, FU088, FU092, HC280?
38. 9650 (1858), Channel No. 6 flora, (5 m; channel la): FU001, FU007, FU016, FU031, FU059, FU061
39. 9651 (1859), Channel No. 6 ironstone, (5 m; channel la): FU001, FU007, FU016, FU031, FU037
40. 9855 (2216), Garrison's Pod, (5 m; pond vb): FU098
41. 9410 (1198), Kirk's Paleocene Channel, (5 m; channel x): FU031
42. 9829 (2206), Robsue Butte, (4 m; pond vb): FU007, FU016, FU023, FU026, FU031, FU059, FU082, HC123, HC231
43. 9830 (2207), The Honeymoon, (4 m; pond vb): FU029, FU031
44. 9832 (2209), Breezy Bench, (3.6 m; pond): FU004, FU007, FU016, FU026, FU057, FU059, FU082, HC123
45. 9843 (2211), Cedar City, (3.3 m; pond): FU007, FU016, FU023, FU026, HC123
46. 9841 (2210), Jim's Awl, (3.25 m; pond): FU004, FU007, FU026, FU059
47. 86149 (3 m; pond ls): FU007, FU029, FU031, FU059
48. 9858 (2217), Das Goods, (2.5 m; pond): FU001, FU002, FU004, FU007, FU009, FU016, FU026, FU029, FU031, FU035, FU036, FU037, FU059, FU061, FU082, FU083, FU084, FU085, FU086, FU087, FU088, HC123
49. 86163 (2 m; channel la): FU031
50. 86107 (427), Pyramid Butte, (1.15 m; pond): FU001, FU007, FU016, FU019, FU029, FU031, FU035, FU036, FU037, FU038, FU039, FU040, FU049, FU059, FU060, FU061, FU061?, FU062, FU063, FU064, FU065, FU069, FU070, FU072, FU073, FU074, FU093, FU094, HC140
51. 9136 (429), Mud Buttes Paleocene, (1 m, pond vb): FU001, FU002, FU007, FU016, FU029, FU031, FU062
52. 86162 (1 m; pond): FU001, FU016

FU0 MEGAFLORAL ZONE

53. 86155 (0.7 m; pond): FU025, FU029, FU031?, FU042
54. 9828 (2205), Steph's Scorpion, (0.56 m; carb splay): FU004, FU007, FU016, FU026, FU031, FU043, FU081, HC071
55. 9145 (438), Cannonball FUI, (0.5 m; carb shale): FU031
56. 9752 (2103), Dig It Again Dennis, (0.34 m; carb shale): FU007, FU059
57. 9844 (2212), Instant Gratification, (0.33 m; carb splay): FU005, FU007, FU016, FU026, FU029, FU031, FU043, FU044, FU059, FU081, FU095, FU096, HC049
58. 9754 (2100), Facet Butte, (0.33 m; carb shale): FU016, FU081

(*continued*)

TABLE A4. MEGAFLORAL MORPHOTYPE OCCURRENCE BY LOCALITY AND MEGAFLORAL ZONE (continued)

HCIII MEGAFLORAL ZONE

59. 9262 (573), Bobcat HCIII, (−1.39 m; channel la): HC001, HC103, HC106, HC111, HC166
60. 9249 (561), Pretty Butte North, (−1.8 m; splay): HC043, HC049, HC105, HC166, HC248, HC261, HC292, HC293
61. 88128 (440), Marshall's Turtle, (−2.2 m; splay): FU037, HC002, HC004, HC004?, HC014, HC043, HC105, HC106, HC111, HC162, HC163, HC164, HC165, HC166, HC175, HC179, HC179?, HC257, HC261, HC275, HC280, HC280?, HC321, HC337
62. 9904 (2308), Dinosaur Nodule, (−2.7 m; splay): FU002, FU037, FU062, HC036, HC044, HC105, HC137, HC179
63. 9322 (900), Battleship, (−2.8 m; channel la): FU009, FU037, FU043?, FU061, HC001, HC002, HC004, HC027, HC043, HC047, HC049, HC057, HC062, HC065, HC103, HC105, HC106, HC111, HC137, HC162, HC164, HC166, HC174, HC175, HC200?, HC208, HC261, HC263, HC264, HC265, HC277, HC278, HC280, HC284, HC295, HC328, HC329, HC330, HC331, HC332, HC333, HC334, HC335, HC336
64. 88107 (−3.1 m; splay): HC103, HC111
65. 9747 (2098), No Leaf Minors, (−3.45 m; pond): FU002, FU037, FU081, HC002, HC036, HC044, HC049?, HC054, HC127?, HC179, HC262, HC263, HC266, HC279, HC169, HC323, HC324, HC325, HC326, HC327
66. 9643 (1854), Tritriceratops, (−5.4 m; splay): FU037, HC001, HC004, HC070, HC071, HC103, HC103?, HC106, HC111, HC174, HC175?, HC199, HC263, HC281?
67. 87151 (−6.00 m; floodplain?): FU037
68. 9146 (439), Hwy. 16 HCIII, (−6.7 m; channel la): FU031, HC001, HC004, HC044, HC106, HC165
69. 87110 (−7.7 m; splay): FU037, FU081, HC001, HC002, HC003, HC004, HC007, HC028?, HC035, HC036, HC043, HC044, HC049?, HC057, HC066, HC070, HC103, HC105, HC106, HC111, HC162, HC163, HC166, HC168, HC169, HC171, HC172
70. 87152 (−7.8 m; channel x): HC105, HC106?, HC111
71. 9736 (2091), Kevin's First Palaeoaster, (−8.95 m; floodplain soil): FU037, FU059, HC001, HC007, HC043, HC044, HC070, HC086?, HC103, HC105, HC106, HC162, HC165, HC179, HC280, HC285, HC287, HC293
72. 88123 (574), Pretty Butte Shoulder, (−9 m; channel la): FU037, FU062, FU076?, HC001, HC002, HC003, HC004, HC036, HC043, HC044, HC049?, HC105, HC106, HC111, HC162, HC166, HC168?, HC174, HC261, HC265
73. 9611 (1782), The X-Butte, (−9 m; channel x): HC001, HC002, HC004, HC209, HC036, HC070, HC103, HC105, HC106, HC111, HC137, HC162, HC174, HC200, HC263, FU062
74. 9739 (2093), Torosaurus Leaves, (−9.5 m; channel x): FU037, HC043?, HC049, HC052, HC057, HC062, HC105, HC106, HC162, HC166, HC280?, HC329
75. 9740 (2094), Michael's Noooooooose, (−10.61 m; channel la): FU002, FU029?, HC002, HC036, HC049, HC103, HC105, HC111, HC162, HC163, HC166, HC174, HC248?, HC250?, HC254?, HC293, HC321
76. 88124 (−11.80 m; floodplain soil): HC105, HC106?, HC108?, HC163
77. 9729 (2089), Above Acronym Boy, (−11.85 m; channel la): HC105
78. 88105 (−12.40; channel la): HC103?, HC105, HC106, HC176
79. 9907 (2306), The Wall, (−12.4 m; channel la): FU029?, HC002, HC036, HC049, HC106, HC111, HC165, HC174, HC179, HC340, HC369
80. 88112 (−13 m; channel): HC105, HC175
81. 88127 (2099), Hastata Heaven, (−14.48 m; channel la): FU059, HC001, HC012, HC028?, HC043, HC049, HC057, HC062, HC105, HC106, HC111, HC162, HC165, HC315
82. 88102 (428), Dean Street, (−15 m; channel x): FU016, FU026, FU029, FU037, FU047, FU062, FU081, HC001, HC002, HC003, HC004, HC004?, HC014, HC036, HC043, HC044, HC047, HC049, HC062, HC062?, HC063, HC066, HC103, HC103?, HC105, HC106, HC111, HC115, HC162, HC165, HC166, HC169, HC169?, HC174, HC175, HC179, HC180, HC182, HC182?, HC183, HC185, HC186, HC187, HC189, HC199, HC200, HC201, HC202, HC203, HC204, FU061, HC207, HC212, HC248, HC254, HC261, HC261?, HC264, HC265, HC275, HC276, HC280, HC286, HC292, HC293, HC294, HC295, HC321, HC340, HC341, HC342, HC343, HC344, HC345, HC346, HC347, HC348, HC349, HC350, HC351, HC352, HC353, HC354, HC355, HC356, HC357, HC358, HC359, HC360, HC362, HC363, HC364
83. 88104 (−15 m; channel x): HC004
84. 87147a (−15.3 m; channel x): FU007?, HC001?, HC002?, HC003?, HC044?, HC066, HC103, HC105, HC106, HC111, HC209
86. 9649a (1857), Dean's Hadrosaur Section, (−15.67 m; channel x): HC002, HC105?
87. 86158 (−16 m; channel x): FU037, HC001, HC002, HC004, HC005, HC036, HC043, HC044, HC047, HC099, HC105, HC106, HC111, HC187
89. 87153 (1250), Palaeoaster Pass, (−17 m; channel la): FU007?, FU059, HC001, HC209, HC007, HC008, HC009, HC011, HC036, HC044, HC049, HC052, HC057?, HC094?, HC103, HC105, HC106, HC151, HC162, HC204, HC247, HC287
90. 9331 (906), New HCIII by Dean's T-rex, (−19 m; channel x): HC004, HC105, HC106, HC111, HC162, HC165, HC166
88. 9239 (517), Stan T-rex Site, (−16.2 m; channel la): FU002, FU038, FU037, FU062, HC001, HC002, HC007, HC032, HC043, HC044, HC062, HC070, HC094?, HC105, HC106, HC162, HC165, HC174, HC200, HC263?, HC280, HC318
91. 9849 (2214), Above Dean's Hadrosaur, (−19 m; channel la): HC044, HC049, HC057?, HC062, HC070
93. 9645 (1855), Triceps, (−20.00 m; channel la): FU047?, HC014, HC049, HC062?, HC094?, HC105, HC179, HC248, HC249, HC250, HC253, HC254, HC255, HC319, HC320
94. 9610 (1781), Dean's Hadrosaur, (−21.3 m; channel la): FU002, HC004, HC014, HC043, HC044, HC054, HC070, HC105, HC111, HC162, HC165, HC200, HC240, HC242, HC254, HC256, HC257, HC258, HC276, HC286, HC312, HC313, HC314, HC315, HC316, HC317, HC338
137. 9257 (569), Steph's Locality (−65 m; channel la): FU037, HC001, HC004, HC049, HC103, HC105, HC106, HC200, HC237, HC238, HC239

(continued)

TABLE A4. MEGAFLORAL MORPHOTYPE OCCURRENCE BY LOCALITY AND MEGAFLORAL ZONE (continued)

HCIIb MEGAFLORAL ZONE

85. 88117 (910), Turtle Grape, (−15.4 m; channel la): HC014
92. 87103 (−19.5 m; channel la): HC049, HC108
95. 88121 (908), Whiterock Hilltop, (−24 m; channel la/fill): HC014, HC049, HC062, HC100, HC188, HC189, HC190, HC311
97. 87102 (−26.90 m; channel x): HC108
102. 9323 (901), Roadside HCIIb, (−30 m; channel la): FU037, FU046, HC011?, HC014, HC044, HC049, HC108
115. 87105 (2088), Ken and Joanne, (−36 m; channel la): HC009, HC014, HC015, HC017, HC049, HC062, HC094, HC212
116. 9811 (2195), Bad Bug Spray, (−40 m; channel x): HC004?, HC108
117. 9812 (2196), Backpack Butte, (−40 m; channel la): HC108
118. 9813 (2197), Right Nipple Nose, (−40 m; channel la): HC049, HC086, HC162, HC297

HCIIa MEGAFLORAL ZONE

96. 86115 (2215) (−25.5 m; channel la): HC049, HC052, HC054, HC055, HC056, HC057, HC058, HC062
98. 88110 (−28.5 m; channel la): HC049, HC057, HC080?, HC190, HC268
99. 9309 (896), Below Stan, (−28.7 m; pond): FU002
100. 9745 (2097), Terry's HCIIa site, (−28.8 m; channel la): HC032, HC049, HC057, HC062, HC064, HC066, HC077, HC086, HC103, HC151, HC254, HC260, HC267, HC286?, HC289, HC290, HC304, HC305, HC306, HC307, HC308, HC309, HC310
101. 9138 (431), Sunset Butte #1, (−29 m; channel la): HC049, HC057, HC064, HC303
103. 87131 (2096), Ironstone Flora, (−30.6 m; floodplain soil): HC009, HC018, HC020, HC021, HC022, HC023, HC049, HC057, HC241
104. 9139 (432), Sunset Butte #2, (−31.5 m; channel la): HC027, HC057, HC062, HC066?, HC086, HC283
105. 9529 (1489), Wet Butt, (−32 m; channel la): FU002, FU031, FU038, FU038?, FU062, HC035, HC049, HC057, HC062, HC063, HC065, HC066, HC071, HC076, HC086, HC103, HC103?, HC114, HC129, HC162?, HC166?, HC204?, HC242, HC254, HC268, HC269, HC270, HC272, HC273, HC274, HC281, HC283, HC285, HC290, HC291
106. 9530 (1490), Wet Butt Too, (−32 m; channel la): HC283
107. 87146 (−33.3 m; channel la): HC043, HC049, HC057, HC064, HC066
108. 88111 (566), Ginkgo Salad, (−33.5 m; floodplain soil): FU003, FU059, HC002, HC032, HC049, HC057, HC060, HC062, HC064, HC065, HC066, HC071, HC077, HC086, HC103, HC114, HC164, HC168?, HC169, HC179, HC189, HC193, HC194, HC291
109. 9731 (2090), Ode to an Ice Cream Truck, (−33.6 m; channel la): HC049, HC057, HC071, HC103
110. 86153 (568), Allison's Attachment, (−33.8 m; channel x): FU063, HC003?, HC009, HC024, HC025, HC027, HC028, HC029, HC031, HC032, HC034, HC036, HC043, HC049, HC054, HC057, HC060, HC062, HC063, HC064, HC065, HC066, HC067, HC068, HC071, HC081, HC086, HC103, HC105, HC114, HC123, HC137, HC151, HC261, HC282, HC283, HC291?
111. 86103 (−33.8 m; channel x): HC002, HC035, HC036, HC049, HC103
113. 9307 (895), Dean's T-rex Pond, (−35 m; pond): FU002, HC077, HC108, HC162, HC196, HC298
114. 87109 (−35.6 m; channel la): HC035, HC049, HC057, HC062, HC064
119. 87136 (565), Dragonfly, (−42 m; splay): HC027, HC049, HC147, HC268?
120. 87129 (425), Roadside, (−42.70 m; channel x): FU035?, FU035?, FU059, HC002, HC036, HC040, HC049, HC052, HC057, HC062, HC064, HC066, HC068?, HC071, HC073, HC077, HC086, HC094, HC099, HC103, HC114, HC209
121. 88114 (−43.4 m; channel la): HC032, HC049, HC057, HC064, HC086
122. 9737 (2092), The Rock Game, (−45.77 m; pond): FU026, HC014, HC036, HC049, HC049?, HC062, HC071, HC129, HC196, HC262, HC271
123. 9312 (897), New Bobcat Channel, (−47 m; channel la): FU003, HC002, HC049, HC057, HC064, HC070, HC071, HC073, HC094, HC098, HC103, HC124, HC129, HC212, HC262, HC296
124. 86102 (2095), Nowhere Hole, (−47 m; channel la): FU037, HC049, HC057, HC064, HC070, HC071, HC072, HC073, HC074, HC075, HC076, HC094, HC098, HC103, HC122, HC123, HC124, HC126, HC127, HC128, HC268
126. 86104 (−52 m; pond): HC077, HC078, HC129

HCIb MEGAFLORAL ZONE

112. 9727 (2087), Acronym Boy, (−34.4 m; channel la): FU005, FU036, FU043, HC014, HC027, HC049, HC054, HC062, HC080, HC083, HC084, HC090, HC092, HC093, HC094, HC114, HC124, HC135, HC212, HC225, HC226, HC241, HC242, HC243, HC244, HC245, HC246, HC247, HC251, HC252, HC288, HC299, HC300, HC301, HC302, HC370
127. 9634 (1860), Little Mound of Dickey Draw, (−55 m; channel la): HC092
128. 86108 (−55.00 m; channel la): HCIb: FU004?, HC241, HC035, HC049?, HC073, HC080, HC081, HC086, HC092, HC093, HC094, HC096, HC099?, HC131, HC140
129. Dynamite Coulee (−55m): HC070, HC073, HC083, HC086, HC092, HC094, HC209, HC212
130. 86100 (567), Marijuana, (−56 m; channel la): HC027, HC035, HC049, HC054, HC070, HC073, HC080, HC081, HC083, HC084, HC085, HC086, HC087, HC088, HC090, HC091, HC092, HC093, HC094, HC099, HC123, HC130, HC131, HC132, HC135, HC136, HC137, HC138, HC139, HC140, HC141, HC142, HC212, HC224, HC225, HC226, HC227, HC240, HC300
131. 9535 (1491), Madeline's Bank, (−56 m; channel la): HC057, HC070, HC071, HC083, HC086, HC090, HC092, HC094, HC135, HC137, HC211, HC212, HC213, HC214, HC215, HC216, HC217, HC218, HC219, HC220, HC222
132. 9822 (2202), Rusty Girls, (−56 m; channel la): HC090, HC092, HC094, HC212, HC223
133. 9824 (2203), Somebody's Garden, (−56 m; channel la): HC014, HC081, HC084, HC090, HC092, HC094, HC136, HC139, HC212, HC224, HC226, HC227, HC228, HC229, HC230, HC231, HC232, HC233, HC234
134. 9825 (2204), Mad Butt, (−56 m; channel la): HC092, HC094, HC212

(*continued*)

TABLE A4. MEGAFLORAL MORPHOTYPE OCCURRENCE BY LOCALITY AND MEGAFLORAL ZONE (continued)

HCIa MEGAFLORAL ZONE

125. 9536 (1492), Triebold's Sandy Site, (−50 m; channel x): FU043, HC044, HC071, HC086, HC100, HC103, HC236
135. 87134 (572), Butte Base, (−60 m; channel la): FU003, FU005, HC014, HC035, HC049, HC057, HC064, HC066?, HC083, HC086, HC099, HC099?, HC108, HC114, HC124, HC137, HC151, HC152, HC154, HC217, HC224?, HC226, HC286, HC291?
136. 9911 (2309), Confusion Butte, (−60 m; channel la): HC014, HC094, HC099, HC212
138. 86105 (571), Little Mound of Glee/Mosquito Butte (−65 m; channel la): FU003, FU004, FU059, HC027, HC035, HC049, HC057, HC063, HC064, HC066, HC071, HC073, HC077, HC086, HC098, HC099, HC100, HC103, HC106, HC108, HC114, HC148, HC150, HC151, HC164, HC209, HC210, HC212, HC339
139. 86119 (−65 m; channel la): HC049, HC086, HC098
140. 9910 (2310), Delores Leaves, (−65 m; channel la): HC002, HC028, HC049, HC057, HC058, HC086, HC087, HC340?, HC368
141. 87133 (−74 m; floodplain soil): HC057, HC094?, HC103?, HC137
142. 87111 (430)(−75 m; channel la): FU060, HC014, HC027, HC049, HC055, HC057, HC063, HC064, HC066, HC073, HC077, HC086, HC099, HC103, HC124, HC151, HC155, HC212, HC268
143. 86142 (1852), Luten's 4H Hadrosaur, (−75 m; channel la): FU003, HC043, HC049, HC057, HC064, HC066, HC073, HC086, HC092, HC094, HC103, HC124?, HC151, HC155, HC161, HC212, HC365, HC366
144. 9733 (2102), Doaks Butte ash bed, (−75 m; volcanic ash): HC049, HC057, HC064, HC092
145. 86143 (−75 m; pond): HC049
146. 86144 (−75 m; channel/pond): HC077
147. 86145 (−75 m; channel/pond): HC049, HC073, HC086, HC103
148. 87112 (−75 m; channel la): HC049, HC056, HC073?, HC086, HC103?, HC123, HC124, HC158, HC159, HC241
149. 87132 (−76.8 m; channel la): HC014?, HC049
150. 9630 (1850) Steep Sandy (−80 m; channel la): HC086, HC099, HC340
151. 9631 (1851) Tapir Point (−80 m; channel la): HC057, HC064, HC066, HC086, HC092, HC094, HC151, HC212, HC337
152. 88139 (−80 m; channel la): HC094
153. 88109 (−83 m; pond): FU026, HC036, HC077, HC196
154. 88108 (−85.1 m; splay): HC014, HC164
155. 9141 (434), Doaks Butte #2, (−88.5 m; channel la): HC032, HC057, HC092, HC137, HC209
156. 9140 (433), Doaks Butte #1, (−91 m; channel la): HC049, HC212, HC241?, HC253, HC280?, HC318, HC367
157. 87113 (−102 m; Colgate channel): HC049
158. 88131 (−103 m; Colgate channel): FU005, FU043, HC049, HC094, HC137, HC165, HC197, HC198

Notes: Localities cited by field number only (those beginning with the numbers 86–88) are Yale Peabody Museum of Natural History localities. Localities cited by field numbers beginning with 91–99 are Denver Museum of Nature & Science localities (followed by DMNH locality number in parentheses). Localities with field numbers beginning with 86–88 (followed by DMNH locality number in parentheses) are localities that have representative collections at both YPM and DMNH. The parentheses following the locality number show the stratigraphic position relative to the Hell Creek–Fort Union formational contact datum and the sedimentary facies.

rent interpretations (Lillegraven and Ostresh, 1990), but is well supported by their data and recent work by Kennedy et al. (1998). Thus, the Hell Creek Formation of western North Dakota was located about 200 km west of the shoreline. Sedimentology of the Hell Creek (Fastovsky, 1987) shows an abundance of medium-sized meandering river channels coursing across an aggrading coastal plain. Ponded water in this landscape occurred almost exclusively in oxbow lakes. Mires and swamps were present but extremely rare and may have been related to sea-level fluctuations. Fastovsky (1987), Fastovsky and McSweeney (1987), and Retallack (1994) presented paleosol data suggesting that Hell Creek soils were wet but not flooded.

There seems to be very little evidence that the Sheridan delta of Gill and Cobban (1973) was truly deltaic in the Marmarth area. Cherven and Jacob (1985, their Fig. 25) illustrated a compelling paleogeographic map for K-T boundary time, but stream trends in southwestern North Dakota are shown trending to the east. Fastovsky (1987) documented that actual Hell Creek paleocurrent indicators in the region trend south and southeast. Thinning of the Hell Creek Formation over the Cedar Creek anticline near Glendive, Montana, has been presented as evidence that the anticline was active in the Maastrichtian (Cherven and Jacob, 1985; Vuke-Foster et al., 1986). It is interesting to note that the Hell Creek paleocurrent directions of Fastovsky (1987) parallel the structural trend of the anticline, and then to speculate that the Hell Creek rivers may have been structurally controlled en route to their debouchment into the Western Interior seaway.

Fort Union landscapes

The marked facies change at the Hell Creek–Fort Union formational contact was probably caused by the damming of groundwater and the subsequent flooding of the landscape as a response to the transgression of the Cannonball seaway. Marine or brackish tongues within the Ludlow Member of the Fort Union Formation begin ~50 m above the base of the Ludlow and the units become increasingly deltaic upsection (Belt et al., 1984; Cherven and Jacob, 1985; Hartman, 1999). The result is a great expansion of mires and ponded water deposits. The dry-foot forests of the Hell Creek are replaced by taxodiaceous swamps and low-diversity levee forests. Stream channels still show meandering patterns in the basal Fort Union but become straighter and more like distributary channels upsection. In terms of specific Fort Union landscapes and vegetation, the variegated mudstone facies that represents widespread ponds is most common in the basal 20 m of the Fort Union, and this corresponds with the occurrence of palms and a slightly more diverse forest. Much more work needs to be done in the upper part of the Ludlow Member to understand trends of floral diversity associated with the nearing of the shoreline.

ACKNOWLEDGMENTS

I thank the following volunteers of the Denver Museum of Nature & Science (formerly Denver Museum of Natural History, DMNH) and the Pioneer Trails Regional Museum, who provided valuable assistance in field collection, specimen preparation, curation, and accessioning of the fossil leaf material in this study: Rich Barclay, Bill Bateman, Kristian Benson, Stephanie Bowman, Peter Buckman, Laura Carsten, Jon Christians, Tom Collins, Regan Dunn, Dolores Eckridge, Beth Ellis, Tim Farnham, John Gurche, Madeline Harrell, Biz Hermsen, Dennis Hopkins, Beth Kaminsky, Tyler Lyson, Karen McMeekin, Dena Meade-Hunter, Doug Nichols, Spiny Nichols, Laurie Oakland, Marianne O'Connell, Michael O'Connell, Susan Passmore, Dean Pearson, Michele Reynolds, Nancy Schaefer, Terry Schaefer, Pam Schmidt, Gary Staab, Erin Steele, Patty Stoiko, Allison Sundine, John van Regenmorter, Kevin Werth, the students of the DMNH 1998 field school, and the DMNH Leaf Whackers. Peter Wilf and Paul Harnick developed the protocol for digitizing the morphotypes. Adrian Kropp, Erin Steele, Doug Nichols, and Michele Reynolds helped with the preparation of the manuscript. Conversations with Leo Hickey, Peter Wilf, Scott Wing, Jason Hicks, Doug Nichols, Steven D'Hondt, David Fastovsky, and Dean Pearson were most valuable.

Property access was allowed by the Little Missouri National Grassland, the Bureau of Land Management, the East Marmarth Pasture Association, the Horse Creek Grazing Association, Scott and Henry Bradac, John and Bud Brown, Mike Sonsella, Lyle and Sharon Gundersen, John Niemi, Danny Flor, and Merle Clark. Support for this research originated from the Denver Museum of Nature & Science (formerly Denver Museum of Natural History). Additional support was supplied by the National Science Foundation (grant EAR-9805474) and the North Dakota Academy of Science.

REFERENCES CITED

Alvarez, L.W., Alvarez, W., Asaro, F., and Michel, H.V., 1980, Extraterrestrial cause for the Cretaceous-Tertiary extinctions: Science, v. 208, p. 1095–1108.

Archibald, J.D., 1996, Dinosaur extinction and the end of an era: What the fossils say: Critical Moments in Paleobiology and Earth History Series: New York, Columbia University Press, 237 p.

Archibald, J.D., and Bryant, L.J., 1990, Differential Cretaceous/Tertiary extinctions of nonmarine vertebrates; evidence from northeastern Montana: Geological Society of America Special Paper 247, p. 549–562.

Bauer, C.M., 1924, The Ekalaka lignite field southwestern Montana: U.S. Geological Survey Bulletin, v. 751-F, p. 231–267.

Belt, E.S., Flores, R.M., Warwick, P.D., Conway, K.M., Johnson, K.R., and Waskowitz, R.S., 1984, Relationship of fluviodeltaic facies to coal deposition in the lower Fort Union Formation (Palaeocene), south-western North Dakota, *in* Rahmani, R.A., and Flores, R.M., eds., Sedimentology of coal and coal-bearing sequences: International Association of Sedimentologists Special Publication 7, p. 177–195.

Belt, E.S., Hicks, J.F., and Murphy, D.A., 1997, A pre-Lancian regional unconformity and its relationship to Hell Creek paleogeography in southeastern Montana: University of Wyoming Contributions to Geology, v. 31, p. 1–26.

Berry, E.W., 1934, A lower Lance florule from Harding County, South Dakota: U.S. Geological Survey Professional Paper, v. 185F, p. 127–133.

Bishop, G.A., 1965, The paleontology and stratigraphy of the Upper Cretaceous Upper Pierre Shale on Cedar Creek anticline, Montana: South Dakota Academy of Science Proceedings, v. 44, p. 91–96.

Bishop, G.A., 1973, Geology, stratigraphy, and biostratigraphy of the north end of Cedar Creek anticline, Dawson County, Montana: Montana Bureau of Mines and Geology Special Publication 61, 1 p.

Bowen, C.F., 1912, The Baker lignite field, Custer County, Montana: U.S. Geological Survey Bulletin, v. 471, p. 202–226.

Brown, B., 1907, The Hell Creek beds of the Upper Cretaceous of Montana: American Museum Natural History Bulletin, v. 23, p. 823–845.

Brown, R.W., 1938, The Cretaceous-Eocene boundary in Montana and North Dakota: Washington Academy of Science Journal, v. 28, p. 421–422.

Brown, R.W., 1939, Fossil plants from the Colgate Member of the Fox Hills Sandstone and adjacent strata: U.S. Geological Survey Professional Paper 189-I, p. 239–275.

Brown, R.W., 1952, Tertiary strata in eastern Montana and western North and South Dakota, *in* Sonnenberg, F.P., ed., Billings Geological Society Guidebook: Billings, Montana, Billings Geological Society, Third Annual Field Conference, p. 89–92.

Brown, R.W., 1962, Paleocene flora of the Rocky Mountains and Great Plains: U.S. Geological Survey Professional Paper 375, 119 p.

Burnham, R.J., 1989, Relationships between standing vegetation and leaf litter in a paratropical forest: implications for paleobotany: Review of Palaeobotany and Palynology, v. 58, p. 5–32.

Burnham, R.J., Wing, S.L., and Parker, G.G., 1992, The reflection of deciduous forest communities in leaf litter: Implications for autochthonous litter assemblages from the fossil record: Paleobiology, v. 18, p. 30–49.

Carlson, C.G., 1979, Geology of Adams and Bowman Counties, North Dakota: North Dakota Geological Survey Bulletin 65, 29 p.

Carlson, C.G., 1983, Geology of Billings, Golden Valley, and Slope Counties, North Dakota: North Dakota Geological Survey Bulletin 76, 40 p.

Carpenter, K., 1987, Paleoecological significance of droughts during the Late Cretaceous of the Western Interior: Fourth Symposium on Mesozoic Terrestrial Ecosytems, *in* Currie, P.J., and Koster, E.H., eds., Occasional paper of the Tyrrell Museum of Palaeontology #3: Drumheller, Alberta, Royal Tyrrell Museum of Palaeontology p. 42–47.

Cherven, V.B., and Jacob, A.R., 1985, Evolution of Paleogene depositional systems, Williston Basin in response to global sea level changes, *in* Flores, R.M., and Kaplan, S.S., eds., Cenozoic paleogeography of the west-central United States: Society of Economic Paleontologists and Mineralogists, Rocky Mountain Symposium 3, p. 127–170.

Christophel, D.C., and Johnson, K.R., 1990, The effect of mosaic rainforest distribution and taphonomy on the physiognomy of modern leaf deposits: Implications for Cretaceous and Paleogene paleoclimatic reconstructions: Geological Society of America Abstracts with Programs, v. 22, no. 7, p. 35.

Colagrande, J., and Felder, L., 2000, In the presence of dinosaurs: Alexandria, Virginia, Time-Life Books, 189 p.

Corbin, R.S., 1938, Stratigraphy and paleontology of the Pierre, Fox Hills, and Lance Formations in southwestern North Dakota [B.A. thesis]: Princeton, New Jersey, Princeton University, 141 p.

Cvancara, A.M., 1976, Geology of the Fox Hills Formation (Late Cretaceous) in the Williston Basin of North Dakota, with reference to uranium potential: North Dakota Geological Survey Report of Investigations, v. 55, 16 p.

Cvancara, A.M., and Hoganson, J.W., 1993, Vertebrates of the Cannonball

Formation (Paleocene) in North and South Dakota: Journal of Vertebrate Paleontology, v. 13, p. 1–23.

Derstler, K., 1994, Dinosaurs of the Lance Formation in Eastern Wyoming: Wyoming Geological Association 44th Annual Field Conference Guidebook, p. 127–146.

Dobbin, C.E., and Larsen, R.M., 1934, Geologic and structure contour map of the southern half of the Cedar Creek anticline, Fallon County, Montana, and Bowman County, North Dakota: U.S. Geological Survey map, scale 1:63360.

Dobbin, C.E., and Reeside, J.B., Jr., 1929, The contact of the Fox Hills and Lance Formations: U.S. Geological Survey Professional Paper, v. 158B, p. 9–25.

Dorf, E., 1940, Relationship between floras of the type Lance and Fort Union Formations: Geological Society America Bulletin, v. 51, p. 213–236.

Dorf, E., 1942, Upper Cretaceous floras of the Rocky Mountain region: II: Flora of the Lance Formation at its type locality, Niobrara County, Wyoming: Carnegie Institution of Washington Publication 508, p. 83–168.

Ellis, M.S., and Colton, R.B., 1994, Geologic map of the Powder River Basin and surrounding area, Wyoming, Montana, South Dakota, North Dakota, and Nebraska: U.S. Geological Survey Miscellaneous Investigations Series 2298, scale 1:500000.

Erickson, H.D., 1956a, Areal geology of the Buffalo Quadrangle: Vermillion, South Dakota Geological Survey, scale 1:62500.

Erickson, H.D., 1956b, Areal geology of the Willett and Midland #1 Quadrangle: Vermillion, South Dakota Geological Survey, scale 1:62500.

Fastovsky, D.E., 1986, Paleoenvironments of vertebrate-bearing strata at the Cretaceous-Paleogene boundary in northeastern Montana and southwestern North Dakota [Ph.D. thesis]: Madison, University of Wisconsin, 301 p.

Fastovsky, D.E., 1987, Paleoenvironments of vertebrate-bearing strata during the Cretaceous-Paleogene transition, eastern Montana and western North Dakota: Palaios, v. 2, p. 282–295.

Fastovsky, D.E., and McSweeney, K., 1987, Paleosols spanning the Cretaceous-Paleogene transition, eastern Montana and western North Dakota: Geological Society of America Bulletin, v. 99, p. 66–77.

Fastovsky, D.E., and Weishampel, D.B., 1996, The evolution and extinction of the dinosaurs: Cambridge, Cambridge University Press, 460 p.

Feldmann, R.M., 1972, Stratigraphy and paleoecology of the Fox Hills Formation (Upper Cretaceous) of North Dakota: North Dakota Geological Survey Bulletin 61, 65 p.

Fort Union Coal Assessment Team, 1999, 1999 Resource assessment of selected Tertiary coal beds and zones in the northern Rocky Mountains and Great Plains region: U.S. Geological Survey Professional Paper 1625-A (CD-ROM).

Fox, S.K., Jr., and Olsson, R.K., 1969, Danian planktonic foraminifera from the Cannonball Formation in North Dakota: Journal of Paleontology, v. 43, p. 1397–1404.

Fox, S.K., Jr., and Ross, R.J., Jr., 1942, Foraminiferal evidence for the Midway (Paleocene) age of the Cannonball Formation in North Dakota: Journal of Paleontology, v. 16, p. 660–673.

Frye, C.I., 1967, The Hell Creek Formation in North Dakota [Ph.D. thesis]: Grand Forks, University of North Dakota, 411 p.

Frye, C.I., 1969, Stratigraphy of the Hell Creek Formation in North Dakota: North Dakota Geological Survey Bulletin 54, 65 p.

Gill, J.R., and Cobban, W.A., 1973, Stratigraphy and geologic history of the Montana Group and equivalent rocks, Montana, Wyoming, and North and South Dakota: U.S. Geological Survey Professional Paper 776, 36 p.

Hares, C.J., 1928, Geology and lignite resources of the Marmarth field, southwestern North Dakota: U.S. Geological Survey Bulletin 775, 110 p.

Hartman, J.H., 1995, Slope County bed names: an annotated list of named beds in Slope County, North Dakota: Final report to the U.S. Geological Survey, National Coal Resources Data System (Agreement No. 1434–94-A-1207) Evaluation and entry of North Dakota lignite data into the NCRDS program: Grand Forks, North Dakota, EERC Publication, 36 p.

Hartman, J.H., 1999, Western exploration along the Missouri River and the first paleontological studies in the Williston Basin, North Dakota and Montana: North Dakota Academy of Sciences Proceedings, v. 53, p. 158–164.

Henderson, D., 2000, Asteroid impact: New York, Penguin Putnam Inc., 40 p.

Hickey, L.J., 1973, Classification of the architecture of dicotyledonous leaves: American Journal of Botany, v. 60, p. 17–33.

Hickey, L.J., 1979, A revised classification of the architecture of dicotyledonous leaves, *in* Metcalfe, C.R., and Chalk, L., eds., Anatomy of the dicotyledons, Second Edition: Oxford, Clarendon Press, p. 25–39.

Hickey, L.J., 1980, Paleocene stratigraphy and flora of the Clark's Fork Basin, *in* Gingerich, P.D., ed., Early Cenozoic paleontology and stratigraphy of the Bighorn Basin, Wyoming: University of Michigan Papers in Paleontology 24, p. 33–49.

Hickey, L.J., 1981, Land plant evidence compatible with gradual, not catastrophic, change at the end of the Cretaceous: Nature, v. 292, p. 529–531.

Hickey, L.J., 1984, Changes in the angiosperm flora across the Cretaceous-Tertiary boundary, *in* Berggren, W.A., and Van Couvering, J.A., eds., Catastrophes in earth history; the new uniformitarianism: Princeton, New Jersey, Princeton University Press, p. 279–313.

Hickey, L.J., and Wolfe, J.A., 1975, The bases of angiosperm phylogeny: Vegetative morphology: Missouri Botanical Gardens Annals, v. 62, p. 538–589.

Hicks, J.F., 1993, Chronostratigraphic analysis of the foreland basin sediments of the latest Cretaceous, Wyoming, U.S.A. [Ph.D. thesis]: New Haven, Connecticut, Yale University, 251 p.

Hicks, J.F., Obradovich, J.D., and Tauxe, L., 1999, Magnetostratigraphy, isotopic age calibration and intercontinental correlation of the Red Bird section of the Pierre Shale, Niobrara County, Wyoming, USA: Cretaceous Research, v. 20, p. 1–27.

Hunter, J.P., 1999, The radiation of Paleocene mammals with the demise of the dinosaurs: Evidence from southwestern North Dakota: North Dakota Academy of Sciences Proceedings, v. 53, p. 141–144.

Johnson, K.R., 1982, Paleobotany and sedimentology of some lower Ludlow strata, Fort Union Formation (Paleocene), Slope County, North Dakota [thesis]: Amherst, Massachusetts, Amherst College, 159 p.

Johnson, K.R., 1988, Quantitative plant taphonomy of the Wharton Brook flood plain, North Haven, Connecticut: American Journal of Botany v. 75, p. 111.

Johnson, K.R., 1989, A high-resolution megafloral biostratigraphy spanning the Cretaceous-Tertiary boundary in the northern Great Plains [Ph.D. thesis]: New Haven, Connecticut Yale University, 556 p.

Johnson, K.R., 1992, Leaf-fossil evidence for extensive floral extinction at the Cretaceous-Tertiary boundary, North Dakota, USA: Cretaceous Research, v. 13, p. 91–117.

Johnson, K.R., 1993, Time resolution and the study of Late Cretaceous and early Tertiary megafloras, *in* Kidwell, S.M., and Behrensmeyer, A.K., eds., Taphonomic approaches to time resolution in fossil assemblages: Short Courses in Paleontology: Knoxville, Tennessee, The Paleontological Society, p. 210–227.

Johnson, K.R., 1996, Description of seven common plant megafossils from the Hell Creek Formation (Late Cretaceous: late Maastrichtian), North Dakota, South Dakota, and Montana): Denver Museum of Natural History Proceedings, ser. 3, v. 3, p. 1–48.

Johnson, K.R., 1999, The reconstruction of ancient landscapes: An example from the Late Cretaceous Hell Creek Formation of North Dakota: North Dakota Academy of Sciences Proceedings, v. 53, p. 134–140.

Johnson, K.R., and Hickey, L.J., 1990, Megafloral change across the Cretaceous-Tertiary boundary in the northern Great Plains and Rocky Mountains, U.S.A., *in* Sharpton, V. L., and Ward, P. D., eds., Global catastrophies in Earth history: An interdisciplinary conference on impacts, volcanism, and mass mortality: Boulder, Colorado, Geological Society of America Special Paper 247 p. 433–444.

Johnson, K.R., and Stucky, R.K., 1995, Prehistoric journey: A history of life on Earth: Boulder, Colorado, Roberts Rinehart, 144 p.

Johnson, K.R., and Wilf, P., 1996, Paleobotanical temperature curve for the Cretaceous-Tertiary boundary interval and a scenario for the revegetation of North America after the terminal Cretaceous event: Geological Society of America Abstracts with Programs, v. 28, no. 7, p. 225.

Johnson, K.R., Nichols, D.J., Attrep, M., Jr., and Orth, C.J., 1989, High-resolution leaf-fossil record spanning the Cretaceous/Tertiary boundary: Nature, v. 340, p. 708–711.

Kennedy, W.J., Landman, N.H., Christensen, W.K., Cobban, W.A., and Hancock, J.M., 1998, Marine connections in North America during the late Maastrichtian: Paleogeographic and paleobiogeographic significance of *Jeletzkytes nebrascensis* Zone cephalopod fauna from the Elk Butte Member of the Pierre Shale, SE South Dakota and NE Nebraska: Cretaceous Research, v. 19, p. 745–775.

Laird, W.M., and Mitchell, R.H., 1942, The geology of the southern part of Morton County, North Dakota: North Dakota Geological Survey Bulletin 14, 42 p.

Leaf Architecture Working Group, 1999, Manual of Leaf Architecture—Morphological description and categorization of dicotyledonous and net-veined monocotyledonous angiosperms: Washington, D.C., Smithsonian Institution, 65 p.

Leffingwell, H.A., 1970, Palynology of Lance (Late Cretaceous) and Fort Union (Paleocene) Formations of the type Lance area, Wyoming, *in* Kosanke, R.M., and Cross, A.T., eds., Symposium on palynology of the Late Cretaceous and early Tertiary: Geological Society of America Special Paper 127, p. 1–64.

Leggitt, M.C., and Johnson, K.R., 1999, From fossil plants to scientifically-accurate dioramas: the fabrication of prehistoric ecosystems: Journal of Natural Science Illustration, v. 3, no. 1, p. 3–6.

Leonard, A.G., 1908, Geology of southwestern North Dakota with special references to the coal: North Dakota Geological Survey 5th Biennial Report, p. 29–114.

Leonard, A.G., 1911, The Cretaceous and Tertiary Formations of western North Dakota and eastern Montana: Journal of Geology, v. 19, p. 507–547.

Lillegraven, J.A., and Ostresh, L.M., Jr., 1990, Late Cretaceous (earliest Campanian/Maastrichtian) evolution of western shorelines of the North American western interior seaway in relation to known mammalian faunas, *in* Bown, T.M., and Rose, K.D., eds., Dawn of the age of mammals in the northern part of the Rocky Mountain Interior, North America: Geological Society of America Special Paper 243, p. 1–30.

Lloyd, E.R., and Hares, C.J., 1915, The Cannonball marine member of the Lance Formation of North and South Dakota and its bearing on the Lance-Laramie problem: Journal of Geology, v. 23, p. 523–547.

Moore, W., 1976, The stratigraphy and environments of the Cretaceous Hell Creek Formation (reconnaissance) and the Paleocene Ludlow Formation (detailed), southwestern North Dakota: North Dakota Geological Survey Report of Investigations, v. 56, 40 p.

Murphy, E.C., Nichols, D.J., Hoganson, J.W., and Forsman, N.F., 1995, The Cretaceous-Tertiary boundary in south-central North Dakota: North Dakota Geological Survey Report of Investigation 98, p. 1–41.

Nichols, D.J., Murphy, E.C., Johnson, K.R., and Betterton, W.J., 2000, A second K-T boundary locality in North Dakota verified by palynostratigraphy and shocked quartz: Geological Society of America Abstracts with Programs, v. 32, no. 7, p. 130.

Obradovich, J.D., 1993, A Cretaceous time scale, *in* Caldwell, W.G.E., and Kauffman, E.G., eds., Evolution of the Western Interior Basin: Geological Association of Canada Special Paper 39, p. 379–396.

Orth, C.J., Gilmore, J.S., Knight, J.D., Pillmore, C.L., Tschudy, R.H., and Fassett, J.E., 1981, An iridium abundance anomaly at the palynological Cretaceous-Tertiary boundary in northern New Mexico: Science, v. 214, p. 1341–1343.

Pearson, D.A., Schaefer, T., Johnson, K.R., and Nichols, D.J., 2001, A gap in the record of all vertebrates at the K-T boundary in North Dakota supports abrupt dinosaur extinction: Geology, v. 29, p. 39–42.

Petsch, B.C., 1956, Areal geology of the Ladner Quadrangle: Vermillion, South Dakota Geological Survey, scale 1:62 500.

Pipiringos, G.N., Chisholm, W.A., and Kepferle, R.C., 1965, Geology and uranium deposits in the Cave Hills area, Harding County, South Dakota: U.S. Geological Survey Professional Paper 476-A, 64 p.

Relf, P., 2000, A dinosaur named Sue: The story of the colossal fossil: New York, Scholastic Inc., 64 p.

Retallack, G.J., 1994, A pedotype approach to latest Cretaceous and earliest Tertiary paleosols in eastern Montana: Geological Society of America Bulletin, v. 106, p. 1377–1397.

Rigby, J.K., Jr., 1987, The last of the North American dinosaurs, *in* Czerkas, S.J., and Olson, E.C., eds., Dinosaurs past and present; Volume II: Seattle, University of Washington Press, p. 118–135.

Schulte, J.J., 1957, Areal geology of the Camp Crook and Midland #4 Quadrangle: Vermillion, South Dakota Geological Survey, scale 1:62 500.

Sheehan, P.M., and Fastovsky, D.E., 1993, Major extinctions of land-dwelling vertebrates at the Cretaceous-Tertiary boundary, eastern Montana: Geology, v. 20, p. 556–560.

Sheehan, P.M., Fastovsky, D.E., Barreto, C., and Hoffman, R.G., 2000, Dinosaur abundance was not declining in a "3 m gap" at the top of the Hell Creek Formation, Montana and North Dakota: Geology, v. 28, p. 523–526.

Shoemaker, R.E., 1966, Fossil leaves of the Hell Creek and Tullock Formations of eastern Montana: Palaeontographica, Abteilung B, v. 119, p. 54–75.

Sloan, R.E., and Van Valen, L., 1965, Cretaceous mammals from Montana: Science, v. 148, p. 220–227.

Spicer, R.A., and Herman, A.B., 1996, *Nilssoniocladus* in the Cretaceous Arctic: New species and biological insights: Review of Palaeobotany and Palynology, v. 92, p. 229–245.

Stanley, E.A., 1965, Upper Cretaceous and Paleocene plant microfossils and Paleocene Dinoflagellates and Hystrichosphaerids from northwestern South Dakota: of American Paleontology Bulletin, v. 49, p. 179–379.

Stanton, T.W., 1910, Fox Hills Sandstone and Lance Formation ("Ceratops") in South Dakota, North Dakota, and eastern Wyoming: American Journal of Science, v. 30, p. 172–188.

Stanton, T.W., 1921, The fauna of the Cannonball marine member of the Lance Formation: U.S. Geological Survey Professional Paper 128-A, 60 p.

Stewart, W.N., and Rothwell, G.W., 1993, Paleobotany and the evolution of plants: Cambridge, Cambridge University Press, 521 p.

Sweet, A.R., and Braman, D.R., 1992, The K-T boundary and contiguous strata in Western Canada; interactions between paleoenvironments and palynological assemblages: Cretaceous Research, v. 13, p. 31–79.

Sweet, A.R., Braman, D.R., and Lerbekmo, J.F., 1990, Palynofloral response to K/T boundary events; a transitory interruption within a dynamic system, *in* Sharpton, V.L., and Ward, P.D., eds., Global catastrophies in Earth history: An interdisciplinary conference on impacts, volcanism, and mass mortality: Boulder, Colorado, Geological Society of America Special Paper 247, p. 457–469.

Swisher, C.C., III, Dingus, L., and Butler, R.F., 1993, $^{40}Ar/^{39}Ar$ dating and magnetostratigraphic correlation of the terrestrial Cretaceous-Paleogene boundary and Puercan Mammal Age, Hell Creek-Tullock Formations, eastern Montana: Canadian Journal of Earth Sciences, v. 30, p. 1981–1996.

Takhtajan, A.L., 1980, Outline of the classification of flowering plants (Magnoliophyta): Botanical Review, v. 46, no. 3, p. 1–225.

Thom, W.T., Jr., and Dobbin, C.E., 1924, Stratigraphy of Cretaceous-Eocene transition beds in eastern Montana and the Dakotas. Geological Society of America Bulletin, v. 35, p. 481–506.

Tschudy, R.H., 1970, Palynology of the Cretaceous-Tertiary boundary in the northern Rocky Mountain and Mississippi Embayment regions, *in* Kosanke, R.M., and Cross, A.T., eds., Symposium on palynology of the Late Cretaceous and early Tertiary: Geological Society of America Special Paper 127, p. 65–111.

Tschudy, R.H., Pillmore, C.L., Orth, C.J., Gilmore, J.S., and Knight, J.D., 1984, Disruption of the terrestrial plant ecosystem at the Cretaceous-Tertiary boundary, Western Interior: Science, v. 225, p. 1030–1032.

Vuke-Foster, S.M., Coulton, R.B., Stickney, M.C., Wilde, E.M., Robocker, J.E., and Christensen, K.C., 1986, Geology of the Baker and Wibaux 30 × 60-minute quadrangles, eastern Montana and adjacent North Dakota: Montana Bureau of Mines and Geology, Geologic Map 41, scale 1:100000.

Williams, B.L., 1988, Megafloral relationships within a Paleocene river system, southeastern Montana [thesis]: Amherst, Massachusetts, Amherst College, 116 p.

Winchester, D.E., Hares, C.J., Lloyd, E.R., and Parks, E.M., 1916, The lignite field of northwestern South Dakota: U.S. Geological Survey Bulletin 627, 167 p.

Wing, S.L., and Tiffney, B.H., 1987a, The reciprocal interaction of angiosperm evolution and tetrapod herbivory: Review of Palaeobotony and Palynology, v. 50, p. 179–210.

Wing, S.L., and Tiffney, B.H., 1987b, Interactions of angiosperms and herbivorous tetrapods through time, *in* Friis, E.M., et al., eds., The origin of angiosperms and their biological consequences: Cambridge, Cambridge University Press, p. 203–224.

Wing, S.L., Hickey, L.J., and Swisher, C.C., 1993, Implications of an exceptional fossil flora for Late Cretaceous vegetation: Nature, v. 363, p. 342–344.

Wolfe, J.A., 1979, Temperature parameters of humid to mesic forests of eastern Asia and relation to forests of other regions in the Northern Hemisphere: U.S. Geological Survey Professional Paper 1106, 37 p.

Manuscript Accepted by the Society September 24, 2001

Printed in the U.S.A.

Geological Society of America
Special Paper 361
2002

Palynology and palynostratigraphy of the Hell Creek Formation in North Dakota: A microfossil record of plants at the end of Cretaceous time

Douglas J. Nichols*
U.S. Geological Survey, MS 939, Box 25046, Denver Federal Center, Denver, Colorado 80225, USA

ABSTRACT

The Hell Creek Formation of North Dakota preserves a rich and varied palynoflora including pollen, spores, and algal cysts; this palynoflora provides insight into the nature of plant life in the region in late Maastrichtian time. This palynological record supplements extensive data from the same ancient vegetation preserved in the megafloral record, and it provides background biostratigraphic data necessary to determine the nature of the Cretaceous-Tertiary (K-T) boundary event.

This study of the Hell Creek palynoflora is based upon occurrences of palynomorphs in 71 samples from 20 measured sections in Bowman and Slope Counties, southwestern North Dakota. The palynoflora documented includes 115 or more palynomorph species, 98 of which are described. Three new genera are proposed, three new species are described, and five new nomenclatural combinations are made. The new taxa are *Palaeoisoetes* gen. nov., *Styxpollenites* gen. nov., *Tschudypollis* gen. nov., *Aquilapollenites marmarthensis* sp. nov., *Osmundacidites stanleyi* sp. nov., *Styxpollenites calamitas* sp. nov., *Polyatriopollenites levis* (Stanley 1965) comb. nov., *Palaeoisoetes subengelmannii* (Elsik 1968) comb. nov., *Triporopollenites triplicatus* (Anderson 1960) comb. nov., *Tschudypollis retusus* (Anderson 1960) comb. nov., and *Tschudypollis thalmannii* (Anderson 1960) comb. nov. The palynofloral taxa include 67 angiosperms, eight gymnosperms, 19 pteridophytes, and three bryophytes. Of these, 33 are regarded as characteristically uppermost Cretaceous taxa (K taxa). A few of the K taxa do not range all the way to the K-T boundary. Extinction of the Hell Creek palynoflora at the K-T boundary is ~30%. The Hell Creek Formation is within the *Wodehouseia spinata* Assemblage Zone.

INTRODUCTION

Purpose of this study

The Hell Creek Formation in North Dakota and adjacent parts of South Dakota and Montana carries an extraordinarily varied and well-preserved microfossil flora of pollen, spores, and algal cysts that provides valuable insight into the nature of plant life in the region in late Maastrichtian time. This palynological record supplements the extensive data from the same rocks and the same ancient vegetation preserved in the megafloral record (Johnson and Hickey, 1990; Johnson, 1992, this volume), and it provides the background biostratigraphic data required to precisely determine the position of the Cretaceous-

*E-mail: nichols@usgs.gov

Nichols, D.J., 2002, Palynology and palynostratigraphy of the Hell Creek Formation in North Dakota: A microfossil record of plants at the end of Cretaceous time, *in* Hartman, J.H., Johnson, K.R., and Nichols, D.J., eds., The Hell Creek Formation and the Cretaceous-Tertiary boundary in the northern Great Plains: An integrated continental record of the end of the Cretaceous: Boulder, Colorado, Geological Society of America Special Paper 361, p. 393–456.

Tertiary (K-T) boundary on the basis of palynology (see Nichols and Johnson, this volume). Despite its importance to palynology, paleobotany, biostratigraphy, and the nature of the terminal Cretaceous extinction event, this palynoflora has not been formally described previously. In this study a reference section extending from the base to the top of the Hell Creek Formation and shorter sections at numerous other stratigraphically well controlled localities provide a database documenting the occurrence of species that constitute the *Wodehouseia spinata* Assemblage Zone, along with associated microfossil taxa. The purpose of this report is to describe, illustrate, and biostratigraphically interpret the palynoflora of the Hell Creek Formation in North Dakota.

Material

Occurrences of palynomorphs in 71 samples collected from 20 measured sections in Bowman and Slope Counties, southwestern North Dakota, are the basis of this study. Samples were collected throughout the stratigraphic thickness of the formation, but they are not evenly distributed. Because of special interest in the K-T boundary, more samples were collected within 1 m of the top of the formation than elsewhere. Table 1 shows the distribution of samples within 10 m intervals from the base to the top of the formation; the uppermost 10 m is further subdivided. Table 2 shows the distribution of samples by lithofacies.

I discuss 98 taxa herein, 14 of which are listed as "(genus name) spp." Each of these categories includes two or more species that are not differentiated in this study for reasons of expediency. More individual species than the 100–115 or so effectively covered here may be present in the Hell Creek palynoflora; Hotton (1988) identified 291 species, exclusive of algae and fungi, in her study of the Hell Creek palynoflora in Montana. The taxa selected for description and analysis in this study include only pollen and spores that are consistently recognizable, reasonably common in occurrence, or deemed to have biostratigraphic or paleoecologic significance. Others were ignored or relegated to an informal category for future study. Those that are discussed here represent the vast majority of the important palynologic taxa that characterize the Hell Creek Formation in North Dakota.

TABLE 1. STRATIGRAPHIC DISTRIBUTION OF SAMPLES

Meters above base	Number of samples
0–10	2
10–20	2
20–30	2
30–40	0
40–50	5
50–60	2
60–70	5
70–80	13
80–90	3
90–100	32
90–95	2
95–99	6
99–100	24

TABLE 2. DISTRIBUTION OF SAMPLES BY LITHOFACIES

Lithofacies	Number of samples
Mudstone	37
Claystone	12
Shale	5
Carbonaceous shale	5
Lignite	6

It is well known that fossil palynomorphs can be extremely abundant in some sedimentary facies. In this study, more than 110 000 specimens were observed during the scanning of microscope slides and 5000 specimens were counted to determine percentage abundances of individual taxa. These figures include only those samples used in the study of the Hell Creek palynoflora and do not consider samples from the Fort Union Formation also analyzed. To provide a basis for comparison with the Hell Creek palynoflora, Fort Union samples were studied; in the course of that investigation, more than 330 000 specimens were scanned and more than 20 000 were counted. Thus, interpretations of presence, absence, and relative abundance presented here are derived from a large database.

Study area

The study area extends in a north-south outcrop belt along the Little Missouri River from T.135N. in Slope County to T.129N. in Bowman County (Fig. 1). Kirk R. Johnson measured numerous reference sections in this area from 1987 to 1999, including intervals within the Hell Creek Formation and across the Hell Creek–Fort Union contact and into the Fort Union Formation. Johnson and I collected hundreds of samples from these sections. Only those from the Hell Creek Formation that yielded abundant plant microfossils are referred to in this study; some other localities are discussed elsewhere in this volume (see Nichols and Johnson; Pearson et al.). Additional reference sections used in this study were measured by D.E. Fastovsky (Sunset Butte) and E.C. Murphy (Pretty Butte) and sampled by me.

The study area includes the most extensive outcrops of the Hell Creek Formation in North Dakota; the more restricted outcrop belt along the Missouri River in the south-central part of the state was the subject of a previous study involving palynology (Murphy et al., 1995). The badlands of southwestern North Dakota afford excellent, often continuous outcrops of the formation extending from the underlying Fox Hills Sandstone through to the overlying Ludlow Member of the Fort Union Formation. However, only one section provides access to the entire thickness of the formation in outcrop. That section is at and near Sunset Butte in Bowman County. The study area also includes several measured sections in the vicinity of Marmarth, North Dakota. Among these are Pretty Butte, a locally well known geomorphological feature in Slope County north of

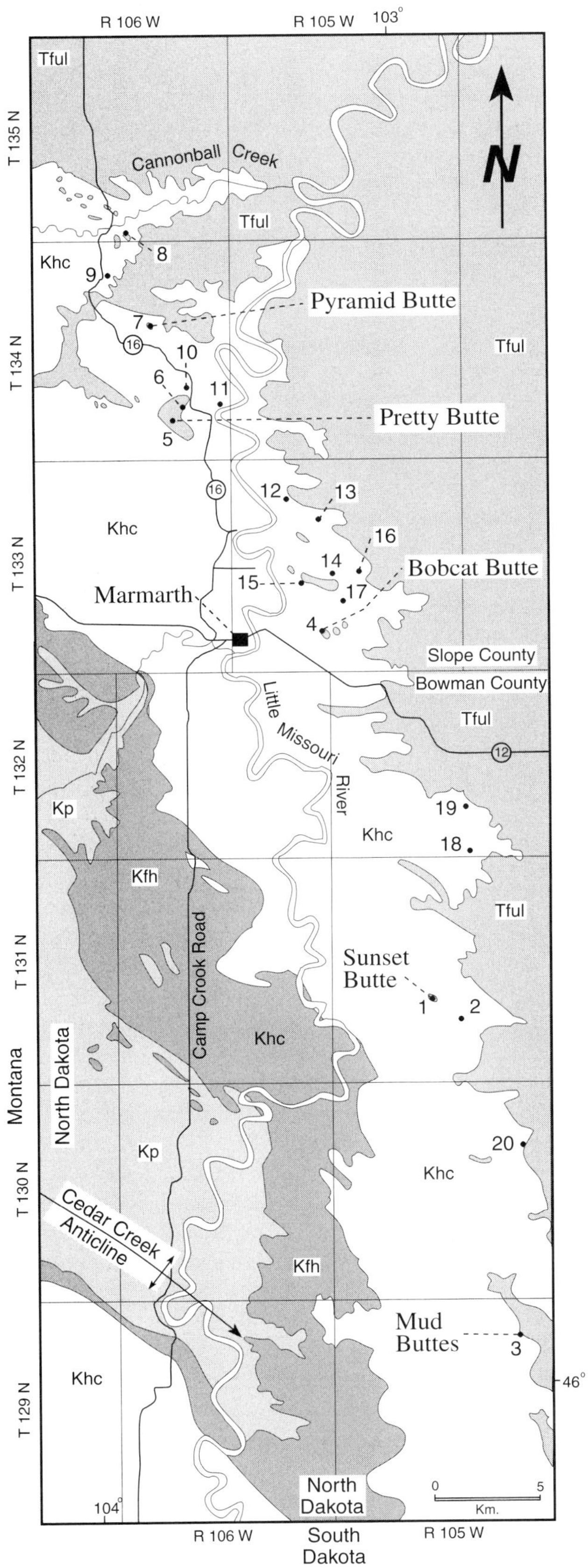

Figure 1. Geologic map of study area in southwestern North Dakota showing major features and locations of measured sections from which Hell Creek Formation samples were collected. Geologic units: Kp, Pierre Shale; Kfh, Fox Hills Sandstone; Khc, Hell Creek Formation; Tful, Ludlow Member of Fort Union Formation. Measured sections: 1, Sunset Butte; 2, Sunset Patch; 3, Mud Buttes; 4, Bobcat Butte; 5, Pretty Butte; 6, Pretty Butte North; 7, Pyramid Butte; 8, Facet Butte; 9, Badland Draw; 10, Golva Road; 11, Hell Creek Coal; 12, River Section; 13, No Leaf Minors; 14, Battleship; 15, Dean's NM1; 16, Barbecue; 17, Roadside HCIIb; 18, Swanke's Nose; 19, Swanke's Coal; 20, Torosaurus Section.

Marmarth, and Pyramid Butte, a key reference section at which the K-T boundary—complete with an anomalous concentration of iridium and the presence of shock-metamorphosed mineral grains—was discovered (Johnson et al., 1989). At the southern end of the outcrop belt is Mud Buttes, a second section in which debris from the Chicxulub impact is preserved at the K-T boundary (Nichols et al., 2000).

Stratigraphy

A detailed discussion of the stratigraphy of the Hell Creek Formation is omitted from this report because that topic is better covered by other chapters in this volume (see especially Murphy et al.; Hicks et al.). A brief summary is given here. Stratigraphic nomenclature is summarized in Figure 2.

The Hell Creek Formation (Upper Cretaceous: Maastrichtian) is ~100 m thick. It overlies either the Colgate Member of the Fox Hills Sandstone (Maastrichtian) or the main body of the Fox Hills Sandstone, and it underlies the Ludlow Member of the Fort Union Formation (uppermost Maastrichtian and/or lower Paleocene). It consists of poorly lithified sandstone, siltstone, mudstone, shale, carbonaceous shale, and rare lignite

	North Dakota South Dakota	Montana	Wyoming	south-central Saskatchewan
Paleocene	Fort Union Formation	Fort Union Formation	Fort Union Formation	Ravenscrag Formation
Maastrichtian	**Hell Creek Formation**	Hell Creek Formation	Lance Formation	Frenchman Formation
	Fox Hills Formation	Fox Hills Sandstone	Fox Hills Sandstone	Whitemud Formation
				Eastend Formation

Figure 2. Chart showing stratigraphic nomenclature for uppermost Cretaceous and lower Tertiary in study area. Basic stratigraphic relations and correlations of Hell Creek Formation are indicated. North Dakota Geological Survey recognizes Fort Union as group, but U.S. Geological Survey nomenclature is followed here. Position of Maastrichtian-Paleocene boundary is approximate; see discussion in text.

beds deposited in fluvial environments at the end of Cretaceous time, during the regression of the Cretaceous seaway. No members are recognized within the formation in the study area (see Murphy et al., this volume). In the field the Hell Creek Formation often is distinguished from the Fort Union Formation by its characteristic "popcorn" weathering appearance, which results in the surface being covered with soft, rounded pellets and is caused by the presence of bentonitic clay, and its generally dark color (early descriptions of the Hell Creek beds refer to them as somber). The Fort Union Formation contains common lignite beds and tends to be finer grained and lighter in color. The formational contact is conformable and gradational, and it may be difficult to pick the contact at some localities if the field-mapper's guide—a basal Fort Union lignite bed—is absent. The Hell Creek Formation typically contains abundant dinosaur bones and fossil leaves of Maastrichtian age.

The contact between the Hell Creek and Fort Union Formations is often taken to mark the K-T boundary. This may be a useful rule-of-thumb in some field studies, but it is not correct in detail at most localities (as discussed more completely by Nichols and Johnson, this volume). At most localities investigated in that report and herein, the Hell Creek palynoflora, which is of latest Cretaceous age, also occurs in the lowermost part of the overlying Ludlow Member of the Fort Union Formation. At one locality near Ekalaka, in eastern Montana, investigated in the course of this study but not discussed herein, the K-T boundary occurs well below the Hell Creek–Fort Union formational contact. Murphy et al. (1995) found the K-T boundary to be near but not coincident with the formational contact in south-central North Dakota. By definition, the formational contact is lithostratigraphic, whereas the K-T boundary is geochronologic.

PREVIOUS STUDIES

Palynological study of the Hell Creek Formation dates from the work of Stanley (1961a, 1961b, 1965), who investigated outcrops of the formation at Crow Butte, Harding County, South Dakota. Stanley (1965) described 46 taxa of plant microfossils, mostly pollen and spores, including 24 that were described as new. Many of Stanley's species are still recognized today, although others have undergone taxonomic and nomenclatural revisions. In his study he named 27 species believed to be restricted to the Hell Creek Formation; the others also occurred in the overlying Fort Union Formation. The current study changes many of those interpretations on the basis of more extensive data. For example, ten species Stanley (1965) indicated as restricted to the Hell Creek Formation are now also recorded in the basal part of the Fort Union Formation, and eight species indicated as restricted to the Fort Union are now known from the Hell Creek.

Norton (1965), Hall and Norton (1967), Norton and Hall (1967, 1969), and Oltz (1969, 1971) conducted a series of investigations of the Hell Creek palynoflora in the Hell Creek area of eastern Montana. Among these, the most important for present purposes are those of Norton and Hall (1967, 1969). In their first paper, these authors listed eight palynomorph species that they considered to be guides to the Hell Creek Formation, i.e., characteristic species. Three of these were species previously described by Stanley (1961a, 1965) and two were described as new. None are restricted to the formation. Norton and Hall (1969) defined four groups of palynomorph taxa: 58 that occurred only in the Hell Creek Formation, 28 that occurred in a so-called transition assemblage that included taxa found in a zone near the Hell Creek–Fort Union formational contact, 22 that occurred only in the Fort Union, and 30 long-ranging species that occurred in both the Hell Creek and Fort Union Formations. The transition flora of Norton and Hall lacks characteristic Maastrichtian species; their definition of this assemblage suggests that they may have had problems picking the formational contact in the field or with correlation of sampled sections. They described and illustrated 116 species-level taxa from the Hell Creek Formation, a number close to that recorded in this study and much less than that reported by Hotton (1988) from the same area.

In their landmark publication on the stratigraphic distribution of species of the characteristic Upper Cretaceous genus *Aquilapollenites,* Tschudy and Leopold (1970) described, illustrated, and determined the stratigraphic ranges of ten species of the genus that occur in the Hell Creek Formation. All of their records were from Montana. Their taxonomic and nomenclatural concepts of these species are followed herein. New data presented here supplement their work.

Tschudy (1970) discussed characteristic species of the Hell Creek and Fort Union Formations in a major study of the K-T boundary in the Hell Creek, Seven Blackfoot Creek, Brockton, and Glendive areas in Montana. Especially important in his study were the measured sections in the Seven Blackfoot Creek and Glendive areas. Tschudy analyzed 26 samples collected through the entire thickness of the Hell Creek Formation at these localities. He designated 19 species as characteristic of the Hell Creek palynoflora. This assemblage differs strongly from its correlative equivalent in the Mississippi Embayment region discussed by Tschudy. Tschudy (1970) reported no species in common in the Maastrichtian assemblages from these two regions.

Robertson (1973) described a new palynomorph species from the Hell Creek Formation in Montana, interpreting it to be a unique fern spore. Robertson and Elsik (1978) reinterpreted the botanical affinity of this species, determining it to be angiosperm pollen, and stated that it is an index fossil for the Maastrichtian in Montana and the adjacent part of Saskatchewan, Canada.

Bergad (1973) described species of Cretaceous megaspores he assigned to the genus *Balmeisporites,* including four based on specimens from the Hell Creek Formation in North Dakota and Montana. (Similar specimens are herein assigned to the genus *Ghoshispora.*)

Hotton (1988) conducted a detailed study of the Hell Creek palynoflora near the K-T boundary at six localities in the Hell Creek area of Montana (see also Hotton, this volume). Her analyses involved 52 samples from the Hell Creek Formation. Hotton described 313 palynomorph taxa, including 291 species of pollen and spores. She recognized four groups of taxa: (1) K taxa, those that are exclusively Cretaceous or that underwent a highly significant reduction in the Tertiary; (2) T taxa, those that underwent a highly significant increase in the Tertiary; (3) K-T taxa, those that show no significant change across the K-T boundary; and (4) those that are too rare to be significant. Her results provided strong support for the extraterrestrial-impact theory of the K-T extinctions (Hotton, 1984). Hotton's current interpretation (this volume) of the nature and significance of the Hell Creek palynoflora in Montana provides an important comparison with the analyses and interpretations presented in this chapter. Only about one-third as many taxa are discussed here—those considered to represent the most important species in the palynoflora in North Dakota—but the assemblages from eastern Montana and western North Dakota are essentially the same. There may be minor differences in taxonomy and nomenclature.

Johnson et al. (1989) briefly discussed the palynology of the K-T boundary in southwestern North Dakota. Only one sample from the Hell Creek Formation was used in that critical analysis, which also relied on extensive data on the Hell Creek megaflora and on iridium analyses. The Hell Creek palynomorph assemblage mentioned included 14 taxa (20 or more individual species grouped for convenience), all of which were well known on the basis of previous studies of uppermost Cretaceous rocks in Montana and Wyoming. The locality—Pyramid Butte—is a key reference section in this study and for others in this volume (see especially Nichols and Johnson; Pearson et al.; Hicks et al.).

Nichols and Sweet's (1993) comprehensive analysis of Upper Cretaceous nonmarine palynofloras in the western United States and Canada included those of the Hell Creek Formation and correlative equivalents from the Yukon Territory to New Mexico. They expanded the definition of the *Wodehouseia spinata* assemblage, which characterizes the upper Maastrichtian *Wodehouseia spinata* Assemblage Zone. Nichols and Sweet named 40 taxa as characteristic of the *Wodehouseia spinata* assemblage; among these, 33 are present in the Hell Creek Formation in North Dakota. Nichols and Sweet noted that the existence of the *Wodehouseia spinata* Assemblage Zone in western North America was terminated by the K-T boundary event, although some individual species in the assemblage survived.

In the only previously published study involving the Hell Creek palynoflora in North Dakota, Murphy et al. (1995) discussed and illustrated 24 species of pollen and spores, seven of which are restricted to the Cretaceous. These species were used to define the stratigraphic position of the K-T boundary in south-central North Dakota. Murphy et al. determined the position of the K-T boundary at 12 localities in that area. The boundary was found to be within 3 m of the Hell Creek–Fort Union formational contact, either above or below it, but not necessarily coincident with the contact.

Six previously published studies of correlative equivalents of the Hell Creek Formation proved to be useful for this study. The relevance of each is discussed briefly here.

Leffingwell (1970) provided detailed descriptions and illustrations of palynomorphs from the Lance Formation in Wyoming. His study is the most detailed taxonomic analysis of the upper Maastrichtian assemblage in the region previously published. Leffingwell discussed 16 taxa that characterize his assemblage A from the uppermost part of the Lance Formation. Six of these taxa were found to be restricted to the Lance Formation, six were found to be common in the Lance and rare in the overlying Tullock Member of the Fort Union Formation, and the other four were about equally common in both stratigraphic units. Leffingwell discussed the stratigraphic ranges and relative abundances of 40 taxa that occur in the Lance Formation (13 samples were used in this compilation). Leffingwell defined some of his taxonomic categories broadly; they include several possibly distinguishable species. He described seven new species and revised three previously published species that are identified in the Hell Creek Formation in this study. His report is well illustrated through a variety of techniques of transmitted-light photomicrography.

Farabee and Canright (1986) also investigated the palynoflora of the Lance Formation in Wyoming. They described and illustrated 146 species of pollen and spores, including two new species; they proposed nine new nomenclatural combinations. Only ~40 of the species identified in the Lance palynoflora by Farabee and Canright are identified in the Hell Creek palynoflora in this study, however, and several of these are given different names here. Some differences between Farabee and Canright's species list and that herein may be attributable to the former concentrating on the lower part of the Lance Formation; others may be due to differences in the palynofloras related to the different geographic positions of the respective study areas. Farabee and Canright (1986) used 30 samples in their study; their report is well illustrated with transmitted-light and scanning-electron photomicrographs.

Jarzen (1978) studied the Frenchman Formation of Saskatchewan, the correlative equivalent of the Hell Creek Formation. The study involved five samples and 15 taxa. The taxa included algal cysts as well as pollen and spores; all were identified and interpreted in terms of their presumed modern affinities and paleoecological significance. No biostratigraphic information was developed in that study other than the coincidental documentation of the presence of these taxa in the Frenchman Formation. Of the taxa discussed by Jarzen, ten can be recognized among the pollen and spores described herein.

The palynoflora of the Frenchman Formation also received attention in the study by Nichols et al. (1986) on the K-T boundary in Saskatchewan. Their analysis summarized the occurrences of 27 pollen and spore taxa in four samples from the

Frenchman below the K-T boundary. It documented local extinction of ~30% of the palynoflora at the stratigraphic level of an iridium anomaly and shock-metamorphosed minerals. The position of the K-T boundary at this locality coincides with the contact between the Frenchman Formation and the overlying Ravenscrag Formation. All of the taxa reported by Nichols et al. (1986) in the palynoflora of the Frenchman Formation in Saskatchewan are present in the Hell Creek Formation in North Dakota.

The palynoflora of the Frenchman Formation in Saskatchewan and Alberta and its correlative equivalent, the Boissevain Formation in Manitoba, have been the subject of detailed investigations by Braman and Sweet (1999) and Sweet et al. (1999). Braman and Sweet (1999) determined stratigraphic ranges of key taxa, many of which are the same as those in the Hell Creek Formation, in the upper Maastrichtian in three core holes in southern Canada. The biostratigraphic zonation proposed by Braman and Sweet (1999) is compared with that for the Hell Creek Formation herein. Sweet et al. (1999) determined ranges and relative abundances of many taxa in the interval including the K-T boundary in core hole and outcrop sections, especially in Saskatchewan. That study also included similar analyses of a section previously studied by Hotton (1988) in Montana that included the uppermost part of the Hell Creek Formation, and new interpretations of data published by Wolfe (1991) and Nichols et al. (1992a) from their respective K-T boundary localities in Wyoming (which included the uppermost part of the Lance Formation, the Hell Creek equivalent in Wyoming).

METHODS

Field and laboratory procedures

Samples discussed in this chapter were collected from carefully measured stratigraphic sections. K.R. Johnson gave each of the measured sections a unique field name, by which they are referred to here. Intervals sampled varied in thickness from 2.5 to 5.0 cm. Stratigraphic positions of all samples were determined relative to the Hell Creek–Fort Union formational contact and are cited in this report as "-(number) cm" to indicate the number of centimeters below the formational contact. To afford maximum accuracy in listing sample positions in this report, some sampled intervals are given as the entire thickness of a unit in the measured section, where the precise level within that unit from which the sample was collected was not recorded in the field; thus, apparent sample intervals of as much as 82 cm are indicated. In no case was the interval actually sampled more than 2.5–5.0 cm, however. Where only a single number is given instead of an interval, the sample was from a 2.5–5.0 cm interval centered on the number given. Great care was taken to avoid cross-contamination of sample material during collection. The samples were stored in heavy-weight plastic bags labeled with field numbers.

Field numbers were eventually replaced by U.S. Geological Survey (USGS) laboratory numbers on file at the USGS palynology lab in Denver, Colorado; these are unique four-digit numbers preceded by the letter D and followed by a hyphen and one or two additional letters to distinguish individual samples, unless a single sample was collected at the locality. Separate collections made at different times from a single measured section may have been assigned different D numbers, which were then used to label all laboratory preparations, microscope slides, unmounted material, and archival splits of unprocessed samples. The repository for all microscope slides, unmounted material, and splits of samples discussed in this chapter is the Central Energy Resources palynology laboratory at the USGS in Denver, Colorado.

Standard palynological laboratory procedures were used to prepare samples for analysis (Doher, 1980; Traverse, 1988). The success rate in preparation was greater than 70%; i.e., more than 70% of samples collected from the Hell Creek Formation in this study yielded abundant, well-preserved palynomorphs when prepared. Preparations that were barren of palynomorphs or that resulted in only sparse recovery of palynomorphs were regarded as unsuccessful.

Microscopic analyses were routine in nature except for the procedure adopted to provide a comprehensive and representative compilation of the diversity and abundance of palynomorphs present in each sample. The standard chosen was to count 200 specimens per slide to determine relative abundance in percent and then to continue to scan the slide until a minimum of 2000 specimens were observed. The additional scanning inevitably revealed the presence of additional rare species. Their presence was recorded but did not change the percentage numbers already determined. This procedure, which tends to emphasize presence or absence data over relative abundance, was adopted because it became evident early in the study that biostratigraphically important species, e.g., members of the *Wodehouseia spinata* assemblage, might not be common in certain samples, yet their presence might be critical to interpretation of the sample. Scanning 2000 or more specimens was required to determine the presence of other species in samples that were numerically dominated by one or a few species. Percentage abundances of individual taxa were not determined for all samples studied; for samples not counted, presence or absence data were recorded based on scanning.

Reference sections and other sample localities

The composite measured section at Sunset Butte is the primary palynological reference section for the Hell Creek Formation in North Dakota. Sunset Butte is a prominent geomorphological feature in southwestern North Dakota, in Bowman County ~16 km southeast of Marmarth (Fig. 1). The entire thickness of the formation is exposed in the butte and in slopes extending down from it to the Little Missouri River. Numerous individual sample localities in the composite measured section were collected, each being assigned a different D number (see Appendix 1). In addition to the palynological study of this measured section discussed here and in the report by Nichols and Johnson (this volume), there is a study of paleomagnetic stratigraphy at Sunset Butte (Hicks et al., this volume). Much of the uppermost part of the section at Sunset Butte consists of lithologies that cannot be successfully sampled for palynology. To compensate for this gap in data, a supplementary section called the Sunset Patch section was measured and sampled nearby (Fig. 1; Appendix 1).

The Mud Buttes locality is ~32 km southeast of Marmarth, at the southernmost extent of the study area (Fig. 1). The measured section at Mud Buttes includes six samples from the Hell Creek Formation (see Appendix 1). This section is especially important because the K-T boundary has been precisely located in this section, both palynologically and by the presence of shocked quartz (Nichols et al., 2000). Numerous productive samples were collected from the overlying Fort Union Formation in this section, but are not discussed herein (see Nichols and Johnson, this volume; see also the discussion of magnetostratigraphy at Mud Buttes; Hicks et al., this volume).

The Bobcat Butte locality is ~3 km east of Marmarth, in about the middle of the study area (Fig. 1). Collections were made from two intervals within the measured section that provide important stratigraphic coverage for the Hell Creek palynoflora. The intervals sampled are one near the top of the Hell Creek Formation and a second ~22–50 m below the top (see Appendix 1). Samples were collected at closely spaced intervals within the lowermost 1 m of the overlying Fort Union Formation at this locality, but only the Hell Creek samples are discussed herein. This measured section was also sampled for magnetostratigraphic and chemostratigraphic studies (see Hicks et al., this volume; Ahrens and Jahren, this volume).

Pretty Butte is ~13 km north of Marmarth, toward the northern extent of the study area (Fig. 1). Pretty Butte affords good exposures of the upper part of the Hell Creek Formation and lower part of the Fort Union Formation, although for this study only a few samples were collected from the Hell Creek (see Appendix 1). The Fort Union was also extensively sampled at Pretty Butte, but those samples are not discussed in this report. A short measured section nearby called Pretty Butte North supplements the primary measured section at Pretty Butte with additional samples collected near the top of the Hell Creek Formation (see Appendix 1).

Several measured sections generally north of Pretty Butte and in the vicinity of Cannonball Creek in Slope County (Fig. 1) each provided small numbers of samples from the Hell Creek Formation that were used in this study (see Appendix 1). The Pyramid Butte section was also the subject of a previously published study of the K-T boundary in North Dakota (Johnson et al., 1989); only the lowermost sample analyzed at Pyramid Butte was from the Hell Creek Formation. Facet Butte is the northernmost measured section in the study area; a single sample was collected from the top of the Hell Creek. The Badland Draw measured section in Slope County brackets the K-T boundary; a single sample was collected there from the top of the Hell Creek. A sample collected at the Golva Road locality is from the middle part of the formation; it provides additional stratigraphic coverage for the Hell Creek palynoflora. One of the relatively rare lignite beds in the Hell Creek Formation is exposed in the cut bank of the Little Missouri River east of Pretty Butte in the section designated Hell Creek Coal; four samples were collected from this interval, which is stratigraphically about in the middle of the formation.

Several measured sections located generally north of Bobcat Butte (Fig. 1) each provided small numbers of samples used in this study (see Appendix 1). The River Section is ~8 km northeast of Marmarth; samples from this measured section include four at considerable stratigraphic distance below the top of the Hell Creek and a fifth near the top of the formation (samples collected from the Fort Union Formation in the River Section are not discussed herein). A single Hell Creek sample was analyzed from the measured section called No Leaf Minors; it is at the top of the formation. A single sample from the top of the Hell Creek was analyzed from the measured section called Battleship; the sample is from just below a Fort Union lignite bed. The measured section designated Dean's NM1 includes a vertebrate microsite (a locality from which vertebrate microfossils were collected; see Pearson et al., this volume); two Hell Creek samples collected from this section are from near the top of the formation. In the Barbecue measured section, a thick lignite bed marking the base of the Fort Union Formation overlies the Hell Creek Formation; the section includes three Hell Creek samples. Two samples were collected at the locality designated Roadside HCIIb; the samples bracket a bed containing fossil leaves collected by K.R. Johnson and identified by him as belonging to his megafloral zone HCIIb, thus making possible a comparison of plant megafossil and microfossil assemblages that characterize this part of the Hell Creek Formation.

Three measured sections generally southeast of Marmarth and toward the southern part of the study area in Bowman County (Fig. 1) each provided small numbers of samples of the Hell Creek Formation (see Appendix 1). The measured section called Swanke's Nose spans the Hell Creek–Fort Union formational contact at a location where there is no basal Fort Union lignite; two Hell Creek samples were analyzed from this locality. Swanke's Coal is another of the relatively rare lignite beds in the Hell Creek; this lignite bed is stratigraphically above the one at the Hell Creek Coal locality. The K-T boundary is closely bracketed at the top of the Torosaurus Section, where it is coincident with the Hell Creek–Fort Union formational contact; a dinosaur skeleton was collected from this section by the

Pioneer Trails Regional Museum in Bowman, North Dakota (see Pearson et al., this volume).

COMPOSITION OF THE PALYNOFLORA

Overview

The palynoflora of the Hell Creek Formation is numerically dominated by angiosperm pollen, although it includes significant percentages of gymnosperm pollen and spores of ferns and other cryptogams; algal cysts are notably common in some samples, reflecting the origin of certain facies in freshwater depositional environments. In this study the presence of algal cysts, of which there are five or more species, was noted, but they were not counted in percentage abundance calculations. The list of taxa identified and tabulated in this study is given in Table 3. The numbers used to designate taxa in Table 3 are also used in Tables 4 and 5 and Figures 3 and 4.

The most commonly occurring taxa, those recorded at 70% or more of the localities sampled, are *Laevigatosporites* spp., *Wodehouseia spinata, Arecipites* spp., *Gleicheniidites* spp., *Pityosporites* spp., *Triporopollenites* spp., *Ulmipollenites krempii, Liliacidites complexus, Stereisporites* spp., *Aquilapollenites quadrilobus, Aquilapollenites reticulatus, Tschudypollis* spp., *Taxodiaceaepollenites hiatus, Kurtzipites trispissatus, Reticuloidosporites pseudomurii, Cyathidites* spp., and *Discoidites parvistriatus.* Because of the well-known differential productivity of pollen and spores by different species of plants, this list cannot represent the absolute abundance of plants in the Hell Creek flora, but it offers some insights. The list includes ten taxa of angiosperm pollen, two of gymnosperm pollen, four of pteridophyte spores, and one of bryophyte spores. Note that six of these taxa are defined at the genus level and include more than one species (spp.). The common occurrence of these taxa suggests that, although angiosperms dominated the flora numerically in terms of diversity, certain gymnosperms were well represented and bryophytes and ferns were present at most locations. It is possible that the bryophytes and ferns constituted much of the vegetational cover in Hell Creek environments. Wing et al. (1993) showed that in the Meeteetse Formation (lower Maastrichtian, central and northern Wyoming), pteridophytes accounted for almost 50% of the ground cover, although they were low in diversity and amounted to <20% of the species in the flora.

With the caveat of differential productivity, other general statements can be made on the basis of this short list of commonly occurring palynomorph taxa. Palms (*Arecipites* spp.) appear to have been prominent members of the flora. Two other taxa (*Ulmipollenites krempii* and *Triporopollenites* spp.) likely were also produced by arboreal angiosperms. Two taxa (*Pityosporites* spp. and *Taxodiaceaepollenites hiatus*) likely were produced by arboreal gymnosperms. The common occurrence of bryophytes (*Stereisporites* spp.) and pteridophytes (e.g., *Cyathidites* spp., *Laevigatosporites* spp., *Gleicheniidites* spp., and *Reticuloidosporites pseudomurii*) in the samples analyzed may reflect an inevitable sampling bias toward aquatic environments in which fine-grained sediments were deposited, including coal-forming mires, but fern spores of the genus *Laevigatosporites* were present in all of the samples analyzed, indicating the wide distribution of ferns in the Hell Creek flora (cf. the Meeteetse flora; Wing et al., 1993). The list includes five taxa (*Wodehouseia spinata, Liliacidites complexus, Tschudypollis* spp., *Aquilapollenites quadrilobus,* and *Aquilapollenites reticulatus*) generally regarded as characteristically Maastrichtian; none is restricted to the Hell Creek Formation, however, as discussed later. The taxa generally regarded as characteristically Maastrichtian contribute to the unique character of the Hell Creek palynoflora.

If taxa occurring at 50% or more of the localities studied are considered, nine are added to the list just given. Among these are six of angiosperm pollen and three of pteridophyte spores. Among the angiosperms added are three species that are characteristically Maastrichtian: *Aquilapollenites collaris, Myrtipites scabratus,* and *Striatellipollis striatellus.* The additional taxa tend to reinforce the general nature of the Hell Creek palynoflora revealed by the first list, i.e., it is dominated by angiosperm pollen but includes a variety of cryptogam spores, and it includes a high percentage of what can be referred to as K taxa. The number of K taxa in Hell Creek assemblages analyzed in this study ranges from three to 17 and averages ten ($n = 35$); the total percentage abundance of K taxa in Hell Creek assemblages averages 26%.

The palynofloral list for the Hell Creek Formation (Table 3) includes 68 angiosperm taxa, eight gymnosperm taxa, 19 pteridophyte taxa, and three bryophyte taxa. Table 4 lists the taxa according to botanical categories, and Table 5 lists them in morphological categories. Of the 98 individual taxa, 33 are regarded as characteristically Upper Cretaceous (K taxa). Only six K taxa are restricted to the Hell Creek Formation, however; all the others were also recorded in Fort Union Formation strata of Maastrichtian age in the course of this study (see Nichols and Johnson, this volume). The K taxa that have restricted ranges within the Hell Creek Formation are the basis for palynostratigraphic zonation of the Hell Creek Formation, as discussed later. Many K taxa range throughout the formation, and in this study, 22 of them are recorded within 10 cm below the top of the formation.

In the brief discussions that follow, individual taxa are sorted into the major groups to which they belong—spores, gymnosperm pollen, and angiosperm pollen—and listed alphabetically (see Table 5 for morphological categories). Data on occurrence and interpretations of botanical affinity and paleoecologic significance are included; synonymies are limited to major changes in taxonomy and/or nomenclature. The treatments are brief because of space limitations and are informal in style, except in cases where new taxa are introduced. Morphological terminology follows Punt et al. (1994). In discussions of occurrence, the term common (or variants such as rare)

TABLE 3. PALYNOMORPH TAXA IN THE HELL CREEK PALYNOFLORA

Number	Taxon
1.	*Anacolosidites rotundus* Stanley 1965
2.	*Aquilapollenites attenuatus* Funkhouser 1961
3.	*Aquilapollenites bertillonites* Funkhouser 1961
4.	*Aquilapollenites catenireticulatus* Srivastava 1968
5.	*Aquilapollenites collaris* (Tschudy and Leopold 1971) Nichols 1994
6.	*Aquilapollenites conatus* Norton 1965
7.	*Aquilapollenites delicatus* Stanley 1961
8.	*Aquilapollenites marmarthensis* sp. nov.
9.	*Aquilapollenites pyriformis* Norton 1965
10.	*Aquilapollenites quadricretaeus* Chlonova 1961
11.	*Aquilapollenites quadrilobus* Rouse 1957
12.	*Aquilapollenites reductus* Norton 1965
13.	*Aquilapollenites reticulatus* (Mchedlishvili 1961) Tschudy and Leopold 1971
14.	*Aquilapollenites senonicus* (Mchedlishvili 1961) Tschudy and Leopold 1971
15.	*Aquilapollenites turbidus* Tschudy and Leopold 1971
16.	*Aquilapollenites vulsus* Sweet 1986
17.	*Arecipites columellus* Leffingwell 1971
18.	*Arecipites tenuiexinous* Leffingwell 1971
19.	*Arecipites* spp. (undifferentiated)
20.	*Azolla circinata* Oltz and Hall in Hall 1968
21.	*Azolla cretacea* Stanley 1965
22.	*Boehlensipollis minimus* (Leffingwell 1971) Pocknall and Nichols 1996
23.	*Cercidiphyllites* sp.
24.	*Chenopodipollis* sp.
25.	*Cicatricosisporites* spp.
26.	*Corollina* sp.
27.	*Cranwellia rumseyensis* Srivastava 1967
28.	*Cranwellia/Striatellipollis* (undifferentiated)
29.	*Cupanieidites inaequalis* Leffingwell 1971
30.	*Cupuliferoidaepollenites* spp.
31.	*Cupuliferoipollenites* sp.
32.	*Cyathidites diaphana* (Wilson and Webster 1946) Nichols and Brown 1992
33.	*Cyathidites* spp.
34.	*Deltoidospora* sp.
35.	*Discoidites parvistriatus* (Norton in Norton and Hall 1967) Nichols and Brown 1992
36.	*Dyadonapites reticulatus* Tschudy 1973
37.	*Echitricolpites* sp.
38.	*Ephedripites jansonii* (Pocock 1964) Muller 1968
39.	*Ephedripites multipartitus* (Chlonova 1961) Yu, Guo and Mao, 1981
40.	*Ephedripites ovatus* (Pierce 1961) Norton in Norton and Hall 1969
41.	*Erdtmanipollis cretaceus* (Stanley 1965) Norton in Norton and Hall 1969
42.	*Foraminisporis undulatus* Leffingwell 1971
43.	*Fraxinoipollenites variabilis* Stanley 1965
44.	*Ghoshispora bella* (Kondinskaya 1966) Srivastava 1978
45.	*Gleicheniidites* spp.
46.	*Hamulatisporis amplus* Stanley 1965
47.	*Hamulatisporis hamulatis* Krutzsch 1959
48.	*Inaperturopollenites* sp. G
49.	*Inaperturotetradites scabratus* Tschudy 1973
50.	*Kurtzipites circularis* (Norton in Norton and Hall 1969) Srivastava 1981
51.	*Kurtzipites trispissatus* Anderson 1960
52.	*Laevigatosporites* spp.
53.	*Lanagiopollis lihoka* (Srivastava 1972) Frederiksen 1988
54.	*Leptolepidites tenuis* Stanley 1965
55.	*Leptopecopites pocockii* (Srivastava 1967) Srivastava 1978
56.	*Libopollis jarzenii* Farabee et al. 1984
57.	*Liliacidites altimurus* Leffingwell 1971
58.	*Liliacidites complexus* (Stanley 1965) Leffingwell 1971
59.	*Liliacidites* spp.
60.	*Lygodiumsporites* sp.
61.	*Marsypiletes cretacea* Robertson 1973 emend. Robertson and Elsik 1978
62.	*Momipites inaequalis* Anderson 1960
63.	*Monocolpopollenites* sp. cf. *M. scabratus* Stanley 1965
64.	*Myrtipites scabratus* Norton in Norton and Hall 1969
65.	*Nyssapollenites* spp.
66.	*Orbiculapollis lucidus* Chlonova 1961
67.	*Osmundacidites stanleyi* sp. nov.
68.	*Palaeoisoetes subengelmannii* (Elsik 1968) gen. nov., comb. nov.

(*continued*)

TABLE 3. PALYNOMORPH TAXA IN THE HELL CREEK PALYNOFLORA (continued)

Number	Taxon
69.	*Pandaniidites typicus* (Norton in Norton and Hall 1969) Sweet 1986
70.	*Penetetrapites inconspicuus* Sweet 1986
71.	*Pityosporites* spp.
72.	*Polyatriopollenites levis* (Stanley 1965) comb. nov.
73.	*Polycingulatisporites* sp.
74.	*Porosipollis porosus* (Mchedlishvili in Samoilovich and Mchedlishvili 1961) Krutzsch 1969
75.	*Racemonocolpites formosus* Sweet 1986
76.	*Retibrevitricolporites beccus* Sweet 1986
77.	*Reticuloidosporites pseudomurii* Elsik 1968
78.	*Retitrescolpites anguloluminosus* (Anderson 1960) Frederiksen 1979
79.	*Retitriletes* sp.
80.	*Rhoipites* spp.
81.	*Simplicepollis rallus* (Stanley 1965) Nichols and Brown 1992
82.	*Sparganiaceaepollenites* sp.
83.	*Stereisporites* spp.
84.	*Striatellipollis striatellus* (Mchedlishvili in Samoilovich and Mchedlishvili 1961) Krutzsch 1969
85.	*Striatopollis tectatus* Leffingwell 1971
86.	*Styxpollenites calamitas* gen. et sp. nov.
87.	*Taxodiaceaepollenites hiatus* (Potonié 1931) Kremp 1949
88.	*Toroisporis major* (Pflug in Thomson and Pflug 1953) Stanley 1965
89.	*Triatriopollenites granulatus* (Simpson 1961) Leffingwell 1971
90.	*Tricolpites hians* Stanley 1965
91.	*Tricolpites microreticulatus* Belsky, Boltenhagen, and Potonié 1965
92.	*Tricolpites* spp.
93.	*Triporopollenites plektosus* Anderson 1960
94.	*Triporopollenites triplicatus* (Anderson 1960) comb. nov.
95.	*Triporopollenites* spp.
96.	*Tschudypollis retusus* (Anderson 1960) gen. et comb. nov.
97.	*Tschudypollis* spp.
98.	*Ulmipollenites krempii* (Anderson 1960) Frederiksen 1979
99.	*Wodehouseia spinata* Stanley 1961
100.	*Zlivisporis novomexicanum* (Anderson 1960) Leffingwell 1971

Note: The taxa *Arecipites* spp. (undifferentiated) and *Cranwellia/Striatellipollis* (undifferentiated) are not separate taxa but categories of convenience used when scanning slides. The first represents two or more species of the genus that were not distinguished in some of the earlier work in this study, and the second represents specimens that belong to either one or the other of these two genera (these genera are difficult to distinguish where preservation is not ideal).

refers to the number of localities at which a taxon was recorded, and the term abundant (or variants) refers to the number or percentage of specimens of the taxon in samples from certain localities.

Spores

Genus *Azolla* Lam.
Azolla circinata Oltz and Hall in Hall 1968
Plate 1, fig. 1

Azolla circinata Oltz and Hall in Hall, 1968, p. 79, 81; Plate 10, fig. 1.

Description. This fossil species encompasses massulae of the living aquatic fern genus *Azolla* that have circinate or fiddlehead-shaped glochidia. It is readily distinguished from *Azolla cretacea* by the distinctive shape of the glochidia. See Hall (1968) for a more detailed description and discussion.

Occurrence. *Azolla circinata* is rare in the Hell Creek palynoflora. It was found at only three localities.

Comments. Hall (1968) first described this species from the Judith River Formation (Campanian) in Montana. This species is not regarded as having particular biostratigraphic significance. Living species of the modern genus *Azolla* (Salviniaceae) are floating aquatic ferns, thus the presence of this species in the Hell Creek Formation indicates the presence of open water in the depositional environment. Microspores of *Azolla* not in association with a massula are illustrated here in Plate 1, fig. 3. They belong to either *A. circinata* or *A. cretacea;* the sample in which they occurred contained both species.

Azolla cretacea Stanley 1965
Plate 1, fig. 2

Azolla cretacea Stanley, 1965, p. 256–257, Plate 33, figs. 1–5.

Description. This fossil species encompasses massulae of the living aquatic fern genus *Azolla* that have anchor-shaped glochidia. It is readily distinguished from *Azolla circinata* by the distinctive shape of the glochidia. See Stanley (1965) for a more detailed description and diagnosis.

Occurrence. *Azolla cretacea* is more common in the Hell Creek palynoflora than is *A. circinata.* It was found at one-third of the localities. It also occurs in the Fort Union Formation, where it is about as common as in the Hell Creek Formation.

TABLE 4. PALYNOMORPH TAXA IN BOTANICAL GROUPS

Group	Taxon	Number
Bryophytes	*Foraminisporis undulatus*	(42)
	Stereisporites spp.	(83)
	Zlivisporis novomexicanum	(100)
Quillworts	*Palaeoisoetes subengelmannii*	(68)
Lycopods	*Hamulatisporis amplus*	(46)
	Hamulatisporis hamulatis	(47)
	Retitriletes sp.	(79)
Ferns	*Azolla circinata*	(20)
	Azolla cretacea	(21)
	Cicatricosisporites spp.	(25)
	Cyathidites diaphana	(32)
	Cyathidites spp.	(33)
	Deltoidospora sp.	(34)
	Ghoshispora bella	(44)
	Gleicheniidites spp.	(45)
	Laevigatosporites spp.	(52)
	Leptolepidites tenuis	(54)
	Lygodiumsporites sp.	(60)
	Osmundacidites stanleyi	(67)
	Polycingulatisporites sp.	(73)
	Reticuloidosporites pseudomurii	(77)
	Toroisporis major	(88)
Gymnosperms	*Corollina* sp.	(26)
	Ephedripites jansonii	(38)
	Ephedripites multipartitus	(39)
	Ephedripites ovatus	(40)
	Inaperturopollenites sp. G	(48)
	Monocolpopollenites sp. cf. *M. scabratus*	(63)
	Pityosporites spp.	(71)
	Taxodiaceaepollenites hiatus	(87)
Angiosperms (Monocots)	*Arecipites columellus*	(17)
	Arecipites tenuiexinous	(18)
	Dyadonapites reticulatus	(36)
	Liliacidites altimurus	(57)
	Liliacidites complexus	(58)
	Liliacidites spp.	(59)
	Marsypiletes cretacea	(61)
	Pandaniidites typicus	(69)
	Penetetrapites inconspicuus	(70)
	Racemonocolpites formosus	(75)
	Sparganiaceaepollenites sp.	(82)
Angiosperms (Dicots)	*Anacolosidites rotundus*	(1)
	Aquilapollenites attenuatus	(2)
	Aquilapollenites bertillonites	(3)
	Aquilapollenites catenireticulatus	(4)
	Aquilapollenites collaris	(5)
	Aquilapollenites conatus	(6)
	Aquilapollenites delicatus	(7)
	Aquilapollenites marmarthensis	(8)
	Aquilapollenites pyriformis	(9)
	Aquilapollenites quadricretaeus	(10)
	Aquilapollenites quadrilobus	(11)
	Aquilapollenites reductus	(12)
	Aquilapollenites reticulatus	(13)
	Aquilapollenites senonicus	(14)
	Aquilapollenites turbidus	(15)
	Aquilapollenites vulsus	(16)

(*continued*)

Comments. Stanley (1965) first described the fossil species from the Hell Creek Formation in South Dakota. It is not regarded as having particular biostratigraphic significance. Living species of *Azolla* (Salviniaceae) are floating aquatic ferns; the glochidia of this species are quite similar to those of the modern species *Azolla filiculoides* (see Tryon and Tryon, 1982, Fig. 122.10). The presence of *A. cretacea* in the Hell Creek Formation indicates the presence of open water in the depositional environment. Microspores of *Azolla* not in association with a massula are illustrated here in Plate 1, fig. 3. They belong to either *A. cretacea* or *A. circinata;* the sample in which they occurred contained both species.

TABLE 4. PALYNOMORPH TAXA IN BOTANICAL GROUPS (continued)

Group	Taxon	Number
Angiosperms (Dicots) (*continued*)	*Boehlensipollis minimus*	(22)
	Cercidiphyllites sp.	(23)
	Chenopodipollis sp.	(24)
	Cranwellia rumseyensis	(27)
	Cupanieidites inaequalis	(29)
	Cupuliferoidaepollenites spp.	(30)
	Cupuliferoipollenites sp.	(31)
	Discoidites parvistriatus	(35)
	Echitricolpites sp.	(37)
	Erdtmanipollis cretaceus	(41)
	Fraxinoipollenites variabilis	(43)
	Inaperturotetradites scabratus	(49)
	Kurtzipites circularis	(50)
	Kurtzipites trispissatus	(51)
	Lanagiopollis lihoka	(53)
	Leptopecopites pocockii	(55)
	Libopollis jarzenii	(56)
	Momipites inaequalis	(62)
	Myrtipites scabratus	(64)
	Nyssapollenites spp.	(65)
	Orbiculapollis lucidus	(66)
	Polyatriopollenites levis	(72)
	Porosipollis porosus	(74)
	Retibrevitricolporites beccus	(76)
	Retitrescolpites anguloluminosus	(78)
	Rhoipites spp.	(80)
	Simplicepollis rallus	(81)
	Striatellipollis striatellus	(84)
	Striatopollis tectatus	(85)
	Styxpollenites calamitas	(86)
	Triatriopollenites granulatus	(89)
	Tricolpites hians	(90)
	Tricolpites microreticulatus	(91)
	Tricolpites spp.	(92)
	Triporopollenites plektosus	(93)
	Triporopollenites triplicatus	(94)
	Triporopollenites spp.	(95)
	Tschudypollis retusus	(96)
	Tschudypollis spp.	(97)
	Ulmipollenites krempii	(98)
	Wodehouseia spinata	(99)

Note: Numbers in parentheses refer to Table 3 and identify taxa in Figures 3 and 4.

Genus *Cicatricosisporites* Potonié and Gelletich 1933
Cicatricosisporites spp.
Plate 1, figs. 4a and 4b

Description. This taxon, which may include one or more species in the Hell Creek Formation, encompasses striate (cicatricose) trilete spores. A typical specimen is illustrated.

Occurrence. Spores assigned to *Cicatricosisporites* spp. were present at 30% of the localities studied. They show no particular pattern of stratigraphic distribution, being present in low numbers throughout the formation. Kroeger (this volume) found *Cicatricosisporites* spores to be more common in point-bar deposits than in other sedimentary environments within the Hell Creek Formation in South Dakota.

Comments. Striate (cicatricose) trilete spores of the genus *Cicatricosisporites* occur in the Hell Creek palynoflora. It is possible that only one species is present, which has undulatory muri about as wide as the spaces between them (illustrated here in Plate 1, figs. 4a and 4b), but no attempt was made to distinguish separate species or to describe these spores in detail in this study. Ferns of the modern family Schizaeaceae, especially those of the genus *Anemia,* produce spores having this type of sculpture (e.g., see Tryon and Tryon, 1982, figs. 6.26–6.28).

Genus *Cyathidites* Couper 1953
Cyathidites diaphana (Wilson and Webster, 1946) Nichols and Brown 1992
Plate 1, figs. 5a and 5b

Deltoidospora diaphana Wilson and Webster, 1946, p. 273, fig. 3.

Cardioangulina diaphana (Wilson and Webster) Stanley, 1965, p. 248, Plate 30, figs. 17–21.

Cyathidites diaphana (Wilson and Webster) Nichols and Brown, 1992, p. 14–15, Plate 3, figs. 3–5.

Description. This species encompasses psilate trilete

TABLE 5. PALYNOMORPH TAXA IN MORPHOLOGICAL CATEGORIES

Category	Taxon	Number
Trilete spores	*Azolla circinata*	(20)
	Azolla cretacea	(21)
	Cicatricosisporites spp.	(25)
	Cyathidites diaphana	(32)
	Cyathidites spp.	(33)
	Deltoidospora sp.	(34)
	Foraminisporis undulatus	(42)
	Ghoshispora bella	(44)
	Gleicheniidites spp.	(45)
	Hamulatisporis amplus	(46)
	Hamulatisporis hamulatis	(47)
	Leptolepidites tenuis	(54)
	Lygodiumsporites sp.	(60)
	Osmundacidites stanleyi	(67)
	Polycingulatisporites sp.	(73)
	Retitriletes sp.	(79)
	Stereisporites spp.	(83)
	Toroisporis major	(88)
	Zlivisporis novomexicanum	(100)
Monolete spores	*Laevigatosporites* spp.	(52)
	Palaeoisoetes subengelmannii	(68)
	Reticuloidosporites pseudomurii	(77)
Bisaccate pollen	*Pityosporites* spp.	(71)
Polyplicate pollen	*Ephedripites jansonii*	(38)
	Ephedripites multipartitus	(39)
	Ephedripites ovatus	(40)
Rimulate pollen	*Corollina* sp.	(26)
Inaperturate pollen	*Inaperturopollenites* sp. G	(48)
	Penetetrapites inconspicuus	(70)
	Taxodiaceaepollenites hiatus	(87)
Monosulcate pollen	*Arecipites columellus*	(17)
	Arecipites tenuiexinous	(18)
	Liliacidites altimurus	(57)
	Liliacidites complexus	(58)
	Liliacidites spp.	(59)
	Marsypiletes cretacea	(61)
	Monocolpopollenites sp. cf. *M. scabratus*	(63)
	Racemonocolpites formosus	(75)
Tricolpate pollen	*Cranwellia rumseyensis*	(27)
	Cupuliferoidaepollenites spp.	(30)
	Discoidites parvistriatus	(35)
	Echitricolpites sp.	(37)
	Fraxinoipollenites variabilis	(43)
	Retitrescolpites anguloluminosus	(78)
	Striatellipollis striatellus	(84)
	Striatopollis tectatus	(85)
	Tricolpites hians	(90)
	Tricolpites microreticulatus	(91)
	Tricolpites spp.	(92)
Zonocolpate pollen	*Leptopecopites pocockii*	(55)
Tricolporate pollen	*Cupuliferoipollenites* sp.	(31)
	Kurtzipites circularis	(50)
	Kurtzipites trispissatus	(51)
	Lanagiopollis lihoka	(53)
	Myrtipites scabratus	(64)
	Nyssapollenites spp.	(65)
	Porosipollis porosus	(74)
	Retibrevitricolporites beccus	(76)
	Rhoipites spp.	(80)
	Styxpollenites calamitas	(86)
Syncolporate pollen	*Boehlensipollis minimus*	(22)
	Cupanieidites inaequalis	(29)
Monoporate pollen	*Pandaniidites typicus*	(69)
	Sparganiaceaepollenites sp.	(82)
Triporate pollen	*Cercidiphyllites* sp.	(23)
	Momipites inaequalis	(62)

(*continued*)

TABLE 5. PALYNOMORPH TAXA IN MORPHOLOGICAL CATEGORIES (continued)

Category	Taxon	Number
	Triatriopollenites granulatus	(89)
	Triporopollenites plektosus	(93)
	Triporopollenites triplicatus	(94)
	Triporopollenites spp.	(95)
	Tschudypollis retusus	(96)
	Tschudypollis spp.	(97)
Pantoporate pollen	*Anacolosidites rotundus*	(1)
	Chenopodipollis sp.	(24)
	Erdtmanipollis cretaceus	(41)
	Libopollis jarzenii	(56)
	Polyatriopollenites levis	(72)
	Ulmipollenites krempii	(98)
Triprojectate pollen	*Aquilapollenites attenuatus*	(2)
	Aquilapollenites bertillonites	(3)
	Aquilapollenites catenireticulatus	(4)
	Aquilapollenites collaris	(5)
	Aquilapollenites conatus	(6)
	Aquilapollenites delicatus	(7)
	Aquilapollenites marmarthensis	(8)
	Aquilapollenites pyriformis	(9)
	Aquilapollenites quadricretaeus	(10)
	Aquilapollenites quadrilobus	(11)
	Aquilapollenites reductus	(12)
	Aquilapollenites reticulatus	(13)
	Aquilapollenites senonicus	(14)
	Aquilapollenites turbidus	(15)
	Aquilapollenites vulsus	(16)
	Orbiculapollis lucidus	(66)
Dyads and tetrads	*Dyadonapites reticulatus*	(36)
	Inaperturopollenites scabratus	(49)
	Simplicepollis rallus	(81)
Other pollen	*Wodehouseia spinata*	(99)

Note: Numbers in parentheses refer to Table 3 and identify taxa in Figures 3 and 4.

spores having thin walls and a rounded triangular amb with concave sides. The laesurae extend only about half way to the margin. See Wilson and Webster (1946) and Nichols and Brown (1992) for more detailed descriptions of this species. The genus *Cyathidites* Couper 1953 differs from *Deltoidospora* Miner 1935 emend. Potonié 1956 in having concave rather than straight sides and in having broadly rounded rather than somewhat more angular apices. In addition, Hell Creek specimens assigned to *Cyathidites* have thinner walls than do those assigned to *Deltoidospora.*

Occurrence. Spores definitely identified as *Cyathidites diaphana* were recorded at one-third of the localities studied. The species is consistently present in the upper third of the Hell Creek Formation. It is possible that spores of this species are present at more localities than this but were assigned to the broader taxon *Cyathidites* spp. during scanning of slides. It is equally common in the lower part of the Fort Union Formation.

Comments. Wilson and Webster (1946) first described this species from a Fort Union lignite in west-central Montana. Nichols and Brown (1992) transferred the species to the genus *Cyathidites.* This species was previously reported in upper Maastrichtian rocks in the region by Nichols et al. (1990). The affinity of *Cyathidites diaphana* is not thought to be with the modern tree-fern genus *Cyathea* or the family Cyathaceae. Living species of the Cyathaceae produce trilete spores having various sculpture patterns not characteristic of these fossil spores (see Tryon and Tryon, 1982).

Cyathidites spp. (not illustrated)

Description. Psilate trilete spores having thin walls and rounded triangular ambs with concave to straight sides were assigned to *Cyathidites* spp. (although some of them may have been referable to *Cyathidites diaphana*). Most specimens assigned to *Cyathidites* spp. differ from those assigned to *C. diaphana* in having laesurae extending close to the margin and less markedly lobate ambs.

Occurrence. Spores assigned to this taxon were recorded at almost 75% of the localities studied. They are present throughout the Hell Creek Formation and in the lower part of the Fort Union Formation in the study area.

Comments. This broadly defined taxon almost certainly includes specimens of *Cyathidites diaphana* present in some samples but not critically distinguished during counting and scanning of slides. It may or may not include another distinct species, although Hotton (1988) identified some spores of similar morphology as *Cyathidites* cf. *C. minor* Couper 1953 (a

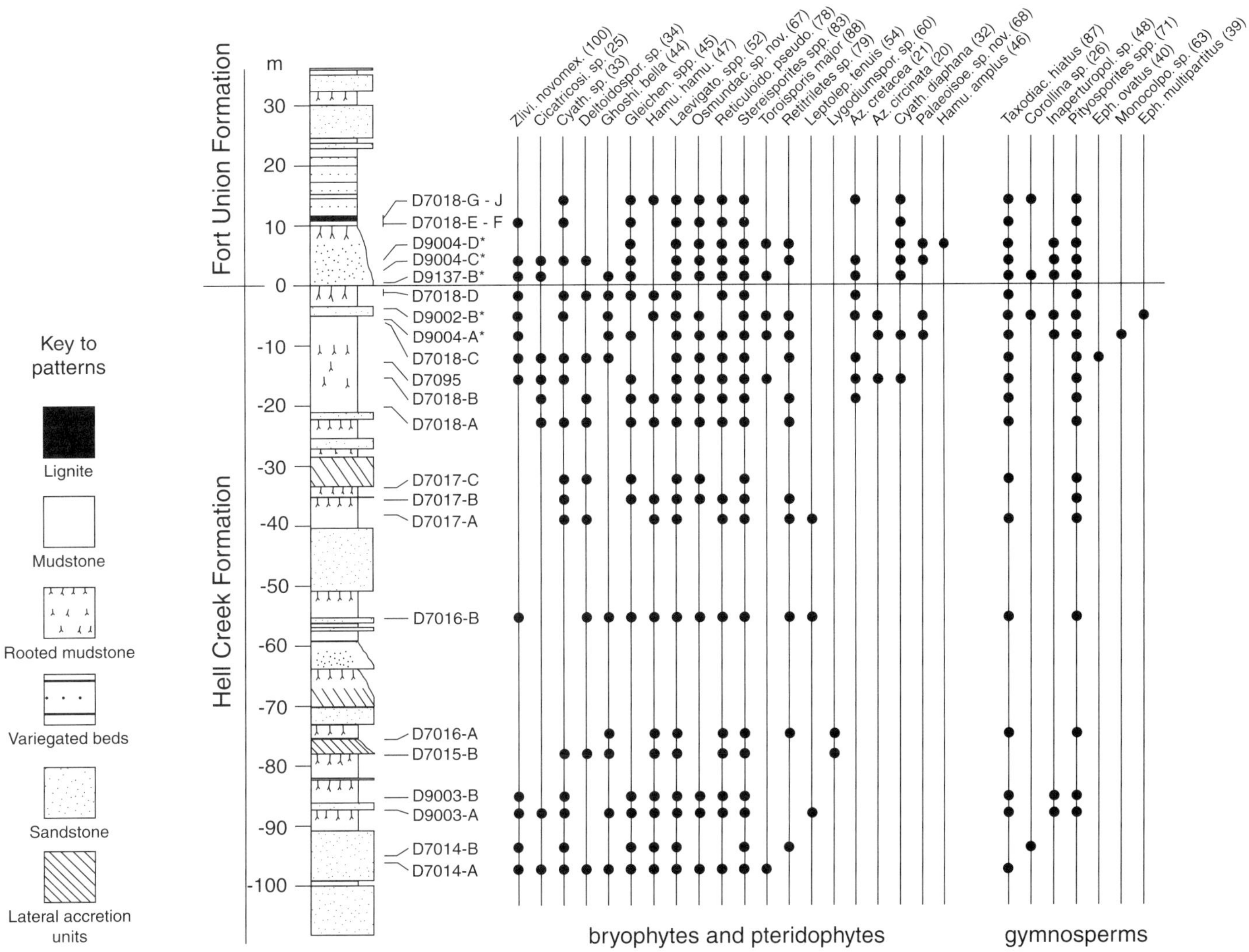

Figure 3. Distribution of bryophyte and pteridophyte spore and gymnosperm pollen taxa in Hell Creek Formation at Sunset Butte, principal reference section for formation in North Dakota. The entire thickness of Hell Creek Formation and lower part of Fort Union Formation are shown; position in measured section is relative to formational contact. Asterisks mark samples from Sunset Patch locality projected into this section. Taxon names are abbreviated; numbers that follow them are keyed to Tables 3–5. Occurrences of all taxa are shown. For practical considerations in drafting, closely spaced samples are composited. Gaps between sample levels are due to lithologies unfavorable for palynology (samples from Sunset Patch locality fill some gaps near top of formation).

species from the Mesozoic of New Zealand) in the Hell Creek palynoflora in Montana.

Genus *Deltoidospora* Miner 1935 emend. Potonié 1956
Deltoidospora sp.
Plate 2, fig. 15

Description. Psilate trilete spores having moderately thick walls, rounded triangular ambs, straight to convex sides, and laesurae extending almost to the margin were assigned to *Deltoidospora* sp. These specimens differ from those herein assigned to *Cyathidites* in wall thickness and/or length of laesurae; most also differ in the shape of the amb. Similar specimens from the Upper Cretaceous of Utah were identified as *Deltoidospora minor* (Couper) Pocock by Nichols (1997, Plate 1, figs. 4–5).

Occurrence. *Deltoidospora* sp. was recorded at more than one-third of the localities studied and is present throughout the Hell Creek Formation. This species is equally common in the Fort Union Formation in the study area.

Comments. The specimen illustrated here (Plate 2, fig. 15) compares with *Intertriletes scrobiculatus* Anderson 1960, but other specimens assigned to *Deltoidospora* sp. in this study lack the structural modification of the exine in the interradial areas exhibited by this specimen and many have somewhat thicker exines. In this study all small, psilate, triangular, trilete spores not assigned to *Cyathidites diaphana* or *Cyathidites* spp. were

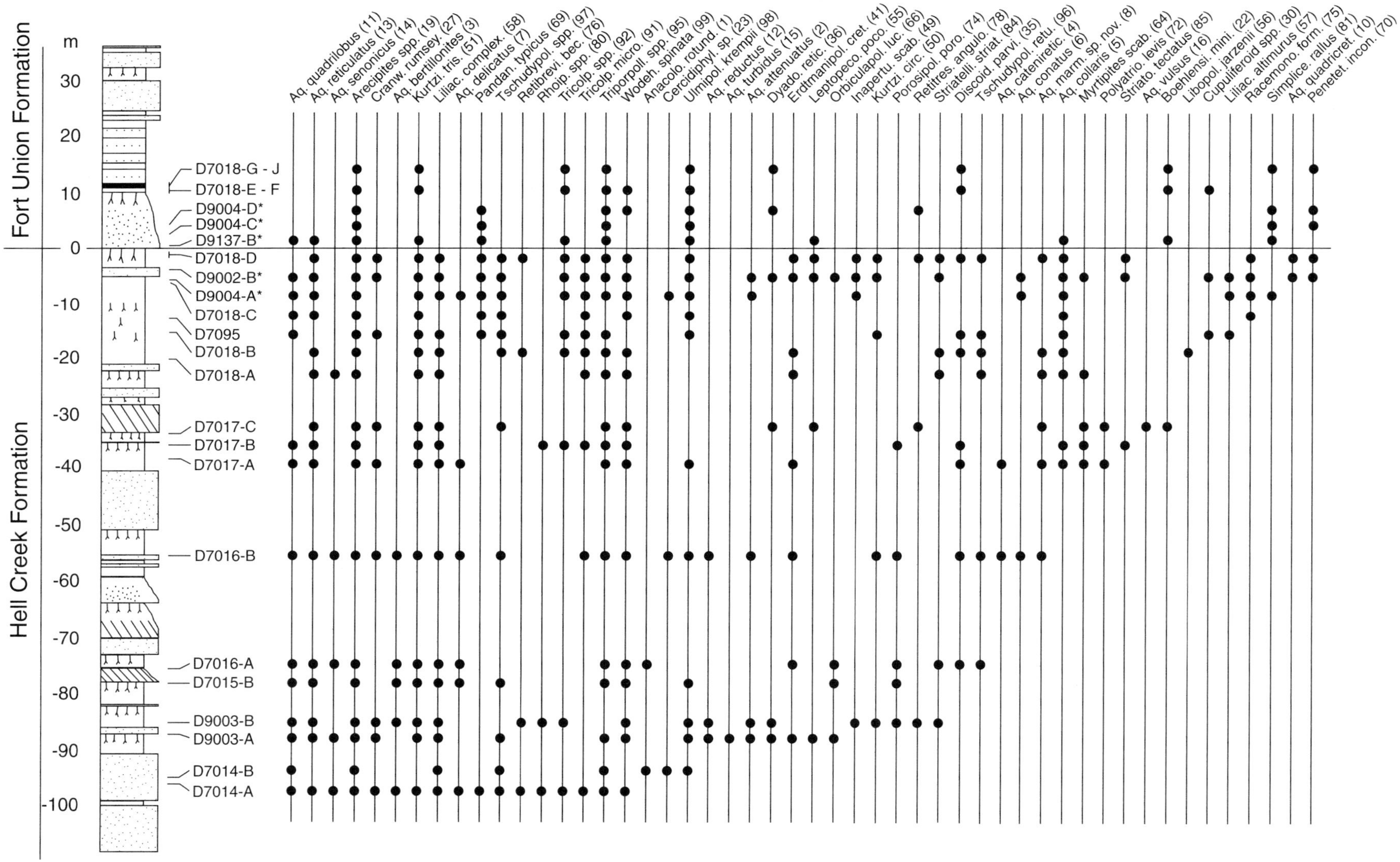

Figure 4. Distribution of angiosperm pollen taxa in Hell Creek Formation at Sunset Butte, principal reference section for formation in North Dakota. Entire thickness of Hell Creek Formation and lower part of Ludlow Member of Fort Union Formation are shown; position in measured section is relative to formational contact. Asterisks mark samples from Sunset Patch locality projected into this section. Lithologic patterns as in Figure 3. Taxon names are abbreviated; numbers that follow them are keyed to Tables 3–5. For practical considerations in drafting, occurrences of only 50 of 63 angiosperm taxa are shown, those most commonly occurring; some individual taxa are included as "spp." Excluded are taxa occurring in only one or two samples, except that all K taxa are shown. Also for practical considerations, closely spaced samples are composited. Gaps between sample levels are due to lithologies unfavorable for palynology (samples from Sunset Patch locality fill some gaps near top of formation).

assigned to *Deltoidospora* sp. Ferns presumably produced these spores. Their morphology is not so distinctive that their affinity with a particular living genus or family can be readily determined.

Genus *Foraminisporis* Krutzsch 1959
Foraminisporis undulatus Leffingwell 1971
Plate 1, fig. 6

Foraminisporis undulatus Leffingwell, 1970, p. 25, Plate 4, figs. 5a and 5b.
Conbaculatisporites undulatus auct. non (Leffingwell), Stone, 1973, p. 67–68, Plate 11, figs. 58–59.

Description. *Foraminisporis undulatus* includes trilete cingulate spores having circular ambs, undulatory to straight laesurae, and scattered sculpture of short spines, verrucae, and/or bacula. See Leffingwell (1970) for a more detailed description and additional illustrations. *Foraminisporis undulatus* differs from *F. wonthaggiensis* Dettmann 1963 in having less densely spaced sculptural elements on the proximal surface and larger sculptural elements of the distal surface. It differs from *Osmundacidites stanleyi* sp. nov. (and all other species of the genus *Osmundacidites*) in having relatively sparse, scattered sculptural elements. Specimens assigned to *Conbaculatisporites undulatus* (Leffingwell) by Stone (1973), who also transferred Leffingwell's species to the genus *Conbaculatisporites,* are not this species because they differ in having more triangular ambs and more robust sculpture.

Occurrence. *Foraminisporis undulatus* is rare in the Hell Creek palynoflora, occurring at fewer than 20% of the localities studied and only in the upper half of the Hell Creek Formation. It is also present but at even fewer localities in the Fort Union Formation.

Comments. Leffingwell (1970) first described this species from the Lance Formation in Wyoming. Botanical affinity is with the bryophyte order Anthocerotales (hornworts); see Krutzsch (1963a).

Although the report by Leffingwell bears the imprint date 1970, it was not actually published until 1971. For nomenclatural purposes the 1971 date is given, but for consistency with procedures in most libraries, the 1970 date is given to identify the publication.

Genus *Ghoshispora* Srivastava 1967
Ghoshispora bella (Kondinskaya 1966) Srivastava 1978
Plate 1, fig. 7

Balmeisporites bellus Kondinskaya, 1966, p. 119, Plate 3, fig. 3.
Styx minor Norton in Norton and Hall, 1967, p. 104, Plate 1, fig. C.
Styx bella (Kondinskaya) Srivastava, 1971a, p. 306, Plate 1, figs. 1–3; Plate 2, figs. 1–5; Plate 3, figs. 1–4; Plate 4, figs. 1–4; Plate 5, figs. 1–6.
Ghoshispora bella (Kondinskaya) Srivastava, 1978a, p. 181–182, Plate 4, figs. 3–13.

Description. This species encompasses relatively large, acrolamellate trilete spores having a distinctive coarse sculpture consisting of muri radiating from conical supratectal processes. See Srivastava (1978a) for a more detailed description.

Occurrence. *Ghoshispora bella* was recorded at more than one-third of the localities studied; it is present throughout the Hell Creek Formation. It was present and somewhat more common in the Fort Union localities analyzed in the course of this study. This species has no particular biostratigraphic significance.

Comments. This species was produced by an aquatic fern (Kondinskaya, 1966). As is true of species of *Azolla,* its occurrence indicates the presence of open water in Hell Creek depositional environments.

Genus *Gleicheniidites* Ross 1949
Gleicheniidites spp.
Plate 2, fig. 1

Description. Specimens assigned to this taxon are psilate, trilete, and have rounded triangular ambs, concave sides, and distinctive interradial thickenings. They differ from other psilate trilete spores in the Hell Creek palynoflora in the presence f interradial thickenings. One or more species may be included. A typical specimen is illustrated.

Occurrence. *Gleicheniidites* spp. are common in the Hell Creek palynoflora. It is present throughout the Hell Creek Formation and was recorded at more than 90% of the localities studied. It is equally common in the Fort Union Formation.

Comments. Stanley (1965) described virtually identical spores as *Gleichenia triangula.* However, Stanley's species and the Hell Creek specimens appear to be indistinguishable from those from the Upper Cretaceous of Europe described by Ross (1949) and Skarby (1964) as *Gleicheniidites senonicus.* No attempt was made to sort out this taxonomic problem in this study, and all specimens resembling either of these species were assigned to *Gleicheniidites* spp. Botanical affinity is commonly thought to be with the living fern family Gleicheniaceae, but spores of modern species of *Gleichenia* illustrated by Tryon and Tryon (1982, Figs. 9.12–9.15) are monolete and bear no resemblance to the fossils from the Hell Creek Formation assigned to *Gleicheniidites* spp.

Genus *Hamulatisporis* Krutzsch 1959
Hamulatisporis amplus Stanley 1965
Plate 1, fig. 8

Hamulatisporis amplus Stanley, 1965, p. 242, Plate 29, figs. 1–6.
Camarozonosporites amplus (Stanley) Dettmann and Playford, 1968, p. 79, Plate 7, figs. 1–3.
Lycopodiacidites amplus (Stanley) Miki, 1972, p. 533, Plate 7, figs. 1–3.

Description. This species encompasses relatively large, circular, cingulate, trilete spores having hamulate sculpture—rugulate muri arranged in a distinctive zigzag pattern. See Stanley (1965) for a more detailed description. *Hamulatisporis amplus*

Plate 1. 1, *Azolla circinata* Hall 1968 (microspore massula); 2, *Azolla cretacea* Stanley 1965 (microspore massula); 3, *Azolla* microspores (species uncertain); 4, *Cicatricosisporites* sp. (two levels of focus on same specimen); 5, *Cyathidites diaphana* (Wilson and Webster 1946) Nichols and Brown 1992 (two levels of focus on same specimen); 6, *Foraminisporis undulatus* Leffingwell 1971; 7, *Ghoshispora bella* (Kondinskaya 1966) Srivastava 1978; 8, *Hamulatisporis amplus* Stanley 1965; 9–12, *Stereisporites* spp.; 13, *Hamulatisporis hamulatis* Krutzsch 1959.

differs from *Hamulatisporis hamulatis* in being larger (diameter 50 μm or more versus ~30 μm). *Hamulatisporis amplus* differs from species of *Camarozonosporites* in lacking interradial thickenings; the distinctive hamulate sculpture characteristic of the genus *Hamulatisporis* differs from the rugulate sculpture of the genus *Lycopodiacidites*.

Occurrence. *Hamulatisporis amplus* was recorded at 30% of the localities studied. It is equally common in the Ludlow Member of the Fort Union Formation.

Comments. Stanley (1965) first described this species from the Hell Creek Formation in South Dakota. Botanical affinity is with the Lycopodiaceae (club mosses or lycopods). *Hamulatisporis amplus* closely resembles *H. hamulatis* except in larger size; because some lycopods are heterosporous, possibly *Hamulatisporis amplus* is the megaspore from the same lycopod species that produced *H. hamulatis*.

Genus *Hamulatisporis hamulatis* Krutzsch 1959
Plate 1, fig. 13

Hamulatisporis hamulatis Krutzsch, 1959, p. 157, Plate 29, figs. 326–328.

Description. This species encompasses small, circular, cingulate, trilete spores having hamulate sculpture—muri arranged in a distinctive zigzag pattern. See Krutzsch (1959) for a more detailed description. *Hamulatisporis hamulatis* differs from *H. amplus* in smaller size (diameter ~30 μm versus 50 μm or more). *Hamulatisporis hamulatis* differs from species of *Camarozonosporites* in lacking interradial thickenings and from species of *Lycopodiacidites* in its distinctive sculpture.

Occurrence. *Hamulatisporis hamulatis* is twice as common in the Hell Creek palynoflora as is *Hamulatisporis amplus*. *Hamulatisporis hamulatis* is present throughout the Hell Creek Formation. It is also present in the Ludlow Member of the Fort Union Formation, where it is somewhat less common than in the Hell Creek.

Comments. Krutzsch (1959) first described this species from the Tertiary of Germany. Botanical affinity is with the Lycopodiaceae (club mosses or lycopods). *Hamulatisporis hamulatis* closely resembles *H. amplus* except in larger size; because some lycopods are heterosporous, possibly *Hamulatisporis hamulatis* is the microspore from the same lycopod species that produced *H. amplus*.

Genus *Laevigatosporites* Ibrahim 1933
emend. Schopf, Wilson, and Bentall 1944
Laevigatosporites spp.
Plate 2, fig. 3

Description. This taxon encompasses psilate monolete spores ranging in size from 25 to 60 μm. At least two and perhaps more than two species are included in this complex that differ only in size, but the size variation appears to be gradational across the entire size range. A typical specimen of the smaller size is illustrated. Spores assigned to *Laevigatosporites* spp. differ from those assigned to *Reticuloidosporites pseudomurii* in lacking sculpture.

Occurrence. *Laevigatosporites* spp. is the most common and abundant taxon in the Hell Creek palynoflora. It was recorded at all localities studied, often in high relative abundance in individual samples; it is present throughout the Hell Creek Formation. These spores are equally common in the Fort Union Formation. Spores of this simple morphology are known throughout the Cretaceous and Tertiary of North America and elsewhere and have no biostratigraphic significance.

Comments. Wilson and Webster (1946) described two species of *Laevigatosporites, L. gracilis* and *L. ovatus,* from the Fort Union Formation in south-central Montana that are similar to some specimens from the Hell Creek Formation in western North Dakota. At least some of the Hell Creek specimens may be assignable to these species, but they are also indistinguishable from specimens from the Tertiary of Europe assigned to *L. haardtii* (Potonié and Venitz) Thomson and Pflug. There seems little practical value in trying to sort out this nomenclatural problem. Although the simple morphology of these spores precludes their affinity being established beyond question, similar spores are produced by living species of the fern family Polypodiaceae (see Tryon and Tryon, 1982). Their ubiquity in the Hell Creek Formation is strong evidence that ferns were a major component of the Maastrichtian vegetation.

Genus *Leptolepidites* Couper 1953 emend. Norris 1968
Leptolepidites tenuis Stanley 1965
Plate 2, figs. 4a and 4b

Leptolepidites tenuis Stanley, 1965, p. 255, Plate 32, figs. 7–11.

Description. This species encompasses distinctive trilete spores with circular ambs and coarsely verrucate sculpture. See Stanley (1965) for a more complete description and additional illustrations. *Leptolepidites tenuis* species differs from all other trilete spores in the Hell Creek palynoflora in having coarse, verrucate sculpture with large, closely packed verrucae.

Occurrence. *Leptolepidites tenuis* is not common in the Hell Creek Formation. It was recorded in about one-fourth of the localities studied and never in great abundance in individual samples. It is only slightly less common in the lower part of the Fort Union Formation.

Comments. Stanley (1965) first described this species from the Hell Creek Formation in South Dakota. Stanley did not report this species from Fort Union samples he studied from the same area.

Genus *Lygodiumsporites* Potonié, Thomson, and Thiergart
1950 ex Potonié, 1956
Lygodiumsporites sp.
(not illustrated)

Description. Spores assigned to this species are relatively large, trilete, and psilate, and have rounded triangular to circular ambs. They differ from other trilete spores in the Hell Creek palynoflora in their large size and shape. See Nichols (1997, Plate 1, fig. 6) for an illustration.

Occurrence. *Lygodiumsporites* sp. is rare in the Hell Creek palynoflora. It was recorded at only ~10% of the localities studied. It is equally rare in the Fort Union Formation.

Comments. These spores were produced by a species of fern, but not necessarily one related to the modern genus *Lygodium,* whose spores differ in form and sculpture (cf. Tryon and Tryon, 1982).

Genus *Osmundacidites* Couper 1953
Osmundacidites stanleyi sp. nov.
Plate 2, figs. 5–7

Osmunda comaumensis (Cookson 1953) Stanley, 1965, p. 250–251, Plate 31, figs. 6–9.

Description. Microspores trilete, oblate spheroidal in form, circular to broadly rounded triangular in outline, thin-walled and tending to fold irregularly, densely sculptured. Sculpture consisting of granulae, verrucae, and short spines; the sculpture is so dense that it may obscure the long, straight, unbordered laesurae; laesurae not extending to margin. Diameter 40.0–63.5 µm (20 specimens). In form and sculpture resembling spores of the modern genus *Osmunda.*

Diagnosis. *Osmundacidites stanleyi* sp. nov. differs from *Baculatisporites comaumensis* (Cookson) Potonié 1956 in having denser sculpture with sculptural elements more varied in form; it differs from all other species of the genus *Baculatisporites* in having finer and denser sculpture. *Osmundacidites stanleyi* sp. nov. differs from species of the genus *Foraminisporis* in having straight laesurae and especially in the density of sculptural elements.

Holotype. Slide D7017-C (1), England Finder coordinate M42/2; Plate 2, fig. 5 herein.

Type locality. Sunset Butte, Bowman County, North Dakota; sample D7017-C, ~33.5 m below top of Hell Creek Formation.

Repository. The slide bearing the type specimen of *Osmundacidites stanleyi* sp. nov. is held in the palynological laboratory of Central Energy Resources, U.S. Geological Survey, Denver, Colorado.

Occurrence. *Osmundacidites stanleyi* sp. nov. is fairly common in the Hell Creek palynoflora. It was recorded at more than 50% of the localities studied; it is present throughout the formation. It is equally common in the lower part of the Fort Union Formation.

Derivation of name. Named for Edward A. Stanley in recognition of his early work on the palynology of the Hell Creek Formation.

Comments. Stanley (1965) originally described and illustrated this species from the Hell Creek Formation in South Dakota as *Osmunda comaumensis* (Cookson 1953) Stanley 1965, which is a species from the Mesozoic of Australia. The species with which he identified his Hell Creek specimens was later reassigned to the genus *Baculatisporites* (see Dettmann, 1963). It seems appropriate to recognize the commonly occurring North American Cretaceous (and lowermost Tertiary) specimens as belonging to a distinct species.

Dettmann (1963) submerged *Baculatisporites* Pflug in Thomson and Pflug 1953 along with *Conosmundacidites* Klaus 1960 and certain fossil species previously assigned to the modern genus *Osmunda* in the genus *Osmundacidites* Couper 1953. Krutzsch (1967) broadened the circumscription of *Baculatisporites,* which was originally described as having widely spaced, cylindrical bacula ~3 µm in length, to include other genera, including *Osmundacidites,* which was originally described as having finer, denser, granulate to papillate sculpture. Norris (1986) reviewed the morphology of spores such as these, many of which resemble those of the modern family Osmundaceae, and rejected Krutzsch's emendation of *Baculatisporites,* proposing to retain *Osmundacidites* as a distinct genus. That recommendation is followed for the Hell Creek specimens, which clearly differ in sculpture from species assigned to *Baculatisporites* (e.g., *Baculatisporites* sp. of Pocknall and Nichols, 1996, Plate 1, (figs. 1–2). Affinity of the Hell Creek specimens is with ferns of the family Osmundaceae (cf. Tryon and Tryon, 1982).

Genus *Palaeoisoetes* gen. nov.

Type species. *Palaeoisoetes subengelmannii* (Elsik 1968) comb. nov.; Elsik, 1968a, p. 288, 290; Plate 5, fig. 7.

Description. Microspores monolete, perinate, ovate. Perine loose, surrounding spore body distally, attached laterally in equatorial region. In form closely resembling microspores of the living genus *Isoëtes* L.

Comments. This new genus is proposed to accommodate isoetalean microspores conforming to the description given here. *Palaeoisoetes* gen. nov. differs from *Permonolites* Erdtman 1947 ex Couper 1953 primarily in closely resembling microspores of the modern genus *Isoëtes* rather than in simply encompassing perinate, monolete spores. This resemblance is particularly noticeable in specimens seen in full lateral view (see Plate 1, figs. 11–13, herein), where the distinctive outline with its proximal extension is observable. Spores of *Palaeoisoetes* also are considerably smaller than those from the Cretaceous of New Zealand assigned to *Permonolites* by Couper (1953, Plate 3, figs. 31–33).

This new microfossil genus is more suitable for the type species, *Palaeoisoetes subengelmannii* comb. nov., than is the genus name of the living plant *Isoëtes,* which circumscribes a much broader concept. Living species of *Isoëtes* often are distinguished by the morphology of their megaspores but not by that of their microspores, all of which are essentially similar in form. The genus *Minerisporites,* of which there are more than 30 species of Cretaceous and Tertiary age, encompasses fossil megaspores of isoetalean affinity.

The type of this new genus is the species *Palaeoisoetes subengelmannii* (Elsik 1968) comb. nov. The holotype specimen of that species (originally described as *Isoetes subengel-*

manni) is indicated in the preceding. Identical specimens from North Dakota are illustrated here (Plate 2, figs. 8–10).

Palaeoisoetes subengelmannii (Elsik 1968) comb. nov.
Plate 2, figs. 8–10

Isoetes subengelmanni Elsik, 1968a, p. 288–290, Plate 5, figs. 5–7.

Peromonolites subengelmanni (Elsik) Jameossanaie, 1987, p. 14–15, fig. 7/15, 7/16.

Basionym. *Isoetes subengelmanni* Elsik, 1968a, p. 288–290, Plate 5, figs. 5–7; holotype, Plate 5, fig. 7.

Description. Spores monolete, perinate, spore body ovate with psilate exine; perine loose, surrounding spore body distally, attached laterally in equatorial region, thin (<0.5 μm), ovate-spheroidal in form and rough in texture (scabrate), often irregularly folded, extending proximally below the general ovate outline in equatorial view at the position of the monolete mark (laesura) on the spore body. In lateral view the perine is seen to extend distally from the equatorial region of the spore body a distance about equal to the polar diameter of the spore and laterally outward; in polar view the perine appears to surround the spore body although it does so only around the distal surface. Size: spore body 10–15 × 20–25 μm; overall dimensions 18–29 × 25–35 μm (10 specimens).

Diagnosis. *Palaeoisoetes subengelmannii* (Elsik 1968) comb. nov. differs from all other monolete spores in the Hell Creek palynoflora and elsewhere in Upper Cretaceous and lower Tertiary rocks in western North America in the presence of the distinctive perine. Specimens of this fossil species closely resemble spores produced by living species of the Isoetaceae (quillworts). Elsik (1968a, p. 290) noted close similarity in morphology of the fossil species with spores of the extant species *Isoëtes engelmannii.*

Occurrence. *Palaeoisoetes subengelmannii* was recorded at about one-fourth of the Hell Creek localities studied. It is present in the Ludlow Member of the Fort Union Formation, where it was recorded at ~30% of the localities studied. I have extensively observed and recorded identical specimens in samples from other Upper Cretaceous formations in the Rocky Mountain region as old as the Frontier Formation (Cenomanian-Turonian). Specimens assigned to *Peromonolites subengelmanni* by Jameossanaie (1987) from the Campanian of New Mexico are *Palaeoisoetes subengelmannii.*

Comments. The distinctive morphology of these spores clearly indicates their affinity with the family Isoetaceae (quillworts) (e.g., see Tryon and Tryon, 1982, Figs. 127.19–127.20). On the basis of presumed similarity in habitat of the Cretaceous and Paleocene species with living relatives, their presence in the Hell Creek Formation is indicative of aquatic depositional environments, including ponds or ephemeral pools.

Genus *Polycingulatisporites* Simoncsics and Kedves 1961
emend. Playford and Dettmann 1965

Polycingulatisporites reduncus (Bolkhovitina 1953)
Playford and Dettmann 1965
Plate 2, fig. 12

Chomotriletes reduncus Bolkhovitina, 1953, p. 35, Plate 3, figs. 23–24.

Taurocusporites reduncus (Bolkhovitina) Stover, 1962, p. 57, Plate 1, figs. 15–21

Polycingulatisporites reduncus (Bolkhovitina) Playford and Dettmann, 1965, p. 144.

Description. Spores from the Hell Creek Formation assigned to *Polycingulatisporites reduncus* are trilete, have a circular amb, are cingulate, and have a distinctive circumpolar ring of thickened exine on the distal surface. Kroeger (this volume, Fig. 7I) illustrates a specimen from the Hell Creek Formation in South Dakota.

Occurrence. *Polycingulatisporites reduncus* is rare in the Hell Creek palynoflora. It was recorded at only two of the localities studied. Kroeger (this volume) found *Polycingulatisporites reduncus* to be more common in point-bar deposits than in other sedimentary environments within the Hell Creek Formation in South Dakota.

Comments. The botanical affinity of *Polycingulatisporites reduncus* is unknown. Presumably it represents some pteridophyte species.

Genus *Reticuloidosporites* Pflug in Thomson and Pflug 1953
Reticuloidosporites pseudomurii Elsik 1968
Plate 2, fig. 2

Reticuloidosporites pseudomurii Elsik, 1968a, p. 290, Plate 7, fig. 2.

Hazaria canadiana Srivastava, 1971b, p. 260–261, Plate 2, fig. 5.

Description. Spores encompassed by this species are monolete and have a distinctive sculpture of elements of low relief forming a coarsely foveolate or negatively reticulate pattern. See Elsik (1968a) for a more detailed description and illustrations. Hell Creek specimens range in length from 35 to 45 μm.

Occurrence. *Reticuloidosporites pseudomurii* is common in the Hell Creek palynoflora. It was recorded at almost 80% of the localities studied. It is present in the Fort Union Formation also, where it is even more common. It is common in other Cretaceous formations in the region as well, including those as old as Coniacian (Nichols, 1994).

Comments. Specimens described by Srivastava (1971b) as *Hazaria canadiana* differ in no significant way from *Reticuloidosporites pseudomurii* and are synonymous. Specimens assigned to *Hazaria* sp. cf. *H. canadiana* by Nichols et al. (1982) are *Reticuloidosporites pseudomurii* and are so reported by Nichols (1994). *Reticuloidosporites pseudomurii* almost certainly was produced by a species of fern, probably one having affinity with the family Polypodiaceae. Several modern genera

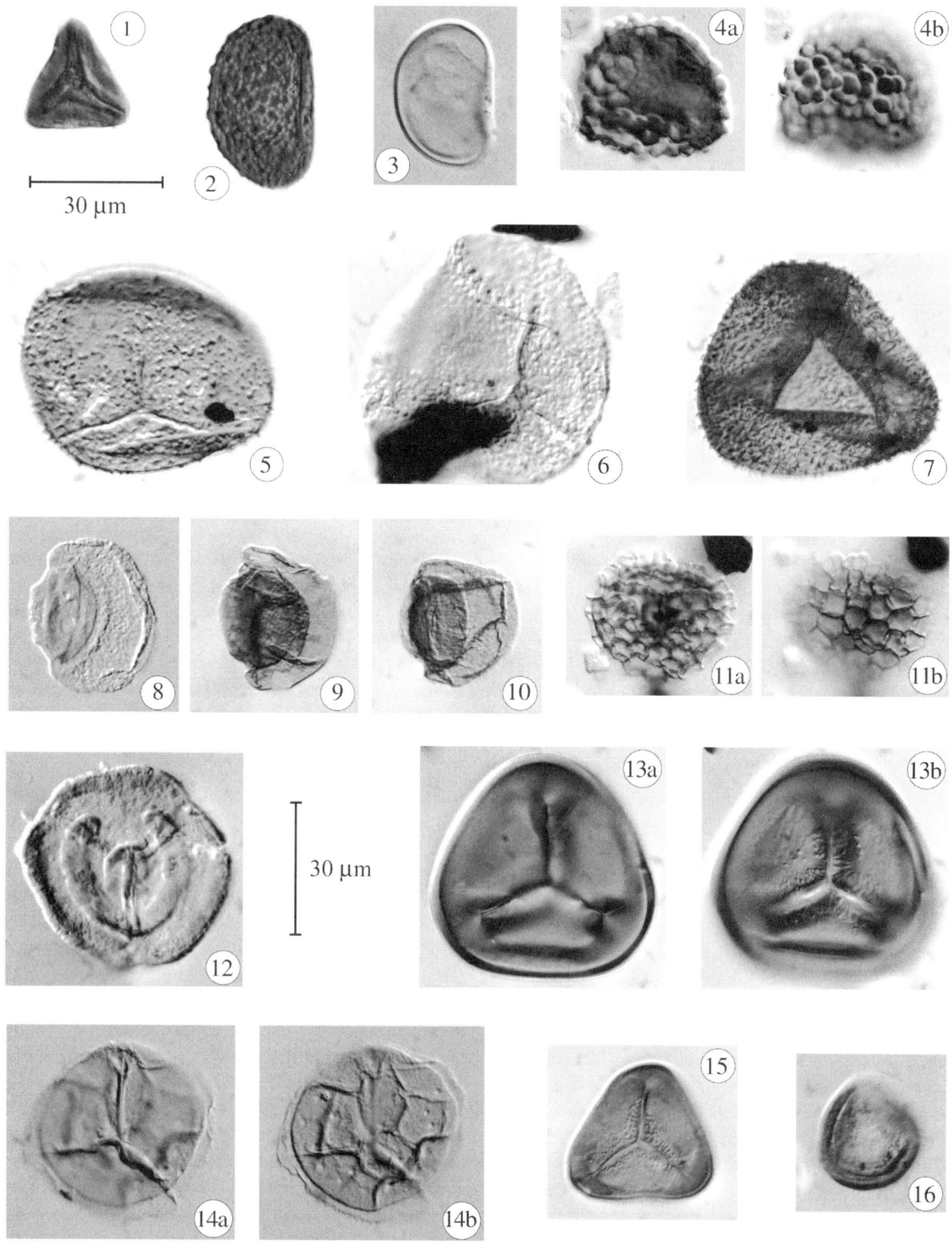

Plate 2. 1, *Gleicheniidites* sp.; 2, *Reticuloidosporites pseudomurii* Elsik 1968; 3, *Laevigatosporites* sp.; 4, *Leptolepidites tenuis* Stanley 1965 (two levels of focus on same specimen); 5, *Osmundacidites stanleyi* sp. nov. (holotype); 6–7, *Osmundacidites stanleyi* sp. nov. (isotypes); 8, *Palaeoisoetes subengelmannii* (Elsik 1968) comb. nov. (genoholotype); 9–10, *Palaeoisoetes subengelmannii* (Elsik 1968) comb. nov.; 11, *Retitriletes* sp. (two levels of focus on same specimen); 12, *Polycingulatisporites reduncus* (Bolkhovitina 1953) Playford and Dettmann 1965; 13, *Toroisporis major* (two levels of focus on same specimen); 14, *Zlivisporis novomexicanum* (Anderson 1960) Leffingwell 1971; 15, *Deltoidospora* sp.; 16, *Corollina* sp.

of this family produce spores having this morphology (see Tryon and Tryon, 1982).

Genus *Retitriletes* Pierce 1961
Retitriletes sp.
Plate 2, figs. 11a and 11b

Description. This species encompasses trilete spores with circular ambs and coarsely reticulate sculpture in which the muri are high and thin.

Occurrence. *Retitriletes* sp. is widespread in the Hell Creek Formation, having been recorded at more than 60% of the localities studied; it is present throughout the formation. It is nowhere abundant, however, and usually is found only while scanning for additional specimens rather than when counting the 200 most abundant specimens in a sample. It is also present in the Fort Union Formation, where it is somewhat less common.

Comments. These spores likely were produced by club mosses (Lycopodiaceae); they are virtually identical with spores produced by some species of the modern genus *Lycopodium.* Some might assign these fossils to *Lycopodiumsporites,* but that name was regarded as a nomen dubium by Krutzsch (1963a).

Genus *Stereisporites* Pflug in Thomson and Pflug 1953
Stereisporites spp.
Plate 1, figs. 9–12

Description. Spores assigned to *Stereisporites* spp. in this study include three or more species. All are trilete and cingulate; they vary in details of structure of the spore wall. Structural features include distal bosses and triradiate thickenings. Small verrucae characterize one species; other species are psilate.

Occurrence. Spores of the several species included in *Stereisporites* spp. are common in the Hell Creek Formation. One or more of the several species included in this taxon was present at almost 90% of the localities studied; they are present throughout the formation. These spores are equally common in the Fort Union Formation. The specimens observed possibly could be assigned to *Stereisporites antiquasporites, S. dakotensis,* and *S. regius,* and perhaps others including species of the *Distancoraesporis* group of Krutzsch (1963b).

Comments. No attempt was made to distinguish and identify separate species of *Stereisporites* in this study. These spores have no biostratigraphic importance, and all species have the same paleoecologic significance. They were produced by Cretaceous species of *Sphagnum* (Bryophyta, sphagnum moss) that inhabited Hell Creek depositional environments, especially the mires in which lignite formed.

Genus *Toroisporis* Krutzsch 1959
Toroisporis major (Pflug in Thomson and Pflug 1953)
Stanley 1965
Plate 2, figs. 13a and 13b

Concavisporites obtusangulus (Potonié) subsp. *major* Pflug in Thomson and Pflug, 1953, p. 50, Plate 1, fig. 42.
Toroisporis (*Toroisporis*) *torus* (Pflug in Thomson and Pflug) subsp. *major* (Pflug in Thomson and Pflug) Krutzsch, 1959, p. 95, Plate 9, figs. 68–69.
Toroisporis major (Pflug in Thomson and Pflug) Stanley, 1965, p. 265, Plate 35, figs. 6–9.

Description. Spores identified as *Toroisporis major* in this study are trilete and have rounded triangular ambs and thick walls. They have kyrtomes in interradial areas. For illustrations see Stanley (1965, Plate 35, figs. 6–9) or Nichols and Brown (1992, Plate 3, figs. 9–11).

Occurrence. *Toroisporis major* spores are not common in the Hell Creek Formation. They were recorded at only about one-fourth of the localities studied, although they are present throughout the formation. They are somewhat more common in the Fort Union Formation.

Comments. The family-level botanical affinity of *Toroisporis major* is unknown. Presumably they were produced by a species of fern.

Genus *Zlivisporis* Pacltová 1961
Zlivisporis novomexicanum (Anderson 1960)
Leffingwell 1971
Plate 2, figs. 14a and 14b

Lycopodium novomexicanum Anderson, 1960, p. 14–15, Plate 1, fig. 2; Plate 8, fig. 1.
Lycopodiumsporites novomexicanum (Anderson) Drugg, 1967, p. 40, Plate 6, fig. 27.
Zlivisporis novamexicanum (Anderson) Leffingwell, 1970, p. 25, Plate 4, figs. 3–4.
Triporoletes novomexicanus (Anderson) Srivastava, 1975, p. 69.

Description. Spores of this species are trilete and have circular ambs; they have a distinctive distal sculpture of thin muri forming an extremely coarse reticulum. See Anderson (1960) for a more detailed description.

Occurrence. *Zlivisporis novomexicanum* is fairly common in the Hell Creek Formation. It is present throughout the formation and was recorded at more than one-third of the localities studied. It is also present in the Fort Union Formation, where it is somewhat more common.

Comments. *Zlivisporis novomexicanum* was inappropriately transferred to the genus *Triporoletes* by Srivastava (1975). The botanical affinity of this species seems to be with the bryophyte order Marchantiales (liverworts), although Krutzsch (1963a) described and illustrated similar species he assigned to the genus *Retitriletes,* whose affinity is with the Lycopodiaceae (club mosses). See also the second comment after *Foraminisporis undulatus* regarding the date of publication of the paper by Leffingwell.

Gymnosperm pollen

Genus *Corollina* Malyavkina 1949
emend. Cornet and Traverse 1975
Corollina sp.
Plate 2, fig. 16

Description. Pollen grains free (not fused in tetrads), oblate spheroidal, with a pore-like opening at one pole and with an equatorial band of fine striations. See Traverse (1988, p. 219–227) for a more detailed description of the morphology of *Corollina* and other circumpolloid pollen.

Occurrence. Specimens of *Corollina* sp. are fairly common in the Hell Creek Formation; they were recorded at 30% of the localities studied. They are nowhere abundant but occur as single specimens usually recorded during scanning rather than during counting. They are clearly present as a minor component of the palynoflora, however, and are present throughout the formation. They are more common in the Fort Union Formation, where they were recorded at almost half of the localities studied.

Comments. The stratigraphic range of *Corollina* (sometimes known by the synonymous name *Classopollis*) is somewhat in dispute. Srivastava (1976) stated that post-Turonian occurrences are most probably redeposited (reworked) pollen. The species here designated *Corollina* sp. occurs consistently although in low numbers in rocks as young as late Paleocene in the western United States, as documented by the Hell Creek specimens discussed here, numerous records from the Ludlow Member of the Fort Union Formation in western North Dakota (Nichols, unpublished data), and records from the Tongue River Member of the Fort Union Formation in Montana and Wyoming (Pocknall and Nichols, 1996). It is also present in rocks as young as Eocene in the Gulf Coast region (Fairchild and Elsik, 1969). The stratigraphic range of this genus is of particular interest because it is pollen of the extinct conifer family Hirmerelliaceae (Cheirolepidiaceae), which has a megafossil record that extends into the Cretaceous. *Corollina* in the Hell Creek Formation is evidence of the existence of that family in the Hell Creek flora.

Genus *Ephedripites* Bolkhovitina 1953 ex Potonié 1958
Ephedripites jansonii (Pocock 1964) Muller 1968
(not illustrated)

Equisetosporites jansonii Pocock, 1964, p. 149–150, Plate 1, figs. 26–27.

Ephedra voluta Stanley, 1965, p. 284–285, Plate 40, figs. 10–11.

Ephedripites jansonii (Pocock) Muller, 1968, p. 9–10, Plate 2, fig. 9.

Gnetaceaepollenites jansonii (Pocock) De Lima, 1980, p. 35, Plate 3, fig. 11.

Description. This species is fusiform and polyplicate (with numerous plicae); the plicae are parallel and are rotated ~45° about the axis extending from one end of the grain to the other. It has ephedroid morphology (it resembles pollen produced by some species of the living gymnosperm genus *Ephedra*), but is unlike pollen produced by any living species of *Ephedra* because none produces polyplicate pollen with the plicae rotated as in *Ephedripites jansonii.* Except for the characteristic 45° rotation of the plicae, however, this species is morphologically similar to some species of *Ephedra,* especially *E. trifurcata. Ephedripites jansonii* differs from *E. ovatus* in having rotated plicae and from *E. multipartitus* in having more numerous plications that are rotated and in lacking colpoid structures in furrows between plicae. See Pocock (1964) or Muller (1968) for a more detailed descriptions and illustrations.

Occurrence. *Ephedripites jansonii* is rare in the Hell Creek Formation. It was recorded at only two of the localities studied. It is somewhat more abundant in the Fort Union Formation than in the Hell Creek.

Comments. The nomenclatural history of the genus name *Ephedripites* Bolkhovitina 1953 ex Potonié 1958 is unusually complex and it is beyond the scope of this study to review it in detail or attempt to resolve it. The name *Ephedripites* is well known and widely used for fossil pollen of ephedroid morphology, but it may not be the correct name. It is used here for convenience, to avoid having to review the history of publication and typification of competing names including *Ephedra, Ephedripites, Equisetosporites, Gnetaceaepollenites,* and perhaps others, and because the name *Ephedripites* is well known.

Pocock (1964) first described the species *Ephedripites jansonii* from the Lower Cretaceous of Saskatchewan. He assigned it to the genus *Equisetosporites,* but Muller (1968) transferred it to *Ephedripites.* Stanley (1965) described a clearly conspecific form from the Hell Creek Formation in South Dakota as *Ephedra voluta.* Stanley's species is here placed in synonymy with *Ephedripites jansonii* (Pocock) Muller. The botanical affinity of *Ephedripites jansonii* is with the gymnosperm family Ephedraceae (Division Gnetophyta, Order Ephedrales).

Ephedripites multipartitus (Chlonova 1961) Yu et al. 1981
Plate 3, figs. 1, 2a, and 2b

Ephedra multipartita Chlonova, 1961, p. 64, Plate 11, fig. 65.

Ephedripites multipartitus (Chlonova) Yu, Guo, and Mao, 1981, p. 53, Plate 22, fig. 16.

Description. *Ephedripites multipartitus* is fusiform and polyplicate (six plicae); it has distinctive, branching, zigzag furrows or colpoid structures between the plicae. It resembles pollen of living species of *Ephedra* such as *E. distachya.* It differs from *Ephedra eocenipites* Wodehouse 1933 in being smaller.

Occurrence. *Ephedripites multipartitus* is not common in the Hell Creek Formation; single specimens were recorded at five of the localities studied in the upper half of the formation. It is somewhat more common in the Fort Union Formation, but there it occurs only in the basal 1 m of the Ludlow Member, in facies bearing an assemblage of Maastrichtian age. Unlike the other species of *Ephedripites* in the Hell Creek palynoflora, *E. multipartitus* is regarded as a K taxon indicative of Maastrichtian age.

Comments. Chlonova (1961) first described this species from the Upper Cretaceous of Siberia. The botanical affinity of *Ephedripites multipartitus* is with the gymnosperm family Ephedraceae (Division Gnetophyta, Order Ephedrales). See also comments about *Ephedripites jansonii.* Nichols and Sweet

(1993) included this species in the *Wodehouseia spinata* assemblage (it is the only species of gymnosperm pollen they included in that assemblage).

Ephedripites ovatus (Pierce 1961) Norton in Norton and Hall 1969
(not illustrated)

Striainaperturites ovatus Pierce, 1961, p. 45, Plate 3, fig. 80.
Equisetosporites sp. cf. *E. ovatus* (Pierce) Singh, 1964, p. 133, Plate 17, fig. 16.
Ephedripites ovatus (Pierce) Norton, in Norton and Hall, 1969, p. 34, Plate 3, fig. 19.

Description. This species encompasses pollen of ephedroid morphology in which the plicae are thin, numerous, and straight, converging at the ends of the grain. It lacks colpoid structures in the furrows between the plicae. See Singh (1964) for a more detailed description. This species is morphologically similar to some species of *Ephedra,* especially *E. trifurcata.* It is distinguished from others of ephedroid morphology in the Hell Creek palynoflora by its straight plications and absence of zigzag colpoid structures. *Ephedripites ovatus* is similar to and possibly conspecific with *Equisetosporites amabilis* Srivastava 1968, which would be a later synonym.

Occurrence. *Ephedripites ovatus* is rare in the Hell Creek Formation. It was recorded at only two of the localities studied. It is somewhat more abundant in the Fort Union Formation than in the Hell Creek.

Comments. Pierce (1961) first described this species from the Lower Cretaceous of Minnesota. Pocock (1964) described a similar species as *Equisetosporites ovatus.* Were Pocock's species to be transferred to the genus *Ephedripites,* it would become a later homonym of *Ephedripites ovatus* (Pierce) Norton in Norton and Hall, and because of its similar morphology, it might be considered a later synonym as well. The botanical affinity of *Ephedripites ovatus* is clearly with the gymnosperm family Ephedraceae (Division Gnetophyta, Order Ephedrales). See also comments about *Ephedripites jansonii.*

Genus *Inaperturopollenites* Pflug and Thomson in Thomson and Pflug 1953
Inaperturopollenites sp. G
Plate 3, figs. 5a and 5b

Description. Pollen assigned to this species is spheroidal, inaperturate, and scabrate; some specimens exhibit a small papilla, similar to pollen of the living gymnosperm genus *Glyptostrobus.* Pollen assigned to *Inaperturopollenites* sp. G differs from that assigned to *Taxodiaceaepollenites hiatus* in this study in not being split open.

Occurrence. *Inaperturopollenites* sp. G is uncommon in the Hell Creek Formation; it was recorded at only seven localities. It is more common in the Fort Union Formation, where it was recorded in more than 40% of the localities analyzed in the course of this study.

Comments. The botanical affinity of this species is with gymnosperms of the Taxodiaceae-Cupressaceae-Taxaceae (TCT) complex, but it is closer to the family Taxodiaceae than to either of the others. Many specimens do not exhibit papillae, but those that do resemble more closely the pollen of the living genus *Glyptostrobus,* and some are reminiscent of pollen of *Sequoia. Inaperturopollenites* sp. G evidently is closely related to *Taxodiaceaepollenites hiatus,* but the two are regarded as separate species because of differences in their patterns of occurrence. They may not be present in the same samples, and *Inaperturopollenites* sp. G is considerably less common in the Hell Creek Formation and disproportionately less common in the Fort Union Formation than is *T. hiatus.* Thus it is evident that the two forms are not morphological variants of the same species but are two distinct taxa. See also *Taxodiaceaepollenites hiatus.*

Genus *Monocolpopollenites* Pflug and Thomson in Thomson and Pflug 1953
Monocolpopollenites sp. cf. *M. scabratus* Stanley 1965
(not illustrated)

Description. Pollen assigned to *Monocolpopollenites* sp. cf. *M. scabratus* is monosulcate, the outline is ovate with tapered ends, and the exine is psilate to scabrate. The sulcus is as long as the grain and may be flared at the ends. These specimens differ from those assigned to species of *Arecipites* in this study in evidently lacking complexly structured exines and in lacking sculpture. *Monocolpopollenites* sp. is similar to and possibly conspecific with *Cycadopites scabratus* Stanley 1965, but the simple morphology and lack of an adequate number of specimens for comparison leave its identification somewhat uncertain. See Stanley (1965, p. 271, Plate 37, figs. 10–15) for a description and illustrations of *M. scabratus.*

Occurrence. *Monocolpopollenites* sp. cf. *M. scabratus* is rare in the Hell Creek Formation. It was recorded at only one locality in this study. It is also present in the Fort Union Formation but is equally uncommon there.

Comments. The botanical affinity of the species appears to be with certain gymnosperms, perhaps the Order Cycadales, but its simple morphology and rare occurrence make this difficult to determine with certainty. It is discussed in this report despite being rare in the Hell Creek palynoflora because it may represent a group of gymnosperms—the cycads—that is not otherwise represented.

Genus *Pityosporites* Seward 1914
Pityosporites spp.
Plate 3, figs. 3a, 4a and 4b

Description. In this study all bisaccate pollen of pinaceoid morphology (resembling pollen of living species of the Pinaceae) is assigned to *Pityosporites* spp. All have reticulate sacci and central bodies with scabrate to finely rugulate caps; their size is variable. All generally resemble pollen produced by modern species of the Pinaceae including *Pinus, Picea,* and *Abies.*

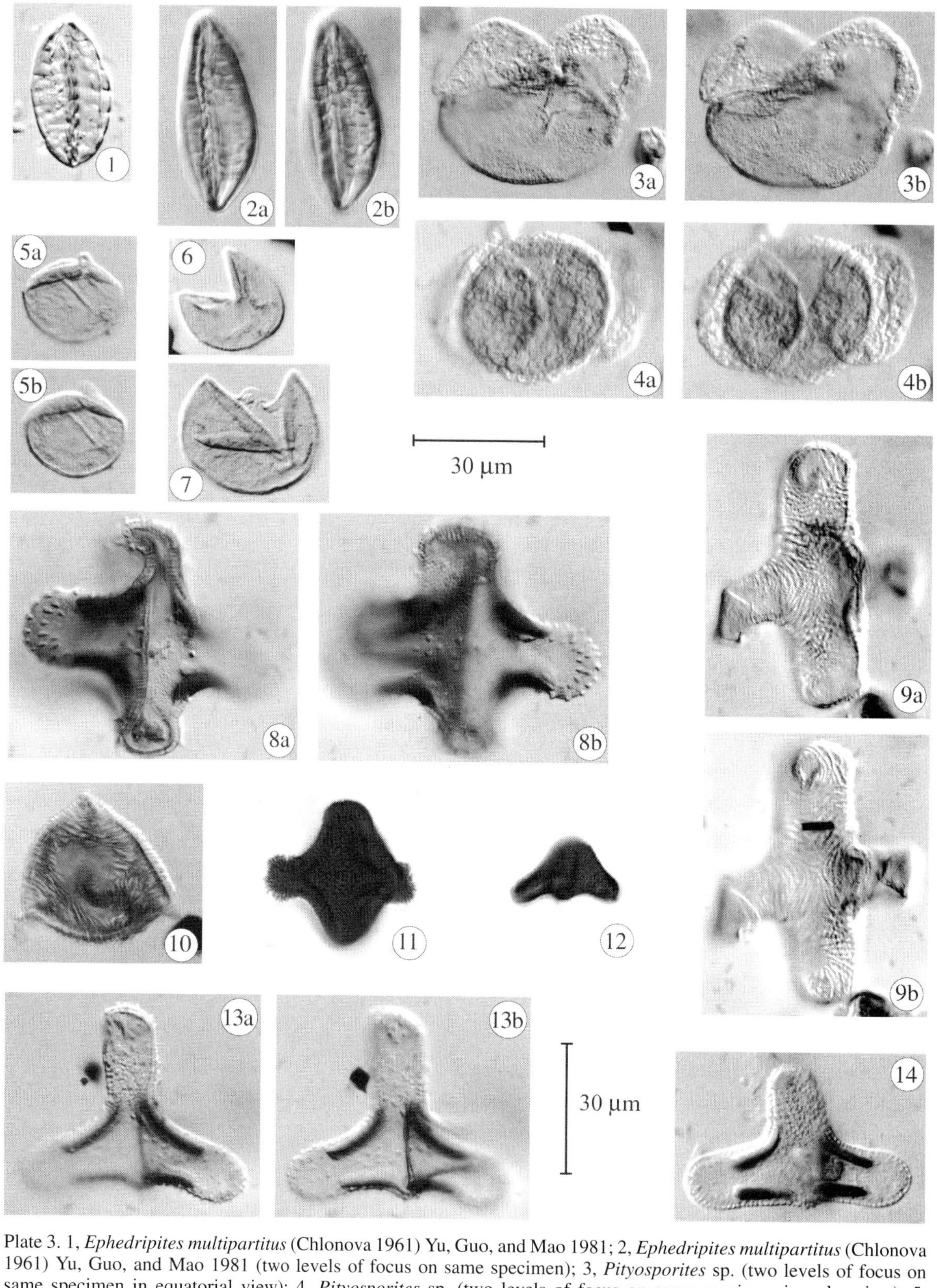

Plate 3. 1, *Ephedripites multipartitus* (Chlonova 1961) Yu, Guo, and Mao 1981; 2, *Ephedripites multipartitus* (Chlonova 1961) Yu, Guo, and Mao 1981 (two levels of focus on same specimen); 3, *Pityosporites* sp. (two levels of focus on same specimen in equatorial view); 4, *Pityosporites* sp. (two levels of focus on same specimen in polar view); 5, *Inaperturopollenites* sp. G (two levels of focus on same specimen); 6–7, *Taxodiaceaepollenites hiatus* (Potonié 1931) Kremp 1949; 8, *Aquilapollenites attenuatus* Funkhouser 1961 (two levels of focus on same specimen); 9, *Aquilapollenites conatus* Norton 1965; 10, *Aquilapollenites bertillonites* Funkhouser 1961 (polar view); 11, *Aquilapollenites catenireticulatus* Srivastava 1968; 12, *Aquilapollenites pyriformis* Norton 1965; 13, *Aquilapollenites collaris* (Tschudy and Leopold 1971) Nichols 1994 (two levels of focus on same specimen); 14, *Aquilapollenites delicatus* Stanley 1961.

Occurrence. As a group, pollen assigned to *Pityosporites* spp. is very common in the Hell Creek palynoflora. One or more species of this group were recorded at more than 90% of the localities studied. It is equally common in the Fort Union Formation.

Comments. The botanical affinity of *Pityosporites* spp. is unquestionably with the Pinaceae. No attempt was made to distinguish separate species because they lack biostratigraphic or paleoecologic significance. Pollen of this morphology is present in the fossil record from the Pennsylvanian to the Holocene. Although some species of bisaccate pollen have more limited stratigraphic ranges, none that do are present in the Hell Creek Formation. Species of living genera of the Pinaceae produce prodigious quantities of pollen, which is dispersed by the wind and may be found far from the forests where it originates. Thus its presence at most localities in the Hell Creek Formation is not necessarily evidence of the proximity or even the presence of pinaceous conifers in the Hell Creek flora. Norton and Hall (1969) postulated that an increased abundance of bisaccate conifer pollen in the Fort Union Formation relative to that in the Hell Creek Formation was an indication of cooling of climate from Cretaceous to Tertiary time, but neither that observation nor that conclusion is substantiated in this study.

Genus *Taxodiaceaepollenites* Kremp 1949 ex Potonié 1958
Taxodiaceaepollenites hiatus (Potonié 1931) Kremp 1949 ex Potonié 1958
Plate 3, figs. 6 and 7

Pollenites hiatus Potonié, 1931, p. 5, fig. 27.
Taxodiaceae-Poll. hiatus (Potonié) Kremp, 1949, p. 59, Plate 5, figs. 31, 37–38.
Taxodioidites hiatus (Potonié) Potonié, Thomson, and Thiergart, 1950, p. 49, Plate A, fig. 23.
Taxodioipollenites hiatus (Potonié) Potonié, 1951, p. 149, Plate 20, fig. 17.
Inaperturopollenites hiatus (Potonié) Thomson and Pflug, 1953, p. 86, Plate 5, figs. 14–20.
Taxodiaceaepollenites hiatus (Potonié) Kremp, Potonié, 1958, p. 78–79, Plate 10, fig. 103.
Thuja? hiatus (Potonié) Stanley, 1965, p. 273–274, Plate 38, figs. 1–3.

Description. Pollen assigned to this species is spheroidal, inaperturate, and scabrate. Although the grains are inaperturate, the exine is always split open in a characteristic manner. They closely resemble pollen produced by living species of the genus *Taxodium.* Pollen assigned to *Taxodiaceaepollenites hiatus* in this study differs primarily from that assigned to *Inaperturopollenites* sp. G in being split and in seldom or never exhibiting a papilla.

Occurrence. *Taxodiaceaepollenites hiatus* is common in the Hell Creek Formation. It was recorded at more than 80% of the localities studied; it is present throughout the formation. It is equally common in the Fort Union Formation. The relative abundance of this species is low at most localities; high productivity of sparse or distant anemophilous parent plants could explain its common occurrence in low numbers.

Comments. Some authors, following Thomson and Pflug (1953), would transfer this species to the genus *Inaperturopollenites,* but there is little or no utility in submerging the simple yet distinctive pollen species *Taxodiaceaepollenites hiatus* in such a broadly circumscribed genus. The botanical affinity of this species is with gymnosperms of the Taxodiaceae-Cupressaceae-Taxaceae (TCT) complex, but it is closer morphologically to the family Taxodiaceae than to either of the others. *Taxodiaceaepollenites hiatus* evidently is closely related to *Inaperturopollenites* sp. G, but the two are regarded as separate species because of differences in their patterns of occurrence. They may not be present in the same samples, and *Taxodiaceaepollenites hiatus* is considerably more common in the Hell Creek Formation and disproportionately more common in the Fort Union Formation than is *Inaperturopollenites* sp. G. Thus it is evident that the two forms are not simply morphological variants of the same species, but are two distinct taxa.

The paleoecological significance of both *Taxodiaceaepollenites hiatus* and *Inaperturopollenites* sp. G is the same—presumably they represent arboreal species that inhabited forested swamps, as do their close living relatives *Taxodium* and *Glyptostrobus.* These species tend to overwhelm assemblages in samples of coal from the Tongue River Member of the Fort Union Formation in Wyoming (Leffingwell, 1970; Pocknall and Nichols, 1996) and North Dakota (Nichols, unpublished data). Given this, these species are less abundant in Hell Creek lignite than might be expected. Six samples of Hell Creek lignite from four different localities were analyzed in this study. In only one of them is taxodiaceous pollen abundant (sample D7644-B, in which the combined total of *Inaperturopollenites* sp. G and *Taxodiaceaepollenites hiatus* is 44%). In three of the others, fern spores (*Laevigatosporites* spp.) are present in higher percentages than is taxodiaceous pollen. These observations suggest that the Taxodiaceae were an important component of the vegetation of some but certainly not all Hell Creek peat-forming mires (ferns were more abundant in the others).

Angiosperm pollen

Genus *Anacolosidites* Cookson and Pike 1954 emend. Potonié 1960
Anacolosidites rotundus Stanley 1965
Plate 5, fig. 1

Anacolosidites rotundus Stanley, 1965, p. 305, Plate 45, figs. 24–28.

Description. *Anacolosidites rotundus* is hexaporate pollen in which the pores occur in three pairs on opposite hemispheres. The shape of the grain is oblate spheroidal, and the amb is circular; the sculpture is reticulate. This species differs from *Libopollis jarzenii* in the absence of furrows connecting pairs of pores across the equator (it has hexaporate rather than "hexaporotricolpate" apertures). See Stanley (1965) for additional illustrations.

Occurrence. *Anacolosidites rotundus* is rare in the Hell Creek Formation; it was recorded at only three localities in this study. It was also recorded in the Ludlow Member of the Fort Union Formation at another locality, but only in the lowermost 30 cm, in a facies bearing a Maastrichtian assemblage. In this study *A. rotundus* is considered to be a K taxon indicative of Maastrichtian age.

Comments. Stanley (1965) first described this species from the Hell Creek Formation in South Dakota. He assigned it to the genus *Anacolosidites.* The genus *Anacolosidites* was proposed by Cookson and Pike (1954) for hexaporate pollen from the Eocene of Australia that has a triangular to rounded-triangular amb. Subsequent descriptions and the designation of a lectotype for the genus by other authors reinforce this description. Although Stanley described *Anacolosidites rotundus* as having sides that are convex to slightly concave, none of the specimens observed in this study nor any of those illustrated by Stanley are other than circular in outline. Furthermore, Stanley (1965, p. 305) noted that the specific epithet *rotundus* was "after the circular shape of this species in polar view." Thus it is evident that this species does not belong in the genus *Anacolosidites.* Although it has much morphologic resemblance to *Libopollis jarzenii* (and may be phylogenetically related to that species), the genus *Libopollis* as described would not accommodate *A. rotundus* because of differences in aperture structure. Too few specimens are available at present on which to base the description of a new genus.

Genus *Aquilapollenites* Rouse 1957 emend. Funkhouser 1961

Comments. The genus *Aquilapollenites* was originally described by Rouse (1957); its description was emended by Funkhouser (1961). Subsequently this genus, whose myriad species are widely distributed in the Upper Cretaceous of western North America and Asia, has been treated taxonomically by many. Some authors (e.g., Mchedlishvili in Samoilovich and Mchedlishvili, 1961; Takahashi, 1964; Srivastava, 1968; Stanley, 1970; Krutzsch, 1970; Farabee, 1990) have split the genus and proposed a variety of names for groups of species that constitute morphologic variants. There are some virtues but many problems associated with each of the nomenclatural schemes proposed in the references cited, the details of which are beyond the scope of this chapter to review. The taxonomic and nomenclatural practice of Tschudy and Leopold (1970) is followed herein, hence only a single genus is recognized. Other generic names are listed in synonymies of individual species of *Aquilapollenites* treated here.

Aquilapollenites attenuatus Funkhouser 1961
Plate 3, figs. 8a and 8b

Aquilapollenites attenuatus Funkhouser, 1961, p. 194, 196; Plate 2, fig. 1a–1c.

Aquilapollenites amplus Stanley, 1961b, p. 342, Plate 1, figs. 1–6, Plate 2, figs. 1–4, Plate 3, figs. 1–5.

Description. Pollen of this species is triprojectate and isopolar to subisopolar; in most specimens the poles are subequal in size. Three equatorial projections extend horizontally; they are broad and flattened latitudinally. Polar and equatorial dimensions are about equal. Colpate apertures are located at the ends of the equatorial projections. Sculpture is echinate, and spines are well developed on the equatorial projections. See Funkhouser (1961) or Srivastava (1969) for a more detailed description. *Aquilapollenites attenuatus* differs from other species of the genus in the Hell Creek palynoflora in size, degree of heteropolarity, and in width and prominence of the equatorial projections. Numerous species described by Srivastava (1969) including *A. accipterus, A. ascriptivus, A. aucellatus, A. augustus, A. drumhellerensis, A. hispidus,* and *A. regalis* are similar and probably conspecific.

Occurrence. *Aquilapollenites attenuatus* is common in the Hell Creek palynoflora. It was recorded at 12 (44%) of the localities studied. It is present throughout the Hell Creek Formation from ~20 m above the base to the top. It is also present in the basal part of the Ludlow Member of the Fort Union Formation at some localities analyzed in this study, but only within the lower 1–1.5 m, in facies that carry a Maastrichtian assemblage. In this study *Aquilapollenites attenuatus* is considered to be a K taxon indicative of Maastrichtian age; it also occurs in rocks as old as late Campanian in Montana and Wyoming (Tschudy and Leopold, 1970).

Comments. Funkhouser (1961) first described *Aquilapollenites attenuatus* from the Lance Formation of Wyoming. Later the same year, Stanley (1961b) described the same species from the Hell Creek Formation in South Dakota as *Aquilapollenites amplus.* The rule of priority of publication applies, and Stanley's species is a later synonym of *Aquilapollenites attenuatus.* The seven species named above described by Srivastava (1969) from the Edmonton Formation of Alberta are probably interspecific variants of the same species and thus would also be synonymous with *Aquilapollenites attenuatus.* Nichols and Sweet (1993) included *Aquilapollenites attenuatus* in the *Wodehouseia spinata* assemblage.

This species is one of 14 species of the genus *Aquilapollenites* identified in the Hell Creek palynoflora. The triprojectate genus *Aquilapollenites* is characteristic of Upper Cretaceous (Campanian through Maastrichtian) rocks in the western North America. The abrupt extinction of most Maastrichtian species of the genus is part of the compelling evidence for continent-wide devastation of plant communities at the end of Cretaceous time—the K–T boundary event. The persistence of some of these species into the lowermost part of the Fort Union Formation does not alter this interpretation. It implies the onset of Fort Union-style deposition (including peat accumulation) shortly before the end of the Cretaceous, and it shows that the lowermost part of the Fort Union Formation is latest Cretaceous in age in some areas.

Aquilapollenites bertillonites Funkhouser 1961
Plate 3, fig. 10

Aquilapollenites bertillonites Funkhouser, 1961, p. 196, Plate 2, figs. 5a–5c.
Aquilapollenites murus Stanley, 1961b, p. 347, Plate 5, figs. 1–9B, Plate 6, figs. 1–9.
Integricorpus bertillonites (Funkhouser) Stanley, 1970, p. 29.
Integricorpus bertillonites (Funkhouser) Krutzsch, 1970, p. 116.

Description. Pollen of this species is triprojectate and isopolar; the equatorial projections are short and slender. The polar dimension is greater than the equatorial dimension. Colpate apertures are located at the ends of the equatorial projections. Sculpture is striate. See Funkhouser (1961) for a more detailed description. The isopolar body and striate sculpture distinguish this species from all others of the genus in the Hell Creek palynoflora.

Occurrence. *Aquilapollenites bertillonites* is restricted to the Hell Creek Formation. It was found only in the lower 42 m of the formation, where it was present in 5 of 7 samples analyzed from that interval (the actual upper limit of its stratigraphic range probably is somewhat higher; a palynologically unproductive interval of ~19 m is above the highest recorded occurrence). Thus, in addition to being one of the K taxa recognized in this study, *A. bertillonites* is a guide fossil for the lower part of the Hell Creek Formation.

Comments. Funkhouser (1961) first described this species from the Lance Formation of Wyoming. Stanley (1961b) described the same species as *Aquilapollenites murus* from the Hell Creek Formation in South Dakota later the same year. The rules of priority apply, and Stanley's species is a later synonym of *Aquilapollenites bertillonites.* Nichols and Sweet (1993) included *Aquilapollenites bertillonites* in the *Wodehouseia spinata* assemblage.

This species is one of 14 species of the genus *Aquilapollenites* identified in the Hell Creek palynoflora. Because its range does not extend to the K–T boundary, it cannot be counted among the species whose extinction is attributed to the K–T boundary event.

Aquilapollenites catenireticulatus Srivastava 1968
Plate 3, fig. 11

Aquilapollenites catenireticulatus Srivastava, 1968, p. 674, Plate 1, figs. 7–9.

Description. The triprojectate species is isopolar to subisopolar, the body of the grain is oval in equatorial view, and the equatorial projections are short and expanded at the ends into knob-like forms. The polar dimension exceeds the equatorial dimension. Sculpture is reticulate to striate and variable in coarseness. See Srivastava (1968) or Tschudy and Leopold (1970) for a more detailed description. The specimen illustrated here in Plate 4 is heavily stained and therefore dark, but the characteristic outline of this species is clearly shown.

Occurrence. This species is rare in the Hell Creek Formation; it was recorded at only three of the localities studied, in about the middle third of the formation. Elsewhere in the South Dakota–Wyoming–Montana region it ranges from upper Campanian to middle Maastrichtian (Tschudy and Leopold, 1970).

Comments. Srivastava (1968) originally described *Aquilapollenites catenireticulatus* from the Edmonton Formation (Maastrichtian) of Alberta. Tschudy and Leopold (1970, p. 154–155) somewhat expanded the circumscription of the species (although they did not formally emend the description) on the basis of numerous specimens in collections of the U.S. Geological Survey, including some from the Hell Creek Formation in Montana. The North Dakota specimens are within the expanded circumscription. *Aquilapollenites catenireticulatus* differs from *A. oblatus* Srivastava 1968 in having larger equatorial projections that are constricted at the base and in the distinctive chain-like pattern in the reticulum (as described by Srivastava, 1968).

Aquilapollenites catenireticulatus is 1 of 14 species of the genus *Aquilapollenites* identified in the Hell Creek palynoflora. It can be considered a K taxon, but because its range does not extend to the K–T boundary, it cannot be counted among the species whose extinctions are attributed to the K–T boundary event.

Aquilapollenites collaris (Tschudy and Leopold 1971)
Nichols 1994
Plate 3, figs. 13a and 13b

Aquilapollenites delicatus Stanley var. *collaris* Tschudy and Leopold, 1970, p. 132–133, 135–138; Plate 3, fig. 6.
Aquilapollenites collaris (Tschudy and Leopold) Nichols, 1994, p. 518–519, fig. 7h.

Description. *Aquilapollenites collaris* is heteropolar with a long, slender major pole and greatly reduced, knob-like minor pole; the equatorial projections are long and slender and extend horizontally. The polar and equatorial dimensions are about equal. There is a distinctive collar-like zone of finely reticulate sculpture separating coarser reticulate sculpture on the major pole. See Tschudy and Leopold (1970) or Nichols (1994) for a more detailed description.

Occurrence. *Aquilapollenites collaris* is common in the Hell Creek Formation; it was recorded at more than 60% of the localities studied and is notably abundant at in some samples. It ranges from the middle of the formation (50 m above the base) to the top and into the overlying Fort Union Formation. In the Fort Union Formation it was recorded in the basal part of the Ludlow Member as much as 2.4 m above the Hell Creek–Fort Union formational contact, but only in facies that otherwise carry an assemblage of Maastrichtian age. It is considered to be a K taxon indicative of Maastrichtian age.

Comments. Tschudy and Leopold (1970) originally described this species as a variety of *Aquilapollenites delicatus* Stanley. It is evident that *A. delicatus* and *A. collaris* are phylogenetically related, the latter evidently having evolved from the former. However, the distinctive morphology and more restricted stratigraphic range of *A. collaris* warrants it being identified as a separate species rather than as a variety. Nichols (1994) raised the variety to species rank. Farabee (1993) cited the name *Aquilapollenites collaris* as a new combination, but he did not formally establish it as such in accordance with the International Code of Botanical Nomenclature. Nichols and Sweet (1993) included this species in the *Wodehouseia spinata* assemblage. (See the second comment about *Aquilapollenites attenuatus.*)

Although the report by Tschudy and Leopold bears the imprint date 1970, it was not actually published until 1971. For nomenclatural purposes the 1971 date is given, but for consistency with procedures in most libraries, the 1970 date is given to identify the publication.

Aquilapollenites conatus Norton 1965
Plate 3, figs. 9a and 9b

Aquilapollenites conatus Norton, 1965, p. 142, Plate 3, figs. 10–11; Plate 4, figs. 12–16.
Triprojectus conatus (Norton) Stanley, 1970, p. 30.
Striatoporites conatus (Norton) Farabee, 1990, p. 346, fig. 19.

Description. *Aquilapollenites conatus* is triprojectate and isopolar; the sculpture is striate. The equatorial projections are moderate in size in proportion to the body of the grain, but the polar and equatorial dimensions are about equal. The ends of the body of the grain are flattened, and they each have three cone-shaped protuberances. See Norton (1965) for a more complete description.

Occurrence. This species is fairly common in the Hell Creek Formation; it was recorded at more than one-third of the localities studied. Stratigraphically it is restricted to the upper half of the Hell Creek and the lowermost 1.5 m of the Fort Union Formation, where it occurs in facies that otherwise carry an assemblage of Maastrichtian age. It is considered to be a K taxon indicative of Maastrichtian age. Tschudy and Leopold (1970) gave its range as middle to upper Maastrichtian in Montana and Wyoming.

Comments. Norton (1965) first described this species from the Hell Creek Formation in eastern Montana. In North Dakota it is a guide fossil for the upper part of the Hell Creek. Nichols and Sweet (1993) included this species in the *Wodehouseia spinata* assemblage. (See the second comment about *Aquilapollenites attenuatus.*)

Aquilapollenites delicatus Stanley 1961
Plate 3, fig. 14

Aquilapollenites delicatus Stanley, 1961b, p. 346–347, Plate 4, figs. 1–12.
Hemicorpus delicatus (Stanley) Krutzsch, 1970, p. 108.

Description. This species is triprojectate and heteropolar; the equatorial projections are large in proportion to the body of the grain; they extend horizontally or at a slight angle downward toward the minor pole. The polar and equatorial dimensions are approximately equal. Sculpture is both reticulate and echinate. See Stanley (1961b) or Tschudy and Leopold (1970) for a more complete description. *Aquilapollenites delicatus* differs from *Aquilapollenites collaris* in having a shorter, broader major polar projection and in lacking the collar of fine sculpture characteristic of the latter species.

Occurrence. *Aquilapollenites delicatus* is fairly common in the Hell Creek Formation; it was recorded at one-third of the localities studied. Stratigraphically it is present throughout the formation, and it was recorded also in the lowermost 1 m of the Fort Union Formation in this study, but only in facies that otherwise carry an assemblage of Maastrichtian age. In this study it is considered to be a K taxon indicative of Maastrichtian age. It ranges as low as the upper Campanian in South Dakota (Tschudy and Leopold, 1970).

Comments. Stanley (1961b) first described this species from the Hell Creek Formation in South Dakota. The cooccurrence of *Aquilapollenites delicatus* with its descendant species, *Aquilapollenites collaris,* in the Hell Creek Formation or elsewhere in no way argues against the latter having evolved from the former, which it almost certainly did. Nichols and Sweet (1993) included this species in the *Wodehouseia spinata* assemblage (see also the second comment about *Aquilapollenites attenuatus*).

Aquilapollenites marmarthensis sp. nov.
Plate 4, figs. 1a–1c, 2, 3a and 3b, 4a and 4b, 5a and 5b, 6a and 6b

Description. Pollen triprojectate, isopolar to subisopolar; major pole broadly rounded, minor pole reduced and similarly rounded but smaller; equatorial projections about as wide as major pole, extending horizontally or at slight angle toward the minor pole; equatorial dimension greater than polar dimension; apertures tricolpate, located at ends of equatorial projections, often obscure; sculpture finely echinate. Exine tectate, thin, ~1 μm or less; sculpture evenly distributed over surface, including at ends of polar projections. Size: 30–35 μm in polar dimension, 35–45 μm in maximum equatorial dimension (20 specimens).

Diagnosis. *Aquilapollenites marmarthensis* sp. nov. is similar to *Aquilapollenites vulsus* Sweet 1986 but differs in having evenly distributed sculpture. It differs from other species of the genus in the Hell Creek palynoflora in small size, polarity and/or heteropolarity, and finely echinate sculpture.

Holotype. Slide D7018-D (2), England Finder coordinate L49/2; 1 in Plate 4, figures 1a–1c herein.

Type Locality. Sunset Butte, Bowman County, North Dakota; sample D7018-D, ~1 m (80–136 cm) below top of Hell Creek Formation.

Repository. The slide bearing the type specimen of *Aquilapollenites marmarthensis* sp. nov. is held in the palynological laboratory of Central Energy Resources, U.S. Geological Survey, Denver, Colorado.

Occurrence. *Aquilapollenites marmarthensis* sp. nov. is fairly common in the Hell Creek Formation; it was recorded at more than one-third of the localities studied. It is present from ~34 m below the top of the formation upward through to the basal part of the Fort Union Formation. In the Fort Union it occurs as high as 2.4 m above the Hell Creek–Fort Union formational contact and is abundant in some samples. In the Fort Union Formation it occurs only in facies that otherwise carry an assemblage of Maastrichtian age. It is considered to be a K taxon indicative of Maastrichtian age.

Comments. This species tends to be more common and abundant in the Maastrichtian basal part of the Ludlow Member of the Fort Union Formation than in the Hell Creek Formation.

Aquilapollenites pyriformis Norton 1965
Plate 3, fig. 12

Aquilapollenites pyriformis Norton, 1965, p. 136, 140; Plate 1, figs. 1–3, Plate 2, fig. 4.
Mancicorpus pyriformis (Norton) Srivastava, 1968, p. 697.
Fibulapollis pyriformis (Norton) Takahashi in Takahashi and Shimono, 1982, p. 81.
Striatocorpus pyriformis (Norton) Farabee, 1990, p. 346, fig. 3.

Description. *Aquilapollenites pyriformis* is triprojectate, heteropolar, and striate. The polar dimension is less than the equatorial dimension, and the equatorial projections extend at an angle toward the minor pole; the outline in equatorial view is roughly triangular (pyramidal). See Norton (1965) for a more detailed description. The specimen illustrated in Plate 3 herein is heavily stained and therefore dark, but the characteristic outline of this species is clearly shown.

Occurrence. This species is rare in the Hell Creek Formation. It was recorded at only one locality in this study, 30 m below the top of the formation. Despite its rare occurrence, it can be considered a K taxon, but because its range does not extend to the K-T boundary, it cannot be counted among the species whose extinction is attributed to the K-T boundary event.

Comments. Norton (1965) first described this species from the Hell Creek Formation in eastern Montana. See also second comment about *Aquilapollenites bertillonites.*

Aquilapollenites quadricretaeus Chlonova 1961
Plate 4, figs. 7a and 7b, 8

Aquilapollenites quadricretaeus Chlonova, 1961, p. 84, Plate 14, fig. 108.

Description. *Aquilapollenites quadricretaeus* is triprojectate and isopolar; sculpture consists of small, scattered verrucae. Polar and equatorial dimensions are about equal, and polar and equatorial projections are about equal in cross-section width. The proportions of the grain and the sculpture distinguish this species from all others of the genus in the Hell Creek palynoflora.

Occurrence. This species is rare in the Hell Creek Formation; it was recorded at only four localities in this study. It was also recorded in the Fort Union Formation, to 2.4 m above the Hell Creek–Fort Union formational contact, but only in facies that otherwise carry an assemblage of Maastrichtian age. It is considered to be a K taxon indicative of Maastrichtian age. Elsewhere in the region it was previously reported in the uppermost Maastrichtian of the Frenchman Formation, the Hell Creek correlative equivalent in southern Saskatchewan (Nichols et al., 1986).

Comments. Chlonova (1961) first described this species from the Maastrichtian of Siberia. The correct spelling of the specific epithet is "*quadricretaeus*" (not "*quadricretaceus*") because that is the way it was spelled in Chlonova (1961); see also Ames and Spackman (1981). Nichols and Sweet (1993) included *Aquilapollenites quadricretaeus* in the *Wodehouseia spinata* assemblage. See the second comment about *Aquilapollenites attenuatus.*

Aquilapollenites quadrilobus Rouse 1957
Plate 4, figs. 9a and 9b, 11, and 12

Designate N2, Radforth and Rouse, 1954, p. 195, Plate 1, fig. 14.
Aquilapollenites quadrilobus Rouse, 1957, p. 371.
Aquilapollenites polaris Funkhouser, 1961, p. 198, Plate 1, figs. 1a–1d, and 2.
Aquilapollenites pulcher Funkhouser, 1961, p. 198, Plate 1, fig. 7a–7c.
Aquilapollenites pulvinus Stanley, 1961b, p. 347–348, Plate 7, figs. 1–12.
Aquilapollenites hirsutus Srivastava, 1969, p. 139, Plate 5, fig. 30.
Aquilapollenites quadrilobus Rouse, Tschudy and Leopold, 1970, p. 125–127, 130–131; Plate 2, figs. 1–6.

Description. *Aquilapollenites quadrilobus* is triprojectate, heteropolar, and echinate. The equatorial projections are large in proportion to the body of the grain, and they extend horizontally or downward at a slight angle toward the minor pole. See the discussion of earlier confusion about the morphology of this species and an extensive and detailed description of the holotype in Tschudy and Leopold (1970). *Aquilapollenites quadrilobus* resembles *A. delicatus* in size and form, but differs in having exclusively echinate, never reticulate, sculpture; *A. senonicus* bears some resemblance but differs in having a more reduced minor pole and in having exclusively reticulate sculpture.

Occurrence. *Aquilapollenites quadrilobus* is common in the Hell Creek Formation and was recorded at 85% of the localities studied. It is also present in the Fort Union Formation at some localities within the basal 2.5 m of the Ludlow Member, but only in facies that otherwise carry an assemblage of Maastrichtian age. In this study it is considered to be a K taxon indicative of Maastrichtian age. Elsewhere in the region it is commonly found in rocks as old as early or middle Campanian as well as in the Maastrichtian in South Dakota, Wyoming, and Montana (Tschudy and Leopold, 1970).

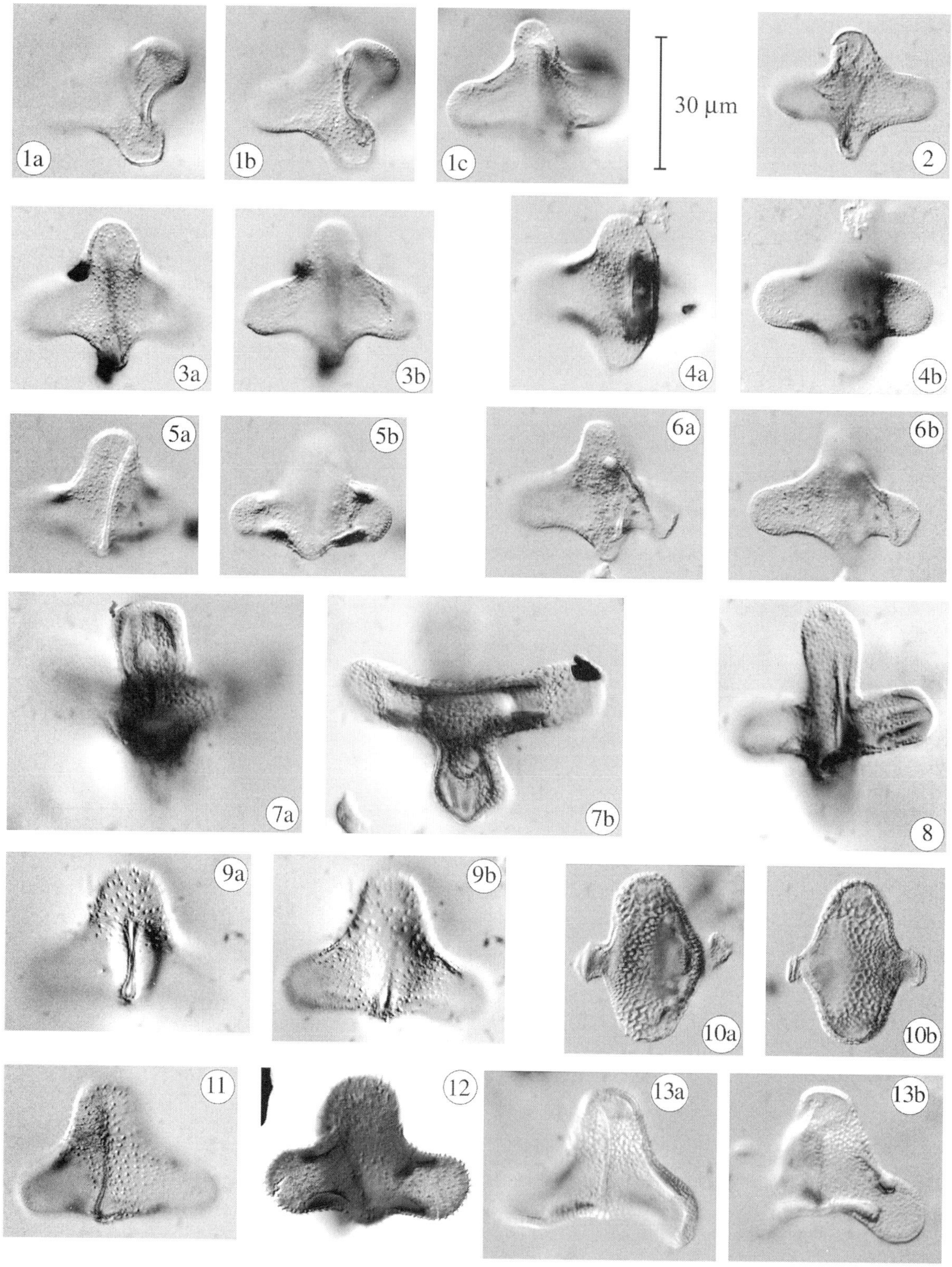

Plate 4. 1, *Aquilapollenites marmarthensis* sp. nov. (holotype; three levels of focus on same specimen); 2, *Aquilapollenites marmarthensis* sp. nov. (isotype); 3, *Aquilapollenites marmarthensis* sp. nov. (isotype; two levels of focus on same specimen); 4, *Aquilapollenites marmarthensis* sp. nov. (isotype; two levels of focus on same specimen); 5, *Aquilapollenites marmarthensis* sp. nov. (isotype; two levels of focus on same specimen); 6, *Aquilapollenites marmarthensis* sp. nov. (isotype; two levels of focus on same specimen); 7, *Aquilapollenites quadricretaeus* Chlonova 1961 (two levels of focus on same specimen); 8, *Aquilapollenites quadricretaeus* Chlonova 1961; 9, *Aquilapollenites quadrilobus* Rouse 1957 (two levels of focus on same specimen); 10, *Aquilapollenites reductus* Norton 1965 (two levels of focus on same specimen); 11–12, *Aquilapollenites quadrilobus* Rouse 1957; 13, *Aquilapollenites senonicus* (Mchedlishvili 1961) Tschudy and Leopold 1971 (two levels of focus on same specimen).

Comments. *Aquilapollenites quadrilobus* is the type species of the genus *Aquilapollenites*. Rouse (1957) originally described but did not illustrate this species from the Upper Cretaceous of western Canada; Radforth and Rouse (1954) had previously illustrated the holotype. Reexamination and reillustration of the holotype specimen by Tschudy and Leopold (1970) did much to correct misconceptions that had arisen about the morphology of this species. Perhaps the most important observation they made was that the holotype is heteropolar—with a greatly reduced minor pole—and not isopolar. This observation renders the genus *Mancicorpus* Mchedlishvili synonymous with *Aquilapollenites*.

In their study of the holotype of *Aquilapollenites quadrilobus,* Tschudy and Leopold (1970) also reexamined the holotype specimens of *A. polaris* Funkhouser, *A. pulcher* Funkhouser, *A. pulvinus* Stanley, and *A. hirsutus* Srivastava. They found all of them to be indistinguishable from the holotype of *A. quadrilobus*. Thus, on the basis of primary evidence, all of these species are later synonyms of *A. quadrilobus*.

Tschudy and Leopold (1970) were able to redescribe and reillustrate the type specimen of *Aquilapollenites quadrilobus* because they borrowed the damaged microscope slide on which it was located, restored the mount, and relocated the specimen. Unfortunately, they eventually returned the slide and now it is lost again, apparently forever. However, in accordance with the International Code of Botanical Nomenclature, the type of the species is the illustration of the holotype specimen. The holotype is illustrated in Radforth and Rouse (1954, Plate 1, fig. 14); the holotype is better illustrated in Tschudy and Leopold (1970, Plate 2, fig. 1, A–B).

Nichols and Sweet (1993) included *Aquilapollenites quadrilobus* in the *Wodehouseia spinata* assemblage although it is also present in older rocks, where it characterizes the *Aquilapollenites quadrilobus* Interval Zone of Nichols et al. (1982). It is one of the most commonly occurring species of the genus *Aquilapollenites* in the Hell Creek Formation.

Aquilapollenites reductus Norton 1965
Plate 4, figs. 10a and 10b

Aquilapollenites reductus Norton, 1965, p. 140, 142; Plate 2, figs. 5–7.
Aquilapollenites funkhouseri Srivastava, 1966, p. 541, Plate 9, figs. 5–8, 10.
Aquilapollenites ceriocorpus Srivastava, 1968, p. 674–676, Plate 2, figs. 1–4.
Aquilapollenites medeis Srivastava, 1968, p. 678–680, Plate 3, fig. 6; Plate 4, figs. 1–2.
Triprojectus reductus (Norton) Stanley, 1970, p. 31.
Integricorpus reductum (Norton) Takahashi in Takahashi and Shimono, 1982, p. 70.

Description. This species is triprojectate and isopolar; the polar dimension is greater than the equatorial dimension. The equatorial projections are small and extend horizontally; they are club-shaped, wider at their extremities than near the body of the grain, and have bands of reticulate sculpture near their extremities. Sculpture is reticulate. *Aquilapollenites reductus* differs from *Aquilapollenites reticulatus* in having club-shaped equatorial projections and generally coarser sculpture; it differs from *A. catenireticulatus* in these features and also in larger size.

Occurrence. *Aquilapollenites reductus* is fairly common in the Hell Creek Formation; it was recorded at one-third of the localities studied. It is also present in the lower part of the Fort Union Formation, where at one locality it was recorded less than 1.4 m above the Hell Creek–Fort Union formational contact in a facies that includes other taxa indicative of Maastrichtian age. In this study it is considered to be a K taxon indicative of Maastrichtian age. Elsewhere in the region it has been recorded in upper Campanian through upper Maastrichtian rocks in South Dakota, Wyoming, and Montana (Tschudy and Leopold, 1970).

Comments. Norton (1965) originally described *Aquilapollenites reductus* from the Hell Creek Formation in eastern Montana. The three species described by Srivastava (1966, 1968) listed in the synonymy for this species differ in no significant way from *Aquilapollenites reductus* and should be regarded as infraspecific variants having the same stratigraphic and geographic ranges. See also the second comment after *Aquilapollenites attenuatus*.

Aquilapollenites reticulatus (Mchedlishvili 1961)
Tschudy and Leopold 1971
Plate 5, fig. 2

Parviprojectus reticulatus Mchedlishvili in Samoilovich and Mchedlishvili, 1961, p. 226–227, Plate 73, figs. 2–3.
Aquilapollenites reticulatus Stanley, 1961b, p. 348–349, Plate 8, figs. 1–12A.
Aquilapollenites mtchedlishvilii Srivastava, 1968, p. 692–693.
Aquilapollenites reticulatus (Mchedlishvili) Tschudy and Leopold, 1970, p. 149–151, Plate 1, fig. 12a–12b.
Integricorpus reticulatus (Mchedlishvili) Stanley, 1970, p. 29.
non *Aquilapollenites reticulatus* Chlonova, 1961, p. 84, Plate 14, fig. 107.

Description. *Aquilapollenites reticulatus* is triprojectate, isopolar, and reticulate; the polar dimension is greater than the equatorial dimension. The equatorial projections are small and extend horizontally, and they taper toward their distal ends. *Aquilapollenites reticulatus* differs from *A. reductus* in being somewhat angular rather than rounded at the poles, in having tapered equatorial projections, and in finer sculpture; it differs from *A. catenireticulatus* in these features and in being larger.

Occurrence. This species is quite common in the Hell Creek Formation; it was recorded at 23 (85%) of the localities studied. It also occurs in the lower part of the Fort Union Formation, where it was recorded at 7 localities to 4.2 m above the Hell Creek–Fort Union formational contact. Elsewhere in the region it has been recorded from upper Campanian through upper Maastrichtian rocks in New Mexico, Colorado, Wyoming, Montana, Saskatchewan, Alberta, and the Northwest Territories (Tschudy and Leopold, 1970; Nichols and Sweet, 1993).

Comments. Mchedlishvili (in Samoilovich and Mchedlishvili, 1961) first described *Aquilapollenites reticulatus* (as *Parviprojectus reticulatus*) from the Upper Cretaceous of Siberia. Later that year, Stanley (1961b) described the same species from the Hell Creek Formation in South Dakota and assigned the name *Aquilapollenites reticulatus* to his specimens. Tschudy and Leopold (1970), recognizing that Mchedlishvili's species and Stanley's are synonymous, transferred Mchedlishvili's species to the genus *Aquilapollenites*. Thus, *A. reticulatus* Stanley is both a later synonym and a homonym of *A. reticulatus* Mchedlishvili. Srivastava (1968), recognizing the homonymy of Mchedlishvili's and Stanley's species, had earlier renamed Mchedlishvili's species as *Aquilapollenites mtchedlishvilii,* but that name is synonymous with *Aquilapollenites reticulatus* (Mchedlishvili) Tschudy and Leopold. *Aquilapollenites reticulatus* Chlonova, coincidentally also first described in 1961 and also from Siberia, is a homonym of *Aquilapollenites reticulatus* (Mchedlishvili) Tschudy and Leopold, but is not a synonym because it differs in morphology (it has an oval outline in equatorial view and very coarsely reticulate sculpture).

Aquilapollenites reticulatus (Mchedlishvili) Tschudy and Leopold is known to occur above the K-T boundary in southern Canada (Jerzykiewicz and Sweet, 1986; Nichols and Sweet, 1993) and in eastern Montana (Hotton, 1988; this volume). Evidently it is one species among the so-called K taxa that survived the K-T boundary extinction event. It is rare in the Fort Union Formation and did not persist long in early Paleocene time. Nichols and Sweet (1993) included this species in the *Wodehouseia spinata* assemblage, despite its longer range than most species in that characteristically Maastrichtian assemblage (this is consistent with the concept of an assemblage that defines a biostratigraphic zone).

Aquilapollenites senonicus (Mchedlishvili 1961) Tschudy and Leopold 1971
Plate 4, figs. 13a and 13b

Mancicorpus senonicum Mchedlishvili in Samoilovich and Mchedlishvili, 1961, p. 224–225, Plate 72, figs. 2–3.
Hemicorpus senonicum (Mchedlishvili) Krutzsch, 1970, p. 108.
Aquilapollenites senonicus (Mchedlishvili) Tschudy and Leopold, 1970, p. 131–133, Plate 1, fig. 2.

Description. *Aquilapollenites senonicus* is triprojectate, heteropolar, and reticulate; the polar and equatorial dimensions are approximately equal. The equatorial projections are large in proportion to the body of the grain and they tend to extend downward at an angle toward the minor pole. This species differs from *Aquilapollenites quadrilobus* and *A. delicatus* in having a more reduced minor pole and in having exclusively reticulate sculpture. It differs from "*Mancicorpus*" *notabile* Mchedlishvili 1961 in having shorter, less slender equatorial projections that, although they extend at an angle toward the minor pole, do so with a less pronounced angle than in the latter species (e.g., cf. specimens illustrated in Nichols and Sweet, 1993, Plate 2, figs. 10 and 32).

Occurrence. *Aquilapollenites senonicus* is rare in the Hell Creek Formation; it was recorded at only five of the localities studied. It was also recorded in the basal part of the Fort Union Formation to 0.5 m above the Hell Creek–Fort Union formational contact, but only in facies that otherwise carry an assemblage of Maastrichtian age. In this study it is considered to be a K taxon indicative of Maastrichtian age. Elsewhere in the region it has been recorded in rocks as old as early Campanian in Montana and Wyoming and in Campanian through upper Maastrichtian rocks in Colorado, Wyoming, and Montana (Tschudy and Leopold, 1970; Farabee and Canright, 1986).

Comments. Mchedlishvili (in Samoilovich and Mchedlishvili, 1961) first described this species from the Upper Cretaceous of Siberia as *Mancicorpus senonicum*. Tschudy and Leopold (1970) transferred Mchedlishvili's species to the genus *Aquilapollenites* because the genus *Mancicorpus* is synonymous with *Aquilapollenites*. (See also the second comment about *Aquilapollenites bertillonites* concerning stratigraphic range and the second comment about *Aquilapollenites collaris* concerning the publication date of the report by Tschudy and Leopold).

Aquilapollenites turbidus Tschudy and Leopold 1971
Plate 5, figs. 3a and 3b

Aquilapollenites turbidus Tschudy and Leopold, 1970, p. 157–159, Plate 2, figs. 7–10.

Description. *Aquilapollenites turbidus* is triprojectate, isopolar, and echinate; the equatorial dimension is greater than the polar dimension. The equatorial projections are longer and wider than the polar projections. See Tschudy and Leopold (1970) for a more detailed description. This species differs from all others in the Hell Creek palynoflora in its small size and in the relative proportions of the equatorial projections and the body of the grain.

Occurrence. This species is rare in the Hell Creek Formation; it was recorded at only two of the localities studied. It was also recorded in the basal part of the Fort Union Formation within the interval 0.5 m above the Hell Creek–Fort Union formational contact, but only in facies that otherwise carry an assemblage of Maastrichtian age. In this study it is considered to be a K taxon indicative of Maastrichtian age. Elsewhere in the region it has been recorded from lower Campanian through upper Maastrichtian rocks in South Dakota, Montana, Wyoming, and Colorado (Tschudy and Leopold, 1970).

Comments. Tschudy and Leopold (1970) originally described this species from the Red Bird Silty Member of the Pierre Shale (middle Campanian) in Wyoming. The records of the species published here are the first from the Hell Creek Formation. The Hell Creek specimen illustrated here departs somewhat from those included in the protologue for the species (Tschudy and Leopold, 1970) in that the equatorial projections are quite wide near the central body and there are some small

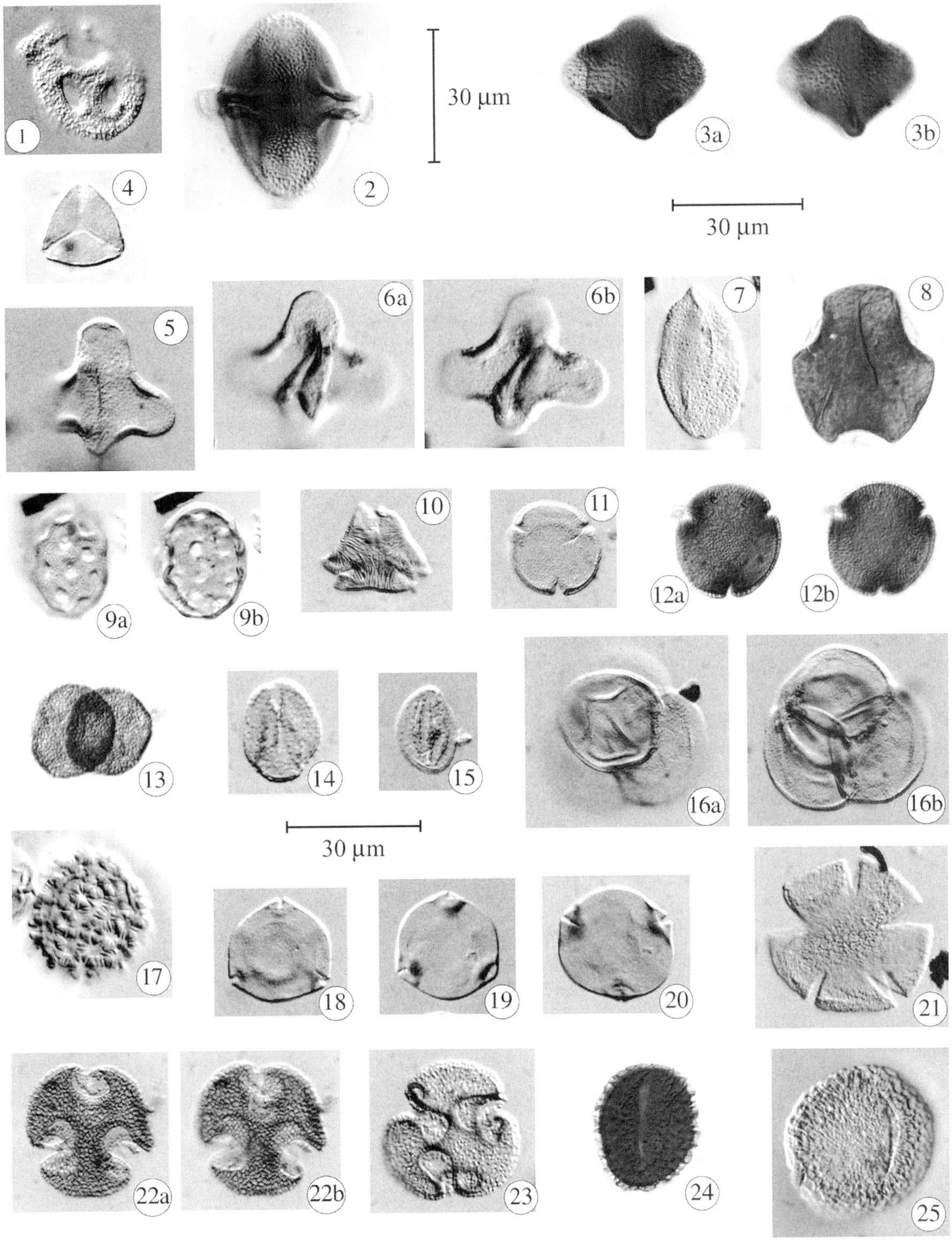

Plate 5. 1, *Anacolosidites rotundus* Stanley 1965; 2, *Aquilapollenites reticulatus* (Mchedlishvili 1961) Tschudy and Leopold 1971; 3, *Aquilapollenites turbidus* Tschudy and Leopold 1971 (two levels of focus on same specimen); 4, *Boehlensipollis minimus* (Leffingwell 1971) Pocknall and Nichols 1996; 5, *Aquilapollenites vulsus* Sweet 1986; 6, *Aquilapollenites vulsus* Sweet 1986 (two levels of focus on same specimen); 7, *Arecipites columellus* Leffingwell 1971; 8, *Cercidiphyllites* sp.; 9, *Chenopodipollis* sp. (two levels of focus on same specimen); 10, *Cranwellia rumseyensis* Srivastava 1967; 11, *Discoidites parvistriatus* (Norton 1967) Nichols and Brown 1992; 12, *Discoidites parvistriatus* (Norton 1967) Nichols and Brown 1992 (two levels of focus on same specimen); 13, *Dyadonapites reticulatus* Tschudy 1973; 14–15, *Echitricolpites* sp.; 16, *Inaperturotetradites scabratus* Tschudy 1973 (two levels of focus on same specimen); 17, *Erdtmanipollis cretaceus* (Stanley 1965) Norton in Norton and Hall 1969; 18, *Kurtzipites circularis* (Norton in Norton and Hall 1969) Srivastava 1981; 19–20, *Kurtzipites trispissatus* Anderson 1960; 21, *Leptopecopites pocockii* (Srivastava 1967) Srivastava 1978; 22, *Libopollis jarzenii* Farabee et al. 1984 (two levels of focus on same specimen); 23, *Libopollis jarzenii* Farabee et al. 1984; 24, *Liliacidites altimurus* Leffingwell 1971, 25, *Liliacidites complexus* (Stanley 1965) Leffingwell 1971.

spines on the polar domes. It conforms to the description of the species, however, in that its equatorial dimension exceeds its polar dimension and in that the presence of reduced spines in some specimens is noted in the description (most typical specimens of *Aquilapollenites turbidus* lack spines on their polar domes).

See also the second comment after *Aquilapollenites collaris* concerning the publication date of the report by Tschudy and Leopold.

Aquilapollenites vulsus Sweet 1986
Plate 5, figs., 5, 6a, and 6b

Aquilapollenites vulsus Sweet, 1986, p. 1378–1379, Plate 1, figs. 3–5.

Description. *Aquilapollenites vulsus* is triprojectate, isopolar to subisopolar, and echinate; the spines are short and blunt, and characteristically are absent from the major polar dome. As noted by Sweet (1986, p. 1379), the absence of sculpture from the polar dome is the feature from which this species derives its name (*vulsus,* Latin for "hairless"). This species differs from *Aquilapollenites marmarthensis* sp. nov. primarily in this characteristic (*A. marmarthensis* is uniformly echinate), and secondarily and subtly in its proportions. In *A. vulsus* the polar dimension is somewhat greater than is the equatorial dimension, but *A. marmarthensis* has equatorial projections that are somewhat greater in dimension than is the polar axis.

Occurrence. This species is rare in the Hell Creek Formation; it was recorded at only two of the localities studied. Elsewhere in the region it has been recorded from the upper Maastrichtian of Alberta and Saskatchewan (Sweet, 1986; Braman and Sweet, 1999). It is considered to be a K taxon indicative of Maastrichtian age.

Comments. Sweet (1986) originally described this species from the uppermost Maastrichtian of Alberta. It is closely similar to *A. marmarthensis,* but can be readily distinguished at high magnification by its psilate polar dome.

Genus *Arecipites* Wodehouse 1933
Arecipites columellus Leffingwell 1971
Plate 5, fig. 7

Arecipites columellus Leffingwell, 1970, p. 140–141, Plate 7, figs. 1–2.

Description. Pollen of this species is monosulcate; the amb is ovate; the exine is tectate and columellate. The sculpture is scabrate in appearance, the result of the tips of the columellae being visible. See Leffingwell (1970) for a more detailed description and illustrations.

Occurrence. *Arecipites columellus* is at least moderately common in the Hell Creek Formation; it was recorded at more than 40% of the localities in this study at which species of *Arecipites* were distinguished. Other records may be included in an undifferentiated category "*Arecipites* spp." used for counting some slides. This category, which includes both *A. columellus* and *A. tenuiexinous* (and possibly other species of the genus not differentiated), was recorded at 96% of the localities studied. On the basis of detailed counts, *Arecipites columellus* is about equally common in the Ludlow Member of the Fort Union Formation.

Comments. Leffingwell (1970) first described the species *Arecipites columellus* from the Lance Formation in Wyoming. Botanical affinity is with the Arecaceae (or Palmae, the palms). Clearly the palms were a major component of the Hell Creek flora, being nearly ubiquitous. Nichols and Sweet (1993) included this species in the *Wodehouseia spinata* assemblage. (See also the second comment about *Foraminisporis undulatus* regarding the date of publication of the paper by Leffingwell.)

Arecipites tenuiexinous Leffingwell 1971
(not illustrated)

Arecipites tenuiexinous Leffingwell, 1970, p. 41, Plate 5, figs. 8–9.

Description. Pollen of this species is monosulcate and the amb is ovate. The exine is thin, and the sculpture is finely reticulate. See Leffingwell (1970) for a more detailed description; see Nichols and Brown (1992, Plate 6, figs. 4–7) for additional illustrations.

Occurrence. *Arecipites tenuiexinous* is at least moderately common in the Hell Creek Formation; it was recorded at more than 40% of the localities in this study in an undifferentiated category "*Arecipites* spp." used for counting some slides. This category, which includes both *A. tenuiexinous* and *A. columellus* (and possibly other species of the genus not differentiated), was recorded at 96% of the localities studied. On the basis of detailed counts, *Arecipites tenuiexinous* is about equally common in the Ludlow Member of the Fort Union Formation.

Comments. Leffingwell (1970) first described the species *Arecipites tenuiexinous* from the Lance Formation in Wyoming. Botanical affinity is with the Arecaceae (Palmae, the palms). Clearly the palms were a major component of the Hell Creek flora, being nearly ubiquitous. (See also the second comment afater *Foraminisporis undulatus* regarding the date of publication of the paper by Leffingwell.)

Genus *Boehlensipollis* Krutzsch 1962
Boehlensipollis minimus (Leffingwell 1971)
Pocknall and Nichols 1996
Plate 5, fig. 4

Syncolporites minimus Leffingwell, 1970, p. 49–50, Plate 9, figs. 6–8.

Boehlensipollis minimus (Leffingwell) Pocknall and Nichols, 1996, p. 21, Plate 5, figs. 9–14.

Description. Specimens of this species are small (<20 μm), heteropolar, and syncolporate to parasyncolporate. See Pocknall and Nichols (1996) for a more detailed description and illustrations; see Nichols and Brown (1992, Plate 8, figs. 3–4) for additional illustrations.

Occurrence. *Boehlensipollis minimus* is rare in the Hell

Creek Formation. It was recorded at only four of the localities studied. The lowest stratigraphic occurrence is ~34 m below the top of the formation. It is considerably more common and abundant in the lower part of the Fort Union Formation than in the Hell Creek.

Comments. Leffingwell (1970) first described this species from the Fort Union Formation in Wyoming. It is also present in the Lance Formation in Wyoming. Pocknall and Nichols (1996) transferred the species to the genus *Boehlensipollis* on the basis of detailed study of the morphology of specimens from the Fort Union. (See also the second comment after *Foraminisporis undulatus* regarding the date of publication of the paper by Leffingwell.)

Genus *Cercidiphyllites* Mchedlishvili in Samoilovich and Mchedlishvili 1961
Cercidiphyllites sp.
Plate 5, fig. 8

Description. Pollen assigned to *Cercidiphyllites* sp. in this study is triporate with unusually large, unbordered apertures. Mchedlishvili (in Samoilovich and Mchedlishvili, 1961) described *Cercidiphyllites brevicolpatus,* type species of the genus, as tricolpate with lalongate apertures, but Hell Creek specimens seen only in polar view appear to be triporate.

Occurrence. *Cercidiphyllites* sp. is rare in the Hell Creek Formation. It was recorded at only four of the localities studied. It is also present in the Fort Union Formation, where it is somewhat less common.

Comments. Mchedlishvili (in Samoilovich and Mchedlishvili, 1961) described the type species of the genus *Cercidiphyllites* from the Upper Cretaceous–Paleocene interval in western Siberia. Specimens closely resemble pollen of the modern genus *Cercidiphyllum* (Cercidiphyllaceae), which is a deciduous tree of eastern Asia.

Genus *Chenopodipollis* Krutzsch 1966
Chenopodipollis sp.
Plate 5, fig. 9a and 9b

Description. Pollen of this species is spheroidal and pantoporate, with numerous, evenly distributed, circular pores. The exine is tectate; the sculpture is granulate to foveolate. It closely resembles pollen produced by species of the modern families Chenopodiaceae and Amaranthaceae. It differs from *Erdtmanipollis cretaceus* in lacking reticulate sculpture with elements arranged in a croton pattern.

Occurrence. *Chenopodipollis* sp. is rare in the Hell Creek Formation. It was recorded at only six of the localities studied. It is also present in the Fort Union Formation, where it is similarly uncommon.

Comments. The modern families Chenopodiaceae and Amaranthaceae cannot be distinguished on the basis of pollen morphology. The Hell Creek specimens could represent ancestral species of either of these families.

Genus *Cranwellia* Srivastava 1961
Cranwellia rumseyensis Srivastava 1967
Plate 5, fig. 10

Cranwellia rumseyensis Srivastava, 1966, p. 538, Plate 11, figs. 3, 7.

Cranwellia striata (Couper) Srivastava, Leffingwell, 1970, p. 37, Plate 8, fig. 13a–13b.

Description. Pollen of this species is tricolporate and oblate with a rounded triangular amb and distinctive striate sculpture. See Srivastava (1966) for a more detailed description. It differs from *Striatellipollis striatellus* in lacking verrucate sculpture at one pole.

Occurrence. *Cranwellia rumseyensis* is fairly common in the Hell Creek Formation. It was recorded at more than 40% of the localities studied. It is distributed stratigraphically throughout the formation, and is also present in the Fort Union Formation, but only within the basal 2 m of the Ludlow Member, in facies that otherwise carry a Maastrichtian assemblage. It is regarded as a K taxon indicative of Maastrichtian age.

Comments. In equatorial view this species is difficult to distinguish from *Striatellipollis striatellus,* to which it is presumably closely related. For this reason, in this study a taxonomic category of convenience in counting was established: "*Cranwellia/Striatellipollis.*" Some records of this taxon are likely to be of, or at least to include, *Cranwellia rumseyensis.* If so, this species is present at more than 50% of the localities studied. Nichols and Sweet (1993) included this species in the *Wodehouseia spinata* assemblage.

Although the report by Srivastava bears the imprint date 1966, it was not actually published until 1967. For nomenclatural purposes the 1967 date is given, but for consistency with procedures in most libraries, the 1966 date is given to identify the publication.

Genus *Cupanieidites* Cookson and Pike 1954 emend. Chmura 1973
Cupanieidites inaequalis Leffingwell 1971
(not illustrated)

Cupanieidites inaequalis Leffingwell, 1970, p. 49, Plate 9, figs. 5, 9, 10.

Description. *Cupanieidites inaequalis* is syncolporate to parasyncolporate and has a rounded triangular amb. The exine is tectate and the sculpture is finely reticulate. See Leffingwell (1970) for a more detailed description and illustrations; see Nichols and Brown (1992, Plate 8, figs. 5–6) for additional illustrations.

Occurrence. This species is rare in the Hell Creek Formation; it was recorded at only one locality. It is also present in the Fort Union Formation, where it is also rare.

Comments. Leffingwell first described this species from the Fort Union Formation in Wyoming. It is also present in the Lance Formation there. (See also the second comment after *Foraminisporis undulatus* regarding the date of publication of the paper by Leffingwell.)

Genus *Cupuliferoidaepollenites* Potonié, Thomson, and Thiergart 1950 ex Potonié 1960
Cupuliferoidaepollenites sp.
(not illustrated)

Description. Pollen assigned to this species in this study is small (<15 μm in polar diameter), prolate, tricolpate, and psilate.

Occurrence. Pollen assigned to *Cupuliferoidaepollenites* sp. is rare in the Hell Creek Formation. It was recorded at only four of the localities studied. It is also present in the Fort Union Formation, where it is more common.

Comments. It is possible that more than one species of *Cupuliferoidaepollenites* could be distinguished among Hell Creek specimens on the basis of size and perhaps other features, but in this study little importance was attached to the occurrence of pollen of this morphology.

Genus *Cupuliferoipollenites* Potonié 1951 ex Potonié 1960
Cupuliferoipollenites sp.
(not illustrated)

Description. Pollen assigned to this species in this study is small (<15 μm in polar diameter), prolate, tricolporate, and psilate. It differs from *Cupuliferoidaepollenites* sp. in being tricolporate rather than tricolpate.

Occurrence. Pollen assigned to *Cupuliferoipollenites* sp. is rare in the Hell Creek Formation. It was recorded at only one of the localities studied. It is also present in the Fort Union Formation, where it is somewhat more common.

Comments. The botanical affinity of this species is unknown (other than that it is pollen of a dicot angiosperm). It has no biostratigraphic significance. It is included here only to record the presence of pollen of this morphology in the Hell Creek palynoflora.

Genus *Discoidites* Muller 1968
Discoidites parvistriatus (Norton 1967)
Nichols and Brown 1992
Plate 5, figs. 11, 12a, and 12b

Tricolpites parvistriatus Norton in Norton and Hall, 1967, p. 105, Plate 1, fig. D.
Tricolpites parvistriatus Norton, Norton and Hall, 1969, p. 45, Plate 7, fig. 2.
Discoidites parvistriatus (Norton) Nichols and Brown, 1992, p. 19–20, Plate 6, figs. 9–11.

Description. *Discoidites parvistriatus* is oblate with a circular amb and brevicolpate apertures; it has supratectal scabrate to granulate sculpture. See Nichols and Brown (1992) for a more detailed description and additional illustrations.

Occurrence. *Discoidites parvistriatus* is common in the Hell Creek Formation; it was recorded at 70% of the localities studied. It is present in the Fort Union Formation, as well, where it is about equally common in occurrence. It is notably abundant in some samples in both formations.

Comments. Norton (in Norton and Hall, 1967) first described this species as *Tricolpites parvistriatus.* The holotype specimen was reillustrated in Norton and Hall (1969). Nichols and Brown (1992) transferred the species to the genus *Discoidites* on the basis of its detailed morphology and because it does not belong in the genus *Tricolpites.* These authors also discussed the possible botanical affinity of this species with the lineage of the modern family Tiliacae. Nichols and Sweet (1993) included this species in the *Wodehouseia spinata* assemblage (as *Tricolpites parvistriatus*), despite its longer range than most species in that characteristically Maastrichtian assemblage.

Genus *Dyadonapites* Erdtman 1947 ex Tschudy 1973
Dyadonapites reticulatus Tschudy 1973
Plate 5, fig. 13

Dyadonapites reticulatus Tschudy, 1973, p. 32, Plate 11, figs. 15–17.

Description. Pollen of this distinctive species occurs as dyads. The individual grains are spheroidal to ovate, finely reticulate, and inaperturate. See Tschudy (1973) for a more detailed description and illustrations; see Nichols and Brown (1992, Plate 9, figs. 16–18) for additional illustrations.

Occurrence. *Dyadonapites reticulatus* is fairly common in the Hell Creek Formation; it was recorded at more than one-third of the localities studied. It is also present in the Fort Union Formation, where it is somewhat more common.

Comments. Tschudy (1973) first described this species from the Judith River Formation (Campanian) in Montana. It is present in Maastrichtian and Paleocene rocks in Montana, Wyoming, and Saskatchewan, as well as in North Dakota. Nichols and Brown (1992) discussed the possible botanical affinity of *Dyadonapites reticulatus* with the modern families Sparganiaceae or Typhaceae. Nichols and Sweet (1993) included this species in the *Wodehouseia spinata* assemblage, despite its having a longer stratigraphic range than most species in that characteristically Maastrichtian assemblage.

Genus *Echitricolpites* Da Silva Pares Regali, Uesugui, and Da Silva Santos 1974
Echitricolpites sp.
Plate 5, figs. 14 and 15

Description. Pollen assigned to this species is tricolpate, spheroidal to prolate, and echinate. The tectate exine bears tiny supratectal spines. This species bears close resemblance to *Echitricolpites supraechinatus* Pocknall and Nichols 1996, although it is smaller in size.

Occurrence. *Echitricolpites* sp. is rare in the Hell Creek Formation and appears to be stratigraphically restricted to the lower half of the formation. It was recorded at only two of the localities studied. Because its range does not extend to the K-T boundary, this species cannot be counted among the species whose extinction is attributed to the K-T boundary event.

Comments. Pocknall and Nichols (1996) described *Echitricolpites supraechinatus* from the upper part of the Fort Union Formation (the Tongue River Member) in Wyoming. The record of a possibly related species in the Hell Creek Formation in North Dakota is intriguing, but the apparent absence of *Echitricolpites* sp. in the lower part of the Fort Union Formation (the Ludlow Member) in North Dakota suggests that the morphologic resemblance may be coincidental. Too few Hell Creek specimens are available on which to base description of a new species.

Genus *Erdtmanipollis* Krutzsch 1962
Erdtmanipollis cretaceus (Stanley 1965) Norton in Norton and Hall 1969
Plate 5, fig. 17

Pachysandra cretacea Stanley, 1965, p. 294–295, Plate 44, figs. 1–9N.
Erdtmanipollis cretaceus (Stanley) Norton, in Norton and Hall, 1969, p. 43, Plate 5, fig. 21.
Erdtmanipollis cretacea (Stanley) Srivastava, 1972a, p. 320–321, Plate 5.

Description. Pollen of this species is spheroidal and pantoporate; it has numerous, evenly distributed, circular pores and distinctive sculpture. The exine is tectate and the sculptural elements are arranged in a croton pattern. It differs from *Chenopodipollis* sp. in the presence of sculpture having the unique croton pattern (crotonoid-reticulate of Krutzsch, 1962). *Erdtmanipollis cretaceus* does not differ in any significant way from *Erdtmanipollis albertensis* Srivastava, which is a later synonym. See Stanley (1965), Norton and Hall (1969), Srivastava (1972a), or Nichols and Brown (1992, Plate 9, figs. 11–12) for illustrations.

Occurrence. *Erdtmanipollis cretaceus* is moderately common in the Hell Creek Formation where it was recorded at more than half of the localities studied. However, it is nowhere abundant—most records are of single specimens found during scanning 2000 or more specimens from a sample. It is about equally common and abundant in the Fort Union Formation.

Comments. Stanley (1965) first described this species from the Hell Creek Formation in South Dakota. He assigned his species to the modern genus *Pachysandra* (Buxaceae). The modern genus *Sarcococca* (Buxaceae) produces pollen of the same morphology. The form genus *Erdtmanipollis* was proposed by Krutzsch (1962) to accommodate fossil pollen of this morphology from the Tertiary of Germany. Norton and Hall (1969) transferred Stanley's species to *Erdtmanipollis.* Srivastava (1972a) reviewed the morphology of the specimens described by Krutzsch and by Stanley and others of his own from the Maastrichtian of Alberta using scanning electron microscopy. He listed the name *Erdtmanipollis cretacea* (sic) as a new combination, but Norton and Hall (1969) made this reassignment three years earlier. Nichols and Sweet (1993) included this species in the *Wodehouseia spinata* assemblage within the taxon *Erdtmanipollis* spp.

Genus *Fraxinoipollenites* Potonié 1951 ex Potonié 1960
Fraxinoipollenites variabilis Stanley 1965 sensu Pocknall and Nichols 1996
(not illustrated)

Fraxinoipollenites variabilis Stanley, 1965, p. 306, Plate 45, figs. 29–35.
Fraxinoipollenites variabilis Stanley sensu Pocknall and Nichols, 1996, p. 15, Plate 4, figs. 15–17.

Description. *Fraxinoipollenites variabilis* is prolate, thin-walled, finely reticulate, tricolpate pollen in which the colpi nearly reach the poles. See Pocknall and Nichols (1996) for more detailed description, discussion, and illustrations; see Nichols and Brown (1992, Plate 6, figs. 12–16) for additional illustrations.

Occurrence. This species is rare in the Hell Creek Formation. It was recorded at only ~10% of the localities studied. It is also present in the Ludlow Member of the Fort Union Formation, where it is equally rare.

Comments. Stanley (1965) first described *Fraxinoipollenites variabilis* from the Fort Union Formation in South Dakota, evidently on the basis of several specimens that are not all conspecific. Pocknall and Nichols (1996) effectively emended the description of this species on the basis of Stanley's original illustrations, specimens illustrated by Leffingwell (1970) from the Fort Union Formation and possibly the Lance Formations in Wyoming, and their own specimens from the Fort Union in Wyoming. The North Dakota specimens are the same as those redescribed by Pocknall and Nichols. There is little or no support for Stanley's suggestion that the botanical affinity of this species is with the modern genus *Fraxinus* (Oleaceae).

Genus *Inaperturotetradites* van Hoeken-Klinkenberg 1964
Inaperturotetradites scabratus Tschudy 1973
Plate 5, figs. 16a and 16b

Inaperturotetradites scabratus Tschudy, 1973, p. 32–33, Plate 11, figs. 18–19.

Description. Pollen of *Inaperturotetradites scabratus* occurs as obligate tetrads of inaperturate grains. Individual grains are spheroidal and scabrate. See Tschudy (1973) for a more detailed description and additional illustrations.

Occurrence. This species is rare in the Hell Creek Formation; it was recorded at fewer than 20% of the localities studied. It also occurs in the Fort Union Formation in the study area, but it is rare in that unit. Elsewhere in the region, this species occurs in rocks as old as Campanian.

Comments. Tschudy first described *Inaperturotetradites scabratus* from the Judith River Formation (Campanian) in Montana. Subsequently it was reported from the Tullock Member of the Fort Union Formation in Montana and Wyoming (Nichols and Brown, 1992). It was also reported from a K-T boundary section in Wyoming by Wolfe (1991), but he did not correctly identify this species and he misinterpreted it as immature tetrads of *Nelumbo*-like pollen. His misidentification was pointed out by Nichols et al. (1992b), who also noted that on that basis and a second misidentification, Wolfe's (1991) scenario for a "June impact winter" at the K-T boundary is effectively discredited. (See also the third comment after *Pandaniidites typicus.*)

Genus *Kurtzipites* Anderson 1960
Kurtzipites circularis (Norton in Norton and Hall, 1969) Srivastava 1981
Plate 5, fig. 18

Momipites circularis Norton in Norton and Hall, 1969, p. 37, Plate 5, fig. 8.
Kurtzipites simplex Leffingwell, 1970, p. 51, Plate 10, figs. 2–4.
Kurtzipites circularis (Norton in Norton and Hall), Srivastava, 1981, p. 874, 876.

Description. *Kurtzipites circularis* is indistinctly tricolporate and brevicolpate; it is oblate and has a rounded triangular amb that is acuminate at the apertures. The supratectal sculpture is microechinate. Different authors have interpreted the apertures differently. Anderson (1960) described the genus *Kurtzipites* as triporate; Norton and Hall (1969) described species they assigned to the genus (and the synonymous species they called *Momipites circularis*) as triporate; Leffingwell (1970), in an emended description of the genus, described the apertures as tricolporate while recognizing that some species are either indistinctly tricolporate or tricolpate (*K. circularis* is such a species); Srivastava (1981) in a revised description referred to the apertures as tricolporate or tricolporoidate; they are here described as indistinctly tricolporate and brevicolpate. *Kurtzipites circularis* differs from *K. trispissatus* in lacking distinctive endexinal thickenings at the apertures.

Occurrence. *Kurtzipites circularis* is fairly common in the Hell Creek Formation. It was recorded at 12 (somewhat less than half) of the localities studied. It is also present in the Ludlow Member of the Fort Union Formation, where it is more common, occurring at more than 50% of the localities analyzed in conjunction with this study.

Comments. Norton (in Norton and Hall, 1969) first described this species from the Hell Creek and Fort Union Formations in eastern Montana, but he assigned it to the genus *Momipites,* to which it clearly does not belong. Srivastava (1981) transferred Norton's species to *Kurtzipites.* Leffingwell's species *Kurtzipites simplex,* which was published after Norton's *Kurtzipites circularis,* is clearly synonymous. Other than that this species was produced by a dicot angiosperm, the botanical affinity is unknown. Nichols and Sweet (1993) included this species in the *Wodehouseia spinata* assemblage, despite its having a longer stratigraphic range than other species in that characteristically Maastrichtian assemblage.

Kurtzipites trispissatus Anderson 1960
Plate 5, figs. 19 and 20

Kurtzipites trispissatus Anderson, 1960, p. 25, Plate 2, figs. 15–17.

Description. *Kurtzipites trispissatus* is tricolporate and brevicolpate; it is oblate and has a rounded triangular amb that is acuminate at the apertures. There are two distinctive pads of thickened endexine at each aperture, located at the ends of the colpi in each hemisphere. The supratectal sculpture is microechinate. *Kurtzipites trispissatus* differs from *K. circularis* in having distinctive endexinal thickenings at the apertures.

Occurrence. *Kurtzipites trispissatus* is common in the Hell Creek Formation. It was recorded at almost 80% of the localities studied, and is considerably more common than is *K. circularis. Kurtzipites trispissatus* is also present in the Ludlow Member of the Fort Union Formation, where it is about as common as it is in the Hell Creek Formation.

Comments. Anderson (1960) originally described *Kurtzipites trispissatus* from the Kirtland Shale (Upper Cretaceous) in New Mexico. This species is widely distributed in uppermost Cretaceous and lowermost Paleocene rocks in western North America. Other than that this species was produced by a dicot angiosperm, the botanical affinity is unknown. McIver et al. (1991) recovered pollen of this species from fossil flowers from the Ravenscrag Formation (Paleocene) in Saskatchewan, but the botanical affinity of the flowers could not be determined. McIver et al. (1991) compared *Kurtzipites trispissatus* with other species of *Kurtzipites* and species of the genera *Fibulapollis, Orbiculapollis,* and *Aquilapollenites* to support their interpretation that *Kurtzipites* is allied with the triprojectate group of fossil pollen. Nichols and Sweet (1993) included this species in the *Wodehouseia spinata* assemblage.

Genus *Lanagiopollis* Morley 1982
Lanagiopollis lihoka (Srivastava 1972) Frederiksen 1988
(not illustrated)

Margocolporites lihokus Srivastava, 1972b, p. 264, Plate 20, figs. 3–5.
Lanagiopollis lihoka (Srivastava) Frederiksen, 1988, p. 56.

Description. Hell Creek specimens assigned to this taxon are relatively large (~45 µm in diameter), tricolporate, and reticulate; the colpi and pores are bordered by margins. Farabee and Canright (1986) identified conspecific specimens from the Lance Formation in Wyoming as *Margocolporites lihokus* Srivastava, 1972. See Srivastava (1972b, Plate 20, figs. 3–5) or Farabee and Canright (1986, Plate 19, figs. 11–12 only) for illustrations.

Occurrence. *Lanagiopollis lihoka* is extremely rare in the Hell Creek Formation; it was recorded at only one locality in this study. It is more common in the Fort Union Formation, although it is not common in that unit either.

Comments. This species is discussed here despite its rare occurrence in the Hell Creek Formation only to record the presence of pollen of this morphology in the palynoflora and to clarify its proper generic assignment. The genus *Margocolporites* Ramanujam 1966, to which Srivastava (1972b) and Farabee and Canright (1986) assigned their specimens of this species, is poorly understood and has been used incorrectly by the au-

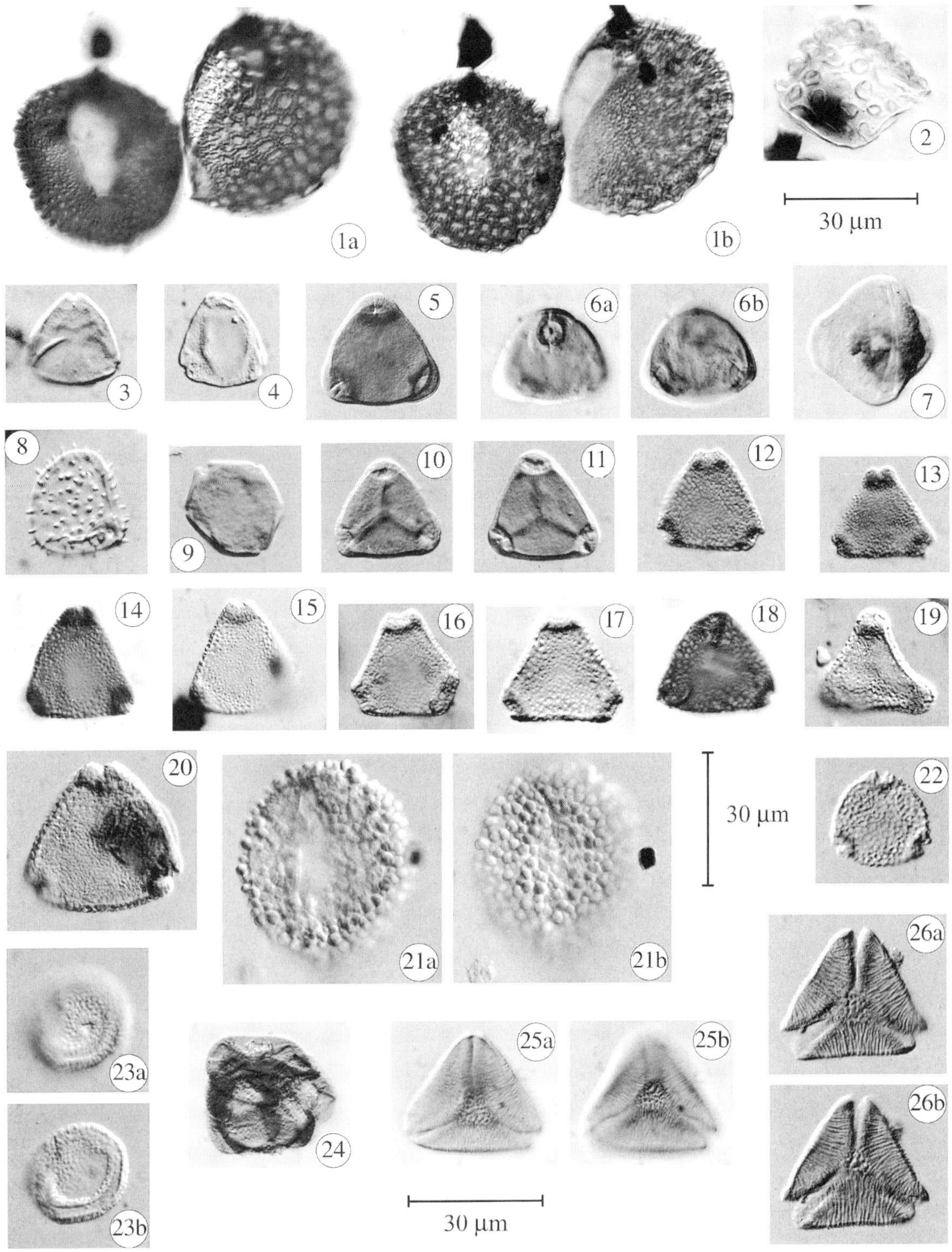

Plate 6. 1, *Liliacidites complexus* (Stanley 1965) Leffingwell 1971 (two levels of focus on same two adjacent specimens); 2, *Marsypiletes cretacea* Robertson 1973 emend. Robertson and Elsik 1978; 3–4, *Momipites inaequalis* Anderson 1960; 5, *Myrtipites scabratus* Norton in Norton and Hall 1970; 6, *Myrtipites scabratus* Norton in Norton and Hall 1970 (two levels of focus on same canted specimen showing one pore in equatorial view in 6a); 7, *Orbiculapollis lucidus* Chlonova 1961 (equatorial view); 8, *Pandaniidites typicus* (Norton in Norton and Hall 1969) Sweet 1986 (monoporate, annulate aperture is visible in lower right quadrant of specimen); 9, *Polyatriopollenites levis* (Stanley 1965) comb. nov.; 10–11, *Porosipollis porosus* (Mchedlishvili in Samoilovich and Mchedlishvili 1961) Krutzsch 1969; 12–18, *Tschudypollis retusus* (Anderson 1960) comb. nov.; 19–20, *Tschudypollis* spp.; 21, *Racemonocolpites formosus* Sweet 1986 (two levels of focus on same specimen); 22, *Retibrevitricolporites beccus* Sweet 1986; 23, *Sparganiaceaepollenites* sp. (two levels of focus on same specimen; monoporate aperture is visible in 23a); 24, *Simplicepollis rallus* (Stanley 1965) Nichols and Brown 1992; 25, *Striatellipollis striatellus* (Mchedlishvili in Samoilovich and Mchedlishvili, 1961) Krutzsch 1969 (two levels of focus on same specimen showing opposite poles); 26, *Striatellipollis striatellus* (Mchedlishvili in Samoilovich and Mchedlishvili, 1961) Krutzsch 1969 (two levels of focus on same specimen showing opposite polar hemispheres).

thors in the references cited. Frederiksen (1988) reassigned this species (along with *Margocolporites cribellatus* Srivastava 1972) to the genus *Lanagiopollis,* which is more appropriate.

Genus *Leptopecopites* Srivastava 1978
Leptopecopites pocockii (Srivastava 1967) Srivastava 1978
Plate 5, fig. 21

Polycolpites pocockii Srivastava, 1966, p. 544, Plate 7, figs. 28–30.
Polycolpites sp., Tschudy, 1970, Plate 4, fig. 24.
Leptopecopites pocockii (Srivastava) Srivastava, 1978b, p. 562, 564, 566; Plate 1, Figs. 1–7, Plate 2, figs. 1–6, Plate 3, figs. 1–8.

Description. *Leptopecopites pocockii* is circular, oblate, and zonocolpate; it has five or six long, unbordered colpi. The exine structure is complex, but the surface sculpture appears to be microstriate or scabrate. See Srivastava (1978b) for a more detailed description, especially of the exine structure. The unique form and apertures of this species distinguish it from all others in the Hell Creek palynoflora.

Occurrence. *Leptopecopites pocockii* is not common in the Hell Creek Formation; it was recorded at 30% of the localities studied. It is also present in the Fort Union Formation, but only within the basal 2 m of the Ludlow Member, in facies that otherwise carry a Maastrichtian assemblage. It is regarded as a K taxon indicative of Maastrichtian age.

Comments. Srivastava (1966) first described this species from the Edmonton Formation in Alberta, and later (Srivastava, 1978b) proposed the genus *Leptopecopites* and made *Leptopecopites pocockii* its type species. Nichols and Sweet (1993) included this species in the *Wodehouseia spinata* assemblage. See also the second comment about *Cranwellia rumseyensis* regarding the date of publication of Srivastava (1966).

Genus *Libopollis* Farabee, Daghlian, Canright, and Oftedahl 1984
Libopollis jarzenii Farabee et al. 1984 Plate 5, figs. 22a, 22b, and 23

Libopollis jarzenii Farabee et al., 1984, p. 152, 154, 156; Plate 1, figs. 1–14; Plate 2, figs. 1–9; Plate 3, figs. 1–5.

Description. This species is circular, oblate, and hexaporotricolpate. The apertures consist of three pairs of pores that are located on opposite hemispheres about midway between the poles and the equator and that are joined by short colpi that cross the equator. The sculpture is homobrochate reticulate. See Farabee et al. (1984) for a more detailed description. The closely related species *L. perferomura* Farabee et al. 1984 differs in having a heterobrochate reticulum. *Libopollis jarzenii* differs from *Anacolosidites rotundus* Stanley 1965 in having colpi connecting the pairs of pores.

Occurrence. *Libopollis jarzenii* is not common in the Hell Creek Formation; it was recorded at only six of the localities studied. It is present at numerous localities in the Fort Union Formation, but only within the basal 2 m of the Ludlow Member, in facies that otherwise carry a Maastrichtian assemblage. It is regarded as a K taxon indicative of Maastrichtian age. Elsewhere it is widespread in Maastrichtian rocks in the Western Interior and the Gulf and Atlantic coastal plains of the United States (Farabee et al., 1984).

Comments. Farabee et al. (1984) described this species from numerous localities. The holotype specimen is from Alabama, and isotype specimens are from Alabama, Arkansas, and Wyoming. This is the first report of the species from North Dakota. Nichols and Sweet (1993) included *Libopollis jarzenii* in the *Wodehouseia spinata* assemblage.

Genus *Liliacidites* Couper 1953
Liliacidites altimurus Leffingwell 1971
Plate 5, fig. 24

Liliacidites altimurus Leffingwell, 1970, p. 41, Plate 5, figs. 6–7.

Description. *Liliacidites altimurus* is monosulcate and coarsely reticulate; the muri of the reticulum are thin and 3–4 μm high. See Leffingwell (1970) for a more detailed description. This species differs from others in the genus in the Hell Creek palynoflora in the nature of the reticulum.

Occurrence. This species is rare in the Hell Creek palynoflora; it was recorded at only four localities. It is restricted to the Hell Creek and does not occur in the Fort Union Formation. It is regarded as a K taxon indicative of Maastrichtian age.

Comments. Leffingwell (1970) originally described this species from the Lance Formation (Maastrichtian) in Wyoming. Nichols and Sweet (1993) included this species in the *Wodehouseia spinata* assemblage. This is the first record of the species from North Dakota. See also the second comment after *Foraminisporis undulatus* regarding the date of publication of the paper by Leffingwell.

Liliacidites complexus (Stanley 1965) Leffingwell 1971
Plate 5, fig. 25; Plate 6, figs. 1a and 1b

Schizosporis complexus Stanley, 1965, p. 267–268, Plate 36, figs. 7–17.
Liliacidites complexus (Stanley) Leffingwell, 1970, p. 41, Plate 6, fig. 3a and 3b.

Description. This species is monosulcate and reticulate; the reticulum is heterobrochate, becoming finer toward the aperture, and has a unique duplicolumellate structure. See Farabee and Canright (1986) for a more complete description.

Occurrence. *Liliacidites complexus* is quite common in the Hell Creek Formation. It was recorded at ~90% of the localities analyzed in this study. It is also present in the Fort Union Formation, but only within the basal 2.5 m of the Ludlow Member, in facies that otherwise carry a Maastrichtian assemblage. It is regarded as a K taxon indicative of Maastrichtian age. It is widely distributed in Maastrichtian strata in western North America (Nichols and Sweet, 1993) and is part of the *Wodehouseia spinata* assemblage.

Comments. Stanley (1965) first described this species from the Hell Creek Formation in South Dakota, but he misinterpreted it as an algal cyst and assigned it to the algal genus *Schizosporis.* This misinterpretation is difficult to understand because *Liliacidites complexus* has a highly complex (as the specific epithet implies) exine structure of a kind only seen in angiosperm pollen and never in algal cysts. Leffingwell (1970) transferred the species to the angiosperm pollen genus *Liliacidites* because it has a clearly defined sulcus rather than a tear, as implied by Stanley's description. Stone (1973) noted that the complexity of the exine structure allied *L. complexus* with the angiosperms. Farabee and Canright (1986) illustrated the duplicolumellate structure with the aid of scanning electron microscopy.

Nichols and Sweet (1993) included *Liliacidites complexus* in the *Wodehouseia spinata* assemblage. See also the second comment after *Foraminisporis undulatus* regarding the date of publication of the paper by Leffingwell.

Liliacidites spp.
(not illustrated)

Description. Specimens assigned to this taxon are monosulcate and reticulate. The coarseness of the reticulum varies among different specimens, and it could be used to distinguish separate species, although no attempt was made to do that in this study because too few specimens were found. In all these specimens the exine structure is simplicolumellate, which distinguishes these species from *Liliacidites complexus,* and in none of them are the muri as high as in *Liliacidites altimurus,* which distinguishes that species from those assigned to *Liliacidites* spp.

Occurrence. Specimens assigned to *Liliacidites* spp. collectively are rare in the Hell Creek Formation; they were recorded at only four localities. This taxon was recorded also in the Fort Union Formation, where it is similarly rare.

Comments. Hotton (1988) distinguished 17 species of *Liliacidites* exclusive of *L. altimurus* and *L. complexus* in her study of the Hell Creek Formation in the Hell Creek area of Montana. There are unlikely to be more than two or three distinguishable species in the Hell Creek in North Dakota that are here assigned to *Liliacidites* spp.

Genus *Marsypiletes* Robertson 1974
emend. Robertson and Elsik 1978
Marsypiletes cretacea Robertson 1973
emend. Robertson and Elsik 1978
Plate 6, fig. 2

New Genus A, Tschudy, 1970, Plate 4, fig. 11.
Marsypiletes cretacea Robertson, 1973, p. 512, 514, figs. 1–15.
Marsypiletes cretacea Robertson emend. Robertson and Elsik (1978, p. 87-88, Plate 1, figs. 1–6).

Description. This species is monosulcate; it has a unique morphology consisting of large protuberances—hollow evaginations—on the distal side. See Robertson and Elsik (1978) for a more detailed description (but note that their term "auriculate" is inappropriate; Robertson's original term evaginations is preferable).

Occurrence. *Marsypiletes cretacea* is rare in the Hell Creek Formation; it was recorded at only two localities in this study. It is also present in the Fort Union Formation, but only within the basal 1.5 m of the Ludlow Member, in facies that otherwise carry a Maastrichtian assemblage. It is regarded as a K taxon indicative of Maastrichtian age.

Comments. Robertson (1973) first described *Marsypiletes cretacea* from the Hell Creek Formation in Montana. In that publication he interpreted this species to be a monolete spore. Jarzen (1976) interpreted the species as fossil spores of the modern bryophyte genus *Plagiopus.* Robertson and Elsik (1978) reinterpreted the species as monosulcate angiosperm pollen. That interpretation is accepted here, although it is recognized that the morphology exhibited by this species is quite unusual and unlike that normally found in angiosperm pollen. Nichols and Sweet (1993) included this species in the *Wodehouseia spinata* assemblage.

Genus *Momipites* Wodehouse 1933 emend. Nichols 1973
Momipites inaequalis Anderson 1960
Plate 6, figs. 3 and 4

Momipites inaequalis Anderson, 1960, p. 25, Plate 6, figs. 7–9; Plate 7, fig. 13.

Description. This species encompasses triporate pollen with juglandaceous morphology that lack modifications of exine at the poles (see Nichols, 1973). See Anderson (1960) for a more detailed description. *Momipites inaequalis* differs from most other species of the genus in having simple exine structure at the poles; it differs from *M. coryloides* Wodehouse 1933 and *M. wyomingensis* Nichols and Ott 1978 in smaller size and in a tendency (not strongly developed) to have sides of unequal length.

Occurrence. *Momipites inaequalis* is uncommon in the Hell Creek Formation; it was recorded at five of the localities analyzed in this study. It is also present in the Fort Union Formation, where it is somewhat more common. It is nowhere abundant at the localities analyzed in conjunction with this study.

Comments. Anderson (1960) first described *Momipites inaequalis* from the Ojo Alamo Sandstone (Paleocene) in New Mexico. Nichols et al. (1990, 1992a) previously reported this species in upper Maastrichtian rocks in the region. Specimens described by Farabee and Canright (1986) as *Momipites wyomingensis* likely are *M. inaequalis.*

Genus *Myrtipites* Norton and Hall 1969
Myrtipites scabratus Norton in Norton and Hall 1969
Plate 6, figs. 5, 6a, and 6b

Myrtipites scabratus Norton in Norton and Hall, 1969, p. 55, Plate 8, fig. 2.

Description. This species is distinctive tricolporate pollen having short colpi, rounded triangular ambs, and scabrate exines. See Norton (in Norton and Hall, 1969) for a more detailed description. *Myrtipites scabratus* differs from *Porosipollis porosus* (Mchedlishvili in Samoilovich and Mchedlishvili 1961) Krutzsch 1969, to which it appears to be closely related, in having short, nonsyncolpate colpi.

Occurrence. *Myrtipites scabratus* is moderately common in the Hell Creek Formation; it was recorded at almost 65% of the localities studied. It is also present in the Fort Union Formation at some localities, but only within the basal 2 m of the Ludlow Member, in facies that otherwise carry a Maastrichtian assemblage. It is regarded as a K taxon indicative of Maastrichtian age.

Comments. Norton (in Norton and Hall, 1969) first described *Myrtipites scabratus* from the Hell Creek Formation in Montana. Nichols and Sweet (1993) included this species in the *Wodehouseia spinata* assemblage.

Genus *Nyssapollenites* Thiergart 1938 ex Potonié 1960
Nyssapollenites spp.
(not illustrated)

Description. Specimens assigned to *Nyssapollenites* spp. in this study are oblate tricolporate pollen having long colpi, triangular ambs, and scabrate exines. Species assigned to this taxon differ from *Myrtipites scabratus* in having long colpi and in details of aperture structure. The morphology is similar to that of pollen of modern species of the genus *Nyssa* (Nyssaceae).

Occurrence. Specimens assigned to *Nyssapollenites* spp. are rare in the Hell Creek Formation; they were recorded at only four localities. They are also present in the Fort Union Formation, where they are similarly rare.

Comments. Too few specimens of *Nyssapollenites* spp. were observed to warrant distinguishing separate species. The botanical affinity of this taxon is with the modern family Nyssaceae.

Genus *Orbiculapollis* Chlonova 1961
Orbiculapollis lucidus Chlonova 1961
Plate 6, fig. 7

Orbiculapollis lucidus Chlonova, 1961, p. 89, Plate 15, fig. 118.
Triporina globosa Chlonova 1957 forma *minuta* Mchedlishvili in Samoilovich and Mchedlishvili, 1961, p. 46, Plate 80, figs. 4–5.
Trialapollis scabratus Stanley, 1965, p. 318, Plate 47, figs. 9–17.

Description. Pollen of this species is prolate, subtly triprojectate, and obscurely tricolpate. The original description (Chlonova, 1961) is in Russian. Stanley (1965) provided a more detailed description of the synonymous species *Trialapollis scabratus* in English, although he regarded this species as inaperturate.

Occurrence. *Orbiculapollis lucidus* is fairly common in the Hell Creek Formation; it was recorded at more than one-fourth of the localities analyzed in this study. It was also recorded in the Fort Union Formation, but only within the basal 2 m of the Ludlow Member, in facies that otherwise carry a Maastrichtian assemblage. It is regarded as a K taxon indicative of Maastrichtian age.

Comments. Chlonova (1961) first described this species from the Upper Cretaceous of Siberia. Stanley (1965) described identical forms from the Hell Creek Formation in South Dakota, but gave them a different name. Nichols and Sweet (1993) included this species in the *Wodehouseia spinata* assemblage.

Genus *Pandaniidites* Elsik 1968
Pandaniidites typicus Norton in (Norton and Hall 1969) Sweet 1986
Plate 6, fig. 8

Spinamonoporites typicus Norton in Norton and Hall, 1969, p. 36, Plate 5, fig. 22.
Pandaniidites radicus Leffingwell, 1970, p. 42–43, Plate 5, figs. 3–5.
Pandaniidites typicus (Norton) Leffingwell et al., 1970, p. 254, 262 (invalid).
Pandaniidites typicus (Norton) Sweet, 1986, p. 1382, Plate 2, fig. 12.

Description. *Pandaniidites typicus* is monoporate pollen with a single annulate pore and echinate sculpture. This species differs from all others in the Hell Creek palynoflora in the unique combination of aperture form and sculpture. See Sweet (1986) for a more detailed description.

Occurrence. *Pandaniidites typicus* is fairly common in the Hell Creek Formation; it was recorded at more than one-third of the localities studied. It is also present in the Fort Union Formation in North Dakota, where it is present at more than half of the localities analyzed in conjunction with this study.

Comments. Norton (in Norton and Hall, 1969) first described this species from the Hell Creek Formation in eastern Montana. He assigned it to a new genus he proposed at that time, *Spinamonoporites.* Prior to that, Elsik (1968a) described a similar species from the Paleocene of Texas, *Pandaniidites texus,* which he assigned to a new genus he proposed at that time, *Pandaniidites.* Leffingwell (1970) described the species *Pandaniidites radicus* from the Lance Formation in Wyoming, which he distinguished from Elsik's species on the basis of *P. radicus* having a thinner exine. It is evident that Leffingwell's species is the same as Norton's, and Leffingwell et al. (1970) proposed the new combination *Pandaniidites typicus;* however, because they did not clearly designate the basionym, that new combination is invalid. Sweet (1986) validly proposed the new combination *Pandaniidites typicus* and redescribed the species, effectively emending it by expanding its circumscription to include all of the specimens described by Norton, Elsik, and Leffingwell. Probably he should have adopted Elsik's name, *Pandaniidites texus,* but he did not, preferring to refer all previously described specimens (and his own from the Maastrichtian of Alberta) to *Pandaniidites typicus.* Sweet is followed here for the North Dakota specimens, which are within the effectively emended circumscription he proposed. Nichols and Sweet (1993) included *Pandaniidites typicus* in the *Wodehouseia spinata* assemblage.

There is some dispute about the botanical affinity of *Pandaniidites typicus.* Elsik (1968a), Norton and Hall (1969), Leffingwell (1970), Leffingwell et al. (1970), Muller (1981), Jarzen (1983), Nichols et al. (1986), Nichols and Brown (1992), and Hotton et al. (1994) all agreed that *Pandaniidites typicus* is pollen of a fossil species of the modern family Pandanaceae. In contrast, Sweet (1986) argued that it is pollen of the modern family Lemnaceae. Either way, its paleoecologic significance is the same with respect to this species being evidence of the presence or proximity of aquatic depositional environments. However, the paleoclimatic significance of this species with respect to events at the K-T boundary is greater if it is related to the Pandanaceae, a tropical family that evidently persisted despite the so-called impact winter (Nichols et al., 1986).

Wolfe (1991) observed but misidentified *Pandaniidites typicus* pollen in a K-T boundary section in Wyoming. Despite their clearly spheroidal form and annulate, monoporate apertures, Wolfe misidentified these specimens as ovoid, monosulcate pollen he called "*Nuphar*-like," implying that they were pollen of pond lilies of the family Nyphaeaceae. Because part of his scenario for a June impact winter at the K-T boundary depended on pond lily pollen being present, the correct identification of the specimens in question as *Pandaniidites typicus* by Nichols (*in* Nichols et al., 1992b) effectively discredits Wolfe's hypothesis. See also the comment after *Inaperturotetradites scabratus.*

Genus *Penetetrapites* Sweet 1986
Penetetrapites inconspicuus Sweet 1986
(not illustrated)

Penetetrapites inconspicuus Sweet, 1986, p. 1382–1383, Plate 1, figs. 16–19.

Description. *Penetetrapites inconspicuus* is spheroidal to oblate and heteropolar; the exine is thin and specimens are easily torn or folded; it has four vaguely defined poroid apertures. See Sweet (1986) for a more detailed description and illustrations; see Nichols and Brown (1992, Plate 9, figs. 13–15) for additional illustrations. The morphology of this species is unlike that of any other in the Hell Creek or Fort Union palynofloras.

Occurrence. This species is rare in the Hell Creek Formation; it was recorded at only four localities in this study. It is present in the Fort Union Formation also, where it is more common, occurring at more than 20% of the localities analyzed in conjunction with this study of the Hell Creek Formation.

Comments. Sweet (1986) first described this species from Maastrichtian strata in Alberta. It is also present in the lower Paleocene in that area. Based on its unusual apertures, the botanical affinity of this species apparently is with the Cyperaceae (Nichols and Brown, 1992).

Genus *Polyatriopollenites* Pflug in Thomson and Pflug 1953
Polyatriopollenites levis (Stanley 1965) comb. nov.
Plate 6, fig. 9

Pterocarya levis Stanley, 1965, p. 303–304, Plate 45, figs. 18–23.

Pterocaryapollenites stellatus auct. non (Potonié) Thiergart, Norton and Hall, 1969, p. 42, Plate 5, fig. 25.

Pterocarya cf.*P. levis* Stanley, Tschudy, 1970, Plate 4, fig. 21.

Basionym. *Pterocarya levis* Stanley, 1965, p. 303–304, Plate 45, figs. 18–23.

Description. Pollen oblate with polygonal amb, zonoporate (five or six apertures), apertures atriate; diameter small <20 µm); exine psilate. See Stanley (1965) for a more detailed description.

Occurrence. *Polyatriopollenites levis* (Stanley 1965) comb. nov. is rare in the Hell Creek Formation. It was recorded at only one locality in this study; it was present at a second locality within the basal 5 cm of the Fort Union Formation, where it occurred in association with species that are of Maastrichtian age. It is regarded as a K taxon indicative of Maastrichtian age. Elsewhere in the region it was reported from the Hell Creek Formation in eastern Montana (Tschudy, 1970).

Comments. Stanley (1965) first described this species from the Hell Creek Formation in South Dakota. He assigned it to the modern genus *Pterocarya* (Juglandaceae) because of its superficial resemblance to pollen of modern species of that genus. It does not closely resemble pollen of *Pterocarya,* however, being considerably smaller in size and having a psilate rather than scabrate exine with evenly spaced fine granules (which is the characteristic sculpture of modern and fossil species of the Juglandaceae; see Whitehead, 1963, 1965; Nichols, 1973; Nichols and Ott, 1978). Furthermore, *Polyatriopollenites levis* (Stanley 1965) comb. nov. became extinct at the K-T boundary and the fossil record of juglandaceous pollen possibly related to *Pterocarya* does not begin until latest Paleocene time. There are no forms of similar morphology that fill this significant gap in geologic time. Thus, the assignment of this species to the genus *Pterocarya* is inappropriate and potentially misleading. Accordingly, this species is hereby transferred to the form genus *Polyatriopollenites.* That form genus includes some species that may be of juglandaceous affinity, but it is in essence a form genus founded on morphologic criteria; it is expressly stated herein that placing the species *Polyatriopollenites levis* (Stanley 1965) comb. nov. in the genus *Polyatriopollenites* does not imply botanical affinity with the Juglandaceae. Norton and Hall (1969) incorrectly identified this species from the Hell Creek Formation in Montana as *Pterocaryapollenites stellatus* (Potonié) Thiergart. That species, which differs in size, sculpture, and apertural structure, is juglandaceous pollen from the Tertiary of Germany. Other than that it is pollen produced by a dicot angiosperm, the botanical affinity of *Polyatriopollenites levis* (Stanley 1965) comb. nov. is unknown.

Genus *Porosipollis* Krutzsch 1969
Porosipollis porosus (Mchedlishvili in Samoilovich and Mchedlishvili 1961) Krutzsch 1969
Plate 6, figs. 10 and 11

Syncolporites porosus Mchedlishvili in Samoilovich and Mchedlishvili, 1961, p. 248, Plate 82, fig. 29.
Porosipollis porosus (Mchedlishvili) Krutzsch, 1969, p. 407.

Description. This species encompasses syncolporate pollen with a rounded triangular amb, well-developed vestibula, and scabrate exine. *Porosipollis porosus* differs from *Myrtipites scabratus,* to which it appears to be closely related, in having long colpi that meet at the poles, giving it a trilete appearance.

Occurrence. *Porosipollis porosus* is rare in the Hell Creek Formation and is stratigraphically restricted to that formation in the study area. It was recorded at four localities analyzed, in samples from the lower part of the formation. In this study it is regarded as a K taxon and as a guide fossil for the lower part of the Hell Creek Formation.

Comments. Mchedlishvili (in Samoilovich and Mchedlishvili, 1961) first described this species from the Upper Cretaceous of Siberia; Krutzsch (1969) transferred it to the genus *Porosipollis.* This species is present in upper Campanian through Maastrichtian strata in the Yukon and Northwest Territories (Nichols and Sweet, 1993), and evidently is at the southern limit of its geographic range in Montana and North Dakota. Because the range of *Porosipollis porosus* does not extend to the K-T boundary in North Dakota, it cannot be counted among the species whose extinction is attributed to the K-T boundary event in the study area. However, it does range up to (but not above) the K-T boundary in Alaska (Frederiksen et al., 1998).

Genus *Racemonocolpites* Sweet 1986
Racemonocolpites formosus Sweet 1986
Plate 6, figs. 21a and 21b

Racemonocolpites formosus Sweet, 1986, p. 1384, Plate 2, figs. 9–11.

Description. This species is monosulcate, ovate, and gemmate. It differs from other monosulcate species in the Hell Creek palynoflora in having gemmate sculpture. See Sweet (1986) for a more detailed description.

Occurrence. *Racemonocolpites formosus* is rare in the Hell Creek Formation. It was recorded at only four localities, where it occurs only within the uppermost 7 m of the formation. It is also present in the basal part of the Fort Union Formation at two localities, but at those localities it occurs within the lowermost 50 cm of the Ludlow Member in a facies that otherwise carries a Maastrichtian assemblage. It is regarded as a K taxon indicative of Maastrichtian age.

Comments. Sweet (1986) first described this species from the Maastrichtian part of the Coalspur Formation in Alberta. Other than that it is pollen of a monocot angiosperm, the botanical affinity of this species is unknown.

Genus *Retibrevitricolporites* Legoux 1978
Retibrevitricolporites beccus Sweet 1986
Plate 6, fig. 22

Retibrevitricolporites beccus Sweet, 1986, p. 1385, Plate 2, fig. 13.

Description. This species encompasses tricolporate (brevicolpate) pollen with circular amb, annulate pores that protrude beak-like, and reticulate sculpture. See Sweet (1986) for a more detailed description. *Retibrevitricolporites beccus* differs from species of *Tschudypollis* in being tricolporate and in having an essentially circular amb.

Occurrence. *Retibrevitricolporites beccus* is uncommon in the Hell Creek Formation; it was recorded at only six localities in this study. It was recorded in the basal Fort Union Formation also, but only within the basal 1 m of the Ludlow Member, in facies otherwise carrying a Maastrichtian assemblage. It is regarded as a K taxon indicative of Maastrichtian age.

Comments. Sweet (1986) first described *Retibrevitricolporites beccus* from Maastrichtian strata in Alberta. Nichols and Sweet (1993) included this species in the *Wodehouseia spinata* assemblage.

Genus *Retitrescolpites* Sah 1967
Retitrescolpites anguloluminosus (Anderson 1960) Frederiksen 1979
(not illustrated)

Tricolpites anguloluminosus Anderson, 1960, p. 26, Plate 6, figs. 15–17; Plate 8, figs. 17–18.
Tricolpites bathyreticulatus Stanley, 1965, p. 320–321, Plate 47, figs. 18–23.
Tricolpopollenites anguloluminosus (Anderson) Elsik, 1968b, p. 624, Plate 24, figs. 15–16; Plate 25, fig. 1.
Retitrescolpites anguloluminosus (Anderson) Frederiksen, 1979, p. 139, Plate 1, fig. 13.

Description. This tricolpate species is distinguished from all others in the Hell Creek palynoflora by its coarsely reticulate sculpture in which the muri are high and thin. See Anderson (1960), Stanley (1965), Elsik (1968b), Leffingwell (1970), Frederiksen (1979), or Nichols and Brown (1992) for illustrations.

Occurrence. *Retitrescolpites anguloluminosus* is fairly common in the Hell Creek Formation; it was recorded at one-third of the localities analyzed. It is also present in the Fort Union Formation, where it was recorded at more than 40% of the localities analyzed in the course of this study.

Comments. Anderson (1960) first described this species (as *Tricolpites anguloluminosus*) from the Ojo Alamo Sandstone (Paleocene) in New Mexico. Frederiksen (1979) transferred Anderson's species to the genus *Retitrescolpites,* which encompasses tricolpate pollen having coarsely rather than finely reticulate sculpture. *Tricolpites bathyreticulatus* Stanley is a later synonym of Anderson's species and differs in no significant way.

Genus *Rhoipites* Wodehouse 1933
Rhoipites spp.
(not illustrated)

Description. In this study, all spheroidal to prolate, reticulate, tricolporate pollen is assigned to this taxon. At least two or three separate species could be identified among these specimens, but no attempt was made to distinguish them in this study because individually they are not common in occurrence and they have little or no biostratigraphic or paleoecologic value.

Occurrence. As a group, *Rhoipites* spp. are fairly common in the Hell Creek Formation. Specimens assigned to this taxon were recorded at one-third of the localities analyzed. *Rhoipites* spp. are also present in the Fort Union Formation, where they were recorded at ~40% of the localities analyzed in the course of this study.

Comments. The genus *Rhoipites* includes many species of reticulate, tricolporate pollen. Several modern families, notably the Anacardiaceae, Burseraceae, Meliaceae, and Rutaceae, produce pollen of this morphology.

Genus *Simplicepollis* Harris 1965
Simplicepollis rallus (Stanley 1965) Nichols and Brown 1992
Plate 6, fig. 24

Ericaceoipollenites rallus Stanley, 1965, p. 296–297, p. 44, figs. 15–18.
Ericipites rallus (Stanley) Farabee and Canright, 1986, p. 48, Plate 17, fig. 10.
Simplicepollis rallus (Stanley) Nichols and Brown, 1992, p. 29, Plate 10, figs. 1–6.

Description. This species occurs as obligate tetrahedral tetrads in which the individual grains are inaperturate to weakly tricolpate. See Nichols and Brown (1992) for a more complete description and illustrations. This species is distinguished from all those in the genus *Ericipites* Wodehouse 1933 or the synonymous genus *Ericaceoipollenites* Potonié 1960, both of which are described as having colporate apertures, in having poorly developed, noncolporate apertures.

Occurrence. *Simplicepollis rallus* is rare in the Hell Creek Formation; it was recorded at only four of the localities studied in detail. It is more common in the Fort Union Formation, where it is present at 50% of the localities analyzed in the course of this study.

Comments. Stanley (1965) first described this species from the Ludlow Member of the Fort Union Formation in South Dakota. Nichols and Brown (1992) transferred it to the genus *Simplicepollis* on the basis of conspecific specimens from the Tullock Member of the Fort Union Formation in Wyoming. Nichols et al. (1990) reported it in upper Maastrichtian rocks in the region (as *Ericaceoipollenites rallus*).

Genus *Sparganiaceaepollenites* Thiergart 1937
Sparganiaceaepollenites sp.
Plate 6, figs. 23a and 23b

Description. Specimens of *Sparganiaceaepollenites* sp. are spheroidal, reticulate, and monoporate. No other species in the Hell Creek palynoflora resembles it. It is readily distinguished from *Pandaniidites typicus* by its reticulate (rather than echinate) sculpture.

Occurrence. *Sparganiaceaepollenites* sp. is rare in the Hell Creek Formation; it was recorded at only two localities studied. It is also present in the Fort Union Formation, where it is somewhat more common.

Comments. The botanical affinity of this species is with the Sparganiaceae or Typhaceae. Either way, it is indicative of marshy, freshwater, depositional environments.

Genus *Striatellipollis* Krutzsch 1969
Striatellipollis striatellus (Mchedlishvili in Samoilovich and Mchedlishvili 1961) Krutzsch 1969
Plate 6, figs. 25a, 25b, 26a, and 26b

Tricolpites striatellus Mchedlishvili in Samoilovich and Mchedlishvili, 1961, p. 247, Plate 79, fig. 6, a–e.
Striatellipollis striatellus (Mchedlishvili) Krutzsch, 1969, p. 411.
Striatellipollis striatella (Mchedlishvili) Krutzsch, Sweet, 1986, p. 1386, Plate 2, figs. 21–23.

Description. *Striatellipollis striatellus* is tricolpate and has a rounded triangular amb with bulging mesocolpia. The sculpture is striate except at one pole where striae are broken into verrucae; striae are straight and parallel within each mesocolpial area. See Sweet (1986) for a more detailed description. *Cranwellia rumseyensis* differs in lacking verrucate exine at the poles but is otherwise closely similar.

Occurrence. *Striatellipollis striatellus* is fairly common in the Hell Creek Formation; it was recorded at ~60% of the localities studied. It is also present in the Fort Union Formation, where it was recorded at several localities, but only within the basal 2 m of the Ludlow Member, in facies otherwise carrying a Maastrichtian assemblage. It is regarded as a K taxon indicative of Maastrichtian age.

Comments. In equatorial view *Striatellipollis striatellus* is difficult to distinguish from *Cranwellia rumseyensis,* to which it presumably is closely related. For this reason, in this study a taxon of convenience in counting was established, "*Cranwellia/Striatellipollis.*" Some records of this taxon are likely to be records of *Striatellipollis striatellus.* Nichols and Sweet (1993) included this species in the *Wodehouseia spinata* assemblage (as *Striatellipollis striatella*).

Genus *Striatopollis* Krutzsch 1959
Striatopollis tectatus Leffingwell 1971
Plate 7, figs. 4 and 5

Striatopollis tectatus Leffingwell, 1970, p. 48, Plate 9, figs. 3–4.

Description. This species is prolate and tricolpate, with long colpi that almost reach the poles; it has distinctive striate sculpture. See Leffingwell (1970) for a more detailed description. The combination of shape and sculpture distinguishes this species from all other tricolpate species in the Hell Creek palynoflora.

Occurrence. *Striatopollis tectatus* is not common in the Hell Creek Formation; it was recorded at only six of the localities studied. It is also present in the Fort Union Formation, where it is similarly uncommon.

Comments. Leffingwell (1970) first described this species from the Fort Union Formation in Wyoming; it is also present in the Lance Formation in that area. Nichols and Sweet (1993) included this species in the *Wodehouseia spinata* assemblage, although it is not restricted to rocks of Maastrichtian age.

See also the second comment after *Foraminisporis undulatus* regarding the date of publication of the paper by Leffingwell.

Genus *Styxpollenites* gen. nov.

Type species. *Styxpollenites calamitas* sp. nov.

Description. Pollen tricolporate, isopolar, oblate, amb rounded triangular; colpi short, extending less than half way from equator to poles; pores atriate; exine two-layered, ectexine thick; sculpture psilate-rugulate, sculptural elements forming a dense mass tending to obscure apertures. Size ~60–75 μm.

Derivation of name. From the River Styx of Greek mythology, one of the rivers of Hades, in reference to Hell Creek.

Repository. The slide bearing the type specimen of the type species of *Styxpollenites* gen. nov. is held in the palynological laboratory of Central Energy Resources, U.S. Geological Survey, Denver, Colorado.

Comments. This genus is characterized by its form, aperture type, and unique, psilate-rugulate sculpture. The sculptural elements in lateral view are seen to be psilate in form with capita that merge forming a rugulate pattern seen in surficial view. No other genus described for fossil pollen in the Hell Creek palynoflora or elsewhere in western North America is similar. *Loranthacites* Mchedlishvili and *Accuratipollis* Chlonova differ in being heteropolar, in having longer colpi of different form, in the shape of the amb, and in sculpture; *Expressipollis* differs strongly in shape of the amb and in having syncolpate apertures.

Styxpollenites calamitas gen. et sp. nov.
Plate 7, figs. 1a–1c, 2a, 2b, and 3

Description. Pollen tricolporate, isopolar, oblate, amb rounded triangular. Colpi short, extending less than half way from equator to poles; pores atriate. Exine two-layered, endexine ~1 μm thick, ectexine ~9 μm thick. Endexine terminates near apertures, forming large endopores; exopores obscure, ~1 μm in diameter; atria ~17 μm wide near equator; colpi unbordered, ~12 μm long on each polar hemisphere, tending to be obscured by sculpture. Sculpture a dense mass of pila with merged capita that form a rugulate pattern, individual sculptural elements ~0.5 μm wide and to 8 μm tall; merged capita (rugulae) aligned in randomly oriented groups that present a striate appearance on the surface of the grains. Size 63–73 μm (15 specimens).

Diagnosis. The combination of shape, aperture type, and the unique structure of the ectexine renders *Styxpollenites calamitas* gen. et sp. nov. unlike any other species in the Hell Creek palynoflora or elsewhere in western North America. Specimens illustrated by McIntyre (1974, Plate 17, figs. 4–7) as *Loranthacites pilatus* Mchedlishvili 1961 and *Loranthacites* sp. 1 and by Nichols and Sweet (1993, Plate 2, fig. 12) as *Anacolosidites* sp. bear superficial resemblance but differ in aperture type, shape of amb, and sculpture.

Holotype. Slide D9007-B (3), England Finder coordinate R23/1; Plate 7, figs. 1a–1c.

Type locality. Swanke's Coal locality, Bowman County, North Dakota; sample D9007-B, 23.9 m below top of Hell Creek Formation.

Repository. The slide bearing the type specimen of *Styxpollenites calamitas* gen. et sp. nov. is in the palynological laboratory of Central Energy Resources, U.S. Geological Survey, Denver, Colorado.

Occurrence. *Styxpollenites calamitas* gen. et sp. nov. is uncommon in the Hell Creek Formation. It was recorded at five localities (~20%) among those studied in detail and at two others. It is also present in the Ludlow Member of the Fort Union Formation at five localities, but only within the basal 1.4 m of the unit, in facies that otherwise carry a Maastrichtian assemblage. It is regarded as a K taxon indicative of Maastrichtian age.

Derivation of name. The specific epithet *calamitas* is Latin for disaster, in reference to the K-T boundary event. Etymologically, "disaster" means "bad star," which in the case of the K-T event is especially appropriate with reference to the extraterrestrial impact theory of the cause of the calamity.

Comments. The sculpture of *Styxpollenites calamitas* gen. et sp. nov. is unique. It is unknown among Upper Cretaceous or lower Tertiary fossil pollen in western North America, or among pollen of familiar modern species. Sculptural elements consist of pilae that stand perpendicular to the surface of the grain with merged capita; they appear to double back on themselves much like the pile of a carpet. The sculptural elements are so thick that they tend to obscure the apertures, especially in specimens that were stained during preparation. A few specimens that inadvertently were prepared unstained provide the best opportunity to observe the structure of the apertures and of the sculpture (see 2 in Plate 7 herein). The botanical affinity of this new species is unknown, other than that it was produced by a dicot angiosperm.

Genus *Triatriopollenites* Pflug in Thomson and Pflug 1953
Triatriopollenites granulatus (Simpson 1961)
Leffingwell 1971
(not illustrated)

Engelhardtia granulata Simpson, 1961, p. 445, Plate 14, figs. 3–4.

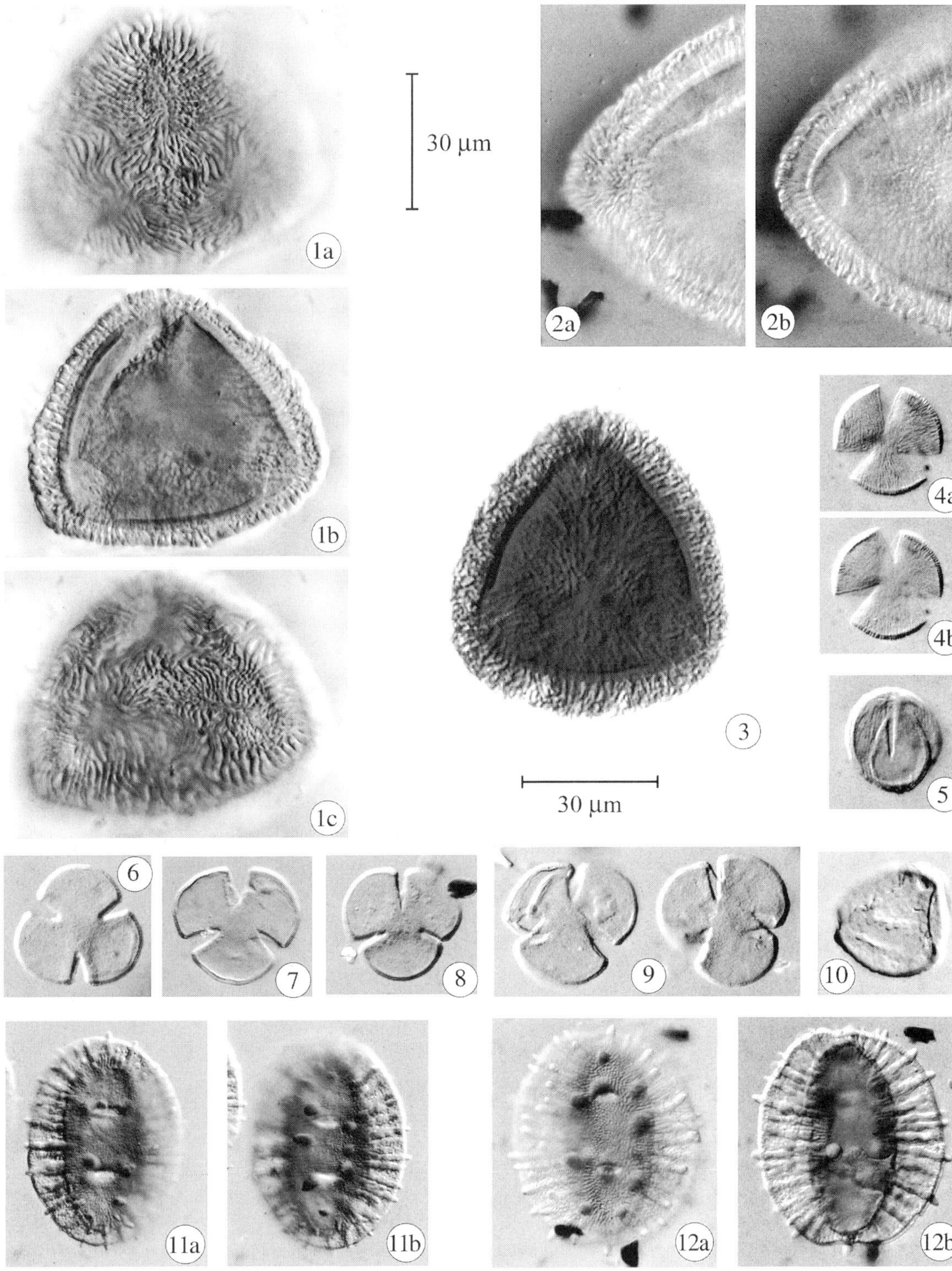

Plate 7. 1, *Styxpollenites calamitas* gen. et sp. nov. (holotype; three levels of focus on same specimen); 2, *Styxpollenites calamitas* gen. et sp. nov. (isotype; two levels of focus on same specimen, enlarger, showing detail of aperture structure); 3, *Styxpollenites calamitas* gen. et sp. nov. (isotype); 4, *Striatopollis tectatus* Leffingwell 1971 (two levels of focus on same specimen in polar view); 5, *Striatopollis tectatus* Leffingwell 1971 (equatorial view); 6–8, *Tricolpites microreticulatus* Belsky, Boltenhagen, and Potonié 1965; 9, *Tricolpites microreticulatus* Belsky, Boltenhagen, and Potonié 1965 (two adjacent specimens); 10, *Ulmipollenites krempii* (Anderson 1960) Frederiksen 1979; 11, *Wodehouseia spinata* Stanley 1961 (two levels of focus on same specimen); 12, *Wodehouseia spinata* Stanley 1961 (two levels of focus on same specimen).

Triatriopollenites granulatus (Simpson) Leffingwell, 1970, p. 51, 54; Plate 10, figs. 8–9.
Engelhardtioidites granulatus (Simpson) Srivastava, 1975, p. 139, Plate 8, fig. 10, Plate 9, figs. 1–3.

Description. Pollen of this species is small, oblate with a circular amb, and triporate; the pores are atriate and the exine is scabrate. There are three crescentic areas of thinner exine on each polar hemisphere between the apertures and the poles (alternatively, there is a triradiate area of thicker exine on each polar hemisphere centered at the poles and directed toward the interporial margins). See Simpson (1961) and/or Leffingwell (1970) for a more detailed description and illustrations; see Nichols and Brown (1992, Plate 8, fig. 20) for an additional illustration.

Occurrence. *Triatriopollenites granulatus* is rare in the Hell Creek Formation; it was recorded at only one locality. It is also present in the Fort Union Formation, where it is equally rare.

Comments. Simpson (1961) first described this species from the Paleocene of Scotland. He assigned the species to the modern genus *Engelhardia* ("*Engelhardtia*"), but it has little in common morphologically with pollen of that genus. Leffingwell transferred it to the form genus *Triatriopollenites.* See also the second comment after *Foraminisporis undulatus* regarding the date of publication of the paper by Leffingwell.

Genus *Tricolpites* Cookson 1947 ex Couper 1953
Tricolpites hians Stanley 1965
(not illustrated)

Tricolpites hians Stanley, 1965, p. 321–322, Plate 47, figs. 24–27.
Tricolpites parvus Stanley, 1965, p. 322, Plate 47, figs. 28–31.
Tricolpites parvus Stanley, Leffingwell, 1970, p. 44, Plate 8, fig. 4.

Description. Specimens identified as this species in this study are tricolpate and finely reticulate. They differ from those assigned to the taxon *Tricolpites* spp. in having finer sculpture, and from those identified as *Retitrescolpites anguloluminosus* in having much finer sculpture. Specimens identified as *Tricolpites parvus* by Stanley (1965) and Leffingwell (1970) are considered to be conspecific. *Fraxinoipollenites variabilis* is larger and more prolate in form and has longer colpi. See Stanley (1965) or Leffingwell (1970) for more detailed descriptions and illustrations; see Nichols and Brown (1992, Plate 7, figs. 5–7) for additional illustrations.

Occurrence. *Tricolpites hians* is moderately common in the Hell Creek Formation; it was recorded at about half or more of the localities analyzed. It is also present in the Fort Union Formation, where it may be somewhat less common, based on records compiled in the course of this study.

Comments. Stanley (1965) first described *Tricolpites hians* from the Ludlow Member of the Fort Union Formation in South Dakota. At the same time he described a similar, virtually indistinguishable, species, *Tricolpites parvus.* That species is here considered to be synonymous. Leffingwell (1970) assigned pollen of this morphology to *T. parvus* and did not mention *T. hians;* presumably his records are of the same species here identified as *T. hians.* Some records of the taxon *Tricolpites* spp. used in this study may include specimens of *T. hians* that were not identified to species level during scanning of slides. In Figure 11, all records of *T. hians* are included in the taxon *Tricolpites* spp.

Tricolpites microreticulatus Belsky, Boltenhagen, and Potonié 1965
Plate 7, figs. 6–10

Tricolpites microreticulata Belsky, Boltenhagen, and Potonié, 1965, p. 75, Plate 12, figs. 8–9.
Tricolpites reticulatus Cookson, Norton and Hall, 1967, p. 102, Plate 1, fig. E.
Tricolpites reticulatus Cookson ex Couper, Norton and Hall, 1969, p. 43–44, Plate 7, fig. 1.
Gunnera microreticulata (Belsky, Boltenhagen, and Potonié) Leffingwell, 1970) p. 37, 40; Plate 7, figs. 7–8.

Description. *Tricolpites microreticulatus* is tricolpate and very finely reticulate but it differs from all other species of the genus, which includes numerous species of this general description, in the distinctive shape of the amb. The amb is broadly rounded triangular with gently protruding intercolpia; in polar view the colpi are positioned interapically. Many specimens also exhibit subtle thickening of exine at the poles. This species closely resembles pollen produced by species of the modern genus *Gunnera.*

Occurrence. *Tricolpites microreticulatus* is common in the Hell Creek Formation; it was recorded at almost 70% of the localities studied. It is also present in the Fort Union Formation, but only within the basal 2 m of the Ludlow Member, in facies otherwise carrying a Maastrichtian assemblage. It is regarded as a K taxon indicative of Maastrichtian age.

Comments. Belsky et al. (1965) first described this species from the Upper Cretaceous of Africa. Norton and Hall (1969) and Leffingwell (1970) called attention to the close similarity of this species and pollen of species of the modern genus *Gunnera* (Gunneraceae); see also Jarzen (1978). Leffingwell (1970) transferred the fossil species to the modern genus on the basis of specimens from the Lance Formation in Wyoming. However, others (e.g., Farabee and Canright, 1986; Hotton, 1988; Nichols and Sweet, 1993) followed Belsky et al. and applied the original name, and that practice is followed here. Nichols and Sweet (1993) included this species in the *Wodehouseia spinata* assemblage.

Tricolpites spp.
(not illustrated)

Description. In this study, specimens assigned to the taxon *Tricolpites* spp. are tricolpate and reticulate. They differ from those identified as *Tricolpites hians* in having coarser sculpture, and from those identified as *Retitrescolpites anguloluminosus* in having finer sculpture. Two or more species may be included but are not distinguished.

Occurrence. Specimens assigned to the taxon *Tricolpites* spp. are moderately common in the Hell Creek Formation; the taxon was recorded at about half of the localities studied. It was also recorded in the Fort Union Formation, where it is somewhat more common, based on records compiled in the course of this study.

Comments. Some records of the taxon *Tricolpites* spp. used in this study may include specimens of *T. hians* that were not identified to species level during scanning of slides. Other species probably are included, but no further attempt was made to distinguish them in this study; this broad taxonomic category was used to simplify the scanning of slides. Hotton (1988) listed 21 species of *Tricolpites* including the two named in this study in her study of the Hell Creek palynoflora of Montana. However, this species complex appears to have little or no biostratigraphic or paleoecologic value that would justify giving it more attention. In Figure 11, all records of *T. hians* are included in this taxon.

Genus *Triporopollenites* Pflug and Thomson in Thomson and Pflug 1953

Triporopollenites plektosus Anderson 1960
(not illustrated)

Triporopollenites plektosus Anderson, 1960, p. 27, Plate 4, fig. 14; Plate 8, fig. 16.

Description. Pollen of this species is spheroidal and triporate; the exine is thin and tends to be folded randomly; the pores are annulate. The thin, randomly folded exine distinguishes *Triporopollenites plektosus* from all other triporate species in the Hell Creek palynoflora. See Anderson (1960) for a more detailed description and illustrations; see Nichols and Brown (1992, Plate 8, figs. 21–23) for additional illustrations.

Occurrence. *Triporopollenites plektosus* is rare in the Hell Creek Formation; it was recorded at only three of the localities studied. It is also present in the Fort Union Formation, where it is similarly rare.

Comments. Anderson (1960) first described this species from the Ojo Alamo Sandstone (Paleocene) of New Mexico. Nichols and Sweet (1993) included this species in the *Wodehouseia spinata* assemblage based on its occurrence in Maastrichtian rocks in western North America.

Triporopollenites triplicatus (Anderson 1960) comb. nov.
(not illustrated)

Basionym. *Paliurus triplicatus* Anderson, 1960, p. 22, Plate 2, fig. 19; Plate 4, figs. 15–18.

Description. *Triporopollenites triplicatus* (Anderson 1960) comb. nov. is triporate pollen with triangular amb and psilate exine. It is distinguished from other triporate species of this general description by distinctive folds in the exine that extend between the pores. See Anderson (1960) for a more detailed description and illustrations.

Occurrence. *Triporopollenites triplicatus* (Anderson 1960) comb. nov. is rare in the Hell Creek Formation; it was recorded at only two localities in this study. It is also present in the Fort Union Formation, where it is similarly rare.

Comments. Anderson (1960) first described this species from the Ojo Alamo Sandstone (Paleocene) and Kirtland Shale (Upper Cretaceous) in New Mexico (as *Paliurus triplicatus*). This is the first record of the species from North Dakota. Anderson assigned this species of fossil pollen to the modern genus *Paliurus* (Rhamnaceae), evidently based on its superficial resemblance to pollen of the modern genus and to fossil specimens from the Cretaceous of Russia similarly assigned by Bolkhovitina (1953). Although affinity with the family Rhamnaceae is possible, based on pollen morphology, definite assignment of this fossil species to the modern genus is here rejected, and the species is reassigned to the form genus *Triporopollenites*.

Triporopollenites spp.
(not illustrated)

Description. Species assigned to *Triporopollenites* spp. in this study are triporate with either annulate or nonannulate pores; sizes range from small to medium; sculpture is subtle, ranging from psilate to scabrate. Three or more species are included that are not distinguished. Excluded from this species complex are *Triporopollenites plektosus* and *T. triplicatus*.

Occurrence. Specimens assigned to *Triporopollenites* spp. as a group are very common in the Hell Creek Formation; they were recorded at more than 90% of the localities analyzed. This broad taxonomic category is also common in the Fort Union Formation, where it was recorded at more than 80% of localities analyzed in the course of this study.

Comments. The broad taxonomic category *Triporopollenites* spp. was used in this study to facilitate scanning of slides. Further study could result in recognition of three or more species, but the distinctions would be subtle, and no attempt was made to distinguish individual species in this study because it appears that the members of this species complex have little or no biostratigraphic or paleoecologic value. The botanical affinity of this group presumably is with certain anemophilous dicot angiosperms.

Genus *Tschudypollis* gen. nov.

Type species. *Tschudypollis retusus* comb. nov.

Description. Pollen triporate, isopolar, oblate, amb triangular with bluntly rounded corners; pores circular to lolongate, annulate; exine two layered, ectexine and endexine subequal in thickness except at pores, which are endexinally thickened; sculpture reticulate, reticulum homobrochate to heterobrochate, where heterobrochate tending to be finer toward the poles. Size ~20–40 μm.

Derivation of name. Patronymic in honor of Robert Hayden Tschudy (1908–1986), pioneer palynologist of the Cretaceous of the western United States.

Repository. The slide bearing the type specimen of the type species of *Tschudypollis* gen. nov. is in the palynological laboratory of Central Energy Resources, U.S. Geological Survey, Denver, Colorado.

Species belonging to this genus. *Tschudypollis retusus* comb. nov. (basionym *Proteacidites retusus* Anderson, 1960, p. 21, Plate 2, figs. 5–7); *Tschudypollis thalmannii* comb. nov. (basionym *Proteacidites thalmannii* Anderson, 1960, p. 21, Plate 2, figs. 1–3, Plate 10, figs. 9–13); undescribed species from the Hell Creek Formation herein assigned to *Tschudypollis* spp.

Comments. Triporate, annulate apertures and reticulate sculpture characterize this genus. The reticulum may be homobrochate or heterobrochate. It bears some resemblance to some fossil and modern species of the family Proteaceae, but this resemblance is superficial (as specified in following). Several authors called attention to this fact (Martin, 1973; Martin and Harris, 1974; Memon, 1983; Frederiksen, 1987). There is also some superficial similarity in shape to some species of the genera *Atlantopollis* Krutzsch and *Pseudatlantopollis* Tschudy of the fossil pollen group Normapolles, but those genera differ in having more complex aperture structure.

The botanical affinity of the North American species attributed to genus *Proteacidites* (now assigned to *Tschudypollis* gen. nov.) has long been uncertain. It has been stated or implied to be with the modern Southern Hemisphere family Proteaceae, dating from the first description of *P. retusus* and *P. thalmannii* by Anderson (1960). From phytogeographic considerations, it is unlikely that it is related to this Southern Hemisphere family, although numerous published records of the North American species have assumed such an affinity, based on superficial resemblance and on the original assignment of Anderson's Cretaceous species to the genus *Proteacidites.* The resemblance of North American "*Proteacidites*" is limited to this pollen being triporate, having a triangular to subangular amb, and in being reticulate (pollen of some but not all species of Proteaceae is reticulate). The extensive discussion of pollen morphology in the modern family Proteaceae by Cookson and Erdtman (in Erdtman, 1966) shows that the North American fossil pollen is morphologically unlike pollen of living species. The endannulate pore structure distinguishes the fossils from most if not all the modern pollen they describe.

It seems more likely that North American species previously assigned to *Proteacidites* belong to the Normapolles group of fossil pollen than to the Proteaceae, although this possible affinity (first mentioned by Tschudy, 1971) remains unproven. Most species of the Normapolles group in North America occur in the Normapolles province in the eastern part of the continent, and very few occur in the *Aquilapollenites* province of the western part.

The suggested affinity of North American species previously assigned to *Proteacidites* with the Symplocaceae (Rouse, 1962; McLeroy, 1971; Frederiksen, 1987) is rejected, at least for those species occurring within the Western Interior. Pollen species herein assigned to *Tschudypollis* gen. nov. differ from pollen of the Symplocaceae in their apertures, which are triporate, never tricolporate. The species discussed by Rouse (1962), McLeroy (1971), and Frederiksen (1987) and assigned to *Proteacidites* occur in exotic terranes that were not part of North America during Cretaceous time (the Continental Margin province of Frederiksen, 1987).

Final determination of the affinity of *Tschudypollis* gen. nov. may await ultrastructural studies involving electron microscopy. It is important now to assign the North American species to a new genus and begin the process of distinguishing these forms from pollen of the modern Southern Hemisphere family.

Tschudypollis retusus (Anderson 1960) comb. nov.
Plate 6, figs. 12–18

Basionym. *Proteacidites retusus* Anderson, 1960, p. 21, Plate 2, figs. 5–7.

Description. Triporate pollen with thickened endexinal annuli at the pores, straight-sided to concave amb, and heterobrochate reticulate sculpture that is finer toward the poles. See Anderson (1960) for a more complete description.

Occurrence. *Tschudypollis retusus* is fairly common in the Hell Creek Formation; it was recorded at 30% of the localities analyzed in this study. It is present in the lowermost Fort Union Formation at one locality, within 5 cm of the base of the formation, in association with other species of Maastrichtian age. It is regarded as a K taxon indicative of Maastrichtian age. Elsewhere in the region it occurs in strata as old as Coniacian (Nichols and Jacobson, 1982).

Comments. Anderson (1960) first described this species from the Kirtland Shale (Upper Cretaceous) in New Mexico (as *Proteacidites retusus*). Additional records of its occurrence in North Dakota may be included in records of "*Tschudypollis* spp." in this study. It is widely distributed both stratigraphically and geographically in western North America, and its abrupt disappearance at the K-T boundary is one of the strongest arguments for a continent-wide extinction event at the end of Cretaceous time based on a single species of pollen. Nichols and Jacobson (1982) reviewed the stratigraphic occurrence of *Tschudypollis retusus* (as *Proteacidites retusus*), and Nichols and Sweet (1993) reviewed its geographic distribution (also as *P. retusus*); it tends to be more common in the southern part of its geographic range. Nichols and Sweet (1993) included this species in the *Wodehouseia spinata* assemblage (as *P. retusus*).

Tschudypollis spp.
Plate 6, figs. 19 and 20

Description. Specimens assigned to this taxon are triporate with thickened endexinal annuli at the pores, rounded triangular ambs that may be straight sided, convex, or concave, and reticulate sculpture. The size and shape of these specimens varies, and the reticulate sculpture, which may be either homobrochate or heterobrochate, varies in coarseness. Several species may be represented, but no attempt was made to consistently distinguish them in this study.

Occurrence. Specimens assigned to *Tschudypollis* spp. are common in the Hell Creek Formation; they were recorded at more than 80% of the localities analyzed. It was also recorded in the Fort Union Formation, but only within the basal 2 m of the Ludlow Member, in facies that otherwise carry a Maastrichtian assemblage. In this study, this taxon, which includes more than one species, is treated as a single K taxon indicative of Maastrichtian age.

Comments. The taxonomic category *Tschudypollis* spp. was used in this study to record the presence of several potentially distinguishable species of the genus *Tschudypollis* in samples being analyzed without taking the time required to separate species on subtle differences in morphology. The illustrated specimens include one that is larger and one that is more concave (retuse) in outline than those herein identified as *Tschudypollis retusus.* Records of this taxon in this study may include some of *Tschudypollis retusus* that were not distinguished during scanning and counting of slides. The botanical affinity of the North American species of *Tschudypollis* is uncertain (see discussion of the genus).

Genus *Ulmipollenites* Wolff 1934 emend. Srivastava 1969
Ulmipollenites krempii (Anderson 1960) Frederiksen 1979
(not illustrated)

Ulmoideipites krempi Anderson, 1960, p. 20, Plate 4, fig. 12; Plate 6, fig. 2; Plate 10, fig. 8.
Ulmoideipites planeraeformis Anderson, 1960, p. 20, Plate 4, fig. 13.
Ulmoideipites tricostatus Anderson, 1960, p. 20–21, Plate 4, figs. 9–11; Plate 6, fig. 4; Plate 7, fig. 8; Plate 8, figs. 8–9.
Carya paleocenica Stanley, 1965, p. 299, Plate 45, figs. 3–7.
Ulmoideipites krempii Anderson emend. Elsik, 1968b, p. 608, Plate 17, figs. 4–7.
Ulmipollenites undulosus auct. non Wolff, Norton and Hall, 1969, p. 41, Plate 5, fig. 23.
Ulmipollenites verrucatus Norton in Norton and Hall, 1969, p. 41–42, Plate 5, fig. 24.
Alnipollenites verus auct. non Potonié, Norton and Hall, 1969, p. 42, Plate 5, fig. 26.
Ulmipollenites sp., Leffingwell, 1970, p. 36, Plate 7, fig. 6.
Ulmipollenites krempii (Anderson) Frederiksen, 1979, p. 154, Plate 3, figs. 6–8.

Description. The species *Ulmipollenites krempii* (Anderson) Frederiksen is zonoporate (three to six pores), oblate, and verrucate; sculpture varies from reduced to robust; arci may be present running between the pores. See Anderson (1960) and Elsik (1968b) for more detailed descriptions. The emended description of Elsik (1968b) is followed here. See any of the references cited in the synonymy for illustrations.

Occurrence. *Ulmipollenites krempii* is common in the Hell Creek Formation; it was recorded at more than 90% of the localities analyzed. It is also present in the Fort Union Formation, where it is about equally common.

Comments. The nomenclatural history of *Ulmipollenites krempii* (Anderson) Frederiksen appears complex, but that is primarily because various authors have tried to distinguish separate species within a single species that exhibits much infraspecific variability. In addition, some misidentifications were made. A summary follows.

Anderson (1960) first described this species from the Ojo Alamo Sandstone (Paleocene) in New Mexico; it also occurs in the Kirtland Shale (Upper Cretaceous) in that area. At the same time he also described two other species that actually are infraspecific variants of the first. He assigned all three species to a genus he proposed at that time, *Ulmoideipites.* That genus is synonymous with *Ulmipollenites* Wolff, however, because it differs in no significant way. Elsik (1968b) synonymized Anderson's three species in an expanded (emended) description of *Ulmipollenites* (al. *Ulmoideipites) krempii;* that species concept is adopted here in accordance with its biostratigraphic and paleoecologic utility. Frederiksen (1979) transferred this species with its broadened circumscription to *Ulmipollenites.*

Stanley (1965) misidentified specimens of *Ulmipollenites krempii* from the Fort Union Formation in South Dakota and placed them in a new species, *Carya paleocenica.* That species—which was inappropriately assigned to the modern genus *Carya* (Juglandaceae) under any circumstances—is synonymous with *U. krempii.* Norton and Hall (1969) misidentified specimens of *U. krempii* from the Hell Creek Formation in Montana as *U. undulosus* Wolff. That species from the Tertiary of Germany differs in size and may be more comparable to certain specimens identified as *Ulmipollenites* sp. by Leffingwell (1970, p. 36, Plate 7, fig. 9 only). Other specimens from the Lance Formation and the lower part of the Fort Union Formations in Wyoming identified as *Ulmipollenites* sp. by Leffingwell (1970, p. 36, Plate 7, fig. 6 only) are *Ulmipollenites krempii.* Norton (in Norton and Hall, 1969) named a new species, *Ulmipollenites verrucatus,* from the Hell Creek Formation in Montana that is synonymous with *U. krempii,* in addition to the misidentified specimens those authors called *U. undulosus.* Lastly, Norton and Hall (1969) also misidentified other Montana Hell Creek specimens as *Alnipollenites verus* Potonié. Those specimens are also *U. krempii;* they are morphologically quite distinct from Potonié's species, which encompasses specimens from the Tertiary of Germany that closely resemble pollen of the modern genus *Alnus* (Betulaceae).

The botanical affinity of *Ulmipollenites krempii* is with the modern family Ulmaceae. It is a species that clearly was unaffected by the K-T boundary extinction event because it is as common and abundant in the Fort Union Formation as in the Hell Creek Formation in North Dakota.

In effect, Nichols and Sweet (1993) included this species in the *Wodehouseia spinata* assemblage, within the taxon *Ulmoideipites* spp. Its stratigraphic range extends both below and above the Maastrichtian, but it is one of the most common species in the Hell Creek palynoflora.

Genus *Wodehouseia* Stanley 1961
Wodehouseia spinata Stanley 1961
Plate 7, figs. 11a, 11b, 12a, and 12b

Wodehouseia spinata Stanley, 1961a, p. 157–158, 160; Plate 1, fig. 1–12.
Wodehouseia stanleyi Srivastava, 1966, p. 548–549, Plate 11, figs. 12, 15.

Description. This morphologically unique species is tetraporate with two pores on each hemisphere of the ovate lenticular body of the grain; it has a prominent, presumably equatorial flange supported by spines, and scattered, echinate sculpture. See Stanley (1961a) for a more detailed description and diagnosis.

Occurrence. *Wodehouseia spinata* is extremely common in the Hell Creek Formation; it was recorded at every locality analyzed in detail in this study. It is present in samples from the base to the top of the formation. It occurs also in the Fort Union Formation at many localities analyzed in conjunction with this study; the highest stratigraphic occurrence of this species in the Fort Union recorded is in a sampled interval 25.53–26.23 m above the base of that formation. Elsewhere in the region this species is present in upper Maastrichtian and younger rocks from Colorado to the Yukon and Northwest Territories (Nichols and Sweet, 1993).

Comments. Despite it having a stratigraphic range that extends well into the lower Paleocene, *Wodehouseia spinata* gives its name to the *Wodehouseia spinata* Assemblage Zone (upper Maastrichtian) of Nichols et al. (1982); see also Nichols (1994) and Nichols and Sweet (1993). This is consistent with the concept and the method of naming an assemblage that defines a biostratigraphic zone. Despite its long stratigraphic range, this species is very common in, and is characteristic of, the Hell Creek Formation, although it cannot be included among the K taxa indicative of Maastrichtian age or be counted among the species whose extinction is attributed to the K-T boundary event.

Stanley (1961a) originally described this species from the Hell Creek Formation in South Dakota. Specimens from the Maastrichtian of Alberta described by Srivastava (1966) as *W. stanleyi* differ in no significant way from *W. spinata* and are synonymous with it.

The botanical affinity of *Wodehouseia spinata* is unknown. It represents a significant kind of plant (almost certainly an angiosperm) that was part of the Hell Creek flora of late Maastrichtian time but that is now extinct and that left no living descendants.

Algal cysts

Five or more species of freshwater algal cysts were recorded during analyses of Hell Creek samples discussed herein. Their significance is limited to being indicators of freshwater in certain depositional environments; therefore, only presence or absence data were recorded and their numbers were not included in relative abundance data. Occurrences of the three most commonly occurring algal cyst taxa are summarized here. None of the Hell Creek algal cysts are illustrated in this report (see Nichols and Brown, 1992, Plate 1, figs. 1–10) for examples.

The most commonly occurring species of algal cysts in the Hell Creek Formation evidently are zygospores or aplanospores related to the modern genus *Spirogyra* (Chlorophyta); see Van Geel (1976) for a discussion of Holocene examples of these fossils and Nichols and Brown (1992) for records from the Tullock Member of the Fort Union Formation in Wyoming. Most Hell Creek specimens are referable to *Schizophacus laevigatus* (Stanley) Nichols and Brown. These algal palynomorphs are present at 30% of the localities studied at various stratigraphic levels throughout the formation. They show no particular pattern of distribution with respect to facies except that they are not present in carbonaceous shale or lignite. These palynomorphs are also present in the Ludlow Member of the Fort Union Formation in North Dakota.

The next most commonly occurring group of algal cysts consists of specimens that are assignable to one or more species of the genus *Tetraporina.* Evidently these are zygospores of a fossil species related to the modern chlorophyte genus *Mougeotia* (Zygnemataceae), based on the work of Van Geel (1976). These palynomorphs are present at ~25% of the localities studied in the Hell Creek Formation. They show no particular pattern of distribution with respect to facies except that they are not present in carbonaceous shale or lignite. They are also present in the Ludlow Member of the Fort Union Formation.

Less common in the Hell Creek Formation are algal coenobia referable to the modern freshwater chlorophyte genus *Pediastrum* (Hydrodictyaceae). These fossils are present at only two localities and appear to be stratigraphically restricted to an interval ~60–70 m above the base of the formation. Living species of the genus are plankton of fluvial and lacustrine environments.

Comparison with the Hell Creek megaflora

Although obviously produced by the same vegetation in Maastrichtian time, the two major kinds of fossil organs—leaves and palynomorphs—give differing pictures of the composition of the Hell Creek flora. These differences are attributable to differential productivity of individual organs by different species, to plant habit, to taphonomy, and to the taxonomic levels these fossils represent.

Differential productivity controls the numbers of potential fossils an individual plant may leave as a record of its existence. Different species of living plants produce variable numbers of potentially fossilizable leaves, and Cretaceous species did the same. This is also true of pollen grains and spores. Factors including the size of an individual plant, its life span, and whether it is deciduous or evergreen will determine how many leaves it can produce that may be preserved as fossils; the mode of reproduction (as determined by botanical affinity) will determine how many pollen grains or spores that same plant will produce. For example, plants that are anemophilous tend to produce prodigious numbers of pollen grains in a season, whereas those that are entomophilous produce far fewer. Counting numbers of specimens of fossil pollen grains in a sample can thus result in anemophilous plants being overrepresented; i.e., they will appear to have been more abundant as individual plants in an ancient community than they actually were, simply because a given number will have produced so many more records of their existence than did the same number of entomophilous plants in the same community. This potential problem for interpretation is not so great when most plants in an ancient community have living relatives with known modes of reproduction that can be taken into account, but it can be a serious problem when many or most species in an ancient flora are extinct or of unknown affinity. This is the case with the Hell Creek flora.

As suggested by D. Braman of the Royal Tyrrell Museum of Palaeontology (2000, written commun.), plant habit (e.g., arboreal versus herbaceous form) may be a factor in accounting for differences in the palynological and paleobotanical records of some species. If a species were arboreal, it might have a greater advantage in interacting with air currents and thus have its pollen spread over a broader area than would a low-growing, herbaceous species. This would be especially likely if the species were anemophilous.

Taphonomy also controls the representation of fossil species in an ancient flora in important ways. Some species produce leaves that are easily fossilized by virtue of their being robust in composition, while others produce frail leaves that seldom survive long enough to be buried and fossilized. It is well known, for example, that herbaceous plants are not well represented in the fossil record because they tend to have thin leaves that are not easily fossilized. In a general way, this is true of pollen grains and spores—different taxa vary in their potential for preservation. It is well understood that entire families of plants are poorly represented in the palynological record because they produce thin-walled pollen grains that are not easily fossilized. Strong differences in taphonomic effects among leaves and pollen are demonstrated well by the Laurales, species of which tend to produce large leaves having thick cuticles that make excellent fossils, but pollen having exceedingly thin exines that are almost never preserved. Taphonomy affects fossil leaves and spores in the opposite way in the case of the ferns, many species of which produce thin, filmy leaves not easily preserved, but robust, thick-walled spores that are well represented in palynofloral assemblages. Added to these is the effect of transport of plant organs prior to fossilization: leaves are not easily transported and tend to be reduced to fragmentary debris within a short distance, but pollen grains and spores can be transported great distances prior to final burial without any noticeable damage; they may end up far from the plants that produced them, perhaps in an entirely different depositional environment.

The final factor influencing interpretations of the nature of a fossil flora provided by paleobotany versus palynology is the taxonomic level represented by the respective fossils. Fossil leaves tend to represent plants at the species level, whereas pollen and spores tend to represent them at the genus level or higher. This is because, as a general rule, a genus of plants that may have numerous species all characterized by differing leaf morphology likely produces only one morphologically distinctive type of pollen grain or spore. The palynological record is one of plants at a high taxonomic level—most palynomorph "species" actually represent botanical genera or even families.

These complexities in determining the nature of a fossil flora like that of the Hell Creek Formation call attention to the value of studying the flora from both the paleobotanical and palynological viewpoints. They also help to explain discrepancies in the fossil record of leaves and pollen (and spores). With these caveats in mind, it is interesting to compare the Hell Creek palynoflora with the Hell Creek megaflora. To avoid problems of overrepresentation or underrepresentation of taxa due to counting numbers of specimens, the summary presented here is based primarily on presence and/or absence data, not relative abundance data.

In his analysis of the Hell Creek megaflora in North Dakota, Johnson (1992) determined that 88.9% of all taxa present are angiosperms (see also Johnson, this volume). In detail, he listed 89% angiosperm species, 6% gymnosperms, 4% pteridophytes, and 1% bryophytes (Johnson, 1992, Table 4). In contrast, the palynoflora of the Hell Creek in North Dakota described here consists of 69% angiosperms, 9% gymnosperms, 18% pteridophytes, and 4% bryophytes. These numbers can be further subdivided and discrepancies in relative abundance of taxonomic groups examined.

Total percentages of gymnosperms and bryophytes in the Hell Creek megaflora and palynoflora differ, but perhaps not significantly, given the different nature of the organs that compose them. The major differences are among the percentages of angiosperms and pteridophytes. It is clear that many more pteridophytes are represented in the Hell Creek palynoflora than in the megaflora, this difference accounting for most of the lower relative number of angiosperm taxa. Among Hell Creek pteridophyte palynomorph taxa, 14% are ferns (the remainder are lycopods and quillworts). In view of the taphonomic factors reviewed here, it is evident that ferns as a group are truly underrepresented in the megaflora. Even allowing that some fern spores may have been transported from distant areas in Maastrichtian time, the relative abundances of some fern-spore taxa in the Hell Creek Formation show that the plants that produced them were indigenous. The most obvious example is *Laevigatosporites* spp., a taxon that includes at least two "species" (genera?) of polypodiaceous ferns. *Laevigatosporites* spp. spores account for 30% or more of the palynomorphs present in six samples and about twice that much in two of them.

Johnson (1992, Table 4) listed only one bryophyte taxon in the Hell Creek megaflora, but three taxa are identified in the palynoflora. The most common of these is *Stereisporites* spp., which includes at least three species of sphagnum moss. *Stereisporites* spp. spores amount to almost 20% of the count in one sample and more than 30% in another. Sphagnum moss was a major component of mire communities in Hell Creek time, but the two samples that have the highest relative abundance of sphagnum spores are not from lignite.

Among angiosperm taxa in the palynoflora, 11% are monocots and 58% are dicots. Among the dicots there are no representatives of the Laurales, although Johnson (1992) identified nine taxa. The reasons for this discrepancy were given here. The most diverse angiosperm taxon in the Hell Creek palynoflora is the genus *Aquilapollenites.* In this study 14 species are identified; whether these palynomorph species represent botanical species or some higher level taxon is unknown because this group is extinct and its botanical affinity is unknown. The answer to the question about which megafloral taxa are represented by *Aquilapollenites* pollen is also unknown. This diverse and locally abundant group may or may not be represented in the megaflora.

PALYNOSTRATIGRAPHY

Wodehouseia spinata Assemblage Zone

The Hell Creek Formation is entirely within the *Wodehouseia spinata* Assemblage Zone, the palynostratigraphic zone that represents late Maastrichtian time in nonmarine (continental) rocks in western North America. This biozone is named for *Wodehouseia spinata,* which as already mentioned is present at all Hell Creek localities analyzed in detail in this study. It was defined by Nichols et al. (1982); see also Nichols and Sweet (1993) and Nichols (1994). The biozone is delimited by the first (lowest stratigraphic) occurrence of *Wodehouseia spinata* and the last (highest stratigraphic) occurrence of *Tschudypollis* spp., and is characterized by the presence of species of the *Wodehouseia spinata* assemblage (Nichols et al., 1982; Nichols and Sweet, 1993; additional data herein). Thus, this zone is an Oppel zone in the sense of Hedberg (1976, p. 57–58; see also North American Commission on Stratigraphic Nomenclature, 1983, Fig. 5B). That *Wodehouseia spinata* ranges above the Hell Creek Formation is in no way inconsistent with either the concept of an assemblage zone or the method of naming an assemblage that defines such a biostratigraphic zone.

As elaborated by Nichols and Sweet (1993), the *Wodehouseia spinata* Assemblage Zone includes 40 palynomorph taxa. Of those taxa, 33 occur in the Hell Creek Formation in North Dakota. On the basis of this study, ten taxa can be added to the list of taxa composing the *Wodehouseia spinata* Assemblage Zone: *Anacolosidites rotundus, Aquilapollenites marmarthensis, A. catenireticulatus, A. reductus, A. senonicus, Cyathidites diaphana, Inaperturopollenites scabratus, Porosipollis porosus, Racemonocolpites formosus,* and *Styxpollenites calamitas.* This additional list brings the total number of species in the *Wodehouseia spinata* assemblage to 50, 43 of which are present in the Hell Creek Formation. Of these, 31 are regarded as K taxa indicative of Maastrichtian age.

The *Wodehouseia spinata* Assemblage Zone is recognized across western North America from New Mexico to the Yukon and Northwest Territories (Nichols and Sweet, 1993). The species present in the assemblage defining the biozone change somewhat with latitude. Thus, it can be considered a composite assemblage zone (see North American Commission on Stratigraphic Nomenclature, 1983, Fig. 5C). The *Wodehouseia spinata* Assemblage Zone is equivalent to the Lancian stage of vertebrate biostratigraphy.

Palynostratigraphic zonation of the Hell Creek Formation

Most palynomorph species present in the *Wodehouseia spinata* Assemblage Zone in North Dakota range throughout the Hell Creek Formation. The few that have more restricted ranges potentially useful for biostratigraphic subdivision of the formation are *Aquilapollenites bertillonites, A. collaris, A. conatus, A. marmarthensis, Liliacidites altimurus, Myrtipites scabratus, Porosipollis porosus, Racemonocolpites formosus,* and *Striatopollis tectatus.* A few others, *Anacolosidites rotundus, Aquilapollenites pyriformis, Echitricolpites* sp., and *Polyatriopollenites levis,* evidently also have restricted ranges, but are so rare that they are of little or no practical value in biostratigraphic correlation. The other stratigraphically restricted taxa provide a basis for subdivision of the assemblage zone. Five informal subzones of the *Wodehouseia spinata* Assemblage Zone in North Dakota are defined, as follows.

In the Sunset Butte reference section, subzone A includes the lower 14 m of the formation; it is characterized by the presence of *Aquilapollenites bertillonites* (*Porosipollis porosus* is absent from this subzone). Subzone B includes the interval from 14 to 42 m above the base of the formation; it is characterized by the presence of both *P. porosus* and *A. bertillonites.* Subzone C includes the interval from 42 to ~70 m above the base of the formation; it is characterized by the presence of *A. collaris, A. conatus, A. marmarthensis, Myrtipites scabratus,* and *Striatopollis tectatus* (*P. porosus* is also present in this subzone). The base of subzone C is the lowest stratigraphic occurrence of *A. conatus* and *A. marmarthensis,* which, in the Sunset Butte reference section, is in the same sample within a sparsely sampled interval (see Fig. 11). Subzone D includes the interval from ~70 to 95 m above the base of the formation; it is characterized by the presence of *Liliacidites altimurus* (all of the species in subzone C are present, except for *P. porosus*). Subzone E includes the uppermost 5 m of the formation. It is characterized by the presence of *Aquilapollenites quadricretaeus* and *Racemonocolpites formosus.* All of the species in subzone D are

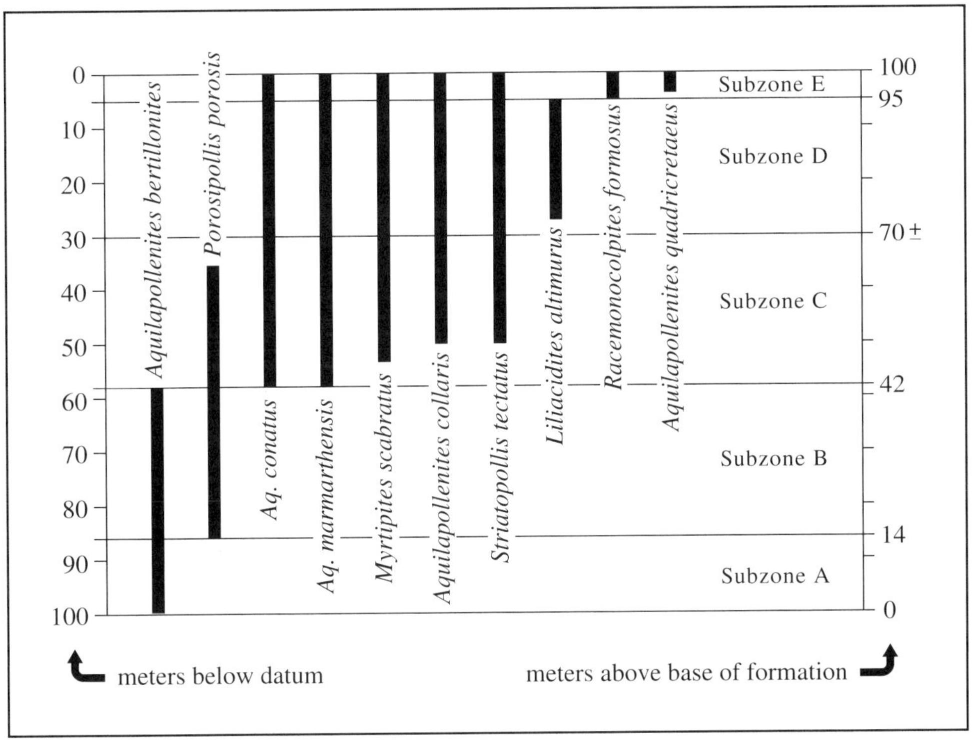

Figure 5. Provisional subzones of *Wodehouseia spinata* Assemblage Zone in western North Dakota. These subzones are interval zones as defined by North American Stratigraphic Code (North American Commission on Stratigraphic Nomenclature, 1983) and include taxon range zone (E), concurrent range zones (B and C), lineage zone (A), and interval zone (D) of International Stratigraphic Guide (Salvador, 1994).

present, except for *L. altimurus*. These data are summarized in Figure 5.

This palynostratigraphic subdivision of the Hell Creek Formation is considered informal because it is based on a limited number of samples primarily from a single stratigraphic section; the upper three subzones are more robustly defined because they are based on a greater number of samples from several localities (see Table 1). Although possibly influenced by sampling density, the occurrence data indicate a rather strong division of the formation into a lower part (subzones A and B) and an upper part (subzones C–E), on the basis of palynostratigraphy. The palynostratigraphic boundary between the lower and upper parts of the Hell Creek is at 42 m above the base of the formation in the Sunset Butte reference section, but an unsampled interval of ~17 m below this horizon makes its exact position uncertain. The boundary between subzones C and D is placed at ~70 m above the base of the formation because of the presence of a similar, unsampled interval of ~16 m that renders the highest stratigraphic occurrence of *Porosipollis porosus* somewhat uncertain. What is not evident from Figure 5 is that the stratigraphic ranges of all the characteristic taxa of subzone E terminate at the K-T boundary. Because the K-T boundary is above the top of the Hell Creek Formation in most sections in the study area (Nichols and Johnson, this volume), subzone E of the *Wodehouseia spinata* Assemblage Zone extends above the top of the Hell Creek Formation.

Comparison of the palynostratigraphic subdivision of the Hell Creek Formation in western North Dakota with the palynological zones of Braman and Sweet (1999) for the Frenchman Formation in southern Alberta and Saskatchewan and the Boissevain Formation of southern Manitoba shows that there are close similarities. Braman and Sweet defined two subzones within the Hell Creek equivalent interval in southern Canada, the *Porosipollis porosus-Aquilapollenites notabile* subzone (lower) and the *Myrtipites scabratus-Aquilapollenites delicatus* var. *collaris* subzone (upper). These two subzones correspond closely with the lower and upper parts of the Hell Creek Formation described herein, specifically with subzones A plus B (lower) and subzones C through E (upper). There are some differences in the palynomorph species selected to define the subzone boundaries in the respective zonations, and in the local ranges of some of those species. These differences result in an offset in the position of subzone boundaries, but the approximate correlation is clear (see Fig. 6).

Braman and Sweet (1999), following Srivastava (1970), identified the *Wodehouseia spinata* zone of Srivastava as encompassing the Frenchman Formation and the upper part of the Boissevain Formation (Hell Creek equivalents) and the lower part of the overlying Ravenscrag Formation and lower part of the Turtle Mountain Formation (Fort Union equivalents). It is important to note that the *W. spinata* zone of Srivastava (1970) is a range zone (not an assemblage zone or Oppel zone, as is the *W. spinata* zone of Nichols et al., 1982). Srivastava defined the top of his *W. spinata* zone by the highest stratigraphic occurrence of *W. spinata;* Braman and Sweet (1999) modified this definition by placing the top of the *W. spinata* zone at the lowest stratigraphic occurrence of *W. fimbriata* (the ranges of these two species overlap). A significant effect of either definition of

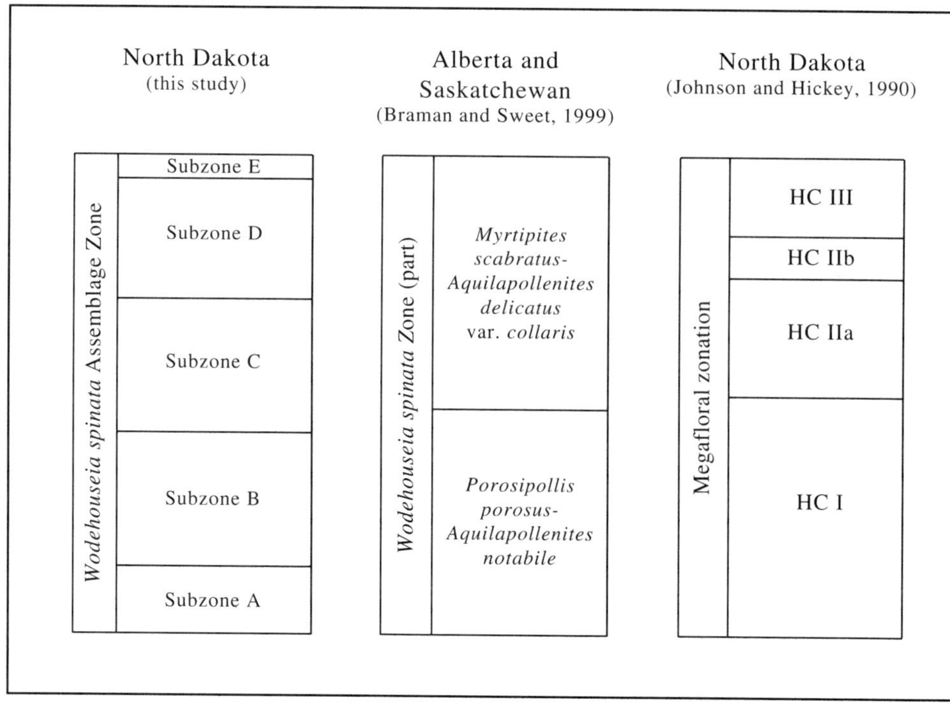

Figure 6. Comparison of palynostratigraphic zonations in North Dakota (this study) and in Alberta and Saskatchewan (Braman and Sweet, 1999) and megafloral zonation in North Dakota (Johnson and Hickey, 1990).

the *W. spinata* range zone in Canada is that it tends to obscure the significance of the range terminations of most K taxa at the K-T boundary (see Nichols and Johnson, this volume). The K-T boundary is within Srivastava's *W. spinata* range zone and is not immediately apparent therein, except at the level of subzonation.

Comparison with megafloral biostratigraphy

As shown in Figure 6, the provisional palynostratigraphic subzones of the *Wodehouseia spinata* Assemblage Zone agree only in part with the subzones of the Hell Creek Formation based on megafloral paleobotany (Johnson and Hickey, 1990; see also Johnson et al., 1990, and Johnson, this volume). Megafloral zone HCI includes the interval from the base to 50 m above the base of the formation, zone HCIIa includes the interval from 50 to 75 m above the base, zone HCIIb includes the interval from 75 to 92 m above the base, and HCIII includes the uppermost 17 m of the formation. Zone HCI corresponds roughly to the lower part of the Hell Creek Formation recognized on the basis of palynostratigraphy in North Dakota (data herein) and also in the equivalent section in southern Canada (Braman and Sweet, 1999). However, the upper part of the Hell Creek is subdivided differently on the basis of plant megafossils and plant microfossils. Megafloral zone HCIIa is roughly equivalent to palynofloral subzone C, but zones and subzones above that level do not correspond well. Explanations for the discrepancies may be in the taxonomic levels of the plant organs on which the palynofloral and megafloral biostratigraphies are based, and the differences in stratigraphic resolution provided by the two kinds of plant fossils. The palynostratigraphic zonation is based primarily on the stratigraphic ranges of higher level botanical taxa (genera or families) rather than on botanical species, and on a limited number of such taxa. Further work may either resolve the differences as well as revealing more about the plant megafossil and microfossil biostratigraphy of the Hell Creek Formation.

CONCLUSIONS

The Hell Creek Formation of North Dakota preserves a rich and varied palynoflora that provides valuable insight into the nature of plant life in the region in late Maastrichtian time. This palynological record supplements the extensive data from the same rocks and the same ancient vegetation preserved in the megafloral record, and it provides background biostratigraphic data necessary to determine the nature of the K-T boundary event. The palynoflora includes 115 or more palynomorph species, 98 of which are described. The palynofloral list for the Hell Creek Formation includes 68 angiosperm taxa, eight gymnosperm taxa, 19 pteridophyte taxa, and three bryophyte taxa. Of the 98 taxa, 33 are regarded as characteristically Upper Cretaceous and are referred to as K taxa, but only six (all angiosperm pollen) are restricted in occurrence to the Hell Creek Formation. All the other taxa were also recorded in the overlying Ludlow Member of the Fort Union Formation at some localities investigated in the course of this study. With the exception of *Wodehouseia spinata* and *Aquilapollenites reticulatus,* the 25 K taxa occurring in the Fort Union Formation are present only in the lowermost 2.5 m or less of the Ludlow

Member, in facies that carry a Maastrichtian assemblage—this interval is termed the Fort Union facies of Maastrichtian age. The level of extinction at the K-T boundary, which is either at or above the top of the Hell Creek Formation in the study area, is ~30%.

ACKNOWLEDGMENTS

I thank the reviewers of the original manuscript, Dennis Braman of the Royal Tyrrell Museum of Palaeontology and Norman Frederiksen of the U.S. Geological Survey, for their comprehensive and constructive critiques. Braman was especially helpful in providing additional synonymies for taxa described and in discussing aspects of palynostratigraphy. Frederiksen offered valuable suggestions on nomenclature and data on stratigraphic ranges. Both contributed appreciably to improvement of the final version.

APPENDIX 1. LOCALITIES AND SAMPLES

Sunset Butte

Composite section incorporating D7014, D7015, D7016, D7017, D7018, D7095, and D9003. Bowman County, North Dakota, Cedar Hills Quadrangle, sec. 30 to sec. 21 and sec. 34, T.131N., R.105W., lat. 46°8′11″–46°8′29″N, long. 103°5′30″–103°4′4″W, elevation 847–879 m. Hell Creek Formation, entire thickness. The complete measured section at Sunset Butte also includes 36 m of Fort Union Formation.

D7018-D	– 80–136 cm	carbonaceous shale
D7018-C	– 660 cm	shale
D7095	– 1205–1387 cm	mudstone
D7018-B	– 1582 cm	carbonaceous shale
D7018-A	– 2000 cm	shale
D7017-C	– 3353 cm	mudstone
D7017-B	– 3566 cm	carbonaceous shale
D7017-A	– 3863 cm	carbonaceous shale
D7016-B	– 5795 cm	shale
D7016-A	– 7500 cm	mudstone
D7015-B	– 7840 cm	carbonaceous shale
D9003-B	– 8591 cm	mudstone
D9003-A	– 8773 cm	mudstone
D7014-B	– 9570 cm	mudstone
D7014-A	– 9965 cm	claystone

Sunset Patch

Composite section incorporating D9002 and D9004. Bowman County, North Dakota, Cedar Hills Quadrangle, sec. 34, T.131N., R.105W., lat. 46°6′12″–46°7′50″N, long. 103°4′14″–103°4′8″W, elevation 908–933 m. Hell Creek Formation, 380–655 cm below top (part of measured section also including 953 cm of Fort Union Formation).

D9002-B	– 380 cm	mudstone
D9004-A	– 492 cm	mudstone
D9002-A	– 655 cm	mudstone

Badland Draw

Locality D7670. Slope County, North Dakota, Badland Draw Quadrangle, sec. 4, T.134N., R.106W., lat. 46°26′24″N, long. 104°0′09″W, elevation 847 m. Hell Creek Formation, 0–1 cm below top (part of measured section including 275 cm of Fort Union Formation).

D7670-A	– 40 cm	claystone

Barbecue

Measured section D8224 (KJ9324). Slope County, North Dakota, Marmarth SE Quadrangle, sec. 22, T.133N., R.106W., lat. 46°19′12″N, long. 103°5′17″W, elevation 873 m. Hell Creek Formation, 0-42 cm below top (part of measured section including 280 cm of Fort Union Formation).

D8224-C	– 0–5 cm	mudstone
D8224-B	– 12–15 cm	mudstone
D8224-A	– 40–42 cm	mudstone

Battleship

Measured section D8149 (KJ9860). Slope County, North Dakota, Marmarth SE Quadrangle, sec. 21, T.133N., R.105W., lat. 46°19′15″–46°20′51″N, long. 103°5′58″–103°5′34″W, elevation 860 m. Hell Creek Formation, 0–2 cm below top (the measured section includes 80 cm of Fort Union Formation).

D8149-C	– 0–2 cm	mudstone

Bobcat Butte

Composite section incorporating D6956, D6957, and D7103 (KJ86148). Slope County, North Dakota, Marmarth Quadrangle, sec. 28, T.133N., R.105W., lat. 46°17′41″–46°17′43″N, long. 103°5′47″–103°5′37″W, elevation 846-861 m. Hell Creek Formation, 0–50 m below top (part of measured section including 37 m of Fort Union Formation).

D7103-A	– 3–5 cm	mudstone
D6956-C	– 10–15 cm	mudstone
D6957-E	– 2197–2237 cm	claystone
D6957-D	– 3392–3474 cm	claystone
D6957-B	– 3963 cm	claystone
D6957-A	– 5000 cm	claystone

Dean's NM1

Locality D7629 (KJ88119). Slope County, North Dakota, Marmarth Quadrangle, sec. 20, T.133N., R.106W., lat. 46°19′6″N, long. 103°5′30″W, elevation 884 m. Hell Creek Formation, 0–47 cm below top (part of measured section including 262 cm of Fort Union Formation).

D7629-C	– 2 cm	claystone
D7629-B	– 17 cm	claystone

Facet Butte

Locality D9097 (KJ9754). Slope County, North Dakota, Pretty Butte Quadrangle, sec. 34, T.135N., R.106W., lat. 46°27′32″–46°20′50″N, long. 103°5′0″W, elevation 835 m. Hell Creek Formation, 0–10 cm below top (part of measured section including 160 cm of Fort Union Formation).

D9097-B	– 0–10 cm	mudstone

Golva Road

Locality D6995 (near KJ87129). Slope County, North Dakota, Pretty Butte Quadrangle, sec. 23, T.134N., R.106W., lat. 46°23′53″N, long. 103°57′2″W, elevation 823 m. Hell Creek Formation, 42.4 m below top.

D6995	– 4240 cm	claystone

Hell Creek Coal

Locality D7644. Slope County, North Dakota, Pretty Butte Quadrangle, sec. 25, T.134N., R.106W., lat. 46°23′10″N, long. 103°56′15″W, elevation 817 m. Hell Creek Formation, 52.6–53.2 m below top.

D7644-D	– 5260 cm	mudstone
D7644-C	– 5280 cm	lignite
D7644-B	– 5300 cm	lignite
D7644-A	– 5320 cm	mudstone

Mud Buttes

Composite section incorporating D7415, D7416, and D7672. Bowman County, North Dakota, Mud Buttes Quadrangle, sec. 1, T.130N., R.105W., lat. 46°1′6″–46°1′8″N, long. 103°4′37″–103°4′31″W, elevation 954–958 m. Hell Creek Formation, 0–4.22 m below top (part of measured section including 839 cm of Fort Union Formation).

D7672-D	– 1 cm	mudstone
D7672-BB	– 8–10 cm	mudstone
D7672-AA	– 20–22 cm	mudstone
D7672-C	– 41 cm	mudstone
D7672-B	– 249 cm	mudstone
D7672-A	– 422 cm	mudstone

No Leaf Minors

Locality D9096 (KJ9750). Slope County, North Dakota, Marmarth SE Quadrangle, sec. 9, T.133N., R.105W., lat. 46°20′49–50″N, long. 103°5′13″W, elevation 851 m. Hell Creek Formation, 0–10 cm below top (part of measured section including 766 cm of Fort Union Formation).

D9096-A	– 0–10 cm	mudstone

Pretty Butte

Composite section incorporating D7656 and D7788 and supplemented by D9094 (KJ9735). Slope County, North Dakota, Pretty Butte Quadrangle, SW¼ SE¼ SE¼ sec. 35, T.134N., R.106W., lat. 46°22′33–35" N, long. 103°56′52″–103°57′8″W, elevation 835–844 m (main section), and sec. 35, T.134N., R.106W., lat. 46°22′41–43″N, long. 103°5′5–52″W, elevation 836–844 m (supplementary section). Hell Creek Formation, 0–26.12 m below top (part of measured section including 68 m of Fort Union Formation).

D9094-D	– 0–10 cm	mudstone
D7788-EE	– 2270–2295 cm	lignite
D9094-A	– 2359 cm	mudstone
D7788-FF	– 2600–2612 cm	mudstone

Pretty Butte North

Section D8025 (KJ9749). Slope County, North Dakota, Pretty Butte Quadrangle, sec. 26, T.134N., R.105W., lat. 46°23′5″–46°22′41″, long. 103°5′7″–103°5′5″W, elevation 890 m. Hell Creek Formation, 0–0.5 m below top (part of measured section including 650 cm of Fort Union Formation).

D8025-D	– 0–5 cm	mudstone
D8025-C	– 5–10 cm	mudstone
D8025-B	– 10–15 cm	mudstone
D8025-A	– 45–50 cm	mudstone

Pyramid Butte

Composite section incorporating D6958 and D6998 (KJ86161). Slope County, North Dakota, Pretty Butte Quadrangle, sec. 15, T.134N., R.106W., lat. 46°25′4″N, long. 103°5′33″W, elevation 857 m. Hell Creek Formation, 0–10 cm below top.

D6958-A	– 0–10 cm	claystone

River Section

Section D7138 (KJ87147). Slope County, North Dakota, Marmarth Quadrangle, sec. 8, T.133N., R.105W., lat. 46°20′57″–46°20′58″N, long. 103°5′49″–103°5′42″W, elevation 851–875 m. Hell Creek Formation, 0–23.60 m below top (part of measured section including 6.8 m of Fort Union Formation).

D7138-E	– 5 cm	mudstone
D7138-D	– 2185 cm	mudstone
D7138-C	– 2341 cm	mudstone
D7138-B	– 2350 cm	lignite
D7138-A	– 2360 cm	mudstone

Roadside HCIIb

Locality D8227 (KJ9323). Slope County, North Dakota, Marmarth Quadrangle, sec. 29, T.133N., R.106W., lat. 46°18′37″N, long. 105°51′42″W, elevation 853 m. Hell Creek Formation, 27–30 m below top.

D8227-B	– 2700 cm	mudstone
D8227-A	– 3000 cm	mudstone

Swanke's Coal

Locality D9007. Bowman County, North Dakota, Cedar Hills Quadrangle, sec. 34, T.132N., R.105W., lat. 46°12′58″N, long. 103°48′4″W, elevation 860 m. Hell Creek Formation, 23.85–24.28 m below top.

D9007-C	– 2385 cm	lignite
D9007-B	– 2392 cm	lignite
D9007-A	– 2428 cm	mudstone

Swanke's Nose

Locality D9006. Bowman County, North Dakota, Cedar Hills Quadrangle, sec. 34, T.132N., R.105W., lat. 46°12′47″–46°23′32″°N, long. 103°4′34″–103°5′37″W, elevation 884 m. Hell Creek Formation, 0–40 cm below top (part of measured section including 40 cm of Fort Union Formation).

D9006-D	– 5–10 cm	shale
D9006-A	– 35–40 cm	shale

Torosaurus Section

Measured section D9095 (KJ9738). Bowman County, North Dakota, Mud Buttes Quadrangle, sec. 12, T.130N., R.105W., lat. 46°22′41″N, long. 103°5′5″W, elevation 922–926 m. Hell Creek Formation, 0–210 cm below top (part of measured section including 3.80 m of Hell Creek and 3.0 m of Fort Union Formations).

D9160-F	– 0–5 cm	claystone
D9160-E	– 5–10 cm	claystone
D9160-D	– 10–15 cm	claystone
D9160-B	– 20–25 cm	claystone
D9095-C	– 30–50 cm	claystone
D9095-B	– 200–210 cm	claystone

REFERENCES CITED

Ames, H.T., and Spackman, W., 1981, Cretaceous spores and pollen from the U.S.A. and the U.S.S.R.: Catalog of Fossil Spores and Pollen, v. 42, 231 p.

Anderson, R.Y., 1960, Cretaceous-Tertiary palynology, eastern side of the San Juan Basin, New Mexico: New Mexico Institute of Mining and Technology Memoir 6, 59 p.

Belsky, C.Y., Boltenhagen, E., and Potonié, R., 1965, Sporae dispersae der Oberen Kreide von Gabun, Aequatoriales Afrika: Paläontologische Zeitschrift, v. 39, p. 72–83.

Bergad, R.D., 1973, North American species of the Cretaceous megaspores *Balmeisporites* and *Monophyllosporites:* Micropaleontology, v. 19, p. 53–67.

Bolkhovitina, N.A., 1953, Spores and pollen characteristic of Cretaceous deposits of central regions of the U.S.S.R.: Trudy Geological Institute Akademia Nauk S.S.S.R., no. 145, Geological Series, no. 61, 184 p. (in Russian).

Braman, D.R., and Sweet, A.R., 1999, Terrestrial palynomorph biostratigraphy of the Cypress Hills, Wood Mountain, and Turtle Mountain areas (Upper Cretaceous–Paleocene) of western Canada: Canadian Journal of Earth Sciences, v. 36, p. 725–741.

Chlonova, A.F., 1961, Spores and pollen of the upper half of the Upper Cretaceous of the eastern part of the West Siberian lowland: Academy of Sciences of the U.S.S.R. (Siberian Branch), Institute of Geology and Geophysics, Trudy, v. 7, 138 p. (in Russian).

Cookson, I.C., and Pike, K.M., 1954, Some dicotyledonous pollen types from Cainozoic deposits in the Australian region: Australian Journal of Botany, v. 2, p. 197–219.

Couper, R.A., 1953, Upper Mesozoic and Cainozoic spores and pollen grains from New Zealand: New Zealand Geological Survey Paleontological Bulletin 22, 77 p.

De Lima, M.R., 1980, Palinologica da Formacao Santana (Cretaceo do Nord Este du Brasil) 3: Descricao sistematica dos polens da Turma Plicates (Subturma Costates): Ameghiniana, v. 17, p. 15–47.

Dettmann, M.E., 1963, Upper Mesozoic microfloras from south-eastern Australia: Royal Society of Victoria Proceedings, v. 77, p. 1–148.

Dettmann, M.E., and Playford, G., 1968, Taxonomy of some Cretaceous spores and pollen grains from eastern Australia: Royal Society of Victoria Proceedings, v. 81, p. 69–94.

Doher, L.I., 1980, Palynomorph preparation procedures currently used in the paleontology and stratigraphy laboratories: U.S. Geological Survey Circular 830, 29 p.

Drugg, W.S., 1967, Palynology of the upper Moreno Formation (Late Cretaceous–Paleocene) Escarpado Canyon, California: Palaeontographica, Abteilung B: Palaophytologie, v. 120, p. 1–71.

Elsik, W.C., 1968a, Palynology of a Paleocene Rockdale lignite, Milam County, Texas. 1. Morphology and taxonomy: Pollen et Spores, v. 10, p. 263–314.

Elsik, W.C., 1968b, Palynology of a Paleocene Rockdale lignite, Milam County, Texas. 2. Morphology and taxonomy (end): Pollen et Spores, v. 10, p. 599–664.

Erdtman, G., 1966, Pollen morphology and plant taxonomy. 1. Angiosperms: An introduction to palynology: New York, Hafner, 553 p.

Fairchild, W.W., and Elsik, W.C., 1969, Characteristic palynomorphs of the lower Tertiary in the Gulf Coast: Palaeontographica, Abteilung B: Palaophytologie, v. 128, p. 81–89.

Farabee, M.J., 1990, Triprojectate fossil pollen genera: Review of Palaeobotany and Palynology, v. 65, p. 341–347.

Farabee, M.J., 1993, Morphology of triprojectate fossil pollen; form and distribution in space and time: The Botanical Review, v. 59, p. 211–249.

Farabee, M.J., and Canright, J.E., 1986, Stratigraphic palynology of the lower part of the Lance Formation (Maestrichtian) of Wyoming: Palaeontographica, Abteilung B: Palaophytologie, v. 199, p. 1–89, pls. 1–26.

Farabee, M.J., Daghlian, C.P., Canright, J.E., and Oftedahl, O., 1984, *Libopollis,* a new genus from the Maestrichtian of North America: Palynology, v. 8, p. 145–163.

Frederiksen, N.O., 1979, Paleogene sporomorph biostratigraphy, northeastern Virginia: Palynology, v. 3, p. 129–167.

Frederiksen, N.O., 1987, Tectonic and paleogeographic setting of a new latest Cretaceous floristic province in North America: Palaios, v. 2, p. 533–542.

Frederiksen, N.O., 1988, Sporomorph biostratigraphy, floral changes, and paleoclimatology, Eocene and earliest Oligocene of the eastern Gulf Coast: U.S. Geological Survey Professional Paper 1448, 68 p.

Frederiksen, N.O., Anderle, V.A.S., Sheehan, T.P., Ager, T.A., Collett, T.S., Fouch, T.D., Franczyk, K.J., and Johnson, M., 1998, Palynological dating of Upper Cretaceous to middle Eocene strata in the Sagavanirktok and Canning Formations, North Slope of Alaska: U.S. Geological Survey Open-File Report 98–471, 44 p.

Funkhouser, J.W., 1961, Pollen of the genus *Aquilapollenites:* Micropaleontology, v. 7, p. 193–198.

Hall, J.W., 1968, A new genus of Salviniaceae and a new species of *Azolla* from the Late Cretaceous: American Fern Journal, v. 58, p. 77–88.

Hall, J.W., and Norton, N.J., 1967, Palynological evidence of floristic change across the Cretaceous–Tertiary boundary in eastern Montana (U.S.A.): Palaeogeography, Palaeoclimatology, Palaeoecology, v. 3, p. 121–131.

Hedberg, H.D., ed., 1976, International stratigraphic guide: New York, John Wiley and Sons, 200 p.

Hotton, C.L., 1984, Palynofloral changes across the Cretaceous-Tertiary boundary in east central Montana, U.S.A.: Sixth International Palynological Conference, Calgary, Alberta, Abstracts, p. 66.

Hotton, C.L., 1988, Palynology of the Cretaceous-Tertiary boundary in central Montana, U.S.A., and its implications for extraterrestrial impact [Ph.D. thesis]: Davis, University of California, 732 p.

Hotton, C.L., Leffingwell, H.A., and Skvarla, J.J., 1994, Pollen ultrastructure of Pandanaceae and the fossil genus *Pandaniidites, in* Kurmann, M.H., and Doyle, J.A., eds., Ultrastructure of fossil spores and pollen: Kew, England, Royal Botanic Gardens, p. 173–191.

Jameossanaie, A., 1987, Palynology and age of South Hospah coal-bearing deposits, McKinley County, New Mexico: New Mexico Bureau of Mines and Mineral Resources Bulletin 122, 65 p.

Jarzen, D.M., 1976, The affinities of *Marsypiletes cretacea* Robertson: Pollen et Spores, v. 18, p. 395–397.

Jarzen, D.M., 1978, Some Maestrichtian palynomorphs and their phytogeographical and paleoecological implications: Palynology, v. 2, p. 29–38.

Jarzen, D.M., 1983, The fossil pollen record of the Pandanaceae: Gardens Bulletin, v. 36, p. 163–175.

Jerzykiewicz, T., and Sweet, A.R., 1986, The Cretaceous-Tertiary boundary in the central Alberta foothills. 1. Stratigraphy: Canadian Journal of Earth Sciences, v. 23, p. 1356–1374.

Johnson, K.R., 1992, Leaf-fossil evidence for extensive floral extinction at the Cretaceous-Tertiary boundary, North Dakota, USA: Cretaceous Research, v. 13, p. 91–117.

Johnson, K.R., and Hickey, L.J., 1990, Megafloral change across the Cretaceous/Tertiary boundary in the northern Great Plains and Rocky Mountains, U.S.A., *in* Sharpton, V.L., and Ward, P.D., eds., Global catastrophes in Earth history: An interdisciplinary conference on impacts, volcanism, and mass mortality: Geological Society of America Special Paper 247, p. 433–444.

Johnson, K.R., Nichols, D.J., Attrep, M., Jr., and Orth, C.J., 1989, High-resolution leaf-fossil record spanning the Cretaceous/Tertiary boundary: Nature, v. 340, p. 708–711.

Johnson, K.R., Nichols, D.J., Tauxe, L., and Clark, D., 1990, Floral zonation and magnetostratigraphy of the Hell Creek (late Maastrichtian) and lower Fort Union (early Paleocene) Formations, North Dakota: Geological Society of America Abstracts with Programs, v. 22, no. 7, p. 323.

Kondinskaya, L.I., 1966, Fossil spores of water ferns in Upper Cretaceous and Paleogene deposits of the west Siberian lowland, *in* Chlonova, A.F., ed.,

Palynology of Siberia: Moscow, Akademia Nauk, p. 116–122 (in Russian).

Kremp, G.O.W., 1949, Pollenanalytische Untersuchung des miozänen Braunkohlenlagers von Konin an der Warthe: Palaeontographica, Abteilung B: Palaophytologie, v. 90, p. 53–93.

Krutzsch, W., 1959, Mikropaläontologische (sporenpaläontologische) Untersuchungen in der Braunkohle des Geiseltals: Geologie, v. 8, no. 21/22, 425 p.

Krutzsch, W., 1962, Stratigraphisch bzw. botanisch wichtige neue Sporen- und Pollenformen aus dem deutschen Tertiär: Geologie, v. 11, p. 265–307.

Krutzsch, W., 1963a, Atlas der mittel- und jungtertiären dispersen Sporen- und Pollen-sowie der Mikroplanktonformen des nördlichen Mitteleuropas, Lieferung II. Die Sporen der Anthocerotaceae und der Lycopodiaceae: Berlin, VEB Verlag der Wissenschaften, 141 p.

Krutzsch, W., 1963b, Atlas der mittel- und jungtertiären dispersen Sporen- und Pollen-sowie der Mikroplanktonformen des nördlichen Mitteleuropas, Lieferung III. Sphagnaceoide und selaginellaceoide Sporenformen: Berlin, VEB Verlag der Wissenschaften, 128 p.

Krutzsch, W., 1967, Atlas der mittel- und jungtertiären dispersen Sporen- und Pollen-sowie der Mikroplanktonformen des nördlichen Mitteleuropas, Lieferung IV/V. Weitere azonotrilete (apiculate, murornate), zonotrilete, monolete und alete Sporenformen: Jena, VEB Gustav Fischer Verlag, 232 p.

Krutzsch, W., 1969, Taxonomie syncolp(or)ater und morphologish Benachbarter Pollengattungen und -arten (Sporae dispersae) aus der Oberkeriede und dem Tertiär, Teil I; Syncolp(or)ate und syncolp(or)atoid: Pollen et Spores, v. 11, p. 397–424.

Krutzsch, W., 1970, Taxonomie syncolp(or)ater und morphologish benachbarter Pollengattungen und -arten (Sporae dispersae) aus der Oberkeriede und dem Tertiär, Teil II; Aquilapolles (= Triprojectacites): Pollen et Spores, v. 12, p. 103–122.

Leffingwell, H.A., 1970, Palynology of the Lance (Late Cretaceous) and Fort Union (Paleocene) Formations of the type Lance area, Wyoming, *in* Kosanke, R.M., and Cross, A.T., eds., Symposium on palynology of the Late Cretaceous and early Tertiary: Geological Society of America Special Paper 127, p. 1–64.

Leffingwell, H.A., Larson, D.A., and Valencia, M.J., 1970, A study of the fossil pollen *Wodehouseia spinata*. 1. Ultrastructure and comparisons to selected modern taxa. 2. Optical microscope recognition of foot layers in differentially stained fossil pollen: Bulletin of Canadian Petroleum Geology, v. 18, p. 238–262.

Martin, A.R.H., 1973, Reappraisal of some palynomorphs of supposed proteaceous affinity. 1. The genus *Beaupreaidites* Cookson ex Couper and the species *Proteacidites hakeoides* Couper, *in* Glover, J.E., and Playford, G., eds., Mesozoic and Cainozoic palynology: Essay in honor of Isabel Cookson: Geological Society of Australia Special Publication 4, p. 73–78.

Martin, A.R.H., and Harris, W.K., 1974, Reappraisal of some palynomorphs of supposed proteaceous affinity. 2. The genus *Proteacidites* Cookson ex Couper: Grana, v. 14, p. 108–113.

McIntyre, D.J., 1974, Palynology of an Upper Cretaceous section, Horton River, District of Mackenzie, N.W.T.: Geological Survey of Canada Paper 74–14, 8 p.

McIver, E.E., Sweet, A.R., and Basinger, J.F., 1991, Sixty-five-million-year-old flowers bearing pollen of the extinct triprojectate complex: A Cretaceous-Tertiary boundary survivor: Review of Palaeobotany and Palynology, v. 70, p. 77–88.

McLeroy, C.A.C., 1971, Late Cretaceous angiosperm pollen from the western San Joaquin Valley, California [abs.]: Geoscience and Man, v. 3, p. 95–96.

Memon, H.R., 1983, Exclusion of North American *Proteacidites* species from the Southern Hemisphere Proteaceae: Sind University Research Journal, Science Series, v. 15, p. 73–84.

Miki, A., 1972, Palynological study of the Kuji Group in northeastern Honshu, Japan: Journal of the Faculty of Science, Hokkaido University, ser. 4, Geology and Mineralogy, v. 15, p. 513–603.

Muller, J., 1968, Palynology of the Pedawan and Plateau Sandstone Formations (Cretaceous-Eocene) in Sarawak, Malaysia: Micropaleontology, v. 14, p. 1–37.

Muller, J., 1981, Fossil pollen records of extant angiosperms: Botanical Review, v. 47, p. 1–142.

Murphy, E.C., Nichols, D.J., Hoganson, J.W., and Forsman, N.F., 1995, The Cretaceous-Tertiary boundary in south-central North Dakota: North Dakota Geological Survey Report of Investigations, v. 98, 73 p.

Nichols, D.J., 1973, North American and European species of *Momipites* ("*Engelhardtia*") and related genera: Geoscience and Man, v. 7, p. 103–117.

Nichols, D.J., 1994, A revised palynostratigraphic zonation of the nonmarine Upper Cretaceous, Rocky Mountain region, United States, *in* Caputo, M.V., et al., eds., Mesozoic systems of the Rocky Mountain region, USA: SEPM (Society for Sedimentary Geology), Rocky Mountain Section, p. 503–521.

Nichols, D.J., 1997, Palynology and ages of some Upper Cretaceous formations in the Markagunt and northwestern Kaiparowits Plateaus, southwestern Utah, *in* Maldonado, F., and Nealey, L.D., eds., Geologic studies in the Basin and Range-Colorado Plateau transition in southeastern Nevada, southwestern Utah, and northwestern Arizona, 1995: U.S. Geological Survey Bulletin 2153, p. 79–105.

Nichols, D.J., and Brown, J.L., 1992, Palynostratigraphy of the Tullock Member (lower Paleocene) of the Fort Union Formation in the Powder River Basin, Montana and Wyoming: U.S. Geological Survey Bulletin 1917-F, 35 p.

Nichols, D.J., and Jacobson, S.R., 1982, Palynostratigraphic framework for the Cretaceous (Albian-Maestrichtian) of the overthrust belt of Utah and Wyoming: Palynology, v. 6, p. 119–147.

Nichols, D.J., and Ott, H.L., 1978, Biostratigraphy and evolution of the *Momipites-Caryapollenites* lineage in the early Tertiary in the Wind River basin, Wyoming: Palynology, v. 2, p. 93–112.

Nichols, D.J., and Sweet, A.R., 1993, Biostratigraphy of Upper Cretaceous nonmarine palynofloras in a north-south transect of the Western Interior Basin, *in* Caldwell, W.G.E., and Kauffman, E.G., eds., Evolution of the Western Interior Basin: Geological Society of Canada Special Paper 39, p. 539–584.

Nichols, D.J., Jacobson, S.R., and Tschudy, R.H., 1982, Cretaceous palynomorph biozones for the central and northern Rocky Mountain region of the United States, *in* Powers, R.B., ed., Geologic studies of the Cordilleran thrust belt, volume 2,: Denver, Colorado, Rocky Mountain Association of Geologists, p. 721–733.

Nichols, D.J., Jarzen, D.M., Orth, C.J., and Oliver, P.Q., 1986, Palynological and iridium anomalies at Cretaceous-Tertiary boundary, south-central Saskatchewan: Science, v. 231, p. 714–717.

Nichols, D.J., Fleming, R.F., and Frederiksen, N.O., 1990, Palynological evidence of effects of the terminal Cretaceous event in western North America, *in* Kauffman, E.G., and Walliser, O.H., eds., Extinction events in Earth history: Lecture Notes in Earth Sciences, Volume 30: New York, Springer-Verlag, p. 351–364.

Nichols, D.J., Brown, J.L., Attrep, M., Jr., and Orth, C.J., 1992a, A new Cretaceous-Tertiary boundary locality in the western Powder River basin, Wyoming: Biological and geological implications: Cretaceous Research, v. 13, p. 3–30.

Nichols, D.J., Hickey, L.J., McWeeney, L.J., and Wolfe, J.A., 1992b, Plants at the K-T boundary: Discussion and reply: Nature, v. 356, no. 6367, p. 295–296.

Nichols, D.J., Murphy, E.C., Johnson, K.R., and Betterton, W.J., 2000, A second K-T boundary locality in North Dakota verified by palynostratigraphy and shocked quartz: Geological Society of America Abstracts and Programs, v. 32, no. 7, p. 130.

Norris, G., 1986, Systematic and stratigraphic palynology of Eocene to Pliocene strata in the Imperial Nuktak C-22 well, Mackenzie Delta region, District of Mackenzie, N.W.T.: Geological Survey of Canada Bulletin 340, 89 p.

North American Commission on Stratigraphic Nomenclature, 1983, North American stratigraphic code: American Association of Petroleum Geologists Bulletin, v. 67, p. 841–875.

Norton, N.J., 1965, Three new species of *Aquilapollenites* from the Hell Creek Formation, Garfield County, Montana: Pollen et Spores, v. 7, p. 135–143.

Norton, N.J., and Hall, J.W., 1967, Guide sporomorphae in the Upper Cretaceous-lower Tertiary of eastern Montana (U.S.A.): Review of Palaeobotany and Palynology, v. 2, p. 99–110.

Norton, N.J., and Hall, J.W., 1969, Palynology of the Upper Cretaceous and lower Tertiary in the type locality of the Hell Creek Formation, Montana, U.S.A.: Palaeontographica, Abteilung B: Palaophytologie, v. 125, p. 1–64.

Oltz, D.F., 1969, Numerical analyses of palynological data from Cretaceous and early Tertiary sediments in east central Montana: Palaeontographica, Abteilung B: Palaophytologie, v. 128, p. 90–166, pls. 39–42.

Oltz, D.F., 1971, Cluster analyses of Late Cretaceous-early Tertiary pollen and spore data: Micropaleontology, v. 17, p. 221–232.

Pierce, R.L., 1961, Lower Upper Cretaceous plant microfossils from Minnesota: Minnesota Geological Survey Bulletin 42, 86 p.

Playford, G., and Dettmann, M.E., 1965, Rhaeto-Liassic plant microfossils from the Leigh Creek coal measures, South Australia: Senckenbergiana Lethaea, v. 46, p. 127–181.

Pocknall, D.T., and Nichols, D.J., 1996, Palynology of coal zones of the Tongue River Member (upper Paleocene) of the Fort Union Formation, Powder River Basin, Montana and Wyoming: American Association of Stratigraphic Palynologists Contributions, ser. 32, 57 p.

Pocock, S.A.J., 1964, Pollen and spores of the Chlamydospermidae and Schizaeaceae from upper Mannville strata of the Saskatoon area of Saskatchewan: Grana Palynologica, v. 5, p. 129–209.

Potonié, R., 1931, Pollenformen aus tertiären Braunkohlen: Jahrbuch Preussischen Geologisches Landesanstalt, v. 52, p. 1–7.

Potonié, R., 1951, Revision stratigraphisch wichtiger Sporomorphen des mitteleuropäischen Tertiärs: Palaeontographica, Abteilung B: Palaophytologie, v. 91, p. 131–151.

Potonié, R., 1958, Synopsis der Gattungen der Sporae dispersae, 2 Teil. Sporites (Nachtrage), Saccites, Aletes, Praecolpates, Polyplicates, Monocolpates: Beihefte zum Geolgischen Jahrbuch, v. 31, 114 p.

Potonié, R., Thomson, P.W., and Thiergart, F., 1950, Zur Nomenklatur und Klassification der neogenen Sporomorphae (Pollen und Sporen): Geologisches Jahrbuch, v. 65, p. 35–70.

Punt, W., Blackmore, S., Nilsson, S., and Le Thomas, A., 1994, Glossary of pollen and spore terminology: Laboratory of Palaeobotany and Palynology Contributions, ser. 1, 71 p.

Radforth, N.W., and Rouse, G.E., 1954, The classification of recently discovered Cretaceous plant microfossils of potential importance to the stratigraphy of western Canadian coals: Canadian Journal of Botany, v. 32, p. 187–201.

Robertson, E.B., 1973, *Marsypiletes cretacea* gen. and sp. nov. from the Hell Creek Formation (Maestrichtian) Montana, U.S.A.: Pollen et Spores, v. 15, p. 511–514.

Robertson, E.B., and Elsik, W.C., 1978, *Marsypiletes* Robertson 1973, emend., an index angiosperm pollen from the Maestrichtian of western North America: Palaeontographica, Abteilung B: Palaophytologie, v. 165, p. 85–88.

Ross, N.E., 1949, On a Cretaceous pollen- and spore-bearing clay deposit of Scania: Uppsala University Geological Institute Bulletin, v. 34, p. 25–43.

Rouse, G.E., 1957, The application of a new nomenclatural approach to Upper Cretaceous plant microfossils from western Canada: Canadian Journal of Botany, v. 35, p. 349–375.

Rouse, G.E., 1962, Plant microfossils from the Burrard Formation of western British Columbia: Micropaleontology, v. 8, p. 187–218.

Salvador, A., ed., 1994, International stratigraphic guide: A guide to stratigraphic classification, terminology, and procedure (second edition): Boulder, Colorado, International Union of Geological Sciences and Geological Society of America, 214 p.

Samoilovich, S.R., and Mchedlishvili, N.D., eds., 1961, Pollen and spores of western Siberia, Jurassic to Paleocene: Moscow, Trudy All-Union Petroleum Scientific-Research and Geological Exploration Institute (VNIGRI), v. 177, 657 p. (in Russian).

Simpson, J.B., 1961, The Tertiary pollen-flora of Mull and Ardnamurchan: Royal Society of Edinburgh Transactions, Earth Sciences, v. 64, p. 421–468.

Singh, C., 1964, Microflora of the Lower Cretaceous Mannville Group, east-central Alberta: Research Council of Alberta Bulletin 15, 239 p.

Skarby, A., 1964, Revision of *Gleicheniidites senonicus* Ross: Stockholm Contributions to Geology, v. 11, p. 59–77.

Srivastava, S.K., 1966, Upper Cretaceous microflora (Maestrichtian) from Scollard, Alberta, Canada: Pollen et Spores, v. 8, p. 497–552.

Srivastava, S.K., 1968, Reticulate species of *Aquilapollenites* and emendation of the genus *Mancicorpus* Mchedlishvili: Pollen et Spores, v. 10, p. 665–699.

Srivastava, S.K., 1969, New spinulose *Aquilapollenites* spp. from the Edmonton Formation (Maestrichtian), Alberta, Canada: Canadian Journal of Earth Sciences, v. 6, p. 133–144.

Srivastava, S.K., 1970, Pollen biostratigraphy and paleoecology of the Edmonton Formation (Maestrichtian), Alberta, Canada: Palaeogeography, Palaeoclimatology, Palaeoecology, v. 7, p. 221–276.

Srivastava, S.K., 1971a, Systematic revision of the genus *Styx* Norton et Hall, 1967: Review of Palaeobotany and Palynology, v. 11, p. 297–309.

Srivastava, S.K., 1971b, Monolete spores from the Edmonton Formation (Maastrichtian), Alberta, Canada: Review of Palaeobotany and Palynology, v. 11, p. 251–265.

Srivastava, S.K., 1972a, Pollen genus *Erdtmanipollis* Krutzsch 1962: Pollen et Spores, v. 14, p. 309–322.

Srivastava, S.K., 1972b, Some spores and pollen from the Paleocene Oak Hill Member of the Naheola Formation, Alabama, U.S.A.: Review of Palaeobotany and Palynology, v. 14, p. 217–285.

Srivastava, S.K., 1975, Microspores from the Fredericksburg Group (Albian) of the southern United States: Paleobiologie Continentale, v. 6, no. 2, p. 1–119.

Srivastava, S.K., 1976, The fossil pollen genus *Classopollis:* Lethaia, v. 9, p. 437–457.

Srivastava, S.K., 1978a, The Cretaceous megaspore genus *Ghoshispora:* Palaeontographica, Abteilung B: Palaophytologie, v. 1167, p. 175–184.

Srivastava, S.K., 1978b, *Leptopecopites,* a new genus from the Canadian Maastrichtian: Pollen et Spores, v. 20, p. 555–568.

Srivastava, S.K., 1981, Fossil pollen genus *Kurtzipites* Anderson: Journal of Paleontology, v. 55, p. 868–879.

Stanley, E.A., 1961a, A new sporomorph genus from South Dakota: Pollen et Spores, v. 3, p. 155–162.

Stanley, E.A., 1961b, The fossil pollen genus *Aquilapollenites:* Pollen et Spores, v. 3, p. 329–352.

Stanley, E.A., 1965, Upper Cretaceous and Paleocene plant microfossils and Paleocene dinoflagellates and hystrichosphaerids from northwestern South Dakota: American Paleontology Bulletins, v. 49, p. 177–384.

Stanley, E.A., 1970, The stratigraphical, biogeographical, paleoecological and evolutionary significance of the fossil pollen group Triprojectacites: Georgia Academy of Science Bulletin, v. 28, p. 1–44.

Stone, J.F., 1973, Palynology of the Almond Formation (Upper Cretaceous), Rock Springs uplift, Wyoming: American Paleontology Bulletins, v. 64, p. 1–135.

Stover, L.E., 1962, *Taurocusporites,* a new trilete spore genus from the Lower Cretaceous of Maryland: Micropaleontology, v. 8, p. 55–58.

Sweet, A.R., 1986, The Cretaceous-Tertiary boundary in the central Alberta Foothills. 2. Miospore and pollen taxonomy: Canadian Journal of Earth Sciences, v. 23, p. 1375–1388.

Sweet, A.R., Braman, D.R., and Lerbekmo, J.F., 1999, Sequential palynological changes across the composite Cretaceous-Tertiary (K-T) boundary claystone and contiguous strata, western Canada and Montana, U.S.A.: Canadian Journal of Earth Sciences, v. 36, p. 743–768.

Takahashi, K., 1964, Sporen und pollen der oberkretazeischen Hakobuchi-Schichtengruppe, Hokkaido: Memoirs of the Faculty of Science, Kyushu University, ser. D, Geology, v. 14, p. 159–271.

Takahashi, K., and Shimono, H., 1982, Maestrichtian microflora of the Miyadani-gawa Formation in the Hida District, central Japan: Bulletin of the Faculty of Liberal Arts, Nagasaki University, Natural Science, v. 22, p. 11–188.

Thomson, P.W., and Pflug, H., 1953, Pollen und Sporen des mitteleuropäischen Tertiärs: Palaeontographica, Abteilung B: Palaophytologie, v. 94, p. 1–138.

Traverse, A., 1988, Paleopalynology: Boston, Unwin Hyman, 600 p.

Tryon, R.M., and Tryon, A.F., 1982, Ferns and allied plants: New York, Springer-Verlag, 857 p.

Tschudy, B.D., 1971, Two new fossil pollen genera from upper Campanian (Cretaceous) rocks of Alaska: U.S. Geological Survey Professional Paper 750-B, p. 53–61.

Tschudy, B.D., 1973, Palynology of the upper Campanian (Cretaceous) Judith River Formation, north-central Montana: U.S. Geological Survey Professional Paper 770, 42 p.

Tschudy, B.D., and Leopold, E.B., 1970, *Aquilapollenites* (Rouse) Funkhouser—Selected Rocky Mountain taxa and their stratigraphic ranges, *in* Kosanke, R.M., and Cross, A.T., eds., Symposium on palynology of the Late Cretaceous and early Tertiary: Geological Society of America Special Paper 127, p. 113–167.

Tschudy, R.H., 1970 [1971], Palynology of the Cretaceous-Tertiary boundary in the northern Rocky Mountain and Mississippi Embayment regions, *in* Kosanke, R.M., and Cross, A.T., eds., Symposium on palynology of the Late Cretaceous and early Tertiary: Geological Society of America Special Paper 127, p. 65–111.

Van Geel, B., 1976, Fossil spores of Zygnemataceae in ditches of a prehistoric settlement in Hoogkarspel (The Netherlands): Review of Palaeobotany and Palynology, v. 22, p. 337–344.

Whitehead, D.R., 1963, Pollen morphology in the Juglandaceae. 1. Pollen size and pore number variation: Arnold Arboretum Journal, v. 44, p. 101–110.

Whitehead, D.R., 1965, Pollen morphology in the Juglandaceae. 2. Survey of the family: Arnold Arboretum Journal, v. 46, p. 369–410.

Wilson, L.R., and Webster, R.M., 1946, Plant microfossils from a Fort Union coal of Montana: American Journal of Botany, v. 33, p. 271–278.

Wing, S.L., Hickey, L.J., and Swisher, C.C., 1993, Implications of an exceptional fossil flora for Late Cretaceous vegetation: Nature, v. 363, p. 342–344.

Wolfe, J.A., 1991, Palaeobotanical evidence for a June 'impact winter' at the Cretaceous-Tertiary boundary: Nature, v. 352, p. 420–423.

Yu, J., Guo, Z., and Mao, S., 1981, Cretaceous palynological assemblages from the district south of the Songhua River, *in* assemblage of Late Cretaceous sporo-pollens in the Songhua Jiang-Liao He Basin: Daqing Oil Field Development Research Institute, Professional Papers on Stratigraphy and Paleontology, p. 1–117 (in Chinese).

Manuscript Accepted by the Society September 24, 2001

Printed in the U.S.A.

Geological Society of America
Special Paper 361
2002

Palynology of the Hell Creek Formation (Upper Cretaceous, Maastrichtian) in northwestern South Dakota: Effects of paleoenvironment on the composition of palynomorph assemblages

Timothy J. Kroeger*
Center for Environmental, Earth, and Space Studies, Bemidji State University, Bemidji, Minnesota 56601, USA

ABSTRACT

Fossil palynomorphs were recovered from fluvial meander-belt and flood-basin sediments of the Maastrichtian Hell Creek Formation. Subenvironments represented within the meander-belt deposits include point-bar, abandoned channel lake and marsh, and point-bar swale. Flood-basin subenvironments include lake, marsh, and crevasse splay. Analyses of the occurrence and relative abundance of 44 palynomorph taxa or groups of taxa allow recognition of numerous taxa that have limited paleoenvironmental distribution. Detrended correspondence analysis demonstrates that the greatest paleoenvironmental distinction occurs among the lacustrine, marsh, swale, and point-bar subenvironments. An additional paleoenvironmental trend, in which taxa occur more commonly in either meander-belt or flood-basin settings, is also evident.

Although many factors affect the relative abundance of palynomorph taxa that are preserved in sediments, comparisons to palynomorph distribution in modern environments suggest that the palynomorph assemblages recovered from marsh deposits were primarily produced by the local flora of the Late Cretaceous marshes. The lake deposits contain palynomorphs that may represent mixtures of local, extra-local, and regional pollen rain. Point-bar sediments probably contain a mixture of locally produced, transported, and reworked palynomorphs. The differences in palynomorph assemblages between meander-belt and flood-basin deposits represent local paleofloral differences due to drier edaphic conditions within the meander-belt paleoenvironments.

INTRODUCTION

Paleoecologic interpretation of palynomorph assemblages is commonly based upon the ecology of modern descendant species, a strategy that is difficult to apply in the Hell Creek Formation because the botanical affinities of many of the palynomorphs are uncertain. In addition, the ecological requirements of the modern descendants may have changed considerably since the Late Cretaceous. Although the taxonomic composition of the abundant, diverse, and well-preserved

*E-mail: tjkroeger@bemidjistate.edu

Kroeger, T.J., 2002, Palynology of the Hell Creek Formation (Upper Cretaceous, Maastrichtian) in northwestern South Dakota: Effects of paleoenvironment on the composition of palynomorph assemblages, *in* Hartman, J.H., Johnson, K.R., and Nichols, D.J., eds., The Hell Creek Formation and the Cretaceous-Tertiary boundary in the northern Great Plains: An integrated continental record of the end of the Cretaceous: Boulder, Colorado, Geological Society of America Special Paper 361, p. 457–472.

palynoflora of the Hell Creek Formation is well documented (see previous works cited in the following), paleoecologic interpretations based upon palynomorphs tend to be generalized.

More specific information on the paleoecology of the palynomorph-producing plants may be uncovered by comparing the relative abundances of palynomorph taxa to the paleoenvironments in which the palynomorphs were deposited. Studies of palynomorph distribution in modern depositional environments (Muller, 1959; Rich and Spackman, 1979; Chmura, 1994) indicate that palynomorph assemblages are distinctive among different depositional environments and commonly reflect the local floral distribution. However, because of vagaries in palynomorph production, preservation, and transport, the relative abundance of palynomorphs does not represent the abundance of plant species within the producing flora.

Objective

The primary objective of this study is to identify the relationship between the relative abundances of palynomorph taxa and depositional environments within the Hell Creek Formation.

Previous studies

Several workers have previously studied the palynology of the Hell Creek Formation and its lateral equivalents. The majority of these studies were devoted to developing a palynostratigraphic framework for Upper Cretaceous and lower Tertiary strata in the northern Western Interior. More recently, attention has focused on microstratigraphy and recognition of the Cretaceous-Tertiary boundary.

Working in northwestern South Dakota, Stanley (1965) completed the first comprehensive palynologic survey of the Hell Creek and lower Fort Union Formations, followed by Norton and Hall (1967, 1969), Oltz (1969, 1971), Robertson (1975), Bergad (1974), and Tschudy (1970), who completed studies in Montana and North Dakota. Studies that addressed, in detail, the palynofloral changes across the Cretaceous-Tertiary boundary include Hotton (1988), who focused on the boundary in northeastern Montana, and Johnson et al. (1989) and Murphy et al. (1995), who examined the boundary in North Dakota.

Previously, workers have identified relationships between paleoenvironment and palynomorph assemblages. Oltz (1969, 1971) completed cluster analysis of palynomorph assemblages from the Hell Creek, Bearpaw, Fox Hills, and lower Fort Union Formations in northeastern Montana, but he did not relate the assemblages that he recognized to the environment of deposition other than in a general manner. Several additional studies of Cretaceous and Paleogene palynomorphs have incorporated multivariate analysis of the palynomorph assemblages with detailed sedimentologic and stratigraphic data to address the relationship between palynomorph assemblages and depositional environments (Farley and Dilcher, 1986; Gennett et al., 1986; Farley, 1989; Fleming, 1990; Kroeger, 1995; Kroeger and Hartman, 1997).

Setting and methods

The rocks sampled for palynomorphs are exposed in northwestern Butte County, South Dakota (Fig. 1). The outcrop area is a small, isolated butte that is ~22 m high and ~60 m in diameter. The butte is located in the NW¼SE¼ sec. 2, T.13N., R.9E, Beverly Creek 7.5 × 15 minute quadrangle. In 1981 and 1982, the first specimen of *Tyrannosaurus rex* Osborn 1905 to be discovered in South Dakota was excavated from the west side of the butte by the Museum of Geology of the South Dakota School of Mines and Technology.

Both Upper Cretaceous and Tertiary rocks are exposed within the area of the study. Most of the land surface within the area of the study consists of exposures of the Hell Creek Formation. The Upper Cretaceous Pierre and Fox Hills Formations are exposed ~6 km to the south within the valley of the South Fork of the Moreau River. Tertiary rocks, consisting of the Fort Union, Golden Valley, Slim Buttes, Chadron, Brule, and Arikaree Formations, are exposed in the Sheep Mountain and Slim Buttes area, ~20 km north of the study site. The Hell Creek–

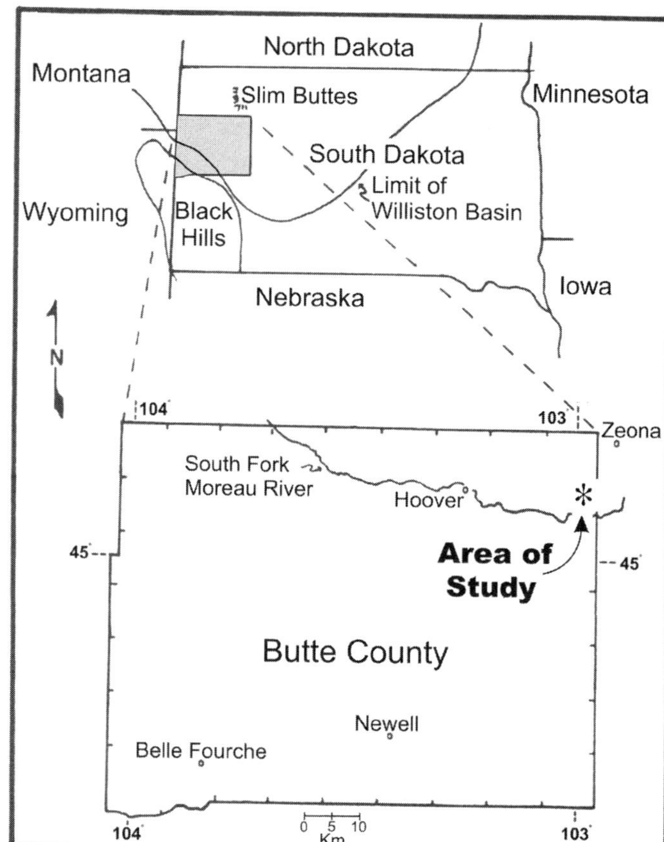

Figure 1. Location of study area within northwestern South Dakota.

Fort Union formational boundary is present in the Sheep Mountain area. The strata dip very gently northward into the Williston Basin.

Unweathered rock samples were collected for palynologic analysis during the measurement of a detailed stratigraphic section. The palynomorphs were extracted through the acid maceration process described by Doher (1980) and Traverse (1988). Palynomorphs were classified according to morphotaxa, in most cases to the species level, although some morphologically similar taxa were lumped into larger taxonomic groups. At least 200 palynomorphs were counted from each sample through nonoverlapping traverses of the microslides.

Sedimentologic interpretations were made by comparison of stratigraphic and sedimentologic features with modern facies models and descriptions of modern fluvial depositional environments (Miall, 1981; Reineck and Singh, 1980). Many lithologic units at the butte are traceable laterally throughout the limited extent of the butte. Both lateral and vertical lithologic changes were utilized in the interpretation of sedimentary environments.

The relative abundance of each palynomorph taxon or groups of similar taxa in each sample was calculated. Rare taxa were not used in the paleoenvironmental analysis; an arbitrarily chosen relative abundance threshold of 1.75% was applied before a taxon was recorded in a sample. In addition, taxa that did not occur above the threshold value in two or more samples were excluded from the analyses. Because rare taxa are excluded, the relative abundance of taxa in each sample does not total 100%. The taxonomic composition of the palynomorph assemblages was related to the depositional environment by use of pollen diagrams and multivariate analysis using detrended correspondence analysis (Hill and Gauch, 1980) using a personal computer version of DECORANA (Hill, 1979).

LOCAL STRATIGRAPHY AND PALEOENVIRONMENTS

Determining the stratigraphic position of the study area is difficult because of the limited exposures. On the basis of the position of the Fox Hills–Hell Creek formational boundary, which is exposed a few kilometers south in the valley of the South Fork of the Moreau River, and the very gentle northward dip of the strata into the Williston Basin, the sample locality is interpreted to be within the lower Hell Creek Formation.

Strata exposed within the study area are divided into lithologic units based upon lithologic character and the sedimentary and biogenic structures present (Fig. 2). The lower half of the butte, lithologic units 1–10, is predominantly claystone and siltstone, whereas the upper half, lithologic units 11–17, is dominated by sandstone.

Within the regressive depositional sequence represented by the Pierre, Fox Hills, and Hell Creek Formations, the strata in this area appear to have been deposited landward of the limit of tidal influence on an upper delta plain or lower alluvial plain. The fluvial depositional model that has been applied to the Hell Creek Formation by previous workers (Frye, 1969; Butler, 1980; Talbot, 1985) allows recognition of both meander-belt and flood-basin facies. Lithologic units 1–10 represent flood-basin depositional environments, and units 11–17 represent a meander-belt setting.

Flood-basin paleoenvironments

Flood basins are the low, poorly drained areas adjacent to active or abandoned stream channels or meander belts. Most undisturbed flood basins are densely vegetated with swamps and marshes, but lakes may also be present. Predominantly fine grained clastic sediments are carried into the flood basins during floods, and autochthonous organic sediments may also accumulate (Reineck and Singh, 1980). Within the study area, flood-basin paleoenvironments include lakes, marshes, and crevasse splays.

Flood-basin lake deposits

In modern delta-plain and alluvial-plain settings, flood-basin lakes commonly form when the rate of sediment compaction and subsidence exceeds the rate of deposition in the flood basin. Generally, the growth of the lakes occurs at the expense of neighboring marshes and swamps, continuing until a stream is diverted into the lake. Coarser grained sediments transported into the lake by the stream accumulate to form lake deltas or crevasse splays (Coleman and Prior, 1982).

The flood-basin lake deposits present in lithologic units 1, 6, 7, and 10 are characterized by laminated claystones that do not contain abundant root traces. Root traces that are present originate within overlying lithologic units. Highly fragmented plant remains are common, but no invertebrate fossils were found. The presence of marsh deposits directly below the lake deposits within lithologic units 6, 7, and 10 and the position of the sandy crevasse-splay deposits above lake deposits in units 1 and 6 fit the depositional scenario cited here. The thickness of the splay deposits indicates that the lakes were quite shallow, probably only a meter or two deep.

Flood-basin marshes

Fine-grained, flood-basin strata that contain ubiquitous and abundant root traces (lithologic units 3, 5, upper part of 6, 8, and 9) are interpreted to represent flood-basin marsh paleoenvironments. In general, these marsh deposits contain more silt and relatively less organic material than is present within the flood-basin lake deposits. Marsh deposits situated directly above sandy crevasse-splay deposits within lithologic units 3 and 6 suggest that progradation of the crevasse splay into the lake decreased the water depth enough for the marshes to become established. No large, tree-sized root traces were noted in

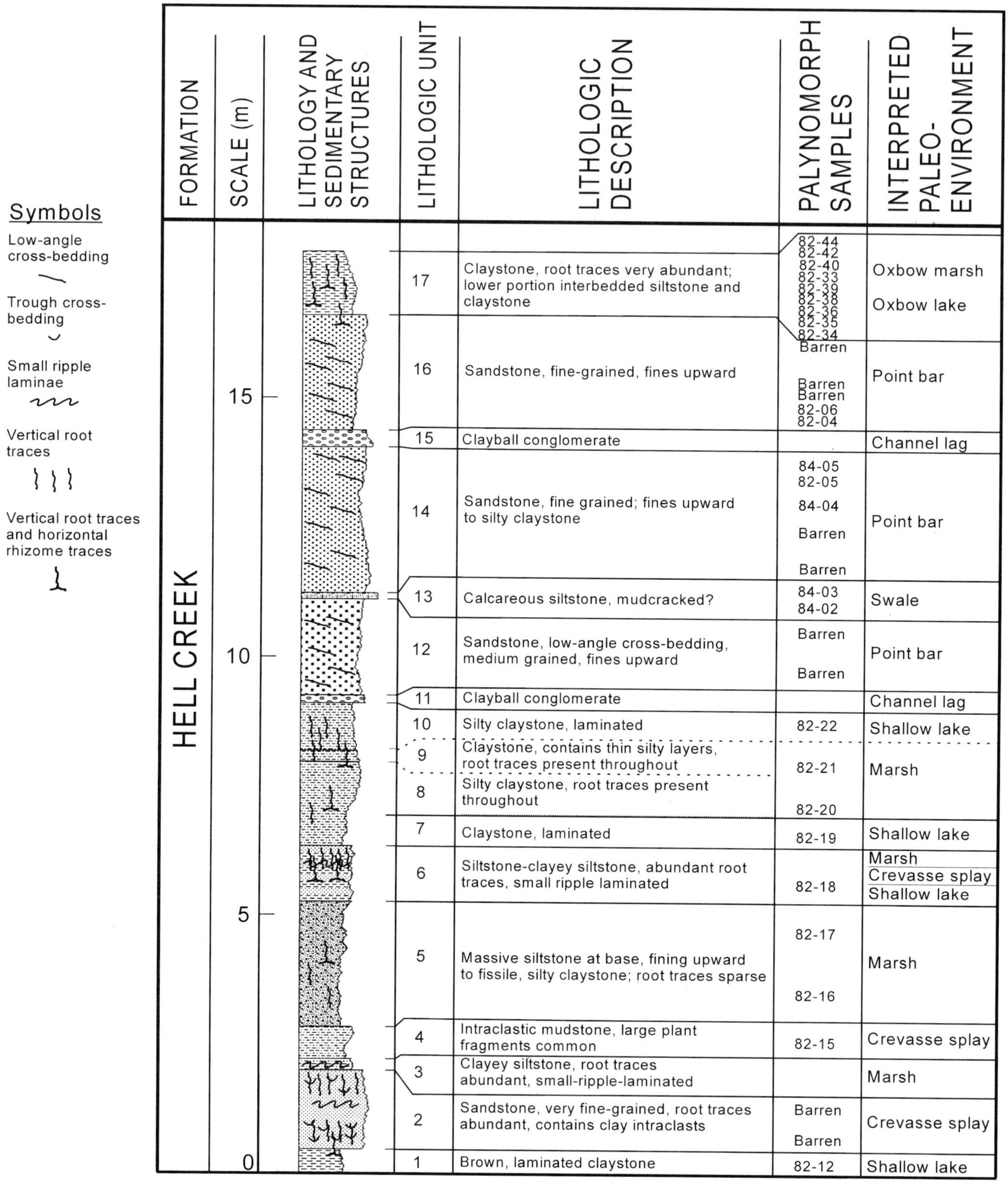

Figure 2. Stratigraphic section of study site. Stratigraphic position and interpreted paleoenvironment of palynomorph sample sites are shown.

these marsh deposits, suggesting that flood-basin forests were not present.

Crevasse-splay deposits

Crevasse splays form when a breach in the natural levee system during flood stage allows river channel sediments to be deposited within the flood basin (Miall, 1981). In this study, relatively thin, sandy or silty beds that are encased within finer grained flood-basin deposits are interpreted as crevasse-splay deposits. The splay deposits are typically very fine grained sand to coarse silt in grain size. Silty splay deposits are characterized by small ripple laminations, whereas the sandy splay deposits tend to have trough cross-bedding.

Meander-belt paleoenvironments

Sediments within fluvial meander-belt environments are deposited during the process of lateral migration of river channels and through vertical accretion in abandoned channels, swales, and upper point bars (Reineck and Singh, 1980).

Channel, point-bar, and swale deposits

Thick, sandstone-rich beds that fine upward in grain size from conglomerate at the base to silty claystone near the top are interpreted to have been deposited in fluvial channels and on fluvial point bars. Low-angle cross-bedding is the dominant inorganic sedimentary structure; root traces become common near the top of the point-bar deposit. A thin bed of calcareous mudstone (lithologic unit 13) that is interbedded within the point-bar deposit is interpreted to represent deposition within a point-bar swale. This type of fining-upward depositional sequence is consistent with the depositional model for the point-bar component of meandering river deposits proposed by Walker and Cant (1984).

The conglomerate at the base of each bed consists of clay-ball clasts to 1.5 cm in diameter. These clasts probably formed when a cutbank was undercut by the river, causing older, clay-rich river-bank deposits to collapse into the channel. The clay-balls were likely transported as bed load during maximum flood stage before they accumulated as a lag deposit at the bottom of the river channel (Reineck and Singh, 1980).

At least two episodes of point-bar deposition occurred, each resulting in a basal clayball conglomerate overlain by sandstone. Most likely, lateral channel migration within the meander belt resulted in partial erosion of an older point-bar deposit and subsequent deposition of a second channel and/or point-bar sequence above the remnant of the older point-bar deposit. The partially articulated specimen of *Tyrannosaurus rex* was recovered from the point-bar sandstone deposits, suggesting that the carcass of the dinosaur had been transported some distance downstream before deposition and subsequent burial.

Abandoned channel deposits

The point-bar deposits are overlain, in sharp contact, by a relatively thick, clay-rich sequence of beds. The basal portion of these beds consists of interbedded siltstone and claystone. Some active current during deposition of the siltstone layers is indicated by small ripple laminae in some of the siltstone beds; other siltstone layers were originally parallel laminated, although they were contorted after deposition. Root traces are sparse in the basal portion, and the traces that are present originate from the overlying layer.

The upper half of the clay-rich beds contains no observable primary bedding structures. Vertical root traces are abundant; most are <5 mm in diameter although a tree-sized root trace was observed.

The clay-rich sequence is interpreted to have been deposited as the result of a sharp decrease in sediment transport energy after the river channel was abandoned by a channel cutoff or by avulsion of the river system to create a new meander belt. The lower, siltier part of the abandoned channel deposit is interpreted to represent a relatively open, abandoned-channel lake paleoenvironment that received some current flow during flood stage. The intensity of rooting in the upper portion of the deposit suggests that vegetation encroached on the open lake environment as the lake filled with sediment, leading to the development of a marsh. The cast of the tree root indicates that eventually the marsh was succeeded by a swamp or lowland forest.

PALYNOLOGICAL RESULTS

I analyzed 34 samples for palynomorph content (Figs. 3 and 4). Of these, 24 contained palynomorphs in sufficient abundance to be included in the paleoenvironmental analysis. Preservation of palynomorphs ranges from excellent to poor. Generally, the finer grained lithologies tend to contain better preserved palynomorphs than the sandier deposits. Overall, 190 palynomorph taxa or groups of morphologically similar taxa were documented. Of these, 44 taxa are present in sufficient abundance to be used in the paleoenvironmental analysis.

Palynomorph taxa in paleoenvironmental analysis

The taxa or groups of taxa used in the paleoenvironmental analysis are listed in the following. In this listing, an abbreviated name, contained in brackets, refers to the abbreviations that are used in the scattergram of the ordination of species (Figs. 5 and 6). Representatives of each taxon or group are illustrated in Figure 7.

Genus *Psiloschizosporis* Jain 1968

Psiloschizosporis spp. (Fig. 7a) [Psiloschiz].

Comments: Several studies (van der Wiel, 1982; van Geel

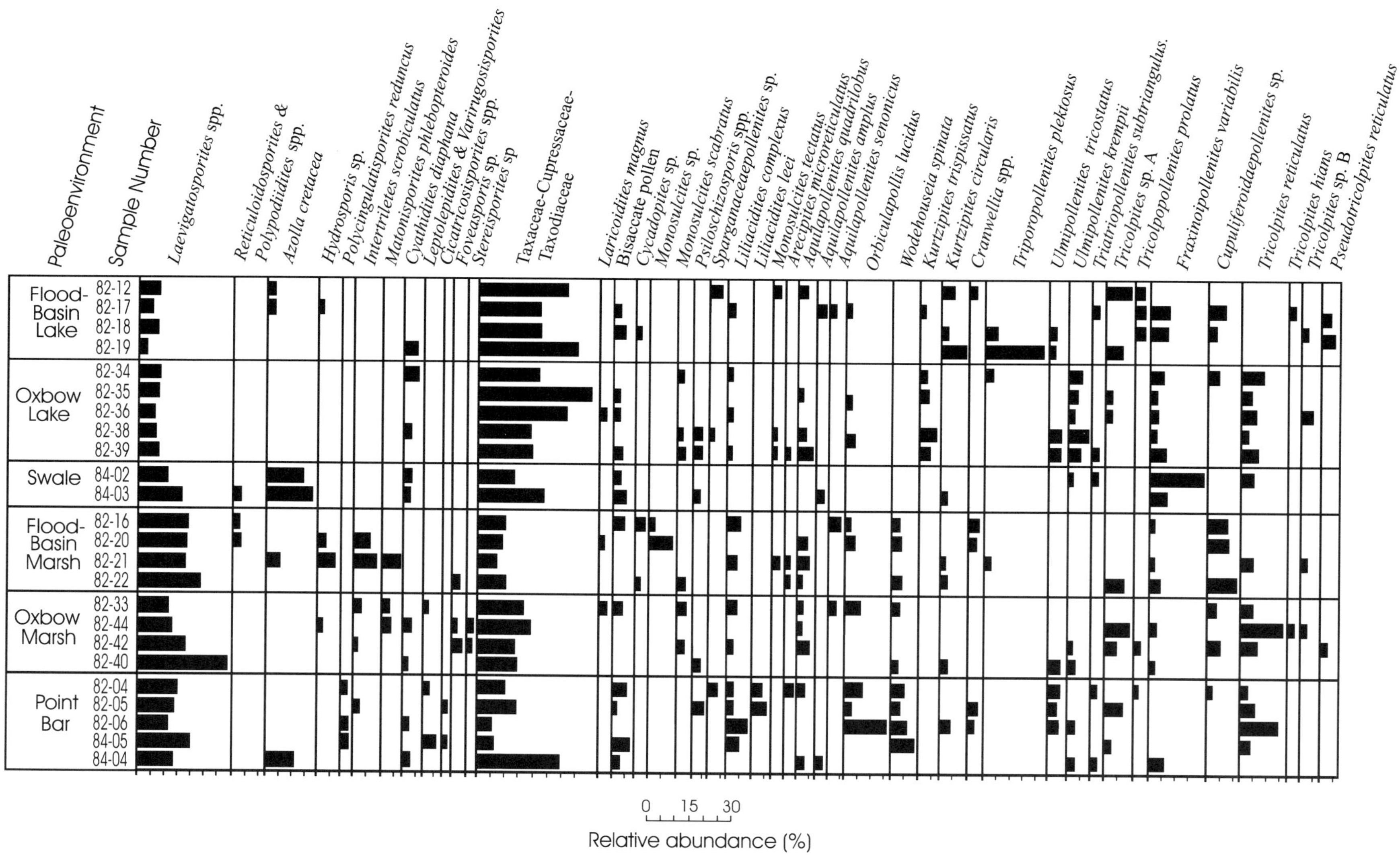

Figure 3. Relative abundance of palynomorph taxa within 24 samples studied. Samples are arranged according to paleoenvironment as interpreted from sedimentologic evidence. Rows represent occurrences of various taxa within individual sample; columns represent relative abundance of stated taxa within samples. Only occurrences in which taxon composes 1.75% or greater of palynomorph assemblage are shown (thus totals do not equal 100%).

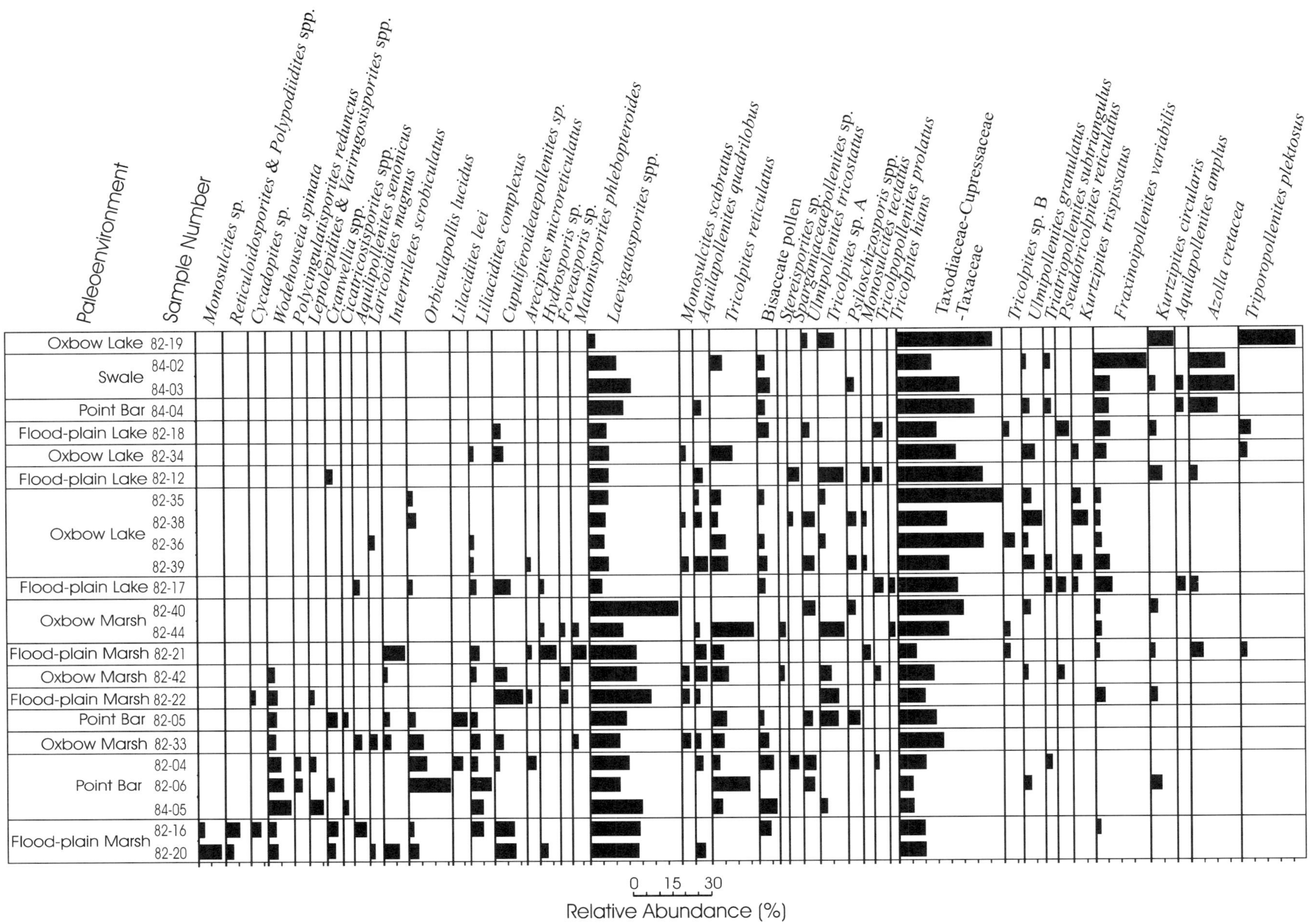

Figure 4. Relative abundance of palynomorph taxa within 24 samples studied. Samples are arranged (from top to bottom) by increasing first axis scores of detrended correspondence analysis (DCA) of samples. Taxa are arranged (from left to right) by decreasing first axis scores of DCA of species. In this arrangement, samples having greatest similarity in palynomorph assemblage are plotted in adjacent rows and species that have most similar occurrences are plotted in adjacent columns. Only occurrences in which taxon composes 1.75% or greater of palynomorph assemblage are shown (thus totals do not equal 100%).

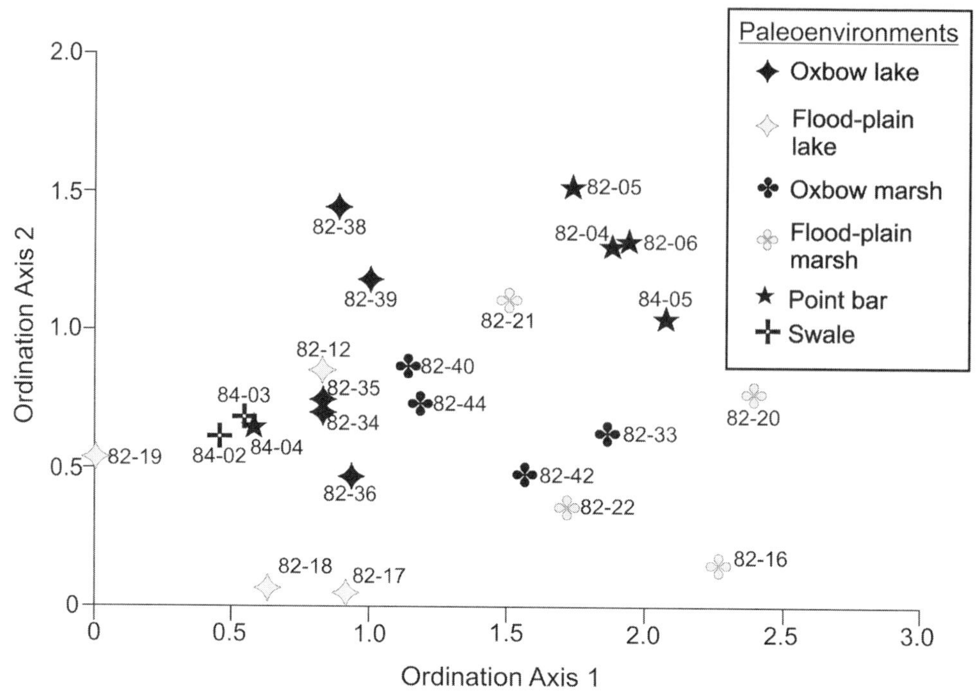

Figure 5. Scatter plot of scores of first two ordination axes produced by detrended correspondence analysis ordination of samples. Units are average standard deviation of species turnover. Numbers at data points are sample numbers.

et al., 1983, 1989; Rich et al., 1982) have associated *Psiloschizosporis*-type spores with the zygospores or aplanospores produced by *Spirogyra* (Zygnemataceae, Chlorophyta). Rich et al. (1982) found this type of spore to be an indicator of deposition in freshwater marshes in Quaternary peat-forming environments. Van Geel et al. (1989) considered similar Quaternary spores to indicate temporary, shallow, open-water environments.

Genus *Stereisporites* Pflug in Thomson and Pflug 1953

Stereisporites sp. (Fig. 7b) [Stereisp].

Genus *Laevigatosporites* Ibrahim 1933 emend. Schopf, Wilson, and Bentall 1944

Laevigatosporites spp. (Fig. 7c) [Laevig].

Genus *Reticuloidosporites* Pflug in Thomson and Pflug 1953

Reticuloidosporites pseudomurii Elsik 1968 (Fig. 7d) [RetPol].

Comments. Spores assignable to *Polypodiidites* Ross 1949 were combined with this species for the paleoenvironmental analysis.

Genus *Azolla* Lamark 1783

Azolla cretacea Stanley 1965 (Fig. 7l) [Azcret].

Comments. These microspore massulae are generally accepted as having been produced by the heterosporous, floating, aquatic fern *Azolla*. *Azolla* megaspores were also noted during this study, but they are rare. Modern *Azolla* thrives in shallow, acidic, fresh water, typically within the tropical or subtropical regions (Jarzen, 1978).

Genus *Hydrosporis* Krutzsch 1962

Hydrosporis sp. (not illustrated) [Hydrosp].

Comments. Spores assignable to *Hydrosporis* are thought to represent isolated microspores of *Azolla*, and the spores are comparable to the microspores that are observable within some *Azolla cretacea* massulae. However, all small, unsculptured, trilete spores having a circular outline are included within *Hydrosporis* regardless of their botanical affinities.

Genus *Polycingulatisporites* Simoncsics and Kedves 1961

Polycingulatisporites reduncus (Bolkhovitina) Playford and Dettmann 1965 (Fig. 7i) [Polycing].

Genus *Intertriletes* Anderson 1960

Intertriletes scrobiculatus Anderson 1960 (Fig. 7f) [Intertri].

Genus *Matonisporites* Couper 1958

Matonisporites phlebopteroides Couper 1958 (Fig. 7e) [Matonisp].

Genus *Cyathidites* Couper 1953

Cyathidites diaphana (Wilson and Webster) Nichols and Brown 1992 (Fig. 7g) [Cyathid].

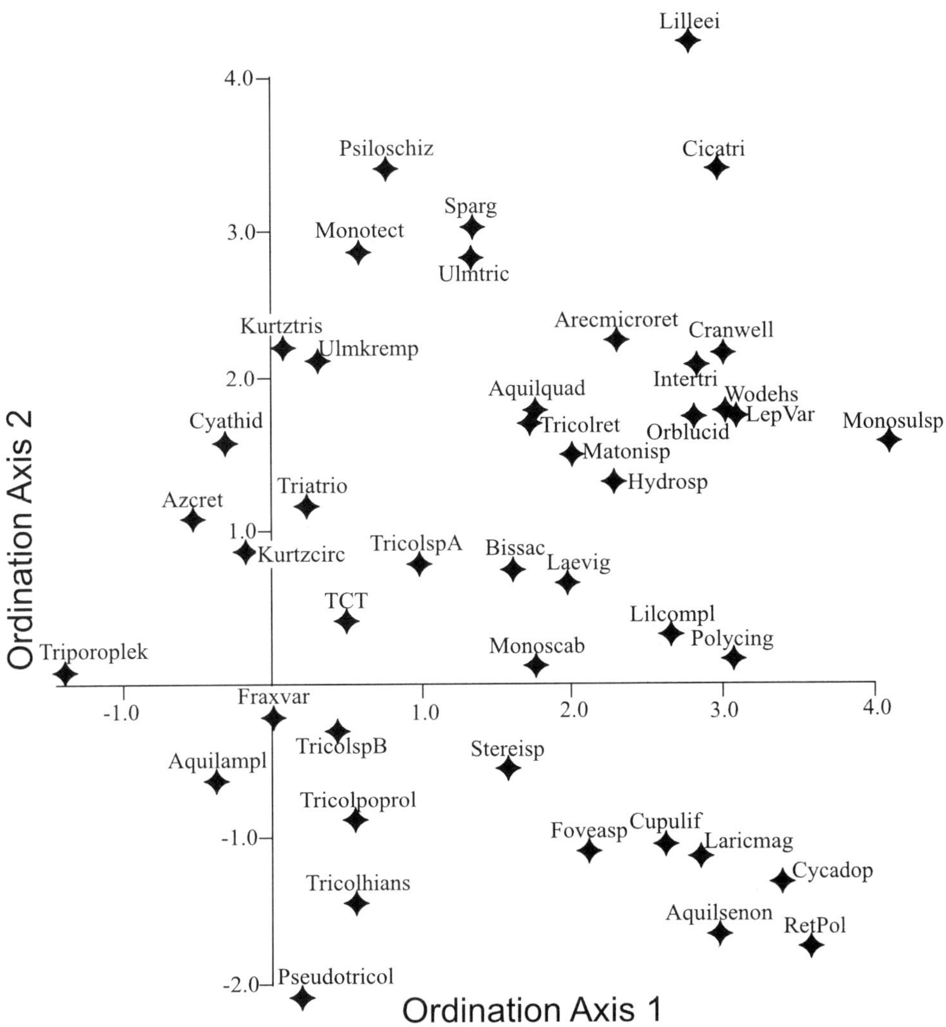

Figure 6. Scatter plot of scores of first two ordination axes produced by detrended correspondence analysis of species. Units are average standard deviation of species turnover. Refer to list of palynomorph taxa in Figure 7 for explanation of abbreviated taxon names.

Genus *Leptolepidites* Couper 1953 emend. Norris 1968

Leptolepidites sp. (not illustrated).

Genus *Varirugosisporites* Döring 1965

Varirugosisporites sp. (Fig. 7j) [LepVar].

Comments. Specimens assignable to *Leptolepidites* and *Varirugosisporites* were combined for the paleoenvironmental analysis.

Genus *Cicatricosisporites* Potonié and Gelletich 1933

Cicatricosisporites spp. (Fig. 7k) [Cicatri].

Comments. Most specimens observed are assignable to *C. dorogensis* Potonié and Gelletich 1933.

Genus *Foveasporis* Krutzsch 1959

Foveasporis sp. (Fig. 7h) [Foveasp].

Genus *Laricoidites* Potonié, Thomson, and Thiergart ex Potonié 1958

Laricoidites magnus (Potonié) Potonié, Thomson, and Thiergart 1950 (Fig. 7o) [Laricmag].

Genus *Cycadopites* Wilson and Webster 1946

Cycadopites sp. (Fig. 7q) [Cycadop].

Pollen of the Taxaceae, Cupressaceae, and Taxodiaceae

Comments. Pollen of these three conifer families (Taxaceae, Cupressaceae, and Taxodiaceae [TCT]) were combined for quantitative purposes (Fig. 7m). Included in this group are inaperaturate to fissurate, scabrate to psilate pollen grains that are generally 20–35 μm in diameter. The most common species within this group is *Taxodiaceaepollenites hiatus* Potonié ex Potonié 1958.

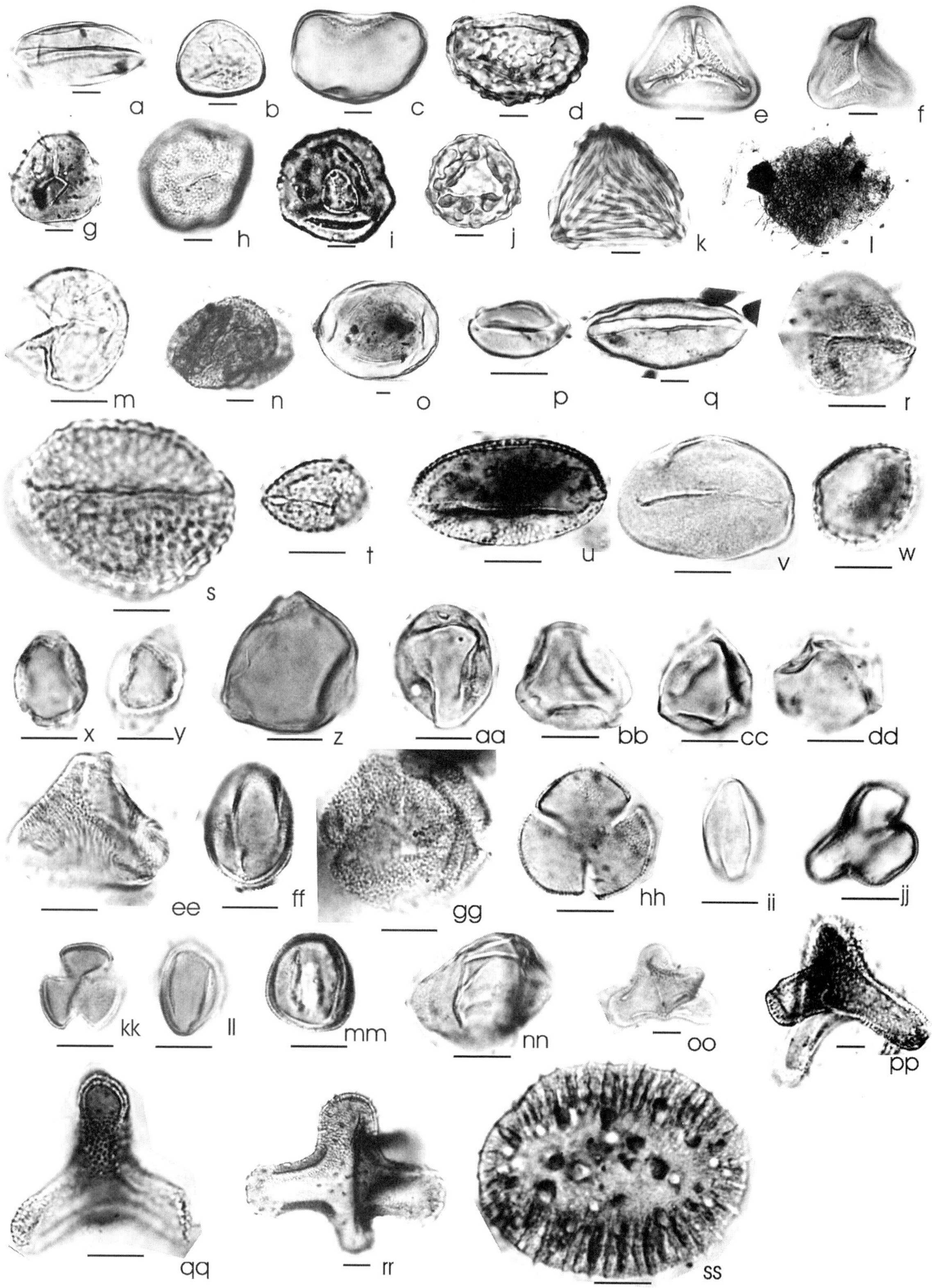

Figure 7. Illustrations of palynomorph taxa used in paleoenvironmental analysis. Scale bars = 10 μm. a, *Psiloschizosporis* sp.; b, *Stereisporites* sp.; c, *Laevigatosporites* sp.; d, *Reticuloidosporites pseudomurii*; e, *Matonisporites phlebopteroides*; f, *Intertriletes scrobiculatus*; g, *Cyathidites diaphana*; h, *Foveasporis* sp.; i, *Polycingulatisporites reduncus*; j, *Varirugosisporites* sp.; k, *Cicatricosisporites* sp.; l, *Azolla cretacea*; m, pollen of Taxodiaceae; Cupressaceae; and Taxaceae; n, Bisaccate pollen; o, *Laricoidites magnus*; p, *Monosulcites* sp.; q, *Cycadopites* sp.; r, *Monosulcites scabratus*; s, *Liliacidites complexus*; t, *Monosulcites tectatus*; u, *Liliacidites leei*; v, *Arecipites microreticulatus*; w, *Sparganaceaepollenites* sp.; x, y, *Ulmipollenites krempii*; equatorial focus and high focus; z, *Triatriopollenites subtriangulus*; aa, *Triporopollenites plektosus*; bb, *Ulmipollenites tricostatus*; cc, *Kurtzipites circularis*; dd, *Kurtzipites trispissatus*; ee, *Cranwellia* sp.; ff, *Fraxinoipollenites variabilis*; gg, *Tricolpites* sp. A; hh, *Tricolpites reticulatus*; ii, *Cupuliferoidaepollenites* sp.; jj, *Tricolpites* sp. B; kk, *Tricolpites hians*; ll, *Tricolpopollenites prolatus*; mm, *Pseudotricolpites reticulatus*; nn, *Orbiculapollis lucidus*; oo, *Aquilapollenites senonicus*; pp, *Aquilapollenites quadrilobus*; qq, *Aquilapollenites delicatus*; rr, *Aquilapollenites amplus*; ss, *Wodehouseia spinata*.

Bisaccate pollen grains [Bissac]

All bisaccate pollen (Fig. 7n), representing conifers, were combined for quantitative purposes. Most of the bisaccate grains are attributable to the Pinaceae.

Genus *Monosulcites* Cookson ex Couper 1953

Monosulcites sp. (Fig. 7p) [Monosulsp].
Monosulcites scabratus (Stanley) Stone 1973 (Fig. 7r) [Monoscab].
Monosulcites tectatus Norton in Norton and Hall 1969 (Fig. 7t) [Monotect].

Genus *Sparganaceaepollenites* Thiergart 1933

Sparganaceaepollenites sp. (Fig. 7w) [Sparg].

Genus *Liliacidites* Couper 1953

Liliacidites leei Anderson 1960 (Fig. 7u) [Lilleei].
Liliacidites complexus (Stanley) Leffingwell 1970 (Fig. 7s) [Lilcompl].

Genus *Arecipites* Wodehouse 1933

Arecipites microreticulatus Anderson 1960 (Fig. 7v) [Arecmicroret].

Genus *Aquilapollenites* Rouse 1957 emend. Funkhouser 1961

Aquilapollenites quadrilobus Rouse 1957 (Fig. 7pp) [Aquilquad].
Aquilapollenites senonicus (Mtchedlishvili) Tschudy and Leopold 1970 (Fig. 7oo) [Aquilsenon].
Aquilapollenites delicatus Stanley 1961 (Fig. 7qq). Specimen counts of *A. delicatus* and *A. senonicus* were combined in the paleoenvironmental analysis.
Aquilapollenites amplus Stanley 1961 (Fig. 7rr) [Aquilampl].

Genus *Orbiculapollis* Chlonova 1961

Orbiculapollis lucidus Chlonova 1961 (Fig. 7nn) [Orblucid].

Genus *Wodehouseia* Stanley 1961

Wodehouseia spinata Stanley 1961 (Fig. 7ss) [Wodehs].

Genus *Kurtzipites* Anderson 1960

Kurtzipites trispissatus Anderson 1960 emend. Srivastava 1981 (Fig. 7dd) [Kurtztris].
Kurtzipites circularis (Norton) Srivastava 1981 (Fig. 7cc) [Kurtzcirc].

Genus *Cranwellia* Srivastava 1961

Cranwellia spp. (Fig. 7ee) [Cranwell].
Comments. Most specimens are assignable to *C. rumseyensis* Srivastava 1966.

Genus *Triporopollenites* Pflug and Thomson in Thomson and Pflug 1953

Triporopollenites plektosus Anderson 1960 (Fig. 7aa) [Triporoplek].

Genus *Ulmipollenites* Wolff 1934 emend. Srivastava 1969

Ulmipollenites krempii (Anderson) Frederiksen 1979 (Fig. 7, x and y) [Ulmkremp].
Ulmipollenites tricostatus (Anderson) Frederiksen 1980 (Fig. 7bb) [Ulmtric].

Genus *Triatriopollenites* Pflug in Pflug and Thomson 1953

Triatriopollenites subtriangulus (Stanley) Frederiksen 1979 (Fig. 7z) [Triatrio].

Genus *Fraxinoipollenites* Potonié 1951

Fraxinoipollenites variabilis Stanley 1965 (Fig. 7ff) [Fraxvar].

Genus *Tricolpites* Cookson ex Couper emend. Jarzen and Dettman 1989

Tricolpites hians Stanley 1965 (Fig. 7kk) [Tricolhians].
Tricolpites reticulatus Cookson 1947 (Fig. 7hh) [Tricolret].
Tricolpites sp. A (Fig. 7gg) [TricolspA].
Tricolpites sp. B (Fig. 7jj) [TricolspB].

Genus *Cupuliferoidaepollenites* Potonié et al. 1950 ex Potonié 1960

Cupuliferoidaepollenites sp. (Fig. 7ii) [Cupulif].

Genus *Tricolpopollenites* Pflug and Thomson in Thomson and Pflug 1953

Tricolpopollenites prolatus (Pierce) Norton, in Norton and Hall, 1969 (Fig. 7ll) [Tricolpoprol].

Genus *Pseudotricolpites* Stanley 1965

Pseudotricolpites reticulatus Stanley 1965 (Fig. 7mm) [Pseudotricol].

Palynologic characteristics and paleoenvironment

The relative abundances of palynomorph taxa in each sample are displayed in Figure 3. Two types of palynomorph association with depositional environment are recognized. Some palynomorph taxa are more common in either meander-belt or flood-basin deposits. Other taxa are more sensitive to the paleoenvironment, such as marshes or lakes.

Flood-basin taxa

The following palynomorph taxa tend to have increased relative abundances or are more common in the fluvial flood-basin deposits: *Hydrosporis* sp., *Cycadopites* sp., *Triporopollenites plektosus, Monosulcites* sp., and *Pseudotricolpites reticulatus.*

Meander-belt taxa

The following taxa tend to have more common occurrence in meander-belt deposits and tend to occur in more than one subenvironment within the meander-belt system: *Cyathidites diaphana, Leptolepidites-Varirugosisporites* spp., *Monosulcites scabratus, Psiloschizosporis* spp., *Ulmipollenites tricostatus, Ulmipollenites krempii, Tricolpites reticulatus*, and *Triatriopollenites subtriangulus.*

Marsh taxa

In general, the marsh deposits are characterized by a high relative abundance of the spores of pteridophytes, bryophytes, and microphyllous vascular plants. Among these spores, *Laevigatosporites* spp. is especially abundant, but *Hydrosporis* sp., *Intertriletes scrobiculatus, Matonisporites phlebopteroides, Stereisporites* sp., and *Foveasporis* sp. all are more common in marsh deposits than they are in other paleoenvironments. *Monosulcites* sp., *Wodehouseia spinata*, and *Cupuliferoidaepollenites* sp. also tend to commonly occur in abundance within the marsh deposits.

Lake taxa

Lake deposits are characterized by the low relative abundance of pteridophyte and bryophyte spores; *Cyathidites diaphana* is the only fern spore that occurs in greater abundance in lake deposits than in any other paleoenvironment. In addition, the lake deposits have high relative abundances of pollen grains produced by the Taxodiaceae, Cupressaceae, and Taxaceae, typically 20%–30% of the palynomorph assemblage within lake deposits. Other taxa that are more abundant in lake deposits are *Monosulcites tectatus, Triporopollenites plektosus*, and *Tricolpopollenites prolatus. Kurtzipites trispissatus* only occurs above the threshold levels within lake deposits and predominantly occurs in meander-belt lakes.

Point-bar taxa

Most of the samples collected from point-bar sandstones were essentially barren of palynomorphs, and those samples that were productive tended to have relatively poorly preserved palynomorphs. Palynomorph assemblages from samples that were adequately preserved typically contain abundant pteridophyte spores, but they are not so abundant as in the marsh deposits. Several taxa have their most conspicuous occurrences in point-bar deposits, including *Polycingulatisporites reduncus, Cicatricosisporites* spp. *Liliacidites leei, Orbiculapollis lucidus*, and *Wodehouseia spinata.*

Swale taxa

The swale deposit is represented by two samples that contain high relative abundances of *Azolla cretacea, Fraxinoipollenites variabilis, Laevigatosporites* sp., and bisaccate pollen.

Application of multivariate analyses

Detrended correspondence analysis (DCA), an ordination technique, was performed on the relative abundance palynomorph data. The ordination results represent the relationships between samples or species on a continuous scale. Samples that have similar assemblages or species that tend to occur within the same samples will plot close together. The first ordination axis accounts for the greatest amount of the variability within the data set and the second and subsequent axes represent progressively lesser amounts of the total variability. Because many factors other than paleoenvironment potentially contribute to the total variability, the underlying cause of the variability that is represented by each axis may not be the same. Thus, the two axes displayed in the scattergrams may vary independently, and each axis should be examined for trends (Gauch, 1982; Kovach, 1988). The relatively limited sequence of strata (~22 m of stratigraphic section) that were examined in this study suggests that variability factors involving very long time scales, such as evolutionary changes and long-term climatic changes, are minimized.

The DCA results are presented by two methods. A pollen diagram (Fig. 4) was constructed in which the samples are arranged (from top to bottom) in order of increasing first axis ordination scores for the sample ordination, and the species are arranged (from left to right) in order of decreasing first axis scores for the species ordination. This arrangement of samples and species plots samples that have the most similar palynomorph assemblages in adjacent rows and species that commonly occur together in adjacent columns. Because only the first ordination axis scores are used, any trends in variability expressed in other ordination axes are not displayed. The DCA results are also plotted as scattergrams of the scores of the first two ordination axes. This method of presentation effectively expresses the degree of similarity between samples or the occurrence of taxa, but it is more difficult to recognize which palynomorph taxa lead to the similarity between samples (Fig. 5) or the samples that contain similar occurrences of a given taxon (Fig. 6).

The results of the ordination analysis indicate that most samples from a specific paleoenvironment are relatively similar. For example, in Figure 4, most of the lake samples and marsh samples plot in adjacent rows, in general indicating that their palynomorph assemblages are more similar to each other than they are to assemblages from other paleoenvironments. However, there does not appear to be a pronounced distinction between assemblages from meander-belt sediments and flood-basin sediments. The scattergram of the DCA ordination of samples (Fig. 5) demonstrates a relatively high degree of similarity along the first axis for the samples from the lake, swale, and point-bar paleoenvironments. The marsh samples, however, demonstrate considerably greater variability on the first axis. Once again, there is no clear distinction between the palynomorph assemblages from flood-basin and meander-belt paleoenvironments.

The results of the DCA of species (Fig. 6) do not demonstrate any discrete groupings of species that can readily be related to paleoenvironment. Because the sample scores are derived by calculating a weighted average of the species scores for that sample, species will plot in the same general region of the species scattergram as the samples that typically bear those species plot on the sample scattergram. For example, species that have high first and second axis scores occur more commonly in samples that also have high first and second axis scores.

Paleofloral interpretations

The abundance of any given palynomorph taxon in a sample is dependent upon a number of biological and physical factors, including the number of producing plants in the given area, the number of palynomorphs produced by each plant, ease of transport, preservability, prevailing wind direction, and the extent of the drainage basin. Not all plant taxa are represented above threshold levels in the palynomorph assemblages, and many plant taxa will be overrepresented in the palynomorph assemblages because of the large quantity of pollen they produce per plant. However, studies of palynomorph distribution in modern continental and coastal environments indicate that palynomorph assemblages generally reflect the local vegetation (Chmura, 1994; Rich, 1978). Paleoenvironments such as lakes that are some distance from the local vegetation will receive higher percentages of palynomorphs that are readily transported by the wind, especially if they are produced by canopy trees (Birks and Birks, 1980).

Lake paleoenvironments

The Hell Creek Formation palynomorph assemblages can be interpreted by analogy to palynomorph distribution in similar modern environments. Palynomorphs that are characteristic of the lake paleoenvironments were probably important in the extralocal and regional pollen rain, especially if the depositional site was some distance from a point where a stream discharged into the lake. The farther from the shoreline the depositional site was located, the more important the contribution of regional and extralocal pollen rain would have been. The high degree of similarity in the first axis ordination scores of the assemblages from lake deposits probably reflects the limited types of palynomorphs that were readily transported into the lake paleoenvironments. The overwhelming abundance of TCT pollen in all paleoenvironments, but especially lake environments, suggests that TCT swamps probably were very common, and may have bordered the lake environments, similar to the abundance of cypress in many modern lowland forests. However, the distribution of bisaccate pollen suggests that the producing plants (probably mostly Pinaceae) were not locally abundant. Pine trees produce prodigious quantities of widely distributed pollen, and pollen spectra from localities near modern pine forests tend to be dominated by pine pollen (Traverse, 1988). Bisaccate pollen grains are never abundant in the Hell Creek palynomorph assemblages, but typically occur in lake, swale, and point-bar paleoenvironments. Their abundance in lakes reflects their transportability by wind, and their occurrence in point-bar and swale environments suggests that they were also commonly transported into the local area by streams.

Differences in the size of lakes may explain differences between the assemblages from the abandoned channel lake and the flood-basin lakes. A narrow, abandoned channel or oxbow lake that was bordered by vegetation would have received pollen and spores produced by understory riparian plants, despite the fact that such plants were generally not significant contributors to the regional pollen rain (Birks and Birks, 1980). In contrast, a large flood-basin lake would have received pollen and spores produced by the understory vegetation only along the shoreline. Elsewhere within the lake, the regional and extralocal pollen rain would have been the most significant contributor to the palynomorph assemblages. The distribution of the two species of *Kurtzipites* may represent this effect. *Kurtzipites trispissatus* tends to occur within the abandoned channel lake, whereas *K. circularis* is more typical of flood-basin lakes.

The distribution of *Cyathidites diaphana* may be different from other trilete spores because of the difference in the stature of the producing plants. This species, which is more common in lake deposits than most other spores, is thought by many workers to have been produced by *Cyathea*, a genus of tree ferns. Because palynomorphs produced by taller plants tend to be more widely dispersed than those produced by short herbaceous plants, spores produced by a tree fern would have been more likely to be transported into a lake than spores produced by a herbaceous fern.

Marsh paleoenvironments

Palynomorph assemblages from the marsh deposits are expected to be more representative of the local flora. In modern floras, plants of low stature typically do not produce palynomorphs that are widely distributed. Many of the palynomorphs that are characteristic of Hell Creek marsh deposits are larger or highly sculptured forms that probably would not be readily transported by wind. The contribution of regional pollen rain to the marsh palynomorph assemblages would have been rendered insignificant in view of high rates of local palynomorph production. Compared to the lake deposits, the marsh deposits exhibit more variability, noted by the wider distribution of the first axis scores in the DCA sample ordination plot (Fig. 5), within their palynomorph assemblages. This variability may reflect changes in floral communities due to differences in the level of the water table or other edaphic conditions from one site to another. The common marsh palynomorphs indicate that spore-producing plants, such as ferns, mosses, and lycopods, were important components of the marsh flora. The as-yet-unidentified producer of the unusual pollen grain, *Wodehouseia*, may also have been a marsh dweller.

Point-bar paleoenvironments

Interpretation of the assemblages from the point-bar deposits demands that palynomorph transport by water be considered (Campbell and Chmura, 1994). In addition to the local floral contribution, palynomorphs produced elsewhere in the drainage basin and palynomorphs reworked from older sediments were transported by the rivers: this may account for the typical occurrence of bisaccate pollen in the point-bar deposits. Differential preservation also plays a more significant role in point-bar deposits, because point-bar palynomorphs are typically poorly preserved. Several of the taxa, notably *Polycingulatisporites reduncus* and *Wodehouseia spinata*, that characterize the point-bar deposits have relatively thick walls and are distinctive enough to be recognizable even when torn or fragmented.

Swale paleoenvironment

Despite the fact that the swale deposit is only ~10 cm thick and is completely encased within point-bar sandstone deposits, the palynomorph assemblages are not at all similar to most of the point-bar assemblages. The absence of a rooted zone below the swale deposit indicates that emergent vegetation was not present in the swale. The abundance of *Azolla cretacea*, produced by a small, floating fern, suggests that a shallow, quiet pond occupied the swale.

Meander-belt versus flood-basin paleoenvironments

The modest differences in palynomorph content between the flood-basin and meander-belt sediments may be due to changes in edaphic conditions. Although many floral components were probably common to flood basins and meander belts, the drier edaphic conditions on the meander belts very likely led to the establishment of different floral communities from those in the flood basins, analogous to floral communities observed in the lower Mississippi River valley (Penfound and Hathaway, 1938). Palynomorphs such as *Cyathidites diaphana, Tricolpites reticulatus, Kurtzipites trispissatus*, and both species of *Ulmipollenites*, may well have been produced by plant taxa that lived only in meander-belt settings.

DISCUSSION

The palynological results, as interpreted here, attribute distinct changes in palynomorph assemblages to changes in paleoenvironment. As data become available for more paleoenvironments and from strata of differing geologic ages, direct application of results such as these will allow more detailed use of palynomorphs as paleoenvironmental indicators. Palynomorphs are especially useful as paleoenvironmental indicators in localities, such as well cores or areas of limited outcrop, where sedimentologic interpretations are hindered by the inability to observe lateral stratigraphic and sedimentologic relationships.

The reliability of biostratigraphic zonations, especially if abundance zones are utilized, will be improved by a better understanding of palynomorph distribution. Results presented here indicate that detailed biostratigraphic zonations should not be based upon samples that are widely separated stratigraphically or collected from limited rock types. Rather, samples should be as closely spaced stratigraphically as feasible, and all lithologies (and paleoenvironments) that are potentially palyniferous should be sampled.

Several of the palynomorph taxa that are reported herein are also considered key indicator taxa in Cretaceous-Tertiary boundary studies (Hotten, 1988; Murphy et al., 1995; Johnson et al., 1989). A number of the taxa that go extinct at the boundary or experience a notable change in abundance across the boundary are paleoenvironmentally sensitive. These taxa include *Liliacidites complexus, Orbiculapollis lucidus, Tricolpites reticulatus, Wodehouseia spinata, Laevigatosporites* spp., and *Cyathidites diaphana*. But other taxa that go extinct at the boundary, notably the three species of *Aquilapollenites*, do not appear to be strongly influenced by paleoenvironment. The use of the paleoenvironmentally sensitive palynomorphs could make determining the position of the boundary less certain if only a few diagnostic taxa are recorded from a sample suite. However, a broad range of taxa is typically used to locate the boundary, including nonsensitive taxa.

Several Cretaceous-Tertiary boundary studies have reported a notable increase in one or more palynomorph taxa just above the iridium anomaly layer. Most of these studies have noticed that fern spores, generally limited to one or few taxa (commonly *Cyathidites diaphana* or *Laevigatosporites*), compose nearly the entire assemblage, leading to a fern spike directly above the boundary (Tschudy et al., 1984; Fleming and Nichols, 1990). In Canada, the assemblages above the boundary are, in places, dominated by the angiosperm taxa, *Kurtzipites, Ulmipopllenites* (as *Ulmoideipites*), or *Syncolporites minimus* (Sweet and Braman, 1992). Both the fern and angiosperm spike assemblages are typically interpreted to have been produced by an early successional flora growing upon a landscape devastated by the effects of the Chicxulub bolide impact. All of the taxa just noted are recognized as paleoenvironmentally sensitive. *Laevigatosporites* tends to be more abundant in marsh deposits, *Kurtzipites* and *Cyathidites* both tend to have higher abundances in lake deposits, and *Ulmipollenites* is more typical of meander-belt deposits. In Paleocene deltaic sediments, *Syncolporites minimus* was associated with marsh, lignite underclay, lake fill, and brackish-water paleoenvironments (Kroeger, 1995). Thus, paleoenvironmental effects must be eliminated before the spike palynofloras can be considered to have resulted from a specific event. Despite the fact that the spike taxa are paleoenvironmentally sensitive, the spike palynomorph assemblages are distinct from any assemblages observed in this study. Even though spore abundances in marsh paleoenvironments are typically high, the sum of all trilete and monolete spores never exceeds 60% of the total assemblage. All of the assemblages are relatively diverse; the minimum number of taxa recorded in a sample (including taxa occurring below threshold levels) is 22. The maximum relative abundance of a given taxon (actually a group of similar taxa) used in this study is 37%. Thus there are distinctive differences between the paleoenvironmental effects observed in the present study and the spike assemblages reported above the Cretaceous-Tertiary boundary. These differences suggest that the spike assemblages represent a distinct paleofloral event rather than paleoenvironmental influence.

ACKNOWLEDGMENTS

I thank Martin B. Farley and Fredrick J. Rich for providing many useful suggestions to help improve this paper. I also thank Mark Fulton for discussions regarding detrended correspondence analysis, W.R. Talbot for assistance in the field, and the Floden family for their hospitality. The Museum of Geology and the Department of Geology and Geological Engineering of the South Dakota School of Mines and Bemidji State University provided logistical support. The editors of this volume provided considerable encouragement.

REFERENCES CITED

Bergad, R.D., 1974, Palynology of the Hell Creek Formation (Maestrichtian) in south-central North Dakota, U.S.A. [Ph.D. thesis]: Minneapolis, University of Minnesota, 454 p.

Birks, H.J.B., and Birks, H.N, 1980, Quaternary paleoecology: London, Edward Arnold, 289 p.

Butler, R.D., 1980, Stratigraphy, sedimentology and depositional environments of the Hell Creek Formation (Late Cretaceous) and adjacent strata, Glendive area, Montana [Ph.D. dissertation]: Grand Forks, University of North Dakota, 398 p.

Campbell, I.D., and Chmura, G.L., 1994, Pollen distribution in the Atchafalaya River, U.S.A.: Palynology, v. 18, p. 55–66.

Chmura, G.L., 1994, Palynomorph distribution in marsh environments in the modern Mississippi Delta plain: Geological Society of America Bulletin, v. 106, p. 705–714.

Coleman, J.M., and Prior, D.B., 1982, Deltaic environments of deposition, *in* Scholle, P.A., and Spearing, D., eds., Sandstone depositional environments: Tulsa, Oklahoma, American Association of Petroleum Geologists, p. 139–178.

Doher, L.I., 1980, Palynomorph preparation procedures currently used in the paleontology and stratigraphy laboratories, U.S. Geological Survey: U.S. Geological Survey Circular 830, 29 p.

Farley, M.B., 1989, Palynological facies fossils in nonmarine environments in the Paleogene of the Bighorn Basin: Palaios, v. 4, p. 565–573.

Farley, M.B., and Dilcher, D.L., 1986, Correlation between miospores and depositional environments of the Dakota Formation (mid-Cretaceous) of north-central Kansas and adjacent Nebraska, U.S.A.: Palynology, v. 10, p. 117–133.

Fleming, R.F., 1990, Palynology of the Cretaceous-Tertiary boundary interval and Paleocene part of the Raton Formation, Colorado and New Mexico [Ph.D. thesis]: Boulder, University of Colorado, 283 p.

Fleming, R.F., and Nichols, D.J., 1990, The fern-spore abundance anomaly at the Cretaceous-Tertiary boundary—A regional bioevent in western North America, *in* Kauffman, E.G., and Walliser, O.H., eds., Extinction events in Earth history: Lecture notes in Earth Sciences Volume 30: New York, Springer-Verlag, p. 347–349.

Frye, C.I., 1969, Stratigraphy of the Hell Creek Formation in North Dakota: North Dakota Geological Survey Bulletin 54, 65 p.

Gauch, H.G., 1982, Multivariate analysis in community ecology: New York, Cambridge University Press, 298 p.

Gennett, J.A., Raymond, A.L., and Parker, W.C., 1986, Changes in floral composition with depositional environment in Texas Jackson Group lignites, *in* Knochemus, D., ed., Continental slope; frontier of the 80's: Gulf Coast Association of Geological Societies Transactions, v. 36, p. 449–456.

Hill, M.O., 1979, DECORANA—A FORTRAN program for detrended correspondence analysis and reciprocal averaging: Ithaca, New York, Cornell University.

Hill, M.O., and Gauch, H.G., Jr., 1980, Detrended correspondence analysis—An improved ordination technique: Vegetatio, v. 42, p. 47–58.

Hotton, C.L., 1988, Palynology of the Cretaceous-Tertiary boundary in central Montana, U.S.A., and its implications for extraterrestrial impact [Ph.D. dissertation]: Davis, University of California, 745 p.

Jarzen, D.M., 1978, Some Maestrichtian palynomorphs and their phytogeographical and paleoecological implications: Palynology, v. 2, p. 29–38.

Johnson, K.R., Nichols, D.J., Attrep, M., Jr., and Orth, C.J., 1989, High-resolution leaf-fossil record spanning the Cretaceous/Tertiary boundary: Nature, v. 340, p. 708–711.

Kovach, W.L., 1988, Multivariate methods of analyzing paleoecological data, *in* DiMichele, W.A., and Wing, S.L., eds., Methods and applications of plant paleoecology: Paleontological Society Special Publication 3, p. 72–104.

Kroeger, T.J., 1995, The paleoecologic significance of Paleocene palynomorph assemblages from the Ludlow, Slope, and Cannonball Formations, southwestern North Dakota [Ph.D. dissertation]: Grand Forks, North Dakota, University of North Dakota, 389 p.

Kroeger, T.J., and Hartman, J.H., 1997, Paleoenvironmental distribution of Paleocene palynomorph assemblages from brackish water deposits in the Ludlow, Slope, and Cannonball Formations, southwestern North Dakota: University of Wyoming Contributions to Geology, v. 32, p. 115–129.

Miall, A.D., 1981, Analysis of fluvial depositional systems: American Association of Petroleum Geologists Education Course Note Series, no. 20, 75 p.

Muller, J., 1959, Palynology of recent Orinoco delta and shelf sediments: Micropaleontology, v. 5, p. 1–2.

Murphy, E.C., Nichols, D.J., Hoganson, J.W., and Forsman, N.F., 1995, The Cretaceous/Tertiary boundary in south-central North Dakota: North Dakota Geological Survey Report of Investigation no. 98, 73 p.

Norton, N.J., and Hall, J.W., 1967, Guide sporomorphae in the Upper Cretaceous–Lower Tertiary of eastern Montana (U.S.A.): Review of Paleobotany and Palynology, v. 2, p. 99–110.

Norton, N.J., and Hall, J.W., 1969, Palynology of the Upper Cretaceous and lower Tertiary in the type locality of the Hell Creek Formation, Montana, U.S.A.: Palaeontographica, Abteilung B, v. 125, p. 1–4.

Oltz, D.F., Jr., 1969, Numerical analysis of palynological data from Cretaceous and early Tertiary sediments in east central Montana: Palaeontographica, Abteilung B, v. 128, no. 3–6, p. 90–164.

Oltz, D.F., Jr., 1971, Cluster analysis of Late Cretaceous–early Tertiary pollen and spore data: Micropaleontology, v. 17, p. 221–232.

Penfound, W.T., and Hathaway, F.S., 1938, Plant communities in the marshlands of southeastern Louisiana: Ecological Monographs, v. 8, p. 1–56.

Reineck, H.E., and Singh, I.B., 1980, Depositional sedimentary environments: Berlin, Springer-Verlag, 549 p.

Rich, F.J., 1978, The origin and development of tree islands in the Okefenokee Swamp, as determined by peat petrography and pollen stratigraphy [Ph.D. dissertation]: University Park, Pennsylvania State University, 301 p.

Rich, F.J., and Spackman, W., 1979, Modern and ancient pollen sedimentation around tree islands in the Okefenokee Swamp: Palynology, v. 3, p. 219–226.

Rich, F.J., Kuehn, D., and Davies, T.D., 1982, The paleoecological significance of *Ovoidites*: Palynology, v. 6, p. 19–28.

Robertson, E.B., 1975, Pollen and spores as stratigraphic indices to the North Dakota Paleocene [Ph.D. dissertation]: Minneapolis, University of Minnesota, 192 p.

Stanley, E.A., 1965, Upper Cretaceous and Paleocene plant microfossils and Paleocene dinoflagellates and hystrichosphaerids from northwestern South Dakota: American Paleontology Bulletins, v. 49, p. 149–384.

Sweet, A.R., and Braman, D.R., 1992, The K-T boundary and contiguous strata in western Canada: Interactions between paleoenvironments and palynological assemblages: Cretaceous Research, v. 13, p. 31–79.

Talbot, W.R., 1985, Sedimentology, depositional environments, and paleomagnetic reconnaissance of the Upper Cretaceous Hell Creek Formation, northeastern Butte County, South Dakota [M.S. thesis]: Rapid City, South Dakota School of Mines and Technology, 141 p.

Traverse, A., 1988, Paleopalynology: Winchester, Massachusetts, Allen and Unwin Inc., 600 p.

Tschudy, R.H., 1970 [1971], Palynology of the Cretaceous-Tertiary boundary in the northern Rocky Mountain and Mississippi Embayment regions, *in* Kosanke, R.M., and Cross, A.T., eds., Symposium on palynology of the Late Cretaceous and early Tertiary: Geological Society of America Special Paper 127, p. 65–112.

Tschudy, R.H., Pillmore, C.L., Orth, C.J., Gilmore, J.S., and Knight, J.D., 1984, Disruption of the terrestrial plant ecosystem at the Cretaceous-Tertiary boundary, Western Interior: Science, v. 225, p. 1030–1032.

van der Wiel, A.M., 1982, A palaeoecological study of a section taken from the foot of the Hazendonk (Zuid–Holland, The Netherlands), based on the analysis of pollen, spores and macroscopic plant remains: Review of Paleobotany and Palynology, v. 38, p. 35–90

van Geel, B., Hallewas, D.P., and Pals, J.P., 1983, A late Holocene deposit under the Westfriese Zeekijk near Enkhuizen (Province of Noord–Holland, The Netherlands)—Palaeoecological and archaeological aspects: Review of Paleobotany and Palynology, v. 38, p. 269–335.

van Geel, B., Coope, G.R., and van der Hammen, T., 1989, Palaeoecology and stratigraphy of the Late Glacial type section at Usselo (The Netherlands): Review of Paleobotany and Palynology, v. 60, p. 25–129.

Walker, R.G., and Cant, D.J., 1984, Sandy fluvial systems, *in* Walker, R.G., ed., Facies models: Geoscience Canada Reprint Series 1, p. 71–89.

Manuscript Accepted by the Society September 24, 2001

Geological Society of America
Special Paper 361
2002

Palynology of the Cretaceous-Tertiary boundary in central Montana: Evidence for extraterrestrial impact as a cause of the terminal Cretaceous extinctions

Carol L. Hotton*
National Center for Biotechnology Information, National Library of Medicine, National Institutes of Health, Building 45, MSC 6510, Bethesda, Maryland 20892-6510 and Department of Paleobiology, MRC 121, National Museum of Natural History, Smithsonian Institution, Washington, D.C. 20560, USA

ABSTRACT

This report details a statistical analysis of palynological change across iridium-enriched claystones defining the Cretaceous-Tertiary (K-T) boundary in the Cretaceous Hell Creek Formation and Tertiary Tullock Member of the Fort Union Formation, eastern Montana, United States. The results strongly support the bolide impact model of terminal Cretaceous extinctions. Chi-square analysis of presence and abundance below and above the iridium datum was used to define characteristic Cretaceous and Tertiary palynomorph species. There is no statistical support for gradual decline in Cretaceous species below the iridium datum, although the presence and abundance of certain Cretaceous species decline somewhat, and the presence of Tertiary species increases slightly, 0–3 m below the iridium datum. An estimated 15%–30% of the Hell Creek palynoflora disappear 0–2 cm above the iridium datum, and another 20%–30% undergoes a significant decline in abundance. Palynoflorules directly overlying the iridium datum are characteristically depauperate. Partial recovery of the palynoflora occurs within 10–20 cm of the boundary, but species richness remains depressed throughout the lower Tullock Member compared to the Hell Creek Formation. Palynoflorules from rare Hell Creek coals contain significantly fewer Cretaceous and significantly more Tertiary species, relative to those of flood-plain deposits, suggesting that perhaps 30%–40% of affected Hell Creek species disappear or decline due to facies change. Greenhouse warming from impact-generated atmospheric CO_2 and consequent increased soil water saturation, combined with still uncertain mass-kill mechanisms, best fit the available data to explain the pattern of change in vegetation at the K-T boundary throughout the Western Interior.

INTRODUCTION

The extinctions that brought the Mesozoic Era to a close have long been recognized as a crisis in the history of life. Among marine invertebrates ~15% of all families went extinct; generic levels of extinction may have been as high as 50% (Sepkoski, 1982). On land, the most important group to go extinct was the dinosaurs; as the dominant vertebrates on land, their disappearance must have profoundly affected the terrestrial ecosystem. The cause or causes of this crisis have long

*E-mail: hotton@ncbi.nlm.nih.gov

Hotton, C.L., 2002, Palynology of the Cretaceous-Tertiary boundary in central Montana: Evidence for extraterrestrial impact as a cause of the terminal Cretaceous extinctions, *in* Hartman, J.H., Johnson, K.R., and Nichols, D.J., eds., The Hell Creek Formation and the Cretaceous-Tertiary boundary in the northern Great Plains: An integrated continental record of the end of the Cretaceous: Boulder, Colorado, Geological Society of America Special Paper 361, p. 473–501.

been in dispute, but little progress was made in resolving the issue until Alvarez et al. (1980) proposed that the terminal Cretaceous extinctions were caused by the impact of an extraterrestrial bolide. Their hypothesis, based on the presence of unusually high concentrations of iridium in a discrete claystone separating the youngest Cretaceous and oldest Tertiary planktonic assemblages at several widely separated sites, offered for the first time a means of testing extinction hypotheses by comparing patterns of biotic change to the iridium marker. The ensuing two decades of lively attempts to refute the impact hypothesis have only strengthened it, by establishing the global occurrence of iridium (Alvarez et al., 1984), its likely extraterrestrial origin (Luck and Turekian, 1983; Kyte, 1998), and its association with uncontested indicators of impact such as shocked quartz (Bohor et al., 1984, 1987a, 1987b; Bohor, 1990) and stishovite (McHone et al., 1989). The probable impact site of the Cretaceous-Tertiary (K-T) bolide has been located off the Yucatan Peninsula of Mexico (Hildebrand et al., 1991; Sharpton et al., 1992). The maximum size of the bolide, inferred from the crater to be ~12 km in diameter, reveals it to have been the largest known impact event in the Phanerozoic (Morgan et al., 1997).

The impact hypothesis spurred paleontologists to reexamine the pattern of biotic change across the K-T boundary in unaccustomed detail, because most existing studies, intended to answer very different kinds of questions, did not offer the required temporal resolution to address such an abrupt short-term event. The recognition that high spore abundance and extinctions were closely associated with an iridium-enriched clay layer, first in New Mexico (Orth et al., 1981; Tschudy et al., 1984) and then throughout the Western Interior of North America (Nichols, 1990; Nichols and Fleming, 1990; Sweet et al., 1999), was the first indication that iridium-enriched horizons associated with extinctions occurred in terrestrial as well as marine environments, thereby bolstering the case for a single global event at the K-T boundary. These observations also highlighted the value of palynology as a high-resolution means of testing the impact hypothesis, by examining in detail the pattern of vegetational change across the K-T boundary and using the iridium datum as a proxy for bolide impact. The sediments of the Hell Creek and Fort Union Formations in the northern Great Plains of North America are especially suitable for this type of study, because the K-T boundary is exposed within a conformable and fossiliferous sequence.

The rationale for this study is to determine whether the pattern of palynofloral change across the iridium datum in the Western Interior is consistent with a hypothesis of catastrophic or gradual change. Assuming a reasonably complete sedimentological record, gradualist and catastrophist hypotheses should produce distinct and contrasting patterns of floral change. Gradual processes such as climatic deterioration predict gradual or stepwise disappearance of characteristic Cretaceous taxa, gradual or stepwise origin of new taxa, and no sharp biostratigraphic boundary. Overall species richness may decline over time, but geologically instantaneous changes would not be predicted. In addition, palynological change might be predicted to track changes in depositional environment. In contrast, catastrophic extinction processes should produce a sharp discontinuity in species richness and floral composition at the event horizon, background levels of extinction below the discontinuity, and a lag in appearance of new species. In addition, change in vegetation might be expected to be independent of changes in depositional environment.

This study attempts to determine which of the two classes of mechanisms better fits the observed pattern of palynofloral change within the study area. The following questions are addressed in turn. (1) What is the magnitude of the floral change at the boundary? (2) Are extinction events at the boundary presaged within the Hell Creek Formation? (3) Can any of the observed floral changes be ascribed to a shift in depositional environment? (4) Do the boundary coals (i.e., coals with an iridium-enriched claystone) record a unique event?

The study described herein is unusual in several respects. First, it has been suggested that biostratigraphic studies tend to focus on stratigraphically important species and that extinction levels may be much lower if the entire flora were sampled, rather than just the most distinctive or most abundant species (Hickey, 1981). In order to overcome this objection, an exceptionally large population (~5000) for each sample was chosen, both to avoid the effects of "swamping" in certain samples caused by very high abundance of a few species at certain levels, and to be reasonably confident that a relatively complete sample of the available flora was obtained. Second, instead of relying on a priori assumptions of what represented stratigraphic markers, an attempt was made to statistically define such stratigraphically important species, as described in the next section. Finally, samples from different sections at a given level were pooled, again in an attempt to define changes statistically over the study area. A sample taken at 10 cm above a datum at one site is not precisely equivalent to a sample taken at 10 cm at another site; however, only by pooling samples from a number of sites can some level of statistical support be achieved. Statistical analyses of paleontological data do not relieve us of the burden of interpreting their biological significance.

MATERIALS AND METHODS

Geological setting

The study area in this report encompasses sediments of the Late Cretaceous Hell Creek Formation and early Tertiary Tullock Member of the Fort Union Formation, in an area of ~70 km in Garfield and western McCone Counties, eastern Montana (Fig. 1). A brief description of the geological setting is provided for this report; for more details, see Fastovsky and Dott (1986), Fastovsky (1987, 1990), Fastovsky and McSweeney (1987), and Retallack (1994). The Hell Creek and Fort Union Forma-

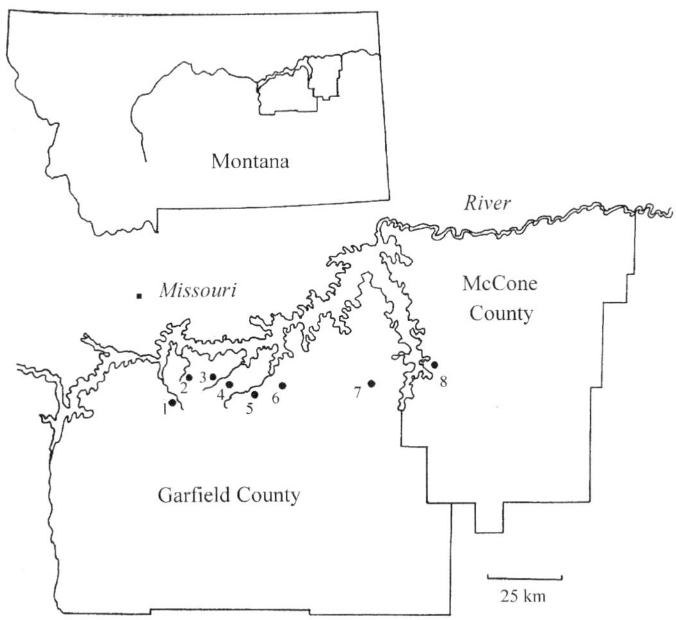

Figure 1. Map of sample localities. 1: Seven Blackfoot (7B83): Garfield County, Montana, Sawmill Creek Quad, center of sec. 32, T. 21 N., R. 34E. 2: Billy Creek (Bi83, BiZ83): Garfield County, Montana, Chalk Butte Quad, NW quarter of sec. 34, T. 22 N., R. 34 E. 3: Herpijunk and Iridium Hill: Garfield County, Montana, Hell Hollow Quad, T. 20 N., R. 35 E. Herpijunk Northeast (HNE82, HNE83), NE quarter of sec. 1; Iridium Hill (IH80, IH81, IH83), NW quarter of sec. 1, on section line. 4: Pearl: Garfield County, Montana, Trumbo Ranch Quad, T. 20 N., R. 37 E., PP82, NW quarter of sec. 6; PL82, SW quarter of sec. 6. 5: Brownie Butte: Garfield County, Montana, Trumbo Ranch Quad, Brownie North (BN82, BN83, PM83) and Brownie East (BEU83, BEL83, BE82), SE quarter of sec. 31, Rick's Place (RPC83), extreme SW corner of SW quarter of sec. 32. 6: Hell Creek State Park overlook and Lerbekmo: Garfield County, Montana, Maloney Hill Quad, Hell Creek State Park Overlook (HC82), T. 20 N., R. 37 E., NE quarter of sec. 1; Lerbekmo (Lb82, LbR82, Lb83), T 21 N., R 37 E., N quarter of sec. 1. 7: Ash Creek localities: Garfield County, Ash Creek West Quad, T. 22 N, R. 41 E., Sandstone Basin (SSB83), sec 14, Thomas Ranch (TR83), SW quarter of sec. 14. 8: Bug Creek localities: McCone County, Montana, Bug Creek Quad, BG-82, SW quarter of sec. 9, GU82, SW quarter of sec. 10.

tions comprise conformable fluvial sediments deposited along the margins of the epicontinental seaway to the east (Frye, 1969; Cherven and Jacob, 1985). The Hell Creek Formation is characterized by banded gray, greenish, and bluish mudstones and siltstones interbedded with cross-stratified fine- to medium-grained sandstones, interpreted as stacked paleosols intercut by channels in an aggrading, meandering fluvial system (Fastovsky, 1987, 1990; Fastovsky and McSweeney, 1987). The Tullock Member of the Fort Union Formation is characterized by subbituminous coals and carbonaceous shales of varying thickness interbedded with planar mudstones and siltstones that are interpreted as shallow ponds or lakes with locally extensive accumulations of peats (Fastovsky, 1987, 1990). Climate was likely subhumid (Retallack, 1994) and mesothermal (mean annual temperature >20 °C, Wolfe and Upchurch, 1987a), with a distinctly higher water table or impeded drainage in the Tullock Member (Fastovsky, 1987, 1990).

Palynomorph samples were collected from a restricted subset of the channel and flood-plain facies recognized by Fastovsky (1987, 1990). Most of the Hell Creek samples are derived from gleyed flood-plain paleosols or thin peat accumulations, and most of the samples from above the iridium datum are derived either from swamp and marsh environments or shallow ponds. Channel and levee facies were generally too coarse grained or oxidized for good palynomorph recovery. Although it would be preferable to compare floras from the same facies above and below the iridium datum to avoid possible confounding facies controls on palynomorph distribution, it was generally not possible to accomplish this within the study area because the biostratigraphic changes are closely associated with a marked change in clastic sedimentation style (but see following).

The K-T boundary in the study area is recognized by the presence of a layered (2–3) pinkish kaolinitic and/or smectitic claystone 1–3 cm thick containing high concentrations of iridium (Alvarez, 1983; Tschudy et al., 1984; Smit and van der Kaars, 1984; Baadsgaard et al., 1988; Fastovsky et al., 1989; Lerbekmo, 1999; F. Asaro, 1984, personal commun.; C.J. Orth, 1984, personal commun.) and shocked quartz (Bohor et al., 1984; Lerbekmo, 1999). These claystones are interpreted as particulate fallout from bolide impact. The palynoflora displays a marked, sharp change in the sediments immediately overlying the iridium-enriched claystone.

The Hell Creek–Fort Union formational boundary is generally placed at the base of the first thick coal in a section (Rogers and Lee, 1923; Archibald, 1982; Swisher et al., 1993). The iridium-enriched claystone usually occurs in thin (2–20 cm), laterally discontinuous coals, generally 1 m to several meters below the first prominent coal, so technically, the K-T boundary occurs in the uppermost Hell Creek Formation. For the sake of clarity, coals containing the iridium-enriched claystone are referred to herein as boundary coals, whether they are Hell Creek or Tullock. Nonboundary coals are termed Hell Creek or Tullock coals, depending on their position below or above the iridium marker and their associated flora. The unique geochemical and mineralogical signature of the iridium-enriched claystone, regardless of origin, marks it as a more reliable isochron across the study area than the traditionally used coal beds.

In western Garfield County the iridium marker typically occurs in the first coal within a section. However, coals are locally developed below the stratigraphic boundary in eastern Garfield and western McCone Counties. Only a minor iridium enrichment of about twice background level has been found in this area (at Bug Creek: Fastovsky and Dott, 1986; Rigby et al., 1987; Smit et al., 1987), and the K-T boundary seems to have been cut out by extensive, crosscutting channel systems (Fastovsky and Dott, 1986; Lofgren, 1995). Thus, the precise relationship of these thin Hell Creek coals to the iridium datum to the west is unknown. Despite this handicap, samples from this area were invaluable to the study because they allowed comparison of peat-forming Cretaceous and Tertiary facies. Sections were measured in this area using the first prominent coal as a local datum; the Hell Creek coals occur between 15 and 20 m below the first thick coal (the so-called Z coal; cf. Swisher et al., 1993) containing early Paleocene pollen florules.

Sampling

I sampled 10 sites across the Hell Creek and Tullock, 6 of which proved to contain a boundary coal with an iridium-enriched claystone (Fig. 2). Trenches were dug across the boundary and sampled at 2 cm intervals through the boundary coal, and at 5 or 10 cm intervals for 1–2 m above and below the boundary. In addition, spot samples were taken from two sections through the Hell Creek Formation and the lower part of the Tullock Member, at Hell Creek State Park Overlook and Billy Creek (Fig. 2). Samples used in the following analysis include 46 samples from the Hell Creek Formation, and 96 samples from the Tullock Member of the Fort Union Formation.

F. Asaro and H.V. Michel (Lawrence Berkeley Laboratory) performed neutron activation analysis of splits of presumed boundary samples for iridium content for all but two sites. Carl J. Orth (Los Alamos National Laboratory) analyzed boundary samples for Billy Creek and Seven Blackfoot Creek (unpublished data). Duplicate samples were processed for pollen in a pollen-free laboratory at the University of California, Davis. Clastic samples were macerated in hydrofluoric acid; most were then briefly (2–4 min) treated with household bleach to remove very fine organic particles. Coal samples were macerated in a warm water bath of dilute potassium hydroxide for 5 min; ~50% of them were then briefly bleached. Most samples were then subjected to brief acetolysis (9:1 acetic anhydride:sulfuric acid) for ~2 min to improve sporopollenin contrast. All residues were subjected to heavy liquid separation and strew-mounted on glass slides for light microscopy.

The samples were first examined to determine the identity of the main components of the palynoflora. A total of 281 pollen and spore species, excluding fungi, algal cysts, dinoflagellates and acritarchs, was identified and characterized. (Note that "species" as applied to dispersed palynomorphs simply refers to a morphospecies or operational taxonomic unit, and has an unknown relationship to actual biological species.) See Appendix 1 for a complete list of taxa recognized in this report and

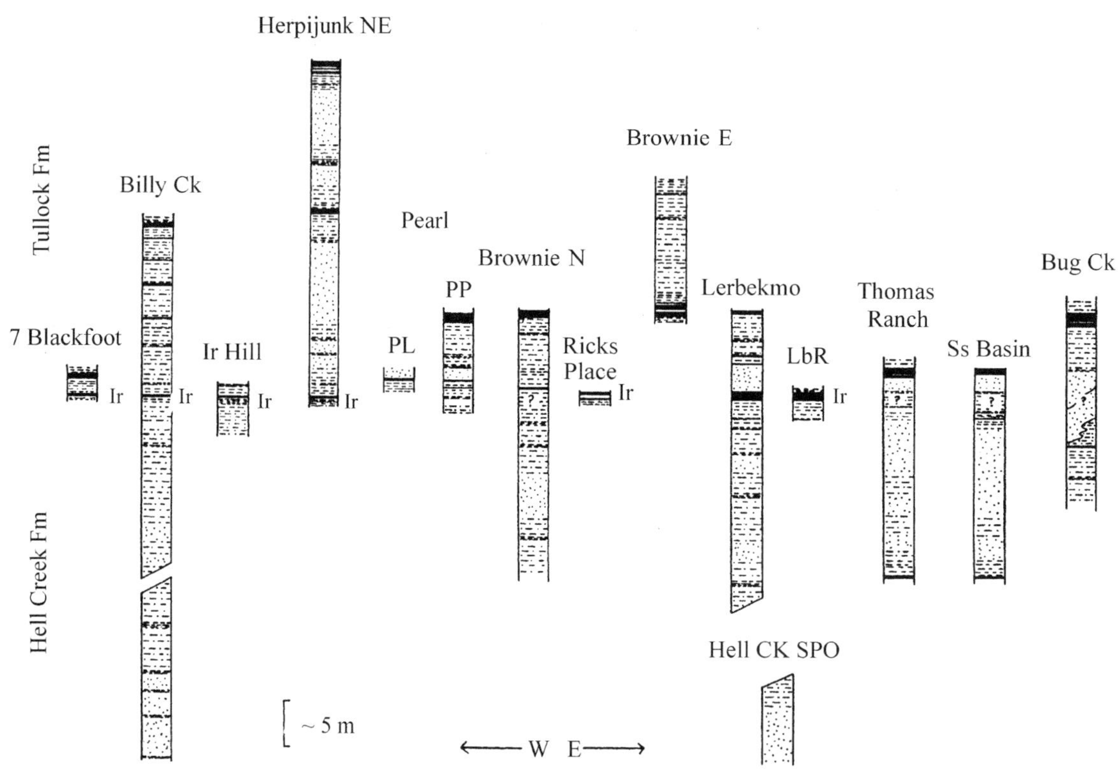

Figure 2. Summary diagrams of sections sampled in this study. The sections are arranged on an east-west axis, but there is no lateral scale. Vertical scale is exaggerated. Ir indicates boundary coals where elevated concentrations of iridium have been measured. Symbols as follows: black = coal or carbonaceous shale; stipple = sandstone; stipple and/or dashed line = siltstone; dashed lines = mudstone and/or shale.

Hotton (1988) for details on taxonomy of the palynoflora. In order to maintain approximately constant sample sizes and to avoid the possibility of swamping by overabundance of a few species, frequency for each species was calculated in an estimated 5000 grain count. From 10 000 to 18 000 grains were estimated for certain boundary samples where there were extremely high abundances of two species. The number of grains per slide was estimated by counting all the grains observed in one vertical and one horizontal scan and multiplying by the number of vertical and horizontal scans per slide. Additional slides were scanned as necessary. Thus, a species that appeared only once in a scan of a given sample was recorded as having a frequency of 0.02%, although the actual frequency could be higher or lower.

Data analysis

Chi-square analysis of species. The first step in this study was to characterize the distribution of palynomorph species above and below the iridium datum. For every recognized species, both presence (i.e., presence or absence in a given sample) and abundance (i.e., number of specimens of a species in a given sample) was recorded. A chi-square analysis was then performed on these data using SAS® Companion for the VMS Operating System (Myrick, 1986).

Species were then classified as belonging to one of four categories. Those species occurring either exclusively below the iridium datum (Appendix Table A1) or undergoing highly significant ($p < 0.004$) reduction in presence above the iridium datum (Appendix Table A2) were termed K species. Those species displaying no significant change in presence or abundance across the boundary were termed K-T species (Appendix Table A4). Those species undergoing significant increase in either presence or frequency in the Tertiary were termed T species (Appendix Table A5). Species that were too rare to generate a significance value with the given sample size were classed as R (rare) species. Rare species were not included in any of the statistical analyses, but are cited in the following discussion.

Included in the K-T category are 33 species that undergo a significant (at $p < 0.05$ level) decline in presence but usually not in frequency above the iridium datum. These were not classified as K species because they are still relatively common and sometimes abundant in the Tullock; they were instead separated into a subcategory of T-decline species (Appendix Table A3). Five species in this category (*Triporoletes tornatilis, Tricolpites parvistriatus, Kurtzipites trispissatus, Triporopollenites plektosus*, and *Ulmipollenites* spp.) actually display a significant decline in both presence and frequency in the Tullock Member, but because they occur in almost every sample, they were classified as K-T taxa. Similarly, several spore species (*Laevigatosporites* spp., *Gleicheniidites senonicus*, and *Stereisporites* spp.) are significantly more abundant in the Tertiary, but because they also are present in virtually every sample, they were classed as K-T species.

K taxa are defined more strictly than K-T or T taxa in order to compensate for the disparity between the two classes of samples (above and below the iridium datum). Samples from the Hell Creek include more spot samples taken from the entire range of the formation, whereas many of the samples from the Tullock come from the boundary coals, which are demonstrably more depauperate than samples taken elsewhere in the section. A further complication comes from the fact that, although the analysis described here assumes independence of each sample, this assumption is violated to some extent, particularly with the Tullock samples. Several samples collected from within the same coal unit or the same trench are more likely to resemble one another in palynological composition than samples separated both vertically and laterally (Hotton, 1988), and more Tullock than Hell Creek samples fall into the former category.

Statistical tests. Samples were split into a number of categories for statistical analysis of distribution over time, between coal and clastic facies, and for analysis of boundary coal composition. For analysis of species change within the Hell Creek, samples were split into four categories: samples from the lower 40 m of the Hell Creek (LHC); from 40 to 3.5 m below the iridium datum (MHC); from 3.5 to 0.04 cm below the iridium datum (UHC); and from 0.04 m below the boundary to the iridium datum (USTK). One-way analysis of variance (ANOVA) was performed on presence of K, T, and K-T species in these four sample categories, and also on presence and abundance for each of 25 common K taxa in each sample category (Appendix Table A6). Pair-wise comparisons of the species richness of K, T, and K-T species were made using the nonparametric Mann-Whitney U-test for differences between median values.

For comparison of palynofloral composition among coal and clastic facies, samples were split into the following categories: Hell Creek coals (KCOAL) and Tullock coals (TCOAL) were compared with clastic sediments from the Hell Creek (KCLAS) and Tullock (TCLAS). No samples from the boundary coals were included in this analysis. For comparison of palynoflorules of boundary coal and other coals, samples were split into the following categories: the basal 2 cm interval in the iridium coals (BASIR); the rest of the iridium coal intervals (UPIR); the basal 2 cm of 5 Tullock coals (TBASC); and samples of Tullock coals with a spore percentage of 50% or greater of the entire (TCOALSP), in addition to the other sample categories listed here. One-way ANOVA on the presence of K, T, and K-T classes, and the presence and abundance of common K and T species, was performed among these sample categories. In addition, mean per sample spore percentage was computed for the boundary coal analysis. Pair-wise tests for differences in K, T, and K-T species richness between sample categories were computed with the Mann-Whitney U-test. See Table 1 for a summary of the sample categories, and Appendix

TABLE 1. EXPLANATION OF ABBREVIATIONS

Category	Number of samples	Stratigraphic position or category
TCLAST	32	Tullock clastics
TCOAL	41	Tullock coals
TCOALSP	22	Subset of Tullock coals with >50% spores
TBASC	5	Basal 2 cm of Tullock coals
UPIR	16	0.02–0.16 m above iridium datum (boundary coal)
BASIR	6	0.0–0.02 m above iridium datum (boundary coal)
USTK	4	0.0–0.04 m below iridium datum
UHC	9	3.5–0.04 m below iridium datum
MHC	16	40.0–3.5 m below iridium datum
LHC	5	Lower 40 m of Hell Creek Formation
KCOAL	10	Hell Creek coals

TABLE 2. DISTRIBUTION OF SPECIES CLASSES IN HELL CREEK AND LOWER TULLOCK

	Number of species
K species	
0.00–0.04 m below iridium datum	34
0.00–3.5 m below iridium datum	52
Total below iridium datum	64
Rare (below iridium datum only)	57
K-T species	
"T decline" species	33
Unchanged across boundary	55
Total K-T species	88
Rare (above and below iridium datum)	35
T species	
T species	16
Rare (above iridium datum only)	21
Total species below iridium datum	256
Total species above iridium datum	193
Total species	281

Note: See Appendix 4 for complete list of and scores for each species; K—Cretaceous; T—Tertiary.

Table A7 for a list of the stratigraphic location and classification of all the samples used in these analyses.

RESULTS

Magnitude of change

A large proportion of the Hell Creek palynoflora is affected at the boundary, not just the stratigraphically important species. Only ~30% of all species in both Hell Creek and Tullock are unchanged in presence and abundance across the iridium datum, and of the other 70%, the large majority decline in abundance or presence, or disappear entirely (Table 2). A precise estimate of the level of apparent extinction and sharp decline is difficult to calculate because of sampling and taphonomic bias, but maximal and minimal levels may be estimated.

Of all pollen and spore species in the Hell Creek Formation, ~47% (121/256, excluding species confined to the Tullock Member) disappear or decline drastically above the iridium datum (Table 2). If the T-decline species are included, another 12% (33/281) undergo significant decline in presence above the iridium datum (Table 2). This yields a maximal estimate of almost 60% of all species in the palynoflora undergoing significant decline or extinction above the boundary. This figure almost certainly exaggerates the magnitude of change at the boundary, because it includes both species that disappear at any point within the Hell Creek as well as all rare species that occur throughout the Hell Creek. If those K species that are present in the uppermost Hell Creek samples (USTK) are included in the calculation, extinction level above the iridium datum drops to ~14% of all Hell Creek species (Table 2). However, this estimate is almost certainly too low. It is likely that most or even all of the common K species that occur within the last 4 m of the Hell Creek extend their range to the iridium claystone (see next section). Furthermore, it is likely that some percentage of the rare species extend their range to the top of the Hell Creek Formation. An estimate that includes the 52 K species that occur in the upper 3.5 m interval, plus the 14 rare species in this interval, yields an estimate of 66/251, or ~26% extinction above the iridium datum.

Compare this number to the background rate of local extinction within the Hell Creek Formation. Only eight species are confined to the lower part of the Hell Creek, and most of these are rare. Excluding rare species, 27 species disappear throughout the ~100 m of the Hell Creek Formation, yielding an extinction rate of ~0.002 species/cm. In contrast, the rate at the iridium datum is at a minimum 34 species/cm. These numbers are certainly not precise, but they illustrate that the level of extinction immediately above the iridium datum is several orders of magnitude greater than background extinctions.

Equally striking is the fact that there are relatively few T species, especially as compared to the number of K species. Only 16/281, or 5.5%, of all species undergo significant increase in presence, while another 5–10 species (depending on species definition) undergo significant increase in abundance above the iridium datum. Only 25 species are confined to the lower Tullock, and 21 of these are rare species. Most T species are present, although rare, in the Hell Creek. Because samples were collected from throughout the Hell Creek Formation, but from a relatively restricted part of the Tullock Member, the number of T species is likely underestimated to some degree. Nevertheless, this disparity between Hell Creek and Tullock palynofloras emphasizes an important aspect of the K-T boundary in Montana (and elsewhere in the Western Interior of North America): the absence of characteristic Cretaceous species, rather than the appearance of new species, defines the biostratigraphic boundary.

Palynofloral extinctions are to some degree selective. Many of the K species belong to a few distinct morphotypes that may comprise clades. Most prominent of these groups is the triprojectates, a morphologically distinctive Late Cretaceous group characteristic of middle and high northern paleolatitudes that includes *Aquilapollenites* sensu lato, *Cranwellia, Striatellipollis, Sindorapollis*, and *Kurtzipites* in the Hell Creek Formation. Although the affinities of the triprojectates are unknown, their highly ornamented and winged morphology clearly mark them as insect pollinated. The only members of this group to survive into the Paleocene (*Kurtzipites* and related taxa) are probably wind pollinated (McIver et al., 1991). North American species of *Proteacidites* sensu lato compose another likely clade. The affinities of this group are obscure, but they are almost certainly not related to the Gondwanan family Proteaceae (Memon, 1983). Their complex pore structure suggests affinities with the Normapolles complex, some members of which belong to the Fagales (Friis, 1983; Sims et al., 1999). Many other K species show no clear morphological similarities to one another, suggesting that the K-T extinctions are selective only to a degree (and also indicating, perhaps, that palynomorphs convey limited information concerning relationships).

Group membership may be one risk factor; could rarity be another? It seems reasonable to postulate that if conditions deteriorate, for whatever reason, then species that consist of few individuals, or are composed of small, scattered populations, or are confined to specialized habitats, would be more susceptible to extinction. To examine this hypothesis, species or species groups displaying high levels of abundance below the iridium datum were examined to determine how relative abundance in the palynoflora affected probability of extinction. (Relative abundance in a dispersed pollen sample is a very rough approximation of ecological importance, because many other factors contribute to representation in the dispersed palynoflora, including, e.g., mode of pollination, stature, phenology, facies type, proximity to catchment area, resistance to degradation, and taphonomic time averaging.)

Species or species groups were ranked by their Hell Creek importance (defined as a mean per-sample frequency of >1.0%, and presence in >65% of samples). The 20 dominant species or species groups recognized below the iridium datum, together compose 50%–75% of all individuals in a given sample (Table 3). Of these groups, 6 include primarily K taxa; together they contribute 10%–30% of all specimens in a typical Hell Creek sample. Certain individual K species, such as *Tricolpites microreticulatus* and *Leptopecopites pocockii*, display high abundance below the iridium datum. Other species that are at high abundance in the Hell Creek, such as *Ulmipollenites* spp., undergo drastic decline, although not extinction, in the Tullock (Table 3). In contrast, while *Aquilapollenites* and *Proteacidites* as groups are important components of the Hell Creek palynoflora, individual species of *Aquilapollenites* and *Proteacidites* often display low abundance, as little as 0.1% of a given sample. Including T-decline species in the tabulation, almost half of those species or species groups that are most abundant in the Hell Creek, which together contribute between 20% and 50% of all specimens in a given sample, are adversely affected at the K-T boundary. This analysis suggests that susceptibility to extinction is not confined to species that are at low abundance in the Hell Creek palynoflora, although low abundance may be a risk factor for certain K species.

TABLE 3. ABUNDANCE OF DOMINANT SPECIES IN HELL CREEK AND LOWER TULLOCK

	Below Ir	Above Ir	Net change
	Mean (SD)	Mean (SD)	
Tricolpites microreticulatus	8.19 (9.93)	—	absence
Aquilapollenites spp.	4.23 (4.45)	—	absence
Arecipites/Liliacidites spp. (K spp.)	3.14 (3.88)	—	absence
Proteacidites s.l. spp.	1.28 (2.05)	—	absence
Leptopecopites pocockii	0.98 (1.33)	—	absence
Cranwellia spp.	0.81 (1.58)	—	absence
Ulmipollenites spp.	10.00 (10.30)	1.18 (2.62)	decrease
Rutihesperipites trochuensis	4.47 (4.22)	0.72 (1.07)	decrease
Tricolpites parvistriatus	1.72 (4.26)	0.20 (0.66)	decrease
Kurtzipites spp.	7.84 (6.53)	10.6 (18.1)	no change
Cyathidites spp.	5.28 (7.06)	4.20 (4.81)	no change
Tricolpites micromunus	3.75 (5.44)	1.60 (1.81)	no change
Tricolpites parvus	2.96 (2.75)	2.37 (3.25)	no change
Dictyophyllidites sp.	1.02 (1.02)	1.92 (2.55)	no change
Taxodiaceae	7.40 (6.37)	33.5 (25.7)	increase
Laevigatosporites spp.	7.22 (6.55)	20.00 (17.30)	increase
Pinaceae (bisaccate pollen)	1.52 (1.81)	3.94 (2.74)	increase
Stereisporites spp. (Sphagnaceae)	0.87 (3.16)	1.82 (4.18)	increase
Gleicheniidites senonicus	0.76 (1.61)	2.79 (3.73)	increase
Reticuloidosporites dentatus	0.12 (3.0)	1.5 (4.06)	increase

Another group of Hell Creek dominants (members of Taxodiaceae, Pinaceae, Sphagnaceae, and leptosporangiate ferns) undergo significant increase above the iridium datum (Table 3). These contribute 10%–25% of a given Hell Creek sample; in the Tullock their contribution jumps to 40%–80% of a given sample. This leaves a relatively small number of species or species groups that remains unchanged across the boundary; together these contribute ~20%–40% of a given Hell Creek or Tullock sample (Table 3). Thus, most of the dominant members of the Hell Creek palynoflora are affected (for better or worse) by the event or events at the iridium datum. Even though the dispersed pollen record is an imperfect reflection of the true structure and diversity of the Hell Creek flora that existed at that time, it seems clear that a significant fraction of the vegetation was affected by the event that produced the iridium clay.

Is the K-T extinction event foreshadowed in the Hell Creek Formation?

The previous discussion makes clear that the iridium datum marks a significant palynofloral change. If these changes are caused entirely or in part by gradual processes (e.g., deteriorating climate) that took place over the course of the Late Cretaceous, they should be presaged within the Hell Creek Formation. Species richness and abundance, especially of K taxa, might be predicted to decrease upsection within the Hell Creek. T taxa might also be expected to increase in abundance or presence upsection within the Hell Creek.

There is little evidence of overall decline in species richness over the course of the Hell Creek. Total per-sample diversity is virtually unchanged from the base of the Hell Creek to the upper 3 m below the iridium datum (Table 4).

The pattern of change in K and T species is more interesting. Species richness of K species per sample appears to decline somewhat upsection. The mean per-sample species richness of K taxa in the lower portion of the Hell Creek is close to significantly higher than in the upper Hell Creek (Tables 4 and 5). T species trend slightly upward in the upper Hell Creek Formation, but the difference is not significant. T species are so rare and at such low abundance in the Hell Creek that differences here are not especially meaningful. The number of K-T taxa per sample remains virtually unchanged between the upper and the lower Hell Creek (Tables 4 and 5).

More detailed analyses of individual K species display no common directional trend. One-way ANOVA of abundance of the 25 most common K taxa over 4 levels of the Hell Creek (LHC, MHC, UHC, USTK) presents a decidedly mixed picture (Appendix Table A6). Only three K species display significant decline in abundance in the upper part of the Hell Creek (Table 6). Of these, *Tricolpites microreticulatus* and "*Nyssa*" *annulata* display a directed decline from the base of the Hell Creek to the top (Appendix Table A7.) These declines are offset by three K taxa that are significantly more abundant in the upper part of the Hell Creek (Table 6). Of these three species, *Trifossapollenites* cf. *T. ellipticus* and *Proteacidites thalmannii* are only abundant in the USTK samples (Appendix Table A6). *Liliacidites complexus* and perhaps *Leptopecopites pocockii* show a monotonic decline from the bottom of the Hell Creek to the iridium datum (Appendix Table A7), but the changes are not statistically significant at any level. Certain other species, such as *Retitriletes muricatus*, *Liliacidites hyalaciniatus*, and *Aquilapollenites quadrilobus*, appear to increase in abundance from the bottom of the Hell Creek to the top, but these changes are not statistically significant at any level. For the rest of the species analyzed, variation in abundance displays no clear trend in either direction (Appendix Table A6). These data suggest that there is little statistical support for progressive decline in K species approaching the iridium datum.

TABLE 4. MEDIAN SPECIES RICHNESS OF CRETACEOUS, TERTIARY, AND CRETACEOUS-TERTIARY SPECIES IN LOWER AND UPPER HELL CREEK

	N	K*	T*	K/T*	Total
UHC	27	29	2	44	72
LHC	5	34	1	43	76
p-value		0.0617	0.4517	0.9586	—

*Values represent medians (Mann-Whitney U-test for differences between median per-sample presence)

TABLE 5. PER-SAMPLE DIVERSITY OF CRETACEOUS, TERTIARY, AND CRETACEOUS-TERTIARY SPECIES IN HELL CREEK AND TULLOCK, BY CATEGORY

Category	N	Mean	SD
K Species			
TCLAST	32	0.78	0.87
TCOAL	41	0.41	0.63
UPIR	16	0.19	0.40
BASIR	6	1.50	1.52
USTK	4	20.75	8.06
UHC	27	27.67	6.43
KCOAL	10	16.70	5.81
LHC	5	33.80	6.87
T Species			
TCLAST	32	5.44	2.18
TCOAL	41	5.63	2.16
UPIR	16	5.44	1.67
BASIR	6	3.50	2.43
USTK	4	3.00	2.45
UHC	27	2.07	1.61
KCOAL	10	3.30	1.57
LHC	5	1.40	1.14
K-T Species			
TCLAST	32	34.13	7.86
TCOAL	41	28.56	7.18
UPIR	16	24.75	5.04
BASIR	6	18.50	5.68
USTK	4	33.25	10.78
UHC	27	42.44	6.89
KCOAL	10	38.90	6.30
LHC	5	43.20	3.37

Last 4 m of the Cretaceous. Dinosaurs and other vertebrates are apparently rare or absent in the upper 3–4 m interval beneath the iridium-enriched clay within the study area. This observation has been interpreted as evidence that their extinction predated the event that produced the iridium enrichment (Clemens and Archibald, 1981; Williams, 1994; MacLeod et al., 1997). As a consequence, this interval, especially the 4 cm below the iridium datum (USTK), was examined in detail.

Mean species richness of K species declines from 34 in the lower Hell Creek to 27.6 in the 0.04–3.5 m interval below the iridium datum. There is a further decline in K species diversity to a mean of 20 in the USTK samples (Table 6). This decline is not statistically significant, however, even when compared to the basal Hell Creek class of samples ($p = 0.1329$). T taxa display a slight upward trend, from a mean of two in the UHC samples to a mean of three in the USTK samples (Table 5): this difference is not statistically significant.

TABLE 6. CHANGE IN ABUNDANCE OF SELECTED CRETACEOUS SPECIES IN USTK SAMPLES

	*p-value**
Significant increase in USTK	
54. cf. *Trifossapollenites ellipticus*	**0.0219**
69. *Liliacidites* sp. 1	**0.0200**
268. *Proteacidites thalmannii*	**0.0373**
Significant decline in USTK	
78. echinoreticulate monosulcate	**0.0455**
122. *Tricolpites microreticulatus*	**0.0200**
184. "*Nyssa*" *annulata*	**0.0200**

*Pair-wise comparisons of LHC and USTK samples; significant at $p < 0.05$ (in bold)

Of the most abundant K species, only *Tricolpites microreticulatus* declines significantly in abundance in the USTK samples (Table 7). *Leptopecopites pocockii* and certain species of *Aquilapollenites* appear somewhat less abundant in the USTK samples, although not significantly so. Abundance of *Proteacidites* and Cretaceous species of *Liliacidites* and *Arecipites* appear somewhat higher in the USTK samples, but not significantly so (Table 7).

The apparent decline in species richness in both the UHC and USTK samples is likely due in part to poorer preservation beneath the boundary coals, and the consequent small sample size in this interval. Most of the boundary coals are underlain by an extensive leached or oxidized zone marked by a color change in the sediments to yellow or almost white, so despite the fact that these sections were heavily sampled, palynomorph recovery was very low. That K-T species decline in tandem with K species in the USTK samples supports this hypothesis (Table 6). The two best-preserved USTK samples (RPC83-1, 7B83-14.5, Appendix Table A7) display mean K species richness very close to the upper Hell Creek mean, lending further support to the hypothesis that the observed decline in K taxa in uppermost Cretaceous samples is due, at least in part, to taphonomic bias.

Leaving aside the question of preservation, there is also the matter of sampling error: any subset of samples would not be expected to include the entire Hell Creek palynoflora. About 50% of all K species occur in the USTK samples, compared to 76% of all K-T taxa (Table 8). This number rises to ~80% of all K taxa in the UHC samples (excluding the USTK samples) compared to 90% of all K-T species in this interval (Table 8). Is this evidence of decline in K species at the very top of the Hell Creek? In the case of certain species, such as *Tricolpites microreticulatus* and "*Nyssa*" *annulata*, the answer is probably yes. However, because individual K species tend to display lower abundance than many K-T species and are more likely to be missed in a small sample, some of the apparent decline may be due to sampling error.

TABLE 7. MEAN FREQUENCY OF DOMINANT CRETACEOUS TAXA IN USTK AND OTHER HELL CREEK SAMPLES

	USTK (N = 4)		Hell Creek (N = 32)		
	Mean	SD	Mean	SD	p-value
Aquilapollenites spp.	3.54	2.39	4.08	4.63	0.880
Cranwellia/Striatellipollis	1.88	2.84	0.66	1.36	1.000
Leptopecopites pocockii	0.62	0.63	1.39	1.61	0.580
Liliacidites/Arecipites (K spp.)	4.53	1.55	2.80	2.08	0.092
Proteacidites spp.	3.66	3.87	0.99	1.54	0.119
Tricolpites microreticulatus	0.50	0.70	9.62	9.81	**0.006**

TABLE 8. PERCENTAGE OF CRETACEOUS AND CRETACEOUS-TERTIARY TAXA IN USTK AND UPPER 4 m OF HELL CREEK FORMATION

	USTK (N = 4)	Upper 4 m (N = 11)	Total Hell Creek (N = 46)
K taxa	34/64 (53%)	52/64 (81%)	64/64
K-T taxa	68/88 (76%)	80/88 (90%)	88/88

Overall, the Hell Creek palynoflora presents a picture of dynamic turnover rather than progressive decline. In particular, samples immediately underlying the boundary clay have a diverse assemblage of K species, demonstrating that the 3 m barren zone beneath the boundary clay is taphonomic rather than evidence of preiridium event extinction. The palynological data corroborate evidence from intensive sampling of vertebrate fossils in the Hell Creek around Glendive, Montana, and North Dakota that suggests that there is no evidence for vertebrate decline within the 3–4 m interval beneath the iridium datum (Sheehan and Fastovsky, 1992; White et al., 1998; Sheehan et al., 2000). The 3–4 m barren zone within the study area seems to be a local taphonomic artifact.

Nevertheless, diversity and abundance of K species appear to decline somewhat, and diversity of T species appears to increase slightly, in the upper 3–4 m of the iridium datum, and especially in the 4 cm interval immediately underlying the iridium clay. The USTK palynoflorules in particular seem to foreshadow, to some degree, the kinds of changes that occur above the K-T boundary. Nothing in the Hell Creek palynoflora, however, presages the magnitude of the change that takes place immediately above the iridium datum.

Change in depositional environment across the K-T boundary: A possible confounding factor

The iridium-enriched clay with associated palynofloral changes always occurs at the base of the first thin coal in the western part of the study area. This close association of depositional and biotic change raises the very real possibility that some or even most biotic change could be tracking local or regional shifts in depositional environment. That clastic sediments in the Tullock represent rather different facies compared to those of the Hell Creek means that any facies effects on distribution of species above and below the iridium datum may be difficult to distinguish from other possible controls on distribution.

Some thin local coals clearly overlain by Cretaceous vertebrate faunas and palynofloras are present in the eastern part of the study area. Despite the absence of iridium-based correlation with the western part of the study area, these Hell Creek coals provide an opportunity to determine the effects of facies from other influences on the pattern of palynofloral change. If facies change plays a part in the palynofloral change at the K-T boundary, K species should similarly display lower presence or abundance and T species should display higher presence or abundance in Hell Creek coals.

Overall per-sample species richness drops significantly in Hell Creek coals, compared to Hell Creek clastics (Table 9); this is not remarkable, because coals, representing more local sample of vegetation, are often less diverse. However, it is notable that loss of K species contributes to most of the diversity

TABLE 9. MEDIAN DIVERSITY OF CRETACEOUS, TERTIARY, AND CRETACEOUS-TERTIARY SPECIES IN HELL CREEK COALS AND HELL CREEK CLASTICS

	N	K*	T*	K-T*
UHC	27	29	2	44
KCOAL	10	15	5	38
p-value		**0.0005**	**0.0402**	0.1509

*values represent medians; significant at $p < 0.05$ (in bold)

decline. Presence of K species is significantly lower in coals than in clastics (Table 9). Furthermore, even though the increase in T species is not large, it is significant in comparison to Hell Creek clastics (Table 9). K-T species, in contrast, display some drop in species richness in Hell Creek coals, but the difference is not significant (Table 9).

Comparison of abundance patterns of individual K species reveals a similar pattern: ~28% of K species are significantly less common in or entirely absent from Hell Creek coals, including some of the most common and abundant species, such as *Tricolpites microreticulatus* and *Leptopecopites pocockii* (Table 10). This estimate is almost certainly too low, because a number of K species are uncommon or absent from coals but are too rare overall to display a significant difference. If those K species that are completely absent from coal are included in the count, regardless of statistical significance, ~40% of all K species are negatively correlated with coal.

Some of the species that appear to be negatively correlated with coals are unexpected, e.g., species of *Azolla*, extinct representatives of the extant water fern. *Azolla* also displays significantly lower abundance above the iridium datum, which at first glance seems contradictory, given that the Tullock has many more water-saturated and ponded environments than the Hell Creek. However, both Hell Creek and Tullock coals contain an abundance of taxodiaceous pollen, suggesting that these represent woody swamp environments, whereas modern *Azolla* occupies open-water habitats.

About 50% of all K taxa, including four species of *Aquilapollenites*, several species of *Cranwellia/Striatellipollis*, several species of *Liliacidites*, and all species of *Proteacidites*, seem as likely to occur in coals as in clastics (Table 10). A few K species are positively correlated with coals, including *Sindorapollis* spp. and *Aquilapollenites turbidus*. Rare taxa as previously defined were excluded from this analysis, but it is notable that very few rare taxa occur in either Hell Creek or Tullock coals.

Examined by group, the most abundant K species display a similarly mixed picture. *Tricolpites microreticulatus* and *Leptopecopites pocockii* are nearly absent from coals (Table 11). However, species of *Aquilapollenites, Cranwellia/Striatellipollis*, and *Proteacidites* all display about the same frequency in either coals or clastics (Table 11). K species of *Arecipites* and *Liliacidites* are likely to be more abundant in coals, although the difference is not significant.

TABLE 10. PRESENCE OF K (CRETACEOUS) TAXA IN HELL CREEK COALS AND CLASTICS

	K Coals (N = 10)	K Clastics (N = 36)	p-value
Coal positive			
155. tricolporate sp. 2	6	3	0.001
195. *Rhoipites* sp. 8	6	3	0.001
246. *Aquilapollenites turbidus*	6	5	0.006
263. *Sindorapollis granulatus*	9	8	0.000
Coal negative			
15. *Gabonisporis cristata*	0	12	0.044
16. *Ghoshispora bella*	2	23	0.028
69. *Liliacidites* sp. 1	1	21	0.011
71. cf. *Arecipites tenuiexinus*	3	31	0.001
75. *Liliacidites altimurus*	2	25	0.009
78. echinoreticulate monosulcate	2	30	0.000
94. *Liliacidites* sp. 12	0	14	0.020
101. *Marsypiletes cretacea*	0	17	0.008
122. *Tricolpites microreticulatus*	6	35	0.006
204. tricolporate sp. 14	0	18	0.004
235. *Aquilapollenites amplus*	0	23	0.001
242. *Aquilapollenites quadricretaceus*	0	12	0.044
243. *Aquilapollenites quadrilobus*	0	32	0.000
275. polyoperculate	0	17	0.008
287. *Leptopecopites pocockii*	0	29	0.000
302. *Inaperturotetradites scabratus*	0	19	0.003
Coal neutral			
6. *Cicatricosisporites* sp. 1	1	6	1.000
8. *Concavissimisporites* sp.	5	24	0.462
18. *Heliosporites* sp.	0	7	0.319
22. *Liburnisporis adnacus*	0	6	0.315
23. cf. *Nevesisporites radiatus*	0	7	0.319
25. *Perotrilites undulosus*	0	8	0.171
29. *Retitriletes muricatus*	3	18	0.306
43. *Sindorapollis pilatus*	3	2	0.061
52. *Equisetosporites amabilis*	2	10	1.000
54. cf. *Trifossapollenites ellipticus*	4	20	0.484
55. *Ephedripites eocenipites*	2	19	0.084
61. cf. *Pristinuspollenites inchoatus*	4	21	0.475
76. *Liliacidites complexus*	8	28	1.000
79. *Liliacidites hyalaciniatus*	6	26	0.465
93. cf. *Liliacidites creticus*	0	10	0.089
99. *Stellatopollis* sp.	0	9	0.172
100. *Luminidites* sp.	0	6	0.282
130. *Foveotricolpites* sp. 1	0	5	0.570
167. *Foveotricolporites* sp. 1	5	13	0.480
184. *"Nyssa" annulata*	5	21	0.726
189. *Rhoipites* sp. 2	0	5	0.570
227. *Tricolporopollenites* sp. 11	1	9	0.420
237. *Aquilapollenites conatus*	4	12	0.720
240. *Aquilapollenites delicatus* var. *collaris*	4	12	0.720
244. *Aquilapollenites reductus*	1	13	0.143
245. *Aquilapollenites reticulatus*	5	9	0.242
252. *Cranwellia rumseyensis*	7	31	0.344
253. *Cranwellia baccata*	3	5	0.344
255. *Accuratipollis macrosolenoides*	2	3	0.294
256. *Striatellipollis radiata*	3	5	0.344
259. *Grewipollenites radiatus*	0	5	0.570
268. *Proteacidites thalmannii*	8	32	0.598
269. *Proteacidites marginus*	2	16	0.274
270. *Proteacidites retusus*	1	11	0.252
272. *Retibrevicolporites beccus*	4	6	0.189
292. *Libopollis jarzenii*	3	5	0.344
299. stephanoporate	0	10	0.089

TABLE 11. MEAN FREQUENCY OF IMPORTANT CRETACEOUS SPECIES IN HELL CREEK COALS AND CLASTICS

Species	K clastics (N = 36) Mean	SD	K coals (N = 10) Mean	SD	p-value
Aquilapollenites spp.	4.02	4.24	5.16	7.18	0.464
Liliacidites/Arecipites spp.	2.91	0.95	3.88	7.11	0.107
Proteacidites spp.	1.29	2.03	0.68	0.66	0.790
Cranwellia/Striatellipollis spp.	0.80	1.57	0.75	1.09	0.790
Tricolpites microreticulatus	8.60	9.68	0.56	0.54	**0.000**
Leptopecopites pocockii	1.30	1.54	0.00	0.00	**0.000**

Most K-T taxa are neutral toward depositional environment (Table 12). Of eight K-T taxa that are less common in coals, seven are actually T-decline species, i.e., species that undergo a significant decline in frequency above the iridium datum.

Although T species as a class appear more common in Hell Creek coals, variation in abundance of individual T species does not display as clear a pattern as the K species. Most T species do not display a pronounced affinity for coals; only two T species, *Brevicolporites colpella* and *Rhoipites pisinnus* (of 11 that were sufficiently common to analyze) were positively correlated with Hell Creek coals (Table 13). None displayed a significant negative correlation to coal.

The overall low abundance and presence of T species in the Hell Creek could explain why they do not present as strong a signal of facies preference. Because T species are both common and abundant above the iridium datum, their occurrence in Tullock coals and clastics was compared to determine facies preferences. Only one species, *Retitriletes muricatus*, displayed a positive association with coal (Table 14). Notably, neither *Brevicolporites colpella* nor *Rhoipites pisinnus* appeared to be significantly more common in Tullock coals, in contrast to their apparent affinity for Hell Creek coals.

These data suggest that a significant percentage of K species may go extinct at the K-T boundary due to changes in depositional environment, e.g., the disappearance of common coal-negative species such as *Tricolpites microreticulatus*, *Aquilapollenites amplus*, and *Leptopecopites pocockii*. Furthermore, at least some of the T-decline species may become less abundant in the Tullock for the same reason. Likewise, many K-T species may cross the boundary unchanged precisely because they are not confined to particular facies. The reduced diversity (species richness) of K species in the 3–4 m interval underlying the iridium datum, and especially of the USTK samples, suggests that the water table was rising, or soil water saturation was increasing, within this interval. Such seems to be the case at Rick's Place (Brownie Butte), where the boundary clay is underlain by several centimeters of brown carbonaceous shales. The observation that palynoflorules immediately underlying coals resemble those within coals would be expected as part of a normal facies progression toward wetter conditions.

Not all of the extinctions of K species can be explained by facies differences. For example, "*Nyssa*" *annulata, Pristinus-*

TABLE 12. PRESENCE OF K-T (CRETACEOUS-TERTIARY) TAXA IN HELL CREEK COALS AND CLASTICS

	K Coals (N = 10)	K Clastics (N = 36)	p-value
Coal positive			
4. *Hamulatisporis amplus*	6	4	0.003
5. *Camarazonosporites concinnus*	9	14	0.010
7. *Cicatricosisporites* sp. 2	4	3	0.031
14. cf. *Retitriletes saxatilis*	5	4	0.015
123. cf. *Tricolpopollenites microscabratus*	9	14	0.010
251. *Polotricolporites rotundus*	5	6	0.043
Coal negative			
1. *Azolla* spp.	2	25	0.009
39. *Triporoletes tornatilis*	0	25	0.000
97. *Pandaniidites typicus*	0	27	0.000
105. *Penetetrapites inconspicuus*	3	24	0.067
120. *Echitricolpites* sp.	1	20	0.013
162. *Cupanieidites carlquistii*	1	19	0.028
281. *Triporopollenites plektosus*	4	31	0.006
300. *Dyadonapites reticulatus*	2	26	0.008
Coal neutral			
2. *Baculatisporites comaumensis*	5	9	0.242
9. *Concavisporites jurienensis*	6	15	0.475
11. *Cyathidites* spp.	8	35	0.115
12. *Dictyophyllidites* sp.	9	33	1.000
17. *Gleicheniidites senonicus*	8	24	0.699
19. *Laevigatosporites* spp.	9	34	0.530
20. *Leptolepidites* sp.	3	14	0.723
21. *Leptolepidites tenuis*	2	15	0.282
28. *Retitriletes austroclavatidites*	8	21	0.282
33. *Stereisporites antiquasporites*	8	15	0.071
36. *Cingulatisporites regius*	4	8	0.416
38. *Seductisporites barbatus*	1	9	0.420
47. *Alisporites grandis*	5	23	0.480
49. *Cycadopites scabratus/C. follicularis*	5	15	0.726
56. taxodiaceous pollen	9	34	0.530
59. *Pinuspollenites semicircularis*	7	31	0.344
60. *Pinuspollenites elongatus*	6	10	0.074
62. *Podocarpidites maximus*	2	14	0.455
70. cf. *Calamuspollenites pertusus*	2	18	0.150
80. cf. *Liliacidites ligniticus*	1	8	0.659
81. *Liliacidites* sp. 3	8	22	0.455
85. *Liliacidites* sp. 6	4	8	0.416
107. *Phimopollenites* sp. 1	1	12	0.240
108. *Phimopollenites* sp. 2	2	7	1.000
113. *Retitrescolpites anguloluminosus*	3	8	0.682
114. *Tricolpites bathyreticulatus*	3	6	0.384
115. *Tricolpopollenites clavireticulatus*	3	11	1.000
Coal positive			
121. *Tricolpites micromunus*	7	29	0.666
124. cf. *Tricolpopollenites obliquimuratus*	2	5	0.634
125. *Tricolpites parvistriatus*	7	27	0.706
126. *Tricolpites parvus*	7	33	0.107
127. cf. *Tricolpites plicatilis*	5	11	0.283
137. *Satishia compacta*	7	30	0.384
157. *Dryadopollis* sp. 2	4	6	0.189
164. *Myrtaceidites solidus*	1	15	0.067
173. cf. *Myrtipites scabratus*	0	5	0.570
175. *Myrtipites* sp. 2	3	8	0.682
178. *Tricolporopollenites incomptus*	6	11	0.139
179. *Dryadopollis* sp. 3	2	3	0.295
192. *Rutihesperipites trochuensis*	8	33	0.295
194. *Simpsonipollis* sp. 2	3	10	1.000
212. *Tricolporites rhomboides*	3	4	0.163

(*continued*)

TABLE 12. PRESENCE OF K-T (CRETACEOUS-TERTIARY) TAXA IN HELL CREEK COALS AND CLASTICS (continued)

	K Coals (N = 10)	K Clastics (N = 36)	p-value
234. *Tricolpopollenites parvicolpatus*	2	11	0.700
260. *Kurtzipites annulatus/simplex*	7	33	0.107
261. *Kurtzipites trispissatus*	6	26	0.465
266. *Momipites* spp.	1	5	1.000
267. *Paraalnipollenites confusus*	0	8	0.171
277. *Triatriopollenites costatus*	8	30	1.000
285. *Ulmipollenites* spp.	8	33	0.295
286. *Jarzenipollenites* spp.	4	27	0.057
288. *Wodehouseia spinata*	9	27	0.420
293. *Erdtmanipollis procumbentiformis*	6	21	1.000

pollenites inchoatus, and *Liliacidites complexus*, all of which appear coal neutral in this analysis, decline significantly or trend downward in USTK samples. The data are more ambiguous with respect to T species, but do not strongly support facies as a cause for increase of T species above the K-T boundary. Given the small sample sizes involved and the likelihood that Hell Creek and Tullock coals do not represent precisely equivalent environments, such apparently contradictory results may not be particularly meaningful. However, they do suggest that change in depositional environment is important, but by no means the only factor in the dramatic palynofloral changes at the K-T boundary.

Are the boundary coals unique?

High spore abundance and otherwise depauperate palynoflorules immediately overlying an iridium enrichment is a widespread phenomenon across the Western Interior (Nichols, 1990; Fleming and Nichols, 1990). The spore spike has been widely interpreted as representing an early successional flora developed in the immediate aftermath of impact. However, coals represent wet habitats conducive to spore-producing plants such as ferns and bryophytes. Furthermore, peats tend to have a much more limited catchment area, so pollen floras in coals tend to more strongly reflect the local vegetation and are therefore less diverse. Could the spore spike and depauperate florules be explained more simply by these factors? To test these contrasting hypotheses, boundary coals were compared with other subsets of samples with respect to their spore percentage, species diversity of K, T, and K-T species, overall species richness, and abundance of individual species.

The spore percentage in the basal 2 cm of the boundary coals ranges from 50% to 98%, and is significantly higher in the boundary coals compared to both Hell Creek and other Tullock coals (Table 15). Samples in the basal 2 cm of the boundary coal (BASIR) also have significantly higher spore content than the basal 2 cm of Tullock coals (TBASC, Table 15). However, spore content varies widely in Tullock coals, ranging as high as 92% at certain levels within a coal. A subsample of Tullock coals with spore percentage >50% is not significantly different

TABLE 13. PRESENCE OF T (TERTIARY) TAXA IN HELL CREEK COALS AND CLASTICS

	Coals (N = 10)	Clastics (N = 36)	p-value
Coal positive			
160. *Brevicolporites colpella*	7	1	0.000
171. cf. *Rhoipites pisinnus*	7	6	0.002
Coal neutral			
27. *Reticuloidosporites dentatus*	5	15	0.746
30. *Retitriletes* cf. *muricatus*	2	9	1.000
34. *Cingulatisporites dakotaensis*	2	5	0.636
145. *Tricolpites* sp. 6	2	12	0.699
193. *Rutihesperipites* sp.	0	6	0.315
213. cf. *Tricolpopollenites levitas*	4	4	0.055
279. "*Betula*" *infrequens*	2	3	0.294
280. "*Carpinus*" *subtriangulus*	3	9	0.706
306. *Verrutricolpites* sp.	0	8	0.171

TABLE 14. PRESENCE OF T (TERTIARY) TAXA IN TULLOCK COALS AND CLASTICS

	Coal (N = 71)	Clastics (N = 32)	p-value
Coal positive			
30. cf. *Retitriletes muricatus*	27	4	0.010
Coal negative			
279. "*Betula*" *infrequens*	22	18	0.018
280. "*Carpinus*" *subtriangulus*	41	30	0.000
Coal neutral			
27. *Reticuloidosporites dentatus*	57	22	0.216
34. *Cingulatisporites dakotaensis*	27	13	0.830
72. *Arecipites* sp. 1	10	6	0.532
145. *Tricolpites* sp. 6	28	18	1.000
160. *Brevicolporites colpella*	23	11	1.000
171. cf. *Rhoipites pisinnus*	45	18	0.519
193. *Rutihesperipites* sp.	19	5	0.314
201. cf. *Tricolporopollenites inductorius*	15	6	1.000
213. cf. *Tricolpopollenites levitas*	28	7	0.115
306. *Verrutricolpites* sp.	23	6	0.236

TABLE 15. MEAN SPORE PERCENTAGE IN HELL CREEK AND TULLOCK COALS

BASIR (N = 6)	KCOAL (N = 10)	TCOAL (N = 41)	TBASC (N = 5)	TCOALSP (N = 22)
73.8 (18.4)*	25.8 (11.8)	48.5 (23.3)	30.5 (20.6)	65.9 (12.4)
p-value†	**0.002**	**0.018**	**0.028**	0.327

Note: See Table 1 for abbreviations.
*Mean (SD)
†Difference from BASIR; significant at $p < 0.05$ (in bold).

from BASIR samples (Table 15). Thus high percentage of spores alone is not necessarily a unique characteristic of boundary coals.

Stratigraphic position of the spore spike also varies within both boundary and nonboundary coals. Spore percentage tends to be at or near the boundary clays. In contrast, spore abundance tends to be highest within the upper levels of Tullock coals.

The most striking feature of boundary coals is the dramatic drop in presence and abundance of K species, from a mean per-sample species richness of 21 in the USTK samples, to 1.5 species in the basal 2 cm of the boundary coals (Table 5). K species virtually disappear in the higher levels of the boundary coals, then rebound slightly to a mean of ~0.8 in the lower Tullock, and remain very low up to the top of the sampled sections (Table 5).

The palynofloral shift is extremely sharp, but there are hints that this change is not instantaneous. One suggestive observation is that K species are somewhat more common in the BASIR samples compared to the UPIR samples, although not significantly so (Table 16). T species are also at Hell Creek levels of species richness at the base of the boundary coal, and then rise to Tullock levels in the upper levels of the boundary coal (Table 5). However, this difference is not significant (Table 16). Another hint that there is a slight lag in the shift from Cretaceous to Tertiary type palynofloras comes from the one sample of iridium clay processed that yielded a palynoflorule, at the Herpijunk NE boundary coal (Appendix Table A6, HNE83-11). The species richness of this palynoflorule is low, but it contains 7 K species, all at low abundance. Furthermore, *Triporoletes tornatilis*, a T-decline species that is otherwise at low abundance in the Tullock, comprises 45% of all specimens in this sample.

Boundary coals are also notable in their low overall species richness. This effect is most noticeable in the basal 2 cm of the boundary coal. Mean species richness is 23 species at the base, significantly lower than overall species richness of all but the TBASC class of samples (Table 17). (TBASC samples probably do not display significant difference from BASIR samples because of the small sample size, because their mean species richness is close to that of Tullock coals.) Because the low species richness of the BASIR samples could be attributed solely to dilution of other species by the overabundance of a few species, it is especially notable that Tullock coals with >50% spores display significantly higher species richness. This, plus the very large per-sample counts to counteract swamping by overabundance of one or two species, suggests that factors other than dilution are responsible for the low species richness at the base of the boundary coals.

Palynofloral composition in the boundary coals also differs. Both Hell Creek and Tullock coals tend to have high percentages of taxodiaceous pollen, suggesting that peat-forming environments both below and above the boundary were swamps. Cretaceous spore species such as *Ghoshispora bella, Gabonisporis cristata*, and *Heliosporites* spp., as well as local abundances of certain K angiosperms such as *Sindorapollis* spp., tend to be important constituents of Hell Creek coals. The primary spore components of nonboundary Tullock coals are *Laevigatosporites, Reticuloidosporites dentatus*, and Sphagnaceae (*Cingulatisporites/Stereisporites*). In contrast, boundary coals are generally dominated by species of *Cyathidites* and *Dictyophyllidites*, especially at their base. Species of *Cyathidites*, although not *Dictyophyllidites*, are significantly more abundant at the base of the boundary coal compared to any other class of samples (Table 18). *Ulmipollenites* are also present at higher than Tullock abundances (1%–20%). *Cyathidites* declines within 2–4 cm of the boundary layer, to be replaced by *Laevigatosporites* and in places to as much as 40% *Kurtzipites* spp. Spore percentage tends to remain high throughout the boundary coals, although the taxonomic composition changes.

Most other species that are present in the basal 2 cm of the

TABLE 16. MEDIAN SPECIES RICHNESS OF CRETACEOUS, TERTIARY, AND CRETACEOUS-TERTIARY SPECIES IN BASAL AND HIGHER LEVELS OF THE BOUNDARY COALS

	N	K*	T*	K/T*
UPIR†	16	0.0	6	24
BASIR	6	1.5	3.5	18
p-value		0.0768	0.0972	0.0507

*Values represent medians.
†See Table 1 for abbreviations.

TABLE 17. PER-SAMPLE SPECIES RICHNESS, BY SAMPLE CATEGORY

	BASIR N = 6	KCLAS N = 36	KCOAL N = 10	TCOAL N = 41	TBASC N = 5	TCOALSP N = 22	TCLAS N = 32
	23 (7.8)*	72 (13.5)*	59 (13.4)*	35 (8.8)*	33 (11.3)*	32 (8.8)*	40 (9.7)*
p-value†		**0.000**	**0.001**	**0.006**	0.171	**0.038**	**0.000**

Note: See Table 1 for abbreviations.
*Mean (SD).
†difference from BASIR; significant at $p < 0.05$ (in bold)

TABLE 18. CHANGES IN ABUNDANCE OF SELECTED TAXA IN BOUNDARY AND OTHER COALS

	BASIR N = 6	KCOAL N = 4	TCOAL N = 15	TCOALSP N = 12	TBASC N = 5
Taxodiaceae					
X	2.17	16.5	21.6	15.62	27.80
SD	5.06	13.5	12.5	7.85	15.40
p-value	—	**0.025**	**0.002**	**0.004**	**0.008**
Pinaceae					
X	0.01	3.07	3.19	3.20	1.54
SD	0.02	2.13	2.06	2.25	1.72
p-value	—	**0.014**	**0.000**	**0.001**	**0.008**
Sphagnaceae spp.					
X	0.01	1.50	3.56	5.60	1.39
SD	0.03	1.41	5.94	9.70	1.97
p-value	—	**0.020**	**0.002**	**0.004**	0.060
Gleicheniidites senonicus					
X	0.00	1.42	3.06	5.17	1.40
SD	0.00	1.47	3.14	7.52	1.78
p-value	—	0.111	**0.008**	**0.018**	0.076
Cyathidites spp.					
X	31.0	1.34	3.08	2.35	3.40
SD	23.6	0.93	1.86	1.44	3.91
p-value	—	**0.020**	**0.013**	**0.008**	0.060
Dictyophyllidites sp.					
X	19.00	1.77	3.07	3.54	1.40
SD	22.00	1.42	3.22	3.58	1.78
p-value	—	0.12	0.22	0.22	0.17

boundary coal are markedly depressed in abundance, compared to their abundance in other Hell Creek and Tullock samples. Pollen of Taxodiaceae and Pinaceae are virtually absent from the base of the boundary coal, a difference that is highly significant compared to any subset of samples (Table 18). Both pollen types rise to lower Tullock levels near the top of the boundary coals. Among other taxa, spores of Sphagnaceae and *Gleicheniidites senonicus* are significantly or nearly significantly lower in abundance in the basal 2 cm of the iridium coal relative to other subsets of samples (Table 18). Other K-T taxa display a similar pattern of lower abundance in BASIR samples compared to other subsets of samples, although none of the differences are statistically significant compared to other sample categories.

Are any of these changes presaged immediately beneath the boundary clay? The USTK samples immediately underlying the iridium claystone do not even hint at the dramatic changes in abundance of K-T species in the basal 2 cm of the boundary coals (Table 19). Taxodiaceae, Pinaceae, Sphagnaceae, *Gleicheniidites, Cyathidites*, and *Dictyophyllidites* are all at normal Hell Creek abundance in the upper 4 cm of the Hell Creek Formation.

Partial recovery of the palynoflora takes place within the boundary coal. Mean overall species richness increases steadily from the base of the boundary coal to ~34 at the top, which is very close to the TCOAL average (Table 20). The mean species richness of both K-T and T species moves steadily upward to reach close to lower Tullock levels by the top of the boundary coal (Table 5). Together the pooled BASIR samples have only 62 species, including nine T species, but >80% of the total lower Tullock flora occurs within the top of the boundary coals, including virtually all of the T species.

In spite of an apparently rapid recovery, the lower Tullock palynoflora appears depauperate compared to the Hell Creek palynoflora. The average per-sample species richness is considerably lower than one would expect from simple removal of K species and addition of T species. The number of characteristic Cretaceous species plummets from an average of 26 species/sample to 0.5/sample, but the number of T species does not expand to compensate for their loss. The mean per-sample number of T species increases from 2.0 in the upper Hell Creek to 5.4 in Tullock clastics and from 3.3 in Hell Creek coals to 5.6 in Tullock coals (Table 5). If the decline in average species richness is due entirely to loss of K species, then the expected average value for Tullock clastics is 50.5 species (72 species [average value for Hell Creek clastics] − 25 K species + 3.5 net gain in T species). The actual average value for Tullock clastics is 40 species, or ~10 fewer species (Table 5). A similar calculation can be made for Tullock coals. The expected per-sample species richness would be ~45 species (59 species [mean species richness for Hell Creek coals] − 16.7 K species [mean species richness of K species] + 2.5 net gain in T species), but the actual mean species richness is 35 species, or ~10 fewer species than expected. A small part of the deficit (about one species/sample) is attributable to the decline of rare species, of which there are about half as many in the Tullock as in the Hell Creek. However, most of the difference may be attributable to the lower frequency of the surviving K-T species within the lower Tullock; the average number of K-T species per sample is ~34 in clastics, compared to 43 per sample for Hell Creek clastics.

From the previous points it is apparent that the boundary coals differ from other coals both above and below the boundary in the following respects. (1) Characteristic K taxa are either absent or at very low abundance. (2) Smooth trilete spores (*Cyathidites* and *Dictyophyllidites* spp.) are significantly more abundant in boundary coals, especially in the basal 2 cm, relative to other coals. (3) The species richness of the basal 2 cm of the boundary coals is significantly lower than in other coals. (4) There is a statistically significantly lower abundance of taxodiaceous and bisaccate pollen, and spores of Sphagnaceae relative to other subsets of samples, in the boundary coals.

TABLE 19. MEAN HELL CREEK AND USTK ABUNDANCE OF SELECTED TAXA

	USTK (N = 4)		Hell Creek (N = 32)		
	Mean	*SD*	*Mean*	*SD*	*p-value*
Cyathidites spp.	3.39	4.71	6.02	7.58	0.326
Dictyophyllidites spp.	1.38	1.60	0.89	0.88	1.000
Gleicheniidites senonicus	0.37	0.48	0.73	1.72	0.880
Sphagnaceae	3.89	7.41	0.34	0.89	0.597
Pinaceae	0.78	0.25	1.43	1.80	0.513
Taxodiaceae	5.75	5.07	6.12	4.26	0.782

TABLE 20. MEAN PER-SAMPLE DIVERSITY THROUGH IRIDIUM COAL

cm above iridium	N	Mean	SD
8–10	5	34.00	5.15
6–8	4	32.25	4.35
4–6	4	26.00	8.12
2–4	4	31.00	4.00
0–2	6	23.00	7.82

DISCUSSION

The palynological data presented here offer strong support for catastrophic extinction at the K-T boundary in east-central Montana. High-diversity palynoflorules dominated by typical Cretaceous species are replaced, within the interval of the intervening iridium-enriched boundary clay, by low-diversity assemblages dominated by herbaceous plants, primarily ferns. The iridium-enriched clay also marks the level at which a minimum of 15% and a maximum of almost 50% of all Hell Creek palynomorph species disappear. Species that dominate the low-diversity boundary coals represent survivors from the Hell Creek palynoflora, rather than new species. Partial recovery of the palynoflora occurs within 10–20 cm of the boundary clay, but assemblages remain distinctly lower in diversity throughout the lower Tullock Member. There is little evidence for gradual decline of the typical Cretaceous palynoflora within the Hell Creek, although some of the palynological changes observed above the boundary claystone appear to be initiated within the upper Hell Creek.

These results largely mirror the K-T palynological record throughout the Western Interior of North America, from New Mexico (Tschudy et al., 1984; Fleming, 1990; Fleming and Nichols, 1990) to Wyoming (Bohor et al., 1987b; Wolfe, 1991, Nichols et al., 1992), Colorado (Tschudy et al., 1984), and Alberta and Saskatchewan (Nichols et al., 1986; Jerzykiewicz and Sweet, 1986; Lerbekmo et al., 1987; Sweet et al., 1990, 1999; Sweet and Braman, 1992). The Canadian sites differ in certain respects from those to the south, especially in the replacement of typical Late Cretaceous palynofloras by lower diversity assemblages several centimeters below the boundary clay, and by the survival of certain Cretaceous species for a short distance into the Tertiary that elsewhere go extinct at the boundary (Sweet et al., 1999). However, the replacement of typical Cretaceous assemblages by a depauperate, fern-dominated assemblage overlying the boundary clay, and an estimated loss of 25%–30% of palynomorph species, is seen throughout the Western Interior of North America (Johnson et al., 1989; Nichols, 1990; Nichols et al., 1992).

Sweet et al. (1999) offer an extraordinarily detailed view of postimpact floristic changes throughout the Western Interior, based on microstratigraphic palynological analysis of boundary claystones and boundary coals with an estimated temporal resolution of year or decades. A key element of their model is the assumption that palynoflorules within the boundary claystones represent autochthonous vegetation that continued growth and reproduction throughout deposition of impact debris, suggesting that photosynthesis never completely ceased after impact. Cretaceous species disappear abruptly within the boundary claystone at a level hypothesized to correspond to eject curtain fallout from the initial impact (Alvarez et al., 1995). These Cretaceous assemblages are replaced by a succession of very low-diversity assemblages composed of a few species of ferns and angiosperms that were presumably tolerant of adverse conditions in the immediate aftermath of impact. Angiosperm dominance and appearance of typical Paleocene assemblages usually occurs within 10–40 cm of the base of a given boundary coal, which is interpreted as a return to full forest canopy (Sweet et al., 1999). A negative carbon isotope excursion of ~130 k.y. ± 5 k.y. duration measured at several sites in eastern Montana and North Dakota probably brackets this interval from the base of the boundary coal to palynofloral recovery (Arens and Jahren, 2000, this volume).

The megafloral record, based on both leaves and dispersed cuticle, largely parallels that of the palynoflora throughout the Western Interior; in addition, it provides valuable taxonomic and paleoclimatic data not available from the palynofloral record. In New Mexico and Colorado, diverse Late Cretaceous megafloras are abruptly replaced by fern-dominated assemblages, which are then succeeded by low-diversity, angiosperm-dominated assemblages appearing 1 m to several meters above the iridium datum (Wolfe and Upchurch, 1986, 1987a, 1987b). Extinctions are primarily of broad-leaved evergreen taxa, and include many members of the Laurales, Illiciales, and Cyclanthaceae that are important constituents of Late Cretaceous megafloras; surviving taxa are often deciduous (Wolfe and Upchurch, 1986, 1987a). Leaf physiognomic analysis suggests a marked increase in both precipitation (Wolfe and Upchurch, 1986, 1987a, 1987b) and temperature (Wolfe, 1990) in the early Paleocene.

The megafloral record differs in one notable respect from the palynofloral record in the Western Interior: it records a much higher level of extinction at the K-T boundary. In North Dakota, Late Cretaceous leaf assemblages are almost entirely replaced by characteristic Paleocene assemblages, with an estimated extinction level of 70%–90% (Johnson et al., 1989; Johnson and Hickey, 1990; Johnson, 1992). Similar high levels of extinction are estimated from leaf floras and dispersed cuticle in New Mexico and Colorado (Wolfe and Upchurch, 1986, 1987a, 1987b). As this chapter demonstrates, estimates of palynofloral extinction have a wide margin of error, but nowhere in the Western Interior of North America does the palynoflora display such a high level of extinction. This disparity has been ascribed to the higher taxonomic resolution afforded by the leaf record, which reflects species-level extinctions, in contrast to the palynofloral record, which reflects extinction at generic and higher taxonomic levels (Wolfe and Upchurch, 1987a; Johnson, 1992). In addition, certain groups that underwent high levels of extinction in the leaf record across the boundary, e.g., the Laurales, are entirely absent from the palynofloral record due to their thin-walled pollen (Wolfe and Upchurch, 1987a, 1987b; Johnson, 1992). Despite the deficiency of the palynofloral record in these respects, given the much greater ease of dispersal of spores and pollen, the dispersed palynological record may more accurately reflect the true magnitude of extinction in the regional flora.

The pattern of replacement above the boundary clay mimics ecological succession, albeit on a longer time scale. The fern-dominated florules immediately overlying the iridium have been interpreted as early successional assemblages colonizing a landscape denuded of vegetation in the immediate aftermath of the impact, analogous to the pioneer role of many modern ferns on bare soil after fire or volcanic eruption (Tschudy and Tschudy, 1986; Wolfe and Upchurch, 1987a, 1987b). Estimates of recovery time after this initial reestablishment of vegetation vary, in part depending on whether the data are derived from megafloras or palynofloras. Sweet et al. (1999) recognized recovery to preimpact assemblages (minus Cretaceous species) within 10–40 cm of the boundary layer. Similarly, in eastern Montana, palynofloras attain Tullock levels of diversity within 10–20 cm of the boundary layer. The megafloral record suggests that complete recovery of the flora may have taken considerably longer. Wolfe and Upchurch (1986, 1987a) noted that the megaflora retains physiognomic characters of early successional plants several meters above the iridium datum; full recovery occurs ~200 m above the iridium datum, or ~0.5–1.0 m.y. after impact. Similarly, megafloras remain depauperate up to the late Paleocene in the northern Great Plains (Gemmill and Johnson, 1997). In eastern Montana, the palynoflora remains markedly less diverse throughout the lower Tullock Member in comparison to the Hell Creek palynoflora. In particular, the lower Tullock palynoflora contains ~75% fewer rare species, suggesting a widespread loss of species that included those far removed from catchment areas. In all of these cases, postimpact recovery occurred on a time scale of 10–100 k.y., considerably longer than typical succession after a local perturbation.

Comparison of Hell Creek coal and clastic palynofloras described in this study reveals a significant facies overprint to the pattern of extinction. From 25% to 40% of the palynofloral extinctions at the boundary are potentially explainable by observed facies changes from the Hell Creek to the Tullock. Reduced presence and abundance of Cretaceous species in the 3–4 m interval below the iridium claystone, and especially in the USTK samples, suggest a rising water table within this interval, with consequent loss of favorable habitats for at least the coal-averse species. Interpreting this interval is complicated by the fact that it represents a palynological barren zone, displaying much poorer palynomorph recovery despite heavy sampling than elsewhere in the Hell Creek Formation. Thus, the apparent reduction in diversity of Cretaceous species may be due to sampling error: if on average Cretaceous species are less abundant than K-T species, they are more likely to be missing from a small sample. Because even the well-preserved palynoflorules immediately underlying the boundary claystone contain a lower proportion of Cretaceous species, however, this deficiency is probably real.

The palynoflorules directly underlying the boundary clay in eastern Montana bear some resemblance to the lower diversity, Taxodiaceae- and angiosperm-dominated assemblages found 5–10 cm below the boundary coal in Alberta and Saskatchewan (Sweet et al., 1990, 1999; Sweet and Braman, 1992). Although Sweet et al. (1999) considered the Rick's Place (Brownie Butte) USTK palynoflorule to be a typical Late Cretaceous assemblage, the analysis presented here suggests that it, like the other palynoflorules directly beneath the boundary clay, contains a lower diversity of Cretaceous species in comparison to palynoflorules several meters below it. These data, combined with sporadic regional development of coal below the K-T boundary, indicate that a shift to wetter conditions had begun before the impact event in eastern Montana, and elsewhere in the Western Interior (Sweet and Braman, 1992; Johnson, 1992; Sweet et al., 1999; Arens and Jahren, 2000).

Despite the facies overprint, the unique floristic pattern above the K-T boundary layer clearly cannot be explained by facies change. The clearest evidence of this comes from two sites in Canada (Sweet and Cameron, 1991; McIver, 1999) and New Mexico (Pillmore et al., 1984; Nichols, 1990), where the iridium clay occurs in the middle of a coal. At these sites, depauperate, fern-dominated florules abruptly replace a diverse Cretaceous assemblage at the level of the iridium clay within the coals, thus removing any possibility that a shift to peat-forming environments accounts for the change. This unique floristic signature, so closely coupled to the iridium clay that two decades of accumulated evidence demonstrates is fallout from a large bolide impact, strongly supports a causal connection between the two events.

The consequences of a 12 km bolide impact on Earth are still poorly understood, but they are likely to have caused severe disruption in Earth's ecosystems. Most impact models have considered short-term effects as the primary agents of mass extinction. Proposed agents of mass kill include one or a combination of the following: global darkening and cessation of photosynthesis for several months (Alvarez et al., 1980) causing prolonged freezing (Pollack et al., 1983) from impact-generated dust, or sulfuric acid aerosols (Siggurdson et al., 1992; Pope et al., 1994, 1997), greenhouse warming from impact-generated CO_2 (O'Keefe and Ahrens, 1989), acid rain (Prinn and Fegley, 1987), coupled with ozone depletion (Brett, 1992), or intense thermal radiation (Melosh et al., 1990) combined with global wildfires (Wolbach et al., 1985). Marine ecosystems appear to have been severely disrupted by one or more of these predicted effects, including a brief cessation of primary productivity (Hsü et al., 1982; Stott and Kennett, 1989). In contrast, terrestrial ecosystems do not seem to have been as severely affected, at least not globally. Southern Hemisphere floras display little evidence of extinction or perturbation at the K-T boundary, even in association with iridium enrichments (Askin, 1990; Johnson, 1993; Elliott et al., 1994). These observations define the severity of impact-generated effects on terrestrial ecosystems, and suggest that the greatest disruption may have occurred within North America because of its proximity to the site of impact in the Yucatan. Even in North America, cold-sensitive organisms such as crocodilians survived into the Paleocene as far north as Canada (Archibald and Bryant, 1990), suggesting that any freeze following the impact was either brief or not severe (Wolfe, 1991).

Any model of mass extinction must explain the disparate effects between marine and terrestrial ecosystems and between Northern and Southern Hemispheres. Many of the most severe scenarios seem less plausible just because they are global in effect. The likeliest models of extinction include postimpact greenhouse warming as an important, or perhaps leading agent of extinction. Several independent lines of evidence suggest that significant warming occurred in the early Paleocene (Hsü et al., 1982; Stott and Kennett, 1989; Wolfe, 1990; Kaiho et al., 1999), as a consequence of impact injection of carbonates into the atmosphere and loss of marine CO_2 sinks (O'Keefe and Ahrens, 1989; D'Hondt et al., 1998). Lomax et al. (2000) modeled the predicted effects of increasing CO_2 and precipitation on Late Cretaceous terrestrial ecosystems, using a combined vegetation-biogeochemical general circulation model, with results that show a good fit to the data. High atmospheric CO_2 and increased precipitation increase primary productivity, leaf area index, and soil carbon reservoirs, especially in middle and high latitudes of North America. Less of a greenhouse effect is seen at low latitudes and in the Southern Hemisphere. Their model also suggests that widespread coal deposits and higher water tables in the Western Interior could be a consequence, at least in part, of the postimpact greenhouse world. High atmospheric CO_2 is predicted to reduce rates of soil decomposition and may also inhibit transpiration through stomatal closure, which would cause increased soil saturation; both waterlogging and slower decomposition would lead to coal formation (Lomax et al., 2000). If this scenario is correct, many of the floral extinctions seen throughout the Western Interior could be a consequence of postimpact greenhouse climate change, rather than a direct consequence of impact. An additional mechanism of extinction is suggested by the observation that artificial CO_2 enrichment in experimental plots markedly increases both growth and allocation to reproduction; therefore, in high CO_2 conditions, fast-growing species may have a competitive advantage over slower growing species (LaDeau and Clark, 2001). This could have the paradoxical effect of reducing species richness over time, and could perhaps explain some of the persistent low diversity in early Paleocene floras.

Modeling the effects of bolide impact on global ecosystem remains an enormously complex problem. Impact effects operated at a variety of levels and at both short- and long-term temporal scales, further complicated by biotic response both affecting and being affected by changes in biogeochemical and geochemical cycles and climate. The postimpact greenhouse model seems to be particularly fruitful for directing future research. If we are fortunate, we will never have the opportunity to examine at first hand the effects of a large body impact on Earth, but the following decades will provide ample opportunity to refine our understanding of the environmental effects of doubling atmospheric CO_2 as we face an anthropogenic greenhouse world. Perhaps one of the benefits of this global experiment will be an enhanced understanding of the early Paleocene world.

ACKNOWLEDGMENTS

I thank J.H. Hartman, D.J. Nichols, and K. Johnson for their invitation to contribute to this volume and for their encouragement and infinite patience as many a deadline slipped away, and R. Askin and an anonymous reviewer for critical reading of the manuscript. Special thanks to J.A. Doyle and W.A. Clemens, whose critical and sometimes skeptical probing of every aspect of this work immeasurably improved the final product. I also thank the following individuals for their various contributions to this study: W. Alvarez, J.D. Archibald, F. Asaro, L. Bryant, L. Dingus, D.E. Fastovsky, L.J. Hickey, H. Hutchison, S. Margolis, D. Russell, E. Shoemaker, R. Tschudy, G.R. Upchurch Jr., J.V. Ward, S. Wing, and J.A. Wolfe. M. Farrabee, F. Fleming, and D.J. Nichols provided helpful palynological advice, F. Asaro, H.V. Michel, and C.J. Orth provided geochemical analyses, and L. Delwiche, M. Rejmanek, M. Stanton, and N. Willits advised me on statistics. I thank D. Clemens, M. Greenwald, N. Greenwald, and P. Morris for help in the field, the Engdahl family for their generous support of University of California Berkeley field crews over the years, and Gonzo (a.k.a. Dustclot) and Pinto for admirably filling their roles as camp kittens. National Science Foundation grant BSR-85-13253 to W.A. Clemens is acknowledged.

DEDICATION

This paper is dedicated to the memory of my parents, Ruth Olmstead Hotton (1923–1986) and Nicholas Hotton III (1921–1999).

APPENDIX 1. SPECIES RECOGNIZED IN THIS STUDY

The number preceding each taxon represents a unique identifier corresponding to a taxon recognized by Hotton (1988), because the names of many of the species have been changed in this chapter. Numerical order corresponds roughly to morphological categories (spores, gymnosperm pollen, monosulcates, tricolpates, and derived types).

1. *Azolla* spp.
2. *Baculatisporites comaumensis* (Cookson) Potonie 1956
3. Spore forma A
4. *Hamulatisporis amplus* Stanley 1965
5. *Camarazonosporites concinnus* Srivastava 1972
6. *Cicatricosisporites* sp. 1
7. *Cicatricosisporites* sp. 2
8. *Concavissimisporites variverrucatus* Couper 1958 sensu Srivastava 1972
9. *Concavisporites jurienensis* Balme 1957
10. Spore forma B
11. *Cyathidites* cf. *C. minor* Couper 1953
11. *Deltoidospora diaphana* Wilson and Webster 1946
12. *Dictyophyllidites* sp.
13. *Foraminisporis undulatus* Leffingwell 1971
14. *Retitriletes* cf. *R. saxatilis* Srivastava 1972
15. *Gabonisporis cristata* (Stanley) Sweet 1986
16. *Ghoshispora bella* (Kondinskaya) Srivastava 1978
17. *Gleicheniidites senonicus* Ross 1949
18. *Heliosporites* sp.
19. *Laevigatosporites haardtii* (Potonie and Venitz) Thomson and Pflug 1953
20. *Leptolepidites* sp.
21. *Leptolepidites tenuis* Stanley 1965
22. *Liburnisporis adnacus* Srivastava 1972
23. *Nevesisporites* cf. *N. radiatus* Srivastava
24. *Baculatisporites* sp.
25. *Perotrilites undulosus* Oltz 1969
26. *Polycingulatisporites* sp.
27. *Reticuloidosporites dentatus* Pflug in Thomson and Pflug 1953
28. *Retitriletes austroclavatidites* (Cookson) Krutzsch 1963
29. *Retitriletes muricatus* Srivastava 1972
30. *Retitriletes* cf. *R. muricatus* Srivastava 1972
31. *Cicatricososporites norrisi* Srivastava 1971
32. *Heliosporites kemensis* (Chlonova) Srivastava 1972

33. *Stereisporites antiquasporites* Wilson and Webster 1946
34. *Cingulatisporites dakotaensis* Stanley 1965
36. *Stereisporites regius* (Drozhastichich) Felix and Burbridge 1973
37. *Lusatisporis dettmannae* (Drugg) Srivastava 1972
38. *Seductisporites barbatus* Tschudy 1973
39. *Triporoletes tornatilis* Srivastava 1972
40. *Polypodiidites* sp.
41. Spore forma C
42. Spore forma D
43. *Sindorapollis pilatus* (Mchedlishvili) Krutzsch 1969
44. Spore forma E
45. *Concavissimisporites* sp. 2
46. *Matonisporites phlebopteroides*
47. *Alisporites grandis* (Cookson) Dettman 1963
49. *Cycadopites scabratus* Stanley 1965
50. *Cycadopites follicularis* Wilson and Webster 1946
52. *Equisetosporites amabilis* Srivastava 1968
53. "*Ephedra*" *voluta* Stanley 1965
54. *Trifossapollenites* cf. *T. ellipticus* Rouse 1957
55. *Ephedripites eocenipites* (Wodehouse) Krutzsch 1970
56. *Sequoipollenites paleocenicus* Stanley 1965
56. *Taxodiaceaepollenites hiatus* Potonie 1931
57. *Sequoipollenites* sp.
59. *Pinuspollenites semicircularis* (Stanley) Wilson 1978
60. *Pinuspollenites elongatus* Norton in Norton and Hall 1969
61. *Pristinuspollenites* cf.*P. inchoatus* (Pierce) Tschudy 1973
62. *Podocarpidites maximus* (Stanley) Norton in Norton and Hall 1969
64. *Rugubivesiculites* sp.
65. *Inaperturopollenites dubius* Potonie and Venitz ex Thomson and Pflug 1953
66. *Callialasporites* cf. *C. segmentatus* (Balme) Srivastava 1963
68. *Singhia urwashii* Srivastava 1968
69. *Liliacidites* sp. 1
70. *Calamuspollenites* cf.*C. pertusus* Elsik 1966
71. *Arecipites* cf. *A. tenuiexinus* Leffingwell 1971 [as "*tenuiexinous*"]
72. *Arecipites* sp. 1
73. *Cycadopites* sp.
74. *Arecipites* sp. 2
75. *Liliacidites altimurus* Leffingwell 1971
76. *Liliacidites complexus* (Stanley) Leffingwell 1971
77. *Liliacidites* sp. 2
78. Echinoreticulate monosulcate
79. *Liliacidites hyalaciniatus* Anderson 1960
80. *Liliacidites* cf. *L. ligniticus* Elsik 1968
81. *Liliacidites* sp. 3
82. *Monocolpopollenites* sp.
83. *Liliacidites* sp. 4
84. *Liliacidites* sp. 5
85. *Liliacidites* sp. 6
86. Sparganiaceae type monoporate
87. *Liliacidites* sp. 7
88. *Liliacidites* sp. 8
89. *Liliacidites* sp. 9
90. *Liliacidites* sp. 10
91. *Liliacidites* sp. 11
92. *Retiinaperturites* sp.
93. *Liliacidites* cf. *L. creticus* Mchedlishvili 1961
94. *Liliacidites* sp. 12
95. *Liliacidites* sp. 13
96. *Liliacidites* sp. 14
97. *Pandaniidites typicus* (Norton) Sweet 1986
98. *Echimonocolpites* sp.
99. *Stellatopollis* sp.
100. *Luminidites* sp.
101. *Marsypiletes cretacea* Robertson 1974
102. *Catinipollis* cf. *C. geiseltalensis* Krutzsch 1966
105. *Penetetrapites inconspicuus* Sweet 1986
106. *Cercidiphyllites* sp.
107. *Phimopollenites* sp. 1
108. *Phimopollenites* sp. 2
110. *Australopollis* sp.
111. *Tricolpites* sp. 1
112. *Fraxinoipollenites variabilis* Stanley 1965
113. *Retitrescolpites anguloluminosus* (Anderson) Frederiksen 1979
114. *Tricolpites bathyreticulatus* Stanley 1965
115. *Tricolpopollenites* cf. *T. clavireticulatus* Norton in Norton and Hall 1969
116. *Tricolpites hyphantos* Farabee and Canright 1986
117. *Tricolpites* sp. 14
119. *Tricolpites hians* Stanley 1965
120. *Echitricolpites* sp.
121. *Tricolpites micromunus* (Groot and Penny) Singh 1971
122. *Tricolpites microreticulatus* Belsky, Boltenhagen and Potonie 1965
123. *Tricolpopollenites microscabratus* Norton in Norton and Hall 1969
124. *Tricolpopollenites obliquimuratus* Chmura 1973
125. *Tricolpites parvistriatus* Norton 1967
126. *Tricolpites parvus* Stanley 1965
127. *Tricolpites* cf. *T. plicatilis* Chmura 1973
128. *Satishia* sp. 1
129. *Albertipollenites* sp. 1
129. *Albertipollenites* sp. 2
129. *Albertipollenites* sp. 3
130. *Foveotricolpites* sp. 1
131. *Retitrescolpites* sp. 1
132. *Tricolpites* sp. 2
133. *Tricolpites* sp. 3
134. *Tricolpites* sp. 4
135. *Tricolpopollenites* cf. *T. opiparus* Chmura 1973
136. *Rousea* sp. 1
137. *Satishia compacta* (Norton) Ward 1986
138. *Tricolpites* cf. *T. psilascabratus* Norton in Norton and Hall 1969
139. *Tricolpites* sp. 5
140. *Rousea* sp. 7
141. *Foveotricolpites* sp. 2
142. *Satishia* sp. 2
143. *Satishia* sp. 3
144. *Dryadopollis* sp. 1
147. *Rousea* sp. 2
148. *Phimopollenites* sp. 3
145. *Tricolpites* sp. 6
149. *Tricolpites* sp. 7
150. *Tricolpites* sp. 1
151. *Tricolpites* sp. 8
152. *Tricolpites* sp. 9
153. *Tricolpites* sp. 10
154. *Tricolpites* sp. 11
155. Tricolporate sp. 2
156. *Tricolpites* sp. 12
157. *Dryadopollis* sp. 2
157. aff. Dryadopollis sp. 2
158. *Satishia* sp. 4
159. *Rousea* sp. 3
160. *Brevicolporites colpella* Anderson 1960
161. *Cupuliferoipollenites pusillus* Potonie 1960

162. *Cupanieidites carlquistii* (Drugg) Chmura 1973
163. *Cupanieidites conspectipolus* Chmura 1973
164. *Myrtaceidites solidus* Sweet 1986
165. *Cupanieidites* sp.
166. *Myrtaceidites* sp. 2
167. *Foveotricolporites* sp. 1
168. *Margocolporites* cf. *M. cribellatus* Srivastava 1972
169. *Margocolporites* sp.
170. Tricolporate sp. 3
171. *Rhoipites* cf. *R. pisinnus* Stanley 1965
173. *Myrtipites* cf. *M. scabratus* Norton in Norton and Hall 1969
174. *Myrtipites* sp. 1
175. *Myrtipites* sp. 2
176. *Myrtipites* sp. 3
177. *Nothopollenites* sp.
178. *Tricolporopollenites incomptus* Chmura 1973
179. *Dryadopollis* sp. 3
180. *Tricolporopollenites* cf. *T. inusitatis* Chmura 1973
181. *Simpsonipollis* sp.
182. *Ilexpollenites* sp. 1
183. *Oacolpopollenites* sp.
184. "*Nyssa*" *annulata* Chmura 1973 [as "*anulata*"]
185. *Rhoipites* cf. *R. cryptoporus* Srivastava 1972
186. *Wilsonipites* sp.
187. *Cupuliferoipollenites* sp.
188. *Rhoipites* sp. 1
189. *Rhoipites* sp. 2
190. *Rhoipites* sp. 3
191. *Tricolporopollenites* cf. *T. striatus* Norton in Norton and Hall 1969
192. *Rutihesperipites trochuensis* Srivastava 1977
193. *Rutihesperipites* sp.
194. *Simpsonipollis* sp.
195. *Rhoipites* sp. 8
196. *Tricolporopollenites* cf. *T. desultorius* Chmura 1973
197. *Rousea* sp. 4
198. *Rousea* sp. 5
199. *Rousea* sp. 4
200. *Rousea* sp. 6
201. *Tricolporopollenites* cf. *T. inductorius* Chmura 1973
202. *Tricolpites mutabilis* Leffingwell 1971
203. *Tricolpopollenites tesselatus* Tschudy 1973
204. Tricolporate sp. 14
205. Tricolporate sp. 4
206. Tricolporate sp. 5
207. *Rhoipites* sp. 5
208. *Ilexpollenites* sp. 2
209. *Tricolporopollenites* cf. *T. viriosus* Chmura 1973
210. Tricolporate sp. 6
211. *Tricolpites* sp. 13
212. *Tricolporites rhomboides* Anderson 1960
213. *Tricolpopollenites* cf. *T. levitas* Tschudy
214. Tricolporate sp. 7
215. Tricolporate sp. 8
217. *Foveotricolporites* sp. 2
219. Tricolporate sp. 9
220. *Foveotricolporites* sp. 3
221. *Rhoipites* sp. 6
223. *Horniella* sp.
226. Tricolporate sp. 10
227. Tricolporate sp. 11
229. Tricolporate sp. 12
228. *Rhoipites* sp. 7
230. *Ailanthipites cupidineus* Srivastava 1969
231. *Porocolpopollenites* cf. *P. sepositus* Chmura 1973
232. *Simpsonipollis* sp. 1
233. Tricolporate sp. 13
234. *Tricolpopollenites parvicolpatus* Norton in Norton and Hall 1969
235. *Aquilapollenites amplus* Stanley 1901
236. *Aquilapollenites amicus* Srivastava 1968
237. *Aquilapollenites conatus* Norton 1965
238. aff. *Aquilapollenites conatus* Norton 1965
239. *Aquilapollenites delicatus* Stanley 1961
240. *Aquilapollenites delicatus var. collaris* Tschudy and Leopold 1970
241. *Aquilapollenites pyriformis* Norton 1965
242. *Aquilapollenites quadricretaeus* Chlonova 1961
243. *Aquilapollenites quadrilobus* Rouse 1957
244. *Aquilapollenites reductus* Norton 1965
245. *Aquilapollenites reticulatus* Mchedlishvili 1961
246. *Aquilapollenites turbidus* Tschudy and Leopold 1970
247. *Aquilapollenites* sp. 1
248. *Mancicorpus* cf. *M. rostratus* Srivastava 1968
249. *Aquilapollenites catenreticulatus* Srivastava 1968
250. *Aquilapollenites* sp. 2
251. *Polotricolporites rotundus* Sweet 1986
252. *Cranwellia rumseyensis* Srivastava 1966
253. *Cranwellia baccata* Srivastava 1969
254. *Cranwellia asperata* Srivastava 1969
255. *Accuratipollis macrosolenoides* (Mchedlishvili) Krutzsch 1969
256. *Striatellipollis radiata* Sweet 1986
257. *Porosipollis porosus* (Mchedlishvili) Krutzsch 1969
258. *Aquilapollenites* sp. 3
259. *Grewipollenites radiatus* Tschudy 1973
260. *Kurtzipites annulatus* Norton in Norton and Hall 1969
260. *Kurtzipites circularis* (Norton) Srivastava 1981
261. *Kurtzipites trispissatus* Anderson 1960
263. *Sindorapollis granulatus* Tschudy 1973
264. *Jarzenipollenites* spp.
266. *Momipites tenuipolis* Anderson 1960
266. *Momipites* cf. *M. triletipollenites* (Rouse) Nichols 1973
267. *Paraalnipollenites confusus* (Zaklinskaya) Hills and Wallace 1969
268. *Proteacidites thalmannii* Anderson 1960
269. *Proteacidites marginus* Rouse 1962
270. *Proteacidites retusus* Anderson 1960
271. *Proteacidites* sensu lato sp. 1
273. *Proteacidites* sensu lato sp. 2
272. *Retibrevitricolporites beccus* Sweet
274. *Polyatriopollenites* sp.
275. polyoperculate
276. "*Tilia*" *wodehousei* Anderson 1960
277. *Triatriopollenites costatus* Norton in Norton and Hall 1969
278. *Triatriopollenites granilabratus* (Stanley) Norton in Norton and Hall 1969
279. "*Betula*" *infrequens* Stanley 1965
280. "*Carpinus*" *subtriangulus* Stanley 1965
281. *Triporopollenites plektosus* Anderson 1960
282. Triporate 1
283. Triporate 2
284. ?*Compositoipollenites* sp.
285. *Ulmipollenites* spp.
286. *Jarzenipollenites trinus* (Stanley) Kedves 1980
287. *Leptopecopites pocockii* Srivastava 1978
288. *Wodehouseia spinata* Stanley 1961
289. *Wodehouseia fimbriata* Stanley 1961
290. *Anacolosidites rotundus* Stanley 1965
292. *Libopollis jarzenii* Farabee et al. 1984

293. *Erdtmanipollis procumbentiformis* (Samoilovich) Krutzsch 1962
294. "*Polyporina*" *cribaria* Srivastava 1969
295. *Disulcipollis* sp.
296. "*Schizosporis*" *scabratus* Stanley 1965
297. ?*Stelckia* sp.
299. stephanoporate
300. *Dyadonapites reticulatus* Erdtman ex Tschudy 1973
301. *Inaperturotetradites* sp.
302. *Inaperturotetradites scabratus* Tschudy 1973
303. *Foveotetradites* sp.
304. *Dicotetradites granulatus* Norton in Norton and Hall 1969
305. Tetrad sp.
306. *Verrutricolpites* sp.
307. *Retiinaperturites* sp. 2
308. *Proteacidites* sensu lato sp. 3
309. polyad

Tables A1–A5 show classification of species according to presence above and below the iridium-enriched claystone.

TABLE A1. K (CRETACEOUS) SPECIES (THOSE APPEARING ONLY BELOW IRIDIUM DATUM)

Species	Below Ir (N = 46)	Above Ir (N = 103)	p-value
15. *Gabonisporis cristata*	16	0	0.000
18. *Heliosporites* spp.	9	0	0.000
22. *Liburnisporis adnacus*	6	0	0.001
23. cf. *Nevesisporites radiatus*	7	0	0.000
25. *Perotrilites undulosus*	9	0	0.000
43. *Sindorapollis pilatus*	5	0	0.001
44. Spore forma E	6	0	0.001
52. *Equisetosporites amabilis*	12	0	0.000
55. *Ephedripites eocenicus*	21	0	0.000
61. cf. *Pristinuspollenites inchoatus*	24	0	0.000
73. *Cycadopites* sp.	6	0	0.002
76. *Liliacidites complexus*	30	0	0.000
93. cf. *Liliacidites creticus*	10	0	0.000
94. *Liliacidites* sp. 12	24	0	0.000
99. *Stellatopollis* sp.	9	0	0.000
101. *Marsypiletes cretacea*	17	0	0.000
130. *Foveotricolpites* sp. 1	5	0	0.002
155. Tricolporate sp. 2	9	0	0.000
189. *Rhoipites* sp. 2	5	0	0.002
195. *Rhoipites* sp. 8	9	0	0.000
227. Tricolporate sp. 11	11	0	0.000
235. *Aquilapollenites amplus*	23	0	0.000
237. *Aquilapollenites conatus*	16	0	0.000
239. *Aquilapollenites delicates*	6	0	0.001
240. *Aquilapollenites delicatus var. collaris*	39	0	0.000
241. *Aquilapollenites pyriformis*	6	0	0.001
242. *Aquilapollenites quadricretaeus*	12	0	0.000
255. *Accuratipollis macrosolenoides*	11	0	0.000
256. *Striatellipollis radiata*	8	0	0.000
257. *Porosipollis porosus*	6	0	0.001
259. *Grewipollenites radiatus*	5	0	0.002
263. *Sindorapollis granulatus*	15	0	0.000
268. *Proteacidites thalmanii*	41	0	0.000
269. *Proteacidites marginus*	18	0	0.000
270. *Proteacidites retusus*	12	0	0.000
272. *Retibrevicolporites beccus*	10	0	0.000
273. *Proteacidites* sp. 2	8	0	0.000
275. polyoperculate	17	0	0.000
287. *Leptopecopites pocockii*	29	0	0.000
292. *Libopollis jarzenii*	8	0	0.000
299. *Stephanoporopollenites* sp.*	10	0	0.000

*Plus 57 rare species confined to the Hell Creek.

TABLE A2. K SPECIES WITH RARE OCCURRENCES ABOVE IRIDIUM

Species	Below Ir (N = 46)	Above Ir (N = 103)	p-value
6. *Cicatricosisporites* sp. 1	10	1	0.000
8. *Concavissimisporites variverrucatus*	29	1	0.000
13. *Foraminisporis undulates*	27	1	0.000
16. *Ghoshispora bella*	25	2	0.000
29. *Retitriletes muricatus*	20	2	0.000
54. cf. *Trifossapollenites ellipticus*	23	5	0.000
69. *Liliacidites* sp. 1.	23	1	0.000
71. cf. *Arecipites tenuiexinus*	34	1	0.000
75. *Liliacidites altimurus*	27	1	0.000
78. echinoreticulate monosulcate	32	9	0.000
79. *Liliacidites hyalaciniatus*	32	1	0.000
100. *Luminidites* sp.	6	1	0.004
122. *Tricolpites microreticulatus*	41	2	0.000
167. *Foveotricolporites* sp. 1	18	1	0.000
184. "*Nyssa*" *annulata*	26	3	0.000
204. tricolporate sp. 14	19	5	0.000
243. *Aquilapollenites quadrilobus*	32	2	0.000
244. *Aquilapollenites reductus*	15	2	0.000
245. *Aquilapollenites reticulates*	14	11	0.004
246. *Aquilapollenites turbidus*	13	4	0.000
252. *Cranwellia rumseyensis*	38	8	0.000
253. *Cranwellia baccata*	8	1	0.000
302. *Inaperturotetradites scabratus*	19	4	0.000

TABLE A3. K-T SPECIES (T-DECLINE SUBSET) DISPLAYING SIGNIFICANT DECLINE IN PRESENCE ABOVE IRIDIUM

Species	Below Ir (N = 46)	Above Ir (N = 103)	p-value
1. *Azolla* spp.	27	22	0.000
5. *Camarazonosporites concinnus*	23	16	0.000
14. cf. *Retitriletes saxatilis*	9	3	0.001
20. *Leptolepidites* sp.	17	12	0.001
21. *Leptolepidites tenuis*	21	21	0.003
28. *Retitriletes austroclavatidites*	29	13	0.000
38. *Seductosporites barbatus*	10	4	0.000
39. *Triporoletes tornatilis*	27	10	0.000
40. *Polypodiidites* sp.	10	6	0.008
47. *Alisporites grandis*	28	33	0.001
70. cf. *Calamuspollenites pertusus*	20	14	0.000
77. *Liliacidites* sp. 2	8	5	0.023
85. *Liliacidites* sp. 6	12	8	0.002
86. Sparganiaceae-type monoporate	10	5	0.003
97. *Pandaniidites typicus*	27	39	0.021
105. *Penetetrapites inconspicuus*	27	11	0.000
108. *Phimopollenites* sp. 2	9	6	0.016
115. *Tricolpopollenites clavireticulatus*	14	13	0.012
120. *Echitricolpites* sp.	21	19	0.001
125. *Tricolpites parvistriatus*	36	26	0.000
164. *Myrtaceidites solidus*	17	20	0.026
178. *Tricolporopollenites incomptus*	19	15	0.001
194. *Simpsonipollis* sp. 2	13	11	0.014
212. cf. *Tricolporites rhomboides*	7	2	0.000
234. cf. *Tricolpopollenites parvicolpatus*	13	11	0.014
261. *Kurtzipites trispissatus*	36	27	0.000
281. *Triporopollenites plektosus*	36	46	0.000
285. *Ulmipollenites* spp.	41	74	0.021
286. *Jarzenipollis* spp.	31	44	0.008
288. *Wodehouseia spinata*	36	25	0.000
293. *Erdtmanipollis procumbentiformis*	27	27	0.000
300. *Dyadonapites reticulatus*	28	40	0.020
304. *Dicotetradites granulata*	38	64	0.014

TABLE A4. K-T SPECIES (THOSE SHOWING NO SIGNIFICANT CHANGE IN PRESENCE ABOVE IRIDIUM DATUM)

Species	Below Ir (N = 46)	Above Ir (N = 103)	p-value
2. *Baculatisporites comaumensis*	14	44	0.203
4. *Hamulatisporis amplus*	10	25	0.836
7. *Cicatricosisporites* sp. 2	6	21	0.263
9. *Concavisporites jurienensis*	26	44	0.155
11. *Cyathidites* spp.	41	93	0.777
12. *Dictyophyllidites* sp.	41	96	0.515
17. *Gleicheniidites senonicus*	32	71	1.000
19. *Laevigatosporites* spp.	43	96	1.000
33. *Stereisporites antiquasporites*	27	68	0.461
36. *Stereisporites regius*	12	41	0.138
49. *Cycadopites* spp.	20	27	0.055
56. Taxodiaceous pollen	43	100	0.373
59. *Pinuspollenites semicircularis*	36	83	0.826
60. *Pinuspollenites elongatus*	16	41	0.589
62. *Podocarpidites maximus*	17	41	0.856
65. *Inaperturopollenites dubius*	8	20	0.825
74. *Arecipites* sp. 2	4	9	1.000
80. cf. *Liliacidites ligniticus*	9	12	0.211
81. *Liliacidites* sp. 3	30	74	0.444
107. *Phimopollenites* sp. 1	13	45	0.098
112. *Fraxinoipollenites variabilis*	14	33	1.000
113. *Retitrescolpites anguloluminosus*	11	13	0.095
114. *Tricolpites bathyreticulatus*	9	13	0.319
117. *Tricolpites* sp. 14	2	8	0.724
121. *Tricolpites micromunus*	36	63	0.060
123. *Cupuliferoidaepollenites microscabratus*	25	56	1.000
124. *Tricolpites obliquimuratus*	7	15	1.000
126. *Tricolpites parvus*	41	82	0.242
127. cf. *Tricolpites plicatilis*	16	35	1.000
137. *Satishia compacta*	37	71	0.168
143. *Satishia* sp. 3	3	7	1.000
157. *Dryadopollis* sp. 2	10	22	1.000
158. *Satishia* sp. 4	5	17	0.459
161. *Cupuliferoipollenites pusillus*	6	10	0.573
162. *Cupanieidites carlquistii*	20	48	0.859
165. *Cupanieidites* sp. 1	3	13	0.392
168. cf. *Margocolporites cribellatus*	1	8	0.068
173. *Myrtipites scabratus*	5	6	0.316
175. *Myrtipites* sp. 2	11	27	0.841
179. *Dryadopollis* sp. 3	5	18	0.340
181. cf. *Simpsonipollis* sp. 1	5	15	0.613
182. cf. *Ilexpollenites* sp. 1	3	8	1.000
185. *Rhoipites cryptoporus*	9	3	0.319
186. *Wilsonipites* sp.	3	7	1.000
187. *Cupuliferoipollenites* sp.	13	26	0.692
192. *Rutihesperipites trochuensis*	41	79	0.115
199. *Rousea* sp. 4	1	9	0.175
203. *Tricolpites tesselatus*	13	45	0.101
217. *Foveotricolporites* sp. 2	1	12	0.066
251. *Polotricolporites rotundus*	13	39	0.272
260. *Kurtzipites circularis/K. annulatus*	41	76	0.051
266. *Momipites* spp.	6	19	0.298
267. *Paraalnipollenites confusus*	8	27	0.298
277. *Triatriopollenites costatus*	38	72	0.112
294. "*Polyporina*" *cribarius**	2	2	0.228

*Plus 35 rare taxa

TABLE A5. T (TERTIARY) SPECIES (THOSE UNDERGOING SIGNIFICANT INCREASE IN EITHER PRESENCE OR FREQUENCY ABOVE IRIDIUM)

Species	Below Ir (N = 46)	Above Ir (N = 103)	p-value
27. *Reticuloidosporites dentatus*	20	82	0.000
30. aff. *Retitriletes muricatus*	11	31	0.555
34. *Stereisporites dakotaensis*	7	38	0.007
72. *Arecipites* sp. 1	0	17	0.002
145. *Tricolpites* sp. 6	14	49	0.074
160. *Brevicolporites colpella*	8	34	0.075
171. cf. *Rhoipites pisinnus*	13	63	0.000
193. *Rutihesperipites* sp.	6	25	0.132
201. *Tricolporopollenites cf. T. inductorius*	1	20	0.002
213. cf. *Tricolpites levitas*	7	36	0.018
279. "*Betula*" *infrequens*	5	37	0.002
280. "*Carpinus*" *subtriangulus*	12	71	0.000
306. *Verrutricolpites* sp.	8	30	0.157
141. *Foveotricolporites* sp. 2	0	8	0.059
229. *Tricolporopollenites* sp. 12	0	6	0.178
289. *Wodehouseia fimbriata**	0	6	0.178

*Plus 21 rare taxa.

TABLE A6. ONE-WAY ANOVA OF ABUNDANCE OF SELECTED K SPECIES THROUGH HELL CREEK*

8. *Concavissimisporites variverrucatus*

Source	DF	SS	MS	F
Factor	3	0.333	0.111	0.54
Error	30	6.130	0.204	—
Total	33	6.463	—	—
Sample category	N	Mean	SD	
USTK	4	0.0175	0.0236	
UHC	9	0.3400	0.6578	
MHC	16	0.1981	0.3982	
LHC	5	0.3040	0.2685	

Pooled SD = 0.4520

13. *Foraminisporis undulates*

Source	DF	SS	MS	F
Factor	3	0.307	0.102	0.30
Error	30	10.079	0.336	—
Total	33	0.386	—	—
Sample category	N	Mean	SD	
USTK	4	0.1400	0.2406	
UHC	9	0.1189	0.2167	
MHC	16	0.3313	0.7860	
LHC	5	0.2200	0.2561	

Pooled SD = 0.5796

16. *Ghoshispora bella*

Source	DF	SS	MS	F
Factor	3	8.78	2.93	0.86
Error	30	102.16	3.41	—
Total	33	110.94	—	—
Sample category	N	Mean	SD	
USTK	4	0.0100	0.0115	
UHC	9	0.0633	0.1640	
MHC	16	1.1062	2.6036	
LHC	5	0.2200	0.2561	

Pooled SD = 1.8454

29. *Retitriletes muricatus*

Source	DF	SS	MS	F
Factor	3	0.1368	0.0456	0.56
Error	30	2.4516	0.0817	—
Total	33	2.5884	—	—
Sample category	N	Mean	SD	
USTK	4	0.2500	0.5000	
UHC	9	0.1800	0.2402	
MHC	16	0.1287	0.2871	
LHC	5	0.0200	0.0283	

Pooled SD = 0.2859

54. *Trifossapollenites cf. T ellipticus*

Source	DF	SS	MS	F
Factor	3	6.325	2.108	8.37
Error	29	7.306	0.252	—
Total	32	13.630	—	—
Sample category	N	Mean	SD	
USTK	4	1.3875	1.5332	
UHC	8	0.0525	0.0623	
MHC	16	0.0550	0.1227	
LHC	5	0.0120	0.0109	

Pooled SD = 0.50191

(continued)

TABLE A6. ONE-WAY ANOVA OF ABUNDANCE OF SELECTED K SPECIES THROUGH HELL CREEK* (continued)

55. *Ephedripites eocenicus*

Source	DF	SS	MS	F
Factor	3	0.4211	0.1404	1.79
Error	30	2.3496	0.0783	—
Total	33	2.7706	—	—
Sample category	N	Mean	SD	
USTK	4	0.3875	0.7420	
UHC	9	0.0911	0.1588	
MHC	16	0.0500	0.1241	
LHC	5	0.2200	0.2573	

Pooled SD = 0.2798

61. *Pristinuspollenites cf. P. inchoatus*

Source	DF	SS	MS	F
Factor	3	3.500	1.167	2.92
Error	30	11.994	0.400	—
Total	33	15.494	—	—
Sample category	N	Mean	SD	
USTK	4	0.0150	0.0191	
UHC	9	0.0222	0.0524	
MHC	16	0.0237	0.0328	
LHC	5	0.9280	1.7288	

Pooled SD = 0.6323

69. *Liliacidites* sp. 1

Source	DF	SS	MS	F
Factor	3	4.192	1.397	5.90
Error	30	7.105	0.237	—
Total	33	11.296	—	—
Sample category	N	Mean	SD	
USTK	4	1.1750	1.3841	
UHC	9	0.0867	0.1949	
MHC	16	0.1231	0.2650	
LHC	5	0.0040	0.0089	

Pooled SD = 0.48665

71. *Arecipites cf. A. tenuiexinus*

Source	DF	SS	MS	F
Factor	3	13.66	4.55	3.35
Error	30	40.78	0.36	—
Total	33	4.44	—	—
Sample category	N	Mean	SD	
USTK	4	1.8750	0.6292	
UHC	8	1.8975	1.9229	
MHC	17	0.7412	0.8985	
LHC	5	0.2040	0.4451	

Pooled SD = 1.1659

75. *Liliacidites altimurus*

Source	DF	SS	MS	F
Factor	3	0.794	0.265	1.46
Error	30	5.431	0.181	—
Total	33	6.225	—	—
Sample category	N	Mean	SD	
USTK	4	0.1700	0.2312	
UHC	9	0.2111	0.3382	
MHC	16	0.4287	0.5388	
LHC	5	0.0160	0.0167	

Pooled SD = 0.4255

TABLE A6. ONE-WAY ANOVA OF ABUNDANCE OF SELECTED K SPECIES THROUGH HELL CREEK* (continued)

76. *Liliacidites complexus*

Source	DF	SS	MS	F
Factor	3	0.93	0.31	0.21
Error	30	44.10	1.47	—
Total	33	45.03	—	—
Sample category	N	Mean	SD	
USTK	4	0.1350	0.2441	
UHC	9	0.4533	1.1540	
MHC	16	0.5606	1.3858	
LHC	5	0.7520	1.0565	

Pooled SD = 1.2125

78. *Echinoreticulate monosulcate*

Source	DF	SS	MS	F
Factor	3	2.815	0.938	3.07
Error	30	9.161	0.305	—
Total	33	11.977	—	—
Sample category	N	Mean	SD	
USTK	4	0.2500	0.2887	
UHC	9	0.2178	0.3491	
MHC	16	0.4313	0.5670	
LHC	5	1.1080	0.8822	

Pooled SD = 0.5526

79. *Liliacidites hyalaciniatus*

Source	DF	SS	MS	F
Factor	3	1.563	0.521	2.16
Error	30	7.251	0.242	—
Total	33	8.814	—	—
Sample category	N	Mean	SD	
USTK	4	0.8875	0.8370	
UHC	9	0.5356	0.5261	
MHC	16	0.3081	0.4293	
LHC	5	0.1480	0.2062	

Pooled SD = 0.4916

122. *Tricolpites microreticulatus*

Source	DF	SS	MS	F
Factor	3	531.6	177.2	2.02
Error	30	2630.9	7.7	—
Total	33	3162.5	—	—
Sample category	N	Mean	SD	
USTK	4	0.5050	0.7024	
UHC	9	6.2533	4.5278	
MHC	16	11.6250	11.4623	
LHC	5	12.6000	11.1209	

Pooled SD = 9.3647

167. *Foveotricolporites* sp. 1

Source	DF	SS	MS	F
Factor	3	0.0178	0.0059	0.40
Error	30	0.4446	0.0148	—
Total	33	0.4624	—	—
Sample category	N	Mean	SD	
USTK	4	0.0050	0.0100	
UHC	9	0.0667	0.1631	
MHC	16	0.0556	0.1241	
LHC	5	0.0120	0.0109	

Pooled SD = 0.1217

(*continued*)

TABLE A6. ONE-WAY ANOVA OF ABUNDANCE OF SELECTED K SPECIES THROUGH HELL CREEK* (continued)

184. *"Nyssa" annulata*

Source	DF	SS	MS	F
Factor	3	2.352	0.784	2.28
Error	30	10.296	0.343	—
Total	33	12.648	—	—
Sample category	N	Mean	SD	
USTK	4	0.0000	0.0000	
UHC	9	0.0340	0.0313	
MHC	16	1.610	4.9300	
LHC	5	8.200	1.2900	

Pooled SD = 0.5858

204. Tricolporate sp. 14

Source	DF	SS	MS	F
Factor	3	0.603	0.201	0.42
Error	30	14.219	0.474	—
Total	33	14.821	—	—
Sample category	N	Mean	SD	
USTK	4	0.0050	0.0100	
UHC	9	0.4600	0.5142	
MHC	16	0.2675	0.8711	
LHC	5	0.2440	0.4241	

Pooled SD = 0.6884

235. *Aquilapollenites amplus*

Source	DF	SS	MS	F
Factor	3	0.4016	0.1339	2.30
Error	30	1.7484	0.0583	—
Total	33	2.1500	—	—
Sample category	N	Mean	SD	
USTK	4	0.0250	0.0300	
UHC	9	0.0611	0.0526	
MHC	6	0.0372	0.0575	
LHC	5	0.3480	0.6467	

Pooled SD = 0.2414

239/240. *Aquilapollenites delicatus*

Source	DF	SS	MS	F
Factor	3	15.47	5.16	1.00
Error	30	154.9	9.5.17	—
Total	33	170.46	—	—
Sample category	N	Mean	SD	
USTK	4	2.1250	3.2755	
UHC	9	1.0978	1.2073	
MHC	16	2.3550	2.6775	
LHC	5	0.7280	0.9500	

Pooled SD = 2.2730

243. *Aquilapollenites quadrilobus*

Source	DF	SS	MS	F
Factor	3	2.46	0.82	0.47
Error	30	52.90	1.76	
Total	33	55.36		
Sample category	N	Mean	SD	
USTK	4	1.1300	1.3093	
UHC	9	0.7222	0.8271	
MHC	16	1.0325	1.6735	
LHC	5	0.3080	0.2629	

Pooled SD = 1.3279

TABLE A6. ONE-WAY ANOVA OF ABUNDANCE OF SELECTED K SPECIES THROUGH HELL CREEK* (continued)

252. *Cranwellia rumseyensis*

Source	DF	SS	MS	F
Factor	3	5.59	1.86	0.98
Error	30	57.12	1.90	
Total	33	62.72		

Sample category	N	Mean	SD
USTK	4	1.8800	2.8351
UHC	9	0.4689	0.5077
MHC	16	0.7988	1.1122
LHC	5	0.8520	1.7603

Pooled SD = 1.3279

268. *Proteacidites thalmanii*

Source	DF	SS	MS	F
Factor	3	36.11	12.04	5.53
Error	30	65.35	2.18	—
Total	33	101.46	—	—

Sample category	N	Mean	SD
USTK	4	3.6500	3.8717
UHC	9	0.9222	1.2684
MHC	16	0.5331	0.7071
LHC	5	0.0360	0.0590

Pooled SD = 1.4756

275. *Polyporate*

Source	DF	SS	MS	F
Factor	3	0.676	0.225	1.26
Error	30	5.357	0.179	—
Total	33	6.033	—	—

Sample category	N	Mean	SD
USTK	4	0.3750	0.4787
UHC	9	0.1189	0.3307
MHC	16	0.1575	0.3961
LHC	5	0.5120	0.6003

Pooled SD = 0.4226

287. *Leptopecopites pocockii*

Source	DF	SS	MS	F
Factor	3	11.07	3.69	2.56
Error	30	43.24	1.44	—
Total	33	54.31	—	—

Sample category	N	Mean	SD
USTK	4	0.6250	0.6292
UHC	9	1.5711	1.5732
MHC	16	0.5300	0.9147
LHC	5	1.9000	1.5572

Pooled SD = 1.2005

302. *Inaperturotetradites scabratus*

Source	DF	SS	MS	F
Factor	3	0.3951	0.1317	1.36
Error	30	2.9028	0.0968	—
Total	33	3.2979	—	—

Sample category	N	Mean	SD
USTK	4	0.1250	0.2500
UHC	9	0.2867	0.5604
MHC	16	0.0250	0.0423
LHC	5	0.1320	0.2100

Pooled SD = 0.3111

Note: See Table 1 for sample category abbreviations.
*DF = degrees of freedom; SS = sum of squares; MS = mean squares; F = F-value (sample variance ratio).

REFERENCES CITED

Alvarez, L.W., 1983, Experimental evidence that an asteroid impact led to the extinction of many species 65 million years ago: National Academy of Sciences Proceedings, v. 80, p. 627–642.

Alvarez, L.W., Alvarez, W., Asaro, F., and Michel, H.V., 1980, Extraterrestrial cause for the Cretaceous-Tertiary extinction: Science, v. 208, p. 1095–1108.

Alvarez, W., Alvarez, L.W., Asaro, F., and Michel, H.V., 1984, The end of the Cretaceous: Sharp boundary or gradual transition?: Science, v. 223, p. 1183–1186.

Alvarez, W., Caeys, P., and Kieffer, S.W., 1995, Emplacement of Cretaceous-Tertiary boundary shocked quartz from Chicxulub crater: Science, v. 269, p. 930–935.

Archibald, J.D., 1982, A study of Mammalia and geology across the Cretaceous-Tertiary boundary in Garfield County, Montana: University of California Publications in Geological Sciences, v. 122, 286 p.

Archibald, J.D., and Bryant, L.J., 1990, Differential Cretaceous/Tertiary extinctions of nonmarine vertebrates; evidence from northeastern Montana, *in* Sharpton, V.L., and Ward, P.D., eds., Global catastrophes in Earth history: Boulder, Colorado, Geological Society of America Special Paper 247, p. 549–562.

Arens, N.C., and Jahren, A.H., 2000, Carbon isotope excursion in atmospheric CO_2 at the Cretaceous-Tertiary boundary: evidence from terrestrial sediments: Palaios, v. 15, p. 314–322.

Askin, R.A., 1990, Campanian to Paleocene spore and pollen assemblages of Seymour Island, Antarctica: Review of Palaeobotany and Palynology, v. 65, p. 105–113.

Baadsgaard, H., Lerbekmo, J.F., and McDougall, I., 1988, A radiometric age for the Cretaceous-Tertiary boundary based on K-Ar, Rb-Sr, and U-Pb ages of bentonites from Alberta, Saskatchewan and Montana: Canadian Journal of Earth Sciences, v. 25, p. 1088–1097.

Bohor, B.F., 1990, Shocked quartz and more; Impact signatures in Cretceous/Tertiary boundary clays, *in* Sharpton, V.L., and Ward, P.D., eds., Global catastrophes in Earth history: Boulder, Colorado, Geological Society of America Special Paper 247, p. 335–343.

Bohor, B.F., Foord, E.E., Modreski, P.J., and Triplehorn, D.M., 1984, Mineralogic evidence for an impact event at the Cretaceous-Tertiary boundary: Science, v. 224, p. 867–869.

Bohor, B.F., Modreski, P.J., and Foord, E.E., 1987a, Shocked quartz in the Cretaceous-Tertiary boundary clays: Evidence for a global distribution: Science, v. 236, p. 705-709.

Bohor, B.F., Triplehorn, D.M., Nichols, D.J., and Millard, H.T., 1987b, Dinosaurs, spherules, and the magic layer—A new K-T boundary clay site in Wyoming: Geology, v. 15, p. 896–899.

Brett, R., 1992, The Cretaceous-Tertiary extinction—A lethal mechanism involving anhydrite target rocks: Geochimica et Cosmochimica Acta, v. 56, p. 3603–3606.

Cherven, V.B., and Jacob, A.F., 1985, Evolution of Paleogene depositional systems, Williston basin, in response to global sea level changes: *in* Flores, R.M., and Kaplan, S.S., eds., Cenozoic paleogeography of the west central United States: Rocky Mountain Symposium No. 3: Society of Economic Paleontologists and Mineralogists, Rocky Mountain Section, Denver, p. 127–170.

Clemens, W.A., and Archibald, J.D., 1981, Out with a whimper not a bang: Paleobiology, v. 7, p. 293–298.

D'Hondt, S., Donaghay, P., Zachos, J.C., Luttenberg, D., and Lindinger, M., 1998, Organic carbon fluxes and ecological recovery from the Cretaceous-Tertiary mass extinction: Science, v. 282, p. 276–279.

Elliot, D.H., Askin, R.A., Kyte, F.T., and Zinsmeister, W.J., 1994, Iridium and dinocysts at the Cretaceous/Tertiary boundary on Seymour Island, Antarctica: Implications for the K-T event: Geology, v. 22, p. 675–678.

Fastovsky, D.E., 1987, Paleoenvironments of vertebrate-bearing strata during

TABLE A7. LOCATION, CATEGORY AND PROPORTION OF K, T, AND K-T SPECIES IN EACH SAMPLE

Section* and stratigraphic position	Sample category†	Sample number	K	T	K-T	Total
Garfield County						
Billy Creek						
19.00 m above Ir	TCOALSP	BiP83-13	1	6	24	31
0.16–0.18 m above Ir	UPIR	BiZ83-12-14	2	5	33	40
0.12–0.14 m above Ir	UPIR	BiZ83-16-18	1	6	29	36
0.10–0.12 m above Ir	UPIR	BiZ83-18-20	0	6	30	36
0.06–0.10 m above Ir	UPIR	BiZ83-20-24	0	7	24	31
0.02–0.04 m above Ir	UPIR	BiZ83-26-28	0	7	24	30
0.00–0.02 m above Ir	BASIR	BiZ83-28-30	4	3	24	31
0.02–0.04 m below Ir	USTK	BiZ83-34-36	16	1	23	40
14.0 m below Ir	MHC	Bi83-15	31	0	37	68
42.0 m below Ir	LHC	Bi83-10	34	1	42	77
52.0 m below Ir	LHC	Bi83-7	34	0	39	73
70.0 m below Ir	LHC	Bi83-4	23	3	43	71
Brownie East#						
~19.0 m above Ir	TCLAS	BE82-10	0	9	46	55
~18.7 m above Ir	TCLAS	BE83-4	0	3	30	33
~14.1 m above Ir	TCLAS	BE82-9a	0	6	40	46
~13.9 m above Ir	TCOALSP	BEU83-30-40A	0	5	32	37
~13.8 m above Ir	TCLAS	BEU83-50-60	1	7	32	40
~13.7 m above Ir	TBASC	BEU83-60-66	0	9	31	40
~13.4 m above Ir	TCOAL	BEU83-70-74	1	10	36	47
~10.7 m above Ir	TCLAS	BEL83-22-28	0	7	32	42
~10.5 m above Ir	TCLAS	BEL83-31-34	0	5	31	36
~10.5 m above Ir	TCLAS	BEL83-45-47	0	6	32	38
~10.45 m above Ir	TCOALSP	BEL83-51-54	0	3	12	15
~10.20 m above Ir	TCLAS	BEL83-74-77	1	6	21	28
~9.90 m above Ir	TCOALSP	BEL83-106-108	0	9	34	43
~9.85 m above Ir	TCOALSP	BEL83-108-10	0	3	19	22
~9.80 m above Ir	TCOAL	BEL83-110-12	0	5	30	36
~9.75 m above Ir	TCOAL	BEL83-113-15	1	4	29	34
~9.50 m above Ir	TCOAL	BEL83-140-42	0	3	35	38
~9.45 m above Ir	TBASC	BEL83-145-50	2	6	25	33
~9.35 m above Ir	TCLAS	BEL83-150-54	0	3	31	34
Brownie North#						
~7.9 m above Ir	TCLAS	BN82-10	0	5	31	36
~3.0 m above Ir	TCLAS	BN82-60-70A	1	4	27	32
~2.50 m above Ir	TCLAS	BN82-20-30A	1	7	32	40
~2.30 m above Ir	TCLAS	BN82-0-10A	1	4	19	24
~2.20 m above Ir	TCOALSP	BN82-0-2	0	7	34	41
~2.16 m above Ir	TCOALSP	BN82-4-6	0	3	28	31
~2.14 m above Ir	TCOALSP	BN82-6-8	1	3	15	19
~2.12 m above Ir	TBASC, SP	BN82-8-10	0	3	19	22
~2.10 m above Ir	TCLAS	BN82-11-15	0	2	23	25
~2.05 m above Ir	TBASC	BN82-23-25	1	7	41	49
~2.00 m above Ir	TCLAS	BN82-10-20B	2	8	34	44
Browne Butte (Rick's Place)						
0.08–0.10 m above Ir	UPIR	RPI83-4	1	5	26	32
0.06–0.08 m above Ir	UPIR	RPC83-6	1	4	23	28
0.04–0.06 m above Ir	UPIR	RPC83-5	0	4	30	34
0.02–0.04 m above Ir	UPIR	RPC83-4	0	7	24	33
0.00–0.02 m above Ir	BASIR	RPC83-3	1	7	25	33
0.00–0.02 m below Ir	USTK	RPC83-1	29	4	51	83
1.30 m below Ir	UHC	BN82-4	23	1	36	60
3.60 m below Ir	MHC	PM83-2	22	1	38	61
Herpijunk North						
33.75 m above Ir	TCOALSP	HNE83-45-50	0	8	28	36
33.60 m above Ir	TCOAL	HNE83-65-74	0	7	24	34
33.15 m above Ir	TCOAL	HNE83-118-29	1	9	34	44
32.80 m above Ir	TCOALSP	HNE83-Y3	0	4	20	24
26.05 m above Ir	TCLAS	HNE83-15	2	8	33	43
25.35 m above Ir	TCLAS	HNE83-10	1	4	35	35
19.30 m above Ir	TCOALSP	HNE83-9	1	6	31	38

(*continued*)

TABLE A7. LOCATION, CATEGORY AND PROPORTION OF K, T, AND K-T SPECIES IN EACH SAMPLE (continued)

Section* and stratigraphic position	Sample category†	Sample number	K	T	K-T	Total
Herpijunk North (cont.)						
15.70 m above Ir	TCLAS	HNE83-6	0	1	35	36
5.00 m above Ir	TCOAL	HNE83-3	0	6	38	44
0.40–0.50 m above Ir	TCLAS	HNE82-30-40B	0	4	20	24
0.30–0.40 m above Ir	TCLAS	HNE82-10-20B	0	3	21	24
0.15 m above Ir	TCLAS	HNE82-16	1	6	34	41
0.08–0.10 m above Ir	TCLAS	HNE83-0-2	0	4	30	34
0.06–0.08 m above Ir	UPIR	HNE83-2-4	0	8	21	29
0.04–0.06 m above Ir	UPIR	HNE83-4-6	0	4	15	19
0.02–0.04 m above Ir	UPIR	HNE83-6-8	0	6	28	34
0.00–0.02 m above Ir	BASIR	HNE83-8-10	2	4	21	27
0.00–0.02 m below Ir	USTK	HNE82-10	12	1	30	43
iridium claystone	unclassified	HNE83-11	7	2	25	34
0.60–0.70 m below Ir	UHC	HNE82-60-70A	29	2	51	83
1.00–1.10 m below Ir	UHC	HNE82-100	32	2	41	75
1.50–1.60 m below Ir	UHC	HNE82-150	21	0	34	55
Iridium Hill						
0.02–0.04 m above Ir	UPIR	IH81-422-24	0	7	36	43
0.00–0.02 m above Ir	BASIR	IH83-2	2	0	11	13
3.40–3.50 m below Ir	UHC	IH80-70-80	30	3	52	85
3.70–8.80 m below Ir	MHC	IH80-40-50	38	3	43	84
3.90–4.00 m below Ir	MHC	IH80-20-30	29	2	29	73
Lerbekmo						
8.20 m above Ir	TCLAS	Lb82-36	0	3	43	46
7.00 m above Ir	TCLAS	Lb82-35	2	3	43	48
6.10 m above Ir	TCLAS	Lb82-31	0	6	36	41
0.24 m above Ir	TCLAS	LbR82-24	3	3	30	36
0.20 m above Ir	TCOAL	LbR82-8	0	7	26	37
0.00–0.02 m above Ir	BASIR	LbR83-15b	0	5	16	21
1.30 m below Ir	UHC	Lb82-27	42	1	56	102
1.40–1.50 m below Ir	UHC	LbR82-150	21	2	32	55
2.40–2.50 m below Ir	UHC	LbR82-250	27	2	47	76
5.50 m below Ir	MHC	Lb82-23	24	2	45	71
16.6 m below Ir	MHC	Lb82-13	29	2	46	77
20.0 m below Ir	MHC	Lb82-8	23	1	45	64
23.5 m below Ir	MHC	Lb82-3	22	3	46	70
~80.0 m below Ir	LHC	HC82-2	42	1	48	91
~82.0 m below Ir	LHC	HC82-1	36	2	44	84
Pearl#						
0.0–0.1 m below top Z coal	TCOALSP	PPD82-0-10	2	7	26	33
0.2–0.3 m below top Z coal	TCOALSP	PPD82-20-30	0	6	21	27
0.4–0.5 m below top Z coal	TCOALSP	PPD82-40-50	2	9	31	42
0.5–0.6 m below top Z coal	TCOALSP	PPD82-50-60	1	8	34	42
0.7–0.8 m below top Z coal	TCOAL	PPD82-70-80	0	7	34	44
0.8–0.9 m below top Z coal	TCOAL	PPD82-80-90	1	8	43	55
0.9–1.0 m below top Z coal	TCOAL	PPD82-90-100	0	5	34	39
1.1–1.2 m below top Z coal	TCLAS	PPD82-110-20	2	5	39	47
6.3 m below top Z coal	TCOALSP	PPB82-10-20	1	4	38	43
6.4 m below top Z coal	TBASC	PPB82-0-10	0	3	15	18
6.5 m below top Z coal	TCOAL	PL83-4	0	6	38	44
6.55 m below top Z coal	TCOAL	PL83-5	0	6	32	38
6.6 m below top Z coal	TCLAS	PL83-6	0	6	27	30
6.65 m below top Z coal	TBASC	PL83-8	0	1	27	28
7.3 m below top Z coal	TCOAL	PPE82-2-10	3	8	50	61
7.35 m below top Z coal	TCLAS	PPE82-27-30	1	10	51	62
Seven Blackfoot						
0.08–0.10 m above Ir	UPIR	7B83-0-2	0	4	22	26
0.06–0.08 m above Ir	UPIR	7B83-2-4	0	7	29	36
0.04–0.06 m above Ir	UPIR	7B83-6-8	1	3	15	19
0.02–0.04 m above Ir	UPIR	7B83-8-10	0	3	22	25
0.00–0.02 m above Ir	BASIR	7B83-10-12	0	2	14	16
0.00–0.02 m below Ir	USTK	7B83-14.5	26	6	44	76

(*continued*)

TABLE A7. LOCATION, CATEGORY AND PROPORTION OF K, T, AND K-T SPECIES IN EACH SAMPLE (continued)

Section* and stratigraphic position	Sample category†	Sample number	K	T	K-T	Total
Sandstone Basin§						
Top of Z coal	TCOALSP	SSB83-17b	0	5	35	39
3.1 m below top of Z coal	TCLAS	SSB83-14b	1	7	42	50
3.2 m below top Z coal	TCLAS	SSB83-14a	1	5	28	34
3.3 m below top Z coal	TCOALSP	SSB83-13	0	5	28	33
4.2 m below top Z coal	MHC	SSB83-9	24	2	39	65
5.2 m below top Z coal	MHC	SSB83-8	26	5	44	75
5.4 m below top Z coal	MHC	SSB83-4	36	4	53	93
5.6 m below top Z coal	KCOAL	SSB83-6	12	3	32	47
Thomas Ranch§						
0.3 m below top Z coal	TCOALSP	TR83-5c	2	7	32	46
0.7 m below top Z coal	TCLAS	TR83-5b	0	7	27	34
1.2 m below top Z coal	TCOALSP	TR83-5a	1	3	29	33
12.1 m below top Z coal	MHC	TR83-4	26	1	39	66
19.0 m below top Z coal	MHC	TR83-2a	22	5	45	72
20.7 m below top Z coal	MHC	TR83-0	35	5	50	94
20.75 m below top Z coal	KCOAL	TR83-2-4	17	5	38	60
20.80 m below top Z coal	KCOAL	TR83-6-8	24	5	47	76
20.82 m below top Z coal	KCOAL	TR83-10-12	13	2	38	53
20.84 m below top Z coal	KCOAL	TR83-12-14	14	1	37	52
20.86 m below top Z coal	KCOAL	TR83-14-16	17	3	39	59
20.88 m below top Z coal	MHC	TR83-16-17	12	2	33	47
McCone County						
Bug Creek§						
Top of Z coal	TCOAL	DR83-A1	0	2	20	22
11.0 m below base of Z coal	MHC	BG83-cl	24	0	37	61
14.2 m below base of Z coal	MHC	GU82-3	25	2	34	71
18.30 m below base of Z coal	KCOAL	BG82-A6	18	3	42	63
18.32 m below base of Z coal	KCOAL	BG82-A7	11	2	33	46
18.34 m below base of Z coal	KCOAL	BG82-A8	11	3	32	47
18.38 m below base of Z coal	KCOAL	BG82-A9	29	6	51	87
19.9 m below base of Z coal	MHC	BG82-B17	29	0	42	71
20.1 m below base of Z coal	MHC	BG82-C22	22	0	34	56

*See Figure 1 for location of sections.
†See Table 1 for explanation of sample categories.
§Iridium datum missing in this section, but know within ~1.0 m by correlation with nearby sections.
#Iridium datum unknown; measurement taken from top of first thick coal (Z coal) with Tertiary age palynoflora.

the Cretaceous-Paleogene transition, eastern Montana and western North Dakota: Palaios, v. 2, p. 282–295.

Fastovsky, D.E., 1990, Rocks, resolution and the record: A review of depositional constraints on fossil vertebrate assemblages at the terrestrial Cretaceous/Paleogene boundary, eastern Montana and western North Dakota: *in* Sharpton, V.L., and Ward, P.D., eds., Global catastrophes in earth history: Geological Society of America Special Paper 247, p. 541–562.

Fastovsky, D.E., and Dott, R.H., Jr., 1986, Sedimentology, stratigraphy, and extinctions during the Cretaceous-Paleogene transition at Bug Creek, Montana: Geology, v. 14, p. 279–282.

Fastovsky, D.E., and McSweeney, K., 1987, Paleosols spanning the Cretaceous-Paleogene transition, eastern Montana and western North Dakota: Geological Society of America Bulletin, v. 99, p. 66–77.

Fastovsky, D.E., McSweeney, K., and Norton, L.D., 1989, Pedogenic development at the Cretaceous-Tertiary boundary, Garfield County, Montana: Journal of Sedimentary Petrology, v. 59, p. 758–767.

Fleming, R.F., 1990, Palynology of the Cretaceous-Tertiary boundary interval and Paleocene part of the Raton Formation, Colorado and New Mexico [Ph.D. thesis]: Boulder, Colorado, University of Colorado, 283 p.

Fleming, R.F., and Nichols, D.J., 1990, The fern-spore abundance anomaly at the Cretaceous-Tertiary boundary: a regional bioevent in western North America: *in* Kauffman, E.G., and Walliser, O.H., eds., Extinction events in earth history: Lecture notes in the earth sciences 30: New York, Springer-Verlag, p. 347–349.

Friis, E.-M., 1983, Upper Cretaceous (Senonian) floral structures of Juglandalean affinity containing Normapolles pollen. Review of Palaeobotany and Palynology 39: 161–188.

Frye, C., 1969, Stratigraphy of the Hell Creek Formation in North Dakota, North Dakota Geological Survey Bulletin 54, 65 p.

Gemmill, C.E.C., and Johnson, K.R., 1997, Paleoecology of a late Paleocene (Tiffanian) megaflora from the northern Great Divide Basin, Wyoming: Palaios, v. 12, p. 439–448.

Hickey, L.J., 1981, Land plant evidence compatible with gradual, not catastrophic, change at the end of the Cretaceous: Nature, v. 292, p. 529–531.

Hildebrand, A.R., Penfield, G.T., Kring, D.A., Pilkington, M., Camargo, A., Jacobsen, S.B., and Boynton, W.V., 1991, Chicxulub crater: A possible Cretaceous Tertiary boundary impact crater on the Yucatan peninsula, Mexico: Geology, v. 19, p. 867–871.

Hotton, C.L., 1988, Palynology of the Cretaceous-Tertiary boundary in central Montana, U.S.A., and its implications for extraterrestrial impact [Ph.D. thesis]: Davis, University of California, 731 p.

Hsü, K.J., He, Q.,McKenzie, J.A., Weissert, H., Perch-Nielson, K., Oberhaensli, H., Kelts, K., LaBrecque, J., Tauxe, L., Kraehenbuehl, U., Percival, S.F., Jr., Wright, R., Karpoff, A.M., Petersen, N., Tucker, P., Poore, R.Z., Gombos, A.M., Pisciotto, K.A., Carman, M.F., Jr., and Schrieber, E., 1982, Mass mortality and its environmental and evolutionary consequences: Science, v. 216, p. 249–256.

Jerzykiewicz, T., and Sweet, A.R., 1986, The Cretaceous-Tertiary boundary in the central Alberta foothills. I. Stratigraphy: Canadian Journal of Earth Sciences, v. 23, p. 1356–1374.

Johnson, K.R., 1992, Leaf fossil evidence for extensive floral extinction at the Cretaceous-Tertiary boundary, North Dakota, U.S.A.: Cretaceous Research, v. 13, p. 91–117.

Johnson, K.R., 1993, Extinctions at the antipodes: Nature, v. 366, p. 511–512.

Johnson, K.R., and Hickey, L.J., 1990, Megafloral change across the Cretaceous/Tertiary boundary in the northern Great Plains and Rocky Mountains, U.S.A., *in* Sharpton, V.L., and Ward, P.D., eds., Global catastrophes in earth history: Boulder, Colorado, Geological Society of America Special Paper 247, p. 433–444.

Johnson K.R., Nichols, D.J., Attrep, M., and Orth, C.J., 1989, High-resolution leaf fossil record spanning the Cretaceous Tertiary boundary: Nature, v. 340, p. 708–711.

Kaiho, K., Kajiwara, Y., Tazaki, K., Ueshima, M., Takeda, N., Kawahata, H., Arinobu, T., Ishiwatari, R., Hirai, A., and Lamolda, M.A., 1999, Oceanic primary productivity and dissolved oxygen levels at the Cretaceous-Tertiary boundary: Their decrease, subsequent warming, and recovery: Paleoceanography, v. 14, p. 511–524.

Kyte, F.T., 1998, A meteorite from the Cretaceous/Tertiary boundary: Nature, v. 396, p. 237–239.

LaDeau, S.L., and Clark, J.S., 2001, Rising CO_2 levels and the fecundity of forest trees: Science, v. 292, p. 95–98.

Lerbekmo, J.F., 1999, Magnetostratigraphy of the Canadian Continental Drilling Project Cretaceous-Tertiary (K-T) Boundary Project cores holes, western Canada: Canadian Journal of Earth Sciences, v. 36, p. 705–715.

Lerbekmo, J.F., Sweet, A.R., and St. Louis, R.M., 1987, The relationship between the iridium anomaly and palynological floral events at three Cretaceous-Tertiary boundary localities in western Canada: Geological Society of America Bulletin 99, p. 325–330.

Lofgren, D.L., 1995, The Bug Creek problem and the Cretaceous-Tertiary transition at McGuire Creek, Montana: University of California Publications in the Geological Sciences, v. 140, 185 p.

Lomax, B.H., Beerling, D.J., Upchurch, G.R. Jr., and Otto-Bliesner, B.L., 2000, Terrestrial ecosystem responses to global environmental change across the Cretaceous-Tertiary boundary: Geophysical Research Letters, v. 27, p. 2149–2152.

Luck, J.M., and Turekian, K.K., 1983, Os^{187}/Os^{186} in Mn nodules and the Cretaceous-Tertiary boundary: Science, v. 222, p. 613–615.

MacLeod, N., Rawson, P.F., Forey, P.L., Banner, F.T., Boudagher-Fadel, M.K., Bown, P. R., Burnett, J.A., Chambers, P., Culver, S., Evans, S.E., Jeffery, C., Kaminski, M.A., Lord, A. R., Milner, A.C., Milner, A.R., Morris, N., Owen, E., Rosen, B. R., Smith, A.B., Taylor, P.D., Urquhart, E., and Young, J.R., 1997, The Cretaceous-Tertiary biotic transition: Geological Society [London] Journal, v. 154, p. 265–292.

McHone, J.F., Nieman, R.A., Lewis, C.F., and Yates, A.M., 1989, Stishovite at the Cretaceous-Tertiary boundary, Raton, New Mexico: Science, v. 234, p. 1182–1184.

McIver, E.E., 1999, Paleobotanical evidence for ecosystem disruption at the Cretaceous-Tertiary boundary from Wood Mountain, Saskatchewan, Canada: Canadian Journal of Earth Sciences, v. 36, p. 775–789.

McIver, E.E., Sweet, A.R., and Basinger, J.F., 1991, 65000000-year-old flowers bearing pollen of the extinct triprojectate complex—A Cretaceous/Tertiary boundary survivor: Review of Palaeobotany and Palynology, v. 70, p. 77–88.

Melosh, H.J., Schneider, N.M., Zahnel, K.J. and Latham, D., 1990, Ignition of global wildfires at the Cretaceous/Tertiary boundary: Nature, v. 343, p. 251–254.

Memon, H.R., 1983, Exclusion of North American *Proteacidites* spp. from the Southern Hemisphere Proteaceae: Sind University Research Journal, Science Series, v. 15, p. 73–84.

Morgan, J., Warner, M., Brittan, J., Buffler, R., Camargo, A., Christeson, G., Denton, P. Hildebrand, A., Hobbs, R., Macintyre, H., Mackenzie, G., Maguire, P., Marin, L.E., Nakamura, Y., Pilkington, M., Sharpton, V.L., Snyder, D., Suarez, G., and Trejo, A., 1997, Size and morphology of the Chicxulub impact crater: Nature, v. 390, p. 472–476.

Myrick, C.W., 1986, SAS® companion for the VMS operating system: Cary, North Carolina, SAS Institute, Inc.

Nichols, D.J., 1990, Geologic and biostratigraphic framework of the non-marine Cretaceous-Tertiary boundary interval in western North America: Review of Palaeobotany and Palynology, v. 65, p. 75–84.

Nichols, D.J., Jarzen, D.M, Orth, C.J., and Oliver, P.Q., 1986, Palynological and iridium anomalies at the Cretaceous-Tertiary boundary, south central Saskatchewan: Science, v. 231, p. 714–717.

Nichols, D.J., and R.F. Fleming, 1990, Plant microfossil record of the terminal Cretaceous event in the western United States and Canada, *in* Sharpton, V.L., and Ward, P.D., eds., Global catastrophes in earth history: Boulder, Colorado, Geological Society of America Special Paper 247, p. 445–456.

Nichols D.J., Brown, J.L., Attrep, M., and Orth, C.J., 1992, A new Cretaceous Tertiary boundary locality in the western Powder River Basin, Wyoming—biological and geological implications: Cretaceous Research, v. 13, p. 3–30.

O'Keefe, J.D. and Ahrens, T.J., 1989, Impact production of CO_2 by the Cretaceous/Tertiary extinction bolide and the resultant heating of the Earth: Nature, v. 338, p. 247–249.

Orth, C.L., Gilmore, J.S., Knight, J.D., Pillmore, C.L., Tschudy, R.H., and Fassett, J.E., 1981, An iridium abundance anomaly in the palynological Cretaceous-Tertiary boundary in northern New Mexico: Science, v. 214, p. 1341–1343.

Pillmore, C.L., Tschudy, R.H., Orth, C.J., Gilmore, J.S., and Knight, J.D. 1984, Geologic framework of non-marine Cretaceous-Tertiary boundary sites, Raton Basin, New Mexico and Colorado: Science, v. 223, p. 1180–1183.

Pollack, J.B., Toon, O.B., Ackerman, T.P., McKay, C.P., and Turco, R.P., 1983, Environmental effects of an impact-generated dust cloud: implications for the Cretaceous-Tertiary extinctions: Science, v. 219, p. 287–289.

Pope, K.O., Baines, K.H., Ocampo, A.C. and Ivanov, B.A., 1994, Impact winter and the Cretaceous/Tertiary extinctions: results of a Chicxulub asteroid impact model: Earth and Planetary Science Letters, v. 128, p. 719–725.

Pope, K.O., Baines, K.H., Ocampo, A.C. and Ivanov, B.A., 1997, Energy, volatile production, and climatic effects of the Chicxulub Cretaceous/Tertiary impact: Journal of Geophysical Research, E, Planets, v. 102, p. 21645–21664.

Prinn, R.G., and Fegley, B., 1987, Bolide impacts, acid rain, and biospheric traumas at the Cretaceous-Tertiary boundary: Earth and Planetary Science Letters, v. 83, p. 1–15.

Retallack, G.J., 1994, A pedotype approach to latest Cretaceous and earliest Tertiary paleosols in eastern Montana, Geological Society of America Bulletin 106, p. 1377–1397.

Rigby, J.K., Jr., Newman, K.R., Smit, J., van der Kaars, S., and Sloan, R.E., 1987, Dinosaurs from the Paleocene part of the Hell Creek Formation, McCone County, Montana: Palaios, v. 2, p. 296–302.

Rogers, G.S. and Lee, W., 1923, Geology of the Tullock Creek coal field, Rosebud and Big Horn Counties, Montana: U.S. Geological Survey Bulletin 749, p. 1–181.

Sepkoski, J.J., Jr., 1982, Mass extinctions in the Phanerozoic oceans: a review, *in* Silver, L.T., and Schultz, P.H., eds., Geological implications of impacts of large asteroids and comets on the Earth: Boulder, Colorado, Geological Society of America Special Paper 190, p. 283–290.

Sharpton, V.L., Dalyrumple, G.B., Marin, L.E., Ryder, G., Schuraytz, B.C., and Urrutiafucugauchi, J., 1992, New links between the Chicxulub impact structure and the Cretaceous-Tertiary boundary: Nature, v. 359, p. 819–821.

Sheehan P.M., and Fastovsky, D.E., 1992, Major extinctions of land-dwelling vertebrates at the Cretaceous-Tertiary boundary, eastern Montana: Geology, v. 20, p. 556–560.

Sheehan P.M., Fastovsky, D.E., Barreto, C., and Hoffmann, R.G., 2000, Dinosaur abundance was not declining in a "3 m gap" at the top of the Hell Creek Formation, Montana and North Dakota: Geology, v. 28, p. 523–526.

Sigurdsson, H., D'Hondt, S., and Carey, S., 1992, The impact of the Cretaceous/Tertiary bolide on evaporite terrane and generation of major sulfuric acid aerosol, Earth and Planetary Science Letters, v. 109, p. 543–559.

Sims, H.J., Herendeen, P., Lupia, R., Christopher, R., and Crane, P.R., 1999, Fossil flowers with Normapolles pollen from the Upper Cretaceous of southeastern North America: Review of Palaeobotany and Palynology, v. 106, p. 131–151.

Smit, J., and van der Kaars, W.A., 1984, Terminal Cretaceous extinctions in the Hell Creek area, Montana: compatible with catastrophic extinctions: Science, v. 223, p. 1177–1179.

Smit, J., van der Kaars, W.A., and Rigby, J.K., Jr., 1987, Stratigraphic aspects of the Cretaceous-Tertiary boundary in the Bug Creek area of eastern Montana, U.S.A.: Memoire de la Societé Géologique de France, new series, 150, p. 53–73.

Stott, L.D., and Kennett, J.P., 1989, New constraints on early Tertiary palaeoproductivity from carbon isotopes in foraminifera: Nature, v. 342, p. 526–529.

Sweet, A.R., Braman, D.R., and Lerbekmo, J.F., 1990, Palynofloral response to K-T boundary events: a transitory interruption within a dynamic system: *in* Sharpton, V.L., and Ward, P.D., eds., Global catastrophes in earth history: Boulder, Colorado, Geological Society of America Special Paper 247, p. 457–470.

Sweet, A.R., and Cameron, A.R., 1991, Palynofacies, coal petrographic facies and depositional environments: Amphitheatre Formation (Eocene to Oligocene) and Ravenscrag Formation (Maastrichtian to Paleocene), Canada: International Journal of Coal Geology, v. 19: 121–144.

Sweet, A.R., and Braman, D.R., 1992, The K-T boundary and contiguous strata in western Canada—Interactions between paleoenvironments and palynological assemblages: Cretaceous Research, v. 13, p. 31–79.

Sweet, A.R., Braman, D.R., and Lerbekmo, J.F., 1999, Sequential palynological changes across the composite Cretaceous-Tertiary (K-T) boundary claystone and contiguous strata, western Canada and Montana, U.S.A.: Canadian Journal of Earth Sciences, v. 36, p. 743–768.

Swisher, C.C., III, Dingus, L., and Butler, R.F., 1993, $^{40}Ar/^{39}Ar$ dating and magnetostratigraphic correlation of the terrestrial Cretaceous-Paleogene boundary and Puercan Mammal Age, Hell Creek–Tullock formations, eastern Montana: Canadian Journal of Earth Sciences, v. 30, p. 1981–1995.

Tschudy, R.H., and Tschudy, B.D., 1986, Extinction and survival of plant life following the Cretaceous-Tertiary boundary event, Western Interior, North America: Geology, v. 14, p. 667–670.

Tschudy, R.H., Pillmore, C.L., Orth, C.J., Gilmore, J.S., and Knight, J.D., 1984, Disruption of the terrestrial plant ecosystem at the Cretaceous-Tertiary boundary, Western Interior: Science, v. 225, p. 1030–1032.

White, P.D., Fastovsky, D.E., and Sheehan, P.M., 1998, Taphonomy and suggested structure of the dinosaurian assemblage of the Hell Creek Formation (Maastrichtian), eastern Montana and western North Dakota: Palaios, v. 13, p. 41–51.

Williams, M.E., 1994, Catastrophic versus noncatastrophic extinction of the dinosaurs: testing, falsifiability and the burden of proof: Journal of Paleontology, v., 68, p. 183–190.

Wolbach, W.S., Lewis, R.S., and Anders, E. 1985, Cretaceous extinctions: evidence for wildfires and search for meteoritic material: Science, v. 230, p. 167–170.

Wolfe, J.A., 1990, Paleobotanical evidence for a marked temperature increase following the Cretaceous/Tertiary boundary: Nature, v. 343, p. 153–156.

Wolfe, J.A., 1991, Palaeobotanical evidence for a June "impact winter" at the Cretaceous/Tertiary boundary: Nature, v. 352, p. 420–423.

Wolfe, J.A., and Upchurch, G.R. Jr., 1986, Vegetation, climatic and floral changes at the Cretaceous-Tertiary boundary: Nature, v. 324, p. 148–152.

Wolfe, J.A., and Upchurch, G.R. Jr., 1987a, Leaf assemblages across the Cretaceous-Tertiary boundary in the Raton Basin, New Mexico and Colorado: National Academy of Science Proceedings, v. 84, p. 5096–5100.

Wolfe, J.A., and Upchurch, G.R. Jr., 1987b, North American non-marine climates and vegetation during the Late Cretaceous, Palaeogeography Palaeoclimatology Palaeoecology, v. 61, p. 33–77.

Manuscript Accepted by the Society September 24, 2001

Printed in the U.S.A.

Geological Society of America
Special Paper 361
2002

Hell Creek Formation: A 2001 synthesis

Kirk R. Johnson*
Department of Earth Sciences, Denver Museum of Nature & Science, 2001 Colorado Boulevard, Denver, Colorado 80205, USA
Douglas J. Nichols
U.S. Geological Survey, MS 939, Box 25046, Denver Federal Center, Denver, Colorado 80225, USA
Joseph H. Hartman
Department of Geology and Geological Engineering and Energy & Environmental Research Center, University of North Dakota, Grand Forks, North Dakota 58202, USA

INTRODUCTION

In the 94 years since Barnum Brown (1907) described the Hell Creek Formation, this rock unit has yielded magnificent dinosaurs and diverse plants and animals from the last Cretaceous terrestrial ecosystem. Fossils collected from near the Cretaceous-Tertiary (K-T) boundary in the Hell Creek Formation have been used to argue catastrophic extinction, gradual extinction, and even the survival of dinosaurs into the Tertiary. The contributors to this volume have attempted to carry the understanding of these uppermost Cretaceous strata to the next level by exploring the lesser studied eastern half of the outcrop area, utilizing new techniques, attempting to study the entirety of the fossilized biota, and applying an integrated approach wherever possible.

From east to west, the preserved remnants of the Hell Creek Formation extend nearly 700 km (Fig. 1). Most of the previous published work on the formation and its fossils was based on outcrops in Garfield and McCone Counties, Montana, around the margins of the Fort Peck Reservoir, and around the margins of the Cedar Creek Anticline near the Montana–North Dakota border. The exposures in North Dakota received some attention in the late 1960s, but interdisciplinary efforts began there primarily in the late 1980s (Johnson et al., 1989). The Hell Creek Formation in South Dakota has yielded spectacular dinosaurs such as the *Tyrannosaurus rex* "Sue," but very little scientific effort has been applied there, and few publications are available. In this summary chapter, we attempt to synthesize and integrate the major findings of the other chapters in this volume.

STRATIGRAPHY

The Hell Creek Formation is exposed in Montana, North Dakota, and South Dakota. Equivalent and contiguous strata in Wyoming are known as the Lance Formation and in Canada as the Frenchman and Scollard Formations. In this volume, authors have focused on the areas that bear the Hell Creek name, primarily those on the margin of the Williston Basin in Montana and the Dakotas (Fig. 1).

The complete thickness of the Hell Creek Formation has been measured at only a few locations. Several factors have contributed to this apparent oversight. The formation is mostly horizontal and is usually thicker than the topographic relief that has been incised into it. This, combined with the lateral discontinuity of beds, a bentonitic surface weathering that obscures bedding, and an upper contact that in places is difficult to identify, has created a situation where it is difficult to make accurate measurements. For these reasons, many have estimated the thickness of the formation rather than directly measuring it. Estimates in the Fort Peck Reservoir area have ranged from a maximum of 170 m in Garfield County, Montana (Brown, 1907) to a minimum of 41 m in McCone County, Montana (Collier and Knechtel, 1939). The most recent work in McCone County reported thicknesses of 51–85 m (Rigby and Rigby, 1990). Murphy et al. (this volume) report six localities in North Dakota where they measured the entire thickness of the formation. At two sites in southwestern North Dakota, the formation is 100 m thick and at four sites in the central North Dakota the formation ranges from 50 to 70 m thick.

*E-mail: kjohnson@dmns.org

Johnson, K.R., Nichols, D.J., and Hartman, J.H., 2002, Hell Creek Formation: A 2001 synthesis, *in* Hartman, J.H., Johnson, K.R., and Nichols, D.J., eds., The Hell Creek Formation and the Cretaceous-Tertiary boundary in the northern Great Plains: An integrated continental record of the end of the Cretaceous: Boulder, Colorado, Geological Society of America Special Paper 361, p. 503–510.

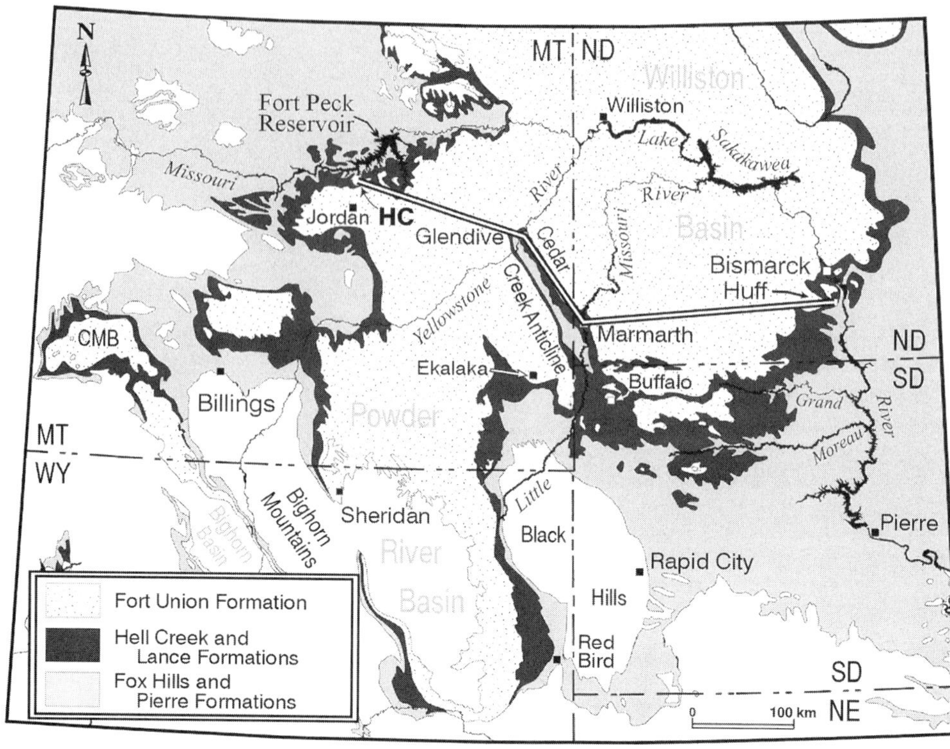

Figure 1. Map of northern Great Plains showing outcrop of Hell Creek and equivalent and adjacent strata. Type area of Hell Creek Formation is marked as HC. Line of cross section for Figures 2 and 3 is marked.

The greatest degree of integration of diverse Hell Creek data has been achieved in sections in southwestern North Dakota. These sections were used to provide a lithostratigraphic framework for magnetostratigraphy, geochronology, chemostratigraphy, palynology, megafloral paleobotany, paleoentomology, and vertebrate paleontology. Patterns revealed by this integration provide new hypotheses that can be tested across the entire outcrop of the formation.

A cross section of the formation presented in Figure 2 extends from the type area of the Hell Creek on the south bank of the Fort Peck Reservoir, to the north end of the Cedar Creek Anticline at Glendive, Montana, to the south end of the anticline near Marmarth, North Dakota, to the Missouri River south of Bismarck, North Dakota. Along this transect, the formation thins, ranging from 80–120 m in the west to 50–70 m in the east. Over the 500 km course of this transect, the top of the formation is always within a few meters, above or below, the K-T boundary, and as such, it can be roughly viewed as diachronous on a local fine scale, but essentially isochronous on the regional scale (Nichols and Johnson, this volume; Hotton, this volume; Fastovsky, 1987).

The easternmost outcrops of the Hell Creek Formation document that the formation interfingers with marine strata. Murphy et al. (this volume) place the marine sediments of the Breien Member of the Hell Creek Formation into stratigraphic context relative to the top of the formation. They also name a new marine deposit, the Cantapeta advance, a unit 10 m below the top of the formation, that contains *Ophiomorpha* burrows and was clearly deposited under the influence of marine conditions. This is the youngest evidence of Cretaceous marine deposition in the northern Great Plains. It is significant because it reduces the stratigraphic separation of the highest Cretaceous marine beds and the lowest Paleocene marine beds to a mere 15 m and raises the probability that a marine K-T boundary may be in the subsurface in the eastern Williston Basin. Hoganson and Murphy (this volume) describe the dimensions and vertebrate fauna of the Breien Member of the Hell Creek Formation. Hartman and Kirkland (this volume) document the previously unpublished invertebrate fauna collected by Charles Frye in the 1960s and interpret variations in salinity based primarily on molluscan associations through the end of the Cretaceous as represented by the Fox Hills and Hell Creek Formations. Collectively, these papers show that the Breien Member was marine, and its fauna was closely related to the marine fauna of the underlying Fox Hills Formation. In addition, brackish and marine mollusks from Hell Creek strata overlying the Breien Member indicate a younger and more complex history of the Western Interior Seaway near the end of the Cretaceous.

MAGNETOSTRATIGRAPHY

Both Lund et al. (this volume) and Hicks et al. (this volume) present papers on the magnetostratigraphy of the Hell Creek Formation. Lund et al. compare two sections at Glendive, Montana, with two sections south of Bismarck, a separation of more than 300 km. Hicks et al. studied six sections along

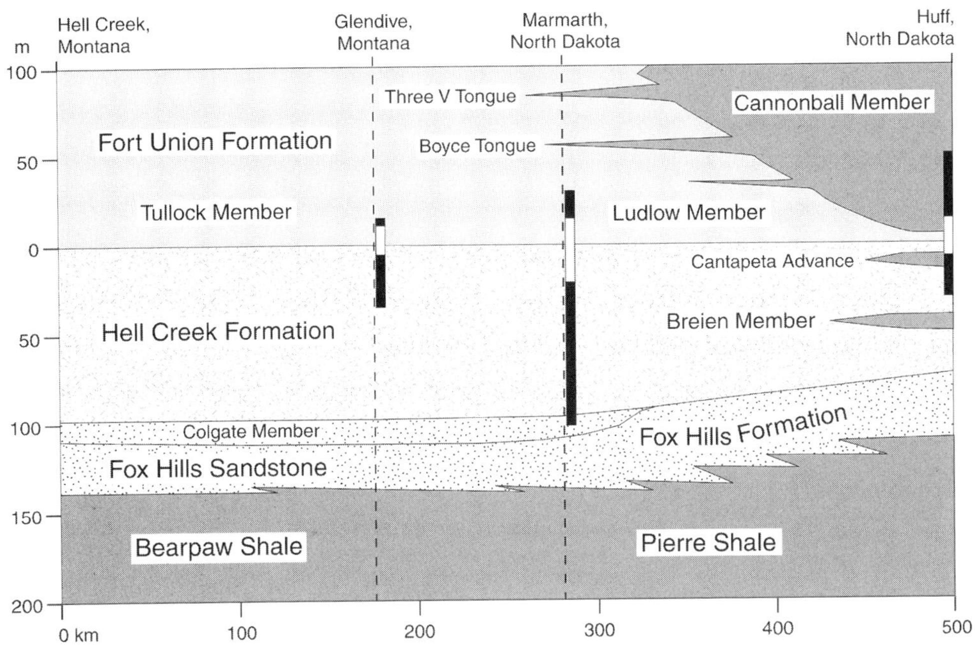

Figure 2. Stratigraphic cross section from Hell Creek to Glendive, Montana, to Marmarth, to Huff, North Dakota showing thickness of Hell Creek and adjacent strata. Black and white bars show position and results of paleomagnetic sections of Hicks et al. (this volume) near Marmarth and Lund et al. (this volume) near Glendive, Montana, and Huff, North Dakota. Datum is top of Hell Creek Formation. The Firesteel lignite seam divides the lower and upper Hell Creek Formation in South Dakota, south of the line of section.

a 45 km length of the Little Missouri River. Both groups identified three polarity subchrons that they interpret as C29n, C29r, and C30n. In all 10 sections, the palynologically defined K-T boundary occurs in a reversed interval interpreted as C29r. The thickness of C29r and the position of the K-T boundary relative to the base of the subchron are variable, but in general, the K-T boundary occurs in the basal third of the subchron. Lund et al. interpret this variation to represent a diachroneity of the top of the Hell Creek Formation and argue that the top of the formation is older to the east than to the west. Hicks et al. see this same amount of variation in the position of the K-T boundary within their local transect, but argue that local diastems within the Hell Creek are responsible for the variation. In the most complete marine sections, the K-T boundary occurs slightly above the middle of C29r (Smit, 1999).

Obradovich (*in* Hicks et al., this volume) recalibrates the age of the K-T boundary to 65.51 Ma by normalizing many recently determined radiometric ages to a uniform standard. The K-T boundary and the beginning and end of magnetic polarity subchron C29r represent time lines. Using estimated durations for the Cretaceous and Paleocene portions of C29r (D'Hondt et al., 1996), the age of the K-T boundary (Hicks et al., this volume), and the thickness of rock between those time lines, it is possible to estimate a sediment accumulation rate for the Hell Creek Formation. Both Lund et al. and Hicks et al. use variants of this method and the assumption of a uniform sediment accumulation rate to make estimates of the duration of the entire formation. Lund et al. estimate a duration of 2.5 m.y., whereas Hicks et al. estimate 1.36 m.y. The difference in these estimates is due to three variables: (1) variable formation thickness, (2) different estimations for the duration of C29r, and (3) different thicknesses or accumulation rates of the Cretaceous portion of C29r. The shorter duration is supported by the absence of C30r in two tightly sampled sections of the lower Hell Creek Formation in Bowman County, North Dakota. Nonetheless, the difference between the two estimates illustrates the need for better age control at the base of the formation.

CHEMOSTRATIGRAPHY

Arens and Jahren (this volume) present analyses of stable carbon isotopes from sedimentary rock samples in four sections in southwestern North Dakota. The $\delta^{13}C$ shift of 1.5‰–2‰ that marks the K-T boundary in marine sections was detected in all four sections, but not always in precise correlation with the palynologically defined K-T boundary. Additional variations of $\delta^{13}C$ throughout the formation have the potential to define an isotope stratigraphy, allowing for independent correlation between sections of the Hell Creek Formation. The differences between these results and the palynological record suggest that chemostratigraphy needs to be refined at more sites where the K-T boundary clay layer is present.

PALYNOLOGY AND PALEOBOTANY

Nichols (this volume) documents the palynology of the entire Hell Creek Formation, and Nichols and Johnson (this volume) use palynology to locate the K-T boundary in 17 sections. In several sections north of Marmarth, North Dakota, they identify a zone as thick as 3 m at the base of the Ludlow Member of the Fort Union Formation that contains Cretaceous palynomorphs. They name this the Fort Union strata of Cretaceous age. Pearson et al. (this volume) have discovered a partial ceratopsian skeleton in this unit. Hotton (this volume) addresses the K-T boundary in the Fort Peck Reservoir region and argues that dramatic changes in the palynoflora at the K-T boundary support the impact hypothesis. Kroeger (this volume) uses palynology to discuss sedimentological facies within the Hell Creek.

Johnson (this volume) describes the megafloral paleobotany of the Hell Creek and lower Fort Union Formations and recognizes 380 megafloral morphotypes. Of this total, 328 morphotypes occur in the Hell Creek Formation. This is more than three times the number of palynofloral taxa described by Nichols from the same beds, despite the huge difference in sample size (Nichols observed ~450 000 playnomorphs and Johnson saw ~30 000 megafossils). This indicates that megafossils sample the flora at a higher taxonomic resolution than palynomorphs (species rather than genera or families). The vegetation of the Hell Creek is dominated by arboreal angiosperms, whereas ferns and conifers are more abundant in the palynoflora than the megaflora. Hotton's statistical analyses are based on a Hell Creek palynoflora composed of about three times as many species as recognized by Nichols, because these authors took somewhat different taxonomic approaches. It is significant that both authors conclude that about one-third of the palynoflora became extinct at the K-T boundary. Johnson, whose fossil data set has higher taxonomic resolution, concludes that ~80% of the Hell Creek flora did not survive the terminal Cretaceous extinction event. These results are not in disagreement, but instead result from sampling a biotic record at different taxonomic levels.

The depth of understanding that comes from combining the strengths of palynology and megafloral paleobotany is impressive. Because pollen and leaves are derived from the same vegetation, interpretations based on one organ can be extrapolated to the other. The high stratigraphic resolution of palynology allows for the effects of the Cannonball transgression (the initiation of lignite deposition) to be separated from the effects of the impact event that deposited iridium and shocked minerals on the landscape. Megafossils lack that stratigraphic resolution, but offer species-level taxonomic resolution, which allows for a more precise assessment of the magnitude of the extinction. In addition, the paleoclimatic abilities of foliar physiognomy allow for the identification of Late Cretaceous climate trends. A dramatic warming, recorded by fossil leaves that first occur ~20 m below the top of the Hell Creek Formation (Johnson and Wilf, 1996), also is recorded in the record of marine foraminifera near Antarctica (Stott and Kennett, 1990).

VERTEBRATE PALEONTOLOGY

Russell and Manabe (this volume) provide a brief overview of the dinosaurian fauna of the Hell Creek Formation. They discuss the much-collected but rarely mentioned monotypic *Edmontosaurus* dinosaur bone beds found in the lower part of the formation in central South Dakota and adjacent North Dakota (Christians, 1992). These bone beds occur above the Fox Hills–Hell Creek formational contact but generally below the Firesteel lignite zone, which marks the contact between the lower and upper Hell Creek Formation in central South Dakota. The presence and nature of these bone beds are reminiscent of monotypic ceratopsian bone beds described by Eberth (1998, written commun.) from the Dinosaur Park Formation of Alberta. He suggests that these deposits were formed by storm surge across a low-relief coastline that provided both the mechanism of death by drowning and the mode of burial.

Pearson et al. (this volume) combine vertebrate microsites and isolated dinosaur occurrences to plot the distribution of 61 vertebrate taxa relative to the K-T boundary in westernmost North Dakota and South Dakota. All common taxa occur throughout the formation and dinosaurs are present in all localities. There is a zone ~2 m thick at the top of the formation in which no bones have been found, and Pearson et al. interpret this to be related to a change in the sedimentological nature of the interval containing the Hell Creek–Fort Union formational contact. This gap has been long recognized (Clemens, this volume) and remains controversial, but Pearson et al. have clarified the issues by providing the first documentation of a large data set precisely located in reference to both the formational contact datum and the K-T boundary.

Hunter and Archibald (this volume) document the mammalian fauna of the Hell Creek Formation in southwestern North Dakota and show it to be a typical Lancian fauna. Holroyd and Hutchison (this volume) compare the turtle fauna of southwestern North Dakota with that of the type Hell Creek in Montana and other Lancian sites elsewhere and document some regional differences in the turtle faunas suggestive of latitudinal gradients.

Hoganson and Murphy document the vertebrate fauna of the marine Breien Member of the Hell Creek Formation and conclude that it more closely resembles the fauna of the underlying Fox Hills Formation than the overlying fauna of the marine Cannonball Member of the Fort Union Formation. Pearson et al. also noted the presence of a number of fish taxa in the Hell Creek Formation in southwestern North Dakota that indicate proximity of brackish or marine water.

Clemens (this volume) provides a detailed review of the recent and historical work in the Hell Creek type area and addresses the role of fossil mammals for understanding the terminal Cretaceous event. He concludes that immigration played a major role in the turnover of mammalian faunas before the K-T boundary. He discusses Paleocene vertebrate sites in the uppermost Hell Creek section in Garfield County, Montana. These observations appear to be in agreement with the work of Nichols and Johnson and Lund et al., who used palynology and magnetostratigraphy, respectively, to argue that the upper contact of the Hell Creek Formation is diachronous.

INVERTEBRATE PALEONTOLOGY

Hartman and Kirkland (this volume) document the previously unpublished invertebrate fauna collected by Charles Frye in the 1960s and interpret possible variations in salinity based primarily on molluscan associations from the Fox Hills Formation through the end of the Cretaceous. This fauna was derived primarily from the Breien Member of the Hell Creek but also includes specimens from higher and lower in the formation. Many of these species are interpreted to have inhabited brackish or marine waters, suggesting that much of the lower Hell Creek in the Missouri River Valley was deposited close to the remnant Cretaceous Western Interior Seaway. Hoganson and Murphy (this volume) present ichnofossil evidence of marine conditions in the Breien Member and also document the existence of a poorly preserved specimen of the zonal ammonite *Jeletzkytes* cf. *J. nebrascensis. Jeletzkytes nebrascensis* represents the youngest zone of the Western Interior ammonite zonation. Examples of this ammonite are known from the Fox Hills Formation and the Elk Butte Member of the Pierre Shale (Kennedy et al., 1998). The poor preservation of the Hell Creek specimen ammonite is unfortunate considering its potential biostratigraphic significance.

PALEOENTOMOLOGY

Labandeira et al. (this volume) apply a novel approach to understanding the insect fauna of the Hell Creek and Fort Union Formations by studying the insect damage found on the fossil leaves described by Johnson (this volume). They document 49 types of insect damage, including foliage feeding, mining, galling, and piercing and sucking. Only 22 types of insect damage cross the K-T boundary extinction event, and of these, 55% were the more generalized feeding types. They interpret this to suggest the extinction of specialist herbivores at the K-T boundary.

AGE OF THE HELL CREEK FORMATION

Age of the upper contact of the Hell Creek Formation

The age and duration of the Hell Creek Formation are longstanding and still partially unanswered questions. The upper contact of the formation was recognized to approximate the K-T boundary by Dorf (1940, 1942) and Brown (1952), although it was not directly correlated to the marine K-T boundary. Ross (1940), Fox and Ross (1942), and Fox and Olsson (1969) described the early Paleocene (Danian) foraminifera of the Cannonball Member of the Fort Union Formation and showed that the terrestrial K-T boundary could be roughly correlated to the marine K-T boundary. More precise correlation followed with the use of palynology to identify the K-T boundary and the discovery of the iridium anomaly at a level very near the top of the Hell Creek Formation in the type area by Alvarez (1983) and Bohor et al. (1984), and in southwestern North Dakota by Johnson et al. (1989) and Nichols et al. (2000). Magnetostratigraphy in the Hell Creek type area (Archibald et al., 1982; Swisher et al., 1993), near Glendive (Lund et al., this volume), Ekalaka (J. Archibald, 1988, written commun.), and Marmarth (Hicks et al., this volume), places the palynological K-T boundary in a zone of reversed polarity interpreted as subchron C29r. Lund et al. (this volume) show two sections where the top of the Hell Creek Formation is located very close to the base of C29r, which suggests to them that the Hell Creek deposition terminated locally before the end of the Cretaceous. Nichols and Johnson (this volume) use palynology to show that the K-T boundary and the formation contact are separated by 0–270 cm in southwestern North Dakota. This result is similar to work near Ekalaka, Montana; Buffalo, South Dakota (Nichols, 1999, written communication); the Hell Creek type area (Hotton, this volume), and central North Dakota (Murphy et al., 1995). Thus the upper contact of the Hell Creek over most of its outcrop area is roughly the age of the K-T boundary plus or minus the time taken to deposit as much as 3 m of sediment. On the basis of the thickness of C29r, 3 m of Hell Creek sediment would likely have accumulated in 30–100 k.y. Thus, over an east-west transect of the Hell Creek Formation, the upper contact is slightly diachronous.

Age of the lower contact of the Hell Creek Formation

Estimates of the age of the lower contact of the Hell Creek Formation are few and poorly delimited. Hicks et al. (this volume) attempted to date a volcanic ash from Doaks Butte near the base of the formation on the South Dakota-North Dakota border, but did not achieve a tight-enough spread of dates to be significant. The age of the youngest dated ammonite below the Hell Creek is the 70.15 Ma *Baculites grandis* zone that occurs in the uppermost Bearpaw Shale in the Fort Peck Reservoir area, and in the uppermost Pierre Shale in the Cedar Creek Anticline. The 69.57 Ma *Baculites clinolobatus* zone underlies the formation in central North Dakota, as do the undated *Hoploscaphites birkelundi, Hoploscaphites nicolletti*, and *Jeletzkytes nebrascensis* zones. Hicks et al. (this volume) and Lund et al. (this volume) provide estimates of the age of the base of the Hell Creek Formation based on extrapolation of sediment accumulation rates: their estimates are 1.36 Ma and 2.5 Ma, respectively. There is a distinct possibility that the base of the formation is significantly diachronous. The oldest ammonite zone in the uppermost portion of the Pierre Shale below the Hell Creek Formation is the *Baculites cuneatus* zone, which occurs near Ekalaka, Montana. At nearly 73 Ma, this would represent the absolute maximum age for the base of the formation.

Basal Hell Creek unconformity

The nature of the basal contact of the Hell Creek Formation is germane to the discussion of the age of its base. In his original description, Brown (1907) discussed the erosional basal contact of the Hell Creek on the underlying Fox Hills Formation. Subsequent work in the Fort Peck Reservoir region identified basal Hell Creek channels containing quartzite cobbles up to 10 cm

(Bauer, 1925; Jensen and Varnes, 1964). The uppermost unit of the Fox Hills Formation in the Fort Peck Reservoir and Cedar Creek Anticline areas is the Colgate Member, a stark white, leached, 10–20-m-thick sandstone with an eroded upper surface. There has been continuing debate over the extent and magnitude of the basal Hell Creek unconformity. Dobbin and Reeside (1929) studied this issue carefully over the entire Rocky Mountain region and concluded that there was not a significant regional unconformity. Belt et al. (1997) combined sedimentological and magnetostratigraphic data to argue that this unconformity was regional and represents a significant period of time between the deposition of the Fox Hills and the Hell Creek Formations. Several observations cause us to regard their hypothesis favorably, as seen in the accompanying Wheeler diagram (Fig. 3). Evidence for an unconformity includes the extrabasinal conglomerate lag at the base of the Hell Creek, the fact that none of the magnetostratigraphic studies have identified subchrons C30r or C31r, and the absence of lignite beds in the base of the Hell Creek Formation. Lignite beds are commonly present in a regression sequence as the beds shoreward of the strand. Such a situation occurs in the Denver Basin, where the Laramie Formation, the unit immediately above the Fox Hills Formation, is rich in coal seams. Figure 3 represents a fairly radical view of the genesis of the Hell Creek Formation. It implies that the formation was deposited in response to the early inception of the Cannonball transgression rather than simply representing the landward component of the Fox Hills regression, as has long been the consensus.

PROXIMITY OF THE WESTERN INTERIOR SEAWAY

Analysis of the easternmost Hell Creek Formation outcrops shows that the Cretaceous Western Interior Seaway, once thought to be absent in Hell Creek time, was probably present

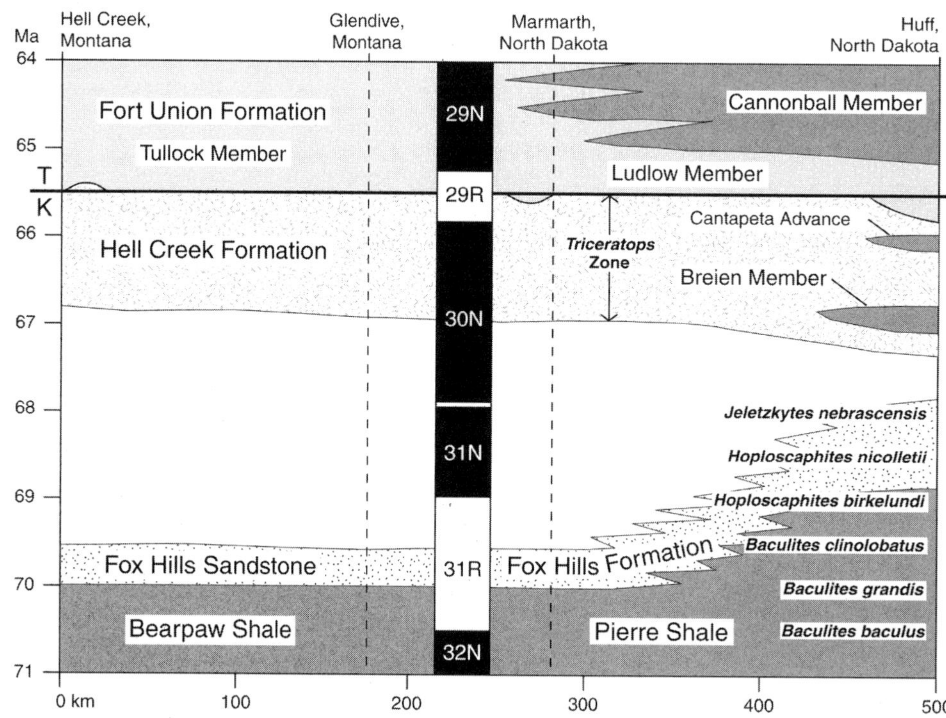

Figure 3. Wheeler diagram based on Figure 2 using Cretaceous-Tertiary (K-T) boundary as datum. In this interpretation, there is large unconformity between base of Hell Creek Formation and top of Fox Hills Formation. This unconformity is based on (1) Hicks et al. (1999) estimate for duration of 1.36 m.y. for Hell Creek Formation near Marmarth; (2) absence of magnetic polarity subchrons C30r, C31n, and C31r in well-sampled base of Hell Creek near Doaks Butte and Sunset Butte, North Dakota; (3) the historically well documented scoured contact of Hell Creek Formation on Fox Hills Formation; and (4) absence of lignite beds in base of Hell Creek Formation, a situation that would be expected if the Hell Creek represented a regressional deposit laterally and temporally equivalent to the Fox Hills Formation. This diagram represents end-member minimum-duration interpretation for Hell Creek Formation. Patterns are only to distinguish units and have no lithologic implications. White space indicates temporal hiatus. Single and partial ammonite specimen identified as *Jeletzkytes* cf. *J. nebrascensis* was found in Breien Member (Hoganson and Murphy, this volume). Highest ammonite zones that have reliable radiometric dates are middle of *Baculites grandis* zone (70.15 ± 0.65 Ma) and top of *Baculites clinolobatus* zone (69.57 ± 0.37 Ma), both from Red Bird, Wyoming (Hicks et al., 1999).

somewhere relatively close to the present eastern erosion-controlled outcrop margin. Evidence of the presence of the seaway exists in the lithologies and ichnofossils described by Murphy et al. (this volume) and Hoganson and Murphy (this volume), the vertebrate fossils described by Pearson et al. (this volume) and Hoganson and Murphy (this volume), and the brackish and marine invertebrate faunas described by Hoganson and Murphy (this volume) and Hartman and Kirkland (this volume). If the monotypic dinosaur bone beds of central South Dakota represent storm-surge–related mass drowning and burials, then they are one more argument for a near-coastal setting for the Hell Creek Formation. Correlation and interpretation of electric logs by Cherven and Jacob (1985) indicated that uppermost Cretaceous marine deposits are present in the subsurface of the Williston Basin, suggesting that the entire thickness of Hell Creek in North Dakota and South Dakota was subject to some marine influence.

In this nontraditional interpretation, the regression of the Cretaceous seaway, represented by the Fox Hills Formation, was followed by a fall in base level that caused sediment bypass and erosion in the area that would later be covered by the Hell Creek Formation. Eventually regression ceased, and the base level stabilized and then began to rise. It was during this time that the deposition of the Hell Creek Formation began. Small-scale transgressions during Hell Creek time resulted in the deposition of the Breien Member and the Cantapeta advance (Murphy et al., this volume). The distal effects of these or similar transgressions were the few lignite beds known in the Hell Creek, most notably the Firesteel lignite in central South Dakota.

The Cannonball transgression began shortly before the K-T boundary. The abundant lignite beds of the Ludlow and Tullock Members of the Fort Union Formation are a testimony to the rise in base level and increase in coastal aggradation caused by a rapidly transgressing Paleocene seaway. That the Hell Creek–Fort Union formational contact, which is defined most often by the inception of peat deposition, is nearly isochronous over a depositional slope of more than 500 km argues for the rapidity of this event and the flatness of the terrain.

K-T BOUNDARY

Data presented in this volume represent the first integrated assessment of a complete terminal Cretaceous terrestrial biota. Never before have plants, vertebrates, and insects been observed together in a well-dated stratigraphic context that contains the physical evidence of the K-T impact event and its associated shift in carbon isotopes. In addition, many of these chapters also assess the biota of the overlying Paleocene beds. The debate concerning the biotic effects of the K-T impact event has been an enduring one. Some workers even question the existence of a major K-T extinction of terrestrial biota (Sarjeant and Currie, 2001). One of the reasons that this controversy persists is that there have been very few integrated studies with adequate sample density to refute the apparent gradualism resulting from small samples. A second reason is that most arguments have relied on individual groups rather than entire biotas. The pattern that emerges in this volume is one of substantial, abrupt, biota-wide devastation at the K-T boundary. Furthermore, high-resolution stratigraphy allows for the effects of the K-T event to be separated from those are attributed to Late Cretaceous climate change and transgression of the Cannonball Seaway.

FUTURE QUESTIONS

This volume has raised and refined a number of questions. Perhaps the most obvious is the unresolved age of the base of the Hell Creek Formation and its implication for the duration of Hell Creek deposition. Does the formation have a duration of 1.36 m.y., as suggested by Hicks et al. (this volume), or does it fill all of the time between the highest dated ammonite and the K-T boundary, a span of 4.6 m.y.? Is there really a major unconformity between the Fox Hills and Hell Creek Formations? The answers to these questions have tremendous implications for our ability to assess the rate of biotic change before the K-T boundary event. Despite the presence of bentonites and the efforts of Hicks et al., there have been no precise radiometric ages derived from the Hell Creek Formation.

The issue of the vertebrate fossil-free gap in the uppermost 2 m of the formation continues to beg an explanation. It may be related to the speed at which the Cannonball Sea transgressed the flat terrain, but the sedimentological details of this formational transition still await insightful analysis. A good start would be the publication of the stratigraphic positions of the huge collections of vertebrate fossils from the Fort Peck Reservoir area that are in the collections at the University of California at Berkeley. More work needs to be done to correlate palynofloral and megafloral taxa, and the megaflora of the Hell Creek type area should be collected and studied. Insights into the facies patterns of the Hell Creek and its relationship to marine strata in the subsurface of the Williston Basin should be sought by interpretation and correlation of electric logs. The possibility that a marine K-T boundary section is beneath the plains of the eastern Williston Basin is tantalizing. Its discovery would be a splendid solution to the ongoing problem of correlating the marine and terrestrial K-T boundary records.

Note added in proof: Recent palynological and paleomagnetic analyses of the Kiowa Core in the Denver Basin (Nichols and Fleming, 2002) document the *Wodehouseia spinata* Assemblage Zone in polarity subchrons C29r and C30n and the underlying *Aquilapollenites striatus* Interval Zone in subchrons C31n and C31r within the Laramie and Fox Hills Formations. Since the base of the Hell Creek section in southwestern North Dakota contains palynomorphs indicative of the *W. spinata* Assemblage Zone and not the older *A. striatus* Interval Zone, there is now further evidence for the Fox Hills–Hell Creek unconformity discussed in this chapter.

REFERENCES CITED

Alvarez, L.W., 1983, Experimental evidence that an oasteroid impact led to the extinction of many species 65 million years ago: National Academy Sciences Proceedings, v. 80, p. 627–642.

Archibald, J.D., Butler, R.F., Lindsay, E.H., Clemens, W.A., and Dingus, L., 1982, Upper Cretaceous–Paleocene biostratigraphy and magnetostratigraphy, Hell Creek and Tullock Formations, northeastern Montana: Geology, v. 10, p. 153–159.

Bauer, C.M., 1925, Quartzite pebbles at the base of the Lance Formation in Montana: American Association of Petroleum Geologists Bulletin, v. 9, p. 344–346.

Belt, E.S., Hicks, J.F., and Murphy, D.A., 1997, A pre-Lancian regional unconformity and its relationship to Hell Creek paleogeography in southeastern Montana: University of Wyoming, Contributions to Geology, v. 31, p. 1–26.

Bohor, B.F., Foord, E.E., Modreski, P.J., and Triplehorn, D., 1984, Mineralogic evidence for an impact event at the Cretaceous-Tertiary boundary: Science, v. 244, p. 867–869.

Brown, B., 1907, The Hell Creek beds of the Upper Cretaceous of Montana: American Museum Natural History Bulletin, v. 23, p. 823–845.

Brown, R.W., 1952, Tertiary strata in eastern Montana and western North and South Dakota: *in* Sonnenberg, F.P., ed., Billings Geological Society Guidebook: Billings, Montana, Billings Geological Society, Third Annual Field Conference, p. 89–92.

Cherven, V.B., and Jacob, A.R., 1985, Evolution of Paleogene depositional systems, Williston Basin in response to global sea level changes, *in* Flores, R.M., and Kaplan, S.S., eds., Cenozoic paleogeography of the west-central United States: Society of Economic Paleontologists and Mineralogists, Rocky Mountain Symposium 3, p. 127–170.

Christians, J.P., 1992, Taphonomy and sedimentology of the Mason Dinosaur Quarry, Hell Creek Formation (Upper Cretaceous), South Dakota [M.S. thesis]: Madison, University of Wisconsin, 91 p.

Collier, A.J., and Knechtel, M.M., 1939, The coal resources of McCone County, Montana: U.S. Geological Survey Bulletin 905, 77 p.

D'Hondt, S., Herbert, T.D., King, J., and Gibson, C., 1996, Planktic foraminifera, asteroids, and marine production: Death and recovery at the Cretaceous-Tertiary boundary, *in* Ryder, G., Fastovsky, D., and Gartner, S., eds., The Cretaceous-Tertiary event and other catastrophes in Earth history: Geological Society of America Special Paper 307, p. 303–317.

Dobbin, C.E., and Reeside, J.B., Jr., 1929, The contact of the Fox Hills and Lance Formations: U.S. Geological Survey Professional Paper 158B, p. 9–25.

Dorf, E., 1940, Relationship between floras of the type Lance and Fort Union Formations: Geological Society of America Bulletin, v. 51, p. 213–236.

Dorf, E., 1942, Upper Cretaceous floras of the Rocky Mountain region, 2. Flora of the Lance Formation at its type locality, Niobrara County, Wyoming: Carnegie Institution of Washington Publication 508, p. 83–168.

Fastovsky, D.E., 1987, Paleoenvironments of vertebrate-bearing strata during the Cretaceous-Paleogene transition, eastern Montana and western North Dakota: Palaios, v. 2, p. 282–295.

Fox, S.K., Jr., and Ross, R.J., Jr., 1942, Foraminiferal evidence for the Midway (Paleocene) age of the Cannonball Formation in North Dakota: Journal of Paleontology, v. 16, p. 660–673.

Fox, S.K., Jr., and Olsson, R.K., 1969, Danian planktonic foraminifera from the Cannonball Formation in North Dakota: Journal of Paleontology, v. 43, p. 1397–1404.

Hicks, J.F., Obradovich, J.D., and Tauxe, L., 1999, Magnetostratigraphy, isotopic age calibration and intercontinental correlation of the Red Bird section of the Pierre Shale, Niobrara County, Wyoming, USA: Cretaceous Research, v. 20, p. 1–27.

Jensen, F.S., and Varnes, H.D., 1964, Geology of the Fort Peck Area, Garfield, McCone, and Valley Counties, Montana: U.S. Geological Survey Professional Paper 414F, 49 p.

Johnson, K.R., Nichols, D.J., Attrep, M., Jr., and Orth, C.J., 1989, High-resolution leaf-fossil record spanning the Cretaceous/Tertiary boundary: Nature v. 340, p. 708–711.

Johnson, K.R., and Wilf, P., 1996, Paleobotanical temperature curve for the Cretaceous-Tertiary boundary interval and a scenario for the revegetation of North America after the terminal Cretaceous event: Geological Society of America Abstracts with Programs, v. 28, no. 7, p. 225.

Kennedy, W.J., Landman, N.H., Christensen, W.K., Cobban, W.A., and Hancock, J.M., 1998, Marine connections in North America during the late Maastrichtian: Paleogeographic and paleobiogeographic significance of *Jeletzkytes nebrascensis* Zone cephalopod fauna from the Elk Butte Member of the Pierre Shale, SE South Dakota and NE Nebraska: Cretaceous Research, v. 19, p. 745–775.

Murphy, E.C., Nichols, D.J., Hoganson, J.W., and Forsman, N.F., 1995, The Cretaceous/Tertiary boundary in south-central North Dakota: North Dakota Geological Survey Report of Investigations 98, 74 p.

Nichols, D.J., and Fleming, F., 2002, Palynology and palynostratigraphy of Maastrichtian, Paleocene, and Eocene strata in the Denver Basin, Colorado: Rocky Mountain Geology, v. 37, no. 2 (in press).

Nichols, D.J., Murphy, E.C., Johnson, K.R., and Betterton, W.J., 2000, A second K-T boundary locality in North Dakota verified by palynostratigraphy and shocked quartz: Geological Society of America Abstracts with Program v. 32, no. 7, p. 130.

Rigby, J.K., and Rigby, J.K., Jr., 1990, Geology of the Sand Arroyo and Bug Creek Quadrangles, McCone County, Montana: Geology Studies, v. 36, p. 69–134.

Ross, R.J., Jr., 1940, The foraminiferal fauna of the Cannonball Formation in North Dakota and its bearing on the problem of the Mesozoic-Cenozoic boundary [B.S. thesis]: Princeton, New Jersey, Princeton University, 84 p.

Sarjeant, W.A.S., and Currie, P.J., 2001, The "Great Extinction" that never happened: The demise of the dinosaurs considered: Canadian Journal of Earth Sciences, v. 38, p. 239–247.

Smit, J., 1999, The global stratigraphy of the Cretaceous-Tertiary boundary impact ejecta: Annual Review of Earth and Planetary Sciences, v. 27, p. 75–113.

Stott, L.D., and Kennett, J.P., 1990, The paleoceanographic and paleoclimatic signature of the Cretaceous-Paleogene boundary in the Antarctic: Stable isotopic results from ODP LEG 113, *in* Proceedings, Ocean Drilling Program, Leg 113: College Station, Texas, Ocean Drilling Program, p. 829–848.

Swisher, C.C., III, Dingus, L., and Butler, R.F., 1993, $^{40}Ar/^{39}Ar$ dating and magnetostratigraphic correlation of the terrestrial Cretaceous-Paleogene boundary and Puercan Mammal Age, Hell Creek-Tullock Formations, eastern Montana: Canadian Journal of Earth Sciences, v. 30, p. 1981–1996.

Manuscript Accepted by the Society September 24, 2001

Index

M

N

T

U